This impressive work is the first comprehensive account of the vegetation of southern Africa. The region contains a remarkable juxtaposition of different ecosystems, yet it forms a cohesive ecological unit with exceptionally high endemism.

The book is divided into three major parts: Part 1 provides the physiographic, climatic, biogeographic and historical background essential for understanding contemporary vegetation patterns and processes. Part 2 includes systematic descriptions of the characteristics and determinants of major vegetation units (the major terrestrial biomes, coastal vegetation, freshwater wetlands and marine vegetation). Part 3 elaborates on selected ecological themes of particular importance, including grazing, fire, alien plant invasions, conservation and human use of plants. These are discussed in the context of prevailing paradigms in the international literature.

Vegetation of Southern Africa

Vegetation of

Southern Africa

Edited by **R M Cowling, D M Richardson & S M Pierce**
Institute for Plant Conservation, University of Cape Town, South Africa

CAMBRIDGE
UNIVERSITY PRESS

PUBLISHED BY THE PRESS SYNDICATE OF THE UNIVERSITY OF CAMBRIDGE
The Pitt Building, Trumpington Street, Cambridge, United Kingdom

CAMBRIDGE UNIVERSITY PRESS
The Edinburgh Building, Cambridge CB2 2RU, UK
40 West 20th Street, New York NY 10011–4211, USA
477 Williamstown Road, Port Melbourne, VIC 3207, Australia
Ruiz de Alarcón 13, 28014 Madrid, Spain
Dock House, The Waterfront, Cape Town 8001, South Africa

http://www.cambridge.org

First published 1997
First paperback edition 2003

Typeset in 8.5/11.5 pt Swift and 6/8 pt Univers

A catalogue record for this book is available from the British Library

Library of Congress Cataloguing in Publication data

Vegetation of southern Africa / edited by R.M. Cowling,
D.M. Richardson & S.M. Pierce.
0 521 57142 1 (hardback)
Includes bibliographical references.
Plant ecology – Africa, Southern.
Plant communities – Africa, Southern.
Phytogeography – Africa, Southern.
Cowling, R.M. (Richard M.)
Richardson, D.M. (David M.), 1958–
Pierce, Shirley M.
QK394 .V45 1997
581.5/0968 20

ISBN 0 521 57142 1 hardback
ISBN 0 521 54801 2 paperback

This volume is dedicated to the memory of
J.P.H. Acocks (1911–1979) whose seminal work, *Veld Types of South Africa*,
inspired two generations of southern African plant ecologists.

Contents

Part three Ecological themes

Contributors [chapter numbers in brackets]

Allsopp, N. Range and Forage Institute, Botany Department, University of the Western Cape, Bellville 7535, South Africa. [16]

Anderson, H.M. National Botanical Institute, Private Bag X101, Pretoria, 0001, South Africa. [4]

Anderson, J.M. National Botanical Institute, Private Bag X101, Pretoria, 0001, South Africa. [4]

Anderson, R.J. Seaweed Unit, Sea Fisheries Research Institute, Private Bag X2, Rogge Bay 8012, South Africa. [15]

Avis, A.M. Department of Botany, Rhodes University, P.O. Box 94, Grahamstown 6140, South Africa. [13]

Bolton, J.J. Botany Department, University of Cape Town, Private Bag, Rondebosch 7700, South Africa. [15]

Bond, W.J. Botany Department, University of Cape Town, Private Bag, Rondebosch 7700, South Africa. [18]

Boucher, C. Department of Botany, University of Stellenbosch, Stellenbosch 7600, South Africa. [13]

Bredenkamp, G.J. Department of Botany, University of Pretoria, Pretoria 0002, South Africa. [10]

Burke, A. Botany Department, University of Namibia, Private Bag 13301, Windhoek, Namibia. [9]

Cowling, R.M. Institute for Plant Conservation, Botany Department, University of Cape Town, Private Bag, Rondebosch 7700, South Africa. [Editor, 3,6,12,19]

Cunningham, A.B. WWF/UNESCO/Kew People and Plants Initiative, P.O. Box 42, Betty's Bay, 7141, South Africa. [20]

Danckwerts, J.E. Roodeplaat Grassland Institute, P.O. Box 94, Adelaide 5760, South Africa. [17]

Davis, G.W. National Botanical Institute, Private Bag X7, Claremont 7735, South Africa. [20]

Dean, W.R.J. FitzPatrick Institute, University of Cape Town, Rondebosch 7700, South Africa. [7]

Henderson, L. National Botanical Institute, Private Bag X101, Pretoria 0001, South Africa. [22]

Hilton-Taylor, C. Conservation Biology Research Unit, National Botanical Institute, Private Bag X7, Claremont 7735, South Africa. [3,19]

Hoffmann, J.H. Zoology Department, University of Cape Town, Private Bag, Rondebosch 7700, South Africa. [22]

Hoffman, M.T. National Botanical Institute, Private Bag X7, Claremont 7735, South Africa. [8,19,21]

Huntley, B.J. National Botanical Institute, Private Bag X7, Claremont 7735, South Africa. [Foreword]

Jacobson, K.M. Desert Ecological Research Unit of Namibia, P.O. Box 1592, Swakopmund, Namibia. [9]

Jürgens, N. Botanical Institute, Albertus Magnus University of Cologne, D-50923 Köln, Germany. [9]

Lubke, R.A. Department of Botany, Rhodes University, P.O. Box 94, Grahamstown 6140, South Africa. [13]

Macdonald, I.A.W. WWF South Africa, P.O. Box 456, Stellenbosch 7599, South Africa. [22]

Midgley, J.J. Botany Department, University of Cape Town, Private Bag, Rondebosch 7700, South Africa. [12,19]

Milton, S.J. FitzPatrick Institute, University of Cape Town, Rondebosch 7700, South Africa. [7]

Mustart, P.J. Institute for Plant Conservation, Botany Department, University of Cape Town, Private Bag, Rondebosch 7700, South Africa. [6]

O'Connor, T.G. Döhne Agricultural Development Institute, Private Bag X15, Stutterheim 4930, South Africa. [10]
Current address: Department of Grassland Science, Unversity of Natal, P. Bag X01, Scottsville 3209, South Africa.

Owen-Smith, N. Department of Zoology, University of the Witwatersrand, P.O. Wits 2050, South Africa. [17]

Palmer, A.R. Agricultural Research Council – Range and Forage Institute, P.O. Box 101, Grahamstown, 6140, South Africa. [8]

Partridge, T.C. Climatology Research Group, University of the Witwatersrand, 12 Cluny Road, Forest Town 2193, South Africa. [1]

Pierce, S.M. Institute for Plant Conservation, Botany Department, University of Cape Town, Private Bag, Rondebosch 7700, South Africa. [Editor]

Rebelo, A.G. National Botanical Institute, Private Bag X7, Claremont 7735, South Africa. [23]

Richardson, D.M. Institute for Plant Conservation, Botany Department, University of Cape Town, Private Bag, Rondebosch 7700, South Africa. [Editor, 6,19,22]

Rogers, K.H. Centre for Water in the Environment, Department of Botany, University of the Witwatersrand, Private Bag 3, Wits 2050, South Africa. [14]

Rutherford, M.C. National Botanical Institute, Private Bag X7, Claremont 7735, South Africa. [5]

Scholes, R.J. Division of Water, Environment and Forestry Technology, CSIR, P.O. Box 395, Pretoria 0001, South Africa. [11]

Schulze, R.E. Department of Agricultural Engineering, Faculty of Engineering, University of Natal, P.O. Box 375, Pietermaritzburg, South Africa. [2,19]

Scott, L. Department of Botany and Genetics, University of the Orange Free State, P.O. Box 339, Bloemfontein 9300, South Africa. [4]

Seely, M.K. Desert Ecological Research Unit of Namibia, P.O. Box 1592, Swakopmund, Namibia. [9]

Seydack, A.H.W. Department of Water Affairs and Forestry, Private Bag X7, Knysna 6570, South Africa. [12]

Steinke, T.D. Estuarine and Marine Group, Department of Botany, University of Durban-Westville, Private Bag X 54001, Durban 4000, South Africa. [13]

Stock, W.D. Botany Department, University of Cape Town, Private Bag, Rondebosch 7700, South Africa. [16]

van der Heyden, F. Range and Forage Institute, Botany Department, University of the Western Cape, Bellville 7535, South Africa. [16]
Current address: Division of Water, Environment and Forestry, CSIR, P.O. Box 320, Stellenbosch 7599, South Africa.

van Wyk, G.F. Division of Forest Science and Technology, CSIR, Futululu Research Centre, Private Bag, Mtubatuba 3935, South Africa. [12]

Vlok, J.H.J. Cape Nature Conservation, P.O. Box 123, De Rust 6650, South Africa. [7]

Witkowski, E.T.F. Department of Botany, University of the Witwatersrand, P.O. Wits 2050, South Africa. [16]

Yeaton, R.I. Botany Department, University of Natal, P.O. Box 375, Pietermaritzburg 3200, South Africa. [7]

Foreword

Southern Africa is endowed with an unusually rich flora with high levels of endemism, assembled within vegetation types ranging from desert to rainforest. Not only is the flora exceptionally rich, with over 22 000 species of indigenous ferns, gymnosperms and angiosperms, but the plants themselves are of great beauty, having attracted the attention of botanists since the seventeenth century. This interest resulted in a long history of botanical exploration, collection and description, and in the commercial exploitation of numerous southern African species, some of which today form the basis of a multi-billion dollar horticultural industry in Europe.

The study of southern Africa's flora and vegetation has been built on a strong tradition of fieldwork. The explorations in the late eighteenth century of C.P. Thunberg, A. Sparrman, F. Masson, C.F. Ecklon, C.L.P. Zeyher, etc. led to the publication of the *Flora Capensis* and many regional floras. The type specimens of much of this flora are housed in Kew, Paris, Berlin and other European institutions, but from the mid-nineteenth century, herbaria were established in Cape Town, Durban, Grahamstown, Pretoria, Windhoek and many other centres. These collections served the taxonomists of the late nineteenth and early twentieth century – J. Medley Wood, H. Bolus, J. Burtt-Davy, E.P. Phillips, R.A. Dyer – prolific workers who generated a momentum which continues to this day. In turn, these herbaria and taxonomists provided the essential foundation for the first wave of plant ecologists who described and mapped our vegetation – H.W.R. Marloth, R.S. Adamson, J.F.V. Phillips, J.W. Bews, I.B. Pole Evans, J.A. Pentz, M.R.B. Levyns, C.L. Wicht – the pioneers of southern African vegetation science.

The broad sweep of subject matter, and of geographic cover, addressed by these early workers is both inspiring and intimidating to the student of the 1990s. The era of such ambitious, single-authoured treatises ended in southern Africa with J.P.H. Acocks' classic *Veld Types of South Africa*, published in 1953. Acocks' map and descriptive memoir were prepared under duress – as a perfectionist he regarded his work as being both rushed and preliminary. Yet to this day, the work is the most valuable reference on the composition and distribution of South Africa's vegetation types. His interpretation of vegetation relationships and long-term dynamics has provided the platform for many subsequent studies and for continuing debate. His exhaustive and meticulously recorded field data, still unpublished, offer exciting prospects for the extension of his work through computerized analysis and modelling. Throughout his professional career, he challenged dogma and presented new insights into the workings of southern African vegetation. It is therefore appropriate that this volume be dedicated to his memory.

Acocks' *Veld Types* stands as a benchmark in southern African vegetation studies. It was followed in the 1960s and 1970s by many excellent regional and local vegetation surveys, some purely descriptive, some following the Braun–Blanquet phytosociological tradition, others testing the then new computer-based classification and ordination techniques. These works added depth and breadth to our knowledge of the distribution and composition of our vegetation. But they tell little of the evolution and functioning of the dynamic mix of landscapes, floras, ecosystems and human impacts that characterize southern Africa.

This volume provides a new synthesis of the wealth of information, understanding and opinion on the subcontinent's vegetation. Most particularly it builds on the results of the renaissance of South African plant ecology which occurred from the mid 1970s to the end of the 1980s.

The new wave of ecosystem research was a direct result of the global upsurge of environmental awareness, which reached its scientific expression in the International Biological Programme, 1967–1972. South Africa played a minor role in the IBP, but the approach and philosophy of large, interdisciplinary and multi-organizational research programmes captured the imagination of South African ecologists and science administrators. A consequence was the initiation of a series of cooperative scientific programmes, focusing on biome-level studies on the structure and functioning of South African ecosystems: these comprised the Savanna Ecosystem Project (initiated in 1973), Fynbos Biome Project (1977), Grassland Biome Project (1982), Forest Biome Project (1985) and Karoo Biome Project (1986). The impetus that developed through these studies extended into the marine environment with the establishment of the Kelp Bed Project, followed by the massive research programme on the Benguela upwelling system, while research at the Desert Ecological Research Unit at Gobabeb in the Namib Desert, and at the Okavango Swamps in Botswana, was stimulated and supported through South Africa's Council for Scientific and Industrial Research and various national and international non-governmental organizations.

The majority of the authors of this volume are products of these programmes. For the first time in southern Africa's history, 'soft money' was available in the 1970s and 1980s for young researchers to commence exciting, if risky, careers in ecology. This new generation of researchers succeeded in breaking out of the existing academic and institutional hierarchies, creating a vibrant and exciting invisible college of environmental scientists. Never before on the subcontinent had such a mix of disciplines and philosophies meshed together in a network of common interest. As a consequence, new dimensions emerged. Palaeoscientists gave a time perspective which, with geomorphologists, pedologists and systematists, presented new opportunities to unravel the biogeography and evolution of both flora and vegetation. The rapidly moving debate on the nature of communities, equilibrium theory and conservation biology provided intellectual challenges to be tested in the field. Significant contributions to the international literature in the form of original papers and synthesis volumes were made. These were heady days in southern African ecology, where a genuine spirit of sharing and testing ideas, field data and infrastructural resources reigned supreme. Sadly, this era emerged at a time of unprecedented political oppression, violence and open warfare in South Africa and Namibia.

Today, largely as a result of changing science policy, the culture of cooperative multi-organizational research is a thing of the past. However, the new political dispensation throughout the subcontinent offers unprecedented opportunities for a new wave of ecological endeavour. Never before have the conditions for a regional programme of ecological research and application been better. The solid intellectual foundation, reflected in this volume, provides a springboard for collaboration between vegetation scientists from throughout the region. The social need for the development of environmental restoration and upliftment programmes to heal the land is critically urgent. The political will to support such programmes has been demonstrated repeatedly by leaders of all the countries in the region. Financial mechanisms to provide support for capacity building, infrastructural development and implementation projects have never been more substantial or more accessible to southern Africa.

The editors and authors have provided this volume as the first step towards launching a new era of vegetation science and its application towards achieving social benefits throughout southern Africa. May their lead inspire others to join together and motivate a new network of southern African ecologists in service of society.

B. J. Huntley
Chief Director, National Botanical Institute
Kirstenbosch, June 1995

Preface

Why a book on the vegetation of southern Africa? We can think of at least three answers to this question. The short and simple answer is that the publishers approached us and, perhaps against our better judgement, we could not resist the temptation. More seriously, the extremely varied vegetation of the subcontinent provides an opportunity for a regional synthesis that includes representative examples of many of the world's major ecosystems. The popular tourist slogan for South Africa, 'a world in one country', is no hyperbole. It is possible, over relatively short distances, to move from hyper-arid desert to mesic savanna, or from tropical rainforest to alpine heathland. Not only is there a remarkable juxtaposition of different ecosystems, but there is also great variation in the forces that determine their structure and functioning: the large herds of ungulates that tracked the pulses of productivity on the vast karoo plains grazed only a short distance beneath the apparently lifeless and fire-driven fynbos on the sterile Cape mountains. It seemed to us that there were fascinating stories in all of this and that an ideal medium for telling them would be an account of the vegetation of the region.

But have these stories not been told before? This brings us to the third reason for producing this book: that a comprehensive volume on the vegetation of southern Africa is overdue. Pole Evans (1936) produced the first map and accompanying memoir of the vegetation of the subcontinent (south of latitude 22°), at a time when ecologists were bold and broad enough to tackle syntheses of this scope. This was followed by Adamson's (1938) pioneering book, *The Vegetation of South Africa*, published under the auspices of the British Empire Vegetation Committee. Although invaluable in their time, these early studies have become dated, in

terms of both their conceptual framework and their information content. They are seldom referred to today.

In contrast, Acocks' (1953) *Veld Types of South Africa*, comprising a richly worded memoir and detailed vegetation map (Fig. x.1*a*), remains a standard text to this day. John Acocks, born and bred in South Africa, was a maverick botanist who pursued his remarkable career on a remote research station amidst the sweeping plains and ancient koppies of the eastern karoo veld that he so dearly loved. Although firmly rooted in the Clementsian paradigm of the time, Acocks provided a uniquely South African perspective of vegetation patterns and dynamics, stressing in particular the role of fire and grazing in shaping the plant cover and composition of the country. His knowledge of the flora, one of the richest in the world, was extraordinary and will probably never again reside in any one individual. It is true, and remarkably so, that with the exception of some remote mountain fynbos areas, Acocks was able, towards the end of his career, to identify perhaps 90% of the plant species observed during a ramble through any of South Africa's 70 Veld Types.

The publication in 1978 of *Biogeography and Ecology of Southern Africa*, edited by Marinus Werger (a plant ecologist 'on loan' from the Netherlands), which included contributions from local and international scholars, was a major landmark in that it internationalized southern African vegetation science. This volume was also significant in that it included material from countries in south-central Africa that, for obvious political reasons, were rapidly distancing themselves from South Africa. Fortunately, Frank White's (1983) *Vegetation of Africa*, comprising a map (Fig. x.1*b*) and descriptive memoir of the vegetation of the entire continent, provided an opportunity for the younger generation of South African plant ecol-

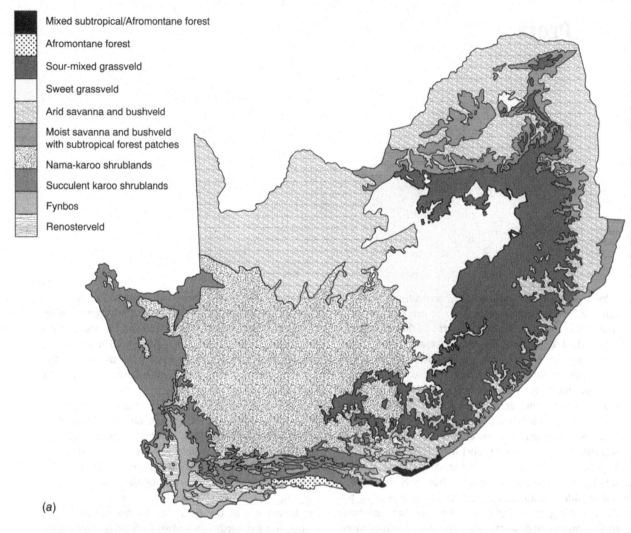

Mixed subtropical/Afromontane forest

Afromontane forest

Sour-mixed grassveld

Sweet grassveld

Arid savanna and bushveld

Moist savanna and bushveld
with subtropical forest patches

Nama-karoo shrublands

Succulent karoo shrublands

Fynbos

Renosterveld

(a)

Figure x.1 **Examples of vegetation maps for southern Africa:**
(a): Simplified version of Acocks' (1953) map of *Veld Types of
South Africa* (R.J. Scholes, unpubl. data). (b) White's (1983)
vegetation map based on phytochoria (see Chap. 3, this volume).

ogists, who had little field experience outside the subcontinent, to view their ecosystems in an African-wide
context.

Southern African vegetation science has made great
strides since the mid 1970s, largely as a result of the
positive impact of National Programmes for Ecosystem
Research initiated by the Council for Scientific and
Industrial Research. These interdisciplinary and collaborative research programmes, organized as a series of
biome-level projects, fostered a productive and imaginative corps of researchers who greatly advanced our understanding of southern African terrestrial ecosystems
(Huntley 1987, 1992). Although there have been two
biome-level syntheses of research carried out during this

era (Cowling 1992; Scholes & Walker 1993), with another
on the arid biomes in preparation, there was a definite
need for an updated subcontinent-wide synthesis of vegetation research.

We decided that the geographic scope of the book
should be limited to southern Africa, defined so as to
include Botswana, Lesotho, Namibia, South Africa and
Swaziland, an area of 2 675 425 km² (Fig. x.2). This region
forms a cohesive ecological and phytogeographical unit
with exceptionally high endemism at the taxic, vegetation type and phytochorion levels (Goldblatt 1978;
Gibbs Russell 1985; Rutherford & Westfall 1986; Cowling & Hilton-Taylor 1994; Chaps 3 and 5, this volume).
However, the shortcomings of this delimitation became

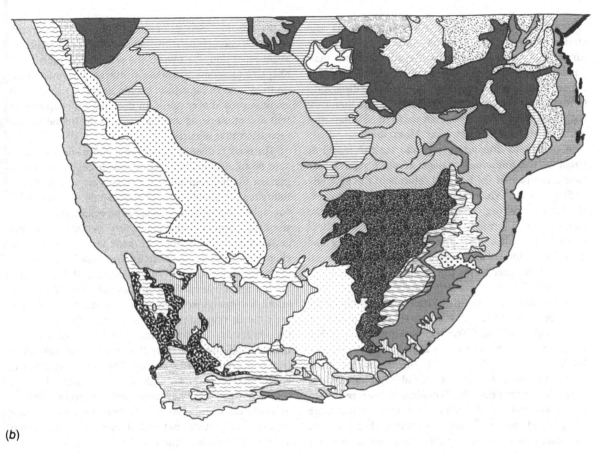

(b)

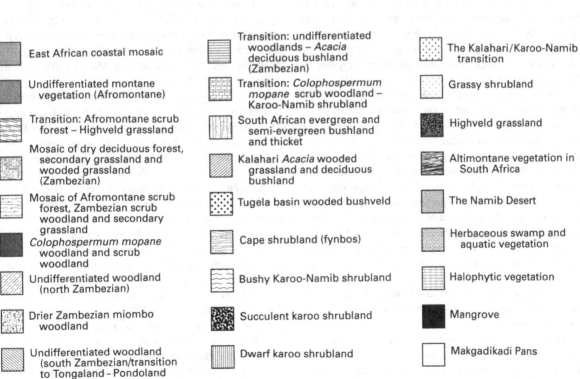

East African coastal mosaic

Undifferentiated montane vegetation (Afromontane)

Transition: Afromontane scrub forest – Highveld grassland

Mosaic of dry deciduous forest, secondary grassland and wooded grassland (Zambezian)

Mosaic of Afromontane scrub forest, Zambezian scrub woodland and secondary grassland

Colophospermum mopane woodland and scrub woodland

Undifferentiated woodland (north Zambezian)

Drier Zambezian miombo woodland

Undifferentiated woodland (south Zambezian/transition to Tongaland – Pondoland bushland)

Transition: undifferentiated woodlands – Acacia deciduous bushland (Zambezian)

Transition: Colophospermum mopane scrub woodland – Karoo-Namib shrubland

South African evergreen and semi-evergreen bushland and thicket

Kalahari Acacia wooded grassland and deciduous bushland

Tugela basin wooded bushveld

Cape shrubland (fynbos)

Bushy Karoo-Namib shrubland

Succulent karoo shrubland

Dwarf karoo shrubland

The Kalahari/Karoo-Namib transition

Grassy shrubland

Highveld grassland

Altimontane vegetation in South Africa

The Namib Desert

Herbaceous swamp and aquatic vegetation

Halophytic vegetation

Mangrove

Makgadikadi Pans

evident in contributions that dealt with savanna vegetation, which extends from its centre in south-central Africa, as a series of somewhat depauperate and marginal communities into the northern reaches of our region. It made sense, therefore, to encourage discussion in these chapters (Chaps 11, 17 and 20) of the literature dealing with the more typical savannas of neighbouring countries such as Zimbabwe and Mozambique.

A potentially frustrating problem for the reader is posed by the numerous changes of place names that have accompanied the political reforms in the Republic of South Africa. While this volume was being prepared, the 'homelands' (e.g. Transkei) were abolished and the country, which formerly comprised four provinces, was divided into nine new provinces incorporating the 'homelands' and subdividing the old Cape and Transvaal provinces. We present for the reader's convenience two maps, one showing the old provincial boundaries (Fig. x.2a) and the other showing the new ones (Fig. x.2b). Province names from both schemes are used throughout the volume. Other than Transkei (location shown in Fig. x.2a), no reference is made to any of the 'homelands'. Where possible, we have tried to use the names of the new provinces. In the case of the old Natal and Orange Free State provinces, the boundaries have remained unchanged and the editors' task was simple. It was more complicated for the other provinces, because, for example, it was not always clear how closely 'western Cape' or 'northern Transvaal' in older publications conformed with the boundaries of the new Western Cape or Northern Province. Where ambiguity could arise, our approach has been to use the old province names, in which case the geographical modifier always appears in lower case (e.g. eastern Cape, eastern Transvaal).

The organization of this book provided us with a few major problems. The first of these was the definition of vegetation science (cf. Wilson 1991). The definition of vegetation as a discreet unit for analysis is fraught with the same conceptual problems associated with the delineation of communities. We adopted a pragmatic and holistic approach (Keddy 1993), defining vegetation science as the study of the patterns and processes that determine the plant cover and composition of a given area. Thus, rather than focus exclusively on the description of various units of vegetation, we have encouraged analysis of the historical and contemporary factors that govern the distribution, structure and functioning not only of vegetation units (or communities) but also of lower order (organisms, populations) and emergent entities (ecosystems).

Since there were no volunteers willing or foolish enough to take on the task of writing the entire volume, we opted for a multi-authored book and recruited authors from across the subcontinent. It is important, in a book comprising contributions on related topics, to avoid repetition. We invited the principal authors of all chapters to attend a workshop during November 1993 on the shores of Langebaan lagoon near Cape Town. This exercise proved very useful in deciding on chapter content and minimizing overlap. We take the blame for any repetition that persists.

The book is divided into three major parts, each prefaced by an introductory text to provide context and integration for the individual chapters. Part 1, comprising four chapters, provides the physiographic, climatic, biogeographic and historical background essential for understanding contemporary vegetation patterns and processes. Part 2, the core of the book and comprising 10 chapters, is the encyclopaedic and systematic description of the characteristics and determinants of the major vegetation units. We chose biomes as the units of organization, for reasons outlined in the introduction to Part 2. Included in this part are three chapters that deal with the vegetation of non-biome-scale units, namely the coastal zone, freshwater bodies and the marine intertidal and subtidal environments. The final section of the book, Part 3, comprises eight chapters, each of which covers a topic that has been well researched and is uniquely interesting. We believe that these cross-biome reviews will provide material of interest to vegetation scientists throughout the world.

Nomenclature for plants listed in the text is according to Arnold & De Wet (1993) except for algae where nomenclature follows Seagrief (1984); occasional exceptions are qualified by author citation. Nomenclature for birds and mammals follows Maclean (1993) and Skinner & Smithers (1990), respectively.

It is a sad reflection of the history of the region that of all contributors to this volume none is black and only eight are women. There are several reasons for this. Foremost among these is the long history of oppression and official denial of access to quality education and employment prospects suffered by black Namibians and South Africans. Furthermore, the contributors are overwhelmingly South African: this country supports by far the strongest research personnel and infrastructure in the subcontinent. We hope that the next volume on the vegetation of southern Africa, which will probably be needed in about 20 years' time, will have a list of contributors more representative of southern Africa.

Finally, we would like to acknowledge the skills of Leslie Shackleton of Research Facilitation Services, Cape Town, for facilitating at the Langebaan workshop; the Foundation for Research Development, Pretoria, and the Institute for Plant Conservation, University of Cape Town, for financial support; Colin Paterson-Jones, the

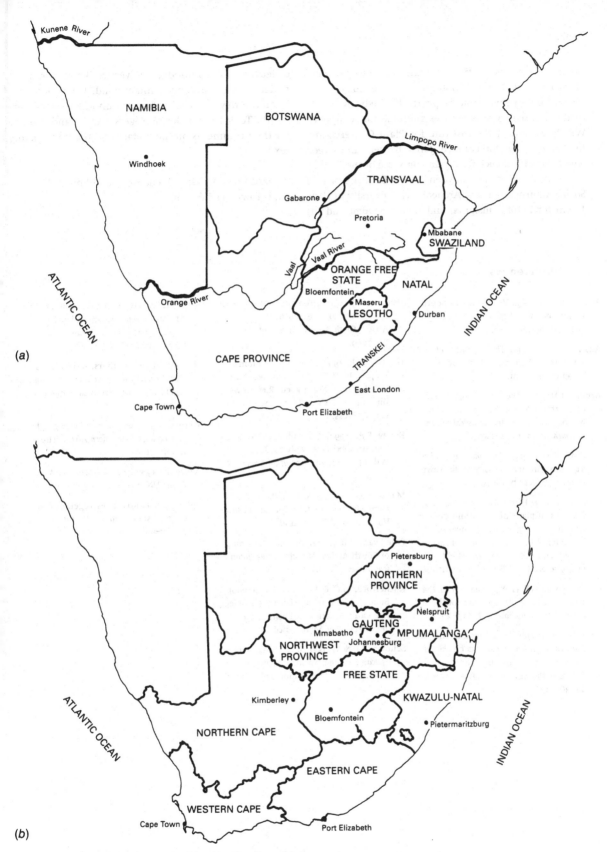

(a)

(b)

Figure x.2 **Southern Africa, as defined in this volume.** (*a*) shows
the old provincial boundaries in the Republic of South Africa
(pre-1992); (*b*) shows the boundaries of the nine new provinces
in the Republic of South Africa.

National Botanical Institute at Kirstenbosch, and Fred Ellery for access to their photographic collections; Corlia Richardson for invaluable help with final editing, proof-reading, miscellaneous services and emotional support; Wendy Paisley, at the Institute for Plant Conservation, for her cheerful forbearance; Janet Barker, Ann Green and Wendy Hitchcock for producing some of the figures; Emile Boonzaaier and Royden Yates (Departments of Social Anthropology and Archeology, University of Cape Town) for helpful insights; and Alan Crowden and his colleagues at Cambridge University Press for their patience. We also thank William Bond, Timm Hoffman, Brian Huntley, Roy Lubke, Tim O'Connor, Norman Owen-Smith, Tony Palmer, Kevin Rogers and Roland Schulze for their comments on the Preface and the introductory texts.

Richard Cowling, Dave Richardson and Shirley Pierce
Cape Town, 31 July 1995

References

Acocks, J.P.H. (1953). Veld Types of South Africa. *Memoirs of the Botanical Survey of South Africa*, **28**, 1–192.

Adamson, R.S. (1938). *The Vegetation of South Africa*. London: British Empire Vegetation Committee.

Arnold, T.H. & De Wet, B.C. (1993). Plants of southern Africa: Names and distribution. *Memoirs of the Botanical Survey of South Africa*, **62**, 1–825.

Cowling, R.M. (1992). *The Ecology of Fynbos. Nutrients, Fire and Diversity*. Cape Town: Oxford University Press.

Cowling, R.M. & Hilton-Taylor, C. (1994). Patterns of plant diversity and endemism in southern Africa: an overview. In *Botanical Diversity in Southern Africa*, ed. B.J. Huntley, pp. 31–52. *Strelitzia* 1. Pretoria: National Botanical Institute.

Gibbs Russell, G.E. (1985). Analysis of the size and composition of the southern African flora. *Bothalia*, **17**, 213–27.

Goldblatt, P. (1978). An analysis of the flora of southern Africa: its characteristics, relationships and origins. *Annals of the Missouri Botanical Garden*, **65**, 369–436.

Huntley, B.J. (1987). Ten years of cooperative ecological research in South Africa. *South African Journal of Science*, **83**, 72–9.

Huntley, B.J. (1992). The Fynbos Biome Project. In *The Ecology of Fynbos. Nutrients, Fire and Diversity*, ed. R.M. Cowling, pp. 1–5. Cape Town: Oxford University Press.

Keddy, P.A. (1993). Do ecological communities exist? A reply to Bastow Wilson. *Journal of Vegetation Science*, **4**, 135–6.

Maclean, G.L. (1993). *Roberts' Birds of Southern Africa*. 6th edn. Cape Town: John Voelcker Bird Book Fund.

Pole Evans, I.B. (1936). A vegetation map of South Africa, *Memoirs of the Botanical Survey of South Africa*, **15**, 1–23.

Rutherford, M.C. & Westfall, R.H. (1986). Biomes of southern Africa – an objective categorization. *Memoirs of the Botanical Survey of South Africa*, **54**, 1–98.

Scholes, R.J. & Walker, B.H. (1993). *An African Savanna. Synthesis of the Nylsvley Study*. Cambridge: Cambridge University Press.

Seagrief, S.C. (1984). A catalogue of South African green, brown and red marine algae. *Memoirs of the Botanical Survey of South Africa*, **47**, 1–72.

Skinner, J.D. & Smithers, R.H.N. (1990). *The Mammals of the Southern African Subregion*. 2nd edn. Pretoria: University of Pretoria.

Werger, M.J.A. (ed.) (1978). *Biogeography and Ecology of Southern Africa*. The Hague: Junk.

White, F. (1983). *The Vegetation of Africa*. Paris: UNESCO.

Wilson, J.B. (1991). Does vegetation science exist? *Journal of Vegetation Science*, **2**, 287–90.

Part 1 of this volume provides the physical, historical and biogeographical background that is necessary for a fuller appreciation of vegetation patterns and processes in southern Africa. It comprises four chapters, namely landscape evolution (Chap. 1), climate (Chap. 2), phytogeography (Chap. 3) and vegetation history (Chap. 4). Each chapter provides a different perspective towards understanding the enormous vegetational diversity of the subcontinent.

In Chap. 1, Partridge discusses the evolution of southern African landscapes. In common with other southern continents, many of these landscapes are extremely ancient. However, because of southern Africa's central location in Gondwanaland, it inherited more relief than the other fragments of the super-continent. A major feature of its geomorphic history was the erosion from the coast towards the interior, during the Cretaceous, of the ancient Gondwanan surface. This recession resulted in the formation of the Great Escarpment, a continuous arc that separates an elevated inland plateau from the dissected coastal margin, and a feature of great phytogeographical significance. Remnants of the pre-African surface persist as elevated areas such as the Cape Folded Belt and the Drakensberg uplands, both supporting very diverse contemporary floras. The planation of the African surface was associated with the formation of massive duricrusts of laterite, silcrete and calcrete. The shallow and sometimes infertile soils developed on these crusts provided a substratum suitable for the early diversification of xerophytic and sclerophyllous angiosperms, now such an overwhelming component of the southern African flora. Tertiary geomorphic history was characterized by successive cycles of uplift and erosion of this African surface. As a consequence, ancient and younger surfaces formed a complex mosaic of substrata. This geomorphic and associated edaphic complexity has undoubtedly played an important role in the development of southern Africa's vegetational diversity.

Schulze (Chap. 2) describes the contemporary climate of southern Africa (excluding Botswana and Namibia) by way of a wide-ranging series of computer-generated maps based on impressive databases of primary climatic and physiographic variables, which are used in conjunction with water and energy-budget simulation models. His analysis clearly shows the semi-arid nature of the subcontinent: less than 5% of the region receives an annual rainfall of greater than 800 mm and, on average, more than 90% of rainfall is returned to the atmosphere by evaporative losses. Biologically, the region is characterized by great uncertainty with most of southern Africa having an inter-annual variability in primary production in excess of 50%. These generalizations, however, tend to mask the high climatic diversity and steep climatic

Part one

Physiography and history

gradients of the subcontinent: from the hyper-arid and foggy Namib desert along the west coast to the hot and humid climate of the east coast; from the semi-arid and drought-prone Little Karoo basin to the cool and rain-soaked summits of the Cape mountains, only 1000 m higher. These patterns are largely the result of the region's transitional location with regard to tropical and temperate circulation systems; the juxtaposition of cold and warm ocean currents; and a varied topography. Thus, climatic diversity acts in concert with edaphic diversity to produce the complex vegetation patterns of southern Africa.

Cowling and Hilton-Taylor (Chap. 3) build on these themes of physiographic diversity to explain the exceptional concentration of phytogeographic units in southern Africa. The striking features of the region's flora are the high diversity and endemism at all taxonomic levels. Thus, southern Africa forms a discrete phytogeographic entity: a southern temperate flora with some characteristics more typical of an oceanic island than a continental landmass. The endemic component is not uniformly distributed across the subcontinent but is concentrated in an almost continuous arc below, and including, sectors of the Great Escarpment and elevated pre-African surfaces. Moreover, endemic species are overwhelmingly associated with a limited number of genera characteristic of the southern (Cape and Karoo–Namib) phytochoria; they are biologically uniform; and are the product of relatively recent diversification. Cowling and Hilton-Taylor suggest that the unusual richness of the subcontinent's flora is more a function of massive and incidental diversification within biologically peculiar lineages, than adaptive radiation in a spatially and temporally heterogeneous environment.

Scott, Anderson and Anderson (Chap. 4) conclude this part with an account of southern Africa's vegetation history, from the origin of land plants to the mid-Holocene. There is nothing in this account to suggest that, relative to other southern continents, there is anything unusual or exceptional about the region's history. Modern vegetation types first appeared in the Cainozoic and expanded, along with the extinction of archaic lineages, with accelerated climatic deterioration from the Miocene onwards. The ever-present saw-tooth fluctuations in climate over the last few million years have resulted in a long series of substantial changes in vegetation boundaries: glacial events favoured the temperate or 'southern' flora, whereas interglacials witnessed the expansion of subtropical savannas and forests. Scott et al. invoke climate change as the major driving force behind these shifts in vegetation boundaries and conclude that the timing and magnitude of these climatic cycles were sufficiently unremarkable to explain the region's exceptional plant diversity.

Why then, when compared to the other southern subcontinents at similar latitude, is southern Africa so rich in plant species and, consequently, floristically characterized vegetation types? Is it because there is something unusual about the region's biota or physiography, or both of these? These four chapters go some way to answering this question, but more comparative biogeographical and evolutionary studies are required to solve this enigma.

Evolution of landscapes

1

T.C. Partridge

1.1 Introduction

Large parts of Africa, like Australia and South America, differ from most northern hemisphere land masses in preserving extensive tracts of ancient landscapes. These have exerted a profound influence on the evolution and distribution of vegetation. An understanding of landscape history is therefore as important as the chronicle of changing climates for comprehending the development of southern African plant communities and their current distributions.

Southern Africa is, in many ways, typical of the face of ancient Africa. It is characterized by a high interior plateau, bounded on three sides by the horseshoe rampart of the Great Escarpment. Above the plateau stand several elevated mountain massifs, among them the highlands of Lesotho, which exceed 3000 m in places. Other parts of the interior are dished into large basins, such as the Kalahari and Transvaal Bushveld. Away from the foot of the Great Escarpment, a plinth 50 to 200 km wide slopes to the coast; its undulating surface is cleft by deep gorges, through which rivers find their way from the interior to the ocean.

The broad physiographic symmetry of southern Africa contrasts with the meridional distribution of its principal climatic belts. These are so aligned that much of the subcontinent grades, west to east, from hyper-arid to humid (Chap. 2, this volume). The imprint of this climatic gradient is evident in both recent landforms and soils, but other relict features indicate that the contrast between the western and eastern seaboards is not of particularly great antiquity. The legacy of past changes in climate is, indeed, readily apparent in the landscapes of the now arid west of southern Africa.

1.2 Early history: Gondwanaland and its fragmentation

Although most of the present landscape of southern Africa postdates the break-up of the Gondwanaland super-continent, some major elements of its architecture have been inherited from earlier periods. From Proterozoic times much of Africa has been characterized by a gross structure of interior basins separated by upwarped zones or 'swells'. Cahen et al. (1984) have shown that these zones of doming and tensional stress, first formed during the 'Pan-African' episode of uplift within mobile belts which ended about 560 million years ago (Myr), were rejuvenated repeatedly during the Phanerozoic. In the course of these uplifts the intervening stable shield areas, or cratons, lagged behind to form large basins. In southern Africa the Kalahari Basin has been the principal focus of continental sedimentation since the late Cretaceous (Partridge & Maud 1987). Although influenced in its form and location by very much earlier geological events, the present Bushveld Basin is considerably more recent, having been lowered along marginal faults as recently as the Pliocene (3–5 Myr) (Partridge, Wood & de Menocal 1995b).

The present configuration of the southern African landscape is overwhelmingly the product of a sequence of events set in motion by the break-up of Gondwanaland. Recent reconstructions confirm the central position occupied by Africa in the assemblage of southern hemisphere continents (Dingle, Siesser & Newton 1983; Fig. 1.1). This central location was, in large measure, responsible for the high altitude of southern and eastern Africa before the fragmentation of Gondwanaland in late Jurassic and early Cretaceous times. Reconstructions

Figure 1.1 **The configuration of the southern hemisphere continents at the time of the break-up of Gondwanaland (after Dingle et al. 1983).**

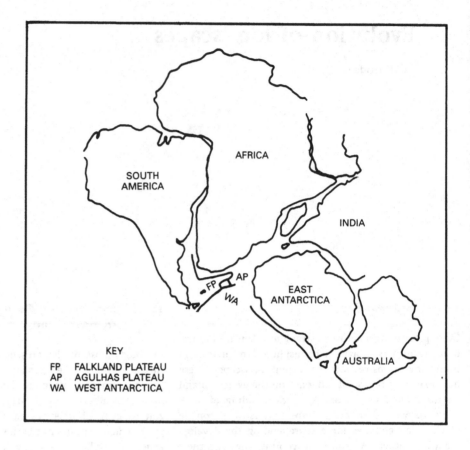

KEY

FP   FALKLAND PLATEAU
AP   AGULHAS PLATEAU
WA   WEST ANTARCTICA

based on the summit elevations of little-lowered remnants of the pre-rifting landsurface (e.g. the crests of the Lesotho Highlands), and on the extent of subsequent erosion as determined from the morphology of kimberlite pipes (Hawthorne 1975; Partridge & Maud 1987), indicate that elevations on this surface probably ranged from about 1800 m in the interior to about 2350 m in zones of upwarping adjacent to the rift margins. The amplitude of marginal downwarping and scarp formation during rifting must therefore have been of considerable magnitude. Their legacy has been preserved during the subsequent evolution of southern Africa in the Great Escarpment, now driven back by erosion between 50 and 200 km from the coastline, which forms a continuous arc separating the inland plateau from the more deeply dissected coastal margins. The interior plateau now ranges in elevation between about 1000 and 1700 m.

These inferences on the extent of erosion since the disruption of Gondwanaland are supported by computations of the volumes of terrigenous marine sediments which have accumulated since the ocean basins surrounding southern Africa were first formed (Rust & Summerfield 1990). Independent confirmation comes from the measurement of fission tracks in apatite crystals in

rocks now exposed on the surface (Brown et al. 1990); these indicate the thickness of overlying material removed since the occurrence of particular geological events. Both sources show that between 1800 and 3000 m has been removed from different parts of the interior since continental separation.

Although there is good evidence that several kilometres of material have been eroded from southern Africa since the fragmentation of Gondwanaland, a few upland areas of significant extent are preserved locally above the new landsurfaces which resulted from this denudation. These include the Namaqua Highlands, the Cape Fold Mountains, the Lesotho Highlands and the Transvaal Drakensberg (for a key to areas and localities referred to in this chapter see Fig. 1.2). Of these the Lesotho Highlands are the highest, exceeding 3000 m in places. Examination of the kimberlite pipes at Letseng Le Terai, near the crest of the Lesotho mountains, suggests that some 300 m of material have been eroded since the emplacement of these pipes (Hawthorne 1975) which, according to Davis (1977), occurred some 87 Myr. It is thus clear that the pre-rifting Gondwana landsurface is nowhere preserved within the present landscapes of the subcontinent.

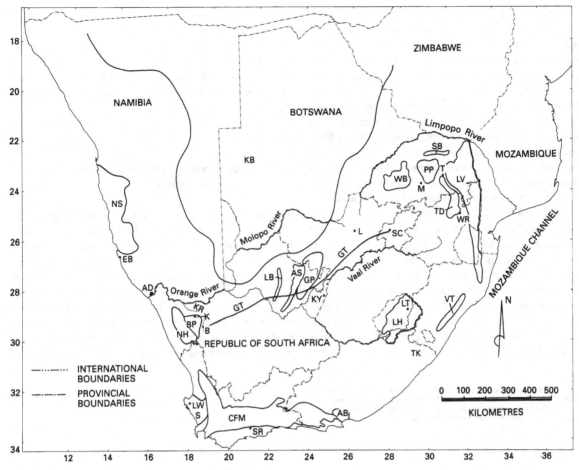

Figure 1.2 **Areas and localities referred to in this chapter. AB, Algoa Basin; AD, Arrisdrift; AS, Asbesberge; B, Bosluispan; BP, Bushmanland Plain; CFM, Cape Fold Mountains; EB, Elisabeth Bay; GP, Ghaap Plateau; GT, Griqualand–Transvaal Arch; K, Kangnas; KB, Kalahari Basin; KR, Koa River; KY, Kimberley; L, Lichtenburg; LB, Langeberge; LH, Lesotho Highlands; LT, Letseng Leterai; LV, Lowveld; LW, Langebaanweg; M, Makapansgat; NH, Namaqua Highlands; NS, Namib sand sea; PP, Pietersburg Plain; S, Swartland; SB, Soutpansberge; SC, Sterkfontein Caves; SR, Spiegel River; T, Tzaneen; TD, Transvaal Drakensberg; TK, Transkei; VT, Valley of a Thousand Hills; WB, Waterberge; WR, White River.**

1.3   Erosion and recession of the Great Escarpment during the Cretaceous

Because of the high elevation of much of southern Africa, a substantial marginal escarpment was created as the adjoining continental masses of South America and Antarctica drifted away from Africa. Rivers operating to the newly created local oceanic base level soon produced an erosional face. Although, during the Cretaceous, southern Africa occupied a position at least 14° further south than at present (Smith & Briden 1982), the subcontinent evidently enjoyed a humid, tropical cli-

mate, and offshore sedimentation rates were high, in keeping with the newly created energy potential, enhanced weathering and high runoff rates (Dingle *et al.* 1983; Partridge 1990). A comparison of the size distribution of basalt clasts of the Drakensberg Formation encountered today in river channels of the southeastern hinterland with those preserved in coastal marine sediments indicates that, by the mid-Cretaceous about 100 Myr, the Great Escarpment in this area had receded some 100 km from the coast; by the end of the Cretaceous this distance had increased to approximately 120 km – i.e. to a line not far seaward of its present position (Matthews 1978; Partridge & Maud 1987; Fig. 1.3). In short, as is reflected by sedimentation rates on the continental shelf, the sculpturing of the marginal zone below the Great Escarpment was largely a product of Cretaceous erosion (i.e. erosion which took place before about 65 Myr).

Simultaneous with denudation in the coastal hinterland was the removal of large volumes of material from the continental interior, as already noted. The bulk of these sediments was transported to the sea via the Limpopo and Orange/Vaal river systems, whose early exits through elevated rift shoulders are confirmed by the presence of Cretaceous sediments within their lower valleys (Partridge & Maud 1987). Although erosion has proceeded simultaneously above and below the Great Escarpment throughout the Cretaceous and the Cainozoic, the base levels which controlled this denudation differed for each zone: that of the interior plateau has been determined by the elevation of the exit points for the major river systems through the Great Escarpment (Partridge & Maud 1987). The resulting erosion surfaces, although formed over the same period, thus occur at different levels above and below the Great Escarpment

(Fig. 1.4). This dichotomy has, in fact, been maintained throughout the geomorphic history of southern Africa. Attempts to correlate surfaces altimetrically between the coastal and inland areas have been at the root of many of the controversies which are recorded in an extensive literature (for a summary see Appendix A in Partridge & Maud 1987).

1.4 The African planation surface and its distribution

The massive removal of material from most parts of the subcontinent was overwhelmingly the work of Cretaceous erosion. Not only was the Great Escarpment driven back almost to its present position during this period, but the extensive African surface was planed to its ultimate level before the beginning of the Cainozoic, some 65 Myr. Although it has been lowered and dissected over large areas by subsequent erosion, sufficient remnants of this surface remain for its previous extent to be reconstructed and its elevation over much of the sub-

Figure 1.3 **The Great Escarpment in the eastern hinterland of South Africa. Here the quartzitic cliffs of the Transvaal Drakensberg overlook the rolling savanna country of the lowveld, which belongs to the Post-African I erosion cycle** (Photo: T.C. Partridge).

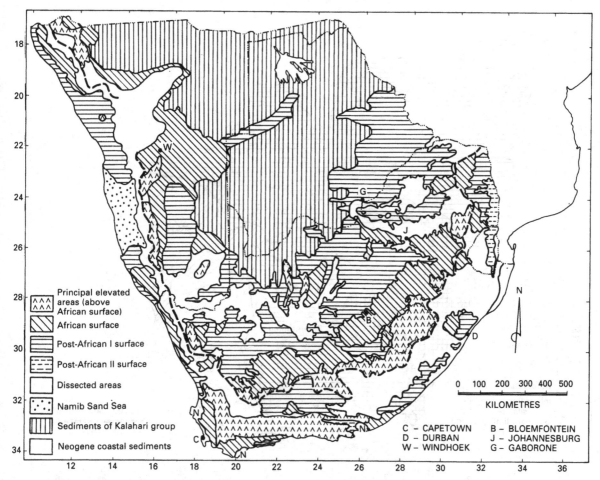

Figure 1.4 **Simplified map of the principal cyclic landsurfaces of southern Africa (modified from Partridge & Maud 1987). B, Bloemfontein; C, Cape Town; D, Durban; G, Gaborone; J, Johannesburg; W, Windhoek. Bold dashed lines indicate the Great Escarpment.**

Legend:
- ^^^^ Principal elevated areas (above African surface)
- African surface
- Post-African I surface
- Post-African II surface
- Dissected areas
- Namib Sand Sea
- Sediments of Kalahari group
- Neogene coastal sediments

C – CAPETOWN   B – BLOEMFONTEIN
D – DURBAN   J – JOHANNESBURG
W – WINDHOEK   G – GABORONE

0 100 200 300 400 500
KILOMETRES

continent to be contoured (Partridge & Maud 1987). Capping all of these pristine remnants are well-developed duricrusts. To the east of a line extending roughly from the Transkei to southeastern Botswana, these are of laterite; in the southern and western coastal hinterland they are preponderantly of silcrete, and in the semi-arid interior they are chiefly calcrete (caliche). This distribution is reflected in Fig. 1.5, but it is important to note that this map depicts the occurrence of pedocretes of all ages and stages of development, whereas those capping African remnants are generally mature (exceeding 2 m in thickness in many cases), frequently themselves weathered, and almost invariably underlain by deeply kaolinized regolith, which in places exceeds 30 m in thickness. These mature duricrust remnants, underlain by deep weathering profiles, serve to identify the African

surface unambiguously in many highly dissected areas where other evidence is lacking (Fig. 1.6).

The age of overlying deposits is obviously crucial in defining the duration of the African erosion cycle. In the coastal zone silcrete caps are locally overlain by sediments which range in age, on palaeontological grounds, from early Eocene to Miocene; one such remnant overlies a weathered olivine melilitite pipe at Spiegel River in the southern Cape Province, which is part of a cluster of more than 270 whose K/Ar ages lie in the range 68–63 Myr (Moore 1979). In the interior the same surface is associated at Kangnas, on the northern edge of the Bushmanland Plain, with Cretaceous sediments containing dinosaur remains (Haughton 1915; Rogers 1915); further inland, on the Ghaap Plateau north of Kimberley, it is crossed by channel deposits containing silicified logs

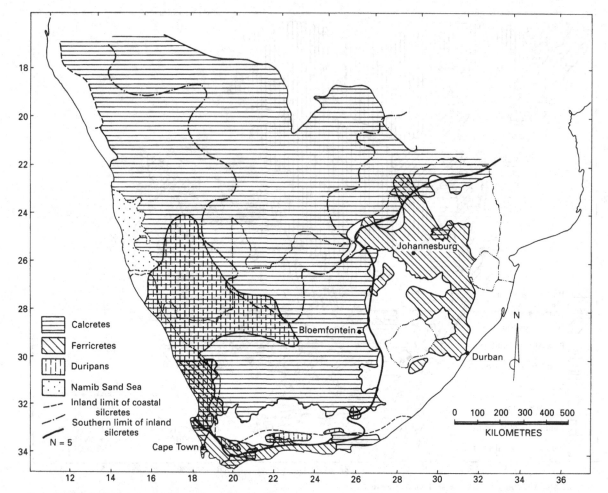

Figure 1.5 Distribution of common occurrences of pedocretes in southern Africa (modified from Du Toit 1954; Weinert 1980; Ellis & Schloms 1984 and Schloms & Ellis 1984). N, Weathering Index (*sensu* Weinert 1974): east of the line, where N < 5, the moisture regime is such that chemical decomposition of rock and soil predominates over mechanical disintergration; west of the line, where N > 5, mechanical disintegration predominates.

of the Upper Cretaceous (80–75 Myr) (T.C. Partridge, M.C.J. de Wit, J.D. Ward and M. Zavada, unpubl. data); and near Lichtenburg in the western Transvaal potholes within analogous palaeo-channel remnants contain late Cretaceous pollens (R.M.H. Smith, pers. comm.). The significance of this evidence is that, at widely separated localities within the Orange/Vaal catchment, which drains much of the interior of southern Africa, the African surface is overlain by deposits of Cretaceous (mostly late Cretaceous) age. On the basis of this and various other evidence it can, in fact, be deduced that the duricrusts which armour the African surface are of end-Cretaceous to Palaeocene age (Partridge & Maud 1989), and that erosion within the African cycle was largely complete by the end of the Cretaceous.

This accords well with the marine evidence (Dingle

et al. 1983), but the record of significant peaks in sedimentation during the Cretaceous suggests that several sub-cycles, probably initiated by local tectonic disturbances, may have occurred during the interval of more than 60 million years, during which the African cycle of erosion was current. However, the great antiquity of these events has resulted in the merging of any separate local responses to form a surface which, for practical purposes, must be regarded as a single unit within the southern African landscape. Contrasting interpretations to the effect that significant denudation persisted into the Palaeogene (Rust & Summerfield 1990) must be rejected in the light of both the geomorphic and terrestrial palaeontological evidence (Partridge *et al.* 1995b).

As indicated in Fig. 1.4, remnants of the African sur-

Figure 1.6 **Silcrete duricrust overlying deeply kaolinized granite: the Kookoppies on the Bushmanland plain in the Nama-karoo biome. The upper surface is the African surface** *(Photo: T.C. Partridge).*

face are widespread in southern Africa, both above and below the Great Escarpment. In the coastal hinterlands, the early Cainozoic landscape was dominated by a gentle, multi-concave pediplain, linking the foot of the escarpment to the coast, and punctuated by isolated coastal ranges in KwaZulu-Natal and the Eastern Cape (e.g. the Amatole Mountains), and by high-standing ridges of the Cape Fold Mountains further to the west. Above the Great Escarpment, elevations on the African surface were probably of the order of 600 m, increasing somewhat in the upwarped areas inland of the present scarp front and adjacent to inland massifs such as the Lesotho Highlands, the Langeberge and Asbesberge of the Northern Cape, and the Waterberge and Soutpansberge of the northern Transvaal. Over most intervening areas a subdued, duricrust-armoured pediplain characterized the interior plateau; but in a zone extending about 800 km southwestward from the western border of Lesotho (Fig. 1.4), where erosion in the African cycle had not consumed all upland remnants, numerous flat-topped koppies and *tafelberge* (table mountains), formed by resistant strata of the Karoo Supergroup, imparted considerable relief. Soils were correspondingly less mature, with a strong colluvial component and a lesser degree of deep weathering and duricrust development.

The humid, tropical conditions which evidently prevailed in southern Africa during the Cretaceous (Partridge 1990) were undoubtedly responsible for the development of the deeply kaolinized weathering profiles which characterize most surviving remnants of the

African surface. The genesis of the lateritic cappings that characterized the eastern areas of the subcontinent, by processes of relative enrichment of iron and aluminium oxides, can be readily comprehended in the context of tropical weathering processes; the nature of the genetic link between kaolinization and the formation of silcrete cappings is, however, more ambiguous. Most authors believe that these mature silcretes formed both in Australia and in southern Africa under humid conditions and low soil pH (e.g. Beckmann 1983; Summerfield 1983a,b; Twidale & Hutton 1986). At a pH of 4 or less, both silica and titanium (which is significantly enriched in silcretes associated with deep weathering profiles) are reasonably soluble (Summerfield 1983a,b); silica also shows enhanced solubility at very high pH. Partridge & Maud (1989) have argued that, although advanced weathering and uptake of silica into solution would undoubtedly have been favoured by the torrid Cretaceous climates, there is abundant evidence for cooling and desiccation at the end of the Cretaceous, which were evidently associated with an increase in pH. This would have led, inevitably, to silica precipitation. Indeed, many of the southern African silcretes overlying deeply weathered profiles display relict columnar or prismatic structures indicative of a sodic soil environment. In the semi-arid to arid western parts of the country an incipient form of silcrete known as 'dorbank' (red–brown hardpan) has formed more recently on lowered remnants of the African surface and on the Post-African I surface through cementation of soil particles by silica and iron oxides under eutrophic conditions. It is considered significant by Partridge & Maud (1989) that 'dorbank' has much the same distribution as silcrete in South Africa (Fig. 1.5). There are, therefore, good grounds for arguing that silcrete formation was linked to a major cooling event that was initiated at the end of the Cretaceous; it would be surprising if this event, whose legacy is preserved in duricrusts as far afield as Australia, southern Africa and even the western United States (Gassaway 1988), were not directly linked to the major upheavals that caused mass extinctions at the Cretaceous–Tertiary boundary (Partridge & Maud 1989).

The implications of these events for the early development of the southern African flora are considerable. Although the wet, torrid climates of the Cretaceous would have favoured the widespread occurrence of forest communities (grading into woodland in the somewhat drier interior), cooling and desiccation at the end of this period and in the early Palaeogene, combined with the nutrient-poor and often shallow substrates occasioned by the ubiquitous development of pedocretes, would have led to very different plant associations. Probably containing a significant proportion of sclerophyllous and shrubby elements, these are likely to have included a fair number of early flowering plants. The continuity of the African surface over very large parts of the subcontinent would also have facilitated species interchange over wide areas. Supporting palaeobotanical evidence is unfortunately almost entirely absent. It is important to realize that, after planation of the African surface was complete, the pattern of landscape development in southern Africa was one of progressive fragmentation of habitats as the effects of successive pulses of uplift and dissection were transmitted into the continental interior. Today, most remnants of the African surface are covered by grassland, except in the arid west where xeric scrub of the Nama and succulent karoo biomes predominate.

1.5 Mid-Tertiary disturbances and incision of the African surface

Although planation within the long African cycle was essentially complete in most areas by the end of the Cretaceous, conditions of relative tectonic stability apparently persisted until the early Miocene. The Palaeogene history of the African cycle was characterized by widespread duricrust armouring and low rates of offshore sedimentation. There is evidence that significant parts of the Palaeocene and the early Eocene were arid, at least in the western areas (Partridge 1993), and this desic-

cation undoubtedly contributed to the slow tempo of landscape development during this interval.

The African cycle was brought to an end by epeirogenic uplift of the subcontinent during the early Miocene, shortly after about 20 Myr. The amplitude of the uplift was greatest in the eastern part of southern Africa, where it approached 300 m and imparted a small but significant westward tilt to the subcontinent. In contrast, the Kalahari Basin apparently lagged behind surrounding areas, and this deepening led to renewed sedimentation within it. Elsewhere deformation of the pre-existing surface and increases in channel gradients initiated a new Post-African I cycle of erosion (Fig. 1.7). Although the resulting offshore sedimentation was small in comparison with that of the Cretaceous, there is nonetheless firm evidence for these movements both in the marine sedimentary record and in the onshore distribution of fossiliferous marine deposits (Partridge & Maud 1987). The driving force for these and later movements can now be linked to large-scale thermal anomalies within the upper mantle of the earth; detailed analyses are offered by Partridge et al. (1995a).

The new Post-African I surface, cut no more than about 250 m (and usually considerably less) below remnants of the African surface, is overlain locally by both marine and terrestrial sediments of early middle Miocene to Pliocene age (Partridge & Maud 1987). Because of its relatively short duration (less than about 15 million years), advanced planation was achieved in the Post-African I cycle in only a few areas; over most of its extent landscape development was limited to the removal of the deep weathering mantles of the African surface and the development of a rolling landscape cut into underlying lithologies. In the coastal areas of the southern and western Cape, which experienced minimal uplift, the full thickness of the earlier kaolinized saprolite was not, in fact, everywhere removed.

The imperfectly planed Post-African landscape is widely represented in southern Africa and gives special character to much of its interior through the undulating topography, with frequent bouldery koppies, which it has produced. Above the Great Escarpment and in the western coastal hinterland it is difficult to distinguish between incision referable to the early Miocene Post-African I cycle and its Pliocene Post-African II counterpart. Because the Pliocene movements were concentrated a short distance inland of the southeast coast, incision within the two cycles is more readily distinguishable in valley cross-profiles of this area, and a separate Post-African II surface has been cut locally across weak lithologies. Over large areas, however, multi-cyclic incision has produced deeply dissected landscapes,

Figure 1.7 **Flat-topped koppies capped by resistant dolerite characterize much of the Nama-karoo biome of the interior of South Africa. The erosion cycle is Post-African I** *(Photo: T.C. Partridge)*.

in which the influence of geological structure and the contrasting resistance of different rock types to erosion are dominant. Such areas are separately delineated in the simplified geomorphological map reproduced in Fig. 1.4.

As is discussed more fully below, soil distributions within the Post-African I landscape frequently owe more to climate and lithology than to geomorphological influences. A much more complex mosaic of soil types is present than is typical of remnants of the African surface. As a broad generalization, it may be noted that the Post-African I surface of the interior plateau is today host to much of the savanna and karroid vegetation of southern Africa. Below the Great Escarpment, vegetation communities are more diverse, owing chiefly to local topographic and climatic influences. It is important to place on record, however, that these communities were not characteristic of the Post-African I landscape over the entire span of 18 million years or so that have elapsed since its formation. Although its inception coincided broadly with the rapid spread of grasses, several important climatic shifts have been reflected in a succession of

vegetation communities over part or all of the interior. Particularly important was a return to more humid conditions in the early mid-Miocene (about 16 Myr) (Dingle & Hendey 1984; de Wit 1993), in contrast to the aridity that apparently characterized the early Tertiary. This wetter period saw the development of an extensive, integrated drainage network over the now arid interior of Bushmanland and Gordonia. Palaeontological remains recovered from the axial channel of this network (the Koa River valley) at Bosluispan indicate the existence of a warm, mesic, woodland environment, whose regional extent is confirmed by similar evidence from Arrisdrift on the lower Orange River and Elisabeth Bay on the now hyper-arid coast of Namibia (Hendey 1978; Dingle & Hendey 1984; Partridge 1985, 1993). It is clear, however, that these equable conditions did not persist for more than about 7 or 8 million years, and by the latter part of the Miocene, flow within the Koa River had effectively ceased (de Wit 1993).

1.6 Late Neogene tectonism and the rise of the southeastern hinterland

The Post-African I cycle was brought to an end during the Pliocene by massive uplift, concentrated chiefly in the southeastern hinterland of the subcontinent. The precise timing of this uplift is as yet uncertain, but it probably dates to between 5 and 3 Myr. Chronological control and evidence on the amplitude of the movement comes from the occurrence of anomalously high Neogene marine sediments on the flanks of the uplift axis, obviously steepened gradients across local remnants of the Post-African I surface, the anomalous convex-upward profiles of rivers crossing the axis, and the tilting of remnants of marine terraces corresponding with sea level high stands of known age (Partridge et al. 1996). These authors discuss, also, the likely mechanisms of uplift, in particular the occurrence of anomalously hot mantle material below the southeastern hinterland of the subcontinent which is considered to be the result, in part, of the migration of a 'hot-spot' from the Mozambique Channel to a present position beneath the Lesotho Highlands (Hartnady 1985, 1990). The principal axis of uplift was located some 80 km inland of the southeastern coastline; in this area vertical movements ranged from 600–900 m (Partridge & Maud 1987). In the interior lesser movements (< 100 m) were concentrated largely along discrete axes such as the Griqualand–Transvaal arch. Elsewhere the uplift was substantially smaller, amounting to no more than a few tens of metres along the west coast. The overall effect was thus to increase the westward gradient of the subcontinent. In the southeastern coastal hinterland uplift was accompanied by outward (monoclinal) flexing, which is manifested in the high elevations of remnants of the African and Post-African I surfaces and the presence of significantly increased coastward gradients on them. A major resurgence of sedimentation offshore of major east coast rivers followed this event (Dingle et al. 1983; Martin 1987).

There can be little doubt that the magnitude of this uplift contributed significantly to receipts of orographic precipitation within the southeastern hinterland of the subcontinent. Simultaneously eastward-draining rivers began to cut deep coastal gorges, thus initiating the Post-African II cycle of erosion. In some areas, such as the Valley of a Thousand Hills in KwaZulu-Natal, dissection has occurred in this cycle to depths of up to 500 m (Fig. 1.8). Only on susceptible lithologies in the Lowveld areas of the Eastern Transvaal and Swaziland and in the hinterland of the Algoa Basin on the southern Cape coast has any degree of planation occurred in the Post-African II cycle (Fig. 1.4). Because of this almost ubiquitous dissection, soils on the Post-African II surface are strongly influenced by underlying rock type and are mostly skeletal; in the limited areas of Post-African II planation, soils are similar to those of the Post-African I landscape.

The present east–west climatic gradient (Chap. 2, this volume), probably first established across the subcontinent during the latter part of the Miocene, seems to have persisted through much of the Pliocene. The existence of arid conditions in the Kalahari during this interval is demonstrated by the presence, within the Kalahari Group sediments, of intercalated dolomitic calcretes and silcretes of a type developed characteristically around alkaline lakes or pans (the 'inland' silcretes of Fig. 1.5) (Watts 1980; Summerfield 1983a). That hyper-arid conditions did not, however, prevail continuously throughout this period is clear from the establishment of the well-integrated drainage network of the Molopo River system across the sedimentary filling of the basin during this time; indeed, such was the erosive power of the Molopo that it succeeded in breaching the southern rim of the basin. Although direct evidence is lacking, there are grounds for inferring that this climatic amelioration may have dated to the periods of Pliocene warming that occurred between 5.1 and 4.2 Myr and 3.5 and 2.8 Myr. It is likely that these events produced the Pliocene high sea levels which are documented by marine terraces at elevations of 60–110 m around the southern African coast (Partridge 1990). These elevations are due, in part, to tectonic effects that account also for the greater height of the terraces along the southeastern coast. Inferences drawn from the species composition of the antelope fauna preserved in the hominid-bearing cave deposits of the Transvaal indicate the occurrence, during the latter warm interval, of wetter conditions with a greater bush cover than at present (Vrba 1985). If the earlier Pliocene deposits at Langebaanweg on the western Cape coast relate to the first of these episodes of elevated sea level, as seems likely, then a rather different set of climatic circumstances appears to have prevailed in the extreme southwestern part of the subcontinent: peats within the Varswater Formation have yielded abundant shrubland elements characteristic of the present fynbos vegetation of the area, together with some grassland and forest taxa (Coetzee & Rogers 1982). These findings, together with the evidence of the accompanying vertebrate fauna (Hendey 1983), indicate the existence of a strongly seasonal rainfall regime that contrasts with the more mesic conditions characteristic of the west coast of southern Africa (including the Namib) during the early to mid-Miocene. It seems significant that increased rainfall seasonality and desiccation apparently recurred in the southwestern areas of the subcontinent during subsequent warming episodes (Partridge 1990).

Figure 1.8 **The Valley of a Thousand Hills in the interior of KwaZulu-Natal. Dissection in the Post-African II cycle has caused incision of up to 500 m below laterite-capped remnants of the African cycle** *(Photo: E.J. Moll).*

The global warming that characterized the late Pliocene was brought to an end by a major decline in temperature around 2.8 Myr, which saw the re-establishment of the East Antarctic Ice Sheet (Denton 1985; Harwood 1985), and the onset of ice rafting in the North Atlantic (Shackleton *et al.* 1984). It was at this time that middle and low latitude climates worldwide became dependent, for the first time, on the growth of ice-sheets in the northern hemisphere, and on subsequent fluctuations in their size in response to orbitally induced variations in receipts of incoming solar radiation (de Menocal & Bloemendal 1995). Concomitant with these events were the beginning of the loess sequence in China and a major episode of climatic deterioration (accompanied by the spread of grasslands and extensive faunal turnover) recorded in terrestrial deposits of both southern and eastern Africa. The Cape winter rainfall regime and the fynbos biome, as we know it today, probably became fully established at this time. These changes heralded the onset of repeated glacial–interglacial couplets that characterized the Quaternary, particularly after about

1.0 Myr. In southern Africa the climatic response to late Pliocene cooling was overwhelmingly in the direction of greater aridity. Although earlier episodes of dune building may have occurred in the Kalahari, there is little doubt that the maximum extension of the red Kalahari sand occurred after this event, which resulted in the spread of dunes and sand blankets in southern Africa far beyond the limits of the Kalahari Basin into Zaire, Angola, Zimbabwe, the western Transvaal and Free State, the Northern Cape and parts of the Karoo extending as far south as the Cape Fold Mountains (Partridge & Maud 1987). The encroachment of dunes during this period caused the lower course of the Molopo River system to become defunct. The oldest terraces of the Vaal River are thought to be probably of late Pliocene age on the basis of limited faunal evidence; their extensive course alluviation is essentially a semi-arid phenomenon (Partridge & Brink 1967). Further evidence is forthcoming from the hominid-bearing cave deposits of Makapansgat and Sterkfontein, in which a change in style of sedimentation, from predominantly fine materials, deposited, at least in part, under subaqueous conditions, to more gravelly debris and sheetflood accumulations, occurred around 2.8 to 2.5 Myr (Partridge 1985). Of added significance is the fact that approximately concurrent faunal changes indicate a shift to more open vegetation (Vrba 1985); Tobias (1985, 1986) has pointed to a major nodal episode in hominid speciation at about this time, and there is a wealth of evidence that these evolutionary events were a response to climatic forcing.

Well-calibrated palaeoenvironmental data for the early and middle Pleistocene of southern Africa are sparse. Lancaster (1984) has adduced evidence for progressive desiccation of the Namib region during the Quaternary, although low amplitude shifts towards more mesic conditions are evident locally. In an earlier study (Lancaster 1981) he noted that extensive systems of parallel linear dunes of the Kalahari, now stabilized by vegetation, form an arc that corresponds approximately to the pattern of winds around the southern African anticyclone. Today the northern part of this area has an annual rainfall of 450–650 mm; in contrast, active dunes are restricted to areas receiving 150–200 mm, and it is clear that the 200-mm isohyet lay some 1200 km northeast of its present position when the dunes were formed in northern Botswana; a lesser shift of some 250–300 km can be deduced for the southern Kalahari. The inferred extent of these shifts has been revised (Lancaster 1988), but there is little doubt that significant latitudinal migration of major circulation patterns occurred during glacial–interglacial cycles. It is, indeed, likely that, since the late Pliocene, the Kalahari dune field has been repeatedly re-activated during the dry periods. Abundant evidence, both global and local, points to the occurrence of the greatest aridity during the cool glacial maxima, and there is little doubt that the main periods of dune formation within the Kalahari Sand Sea were during these intervals, in contrast to the apparent persistence of mobile sands in the Namib Desert of the west coast since the later part of the Miocene (Partridge 1990). Under present climatic conditions and a savanna vegetation the Kalahari sands have undergone noticeable weathering and pedogenesis, and catenal sequences relating the moisture regimes within and between dunes are present.

1.7   Soils

1.7.1   High ground above the African surface

Although African planation was more pervasive than that of any subsequent erosion cycle, residual upland areas remained in several important areas. The Cape Fold Mountains and Lesotho Highlands are the largest. On them soils are sporadically developed, immature and litholic. Upon the chiefly siliceous rocks of the Cape ranges, with their predominantly winter rainfall regime, inceptisols and spodosols are widespread (Partridge et al. 1996). The character of the important fynbos flora is believed to owe much to the influence of these poor substrates (Deacon, Hendey & Lambrechts 1983; Chap. 6, this volume). Soils of the Lesotho Highlands differ in conformity, with the more basic nature of the host rock (basalt) and the influence of freeze-and-thaw processes, and are dominated by mollisols. The immature and discontinuous soil cover of these upland areas is in marked contrast to the advanced weathering and pedogenesis that characterized the African surface.

1.7.2   African surface

Pristine remnants of the African surface are poorly preserved within the landscape of southern Africa, owing to the ravages of subsequent cycles of erosion acting upon a relatively high continental interior. Where present, these residuals are capped by duricrusts. The ferricrete cappings of those that have survived in the more humid eastern part of the subcontinent are relatively thin in comparison with true laterites of the tropics and have frequently undergone weathering to produce a ferruginous gravel. Yet, despite the destruction of their protective cappings, these relics of the Cretaceous surface have seldom been consumed by gullying or large-scale stripping, but have rather undergone gentle lowering through the slow removal of the kaolinitic weathering mantle. Many such lowered remnants of the African surface are preserved throughout southern Africa; larger examples include the Highveld of the Eastern Transvaal and North-West Province, the Ghaap Plateau of the Northern Cape, the Pietersburg Plain in the Northern Transvaal, the White River–Tzaneen Bench below the Great Escarpment and the Swartland of the Western Cape. In most respects they conform to the concept of the etchplain, formed by small-scale lowering of the landsurface through the removal of the weathered rock or saprolite, as proposed by Wayland (1934) and developed by Twidale (1988). The present characteristics of soils in all these areas owe much to the survival of remains of the original deep weathering mantles, whose properties are frequently incompatible with the more arid climates of today. It is only in the area in Fig. 1.4, where planation in the African cycle was generally incomplete, that ancient saprolites have not been inherited to any marked extent.

The limited extent to which the African surface has been lowered, and the preservation of remnants under relatively thin duricrusts over a period of more than 60 million years, during which considerable volumes have been removed by erosion in marginal areas, bears graphic testimony to the dominance of backwearing over downwearing in the geomorphic evolution of southern Africa.

The presence of ancient saprolitic parent material has been important in imparting an 'old' character to soil profiles within upland plateau zones of the landscape, but local factors have also been significant in the evolution of soil mantles in such etchplain environments. The upper part of most profiles has been extensively modified by wind action, sheetwash and termite activity during many climatic cycles to which the landscape has been subject. Lithological and regional climatic influences have also been of importance, as in the development of manganese-rich saprolites over dolomitic rocks of the Transvaal and Griqualand West Supergroups. Elsewhere, aridification since the later part of the Miocene has led to the development of calcrete (caliche); on the Ghaap Plateau, dolomitic rocks are overlain by well-developed calcrete duricrusts, and mature calcretes are also present on areas of the Post-African I landsurface in Bushmanland and around the margins of the Kalahari Basin. Less well-developed calcareous horizons characterize profiles on the lowered African surface of the Swartland in the Western Cape (Fig. 1.4). In contrast, in the eastern areas of the subcontinent, particularly those that continued to enjoy an annual moisture surplus, ferricrete dominates most profiles. It is particularly in these areas that important catenal sequences have developed on the deeply weathered saprolites of the lowered African surface (Partridge *et al.* 1996).

1.7.3 Post-African planation and dissection

By far the most extensive areas occupied by the Post-African I surface lie within those parts of the subcontinent that are at present arid to semi-arid (Fig. 1.4). As has been discussed previously, there is considerable evidence to suggest that, apart from mesic intervals in the mid-Miocene and during one or more warm periods in the Pliocene, fairly dry climatic conditions have, in fact, characterized these areas during most of the time since the beginning of the Post-African I cycle in the early Miocene. Siesser (1978, 1980) has drawn attention to evidence, from marine sediments on the Walvis Ridge off Namibia, that upwelling intensified significantly in the late Miocene sometime after about 10 Myr, giving rise to the cold Benguela Current system with which the hyperarid climatic regime of the Namib Desert is intimately associated. This episode of aridification conforms with the global pattern of step-wise cooling, accompanied by desiccation, which characterized the Tertiary (Miller & Fairbanks 1985). It is therefore scarcely surprising that,

apart from the formation of calcrete duricrusts and rubefication of aeolian cover sands, there is little evidence of significant pedogenesis in the soils of the Post-African surfaces. In the more humid areas of the eastern Bushveld Basin and the lowveld areas of the Transvaal, Swaziland and northern KwaZulu-Natal, weathering and soil formation on the Post-African surfaces are predictably more advanced; red colours are common in well-drained situations, but dark vertisols are also well represented in areas of low relief, particularly where basic igneous rocks are present (e.g. those of the Bushveld Igneous Complex). Catenal sequences are generally not as spectacularly developed as on lowered remnants of the African surface.

Where dissection has been profound, as in the southeastern coastal hinterland, soils are mostly immature and strongly influenced by bedrock lithology. In the coastal zone, however, intense weathering during the last one or two interglacial cycles has produced red, clay-rich profiles, particularly on the older coastal dune sand and river terraces (Maud 1968). The absence of similar profiles elsewhere in this area implies the occurrence of a major phase of erosion as the climate deteriorated, following the onset of global cooling in the last glacial cycle.

1.7.4 The Kalahari Basin

No account of major soil distributions within southern Africa would be complete without reference to the single major area of the subcontinent in which deposition has occurred, probably since the late Cretaceous, in contrast to the all-pervasive denudational processes that sculpted the remainder of the subcontinent. The Kalahari Basin at present lies within the semi-arid zone of the western interior of the subcontinent and the sands that form the uppermost unit of its sedimentary filling are largely stabilized by vegetation; that this was not so during most xeric episodes in the past is evident from the preservation of extensive systems of linear dunes within this sand mantle, which extends continuously into Angola, Zimbabwe and Zaire and which constitutes the single largest body of aeolian sand in the world (Fig. 1.9). Of significance is the fact that discontinuous remnants of an even more extensive sand cover are preserved beyond its present limits.

As has been indicated earlier, the initial distribution of this sand blanket was probably the result of global cooling and aridification about 2.8 Myr. Since then it has been subject to recurrent episodes of aeolian redistribution, interrupted by interludes of weathering under a stabilizing vegetation cover, as local climates have responded to the repeated cycles through which ice-sheets and glaciers waxed and waned in high latitudes

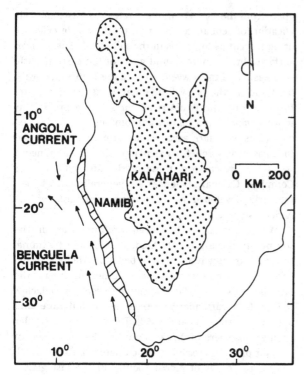

Figure 1.9 **The Namib and Kalahari areas of southern Africa. Active dunes occur today in the central part of the Namib, and the Kalahari is mantled by windblown sand, almost all of which is stabilized by vegetation.**

over previous dune topography bears testimony to the widespread dryness which must have characterized large areas in the western hinterland of Africa, extending to the equator and beyond, during at least part of the late Pliocene and Pleistocene.

1.8 Conclusion

There can be no doubt that the interplay between patterns of landscape development and changing climates since the Cretaceous has fundamentally influenced present soil and vegetation distributions in southern Africa (see also Chap. 2, this volume). In the broadest terms the trend over the last 65 million years has been towards increasing fragmentation of the landscape, as the great plains of the African surface were progressively dissected in the successive cycles of erosion that followed episodes of tectonic uplift within the subcontinent. The most recent of these uplifts was of sufficient magnitude to have a major influence on patterns of temperature and precipitation in the eastern hinterland. Over the same period climates changed dramatically in a series of step-like shifts towards cooler and drier conditions; these separated a few, relatively short-lived intervals during which there was a return to greater humidity. As climates became drier, so the present east–west climatic gradient became firmly established across southern Africa. These events culminated in major aridification about 2.8 Myr, concurrent with the first development of large, high-latitude ice-sheets in the northern hemisphere. This ushered in the Pleistocene with its recurrent glacial–interglacial cycles, adaptions to which have left their mark on the composition and distribution of vegetation communities across the globe.

during the Pleistocene. That some intervals were decidedly moister than at present is indicated by hydromorphic influences in interdune areas, which have led to the development of local catenal sequences. Today all but some dune crests in the most arid areas are stabilized by vegetation (an arid *Acacia* savanna in the core area of the dry Kalahari; Chap. 11, this volume), but elsewhere, as in Angola and Zaire, the existence of a forest cover

1.9 References

Beckman, G.G. (1983). Development of old landscapes and soils. In *Soils: an Australian Viewpoint*, pp. 51–72. Canberra: CSIRO/Academic Press.

Brown, R.W., Rust, D.J., Summerfield, M.A., Gleadow, A.J.W. & de Wit, M.C.J. (1990). An early Cretaceous phase of accelerated erosion on the south-western margin of Africa: evidence from apatite fission track analysis and the offshore sedimentary record. *Nuclear tracks and radiation measurements. International Journal of Radiation and Applied Instrumentation*, Part D, **17**, 339–50.

Cahen, L., Snelling, N.J., Delhal, J. & Vail, J.R. (1984). *Geochronology and Evolution of Africa*. Oxford: Clarendon Press.

Coetzee, J.A. & Rogers, J. (1982). Palynological and lithological evidence for the Miocene palaeoenvironment in the Saldanha region (South Africa). *Paleogeography, Palaeoclimatology, Palaeoecology*, **39**, 71–85.

Davis, G.L. (1977). The ages and uranium contents of zircons from kimberlites and associated rocks. *Extended Abstracts, Second International Kimberlite Conference*, Santa Fe, New Mexico (unpaginated).

Deacon, H.J., Hendey, Q.B. & Lambrechts, J.J.N. (eds.) (1983). *Fynbos Palaeoecology: A Preliminary Synthesis. South African National Research Programmes Report 75.* Pretoria: CSIR.

de Menocal, P.B. & Bloemendal, J. (1995). Plio-Pleistocene subtropical African climate variability and the paleoenvironment of hominid evolution: a combined data-model approach. In *Palaeoclimate and Evolution with Emphasis on Human Origins*, eds. E.S. Vrba, G.H. Denton, L.H. Burckle & T.C. Partridge, pp. 262–88. New Haven: Yale University Press.

Denton, G.H. (1985). Did the Antarctic ice sheet influence Late Cainozoic climate and evolution in the southern hemisphere? *South African Journal of Science*, **81**, 224–9.

de Wit, M.C.J. (1993). *Cainozoic Evolution of Drainage Systems in the North-Western Cape*. PhD thesis. Cape Town: University of Cape Town.

Dingle, R.V. and Hendey, Q.B. (1984). Late Mesozoic and Tertiary sediment input into the eastern Cape basin (S.E. Atlantic) and palaeodrainage systems in south western Africa. *Marine Geology*, **56**, 13–26.

Dingle, R.V., Siesser, W.G. & Newton, A.R. (1983). *Mesozoic and Tertiary Geology of Southern Africa*. Rotterdam: A.A. Balkema.

du Toit, A.L. (1954). *The Geology of South Africa*, 3rd edn. Edinburgh: Oliver and Boyd.

Ellis, F. & Schloms, B.H.A. (1984). Distribution and properties of dorbanks (duripans) in South and South-Western South Africa. Poster at *12th Congress of Soil Science Society of South Africa*, Bloemfontein.

Gassaway, J.S. (1988). Silcrete-Palaeocene marker beds in the Western Interior. *Geological Society of America Abstracts*, **21**, 379.

Hartnady, C.J.H. (1985). Uplift, faulting, seismicity, thermal spring and possible incipient volcanic activity in the Lesotho-Natal region, SE Africa: the Quathlamba hotspot hypothesis. *Tectonics*, **4**, 371–7.

Hartnady, C.J.H. (1990). Seismicity and plate boundary evolution in southeastern Africa. *South African Journal of Geology*, **93**, 473–84.

Harwood, D.M. (1985). Late Neogene climatic fluctuations in the southern high-latitudes: implications of a warm Pliocene and deglaciated Antarctic continent. *South African Journal of Science*, **82**, 239–41.

Haughton, S.H. (1915). On some dinosaur remains from Bushmanland. *Transactions of Royal Society of South Africa*, **5**, 259–64.

Hawthorne, J.B. (1975). Model of kimberlite pipe. In *Physics and Chemistry of the Earth*, vol. 9, ed. L.H. Ahrens, J.B. Dawson, A.R. Duncan & A.J. Erlank, pp. 1–15. Oxford: Pergamon.

Hendey, Q.B. (1978). Preliminary report on the Miocene vertebrates from Arrisdrift, South West Africa. *Annals of the South African Museum*, **76**, 1–41.

Hendey, Q.B. (1983). Palaeoenvironmental implications of the late Tertiary vertebrate faunas of the Fynbos region. In *Fynbos Palaeoecology: a Preliminary Synthesis*, ed. H.J. Deacon, Q.B. Hendey & J.N.N. Lambrechts, pp. 100–15. *South African National Programmes Report* 75. Pretoria: CSIR.

Lancaster, I.N. (1981). Palaeoenvironmental implications of fixed dune systems in southern Africa. *Palaeogeograhy, Palaeoclimatology, Palaeoecology*, **33**, 327–46.

Lancaster, I.N. (1984). Aridity in southern Africa: age, origins and expression in landforms and sediments. In *Late Cainozoic Palaeoclimates of the Southern Hemisphere*, ed. J.C. Vogel, pp. 433–44. Rotterdam: A.A. Balkema.

Lancaster, I.N. (1988). Development of linear dunes in the south western Kalahari, Southern Africa. *Journal of Arid Environments*, **14**, 233–44.

Martin, A.K. (1987). Comparison of sedimentation rates in the Natal Valley, south-western Indian Ocean, with modern sediment yields in east coast rivers of Southern Africa. *South African Journal of Science*, **83**, 716–24.

Matthews, P.E. (1978). An examination of Sternberg's Law using river-pebble measurements from the Natal Drakensberg. *Petros*, **8**, 26–33.

Maud, R.R. (1968). Quaternary geomorphology and soil formation in coastal Natal. *Zeitschrift für Geomorphologie, N.F., Suppl-Bd*, **7**, 155–99.

Miller, K.G. & Fairbanks, R.G. (1985). Cainozoic ^{18}O record of climate and sea level. *South African Journal of Science*, **81**, 248–9.

Moore, A.E. (1979). *The Geochemistry of the Olivine Melilitites and Related Rocks of Namaqualand/Bushmanland, South Africa*, PhD thesis. Cape Town: University of Cape Town.

Partridge, T.C. (1985). The palaeoclimatic significance of Cainozoic terrestrial stratigraphic and tectonic evidence from southern Africa: a review. *South African Journal of Science*, **81**, 245–7.

Partridge, T.C. (1990). Cainozoic environmental changes in southern Africa. *South African Journal of Science*, **86**, 315–17.

Partridge, T.C. (1993). The evidence for Cainozoic aridification in southern Africa. *Quaternary International*, **17**, 105–10.

Partridge, T.C., Bond, G.C., Hartnady, C.J.H., de Menocal, P.B. & Ruddiman, W.F. (1995a) Climatic effects of late Neogene tectonism and volcanism. In *Palaeoclimate and Evolution with Emphasis on Human Origins*, ed. E.S. Vrba, G.H. Denton, L.H. Burckle & T.C. Partridge, pp. 8–23. New Haven: Yale University Press.

Partridge, T.C. & Brink, A.B.A. (1967). Gravels and terraces of the lower Vaal River basin. *South African Geographical Journal*, **49**, 21–38.

Partridge, T.C., de Villiers, J.M., Fitzpatrick, R.W. & Maud, R.R. (1996) Soil-landscape–climate relations in southern Africa. *Catena* (in press).

Partridge, T.C. & Maud, R.R. (1987). Geomorphic evolution of southern Africa since the Mesozoic. *South African Journal of Geology*, **90**, 179–208.

Partridge, T.C. & Maud, R.R. (1989). The end-Cretaceous event: new evidence from the southern hemisphere. *South African Journal of Science*, **85**, 428–30.

Partridge, T.C., Wood, B.A. & de Menocal, P.B. (1995b). Influence of global climatic change and regional uplift on large mammal evolution in East and Southern Africa. In *Palaeoclimate and Evolution with Emphasis on Human Origins*, ed. E.S. Vrba, G.H. Denton, L.H. Burckle & T.C. Partridge, pp. 331–655. New Haven: Yale University Press.

Rogers, A.W. (1915). The occurrence of dinosaurs in Bushmanland. *Transactions of the Royal Society of South Africa*, **5**, 265–72.

Rust, D.J. & Summerfield, M.A. (1990). Isopach and borehole data as indicators of rifted margin evolution in southwestern Africa. *Marine and Petroleum Geology*, **7**, 277–87.

Schloms, B.H.A & Ellis, F. (1984). Distribution of silcretes and properties of some soils associated with silcretes in Cape Province, South Africa. *Proceedings of 12th Congress of Soil Science Society of South Africa*, Bloemfontein.

Shackelton, N.J., Backman, J., Zimmerman, H., Kent, D.V., Hall, M.A., Roberts, D.G., Schnitker, D., Baldaur, J.G., Desprairies, A., Hormrighausen, R., Huddeleston, P., Keene, J.B., Kaltenbuck, A.J., Krumsiek, K.A.O., Morton, A.C., Murray, J.W. & Westberg-Smith, J. (1984). Oxygen isotope calibration of the onset of ice rafting and history of glaciation in the North Atlantic region. *Nature*, **307**, 620–3.

Siesser, W.G. (1978). Aridification of the Namib Desert, evidence from oceanic cores. In *Atlantic Glacial History and World Palaeoenvironments*, ed. E.M. van Zinderen Bakker, pp. 105–13. Rotterdam: A.A. Balkema.

Siesser, W.G. (1980). Late Miocene origin of the Benguela upwelling system off northern Namibia. *Science*, **208**, 283–5.

Smith, A.G. & Briden, J.C. (1982). *Mesozoic and Cenozoic Palaeocontinental Maps.* Cambridge: Cambridge University Press.

Summerfield, M.A. (1983a). Silcrete as a palaeoclimatic indicator: evidence from southern Africa. *Palaeogeography, Palaeoclimatology, Palaeoecology*, **41**, 65–79.

Summerfield, M.A. (1983b). Silcrete. In *Chemical Sediments and Geomorphology*, ed. A.S. Goudie & K. Pye, pp. 59–91. London: Academic Press.

Tobias, P.V. (1985). Ten climacteric events in hominid evolution. *South African Journal of Science*, **81**, 271–2.

Tobias, P.V. (1986). Delineation and dating of some major phases in hominidization and hominization since the Middle Miocene. *South African Journal of Science*, **82**, 92–4.

Twidale, C.R. (1988). The missing link: planation surfaces and etch forms in southern Africa. In *Geomorphological Studies in Southern Africa*, ed. G.F. Dardis & B.P. Moon, pp. 31–46. Rotterdam: A.A. Balkema.

Twidale, C.R. & Hutton, J.T. (1986). Silcrete as a climatic indicator: discussion. *Palaeogeography, Palaeoclimatology, Palaeoecology*, **52**, 351–60.

Vrba, E.S. (1985). Early hominids in southern Africa: updated observations on chronological and ecological background. In *Hominid Evolution: Past, Present and Future*, ed. P.V. Tobias, pp. 195–200. New York: Alan R. Liss.

Watts, N.L. (1980). Quaternary pedogenic calcretes from the Kalahari (southern Africa), mineralogy, genesis and diagenesis. *Sedimentology*, **27**, 661–86.

Wayland, E.J. (1934). Peneplains and some erosional landforms. *Geological Survey of Uganda, Annual Report and Bulletin*, **1**, 77–9.

Weinert, H.H. (1974). A climate index of weathering and its application in road construction. *Geotechnique*, **24**, 457–88.

Weinert, H.H. (1980). *The Natural Road Construction Materials of Southern Africa.* Cape Town: Academia.

Climate

2

R.E. Schulze

2.1 Introduction: the climate–vegetation link

Of the three great natural patterns that dominate the earth's environment, *viz.* patterns of climate, vegetation and soil, climate is inevitably perceived as the principal dynamic component and the obvious independent variable shaping the other two, especially vegetation, on both micro and sub-continental scales (Akin 1991). The term climate is a composite concept and may be defined as the 'long range pattern of weather' (Kendrew 1949), taking account not only of prevailing weather conditions, but also of dynamic and intricate variations occurring diurnally, daily, monthly, seasonally and annually and in addition allowing for the probability for climate to vary from the norm. Climate is thus more than 'average weather'.

Woodward (1986) suggests a close link between the time for a plant's life cycle to be completed and the temporal variation of the controlling climatic processes. Examples of differential life-cycle sensitivity to climate include rate of germination, rate in amount of die-off/survival, biomass production related to temperature/precipitation distributions, intensity of flowering and reproductive capacity. For certain species, in one phase of the life cycle, climate may be critical for survival. For another species, no particular phase is critical, but it is rather that climatic constraints cause progressive reductions in probabilities of survival or production at a number of stages in the life cycle. Ecologists should take care in attempting to correlate the distribution of plant species with climatic thresholds. The identification of these climatic thresholds for plant distribution are important for predicting the impacts of climate change.

The climatic factors of greatest importance in vegetation development are light, temperature and moisture, all of which vary sub-continentally as well as on a meso- and micro-scale. In much of the earlier literature individual climatic parameters have been treated separately in their relation to vegetation distribution, with little attention being paid to the way in which they might interact. Today, however, there is a greater focus on climatic interactions and the fact that climatic parameters operate in combination to produce relatively homogeneous environments in which certain plant communities can attain importance.

After a brief introduction on techniques of climate mapping and approaches to deriving climate-related variables in southern Africa, major sections follow on solar radiation patterns including the influence of topography on solar radiation, temperature and temperature parameters including frost, precipitation including long-term patterns, variability and seasonality, and reference potential evaporation. Thereafter, climatic interactions are discussed, assessing first moisture availability to plants and patterns of soil-moisture stress over southern Africa, followed by a section on distributions and inter-annual variability of primary production.

Southern Africa in the context of this chapter comprises South Africa, Lesotho and Swaziland, but excludes Namibia and Botswana, for which the relevant climatic and other detailed information was not readily available to the author.

2.2 Information sources, approaches to climate analysis and mapping techniques

To map climate parameters with sufficient detail over southern Africa, easily determinable and readily available causative factors influencing climate in a defined region were sought, e.g. altitude, latitude, continentality, aspect, daylength/extraterrestrial radiation and lapse rates. From these, regional and temporal (e.g. monthly) multiple regression equations for various climate parameters were developed for application with gridded point information for mapping. For this purpose, a grid of altitudes one minute by one minute of a degree latitude and longitude (i.e. $1' \times 1'$, spaced at approximately 1.6×1.6 km) consisting of over 437 000 points was developed for southern Africa (Dent, Lynch & Schulze 1989 and subsequently refined).

A second method of climatic regionalization was the delineation of southern Africa into 2136 relatively homogeneous zones of climatic response (Dent, Schulze & Angus 1989 and subsequently refined). To each of these zones a representative climate station with a daily precipitation record for 30–100 years was assigned, as were values for other climate variables, soils and natural vegetation (Schulze & Lynch 1992). This information base, on geographic information systems (GIS), was coupled to the ACRU agrohydrological modelling system – a multi-purpose and multi-level daily soil-water budgeting model (Schulze 1995) which predicts, *inter alia*, soil-water responses and primary production as well as standard hydrological output (Fig. 2.1).

2.3 Light (solar radiation)

The energy resources of all ecosystems are ultimately dependent on the quantity of incoming solar radiation intercepted and its seasonal variation. In southern Africa solar radiation is seldom a limiting factor to vegetation

Figure 2.1 **The structure of the ACRU agrohydrological modelling system** *(Schulze 1995).*

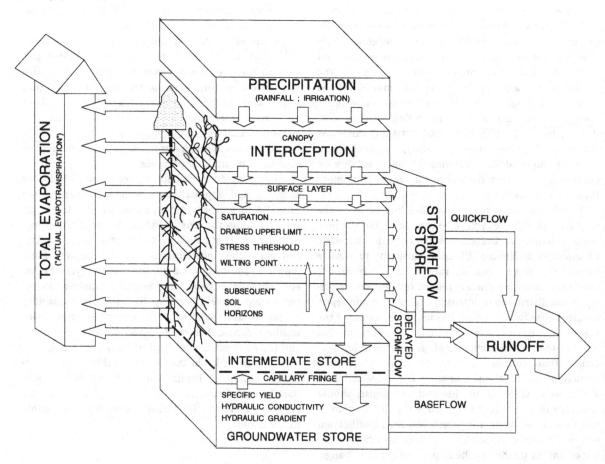

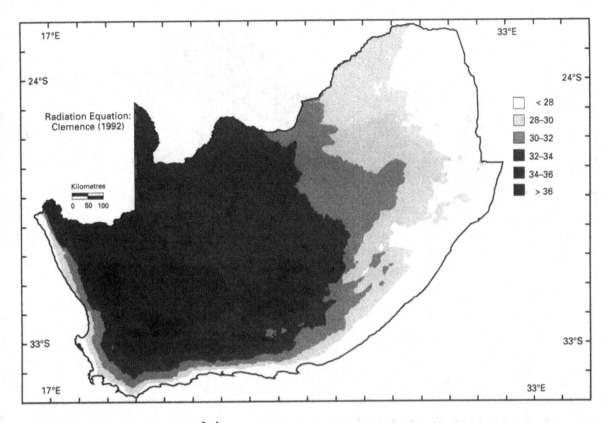

Figure 2.2 **Mean daily solar radiation (MJ m⁻² d⁻¹) over southern Africa for January.**

development on a macro-scale. However, on the meso- and micro-scale it frequently is. Ecologically, it is therefore meaningful to view seasonal patterns of solar radiation and also the effects which varying topography can have on it.

2.3.1 The seasonal distribution of solar radiation over southern Africa

Solar radiation patterns over southern Africa were determined by the Clemence (1992) equation, which uses maximum and minimum temperatures, T_{max} and T_{min}, together with extraterrestrial radiation, R_a (obtained from latitude and date). With the use of the $1' \times 1'$ gridded T_{max} and T_{min} values generated for southern Africa (Schulze & Maharaj 1994; Figs. 3.6 and 3.7) and latitude/date-derived R_a estimates (Chidley & Pike 1970), patterns of mean daily solar radiation were mapped for January and July, representing a mid-summer and a mid-winter month, respectively.

Patterns of solar radiation in summer (Fig. 2.2) vary from <28 MJ m⁻² d⁻¹ along the east and south coast zones to highs exceeding 34 MJ m⁻² d⁻¹ in the arid western Cape interior. Winter distributions (Fig. 2.3) follow latitudinal zonation closely, with values over 18 MJ m⁻² d⁻¹ in the

north decreasing to less than 14 MJ m⁻² d⁻¹ in the winter-rainfall dominated southern Cape.

2.3.2 The influence of topography on solar radiation

Permanent differences in amounts of radiant energy intercepted on contrasting exposures may cause marked variations in distribution of plant communities (as well as in geomorphology, soil moisture, snowmelt rates and streamflow). While many plant ecological studies worldwide have drawn attention to such existing relationships, most studies remain relatively qualitative. Granger & Schulze (1977) have, however, produced maps of incoming radiation patterns in a protected research catchment at Cathedral Peak in South Africa to explain temporal changes in vegetation, physiognomy and species composition. They furthermore ascribe the presence of mature *Podocarpus latifolius* forest on a cool slope versus *Protea* savanna and subtropical grasses on the corresponding warm slope to differences in the energy budget, cool slopes receiving winter incoming radiation loadings of 9 MJ m⁻² d⁻¹ versus 22 MJ m⁻² d⁻¹ on the warm slopes.

Daily incoming radiant flux densities on sloping ter-

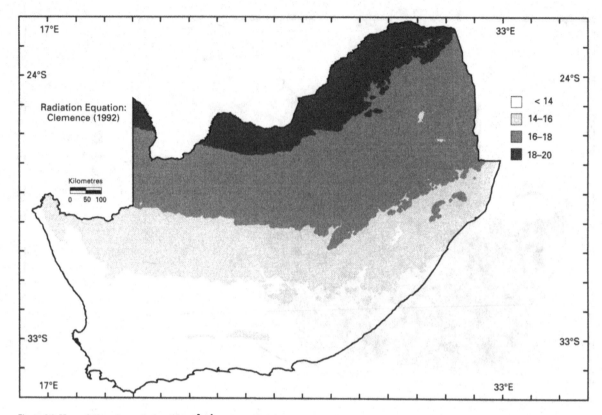

Figure 2.3 **Mean daily solar radiation (MJ m⁻² d⁻¹) over southern Africa for July.**

rain as a function of slope, aspect and season have been presented for cloudless days in southern Africa for the latitudinal range 20° S–35° S by Schulze (1975). Results are based on equations developed by Archer (1964) and Schulze (1975) using radiation data from southern African stations. Seasonal variations of radiation income on slopes are illustrated in Fig. 2.4 using December 22, March/September 22 and June 21 as representing mid-summer, the equinoxes and mid-winter clear sky conditions, respectively.

In mid-summer on cloudless days radiation flux densities in southern Africa generally exceed 25 MJ m⁻² d⁻¹, with higher values occurring on flatter slopes. Solar radiation loadings increase with latitude for north, northeast/northwest and for east/west aspects (Fig. 2.4). There is little variation with latitude on south-east/southwest facing slopes and only on southern aspects is a decrease of radiation evident with increasing latitude, especially on the steeper slopes. On cloudless days at the equinoxes, slopes intercept about 25 MJ m⁻² d⁻¹ on northern aspects, decreasing to 17–21 MJ m⁻² d⁻¹ on southern aspects. The northern and northeast/northwest facing slopes exhibit relatively little variation in interception of radiant energy with different

gradients and with latitude. On the northern aspect the steeper slopes receive most energy, but the more usual decrease of radiation as slope increases prevails on aspects deviating 90° or more from north. Around the equinoxes the influence of slope is most marked on the steeper southern aspects and this influence is further accentuated as latitude increases (Fig. 2.4). The mid-winter influence of latitude on incoming radiation fluxes on slopes is most noticeable. The lowest differences due to gradient throughout the latitudinal range graphed are found on east/west aspects. The most extreme effects of latitude and aspect again occur on the steep slopes, which intercept considerably more radiation than flat slopes on northern aspects and considerably less on southern aspects.

A more recent verified model for estimating radiation budgets on sloping terrain, RADSLOPE (Schulze & Lambson 1986), incorporates many features frequently omitted by other authors, including atmospheric transmissivity, attenuation of shortwave radiation by different cloud types, shading effects of surrounding topography, the contribution of reflected radiation from surrounding topography and the incorporation of equations for the estimation of surface temperature and potential evapor-

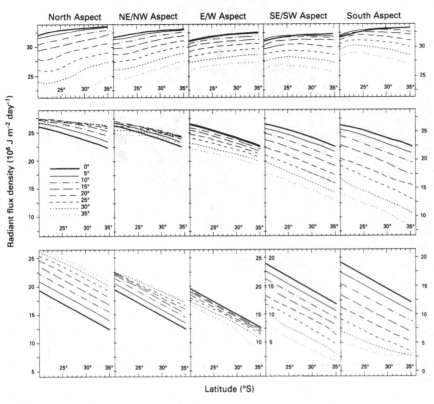

Figure 2.4 **Seasonal influences of slope and aspect on clear-sky solar radiation loadings in southern Africa. Top, midsummer; middle, equinoxes; bottom, midwinter. After Schulze (1975).**

ation differences on sloping terrain. This model is available from the author.

2.4 Temperature

Temperature is a basic climatological variable frequently used as an index to the energy status of the environment (Schulze 1982). Consequently, temperature parameters are vital controls by which the distribution of vegetation is frequently limited.

2.4.1 Distributions of means, maxima, minima and ranges of temperature over southern Africa

The isotherms of mean annual temperatures (MAT), depicted in Fig. 2.5, display several major characteristics, *viz.* high values exceeding 20 °C along the northeastern coastal belt and the eastern border of South Africa and to a lesser extent along the northern border. In addition, temperature irregularities are induced by topographic variation on the subcontinent: for instance the lower MAT along the escarpment along the perimeter of southern Africa and particularly the low MAT over most of Lesotho.

Fig. 2.6 shows that in January the highest means of daily maximum temperature occur over the Kalahari, with the lowest mid-summer maxima along the west coast (caused by the cold Benguela Current), the south coast (latitudinal effects) and along the Drakensberg escarpment (altitude effects).

Monthly means of daily minimum temperatures for the coldest month of the year, July, are illustrated in Fig. 2.7. Cold night-time temperatures averaging 0 °C and below, characterize the interior plateau regions (influence of altitude and continentality), while, under the influence of the warm Mozambique Current, the July minima of the east coast of KwaZulu-Natal still average above 8 °C.

Diurnal temperature ranges (T_r) between daytime maxima and night-time minima in mid-summer (January) display an essentially longitudinal trend, with T_r increasing from east to west generally in phase with rainfall distribution and thus reflecting high humidities and high degrees of cloudiness, both of which suppress the temperature range (Fig. 2.8). In contrast, July T_r displays a far greater latitudinal trend (Fig. 2.9), with highest ranges between daytime maxima and night-time minima generally decreasing from north (>18 °C) to south (generally <13 °C), but with the entire coastline

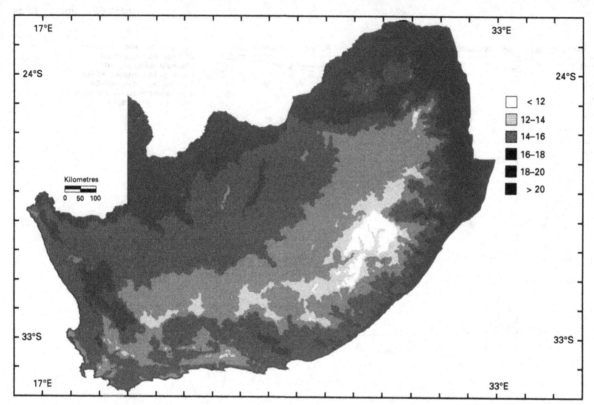

Figure 2.5 **Mean annual temperature (°C) over southern Africa.**

Figure 2.6 **Means of daily maximum temperature (°C) over southern Africa in January.**

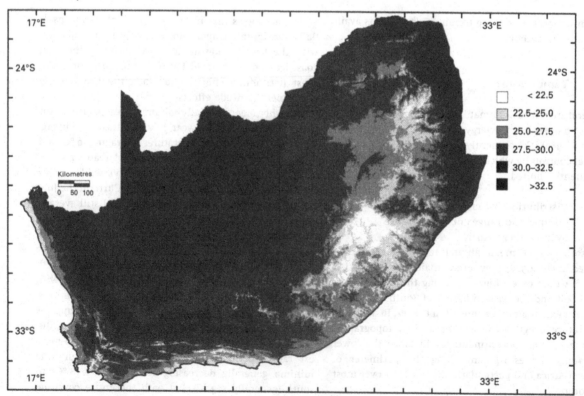

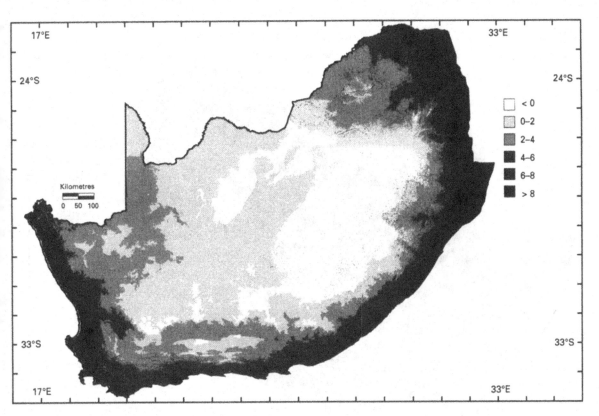

Figure 2.7 **Means of daily minimum temperature (°C) over southern Africa in July.**

Figure 2.8 **Diurnal temperature ranges (°C) over southern Africa in January.**

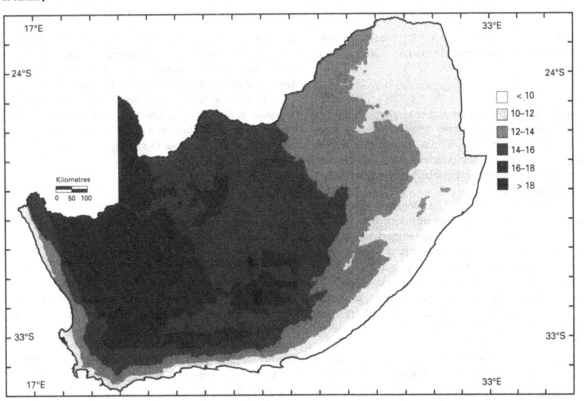

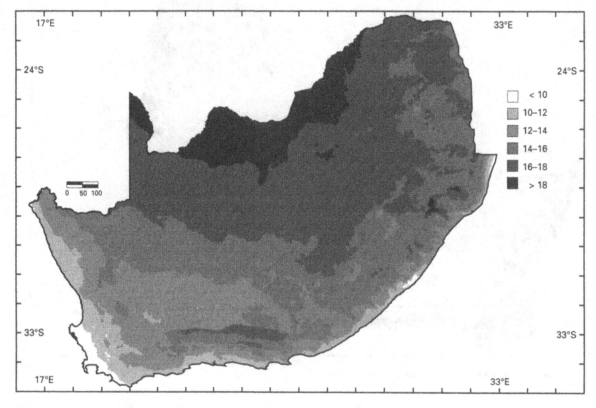

Figure 2.9 **Diurnal temperature ranges (°C) over southern Africa in July.**

being under oceanic influences and having small diurnal temperature ranges of approximately 12 °C and lower in winter.

2.4.2 Frost

Although plants possess a number of physical and biological mechanisms that serve to avoid freezing, none provides complete protection from below zero minimum temperatures. Low temperatures and frost are, therefore, often critical in determining plant survival and hence their distribution.

The information given for 244 southern African stations by Kotze (1980) on probabilities of first and last dates of frost-related minimum air temperatures of varying severity were used by Kunz (1994) to develop linear regression equations for the average (50th percentile) first date of frost and the average duration of the frost-prone period ($r^2 = 0.81$). These equations were used with the $1' \times 1'$ grid point data sets, assuming light frosts to occur at minimum screen-height air temperatures of 3 °C.

Fig. 2.10 illustrates that the occurrence of frost, as delineated by the technique described above, is a feature of the high-altitude interior areas of southern Africa.

Average first frost occurrence is in the first half of March already in the high-lying Drakensberg and Maluti mountains and average commencement dates progressively later in a northwesterly direction. The duration of the period during which frost can occur with a 50% annual probability exceeds 225 days in the highest altitude zones of the Great Escarpment and in parts of Lesotho (Fig. 2.11), with a decrease to 150 days, again in a northwesterly direction.

2.5 Precipitation

2.5.1 The importance of precipitation and moisture in vegetation studies

Among the various individual climatic parameters that influence vegetation differences on earth, Walter (1972) considers water to be the most important. Limitations in water availability are frequently a restrictive factor in plant development, and water is essential for the maintenance of physiological and chemical processes within the plant and in the exchange of energy and transport of soluble nutrients.

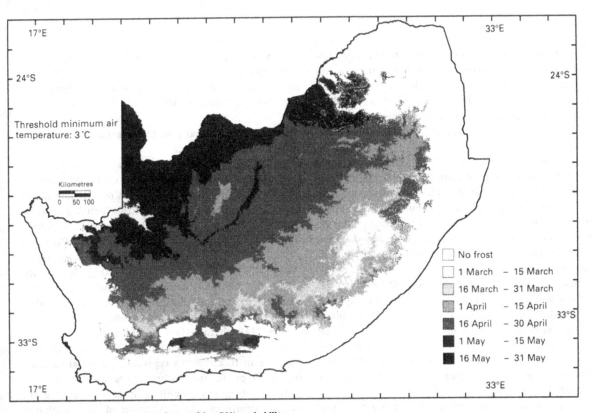

Figure 2.10 **Average commencement dates, with a 50% probability of occurrence, of first frosts over southern Africa.**

Figure 2.11 **Average duration (days) over which frost can occur over southern Africa.**

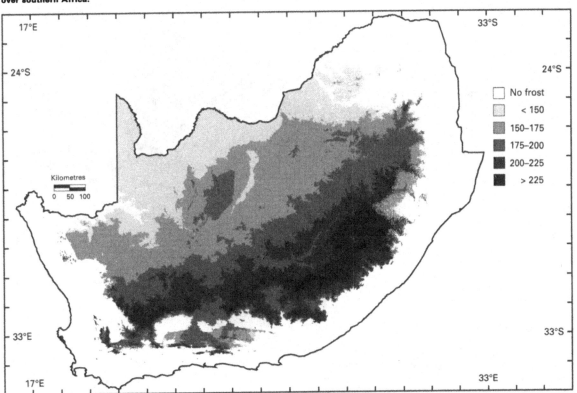

Invariably, ecologists focus on the patterns of precipitation in time and space, i.e. how much it rains (its magnitude), where it rains (its spatial distribution) and when it rains (its seasonal distribution). However, their analyses of precipitation are extended, since they need to consider the variability and concentration of precipitation as well.

The reservoir from which land plants draw their moisture supply is derived from precipitation mainly in the form of rainfall, fog and snow, of which the first two are considered important in South Africa. Not all precipitation is, however, freely available to the vegetation through the soil: some is intercepted by the plant before reaching the soil; some runs into streams as stormflow after rainfall events without being utilized by plants; some percolates into the deep soil layers beyond the root zones; and a portion is evaporated directly from the ground without being transpired through the plant.

Most major subdivisions of vegetation formations on a subcontinental scale reflect annual and seasonal soil-moisture balances rather than gross precipitation. Aspects of the moisture balance of southern Africa are evaluated in a subsequent section. However, as background, the overall patterns and variability of annual precipitation and rainfall seasonality and concentration are discussed in an ecological context with short sections on rainfall interception, fog, snow and lightning incidence over southern Africa.

2.5.2 Mean annual precipitation patterns over southern Africa

Dent, Lynch & Schulze (1989) delineated southern Africa into 34 regions, each of which was considered relatively homogeneous in relation to controls on distributions of seasonal precipitation. These controls included altitude, distance from sea, aspect, physiographic roughness, direction of prevailing rain-bearing winds and other variables. Equations for determining mean annual precipitation (MAP in mm) and median monthly precipitation (MMP in mm) were developed for each region, from which $1' \times 1'$ of a degree grid surface MAP and MMP maps were generated (Dent *et al.* 1989).

The MAP characterizes the long-term quantity of water that is available to a region. Therefore, it gives an upper limit to a region's vegetative potential in terms of biomass production if other factors such as light, temperature, topography or soils are in no way limiting. The overall feature of the distribution of MAP is that it

Figure 2.12 **Mean annual precipitation (mm) over southern Africa. After Dent, Lynch & Schulze (1989).**

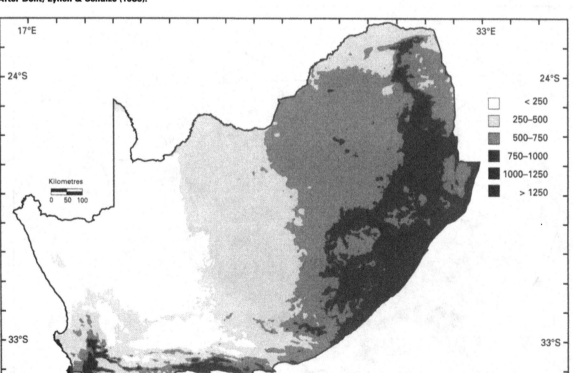

decreases fairly uniformly westwards from the escarpment across the plateau (Fig. 2.12). Between the escarpment and the sea in both the southern and the eastern coastal margins there is the expected complexity induced by topographic regularities. About one-quarter of southern Africa receives less than 250 mm yr⁻¹ and only about 8% has a MAP exceeding 750 mm.

2.5.3 Variability of annual precipitation over southern Africa

The average amount of precipitation need not necessarily be a constraint to the distribution of vegetation. Averaging procedures, however, eliminate the year-to-year natural variability of precipitation. For this reason the coefficient of variation of inter-annual precipitation, V_r, has been mapped for southern Africa using inverse relations between MAP and V_r (Schulze 1996), which had previously been shown to hold worldwide (Conrad 1941; Woodward 1986) and in South Africa (Schulze 1982).

The distributions of V_r (Fig. 2.13) depict a westward increase over southern Africa from less than 25% in the east to 40% in the west. Seasonal and monthly coefficients of variation (not given in this chapter) are even higher (e.g. Schulze 1982).

2.5.4 Concentration and seasonality indices of precipitation over southern Africa

The distribution and growth of vegetation depends strongly on the season in which precipitation falls and the degree to which the precipitation is concentrated over a short or longer period within that season. In order to obtain an index of concentration of precipitation in southern Africa, Markham's vector technique (Markham 1970; Schulze 1982) was applied; the value of the concentration index, P_c, ranges from 100% (if all the precipitation were to occur during a single month) to 0% (if the precipitation were the same for every month).

Precipitation concentration is not to be confused with the magnitude of precipitation shown in Fig. 2.12, nor its seasonality shown in Fig. 2.15. Fig. 2.14 indicates that the highest intra-annual concentrations of precipitation, when P_c is >65%, occur in the far north and northwest, whereas the lowest values, as expected, are found in the all-year rainfall region. A further band of low concentrations of precipitation is found along the east coast and at the transitions between the all-year and summer- as well as the winter/summer-rainfall regions.

To delineate southern Africa into precipitation-seasonality regions, a combination of Markham's concen-

Figure 2.13 **Coefficient of variation (%) of annual precipitation over southern Africa. After Schulze (1996).**

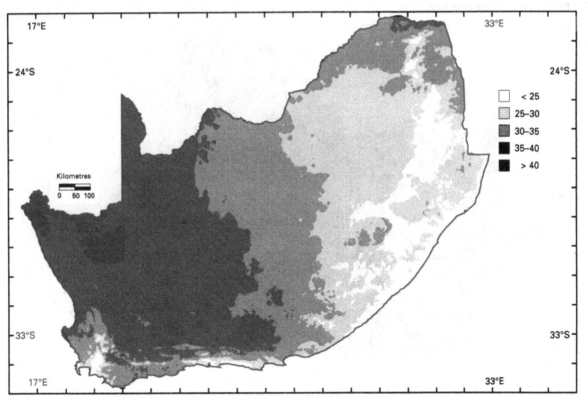

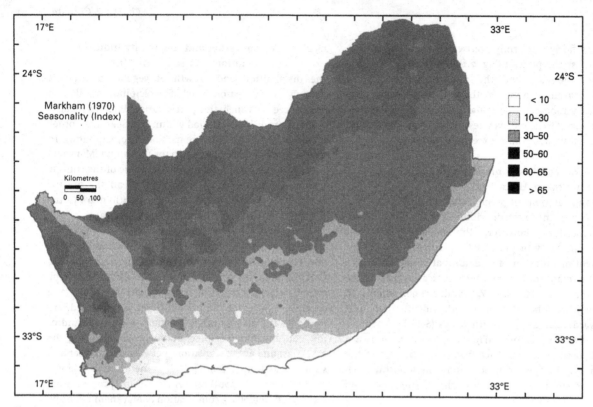

Figure 2.14 **Precipitation concentration (%) over southern Africa described by the Markham (1970) technique (see text).**

Figure 2.15 **Rainfall seasonality over southern Africa.**

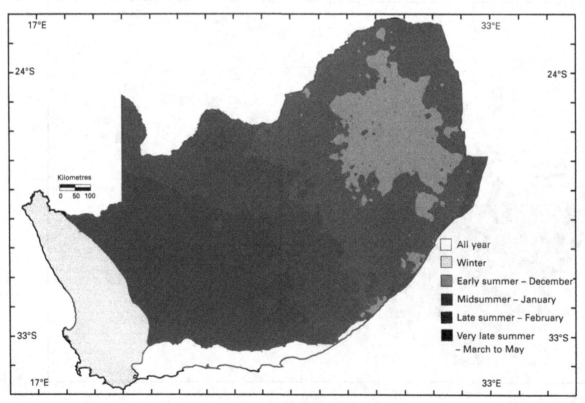

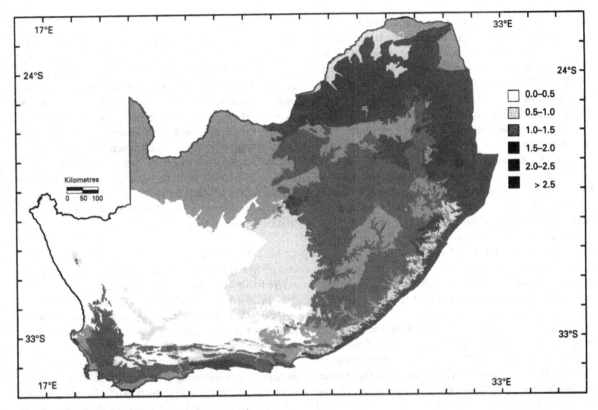

Figure 2.16 **Estimates of interception loss (mm rainday⁻¹) for natural vegetation in southern Africa. After Schulze (1981).**

tration index, P_c, and a smoothed percentage technique (Schulze 1996) was used. The dominant summer-rainfall region was subdivided into an early-summer rainfall region (peaking December or earlier), a mid-summer (January peak), late-summer (February) and a very-late-summer rainfall region where precipitation peaks occurred between March and May. Distributions of rainfall seasonality in southern Africa are shown in Fig. 2.15.

2.5.5 Precipitation interception by vegetation
The interception of rainfall by plant foliage and plant litter is frequently one of the least considered aspects of vegetation studies at any scale. Interception is the process by which precipitation is 'caught' by the vegetative canopy, stored temporarily on the canopy surfaces as interception storage and then redistributed. Important in vegetation studies is the interception loss, i.e. the portion of precipitation which after interception does not reach the ground, because it is retained by the aerial portion of the vegetation to be either absorbed by it or (more likely) returned to the atmosphere by evaporation. Interception of precipitation is a function of vegetation characteristics (canopy density, structure and number of

storeys) as well as climate characteristics (mainly precipitation amount, intensity, duration and frequency and also evaporation rates). In the southern African context, Whitmore (1971) estimated that interception by vegetation would amount to 11.6% of South Africa's rainfall and 18.3% of the rainfall in KwaZulu-Natal. Estimates of interception loss per rain day for Acocks' Veld Types were derived by Schulze (1981) from information extracted from the work by de Villiers (1975). Distributions of interception loss per rain day are shown in Fig. 2.16.

2.5.6 Fog
On foggy days and/or in zones where advective sea fog is prevalent, moisture may be intercepted by the vegetation even though standard rain gauges do not necessarily record any precipitation during the same period. This deposition of fog droplets on the foliage is generally accepted to be beneficial and it may have a profound effect on the growth, development and distribution of plants (Kerfoot 1968). Observations in southern Africa and elsewhere have shown that a relationship exists between fog incidence and the presence of vascular epi-

phytes. The *Welwitschia* zone in the desert biome of Angola and Namibia, for instance, is near the sea and it has been suggested that, although fog is not entirely a limiting factor, the bulk of moisture required for current growth could be derived from sea fogs (Airy Shaw 1947). Since the amount of fog is directly proportional to the liquid water content of the air, fog frequencies and amounts in the winter-rainfall regions of South Africa are highest in winter. Similarly, summer fog predominates in the summer-rainfall areas.

The systematic fog precipitation measurements of Nagel (1962) in the mountainous areas of winter-rainfall area have shown, *inter alia*, that over a five-year period at the highest point of Table Mountain (Cape Town) the 5664 mm of fog intercepted yr^{-1} was three times as high as the annual rainfall, and that no month recorded less than 311 mm of fog. These may be extreme results, but in the Jonkershoek Mountains near Stellenbosch orographically induced moisture from fog (not recorded by standard gauges) exceeded 600 mm yr^{-1}.

Along the west coast of South Africa, radiation and advection fog is formed when surface water of the warm inshore mixes with upwelling cold water of the Benguela Current. The afternoon sea breeze blows this fog inshore. For Swakopmund in Namibia, Nagel (1962) cites 121 fog days yr^{-1} with an amount intercepted in 1958 equivalent to 130 mm of rainfall – more than seven times the mean annual rainfall! Nagel estimates that along a 3 km coastal strip of the west coast around Swakopmund, fog precipitation equivalent to 150 mm may be intercepted by plants. Nagel (1962) also asserts that in the block of land in the western Cape formed by the ocean, latitude 32° S and longitude 20° E (i.e. an area of some 50 000 km²), the fog precipitation is estimated to be equivalent to 300 mm yr^{-1}.

Observations of the contribution of fog to precipitation in the summer-rainfall area are somewhat less scientific than those of Nagel's (1962). Results from 18 standard 'fog gauges' in northern and eastern Transvaal have yielded precipitation figures in excess of those recorded by standard gauges of between 105.2 and 280.1% (Fabricius 1969). In the KwaZulu-Natal midlands at Cedara, fog catch, over and above measured precipitation in standard gauges, ranges from 100 mm to over 1400 mm yr^{-1} at higher altitudes above a local condensation level of 1250 m (Schmidt & Schulze 1989).

This brief review emphasizes the importance of fog as an ecological agent in southern Africa, indicating that considerable amounts of moisture not recorded conventionally may in fact be intercepted and utilized, directly or indirectly, by vegetation cover.

2.5.7 Snow

Snow occurs only spasmodically in southern Africa, mainly on the higher mountain ranges of the Great Escarpment (Schulze 1965), falling most often on the ranges of the Western Cape (with an annual frequency of 5.4 snowfalls, 3.2 of which may be expected from June to August), and, on account of the greater elevation, along the Drakensberg escarpment and in Lesotho (annual frequency 8.3 with an average of 5.2 falls from May to August and snowfalls having been recorded in every month of the year). Snowmelt is usually completed within days. Snowfalls are rare north of 23° S. The effects of snow on vegetation are therefore thought to be minimal in southern Africa.

2.5.8 Lightning

In a review by Edwards (1984), lightning is generally considered the most significant of natural causes of veld fires in South Africa. Although lightning is accepted as an ignition source, opinions differ as to the frequency and importance of lightning-induced fires in the natural ecosystems of southern Africa.

Lightning ground-flash densities (LGFD) have been recorded from lightning counters by the CSIR since the early 1970s and Fig. 2.17 depicts LGFD (flashes $km^{-2}\,yr^{-1}$) over southern Africa. It is noteworthy that the southern Cape records fewer than two flashes $km^{-2}\,yr^{-1}$ against the more than eight over most of the Drakensberg escarpment, with highs of 12 to 14 flashes $km^{-2}\,yr^{-1}$.

Although Edwards (1984) and Le Roux (1979) show that lightning causes approximately 25–30% of fires in state forest land and fynbos, respectively, the frequency of lightning-induced fires is low, with the equivalent of about 1 successful lightning-induced veld fire per 500 ground-flashes $km^{-2}\,yr^{-1}$. Data from Horne (1981) suggest that the southern Cape, despite its low LGFD, experiences a far higher incidence of lightning fires than elsewhere, especially in the areas of high lightning-flash density in the eastern half of southern Africa. The reason may lie in the highly combustible fuel associated with the dry summers of the southern Cape, whereas in the summer-rainfall regions, lightning-induced fires may soon be extinguished by any accompanying rain and also by green, actively growing vegetation being struck, and thus may go largely undetected (Granger 1984).

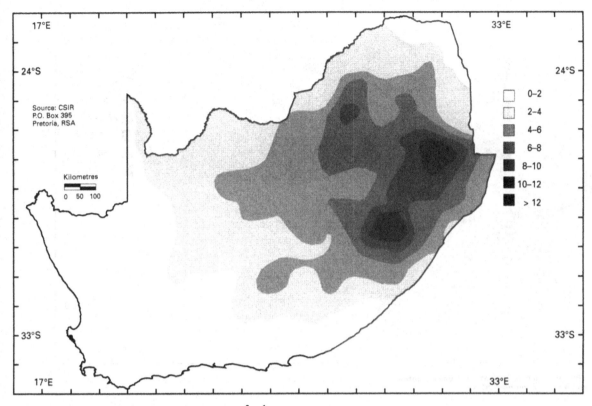

Figure 2.17 **Lightning ground-flash densities (flashes km⁻² yr⁻¹) over southern Africa. After CSIR (1994).**

2.6 Potential evaporation

2.6.1 The importance of evaporation in vegetation studies

The amount of water 'consumed' by a vegetated surface, termed total evaporation (formerly 'actual evapotranspiration') is the combination of transpiration from the leaves of growing plants and evaporation from the soil surface and the plant surface (i.e. from the intercepted water after rainfall). Whether soil-water evaporation or transpiration are calculated from either dry or wet soil conditions, the forcing function of all evaporative losses is potential evaporation, E_p. Since evaporation is, by definition, the conversion of liquid water to vapour at an evaporating surface and the vertical transport of vapour into the atmospheric boundary layer, E_p may be considered an 'atmospheric demand', determined by climatic variables such as net radiation, wind and vapour pressure deficits, or their surrogates. The accurate estimation of E_p is vital, particularly in regions such as southern Africa, where overall an estimated 91% of the MAP is returned to the atmosphere by evaporative losses (Whitmore 1971), as against a global average of 65%.

2.6.2 Reference potential evaporation and its estimation in southern Africa

There are many methods of estimating E_p, ranging from complex physically based equations to simple surrogates based on single variables such as temperature. These methods all yield different answers under different climatic conditions, and a reference potential evaporation, E_r (with its inherent advantages and defects), therefore has to be selected, against which other estimates must be adjusted appropriately. A-pan equivalent evaporation was selected as a reference potential evaporation for southern Africa because of an abundance of data and for other reasons outlined by Schulze & Maharaj (1991).

Because of the geographic unrepresentativeness of the A-pan station network in southern Africa and many other problems associated with A-pan measurements (as reviewed by Schulze & Maharaj, 1991), A-pan equivalent evaporation has been estimated by multiple-regression techniques on a month-by-month basis from 13 defined evaporation regions in southern Africa using maximum temperature (as a surrogate for net radiation), extra-terrestrial radiation (to correct the maximum temperature for the relative influence of daylength), altitude (to

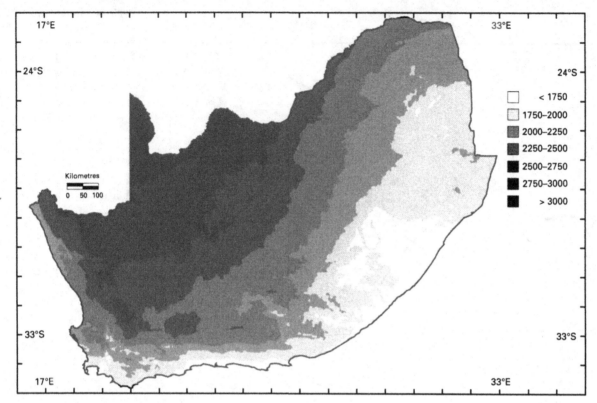

Figure 2.18 **Annual A-pan equivalent reference potential evaporation (mm) over southern Africa.**

Figure 2.19 **January A-pan equivalent reference potential evaporation (mm) over southern Africa.**

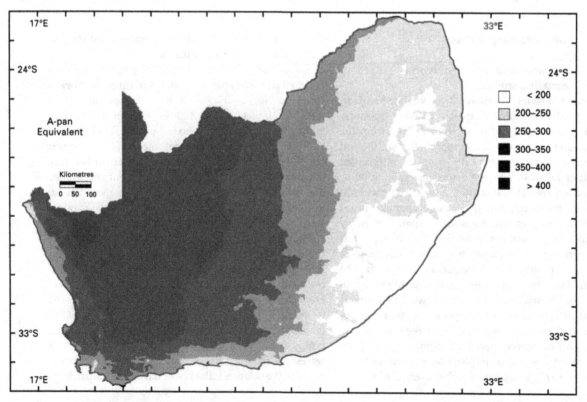

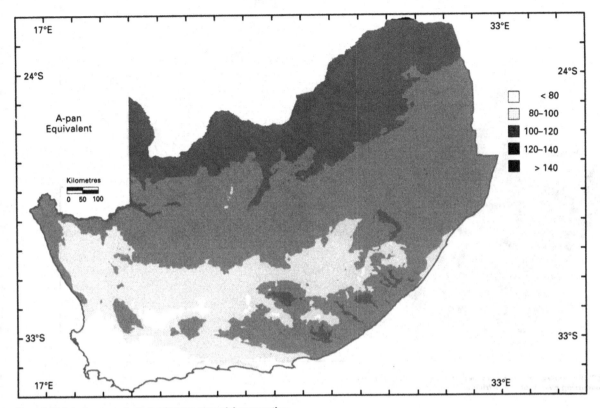

Figure 2.20 **July A-pan equivalent reference potential evaporation (mm) over southern Africa.**

account for the atmospheric pressure influence on the vaporization process) and median monthly rainfall (as a simple determinant of intra-regional differences of cloudiness and vapour-pressure deficits).

For all these variables, values either exist (Schulze 1996) or can be computed easily for the 437 000 points which make up the $1' \times 1'$ grid of southern Africa.

2.6.3 Annual and seasonal distributions of reference potential evaporation over southern Africa

Distributions of annual E_r are shown in Fig. 2.18. Values of less than 1750 mm occur in the east and south, with a general southeast–northwest increasing trend culminating in highs exceeding 3000 mm A-pan equivalent evaporation yr^{-1} in the northwest and along the Orange River. Evaporation patterns for January (Fig. 2.19) display a similar trend, with a longitudinal alignment of relative low evaporation in the east (<200 mm in January) to relatively high evaporation in the west (>400 mm for January). Winter potential evaporation, typified by July patterns (Fig. 2.20), has a distinct latitudinal alignment, with the western Cape region, subject to low temperatures and high incidence of cloudiness/rain days in July, having monthly E_r values of <80 mm.

2.7 Moisture availability to plants

2.7.1 The importance of the soil–water budget in vegetation studies

Climatic interactions and the fact that climate parameters operate in combination with one another and through the soil to produce the environment in which the plant has to grow have been alluded to in the introduction of this chapter. Water constitutes a major portion of plant tissues, is a major reagent in photosynthesis and other chemical processes involved in plant metabolism, transports nutrients from the soil to the plant, is essential in maintaining turgidity, is transpired from the leaves (with its consequent cooling effect), has a strong influence on growth through leaf extension and hence, leaf area index (LAI), biomass and dry matter production, and exerts a strong control over leaf abscission (Woodward 1986).

Soil-water availability is determined from the local hydrological budget which includes local soil characteristics and can yield meaningful indices of, for example, the duration or probability of plant stress of a given severity. The ACRU model was used for this purpose.

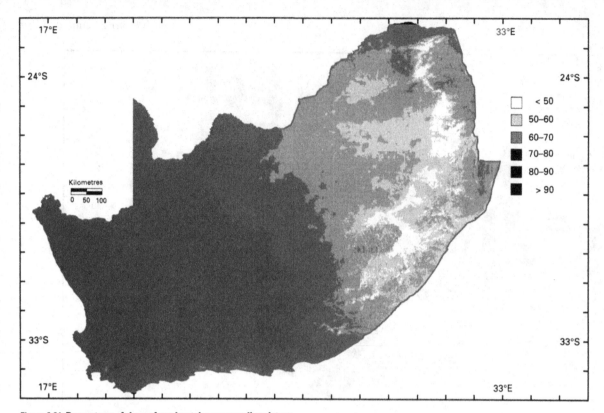

Figure 2.21 **Percentage of days of moderately severe soil moisture stress occurring over southern Africa in January.**

Figure 2.22 **Percentage of days of moderately severe soil moisture stress occurring over southern Africa in July.**

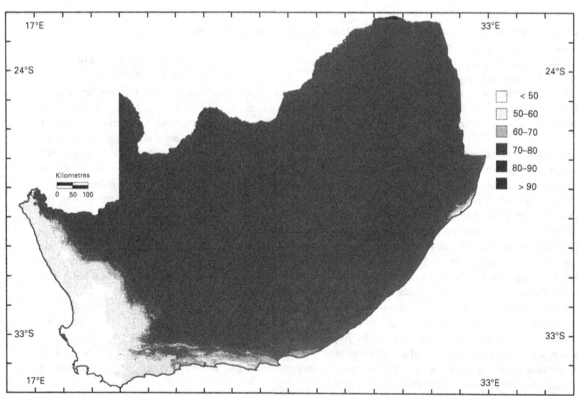

2.7.2 Soil moisture stress patterns over southern Africa

For the purposes of this study, moderately severe soil moisture stress was assumed to occur when, on a given day, the total evaporation (i.e. 'actual evapotranspiration') was half or less than half the maximum evaporation (i.e. 'potential evapotranspiration') in that upper portion of the soil profile where 80% of the active roots are found. The assumption was made in the model that soil moisture stress commences when the soil moisture content drops to below 40% of the water available to the plants in the profile, a typical value for many plants. During periods of stress the effective LAI of the vegetation declines exponentially and with rainfall and the wetting of the soil, a 'recovery' of LAI takes place, dependent on the degree of previous stress and also on the prevailing temperature conditions (with a slower recovery of LAI in winter than in summer).

The percentage of days in January experiencing the moderately severe stress levels as defined above is shown in Fig. 2.21. As expected, the highest percentages, in places exceeding 90%, are in the more arid western half and in the winter-rainfall regions of southern Africa, decreasing to <50% in the wetter east. A reversal of these patterns occurs in July (Fig. 2.22) when the winter-rainfall regions display the lower probabilities of soil moisture stress.

The above examples have illustrated clearly that plants in southern Africa live under water stress for much of the year. There is little doubt that this has a most important influence on plant form and structure.

2.8 Primary production

2.8.1 Concepts

Primary production is a quantitative expression of vegetative matter (e.g. harvestable yield) which can be produced (e.g. in tonnes) by the natural environment at a location per unit area (e.g. per hectare) over a given period of time (e.g. in a season or a year). It may be conceptualized as a generalized expression of sustainable production and is a quantification of the rain-fed environmental status of a location. The concept is particularly useful and objective in assessing the intrinsic environmental biomass production capability and in comparing the environmental resource potential of one location with others.

2.8.2 Estimation of net above-ground primary production in southern Africa

There are many approaches to and techniques of estimating primary production. However, from the foregoing, a number of characteristics were sought for an objective model of primary production suitable for detailed mapping on a subcontinental scale, bearing in mind the diversity of soils, climate and growth responses found in southern Africa. For this reason, Rosenzweig's (1968) approach was selected, in which net above-ground production is related logarithmically to total evaporation, E, in which rainfall, energy status, vegetative and soil characteristics interact with one another on a day-by-day basis.

In a natural environment E synthesises the two most variable photosynthetic resources, *viz.* water and energy. It is a measure of the simultaneous availability of soil water and atmospheric evaporative demand of a plant community at a given stage of growth in a given period of time and it is the amount of water actually entering the atmosphere through the soil/vegetation complex. Being a quantifiable measure of the energy flow in a plant community, it is fundamentally a useful predictor of biomass production. Although it is not without some conceptual limitations, the Rosenzweig model was 'imbedded' in the daily ACRU modelling system as an option that could be requested in a simulation.

2.8.3 Distributions of mean annual net above-ground primary production and its inter-annual variability over southern Africa

Daily values of total evaporation (E) generated by ACRU were accumulated over a July 1–June 30 growing season to yield annual values of primary production. Season-by-season values of primary production were then generated for the 2136 modelling units identified in southern Africa and the risk analyses and other statistics were stored for further interpretation. A primary production value was then assigned to each of the over 437 000 pixels (one minute by one minute of a degree in size) covering southern Africa to make up a raster image of primary production (Schulze *et al.* 1990; Schulze & Lynch 1992).

Mean annual primary production estimated by this technique ranges in southern Africa from <2 t ha^{-1} season^{-1}, and even <1 t ha^{-1}, in the west, to over 12 t ha^{-1} season^{-1}, and even exceeding 23 t ha^{-1} in places along the east coast (Fig. 2.23). Regions of highest natural environmental potential are, clearly, in the eastern areas. Primary production changes most rapidly in the southern and southwestern coastal areas and along the eastern seaboard regions, mainly in phase with rainfall gradients.

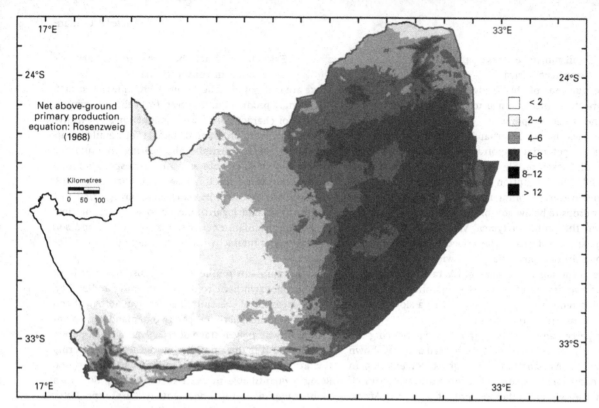

Figure 2.23 **Mean annual net above-ground primary production**
(t ha^{-1} season^{-1}) over southern Africa. After Schulze & Lynch
(1992).

Figure 2.24 **Inter-annual variability (%) of net above-ground primary**
production over southern Africa. After Schulze & Lynch (1992).

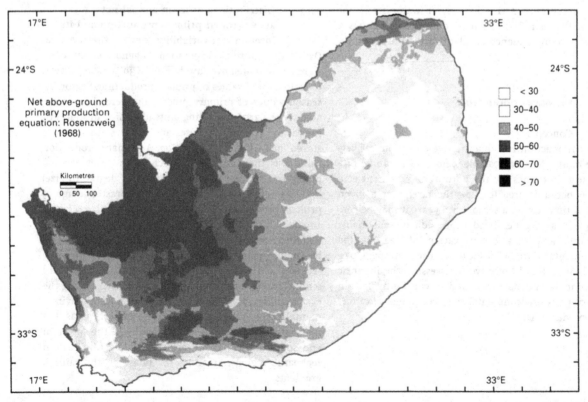

The high-risk natural environment of southern Africa is reflected in the map of inter-annual variability of primary production, expressed by its percentage coefficient of variation, CV (Fig. 2.24). Nearly half the region displays CVs exceeding 50%, mainly in the arid west and also along the extreme northern boundary with Zimbabwe, while the northwestern areas have inter-annual CVs above 100%. Lowest risk (<30% CV) is found in those areas in the east which, today, are largely under intensive commercial agriculture and forestry.

2.9 Conclusions

Climate affects the vegetation of southern Africa both directly and indirectly: directly through factors such as solar radiation, temperature and moisture (independently and in some combinations) that determine the distribution of species; and indirectly through the influence of climate on soil conditions and fire regime (Chap. 18, this volume). Furthermore, and considered in this chapter, the reciprocal influence of vegetation on the micro-climate may in turn influence the distribution of certain species.

Treatment of climatic factors in this review has been from both the individual parameter and the 'combined index' point of view, but has been by no means exhaustive. Thus, for instance, the effect of wind has not been dealt with because of a lack of available data and as it is considered to be largely local.

In conclusion it should be emphasized again that the degree to which the distribution of vegetation can be explained on the basis of climatic parameters and indices depends largely on the proper selection of the climatic factors. In this review it is therefore considered that for southern Africa no one factor or index should be used to the exclusion of others, although some indices dealt with are inherently more suitable than others. Thus, temperature and precipitation by themselves describe climate poorly, as precipitation does not really indicate whether a climate is moist or dry unless one is able to compare it with the plant water needs of a location, and the temperature *per se* does not really reveal the energy that is useful for plant development unless the moisture condition of the soil is also known at the time. However, both temperature and precipitation serve as most important inputs in determinations of more active factors of climate in relation to vegetation that find expression through the local water budget.

To this end, the application of a physical conceptual and daily multi-soil-layer water budgeting model such as ACRU becomes a powerful tool in understanding soil–vegetation–climate transfer functions and their relationships with vegetation distributions. Similarly, the gridded climatic information bases facilitate a powerful interplay of climate variables significant to plant distributions.

2.10 Acknowledgements

Over a number of years and through a series of projects, the Water Research Commission has been the major funding agency through whose generosity this work has been made possible. Their financial support is acknowledged gratefully. Furthermore, tribute is paid to the many colleagues and postgraduate students who have contributed to the development of the ACRU model. Particular personal thanks, however, are due to present and erstwhile Departmental colleagues Richard Kunz (for his input in mapping climate parameters), Steven Lynch (for his contributions to original precipitation mapping), Mrs Manju Maharaj (for her patient and meticulous development of temperature and evaporation equations), Mrs Cynthia O'Mahoney (for preparation of this chapter and previous documents on which this work was based), and Dr Mark Dent of the Computing Centre for Water Research (for GIS and computing facilities and for his original contributions to precipitation mapping and gridded/zonal information bases).

2.11 References

Airy Shaw, J.P.H. (1947). The vegetation of Angola. *Journal of Ecology*, **35**, 23–48.

Akin, W.E. (1991). *Global Patterns: Climate, Vegetation and Soils*. Norman: University of Oklahoma Press.

Archer, C.B. (1964). The relationship between radiation and solar altitude in southern Africa. *Notos*, **13**, 21–4.

Chidley, T.R.E. & Pike, J.G. (1970). A generalised program for the solution of the Penman equation for evapotranspiration. *Journal of Hydrology*, **10**, 75–89.

Clemence, B.S.E. (1992). An attempt at estimating solar radiation at South African sites which measure air temperature only. *South African Journal of Plant and Soil*, **9**, 40–2.

Conrad, V. (1941). The variability of precipitation. *Monthly Weather Review*, **69**, 5–11.

CSIR (1994). Lightning ground-flash density map. Pretoria: CSIR.

Dent, M.C., Lynch, S.D. & Schulze, R.E. (1989). *Mapping Mean Annual and Other Rainfall Statistics over Southern Africa.* WRC Report 109/1/89. Pretoria: Water Research Commission.

Dent, M.C., Schulze, R.E. & Angus, G.R. (1988). *Crop Water Requirements, Deficits and Water Yield for Irrigation Planning in Southern Africa.* WRC Report 118/1/88. Pretoria: Water Research Commission.

de Villiers, G. du T. (1975). *Reënvalonderskeppingsverliese in die Republiek van Suid-Afrika – 'n Streekstudie.* PhD Thesis. Bloemfontein: University of the Orange Free State.

Edwards, D. (1984). Fire regimes in the biomes of South Africa. In *Ecological Effects of Fire in South African Ecosystems,* ed. P. de V. Booysen & N.M. Tainton, pp. 19–38. Heidelberg: Springer-Verlag.

Fabricius, A.F. (1969). Die onttrekking van water uit die newellaag en lae wolke by Mariepskop. In *Technical Note 20,* ed. J.M. Schutte, pp. 1–21. Pretoria: Department of Water Affairs, Division of Hydrological Research.

Granger, J.E. (1984). Fire in forest. In *Ecological Effects of Fire in South African Ecosystems,* ed. P. de V. Booysen & N.M. Tainton, pp. 177–98. Heidelberg: Springer-Verlag.

Granger, J.E. & Schulze, R.E. (1977). Incoming solar radiation patterns and vegetation response: examples from the Natal Drakensberg. *Vegetatio,* 35, 47–54.

Horne, I.P. (1981). The frequency of fires in the Groot Swartberg mountain catchment area, Cape Province. *South African Forestry Journal,* 118, 56–60.

Kendrew, W.G. (1949). *Climatology.* London: Oxford University Press.

Kerfoot, O. (1968). Mist precipitation on vegetation. *Forestry Abstracts,* 29, 8–20.

Kotze, A.V. (1980). *Waarskynlike In- en Uittreedatums van Ryp in Suid-Afrika.* Report 167. Pretoria: Department of Agriculture and Fisheries.

Kunz, R.P. (1994). *Techniques to Assess Possible Impacts of Climate Change in Southern Africa.* MSc Thesis. Pietermaritzburg: University of Natal.

Le Roux, P.J. (1979). The occurrence of fires in the southern Cape fynbos. Paper presented at the *Conference on Terrestrial Ecology of the Southern Cape,* George.

Markham, C.G. (1970). Seasonality of precipitation in the United States. *Annals of the Association of American Geographers,* 60, 593–7.

Nagel, J.F. (1962). Fog precipitation measurements of Africa's southwest coast. *Notos,* 11, 51–60.

Rosenzweig, M.L. (1968). Net primary productivity of terrestrial communities: prediction from climatological data. *American Naturalist,* 102, 67–74.

Schmidt, E.J. & Schulze, R.E. (1989). *The Cedara Hydrological Research Catchments 1974–1989.* ACRU Report 34. Pietermaritzburg: Department of Agricultural Engineering, University of Natal.

Schulze, B.R. (1965). *Climate of South Africa. Part 8. General Survey.* SA Weather Bureau Report 28. Pretoria: Government Printer.

Schulze, R.E. (1975). Incoming radiation fluxes on sloping terrain: a general model for use in southern Africa. *Agrochemophysica,* 7, 55–60.

Schulze, R.E. (1981). The land use component in hydrological modelling: an evaluation. In *Workshop on the Effects of Rural Land Use and Catchment on Water Resources,* ed. H. Maaren, pp. 34–61. Technical Report TR 113. Pretoria: Department of Water Affairs, Forestry and Environmental Conservation.

Schulze, R.E. (1982). *Agrohydrology and Climatology of Natal.* Pretoria: Water Research Commission.

Schulze, R.E. (1995). *Hydrology and Agrohydrology.* Pretoria: Water Research Commission.

Schulze, R.E. (1996). *South African Atlas of Agrohydrology and Climate.* Pretoria: Water Research Commission (in press).

Schulze, R.E., Angus, G.R., Lynch, S.D. & Furniss, P.W. (1990). Primary productivity over southern Africa: an example of agricultural resource determination. *Agricultural Engineering in South Africa,* 22, 22–33.

Schulze, R.E. & Lambson, C. (1986). *RADSLOPE – A Model to Simulate Radiant Energy Fluxes and Potential Evaporation on Sloping Terrain: Theory and Background.* UWRL Report Series. Logan: Utah Water Research Laboratory, Utah State University.

Schulze, R.E. & Lynch, S.D. (1992). Distributions and variability of primary productivity over southern Africa as an index of environmental and agricultural resource determination. *ICID International Symposium on Impacts of Climatic Variations and Sustainable Development in Semi-Arid Regions,* Fortaleza, Brazil. *Conference Proceedings,* Vol III, pp. 721–40.

Schulze, R.E. & Maharaj, M. (1991). Mapping A-pan equivalent potential evaporation over southern Africa. *Proceedings, 5th South African National Hydrological Symposium,* Stellenbosch, 4B-4-1 to 4B-4-9.

Schulze, R.E. & Maharaj, M. (1994). *Temperature and Evaporation Mapping Over Southern Africa.* Pietermaritzburg: Department of Agricultural Engineering, University of Natal.

Walter, H. (1972). Der Wasserhaushalt der Pflanzen in kausaler und kybernetischer Betrachtung. *Berichte der Deutschen Botanischen Gesellschaft,* 85, 301–13.

Whitmore, J.S. (1971). The influence of vegetation in South African mountain catchments on water supplies. *South African Journal of Science,* 67, 166–76.

Woodward, F.I. (1986). *Climate and Plant Distribution.* Cambridge: Cambridge University Press.

Phytogeography, flora and endemism

3

R.M. Cowling and C. Hilton-Taylor

3.1 Introduction

The flora of southern Africa is extremely rich in species, the vast majority of which are endemic to the region. These species form the building blocks of the more than 70 major vegetation units (Acocks 1953) that are nested within the subcontinent's six floristically distinct biomes (Rutherford & Westfall 1986; Gibbs Russell 1987; Chap. 5, this volume). Since much of the vegetation diversity of the area is a result of these patterns of richness and endemism, it is appropriate to provide a review of the phytogeography, flora and endemism of southern Africa as a background to this volume.

Extensive reviews of the region's phytogeography (Werger 1978a; White 1983) and flora (Goldblatt 1978; Gibbs Russell 1985) have appeared relatively recently. We therefore review very briefly the development and characteristics of the phytogeographical regions in southern Africa. We also provide a short account of the flora, focusing on a few generalizations regarding its size, composition and affinities. This account is mainly a summary of the recent literature. Readers are referred to the above-mentioned references for details. The bulk of this chapter is devoted to an analysis of the patterns and correlates of species-level endemism in the flora. The underlying structure to our analysis is the general question: are endemics a random assemblage with regard to habitat preferences, biological traits, phylogenetic lineages and age of origin (Kruckeberg & Rabinowitz 1985; Major 1988)? We conclude the chapter by discussing the unique features of the southern African flora in relation to ecological, historical and phylogenetic factors.

3.2 Phytogeographical division

Southern Africa was one of the first areas outside the European sphere to be explored botanically. Between the late eighteenth century and mid-nineteenth century most of the subcontinent was collected, or at least observed, by botanists (Gunn & Codd 1981). This knowledge enabled the delimitation, at a very early stage, of phytogeographical regions. A detailed account of the history of phytogeographical division of the subcontinent is given by Werger (1978a). We provide a brief synopsis below.

Most early attempts at subdivision recognized a southwestern region (more or less equivalent to the contemporary Cape Region), a Karoo region, a Kalahari region and eastern seaboard region (Bolus 1875, 1886, 1905; Rehman 1880; Engler 1882; Marloth 1887). These authors emphasized the difference between the flora of the Cape Region and the remainder of the subcontinent. This practice culminated in the recognition of this area as the Cape Floristic Kingdom, distinct from the Palaeotropical Kingdom, which covers the rest of the subcontinent (Takhtajan 1969; Good 1974).

The next phase in southern African phytogeography involved an increasingly finer subdivision of the extra-Cape regions (Marloth 1908; Pole Evans 1922; Acocks 1953) and the location of the subcontinent's phytogeography in an Africa-wide context (Lebrun 1947; Monod 1957; White 1965, 1971; Troupin 1966; Good 1974). These patterns have been analysed by Werger (1978a), who produced a synthetic phytogeographic map for southern and south-central Africa. The most comprehensive phytogeographical treatment of Africa to date is that of White (1971, 1976, 1983). His non-hierarchical scheme recognizes several phytochoria including regional centres of

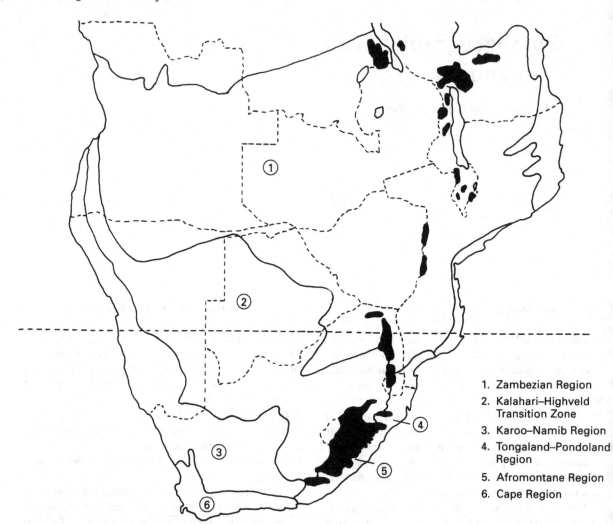

Figure 3.1 **Phytogeographical regions of southern Africa (Botswana, Lesotho, Namibia, South Africa and Swaziland) following White (1976) with changes to the Cape Region according to Goldblatt (1978).**

1. Zambezian Region
2. Kalahari–Highveld Transition Zone
3. Karoo–Namib Region
4. Tongaland–Pondoland Region
5. Afromontane Region
6. Cape Region

endemism, regional transition zones and regional mosaics. We follow Goldblatt (1978) in adopting White's (1976) scheme for the phytogeographical division of southern Africa (Fig. 3.1).

3.2.1 Zambezian Region

This regional centre of endemism (White 1983) occupies about 3.8 million km² of south-central Africa, mainly on the Great African Plateau, from southern Zaire and Tanzania southwards to include, in southern Africa, the northern part of Namibia, northeastern Botswana and the extreme northeastern part of South Africa (Fig. 3.1). The southern African section is relatively depauperate, having only two endemic genera and few endemic spec-

ies (Goldblatt 1978). The climate is one of summer rainfall with mild winters, which become increasingly frost-prone southwards (Chap. 2, this volume). The vegetation is generally some form of wooded grassland dominated by Poaceae and Fabaceae (Werger & Coetzee 1978; White 1983) and commonly referred to as savanna. All of the Zambezian Region in southern Africa is included in the savanna biome (Chaps. 5 and 11, this volume).

3.2.2 Kalahari–Highveld Region

Lying between the tropical Zambezian Region and temperate Karoo–Namib Region, this regional transition zone (White 1983) occupies about 1.2 million km² from southern Angola to the Eastern Cape (Fig. 3.1). Most of the region occupies an area of low relief between 850 and 1000 m on the interior plateau of southern Africa. The predominantly summer rainfall is between 250 and 500 mm yr⁻¹ and winter frost is widespread and severe (Chap. 2, this volume). The vegetation is predominantly grassland (Chap. 10, this volume) with a depauperate Zambezian tree flora (mainly *Acacia* spp.) in the northwest (especially in the Kalahari Desert) (Chap. 11, this volume) and a strong admixture of dwarf shrubs of Karoo–Namib affinity towards the south (Chap. 8, this volume). Despite this being southern Africa's largest phytogeographical region, the flora is relatively poor (c. 3000 species) and has few endemics, exemplifying its transitional nature (Goldblatt 1978; White 1983). This is also borne out by the inclusion of parts of the region in the savanna biome (northwest), the grassland biome (east) and the Nama-karoo biome (southwest) (Chap. 5, this volume).

3.2.3 Karoo–Namib Region

Occupying about 0.7 million km² in an increasingly broad zone from southern Angola to the Eastern Cape, this regional centre of endemism (White 1983) includes the arid west coastal plain and the arid to semi-arid regions of the interior plateau and Great Escarpment (Fig. 3.1). The physiography is extremely diverse including numerous geomorphic provinces such as the Namib Desert, Kaokoveld, Namaqualand, Great Karoo and Upper Karoo (Chap. 1, this volume). Altitude ranges from sea level to 2695 m. Rainfall varies from less than 50 mm yr⁻¹ in the Namib Desert to slightly more than 300 mm in the wetter areas of the Great Escarpment. The southwestern sector from southern Namibia to the fringes of the Cape Region receives most of its rain in winter and fog is frequent on the coastal plain; in the remainder rain may fall at any time of the year or, towards the north and east, predominantly in the warmer summer or autumn months. The coastal belt is frost-free but in the interior the frost period lasts for 5–6 months (Chap. 2, this volume). The vegetation in the eastern sector is dominated by dwarf shrubs, mainly Asteraceae, and a lesser and sometimes ephemeral component of grasses (Chap. 8, this volume). The winter-rainfall southwestern sector is characterized by a dwarf succulent shrubland with numerous succulent species of Mesembryanthemaceae, Asclepiadaceae, Crassulaceae, Zygophyllaceae, Geraniaceae, Liliaceae *sensu lato* and Asteraceae, amounting to the richest succulent flora in the world (Compton 1929; Van Jaarsveld 1987; Hilton-Taylor & Le Roux 1989; Chap. 7, this volume). The Namib Desert is an area of sparse grasslands with a few woody emergents of tropical affinity – it is also home to the interesting genus *Welwitschia*, a relict gymnosperm

(Chap. 9, this volume). The entire flora of the Karoo–Namib region probably exceeds 7000 species, of which up to 50% are endemic (Hilton-Taylor & Le Roux 1989). At least 80 genera are endemic or centred in the region and of these some 50 belong to the Mesembryanthemaceae (Goldblatt 1978). The region is divided into three biomes: the desert biome (central and northern Namib Desert); the succulent karoo biome (winter-rainfall sector including the southern Namib Desert); and the Nama-karoo biome (summer-rainfall sector in the interior) (Chap. 5, this volume). Details on the flora and phytogeography of the Karoo–Namib Region are given by Werger (1978b), Hilton-Taylor (1987, 1996) and Jürgens (1991).

3.2.4 Tongaland–Pondoland Region

This subtropical regional mosaic (White 1983) occupies about 150 000 km² along the eastern seaboard of southern Africa from the mouth of the Limpopo River to Port Elizabeth (Fig. 3.1). It includes the broad, sandy coastal plain of Tongaland in the north; southwards it occupies a very narrow belt where the mountains come close to the sea. Here it abuts on the Afromontane Region, but penetrates inland along river valleys. Under the influence of the warm Mozambique Current, the climate is warm with a moderately high (c. 800–1200 mm yr⁻¹) and well distributed rainfall along the coast, and drier (400–600 mm yr⁻¹) and more extreme conditions inland. The vegetation is a mosaic of subtropical forest (related floristically to the forests of the Zanzibar–Inhambane and Guinea–Congolian Regions), thicket, savanna and grassland (Moll & White 1978; White 1983). The xeric thickets of the southeastern river basins have a strong admixture of succulents of Karoo–Namib affinity (Cowling 1983). The area has about 3000 species and many endemics (40% of the 200 larger woody species are endemic; White 1983), including many taxonomically isolated endemics on the sandstones of southern KwaZulu-Natal and Pondoland (Van Wyk 1990, 1994). There are about 15 endemic genera (Goldblatt 1978). As forest covers a very small portion of the Tongaland–Pondoland Region (Chap. 12, this volume), it is included by Rutherford & Westfall (1986; see also Chap. 5, this volume) in the savanna biome.

3.2.5 Afromontane Region

This is not a discrete region but a series of isolated areas forming an archipelago-like regional centre of endemism (White 1983) distributed from northeastern Africa, along the East African uplands to southeastern Africa and the Cape Peninsula. Significant outliers occur on the high mountains of West Africa and Angola. The region is characterized by temperate and mostly high rainfall (c. 1000 mm yr⁻¹) conditions, although freezing night-time temperatures and relative dryness are experienced in the 'alpine' zone above the forested belt of the higher mountains. In the tropics, Afromontane communities are found only above 2000 m; further south, where latitude compensates for altitude, they descend progressively further, and in the Cape Region pockets of Afromontane forest occur at sea level. In southern Africa, the Afromontane Region is centred in the Lesotho and KwaZulu-Natal Drakensberg with extensions to the north and south along the Great Escarpment (Fig. 3.1). A secondary centre occurs on the coastal platform of the southern Cape which comprises the largest contiguous block of forest in the subcontinent (Chap. 12, this volume). As in the rest of Africa, the predominant vegetation is grassland, usually with a mixture of temperate C$_3$ and tropical C$_4$ species (Vogel, Fuls & Ellis 1978). Other vegetation types are forest, usually dominated by *Podocarpus* spp., and ericaceous shrublands that resemble the fynbos of the Cape Region (Killick 1978, 1994; White 1978, 1983). White (1983) suggests an Africa-wide flora of about 4000 species; this is probably an underestimate, since Hilliard & Burtt (1987) have recorded 1261 species from the southern KwaZulu-Natal Drakensberg alone. Although region-level endemism among forest trees may be in excess of 75%, local endemism is low (White 1978; Chap. 12, this volume). Local endemism is, however, very high among grassland herbs throughout the region (Hilliard & Burtt 1987; Meadows & Linder 1993). The forested parts of the Afromontane Region are included in the forest biome, and the remainder forms part of the grassland biome (Chap. 5, this volume).

3.2.6 Cape Region

This regional centre of endemism occupies about 90 000 km^2 in the extreme southwest of the subcontinent (Fig. 3.1). The landscape is dominated by the parallel ranges of the quartzitic Cape Fold Belt with an average altitude of 1000–1500 m (Chap. 1, this volume). Soils derived from these mountains and those of the sandy coastal fringe are extremely deficient in nutrients. The climate in the western sector is mediterranean with more than 60% of the 250–3000 mm annual rain falling in the winter months (Chap. 2, this volume). Eastwards, an increasing proportion of rain falls in the summer months, although cool season precipitation is a characteristic of the entire region. The prevalent vegetation is fynbos, a fire-prone, sclerophyllous shrubland characterized by the presence of evergreen graminoids (restioids) belonging to the Restionaceae (Chap. 6, this volume). On the heavier soils of the lowlands, fynbos is replaced by renosterveld, an ericoid shrubland that lacks Restionaceae and many other taxa typical of fynbos. The drier intermontane valleys include substantial areas of

succulent karoo shrubland, and patches of depauperate Tongaland–Pondoland thicket occur along the coast. Enclaves of Afromontane forest occur at the base of the wetter, coastal mountains. The flora of the Cape Region is spectacularly rich, comprising some 8600 species, of which 68% are endemic (Bond & Goldblatt 1984). The seven endemic families and 198 endemic genera (19.5% of the total) contribute to the region's status as a floristic kingdom. The Cape Region coincides with the fynbos biome except for the enclaves of karoo and forest, which form part of the succulent karoo and forest biomes, respectively. The flora and phytogeography of the area have been reviewed by Goldblatt (1978), Taylor (1978), Bond & Goldblatt (1984), Cowling & Holmes (1992a), Cowling, Holmes & Rebelo (1992) and Linder, Meadows & Cowling (1992).

3.2.7 Controversies in phytogeographical delimitation

Like any classification, the delimitation of phytogeographical regions in southern Africa has had its share of controversies, some of which have not been resolved. Werger (1978a,b) did not recognize the Kalahari-Highveld Region, preferring to include the southwestern portion of this transitional region, including the southern Kalahari, as part of the Karoo–Namib Region and the remainder as part of the Zambezian Region. Jürgens (1991) has suggested that the succulent, winter-rainfall portion of the Karoo–Namib Region should be included in the Cape Region. Although the floristic links between these two winter-rainfall regions are strong (Acocks 1953; Hilton-Taylor 1987), Gibbs Russell (1987) provides quantitative data which show that links between the succulent karoo and the Nama-karoo floras are marginally stronger than between succulent karoo and fynbos. The recognition of an Afroalpine Region remains problematic: Werger (1978a) upholds the region in southern Africa, assigning to it the vegetation above the forest belt in the southern Drakensberg (see Killick 1978). White (1983), however, includes the Afroalpine as an 'archipelago-like region of extreme floristic impoverishment' within the Afromontane Region, arguing that the flora is too poor, and the definition of the region too vague, to warrant independent phytochorological status. Because of the strong links between the floras of the Cape and Afromontane Regions (Adamson 1938; Weimarck 1941; Levyns 1964; Hedberg 1965; Cowling 1983), both Tinley (1977) and Linder (1990) suggest combining these regions into an Afrotemperate phytochorion. On the other hand, Hilliard & Burtt (1987) argue for the recognition of the Drakensberg Afromontane enclave as a distinct phytochorion on the basis that floristic affinities with adjacent lowlands are stronger than with more distant upland areas. Finally, being a zone of convergence for five phytogeographical areas (Fig. 3.1), the western part of the Eastern Cape poses many problems for accurate delimitation (Goldblatt 1978). Many communities in this area show complex and transitional phytogeographic affinities. Recent research has made much progress in resolving these problems (Cowling 1983; Palmer 1990; Hoffman & Cowling 1991).

3.3 Flora

3.3.1 Size

At 21 137 indigenous species (Arnold & De Wet 1993), the flora of southern Africa is among the richest in the world for similar sized areas, including those in the trop-

ical areas of Africa and elsewhere (Goldblatt 1978; Gibbs Russell 1985; Cowling *et al.* 1989; Cowling & Hilton-Taylor 1994; Fig. 3.2). This is exceptional, given the fact that southern Africa is a predominantly warm temperate, semi-arid region with an overall mean annual rainfall of less than 400 mm (Schulze & McGee 1978; Chap. 2, this volume). Most other species-rich regions include large areas of tropical rainforest or have mediterranean-type climates (Myers 1988, 1990; McNeely et al. 1990). Forests cover less than 1% of southern Africa's land surface (Rutherford & Westfall 1986; Chaps. 5 and 12, this volume) and, even excluding the winter-rainfall Cape Region, the flora of the remaining area is remarkably rich (Gibbs Russell 1985). Richness is not, however, uniformly distributed across the subcontinent. Richer-than-average areas are the southwestern parts of the Cape

Figure 3.2 **Relationship between number of vascular species and area in large (>1 million km²) landmasses distributed across the globe. Tropical Asia includes India, Pakistan, Bangladesh and Burma. Data compiled from Major (1988) and Cowling *et al.* (1989).**

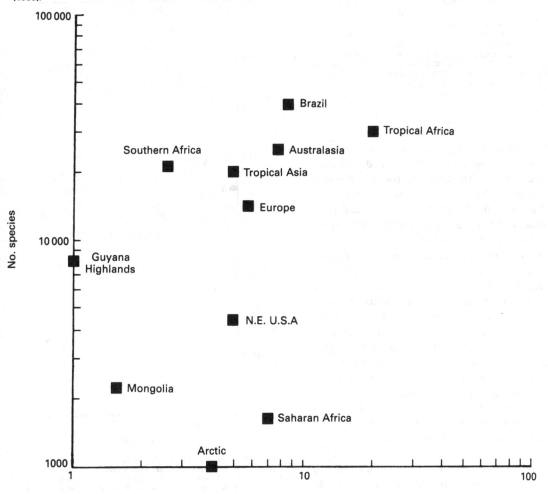

Table 3.1 **Large genera (100 spp. or more) in the southern African flora and their centres of diversity in relation to major phytochoria. Data from Goldblatt (1978) and Gibbs Russell (1985, 1987)**

Genus	Family	No. spp.	Region
Erica	Ericaceae	638	Cape
Ruschia[a]	Mesembryanthemaceae	352	Cape,K-N
Senecio	Asteraceae	309	Cape
Conophytum[a]	Mesembryanthemaceae	301	K-N
Euphorbia	Euphorbiaceae	266	Zamb,T-P
Aspalathus[a]	Fabaceae	256	Cape
Helichrysum	Asteraceae	241	Afro
Lampranthus[a]	Mesembryanthemaceae	218	Cape
Indigofera	Fabaceae	212	Zamb
Pelargonium	Geraniaceae	211	Cape
Oxalis	Oxalidaceae	193	Cape
Thesium	Santalaceae	166	Cape
Delosperma	Mesembryanthemaceae	159	Afr,K-H
Haworthia	Liliaceae[b]	153	Cape
Aloe	Liliaceae[b]	152	Wide
Phylica	Rhamnaceae	147	Cape
Hermannia	Sterculiaceae	146	Cape
Agathosma[a]	Rutaceae	138	Cape
Wahlenbergia	Campanulaceae	135	Wide
Selago	Selaginaceae	127	Cape
Sutera	Scrophulariaceae	125	Wide
Muraltia	Polygalaceae	116	Cape
Restio	Restionaceae	113	Cape
Drosanthemum[a]	Mesembryanthemaceae	111	Cape
Cliffortia	Rosaceae	108	Cape
Sphalmanthus[a]	Mesembryanthemaceae	108	K-N
Gladiolus	Iridaceae	107	Cape
Lotononis	Fabaceae	104	Afr
Cheiridopsis[a]	Mesembryanthemaceae	103	K-N

Afro, Afromontane Region; K–H, Kalahari–Highveld Region; K–N, Karoo–Namib Region; T–P, Tongaland–Pondoland Region; Zamb, Zambezian Region.
[a] Endemic genera.
[b] Sensu lato.

Region, mesic parts the Tongaland–Pondoland Region, and the Afromontane uplands of the east and northeast; relatively depauperate areas include the northwestern Kalahari–Highveld Region, the summer-rainfall and hyper-arid areas of the Karoo–Namib Region and parts of the southern and southeastern coastal belt (Cowling et al. 1989; Chap. 19, this volume).

With 1930 genera in 226 families, the southern African flora is also very rich at supra-species level. For example, the flora of West Tropical Africa includes 1742 genera in an area 1.7 times larger than southern Africa (Goldblatt 1978). An exceptional characteristic of the flora is the existence of numerous large genera: six have more than 250 species and 29 have more than 100 species (Table 3.1). Seventeen of these large genera are centred in the Cape Region. It is not surprising, therefore, that species : genus ratios in southern Africa (10.9) and the Cape Region (8.9) are the highest recorded values in the world, being comparable with data from oceanic islands such as Hawaii (7.5) and New Zealand (7.4) (Goldblatt 1978).

The species density in the Cape Region (8600 species in 90 000 km²) is amongst the highest in the world

(Cowling et al. 1992) and is substantially higher than values from climatically similar (warm temperate to subtropical) regions (Fig. 3.3a). Although the genus : family ratio for the Cape Region is not remarkable (Fig. 3.3b), the number of species per family is considerably higher than in other regions (Fig. 3.3c). Thus, the enormous species-level richness of the Cape Region is not accompanied by unusually high richness at the genus or family levels (Linder et al. 1992).

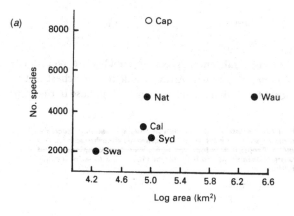

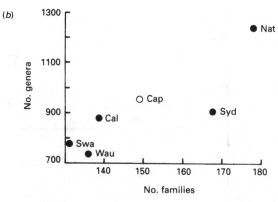

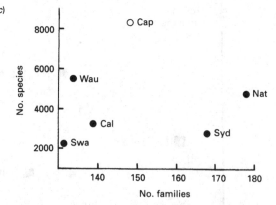

Figure 3.3 **Comparison of taxic richness of the Cape Region (Cap) with Natal (Nat) and Swaziland (Swa) in southern Africa and the climatically similar southwestern Australia (Wau), southern California (Cal) and Sydney (Syd) areas. From Linder et al. (1992).**

Table 3.2 Ten largest families in the southern African flora. Data from Goldblatt (1978) and Gibbs Russell (1985)

Family	No. spp.	Endemic spp. (%)	No. genera	Endemic genera (%)
Asteraceae	2116	86	174	46
Mesembryanthemaceae	2408	98	123	96
Fabaceae	1540	75	115	17
Liliaceae[a]	1066	89	54	39
Iridaceae	858	97	44	61
Ericaceae	804	99	24	79
Poaceae	783	44	167	14
Asclepiadaceae	769	87	60	35
Scrophulariaceae	543	87	51	27
Cyperaceae	464	52	30	23

[a] *Sensu lato.*

Table 3.3 Global patterns of higher plant species endemism. Data from World Conservation Monitoring Centre (1992)

Country	Endemism (%)	Country	Endemism (%)
New Zealand	82	Ecuador	21
Southern Africa	80	United States	21
Australia	80	Costa Rica	15
New Caledonia	80	Greece	15
Madagascar	68	Mexico	14
Indonesia	67	Panama	14
China	56	Algeria	8
Papua New Guinea	55	Mozambique	4
Chile	51	Nigeria	4
Zaire	29	Zambia	4
Sri Lanka	28	Zimbabwe	2
Argentina	25	Germany	<1
Angola	24	Sweden	<1

3.3.2 Composition and endemism

The family-level composition of the southern African flora (Table 3.2) differs markedly from that of tropical African floras. The presence among the ten largest southern African families of Mesembryanthemaceae, Iridaceae, Ericaceae and Asclepiadaceae, and the high rank of Liliaceae *sensu lato* and low rank of Poaceae, is not a feature of other African floras (Gibbs Russell 1985). Another unusual feature of the flora is that the petaloid monocot families – Iridaceae, Liliaceae sensu lato, and Orchidaceae – comprise 11.2% of the flora, a situation surpassed only by the flora of the Cape Region (Goldblatt 1978).

Ten families are endemic to southern Africa (Goldblatt 1978). Seven of these – Bruniaceae, Peneaceae, Stilbaceae, Grubbiaceae, Roridulaceae, Geissolomaceae and Retziaceae – are endemic or nearly so to the Cape Region. The remainder – Stangeriaceae, Greyiaceae and Achariaceae – are restricted to Afromontane and Tongaland-Pondoland parts of the mesic eastern seaboard. Six of these endemic families are mono-, di- or tritypic and the Cape ones are taxonomically isolated, and almost certainly mesophytic relicts (Carlquist 1976). The Mesembryanthemaceae and Selaginaceae, although not strictly endemic, have the vast majority of their taxa in southern Africa (Goldblatt 1978).

Some 560 or 29% of native genera are endemic to southern Africa, an exceptionally high value for a subcontinental landmass (Goldblatt 1978). A further 50 genera are concentrated in the region, with a few species extending locally into tropical Africa. The major centres of generic endemism are the Cape Region (248 genera endemic or concentrated there), the Karoo–Namib Region (80 genera) and the Afromontane Region (20 genera). Families with the largest numbers of endemic genera are the Mesembryanthemaceae (108), Asteraceae (80), Iridaceae (27), Poaceae (23), Liliaceae *sensu lato* (22) and Asclepiadaceae (20) (Goldblatt 1978).

Approximately 80% of the species in the southern African flora are endemic to the region, an unusually high value for a continental region (Golblatt 1978; Cowling & Hilton-Taylor 1994). Similarly high levels are usually associated with islands and large isolated landmasses such as Australia (80%), Madagascar (68%), New Caledonia (80%) and New Zealand (82%) (Table 3.3).

Endemics are not a random assemblage phylogenetically. Significantly higher than average levels of endemism are associated with 20 of the larger families (Table 3.4), most of which are concentrated in the Cape and Karoo–Namib Regions. On a global scale, the southern African flora comprises a distinct phylogenetic assemblage. Of the 26 largest families that have levels of endemism significantly higher or the same as the total flora (as an independent sample), only eight rank among the 30 largest families in the world. In addition to the endemic and near endemic families listed above, southern Africa includes the majority of the world's species of Ericoideae (Ericaceae), Aizoaceae, Amaryllidaceae, Iridaceae and Restionaceae, as well as a high proportion of Geraniaceae, Proteaceae and Rutaceae (Goldblatt 1978; Gibbs Russell 1985). In contrast, families with levels of endemism significantly lower than the average flora are widespread in southern Africa as well as other parts of the globe (Cowling & Hilton-Taylor 1994).

3.3.3 Affinities

The major proportion (76%) of families and about 23% of the genera in the southern African flora are cosmopolitan or pantropic in distribution (Goldblatt 1978). Many of these are represented by a few species associated with tropical savanna and forest vegetation in the extreme north of the region. The palaeotropic element is represented by 255 genera (14% of total) and only five families. Ten families (5%) and 473 genera (25%) are African or African–Madagascan endemics. Many of these genera just extend into the region from the north, but

Table 3.4 **Species endemism in the 30 largest families of flowering plants in the flora of southern Africa. Data from Gibbs Russell (1985).** ***, $P < 0.001$, **, $P < 0.01$, *, $P < 0.05$, NS, not significant

Family	Rank in southern Africa	Rank in world[a]	No. spp. (%)		χ^2
			Endemic	Non-endemic	
Total flora	–	–	16298 (80)	4074 (20)	–
Endemism significantly higher than total flora					
Mesembryanthemaceae	1	–	2360 (98)	48 (2)	552.0***
Asteraceae	2	1	1820 (86)	296 (14)	52.9***
Liliaceae[b]	4	11	943 (89)	123 (11)	49.8***
Iridaceae	5	–	828 (97)	30 (3)	151.4***
Ericaceae	6	21	802 (99)	2 (1)	202.8***
Asclepiadaceae	8	13	668 (87)	101 (13)	23.1***
Scrophulariaceae	9	8	471 (87)	72 (13)	16.5***
Proteaceae	13	–	355 (97)	11 (3)	66.2***
Rutaceae	15	–	274 (94)	17 (6)	36.1***
Restionaceae	16	–	265 (94)	17 (6)	34.0***
Geraniaceae	17	–	257 (96)	10 (4)	43.6***
Campanulaceae	18	–	236 (92)	20 (8)	23.3***
Crassulaceae	20	–	190 (89)	25 (11)	9.0**
Selaginaceae	21	–	208 (97)	10 (3)	31.7***
Polygalaceae	23	–	179 (88)	26 (12)	6.5*
Oxalidaceae	25	–	191 (98)	4 (2)	38.5***
Thymelaeaceae	26	–	180 (95)	9 (5)	26.7***
Aizoaceae	27	–	180 (98)	4 (2)	35.8***
Apiaceae	28	17	162 (90)	14 (10)	9.0**
Santalaceae	29	–	166 (94)	10 (6)	21.8***
Endemism same as total flora					
Euphorbiaceae	11	6	367 (80)	94 (20)	0.0NS
Orchidaceae	12	2	353 (81)	86 (19)	0.0NS
Amaryllidaceae	29	–	162 (82)	36 (18)	0.3NS
Sterculiaceae	30	–	148 (85)	27 (15)	2.0NS
Endemism significantly lower than total flora					
Fabaceae	3	3	1147 (75)	393 (25)	31.4***
Poaceae	7	4	348 (44)	435 (56)	641.2***
Cyperaceae	10	9	240 (52)	224 (48)	235.5***
Acanthaceae	14	15	233 (66)	118 (34)	40.5***
Lamiaceae	19	7	142 (63)	83 (37)	39.5***
Rubiaceae	22	5	108 (52)	99 (48)	99.5***

[a] World rank is among 30 largest families.
[b] Sensu lato.

the majority (76%) are relatively widespread in southern and tropical Africa. Some of these show a disjunct distribution between the arid areas of southwestern and northeastern Africa (De Winter 1971); others are centred in the Cape and extend outside southern Africa mainly in temperate Afromontane areas (Weimarck 1941; Levyns 1964).

Some ten southern African families and 48 genera (2.5%) have an African–Eurasian distribution (Goldblatt 1978). Many of the families are associated with xeric regions in Africa and southwestern Asia but others occur in temperate habitats in Africa (Cape–Afromontane) and Eurasia (Burtt 1971; Linder et al. 1992). African–New World taxa comprise seven families and 23 genera. Some of these occur in the arid areas of southwestern Africa and the New World, suggesting at least pre-Pliocene aridity in the former region (Goldblatt 1978; Ward, Seely & Lancaster 1983; Van Zinderen Bakker & Mercer 1986;

Chap. 1, this volume). Strictly austral or Gondwanan taxa comprise only seven families and 23 genera. There is good evidence that this component was richer in the Palaeogene and Miocene (Coetzee, Scholtz & Deacon 1983) but suffered severe depletion as the climate of the subcontinent deteriorated from the Pliocene onwards (Coetzee 1978; Linder et al. 1992). The origins of Gondwanan taxa have been attributed to: persistence of ancient lineages (gymnosperms) after continental breakup (e.g. *Podocarpus*); derivation from tropical African forest ancestors (e.g. Proteaceae); or dispersal at an early stage across a narrower Indian Ocean to Australasia (see refs. in Goldblatt 1978). Recently, Linder et al. (1992) have proposed a vicariance explanation for the distribution of African Restionaceae, arguing that the time of origin of angiosperms is not known accurately enough to discount the presence of this family on the Gondwanan super-continent prior to its fragmentation.

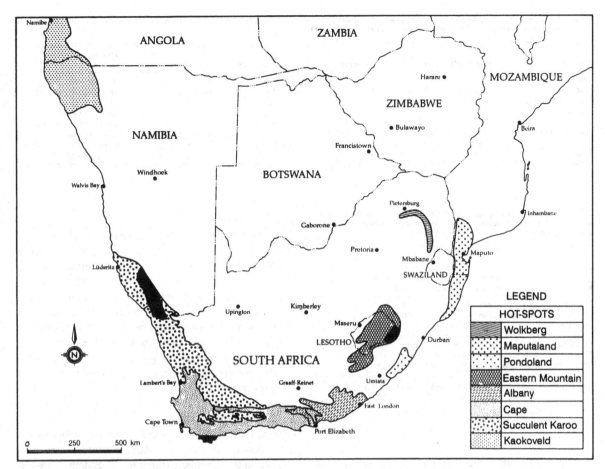

Figure 3.4 **Centres of plant species endemism in southern Africa. Note that the Maputaland and Kaokoveld Centres extend beyond the boundaries of the region. Also shown (in solid shading) is the location of floras used in the analyses of species-level endemism (Fig 3.5; Table 3.7). From Cowling & Hilton-Taylor (1994).**

3.4   Patterns and correlates of species endemism

3.4.1   Patterns

Centres of exceptional species endemism in southern Africa have recently been identified under the auspices of the IUCN Plant Conservation Programme (Davis, Heywood & Hamilton 1994). The formal criteria for the inclusion of sites as centres is that they must both be species-rich and have high levels of endemism. According to these criteria, there are eight such centres in southern Africa [seven centres from Davis *et al.* (1994) and the Wolkberg Centre of Matthews, Van Wyk & Bredenkamp (1993)]. They are not randomly distributed across the subcontinent, but are located in an almost continuous arc below, and including sectors of, the Great Escarpment (Fig. 3.4). Many are associated with

ancient (pre-African surface) elevated areas (Chap. 1, this volume). The centres are as follows: the Wolkberg Centre of the northeastern Transvaal Escarpment (Matthews *et al.* 1993) and the Eastern Mountain Centre of the KwaZulu-Natal Drakensberg and associated uplands (Killick 1994), both in the Afromontane Region; the Maputaland, Pondoland (Van Wyk 1994) and Albany Centres (Davis *et al.* 1994) in the Tongaland–Pondoland Region; the Cape Centre comprising the Cape Region minus enclaves of succulent karoo vegetation (Rebelo 1994); and the Succulent Karoo (Hilton-Taylor 1994a) and Kaokoveld (Hilton-Taylor 1994b) Centres in the Karoo–Namib Region (Fig. 3.4; Table 3.5). These centres encompass diverse environments and vegetation types, ranging from perhumid summer rainfall areas with subtropical rainforests to arid winter-rainfall climates with dwarf succulent shrublands (Table 3.5).

Table 3.5 **Characteristics of southern African centres of plant species endemism. Species number includes vascular plants only. Data from Davis et al. (1994), Matthews et al. (1993) and Hilton-Taylor (1996)**

Centre	Area (km^2)	No. spp.	Endemics (%)	Rainfall (mm/yr^{-1})	Rainfall (season)	Phytogeographical region
Wolkberg	5980	2700	4	500–2000	Summer	Afromontane
Maputaland	26 734	1100	15	600–1200	Summer	Tongaland–Pondoland
Eastern Mountain	40 000	1750	30	1500–2000	Summer	Afromontane
Pondoland	1880	1500	8	1000–2000	Summer	Tongaland–Pondoland
Albany	22500	2000	10	350–750	All year	Tongaland–Pondoland
Succulent Karoo	111 212	4849	40	20–300	Winter to all year	Karoo–Namib
Cape	90 000	7000	80	250–3000	Winter to all year	Cape
Kaokoveld	70 000	952	12	10–300	Summer	Karoo–Namib

Table 3.6 **Numbers of endemic species in southern African centres and other 'hot-spots' (sensu Myers 1988, 1990) of plant diversity and endemism**

Region	Area (km^{-2})	% Earth's area	No. endemic spp.	Endemics as % of total plant spp.
Tropical rainforests[a]	2 428 000	1.6	37235	14.9
Mediterranean-climate	765 400	0.5	14165	5.6
Southern Africa	305 306	0.2	8946	3.6
Total[b]	3 408 706	2.2	54498	21.8

[a] Tropical rainforests include 14 areas and the mediterranean-climate region includes three areas. Data are from Myers (1988, 1990) and Hopper (1992).
[b] The Cape Centre is included in both mediterranean-climate regions and the southern African centres, but duplication has been avoided in the totals.

The eight centres together include a total of 8946 endemic species, some 55% of southern Africa's endemic flora in 12% of the region's area. With some 5600 endemic species, the Cape Centre has southern Africa's and the world's largest concentration of endemic plant species (Myers 1990). The Succulent Karoo Centre, with 1954 endemic plant species, comprises the highest concentration of endemic plants for any semi-arid area in the world (Cowling & Hilton-Taylor 1994; Hilton-Taylor 1996). These two southwestern and essentially winter-rainfall centres harbour some 46% of southern Africa's endemic flora; the remaining centres include 1232 endemic species or only 8% of the total for the subcontinent. The remaining endemic species, some 45% of the total, are more widely distributed across the region.

The southern African centres include about 3.6% of the world's flora in 0.2% of the Earth's surface, a ratio considerably higher than mediterranean-climate and tropical rainforest centres or 'hot-spots' identified by Myers (1988, 1990) (Table 3.6). The southern African centres of endemism, therefore, represent one of the world's most important areas for interventions aimed at reducing rates of species extinctions (Cowling & Hilton-Taylor 1994).

3.4.2   Correlates

Are the endemic floras in these centres random assemblages with regard to taxonomic affinity and biological traits and are they over-represented in certain habitats (Kruckeberg & Rabinowitz 1985; Major 1988; Cowling & Holmes 1992b)? Are the endemics isolated and disjunct relics of ancient lineages or are they the products of recent speciation surrounded by many close taxonomic relatives (Stebbins & Major 1965; Cronk 1992)? In this section we make whatever generalizations are possible on the correlates of species endemism in southern African centres.

3.4.2.1   Habitat aspects

There are few data on the habitat aspects of species-level endemism in the southern African centres. One generalization for the subcontinent is that endemics are relatively more common on nutritionally peculiar or isolated substrata, a phenomenon common in other endemic-rich areas (e.g. Raven & Axelrod 1978; Hopper 1979; Papanicolaou, Babalonas & Kokkinii 1983; Gentry 1986; Brown & Prance 1987).

In the Cape Centre, high levels of endemism are associated with fynbos vegetation on nutrient-poor soils largely derived from ancient, sterile quartzites; levels of endemism in non-fynbos vegetation on more fertile soils are much lower (Dahlgren 1968; Cowling 1983; Cowling & Holmes 1992b). Within fynbos landscapes, each substratum may support a distinct endemic flora (Cowling et al. 1992), and unusual and isolated substrata (e.g. limestone) may support a disproportionately high number of endemics (Cowling & Holmes 1992b; Willis, Cowling & Lombard 1996). In the Wolkberg, 71% of endemics occur on soils derived from nutrient-poor quartzites; the remainder are associated with soils derived from calcium- and magnesium-rich dolomites (Matthews et al. 1993). The Pondoland Centre is exclusively associated with an outcrop of ancient quartzite, which is part of the Cape Supergroup (Van Wyk 1990; 1994). Many of the Maputaland endemics occur in grasslands on infertile, sandy soils (Van Wyk 1994). In the Succulent Karoo, soil type plays an important role in determining vegetation patterns (Jürgens 1986; Esler & Cowling 1993) and local endemics are often associated with unusual substrata and ancient geomorphic surfaces (Van Jaarsveld 1987; Hammer 1993; Hilton-Taylor 1996). Clearly, edaphic speciation (Raven 1964; Stebbins & Major 1965; Kruckeberg 1986) has played an important role in the evolution of the southern African endemic flora (Cowling 1983; Van Wyk 1990; Cowling *et al.* 1992; Matthews *et al.* 1993). Alternatively, unusual geological substrata may also provide a refuge from competition for palaeoendemics (Zedler, Gautier & Jacks 1984; Major 1988) as suggested for Pondoland by Van Wyk (1990, 1994).

In the Eastern Mountain and Wolkberg Centres almost all endemics are associated with grassland habitats (Hilliard & Burtt 1987; Matthews *et al.* 1993). These are drier and more exposed to the elements (including fire) than adjacent forest habitats. In the Albany Centre, endemism is higher in xeric than mesic succulent thicket (Cowling 1983; Hoffman & Cowling 1991). In the Cape Centre, McDonald & Cowling (1995) found that in a mountain fynbos landscape, most endemics occur on the wetter, coastal slopes and, in terms of habitat size, endemics are over-represented in high altitude, wet habitats. A similar pattern is evident on the Cape Peninsula where 36% of all endemics are found in the extensive proteoid shrublands of the drier, lower slopes, whereas, relative to their area, endemics are over-represented in wet fynbos types at higher altitudes (Trinder-Smith, Cowling & Linder 1996).

3.4.2.2 *Phylogenetic aspects*
Endemics in southern African centres are not a random assemblage phylogenetically. This is illustrated in Table 3.7, which shows species endemism in families of flowering plants in selected floras from three centres (Fig. 3.4). In the southern KwaZulu-Natal Drakensberg flora from the Eastern Mountain Centre, families with higher than average levels of endemism include the Asteraceae, Scrophulariaceae and Ericaceae. In most of the larger families, the proportion of endemics is similar to that of the total flora, and endemics are significantly under-represented in Poaceae and Orchidaceae. Similar patterns to these probably exist for the Wolkberg Centre (Matthews *et al.* 1993).

In the flora of the Agulhas Plain, a lowland region in the Cape Centre, endemics are significantly over-represented in six families including typical Cape families such as the Ericaceae, Restionaceae and Proteaceae, but also the largely Karoo–Namib family, Mesembryanthemaceae (Table 3.7). In contrast to the Eastern Moun-

tain flora, Asteraceae and Scrophulariaceae do not have proportionally high levels of endemism; in common with that flora, Poaceae and Orchidaceae are under-represented in terms of endemics. Similar patterns are evident for other floras from this Centre, although regional differences do exist (Cowling *et al.* 1992; McDonald & Cowling 1995; Willis *et al.* 1996; Trinder-Smith *et al.* 1996).

Higher than average endemism in the Gariep flora from the Succulent Karoo Centre is associated with three families, namely the Mesembryanthemaceae, Crassulaceae and Asclepiadaceae (Table 3.7). These are all families of predominantly succulent species which are concentrated in the winter-rainfall areas of the Karoo–Namib Region (Gibbs Russell 1987; Hilton-Taylor 1987, 1994a, 1996; Van Jaarsveld 1987). In an analysis of the entire flora of the Succulent Karoo Centre, significantly higher than average endemism was recorded for the three above-mentioned families as well as the entirely succulent Aloaceae, and four predominantly geophytic families (Amaryllidaceae, Eriospermaceae, Iridaceae and Oxalidaceae) (Hilton-Taylor 1996). Over-representation of endemics in the Mesembryanthemaceae, Crassulaceae and Asclepiadaceae, as well as succulent Euphorbiaceae, is also evident for an Albany Centre flora (Cowling & Holmes 1991). Interestingly, the Mesembryanthemaceae have produced numerous endemics in both dry karroid and wetter, fire-prone fynbos habitats (Hartmann 1991). Highly under-represented families among endemics in the Gariep flora include Asteraceae, Poaceae and Scrophulariaceae (see also Hilton-Taylor 1996).

In conclusion, there are clear phylogenetic correlates of endemism at the family level in the southern African flora, but these show greater differences than similarities in endemic-rich floras from the Afromontane, Cape and Karoo–Namib Regions. Interestingly, the same families often have different proportions of endemics in each region. This is to be expected, given the distinct floras at the generic level (White 1983; Gibbs Russell 1987) and the different ecological conditions in each of the centres (Table 3.5). Nonetheless, large and endemic-rich families such as the Ericaceae (Oliver 1991) and the Mesembryanthemaceae (Hartmann 1991; Ihlenfeldt 1994) have produced numerous endemics in more than one region. The Poaceae and Orchidaceae have consistently low proportions of endemic species.

3.4.2.3 *Biological aspects*
Endemics from southern African centres are not a random assemblage with regard to growth form and other biological attributes. In the southern KwaZulu-Natal Drakensberg flora (Eastern Mountain Centre), forbs and shrubs are more frequent, and other growth forms

Table 3.7 **Species endemism (vascular plants only) in the largest families of flowering plants in three southern African centres of endemism. Data from Cowling & Hilton-Taylor (1994).** *** $P < 0.001$; ** $P < 0.01$; * $P < 0.05$; NS, not significant

Family	No. spp. (%)		χ^2
	Endemic[a]	Non-endemic	
Eastern Mountain Centre: S. KwaZulu-Natal Drakensberg (1115 km^2)			
Total flora	373 (29.3)	898 (70.7)	—
Endemism significantly higher than total flora			
Asteraceae	118 (41.5)	167 (58.5)	35.7***
Scrophulariaceae	36 (45.6)	43 (54.4)	9.9**
Ericaceae	15 (57.6)	11 (42.3)	8.9*
Endemism same as total flora			
Liliaceae[b]	20 (23.8)	64 (76.2)	1.1[NS]
Fabaceae	14 (21.5)	51 (78.5)	1.6[NS]
Iridaceae	20 (30.8)	45 (69.2)	0.0[NS]
Cyperaceae	14 (23.7)	45 (76.3)	0.7[NS]
Asclepiadaceae	13 (29.5)	31 (70.5)	0.0[NS]
Campanulaceae	9 (36.0)	16 (64.0)	0.3[NS]
Apiaceae	7 (33.3)	14 (66.6)	0.0[NS]
Gentianaceae	6 (28.6)	15 (71.4)	0.0[NS]
Crassulaceae	6 (30.0)	14 (70.0)	0.09[NS]
Geraniaceae	5 (25.0)	15 (75.0)	0.0[NS]
Endemism significantly lower than total flora			
Poaceae	19 (17.6)	89 (82.4)	7.3**
Orchidaceae	15 (18.1)	68 (81.9)	4.9*
Cape Centre: Agulhas Plain (1609 km^2)			
Total flora	513 (29.3)	1238 (70.7)	—
Endemism significantly higher than total flora			
Ericaceae	80 (64.5)	44 (35.5)	78.1***
Restionaceae	35 (39.8)	53 (60.2)	4.4*
Mesembryanthemaceae	26 (44.8)	32 (55.2)	6.2*
Proteaceae	39 (67.2)	19 (32.8)	39.8***
Rutaceae	34 (72.3)	13 (27.7)	41.1***
Polygalaceae	19 (54.3)	16 (45.7)	9.6**
Endemism same as total flora			
Asteraceae	52 (25.6)	151 (74.4)	1.3[NS]
Iridaceae	38 (25.5)	111 (74.5)	0.9[NS]
Fabaceae	49 (35.5)	89 (64.5)	2.5[NS]
Campanulaceae	14 (26.9)	38 (73.1)	0.1[NS]
Thymelaeaceae	15 (41.7)	21 (58.3)	2.1[NS]
Endemism significantly lower than total flora			
Poaceae	3 (4.6)	62 (95.4)	19.0***
Liliaceae[b]	12 (16.0)	63 (84.0)	6.0*
Cyperaceae	7 (11.1)	56 (88.9)	9.5**
Scrophulariaceae	8 (14.3)	48 (85.7)	5.6*
Orchidaceae	2 (5.3)	36 (94.7)	9.7**
Succulent Karoo Centre: Gariep (40 931 km^2)			
Total flora	397 (19.8)	1613 (81.2)	—
Endemism significantly higher than total flora			
Mesembryanthemaceae	140 (54.7)	116 (45.3)	229.6***
Crassulaceae	40 (40.8)	58 (59.2)	27.5***
Asclepiadaceae	23 (42.6)	31 (57.4)	16.8***
Endemism same as total flora			
Liliaceae[b]	37 (21.6)	134 (78.4)	2.6[NS]
Fabaceae	17 (18.3)	76 (81.7)	0.1[NS]
Iridaceae	18 (30.0)	42 (70.0)	3.5[NS]
Geraniaceae	15 (26.8)	41 (73.2)	1.4[NS]
Euphorbiaceae	11 (20.0)	44 (80.0)	0.0[NS]
Chenopodiaceae	6 (14.3)	36 (85.7)	0.5[NS]
Acanthaceae	3 (7.9)	35 (92.1)	2.7[NS]
Amaryllidaceae	11 (33.3)	22 (66.6)	3.1[NS]
Zygophyllaceae	2 (6.1)	31 (93.9)	3.1[NS]
Brassicaceae	5 (17.2)	24 (82.8)	0.0[NS]
Endemism significantly lower than total flora			
Asteraceae	29 (9.8)	266 (90.2)	20.7***
Poaceae	9 (7.4)	112 (92.6)	11.5***
Scrophulariaceae	3 (3.4)	86 (96.6)	14.7***
Aizoaceae	1 (1.2)	79 (98.8)	16.8***
Sterculiaceae	1 (3.5)	28 (96.5)	3.9*

[a] Areas of endemism are the entire Eastern Mountain Centre for S. KwaZulu-Natal Drakensberg flora; the South Western Centre (Weimarck 1941) for the Agulhas Plain flora and the Gariep Centre (Hilton-Taylor 1994a) for Gariep flora.
[b] Sensu lato.

less frequent, among endemics than non-endemics (Fig. 3.5a). Despite substantial areas of forest in this and the Wolkberg Centre, there are no trees endemic to these regions: most endemics are forbs and low shrubs associated with grasslands (Hilliard & Burtt 1987; Matthews *et al.* 1993; Killick 1994). This is not the case for the Maputaland and Pondoland Centres, which harbour many tree endemics (Van Wyk 1994). However, most endemics in these areas are also forbs and low shrubs of grassland habitats.

The biological aspects of endemism in the Cape Region have been reasonably well studied (Cowling & Holmes 1992b; Cowling *et al.* 1992; McDonald & Cowling 1995; McDonald *et al.* 1995; Trinder-Smith *et al.* 1996; Willis *et al.* 1996). The generalization from both lowland and mountain floras is that there is a greater than average chance that an endemic will be a fire-sensitive, low to dwarf shrub (Fig. 3.5b) with soil-stored seed banks and short dispersal distance. Both forbs and geophytes are under-represented as endemics and tree endemics are absent. Levels of endemism among graminoids in Cape Centre floras are the highest in southern Africa, owing to the high number of local endemics in the Restionaceae (Linder 1991).

The growth form profile of the endemics in the Gariep flora from the Succulent Karoo Centre is identical to an Albany flora profile (Cowling & Holmes 1991) and similar to those from the Cape Centre except for a reverse of the pattern for graminoids (relatively higher in Cape floras) and geophytes (lower in Cape floras) (Fig. 3.5c; see also Hilton-Taylor 1996). It is interesting that despite southern Africa having the richest geophyte flora in the world (Goldblatt 1978), endemic geophytes are more frequent than widespread species only in semi-arid winter-rainfall areas (Hilton-Taylor 1996; Fig. 3.5). Succulents are massively over-represented as endemics in the Gariep flora, where 60% of endemics are succulents (Cowling & Hilton-Taylor 1994). A similar pattern exists for other Succulent Karoo floras (Hilton-Taylor 1996) as well as an Albany Centre flora (Cowling & Holmes 1991). Tree endemics in southern African arid lands are rare except for the Kaokoveld Centre, where there are a number of taxonomically isolated endemic trees (Goldblatt 1978; Hilton-Taylor 1994b).

In conclusion, biological aspects of endemism show important differences across southern African centres. Endemic forbs are relatively common in moist, summer-rainfall eastern centres; endemic shrubs are common in all centres; endemic succulents are common in the semi-arid winter and non-seasonal rainfall zone; endemic geophytes are common in the semi-arid southwest; and endemic trees are rare everywhere except for a relatively low occurrence in the Maputaland, Pondoland and Kaokoveld centres.

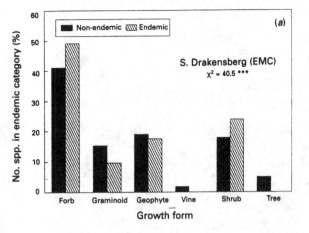

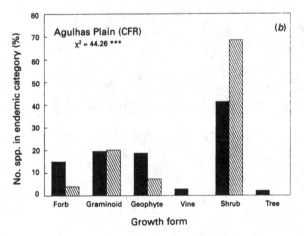

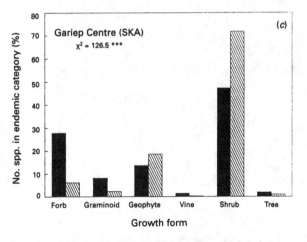

Figure 3.5 **Association between endemism and growth form in floras (higher plants only) from three southern African centres of endemism.** (*a*) S. Drakensberg flora is from the Eastern Mountain Centre; (*b*) Agulhas Plain flora from the Cape Centre; (*c*) Gariep flora from the Succulent Karoo Centre. From Cowling & Hilton-Taylor (1994).

Figure 3.6 **Endemic flora of southern Africa.** (*a*) *Nerine pancratioides* **(Amaryllidaceae), restricted to the Eastern Mountain Centre** (Photo: J.C. Paterson Jones). (*b*) *Encephalartos longifolius* **(Zamiaceae) from the Albany Centre)** (Photo: J.C. Paterson Jones). (*c*) *Erica pageana* **(Ericaceae) from the Cape Centre** (Photo: J.C. Paterson Jones).

(*d*) *Cheiridopsis robusta* **(Mesembryanthemaceae) from the succulent karoo** (Photo: R.M. Cowling). (*e*) **The remarkable** *Welwitschia mirabilis* **(Welwitschiaceae) from the Namib Desert** (Photo: R.M. Cowling). (*f*) *Pachypodium lealii* **(Apocynaceae) from the Kaokoveld Centre** (Photo: R.M. Cowling).

3.4.2.4 Age

It is very difficult without the appropriate phylogenetic and historical data to estimate the age of an endemic flora. There are, however, shortcuts, which are based on taxonomic relatedness, habitat preference and growth form (Cronk 1987, 1992). Following on Cronk's (1992) scheme for the endemic floras of certain Atlantic Ocean islands, we recognize three main types of endemics.

Type 1 endemics are of Miocene age or older. They show extreme taxonomic isolation (low ranking taxa) and are mostly trees of tropical and subtropical forests, relicts of an Africa almost entirely covered in humid forests (Axelrod & Raven 1978; Coetzee & Muller 1984). Many Kaokoveld endemics present a problem, since they belong to taxa that are extremely isolated taxonomically, show continental-scale disjunctions (Goldblatt 1978) but occur in arid to semi-arid environments. This suggests pre-Pliocene aridity along southern Africa's west coast (Goldblatt 1978; Ward *et al.* 1983; Van Zinderen Bakker & Mercer 1986; but see Chap. 4, this volume). We include these (e.g. *Welwitschia mirabilis*) in our Type 1 category (Fig. 3.6). Type 1 endemics, being mainly trees, are not common in the southern African flora, but those that do occur are concentrated in the Kaokoveld and Pondoland centres. The latter region includes one monotypic family and six monotypic genera of forest trees (Van Wyk 1990, 1994).

Type 2 endemics are of Late Miocene or Pliocene age, show moderate taxonomic isolation and are mostly trees and shrubs of tropical or subtropical, seasonally dry vegetation. This group includes trees from all centres as well as some taxonomically isolated shrubs in the Cape Region (e.g. some members of the Penaeaceae, Bruniaceae, etc.).

Type 3 endemics are of Late Pliocene or Pleistocene age, show little taxonomic or geographic disjunction from their nearest relatives, and are usually forbs, geophytes or low shrubs of arid to semi-arid or temperate vegetation (Fig. 3.6). This group includes the overwhelming majority of southern African endemics.

This analysis, albeit superficial, suggests that most southern African endemics are members of relatively young lineages (neoendemics) with numerous close relatives. Many of these endemic species' clusters are ecologically uniform (Bond 1989; Cowling *et al.* 1992). Palaeoendemics are few and concentrated on the eastern seaboard and in the arid to semi-arid west.

3.5 Concluding discussion

Situated on a continental landmass, the southern African flora is remarkable for its high plant richness and endemism. Endemism is pronounced at all taxonomic levels, confirming the region's status as a distinct phytogeographical unit – the southern, temperate element of the African flora (Goldblatt 1978). Richness is particularly pronounced at the species level, and especially so in the Cape Region, where numerous species are associated with few genera and families (Linder *et al.* 1992). The floras of the other regions are more balanced, although highly species-rich genera and families are also a feature of the winter-rainfall portion of the Karoo–Namib Region (Goldblatt 1978; Ihlenfeldt 1994; Hilton-Taylor 1996). Both endemism and richness are considerably higher in southern Africa than in tropical Africa (Goldblatt 1978; Gibbs Russell 1985; Cowling *et al.* 1989), a pattern which contradicts the generalization that plant richness decreases, and range size increases, with increasing latitude and associated climatic extremes (Pianka 1966; Stevens 1989; cf. Cowling & Samways 1996).

With the exception of the Kaokoveld Centre in the northern Namib Desert and parts of the Succulent Karoo Centre, there is a general concordance between high species richness and high endemism in southern Africa. This is not the case for plants in many species-rich tropical regions (Gentry 1992). Endemism (and richness; Chap. 19, this volume) is particularly low in the large, semi-arid to subhumid, interior plateau of the subcontinent. It appears that endemics are associated with a predictable set of habitat factors. However, with the exception of some work in the Cape Region (Cowling *et al.* 1992; McDonald & Cowling 1995; Trinder-Smith *et al.* 1996), little has been done to quantify and model the habitat correlates of plant endemism in southern Africa.

Assemblages of species endemic to particular phytogeographical regions in southern Africa are not randomly organized with regard to phylogenetic lineage and biological traits. The existence of discernible phylogenetic correlates of endemism implies that range size may be an evolutionarily stable character (Jablonski 1987; Ricklefs & Latham 1992; Lawton 1993). These endemic-rich lineages are associated with a predictable suite of biological traits: in the Cape Region these amount to low shrub stature, fire sensitivity and limited dispersal distance (Cowling & Holmes 1992b; McDonald *et al.* 1995; Trinder-Smith *et al.* 1996). By promoting fire-induced population reduction and fragmentation (Cowling & Holmes 1992b), these traits may be the cause rather than the consequence of narrow geographical range (Vrba 1980; Kunin & Gaston 1993) – lineages possessing them (e.g. Ericaceae, Proteaceae, Rutaceae) have a greater potential for diversification and the production of numerous local endemics. In this respect, the identification of the biological aspects of diversification in endemic-rich lineages in the Karoo–Namib and Afromontane Regions remains a major challenge. Ihlenfeldt (1994) discusses some aspects of the recent (post-Miocene) and spectacular diversification in the southwestern Karoo–Namib Region of the succulent family Mesembryanthemaceae. However, more research is required to understand the unrivalled (both for arid regions and higher plants) diversification of this group.

The reasons for the uniquely high species richness and endemism of the southern African flora are varied and complex (Adamson 1938; Weimarck 1941; Levyns 1964; Goldblatt 1978; Gibbs Russell 1985; Cowling et al. 1989). The subcontinent comprises a transition from tropical, summer-rainfall to temperate, winter-rainfall environments (Chap. 2, this volume); these environments have distinct floras at the species- and higher taxon levels (Gibbs Russell 1987) and their overlap in the region certainly contributes to the overall size of the flora (Goldblatt 1978). The high variation in topography, soils and climate at a regional scale (Chaps. 1 and 2, this volume) result in steep ecological gradients, along which many species can be packed (Chap. 19, this volume). Yet more species are packed into southern African landscapes than in topographically and climatically similar landscapes in other parts of the world such as southwestern North America and parts of Australia (Westoby 1988;

Cowling et al. 1989, 1992; Chap. 19, this volume). The pronounced, but non-catastrophic, climatic fluctuations during the Quaternary, and persistence in refugia of low-ranking or relictual taxa, which Goldblatt (1978) cites as factors contributing to the subcontinent's high richness and endemism, are not unique to southern Africa (Axelrod & Raven 1978). The massive post-Miocene diversification within a limited number of lineages, which has produced flocks of closely related and ecologically uniform species and infra-specific taxa, is unparalleled in the world. Processes similar to those which have produced plant richness and endemism of the Cape Region have probably operated in the species- and endemic-rich mediterranean-climate regions of southwestern Australia and California (Cowling et al. 1992; Chap. 19, this volume), but the scale of diversification in the Cape is without equal. It may be that the uniquely rich and endemic flora of southern Africa is largely the result of the presence in the area of lineages that are especially, and incidentally, prone to diversification.

3.6 References

Acocks, J.P.H. (1953). Veld Types of South Africa. Memoirs of the Botanical Survey of South Africa, 28, 1–192.

Adamson, R.S. (1938). The Vegetation of South Africa. London: British Empire Vegetation Committee.

Arnold, T.H. & De Wet, B.C. (1993). Plants of southern Africa: names and distribution. Memoirs of the Botanical Survey of South Africa, 62, 1–825.

Axelrod, D.I. & Raven, P.H. (1978). Late Cretaceous and Tertiary vegetation history of Africa. In Biogeography and Ecology of Southern Africa, ed. M. J. A. Werger, pp. 79–130. The Hague: Junk.

Bolus, H. (1875). Letter from Mr. Bolus to Dr. J. B. Hooker. Journal of the Linnean Society, 14, 482–4.

Bolus, H. (1886). Sketch of the flora of South Africa. In Official Handbook of the Cape of Good Hope. Cape Town: Richards.

Bolus, H. (1905). Sketch of the floral regions of South Africa. In Science in South Africa, ed. W. Flint & J.D.F. Chilchrist, pp. 198–240. Cape Town: Maskew Miller.

Bond, P. & Goldblatt, P. (1984). Plants of the Cape flora. A descriptive catalogue. Journal of South African Botany Supplementary Volume, 13, 1–455.

Bond, W.J. (1989). Describing and conserving biotic diversity. In Biotic Diversity in Southern Africa: Concepts and Conservation, ed. B.J. Huntley, pp. 2–18. Cape Town: Oxford University Press.

Brown, K.S. (Jr.) & Prance, G.T. (1987). Soils and vegetation. In Biogeography and Quaternary History in Tropical America, ed. T.C. Whitmore & G.T. Prance, pp. 19–45. Oxford: Clarendon Press.

Burtt, B.L. (1971). From the south: an African view of the floras of western Asia. In Plant Life of Southwest Asia, ed. P.H. Davis, P.C. Harper & I.C. Hedge, pp. 134–49. Aberdeen: Edinburgh Botanical Society.

Carlquist, S. (1976). Wood anatomy of Roridulaceae: ecological and phylogenetic implications. American Journal of Botany, 63, 1003–8.

Coetzee, J.A. (1978). Climatic and biological changes in southwestern Africa during the late Cainozoic. Palaeoecology of Africa, 10, 13–29.

Coetzee, J.A. & Muller, J. (1984). The phytogeographic significance of some extinct Gondwanan pollen types from the Tertiary of the southwestern Cape (South Africa). Annals of the Missouri Botanical Garden, 70, 1088–99.

Coetzee, J.A., Scholtz, A. & Deacon, H.J. (1983). Palynological studies and vegetation history of the fynbos. In Fynbos Palaeoecology: A Preliminary Synthesis, ed. H. J. Deacon, Q.B. Hendey & J.J.N. Lambrechts, pp. 156–73. South African National Scientific Programmes Report 75. Pretoria: CSIR.

Compton, R.H. (1929). The flora of the Karoo. South African Journal of Science, 26, 160–5.

Cowling, R.M. (1983). Phytochorology and vegetation history in the south eastern Cape, South Africa. *Journal of Biogeography*, **10**, 393–419.

Cowling, R.M., Gibbs Russell, G.E., Hoffman, M.T. & Hilton-Taylor, C. (1989). Patterns of plant species diversity in southern Africa. In *Biotic Diversity in Southern Africa: Concepts and Conservation*, ed. B.J. Huntley, pp. 19–50. Cape Town: Oxford University Press.

Cowling, R. M. & Hilton-Taylor, C. (1994). Patterns of plant diversity and endemism in southern Africa: an overview. In *Botanical Diversity in Southern Africa*, ed. B. J. Huntley, pp. 31–52. *Strelitzia*, 1. Pretoria: National Botanical Institute.

Cowling, R.M. & Holmes, P.M. (1991). Subtropical thicket in the south eastern Cape: a biogeographical perspective. In *Proceedings of the First Valley Bushveld/Subtropical Thicket Symposium*, ed. P.J.K. Zacharias, G.C. Stuart-Hill & J. Midgley, pp. 3–4. Howick: Grassland Society of Southern Africa, Special Publication.

Cowling, R.M. & Holmes, P.M. (1992a). Flora and vegetation. In *The Ecology of Fynbos: Nutrients, Fire and Diversity*, ed. R.M. Cowling, pp. 23–61. Cape Town: Oxford University Press.

Cowling, R.M. & Holmes, P.M. (1992b). Endemism and speciation in a lowland flora from the Cape Floristic Region. *Biological Journal of the Linnean Society*, **47**, 367–83.

Cowling, R.M., Holmes, P.M. & Rebelo, A.G. (1992). Plant diversity and endemism. In *The Ecology of Fynbos: Nutrients, Fire and Diversity*, ed. R. M. Cowling, pp. 62–112. Cape Town: Oxford University Press.

Cowling, R.M. & Samways, M.J. (1996). Predicting global patterns of endemic plant species richness. *Biodiversity Letters*, **2**, 17–21.

Cronk, Q.C.B. (1987). The history of the endemic flora of St. Helena: a relictual series. *New Phytologist*, **105**, 509–20.

Cronk, Q.C.B. (1992). Relict floras of Atlantic islands: patterns assessed. *Biological Journal of the Linnean Society*, **46**, 91–103.

Dahlgren, R. (1968). Distribution and substrate in the South African genus, *Aspalathus*. *Botaniska Notiser*, **116**, 431–72.

Davis, S.D., Heywood, V.H. & Hamilton, A.C. (eds.) (1994). *Centres of Plant Diversity. A Guide and Strategy for their Conservation*. Volume 1 Europe, Africa, South West Asia and the Middle East. Cambridge: IUCN Publications Unit.

De Winter, B. (1971). Floristic relationships between the northern and southern arid areas in Africa. *Mitteilungen der Botanischen Staatssammlung München*, **10**, 424–37.

Engler, A. (1882). *Versuch einer Entwicklungsgeschinchte der Pflanzenwalt insbesondere der Florengebiete seit der Tertiärperiode*, Volume 2. Leipzig: Engelmann.

Esler, K.J. & Cowling, R.M. (1993). Edaphic factors and competition as determinants of pattern in South African Karoo vegetation. *South African Journal of Botany*, **59**, 287–95.

Gentry, A.H. (1986). Endemism in tropical vs. temperate plant communities. In *Conservation Biology*, ed. M. Soulé, pp. 153–81. Sunderland, Massachusetts: Sinauer Press.

Gentry, A.H. (1992). Tropical forest biodiversity: distributional patterns and their conservation significance. *Oikos*, **63**, 19–28.

Gibbs Russell, G.E. (1985). Analysis of the size and composition of the southern African flora. *Bothalia*, **15**, 613–29.

Gibbs Russell, G.E. (1987). Preliminary floristic analysis of the major biomes in southern Africa. *Bothalia*, **17**, 213–27.

Goldblatt, P. (1978). An analysis of the flora of southern Africa: its characteristics, relationships and origins. *Annals of the Missouri Botanical Garden*, **65**, 369–436.

Good, R. (1974). *The Geography of the Flowering Plants*, 4th edn. London: Longmans.

Gunn, M. & Codd, L.E. (1981). *Botanical Exploration of Southern Africa*. Cape Town: Balkema.

Hammer, S.A. (1993). *The Genus Conophytum: a Conograph*. Pretoria: Succulent Plant Productions.

Hartmann, H.E.K. (1991). Mesembryanthema. *Contributions from the Bolus Herbarium*, **13**, 75–157.

Hedberg, O. (1965). Afro-alpine flora elements. *Webbia*, **19**, 519–29.

Hilliard, O.M. & Burtt, B.L. (1987). *The Botany of the Southern Natal Drakensberg*. Cape Town: National Botanic Gardens.

Hilton-Taylor, C. (1987). Phytogeography and origins of the Karoo flora. In *The Karoo Biome: a Preliminary Synthesis. Part 2 –Vegetation and History*, ed. R.M. Cowling & P.W. Roux, pp. 70–95. *South African National Scientific Programmes Report* 142. Pretoria: CSIR.

Hilton-Taylor, C. (1994a). Karoo–Namib Regional Centre of Endemism: CPD Site AF51. Western Cape Domain (Succulent Karoo) Republic of South Africa and Namibia. In *Centres of Plant Diversity: a Guide and Strategy for their Conservation*. Volume 1 Europe, Africa, South West Asia and the Middle East, ed. S.D. Davis, V.H. Heywood & A.C. Hamilton, pp. 204–17. Cambridge: IUCN Publications Unit.

Hilton-Taylor, C. (1994b). Karoo–Namib Regional Centre of Endemism: CPD Site AF50. The Kaokoveld, Namibia and Angola. In *Centres of Plant Diversity: a Guide and Strategy for their Conservation*. Volume 1 Europe, Africa, South West Asia and the Middle East, ed. S.D. Davis, V.H. Heywood & A.C. Hamilton, pp. 201–3. Cambridge: IUCN Publications Unit.

Hilton-Taylor, C. (1996). Patterns and characteristics of the flora of the succulent karoo biome, southern Africa. In *The Biodiversity of African Plants*, ed. L.J.G. van der Maesen, X.M. van der Burgt & J.M. van Medenbach de Rooy, pp. 58–72. Dordrecht: Kluwer Academic Publishers.

Hilton-Taylor, C. & Le Roux, A. (1989). Conservation status of the fynbos and karoo biomes. In *Biotic Diversity in Southern Africa: Concepts and Conservation*, ed. B.J. Huntley, pp. 202–23. Cape Town: Oxford University Press.

Hoffman, M.T. & Cowling, R.M. (1991). Phytochorology and endemism along aridity and grazing gradients in the lower Sundays River Valley, South Africa: implications for vegetation history. *Journal of Biogeography*, **18**, 189–201.

Hopper, S.D. (1979). Biogeographical aspects of speciation in the south-west Australian flora. *Annual Review of Ecology & Systematics*, 10, 399-422.

Hopper, S.D. (1992). Patterns of plant diversity at the population and species level in south-west Australian mediterranean ecosystems. In *Biodiversity in Mediterranean Ecosystems in Australia*, ed. R.J. Hobbs, pp. 27-46. Norton, New South Wales: Surrey, Beatty & Sons.

Ihlenfeldt, H.D. (1994). Diversification in an arid world – the Mesembryanthemaceae. *Annual Review of Ecology & Systematics*, 25, 521-46.

Jablonski, D. (1987). Heritability at the species level: analysis of geographic ranges of Cretaceous molluscs. *Science*, 238, 360-3.

Jürgens, N. (1986). Untersuchungen zur Ökologie sukkulenter Pflanzen des südlichen Afrika. *Mitteilungen aus dem Institut für Allgemeine Botanik Hamburg*, 21, 139-365.

Jürgens, N. (1991). A new approach to the Namib Region. I: Phytogeographic sub-division. *Vegetatio*, 97, 21-38.

Killick, D.J.B. (1978). The Afro-Alpine Region. In *Biogeography and Ecology of Southern Africa*, ed. M.J.A. Werger, pp. 515-42. The Hague: Junk.

Killick, D.J.B. (1994). Drakensberg Alpine Region: CPD Site AF82. Drakensberg Alpine Region, Lesotho and South Africa. In *Centres of Plant Diversity: a Guide and Strategy for their Conservation*. Volume 1 Europe, Africa, South West Asia and the Middle East, ed. S.D. Davis, V.H. Heywood & A.C. Hamilton, pp. 257-60. Cambridge: IUCN Publications Unit.

Kruckeberg, A.R. (1986). An essay: the stimulus of unusual geologies for plant speciation. *Systematic Botany*, 11, 455-63.

Kruckeberg, A.R. & Rabinowitz, D. (1985). Biological aspects of endemism in higher plants. *Annual Review of Ecology & Systematics*, 16, 449-79.

Kunin, W.E. & Gaston, K.J. (1993). The biology of rarity: patterns, causes and consequences. *Trends in Ecology and Evolution*, 8, 298-301.

Lawton, J.H. (1993). Range, population abundance and conservation. *Trends in Ecology and Evolution*, 8, 409-13.

Lebrun, J. (1947). La végétation de la plaine alluviale au sud du lac Édouard. Exploration des Parcs National Albert. Fasc. 1, 1-800. Brussels: Parcs National Congo Belge.

Levyns, M.R. (1964). Migrations and origins of the Cape flora. *Transactions of the Royal Society of South Africa*, 37, 85-107.

Linder, H.P. (1990). On the relationship between the vegetation and floras of the Afromontane and the Cape regions of Africa. *Mitteilungen aus dem Institut für Allgemeine Botanik Hamburg*, 23b, 777-90.

Linder, H.P. (1991). A review of the African Restionaceae. *Contributions from the Bolus Herbarium*, 13, 209-64.

Linder, H.P., Meadows, M.E. & Cowling, R.M. (1992). History of the Cape flora. In *The Ecology of Fynbos: Nutrients, Fire and Diversity*, ed. R.M. Cowling, pp. 113-34. Cape Town: Oxford University Press.

Major, J. (1988). Endemism: a botanical perspective. In *Analytical Biogeography. An Integrated Approach to the Study of Animal and Plant Distributions*, ed. A.A. Myers & P.S. Giller, pp. 117-46. New York: Chapman & Hall.

Marloth, R. (1887). Das südöstliche Kalahari-Gebiet. Ein Beitrag zur Pflanzengeographie Süd Afrikas. *Botanische Jahrbücher*, 8, 247-60.

Marloth, R. (1908). *Das Kapland*. Jena: Fischer.

Matthews, W.S., Van Wyk, A.E. & Bredenkamp, G.J. (1993). Endemic flora of the north-eastern Transvaal escarpment, South Africa. *Biological Conservation*, 63, 83-94.

McDonald, D.J. & Cowling, R.M. (1995). Towards a profile of an endemic mountain fynbos flora: Implications for conservation. *Biological Conservation*, 72: 1-12.

McDonald, D.J., Juritz, J.M., Cowling, R.M. & Knottenbelt, W.J. (1995). Modelling the biological aspects of local endemism in South African fynbos. *Plant Systematics and Evolution*, 195, 137-47.

McNeely, J.A., Miller, K.R., Reid, W.V., Mittermeier, R.A. & Werner, T.B. (1990). *Conserving the World's Biological Diversity*. Gland, Switzerland and Washington, DC: International Union for the Conservation of Nature.

Meadows, M.E. & Linder, H.P. (1993). A palaeoecological perspective on the origin of Afromontane grasslands. *Journal of Biogeography*, 20, 345-55.

Moll, E.J. & White, F. (1978). The Indian Ocean coastal belt. In *Biogeography and Ecology of Southern Africa*, ed. M.J.A. Werger, pp. 561-98. The Hague: Junk.

Monod, T. (1957). Les grands divisions chorologiques de l'Afrique. *C.S.A./C.C.T.A. Publ.* No. 24, 1-150. London: C.S.A./C.C.T.A.

Myers, N. (1988). Threatened biotas: 'Hotspots' in tropical forests. *The Environmentalist*, 8, 1-20.

Myers, N. (1990). The biodiversity challenge: expanded hot-spots analysis. *The Environmentalist*, 10, 243-55.

Oliver, E.G.H. (1991). The Ericoideae (Ericaceae) – a review. *Contributions from the Bolus Herbarium*, 13, 158-208.

Palmer, A.R. (1990). A qualitative model of vegetation history in the eastern Cape midlands, South Africa. *Journal of Biogeography*, 17, 35-46.

Papanicolaou, K., Babalonas, D. & Kokkinii, S. (1983). Distribution patterns of some Greek mountain endemic plants in relation to geological substrate. *Flora*, 174, 405-37.

Pianka, E.R. (1966). Latitudinal gradients in species diversity: a review of concepts. *The American Naturalist*, 100, 33-46.

Pole Evans, I.B. (1922). The main botanical regions of South Africa. *Memoirs of the Botanical Survey of South Africa*, 4, 49-53.

Raven, P.H. (1964). Catastrophic selection and edaphic endemism. *Evolution*, 18, 336-8.

Raven, P.H. & Axelrod, D.I. (1978). Origin and relationships of the California flora. *University of Californian Publication in Botany*, 72, 1-134.

Rebelo, A.G. (1994). Cape Regional Centre of Endemism: CPD Site AF53. Cape Floristic Region, Republic of South Africa. In *Centres of Plant Diversity: a Guide and Strategy for their Conservation*. Volume 1 Europe, Africa, South West Asia and the Middle East, ed. S. D. Davis, V.H. Heywood & A.C. Hamilton, pp. 218–24. Cambridge: IUCN Publications Unit.

Rehman, A. (1880). Geo-botaniczne stosunki potudniowéj Afryki. *Pamietnik Akademii Umiejetnosci, Wydzial Matematyczno-Przyrodniczy*, 5, 28–96.

Ricklefs, R.E. & Latham, R.E. (1992). Intercontinental correlation of geographic ranges suggests stasis in ecological traits of relict genera of temperate perennial herbs. *The American Naturalist*, 142, 1–16.

Rutherford, M.C. & Westfall, R.H. (1986). Biomes of southern Africa – an objective categorization. *Memoirs of the Botanical Survey of South Africa*, 54, 1–98.

Schulze, R.E. & McGee, O.S. (1978). Climatic indices and classifications in relation to the biogeography of southern Africa. In *Biogeography and Ecology of Southern Africa*, ed. M.J.A. Werger, pp. 19–52. The Hague: Junk.

Stebbins, G.L. & Major, J. (1965). Endemism and speciation in the California flora. *Ecological Monographs*, 35, 1–31.

Stevens, G.C. (1989). The latitudinal gradient in geographical range: how so many species coexist in the tropics. *The American Naturalist*, 133, 240–56.

Takhtajan, A. (1969). *Flowering Plants: Origin and Dispersal*. Edinburgh: Oliver & Boyd.

Taylor, H.C. (1978). Capensis. In *Biogeography and Ecology of Southern Africa*, ed. M. J. A. Werger, pp. 171–229. The Hague: Junk.

Tinley, K.L. (1977). *Framework of the Gorongosa Ecosystem*. DSc Thesis. Pretora: University of Pretoria.

Trinder-Smith, T.H., Cowling, R.M. & Linder, H.P. (1996). Profiling a besieged flora: rare and endemic plants of the Cape Peninsula, South Africa. *Biodiversity and Conservation* 5, 575–89.

Troupin, G. (1966). Étude phytocénologique du Parc National de l'Akagera et du Rwanda oriental. *Publications d'Institut National de Recherches Scientifiques Butare, Rwanda*, 2, 1–293.

van Jaarsveld, E.J. (1987). The succulent riches of South Africa and Namibia. *Aloe*, 24, 45–92.

van Wyk, A.E. (1990). The sandstone regions of Natal and Pondoland: remarkable centres of endemism. *Palaeoecology of Africa*, 21, 243–57.

van Wyk, A.E. (1994). Indian Ocean Coastal Belt: CPD Site AF59. Maputaland-Pondoland Region, South Africa, Swaziland and Moçambique. In *Centres of Plant Diversity: a Guide and Strategy for their Conservation*. Volume 1 Europe, Africa, South West Asia and the Middle East, ed. S.D. Davis, V.H. Heywood & A.C. Hamilton, pp. 227–35. Cambridge: IUCN Publications Unit.

van Zinderen Bakker, E.M. & Mercer, J.H. (1986). Major late Cainozoic climatic events and palaeoenvironmental changes in Africa viewed in a world context. *Palaeogeography, Palaeoclimatology and Palaeoecology*, 56, 217–35.

Vogel, J.C., Fuls, A. & Ellis, R.P. (1978). The geographical distribution of Krantz grasses in South Africa. *South African Journal of Science*, 74, 9–15.

Vrba, E.S. (1980). Evolution, species and fossils: how does life evolve. *South African Journal of Science*, 76, 61–84.

Ward, J.D., Seely, M.K. & Lancaster, N. (1983). On the antiquity of the Namib. *South African Journal of Science*, 79, 175–83.

Weimarck, H. (1941). Phytogeographical groups, centres and intervals within the Cape Flora. *Lunds Universitets Årsskrift N.F. Avd. 2*, 37(5), 1–143.

Werger, M.J.A. (1978a). Biogeographical divisions of southern Africa. In *Biogeography and Ecology of Southern Africa*, ed. M.J.A. Werger, pp. 145–70. The Hague: Junk.

Werger, M.J.A. (1978b). The Karoo-Namib Region. In *Biogeography and Ecology of Southern Africa*, ed. M.J.A. Werger, pp. 233–99. The Hague: Junk.

Werger, M.J.A. & Coetzee, B.J. (1978). The Sudano-Zambezian Region. In *Biogeography and Ecology of Southern Africa*, ed. M.J.A. Werger, pp. 301–462. The Hague: Junk.

Westoby, M. (1988). Comparing Australian ecosystems to those elsewhere. *Bioscience*, 38, 549–56.

White, F. (1965). The savanna woodlands of the Zambezian and Sudanian Domains: an ecological and phytogeographical comparison. *Webbia*, 19, 651–81.

White, F. (1971). The taxonomic and ecological basis of chorology. *Mitteilungen der Botanischen Staatssammlung München*, 10, 91–112.

White, F. (1976). The vegetation map of Africa – the history of a completed project. *Boissiera*, 24, 659–66.

White, F. (1978). The Afromontane Region. In *Biogeography and Ecology of Southern Africa*, ed. M.J.A. Werger, pp. 463–513. The Hague: Junk.

White, F. (1983). *The Vegetation of Africa*. Paris: Unesco.

Willis, C.K., Cowling, R.M. & Lombard, A.T. (1996). Patterns of endemism in the limestone flora of South African lowland fynbos. *Biodiversity and Conservation*, 5, 55–73.

World Conservation Monitoring Centre (1992). *Global Biodiversity: Status of the Earth's Living Resources*. London: Chapman & Hall.

Zedler, P.H., Gautier, C.R. & Jacks, P. (1984). Edaphic restriction of *Cupressus forbesii* (tectate cypress) in southern California, USA – a hypothesis. In *Being Alive on Land*, ed. N.S. Margaris, M. Arianoustou-Farragitaki & W.C. Oechel, pp. 237–43. The Hague: Junk.

Vegetation history

4

L. Scott, H.M. Anderson and J.M. Anderson

4.1   Introduction

Having produced some of the oldest prokaryotic fossils on earth of more than 3 billion years of age, the southern African region is important in answering the question of the origin of life (Knoll 1984). Valuable insights into Palaeozoic life are provided by its richness in early fossil land plants, which appeared in the Devonian period (c. 400 Myr). These ferns and gymnosperms eventually gave rise to the flowering plants in the Cretaceous, c. 100 Myr (Plumstead 1969; Anderson & Anderson 1983). The subsequent evolution of the angiosperms during the rest of the late Cretaceous and Cainozoic phases in southern Africa is less well documented, but palynological research has provided several clues regarding this history. In the later part of the Cainozoic (Neogene), modern plant communities became established. Long-term Quaternary studies, including palynology, indicate that their composition and boundaries were constantly shaped by regular shifts in environmental conditions.

During recent centuries, there have been distinct changes in the vegetation of southern Africa, such as the spread of karoo plants and the effects of human settlement (Acocks 1953). These changes are dealt with in Chap. 21 (this volume), whereas this chapter focuses on the distant history of land plants. Several previous investigations on the origins of vegetation in southern Africa leaned on the biogeographical distribution of the modern flora (Levyns 1952, 1964; Axelrod & Raven 1978; Goldblatt 1978; Linder, Meadows & Cowling 1992). This chapter, instead, is confined to fossil evidence including macro- and microfossil studies in the subcontinent. The descriptions and classification of mainly plant macrofossils and, to a lesser extent, fossil pollen and spores from old Mesozoic and Palaeozoic rocks (e.g. Seward 1903; Plumstead 1969; Anderson 1977; Anderson & Anderson 1983) are the basis for the reconstruction of the earliest vegetation history. The Tertiary and Quaternary histories are generally derived from research aimed at elucidating changing palaeoenvironmental conditions, and rely largely on fossil pollen research (Van Zinderen Bakker 1957; Coetzee 1967; Scott 1982a,b, 1989a, 1994, 1995). Faunal (Klein 1984), microfaunal (Avery 1990, 1991), archaeological (Deacon & Lancaster 1988) and isotopic studies (Vogel 1983) of the Quaternary also make a contribution to the vegetation history of southern Africa during the Late Cainozoic.

Environmental factors, such as atmospheric and crustal changes, which influenced the development of vegetation during different periods, are discussed in Chap. 1 (this volume). In the present chapter the emerging picture of Cainozoic vegetation change is considered briefly in the global context of African climates responding to orbital forcing and other factors (Shackleton & Kennet 1975; Siesser 1978; Wright et al. 1993).

4.2   The Pangaeic phase

4.2.1   The fossil record

Well-preserved megafloras (plant macrofossils) representing plants that evolved during the span of roughly 300 million years from the Lower Devonian to the Lower Cretaceous, are found in southern Africa (Fig. 4.1). They document the early appearance of vascular land plants to that interval just before the dramatic rise of flowering

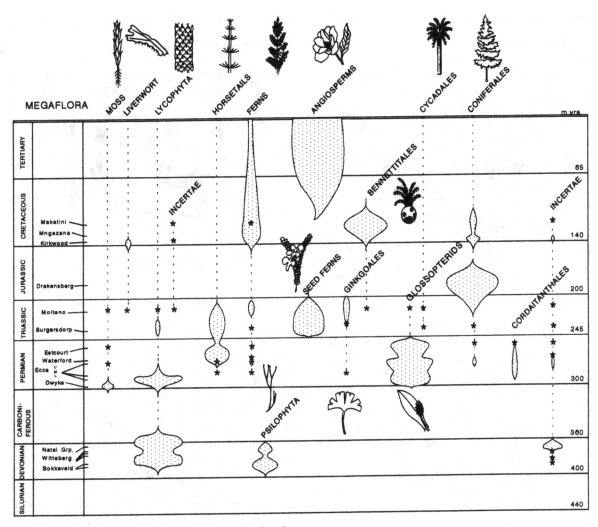

Figure 4.1 **Evolution of Southern African megaplant groups based on macrofossils, adapted after Anderson & Anderson (1985). Asterisk, 1% or less of megaflora.**

plants (Plumstead 1969; Anderson & Anderson 1983, 1985). These developments coincide broadly with that phase when all the earth's landmasses formed a single super-continent: Pangaea.

The collection of fossil megaplants in South Africa has continued with increasing momentum for close on 150 years, since Richard Nathaniel Rubidge in 1845 and Andrew Geddes Bain in c. 1846 made the earliest recorded finds in the Lower Cretaceous and Devonian, respectively. Although there remains considerable scope for further sampling, a good record of the floras is now available for the Devonian, Permian, Triassic and Lower Cretaceous periods. The notable exceptions are the Carboniferous, where deposits are lacking and the Jurassic, where the strata are unsuitable for plant preservation.

Since productive strata elsewhere in Africa appear remarkably limited, the southern African sequence is likely to remain representative for this whole super-region of Gondwana.

Palynology and the study of fossil wood offers the potential to expand the early vegetation history substantially, but the systematic integration of data remains to be done. Well-preserved spore and pollen assemblages are available from the Permian (Middle Ecca in particular), Triassic (Molteno Formation) and Lower Cretaceous (e.g. Scott 1976a; Anderson 1977; McLachlan & Pieterse 1978). A brief survey of the more important fossil floras and vegetation is given for each successive geological period. The synthesis outlined here for the Palaeozoic and Mesozoic is drawn principally from data

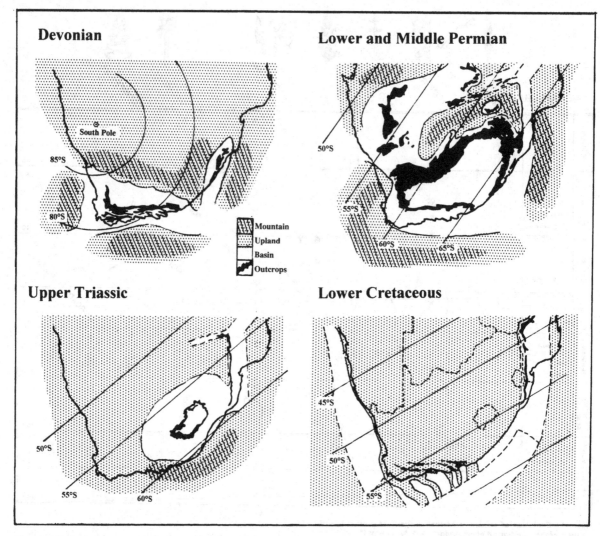

Figure 4.2 **Evolution of depositional basins from the Devonian to the lower Cretaceous. Outcrops of sediments with plant fossils.** After Anderson & Anderson (1985).

in Anderson & Anderson (1983, 1985), but is supplemented for the Devonian through an excellent new locality near Grahamstown (Anderson, Hiller & Gess 1995), and for the Molteno Formation through additional collections made over the past decade (J.M. Anderson & H.M. Anderson, unpubl. data).

4.2.2 The early vascular plants (c. 390–360 Myr)
The earliest, undoubted vascular plants (small herbaceous psilophytes) of the Cape Devonian flora colonized land in the Late Silurian (c. 420 Myr); but the first records in southern Africa occur only in the Middle

to Late Devonian in the Cape Fold Mountains. By this interval floras had evolved to include definite arboreal elements. The Devonian palaeoenvironment (Fig. 4.2) in which the preserved Devonian floras occurred was the littoral belt (deltas, lagoons, swamps, estuaries) of an east–west seaway coinciding with the present Cape Fold Belt. The available sample includes around 2500 identifiable specimens from 53 assemblages scattered widely over 46 localities through the fold belt. This restricted, primitive land flora comprised 19 species (ten genera) and was clearly dominated in abundance and diversity by lycopods (*Haplostigma* and *Archaeosigillaria*) and to a far lesser extent by psilophytes. Rare pro-gymnosperms and unassigned forms also occur. Plant associations were generally monospecific or of very low diversity. Typical examples included *Dutoitia*, a ground cover of very slender, herbaceous psilophytes fringing shallow, quiet lagoons or interdistributary bays. *Archaeosigillaria* constituted a monospecific cover of small herbaceous lycopods along distributary banks in deltaic plains. *Haplostigma* formed monospecific herbaceous to semi-arboreal lycopod reed stands in the shallow waters of lagoons or deltas. *Leptophloeum* formed a monospecific arboreal lycopod forest on higher ground or levee banks in upstream reaches of the deltas. The only other known tree genus of this ancient Cape flora was the pro-gymnosperm *Archaeopteris*.

4.2.3 The coal-forming Glossopterid forests (c. 280 Myr)

South Africa's rich and extensive coal deposits owe their existence to the luxuriant forests of the late Lower Permian (the Middle Ecca flora) that are characterized by a diversity of *Glossopteris* species. An equally important and better sampled flora of the Upper Permian, with 25 assemblages from 21 localities, derives from the Estcourt Formation of KwaZulu-Natal. Glossopterid species still dominated these later floras, but there was a major shift from the prominent occurrence of lycopods in the Ecca to that of horsetails.

With the melting of the Gondwana-wide glaciers in the earliest Permian, a massive non-marine epicontinental sea developed in southern Africa during the Lower and Middle Permian (Fig. 4.2). The coal-forming forests of the Middle Ecca flourished on the deltas, and swamps developed around the northern margin of this sea. The available sample of the flora includes around 20 000 identifiable specimens from five well-sampled localities. The flora of 43 known species (24 genera) is dominated by various glossopterid (seed fern) and, to a lesser extent, lycopod genera. Rare forms include horsetails, ferns, ginkgos and conifers. Two characteristic plant associations are evident: medium to diverse glossopterid forest and woodland along river banks (levees) and other elevated ground; and dense monospecific lycopod stands fringing interdistributary pans and swamps (the habitat occupied by the horsetails in the later Permian).

4.2.4 A clear biodiversity peak (c. 210 Myr)

The Late Triassic Molteno flora is clearly the richest and most comprehensively sampled fossil flora (pre-angiosperm) in Africa, and from any Triassic sequence worldwide. The Late Triassic plant communities define the ecosystems in which the earliest dinosaurs, mammals and possibly the flowering plants appeared. Much

of the extant spectrum of insect orders also appeared at this time (Anderson & Anderson 1993). The Molteno Formation was deposited in an extensive low-lying riverine basin, at 55–60° S, well within the Upper Triassic Gondwana landmass (Fig. 4.2). The braided river systems traversing the basin drained the actively folding Cape Fold Belt to the south. This formation, which yields numerous sampling localities, is widely exposed in a belt of terraced landscape surrounding the Lesotho highlands. Over the past 30 years some 30 000 fossiliferous slabs from 100 assemblages have been collected. The 204 species (54 genera) attest to a rich biodiversity. The taxa are fairly evenly split between the pteridophytes and gymnosperms. The abundant and diverse seed-fern *Dicroidium* (Fig. 4.3a and b) particularly characterized the flora. Conifers are abundant although not diverse, whereas the horsetails are abundant and relatively diverse. The ferns, cycads (Fig. 4.3c) and ginkgos are relatively rare, yet diverse. Mosses, liverworts and lycopods are rare. Six broad vegetation types are recognized: (1) *Dicroidium* high-diversity, multi-storeyed, riparian forest, on the old eroded landscape; (2) *Dicroidium* medium-diversity single-storeyed riparian forest on the young floodplain landscape; (3) *Dicroidium* low- to medium-diversity woodland on the open floodplain; (4) *Sphenobaiera* (seed fern; Fig. 4.3d) medium-diversity woodland on lake margins on the floodplain; (5) *Heidiphyllum* monospecific thicket of herbaceous conifers in depressions on the floodplain; and (6) *Equisetum* (horsetail) monospecific thicket on sandbanks within the braided river or in wetland areas on the floodplain.

4.2.5 The Bennettitalean heyday (140–120 Myr)

The Lower Cretaceous flora of the sea board records the last flurry of pre-angiosperm floras which were strongly characterized by the extinct, cycad-like Bennettitales. This development coincided with the fragmentation of Gondwana during the Lower Cretaceous. The early angiosperms had probably already evolved, but do not appear in these floras. Their rise to prominence globally begins around the mid-Cretaceous. The Lower Cretaceous deposits and floras around the periphery of southern Africa relate directly to rift valleys of the Gondwana break-up (Fig. 4.2). Conglomeratic, riverine, lacustrine, estuarine and marine strata successively filled the rifts that opened up around the (extant) African continent. Around 2000 identifiable specimens from 16 assemblages (14 localities) are available. The localities occur primarily in the coastal plains of northern KwaZulu-Natal and a string of intermontane basins within the Cape Fold Belt. This relatively sparse flora of 32 species (15 genera), was dominated by the widely occurring Bennettitales and to a lesser extent by ferns, cycads and

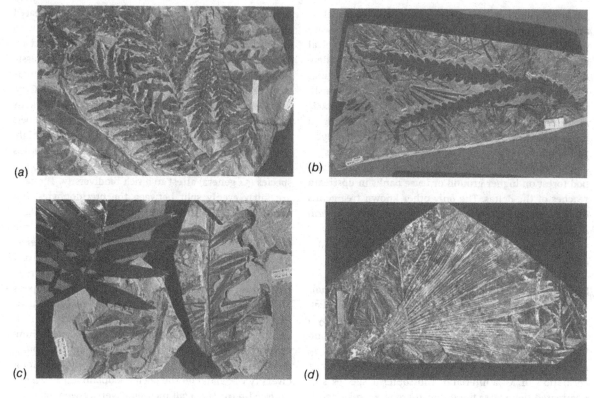

Figure 4.3 **Fossil plants from the Molteno Formation (Upper Triassic) in South Africa. (a and b) Species of** *Dicroidium*, **an abundant and diverse genus of seed-ferns. (c) A cycad shown with** a contemporary representative of *Encephalartos* sp. (d) *Sphenobaiera*, a seed fern *(Photos: H.M. Anderson)*.

conifers. Liverworts were rare. Different species of *Zamites* (Bennettitales) dominated the most frequently occurring low-diversity woodland associations. Despite the dominance of Bennettitales in the megaflora, the Lower Cretaceous pollen flora was dominated by the conifer genus *Classopolis* (Cheirolepidaceae) (Scott 1976a).

4.3 The spread of flowering plants

4.3.1 The mid-Cretaceous to early Tertiary environment (c. 120–60 Myr)

Through continental drift, Africa was isolated geographically and floristically from other Gondwana landmasses by mid- and late Cretaceous times (Smith & Briden 1977; Axelrod & Raven 1978). Global climates at this stage were relatively warm (Shackleton & Kennet 1975). Angiosperms were well established worldwide by the end of the Cretaceous and showed considerable diversity (Penny 1969; Axelrod & Raven 1978; Traverse 1988; Coetzee 1993). Information on mid- and late Cretaceous palaeobotany in southern Africa is scarce, but published and unpublished palynological studies of offshore wells by

the Southern Oil Exploration Corporation suggest that the microfloras from the mid-Cretaceous period conform well with those from other southern hemisphere continents (Scott 1976a; McLachlan & Pieterse, 1978). As found elsewhere in the world, *Clavatipollenites* (affinity with Chloranthaceae) is recorded as the first angiosperm member (Kemp 1968; Scott 1976a). By the Cretaceous–Tertiary transition the African vegetation had developed distinctive characteristics (Belsky, Boltenhagen & Potonié 1965; Partridge 1978; Scholtz 1985; A. Scholtz, unpubl. data), lacking the *Nothofagus* component that developed in southerly sections of Gondwana landmasses (Cranwell 1964; Muller 1981).

4.3.2 Macrofossils

Several types of gymnosperms and primitive angiosperm fossil drift-wood have been reported from the late Cretaceous, Umzamba beds, of the Eastern Cape region (Fig. 4.4), including Monimiaceae and Euphorbiaceae (Mädel 1960, 1962; Klinger & Kennedy 1980). The presence of Monimiaceae, which have also recently been identified in the Mahura Muthla sediments from the interior Ghaap plateau (Fig. 4.4), suggests that this family was a common element in the Upper Cretaceous flora of Africa

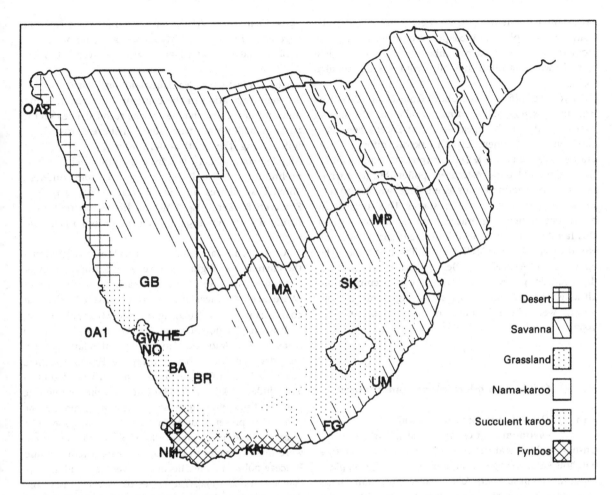

Figure 4.4 **Localities with evidence for vegetation during the Tertiary. BA, Banke; BR, Brandvlei; FG, Fort Grey; GB, Gross Brukkaros; GW, Graafwater; HE, Henkries; KN, Knysna; LB, Langebaanweg; MA, Mahura Muthla; MP, Makapansgat; NH, Noordhoek; NO, Noup; OA1, OA2, Offshore areas 1 and 2; SK, Sterkfontein and Kromdraai; UM, Umzamba.**

(T. C. Partridge and M. S. Zavada, pers. comm.). Plant remains from lake deposits in the Gross Brukkaros crater, southern Namibia (Fig. 4.4), include horsetails, ferns, gymnosperms and possible angiosperms, suggesting a Cretaceous or post-Cretaceous age (Kelber *et al.* 1993). At Fort Grey (Fig. 4.4), fossil plants attributable to *Podocarpus*, *Widdringtonia* and various mono- and dicotyledons, including *Curtisia*, were recorded in silcrete deposits (Adamson 1934). Since widespread formation of silcretes in South Africa occurred no later than the Eocene, the Fort Grey fossils are probably not younger than this (Chap. 1, this volume).

4.3.3 Pollen and spores

The terminal Cretaceous to early Tertiary vegetation in southern Africa is best revealed by well-preserved fossil pollen remains from crater deposits overlying the Arnot kimberlite pipe at Banke in Namaqualand (Fig. 4.4) (Scholtz 1985). On the basis of radiometric dating of a cluster of similar kimberlite volcanoes in the vicinity, the latter microflora is likely to be 71–64 million years old. Although the life forms of parent plants of ancient pollen form-genera may have differed from modern relatives, the best apparent analogies for the palaeovegetation at Banke are some of the extant drier forest types occurring in East Africa and Australia. Taxonomic affinities suggest that the vegetation contained mainly trees but also lianas, epiphytes, forest floor mosses and forest margin elements. Early representatives of taxa that are today characteristic of fynbos, such as Proteaceae, Ericaceae and Restionaceae, were also present and could have been part of the forest understorey vegetation. These proto-fynbos elements at Banke were present under climatic, topographic and edaphic conditions differing from those of the modern fynbos environment which include winter rainfall and oligotrophic soils (Scholtz 1985).

A younger offshore microflora from Noup (OA1, Fig. 4.4), in Paleogene shelf sediments offshore of the southern Namib, which appears to be of Eocene age, indicates the development of a markedly different vegetation (Scott, Cadman & Corbett 1993; A. Cadman and I. Corbett, unpubl. data). Asteraceae pollen (Mutisiae tribe, *Gerbera* type), dominates this assemblage, which also contains a high proportion of fern spores. This suggests an earlier occurrence of Asteraceae than the Miocene, when this family is generally believed to have spread over the world (Leopold 1969; Muller 1981; Traverse 1988). The abundance of Asteraceae pollen indicates that relatively dry conditions prevailed at the time of accumulation, although the fern spores suggest that high moisture conditions were either localized or seasonal (Scott *et al.* 1993).

4.4   Origin of modern plant communities

4.4.1   The Neogene fossil record and environmental change (c. 20–2 Myr)

Onshore palaeobotanical research and offshore marine borehole studies suggest that, during the Neogene, plant communities in southern Africa evolved into equivalents of modern biomes of the subcontinent. Insights in the Neogene vegetation history have been gained despite the scarcity of dateable fossils and material for chronometric dating. Palaeodata, especially palynology from the southern and southwestern Cape (Coetzee 1978a,b), are most abundant in coastal and offshore areas where anaerobic conditions prevailed in depositional basins. The erosional nature and climate in the central interior of the subcontinent were generally not favourable for accumulation and long-term preservation of organic material (Scott 1995).

During the Neogene, southern Africa experienced considerable climate change related to a progressive cooling of ocean waters, the growth of the Antarctic Ice Sheet and development of the circum-Antarctic current in the southern ocean (Shackleton & Kennet 1975; Van Zinderen Bakker 1975; Coetzee 1978a,b; Siesser 1978). Under the influence of global environmental changes angiosperms continued to diversify and plant communities evolved together with modern antelopes and, in the African interior, hominid genera appeared (Coetzee, Scholtz & Deacon 1983; Coetzee & Muller 1984; Hendey

1984; Klein 1984; Vrba 1985; Scott 1995). The Neogene vegetation history of southern Africa is summarized in Table 4.1 (after Scott 1995). Details are given below.

4.4.2   West coast, Namaqualand and Karoo

Neogene vegetation in the contemporary desert, Nama- and succulent karoo biomes can be reconstructed on the basis of pollen data from some offshore marine boreholes along the southwestern African and southern Angolan coast, fossil wood from the Brandvlei area, and preliminary undated pollen spectra from different alluvial sequences in Namaqualand (Fig. 4.4) (Van Zinderen Bakker 1984; Sancetta, Heusser & Hall 1992; Bamford & De Wit 1993; De Wit & Bamford 1993; Scott 1995).

Geomorphic data from the Namib Desert suggest that the current arid regime is a Neogene feature, although dry conditions have occurred as early as the Cretaceous (Ward, Seely & Lancaster 1983). The results from the west coast offshore areas (OA2, Fig. 4.4) show that pollen types of Poaceae, Chenopodiaceae and Asteraceae dominated since the Late Miocene, confirming that relatively open and dry conditions had already developed by then. Lowest values of Chenopodiaceae pollen were found in Van Zinderen Bakker's (1984) Late Miocene sample (the oldest of the sequence), possibly indicating a more mesic phase. The presence of palm and other tree pollen like Combretaceae may suggest a tropical climate. There were marked cycles in Chenopodiaceae, Asteraceae and Poaceae pollen in the Pliocene between 2.6 and 2.2 Myr (Sancetta *et al.* 1992). The cycles of wetter, grassy conditions can possibly be related to the earth's orbital movements, because they seem to follow the 41 000-yr obliquity pattern. High values of Asteraceae pollen between 2.32 and 2.4 Myr were attributed to colder winter rainfall conditions.

A fauna in deposits along the Orange River at Arrisdrift suggests that woodland vegetation occurred during the Miocene (Hendey 1984). Mid-Miocene fossil plant remains from the Brandvlei area of Bushmanland include Dipterocarpaceae, Fagaceae, Myrtaceae, Oleaceae and Rutaceae, indicating subtropical and wetter climates in this region where dry Nama-karoo shrubland currently occurs (Fig. 4.4) (Bamford & De Wit 1993; De Wit & Bamford 1993). In Namaqualand, pollen in channel clays from Noup (Fig 4.4) suggests a moist climate with woodland and forest elements, differing entirely from the succulent karoo vegetation in the presently arid area (Scott 1995). Pollen of *Podocarpus*, *Olea*, Proteaceae and Myrtaceae, together with fern spores and Asteraceae types (of both the Mutisiae type and typical long-spine groups), suggest similarities with the southern and southwestern Cape pollen spectra (discussed below), and point to a

Table 4.1 **Indications of vegetation in parts of southern Africa during the Neogene, based on pollen data (after Scott 1995)**

	Southern and southwestern Cape	Namaqualand
Quaternary	Fynbos	Succulent rich dwarf shrubland or grassland
Pliocene	Fynbos	
Late Miocene/Pliocene	Transition from subtropical woodland to fynbos	Karroid shrubland with fynbos and woodland elements
Miocene	Subtropical woodland	Subhumid subtropical woodland

probable Miocene age. An alluvial deposit at Graafwater (Fig. 4.4), also in Namaqualand, contains mainly Asteraceae pollen (long-spine, *Pentzia* and *Pacourina* types), Cyperaceae, but also some fynbos elements (*Passerina, Cliffortia*), and arboreal pollen (*Podocarpus, Olea, Rhus, Celtis* and Capparaceae) (Scott 1995).

Diatomaceous clays from Henkries (Fig. 4.4), which are apparently younger, contain mainly pollen of grasses, Cyperaceae, long-spine Asteraceae, aquatics and others, indicating conditions comparable to those of the present Nama-karoo or highveld grassland regions of the interior (Scott 1995).

4.4.3 Southern and southwestern coasts

In the winter-rainfall region of the contemporary fynbos biome, Neogene vegetation can be reconstructed mainly from pollen sequences from Noordhoek and Langebaanweg in the southwestern Cape (Fig. 4.4), although precise dating is not possible (Coetzee 1978a,b; Coetzee & Rogers 1982; Coetzee *et al.* 1983; Coetzee & Muller 1984; Scott 1995). The pollen data show that a markedly different vegetation with subtropical forests, including palms, was present during the early Neogene in this region. The similarity of these, apparently Miocene, pollen sequences to assemblages from Knysna in the southern Cape (Fig. 4.4), suggests that they are possibly of a comparable age (Thiergart, Frantz & Raukopf 1962; Helgren & Butzer 1977; Coetzee *et al.* 1983; Maud & Partridge 1987; Partridge & Maud 1987; Scott 1995). The well-known fauna from the Early Pliocene Varswater Formation at Langebaanweg (Hendey 1984) is associated with open fynbos vegetation (Scott 1995), postdating subtropical forest in the Miocene (Coetzee & Rogers 1982). The changes in pollen composition confirm the faunal indications that, although woody elements were present, the vegetation was relatively open (Hendey 1984). This transition from subtropical forest to fynbos vegetation has been linked to developments in the southern ocean, with the [18]O chronology from the southeastern Indian Ocean, global sea-level fluctuations (Shackleton & Kennet 1975; Vail & Hardenbol 1979) and the development of the cool Benguela Current (Van Zinderen Bakker 1975; Siesser 1978).

Mutisiae tribe pollen was still prominent in older Miocene levels in the southwestern Cape. Precise data are not available, but illustrations of Asteraceae from the Miocene in the Elands Bay Formation (Coetzee & Rogers 1982), Noordhoek (Coetzee 1978a) and late Miocene in offshore deposits in the Angola Basin (Partridge 1978) all suggest *Gerbera*-like forms. Major evolution in the Asteraceae apparently took place after the Miocene with the development of more diverse long-spine and other forms. The same pattern of evolution can be observed in Argentina (Barreda 1993).

4.4.4 Interior plateau

From the highveld summer-rainfall plateau, there are faunal indications of vegetation from hominid-bearing breccias of 3 million years of age and younger. A change is reported from dense woodland to more open vegetation around Plio/Pleistocene times (Vrba 1985). Pollen in deposits from Kromdraai and Sterkfontein suggest that open *Protea* savanna occurred after this change and that the setting began to resemble the contemporary one (Scott & Bonnefille 1986; Scott 1995). However, caveats have been identified with palynological research on these calcareous hominid breccias, which lack adequate pollen concentrations and are susceptible to modern contamination (Scott 1982c; Scott & Bonnefille 1986; Scott 1995). On this basis, pollen indications at Sterkfontein (Horowitz 1975) and Makapansgat caves (Cadman & Rayner 1989; Rayner, Moon & Masters 1993; Zavada & Cadman 1993) (Fig. 4.4), can be questioned (Scott 1995).

4.5 Quaternary vegetation changes

4.5.1 Cyclic change during the Quaternary (c. 1.8 Myr)

During the Quaternary, marked shifts in modern phytochoria occurred in response to glacial–interglacial changes in climate. No single detailed palaeovegetation record for the whole Quaternary exists, but shorter pollen profiles confirm this pattern. The oldest ones from the Pretoria Saltpan and Port Durnford suggest that, although modern biomes were well established during the Quaternary, and probably before, marked cycles of vegetation change occurred during this period which resulted in wide shifts in biome composition and boundaries. These were apparently driven by fluctuations in temperature, precipitation and seasonal distribution patterns of moisture. The climatic changes can be related to orbital forcing, the mechanisms of which are considered briefly at the end of this chapter.

Broad trends of vegetation change can be identified for the Late Pleistocene, although our understanding of conditions during the warmest phase of the Last Interglacial (c. 120 000 yr BP) is still poor. Global palaeoenvironmental evidence for this warm period, however, suggests that it resembled the Holocene. Where pollen data in different parts of the region are available from deposits between roughly 40 000 and 75 000 yr BP of age, they provide indications that forests were more widespread under much improved moisture conditions. During the Last Glacial Maximum period, c. 18 000 yr BP, vegetation belts were generally lowered by c. 1000 m in altitude, responding to a drop in temperature of c. 5 °C. Since fluctuations in temperature and precipitation could have had similar effects on fossil pollen spectra, their interpretations are not always firm. Nevertheless, regional palaeobotanical evidence from the end of the Pleistocene (c. 25 000–10 000 yr BP) in southern Africa, suggests that precipitation rates fluctuated considerably, reaching low levels at c. 18 000 yr BP. However, evaporation rates declined with cooler conditions and resulted in a greater effectiveness of the available moisture. In general temperatures progressively ameliorated since c. 14 000, while precipitation increased. Rainfall declined markedly at the start of the Holocene, but c. 7000 yr BP the biomes began to reflect modern conditions, although smaller fluctuations continued until recently. Details of vegetation change in the different biomes are presented below.

4.5.2   Desert biome
There are very few data on vegetation changes from this biome for the Quaternary. The only long-term palynological sequences are derived from offshore marine sediments, but most of these are of Pliocene rather than Pleistocene age (Van Zinderen Bakker 1984; Sancetta *et al.* 1992). Macrofossil (reeds, roots and wood) in the Namib sand sea (Vogel 1989; Vogel & Rust 1987) and hyrax middens (Scott 1996) provide evidence for the more recent Holocene vegetation history. Palynological reconstructions from pollen in pan sediments at Sossus-Vlei (Fig. 4.5), was problematic, owing to poor preservation of organic material in the oxic sediments and difficulties with dating (Van Zinderen Bakker & Muller 1987).

Van Zinderen Bakker & Muller (1987) provided pollen data from two samples from the offshore borehole RC13-229 (Fig. 4.5), estimated to be roughly 250 000 and 550 000 years old. Both contain relatively high proportions of Asteraceae, together with environmental indicators of either upland or southerly areas, such as *Myrica*, Ericaceae, Proteaceae and Restionaceae pollen. This is compatible with wetter conditions, in association with either cooling or a seasonal shift to more winter rain. The two pollen spectra should be viewed in the light of the record of Sancetta *et al.* (1992), which suggests that a cyclic pattern of vegetation change was well established by the beginning of the Quaternary. In view of the length of recorded cycles, the two spectra from RC13-229 represent different climatic pulses and the period which separate them probably represents several major climatic variations.

Data on vegetation in the Namib Desert during the Holocene are scarce. It has been inferred from micromammalian evidence from Mirabeb Shelter (Fig. 4.5) that the desert was more grassy c. 6500 yr BP (Brain & Brain 1977). Pollen in hyrax middens from the Kuiseb River Basin (Scott 1996) suggests that low-intensity short-term cycles of c. 200 years occurred, with the present environment being relatively warm and dry. These cycles are obviously less intense relative to longer-term Pleistocene climatic fluctuations, affecting only the proportions of existing taxa in the Kuiseb River basin, and not large-scale migration. Cyclic enrichment of ^{13}C in the hyrax middens corresponds partly with this pattern, suggesting periods of greater C_4 grass incorporation in hyrax diet during relatively wet periods.

4.5.3   Succulent karoo biome
Preliminary palynological data from spring deposits at Eksteenfontein in the Richtersveld (Fig. 4.5) show a marked change from assemblages of predominantly Chenopodiaceae–Amaranthaceae and different Asteraceae (including *Stoebe*-type) pollen in the Late Pleistocene (before 10 700 yr BP), to assemblages totally dominated by succulent pollen of the Aizoaceae–Mesembryanthemaceae type in the early Holocene (10 700–8450 yr BP) (Scott, Steenkamp & Beaumont 1995). A tentative interpretation of this change is that it represents a transition from cooler to warmer conditions in a constantly dry environment. Poaceae are present in low numbers in this pollen sequence, declining even more towards the top of the profile in the early Holocene. The low grass values may explain a very slight rise in C_3 plants in the diet of Late Pleistocene animals whose bones have been reported in nearby Apollo Cave in Namibia (Vogel 1983). The suggestion by Avery (1990), based on micromammalian data, that the succulent karoo boundary may have moved southward c. 8000 yr BP, can be supported by the hyrax pollen data from the Cederberg, showing that more succulents (Aizoaceae–Mesembryanthemaceae-type) occupied the northern fynbos zone (Scott 1994).

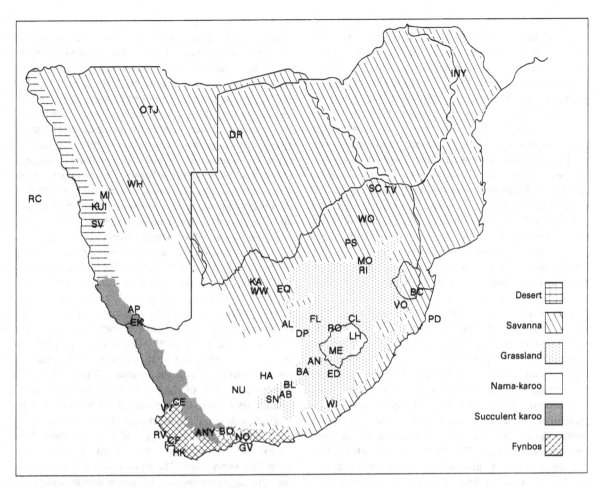

Figure 4.5 **Localities with evidence for vegetation during the Quaternary.** AB, Abbot's Cave; AL, Alexandersfontein Pan; AN, Aliwal North; ANY, Anysberg; AP, Apollo Cave; BA, Badsfontein; BC, Border Cave; BL, Blydefontein; BO, Boomplaas; CE, Cederberg; CF, Cape Flats; CL, Clarens; DP, Deelpan; DR, Drotsky's Cave; ED, eastern Cape Drakensberg; EK, Eksteenfontein; EQ, Equus Cave; FL, Florisbad; GV, Groenvlei; HA, Haaskraal Pan; HK, Hangklip; INY, Inyanga Mountains; KA, Kathu Pan; KUI, Kuiseb River; LH, Lesotho highlands (Malutis); ME, Melikane Cave; MI, Mirabeb; MO, Moreletta River; NO, Norga; NU, Nuweveldberge; OTJ, Otjikoto Lake; PD, Port Durnford; PS, Pretoria Saltpan; RC, borehole RC13-229; RI, Rietvlei Dam; RO, Rose Cottage Cave; RV, Rietvlei; SC; Scot; SN, Sneeuberg; SV, Sossus Vlei; TV, Tate Vondo; VO, Voordrag; VV, Verlorenvlei; WH, Windhoek; WI, Winterberg; WO, Wonderkrater; WW, Wonderwerk Cave.

4.5.4 Nama-karoo biome

Just as in the desert biome, environmental conditions in the Nama-karoo biome are generally not favourable for the preservation of organic material and consequently few Quaternary fossil records are available. The pollen sites from Blydefontein, Winterberg and Sneeuberg are broadly situated in the Nama-karoo region (Fig. 4.5), but because of higher elevations, they lie in grassland and are also referred to in the section on the grassland biome.

Old Pleistocene deposits from Haaskraal Pan lacked fossil pollen, and were contaminated by recent mud-cracks (Partridge & Dalbey 1986; Scott & Brink 1992). Therefore, direct data pertaining to the Pleistocene vegetation in the karoo are lacking. However, by the end of the Pleistocene karroid shrublands seemed to have had a much wider distribution than at present, being recorded in the southern Kalahari, at Equus and Wonderwerk Caves at c. 10 400 yr BP (Scott 1987b; Beaumont, Miller & Vogel 1992; Scott et al. 1996). The pollen spectra include *Passerina*, now largely associated with temperate Cape and Afromontane shrublands, which indicates that the vegetation was probably not the result of dry conditions, but of cooler and possibly seasonally altered climates. At Aliwal North (Fig. 4.5) in an area bordering on grassland, spring deposits provided evidence for cyclic alteration between grassy and karroid veld at the end of the Pleistocene (Coetzee 1967). In general, however, the whole sequence probably represented more grassy vegetation in comparison with that implied by pollen trapped in similar, but younger, Holocene spring deposits at Badsfontein (Scott & Cooremans 1990). The early Holocene alluvial deposits from Blydefontein was characterized by relatively low proportions of grass pollen, but grasses in the vegetation returned by the middle Holocene (Bousman *et al.* 1988; Scott 1993). The present karroid vegetation at Deelpan and Alexandersfontein appeared to have developed only after c. 1000 yr BP when the grass component in the vegetation declined (Scott 1988; Scott & Brink 1992). Vlei pollen from the Nuweveldberge (Sugden & Meadows 1989) suggests relatively stable vegetation during the last 760 yr BP, with shifts indicating changes between mountain grassland and various karroid veld types (Fig. 4.5). In contrast, two hyrax midden pollen sequences from Blydefontein (east of the Nuweveldberge) point to a progressive replacement of grass with shrubby Asteraceae during the last 300 to 400 years (Scott & Bousman 1990; Bousman & Scott 1994). This deterioration started before the advent of modern grazing practices and was initiated by climatic change; however, overgrazing may have exacerbated the process. Although the deterioration is mainly the result of mismanagement by European farmers, the possibility of earlier alteration of vegetation by grazing of stock owned by Khoikhoi has been mentioned (Sampson 1986; Chap. 21, this volume). Inferences about vegetation cover in the Nama-karoo biome during the last millennium on the basis of micromammal remains in archaeological deposits from Abbot's Cave (Avery 1991) do not support the palynological interpretation strongly. Potential discrepancies could in future be explained by further investigations into both methods.

In general, palynological evidence suggests that karroid shrubland elements in the interior of South Africa were well established and more widespread during the late Pleistocene. More grass occurred in the Nama-karoo region since the middle Holocene, but the grass and shrub ratios continued to alternate. Before the arrival of farmers with domestic stock, these shifts were caused by climatic change, whereas contemporary changes are influenced by both climate and land management (Chap. 21, this volume).

4.5.5 Fynbos biome

Progress with Quaternary vegetation reconstruction of the fynbos biome has been slow despite the occurrence of potential polleniferous sediments in extensive mountains and valleys of the southern Cape area. The fossil pollen data suggest that marked changes in the fynbos vegetation took place during the Quaternary period, but the biome remained essentially intact despite climatic change. Although data on boundary shifts are lacking, there is evidence for forest and woodland expansion within the biome at times before the Last Glacial Maximum phase. In turn this phase was characterized by lowered vegetation zones as a result of cooler temperatures and the spread of asteraceous shrubland types such as renosterveld (see Chap. 6, this volume).

In the southwestern Cape, Pleistocene pollen in cores of lagoon sediments from Rietvlei (Fig. 4.5) show spectra that do not differ markedly from modern ones in the region (Schalke 1973). Schalke's pollen chronology of the Rietvlei sediments (reviewed by Scott 1984a) cannot be relied upon, because several radiocarbon datings gave ages of around 40 000 yr BP; these are probably beyond the limits of reliability of available methods. From an intermontane valley in the southern Cape, a long Late Pleistocene record of change in the fynbos during the Quaternary is given by charcoal and pollen from archaeological sediments in Boomplaas Cave (Deacon *et al.* 1984; Scholtz 1986; Fig. 4.5), and is supplemented by micro- and macrofaunal evidence (Avery 1982; Klein 1983; Thackeray 1987). The contemporary vegetation at the site is characteristic of nutrient-rich areas and includes riverine woodland and renosterveld, with few fynbos elements. The two sources of palaeobotanical evidence at Boomplaas Cave, charcoal (Scholtz 1986) and pollen assemblages (Deacon & Lancaster 1988), show divergent patterns attributable to different trapping mechanisms. Charcoal data are biased because of selection of firewood by prehistoric people and pollen by different production and transport qualities of various taxa. Scholtz's (1986) palaeobotanical reconstruction of Boomplaas Cave relies on relative abundance of morphological types of charcoal, as well as his proposed new methods of ecologically diagnostic xylem analysis, and practical and cultural factors of firewood use. His results show distinct changes in vegetation over the sequence, the strongest of which occurred at c. 14 000 yr BP following the amelioration of climate after the Last Glacial Maximum. ^{18}O values from the nearby Cango Caves (Talma & Vogel 1992) indicate a strong temperature increase of c. 5 °C which corresponds to the change in plant cover between Pleistocene and Holocene times. The vegetation transition comprised a change from open asteraceous shrubland to woodland with *Olea, Dodonea*,

Rhus, Ebenaceae, *Buddleja*, *Protea*, *Acacia*, etc. (Scholtz 1986). At Boomplaas Cave, the Late Pleistocene open vegetation phase between c. 55 000 yr BP and 18 000 yr BP was generally characterized mainly by various Asteraceae elements (e.g. *Elytropappus* and *Euryops*) and Ericaceae. The vegetation represented in levels of c. 45 000 yr BP and older contained some woodland charcoal of, for example, *Olea* and *Dodonea* types, whereas that of the cold phase around 18 000 to 21 000 yr BP was strongly dominated by Asteraceae, suggesting that this was the driest phase in the Boomplaas Cave sequence.

In the northwestern sector of the biome a palynological study of peat deposits in a high-altitude marsh site on the fynbos-clad Cederberg Mountains (Fig. 4.5) shows very little vegetation change after 14 500 yr BP (Meadows & Sugden 1990, 1991). The low degree of change in the fynbos with, Ericaceae, Restionaceae, etc., during the terminal Pleistocene (c. 14 000 yr BP), is surprising in the light of worldwide patterns and new pollen evidence from a 20 000 yr BP hyrax midden sequence in the same mountain range (Scott 1994). The middens occurring in a cave at a lower altitude than the marsh site and on the interior succulent karoo-fringed side of the Cederberg range indicate considerable change at c. 14 000 yr BP, corresponding with the transition in the Boomplaas Cave sequence (Fig. 4.5). The pollen spectra of the cold phase (c. 20 000 to 14 000 yr BP) indicate fynbos vegetation with more *Stoebe*-type (*Elytropappus*) plants, similar to present high-altitude types, whereas the younger spectra are characterized by woody elements such as *Dodonea*, *Euclea* and *Olea*.

Evidence for the development of coastal fynbos during the Holocene was obtained from palynological records at Hangklip (Schalke 1973), Groenvlei (Martin 1968) and Verlorenvlei (Meadows, Baxter & Adams 1994; Baxter & Meadows 1994) (Fig. 4.5). The Groenvlei area was covered by fynbos before 7000 yr BP, but at present the area supports temperate (Afromontane) forest (Martin 1968). The oldest samples in the peat sequence of Hangklip, c. 7000 yr BP, also indicate the occurrence of fynbos at the time. No data for the period between 6000 and 2000 yr BP were recorded at Hangklip, but fynbos occurred in the region during the last 2000 years, showing some variation in composition (Schalke 1973). Late Holocene pollen data from Verlorenvlei, a coastal lake, some 180 km north of Cape Town, show that local conditions changed from an open saltmarsh environment associated with a raised sea level, to a freshwater one in a closed lake system some time after c. 4000 yr BP (Meadows *et al.* 1994). Charcoal pieces from archaeological sites at Elands Bay, near the mouth of the Verlorenvlei basin, suggest that the same species (e.g. *Ruschia*, *Zygophyllum* and *Euclea*) were collected for firewood in the area for 4000 years (February 1992), but more negative ^{13}C isotope values around 3000 yr BP and 1620 yr BP may be related to relatively wetter conditions (February & Van der Merwe 1992). It has been inferred that anthropogenic disturbance since c. AD 1700 is responsible for a series of smaller changes in the youngest levels of the Verlorenvlei sedimentary and pollen sequence (Baxter & Meadows 1994).

In the intermontane valley at Boomplaas Cave, the composition of woody vegetation changed during the Late Holocene, c. 2000 yr BP, with the replacement of more diverse thicket vegetation by *Acacia karroo*. Possibly related to this is the maximum value of a ^{13}C curve from a stalagmite in the nearby Cango Cave, implying an increase in the abundance of C_4 plants (Talma & Vogel 1992). The vegetation reconstructed from pollen in hyrax middens on the dry succulent karoo side of the Cederberg (Scott 1994), also shows variation during the Holocene. This is in contrast to the very low degree of change in the pollen record of fynbos of the higher-lying swamp site (Meadows & Sugden 1990, 1991), where the most marked change is a very gradual decline in pollen of the locally endemic tree *Widdringtonia cedarbergensis* (Cupressaceae). The hyrax midden pollen from the Cederberg suggests that before 8000 yr BP Restionaceae, Cyperaceae and succulent Aizoaceae–Mesembryanthemaceae-types were prominent, but *Dodonea* and Asteraceae (including *Stoebe*-type or *Elytropappus*) pollen was relatively less important. Gradually by 5000 to 4000 yr BP this situation was reversed. Pollen from a hyrax midden at Keerkloof in the Anysberg Nature Reserve (Fig. 4.5), more than 200 km to the southeast, also shows replacement of dominant Restionaceae pollen, c. 8400 yr BP by more Asteraceae, including *Stoebe*-type, around 7500 yr BP (L. Scott, unpubl. data). Some time after 1400 yr BP, thickets of *Dodonea* and *Euclea* at the Cederberg hyrax site show a sharp decline, possibly as a result of human influence on the vegetation (Fig. 4.6) (Scott 1994). This decline in woody elements is more sudden than the very gradual and longer-term pattern reported for *Widdringtonia cedarbergensis* (Meadows & Sugden 1990, 1991).

With the exception of the Cederberg, well-dated long-term records are scarce in fynbos areas. Therefore, our knowledge of changes in other mountains and the lowlands of the biome is still rather limited.

4.5.6 Grassland biome

Grassland is widespread in the interior plateau, and includes fynbos-like vegetation in moist higher-altitude areas (Chap. 10, this volume). Isotope studies of bone suggest that C3 plants became more prominent in the diet of animals living at Melikane Cave, Lesotho during the

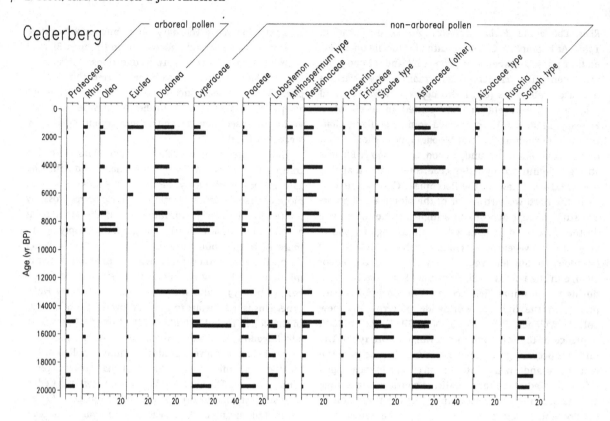

Figure 4.6 **Pollen diagram of hyrax middens in a rock shelter in the Pakhuis Pass, Cederberg, adapted after Scott (1994). Scroph type, Scrophulariaceae type.**

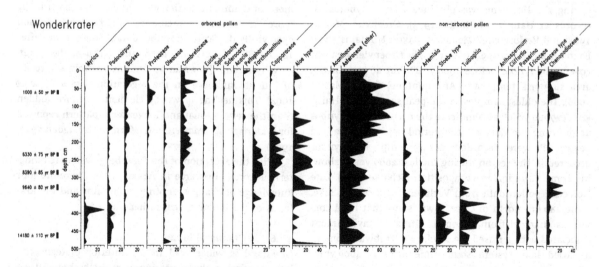

Figure 4.7 **Diagram of selected pollen taxa from the Wonderkrater spring deposit. After Scott (1982b).**

Late Pleistocene (Vogel 1983), and this could be related to the lowering of upland grassland. Pleistocene pollen records of the highland mountain fynbos of Lesotho are not available, but the studies at Clarens suggest that this vegetation type probably migrated to lower elevations during the Last Glacial Maximum (Scott 1989b). The presence of shrubby fynbos elements among grass pollen conforms with the suggestions of the spread of so-called "cold" grassland during glacial periods (Van Zinderen Bakker & Werger 1974; Van Zinderen Bakker 1983). However, details of such changes are not always detectable by pollen analysis, as grass pollen can generally not be distinguished below family level.

Because of downward shifts of zones during glacial events, highveld grassland types expanded regularly during the Quaternary at the expense of woody vegetation. Pleistocene pollen records from woodland outside the present grassland biome, at Wonderkrater (Fig. 4.7), the Pretoria Saltpan and Equus Cave (Fig. 4.5), suggest that grassy vegetation occupied a much greater area to the north during past cooler episodes (Scott 1982b, 1987b; Partridge *et al.* 1993). Sites inside the contemporary grassland biome, such as the spring at Florisbad (Van Zinderen Bakker 1989; Scott & Brink 1992) and swamp accumulations at Clarens (Scott 1989b), indicate cyclic changes in this vegetation during the Pleistocene. In the pollen record of the pan (playa) environment of Florisbad (Fig. 4.5) in the central plains, the main fluctuations comprise the alternation of grass with local halophytic or spring plants, presumably flourishing in response to climatic change. In the higher-lying grassland at Clarens (2000 m), the downward spread of relatively dry mountain fynbos elements is indicated during the cold Last Glacial Maximum, although grass remained the most important component of the vegetation in the Late Pleistocene (Scott 1989b). This is supported by charcoal analysis from Rose Cottage Cave (Fig. 4.5) near Ladybrand (Wadley, Esterhuysen & Janneret 1992) and pollen analysis from Voordrag, KwaZulu-Natal (Botha *et al.* 1992). The charcoal study from Rose Cottage Cave suggests that *Protea* and *Cliffortia*, typical of upland vegetation, were more available as firewood during the Late Pleistocene than during the Holocene when plants like *Buddleja* and *Maytenus*, common elements of the subtropical thicket and dry forest, were more prominent.

Evidence of vegetation during the Holocene in the contemporary grassland biome has been found at various sites from as far north as, for example, the Inyanga Mountains in Zimbabwe (Tomlinson 1974) and the highveld, to the KwaZulu-Natal and Lesotho Drakensberg range (Van Zinderen Bakker 1955), the central Free State (Van Zinderen Bakker 1989) and the Eastern Cape (Bousman *et al.* 1988; Meadows & Meadows 1988). Sites

from the last-mentioned region can be considered together with karoo vegetation records, because they represent upland grassland zones in the proximity of Nama-karoo biome communities (Fig. 4.5). Data from different areas suggest that the grassland biome occupied roughly the same area during the Holocene as it does today, but some movement of its boundaries is indicated. Around 7790 yr BP, shrubby karroid vegetation, characterized by a high incidence of Asteraceae pollen, such as *Stoebe* (or *Elytropappus*)-type, seemed to have occurred in the Kikvorsberg at Blydefontein near Noupoort. This apparently reflects slightly cooler or drier conditions with less marked seasonal rainfall patterns (Bousman *et al.* 1988; Bousman 1991). Dry conditions before c. 7000 yr BP in the southernmost Drakensberg (Ravenscraig, Colwinton and Bonawe sites) are also indicated by charcoal analysis (Tusenius 1989). During a "climatic optimum" in the middle Holocene around 7000 yr BP, on the northern boundary of the main Highveld region, bushveld vegetation temporarily spread southwards over the edge of the plateau at Rietvlei Dam (Scott & Vogel 1983). Late Holocene pollen records from the present transition of the Nama-karoo biome to grassland are furthering understanding of the recent vegetation on the western limits of the grassland biome. Under favourable conditions in the Late Holocene around 4000–1000 yr BP, the grassland was very well developed in the transitional area at Deelpan and Alexandersfontein (Uitzigt), Free State (Scott 1976b, 1988; Scott & Brink 1992), suggesting that the grassland boundary must have been further west. At the same time the upland grassland at Sneeuberg and Blydefontein contained less karroid Asteraceae (Meadows & Sugden 1988; Scott & Bousman 1990). Relatively moist conditions are also indicated around 2400 yr BP, by charcoal analysis of archaeological sediments in the Drakensberg (February 1994).

Apart from temporary boundary and composition shifts (Scott & Vogel 1983), pollen data from several sites support the view that grasslands have essentially been in place throughout the Holocene, and that they were often more widespread during the Pleistocene. This does not support the widely held view (e.g. Acocks 1953) that Afromontane grassland is the result of forest clearance in the recent past (Meadows & Linder 1993). Human influence on grassland as a result of agricultural activities in South Africa is probably restricted to the Late Holocene (Chap. 21, this volume). Climatic changes, therefore, generally influenced the long-term history of grasslands, e.g., lower temperatures and less marked seasonal rainfall patterns allowed the downward spread of Afromontane fynbos and C_3 grasses.

4.5.7 Savanna biome

Evidence of Quaternary vegetation history in the southern African savanna biome has been obtained from pollen and charcoal analysis of spring, swamp, lake and cave deposits and indirectly from charcoal, faunal, microfaunal and isotopic analyses of archaeological sites.

Long pollen sequences from organic deposits at a warm spring at Wonderkrater (Scott 1982a,b) and lake deposits from the Pretoria Saltpan crater (Partridge *et al.* 1993) show strong cyclic patterns during the Late and Middle Pleistocene. Although the 90-m palynological record from the Pretoria Saltpan crater (Fig. 4.5) is interrupted by wide barren sections, it provides new evidence for environmental conditions in the interior as far back as 180 000 yr BP, and gives further insights into the problematic aspects of the chronology in part of the Wonderkrater pollen sequence (Scott 1982b). The data suggest that vegetation alternated markedly, ranging from woodland savanna during warm interglacial phases, to cool open upland grassland (including fynbos elements) during glacial maxima, and to mesic woodland with *Podocarpus* forests during some intermediate phases. Support for a cyclic pattern of vegetation change of this scale can be found in faunal and microfaunal data from Border Cave on the Swaziland/KwaZulu-Natal border (Klein 1977; Avery 1993), but precise chronological correlations with the pollen sequences have not been established. Microscopic charcoal in the Pretoria Saltpan sequence indicates that phases of more intense fires occurred in the bushveld environment in the past (Scott 1995). Since the Middle Pleistocene denser bush cover did not always appear to coincide with charcoal abundance, suggesting that fire frequency is not dependent on vegetation type, but presumably on the activities of humans in episodes of occupation of the area.

At the end of the Pleistocene, there is evidence for the dramatic change in the savanna biome from upland grassy fynbos with Ericaceae and patches of *Podocarpus* forest to semi-arid savanna (similar to that currently occurring in the southwestern Kalahari) in the earliest Holocene, c. 8000 yr BP. Broad-leaved savanna was established consequently, c. 7000 yr BP. Relatively more mesic and presumably cooler conditions with a partial return of *Podocarpus* forests in the vicinity is indicated later in the Holocene. This pattern of vegetation change since the Pleistocene is best recorded in the Wonderkrater sequence (Scott 1982b), in which the dating reliability has been supported by new radiocarbon measurements (Scott *et al.* 1995). It has also been found in the uppermost section of the Pretoria Saltpan lake sequence, shorter spring sequences from the Soutpansberg region (Scot and Tate Vondo sites), and the Moreletta River, Pre-

toria (Fig. 4.5), as well as in archaeological sediments with charcoal evidence (Scott 1982d, 1984b, 1987a; Prior & Price Williams 1985; Dowson 1988; Partridge *et al.* 1993).

The Late Pleistocene and Holocene history of the thornveld vegetation of the southern Kalahari margin is elucidated by pollen analysis of sediments in Equus Cave (Fig. 4.5) (Scott 1987b). The sequence which spans the last 30 000 years (Beaumont *et al.* 1992) shows a transition from karroid grassland in the Late Pleistocene to Kalahari thornveld in the Holocene. This tendency is supported by new results from a terminal Pleistocene porcupine midden at Wonderwerk Cave (Scott *et al.* 1995). The results do not support Van Zinderen Bakker's (1982) proposal of bee contamination as the explanation for high Asteraceae values in early Holocene layers from Wonderwerk Cave, because the hyaena coprolites at Equus Cave and porcupine middens at Wonderwerk (Scott 1987b; Scott *et al.* 1995) also show an increase in Asteraceae pollen; it is very unlikely that these materials could have been contaminated by bee burrows. In contrast to the pollen sequences in the southern Kalahari area, data from a speleothem from Drotsky's Cave, Botswana, in the northern Kalahari, do not show markedly increased Asteraceae pollen in the early Holocene (Burney, Brook & Cowart 1994). Avery's (1981) and Van Zinderen Bakker's (1982) interpretations of microfaunal and pollen data from Wonderwerk Cave show both agreement and disagreement in terms of minor changes in environmental conditions during the Holocene. Details of pollen analysis from Kathu Pan from this area remain unpublished, but a preliminary study suggests a locally drier environment between c. 4400 and 3000 yr BP (Beaumont, Van Zinderen Bakker & Vogel 1984).

Further evidence for change in savanna vegetation during the Holocene is presented by pollen data from spring accumulations, and cave bat guano from central Namibia at Windhoek (Scott *et al.* 1991; L. Scott and E. Marais unpubl. data). The data suggest that relatively moist grassy conditions developed soon after c. 7000 yr BP, but deterioration followed c. 5600 yr BP. Furthermore, a 1-m core in sediments of a karstic sinkhole, Lake Otjikoto (Fig. 4.5) in northern Namibia which formed under 50 m of water, shows fluctuations in the pollen composition during the late Holocene (Scott *et al.* 1991). The evidence includes indications for a drier savanna c. 1000 yr BP, apparently corresponding with pollen data from hyrax middens in the Kuiseb River which have been described above (Scott 1996).

It is concluded that although the savanna region provided the most detailed record of vegetation change during the Quaternary, there are still wide spatial and temporal gaps in the fossil pollen record. The latter is demonstrated by the Pretoria Saltpan sequence, which lacks pollen data from levels dating to the Last Interglacial period. It is clear, however, that savanna vegetation underwent considerable changes during the Quaternary. During cold glacial phases the southern boundaries of savanna communities shifted far northwards, whereas during periods with more mesic conditions forests colonized parts of the mountain slopes currently occupied by savanna vegetation. Although broad-leaved savannas developed as early as 7000 yr BP, the contemporary savanna structure probably only came into being around 1000 yr BP.

4.5.8 Forest biome

Dated, long-term Pleistocene records of vegetation change are not available from the Knysna region where the main body of the forest biome is located, although Holocene records were provided by Martin (1968) and Scholtz (1986). Outside this biome, fossil pollen spectra in a palaeosol from the Cape Flats (Fig. 4.5), dating to c. 41 500 yr BP, show increased (up to 40%) *Podocarpus* pollen (Schalke 1973), which may indicate the expansion of forests across the southwestern Cape during the Pleistocene. The dating of the palaeosol is possibly beyond the limits of reliability of radiocarbon dating and could be much older, representing a relatively warm episode as is indicated by tropical *Alchornea* pollen, a taxon not presently found in the area (Scott 1984a).

The Holocene history of the forest biome is elucidated by Martin's (1968) detailed pollen analysis of sediments of Groenvlei (Fig. 4.5), a coastal lake, and for the later part of the Holocene it can be complemented by a similar study by Scholtz (1986) of the peat deposits at Norga near George. The Groenvlei pollen data suggest that forests including species of *Podocarpus* and *Olea* were present throughout the Holocene. However, at the start of this period, fynbos was much more prominent relative to forest vegetation. By c. 7000 yr BP, there was a climatic amelioration with an increase in the extent of forest. The middle part of the sequence is characterized by a marine transgression, which complicates interpretation of the pollen spectra (Martin 1968). Before 1905 yr BP, however, a change in arboreal pollen composition took place, comprising a decline in *Podocarpus* in favour of other woody elements. After 1905 yr BP *Podocarpus* pollen percentages recovered markedly, presumably representing a spread of *P. falcatus* on the coastal dunes around Groenvlei, together with dry forest elements. The pollen sequence from Norga (Scholtz 1986), on the periphery of the forest biome, suggests a decrease in forest cover (mainly *Podocarpus* and *Olea*) c. 2570 to 1810 yr BP, which corresponds roughly with the decline in *Podocarpus* pollen (and the increase of other woody elements) in Groenvlei.

Records of forest history in the interior, north of the southern forest block, provide evidence for much wider distributions of forest patches during the Pleistocene. Pollen in lignite deposits at Port Durnford along the Zululand coast (Fig. 4.5) suggests that after the Last Interglacial phase, *Podocarpus* was very prominent (Scott, Cooremans & Maud 1992). This is in contrast with the situation in contemporary forests where the numbers of gymnosperms are low (Van Daalen *et al.* 1986; cf. Chap. 12, this volume). Increased *Podocarpus* pollen, which have airbags and are thus well distributed over long distances, have been used to identify times during which generally more widespread forest patches existed in the interior plateau. On this basis, certain periods of favourable climate with increased rainfall and cloudiness, in the Late and Middle Pleistocene, have been inferred at sites such as Wonderkrater and the Pretoria Saltpan in the interior (Fig. 4.5) (Scott 1982b; Partridge *et al.* 1993). Forests probably occupied extensive areas on mountains like the Waterberg and even occurred on lower hills like the rim of the Pretoria Saltpan crater. During the last 180 000 years the size of the forest patches fluctuated considerably as indicated by variations in *Podocarpus* pollen numbers (Partridge *et al.* 1993).

During the Last Glacial Maximum, forest patches survived in protected mountain valleys on the iterior plateau. Pollen data from the Vryheid region at Voordrag (Fig. 4.5) suggest that *Podocarpus* forests in the KwaZulu-Natal Midland were formerly well developed, during and before the Last Glacial Maximum, in an area were forests are absent today (Botha *et al.* 1992). In the high Drakensberg, however, temperatures were probably too low for their survival and virtually no long-distance *Podocarpus* pollen reached the Clarens area (Fig. 4.5) at Elim (Scott 1989b). Pollen in younger alluvial sediments at Graigrossie, Clarens (Scott 1989b) suggest that with amelioration of temperatures and moisture after the last Glacial Maximum, *Podocarpus* forests must have spread up the lower mountain ravines of the Drakensberg Range and Malutis, Lesotho (Fig. 4.5), and must have been temporarily more widespread than today.

At Wonderkrater in the savanna biome, *Podocarpus* pollen declined markedly during the early Holocene (c. 8000 yr BP) (Fig. 4.7), implying that forests in mountain ravines were reduced as a result of aridification (Scott 1982b). Over this time, however, they persisted in more favourable environments such as Tate Vondo (Fig. 4.5), Zoutpansberg region (Scott 1987a). By the middle and late Holocene, *Podocarpus* in the interior (Wonderkrater and the Pretoria Saltpan) spread again, but never reached the same proportions as during some phases of the Late and Middle Pleistocene or even the Last Glacial Maximum. Finally by c. 1500 yr BP, forests at Tate Vondo in the Zoutpansberg declined, coinciding with the arrival of Iron Age people (Scott 1987a).

4.6 Evolution of vegetation and global palaeoclimate

From the earliest times the preservation of the first fossil land plants of the Palaeozoic and Mesozoic in southern Africa was controlled by both the local depositional environment and regional climatic conditions. Where vegetation was not prevented by glaciers or desert conditions, plant remains were preserved in local swampy basins in oxygen-free sediments, or were destroyed by oxidation and decomposition. It is possible to draw only relatively precise conclusions about climatic change from taxonomic relationships rather than geological data in the case of relatively modern fossil material from the Tertiary and Quaternary periods. Palaeobotanical data on earlier vegetation patterns related to moisture and temperature changes suggest that the overall magnitude of environmental changes during the Neogene and the Quaternary in southern Africa generally conforms with that anticipated in global simulation models and palaeoclimatic data from other continents and the rest of Africa. Therefore, palaeoclimate can probably not be considered to be an unusual factor in the evolution of modern plant communities in the region.

The long-term fossil pollen record during the Neogene from the southwestern Cape (Coetzee 1978a,b) suggests that since the beginning of the Cainozoic, evolution of the vegetation is strongly related to gradual global cooling (Shackleton & Kennet 1975). The reason for cooling is not well understood, but complex factors could be involved: e.g. energy loss through albedo feedback following large-scale tectonic uplifting of landmasses such as the Tibetan Plateau (Ruddiman & Kutzbach 1991); changes in atmospheric greenhouse gasses through interactions of the atmosphere, the biosphere and oceans; or even volcanic activity (Williams et al. 1993). Continental movements must have influenced global climates as well, and the establishment of the Antarctic Circumpolar Current certainly had an effect on Africa (Shackleton & Kennet 1975; Van Zinderen Bakker 1975). Although possible external effects such as orbital variations are also a possible cause for long-term cooling through the Cainozoic, their role is better established in the case of the relatively young Quaternary environments (COHMAP 1988; Wright et al. 1993). Good correlations between some palaeoclimatic proxy data and orbital movements do exist, highlighting the role of past

climates in the distribution and development of vegetation. Although correlation is possible only in relatively young Quaternary deposits, it can be assumed that the effects of orbital changes prevailed for a much longer period, causing numerous and regular cycles of change in vegetation through the Pleistocene and Tertiary period and probably beyond. Numerous constant cycles of global change have been recorded in an isotopic curve of Pacific ocean sediments across the Miocene/Pliocene boundary (Shackleton 1995) and a similar pattern occurs in the pollen composition of Colombian lake sediments of at least 3 Myr (Hooghiemstra 1995).

We are well acquainted with the present interglacial scenario forming the last peak on the saw-tooth pattern of the palaeoclimate curve, but not with glacial and interstadial events that are part of the long-term picture of a typical African vegetation. Both recurring glacial and interglacial scenarios must influence evolutionary processes in plant communities and contribute to determine the nature of the modern vegetation.

The last full glacial cycle of the southern hemisphere, corresponding with the 100 000-year orbital eccentricity cycle, is well recorded in the Vostoc Ice core from Antarctica. The long pollen record from the Pretoria Saltpan (Partridge et al. 1993; L. Scott, unpubl. data) which goes beyond this cycle is unfortunately not continuous, lacking direct and well-dated evidence for the vegetation in southern Africa during the hottest part of the last Interglacial, c. 125 000 years ago. Long pollen records from the southern hemisphere, which can be compared with the southern African data, are very scarce. The most detailed records over the last full glacial cycle were obtained in Australia. Pollen profiles from the Atherton Tableland and Western Plains, Australia (Kershaw & Nanson 1993) show broad similarities with the available pattern of vegetation changes in the interior of South Africa over the last full glacial cycle. In both continents it is suggested that temperature and moisture changes were not in phase during the Pleistocene. The lack of correlation between temperature and moisture can be expected if the complex influences of orbital forcing on the earth's climate is considered. Past vegetation patterns during the last full glacial cycle, which are a reflection of these complex changes, can therefore be expected to include a relatively wide diversity of community types. The extent to which upland vegetation migrated northwards during the last Glacial Maximum period, i.e. c. 500 km in the case of the highveld grassland (Dowson 1988), is comparable with similar trends elsewhere on the globe and also in Africa (Coetzee 1967; Livingstone 1971; Vincens et al. 1993).

Since past moisture changes following the 22 000-year orbital cycle of precession did not affect the globe uniformly, monsoonal precipitation increased significantly in the northern hemisphere tropics during the early Holocene (especially in Africa), but not in the southern hemisphere where the summer monsoon was weaker. This pattern is shown by climatic model simulations and data on palaeo-lake levels (Kutzbach 1981; COHMAP 1988; Street-Perrot & Perrot 1993), and in southern Africa is reflected in pollen data from the interior region which suggest that Kalahari thornveld elements, *Tarchonanthus* and Capparaceae, were more prominent (Street-Perrot & Perrot 1993; Scott 1989a). Modelling experiments further suggest that the seasonality in the early Holocene was different in the southern hemisphere (Kutzbach *et al.* 1993), and this is in keeping with pollen data from the southern Kalahari, the Free State grasslands and the Nama-karoo, which suggest less grass and more Asteraceae pollen (including *Stoebe* or *Elytropappus* type) (Scott 1987b, 1989b, 1993; Bousman *et al.* 1988). The existing data and comparisons with other continents, therefore, suggest that the Late Quaternary vegetation of the subcontinent responded according to patterns of global change. More palaeobotanical research is necessary to refine the Late Pleistocene and Holocene vegetation history in southern Africa and to tie the data in with global simulation models like that of the COHMAP group (Wright *et al.* 1993).

Concerning the origins of biomes, it can further be concluded that early types of fynbos and semi-arid, succulent shrublands and desert vegetation of the southwestern Cape, the west coast and the Karoo only became established by the end of the Miocene, i.e. more or less at the same time as modern communities on other continents. Although desertic conditions prevailed in the Namib region much earlier (Ward *et al.* 1983),

there is no evidence that contemporary desert plant types were present in the Palaeogene. Therefore, if the age of the desert and fynbos communities is not exceptionally great in the Cainozoic framework, this cannot be identified as the prime cause for the enormous species diversity in these biomes. Although rhythmic Cainozoic climatic cycles probably played an important role in the development of the unusual species richness of the southern African region, they are probably of a secondary nature, because they are not exclusive to this region. An explanation of species richness in southern Africa is more likely to be found among factors such as the development of physiographic characteristics, including crustal movements, substrata, oceanic currents and climate. In this regard, the dead-end of migration routes in the southern tip of the continent, and barriers between the southern biomes and the rest of Africa, are possibly also of key importance. However, to provide answers to the question of the evolution and speciation within the modern biomes, palaeobotanical data still have to be integrated effectively with biogeographical and genetic studies.

4.7 Acknowledgements

Drs Ann Cadman and Ian Corbett are thanked for providing information concerning west coast offshore deposits and Prof T. C. Partridge for supplying information about the Mahuru Muthla formation. Dr M. Smith (Steenkamp) is responsible for pollen analyses of the Anysberg hyrax middens, and Mr F. Scott for preparation of graphic art work.

4.8 References

Acocks, J.P.H. (1953). Veld Types of South Africa. *Memoirs of the Botanical Survey of South Africa*, **28**, 1–192.

Adamson, R.S. (1934). Fossil plants from Fort Grey near East London. *Annals of the South African Museum*, **31**, 67–96.

Anderson, H.M., Hiller, N. & Gess, R.W. (1995). *Archaeopteris* (Progymnospermopsida) from the Devonian of Southern Africa. *Botanical Journal of the Linnean Society*, **117**, 305–20.

Anderson, J.M. (1977). The biostratigraphy of the Permian and Triassic. *Memoirs of the Botanical Survey of South Africa*, **41**, 1–132.

Anderson, J.M. & Anderson, H.M. (1983). Vascular plants from the Devonian to Lower Cretaceous in Southern Africa. *Bothalia*, **14**, 337–44.

Anderson, J.M. & Anderson, H.M. (1985). *Palaeoflora of Southern Africa. Prodromus of South African megafloras Devonian to Lower Cretaceous.* Rotterdam: Balkema.

Anderson, J.M. & Anderson, H.M. (1993). Terrestrial flora and fauna of the Gondwana Triassic: Part 2 – Co-evolution. In *The Non-Marine Triassic*, ed. S.G. Lucas & M. Morales, pp. 13–25. New Mexico Museum of Natural History & Science Bulletin 3. Albuquerque: New Mexico Museum of Natural History & Science.

Avery, D.M. (1981). Holocene micromammalian faunas from the Northern Cape, South Africa. *South African Journal of Science*, **77**, 265–73.

Avery, D.M. (1982). Micromammals as palaeoenvironmental indicators and an interpretation of the Late Quaternary in the Southern Cape Province, South Africa. *Annals of the South African Museum*, 85, 183–374.

Avery, D.M. (1990). Late Quaternary environmental change in southern Africa based on micromammalian evidence: a synopsis. *Palaeoecology of Africa*, 21, 131–42.

Avery, D.M. (1991). Micromammals, owls and vegetation change in the eastern Cape Midlands, South Africa, during the last millennium. *Journal of Arid Environments*, 20, 357–67.

Avery, D.M. (1993). Last Interglacial and Holocene altithermal environments in South Africa and Namibia: micromammalian evidence. *Palaeogeography, Palaeoclimatology, Palaeoecology*, 101, 221–8.

Axelrod, D.I. & Raven, P.H. (1978). Late Cretaceous and Tertiary vegetation history of Africa. In *Biogeography and Ecology of Southern Africa*, ed. M.J.A. Werger, pp. 77–130. The Hague: Junk.

Bamford, M.K. & De Wit, M.C.J. (1993). Taxonomic descriptions of fossil wood from Cainozoic Sak River terraces, near Brandvlei, South Africa. *Palaeontologia Africana*, 30, 71–80.

Barreda, V.D. (1993). Late Oligocene?–Miocene pollen of the families Compositae, Malvaceae and Polygonaceae from the Chenque Formation, Golfo San Jorge basin, Southeastern Argentina. *Palynology*, 17, 169–86.

Baxter, A.J. & Meadows, M.E. (1994). Palynological evidence for the impact of colonial settlement within lowland fynbos: a high-resolution study from the Verlorenvlei, Southwestern Cape Province, South Africa. *Historical Biology*, 9, 61–70.

Beaumont, P.B., Miller, G.H. & Vogel, J. C. (1992). Contemplating old clues to the impact of future greenhouse climates in South Africa. *South African Journal of Science*, 88, 490–8.

Beaumont, P.B., van Zinderen Bakker, E.M. & Vogel, J.C. (1984). Environmental changes since 32 000 BP at Kathu Pan, northern Cape. In *Late Cainozoic Palaeoclimates of the Southern Hemisphere*, ed. J.C. Vogel, pp. 329–38. Rotterdam: Balkema.

Belsky, C.Y., Boltenhagen, E. & Potonié R. (1965). Sporae dispersae der Oberen Kreide von Gabun, äquatoriales Afrika. *Paläontologische Zeitschrift*, 39, 72–83.

Botha, G.A., Scott L., Vogel J.C. & von Brunn, V. (1992). Palaeosols and palaeoenvironments during the Late Pleistocene Hypothermal in Northern Natal. *South African Journal of Science*, 88, 508–12.

Bousman, C.B. (1991). *Holocene paleoecology and Later Stone Age hunter–gatherer adaptations in the South African interior plateau*, PhD thesis. Dallas: Southern Methodist University.

Bousman, C.B., Partridge, T.C., Scott, L., Metcalfe, S.E., Vogel, J.C., Seaman, M. & Brink, J.S. (1988). Palaeoenvironmental implications of Late Pleistocene and Holocene valley fills in Blydefontein basin, Noupoort, C. P., South Africa. *Palaeoecology of Africa*, 19, 43–67.

Bousman, C. B. & Scott, L. (1994). Climate or overgrazing? the palynological evidence for vegetation change in the eastern Karoo. *South African Journal of Science*, 90, 575–8.

Brain, C.K. & Brain, V. (1977). Microfaunal remains from Mirabeb: some evidence of palaeoecological changes in the Namib. *Madoqua*, 10, 285–305.

Burney, D.A., Brook, G.A. & Cowart, J.B. (1994). A Holocene pollen record for the Kalahari Desert of Botswana from a U-series dated speleothem. *The Holocene*, 4, 225–32.

Cadman, A. & Rayner, R.J. (1989). Climate change and the appearance of *Australopithecus africanus* in the Makapansgat sediments. *Journal of Human Evolution*, 18, 107–13.

Coetzee, J.A. (1967). Pollen analytical studies in East and Southern Africa. *Palaeoecology of Africa*, 3, 1–146.

Coetzee, J.A. (1978a). Late Cainozoic palaeoenvironments of southern Africa. In *Antarctic Glacial History and World Palaeoenvironments*, ed. E.M. van Zinderen Bakker, pp. 115–27. Rotterdam: Balkema.

Coetzee, J.A. (1978b). Climate and biological changes in south-western Africa during the Late Cainozoic. *Palaeoecology of Africa*, 10, 13–29.

Coetzee, J.A. (1993). African flora since the terminal Jurassic. In *Biological Relationships between Africa and South America*, ed. P. Goldblatt, pp. 37–61. New Haven: Yale University Press.

Coetzee, J.A. & Muller, J. (1984). The phytogeographic significance of some extinct Gondwana pollen types from the Tertiary of the southwestern Cape (South Africa). *Annals of the Missouri Botanical Gardens*, 71, 1088–99.

Coetzee, J.A. & Rogers, J. (1982). Palynological and lithological evidence for the Miocene palaeoenvironment in the Saldanha region (South Africa). *Palaeogeography, Palaeoclimatology, Palaeoecology*, 39, 71–85.

Coetzee, J.A., Scholtz, A. & Deacon, H.J. (1983). Palynological studies and the vegetation history of the fynbos. In *Fynbos Palaeoecology. A Preliminary Synthesis*, ed. H.J. Deacon, Q.B. Hendey & J.J.N. Lambrechts, pp. 156–73. *South African National Scientific Programmes Report 75*. Pretoria: CSIR.

COHMAP (1988). Climatic change of the last 18 000 years: observations and model simulations. *Science*, 241, 1043–52.

Cranwell, L.M. (1964). Antarctica: cradle or grave for its *Nothofagus*? In *Ancient Pacific Floras. The Pollen Story*, ed. L.M. Cranwell, pp. 87–93. Honolulu: University of Hawaii Press.

Deacon, H.J., Deacon, J., Scholtz, A., Thackeray, J.F. & Brink, J.S. (1984). Correlation of palaeoenvironmental data from the Late Pleistocene and Holocene deposits at Boomplaas cave, Southern Cape. In *Late Cainozoic Palaeoclimates of the Southern Hemisphere*, ed. J.C. Vogel, pp. 339–51. Rotterdam: Balkema.

Deacon, J. & Lancaster, N. (1988). *Late Quaternary Palaeoenvironments of Southern Africa*. Oxford: Clarendon Press.

De Wit, M.C.J. & Bamford, M.K. (1993). Fossil wood from the Brandvlei area, Bushmanland as an indication of palaeoenvironmental changes during the Cainozoic. *Palaeontologia Africana*, 30, 81–9.

Dowson, T.A. (1988). Shifting vegetation zones in the Holocene and later Pleistocene: preliminary charcoal evidence from Jubilee Shelter, Magaliesberg, Southern Transvaal. *Palaeoecology of Africa*, 19, 233–9.

February, E. (1992). Archaeological charcoals as indicators of vegetation change and human fuel choice in the late Holocene at Elands Bay, western Cape Province, South Africa. *Journal of Archaeological Science*, **19**, 347–54.

February, E. (1994). Rainfall reconstruction using wood charcoal from two archaeological sites in South Africa. *Quaternary Research*, **42**, 100–7.

February, E. & van der Merwe, N.J. (1992). Stable carbon isotope ratios of wood charcoal during the past 4000 years: anthropogenic and climatic influences. *South African Journal of Science*, **88**, 291–2.

Goldblatt, P. (1978). An analysis of the flora of southern Africa: its characteristics, relationships and origins. *Annals of the Missouri Botanical Garden*, **65**, 369–436.

Helgren, D.M. & Butzer, K.W. (1977). Palaeosols of the southern Cape coast, South Africa: implications for laterite definition, genesis, and age. *Geographical Review*, **67**, 430–45.

Hendey, Q.B. (1984). Southern African late Tertiary vertebrates. In *Southern African Prehistory and Palaeoenvironments*, ed. R.G. Klein, pp. 81–106. Rotterdam: Balkema.

Hooghiemstra, H. (1995). Pliocene–Quaternary paleoclimatic change and evolution of northern Andean montane forests and Paramo ecosystems. In *Palaeoclimate and Evolution with Emphasis on Human Origins*, ed. E.S. Vrba, G.H. Denton, T.C. Partridge & L.H. Burckle, pp. 249–61. New Haven: Yale University Press.

Horowitz, A. (1975). Preliminary palynological implications of pollen analysis of Middle Breccia from Sterkfontein. *Nature*, **258**, 417–18.

Kelber, K.-P., Franz, L., Stachel, T., Lorenz, V. & Okrusch, M. (1993). Plant fossils from Gross Brukkaros (Namibia) and their biostratigraphical significance. *Communications of the Geological Survey of Namibia*, **8**, 57–66.

Kemp, E.M. (1968). Probable angiosperm pollen from the British Barremian to Albian strata. *Palaeontology*, **11**, 421–34.

Kershaw, A.P. & Nanson, G.C. (1993). The last full glacial cycle in the Australian region. *Global Planetery Change*, **7**, 1–9.

Klein, R.G. (1977). The mammalian fauna of Middle and Later Stone Age (later Pleistocene) levels of Border Cave, Natal Province, South Africa. *South African Archaeological Bulletin*, **32**, 14–27.

Klein, R.G. (1983). Palaeoenvironmental implications of Quaternary large mammals in the Fynbos region. In *Fynbos Palaeoecology. A Preliminary Synthesis*, ed. H.J. Deacon, Q.B. Hendey & J.J.N. Lambrechts, pp. 116–38. *South African National Scientific Programmes Report 75*. Pretoria: CSIR.

Klein, R.G. (1984). The large mammals of southern Africa: Late Pliocene to recent. In *Southern African Prehistory and Palaeoenvironments*, ed. R.G. Klein, pp. 107–46. Rotterdam: Balkema.

Klinger, H.C. & Kennedy, W.J. (1980). The Umzamba Formation at isotope section Umzamba Estuary (Pondoland, Transkei), the ammonite content and palaeogeographical distribution. *Annals of the South African Museum*, **81**, 207–22.

Knoll, A.H. (1984). Africa and Precambrian biological evolution. *Bothalia*, **14**, 329–36.

Kutzbach, J.E. (1981). Monsoon climate of the early Holocene climate experiment using the earth's orbital parameters for 9000 years ago. *Science*, **214**, 59–61.

Kutzbach, J.E., Guetter, P.J., Behling, P.J. & Selin, R. (1993). Simulated climatic changes: results of the COHMAP climate-model experiments. In *Global Climates since the Last Glacial Maximum*, ed. H.E. Wright, J.E. Kutzbach, T. Webb, W.E. Ruddiman, F.A. Street-Perrot & P.J. Bartlein, pp. 24–93. Minneapolis: University of Minnesota Press.

Leopold, E.B. (1969). Late Cainozoic Palynology. In *Aspects of Palynology*, ed. R.H. Tschudy & R.A. Scott, pp. 377–438. New York: Wiley-Interscience.

Levyns, M.R. (1952). Clues to the past in the Cape Flora of today. *South African Journal of Science*, **49**, 155–64.

Levyns, M.R. (1964). Migrations and origin of the Cape Flora. *Transactions of the Royal Society of South Africa*, **39**, 85–107.

Linder, H.P., Meadows, M.E. & Cowling, R.M. (1992). History of the Cape flora. In *The Ecology of Fynbos. Nutrients, Fire and Diversity*, ed. R.M. Cowling, pp. 113–34. Cape Town: Oxford University Press.

Livingstone, D.A. (1971). A 22,000-year pollen record from the plateau of Zambia. *Limnology and Oceanography*, **16**, 349–56.

Mädel, E. (1960). Monimiaceen-Hölzer aus den oberkretazishen Umzamba-Schichten von Ost-Pondoland (S-Afrika). *Senckenbergia Lethaea*, **41**, 331–91.

Mädel, E. (1962). Die fossilen Euphorbiaceae-Hölzer mit besonderer Berücksichtigung neuer Funde aus der Oberkreide Süd-Afrikas. *Senkenbergia Lethaea*, **43**, 283–321.

Martin, A.R.H. (1968). Pollen analysis of Groenvlei lake sediments, Knysna (South Africa). *Review of Palaeobotany and Palynology*, **7**, 107–44.

Maud, R.R. & Partridge, T.C. (1987). Regional evidence for climate change in southern Africa since the Mesozoic. *Palaeoecology of Africa*, **18**, 337–48.

McLachlan, I.R. & Pieterse, E. (1978). Preliminary palynological results: site 361, leg 40. *Initial reports of the Deep See Drilling Project*, **40**, 857–81.

Meadows, M.E., Baxter, A.J. & Adams, T. (1994). The late Holocene vegetation history of lowland fynbos, Verlorenvlei, southwestern Cape Province, South Africa. *Historical Biology*, **9**, 47–59.

Meadows, M.E. & Linder, H.P. (1993). A palaeoecological perspective on the origin of Afromontane grasslands. *Journal of Biogeography*, **20**, 345–55.

Meadows, M.E. & Meadows, K.F. (1988). Late Quaternary vegetation history of the Winterberg Mountains, eastern Cape, South Africa. *South African Journal of Science*, **84**, 253–9.

Meadows, M.E. & Sugden, J.M. (1988). Late Quaternary environmental changes in the karoo, South Africa. In *Geomorphological Studies in Southern Africa*, ed. G.F. Dardis & B.P. Moon, pp. 337–53. Rotterdam: Balkema.

Meadows, M.E. & Sugden, J.M. (1990). Late Quaternary vegetation history of the Cederberg, south-western Cape. *Palaeoecology of Africa*, **21**, 269–82.

Meadows, M.E. & Sugden, J.M. (1991). A vegetation history of the last 14 000 years on the Cederberg, south-western Cape Province. *South African Journal of Science*, **87**, 34–43.

Muller, J. (1981). Fossil pollen records of extant angiosperms. *Botanical Review*, 47, 1–142.

Partridge, A.D. (1978). Palynology of the Late Tertiary sequence at site 365, leg 40. *Initial Reports of the Deep Sea Drilling Project*, 40, 953–61.

Partridge, T.C. & Dalbey, T. (1986). Geoarchaeology of the Haaskraal Pan: a preliminary palaeoenvironmental model. *Palaeoecology of Africa*, 17, 69–78.

Partridge, T.C., Kerr, S.J., Metcalfe, S.E., Scott L., Talma, A.S. & Voge,l J.C. (1993). The Pretoria Saltpan: a 200 000 year southern African lacustrine sequence. *Palaeogeography, Palaeoclimatology, Palaeoecology*, 101, 317–37.

Partridge, T.C. & Maud, R.R. (1987). Geomorphic evolution of southern Africa since the Mesozoic. *South African Journal of Geology*, 902, 179–208.

Penny, J.S. (1969). Late Cretaceous and early Tertiary Palynology. In *Aspects of Palynology*, ed. R.H. Tschudy & R.A. Scott, pp. 331–76. New York: Wiley-Interscience.

Plumstead, E.P. (1969). Three thousand million years of plant life in Africa. *Geological Society of South Africa Annexure*, 72, 1–72.

Prior, J. & Price Williams, D. (1985). An investigation of climatic change in the Holocene Epoch using archaeological charcoal from Swaziland, southern Africa. *Journal of Archaeological Science*, 12, 457–75.

Rayner, R.J., Moon, B.P. & Masters, J.C. (1993). The Makapansgat australopithecine environment. *Journal of Human Evolution*, 24, 219–31.

Ruddiman, W.F. & Kutzbach, J.E. (1991). Plateau uplift and climatic change. *Scientific American*, 264(3), 42–50.

Sampson, C.G. (1986). Veld damage in the karoo caused by its pre-trekboer inhabitants: preliminary observations in the Seacow Valley. *The Naturalist*, 30, 37–42.

Sancetta, C., Heusser, L. & Hall, M.A. (1992). Late Pliocene climate in the southeast Atlantic: preliminary results from a multi-disciplinary study of DSDP Site 532. *Marine Micropaleontology*, 20, 59–75.

Schalke, H.J.W.G. (1973). The upper Quaternary of the Cape Flats area (Cape Province, South Africa). *Scripta Geologica*, 15, 1–57.

Scholtz, A. (1985). Palynology of the Upper Cretaceous lacustrine sediments of the Arnot pipe, Banke, Namaqualand. *Annals of the South African Museum*, 95, 1–109.

Scholtz, A. (1986). *Palynological and Palaeobotanical Studies in the Southern Cape*. MSc thesis. Stellenbosch: Stellenbosch University.

Scott, L. (1976a). Palynology of Lower Cretaceous deposits from the Algoa Basin (Republic of South Africa). *Pollen et Spores*, 18, 563–609.

Scott, L. (1976b). Preliminary palynological results from the Alexandersfontein Basin near Kimberley. *Annals of the South African Museum*, 71, 193–9.

Scott, L. (1982a). Late Quaternary fossil pollen grains from the Transvaal, South Africa. *Review of Palaeobotany and Palynology*, 36, 241–78.

Scott, L. (1982b). A Late Quaternary pollen record from the Transvaal bushveld, South Africa. *Quaternary Research*, 17, 339–70.

Scott, L. (1982c). Pollen analyses of Late Cainozoic deposits in the Transvaal, South Africa, and their bearing on palaeoclimates. *Palaeoecology of Africa*, 15, 101–7.

Scott, L. (1982d). A 5000-year old pollen record from spring deposits in the bushveld at the north of the Soutpansberg, South Africa. *Palaeoecology of Africa*, 14, 45–55.

Scott, L. (1984a). Palynological evidence for Quaternary paleoenvironments in southern Africa. In *Southern African Paleoenvironments and Pre-history*, ed. R.G. Klein, pp. 65–80. Rotterdam: Balkema.

Scott, L. (1984b). Reconstruction of Late Quaternary palaeoenvironments in the Transvaal region, South Africa, based on palynological evidence. In *Late Cainozoic Palaeoclimates of the Southern Hemisphere*, ed. J.C. Vogel, pp. 317–27. Rotterdam: Balkema.

Scott, L. (1987a). Late Quaternary forest history in Venda, Southern Africa. *Review of Palaeobotany and Palynology*, 53, 1–10.

Scott, L. (1987b). Pollen analysis of hyena coprolites and sediments from Equus Cave, Taung, Southern Kalahari (S. Africa). *Quaternary Research*, 28, 144–56.

Scott, L. (1988). Holocene environmental change at western Free State pans, South Africa, inferred from pollen analysis. *Palaeoecology of Africa*, 19, 109–18.

Scott, L. (1989a). Climatic conditions in Southern Africa since the Last Glacial Maximum, inferred from pollen analysis. *Palaeogeography, Palaeoclimatology, Palaeoecology*, 70, 345–53.

Scott, L. (1989b). Late Quaternary vegetation history and climatic change in the eastern O.F.S. *South African Journal of Botany*, 55, 107–16.

Scott, L., (1993). Palynological evidence for late Quaternary warming episodes in Southern Africa. *Palaeogeography, Palaeoclimatology, Palaeoecology*, 101, 229–35.

Scott, L. (1994). Palynology of late Pleistocene hyrax middens, south-western Cape Province, South Africa: a preliminary report. *Historical Biology*, 9, 71–81.

Scott, L. (1995). Pollen evidence for vegetation and climate change in Southern Africa during the Neogene and Quaternary. In *Paleoclimate and Evolution with Emphasis on Human Origins*, ed. E.S. Vrba, G.H. Denton, T.C. Partridge & L.H. Burckle, pp. 56–76. New Haven: Yale University Press.

Scott, L. (1996). Palynology of hyrax middens: 2000 years of palaeoenvironmental history in Namibia. *Quaternary International*, 33, 73–9.

Scott, L. & Bonnefille, R. (1986). A search for pollen from the hominid deposits of Kromdraai, Sterkfontein and Swartkrans: some problems and preliminary results. *South African Journal of Science*, 82, 380–2.

Scott, L. & Bousman, C.B. (1990). Palynological analysis of hyrax middens from southern Africa. *Palaeogeography, Palaeoclimatology, Palaeoecology*, 79, 367–79.

Scott, L. & Brink, J. S. (1992). Quaternary palaeoenvironments of pans in central South Africa: palynological and palaeontological evidence. *South African Geographer*, 19, 22–34.

Scott, L., Cadman, A. & Corbett, I. (1993). Development of vegetation in southern Africa since the Eocene. In Conference on *Paleoclimate and Evolution, with Emphasis on Human Origins*, ed. E.S. Vrba, Airlie Conference Center, Virginia.

Scott, L. & Cooremans, B. (1990). Late Quaternary pollen from a hot spring in the upper Orange River Basin, South Africa. *South African Journal of Science*, 86, 154–6.

Scott, L., Cooremans, B., de Wet, J.S. & Vogel., J.C. (1991). Holocene environmental changes in Namibia inferred from pollen analysis of swamp and lake deposits. *The Holocene*, 1, 8–13.

Scott, L., Cooremans, B. & Maud, R.R. (1992). Preliminary palynological evaluation of the Port Durnford Formation at Port Durnford, Natal coast, South Africa. *South African Journal of Science*, 88, 470–4.

Scott, L., Steenkamp, M. & Beaumont, P.B. (1995). Palaeoenvironments in South Africa at the Pleistocene–Holocene transition. *Quaternary Science Reviews*, 14, 937–47.

Scott, L. & Vogel, J.C. (1983). Late Quaternary pollen profile from the Transvaal highveld, South Africa. *South African Journal of Science*, 79, 266–72.

Seward, A.C. (1903). Fossil floras of the Cape Colony. *Annals of the South African Museum*, 4, 1–22.

Shackleton, N.J. (1995). New data on the evolution of Pliocene climatic variability. In *Paleoclimate and Evolution, with Emphasis on Human Origins*, ed. E.S. Vrba, G.H. Denton, T.C. Partridge & L.H. Burckle, pp. 242–8. New Haven: Yale University Press.

Shackleton, N.J. & Kennet, J.P. (1975). Palaeotemperature history of the Cenozoic and the initiation of the Antarctic glaciation: oxygen and carbon isotope analyses in DSDP sites 277, 279 and 281. *Initial reports of the DSDP*, 29, 743–55.

Siesser, W.G. (1978). Aridification of the Namib Desert: evidence from ocean cores. In *Antarctic Glacial History and World Palaeoenvironments*,. ed. E.M. van Zinderen Bakker, pp. 105–13. Rotterdam: Balkema.

Smith, A.G. & Briden, J.C. (1977). *Mesozoic and Cenozoic Paleocontinental Maps*. Cambridge: Cambridge University Press.

Street-Perrot, F.A. & Perrot, R.A. (1993). Holocene vegetation, lake levels, and climate of Africa. In *Global Climates since the Last Glacial Maximum*, ed. H.E. Wright, J.E. Kutzbach, T. Webb, W.E. Ruddiman, F.A. Street-Perrot & P.J. Bartlein, pp. 318–56. Minneapolis: University of Minnesota Press.

Sugden, J.M. & Meadows M.E. (1989). The use of multiple discriminant analysis in reconstructing recent vegetation changes on the Nuweveldberge, South Africa. *Review of Palaeobotany Palynology*, 60, 131–47.

Talma, A.S. & Vogel, J.C. (1992). Late Quaternary paleotemperatures derived from a speleothem from Cango Caves, Cape Province, South Africa. *Quaternary Research*, 37, 203–13.

Thackeray, J.F. (1987). Late Quaternary environmental changes inferred from small mammalian fauna, southern Africa. *Climatic Change*, 10, 285–305.

Thiergart, F., Frantz, U. & Raukopf, K. (1962). Palynologische Untersuchungen von Tertiärkohlen und einer Oberflächenprobe nahe Knysna, Südafrika. *Advancing Frontiers of Plant Sciences*, 4, 151–78.

Tomlinson, R.W. (1974). Preliminary biogeographical studies on the Inyanga Mountains, Rhodesia. *South African Geographical Journal*, 56, 15–26.

Traverse, A. (1988). *Palaeopalynology*. Boston: Unwin Hyman.

Tusenius, M.L. (1989). Charcoal analytical studies in the north-eastern Cape, South Africa. *South African Archaeological Society Goodwin Series*, 6, 77–83.

Vail, P.R. & Hardenbol, J. (1979). Sea-level changes during the Tertiary. *Oceanus*, 22, 71–9.

van Daalen, J.C., Geldenhuis, C.J., Frost, P.G.H. & Moll, E.J. (1986). A rapid survey of forest succession at Mlazi Nature Reserve. *Foundation of Research Development Ecosystem Programmes, Occasional Report*, 11, 1–31.

van Zinderen Bakker, E.M. (1955). A preliminary survey of peat bogs of the Alpine belt of Northern Basutoland. *Acta Geographica*, 14, 413–22.

van Zinderen Bakker, E.M. (1957). A pollen analytical investigation of the Florisbad deposits (South Africa). In *Proceedings of the Third Pan African Congress on Prehistory, Livingstone*, ed. J.D. Clark, pp. 56–67. London: Chatto & Windus.

van Zinderen Bakker, E.M. (1975). The origin and palaeoenvironment of the Namib Desert biome. *Journal of Biogeography*, 2, 65–73.

van Zinderen Bakker, E.M. (1982). Pollen analytical studies of the Wonderwerk Cave, South Africa. *Pollen et Spores*, 24, 235–50.

van Zinderen Bakker, E.M. (1983). The Late Quaternary history of climate and vegetation in East and southern Africa. *Bothalia*, 14, 369–75.

van Zinderen Bakker, E.M. (1984). Palynological evidence for Late Cenozoic arid conditions along the Namibia coast from holes 532 and 530A, leg 75, Deep Sea Drilling Project. *Initial Reports of the Deep Sea Drilling Project*, 75, 763–8.

van Zinderen Bakker, E.M. (1989). Middle Stone Age palaeoenvironments at Florisbad (South Africa). *Palaeoecology of Africa*, 20, 133–54.

van Zinderen Bakker, E.M. & Muller, M. (1987). Pollen studies in the Namib Desert. *Pollen et Spores*, 29, 185–205.

van Zinderen Bakker, E.M. & Werger, M.J.A. (1974). Environment, vegetation and phytogeography of the high altitude bogs of Lesotho. *Vegetatio*, 29, 37–49.

Vincens, A., Chalié, F., Bonnefille, R., Guiot, J. & Tiercelin, A.J. (1993). Pollen-derived rainfall and temperature estimates from Lake Tanganyika and their implications for Late Pleistocene water levels. *Quaternary Research*, 40, 343–50.

Vogel, J.C. (1983). Isotopic evidence for past climates and vegetation of southern Africa. *Bothalia*, 14, 391–4.

Vogel, J.C. (1989). Evidence for past climatic change in the Namib desert. *Palaeogeography, Palaeoclimatology, Palaeoecology*, 70, 355–66.

Vogel, J.C. & Rust, U. (1987). Environmental changes in the Kaokoland Namib Desert during the present millennium. *Madoqua*, 15, 5–16.

Vrba, E.S. (1985). Early hominids in southern Africa: updated observations on chronological and ecological background. In *Hominid Evolution*, ed. P.V. Tobias, pp. 195–200. New York: Alan R. Liss.

Wadley, L., Esterhuysen, A. & Jeanneret, C. (1992). Vegetation changes in the eastern Orange Free State: the Holocene and Later Pleistocene evidence from charcoal studies at Rose Cottage Cave. *South African Journal of Science*, **88**, 558–63.

Ward, J.D., Seely, M.K. and Lancaster, N. (1983). On antiquity of the Namib. *South African Journal of Science*, **79**, 175–83.

Williams, M.A.J., Dunkerley, D.L., De Decker, P., Kershaw, A.P. & Stokes, T. (1993). *Quaternary Environments*. London: Edward Arnold.

Wright, H.E., Kutzbach, J.E., Webb, T. III, Ruddiman, W.F., Street-Perrot, F.A. & Bartlein, P.J. (1993). *Global Climates since the Last Glacial Maximum*. Minneapolis: University of Minnesota Press.

Zavada, M.S. & Cadman, A. (1993). Palynological investigations at the Makapansgat Limeworks: an australopithecine site. *Journal of Human Evolution*, **25**, 337–50.

The second part of this volume is the largest, and comprises ten chapters that describe, in a similar format, pattern and process in southern Africa's major vegetation categories. This section of the book is, in many respects, an inventory, where the reader can obtain information on a particular topic within or across vegetation types. The bulk of the material deals with terrestrial vegetation. However, important aquatic vegetation types such as salt marshes, mangroves, freshwater wetlands and marine algal communities are also included.

The approach adopted for this volume has been to use biomes as major categories for dealing with the terrestrial vegetation of southern Africa. Biomes are the highest order entities in the hierarchy of vegetation units and are characterized in terms of climatic parameters and life-form mix. The use of the biome as the first-order category for this volume is both convenient and appropriate. Firstly, Rutherford (Chap. 5, this volume) has provided a simple and workable characterization of southern Africa's seven biomes and these have been mapped at a 1 : 10 000 000 scale across the entire subcontinent. The only other subcontinental-scale classification is Frank White's vegetation map, where units are organized according to phytochoria (see Chap. 3, this volume). Since the focus of this volume is ecological rather than biogeographical, Rutherford's scheme was clearly preferable. Secondly, terrestrial ecological research in South Africa between the mid-1970s and the late 1980s was organized according to biome-specific research projects under the auspices of the National Programme for Ecosystem Research of the Council for Scientific and Industrial Research (CSIR). These projects explicitly encouraged collaboration between researchers, and between researchers and managers. They were particularly successful in exposing researchers (including young postgraduates) to the practical problems of vegetation management and in exposing managers to the conceptual, theoretical and analytical tools employed in the scientific study of populations, communities and ecosystems. Although these programmes were terminated in the late 1980s when research coordination and funding became the responsibility of the Foundation for Research Development (FRD), the spirit of collaborative and goal-orientated research persisted in some of the biomes. This period, from the mid-1970s onwards, saw a massive upsurge in research on South African terrestrial ecosystems, resulting in the publication of several hundred papers in the primary literature and several synthesis volumes, and the emergence of a well-trained and productive corps of young researchers. Research in each of the biomes developed its own identity as a result of both the particular interests of the researchers and the research questions posed in the descriptive documents for the different biome projects. It would be unwise, for pragmatic and historical reasons, to adopt an organizational approach that emphasized cross-biome patterns and processes. This perspective is, however, given in Part 3 of this volume, where selected themes in vegetation science are discussed for the entire subcontinent.

The structure of each of the biome chapters is broadly similar. The first section of each chapter deals with pattern at different spatial scales, and focuses on the determinants of boundaries at the biome- and community levels. This section also includes a description and the environmental correlates of the major community types. In most cases, these data are summarized in tables. No attempt was made to standardize the community descriptions across biomes: these range from restatements of John Acocks' Veld Types (for succulent karoo), to structurally defined types (fynbos), to formally characterized units arising from a hierarchical classification of many thousands of phytosociological relevés sampled in the Braun–Blanquet tradition (grassland and desert). The next section comprises an analysis of patterns and processes at the organism, population, community and ecosystem levels. It is in this section that the major gaps in research for particular biomes becomes most apparent. The final section deals with (usually) three themes relating to land use and management which were chosen to minimize overlap with the chapters in Part 3 of this volume (especially Chaps 20–23).

We now provide some background to identify and explain the uneven coverage of topics in the various chapters. The chapters (and biomes) are discussed in the order in which they appear in this volume. The Fynbos Biome Project, initiated in 1977, stimulated a concerted and productive research programme that focused on all aspects of vegetation science. Therefore, coverage of this biome (Chap. 6) is more comprehensive than the others. Particular strengths in terms of scope and depth include community description, reproductive ecology, plant–animal interactions, community structure (especially coexistence and diversity) and dynamics (especially in relation to fire regime), the ecology of invasive alien plants, and some aspects of conservation biology. Much less is known about ecophysiological and ecosystem processes, although long-term hydrological experiments in mountain catchments have provided many valuable insights. A great deal of the research has been descriptive, although there has been a trend in recent years for more experimental studies. It is tempting to suggest that the enormous floristic diversity in the fynbos and its international significance as a biological hot-spot have played a role in stimulating imaginative research. However, there has also been a longer history of research in this than any of the other biomes, and the pioneering work in the early part of this century of Rudolf Marloth, Margaret Levyns, R.S. Adamson and others on the systematics, natural history and fire responses of fynbos plants, certainly played a role. The Fynbos Biome Project was highly successful in fostering a spirit of collaboration among researchers and between them and resource managers. Finally, ecological research in the fynbos has benefited enormously from collaboration with researchers from other mediterranean-climate regions of the world. Strong links between fynbos and other regions, particularly Australia, fostered inter-regional exchange and comparison. Starting in 1971, a series of international conferences on mediterranean-type ecosystems (MEDECOS) drew together workers from all regions. These and other meetings produced a wealth of detailed information on topics such as the origin and structure of mediterranean-type shrublands, fire ecology and management, the role of nutrients, ecosystem resilience, plant–water relations, plant–animal interactions, and the links between biodiversity and the functioning of ecosystems. No other southern African biome has enjoyed the benefits of such focused and sustained international collaboration.

The Karoo Biome Project, designed to include both the Nama-karoo and succulent karoo biomes, was short-lived, having been initiated only in 1986. Until that time, most research had been done in the Nama-karoo by agricultural scientists and dealt almost exclusively with the development of protocols for improving range condition. Almost no research (including range management studies) had been done in the succulent karoo. Because the Karoo is remote from all the major universities, a field station was established, in 1987, on the boundary between the Nama-karoo and succulent karoo biomes for university-based research. There followed a brief, but highly productive, phase resulting in numerous insights, mainly for the succulent karoo, on plant reproductive ecology, plant–animal interactions, disturbance ecology and community dynamics (Chap. 7). The emphasis on these themes is probably due to the strong overlap of research leadership between the Fynbos and Karoo Biome Projects. Other strengths of research in the succulent karoo include descriptive biogeography and the ecophysiology of succulents. There are few detailed descriptions of community structure and diversity in this biome and almost nothing is known about ecosystem processes.

Given its size and economic importance, the Nama-karoo is, at least in terms of research coverage, the Cinderella of southern African biomes (Chap. 8). Most research has been done in its more mesic, eastern regions by agricultural scientists based at the Grootfontein Research Institute. Results of this research played a key role in developing recommendations for range management, the implementation of which, although controversial, has resulted in considerable improvement in veld condition over many areas. However, few of these studies have been published in the primary literature. Futhermore, few attempts were made to test rigorously the empirical bases of these management models. There are, therefore, large gaps in almost every aspect of autecological research in the biome. Only recently have there been attempts to analyse the results of grazing experiments using appropriate statistical tools, and to develop community classifications at a level lower than Acocks' Veld Types. Ecosystem-level research has been entirely neglected.

Even less is known about the desert biome (Chap. 9) than the Nama-karoo. This is a result not only of its remoteness but also of the poor research infrastructure in the formerly colonial South West Africa and the now independent Namibia. Some excellent research on the biophysical processes and animal life of the Namib Desert has, however, been carried out at the Desert Ecological Research Unit at Gobabeb in the central Namib. There has been much less plant-related research at Gobabeb, and what has been done has dealt mainly with sources of production for higher trophic levels, phytosociology, the structure of dune grass assemblages and, more recently, seed biology. Over the past decade, Norbert Jürgens and his colleagues have undertaken phytochorological and phytosociological research across the entire biome. D.J. von Willert and his associates have made considerable progress in research on the ecophysiology of desert succulents, especially on the southern fringes of the biome. There is much scope for further research on the plants and vegetation of this, the world's most biologically interesting and ancient desert.

Because of its importance to the livestock industry, the grassland biome (Chap. 10) has a long history of research by agricultural scientists. Indeed, many of the paradigms of vegetation dynamics, which profoundly influenced researchers in other southern African biomes, were developed by grassland scientists such as J.F.V. Phillips, E.R. Roux, J.D. Scott, P. de V. Booysen and N.M. Tainton, working mainly out of the Universities of Natal and the Witwatersrand. Unlike in the Nama-karoo, detailed autecological studies on major grass species provided a mechanistic understanding of the results of grazing and burning trials, some of which were initiated in the early part of this century. This tradition was reinforced by studies initiated after the launch of the Grassland Biome Project in 1982. A major thrust of this programme was the detailed classification of grassland vegetation in the Braun–Blanquet tradition, an activity which has continued to this day and resulted in the most comprehensive phytosociological treatment of all of the biomes. More recently there has been a strong focus on seed biology, disturbance ecology and the dynamics of biome boundaries in relation to global change. Ecosystem-level research is largely confined to hydrological processes in mountain catchments. Overall, the grassland is one of the more comprehensively studied biomes of southern Africa.

Initiated in 1973, the Savanna Ecosystem Project was the first of the biome projects. It is unusual among these projects in that the underlying philosophy of the research programme was strongly influenced by the ecosystem approach inherent in the International Biological Programme. Furthermore, much of the research funded by the project was confined to the Nylsvley Nature Reserve, a broad-leaved savanna site in the Northern Province. It was possible, therefore, to develop an impressive understanding of the structure and function of this ecosystem at a level of detail unparalleled in the other biomes (Chap. 11). The research emanating from the Savanna Ecosystem Project was enormously influential in savanna research worldwide. B.H. Walker, P.G.H. Frost, B.J. Huntley, R.J. Scholes and their colleagues provided a functional classification of savanna types (eutrophic, narrow-leaved versus dystrophic, broad-leaved); a framework for assessing responses to stress and disturbance; and the conceptual and theoretical basis for 'event-driven' dynamics. Other strengths of this project included studies on interactions between large mammals and vegetation, grass–tree interactions, the effects of fire on community and ecosystem processes, and community descriptions. Weaknesses of the project, and southern African savanna vegetation science in general, include a poor knowledge of plant reproductive ecology and plant population dynamics (especially of the tree component).

There has been a long-standing interest by botanists in the forests of southern Africa. John Phillips' study, in the 1920s, on the ecology of the Knysna forests addressed a wide range of topics, including plant reproductive biology, plant–animal interactions and dynamics, that only recently attracted renewed interest by researchers. J.W. Bews carried out some very interesting research on the ecophysiology of forest trees between the two World Wars, and in the 1960s, E.J. Moll and his colleagues provided a number of community-level descriptions of forest vegetation, a tradition that has been extended by many of his students, including C.J. Geldenhuys and S.G. Cawe. The Forest Biome Project was launched in only 1985 and had little impact on research activities by the time it was terminated in the late 1980s. However, the project did foster excellent collaboration between scientists and managers. Recent studies, principally by researchers at the CSIR's Division of Forest Science and Technology (formerly the South African Forestry Research Institute), have focused on the dynamics of forest communities. These studies have shown fundamental differences in the population structure and dynamics of temperate (Afromontane) versus subtropical (Indian Ocean) forests. This research has provided the theoretical basis for decades-old harvesting programmes and will undoubtedly provide direction for new management initiatives. Generally, however, there is still much to be learnt about the forests of the subcontinent.

The vegetation of southern Africa's coastal zone has been very poorly studied (Chap. 13). The CSIR programmes on marine ecosystems concentrated on intertidal, subtidal and pelagic systems rather than coastal habitats. Although there are some comprehensive descriptive studies, including an insightful review by K.L. Tinley, little is known about community dynamics and ecosystem processes of terrestrial coastal habitats. Most research of this nature has been concentrated along the southeastern coastline. This lack of information is unfortunate, given the importance of coastal zone vegetation in the face of rising sea levels and increasing development pressures.

In contrast, the wetland plant communities of the subcontinent are poorly characterized and described, yet some excellent case studies at the population, ecosystem and landscape levels have provided new insights on the development and function of these systems. The description of the dynamics of the Okavango delta system, described in detail in Chap. 14, must rank as one of the most comprehensive and fascinating accounts of its kind.

Like its terrestrial counterpart, the southern African marine environment is extremely varied, ranging from the tropical coral reefs of the northeast, to the cool and highly productive kelp forests of the Cape west coast. The zoological component of these diverse environments has been intensively studied for decades and the zoology departments of several universities have strong marine ecology programmes. However, until the early 1980s, most research on the marine flora of southern Africa was taxonomic. Since then there has been an upsurge in activity, principally in the Western Cape. Research has focused on the environmental tolerances of individual species, community description, community structure in relation to disturbance regime, and the ecology of economically important species. However, large tracts of intertidal shore and almost the entire subtidal environment have yet to be surveyed in a much more intensive manner.

The terrestrial, freshwater and marine environments of southern Africa include a staggering array of plant diversity at all levels of organization. The chapters in this part of the book provide an overview, albeit patchy in some cases, of the vegetation of these habitats. After a very rapid and very promising start in the early part of the twentieth century, fostered by the pioneering and internationally acclaimed studies by R.S. Adamson, J.W. Bews, J.F.V. Phillips and I.B. Pole Evans, there was a prolonged slump in output during the 1950s and 1960s. This hiatus was followed by a period of rigorous and productive research, stimulated by the initiation of the biome projects. It is our sincere wish that future research should strengthen this foundation so that solutions can be found for the developing needs of the subcontinent that will ultimately depend on the wise use of the region's plant resources.

Categorization of biomes 5

M.C. Rutherford

5.1 Introduction

Biomes are broad ecological units that represent major life zones extending over large natural areas (Smith 1974; Godman & Payne 1979; Abercrombie, Hickman & Johnson 1980). A biome forms the highest type of ecological unit and is followed by the community and other levels of detail. Phenomena occur at each level and should be measured at the spatial and temporal scale characteristic of their hierarchical level. Biomes are characterized as weakly organized systems with very poor feedback control (O'Neill *et al.* 1986). It is difficult to predict the effect of varying landscape grain and patchiness on the consistency of emergent properties at the biome scale (Ehleringer & Field 1993).

Vegetation pattern and climate distribution have long been associated in plant ecology. Global correlative schemes from the first half of the twentieth century, such as that of Holdridge (1947), tended to rely on class boundaries that were strictly regular dissections of a climate space defined by two variables (Prentice *et al.* 1992). More recently, global models have attempted to determine biomes through prediction of their functional type components (Box 1981; Prentice *et al.* 1992). The development of a detailed, consistent and explicit classification system of plant functional types is still, however, in its infancy. One of the few well-established schemes that provides functional types which are sufficiently well defined for consistent and universal application is the simple system of plant life forms of Raunkiaer (1934). Global models tend to be severely limited by their reliance on inconsistently derived vegetation units for the world. Consequently, global models have yet to generate results that adequately match the main vegetation patterns within southern Africa.

Only recently have the biomes of southern Africa been determined according to explicit criteria and methodology. This chapter follows the first of these approaches for southern Africa (Rutherford & Westfall 1986), in which a biome is viewed as: (1) the largest land community unit recognized at a continental or subcontinental level; (2) a unit mappable at a scale of no larger than about 1 : 10 million (i.e. only at a broad scale); (3) distinguished from other biomes primarily on the basis of dominant life form(s) in the long term; (4) distinguished from other biomes secondarily on the basis of those major climatic features that most affect the biota; and (5) not an unnatural or anthropogenic system. Life forms of plants are used while the *a posteriori* association of certain zoological component forms with biome units (Irish 1994) is recognized. It is important to stress that biomes are not determined on the basis of floristic or phylogenetic composition and, therefore, do not necessarily coincide with phytochoria (see Chap. 3, this volume).

The plant life forms used in this system are based on the level of protection of the renewal buds in the unfavourable (cold or drought or both) growth season relative to their position within the steep climatic gradient near the ground (Geiger 1965). These represent a modified Raunkiaer-type system (Raunkiaer 1934; Orshan quoted in Zohary 1962) with phanerophytes (trees and larger shrubs), chamaephytes (small shrubs) (as modified by Mueller-Dombois & Ellenberg 1974), hemicryptophytes (mainly perennial grasses), cryptophytes (geophytes) and therophytes (annuals). More detailed growth-form or functional characteristics such as sclerophylly, succulence (Chap. 16, this volume) or photosynthetic response type (Rutherford 1991) are not consistent emergent properties at the biome level.

The method of categorization (Rutherford & Westfall 1994) determined the relative contributions of each main life form to estimates of mean annual production, using a life-form dominance hierarchy in which tree production is weighted more heavily than the same production of, for example, annuals. Dominance hierarchies are employed as surrogates for expressing unequal interactions between components, in which, for example, woody plants have a greater effect than herbaceous plants. The dominance hierarchy also provided rules that determined limits to the number of co-dominant life forms in a biome. Reduction of information at the hierarchical higher level of the biome not only facilitated interpolation and mapping at a scale of about 1 : 10 million, but also conveniently obviated converting certain data to productivity estimates (Rutherford 1993) where the specific contribution of plant forms were not critical for diagnostic purposes.

5.2　The biome units

Seven biome units were identified for southern Africa (Botswana, Lesotho, Namibia, South Africa and Swaziland) in accordance with dominance or co-dominance of plant life forms at the biome scale. Combinations involving phanerophytes, chamaephytes, hemicryptophytes and therophytes were found sufficient to

Table 5.1 **Surface areas of the biomes of southern Africa shown in Figure 5.2**

Biome	Proportion (%)	Area (km²)
Savanna	53.7	1 435 713
Nama-karoo	22.7	607 235
Grassland	13.1	349 174
Desert	4.2	111 147
Succulent karoo	3.7	100 251
Fynbos	2.7	71 337
Forest	<0.1	568
		2 675 425

differentiate biomes (Fig. 5.1). Cryptophytes were found to intergrade with hemicryptophytes. Uni-dominance of phanerophytes, chamaephytes, hemicryptophytes and therophytes represents the forest, succulent karoo, grassland and desert biomes, respectively. The intersection sets determine the savanna, Nama-karoo and fynbos biomes. The desert biome contains, pragmatically, a few areas of co-dominance with chamaephytes or with hemicryptophytes (Fig. 5.1) especially in areas of transition to the succulent karoo or Nama-karoo biomes, respectively. No co-dominance of only chamaephytes and phanerophytes was found at the biome scale for southern Africa, a combination also apparently lacking extensively elsewhere in the world (Cain & Castro 1959). In terms of the biome definition applied, some widely recognized important units such as the arid, nutrient-rich savanna and moist dystrophic savanna (Huntley 1984; Chap. 11, this volume) do not attain biome status.

The areas (Table 5.1) of the mapped biomes (Fig. 5.2) in southern Africa range widely, and differ from those of Rutherford & Westfall (1994), owing to extension of the region and generally minor boundary changes based on new information.

5.3　Nomenclature

Most of the biome names are universally familiar. The others are dictated by the common practice of assigning locally derived terms to describe more specialized units. Fynbos is a vernacular term (literally meaning fine-leaved bush). It has recently gained acceptance by the international scientific community to distinguish the floristically unique area of the fynbos biome from other sclerophyllous shrublands on nutrient-poor soils elsewhere in Africa, and from structurally similar vegetation in mediterranean-climate regions on other continents.

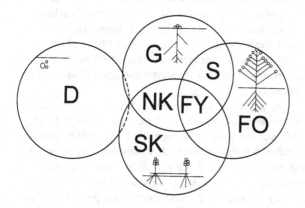

Figure 5.1 **Dominance or co-dominance of four plant life form sets in relation to biomes.** Life forms are schematically represented with phanerophytes on the right, chamaephytes below, hemicryptophytes above and therophytes on the left. Biomes are: FO, forest; G, grassland; SK, succulent karoo; D, desert; NK, Nama-karoo; S, savanna, and FY, fynbos.

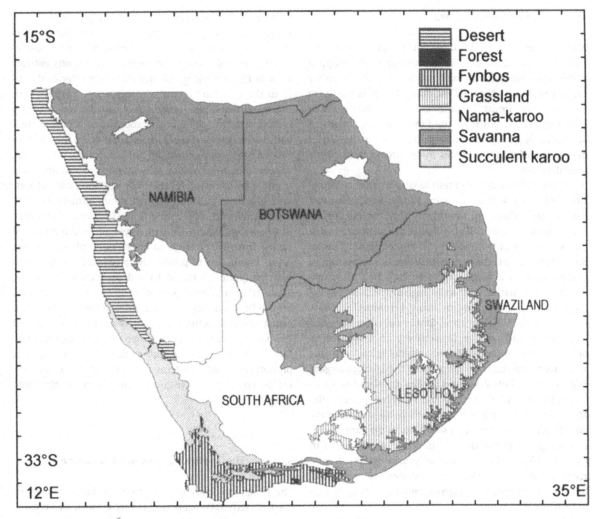

Figure 5.2 **The biomes of southern Africa.**

Nama-karoo is a concatenation of Namaland of southern Namibia and karoo of South Africa. Succulent plants are a recurring feature in most of the area of the succulent karoo biome, but can be rare in places when compared to non-succulent dwarf shrubs. The essential diagnostic feature of the succulent karoo biome is not succulence but the comparatively low abundance of the perennial grass form. To assist in placing the southern African biomes in a perhaps more familiar frame of reference, Table 5.2 lists examples of some corresponding biotic units from the Americas.

Table 5.2 **Southern African biomes with examples of type equivalents from America (North American units after Brown, Lowe & Pase 1979)**

Southern African Biome	American Type Example
Desert	Northern Atacama Desert (northern Chile)
Forest	Warm temperate forests (southeastern North America)
Fynbos	Californian chaparral (warm temperate shrublands, western North America)
Grassland	Plains grassland (cold temperate grasslands, central North America)
Nama-karoo	Great Basin desert-scrub (cold temperate desertlands, western North America)
Savanna	Cerrado (central Brazil)
Succulent karoo	Agave–Bursage ('Vizcaino') Series (Sonoran desert scrub, Baja, California Peninsula, Mexico)

5.4   Important outliers

All the biomes have sub-biomic patches of other biomes within each of their borders. For example, the mapped contiguous Nama-karoo has patches of all six other biomes within its borders, including forest in the form of river gallery forest. The minuscule forest biome's area increases several-fold if all the forest patches elsewhere are included (Chap. 12, this volume). Only the strict observance of scale eliminates these patches from biome consideration.

Outliers illustrate the extent to which genetic composition can vary within the same biome and continent. For example, there are two highly disjunct outliers of the Nama-karoo biome (Fig. 5.2) that are associated with two kinds of aridity condition in higher rainfall areas. Both differ from their relatively mesic surrounding biomes to such a degree as to be classified as Nama-karoo. One situation is that of high-altitudinal physiological drought, resulting in many xeromorphic adaptations (Smith & Geller 1979; Smith 1981) on the high and leeward side of the Drakensberg in Lesotho. The other is the salinity-induced physiological drought associated with the large salt pans of the Madkgadikgadi depression in Botswana and Etosha in Namibia (Irish 1994). In particular, the Sua and Ntwetwe pans of the Madkgadikgadi are associated with belts of arid grassland (Bekker & De Wit 1991) which, as in the case of arid edaphic grasslands of the Namib fringe, are included in the Nama-karoo biome. *Odyssea paucinervis* is often a recurring element in this pan-associated arid grassland, but the chaemophytic drought-adapted *Suaeda* sp. is limited to the pan fringe.

The meeting ground of the three most arid biomes (Nama-karoo, succulent karoo and desert; see Fig. 5.2) contains the main modification of the biome map of Rutherford & Westfall (1994). A newly recognized outlier of the desert biome occurs in the interior of southern Namibia and a small part of South Africa (Irish 1994; see also Chap. 9, this volume).

5.5   Typification limits

Biomes are likely to contain typical and less typical areas. Less typical forms of the savanna biome occur patchily along the eastern seaboard (Feely 1987) and in limited areas of the far south of the subcontinent. In the southern Cape region there are areas of dense thicket that may satisfy the forest definition (Chap. 12, this volume), except perhaps for the criterion of extent. This

area includes several vegetation types which are especially difficult to assign to any of the biomes, e.g. spekboomveld (dominated by *Portulacaria afra*) (Acocks 1988). The arid grassland associated with sandy patches such as those fringing the Namib Desert interior differs from the grassland biome with respect to edaphic limitation, climate and mappability and is consequently incorporated in the Nama-karoo biome (Chap. 8, this volume). In the savanna biome, apart from the salinity-related depressions classified as Nama-karoo, other large depressions in the region are classified as savanna. This is either by virtue of: (1) the close agglomeration of wetland grassland and well-wooded savanna interleaving at biome-scale such as is found in the Okavango Delta area of Botswana (Astle & Graham 1976); or (2) the relatively recent invasion of large sump areas by phanerophytes under conditions of progressively reduced water inflow such as applies in the Mababe depression of Botswana. Some other smaller depressions, such as that of Kazuma on the Botswana–Zimbabwe border, appear to exhibit a Nama-karoo structure, but the *Colophospermum mopane* 'low shrub' component is anomalous in having large surface burls and below-ground systems of phanerophytic proportions. Such systems can change to a savanna biome structure above-ground over a relatively short time.

5.6   The status of gradual transitions

The transition from savanna to Nama-karoo biomes in the southern Kalahari region is very gradual and reflects a changing linear matrix of savanna elements on the dunes and Nama-karoo elements in the inter-dune areas (Werger 1978; Chaps 8 and 11, this volume). Biome transitions can produce limited areas of transitional life-form combinations such as co-dominance of phanerophyte, chamaephyte and hemicryptophyte in parts of the southern Kalahari.

The boundary between the succulent karoo biome and the main block of the desert biome approximates that indicated by the floristic data of Jürgens (1991; Chap. 9, this volume). Except for the outlier of desert biome described above, this boundary does not follow that given by Irish (1994), whose somewhat different definition of desert biome results in an alternative boundary. This contact zone consists of a mix of very abrupt and very gradual transitions between succulent karoo and desert biomes.

5.7 Biome determinants

Currently recognized environmental links should be viewed as primarily correlative and not necessarily predictive at the biome scale. Reasons for this include the following.

- Environmental control at the biome level is likely to be intrinsically weak, owing to the very limited effective feedback loops.
- Independent response and sensitivity of individual components can result in different combinations, especially under changing environmental conditions (Rutherford et al. 1995).
- Existing vegetation may not be in equilibrium with current climate (an intrinsic limitation of equilibrium models – Prentice et al. 1992).
- Combinations predicted at a local scale with real interaction between components do not necessarily apply at broader scales where, with spatial agglomeration, it may be possible for components to avoid direct interaction.
- Simple scaling-up from local results to larger spatial scales does not necessarily take into account the variation in environmental landscape texture (Neilson, King & Koerper 1992).

Attempts at mechanistic explanation and prediction of biomes have tended, in practice, to relate to the biome unit rather than predicting the individual components. Even where components are predicted, it is particularly challenging to predict the effect of environment on the competitive balance between components to produce the resultant composition (e.g. the Walter hypothesis; Chaps 11 and 16, this volume). Prentice et al. (1992) emphasize independent control of biome components (plant functional types) in their ambitious global biome model. However, for southern Africa, their functional type is usually identical to the vegetation type (biome).

Rutherford & Westfall (1994) and Irish (1994) found a generally close association between southern African biomes and a combination of the annual mean proportion of winter rainfall and a mean Summer Aridity Index (SAI) (edaphic and other special biome outliers excepted). These climatic parameters also performed well for areas near the perimeter of given biomes and the SAI was particularly effective in helping to discriminate between the succulent karoo and Nama-karoo biomes. The advantage of the SAI is that it maximizes the number of usable climatic stations where certain effects of temperature are accounted for without the explicit requirement for actual temperature data at each station. However, to discriminate between the grassland biome and the savanna biome, mean minimum temperature of the coldest month needed to be combined with moisture availability. Within the main biome areas, soil types varied greatly and appeared important only in determining several units below the level of biome (e.g. Chap. 6, this volume). Phanerophytes in the southern extension of savanna in the central interior of southern Africa are difficult to explain (see also Chap. 10, this volume) and may undergo some form of persistent cold-hardening under generally more arid conditions than occur in the grassland biome.

In a useful attempt to find biologically meaningful parameters to predict the biome units identified by Rutherford & Westfall (1986), Ellery, Scholes & Mentis (1991) derived an effective algorithm for predicting the biomes (except the desert biome) based on number of days above a certain level of water availability (calculated via Linacre's formula), mean temperature when water was available for growth (growth temperature) and mean temperature when moisture was not available for growth (no-growth temperature) (Fig. 5.3). Although these parameters do relate to physiological growth of the plant, it is unclear how a generalized growth model should discriminate between biomes irrespective of differences in their life-form composition. In effect, the results of Ellery et al. (1991) confirm the environmental correlations of the biomes of Rutherford & Westfall (1986), since the environmental parameters used tend to correlate with the main three used by Rutherford & Westfall (1986; see Fig. 5.4). Unfortunately, owing to the explicit requirement for temperature data, the approach of Ellery et al. (1991) could not be tested with the use of the much larger set of station locations used by Rutherford & Westfall (1986). It is possible that the discrimination of grassland biome from savanna biome by the fact of savanna having a no-growth temperature higher than 17 °C, may be an artifact for the south-central extension of savanna mentioned above, since mean monthly temperatures in this area, with warm winter days, obscure the intensity of frost-induced stress. Consideration of several independent climatic parameters (Fig. 5.4) indicates that several biomes are differentiated by more than one climatic variable and suggests that maximum climatic discrimination may occur between the succulent karoo and grassland biome.

Given all the attempts thus far at prediction of biomes in southern Africa, no adequate mechanistic explanation has yet been derived that will predict changes in the main diagnostic components of biomes with climate change. The selection by Ellery et al. (1991) of a uni-dominant component biome (grassland), to test

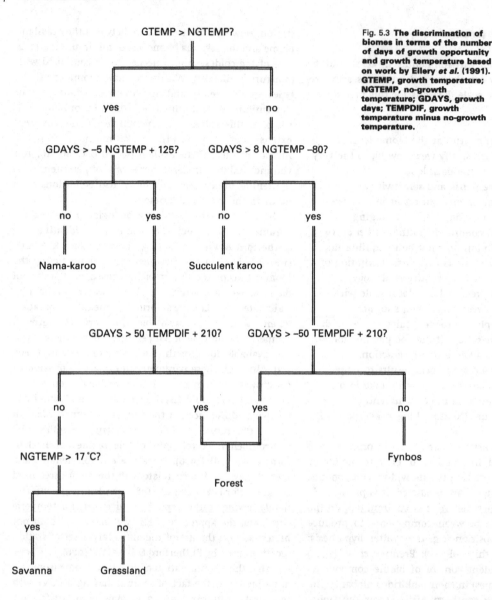

Fig. 5.3 **The discrimination of biomes in terms of the number of days of growth opportunity and growth temperature based on work by Ellery *et al.* (1991). GTEMP, growth temperature; NGTEMP, no-growth temperature; GDAYS, growth days; TEMPDIF, growth temperature minus no-growth temperature.**

Figure 5.4 **Main climatic differences between the major contiguous areas of southern African biomes, modified from Rutherford & Westfall (1986). Climatic factors not in bold are regarded as not independent of other parameters in the same cell. The forest biome is omitted, since its potential climate space is not adequately defined from southern African data. Biome order follows rainfall. Differentiating climate factors are: A, rainfall seasonality; B, Summer Aridity Index (SAI); C, combined rainfall seasonality and SAI; D, mean annual rainfall; E, combined temperature minima and summer rainfall; F, mean minimum temperature for the coldest month. The nomenclature of Ellery *et al.* (1991) is used as follows: GTEMP versus NGTEMP relates to A or C; GDAYS combined with NGTEMP relates to E or F (see Figure 5.3).**

	Grassland	Fynbos	Savanna	Nama-karoo	Succulent karoo
Fynbos	**ACE**				
Savanna	**E**	**C**			
Nama-karoo	**C**	**C**	**C**		
Succulent karoo	**ABCDE**	**C**	**BCE**	**CE**	
Desert	**BCDEF**	**CDE**	**BCDE**	**BCEF**	**C**

the effect of a climate change scenario, was therefore wise. Rutherford & Westfall's (1986) work provides only some indication of environmental control of life-form dominance. The lower limits of dominance or co-dominance by hemicryptophytes tend to be determined by growth season drought (SAI), whereas the lower limits of dominance by phanerophytes and chamaephytes appear to relate more closely to integrated mean annual water availability. Therophytic dominance relates only indirectly to climate and directly to absence of competition from perennial plants.

5.8 Floristic relationships

Biomes are not determined floristically, but do they relate to floristic units? Independent studies (Gibbs Russell 1987; Cowling *et al.* 1989) of floristic species relationships between the main areas of biomes identified by Rutherford & Westfall (1986) showed that each biome (forest was not analyzed) is floristically distinct, except to some extent the Nama-karoo. Grassland and savanna biomes are most closely allied, whereas the two 'karoo biomes' are clearly distinct. Fifty-five per cent of species of succulent karoo is shared with fynbos and only 28% with Nama-karoo. Fifty-five per cent and 32% of species of the Nama-karoo are shared with the savanna and grassland biomes, respectively, but only 28% with succulent karoo. The floristic correspondence with biome units confirms the expectation of structural–floristic convergence at biome level identified by Rutherford & Westfall (1986, Fig. 24). This does not necessarily suggest close agreement with independently determined phytochorological units (Chap. 3, this volume). However, the floristic distinctness of biomes may, in part, explain the success of environmental correlates with biome units irrespective of life-form composition. If true, this could mean that phylogenetic differences within a life form are more important than differences between life forms.

5.9 Conclusions

Relative to its area, southern Africa is particularly rich in biome types. This may relate largely to the orthogonal geographical arrangement of several important climatic determinants of biomes in a region of tropical–temperate convergence.

The correct status of biomes is highly dependent on recognition of the reduction of information at increasingly higher levels in accordance with hierarchy theory. A re-assessment of the status of biomes at global level is needed to reconcile the plethora of different biome units currently in use. Divergent genetic histories and regions of environmental disequilibrium tend to limit the levels of convergence expected from climatically driven predictive models of biomes at the global scale. A re-synthesis and re-survey of the vegetation structure of the globe is needed, with the use of a standardized method and a universally accepted finite set of plant functional types. This is a mammoth task, but is important, not only to avoid the uncertainty caused by the circularity of environment–vegetation correlation inherent in many biome models, but also to provide an essential validation base beyond the much relied upon interim works such as that of Olson, Watts and Allison (1983). In a contrary view, Mueller-Dombois (1984) cautions: 'Because of the limitations inherent in all of the major classifications so far devised, it is probably not possible to devise a general purpose classification of the world's vegetation'.

5.10 Acknowledgements

I thank Rob Westfall, John Irish, Sue Ringrose and Christina Skarpe for information and discussion and Jeanne Hurford for producing the new map on GIS.

5.11 References

Abercrombie, M., Hickman, C.J. & Johnson, M.L. (1980). *The Penguin Dictionary of Biology*. Harmondsworth: Penguin Books.

Acocks, J.P.H. (1988). Veld Types of South Africa, 3rd edn. *Memoirs of the Botanical Survey of South Africa*, 57, 1–146.

Astle, W.L. & Graham, A. (1976). Ecological investigations of the UNDP in the Okavango delta. In *Proceedings of the Symposium on the Okavango Delta and its Future Utilisation*, pp. 81–91. Gabarone: Botswana Society.

Bekker, R.P. & De Wit, P.V. (1991). *Contribution to the Vegetation Classification of Botswana*. Gabarone: Food and Agricultural Organization of the United Nations, United Nations Development Programme, Government of Botswana.

Box, E.O. (1981). *Macroclimate and Plant Forms: An Introduction to Predictive Modeling in Phytogeography*. The Hague: Junk.

Brown, D.E., Lowe, C.H. & Pase, C.P. (1979). A digitized classification system for the biotic communities of North America, with community (series) and association examples for the Southwest. *Journal of the Arizona–Nevada Academy of Science*, **14** (Suppl. 1), 1–16.

Cain, S.A. & Castro, G.M.de O. (1959). *Manual of Vegetation Analysis*. New York: Harper.

Cowling, R.M., Gibbs Russell, G.E., Hoffman, M.T. & Hilton-Taylor, C. (1989). Patterns of plant species diversity in southern Africa. In *Biotic Diversity in Southern Africa: Concepts and Conservation*, ed. B.J. Huntley, pp. 19–50. Cape Town: Oxford University Press.

Ehleringer, J.H. & Field C.B. (eds) (1993). *Scaling Physiological Processes – Leaf to Globe*. San Diego: Academic Press.

Ellery, W.N., Scholes, R.J. & Mentis, M.T. (1991). An initial approach to predicting the sensitivity of the South African grassland biome to climate change. *South African Journal of Science*, **87**, 499–503.

Feely, J.M. (1987). Review. Biomes of southern Africa. *Bulletin of the South African Institute of Ecologists*, **6**(3), 57–61.

Geiger, R. (1965). The Climate Near the Ground. Cambridge, Massachusetts: Harvard University Press.

Gibbs Russell, G.E. (1987). Preliminary floristic analysis of the major biomes in southern Africa. *Bothalia*, **17**, 213–27.

Godman, A. & Payne, E.M.F. (1979). *Longman Dictionary of Scientific Usage*. London: Longman.

Holdridge, L.R. (1947). Determination of the world formations from simple climatic data. *Science*, **105**, 367–8.

Huntley, B.J. (1984). Characteristics of South African biomes. In *Ecological Effects of Fire in South African Ecosystems*, ed. P. de V. Booysen & N.M. Tainton, pp. 1–17. Berlin: Springer.

Irish, J. (1994). The biomes of Namibia, as determined by objective categorisation. *Navorsinge van die Nasionale Museum, Bloemfontein*, **10**, 549–92.

Jürgens, N. (1991). A new approach to the Namib region. I. Phytogeographic subdivision. *Vegetatio*, **97**, 21–38.

Mueller-Dombois, D. (1984). Classification and mapping of plant communities: a review with emphasis on tropical vegetation. In *The Role of Terrestrial Vegetation in the Global Carbon Cycle: Measurement by Remote Sensing*, SCOPE 23, ed. G.M. Woodwell, pp. 21–90. Chichester: John Wiley.

Mueller-Dombois, D. & Ellenberg, H. (1974). *Aims and Methods of Vegetation Ecology*. New York: Wiley.

Neilson, R.P., King, G.A. & Koerper, G. (1992). Toward a rule-based biome model. *Landscape Ecology*, **7**, 27–43.

Olson, J.S., Watts, J.A. & Allison, L.J. (1983). *Carbon in Live Vegetation of Major World Ecosystems*, ORNL-5862. Oak Ridge: Oak Ridge National Laboratory.

O'Neill, R.V., DeAngelis, D.L., Waide, J.B. & Allen, T.F.H. (1986). *A Hierarchical Concept of Ecosystems*. Princeton: Princeton University Press.

Prentice, C., Cramer, W., Harrison, S.P., Leemans, R., Monserud, R.A. & Solomon, A.M. (1992). A global biome model based on plant physiology and dominance, soil properties and climate. *Journal of Biogeography*, **19**, 117–34.

Raunkiaer, C. (1934). *The Life Forms of Plants and Statistical Plant Geography*. Oxford: Oxford University Press.

Rutherford, M.C. (1991). Diversity of photosynthetic responses in the mesic and arid Mediterranean-type climate regions of southern Africa. In *Modern Ecology – Basic and Applied Aspects*, ed. G. Esser & D. Overdieck, pp. 133–60. Amsterdam: Elsevier.

Rutherford, M.C. (1993). Empiricism and the prediction of primary production at the mesoscale: a savanna example. *Ecological Modelling*, **67**, 129–46.

Rutherford, M.C., O'Callaghan, M., Hurford, J.L., Powrie, L.W., Schulze, R.E., Kunz, R.P., Davis, G.W., Hoffman, M.T. & Mack, F. (1995). Realized niche spaces and functional types: a framework for prediction of compositional change. *Journal of Biogeography*, **22**, 523–31.

Rutherford, M.C. & Westfall, R.H. (1986). Biomes of southern Africa: an objective categorization, 1st edn. *Memoirs of the Botanical Survey of South Africa*, **54**, 1–98.

Rutherford, M.C. & Westfall, R.H. (1994). Biomes of southern Africa: an objective categorization, 2nd edn. *Memoirs of the Botanical Survey of South Africa*, **63**, 1–94.

Smith, R.L. (1974). *Ecology and Field Biology*. New York: Harper & Row.

Smith, W.K. (1981). Temperature and water relation patterns in subalpine understory plants. *Oecologia*, **48**, 353–9.

Smith, W.K. & Geller, G.N. (1979). Plant transpiration at high elevations: theory, field measurements, and comparisons with desert plants. *Oecologia*, **41**, 109–22.

Werger, M.J.A. (1978). The Karoo–Namib region. In *Biogeography and Ecology of Southern Africa*, ed. M.J.A. Werger, pp. 231–99. The Hague: Junk.

Zohary, M. (1962). *Plant Life of Palestine, Israel and Jordan*. New York: Ronald.

Fynbos

6

R.M. Cowling, D.M. Richardson and P.J. Mustart

6.1 Introduction

The fynbos biome has been the focus of intensive research over the past two decades (Huntley 1992) and considerable advances have been made in our understanding of the ecology and biogeography of fynbos (see Cowling 1992 for a review). This chapter provides a necessarily brief summary which is by no means comprehensive. Details of fynbos biome ecology not dealt with in depth in this chapter are covered elsewhere in this volume and cover plant biogeography (Chap. 3), sclerophylly (Chap. 16), herbivory (Chap. 17), fire (Chap. 18), regional species richness (Chap. 19) and alien plant invasions (Chap. 22).

The first section of this chapter gives an overview of the major characteristics of the biome. This is followed by an analysis of the determinants of boundaries at the biome, landscape and community levels. We then summarize research on the reproductive and physiological traits of plants, emphasizing the role of fire and nutrient scarcity. This section provides an essential basis for the next, which explores patterns of community structure and the processes that determine them. A discussion of the unusual features of fynbos ecosystems, such as low productivity, slow nutrient cycling and the year-round surplus of soil moisture, provides a process–functional perspective to the chapter. We deal with the conservation of fynbos biome vegetation by discussing three major problems of resource management: alien plant invasions, habitat fragmentation and the harvesting of indigenous plants. In the concluding section, we identify significant gaps in our knowledge of pattern and process in fynbos biome vegetation.

6.2 The fynbos biome in perspective

The fynbos biome (Kruger 1978; Rutherford & Westfall 1986; Chap. 5, this volume) occupies an area of 71 337 km² in the extreme southwestern and southern parts of southern Africa (2.7% of the area) where rain falls mainly in winter (Fig. 6.1). The landscape is an ancient one, dominated by the rugged and steep quartzitic mountains of the Cape Folded Belt, which reach moderate elevations (c. 2000 m). Mountain soils, which are derived from ancient Cape Supergroup (Table Mountain Group and Witteberg Groups) sediments (Fig. 6.1; see also Chap. 1, this volume), are extremely poor in exchangeable bases and extractable phosphorus (Kruger 1979). Soils of the upper slopes and plateaux are mainly shallow, leached sands, which are often podzolized in humid zones; the lower colluvial slopes are mainly red apedal soils, which are finer grained and more fertile. The coastal forelands have moderately fertile and clay-rich residual and duplex soils derived from the pre-Cape Malmesbury Group in the west and the Bokkeveld Group (Cape Supergroup) in the southwest and south. Duplex soils are also associated with exposures of the Cape Granite Suite, which are concentrated in the western lowlands. The coastal margin is mantled with a complex sequence of nutrient-poor acidic and alkaline sands of aeolian and marine origin (including large areas of Mio-Pliocene limestone along the south coast), respectively. Rainfall distribution is typically mediterranean in the west, gradually changing to non-seasonal in the southeast (Fig. 6.1; Chap. 2, this volume). Across the biome annual rainfall has the largest range for southern African biomes: it varies from about 210 mm in the

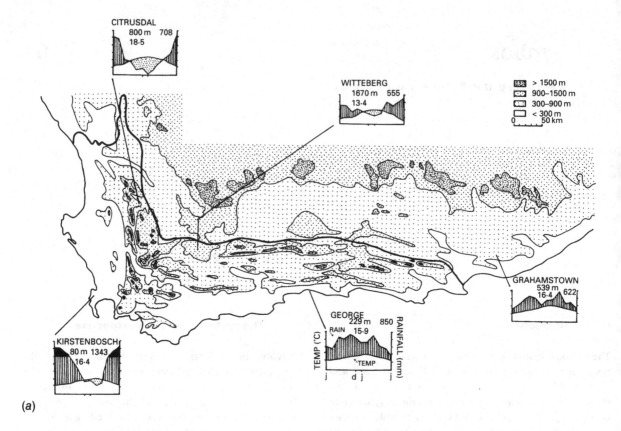

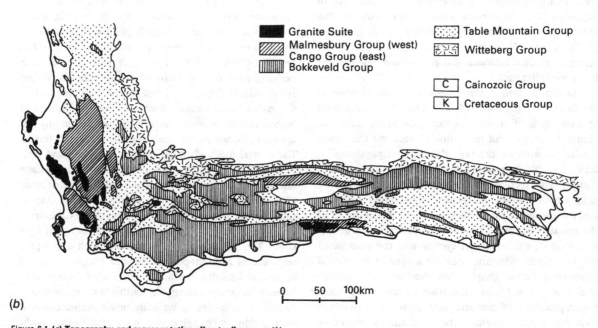

Figure 6.1 (*a*) Topography and representative climate diagrams; (*b*) geology; and (*c*) major vegetation types of the fynbos biome. The bold line in (*a*) and the coverage in (*c*) is the boundary of the Cape Floristic Region.

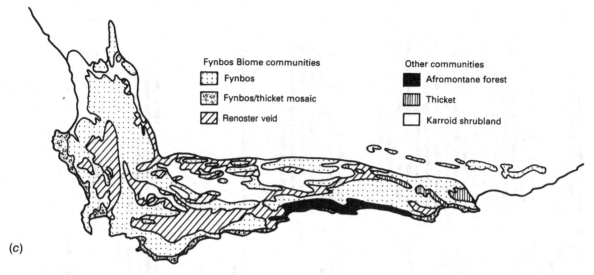

Fynbos Biome communities
- Fynbos
- Fynbos/thicket mosaic
- Renoster veld

Other communities
- Afromontane forest
- Thicket
- Karroid shrubland

(c)

Figure 6.1 **contd.**

inland valleys, to about 400 mm on the broad coastal forelands, and about 800–3000 mm in the mountains where the rainfall gradients are very steep. Temperatures are mild on the coastal forelands and lower slopes (mean annual temperatures of 16–20 °C) where frost is seldom recorded but freezing temperatures and occasional winter snowfalls are widespread at higher altitudes (Linder *et al.* 1993). Further details on the physiography, soils and climate of the fynbos biome are given by Kruger (1978); Campbell (1983); Deacon, Hendey & Lambrechts (1983); Deacon, Jury & Ellis (1992), and Versfeld *et al.* (1992).

Fynbos is a vernacular term (literally meaning fine-leaved bush) used to describe the predominant vegetation type in the biome (Fig. 6.1c; Table 6.1). It is an evergreen, fire-prone shrubland, confined largely to sandy, infertile soils, and characterized structurally by the universal presence of restioids (wiry, evergreen graminoids; Fig. 6.2), a high cover of ericoid shrubs (especially Ericaceae) and the common occurrence of overstorey proteoid shrubs (see Table 6.2 for definitions of growth forms) (Taylor 1978; Kruger 1979; Campbell 1985; Cowling & Holmes 1992). Fynbos has been described as a heathland (Specht 1979; Specht & Moll 1983); in terms of structure and function it resembles quite closely the high-altitude heathlands of tropical Africa (Killick 1979) and the heathy kwongan of the mediterranean-climate region of southwestern Australia (Milewski 1983; Cowling & Witkowski 1994). The similarities between fynbos and the fire-prone shrublands of other mediterranean-climate regions are less pronounced (Cody & Mooney 1978; Cowling & Campbell 1980; Keeley 1992; Le Maitre & Midgley 1992).

Besides fynbos, the biome includes two other major vegetation types, namely renosterveld and a mosaic of fynbos and subtropical thicket (often termed strandveld) (Fig. 6.1; Kruger 1978; Taylor 1978). Renosterveld occupies about 20 000 km² on the moderately fertile, shale-derived soils of the coastal forelands and inland valleys (Fig. 6.3). It is an evergreen, fire-prone vegetation dominated by small-leaved, asteraceous shrubs (especially *Elytropappus rhinocerotis*, the renosterbos or rhinoceros bush) and has an understorey of Poaceae and geophytes (Table 6.1; Boucher & Moll 1980; Campbell 1985). About 60% has been transformed by agriculture (Moll & Bossi 1984). Strandveld is a mosaic of asteraceous or restioid fynbos (Table 6.1) and subtropical thicket (see Chap. 12, this volume) which occupies about 4500 km² on calcareous coastal dunes (see Chap. 13, this volume). The largely evergreen thicket species either are intermingled with fynbos or form distinct patches of dense thicket or low forest (Taylor 1972; Cowling 1984; Fig. 6.4). Thicket shrubs and trees (obligate resprouters in Table 6.2) differ from fynbos plants in that: most also occur in subtropical forests outside the biome (Cowling 1983a); they have low flammability (Van Wilgen, Higgins & Bellstedt 1990); they have fleshy, ornithochorous fruits (Knight 1988); and seedling establishment is confined to the intervals between (rather than shortly after) fires (Manders & Richardson 1992; Cowling *et al.* 1997). Vegetation types adjacent to the fynbos biome include Afromontane forest and subtropical thicket (Fig. 6.5; Chap. 12, this volume), Nama-karoo (Chap. 8, this volume), savanna (Chap. 11, this volume), and succulent karoo (Chap. 7, this volume).

The fynbos biome is roughly coincident with the Cape Floristic Region (Bond & Goldblatt 1984), which is recog-

Table 6.1 Major plant communities in the fynbos biome. Typology is based on Campbell's (1985) structural scheme as modified for the lowlands by Cowling et al. (1988)

Community[a]	Formation[b]	Differentiating features	Floristics	Distribution[c]	Environment
Fynbos[d] Group: Cape fynbos shrubland					
Grassy fynbos	Low, mid-dense to closed leptophyllous (ericaceous[e] or ericoid[f]) shrubland	High grass cover and a relatively high cover of non-proteoid[f] nanophylls and forbs	Variable: *Themeda triandra, Tristachya leucothrix, Restio triticeus, Leucadendron salignum, Phylica axillaris, Erica* spp., *Helichrysum* spp., *Rhus* spp., etc.	Mainly E but also SI and SC on lower mountain slopes and coastal forelands	Easternmost fynbos (high proportion of summer rain) on finer textured and more fertile soils than other fynbos types
Asteraceous fynbos	Low to mid-high, open to mid-dense, leptophyllous shrubland	Low total cover, often high grass and elytropappoid[g] cover and a high cover of non-ericaceous ericoids	*Elytropappus, Pentaschistis, Phylica, Passerina, Agathosma* and other Rutaceae, *Aspalathus, Restio, Felicia, Hypodiscus, Cliffortia, Maytenus oleoides, Protea nitida*	Throughout the biome except for mountains in E	Occupies the driest fynbos sites on a range of substrata, linking fynbos to karroid shrubland and renosterveld; widespread on coastal forelands on calcareous dunes and on shales and silferricretes where rainfall is <550 mm yr^{-1}; in the mountains, rainfall ranges from 450 to 950 mm yr^{-1} and soil depth averages <0.4 m
Restioid fynbos	Dwarf to tall, mid-dense to closed restioland with a sparse shrub stratum	Highest restioid[f] cover (>60%) and the lowest shrub cover (<30%) of all fynbos types; it has the lowest constancy of tall (>1.5 m) shrubs	Restionaceae, *Tetraria, Leucadendron, Pentaschistis, Phylica, Protea, Stoebe,* etc.	Throughout the biome except for the mountains in E; rare in SC	Occurs where conditions are limiting for shrub growth, owing to either excessive waterlogging or drainage; it occupies more mesic sites than asteraceous fynbos; soils may be deep sands or shallow and rocky; in the mountains restioid fynbos is a feature of dry, north slopes
Ericaceous fynbos	Low to mid-high, closed ericaceous shrubland	Like asteraceous fynbos, a leptophyllous shrubland, but has a high cover of restioids and the shrubs are mainly ericaceous ericoids. Shrub cover and total cover are also higher than asteraceous fynbos	Ericaceae, Restionaceae, Bruniaceae, Peneaceae, Grubbiaceae, *Tetraria, Leucadendron*	Mostly SW and SC	Largely confined to south-facing and wet (rainfall average of 1500 mm yr^{-1}) slopes of the coastal mountains; soils have a high carbon content, low pH and high fine-particle fraction
Proteoid fynbos	Low to tall, open to closed, proteoid shrubland	Differs from other types by having >10% cover of mid-high to tall, non-sprouting, proteoid shrubs; included also are some communities on the coastal foreland where the canopy proteoids are <1.5 m	*Leucadendron conicum, L. coniferum, L. eucalyptifolium, L. gandogeri, L. laureolum, L. loeriense, L. meridianum, L. rubrum, L. uliginosum, L. xanthoconus, Protea aurea, P. compacta, P. eximia, P. laurifolia, P. lorifolia, P. mundii, P. neriifolia, P. punctata, P. obtusifolia, P. repens, P. susannae,* Restionaceae, Ericaceae, etc.	Throughout the biome except I; relatively rare in NW and E and most extensive in the SC	Like grassy fynbos, proteoid fynbos occurs at lower altitudes and on more fertile soils than other fynbos types; soils are mostly deep, well-drained and derived from a wide range of substrata including colluvial sands and limestone on the forelands; altitude ranges from sea level to 950 m, and rainfall from 400 to 1100 mm yr^{-1}
Closed-scrub fynbos	Mid-high to tall, open to closed, nano-microphyllous shrubland with an open cover of tall restioids	Similar to forest and thicket in the relatively high cover of mesophyllous, non-proteoid woody plants. Differs by having a high cover (>10%) of restioids and a high frequency of ericaceous ericoids	*Metrosideros angustifolia, Brachylaena neriifolia, Calopsis paniculatus, Salix mucronata, Cannomois virgata, Berzelia intermedia, Empleurum serrulatum, Leucadendron salicifolium, Elegia capensis*	Throughout the biome except in the E and I	Associated with well-drained riparian habitats
Non-fynbos Group: Renosterveld and karroid shrubland					
Renosterveld	Low to mid-high, open to mid-dense leptophyllous shrubland, often with an open, grassy understorey	Generally lacks fynbos characteristics (restioids, proteoids, ericaceous ericoids); high cover of elytropappoids, grasses and fleshy-leaved shrubs	*Elytropappus rhinocerotis, Pteronia* spp., *Anthospermum aethiopicum, Dodonea viscosa, Pentaschistis* spp., *Relhania* spp., *Helichrysum* spp., *Ruschia* spp., *Themeda triandra*	Throughout the fynbos biome on lower mountain slopes, interior valleys and coastal forelands	On fine textured soils usually on the ecotone between fynbos and succulent (karroid) shrubland; very occasionally on quartzite; rainfall mostly <450 mm yr^{-1}

[a] Communities described are at the series level in Campbell's (1985) scheme.
[b] According to Campbell et al. (1981).
[c] In relation to the montane regions of Campbell (1985): C, central; E, eastern; I, interior; NW, northwestern; SC, southern coastal; SI, southern interior.
[d] See text for distinction between fynbos and non-fynbos.
[e] Refers to ericoid shrubs belonging to the Ericaceae.
[f] See Table 6.2 for description of growth forms.
[g] Shrubs with scale-like, pubescent leaves; most are species of *Elytropappus* and *Stoebe* (Asteraceae).

Table 6.2 Structural and reproductive characteristics of major growth forms in the fynbos biome[a]. Taxa and characteristics listed in order of their importance

Growth form	Structural characterization	Number of spp.	Dominant families	Age to maturity (years)[a]	Life span (years)	Reseeder/ resprouter (%)	Pollinators	Seed storage	Seed size	Dispersal	Germination cues
Proteoids	Mid-tall (1.5–5 m) shrubs with nano-mesophyllous, isobilateral leaves	120	Proteaceae	1–5	>15	>80	Insects/birds Rodents Wind	Canopy Soil	Large (>5 mg)	Wind Ants Passive	Cold Fluctuating temperatures Heat
Ericoids	Dwarf-mid (0.5–2 m) shrubs with leptophyllous, rolled leaves	c. 3000	Ericaceae, Asteraceae Fabaceae, Rutaceae Proteaceae	1–4	>6	>60	Insects Wind Birds	Soil Canopy	Fine (<1 mg) –large	Passive Ants Wind	Fluctuating temperatures Heat Cold Smoke
Restioids	Dwarf-mid evergreen graminoids	c. 550	Restionaceae Cyperaceae, Poaceae	1–2	>25	<20	Wind	Soil	Small (1 mg) –large	Ants Passive Wind	Largely unknown Smoke
Geophytes	Evergreen and deciduous herbs with below-ground storage organs	c. 1400	Iridaceae, Liliaceae Orchidaceae Amaryllidaceae	1	<15	0	Insects Birds	Soil No storage	Fine–large	Passive Ants	Unknown Fleshy seeds are non-dormant
Fire ephemerals	Forbs and dwarf-low (0.5–1 m) shrubs	c. 1000	Asteraceae Scrophulariaceae Fabaceae, Poaceae Mesembryanthemaceae	0.5–2	<8	>80	Insects Wind	Soil	Fine–small	Wind Passive	Cold Smoke Unknown
Obligate resprouters	Low-tall shrubs with nano-mesophyllous dorsiventral leaves	102	Anacardiaceae Celastraceae, Ebenaceae	1–5	>>100	0	Insects Wind	No storage	Large	Birds Wind	Mainly fleshy and non-dormant

[a] Data from Johnson (1992a), Le Maitre & Midgley (1992) and Van Wilgen & Forsyth (1992). Additional sources are Brown (1993), Brown et al. (1993), Pierce & Moll (1994), P.J. Mustart (unpubl. data) and R.M. Cowling (unpubl. data).

[b] Primary juvenile period for reseeders and secondary (post-fire) juvenile period for resprouters.

Figure 6.2 **The presence of restioids (wiry, aphyllous graminoids of the Restionaceae) uniquely characterizes fynbos in southern Africa. Shown here is _Elegia filacea_-dominated restioid fynbos on seasonally waterlogged sands in the southwestern mountains** (Photo: J.C. Paterson-Jones).

nized as one of the world's six floristic kingdoms (Good 1974). The biome includes about 7300 species, of which about 80% are endemic (Gibbs Russell 1987). Some 109 large genera (with ten or more taxa) are centred there and the biome is distinguished floristically from other southern African biomes by having large numbers of species belonging to the Ericaceae, Restionaceae, Rutaceae, Polygalaceae, Thymelaceae, Rhamnaceae, Rosaceae and Lobeliaceae (Gibbs Russell 1987). Further details on the flora and phytogeography of the biome are given by Goldblatt (1978), Taylor (1978), Bond & Goldblatt (1984), Cowling & Holmes (1992), Linder, Meadows & Cowling (1992) and Chap. 3 (this volume).

6.3 Determinants of pattern

6.3.1 Biome-scale

According to Rutherford & Westfall (1986), the fynbos biome is characterized climatically by having more than 40% of annual rain falling in the winter months (April to September) and a moderate to low index of summer aridity. It is differentiated from the grassland and Nama-karoo biomes by having a higher proportion of winter rainfall, from the succulent karoo biome by having lower summer aridity, and rather weakly from the forest biome by having higher summer aridity. In terms of life forms, fynbos is uniquely characterized among southern African biomes by the co-dominance of hemicryptophytes, chamaephytes and phanerophytes (Rutherford & Westfall 1986; Chap. 5, this volume).

Ellery, Scholes & Mentis (1991) have differentiated southern African biomes, using the more refined indices of moisture availability, incorporating the following: 'growth days' where rainfall exceeds evaporative demand; 'growth temperatures' or the mean temperature on days when water is available for growth; and 'no-growth' temperatures, the corollary of the above for the dry season. Fynbos climates are distinguished from those of the grassland and Nama-karoo biomes by 'growth temperatures' cooler than 'no-growth' temperatures (i.e. winter rainfall); and from the winter-rainfall succulent karoo by longer growth seasons (higher rainfall), especially where summers are hot. Forest climates in the winter-rainfall region (e.g. in the Knysna enclave of the southern Cape) cannot be differentiated from many fynbos climates on the basis of the duration of the growth season.

Climate is only one of several factors that differentiate the fynbos biome from adjacent biomes. The distribution of fynbos also coincides with nutrient-poor soils (Taylor 1978; Kruger 1979; Cowling & Campbell 1983; Specht & Moll 1983; Campbell 1986). Fynbos biome soils tend to have significantly lower pH values, lower available phosphorus, lower base status and higher clay content than those of adjacent biomes (Cowling 1984; Campbell 1986). The regime of recurrent and intense summer fires at intervals of between four and 40 years – a consequence of nutrient poverty and summer drought (Kruger 1983; Van Wilgen 1984, 1987; Le Maitre & Midgley 1992) – is a unique feature of the fynbos biome and may be responsible for its persistence (Bond & Midgley 1995; Chap. 18, this volume). We discuss below the interaction of climate, soils and fire regime in maintaining vegetation boundaries within fynbos biome landscapes.

Figure 6.3 **The grassy aspect of renosterveld is evident in this fragment in an agricultural landscape on the clay-rich south coastal lowlands near Caledon. Dominant grasses are** *Merxmuellera stricta* **and** *Themeda triandra* *(Photo: D.M. Richardson).*

Figure 6.4 **Mosaic of subtropical thicket (foreground) and asteraceous fynbos on well-drained, calcareous coastal dunes on the southwestern lowlands** *(Photo: J.C. Paterson-Jones).*

Figure 6.5 **Forest species (***Cunonia capensis, Heeria argentea, Maytenus oleoides***) associated with rocky and fire-free scree habitat in a fire-prone fynbos landscape at Jonkershoek in the southwestern mountains** *(Photo: D.M. Richardson).*

6.3.2 Landscape-scale

6.3.2.1 *Fynbos/forest boundary*

The poor differentiation in climate between the fynbos and forest biomes suggests that much of the biome is bioclimatically suitable for forest (Cowling & Holmes 1992; Manders, Richardson & Masson 1992; Euston-Brown 1995). However, forest vegetation (including scrub-forest or thicket; Chap. 12, this volume) has a limited occurrence in the biome and is usually separated from fynbos by a distinct boundary. Forests are invariably associated with deeper (Fig. 6.6) and more fertile soils than fynbos (Cowling & Holmes 1992). However, many studies suggest that forest may develop on soils identical to those which support fynbos and that chemical differences may be plant-induced effects associated with different nutrient cycling processes of litter in the fire-free forest environment (Van Daalen 1981; Cowling 1984; Thwaites & Cowling 1988; Manders 1990a; Manders *et al.* 1992; Euston-Brown 1995; Richardson *et al.* 1995).

A general model that considers the interactions among site conditions, fire regime and the biology of establishment of forest seedlings was proposed by Manders & Richardson (1992) to explain the fynbos/forest boundary. They suggest that forest species are excluded from fynbos because their recruitment is not coupled with fire. Fruits of forest species are dispersed by birds to sites beneath emergent fynbos (usually proteoid) shrubs where reduced light and enhanced nutrient and moisture conditions are optimal for germination and subsequent seedling growth (Bond & Stock 1989; Cowling & Gxaba 1990; Holmes & Cowling 1993; Cowling *et al.* 1997). Seedlings are usually destroyed by fire before they emerge from the canopy and develop fire resistance. However, in the absence of fire, long-lived forest species form nuclei which eventually develop into forest patches (Manders & Richardson 1992). Once established, these patches are fire-resistant (Manders & Richardson 1992), largely due to the low flammability of forest fuel (Van Wilgen *et al.* 1990). Moreover, most forest species can, when mature, coppice after fire (Phillips 1931). It follows that forest should be more widespread in fire-protected areas such as coastal and river margins, rocky scree and talus slopes, and steep valley walls (Geldenhuys 1989, 1994; Cowling *et al.* 1997). Forest development in fire-prone areas should be most rapid in permanently moist, fertile and warm sites, such as occur in the eastern parts of the fynbos biome. This is substantiated by Campbell (1986), who showed that forest in the mountains of the fynbos biome is associated with clay-rich, rock-free and deep soils at lower altitudes, especially in the eastern part of the biome, which experiences more summer rain (Fig. 6.7).

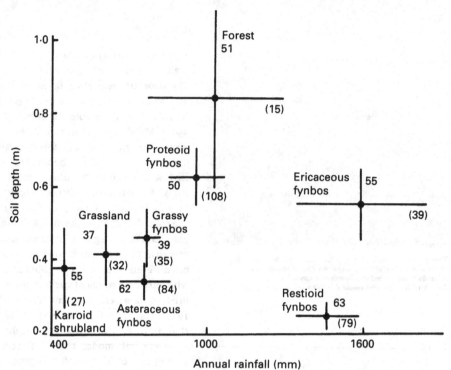

Figure 6.6 **The major plant communities of the mountains of the fynbos biome in relation to soil depth and annual rainfall. Vertical and horizontal lines indicate the 95% confidence intervals for the means of the two variables in each community. The numbers in the figure are winter rainfall (%) and number of plots per community (in parenthesis). Redrawn, with permission, from Campbell & Werger (1988).**

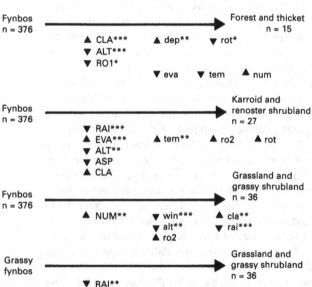

Figure 6.7 **Stepwise discriminant analyses and significance tests of environmental variables between fynbos and non-fynbos vegetation types in the mountains of the fynbos biome. All the variables listed are significantly different ($P \le 0.05$) between groups: *, $P < 0.01$; **, $P < 0.001$; ***, $P < 0.0001$. The variables in upper case are those in the discriminant function. The top variable in each list is the best discriminating variable, the second is the variable that accounts for most of the remaining difference between groups, and so on. The variables in lower case, which are not in the discriminant function, are listed next to the discriminating variables with which they are best correlated. (△, increasing; ▽, decreasing). Abbreviations are as follows: alt, altitude; asp, aspect (hot to cool gradient); cla, % clay content; dep, soil depth; eva, annual pan evaporation; num, transect number (1–22 from northwest to southeast); rai, annual rainfall; rot, total rock cover; ro1, rock (>500 mm in diameter) cover; ro2, rock (50–500 mm) cover; tem, temperature range (summer mean minus winter mean); win, % winter rainfall. Redrawn, with permission, from Campbell (1986).**

6.3.2.2 *Fynbos/karroid and renosterveld boundary*

Campbell (1986) has shown that karroid shrublands and renosterveld in the mountains of the fynbos biome are associated with interior, low-rainfall areas (Fig. 6.7). These shrublands also occur on soils with a higher pH than those of fynbos (Fig. 6.8). Succulent karroid shrubland replaces fynbos on sandy soils below 200–300 mm yr⁻¹ and renosterveld replaces fynbos on clay-rich soils (shale and granite-derived) below 600 mm yr⁻¹ (Kruger 1979; Cowling 1984; Boucher 1987; Euston-Brown 1995). The latter transition is usually attributed to a leaching-induced gradient in soil nutrient status. However, Miller (1982) proposed that soil moisture determined the fynbos/karoo boundary: in dry sites the carbon costs of evergreen leaves cannot be maintained and drought-deciduousness and leaf succulence are favoured (see also Bond 1981; Cowling & Campbell 1983; Campbell & Werger 1988). Boucher (1987) and Cowling et al. (1988) have suggested that post-colonial degradation by burning and grazing have been responsible for the transformation of fynbos, on clay-rich soils of the lowlands, to renosterveld. Experimental work is required

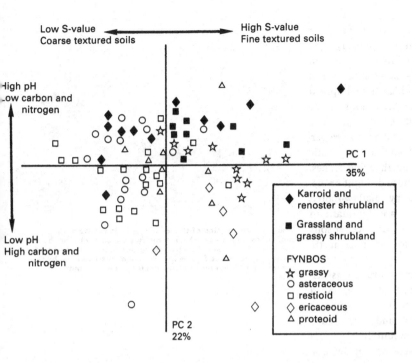

Figure 6.8 **Soil community relationships as indicated by an ordination of 81 plots from the mountains of the fynbos biome. The ordination is a principle components analysis of ten soil variables. Redrawn, with permission, from Campbell (1986).**

to identify the physiological and biotic determinants of the fynbos/karroid and renosterveld boundary. In this respect, Euston-Brown (1995) found that, in experimental transplants, fynbos Proteaceae, but not Restionaceae, grew and survived well in a renosterveld site in the southeastern fynbos biome.

6.3.2.3 *Fynbos/grassland boundary*

Grassy non-fynbos communities are associated with the higher incidence of summer rainfall and/or the finer textured soils of the east (Campbell 1983, 1986; Figs. 6.7, 6.8 and 6.9). Grassy fynbos, which is restricted to the eastern fynbos biome (Table 6.1), is replaced by non-fynbos grassy communities on drier, more inland and north-facing (i.e.

Figure 6.9 **A landscape in the Zuurberg Mountains in the extreme east of the fynbos biome. The planed summits and soft slopes support a mosaic of grassland (mainly north-facing slopes) and grassy fynbos (mainly south-facing slopes); subtropical thicket is widespread in the valley bottoms. Shown in the foreground is *Leucospermum cuneiforme*, a widespread resprouting proteoid, and scattered clumps of *Restio triticeus* in a matrix of subtropical C_4 grasses** (Photo: J.C. Paterson-Jones).

equator-facing) sites, especially under lower rainfall conditions (Figs 6.2 and 6.3; see also Euston-Brown 1995). This increased grassiness, which corresponds mainly to an increase in the cover of summer-growing C_4 grasses, could be due to the generally more fertile soils of the east (Cowling 1984) and the higher growth-season temperatures, resulting from a higher proportion of summer rain (Vogel, Fuls & Ellis 1978). Within eastern landscapes, growth-season temperatures and soil nutrient levels are highest on north-facing slopes (Campbell 1983; Cowling & Campbell 1983). Unlike the uplands of the grassland biome where grazing and fire regimes may determine the boundary between grassland and grassy fynbos (Trollope 1973; Chap. 10, this volume), this boundary appears stable in the fynbos biome (Cowling 1984; Campbell 1985). However, Euston-Brown (1995) found poor vegetation–environment relationships for grassy fynbos communities, suggesting biotic controls on their distribution. Experimental plantings of *Themeda triandra* (a C_4 grass) grew well across a range of fynbos (including grassy fynbos) and non-fynbos sites (including *Themeda* grassland), indicating that this species may be excluded from non-grassy fynbos by competition.

6.3.3 **Community-scale**

6.3.3.1 *Structurally characterized communities*

Campbell's (1985) structural classification of fynbos biome vegetation is used as the basis for vegetation description in this chapter (Table 6.1). The determinants of community change in the mountains are best described in terms of three major environmental gradients: an altitude–aspect gradient; a coast-to-interior gradient; and a west-to-east gradient (Campbell 1983).

Rainfall and soil factors play very important roles in determining community boundaries along altitude and aspect gradients. Proteoid fynbos, which is widespread on the deep and relatively fertile colluvial soils at the base of mountains (Fig. 6.10), is replaced at higher altitudes by ericaceous fynbos (Fig. 6.11) where soils are permanently wet, relatively fine-grained, and rich in organic carbon (Figs 6.6 and 6.8). Warmer north-facing slopes, where soils are both shallower and more drought-prone, support restioid fynbos dominated by shallow-rooted restioids. Deep-rooted ericoid shrubs are dominant in asteraceous fynbos, which is associated with deeper soils at the dry end of the gradient (i.e. low-altitude, north-facing slopes).

The coast-to-interior gradient is most strongly associated with decreasing rainfall (Campbell 1983). As a result, coastal mountains support mainly proteoid and ericaceous fynbos, whereas restioid and asteraceous fynbos predominate in the interior mountains.

The west-to-east gradient is characterized by decreasing summer drought intensity and increasing soil fertility (Campbell 1983), with grassy fynbos replacing proteoid fynbos, grassland replacing restioid fynbos, and grassy shrubland replacing asteraceous fynbos.

Most lowland areas receive less than 600 mm of rain annually, and dry fynbos communities predominate (Cowling *et al.* 1988; Rebelo *et al.* 1991). Asteraceous fynbos occurs on calcareous coastal dunes, on remnant silcrete and ferricrete surfaces, and on duplex, clay-rich soils where annual rainfall exceeds 600 mm yr^{-1}. Restioid fynbos grows on the better developed but more leached soils of inland dunes and on seasonally waterlogged shallow sands on bedrock. Elsewhere, floristically distinct types of proteoid fynbos are associated with different soil types (Fig. 6.12; and see below).

6.3.3.2 Floristically characterized communities

Despite the many detailed phytosociological studies in the fynbos biome (see Cowling 1992 for references), few multivariate vegetation–environmental models have been produced. Richards, Cowling & Stock (1995a) used multivariate direct gradient analysis to explore the vegetation–environment relationships of six lowland fynbos communities (four proteoid, one ericaceous and one restioid) associated with four substrata (limestone, quartzite, acid colluvial sand, neutral colluvial sand) in a 30-ha site. Soil depth, texture, fertility, pH and rock cover were the major factors explaining floristic patterns (Fig. 6.13). This study reinforces the importance of both physical and chemical soil factors in determining the boundaries of fynbos communities (Kruger 1979; Bond

Figure 6.10 **Vegetation of the lower slopes in the western mountains. Shown here is proteoid fynbos, characterized by the reseeding *Protea laurifolia* (left foreground), and a widespread form of asteraceous fynbos (waboomveld) characterized by the resprouting *Protea nitida* (waboom) (right background)** (Photo: J.C. Paterson-Jones).

Figure 6.11 **Ericaceous fynbos, dominated by *Erica inflata*, on a high altitude plateau in the western mountains. Note the geophyte, *Disa uniflora*, on the streambank in the foreground** (Photo: J.C. Paterson-Jones).

Figure 6.12 ***Protea compacta*-dominated proteoid fynbos on deep, acid sand, which blankets much of the southwestern coastal lowlands** (Photo: J.C. Paterson-Jones).

1981; Cowling 1984; Campbell 1986; Thwaites & Cowling 1988; Van Wilgen & McDonald 1992; Fig. 6.14).

In a larger-scale study, covering 1748 km² of the Langeberg mountains – a coastal range in the southern parts of the biome – McDonald, Cowling & Boucher (1996) showed that patterns in community distribution were related to two principle gradients. The main gradient was one of increasing annual rainfall and decreasing winter radiation loads (i.e. wet, south-facing slopes to dry, north-facing slopes); the second gradient ranged from deep, non-rocky and clay-rich to shallow, rocky and sandy soils (see also Bond 1981; Campbell 1986). Soil chemical factors played a subsidiary role in explaining floristic pattern. The multivariate analyses (canonical correspondence analysis) explained approximately 50% of the compositional variation, suggesting that stochastic factors such as fire may influence community structure (Cowling 1987). Nonetheless, McDonald *et al.* (1996) were able effectively to characterize their communities in terms of a limited number of site factors (Fig. 6.14).

Field and laboratory experimental studies on the southwestern coastal lowlands suggest that the distributions of fynbos species (and hence, community boundaries) are a function of edaphic specialization rather than competitive interactions (Newton, Cowling & Lewis 1991; Mustart & Cowling 1993; Richards 1994). However, all the above studies were done in areas of sharp edaphic discontinuities and contrasts (Thwaites & Cowling 1988). Reciprocal transplants of fynbos species along more gradual edaphic gradients in the southeastern mountains showed that most species performed best outside their actual range, thus implying biotic controls of pattern (Euston-Brown 1995).

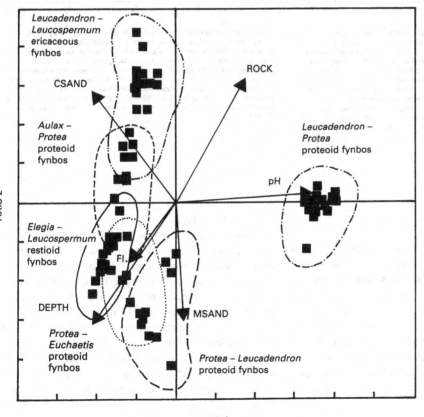

Figure 6.13 **Canonical correspondence analysis (CCA) biplot of site scores for 75 sites and the six most important environmental factors (vectors) on the first two CCA axes for fynbos on the Soetanysberg, Western Cape. CSAND, % coarse sand, DEPTH, soil depth, FI, soil fertility index, MSAND, % medium sand, ROCK, % rock cover. Communities were classified using two-way indicator species analysis. Redrawn, with permission, from Richards, Cowling & Stock (1995a).**

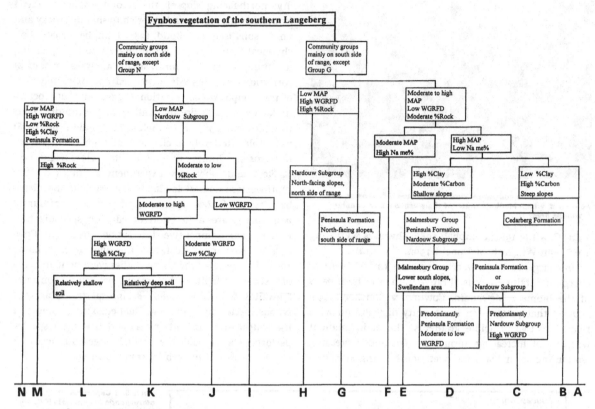

Figure 6.14 **Classification of 14 fynbos community groups from the southern Langeberg mountains in relation to abiotic site factors.** MAP, mean annual precipitation; WGRFD, winter global radiant flux density. Peninsula Formation and Nardouw Subgroups are shale. A, *Protea grandiceps–Erica transparens* ericaceous fynbos; B, *P. grandiceps–Helichrysum oxyphyllum* ericaceous fynbos; C, *Ehrharta rupestris* subsp. *dodii–Platycaulos major* ericaceous fynbos; D, *Blechnum tabulare–Erica atropurpurea* ericaceous fynbos; E, *Erica pubigera–Protea piscina* proteoid fynbos; F, *Centella lanata–Indigofera alopecuroides* ericaceous fynbos;

Hypodiscus aristatus–Erica multumbellifera restioid fynbos; H, *Erica versicolor–Agathosma ovata* (unclassified); I, *Centella glabrata–Pelargonium ovale* proteoid fynbos; J, *Leucadendron eucalyptifolium–Hypodiscus argenteus* proteoid fynbos; K, *Paranomus spathulatus–Osteospermum junceum* asteraceous fynbos; L, *Leucadendron eucalyptifolium–Elegia filacea* proteoid fynbos; M, *Elytropappus rhinocerotis–Passerina obtusifolia* asteraceous fynbos; N, *Erica hispidula–Protea nitida* asteraceous fynbos. Reproduced, with permission, from McDonald *et al.* (1996).

6.4 Plant structure and function

A great deal of research has been carried out on the reproductive ecology of individual plants in the fynbos biome (see Johnson 1992a, Le Maitre & Midgley 1992 for reviews) and the distribution of structural–functional traits along environmental gradients (Cowling & Campbell 1983; Campbell & Werger 1988). There has been much less emphasis on physiological studies (Stock, Van der Heyden & Lewis 1992). In this section we focus on the autecology of fynbos plants in relation to reproductively and structurally characterized plant types or guilds.

6.4.1 Plant types

A striking feature of fynbos biome landscapes is the limited change in growth-form mix over substantial edaphic and climatic gradients (Kruger 1979; Cowling & Campbell 1983; Cowling 1990). Thus, environmentally different sites support the same mix of growth forms but the relative importance of these vary in a predictable way along resource gradients (Campbell & Werger 1988). However, even subtle changes in habitat conditions, which do not result in appreciable differences in the coverage of growth forms, are associated with major floristic changes or high beta diversity (Cowling & Campbell 1984; Cowling 1990; Cowling, Holmes & Rebelo 1992; Chap. 19, this volume).

Some 85% of plant species (6172) in the fynbos biome can be collapsed into just six plant types (Table 6.2). Ericoid shrubs are the largest group, comprising about 40% of the flora. These species show remarkable structural convergence across different taxa, but differ markedly in their reproductive characteristics (Bond 1989; Johnson 1992b). Fire ephemerals, on the other hand, have similar reproductive behaviour (fire-stimulated germination, persistent seed banks, fast growth rates, early reproductive maturities, high reproductive efforts relative to biomass and relatively short lifespans) but include annual and perennial herbs, and erect and sprawling shrubs and semi-shrubs with a diverse array of leaf shapes and sizes (Hoffman, Moll & Boucher 1987; Kruger 1987; Cowling & Pierce 1988; Le Maitre & Midgley 1992; Van Wilgen & Forsyth 1992). Despite their importance in most fynbos vegetation, restioids have been little studied. The geophytic flora of the fynbos biome (Fig. 6.15), totalling about 1400 species, comprises the largest concentration of these species in the world (Goldblatt 1978; Le Maitre & Midgley 1992). However, only recently have these plants been the subject of intensive research (e.g. Le Maitre & Brown 1992; Ruiters, McKenzie & Raitt 1992; Ruiters *et al.* 1993; Johnson & Bond 1994). Obligate resprouting shrubs, which establish seedlings from bird-dispersed propagules in the intervals between fires, are largely forest and thicket species that are capable of persisting in fynbos biome vegetation (Cowling *et al.* 1997); they include relatively few species and generally comprise little cover in most fynbos biome vegetation types (Cowling & Campbell 1983; Campbell 1985). Species belonging to this plant type are a frequent and often dominant component in mediterranean shrublands in California (Keeley 1992), Chile (Hoffmann, Tellier &

Fuentes 1989) and in the Mediterranean Basin (Herrera 1984). Proteoid shrubs, which include only 120 species in the biome but which comprise the bulk of the biomass in many low- to middle-altitude and mesic habitats (Van Wilgen 1984; Campbell 1985; Cowling *et al.* 1988), have been the most intensively studied of all plant types in the region.

6.4.2 Reproductive ecology

Here we discuss flowering phenology, pollination, and seed and seedling ecology of fynbos plants. Where possible, we examine the relationship of these traits to climate, nutrient-poor soils and recurrent fire.

6.4.2.1 *Flowering phenology and pollination*

How seasonal is the flowering of fynbos taxa? Despite predictable winter rainfall throughout the fynbos biome, at least 20% of the flora is in flower at any time of the year, including the dry summer months (Johnson 1992b). This reinforces the finding that fynbos plants do not suffer the severe water stress in summer that is recorded in other mediterranean-type ecosystems (Stock *et al.* 1992), where there is a low incidence of summer flowering. However, a shift from a spring-flowering peak in the western part of the biome (winter rainfall) to an early-summer peak in the eastern part (non-seasonal rainfall) (Johnson 1992b) does suggest some effect of water availability on phenology. Furthermore, many geophytes have uncoupled flowering and growth seasons. Many Amaryllidaceae and Iridaceae grow vegetatively during the wet season and flower in a leafless condition at the end of the dry season. However, fire-lilies (*Cyrtanthus* spp.) flower only after fire, irrespective of season (Le Maitre & Brown 1992). Smoke is the cue for flowering in *Cyrtanthus ventricosus* (Keeley 1993).

Most fynbos plants (c. 83%) are pollinated by insects, with only small proportions by wind (12%), birds (4%) and mammals (<1%) (Johnson 1992a; Table 6.2). Despite its importance in fynbos, insect pollination remained very poorly studied until recently. An interesting finding is the dependence of a guild of unrelated plant species on a single pollinator, the mountain pride butterfly *Meneris tulbaghia* (Johnson & Bond 1994). Fifteen species from four families (Crassulaceae, Amaryllidaceae, Iridaceae and Orchidaceae) rely on this butterfly for pollination, and share several convergent characteristics including large, red flowers with straight, narrow nectar tubes, and a flowering period in late summer – the flight period of the butterfly. Although red flowers are usually associated with bird- and not insect pollination, birds are precluded from feeding on the nectar of these species by the straight and narrow nectar tubes.

Figure 6.15 **Immediate post-fire phase in fynbos succession showing the geophyte, *Pillansia templemanii* (Iridaceae), flowering en masse** *(Photo: J.C. Paterson-Jones).*

Beetles (Coleoptera) are important in pollinating many *Protea* (Coetzee & Giliomee 1985) and *Leucadendron* species, although in the latter genus there is a trend towards wind-pollination, a derived condition (Midgley 1987). Many *Leucadendron* species exhibit leaf dimorphism between the sexes. In particular, the wind pollinated members of this dioecious genus show the greatest differences between leaf size and other correlated features: male plants have more and smaller inflorescences, and smaller leaves and stems, than the females (Bond & Midgley 1988). Since male fitness in anemophilous species is closely related to the number of inflorescences displayed, the selection for large numbers of small, male inflorescences has resulted, as a consequence of allometry, to smaller supporting stems and leaves than in their female counterparts.

Bird pollinators are also attracted to *Protea* species by their bright, conspicuous inflorescences. Despite the low representation of bird-pollinated plants in the fynbos flora, at 4% it is higher than that of the South African flora (2.5%). This may be a consequence of copious nectar production linked to excess carbohydrate production of plants in nitrogen-poor environments (Rebelo 1987). On the other hand, conditions in fynbos such as high wind speeds, frequent mists and rain, and unpalatable, sclerophyllous leaves, are not advantageous to insects and this may have contributed towards a shift to bird pollination (Johnson 1992a). Ground-flowering Proteaceae are pollinated by small, nocturnal rodents that are attracted to their strong, yeasty odour (Wiens *et al.* 1983). In a similar way, non-flying mammals (rodents and marsupials) also pollinate Proteaceae species in the kwongan shrublands of nutrient-poor southwestern Australia (Collins & Rebelo 1987).

Wind is an important pollinating agent in many of the plant types, especially in the restioids (Table 6.2). Contrary to the notion that wind pollination is inefficient and wasteful, a study of *Staberoha banksii* has shown otherwise. The transfer of pollen by wind is efficient and leads to high fertilization levels in this dioecious species (Honig, Linder & Bond 1992).

6.4.2.2 *Seed ecology*

An extremely high proportion of proteoids, ericoids and fire ephemerals are reseeders (Table 6.2). This dependence of so many fynbos species on seeds for survival after fire is unparalleled in other mediterranean regions (Le Maitre & Midgley 1992), and highlights the importance of a full understanding of their seed biology. This includes seed production, seed bank dynamics, dispersal, germination and seedling establishment.

The infertile fynbos soils have led to the selection in some taxa (e.g. Proteaceae) for large, protein-rich seeds (Kuo, Hocking & Pate 1982), thus facilitating the rapid growth of seedling roots (Esler *et al.* 1989; Stock, Pate & Delfs 1990). This also improves moisture acquisition during the dry summer period (Mustart & Cowling 1992a). In *Protea* species there is evidence of a trade-off between seed size and number; larger-seeded species produce fewer seeds than species with smaller seeds (Esler *et al.* 1989; Maze & Bond 1996). Further evidence of what appears to be a nutrient-induced limitation of seed numbers is the finding that increased population density results in lower seed production per plant (Esler & Cowling 1990; Bond, Maze & Desmet 1995). The low seed set (c. 1–10% of florets that develop seed) of Proteaceae has been attributed to nutrient (Collins & Rebelo 1987) or pollinator limitation (Johnson 1992a). The importance of these alternative hypotheses has not been resolved. The determinants of seed production in non-proteoid taxa has not been investigated. However, ericoids do have smaller seeds than proteoids and they produce more seeds annually (Table 6.3), indicating a trade-off of resources.

Many proteoids store their seeds on the canopy in fire-resistant cones (serotiny) which open after scorching by fire (Bond 1985). The valuable, nutrient-rich seeds are released just prior to the wet season. This is favourable for germination. In this way exposure to mammal predation is minimized. *Protea* cones produce a large proportion of infertile seeds, buffering the loss of fertile seeds, since insects feed non-selectively on both (Coetzee & Giliomee 1987; Mustart, Cowling & Wright 1995). Fertile seeds are arranged in small clusters which are scattered among the infertile component of the seed-bearing cone, an arrangement that may further reduce the chance of predation of fertile seeds (Mustart *et al.* 1995). Small-mammal granivory is minimized by the satiation effect of simultaneous, large-scale seed release after fire. Seeds of serotinous proteoids are winged or hairy and are dispersed up to 100 m by wind (Bond 1988). Seeds of non-serotinous proteoids are passively dispersed, or bear fatty elaiosomes that attract ants as dispersal agents over even shorter ranges. Seeds buried in ant nests are safe from predation by small mammals (Bond & Breytenbach 1985).

What are the patterns of seed-bank accumulation in fynbos species? This information is important for gauging the impact of wildflower harvesting on seed reserves. Seed-bank size in serotinous proteoids increases by the nett annual number produced, less those lost to insect granivory (Mustart, Cowling & Dunne 1994) from plant maturity (1–5 years) (Table 6.2) to senescence and death at 30 to 50 years (Bond 1980; Van Wilgen 1981; Table 6.4).

Table 6.3 **Seed biological characteristics of fynbos species (ND, no data available)**

Species	Importance value	Stand age (years)	Post-fire regenera-tion	Seed size (mg)	Seed dispersal	Annual seed crop	Seed bank	Seedlings parent^{-1} (post-fire)	Source
Proteoids									
Paranomus bracteolaris	1 690[a]	16	Reseeder	ND	Ants	3 086[d]	4 630[g]	6.6	Le Maitre (1988)
Protea obtusifolia	1 940[a]	18	Reseeder	23	Wind	819[d]	1 602[g]	20.5	Mustart, Cowling & Dunne (1994)
P. neriifolia	4 400[a]	27	Reseeder	20	Wind	9[d]	44[g]	18.7	Le Maitre (1992)
P. neriifolia	6 800[a]	9	Reseeder	20	Wind	10[d]	19[g]	8.7	Maze & Bond (1996)
P. repens	5 200[a]	9	Reseeder	83	Wind	ND	15[g]	7.2	Maze & Bond (1996)
P. susannae	850[a]	18	Reseeder	36	Wind	753[d]	3 178[g]	13.5	Mustart *et al.* (1994)
Leucadendron coniferum	450[a]	18	Reseeder	12	Wind	9 387[d]	24 024[g]	33.0	Mustart *et al.* (1994)
L. meridianum	1 380[a]	18	Reseeder	9	Wind	965[d]	2 550[g]	16.2	Mustart *et al.* (1994)
L. pubescens	465[a]	16	Reseeder	125	Passive	465[d]	272[g]	3.6	Le Maitre (1988)
Ericoids									
Agathosma apiculata[c]	30[b]	13–14	Reseeder	4.6	Ants	5 438[e]	61.3[h]	ND	Pierce & Cowling (1991)
A. stenopetala[c]	7.5[b]	13–14	Reseeder	2.1	Ants	5 055[e]	18.7[h]	ND	Pierce & Cowling (1991)
Metalasia muricata[c]	18[b]	13–14	Reseeder	0.6	Wind	17 508[e]	56[h]	ND	Pierce & Cowling (1991)
M. muricata	8[b]	11	Reseeder	0.6	Wind	2 853[g]	15.5[h]	ND	Musil (1991)
Muraltia squarrosa[c]	15[b]	13–14	Reseeder	2.5	Ants	202[e]	114[h]	ND	Pierce & Cowling (1991)
Passerina paleacea	19.7[b]	25	Reseeder	0.8	Passive	13 253[e]	441[h]	5.2	Kilian & Cowling (1992)
P. vulgaris[c]	22[b]	13–14	Reseeder	1.3	Passive	44 000[e]	629[h]	ND	Pierce & Cowling (1991)
Phylica ericoides	15.3[b]	25	Reseeder	2.5	Ants	454[e]	278[h]	3.2	Kilian & Cowling (1992)
P. stipularis	20[b]	11	Resprouter	24.0	Ants	771[e]	<1[h]	ND	Musil (1991)
Fire ephemerals									
Felicia echinata[c]	3[b]	13–14	Reseeder	0.4	Wind	0[e,f]	35[h]	ND	Pierce & Cowling (1991)
Restioids									
Thamnochortus erectus	12 300[a]	>40	Resprouter	1.5	Wind	12 489[d]	2 530[h]	0.02	J.M. Ball (unpubl. data)
T. insignis	11 700[a]	8	Reseeder	1.0	Wind	19 412[d]	3 773[h]	4.0	J.M. Ball (unpubl. data)

[a] No. plants ha^{-1}.
[b] % canopy cover at site.
[c] Seed production and seed bank sizes are means of two and three years' data, respectively.
[d] Seeds plant^{-1}.
[e] Seeds m^{-2} canopy.
[f] Plants either dead or senescent in mature stand.
[g] No. seeds plant^{-1}.
[h] No. seeds m^{-2}.

Table 6.4 **Post-fire successional phases in mountain fynbos (after Kruger & Bigalke 1984). Data on the proportion of species that have reached reproductive maturity in each phase are for a** sample of 210 species at Swartboskloof near Stellenbosch (Van Wilgen & Forsyth 1992)

Phase	Period after fire(years)	Characteristics
Immediate post-fire	0–1	Seed germination and vegetative regeneration takes place. Most annuals, many fire ephemerals and some geophytes reproduce only in this phase. About 76% of all species reach reproductive maturity.
Youth	1–5	Restioids, graminoid herbs and sprouting shrubs dominate. Another 13% of species reach reproductive maturity. Canopy cover approaches pre-fire levels and the vegetation becomes flammable.
Transitional	5–10	All plants reach reproductive maturity. Non-sprouting shrubs emerge from the canopy.
Mature	10–30	Tall shrubs reach maximum height and flowering activity. Low shrubs begin to die.
Senescent	30–60?	Mortality of tall shrubs accelerates. Crowns open and litter accumulates.

In some species, seeds retain high germinability during canopy storage (Le Maitre 1990; Mustart & Cowling 1991), whereas in others there is a decline (Van Staden 1978; Bond 1985). Since flower harvesting usually removes the current year's seed crop, these canopy-stored seeds are important in determining post-fire recruitment levels (see 6.7.3).

The seeds of most ericoid species are released each year and dispersed passively, by wind, or by ants. Seeds are rapidly removed from the soil surface by ant dispersal or rodent granivory. Substantial losses are incurred in the soil, due to decay and premature germination (Pierce & Cowling 1991). Annual seed input and losses vary widely among ericoids, leading to unreliable seed banks, and the possibility of population fragmentation and extinction after fire (Cowling *et al.* 1994b).

Fynbos and southwestern Australian kwongan have an unusually high incidence of myrmecochory (seed dispersal by ants): it is found in about 30% of the flora (29 families, 78 genera and about 2500 species) (Bond, Yeaton & Stock 1991; see also Cowling & Witkowski 1994). Why are there so many myrmecochorous species in these two areas? Cowling *et al.* (1994b) suggest that the low nutrient status of their soils has led to the selection of fewer, larger seeds, which are protected from predation by burial in ant nests. The virtual absence of myrmecochores in the moderately fertile renosterveld soils supports this hypothesis. The vulnerability of myrmecochorous lineages, which characteristically have small, transient seed banks, to population fragmentation may have led to the exceptionally high diversification of obligate reseeding lineages (Cowling *et al.* 1994b; Chap. 3, this volume).

Bond (1994) has warned that the many fynbos species that do not resprout after fire and are limited by seed for recruitment face a high risk of extinction if their pollinator and dispersal mutualisms collapse. For example, myrmecochorous Proteaceae are dependent on seed dispersal by ants to escape predation by rodents (Bond & Slingsby 1983). The alien Argentine ant (*Linepithema humile*) displaces native ants, and since the former does not disperse seeds into ant nests, this results in high seed losses (Bond & Slingsby 1984). Fynbos is unique in that it includes many obligate reseeding species, and many of these are reliant on animals for pollination and dispersal (Table 6.2).

6.4.2.3 *Germination*

Since most regeneration occurs after fire, it is reasonable to expect that seeds of fynbos species have germination cues linked to fire, either directly via heat, charcoal and smoke extracts, or indirectly to post-fire conditions such as altered light or temperature regimes as a consequence of removal of the insulating vegetation. Germination trials indicate that seeds of most serotinous Proteaceae are not dormant and germinate readily with adequate moisture at cool, diurnally fluctuating temperatures (10/20 °C for 14/10 h), or at a constant low temperature (5–10 °C) (Van Staden & Brown 1977; Brits 1986a; Mustart & Cowling 1991). These temperatures correlate well with soil surface conditions in the field during the winter season after fire (Brits 1986a; Mustart & Cowling 1993) and explain the large-scale germination that occurs at this time. Proteaceae with soil-stored seeds have dormancy-breaking requirements, and laboratory experiments have shown that elevated oxygen levels, such as those likely to occur in post-fire soils with diminished

root respiration, stimulate germination (Brits 1986b). Experimental desiccation followed by wetting, thought to parallel the fire/rain cycle in the field, results in scarification of their hard seed coats (Brits *et al.* 1993). This may favour the emergence of myrmecochorous species after hot fires (Bond, le Roux & Erntzen 1990; Brits *et al.* 1993).

Most ericoid species store their seeds in the upper layers of the soil. There are many examples of seeds with germination cues indirectly linked to fire, namely the diurnally fluctuating temperatures of the cool, winter months (Musil 1991; Kilian & Cowling 1992; Pierce & Moll 1994). However, germination levels are often low (<50%) and, with few exceptions (Musil 1991; Kilian & Cowling 1992), do not respond to charate or heat treatments, as do many Californian (Keeley 1991) and Mediterranean Basin (Thanos & Georghiou 1988) species. The dependence of the ericoid seed component on fluctuating temperatures in the upper soil layers for germination makes depth of burial an important factor for seedling emergence (Pierce & Moll 1994); this probably contributes to the low and variable field emergence levels. Although smoke extract stimulates germination of many species of Asteraceae, Ericaceae and Restionaceae (Brown 1993; Brown *et al.* 1993), its ecological significance is unclear, with recent research showing that smoke treatment also stimulates germination in taxa from habitats that are not prone to fire (Pierce, Esler & Cowling 1995).

Although there are relatively comprehensive data on the seed ecology of fynbos proteoids and ericoids, little is known about other fynbos plant types, namely the restioids, geophytes and fire-ephemerals. Since many restioids (>80%) and geophytes (100%) are resprouters (Table 6.2), there has been little incentive to investigate their post-fire seeding strategies. However, the economically important thatching reed, *Thamnochortus insignis* (Restionaceae), is a reseeder, and the need to encourage recruitment of this species has spurred research into its seed ecology.

6.4.2.4 *Seedling establishment*

In addition to adequate levels of seed production and germination, the successful establishment of seedlings is of crucial importance in determining replacement of fynbos populations after fire.

The post-fire recruitment of reseeding serotinous proteoids has been studied extensively across the biome by examining parent : seedling ratios in relation to factors such as season and frequency of fire (Bond 1984; Bond, Vlok & Viviers 1984; Midgley 1989). Such studies have shown that winter and spring burns generally lead to poor recruitment (Bond *et al.* 1984; Van Wilgen & Viviers

1985). In contrast, successful establishment after summer and autumn fires is ascribed to the relatively short exposure to granivory by rodents in the period between seed release and the onset of winter germination conditions (Bond 1984). Most of these studies emphasize the importance of exogenous post-burn factors in determining recruitment, but ignore the importance of variation in pre-burn seed numbers according to season of seed maturation (Jordaan 1949) and endogenous, density-dependent effects on per capita reproductive output (Laurie & Cowling 1994; Bond *et al.* 1995). Maze & Bond (1996) have shown that post-fire seedling numbers of two *Protea* species are linearly related to the number of seeds available before the fire. Bond *et al.* (1995) have shown that density-dependent reduction in fecundity may lead to endogenous and chaotic oscillations in the populations of some proteoid species over several fire cycles (see also Chap. 18, this volume). Fire in young fynbos results in poor recruitment of reseeders, owing to the small seed bank size of immature plants (Van Wilgen 1981), and further points to the importance of seed numbers in this regard. Recruitment is also a function of seedling mortality during subsequent growth to maturity. Seedling mortality is generally lower in mountain fynbos (7–25%) (Midgley 1988; Van Hensbergen *et al.* 1992) than in lowland fynbos (10–45%) (Mustart & Cowling 1993; Maze & Bond 1996). Little is known about the causes of this mortality. In order to gain a better understanding of the cycle of population replacement, it will be necessary to assess the relative importance of climate-related and density-dependent causes of mortality in seedlings.

The importance of regeneration stages in determining community structure has been tested in reciprocal seed transplants of two pairs of proteoid species occurring on adjacent, different soil types (Mustart & Cowling 1993). Although sorting did not occur at the seed germination stage, seedlings showed improved growth and survival on their native soil type. This is attributed to differences in root size, as determined by seed size (Mustart & Cowling 1992a), which enables the larger-seeded species to establish on the sandier soil type with its lower moisture-holding capacity.

6.4.3 Physiological traits and processes

Summer drought and nutrient limitations, especially with respect to nitrogen and phosphorus, are the main environmental stresses in the region. Several studies have explored the water relations of different fynbos plant forms. The diurnal and seasonal patterns of leaf conductance and xylem pressure potential (XPP) in response to climatic gradients have been documented at a number of sites ranging from semi-arid to mesic (Stock

et al. 1992; Richards, Stock & Cowling 1995b). These studies show that water stress is much less important as a selective force than in some other mediterranean-climate regions (see Smith *et al.* 1992 for a comparison between fynbos, Californian chaparral and Australian kwongan).

There are, nevertheless, marked differences in water relations between different fynbos growth forms. Shallow-rooted restioids and ericoid shrubs show greater declines in XPP during summer than the deeper-rooted proteoids; the latter experience high XPPs that remain relatively constant throughout the year. The broad classes of rooting depth represented by the different growth forms [restioid (shallow roots), ericoid (medium-depth rooting), and proteoid (deeper roots) (Higgins, Lamb & Van Wilgen 1987)] behave differently with respect to their ability to take up soil water. The deep-rooted proteoids have access to soil water in deep soil horizons and are 'water spenders'. Richards *et al.* (1995b), in a comparative study of the water relations of *Protea compacta* and *P. susannae*, found that seedlings and adults of both species showed different water-use strategies. They suggest that switches in water-use strategy between seedling and adult stages (from water-spending to conservative in *P. compacta* and vice versa in *P. susannae*), together with differences in water relations, particularly at the seedling stage, show that habitat specialization is important in determining the distribution of these species (with *P. compacta* and *P. susannae* being associated with shallow and deep soil, respectively). Restioids and the shallow-rooted ericoids are relatively drought-resistant and use water more conservatively. They maintain high stomatal conductances irrespective of water availability and consequently show highly variable diurnal and seasonal patterns of XPP. These patterns explain the distribution of major growth forms in fynbos landscapes; proteoids occur mainly on deep and relatively fine-grained (water retaining) soils, whereas restioids and ericoids (excluding ericaceous ericoids) predominate on drier sites (Campbell 1986). Restioids can also tolerate water-logged soils, and often dominate such sites (Chap. 16, this volume).

Patterns of water use along aridity gradients in fynbos are complicated by associated nutrient gradients (Stock *et al.* 1992). Drought-deciduous taxa and succulents are better represented in the more arid areas in the west of the biome (Campbell & Werger 1988), but they also occur on richer soils. It is only on these richer sites that water stress appears to act as a selective force for the development of anatomical and morphological traits (see Chap. 16, this volume).

Photosynthetic capacities of fynbos shrubs are low (3–14 μmol m^{-2} s^{-1} at 21 °C), but similar to evergreen shrubs in other mediterranean-climate regions (Mooney et al. 1983). Van der Heyden & Lewis (1989) identified two distinct gas-exchange guilds in fynbos based on photosynthetic rates. Restioid and ericoid species had low photosynthetic rates, whereas proteoid species showed relatively high rates throughout the year. Midday net assimilation rates for the proteoid guild during summer were between 43 and 95% of their spring rates, whereas those for restioids and ericoids were only between 3 and 6%.

Plant water status is not the overriding determinant of photosynthetic capacity in fynbos shrubs. This is shown by poor correlations between XPPs and net photosynthetic rates (Van der Heyden 1988; Von Willert, Herppich & Miller 1989; Richardson & Kruger 1990). Van der Heyden & Lewis (1990) examined the effect of water availability on photosynthetic responses of proteoid, restioid and ericoid species during a late-summer drought period. Photosynthetic rates of proteoids and ericoids were not affected by adding water which suggests that they are not limited by low soil water availability during summer. In contrast, the net photosynthetic rate of a restioid, Thamnochortus lucens, increased by 20–40% when water was added. Van der Heyden & Lewis suggest that the limitations imposed on restioids by water shortage during summer are offset by their ability to take up water quickly when it becomes available, thereby maximizing CO_2 uptake. This capacity, together with various morphological and anatomical features that help restioids avoid desiccation during summer (e.g. heavily lignified culms and non-photosynthetic scale leaves), allows some species to survive and thrive in arid areas where ericoids and proteoids cannot survive (Stock et al. 1992). Some restioids also enhance their water status by trapping mist (Moll & Romoff 1983).

The physiological significance of photosynthetic pathways, sclerophyllous leaves, mycorrhizal relationships and other root structures (e.g. root sheaths in restioids, cluster roots in the Fabaceae and proteoid roots in the proteoids) in fynbos is discussed by Stock et al. (Chap. 16, this volume).

6.5   Community structure and dynamics

Substantial advances have been made over the past decade in the study of fynbos biome communities: the assemblages of interacting species and the various relationships which bind them. Given the importance of fire in shaping community structure (Cowling 1987), the long-standing appreciation of the richness of fynbos

communities (Taylor 1978) and the presence in communities of many closely related species (Cody 1986), it is understandable that much of this research effort has focused on fire effects (Kruger & Bigalke 1984; Van Wilgen et al. 1992b), the maintenance of biodiversity (Cowling et al. 1992) and competition and coexistence of ecologically similar species (Bond, Cowling & Richards 1992). These topics are briefly reviewed in this section.

6.5.1   Fire and succession

Fire drives succession in fynbos. Its effect on species composition, vegetation structure and successional patterns depends on the frequency, intensity and season of the fire, and there is also considerable variation in response between sites (Kruger & Bigalke 1984). Most fynbos species are able to persist under any fire regime, but components of the regime mediate recruitment success in several groups of plants, some of which (e.g. serotinous overstorey proteoids and pyrophytic perennials, including nitrogen-fixing legumes) have a disproportionately large influence on subsequent succession. Most work on post-fire succession has been conducted in mountain fynbos in the wetter parts of the biome (Kruger 1977, 1987; Van Wilgen & Forsyth 1992; Kruger & Bigalke 1994; Table 6.4), and only two comprehensive studies have been done at drier sites in lowland fynbos (Hoffman et al. 1987; Cowling & Pierce 1988). Post-fire succession at all fynbos sites is notable for a rapid initial recovery, owing mainly to growth of fire ephemerals, perennial graminoid herbs and sprouting shrubs (Cowling & Pierce 1988). Immigration of species is not an important process in the early phases of succession, except in some ecotonal communities (Manders 1990b). The generalized scheme of post-fire successional phases derived from results in mountain fynbos (Table 6.4) thus applies to most fynbos communities. The main differences between mountain and lowland sites seems to be that geophytes and annuals persist longer in the lowlands, and that restioids form a greater proportion of the biomass in the mature phase than is the case at most mountain sites. At all sites, species diversity is greatest in the immediate post-fire phase or the first year of the youth phase. In mountain fynbos (and at least one lowland site) there is a marked decrease in species diversity as the cover of overstorey shrubs increases (e.g. Kruger 1987; Cowling & Pierce 1988; Cowling & Gxaba 1990; Fig. 6.16). Hoffman et al. (1987) reported an increase in species number in increasing cover to post-fire ages of up to 19 years; they suggest that growth in the drier lowland sites is much slower and that suppression of understorey species probably starts only after about 30 years, if at all. At many sites, seeds of forest trees are deposited under perch plants (not only fruit-bearing species) during all phases

Figure 6.16 **Senescent phase in fynbos succession showing death and decline of *Protea neriifolia*, ericoids and restioids, and persistence of the resprouting gymnosperm, *Widdringtonia nodiflora*** (Photo: B.W. van Wilgen).

of succession (Manders 1990b; Manders *et al.* 1992). Fire cycles are usually too short to allow forest trees to establish and these waves of immigrants are usually eliminated before they attain permanent residence (see 6.3.2.1).

The diversity of regeneration traits (notably age to maturity, seed production or sprouting ability, seed size, seed storage and dispersal syndromes, germination cues, pollination and breeding systems) within growth forms seems to provide a 'backup' in the event of stochastic population fluctuations induced by fire (see 6.5.2.3). This ensures long-term stability of community-level functional diversity and low variance in ecosystem processes in the absence of alien trees and shrubs (Richardson & Van Wilgen 1992; Richardson *et al.* 1995). Invasive alien plants such as *Hakea sericea* and *Pinus pinaster* initially behave much like the indigenous shrubs in the succession. However, because they are well buffered against fire-induced population crashes (because of their short juvenile periods and large reserves of viable seeds), the non-equilibrium status of natural communities is disrupted, resulting in new depauperate steady states (Richardson & Cowling 1992; Richardson & Van Wilgen 1992).

6.5.2 Diversity

Here we focus on alpha diversity or patterns and maintenance of species richness at the community or within-habitat scale. The interactions between independent diversity components (alpha, beta and gamma diversities) and other factors in determining patterns of regional richness in the fynbos biome are dealt with in Chap. 19 (this volume).

6.5.2.1 *Patterns*

Patterns of alpha diversity have been reasonably thoroughly studied in the fynbos biome (Cowling *et al.* 1992).

Mean species richness for fynbos sites is 15.9 at the 1-m² level and 65.7 at the 1000-m² level: these values do not vary significantly in different geographical regions of the biome. Renosterveld and western succulent karroid shrublands are richer than fynbos at the 0.1-ha scale (mean value 84.2 and 71.0 spp., respectively), while eastern karroid shrublands, subtropical thicket and forest communities have fewer species (mean value 55.9, 58.7 and 53.7 spp., respectively). At a subcontinental level, fynbos biome communities are not particularly rich (Cowling *et al.* 1989; Chap. 19, this volume); by global standards, they are moderately rich, having only half the number of species recorded in tropical rainforest communities at the 0.1-ha scale (Gentry 1988) and double the number in mature Californian chaparral (Bond 1983). Other temperate communities such as certain European grasslands and grazed woodlands in Israel are richer than fynbos at this scale (Naveh & Whittaker 1979; Bond 1983).

6.5.2.2 *Determinants*

Classical niche theory predicts that patterns of diversity are determined by primary differences in the physical environment which, in turn, regulate differences in resource quantity and quality (Whittaker 1977; Brown 1988; Diamond 1988). Diversity models developed for plants predict humped curves with maximum richness at sites with low to moderate soil fertility and productivity (Grime 1979; Huston 1979; Tilman & Pacala 1993). Variation in alpha diversity in fynbos biome communities is not related in any predictable way to soil-nutrient levels (Bond 1983; Cowling 1983b; 1990). However, both Bond (1983) and Kruger (1987) show that diversity of mature fynbos communities is highest at intermediate levels of productivity as measured by soil moisture and biomass. Simmons & Cowling (1996) were unable to demonstrate any significant relationships between richness and several measures of resource availability in a large data set from the Cape Peninsula. However, high richness values were invariably recorded at sites with the lowest resource levels. Temporal variation in resource supply is also an important determinant of diversity. Niche theory predicts that fynbos biome communities would be richest in the non-seasonal rainfall zone, since resource axes for phenological and growth-form differentiation should be longer there than in the strictly winter-rainfall zone (Cody 1989). However, there is no significant difference in the diversity of fynbos communities from the two rainfall zones (Cowling *et al.* 1992). Compared with forest and thicket, growth-form diversity is a relatively poor predictor of fynbos alpha diversity throughout the biome, suggesting high levels of ecological redundancy, at least in terms of structurally characterized plant types (Cowling *et al.* 1994a).

Secondary or biological determinants of diversity consider species interactions and the dynamics of dispersal or extinction in ecological time (Whittaker 1977; Shmida & Wilson 1985). The importance of competition in determining the richness of fynbos communities has not received much attention. The higher richness of fynbos relative to chaparral has been attributed to slower rates of competitive displacement after fire in fynbos, owing to lower soil nutrients (Tilman 1982; Keeley 1992). Environments with relatively high productivity (which favour the development of a dense shrub overstorey in fynbos) support species-poor communities as a result of the competitive suppression of the understorey (Campbell & Van der Muelen 1980; Bond 1983; Cowling & Gxaba 1990; Esler & Cowling 1990; Simmons & Cowling 1996).

In fynbos, fire is a stochastic variable that maintains diversity, in association with nutrient poverty, by preventing competitive exclusion and facilitating the coexistence of species with different life histories (Campbell & Van der Muelen 1980; Cowling 1983b; Kruger 1983). Thus, as in chaparral and other fire-prone shrublands, the highest richness in fynbos is recorded in the early post-fire years when fire ephemerals and longer-lived species coexist (Campbell & Van der Muelen 1980; Cowling & Pierce 1988), but species inhibition over time is a relatively slow process (Kruger 1983). Intermediate fire frequencies maximize diversity in fynbos: short rotation fires (every 3–4 years) eliminate slow-maturing, reseeding shrubs, whereas long fire-free intervals result in the loss of short-lived species (Campbell & Van der Muelen 1980; Kruger 1983; Bond, Midgley & Vlok 1988; Van Wilgen & Forsyth 1992). Fire may also maintain diversity in fynbos by altering competitive hierarchies as a result of fire-induced differential recruitment (Cowling 1987; Cowling and Gxaba 1990; Yeaton & Bond 1991; Kilian & Cowling 1992).

Tertiary or historical determinants of diversity invoke patterns of species origination, arrival and loss over evolutionary time (Brown 1988; Ricklefs & Schluter 1993). The low diversity of upland fynbos communities in the extreme southeastern part of the biome relative to those on environmentally matched sites further west is attributed to insufficient time for migration and speciation of fynbos taxa since Holocene climatic amelioration (Cowling 1983b). This implies a relationship between local and regional species richness (Ricklefs & Schluter 1993; see also Chap. 19, this volume).

In summary, fynbos biome communities are of moderate richness. The primary determinants of fynbos alpha diversity are low-nutrient soils and climatic conditions which promote recurrent fire at relatively long (c. 15 yr) intervals. Within fynbos landscapes, the most species-poor communities occur on the most productive sites. Nothing is known about the determinants of diversity of renosterveld communities.

6.5.2.3 Competition and coexistence

An intriguing feature of fynbos communities is the coexistence of numerous species which belong to the same growth form – the so-called trophically equivalent species of Shmida & Ellner (1984). For example, up to nine proteoid shrubs and more than 20 ericoid shrubs may coexist in small (50–100 m²) patches of fynbos. Contemporary theory has split into two sets of contradictory predictions regarding coexistence in communities of sessile organisms such as plants. Classical theory predicts that the most similar species cannot coexist, whereas a variety of more recent models predict the reverse – the most similar species are likely to coexist. In this section we explore the predictions of these theories using data from assemblages of proteoid and ericoid shrubs in fynbos.

Classical competition theory, as applied to plants, predicts that coexisting potential competitors should exhibit niche differentiation which will be manifest as morphological or growth-form differentiation – the 'growth-form niche' of Cody (1986). Patterns of non-overlapping leaf morphologies of proteoid shrubs at several fynbos sites have been cited by Cody (1986) as evidence for competitive controls on community membership of this guild. Bond et al. (1992) found statistically significant evidence for a shift in leaf morphology (character displacement) of Protea repens at one of two study sites. A major problem with this notion of complementary and co-evolved sets of species is that it assumes a predictable (equilibrium) community structure that is determined largely by species interactions; in reality, fire may induce considerable variation in population sizes of proteoid shrubs (Bond et al. 1984), thus reducing the role of these interactions.

Grubb (1977) made an important contribution to plant community ecology by arguing that coexisting species with similar growth form, habitat and phenological niches could be separated along a regeneration niche axis. Thus, coexistence could be explained by species-specific differences in dispersal, germination and seedling establishment. Yeaton & Bond (1991) demonstrated that the different dispersal modes of two coexisting proteoid shrubs reduced the tendency for exclusion of the poorer competitor, but did not prevent it, over several fire cycles. In fact, the two species they studied are largely segregated by different habitat requirements. Kilian & Cowling (1992) argued that the long-term coexistence of two competitively equivalent ericoid shrubs was mediated by differential regeneration success in response to fire intensity.

Lottery models, which have been developed to explain coexistence of sessile organisms that compete primarily for space, assume that the recruitment success for a given species is proportional to its relative frequency in the pool of propagules (e.g. Chesson & Warner 1981; Fägerström 1988). Two criteria must be met for long-term coexistence in lottery models: first, environmental variation leads to occasional, strong recruitment in each species; second, species must have overlapping generations, so that the effects of good recruitment events can be stored over poor years. In complete contradiction to classical niche theory, lottery models predict that species with similar per capita reproductive output and competitive abilities should coexist (Bond *et al.* 1992).

Many fynbos species violate the condition of overlapping generations: adults are killed by fire, recruitment is limited to the immediate post-fire period and seeds do not survive the intervals between fires. Under these conditions, the inferior competitor would soon be lost from the system and even species with identical per capita reproduction would not coexist indefinitely (Bond *et al.* 1992). To persist indefinitely, populations would have to be replenished from elsewhere (i.e. a spatial storage effect; Chesson 1990) or have a rare species advantage. Laurie & Cowling (1994) have developed a lottery model to explain the coexistence of species with non-overlapping generations. The model assumes density-dependence in reproductive output that involves higher relative per capita population growth of the rarer species in a community of coexisting ecological equivalents (cf. Bond *et al.* 1995). Laurie & Cowling suggest that the rare species advantage could be mediated by lower pre-dispersal seed predation as host shrub density declines (cf. Coetzee & Giliomee 1987; Esler & Cowling 1990; Mustart *et al.* 1994). Candidates for lottery coexistence, which have been studied in some detail, are two reseeding, serotinous proteoids that coexist throughout their range on limestone in the southern Cape (Cowling *et al.* 1994b; Laurie, Mustart & Cowling 1997). These species exhibit symmetric competition, have identical regeneration characteristics and similar reproductive output: they occupy the same niche, no matter how this is defined. Furthermore, spatial patterns in the ratio of species coexistence throughout their environmentally homogeneous distributions are clustered at the 1 : 1 ratio, providing strong inferential evidence for a density-dependent lottery model (Laurie *et al.* 1997).

6.6 Ecosystem processes

Relatively little work has been done on the primary processes that mediate the capture, storage and transfer of energy, CO_2, nutrients and water in fynbos ecosystems. In this section we review what is known about the factors that control three important ecosystem processes: biomass accumulation, catchment hydrology and nutrient cycling.

6.6.1 Biomass

A notable feature of fynbos plant communities is their low height when compared to communities in regions with similar climate and soils on other continents. For example, the vegetation in wetter parts (>800 mm rain yr^{-1}) of mediterranean-climate Western Australia is an order of magnitude taller than that for matched sites in the southern and western Cape (Richardson & Cowling 1992). Whereas trees contribute greatly to the biomass at these high-rainfall sites on deep sands and on young soils on granite in Western Australia, shrubs are the dominant component throughout the Cape landscape, except for isolated forest patches in fire-protected sites.

It is largely as a result of the scarcity of trees that fynbos communities support a lower biomass relative to similar areas on other continents. The above-ground biomass in mature (but not senescent) mountain fynbos ranges from 15 000 to 51 000 kg ha^{-1} (Richardson & Cowling 1992). This is lower than in most structurally similar communities in other mediterranean-climate regions on richer soils but with lower rainfall, and an order of magnitude less than areas within other mediterranean-climate regions with similar rainfall. Most mountain fynbos areas receive more than 800 mm of rainfall per year. Equivalent habitats in the other mediterranean-climate regions support forest with much higher biomass. For example, above-ground biomass in dense pine forests in mediterranean-climate France (1100 mm rain yr^{-1}) is up to two orders of magnitude greater than that of fynbos communities in parts of the Cape with an equivalent annual rainfall. Similarly, in lower rainfall (<800 mm rain yr^{-1}) regions (e.g. dry heaths in Australia, *Ceanothus* chaparral in California and Chilean matorral) above-ground biomass of fynbos communities is markedly lower (references in Richardson & Cowling 1992).

Annual biomass increments of between 1000 and 4000 kg ha^{-1} in the first few years after fire in fynbos are similar to those in Californian chaparral, European garrigue, scrub and heathland, and Australian heath and mallee-broombush communities (Richardson & Cowling 1992).

6.6.2 Hydrology

Studies on the hydrological properties of fynbos catchments have focused on the effects of different fire cycles and afforestation with *Pinus radiata* on streamflow and water quality in the southwestern Cape (see Wicht 1948, 1967 for historical overviews). There have been no studies in catchments in the more arid parts of the biome.

Runoff in most fynbos catchments amounts to between 35 and 55% of rainfall, but some, such as Langrivier and Tierkloof at Jonkershoek, yield up to 70% of rainfall (Bosch & Hewlett 1982); this is amongst the highest in the world for areas with similar rainfall (J.M. Bosch, pers. comm.). One reason for this is the low biomass (see above), which results in low levels of interception (estimated at 5–10%; Versfeld & Van Wilgen 1986) and transpiration per unit area.

There are usually marked pulses of increased streamflow after fire in fynbos catchments (Scott & Van Wyk 1992). Water yield declines by about 1% for every successive year after a fire until the vegetation reaches a postfire age of 15 years; thereafter water consumption levels off (Van der Zel & Kruger 1975; Van Wyk 1977). Experimental burns at intervals of 6 and 12 years in the southwestern Cape caused temporary increases of 700–800 m^3 ha^{-1} in water yield in the first year after fires (Lindley, Bosch & Van Wyk 1988). This means that 6- and 12-year burning cycles could theoretically increase water yield by 600 and 400 m^3 ha^{-1} over 40 years, respectively (Bosch 1984). Changes in water yield from fynbos catchments are directly related to changes in plant biomass. Bosch, Van Wilgen & Bands (1986) suggested that: firstly, the magnitude of hydrological response depends on the pre-fire biomass which determines levels of water loss through interception and transpiration; and secondly, that the rate of decrease in streamflow after fire depends on the relative representation of resprouting plants which influences the rate of recovery of the vegetation to pre-fire levels.

Effects of fire on stormflows, surface runoff, sediment yield and water repellency have been studied in several catchments in the Jonkershoek Valley near Stellenbosch. Results show that only exceptionally severe fires (those with a combination of high fuel load, dry fuel, dry soil and hot, dry weather) have a marked effect on these factors (reviewed by Scott & Van Wyk 1992). Impacts of alien plants on hydrology are discussed later in this chapter.

6.6.3 Nutrient cycling

The largest reserves of carbon, nitrogen and phosphorus in fynbos ecosystems are stored in the soil. As in other vegetation types where these conditions prevail, there is a high incidence of sclerophylly, low tissue nutrient contents and high root/shoot ratios in fynbos plants

(Chap. 16, this volume). Herbivores are less important in nutrient cycling in fynbos than in any other southern African biomes (Chap. 17, this volume). The transfer of nutrients from standing phytomass to the soil between fires therefore depends almost entirely on the processes of litterfall and decomposition. Rates of litterfall increase as the amount of standing phytomass increases, and nitrogen and phosphorus pool sizes vary accordingly. The actual rates of litter production are low (e.g. 78 g m^{-2} yr^{-1} in eight-year-old lowland fynbos; Mitchell *et al.* 1986; and 354–448 g m^{-2} yr^{-1} in 23-year-old mountain fynbos; Mitchell & Coley 1987). The dynamics of nutrient return via litterfall in fynbos have been little studied, but rates are very low. Decomposition of the sclerophyllous litter is also extremely slow and contributes very little to nutrient release (review by Stock & Allsopp 1992).

Fire is the major mineralizing agent in fynbos, returning mineral elements held in above-ground plant material and litter to the soil. There are post-fire flushes of nutrients, with increased availability of nitrogen, phosphorus and cations. The availability of these elements decreases rapidly (after about nine months for nitrogen) as elements are incorporated into plant biomass or immobilized by decomposer organisms. Phosphorus levels increase markedly after fire, but return to pre-fire levels within four months. High concentrations of phosphorus and low concentrations of nitrogen immediately after very hot fires favours certain species, notably a nitrogen-fixing guild comprising mainly *Aspalathus* spp. (Fabaceae). This component is later replaced as fixed nitrogen is returned to the system via either atmospheric deposition or biological nitrogen fixation. The latter source is much more important than the former, since free-living nitrogen-fixing bacteria (Van Reenen, Visser & Loos 1992) and nitrogen-fixing members of the Fabaceae (Cowling & Pierce 1988) often dominate in the early post-fire stage. Catchment-level studies have shown that fire has little effect on overall nutrient capital at this scale (Van Wyk, Lesch & Stock 1992). Less flammable vegetation in riparian zones sometimes acts as a nutrient filter or trap, preventing the loss of nutrients via streamflow. Invasive alien plants have disrupted nutrient-cycling regimes in many parts of the biome (see 6.7.2).

The effects of nutrient enrichment of fynbos have been studied at sites in lowland (Witkowski 1988; Witkowski & Mitchell 1989; Witkowski, Mitchell & Stock 1990) and mountain fynbos (Lamb & Klaussner 1988). These studies examined the effects of single applications of low-level nutrient addition in the form of single nutrient elements and balanced nutrient applications. Nitrogen applications in the lowlands induced slight increases in shoot growth and biomass accumulation, whereas addition of phosphorus reduced growth. The effect was opposite in mountain fynbos: increased growth in response to phosphorus application, but decreased growth when nitrogen was added. Small changes in morphological features, phenology and allocation to reproductive structures were also observed in the proteoid, ericoid and restioid groups (Lamb & Klaussner 1988; Witkowski 1988). Litterfall (especially leaf fall) increased in response to nutrient additions and the nitrogen and phosphorus content of the litter was increased (Witkowski 1989). In reviewing these responses to nutrient additions, Stock & Allsopp (1992) suggested that small nutrient additions (such as those caused through human-induced nutrient enrichment) are unlikely to effect marked changes in productivity or species composition outside the natural amplitude of variation for a particular site.

6.7   Conservation and management

Advances in the understanding of fynbos community and ecosystem structure and functioning have been readily and rapidly incorporated into conservation management strategies and actions (Rebelo 1992; Van Wilgen, Bond & Richardson 1992a; Le Maitre, Van Wilgen & Richardson 1993; Richardson *et al.* 1994; Van Wilgen, Richardson & Seydack 1994). In this section we focus on three themes: the effects of habitat fragmentation; alien plants; and the harvesting of indigenous plant products. More details on conservation of the fynbos biome are given in Chap. 23 (this volume).

6.7.1   Fragmentation
The fragmentation of once continuous habitats is a major problem in the lowlands of the fynbos biome (Rebelo 1992). Some 43% of fynbos biome reserves are smaller than 500 ha and 17% are less than 50 ha. Many of these reserves and other fragments are surrounded by transformed land. What is the likely impact of fragmentation on biodiversity and what are the processes responsible for species loss in small fragments? These questions have only recently been addressed in the fynbos biome.

Bond *et al.* (1988) studied fragmentation effects in the fynbos of the southern Cape by comparing plant species richness on fynbos 'islands' in a 'sea' of forest to similar-sized areas of an adjacent fynbos 'mainland'. Smaller islands supported up to 75% fewer species than equivalent areas of mainland. The island effect disappeared at about 600 ha. Species loss was attributed to a fire-regime effect: the smaller islands apparently burn less fre-

quently, which explains the absence of the fire-ephemeral flora in these fragments. Bond *et al.* (1988) concluded that biodiversity could be maintained in small fragments by applying appropriate fire regimes.

Cowling & Bond (1991) conducted a similar study where the 'islands' were outcrops of limestone in a 'sea' of acid sand. Both substrata support floristically distinct forms of proteoid fynbos, are burnt by the same fires, and harbour similar assemblages of pollination and dispersal vectors. In this study the island effect disappeared at 4–15 ha, suggesting that small reserves may play a useful role in preserving fynbos biodiversity, provided they are subject to fire and are accessible to the pollinator and disperser fauna. Studies are still required to determine the impact of fragmentation in agricultural landscapes where fragments are surrounded by transformed land and subject to grazing, herbicides and pesticides.

6.7.2   Alien plants
Several species of alien plants, notably trees and shrubs of the genera *Acacia*, *Hakea* and *Pinus*, have invaded natural vegetation over large parts of the region (Macdonald & Richardson 1986; Richardson *et al.* 1992; Chap. 22, this volume). Invasion of fynbos by species such as *A. longifolia*, *A. saligna* and *H. sericea* can double or treble the total above-ground plant biomass (Milton & Siegfried 1981; Van Wilgen & Richardson 1985). The biomass in self-sown pine forests can be up to five times that of the pre-invasion fynbos (Versfeld & Van Wilgen 1986). The alien stands suppress indigenous plant species (Richardson, Macdonald & Forsyth 1989), threatening many taxa with extinction: about 750 species are currently at risk (Richardson *et al.* 1992).

Invasive alien plants have totally disrupted nutrient-cycling processes in many parts of the fynbos biome. All the most important invaders have mechanisms for enhancing nutrient acquisition (nitrogen-fixing symbionts, vesicular-arbuscular and sheathing mycorrhizae, or extensive proteoid root production). Proliferation of these taxa causes substantial changes to the nutrient status. The most notable feature in stands of alien *Acacia* and *Pinus* species is the altered distribution of nutrients (Musil & Midgley 1990; Witkowski 1991; Musil 1993). The greater productivity of alien stands relative to fynbos results in increased biomass which influences litterfall dynamics and thus the input of organic matter and nutrients to the soil (Stock & Allsopp 1992). Impacts on soil-nutrient budgets depend on soil type and the invading species involved. On clay-rich soils that support renosterveld, *A. saligna* greatly enhances organic matter and the concentration of total nitrogen, calcium, magnesium and phosphorus in litter and topsoil. These effects are less evident on sandy soils that support fynbos and dune thicket (Low 1988). The enhanced levels of organic matter are due to the higher carbon-gaining capacities of the acacias, whereas increased nitrogen levels presumably result from increased fixation of N_2 by the acacia symbionts. With regular fires and the volatilization of nitrogen and the enrichment of the soil surface with cations and phosphorus, it appears that the alien acacias create an environment that ensures the survival of early-succession nitrogen-fixing species such as themselves (Stock & Allsopp 1992; Richardson et al. 1995). Invading species of Hakea and Pinus do not fix nitrogen, and appear to have a less marked effect on soil properties (organic matter, pH, total nitrogen, total phosphorus) than do A. cyclops or A. saligna (Stock & Allsopp 1992).

Invasion by alien trees and shrubs, which results in marked increases in biomass (Van Wilgen & Richardson 1985; Versfeld & Van Wilgen 1986), has major impacts on hydrology. Studies in catchments planted with Pinus radiata suggest that interception losses due to invasion by alien trees and shrubs is about 20% of total annual rainfall. Judging from catchment experiments (Van Wyk 1987), runoff in catchments completely invaded by aliens could be reduced by between 30 and 70%, depending on the annual rainfall and the age and density of the alien stand. A recent modelling study suggests that invasion could result in an average decrease in water production from fynbos catchments of 347 cm^3 water ha^{-1} yr^{-1} over 100 years, resulting in an average loss of more than 30% of the water supply to the city of Cape Town. In individual years, where large areas are covered by mature trees, losses could be much greater (Le Maitre et al. 1996). Intense fires in dense alien stands (plantations, standing or felled thickets or self-sown invasive trees and shrubs) have had devastating effects, apparently as a result of greatly increased water repellency caused by, among other factors, the very large fuel loads. Effects are particularly severe when dense stands are felled (Richardson & Van Wilgen 1986).

Stands of alien trees and shrubs generally support sparse, species-poor faunas (Breytenbach 1986; Armstrong 1993), although many arboreal birds have invaded the fynbos from adjacent biomes in response to the proliferation of alien trees [reviewed by Macdonald & Richardson (1986)].

Many alien herbs have large ranges in the fynbos biome, but most are confined to severely disturbed sites, at least on nutrient-poor soils (Richardson et al. 1992). Some lowland communities have been invaded by alien herbs. The impacts of such invasions are less obvious than those of alien trees and shrubs. Vlok (1988) has shown that indigenous annuals and geophytes are severely threatened by alien annuals in several parts of the biome. Further research is needed to determine the magnitude of this threat.

Very few alien animals have invaded intact natural vegetation in the fynbos biome, but some species have nevertheless had a marked impact on the natural vegetation. Although mainly confined to disturbed sites, the Argentine ant poses a threat to the continued existence of many of the approximately 1200 fynbos plants with ant-dispersed seeds (see 6.4.2.2). This alien ant displaces native ants that bury seeds (thereby protecting them from granivores). Instead, the Argentine ant consumes elaiosomes on the soil surface, leaving them exposed to predation (Bond & Slingsby 1984).

Biological-control agents have been released and others are about to be released against most of the major alien invasive trees and shrubs in fynbos (Chap. 22, this volume). Mechanical control measures, involving felling stands and burning after 12–18 months, are probably the most intensive undertaken anywhere in the world. The control of invasive alien plants is by far the most expensive and time-consuming task for nature conservation authorities in the biome. Finding more effective and cheaper solutions to this escalating problem is probably the most urgent challenge facing conservationists in the biome.

6.7.3 Plant harvesting

Many proteoids, ericoids and restioids are harvested from the fynbos for a lucrative local and export wildflower trade (Greyling & Davis 1989; Chap. 20, this volume). The wildflower industry provides employment in areas that are otherwise unsuitable for agriculture, owing to low soil fertility and rainfall conditions. It is clearly important that harvesting is practised on a sustainable basis, in order to satisfy both economic and conservation considerations. Since the post-fire establishment of proteoid seedlings is directly related to the pre-fire seed reserves (Maze & Bond 1996), the major consideration when harvesting flowers of obligately reseeding species, is to ensure that enough seeds remain for population replacement after fire. On the basis of patterns of seed accumulation in serotinous Proteaceae, it is recommended that harvesting levels approximate the loss (22 to 51%) which would be incurred by an 'unseasonal' fire (one that occurs before the current crop has matured; Mustart & Cowling 1992b). Maze & Bond (1996) have developed a species-specific model that incorporates density-dependent plant fecundity, and seedling establishment and mortality levels; they suggest that the surplus of seedlings above the optimal adult reproductive densities can be harvested. Application of this model requires an understanding of the factors determining seedling mortality, and further work is required to elucidate these factors (see 6.4.2.4).

Plants with soil-stored seed banks, particularly ericoid taxa, are extremely sensitive to over-harvesting, since seed stores are characteristically small and population replacement relies heavily on seed from the current crop (Pierce & Cowling 1991; Kilian & Cowling 1992). Annual seed production is highly (and unpredictably) variable, and the size of soil-stored seed reserves is difficult and time-consuming to gauge. More research is needed before recommendations can be made for the sustainable harvesting of these taxa.

6.8   Conclusions

The past two decades have witnessed great advances in our understanding of vegetation patterns and processes in the fynbos biome. There are, however, still large gaps in our knowledge. We conclude this chapter by listing some of the more important ones.

(1) There is an inadequate understanding of the causal determinants of vegetation boundaries at all spatial scales. The results of such studies are an essential input for mechanistic models aimed at predicting the impact of global change on vegetation and species distributions (Bond 1996; Euston-Brown 1995).

(2) Because of the scale of the problem, more research is required on developing models that predict the rates of spread, and impacts on biodiversity and ecosystem processes, of invasive alien plants.

(3) More research is required on all aspects of the ecology of renosterveld and the dry fynbos communities of the northwest.

(4) Very little research has been done on the biology of facultative and obligate resprouters in fynbos and renosterveld. Since species in these groups are likely to persist in the face of global change (Bond 1996), it is important to understand their role in ecosystem functioning.

(5) More research is required on the autecology of restioids, ericoids (especially ericaceous ericoids) and fire ephemerals.

(6) Further research is needed to determine the magnitude of the threat posed by invasive alien herbs in lowland communities.

(7) More research is needed in order to make recommendations for sustainable harvesting practices of plants with soil-stored seed banks.

6.9   References

Armstrong, A.J. (1993). Is the 'inhospitable sea' assumption of island biogeographic theory applicable in the context of natural habitat patches within southwestern Cape pine plantations? *Environmental Forum Report*. Pretoria: Foundation for Research Development.

Bond, P. & Goldblatt, P. (1984). Plants of the Cape Flora – a descriptive catalogue. *Journal of South African Botany*, Supplementary Volume, 13, 1–455.

Bond, W. (1980). Fire and senescent fynbos in the Swartberg, Southern Cape. *South African Forestry Journal*, 114, 68–71.

Bond, W.J. (1981). *Vegetation Gradients in Southern Cape Mountains*. MSc Thesis. Cape Town: University of Cape Town.

Bond, W.J. (1983). On alpha diversity and the richness of the Cape flora: a study in southern Cape fynbos. In *Mediterranean-type Ecosystems. The Role of Nutrients*, ed. F.J. Kruger, D.T. Mitchell & J.U.M. Jarvis, pp. 225–43. Berlin: Springer-Verlag.

Bond, W.J. (1984). Fire survival of Cape Proteaceae – influence of fire season and seed predators. *Vegetatio*, 56, 65–74.

Bond, W.J. (1985). Canopy-stored seed reserves (serotiny) in Cape Proteaceae. *South African Journal of Botany*, 51, 181–6.

Bond, W.J. (1988). Proteas as tumbleseeds: wind dispersal through the air and over the soil. *South African Journal of Botany*, 54, 455–60.

Bond, W.J. (1989). Describing and conserving biotic diversity. In *Biotic Diversity in Southern Africa. Concepts and Conservation*, ed. B.J. Huntley, pp. 2–18. Cape Town: Oxford University Press.

Bond, W.J. (1994). Do mutualisms matter? Assessing the impact of pollinator and disperser disruption on plant extinction. *Philosophical Transactions of the Royal Society of London B*, 344, 83–90.

Bond, W.J. (1996). Predicting the impact of global biodiversity change: a case study in Cape Fynbos. In *Functional Types For Predicting The Effects of Global Change*, ed. H.H. Shugart & T.E. Smith. Cambridge: Cambridge University Press (in press).

Bond, W.J. & Breytenbach, G.J. (1985). Ants, rodents and seed predation in Proteaceae. *South African Journal of Zoology*, 20, 150–5.

Bond, W.J., Cowling, R.M. & Richards, M.B. (1992). Competition and coexistence. In *The Ecology of Fynbos. Nutrients, Fire and Diversity*, ed. R.M. Cowling, pp. 206–25. Cape Town: Oxford University Press.

Bond, W.J., Le Roux, D. & Erntzen, R. (1990). Fire intensity and regeneration of myrmecochorous Proteaceae. *South African Journal of Botany*, 56, 326–30.

Bond, W.J., Maze, K. & Desmet, P. (1995). Fire life histories and the seeds of chaos. *EcoScience*, 2, 252–60.

Bond, W.J. & Midgley, J.J. (1988). Allometry and sexual differences in leaf size. *American Naturalist*, 131, 901–10.

Bond, W.J. & Midgley, J.J. (1995). Kill thy neighbour: an individualistic argument for the evolution of flammability. *Oikos*, 73, 79–85.

Bond, W.J., Midgley, J.J. & Vlok, J. (1988). When is an island not an island? Insular effects and their causes in fynbos shrublands. *Oecologia*, 77, 512–21.

Bond, W.J. & Slingsby, P. (1983). Seed dispersal by ants in shrublands of the Cape Province and its evolutionary implications. *South African Journal of Science*, 79, 231–3.

Bond, W.J. & Slingsby, P. (1984). Collapse of an ant–plant mutualism: the Argentine ant (*Iridomyrmex humilis*) and myrmecochorous Proteaceae. *Ecology*, 65, 1031–7.

Bond, W.J. & Stock, W.D. (1989). The costs of leaving home: ants disperse myrmecochorous seeds to low nutrient sites. *Oecologia*, 81, 412–7.

Bond, W.J., Vlok, J. & Viviers, M. (1984). Variation in seedling recruitment in Cape Proteaceae after fire. *Journal of Ecology*, 72, 209–21.

Bond, W.J., Yeaton, R.I. & Stock, W.D. (1991). Myrmecochory in Cape fynbos. In *Ant–Plant Interactions*, ed. C. Huxley & D. Cutler, pp. 448–62. Oxford: Oxford University Press.

Bosch, J.M. (1984). *Water Yield and the Conservation of Fynbos Mountain Catchments*. Information Leaflet 10. Pretoria: Directorate of Forestry.

Bosch, J.M. & Hewlett, J.D. (1982). A review of catchment experiments to determine the effect of vegetation changes on water yield and evapotranspiration. *Journal of Hydrology*, 55, 3–23.

Bosch, J.M., Van Wilgen, B.W. & Bands, D.P. (1986). A model for comparing water yields from fynbos catchments burnt at different intervals. *Water SA*, 12, 191–202.

Boucher, C. (1987). *A Phytosociological Study of Transects Through the Western Cape Foreland, South Africa*. PhD Thesis. Stellenbosch: Stellenbosch University.

Boucher, C. & Moll, E.J. (1980). South African mediterranean shrublands. In *Mediterranean-Type Shrublands*, ed. F. di Castri, D.W. Goodall & R.L. Specht, pp. 233–48. Amsterdam: Elsevier.

Breytenbach, G.J. (1986). Impacts of alien organisms on terrestrial communities with emphasis on the south-western Cape. In *The Ecology and Management of Biological Invasions in Southern Africa*, ed. I.A.W. Macdonald, F.J. Kruger & A.A. Ferrar, pp. 229–38. Cape Town: Oxford University Press.

Brits, G.J. (1986a). Influence of fluctuating temperatures and H_2O_2 treatment on germination of *Leucospermum cordifolium* and *Serruria florida* (Proteaceae) seed. *South African Journal of Botany*, 52, 286–90.

Brits, G.J. (1986b). The effect of hydrogen peroxide treatment on germination in Proteaceae species with serotinous and nut-like achenes. *South African Journal of Botany*, 52, 291–3.

Brits, G.J., Calitz, F.J., Brown, N.A.C. & Manning, J.C. (1993). Desiccation as the active principle in heat-stimulated seed germination of *Leucospermum* R.Br. (Proteaceae) in fynbos. *New Phytologist*, 125, 397–403.

Brown, J.H. (1988). Species diversity. In *Analytical Biogeography. An Integrated Approach to the Study of Animal and Plant Distributions*, ed. A.A. Myers & P.S. Giller, pp. 57–89. London: Chapman & Hall.

Brown, N.A.C. (1993). Promotion of germination of fynbos seeds by plant-derived smoke. *New Phytologist*, 123, 575–83.

Brown, N.A.C., Kotze, G. & Botha, P.A. (1993). The promotion of seed germination of Cape *Erica* species by plant-derived smoke. *Seed Science & Technology*, 21, 573–80.

Campbell, B.M. (1983). Montane plant environments in the fynbos biome. *Bothalia*, 14, 283–98.

Campbell, B.M. (1985). A classification of the mountain vegetation of the fynbos biome. *Memoirs of the Botanical Survey of South Africa*, 50, 1–115.

Campbell, B.M. (1986). Montane plant communities of the fynbos biome. *Vegetatio*, 66, 3–16.

Campbell, B.M., Cowling, R.M., Bond, W. & Kruger, F.J. (1981). *Structural Characterization of Vegetation in the Fynbos Biome. South African National Scientific Programmes Report 52*. Pretoria: CSIR.

Campbell, B.M. & Van der Meulen, F. (1980). Patterns of plant species diversity in fynbos. *Vegetatio*, 43, 43–7.

Campbell, B.M. & Werger, M.J.A. (1988). Plant form in the mountains of the Cape, South Africa. *Journal of Ecology*, 76, 637–53.

Chesson, P.L. (1990). Geometry, heterogeneity and competition in variable environments. *Philosophical Transactions of the Royal Society of London B*, 330, 165–73.

Chesson, P.L. & Warner, R.R. (1981). Environmental variability promotes coexistence in lottery competitive systems. *American Naturalist*, 117, 923–43.

Cody, M.L. (1986). Diversity, rarity and conservation in mediterranean-climate regions. In *Conservation Biology. The Science of Scarcity and Rarity*, ed. M.E. Soulé, pp. 122–52. Sunderland, Massachusetts: Sinauer.

Cody, M.L. (1989). Growth-form diversity and community structure in desert plants. *Journal of Arid Environments*, 17, 199–209.

Cody, M.L. & Mooney, H.A. (1978). Convergence versus non-convergence in mediterranean-climate systems. *Annual Review of Ecology and Systematics*, 9, 265–321.

Coetzee, J.H. & Giliomee, J.H. (1985). Insects in association with the inflorescence of *Protea repens* (L.) (Proteaceae) and their role in pollination. *Journal of the Entomological Society of Southern Africa*, **48**, 303–14.

Coetzee, J.H. & Giliomee, J.H. (1987). Seed predation and survival in the infructescences of *Protea repens* (Proteaceae). *South African Journal of Botany*, **53**, 61–4.

Collins, B.G. & Rebelo, A.G. (1987). Pollination biology of the Proteaceae in Australia and southern Africa. *Australian Journal of Ecology*, **12**, 387–421.

Cowling, R.M. (1983a). Phytochorology and vegetation history in the south eastern Cape, South Africa. *Journal of Biogeography*, **10**, 393–419.

Cowling, R.M. (1983b). Diversity relations in Cape shrublands and other vegetation in the south eastern Cape, South Africa. *Vegetatio*, **45**, 103–27.

Cowling, R.M. (1984). A syntaxonomic and synecological study in the Humansdorp region of the fynbos biome. *Bothalia*, **15**, 175–227.

Cowling, R.M. (1987). Fire and its role in coexistence and speciation in Gondwana shrublands. *South African Journal of Science*, **83**, 106–11.

Cowling, R.M. (1990). Diversity components in a species-rich area of the Cape Floristic Region. *Journal of Vegetation Science*, **1**, 699–710.

Cowling, R.M. (ed.) (1992). *The Ecology of Fynbos: Nutrients, Fire and Diversity*. Cape Town: Oxford University Press.

Cowling, R.M. & Bond, W.J. (1991). How small can reserves be? An empirical approach in Cape fynbos, South Africa. *Biological Conservation*, **58**, 243–56.

Cowling, R.M. & Campbell, B.M. (1980). Convergence in vegetation structure in the mediterranean communities of California, Chile and South Africa. *Vegetatio*, **43**, 191–7.

Cowling, R.M. & Campbell, B.M. (1983). A comparison of fynbos and non-fynbos coenoclines in the lower Gamtoos River Valley, south-eastern Cape, South Africa. *Vegetatio*, **53**, 161–78.

Cowling, R.M. & Campbell, B.M. (1984). Beta diversity along fynbos and non-fynbos coenoclines in the lower Gamtoos River Valley, south-eastern Cape. *South African Journal of Botany*, **50**, 187–9.

Cowling, R.M., Campbell, B.M., Mustart, P., McDonald. D.J. Jarman, M.L. & Moll, E.J. (1988). Vegetation classification in a floristically complex area: the Agulhas Plain. *South African Journal of Botany*, **54**, 290–300.

Cowling, R.M., Gibbs Rusell, G.E., Hoffman, M.T. & Hilton-Taylor, C. (1989). Patterns of plant species diversity in southern Africa. In *Biotic Diversity in Southern Africa. Concepts and Conservation*, ed. B.J. Huntley, pp. 19–50. Cape Town: Oxford University Press.

Cowling, R.M. & Gxaba, T. (1990). Effects of a fynbos overstorey shrub on understorey community structure: implications for the maintenance of community-wide species richness. *South African Journal of Ecology*, **1**, 1–7.

Cowling, R.M. & Holmes, P.M. (1992). Flora and vegetation. In *The Ecology of Fynbos: Nutrients, Fire and Diversity*, ed. R.M. Cowling, pp. 23–61. Cape Town: Oxford University Press.

Cowling, R.M., Holmes, P.M. & Rebelo, A.G. (1992). Plant diversity and endemism. In *The Ecology of Fynbos: Nutrients, Fire and Diversity*, ed. R.M. Cowling, pp. 62–112. Cape Town: Oxford University Press.

Cowling, R.M., Kirkwood, D., Midgley, J.J. & Pierce, S.M. (1997). Invasion and persistence of subtropical thicket in Cape fynbos. *Journal of Vegetation Science* (in press).

Cowling, R.M., Mustart, P.J. Laurie, H. & Richards, M.B. (1994a). Species diversity, functional diversity and functional redundancy in fynbos communities. *South African Journal of Science*, **90**, 333–7.

Cowling, R.M. & Pierce, S.M. (1988). Secondary succession in coastal dune fynbos: variation due to site and disturbance. *Vegetatio*, **76**, 131–9.

Cowling, R.M., Pierce, S.M., Stock, W.D. & Cocks, M. (1994b). Why are there so many myrmecochorous species in the Cape fynbos? In *Plant–Animal Interactions in Mediterranean-Type Ecosystems*, ed. M. Arianoutsou & R.H. Groves, pp. 159–68. Dordrecht: Kluwer.

Cowling, R.M. & Witkowski, E.T.F. (1994). Convergence and non-convergence of plant traits in climatically and edaphically matched sites in Mediterranean Australia and South Africa. *Australian Journal of Ecology*, **19**, 220–32.

Deacon, H.J., Hendey, Q.B. & Lambrechts, J.J.N. (ed.) (1983). *Fynbos Palaeoecology: A Preliminary Synthesis*. South African National Scientific Programmes Report 75. Pretoria: CSIR.

Deacon, H.J., Jury, M.R. & Ellis, F. (1992). Selective regime and time. In *The Ecology of Fynbos: Nutrients, Fire and Diversity*, ed. R.M. Cowling, pp. 6–22. Cape Town: Oxford University Press.

Diamond, J.M. (1988). Factors controlling species diversity: overview and synthesis. *Annals of the Missouri Botanical Garden*, **75**, 117–29.

Ellery, W.N., Scholes, R.J. & Mentis, M.T. (1991). An initial approach to predicting the sensitivity of the South African grassland biome to climate change. *South African Journal of Science*, **87**, 499–503.

Esler, K.J. & Cowling, R.M. (1990). Effects of density on the reproductive output of *Protea lepidocarpodendron*. *South African Journal of Botany*, **56**, 29–33.

Esler, K.J., Cowling, R.M., Witkowski, E.T.F. & Mustart, P.J. (1989). Reproductive traits and accumulation of nitrogen and phosphorus during the development of fruits of *Protea compacta* R.Br. (calcifuge) and *Protea obtusifolia* Beuk. ex Meisn. (calcicole). *New Phytologist*, **112**, 109–15.

Euston-Brown, D. (1995). *Environmental and Dynamic Determinants of Vegetation Distribution in the Kouga and Baviaanskloof Mountains, Eastern Cape*. MSc Thesis. Cape Town: University of Cape Town.

Fägerstrom, T. (1988). Lotteries in communities of sessile organisms. *Trends in Ecology and Evolution*, **3**, 303–6.

Geldenhuys, C.J. (1989). *Environmental and Biogeographic Influences on the Distribution and Composition of the Southern Cape forests (Veld Type 4)*. PhD Thesis. Cape Town: University of Cape Town.

Geldenhuys, C.J. (1994). Bergwind fires and the location of forest patches in the southern Cape landscapes, South Africa. *Journal of Biogeography*, 21, 49–62.

Gentry, A.H. (1988). Change in plant community diversity and floristic composition on environmental and geographic gradients. *Annals of the Missouri Botanic Garden*, 75, 1–34.

Gibbs Russell, G.E. (1987). Preliminary floristic analysis of the major biomes in southern Africa. *Bothalia*, 17, 213–27.

Goldblatt, P. (1978). An analysis of the flora of southern Africa: its characteristics, relationships, and origins. *Annals of the Missouri Botanic Gardens*, 65, 369–436.

Good, R. (1974). *The Geography of Flowering Plants*. London: Longmans.

Greyling, T. & Davis, G.W. (ed.) (1989). *The Wildflower Resource: Commerce, Conservation and Research*. Terrestrial Ecosystems Section, Occasional Report No. 40. Pretoria: CSIR, Ecosystems Programme.

Grime, J.P. (1979). *Plant Strategies and Vegetation Processes*. Chichester: Wiley.

Grubb, P.J. (1977). The maintenance of species richness in plant communities and the importance of the regeneration niche. *Biological Reviews*, 52, 107–45.

Herrera, C.M. (1984). A study of avian frugivores, bird-dispersed plants, and their interaction in Mediterranean scrublands. *American Naturalist*, 140, 421–46.

Higgins, K.B., Lamb, A.J. & Van Wilgen, B.W. (1987). Root systems of selected plant species in mesic fynbos in the Jonkershoek Valley, South Western Cape Province. *South African Journal of Botany*, 53, 249–57.

Hoffman, M.T., Moll, E.J. & Boucher, C. (1987). Post-fire succession at Pella, a South African lowland fynbos site. *South African Journal of Botany*, 53, 370–4.

Hoffmann, A.J., Tellier, S. & Fuentes, E.R. (1989). Fruit and seed characteristics of woody species in mediterranean-type regions of Chile and California. *Revista Chilena de Historia Natural*, 62, 43–60.

Holmes, P.M. & Cowling, R.M. (1993). Effects of shade on seedling growth, morphology and leaf photosynthesis in six subtropical thicket species from the eastern Cape, South Africa. *Forest Ecology and Management*, 61, 199–220.

Honig, M.A., Linder, H.P. & Bond, W.J. (1992). Efficacy of wind pollination: pollen load size and natural gametophyte populations in wind-pollinated *Staberoha banksii* (Restionaceae). *American Journal of Botany*, 79, 443–8.

Huntley, B.J. (1992). The Fynbos Biome Project. In *The Ecology of Fynbos: Nutrients, Fire and Diversity*, ed. R.M. Cowling, pp. 1–5. Cape Town: Oxford University Press.

Huston, M. (1979). A general hypothesis of species diversity. *American Naturalist*, 133, 81–101.

Johnson, S.D. (1992a). Plant–animal relationships. In *The Ecology of Fynbos: Nutrients, Fire and Diversity*, ed. R. M. Cowling, pp. 174–205. Cape Town: Oxford University Press.

Johnson, S.D. (1992b). Climatic and phylogenetic determinants of flowering seasonality in the Cape flora. *Journal of Ecology*, 81, 567–72.

Johnson, S.D. & Bond, W.J. (1994). Red flowers and butterfly pollination in the fynbos of South Africa. In *Plant–Animal Interactions in Mediterranean-type Ecosystems*, ed. M. Arianoutsou & R.H. Groves, pp. 137–48. Dordrecht: Kluwer.

Jordaan, P.G. (1949). Aantekeninge oor die voortplanting en brandperiodes van *Protea mellifera* Thunb. *Journal of South African Botany*, 15, 121–5.

Keeley, J.E. (1991). Seed germination and life history syndromes in the Californian chaparral. *Botanical Review*, 57, 81–116.

Keeley, J.E. (1992). A Californian's view of fynbos. In *The Ecology of Fynbos. Nutrients, Fire and Diversity*, ed. R.M. Cowling, pp. 372–88. Cape Town: Oxford University Press.

Keeley, J. E. (1993). Smoke-induced flowering in the fire-lily *Cyrtanthus ventricosus*. *South African Journal of Botany*, 59, 638.

Kilian, D. & Cowling, R.M. (1992). Comparative seed biology and co-existence of two fynbos shrub species. *Journal of Vegetation Science*, 3, 637–46.

Killick, D.J.B. (1979). African mountain heathlands. In *Heathlands and Related Shrublands of the World*, ed. R.L. Specht, pp. 97–116. Amsterdam: Elsevier.

Knight, R.S. (1988). *Aspects of Plant Dispersal in the South-western Cape with Particular Reference to the Roles of Birds as Dispersal Agents*. PhD Thesis. Cape Town: University of Cape Town.

Kruger, F.J. (1977). Ecology of Cape fynbos in relation to fire. In *Proceedings of the Symposium on the Environmental Consequences of Fire and Fuel Management in Mediterranean Ecosystems*, ed. H.A. Mooney & C.E. Conrad. USDA Forest Service General Technical Report WO-3, 230–44.

Kruger, F.J. (1978). *A Description of the Fynbos Biome Project*. South African National Scientific Programmes Report 28. Pretoria: CSIR.

Kruger, F.J. (1979). South African heathlands. In *Heathlands of the World. A. Descriptive Catalogue*, ed. R.L. Specht, pp. 19–80. Amsterdam: Elsevier.

Kruger, F.J. (1983) Plant community diversity and dynamics in relation to fire. In *Mediterranean-type Ecosystems. The Role of Nutrients*, ed. F.J. Kruger, D.T. Mitchell & J.U.M. Jarvis, pp. 446–72. Berlin: Springer-Verlag.

Kruger, F.J. (1987). *Succession After Fire in Selected Fynbos Communities of the South-Western Cape*. PhD Thesis. Johannesburg: University of the Witwatersrand.

Kruger, F.J. & Bigalke, R.C. (1984). Fire in fynbos. In *Ecological Effects of Fire in South African Ecosystems*, ed. P. de V. Booysen & N.M. Tainton, pp. 69–114. Berlin: Springer-Verlag.

Kuo, J., Hocking, P.J. & Pate, J.S. (1982). Nutrient reserves in seed. of selected Proteaceous species from south western Australia. *Australian Journal of Botany*, 30, 231–49.

amb, A.J. & Klaussner, E. (1988). Response of the fynbos shrubs *Protea repens* and *Erica plukenetii* to low levels of nitrogen and phosphorus applications. *South African Journal of Botany*, 59, 558–64.

aurie, H. & Cowling, R.M. (1994). Lottery coexistence models extended to plants with disjoint generations. *Journal of Vegetation Science*, 5, 161–8.

aurie, H., Mustart, P.J. & Cowling, R.M. (1997). A shared niche? The case of the species pair *Protea obtusifolia* and *Leucadendron meridianum*. *Oikos* (in press).

e Maitre, D.C. (1988). Effects of season of burn on the regeneration of two Proteaceae with soil-stored seed. *South African Journal of Botany*, 54, 575–80.

e Maitre, D.C. (1990). The influence of seed aging on the plant on seed germination in *Protea neriifolia* (Proteaceae). *South African Journal of Botany*, 56, 49–53.

e Maitre, D.C. (1992). The relative advantages of seeding and sprouting in fire-prone environments: a comparison of life histories of *Protea neriifolia* and *Protea nitida*. In *Fire in South African Mountain Fynbos*, ed. B.W. Van Wilgen, D.M. Richardson, F.J. Kruger & H.J. Van Hensbergen, pp. 123–44. Berlin: Springer-Verlag.

e Maitre, D.C. & Brown, P.J. (1992). Life cycles and fire-stimulated flowering in geophytes. In *Fire in South African Mountain Fynbos*, ed. B.W. Van Wilgen, D.M. Richardson, F.J. Kruger & H.J. Van Hensbergen, pp. 145–60. Berlin: Springer-Verlag.

e Maitre, D.C. & Midgley, J.J. (1992). Plant reproductive ecology. In *The Ecology of Fynbos. Nutrients, Fire and Diversity*, ed. R.M. Cowling, pp. 135–74. Cape Town: Oxford University Press.

e Maitre, D.C., Van Wilgen, B.W., Chapman, R.A. & McKelly, D.H. (1996). Invasive plants and water resources in the Western Cape Province, South Africa: Modelling the consequences of a lack of management. *Journal of Applied Ecology*, 33, 161–72.

e Maitre, D.C., Van Wilgen, B.W. & Richardson, D.M. (1993). A computer system for catchment management: background, concepts and development. *Journal of Environmental Management*, 39, 121–42.

Linder, H.P., Meadows, M.E. & Cowling, R.M. (1992). History of the Cape flora. In *The Ecology of Fynbos. Nutrients, Fire and Diversity*, ed. R.M. Cowling, pp. 113–34. Cape Town: Oxford University Press.

Linder, H.P., Vlok, J.H., McDonald, D.J., Oliver, E.G.H., Boucher, C., Van Wyk, B.-E. & Schutte, A. (1993). The high altitude flora and vegetation of the Cape Floristic Region, South Africa. *Opera Botanica*, 121, 247–61.

Lindley, A.J., Bosch, J.M. & Van Wyk, D.B. (1988). Changes in water yield after fire in fynbos catchments. *Water SA*, 14, 7–12.

Low, A.B. (1988). Phytomass and major nutrient pools in an 11-year post-fire coastal fynbos community. *South African Journal of Botany*, 2, 98–104.

Macdonald, I.A.W. & Richardson, D.M. (1986). Alien species in terrestrial ecosystems of the fynbos biome. In *The Ecology and Management of Biological Invasions in Southern Africa*, ed. I.A.W. Macdonald, F.J. Kruger & A.A. Ferrar, pp. 77–91. Cape Town: Oxford University Press.

Manders, P.T. (1990a). Fire and other variables as determinants of forest/fynbos boundaries in the Cape Province. *Journal of Vegetation Science*, 1, 483–90.

Manders, P.T. (1990b). Soil seed banks and post-fire seed deposition across a forest–fynbos ecotone in the Cape Province. *Journal of Vegetation Science*, 1, 491–8.

Manders, P.T. & Richardson, D.M. (1992). Colonization of Cape fynbos communities by forest species. *Forest Ecology and Management*, 48, 277–93.

Manders, P.T., Richardson, D.M. & Masson, P.H. (1992). Is fynbos a stage in succession to forest? Analysis of the perceived ecological distinction between two communities. In *Fire in South African Mountain Fynbos*, ed. B.W. Van Wilgen, D.M. Richardson, F.J. Kruger & H.J. Van Hensbergen, pp. 81–107. Berlin: Springer-Verlag.

Maze, K.E. & Bond, W.J. (1996). Are *Protea* populations seed limited? Implications for wild flower harvesting in Cape fynbos. *Australian Journal of Ecology*, 21, 96–105.

McDonald, D.J., Cowling, R.M. & Boucher, C. (1996). Vegetation-environment relationships on a species-rich mountain range in the fynbos biome (South Africa). *Vegetatio*, 123, 165–82.

Midgley, J.J. (1987). *Aspects of the Evolutionary Ecology of the Proteaceae, with Emphasis on the Genus Leucadendron*. PhD Thesis. Cape Town: University of Cape Town.

Midgley, J.J. (1988). Mortality of Cape Proteaceae seedlings during their first summer. *South African Forestry Journal*, 145, 9–12.

Midgley, J.J. (1989). Season of burn of serotinous Proteaceae: a critical review and further data. *South African Journal of Botany*, 55, 165–70.

Milewski, A.V. (1983). A comparison of ecosystems in mediterranean Australia and southern Africa: nutrient-poor sites at the Barrens and the Caledon coast. *Annual Review of Ecology and Systematics*, 14, 57–76.

Miller, P.C. (1982). Some bioclimatic and pedologic influences on the vegetation in the mediterranean-type region of South Africa. *Ecologia Mediterranea*, 8, 141–56.

Milton, S.J. & Siegfried, W.R. (1981). Aboveground biomass of Australian acacias in the southern Cape, South Africa. *Journal of South African Botany*, 47, 701–16.

Mitchell, D.T. & Coley, P.G.F. (1987). Litter production and decomposition from shrubs of *Protea repens* growing in sand plain lowland and mountain fynbos, south-western Cape. *South African Journal of Botany*, 53, 25–31.

Mitchell, D.T., Coley, P.G.F., Webb, S. & Allsopp, N. (1986). Litterfall and decomposition processes in the coastal fynbos vegetation, south-western Cape, South Africa. *Journal of Ecology*, 74, 977–93.

Moll, E.J. & Bossi, L. (1984). Assessment of the natural vegetation of the fynbos biome of South Africa. *South African Journal of Science*, 80, 355–8.

Moll, E.J. & Romoff, N. (1983). Evidence of mist trapping by *Thamnocortus punctatus* at Pella in the south-western Cape Province. *South African Journal of Science*, 79, 432–5.

Mooney, H.A., Field, C., Gulmon, S.L., Rundel, P. & Kruger, F.J. (1983). Photosynthetic characteristics of South African sclerophylls. *Oecologia*, 58, 398–401.

Musil, C.F. (1991). Seed bank dynamics in sand plain lowland fynbos. *South African Journal of Botany*, 57, 131–42.

Musil, C.F. (1993). Effect of invasive Australian acacias on the regeneration, growth and nutrient chemistry of South African lowland fynbos. *Journal of Applied Ecology*, 30, 361–72.

Musil, C.F. & Midgley, G.F. (1990). The relative impact of invasive Australian acacias, fire and season on the soil chemical status of a sand plain lowland fynbos community. *South African Journal of Botany*, 56, 419–27.

Mustart, P.J. & Cowling, R.M. (1991). Seed germination of four serotinous Agulhas Plain Proteaceae. *South African Journal of Botany*, 57, 310–13.

Mustart, P.J. & Cowling, R.M. (1992a). Seed size: phylogeny and adaptation in two closely related Proteaceae species pairs. *Oecologia*, 91, 292–5.

Mustart, P.J. & Cowling, R.M. (1992b). Impact of flower and cone harvesting on seed banks and seed set of serotinous Agulhas Proteaceae. *South African Journal of Botany*, 58, 337–42.

Mustart, P.J. & Cowling, R.M. (1993). The role of regeneration stages in the distribution of edaphically restricted fynbos Proteaceae. *Ecology*, 74, 1490–9.

Mustart, P.J., Cowling, R.M. & Dunne, T.T. (1994) Reproductive traits of two closely related species pairs on adjacent, different soil types in South African Fynbos. *Vegetatio*, 111, 161–71.

Mustart, P.J., Cowling, R.M. & Wright, M.G. (1995). Clustering of fertile seed in infructescences of serotinous *Protea* species: an anti-predation mechanism? *African Journal of Ecology*, 33, 224–9.

Naveh, Z. & Whittaker, R.H. (1979). Structural and floristic diversity of shrublands and woodlands in northern Israel and other mediterranean areas. *Vegetatio*, 41, 171–90.

Newton, I.P., Cowling, R.M. & Lewis, O.A.M. (1991). Growth of calcicole and calcifuge Agulhas Plain Proteaceae on contrasting soil types, under glasshouse conditions. *South African Journal of Botany*, 57, 319–24.

Phillips, J.F.V. (1931). Forest succession and ecology in the Knysna region. *Memoirs of the Botanical Survey of South Africa*, 14, 1–327.

Pierce, S.M. & Cowling, R.M. (1991). Dynamics of soil-stored seed banks of six shrubs in fire-prone dune fynbos. *Journal of Ecology*, 79, 731–47.

Pierce, S.M., Esler, K. & Cowling, R.M. (1995). Smoke-induced germination of succulents (Mesembryanthemaceae) from fire-prone and fire-free habitats in South Africa. *Oecologia*, 102, 520–2.

Pierce, S.M. & Moll, E.J. (1994). Germination ecology of six shrubs in fire-prone Cape fynbos. *Vegetatio*, 110, 25–41.

Rebelo, A.G. (1987). Bird pollination in the Cape flora. In *A Preliminary Synthesis of Pollination Biology in the Cape Flora*, ed. A.G. Rebelo, pp. 83–108. *South African National Scientific Programmes Report* 141. Pretoria: CSIR.

Rebelo, A.G. (1992). Preservation of biotic diversity. In *The Ecology of Fynbos. Nutrients, Fire and Diversity*, ed. R.M. Cowling, pp. 309–44. Cape Town: Oxford University Press.

Rebelo, A.G., Cowling, R.M., Campbell, B.M. & Meadows, M.E. (1991). Plant communities of the Riversdale Plain. *South African Journal of Botany*, 57, 10–28.

Richards, M.B. (1994). *Soil Factors and Competition as Determinants of Fynbos Plant Species Distribution in the South-western Cape, South Africa*. PhD Thesis. Cape Town: University of Cape Town.

Richards, M.B., Cowling, R.M. & Stock, W.D. (1995a). Fynbos plant communities and vegetation–environment relationships in the Soetanysberg hills, Western Cape. *South African Journal of Botany*, 61, 298–305.

Richards, M.B., Stock, W.D. & Cowling, R.M. (1995b). Water relations of seedlings and adults of two fynbos *Protea* species in relation to their distribution patterns. *Functional Ecology*, 9, 575–83.

Richardson, D.M. & Cowling, R.M. (1992). Why is mountain fynbos invasible and which species invade? In *Fire in South African Mountain Fynbos*, ed. B.W. Van Wilgen, D.M. Richardson, F.J. Kruger & H.J. Van Hensbergen, pp. 161–81. Berlin: Springer-Verlag.

Richardson, D.M., Cowling, R.M., Bond, W.J., Stock, W.D. & Davis, G.D. (1995). Links between biodiversity and ecosystem function in the Cape Floristic Region. In *Mediterranean-type Ecosystems. The Function of Biodiversity*, ed. G.W. Davis & D.M. Richardson, pp. 285–333. Berlin: Springer-Verlag.

Richardson, D.M. & Kruger, F.J. (1990). Water relations and photosynthetic characteristics of selected trees and shrubs of riparian and hillside habitats in the south-western Cape Province, South Africa. *South African Journal of Botany*, 56, 214–25.

Richardson, D.M., Macdonald, I.A.W. & Forsyth, G.G. (1989). Reductions in plant species richness under stands of alien trees and shrubs in the fynbos biome. *South African Forestry Journal*, 149, 1–8.

Richardson, D.M., Macdonald, I.A.W., Holmes, P.M. & Cowling, R.M. (1992). Plant and animal invasions. In *The Ecology of Fynbos: Nutrients, Fire and Diversity*, ed. R.M. Cowling, pp. 271–308. Cape Town: Oxford University Press.

Richardson, D.M. & Van Wilgen, B.W. (1986). The effects of fire in felled *Hakea sericea* and natural fynbos and the implications for weed control in mountain catchments. *South African Forestry Journal*, 139, 4–14.

Richardson, D.M. & Van Wilgen, B.W. (1992). Ecosystem, community and species response to fire in mountain fynbos: conclusions from the Swartboskloof experiment. In *Fire in South African Mountain Fynbos*, ed. B.W. Van Wilgen, D.M. Richardson, F.J. Kruger & H.J. Van Hensbergen, pp. 161–81. Berlin: Springer-Verlag.

Richardson, D.M., Van Wilgen, B.W., Le Maitre, D.C. Higgins, K.B. & Forsyth, G.G. (1994). A computer-based system for fire management in the mountains of the Cape Province, South Africa. *International Journal of Wildland Fire*, 4, 17–32.

icklefs, R.E. & Schluter, D. (1993). Species diversity: regional and historical influences. In *Species Diversity in Ecological Communities. Historical and Geographic Perspectives*, ed. R.E. Ricklefs & D. Schulter, pp. 350–63. Chicago: University of Chicago Press.

uiters, C., McKenzie, B., Aalbers, J. & Raitt, L.M. (1993). Seasonal allocation of biomass and resources in the geophytic species *Haemanthus pubescens* subspecies *pubescens* in lowland coastal fynbos, South Africa. *South African Journal of Botany*, 59, 251–8.

uiters, C., McKenzie, B. & Raitt, L.M. (1992). Ontogenic and demographic studies of *Sparaxis grandiflora* subspecies *fimbriata* (Iridaceae). *South African Journal of Botany*, 58, 182–7.

utherford, M.C. & Westfall, R.H. (1986). Biomes of southern Africa. An objective characterization. *Memoirs of the Botanical Survey of South Africa*, 54, 1–98.

:ott, D.F. & Van Wyk, D.B. (1992). The effects of fire on soil water repellency, catchment sediment yields and streamflow. In *Fire in South African Mountain Fynbos*, ed. B.W. Van Wilgen, D.M. Richardson, F.J. Kruger & H.J. Van Hensbergen, pp. 216–39. Berlin: Springer-Verlag.

imida, A. & Ellner, S. (1984). Coexistence of plant species with similar niches. *Vegetatio*, 58, 29–55.

imida, A. & Wilson, M.V. (1985). Biological determinants of species diversity. *Journal of Biogeography*, 12, 1–20.

mmons, M. & Cowling, R.M. (1996). Why is the Cape Peninsula so rich in plant species? An analysis of the independent diversity components. *Biodiversity and Conservation*, 5, 551–73.

nith, R.E., Van Wilgen, B.W., Forsyth, G.G. & Richardson, D.M. (1992). Coexistence of seeders and sprouters in a fire-prone environment: the role of ecophysiology and soil moisture. In *Fire in South African Mountain Fynbos*, ed. B.W. Van Wilgen, D.M. Richardson, F.J. Kruger & H.J. Van Hensbergen, pp. 108–22. Berlin: Springer-Verlag.

pecht, R.L (1979). Heathlands and related shrublands of the world. In *Heathlands and Related Shrublands of the World*, ed. R.L. Specht, pp. 1–18. Amsterdam: Elsevier.

Specht, R.L. & Moll, E.J. (1983). Heathlands and sclerophyllous shrublands – an overview. In *Mediterranean-Type Ecosystems. The Role of Nutrients*, ed. F.J. Kruger, D.T. Mitchell & J.U.M. Jarvis, pp. 41–65. Berlin: Springer-Verlag.

Stock, W.D. & Allsopp, N. (1992). Functional perspectives of ecosystems. In *The Ecology of Fynbos. Nutrients, Fire and Diversity*, ed. R.M. Cowling, pp. 241–59. Cape Town: Oxford University Press.

Stock, W.D., Pate, J.S. & Delfs, J. (1990). Influence of seed size and quality on seedling development under low nutrient conditions in five Australian and South African members of the Proteaceae. *Journal of Ecology*, 78, 1005–20.

Stock, W.D., Van der Heyden, F. & Lewis, O.A.M. (1992). Plant structure and function. In *The Ecology of Fynbos. Nutrients, Fire and Diversity*, ed. R.M. Cowling, pp. 226–40. Cape Town: Oxford University Press.

Taylor, H.C. (1972). Notes on the vegetation of the Cape Flats. *Bothalia*, 10, 637–76.

Taylor, H.C. (1978) Capensis. In *Biogeography and Ecology of Southern Africa*, ed. M.J.A. Werger, pp. 171–230. The Hague: Junk.

Thanos, C.A. & Georghiou, K. (1988). Ecophysiology of fire-stimulated seed germination in *Cistus incanus* ssp. *creticus* (L.) Heywood and *C. salvifolius* L. *Plant, Cell Environment*, 11, 841–9.

Thwaites, R.N. & Cowling, R.M. (1988). Soil–vegetation relationships on the Agulhas Plain, South Africa. *Catena*, 15, 333–45.

Tilman, D. (1982). *Resource Competition and Community Structure*. Princeton: Princeton University Press.

Tilman, D. & Pacala, S. (1993). The maintenance of species richness in plant communities. In *Species Diversity in Ecological Communities. Historical and Geographic Perspectives*, ed. R.E. Ricklefs & D. Schulter, pp. 13–25. Chicago: University of Chicago Press.

Trollope, W.S.W. (1973). Fire as a method for controlling macchia (fynbos) vegetation on the Amatole mountains of the eastern Cape. *Proceedings of the Grassland Society of Southern Africa*, 8, 35–41.

van Daalen, J.C. (1981). The dynamics of the indigenous forest–fynbos ecotone in the southern Cape. *South African Forestry Journal*, 119, 14–23.

van der Heyden, F. (1988). *An Investigation of Photosynthetic Carbon Fixation in Fynbos Growth Forms and its Variation With Season and Environmental Conditions*. MSc Thesis. Cape Town: University of Cape Town.

van der Heyden, F. & Lewis, O.A.M. (1989). Seasonal variation in photosynthetic capacity with respect to plant water status of five species of the mediterranean climate region of South Africa. *South African Journal of Botany*, 55, 509–15.

van der Heyden, F. & Lewis, O.A.M. (1990). Environmental control of photosynthetic gas exchange characteristics of fynbos species representing three growth forms. *South African Journal of Botany*, 56, 654–8.

van der Zel, D.W. & Kruger, F.J. (1975). Results of the multiple catchment experiments at the Jonkershoek Research Station, South Africa. II. Influence of protection of fynbos on stream discharge in Langrivier. *Forestry in South Africa*, 16, 13–8.

van Hensbergen, H.J. Botha, A.A., Forsyth, G.G. & Le Maitre, D.C. (1992). Do small mammals govern vegetation recovery after fire in fynbos? In *Fire in South African Mountain Fynbos*, ed. B.W. Van Wilgen, D.M. Richardson, F.J. Kruger & H.J. Van Hensbergen, pp. 182–202. Berlin: Springer-Verlag.

van Reenen, C.A., Visser, G.J. & Loos, M.A. (1992). Soil microorganisms and activities in relation to season, soil factors and fire. In *Fire in South African Mountain Fynbos*, ed. B.W. Van Wilgen, D.M. Richardson, F.J. Kruger & H.J. Van Hensbergen, pp. 258–72. Berlin: Springer-Verlag.

van Staden, J. (1978). Seed viability in *Protea neriifolia*. 1. The effects of time of harvesting on seed viability. *Agroplantae*, 10, 65–7.

van Staden, J. & Brown, N.A.C. (1977). Studies on the germination of South African Proteaceae – a review. *Seed Science and Technology*, 5, 633–43.

van Wilgen, B. W. (1981). Some effects of fire frequency on fynbos plant community composition and structure at Jonkershoek, Stellenbosch. *South African Forestry Journal*, 118, 42–55.

van Wilgen, B.W. (1984). Adaptation of the United States Fire Danger Rating System to fynbos conditions. I. A fuel model for fire danger rating in the fynbos biome. *South African Forestry Journal*, 129, 61–5.

van Wilgen, B.W. (1987). Fire regimes in the fynbos biome. In *Disturbance and the Dynamics of Fynbos Biome Communities*, ed. R.M. Cowling, D.C. Le Maitre, B. McKenzie, R.P. Prŷs-Jones & B.W. Van Wilgen, pp. 6–14. *South African National Scientific Programmes Report* 135. Pretoria: CSIR.

van Wilgen, B.W., Bond, W.J. & Richardson, D.M. (1992a). Ecosystem management. In *The Ecology of Fynbos. Nutrients, Fire and Diversity*, ed. R.M. Cowling, pp. 345–71. Cape Town: Oxford University Press.

van Wilgen, B.W. & Forsyth, G.G. (1992). Regeneration strategies in fynbos plants and their influence on the stability of community boundaries after fire. In *Fire in South African Mountain Fynbos*, ed. B.W. Van Wilgen, D.M. Richardson, F.J. Kruger & H.J. Van Hensbergen, pp. 54–80. Berlin: Springer-Verlag.

van Wilgen, B.W., Higgins, K.B. & Bellstedt, D.U. (1990). The role of vegetation structure and fuel chemistry in excluding fire from forest patches in the fire-prone fynbos shrublands of South Africa. *Journal of Ecology*, 78, 210–22.

van Wilgen, B.W. & McDonald, D.J. (1992). The Swartboskloof experimental site. In *Fire in South African Mountain Fynbos*, ed. B.W. Van Wilgen, D.M. Richardson, F.J. Kruger & H.J. Van Hensbergen, pp. 1–20. Berlin: Springer-Verlag.

van Wilgen, B.W. & Richardson, D.M. (1985). The effects of alien shrub invasions on vegetation structure and fire behaviour in South African fynbos shrublands: a simulation study. *Journal of Applied Ecology*, 22, 955–66.

van Wilgen, B.W., Richardson, D.M., Kruger, F.J. & Van Hensbergen, H.J. (eds) (1992b). *Fire in South African Mountain Fynbos*. Berlin: Springer-Verlag.

van Wilgen, B.W., Richardson, D.M. & Seydack, A.H.W. (1994). Managing fynbos for biodiversity: constraints and options in a fire-prone environment. *South African Journal of Science*, 90, 322–9.

van Wilgen, B.W. & Viviers, M. (1985). The effect of season of fire on serotinous Proteaceae in the western Cape and the implications for fynbos management. *South African Forestry Journal*, 133, 49–53.

van Wyk, D.B. (1977). *Die Invloed van Bebossing met Pinus radiata op die Totale Jaarlikse Afvoer van die Jonkershoek Strome*. MSc Thesis. Stellenbosch: Stellenbosch University.

van Wyk, D.B. (1987). Some effects of afforestation on streamflow in the Western Cape Province, South Africa. *Water SA*, 13, 31–6.

van Wyk, D.B., Lesch, W. & Stock, W.D. (1992). Fire and catchment chemical budgets. In *Fire in South African Mountain Fynbos*, ed. B.W. Van Wilgen, D.M. Richardson, F.J. Kruger & H.J. Van Hensbergen, pp. 240–57. Berlin: Springer-Verlag.

Versfeld, D.B., Richardson, D.M., Van Wilgen, B.W., Chapman, R.A. & Forsyth, G.G. (1992). Climate of Swartboskloof. In *Fire in South African Mountain Fynbos*, ed. B.W. Van Wilgen, D.M. Richardson, F.J. Kruger & H.J. Van Hensbergen, pp. 21–36. Berlin: Springer-Verlag.

Versfeld, D.B. & Van Wilgen, B.W. (1986). Impacts of woody aliens on ecosystem properties. In *The Ecology and Management of Biological Invasions in Southern Africa*, ed. I.A.W. Macdonald, F.J. Kruger & A.A. Ferrar, pp. 239–46, Cape Town: Oxford University Press.

Vlok, J.H.J. (1988). Alpha diversity of lowland fynbos herbs at various levels of infestation by alien annuals. *South African Journal of Botany*, 54, 623–7.

Vogel, J.C., Fuls, A. & Ellis, R.P. (1978). The geographical distribution of Krantz grasses in South Africa. *South African Journal of Science*, 74, 9–15.

Von Willert, D.J., Herppich, M. & Miller, J.M. (1989). Photosynthetic characteristics and leaf water relations of mountain fynbos vegetation in the Cedarberg area (South Africa). *South African Journal of Science*, 55, 288–98.

Whittaker, R.H. (1977). Evolution of species diversity in land communities. *Evolutionary Biology*, 10, 1–67.

Wicht, C.L. (1948). Hydrology research in South Africa. *Journal of the South African Forestry Association*, 6, 4–21.

Wicht, C.L. (1967). Forest hydrology research in the South African Republic. In *International Symposium on Forest Hydrology*, ed. W.E. Sopper & W. Lull, pp. 75–84. Oxford: Pergamon.

Wiens, D., Rourke, J.P., Casper, B.B., Rickart, E.A., Lapine, T.R., Peterson, C.J. & Channing, A. (1983). Nonflying mammal pollination of southern African proteas: a coevolved system. *Annals of the Missouri Botanical Garden*, 70, 1–31.

Witkowski, E.T.F. (1988). *Response of a Sand-plain Lowland Fynbos Ecosystem to Nutrient Additions*. PhD Thesis. Cape Town: University of Cape Town.

Witkowski, E.T.F. (1989). Effects of nutrient additions on litter production and nutrient return in a nutrient-poor Cape fynbos ecosystem. *Plant and Soil*, 117, 227–35.

Witkowski, E.T.F. (1991). Effects of invasive alien Acacias on nutrient cycling in the coastal lowlands of the Cape fynbos. *Journal of Applied Ecology*, 28, 1–15.

Witkowski, E.T.F. & Mitchell, D.T. (1989). The effects of nutrient additions on above-ground phytomass and its phosphorus and nitrogen contents of sand-plain lowland fynbos. *South African Journal of Botany*, 55, 243–9.

Witkowski, E.T.F., Mitchell, D.T. & Stock, W.D. (1990). Response of a Cape fynbos ecosystem to nutrient additions: shoot growth and nutrient contents of a proteoid (*Leucospermum parile*) and an ericoid (*Phylica cephalantha*) evergreen shrub. *Acta Oecologica*, 11, 311–26.

Yeaton, R.I. & Bond, W.J. (1991). Competition between two shrub species: dispersal differences and fire promote coexistence. *American Naturalist*, 138, 328–41.

Succulent karoo

7

S.J. Milton, R.I. Yeaton, W.R.J. Dean and J.H.J. Vlok

7.1   Introduction

The succulent karoo with >5000 species in 100 251 km^2 has the highest species richness recorded for semi-arid vegetation, and more than 50% of the plant species are endemic to this biome (Cowling *et al.* 1989; Chap. 3, this volume). Growth-form spectra for the succulent karoo are unusual for semi-arid regions of the world, in the predominance of chamaephytes and geophytes and scarcity of tall shrubs, trees and grasses (Evenari, Noy-Meir & Goodall 1985). Perhaps the most unusual feature of the biome is the enormous concentration of leaf-succulent, low to dwarf shrubs (Werger 1985; Jürgens 1986). Most of the species in the biome's two largest families, Mesembryanthemaceae and Crassulaceae, are leaf-succulents. The former, the richest family in the southern African flora (>2000 spp.), is centred in the succulent karoo (Jürgens 1986), which is also a centre of diversity for African Crassulaceae (Tölken 1985). The region is unusually rich in geophytic and leaf-succulent, petaloid monocots (Goldblatt 1978; Cowling *et al.* 1989; Fig. 7.1).

In this chapter we first discuss the environmental and biotic determinants of boundaries between the succulent karoo and the neighbouring desert, fynbos, Nama-karoo and savanna biomes; between structurally different vegetation types within the biome; and between the plant communities that constitute these vegetation types. The morphological, behavioural (phenology, reproduction) and physiological adaptations of typical succulent karoo growth-forms are discussed in relation to their environment. The section dealing with community diversity and dynamics provides a framework for interpreting the effects of grazing, disturbance and climatic change on the composition and structure of plant

Figure 7.1 **The hallmark of the succulent karoo is the high diversity and strong dominance of dwarf to low, leaf-succulent shrubs, especially Mesembryanthemaceae. Shown here is the mesemb *Astridia velutina* and other succulents in the Richtersveld, near the Namibian border** (Photo: D.M. Richardson).

communities in the succulent karoo. Finally, we discuss current management and conservation issues and identify gaps in the present understanding of the ecology of the succulent karoo.

7.2   Boundaries at the biome, landscape and community scales

7.2.1   Biome-scale boundaries

The succulent karoo comprises coastal plains and intermontane valleys lying along the western and southern edges of the Great Escarpment and mostly <1000 m above sea level (Fig. 7.2a). It is bounded by desert (Chap. 9, this volume), Nama-karoo (Chap. 8, this volume), fynbos (Chap. 6, this volume) and subtropical thicket

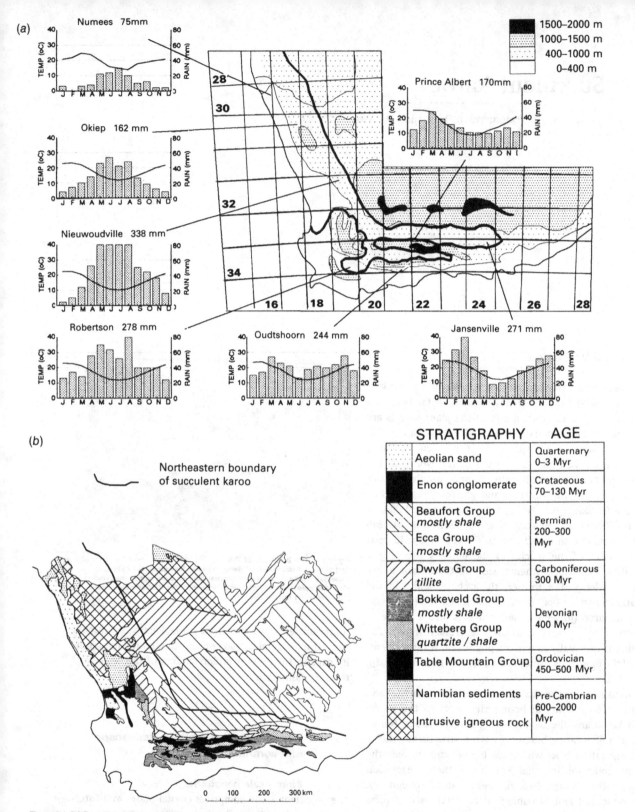

Figure 7.2 (a) Boundary of the succulent karoo biome relative to latitude, altitude and climate. (b) The complex geology of the succulent karoo includes substrata of aeolian, igneous and sedimentary origin ranging in age from 3 Myr to 2000 Myr. After Visser (1986). (c) Vegetation types (Veld Types) of the succulent karoo based on work by Acocks (1953). The northeastern margins of VT33, 39, 28 and 29 lie outside the climatic definition of the succulent karoo.

(c)

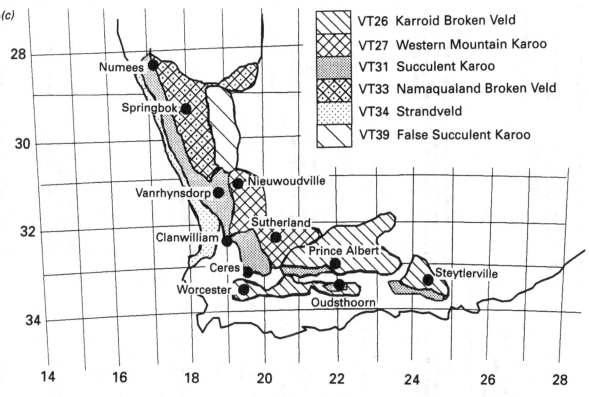

VT26 Karroid Broken Veld
VT27 Western Mountain Karoo
VT31 Succulent Karoo
VT33 Namaqualand Broken Veld
VT34 Strandveld
VT39 False Succulent Karoo

Figure 7.2 (cont.)

(Chap. 12, this volume). Climatically, the biome is characterized by low (20–290 mm yr⁻¹) but fairly reliable (coefficient of variation <50%) annual rainfall, of which >40% falls in the winter half of the year (Fig. 7.2*a*), and by relatively high summer aridity.

The duration and temperature of the growing season clearly separates succulent karoo (short, cool) from fynbos (long, cool), Nama-karoo (short, warm) and sav-

Figure 7.3 **The succulent karoo and desert biomes interdigitate in a complex way in the Namaqualand Broken Veld (VT33) of the Richtersveld. Here the uplands, which receive orographic precipitation associated with winter fronts, support succulent karoo, whereas the drier valleys are desert-like. Note the northward-leaning *Pachypodium namaquanum* on the right**
(Photo: D.M. Richardson).

anna (long, warm) in a climatic model developed by Ellery, Scholes & Mentis (1991). The desert biome to the north has a lower and less reliable rainfall than the succulent karoo. On its eastern boundary the succulent karoo interfaces with subtropical thicket (termed Spekboomveld) dominated by the arborescent succulent *Portulacaria afra*. In the strongly winter-rainfall area of the succulent karoo this thicket vegetation is largely restricted to warmer, north-facing slopes (Levyns 1950; Acocks 1953; Campbell 1985; Fig. 7.3).

The fynbos biome to the south has more rainfall and cooler summers than the succulent karoo. Even within the boundaries of the succulent karoo, on the tops and southern slopes of isolated hills in the Little Karoo (Levyns 1950; Campbell 1985), and on the Kamiesberg inselberg near Okiep (Adamson 1938), lower temperatures and improved moisture availability lead to the development of fynbos and renosterveld. In addition to climatic factors, geology (Fig. 7.2*b*) and soils distinguish the succulent karoo from the fynbos. Whereas nutrient-poor sandstones of the Table Mountain Group underlie much of the fynbos (Chap. 6, this volume), the succulent karoo occurs on intrusive igneous rock and on soils derived from fine-grained sedimentary rocks and on recent alluvial deposits (see Chap. 1, this volume). Soils of the succulent karoo are generally finer-grained, less

Table 7.1 Structural, floristic and environmental characteristics of major vegetation communities in the succulent karoo. Terminology follows Acocks (1953). VT and associated number refer to Acocks' Veld Types

Community vegetation structure	Dominant perennial taxa	Environment	References
VT34 Strandveld Dense to open shrubland; medium height (0.5–1.5 m); sclerophylls, evergreen and deciduous shrubs, grasses and Restionaceae; annual understorey; biomass 3500–8250 g m^{-2}; dry mass 410–1050 g m^{-2}	**Succulents** Aloe, Euphorbia, Ruschia, Tetragonia, Tylecodon **Non-succulents** Chrysanthemoides, Ehrharta, Euclea, Galenia, Grielum, Lebeckia, Lycium, Limonium, Pelargonium, Pteronia, Putterlickia, Nylandtia, Rhus, Salvia, Wiborgia, Willdenowia, Zygophyllum	Alt. 0–200 m; coastal forelands, dunes, limestone and granite outcrops; deep aeolian sand; rainfall 50–300 mm (winter)	Boucher & Jarman (1977) Werger & Morris (1991)
VT31a Succulent Karoo, Namaqualand coastal belt Sparse shrubland; dwarf (<0.4 m); high diversity of succulents, geophytes, therophytes; taller mixed shrubland with grasses on sand; biomass 810 g m^{-2}	**Succulents** Argyroderma, Conophytum, Cotyledon, Crassula, Drosanthemum, Euphorbia, Lampranthus, Monilaria, Oophytum, Rhinephyllum, Ruschia, Sphalmanthus, Sceletium **Non-succulents** Androcymbium, Bulbine, Chaetobromus, Didelta, Ehrharta, Galenia, Hoplophyllum, Lebeckia, Pteronia, Salsola, Zygophyllum	Alt. 0–450 m; flat to hilly; tillite, calcrete metamorphic, quartz; shallow loam/sand pebble pavements; rainfall 50–150 mm (winter)	Jürgens (1986) Milton (1978) Werger & Morris (1991)
VT31b Succulent Karoo, Valley of Tanqua and Doorn rivers Very sparse shrubland; dwarf (<0.3 m); succulents on shallow soil; grass and ephemerals on sandy alluvium	**Succulents** Augea, Cephalophyllum, Crassula, Pleiospilos, Psilocaulon, Hereroa, Rhinephyllum, Ruschia, Sphalmanthus, Sceletium, Tetragonia **Non-succulents** Acacia karroo, Galenia, Hermannia, Lycium, Osteospermum, Pteronia, Salsola, Stipagrostis, Tamarix, Zygophyllum	Alt. 300–450 m; plains, ridges; skeletal soils over tillite, shale and calcrete; alluvium along major rivers; rainfall 100–150 mm; (autumn–winter)	Lane (1978) Jürgens (1986) Boobyer (1989)
VT31c Succulent Karoo, Steytlerville karoo Medium to dense shrubland; low (0.2–0.5 m); plains: succulent shrubs, creeping succulents, non-succulent shrubs, grasses; slopes: trees and tall shrubs (1–3 m)	**Succulents** Augea, Aloe, Crassula, Delosperma, Drosanthemum, Euphorbia, Malephora, Portulacaria, Psilocaulon, Ruschia, Sphalmanthus, Tetragonia **Non-succulents** Eragrostis, Eriocephalus, Euryops, Galenia, Lycium, Pappea, Pentzia, Protasparagus, Salsola, Tragus, Zygophyllum	Alt. 200–600 m; silty alluvium in valley bottoms; ridges of tilliteor Bokkeveld shale; rainfall 150–250 mm (summer to even)	Hoffman (1989) Milton et al. (1995)
VT33 Namaqualand Broken Veld Sparse succulent dwarf shrubland; with emergent trees slopes: tree-succulents; plains: low succulents; drainages: grasses, trees non-succulent shrubs	**Succulents** Aloe dichotoma, Cotyledon, Crassula, Euphorbia, Othonna, Pachypodium namaquanum, Pelargonium, Sarcocaulon, Ruschia, Tylecodon	Alt. 300–1350 m; broken topography; igneous rock and recent sediments; rainfall 150–300 mm (winter or autumn)	Jürgens (1986) Werger (1978) Van Rooyen et al. (1990)

Non-succulents
Acacia karroo, Didelta, Euclea, Ficus, Hermbstaedtia, Maytenus, Microloma, Monechma, Peronia, Rhus, Senecio, Stipagrostis, Sisyndite

Le Roux (1984)

Struck (1994)

VT39 False Succulent Karoo

Sparse, patchy; low to medium (0.2–1.0 m); succulent to non-succulent shrubland on shallow soil; desert grassland on aeolian sand

Succulents
Augea, Brownanthus, Eberlanzia, Drosanthemum, Hereroa, Conicosia, Psilocaulon, Ruschia, Sphalmanthus

Non-succulents
Aptosimum, Salsola, Stipagrostis obtusa, Stipagrostis ciliata, Centropodia, Lycium

Alt. 1000 m; flat to undulating; granite, metamorphic and tillite rocks calcrete, alluvium aeolian sand; rainfall 100–150 mm; (summer to even)

Lloyd (1985)

Lloyd (1989)

VT28b Western Mountain Karoo, Low-altitude form

Dense, fairly homogeneous; low (<50 cm); succulent to non-succulent shrubland on skeletal soil; desert grassland on aeolian sand; biomass 1500 g m^{-2}; dry mass 520 g m^{-2}

Succulents
Brownanthus, Drosanthemum, Euphorbia, Leipoldtia, Malephora, Ruschia, Sphalmanthus, Tylecodon

Non-succulents
Chrysocoma, Eriocephalus, Galenia, Hirpicium, Lycium, Hermannia, Osteospermum, Merxmuellera, Pentzia, Protasparagus, Pteronia, Rosenia, Salsola, Walafrida, Zygophyllum

Alt. 600–1000 m; undulating topography; skeletal soils over Beaufort shale; rainfall 150 mm; winter–non-seasonal

Snijman & Perry

Werger & Morris (1991)

VT26a Karroid Broken Veld, Great Karoo form

Sparse to moderately dense; low to medium (<0.5–1.5 m); mixed shrublands; succulent to non-succulent shrubland on plains; grasses, taller shrubs on ridges; desert grasses, trees in sandy drainages; biomass 500–900 g m^{-2}; drymass 200–400 g m^{-2}

Succulents
Aloe, Brownanthus, Conophytum, Crassula, Cotyledon, Delosperma, Drosanthemum, Euphorbia, Glottiphyllum, Haworthia, Pelargonium, Ruschia, Sphalmanthus

Non-succulents
Aristida, Acacia karroo, Cadaba, Chrysocoma, Digitaria, Diospyros, Enneapogon, Eriocephalus, Galenia, Garuleum, Hermannia, Lycium, Osteospermum, Pentzia, Pteronia, Rhigozum, Rhus, Stipagrostis, Zygophyllum

Alt. 500–900 m; plains and ridges skeletal soil over Tillite, Ecca and Beaufort shales; rainfall 100–150 mm; (autumn–winter)

Compton (1929)

Compton (1931)

Milton (1990a, b)

Milton et al. (1992)

Yeaton & Esler (1990)

VT26b Karroid Broken Veld, Little Karoo form

Dense; medium tall (0.5–1.5 m); succulent to non-succulent shrubland; creeping succulents in valley bottoms; taller shrubs and trees on heuweltjies and hillsides; drymass 765 g m^{-2}

Succulents
Aloe, Conophytum, Crassulaceae, Drosanthemum, Gibbaeum, Euphorbia, Haworthia, Lampranthus, Othonna, Portulacaria, Ruschia, Senecio, Tetragonia

Non-succulents
Ehrharta, Enneapogon, Eriocephalus, Euclea, Galenia, Hirpicium, Osteospermum, Pteronia, Nymania, Pentzia, Rhigozum, Rhus

Alt. 150–250 m; broken topography; soils skeletal to deep and loamy to sandy on Bokkeveld sediment and Enon conglomerate; rainfall 250–300 mm; (winter)

Boshoff (1989)

Olivier (1966)

Joubert (1968)

Rutherford (1978)

Smitheman & Perry (1990)

Stokes (1994)

Wood (1991)

Figure 7.4 **Towards the boundary of the succulent karoo and Nama-karoo biomes, as shown here at Tierberg on the southern fringe of the Great Karoo, the succulent component is subordinate. The dominant non-succulent shrub is *Pteronia pallens* (Asteraceae); *Brownanthus ciliatus* (Mesembryanthemaceae), a dwarf succulent shrub, occupies the interstices. Note the drainage line woodland dominated by *Acacia karroo* (Fabaceae)** (Photo: R.M. Cowling).

leached, with higher pH values (>7) and higher cation exchange capacities than fynbos soils (Ellis & Lambrechts 1986; Chap. 6, this volume).

Higher minimum temperatures and predictable cool-season precipitation distinguish the succulent karoo from the Nama-karoo of the inland plateau, which has frosty winters (Cowling *et al.* 1994). Most of the precipitation that reaches the succulent karoo is brought by fog and predictable cyclonic systems (Schulze & McGee 1978), whereas equally arid sites in the Nama-karoo receive relatively variable convectional rainfall during summer. Thunderstorms (see Chap. 2, this volume) and hail are therefore rare in the succulent karoo (Rutherford & Westfall 1986; Fig. 7.4).

Management-induced changes can confound the separation of succulent and Nama-karoo biomes on physiognomic grounds. The dwarf shrublands of the succulent karoo occur at lower altitudes and on finer, more saline/alkaline soils, and are usually more succulent and less grassy than those of the Nama-karoo. However, both grazing and aridity decrease grassiness and increase the proportion of succulents in karoo shrublands (Acocks 1953; Roux & Vorster 1983; Milton 1990a; Stokes 1994). Soil isotope evidence (Bond, Stock & Hoffman 1994) supports the notion proposed by Acocks (1953) that some karoo shrublands were more grassy before the advent of settled pastoralism. As a result of vegetation changes in response to land use and inter-annual variation in rainfall seasonality (Hoffman, Barr & Cowling 1990), boundaries between the succulent and Nama-karoo biomes are fluid and blurred (Fig. 7.4). In this chapter, the Acocks

(1953) Veld Types, False Succulent Karoo and parts of the Western Mountain Karoo and Karroid Broken Veld have been included here on physiognomic grounds. On climatic grounds, these Veld Types are more strongly linked to the Nama-karoo (Chap. 8, this volume).

7.2.2 Landscape-scale boundaries

Six Veld Types described by Acocks (1953) fall within the climatic definition of the succulent karoo biome, or are characterized by the presence of leaf-succulents (Fig. 7.2c). These vegetation units are separated by physiognomic and floristic characteristics related to differences in soil, rainfall, altitude and topography, as well as geographical position. Details of characteristic physiognomy, genera, environmental factors and references to phytosociological studies are given in Table 7.1.

The most succulent of these dwarf shrublands, Succulent Karoo in the strict sense (Veld Type (VT) 31), occurs only at low altitudes and usually on fine-grained soils with high conductivity. Three forms of Succulent Karoo (VT31a, b and c) are separated on the basis of rainfall seasonality: all are sparse, low and dominated by Mesembryanthemaceae (Figs. 7.1 and 7.3). Abundance of leaf succulents generally decreases eastwards towards the summer rainfall zone and, for this reason, Acocks (1953) considered that the predominance of Mesembryanthemaceae in the Steytlerville area (VT31c) and in the False Succulent Karoo (VT39) was a response to rangeland mismanagement.

Strandveld (VT34) is generally taller than Succulent Karoo (Fig. 7.5) and is restricted to deep, coarse aeolian sediments of the winter-rainfall west coast. Succulence decreases with an increase in altitude (Werger & Morris 1991). Thus, Western Mountain Karoo (VT28) on the central plateau, is a predominantly asteraceous shrubland, with a relatively low diversity of leaf succulents.

Hilly topography and tall shrubs and trees differentiate Broken Veld vegetation types from low shrublands on fairly homogeneous plains (VT26, 31, 34, 39; Fig. 7.6). The Namaqualand Broken Veld (VT33) shows the influence of adjacent desert and arid savanna floras, whereas the Karroid Broken Veld (VT26) in valleys among fynbos-clad mountains, includes many fynbos-centred genera (Levyns 1950; Gibbs Russell 1987; Fig. 7.7). Grassiness in this Veld Type increases from west to east with an increase in the proportion of summer rainfall (Hilton-Taylor 1986), but may also vary in response to inter-annual fluctuations in rainfall seasonality and amount (Hoffman *et al.* 1990).

7.2.3 Community-scale boundaries

Within a landscape, abrupt boundaries between plant communities may occur along edaphic discontinuities,

Figure 7.5 **This patch of Succulent Karoo (VT31), near the Atlantic coastline west of Vanrhynsdorp, is entirely dominated by succulents belonging to many genera including *Drosanthemum, Euphorbia, Othonna, Ruschia* and *Zygophyllum*** *(Photo: D.M. Richardson).*

Figure 7.6 **The shale ridges in the Karroid Broken Veld (VT26) of the Little Karroo support diverse assemblages of leaf-succulent Crassulaceae. Shown here is *Crassula rupestris* and *Cotyledon orbiculata*** *(Photo: D.M. Richardson).*

Figure 7.7 **The eastern fringes of the Karroid Broken Veld (VT26) include a wide array of succulent growth forms. Shown here in the Baviaanskloof valley of the Eastern Cape are *Euphorbia atrispina* (foreground), *Portulacaria afra* (middleground) and *Aloe ferox* (background)** *(Photo: J.C. Paterson-Jones).*

on opposing aspects of slopes, between drainage lines or pans and their watersheds, and at the edges of zoogenically modified soil patches.

Interfaces between the plant communities associated with substrata differing in pH and conductivity are often clearly demarcated. In the Succulent Karoo of the Knersvlakte near Vanrhynsdorp, shallow soils on the steepest slope of a hill have higher conductivity and lower pH values than deeper soils on the hill top or in the valley bottom (Jürgens 1986). Soil zonation is reflected in the distribution of growth-forms (Fig. 7.8). Dwarf succulents are confined to the shallow, saline soils of the slope, whereas shrubby communities develop on deeper soils. There is almost complete compositional change (or high beta diversity) along this edaphic gradient (Cowling *et al.* 1989).

Ordination of vegetation sample plots from Vaalputs on the boundary between Namaqualand Broken Veld and False Succulent Karoo (Fig. 7.9) clustered leaf-succulent assemblages (*Brownanthus, Eberlanzia, Ruschia*) on shallow saline soils, clearly separating them from non-succulent shrublands (*Aptosimum, Salsola*) on sandy soil and calcrete and grasslands (*Stipagrostis*) on aeolian sands (Lloyd 1989).

In the Karroid Broken Veld at Tierberg, a non-succulent shrub (*Pteronia pallens*: Asteraceae) replaced other *Pteronia* species on calcium and nutrient-rich soils (Esler & Cowling 1993). Nearest-neighbour distances between *Pteronia* plants at this site indicate that community boundaries may be maintained by resource- or stress-mediated competition (Esler & Cowling 1993).

Aspect strongly influences plant communities in broken terrain. In the Richtersveld form of Namaqualand Broken Veld, a sparse cover (2%) of succulent shrubs

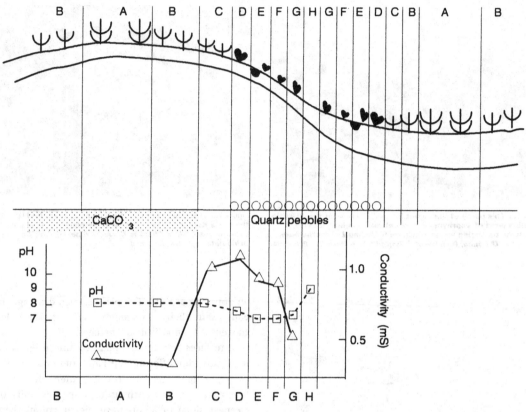

Figure 7.8 **Zonation of plant growth forms along a gradient of diminishing soil depth in the Knersvlakte (VT31a) north of Vanrhynsdorp. Leaf succulents dominate all zones, but differ in growth form as follows: A, tall shrubs; B and C, low shrubs; D, contracted; E and F, reduced and embedded; G, dwarf annuals. After Jürgens (1986).**

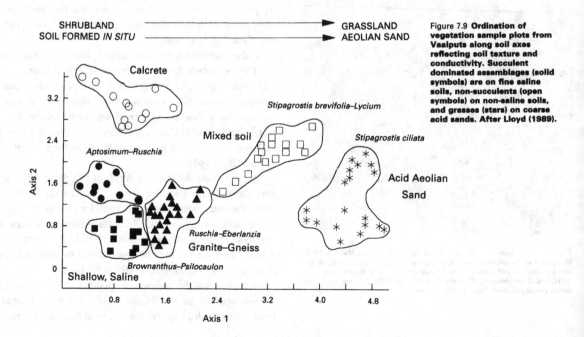

Figure 7.9 **Ordination of vegetation sample plots from Vaalputs along soil axes reflecting soil texture and conductivity. Succulent dominated assemblages (solid symbols) are on fine saline soils, non-succulents (open symbols) on non-saline soils, and grasses (stars) on coarse acid sands. After Lloyd (1989).**

(a)

(b)

Figure 7.10 *Heuweltjies* or Mima-like mounds are a widespread feature of the succulent karoo. (a) In the heavily grazed areas of Succulent Karoo (VT31), as shown here near Vanrhynsdorp, the *heuweltjies* support a sparser plant cover (in this spring aspect, comprising largely annuals) than the surrounding matrix. Note the pattern created by the *heuweltjies* on the slope in the background. (b) Normally, as can be seen in Karroid Broken Veld (VT26) near Worcester, the plant biomass on *heuweltjies* is higher than the matrix *(Photos: J.C. Paterson-Jones).*

occurs on dry, east-facing slopes, whereas west-facing slopes that receive more fog precipitation are better vegetated (20–40%) and support arborescent succulents (Jürgens 1986). Similarly, mesophyllous, non-succulent shrubs and trees occur on shaded south-facing slopes in the Little Karoo, whereas succulents predominate on northern slopes (Levyns 1950; Fig. 7.7).

Timing and duration of water availability differs between drainage lines and the surrounding watershed (Milton 1990a). On the banks of large drainage lines throughout the succulent karoo there is an abrupt change in vegetation structure from dwarf succulent shrubland to riverine woodland (Compton 1929; Acocks 1976). Steep gradients in soil texture, salinity and herbivory coincide with moisture gradients around ephemeral endorheic pans near Prince Albert (Karroid Broken Veld) and at Aggeneys (Namaqualand Broken Veld), leading to the development of specialized plant assemblages (Milton 1991; Milton, Dean & Marincowitz 1992b). At Vaalputs in the False Succulent Karoo, the boundaries between succulent and non-succulent assemblages are defined by the edges of depressions, where salts deposited by runoff water have accumulated (Lloyd 1989).

Heuweltjies ('little hills'), also called mima-like mounds (Fig. 7.10), are widespread in the succulent karoo, from the Richtersveld (Willis 1992) southwards through the Olifants River valley (Knight, Rebelo & Siegfried 1989; Moore & Picker 1991) to the Tanqua Karoo (Lovegrove & Siegfried 1989) and through the Little Karoo (Midgley & Musil 1990) to Prince Albert (Milton & Dean 1990a). They are over-dispersed, 5–30 m in diameter and 0.1–0.5 m high, cover about 3% of the land surface (Lovegrove & Siegfried 1989; Milton, Dean & Kerley 1992a) and differ from their surroundings in soil nutrient status and vegetation composition. *Heuweltjie* soils are more permeable to water (Dean 1992a), having lower bulk densities and more organic matter than soils of the surrounding matrix (Midgley & Musil 1990). High pH, concentrations of Ca, Na, K and organic enrichment were initially attributed to basic rock outcrops within more acidic parent material (Van der Merwe 1962). It is now generally accepted that *heuweltjie* soils have a zoogenic origin, being the soil and frass dumps that overlie colonies of the small harvester termite *Microhodotermes viator* (Coaton & Sheasby 1974; Milton & Dean 1990a; Moore & Picker 1991), which is endemic to the southwestern arid zone of southern Africa and is most abundant in the succulent karoo. *Heuweltjies* generally support taller, denser vegetation with a higher woody or succulent component than surrounding shrublands (Olivier 1979; Knight *et al.* 1989; Marais 1989; Midgley & Musil 1990; Smitheman & Perry 1990; Milton 1990a). This pattern is reversed in overgrazed areas and on shallow soils, where they tend to be barer than their surroundings and dominated by ephemerals (Acocks 1953; Milton 1978; Steinschen, Görne & Milton 1996). Plant community boundaries on *heuweltjies* can be explained in terms of disturbance ecology as well as edaphic factors (see section 7.5.2).

7.3 Organisms

7.3.1 Morphology

In the succulent karoo, succulent and non-succulent chamaephytes, geophytes and therophytes are unusually common relative to trees and grasses (Table 7.2). Succulents occupy exposed habitats on bare soil as well as

Table 7.2 **Growth-form spectra in selected floras from the succulent karoo. Growth forms are nanophanerophytes (NPh), chamaephytes including succulents (Ch), succulents as % of** **chamaephytes (Su), hemicryptophytes (Hem), geophytes (Geo), therophytes (Th) and parasites (Par)**

Locality	Annual rainfall (mm)	Winter % rain	Total species	NPh	Ch (Su)	Hem	Geo	Th	Par	Source
Namaqualand Broken Veld	–	–	1043	9	38 (71)	8	19	30	–	Acocks (undated)
Richtersveld	50	>70	276	13	50 (90)	5	9	22	1	Jürgens (1986)
Succulent Karoo	–	–	863	5	38 (82)	9	17	30	–	Acocks undated
Namaqualand	162	75	582	6	31 (45)	17	16	28	1	Van Rooyen et al. (1990)
Karroid Broken Veld	–	–	1281	6	41 (66)	10	13	25	–	Acocks (undated)
Worcester	275	66	389	2	51 (33)	6	26	13	2	Smitheman & Perry (1990)
Whitehill	240	65	700	9	42 (c.48)	2	18	23	6	Werger (1985)
Tierberg	167	46	189	8	55 (45)	6	7	20	3	Milton et al. (1992b)
Western Mountain Karoo	–	–	1228	3	39 (62)	10	21	29	–	Acocks (undated)
Strandveld	–	–	1061	7	40 (47)	13	20	29	–	Acocks (undated)

shaded habitats beneath non-succulent shrubs, and show remarkable variation in growth-form and life history (Table 7.3).

Evergreen and deciduous leaf-succulents, particularly Mesembryanthemaceae (Fig. 7.11), are most abundant and diverse in the winter rainfall zone, where evapotranspiration during the growing season is low (Jürgens 1986; Cowling *et al.* 1994; Chap. 16, this volume). They tend to be shrubby on deep soils and contracted (leaves at soil surface) or embedded on shallow soils (Jürgens 1986). Stem succulence is correlated with high temperatures during the wet season (Jürgens 1986), being most common in frost-free parts of the summer rainfall region (Fig. 7.12). Fast-growing annual and pauciennial succulents have high water : carbon ratios (Von Willert *et al.* 1992), are tolerant of extremes in salinity and alkalinity and colonize ephemeral pans and disturbed sites (Jürgens 1986; Dean & Milton 1995; Fig. 7.13).

Contracted and reduced succulents grow on shallow soil and in rock crevices, often in very small and isolated populations (Cole 1988; Hammer 1993). In embedded forms, including *Bulbine mesembryanthemoides*, *Haworthia semiviva* and many species of *Opthalmophyllum* and *Lithops*, light reaches internal chloroplasts in the buried leaf through windows of unpigmented cells on the exposed leaf tips (Von Willert *et al.* 1992). Small succulents resemble the soil, stones or dung of their habitats, or the shadows and woody stems of the shrubs that shelter them (Jacobsen & Volk 1960; Wiens 1982). Individuals that escape predation may be remarkably long-lived, up to 95 years in the case of *Lithops* plants (Cole 1988).

The proportion of non-succulents varies with position in the landscape, increasing relative to succulents along a gradient of increasing summer moisture availability (Milton 1990a). Dwarf non-succulent shrubs are slow-growing and their stems split longitudinally with age, each offset retaining its own canopy and root system (Theron 1964). Stem-splitting, also documented among shrubs of the Nama-karoo (Chap. 8, this volume), and the

Negev and Sonoran deserts, is thought to spread the risk of drought mortality among the offsets (Werger 1985). Drought-deciduous shrubs are more frequent in the strictly winter-rainfall zone (80% of species) than at Worcester (16%) and Prince Albert (10%) where there is a higher probability of receiving some rainfall at other seasons (Le Roux, Perry & Kyriacou 1989; Milton 1990a). Dwarf shrubs are replaced by larger shrubs in dry washes, and by trees in rivers beds and among boulders on shaded hill slopes (Milton 1990a). Drainage line assemblages are generally more spinescent than surrounding shrublands (Milton 1991) and include savanna tree and shrub genera at the southwestern limits of their distributions (Acocks 1976). Mistletoes are largely restricted to drainages in the succulent karoo, and their species richness is lower than in mesic savanna but higher than in fynbos or forest (Dean, Midgley & Stock 1994), suggesting that these hemi-parasites require access to ample water and nutrients. The specialized root and canopy holo-parasites of succulents are largely endemic to the succulent karoo (Visser 1981).

Geophytes are well-represented in the succulent karoo (Table 7.3). The petaloid monocot assemblages on deep clay and loam soils around Vanrhynsdorp and Nieuwoudville in the western succulent and mountain karoo are among the richest in the world (Snijman & Perry 1987). A feature of the western part of the succulent karoo are the spring floral displays of winter-growing annuals (Eliovson 1972; Van Rooyen, Theron & Grobbelaar 1979a). Summer-growing annuals and ephemerals with desert affinities are less well represented in the succulent karoo (Leistner 1994).

7.3.2 Phenology

Within the succulent karoo, phenology differs among growth forms, between habitats and along the rainfall seasonality gradient. Perennials, including geophytes, succulent and non-succulent shrubs, generally grow in autumn or winter and flower in spring (Van Rooyen *et*

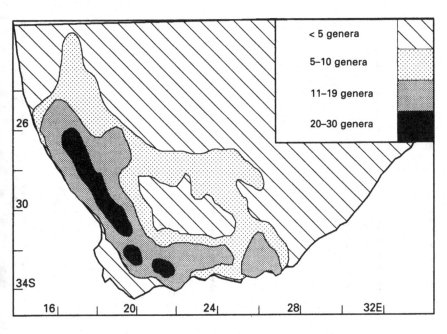

Figure 7.11 **Centres of generic diversity in the Mesembryanthemaceae, based on 107 genera. Isolines show number of genera 100 m^{-2}** areas. After Jürgens (1986).

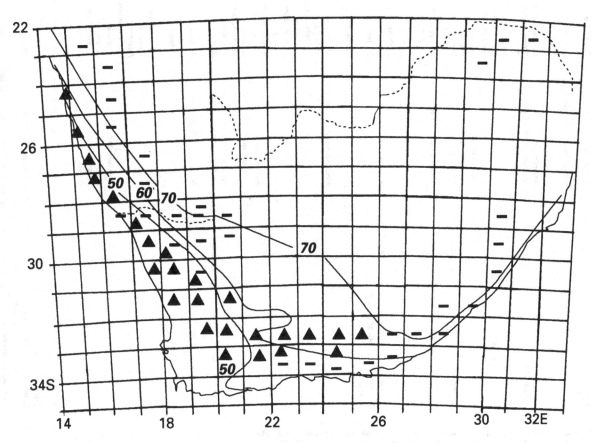

Figure 7.12 **Distribution of leaf (triangle) and stem (rectangle) succulents, after Jürgens (1986). Isolines show summer rainfall (October to March) as percentage of mean annual rainfall.**

Table 7.3 Growth-forms of flowering plants of the succulent karoo. The list excludes crypto-gams. The numbers of taxa within each growth-form category are based on counts of species (not subspecies) in revisions of all the major genera in this biome. Nomenclature and arrangement of families follows Arnold & De Wet (1993)

Growth form	Number of taxa	Main families and genera	Seed attributes	Anti-herbivore defence[a]
Leaf succulent				
Annual	25	Aizoaceae: *Tetragonia* Mesembryanthemaceae: *Dorotheanthus*, *Mesembryanthemum* Crassulaceae: *Crassula* Asteraceae: *Gymnodiscus*	Small dormant endozoochores or large dormant anemochores	Salt, oxalic acid
Pauciennial Usually decumbent	65	Mesembryanthemaceae: *Browanthus*, *Malephora*, *Drosanthemum*, *Sphalmanthus* Crassulaceae: *Crassula* Zygophyllaceae: *Augea* Asphodelaceae: *Bulbine*	Small dormant endozoochores or large dormant anemochores Small hydrochores	Salt, oxalic acid Cryptic
Embedded Most of leaf below soil surface	35 (100)[b]	Asphodelaceae: *Haworthia* Mesembryanthemaceae: *Conophytum*, *Fenestraria*, *Lithops*, *Gibbaeum*, *Opthalmophyllum*	Seed stored in capsules	
Contracted Stemless; all leaves at ground level	400 (600)[b]	Asphodelaceae: *Aloe*, *Gasteria*, *Haworthia* Mesembryanthemaceae: *Argyroderma*, *Cheiridopsis*, *Cephalophyllum*, *Gibbaeum*, *Glottiphyllum*, *Pleiospilos*, *Rhinephyllum* Portulacaceae: *Anacampseros* Crassulaceae: *Crassula*, *Adromischus* Asteraceae: *Othonna*	Small hydrochores Seed stored in capsules or large with wing or pappus	Oxalic acid, cardiac glycosides
Shrubs 0.1–1.0 m Leaves above ground level; deciduous or evergreen	325 (900)[b]	Asphodelaceae: *Bulbine* Mesembryanthemaceae: *Aridaria*, *Delosperma*, *Drosanthemum*, *Hereroa*, *Ruschia*, *Sphalmanthus* Crassulaceae: *Crassula*, *Cotyledon* Zygophyllaceae: *Zygophyllum* Asteraceae: *Othonna*, *Senecio*	Small seed hydrochorous Seed stored in capsules or large, winged seed or vegetative	Oxalate, tannin, glycoside, spines
Stem succulent				
Green stem with reduced leaves or brown stem with deciduous leaves	130	Mesembryanthemaceae: *Psilocaulon*	Large seeds anemochorous or autochorous or small, dust-like seeds	Spines, latex, alkaloid, oxalate, glycoside, cryptic

Growth form	No.	Family	Genera	Seed dispersal	Defence
		Geraniaceae:	*Pelargonium* *Sarcocaulon*		
		Euphorbiaceae:	*Euphorbia*		
		Asclepiadaceae:	*Hoodia* *Caralluma* *Huernia* *Piaranthus* *Sarcostemma*		
		Asteraceae:	*Senecio* *Kleinia*		
Arborescent	5		*Aloe* *Tylecodon* *Pachypodium* *Portulacaria*	Anemochorous dust-like or winged	Spines, oxalate
		Asphodelaceae: Crassulaceae: Apocynaceae: Portulacaceae:			
Geophyte	630	Colchicaceae:	*Androcymbium*	Large seeds anemochorous, autochorous, zoochorous and vegetative reproduction	Cardiac glycoside, oxalic acid, alkaloid, or cryptic
		Asphodelaceae:	*Trachyandra*		
		Eriospermaceae:	*Eriospermum*		
		Hyacinthaceae:	*Albuca* *Lachenalia* *Ornithogalum* *Urginea*		
Monocotyledon		Amaryllidaceae:	*Brunsvigia* *Gethyllis* *Hessea*		
		Hypoxidaceae:	*Spiloxene*		
		Dioscoreaceae:	*Dioscorea*		
		Iridaceae:	*Babiana* *Ixia* *Moraea* *Homeria* *Lapeirousia* *Gladiolus* *Cyanella*		
		Tecophilaeaceae:			
Dicotyledon		Mesembryanthemaceae:	*Conicosia*		
		Crassulaceae:	*Crassula*		
		Geraniaceae:	*Pelargonium*		
		Oxalidaceae:	*Oxalis*		
		Apiaceae:	*Chamarea*		
		Asclepiadaceae:	*Fockea*		
		Cucurbitaceae:	*Kedrostis*		
		Asteraceae:	*Othonna*		
Annual and ephemeral	390	Poaceae:	*Schismus* *Tribolium* *Urochlaena*	Large or dust-like; anemochorous or antitelechorous; usually dormant	Poorly defended transient resource, cyanogenic
		Cyperaceae:	*Isolepis*		
		Aizoaceae:	*Galenia* *Hypertelis*		
		Brassicaceae:	*Heliophila*		
		Rosaceae:	*Grielum*		
		Fabaceae:	*Lotononis*		
		Boraginaceae:	*Anchusa*		
		Scrophulariaceae:	*Diascia* *Manulea*		

Table 7.3 contd.

Growth form	Number of taxa	Main families and genera	Seed attributes	Anti-herbivore defence[a]
		Campanulaceae: Asteraceae:	*Nemesia* *Zaluzianskya* *Wahlenbergia* *Arctotis* *Cotula* *Dimorphotheca* *Osteospermum* *Ursinia*	Spiny, small or poorly defended seasonal resource
Graminoid	90	Poaceae: Cyperaceae:	*Aristida* *Cladoraphis* *Ehrharta* *Enneapogon* *Eragrostis* *Fingerhuthia* *Karroochloa* *Pentaschistis* *Stipagrostis* *Scirpus*	Small endochorous or large anemochorous
Dwarf shrubs Below 0.6 m; evergreen lepto- or nanophyll; or deciduous micro- or mesophyll	560	Urticaceae: Aizoaceae: Chenopodiaceae: Caryophyllaceae: Fabaceae: Geraniaceae: Polygalaceae: Sterculiaceae: Thymelaeaceae: Asclepiadaceae: Scrophulariaceae: Selaginaceae: Asteraceae:	*Forsskaolea* *Galenia* *Limeum* *Salsola* *Dianthus* *Indigofera* *Lessertia* *Lotononis* *Melolobium* *Pelargonium* *Polygala* *Hermannia* *Gnidia* *Microloma* *Aptosimum* *Sutera* *Selago* *Berkheya* *Chrysocoma* *Eriocephalus* *Euryops* *Felicia* *Helichrysum* *Osteospermum* *Pteronia*	Seeds large to medium wings, pappus anemochorous or autochorous	Alkaloids, phenols, cyanogenesis, terpenes, or leaf spines
Mid-high shrubs 0.6–2 m; deep-rooted; generally deciduous	190	Asparagaceae:	*Protasparagus*	Large seeded anemochorous or bird dispersed	Stem spines, high wood to leaf ratio

Family	Genera	Growth form	No.	Seed dispersal	Defence
Fabaceae:	*Lebeckia*				
	Wiborgia				
	Nymania				
	Nylandtia				
Meliaceae:	*Rhus*				
Polygalaceae:					
Anacardiaceae:					
Melianthaceae:	*Melianthus*				
Tiliaceae:	*Grewia*				
Thymelaeaceae:	*Gnidia*				
Oleaceae:	*Menodora*				
Apocynaceae:	*Carissa*				
Asclepiadaceae:	*Asclepias*				
Solanaceae:	*Lycium*				
Bignoniaceae:	*Rhigozum*				
Asteraceae:	*Euryops*				

Trees

Growth form	No.	Family	Genera	Seed dispersal	Defence
Evergreen or deciduous micro- or mesophylls	30	Salicaceae:	*Salix*	Small-seeds dispersed by wind or water; or large seeds in fleshy fruit, dispersed by birds or mammals	Tannin, fibre, phenolics, or spines
		Fabaceae:	*Schotia*		
			Acacia karroo		
		Anacardiaceae:	*Rhus*		
		Celastraceae:	*Maytenus*		
		Sapindaceae:	*Pappea*		
		Tamaricaceae:	*Tamarix*		
		Ebenaceae:	*Diospyros*		
			Euclea		

Parasites

Growth form	No.	Family	Genera	Seed dispersal	Defence
Canopy	30	Loranthaceae:	*Loranthus*	Large seeds fleshy fruit bird-dispersed	Undefended, cryptic
			Moquinella		
			Septulina		
		Viscaceae:	*Viscum*		
		Lauraceae:	*Cassytha*		
		Convolvulaceae:	*Cuscuta*		
		Santalaceae:	*Thesium*	Bird-dispersed, or small seeds dispersed by wind, ants	
Root		Hydnoraceae:	*Hydnora*		
		Scrophulariaceae:	*Hyobanche*		

Vines and lianes

Growth form	No.	Family	Genera	Seed dispersal	Defence
Perennial stems or rootstocks	20	Asparagaceae:	*Protasparagus*	Large seeds, dispersed by wind, birds or mammals	Latex, trichomes, or cryptic
		Asclepiadaceae:	*Myrsiphyllum*		
			Ceropegia		
			Microloma		
		Fumariaceae:	*Cysticapnos*		
		Oleaceae:	*Jasminum*		
		Rubiaceae:	*Galium*		
		Cucurbitaceae:	*Cucumis*		
			Kedrostis		
		Lobeliaceae:	*Cyphia*		

Grand Total 2925 (3765)[b]

[a] Kellerman *et al.* 1988; Van Wyk 1989; Milton, Dean & Siegfried 1994b.

[b] Lower estimate based on recent revisions of Mesembryanthemaceae by H.E.K. Hartmann and colleagues (University of Hamburg); higher estimate if traditional (L. Bolus) species concepts are used for Mesembryanthemaceae.

Figure 7.13 **Growth forms in the succulent karoo.** (*a*) **Arborescent leaf-succulent *Aloe dichotoma* (Asphodelaceae) in Namaqualand Broken Veld (VT33).** (*b*) **Evergreen, sclerophyllous tree *Ozoroa dispar* (Anacardiaceae) growing on a granite dome in Succulent Karoo (VT31).** (*c*) **Dwarf stem-succulent *Sarcocaulon crassicaule* (Geraniaceae) in Succulent Karoo (VT31)** *(Photos: J.C. Paterson-Jones).* (*d*) **Contracted leaf-succulent *Argyroderma delaetii***

(Mesembryanthemaceae) **associated with quartz patches in the Knersvlakte region of the Succulent Karoo (VT31)** *(Photo: D.M. Richardson).* (*e*) **Hysteranthous, autumn-flowering geophyte *Brunsvigia bosmaniae* (Amaryllidaceae) in the Western Mountain Karoo (VT28).** (*f*) **Spring annuals (mainly Asteraceae) and geophytes (petaloid monocots) in heavily grazed Namaqualand Broken Veld (VT33)** *(Photos: J.C. Paterson-Jones).*

al. 1979a; Boshoff 1989; Hoffman 1989b; Le Roux *et al.* 1989; Milton 1990a; Struck 1992). This pattern differs from the winter-rainfall semi-arid area of North America, where both growth and flowering are largely confined to the spring, perhaps because of lower winter temperatures (Esler, Rundel & Cowling 1994). Esler *et al.* (1994) argue that the lower growth-form diversity of the succu-

lent karoo (Cowling *et al.* 1994) could result from conditions that favour winter growth by shallow-rooted leaf-succulents. The spring-growing species of the Mojave desert exploit moisture that has reached greater depths, and deeper rooting results in greater above-ground morphological diversity (Cody 1989).

Differences in the depth and seasonal availability of

water may also explain phenological and structural differences between plant assemblages of plains and drainage lines in the southern Karoo (Boshoff 1989). Plains species, mainly succulents and dwarf shrubs, are active in winter, whereas large shrubs and trees of drainage lines grow and flower throughout the year (Milton 1990a). These patterns appear to be tied to seasonal differences in rainfall intensity: runoff and water storage following intense summer thunder showers benefits deep-rooted plants in drainage lines, whereas in winter, when evaporation rates are low, shallow-rooted assemblages of plains use light drizzle that produces little runoff.

Phenological rhythms appear to be more clearly defined in the winter-rainfall zone than in other parts of the succulent karoo. In the winter-rainfall zone long (dolichoblast) shoots are strictly winter-growing, whereas in the remainder of the succulent karoo they are produced throughout the year with a peak in winter and a trough in summer (Le Roux *et al.* 1989) (Fig. 7.14). Short shoots (brachyblast) are produced throughout the year in both rainfall zones, but summer dormancy was more marked in the winter-rainfall site. In some species, growth is bimodally distributed in autumn and spring when temperatures are moderate (Van Rooyen *et al.* 1979a).

Evergreen, non-succulent shrubs in the succulent karoo lose some leaves when water-stressed and produce new leaves after rain, so that leaf turnover is greater than in sclerophyllous fynbos shrubs (Boshoff 1989; Le Roux *et al.* 1989; Milton 1990a). Drought-deciduous shrubs are leafless in the dry summers of the western region, but grow in both spring and autumn (Van Rooyen *et al.* 1979a; Boshoff 1989). Further east, they respond opportunistically to rain, resprouting and flowering whenever moisture is available (Hoffman 1989b; Milton 1992a). Geophytes with large storage capacity (*Brunsvigia* and *Eriospermum*, *Fockea*, *Kedrostis*) are hysteranthous, flowering at the end of the dry summer, before they produce leaves (Van Rooyen *et al.* 1979a; Hoffman 1989b). However, most geophytes are synanthous, flowering at the end of their winter growing season (Van Rooyen *et al.* 1979a). Exendospermous *Oxalis* species make very brief appearances, flowering, seeding and leafing within a few weeks of rain (Rösch 1977).

Winter-growing annual forbs flower before most perennials. Flower initiation is unrelated to daylength but is stimulated by low temperatures and delayed by moisture

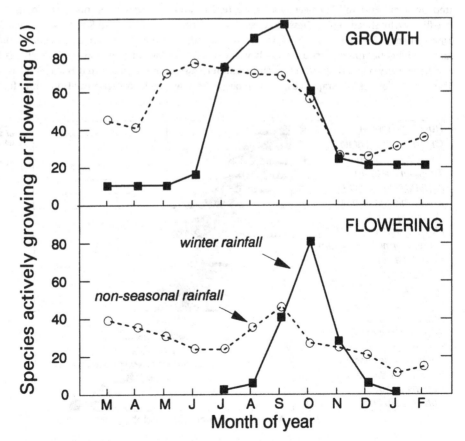

Figure 7.14 Growth and flowering phenology of plant assemblages under winter rainfall (—) at Goegab Nature Reserve near Springbok, and non-seasonal rainfall (- -) at Worcester in the southern Karoo. After Le Roux *et al.* (1989).

stress (Van Rooyen *et al.* 1991). Plants germinating in autumn are slow growing and large relative to late germinators that are short lived and allocate most of their limited resources to reproduction (Van Rooyen *et al.* 1992). A model, based on the relationship between thermal units accumulated during the growing season and the time of flowering, predicts the timing and intensity of Namaqualand floral displays from the date of the first substantial rains (Steyn *et al.* 1994).

7.3.3 Reproductive biology

7.3.3.1 *Floral displays and pollination*

Pollinators are essential for the persistence of most plant species in this biome, because Asclepiadaceae, Asphodelaceae, Mesembryanthemaceae and Oxalidaceae are usually self-sterile (Jacobsen 1960). More than 300 species of insects were identified as floral visitors at Goegap Nature Reserve in Namaqualand, of which 60% were Hymenoptera, 20% Diptera (particularly bee-flies Bombyliidae) and 20% beetles, mainly Hopliinae (Struck 1992). Potential competition for pollinators appears to be reduced by predictable, sequential flowering (Fig. 7.15), despite variability in timing and amount of rainfall (Struck 1992).

Day-flowering succulents of open habitats have showy displays of yellow, orange, purple and pink flowers visited by social and solitary bees, wasps, butterflies and bee-flies (Whitehead 1984; Gess & Gess 1989; Struck 1992), many of which are endemic to the succulent karoo. Night-opening flowers are usually white or yellow and have fruity odors that attract moths (Hammer 1991). The foetid flowers of Stapeliads, which occur at low densities beneath shrubs, are fly-pollinated (Bayer 1978). Other subcanopy (nursed) species display their flowers on long peduncles that emerge through the canopy of the host plants (Von Willert *et al.* 1992). Flowers of drainage-line trees and shrubs are generally small, white and strongly scented, attracting a variety of Diptera and Hymenoptera (Gess & Gess 1989).

Unusual insect attractants and deterrents have evolved in some succulent karoo annuals. The annual *Diascia* spp. (Scrophulariaceae) are pollinated by specialized oil-collecting bees of the genus *Rediviva*, and there is a correlation between flower and bee density (Whitehead & Steiner 1985). Dark outgrowths on ray florets of *Gorteria diffusa* and gaps between the rays of *Ursinia* sp. mimic beetles (Rutelinae, Hopliini) that feed on pollen of annual daisies. This 'occupied' look apparently deters Hopliinid beetles from alighting on the flowers (Midgley 1991).

7.3.3.2 *Dispersal mechanisms and establishment sites*

Whereas most plants of the succulent karoo depend on insects to transport their pollen, seeds are predominantly wind- or water-dispersed (Hoffman & Cowling 1987; Milton 1990a; Van Rooyen *et al.* 1990; Table 7.4). Seeds of most species ripen in spring (Van Rooyen *et al.* 1979a), but many have seed-retention mechanisms that prevent seed dispersal in space, but facilitate dispersal in time (Van Rooyen *et al.* 1990; Esler 1993). Mesembryanthemaceae are hydrochorous, having capsules that open when wet and close again as they dry (Volk 1960), a mechanism that releases seeds when conditions are suitable for germination (Ihlenfeldt 1971). The numerous

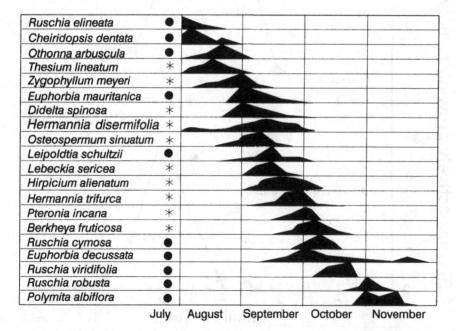

	July	August	September	October	November
Ruschia elineata ●					
Cheiridopsis dentata ●					
Othonna arbuscula ●					
Thesium lineatum ∗					
Zygophyllum meyeri ∗					
Euphorbia mauritanica ●					
Didelta spinosa ∗					
Hermannia disermifolia ∗					
Osteospermum sinuatum ∗					
Leipoldtia schultzii ●					
Lebeckia sericea ∗					
Hirpicium alienatum ∗					
Hermannia trifurca ∗					
Pteronia incana ∗					
Berkheya fruticosa ∗					
Ruschia cymosa ●					
Euphorbia decussata ●					
Ruschia viridifolia ●					
Ruschia robusta ●					
Polymita albiflora ●					

Figure 7.15 **Sequential flowering of 20 perennial succulent (circle) and non-succulent (star) shrubs in Namaqualand. After Struck (1992).**

Table 7.4 **Dispersal spectra (% of species in sample plant communities) for Goegab Nature Reserve, Namaqualand (Van Rooyen *et al.* 1990) and on plains, *heuweltjies* mounds and drainage lines (river) at Tierberg, near Prince Albert (Milton 1990a)**

Site	Wind	Water	Self	Animal ingestion	Animal adhesion	Anti-telochory
Namaqualand[a]	66	26	20	6	8	10
Tierberg plain	41	33	13	8	5	0
Tierberg mound	27	38	0	29	0	6
Tierberg river	35	18	8	31	5	3

[a] Some seeds were allocated to more than one dispersal category.

(30–900) small seeds (Esler 1993) are ejected over short distances (20 to 90 cm) by the impact of raindrops (Volk 1960). In the tribe Dorotheanthinae, capsules in relatively mesic southern Namaqualand endure for some years, but towards the dry northern extreme of their distribution in the Richtersveld, capsules are increasingly flimsy, decaying within a few months (Struck 1989). Durable capsules spread seed release over many rain events, reducing the risk of total seed loss (Ihlenfeldt 1971), and thereby increasing the probability of seedling survival (Esler & Cowling 1995).

Plants that colonize exposed sites (mainly Aizoaceae and Mesembryanthemaceae) usually have small seeds (<1 mm diameter) that are trapped by soil particles (Yeaton & Esler 1990; Milton 1995; Esler, Cowling & Juritz 1996). However, some larger-seeded Asteraceae (*Euryops multifidus*), Brassicaceae (*Lepidium*) and Zygophyllaceae (*Augea capensis*) are also effective pioneers, because their seeds exude a sticky secretion, gluing them to damp surfaces (Van Breda 1939; Rösch 1977).

Species that establish beneath dwarf shrubs on open plains, have seeds that glide in the wind or tumble over the ground (Table 7.3). Soil-stored seed banks at Tierberg were generally larger beneath mat-forming shrubs than on bare soil, but this pattern was particularly marked for seeds of wind-dispersed species (Esler *et al.* 1996). The spongy, globose capsules of shade-loving Mesembryanthemaceae (*Glottiphyllum, Sceletium*) are shed when dry and rolled by the wind until they lodge beneath shrubs or other obstacles (Ihlenfeldt 1971; Hartmann & Gölling 1993). *Euryops, Osteospermum, Lachenalia* and *Zygophyllum*, which have seeds modified for ant dispersal in the fynbos (Bond & Slingsby 1983), have winged seeds in the succulent karoo. Since the ant vectors (mainly Formicines) occur and respond positively to seeds bearing elaiosomes (Breytenbach 1988), we infer that the loss of the mutualism reflects the advantages of wind dispersal over ant dispersal in the succulent karoo.

Seed polymorphism is a common phenomenon in the succulent karoo (Rösch 1977; Von Willert *et al.* 1992). Polymorphisms in seed coat anatomy, diaspore size, morphology or mechanism and timing of release from a multi-seeded capsule (Rösch 1977) allow plants to disperse some of their seeds in space (winged diaspores) and others in time (anti-telochorous diaspores that remain attached to the dead inflorescence). For example, 95% of winged diaspores of *Dimorphotheca sinuata* germinate in their first year, whereas all but 3% of unwinged diaspores are dormant (Beneke *et al.* 1993).

Zoochory in the succulent karoo is largely restricted to moist and nutrient-rich habitats (Table 7.4). Fleshy fruits of trees and canopy parasites are available throughout the year with a peak in autumn and winter (Dean, Williams & Milton 1993). Frugivorous birds (Knight 1988; Dean *et al.* 1993) and mammals, particularly the bat-eared fox (*Otocyon megalotis*), eat fruit in winter when insects are scarce (Kuntzsch & Nel 1992; Stuart 1981;). They appear to disperse tree seeds to *heuweltjies* and river beds (Milton 1990a) that provide the moist, shaded conditions these seedlings require for establishment (Midgley & Cowling 1993). Endozoochory by tortoises, rodents and aardvarks may be an important secondary dispersal mechanism for Aizoaceae, Molluginaceae and Mesembryanthemaceae (Jump 1988; Dean & Yeaton 1992; Milton 1992b). Adaptations for epizoochory are rare in the succulent karoo, but the climber *Galium tomentosum* (Rubiaceae) is effectively dispersed to trees in drainage lines by birds that use its hirsute female inflorescences to construct nests (Dean, Milton & Siegfried 1990). The indehiscent seeds, incorporated into nests, fall to the ground beneath the tree about one year later when the nest falls apart (Dean & Milton 1993).

Succulents are well suited to vegetative reproduction, because their leaves or stems store water and carbohydrates. Leaves of Crassulaceae shed during dry periods (Von Willert *et al.* 1992) are blown or washed in runoff water to suitable establishment sites among litter or beneath shrubs, as are joints of cholla cacti (*Opuntia* spp.) in the Mojave Desert of the USA (Cody 1993). *Gasteria* spp. (Asphodelaceae) survive herbivory because their leaves detach easily and fall to the ground where they root and proliferate (Van Jaarsveld 1994). Stapeliads, and some Mesembryanthemaceae, Asteraceae and Asphodelaceae, multiply by suckers and runners to form clones in favourable sites (Jacobsen 1960).

7.3.3.3 Seed bank dynamics and germination

In variable environments, delayed seed germination is an evolutionarily stable strategy in short-lived but not in long-lived plants (Rees 1994). The abundance of dormant seed in the soil is thus inversely related to the life expectancy of the dominant plant species (Grubb 1988), so that large seed banks may be expected only in frequently disturbed or very arid sites where annuals and pauciennials predominate. Seed-bank dynamics of succulent karoo plant assemblages appear to conform to this general pattern.

Seed production at Tierberg, in shrubland dominated by perennials throughout the year, is about 3000 seeds m^{-2} (Milton & Dean 1990b). Soil-stored seed banks were low by comparison with other winter-rainfall deserts (Esler *et al.* 1996), and fluctuated seasonally, reaching a summer maximum of 430 seeds m^{-2} (Esler *et al.* 1996). At Goegap Nature Reserve in Namaqualand, seed densities in soil were highest (41 000 m^{-2}) in sandy bottomlands dominated by annual plant assemblages, and lowest (5000 m^{-2}) on ridge tops with scattered perennials and few annuals (Van Rooyen & Grobbelaar 1982).

Large-seeded, non-succulent karoo shrubs with long life spans do not appear to maintain dormant seed banks. Seeds of asteraceous shrubs, *Osteospermum sinuatum* and *Pteronia* spp., which are germinable when fresh, were recorded in the seed bank at Tierberg in the summer following seed dispersal, but were absent from soil samples collected in the following winter (Esler *et al.* 1996). All *Pteronia* seeds buried at Tierberg decayed within a year (Esler *et al.* 1996) and seedlings from *O. sinuatum* and *Pteronia* spp. artificially sown in summer emerged only in the first autumn (Milton 1994a). Populations of shrub species that do not maintain soil-stored seed banks may become seed limited (Milton 1995) where sheep annually remove 90% of the seed production (Milton & Dean 1990b). Granivorous ants can take up to 60% of the annual seed production of some plant species, but because they select the most abundant seeds, they

have less effect on vegetation composition than do domestic livestock (Milton & Dean 1993).

The small, tannin-coated seeds of *Galenia fruticosa* (Aizoaceae), which are dormant when fresh, were present in the seed bank throughout the year and 44% of buried seeds survived for two years under field conditions (Esler *et al.* 1996). Most small-seeded Mesembryanthemaceae at Tierberg had short-lived, canopy-stored seed banks (limited to the current year's production) and soil-stored seed banks were equivalent to 1% of annual seed production (Esler, Cowling & Ivey 1992).

Germination of seeds of annual and pauciennial succulents is often staggered (Fig. 7.16), whereas germination of perennial succulents rarely shows innate seed dormancy (Von Willert *et al.* 1992; Esler 1993; Hartmann & Stüber 1993). Some Mesembryanthemaceae have a bimodal or tardy germination pattern in which some seeds germinate 24 h after wetting, and further germination is scattered over the following three weeks (Ihlenfeldt 1971; Hartmann & Gölling 1993; Hartmann & Stüber 1993). Selection for dormancy within a single species may differ between sites (Hartmann 1983).

Laboratory trials indicate that seeds of succulent karoo Mesembryanthemaceae germinate better under cool (10 °C night/20 °C day), than warm (15 °C night/30 °C day) conditions (Esler 1993). In the field, seedlings of most growth-forms emerge in autumn and winter (Van Rooyen *et al.* 1979a; Van Breda & Barnard 1991; Esler 1993; Milton 1995). There was no germination in response to spring (October) irrigation in the Richtersveld, but abundant germination of grasses, succulents and forbs followed March rain (Von Willert *et al.* 1992). On plains at Tierberg, seedlings emerged between March and May in each of six consecutive years (Fig. 7.17), but densities varied with autumn rainfall (Esler 1993; Milton 1995) and were possibly also influenced by seed pro-

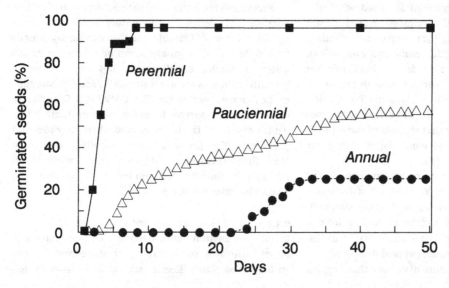

Figure 7.16 **Germination rates of annual (*Mesembryanthemum aitonis*), pauciennial (*Malephora lutea*) and perennial (*Delosperma pergamentaceum*) Mesembryanthemaceae.** After Von Willert *et al.* (1992).

duction in the preceding spring (Wiegand, Milton & Wissel 1995).

7.3.3.4 *Seedling establishment*

Frequency and timing of rainfall influences the relative abundance of annual Mesembryanthemaceae species surviving to maturity: rainfall spread evenly through the winter led to dominance of a Richtersveld plain by *Mesembryanthemum pellitum*, whereas the same area was covered with *Opophytum aquosum* in a year with late winter rain (Von Willert *et al.* 1992). Seedling recruitment is low for long-lived species (1–5% of emergent seedlings survive for 12 months in average years, 20–30% in good years, Milton 1995), but is relatively high for ephemerals, annuals and biennials. In Namaqualand, 47–74% of annuals survived to maturity after emerging at densities of 115–1810 m^{-2} (Rösch 1977; Van Rooyen, Theron & Grobbelaar 1979b).

Survival of perennial succulent and shrub seedlings at Tierberg was highest when rain fell frequently during the three months following emergence, or where a seedling emerged in a gap created by the death of an established shrub (Milton 1995). Shrubs depleted soil moisture 150 mm below the surface within four days, whereas soil in cleared 5 × 5 m plots remained moist for 12 days after rain. In response to moisture availability, seedling survival increased from the perimeters to the centres of cleared plots (Milton 1995). Hence, patchiness in seedling survival is the result of both site heterogeneity and the differing requirements of individual species (Esler 1993).

Seedlings of *Ruschia spinosa* (L. Dehn, Mesembryanthemaceae), which normally establish in open sites, are more drought tolerant than seedlings of non-succulent Asteraceae that establish in the shelter of leaf succulents (Esler & Phillips 1994). Under field conditions, where soil surface temperatures reach 61 °C in summer months (Dean 1992b), heat and drought tolerance may be critical for seedlings of species that pioneer bare ground. The availability of microsites sheltered by living or dead plants sometimes influences the density and recruitment success of wind-dispersed species (Esler 1993; Milton 1994b). Addition of cover to undisturbed and cleared vegetation increased emergence densities of wind-dispersed seedlings, although seedlings survived only in cleared plots (Milton 1995).

7.3.3.5 *Population structure and turnover*

Annual turnover in marked populations of the shrubs

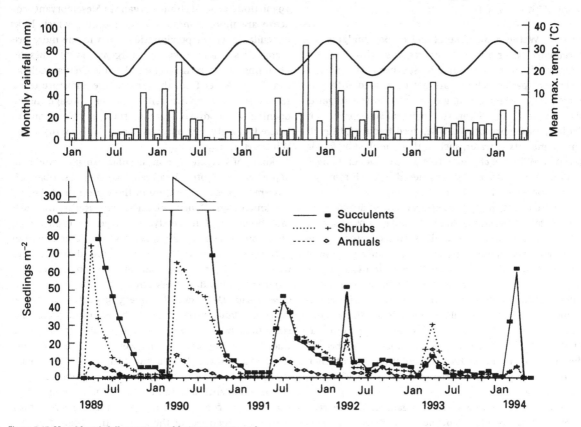

Figure 7.17 **Monthly rainfall, mean monthly temperature and densities of newly germinated and surviving seedlings of shrubs, succulents and annuals in permanent plots at Tierberg in the southern succulent karoo. After Milton & Dean (1995).**

Osteospermum sinuatum, Pteronia empetrifolia and *P. pallens* at Tierberg was 3.0%, 3.4% and 2.2%, respectively, and most of the recorded mortalities (88%) occurred among pre-reproductive plants (Milton 1994b). Assuming that mortality of shrubs is unsynchronized, the proportion of dead individuals in a population can provide an index of turnover and longevity. Esler *et al.* (1992) reported that in Karroid Broken Veld at Tierberg, 8% of the population of a leaf-succulent shrub *Drosanthemum montaguense* was dead as compared with 22% and 24% in contracted and decumbent leaf succulents (*Rhinephyllum macradenium, Brownanthus ciliatus*).

Factors that alter rates of adult mortality or seedling recruitment influence the structure of plant populations. In the succulent karoo, drought (Milton *et al.* 1995), hail (Milton & Collins 1989; Powrie 1993), disease and disturbance (Dean & Yeaton 1992), including herbivory, cause mortality of established plants at local population, patch or individual scales. Florivory by sheep differentially influences seed production, and consequently recruitment, of forage and non-forage species. Relative to moderately grazed rangeland, seedling to adult ratios in rangeland overstocked with sheep were lower in palatable *O. sinuatum*, but higher in toxic *P. pallens* (Milton 1994b).

7.3.4 Water, heat, mineral and carbon budgets

Succulents occur in frost-free regions with low but reliable winter rainfall, and in southern Africa their cover and species richness are not correlated with prolonged water stress (Cowling *et al.* 1989; 1994; Werger & Morris 1991; Chap. 16, this volume). Leaf succulents, particularly Mesembryanthemaceae, need frequent, though small, amounts of water, such as that supplied by fog or dew (Von Willert *et al.* 1990). Tolerance of moderate heat and high evaporation rates is achieved through morphological and physiological features.

Waxy cuticles, glassy papillae, cocoons of dead leaves, fibres or sticky secretions that trap sand (Jacobsen & Volk 1960; Hammer 1993), as well as sunken stomata or densely packed spines, that create a boundary layer of saturated air, reduce water loss (Von Willert *et al.* 1992). Unpigmented cells, which allow light to reach internal chloroplasts, compensate for low photosynthetic area to volume ratios in cylindrical stems and leaves (Jacobsen & Volk 1960; Von Willert *et al.* 1992). Windows and internal photosynthetic surfaces are best developed in embedded dwarf succulents, but reduced evaporation presents thermoregulation problems (Turner & Picker 1993; Chap. 16, this volume).

Orientation of photosynthetic organs influences radiation budgets. Erect leaves of young *Aloe dichotoma* plants absorb 40% less radiation per leaf area than horizontal leaves of older plants (Von Willert *et al.* 1992). Although orientation of cylindrical succulents and American Cactaceae towards the sun is thought to reduce heat loads through self shading (Von Willert *et al.* 1992), Rundel *et al.* (1995) argue that, for *Pachypodium namaquanum* on shaded, south-facing slopes in the winter-rainfall Richtersveld, the phototropic inclination of the terminal rosette of leaves maximizes insolation during the winter-growing period.

The leaves and stems of succulents may use different photosynthetic pathways. For example, the stems of *Pachypodium namaquanum* are CAM (stomata open at night) but the leaves, present only in winter when moisture is available, are C_3 with diurnally active stomata (Von Willert *et al.* 1992). Other succulents, including *Portulacaria afra*, switch from C_3 to CAM pathways during dry periods (Chap. 16, this volume).

Most succulents are probably shallow-rooted. Beukman (1991) found that >40% of *Ruschia* roots and 100% of *Tylecodon wallichii* roots were in the top 0.1 m of the soil where they are well placed to absorb fog, dew and light rain (Chap. 16, this volume). However, roots of *Euphorbia* spp. excavated near Worcester extended 4 m in depth and 2 m beyond the canopy (Scott & Van Breda 1939). Underground storage organs in Mesembryanthemaceae are more common in Nama-karoo, which lacks dependable winter precipitation, than in the more predictable succulent karoo (Jürgens 1986). Besides obtaining moisture from the soil, some CAM succulents can absorb water from the atmosphere. Moisture transfer takes place across cell walls, or through night-opening or modified stomata such as the trichome-fringed hydrathodes of Crassulaceae (Tölken 1977; Von Willert *et al.* 1992).

Roots of long-lived, non-succulent shrubs extend to depths of 8 m (Scott & Van Breda 1937a,b). Less than 10% of *Pteronia pallens* roots were in the top 0.1 m of the soil (Beukman 1991). Rooting characteristics (rather than leaf morphology and longevity, soil type, or exposure to radiation) appear to influence seasonal water potential courses in non-succulent shrubs, and differences in water acquisition rather than in water loss determine differences in water stress among deciduous and evergreen shrubs (Midgley & Bösenberg 1990). Rates of stomatal conductance were found to be high and water use efficiency low, for the asteraceous shrubs *Pentzia incana* and *Eriocephalus ericoides*, in wet soils, but efficiency improved as the soil dried out (Midgley & Moll 1993).

Salt facilitates the storage and acquisition of water. Some succulents accumulate sodium, potassium and other salts at concentrations two to three orders of magnitude greater than in the soil, so that their dry mass comprises up to 50% ash (mostly salts) and 9% sodium

Table 7.5 **Mean species richness and growth-form diversity**[a] **on plains in 25 m**^{2} **plots in desert, succulent and Nama-karoo (Cowling *et al.* 1994) relative to total rainfall and an index of summer aridity**[b]**. Total (cumulative) species richness and succulent species richness in five 25 m**^{2} **plots (125 m**^{2}**) is also shown**

Site	Annual rain (mm)	Summer aridity index	Growth-form diversity (25 m^2)	Species richness (25 m^2)	Total species (125 m^2)	Succulent species (125 m^2)
Desert						
Namib coast	16.7	6.43	0.0	0.8	1	0
Namib inland	87.0	4.72	0.14	1.2	2	0
Namib-karoo						
Amospoortjie	144.7	4.87	1.53	6.4	13	4
Merweville	158.4	4.90	1.14	3.6	7	4
Carnarvon	188.3	4.37	1.48	7.0	12	1
Molteno's pass	240.6	4.23	0.75	7.2	14	1
Succulent karoo						
Tanqua	72.3	6.31	1.10	6.6	12	7
Ladismith	100.8	5.29	1.15	14.0	28	20
Garies	134.0	6.30	1.27	4.6	9	5
Knersvlakte	146.0	6.27	0.38	8.2	12	12
Springbok	162.0	5.86	1.42	7.6	15	10
Tierberg	168.8	4.80	1.21	10.6	20	13
Worcester	266.1	5.42	1.64	14.4	19	16

[a] Shannon–Weiner diversity index.
[b] Defined in Chap. 5, this volume.

(Von Willert *et al.* 1990; Milton, Dean & Siegfried 1994b). Salt-accumulating succulents use less energy (dry matter) to store water than do other succulents (Von Willert *et al.* 1990). Freezing temperatures limit the distribution of succulents (Chap. 16, this volume), but accumulation of salts and acids may serve to elevate the freezing point.

Non-structural carbohydrates, largely fructans and starch, in the evergreen leaf-succulent *Ruschia spinosa* (L. Dehn) are stored mainly in the leaves and stems (Van der Heyden 1992). Carbohydrate reserves show little seasonal fluctuation, probably because growth uses current photosynthates rather than stored reserves. Storage sites for non-structural carbohydrates in shrubby Asteraceae differ between deciduous *Osteospermum sinuatum* and evergreen *Pteronia pallens*. Both species store fructans in vacuoles, thereby increasing their osmotic potential, possibly improving their cold and drought tolerance (Van der Heyden 1992). However, in *P. pallens* carbohydrates are stored in leaves and stems, whereas in *O. sinuatum* they accumulate in roots and shoots. Stored carbohydrates are used to replace tissue removed by defoliation (Van der Heyden 1992).

Cool temperatures during the growing season in the western succulent karoo explain why C_3 grasses make up >90% of the limited grass cover in this area (Vogel, Fuls & Ellis 1978), whereas aridity may account for the lack of malate-forming, C_4 grasses (Werger 1985). Further east, grass cover comprises a mixture of C_3 and C_4 species (Werger 1985). Perennial C_4 aspartate-forming grasses (*Aristida*, *Eragrostis*, *Stipagrostis*) dominate vegetation on sand patches throughout the succulent karoo. The physiology of this phenomenon remains unexplored, but is

probably a result of the inverse texture effect whereby, in arid climates (<300 mm), moisture is more accessible to plants in coarse than in fine soils (Noy-Meir 1973).

7.4 Community structure and dynamics

7.4.1 Diversity

Species richness of perennial plants at the community scale (Table 7.5) is unusually high (4–20 spp. m^{-2}, 6–23 spp. 25 m^{-2}, 32–113 spp. 1000 m^{-2}) in the succulent karoo compared with the Nama-karoo (2–15 spp. m^{-2}, 4–23 spp. 25 m^{-2}, 23–78 spp. 1000 m^{-2}), or with other parts of the world that have a comparable climate (Cowling *et al.* 1989), for example, the Mojave and Sororan Deserts (10–12 spp. 25 m^{-2}, 4–17 spp. 100 m^{-2}, 33–63 spp. 1000 m^{-2}) (Bond 1983; Esler *et al.* 1994; Hunter 1994).

At community level (25 m^2), spatially heterogeneous slopes were not richer in species or more diverse in structure than the adjacent homogeneous plains. However, at the site scale (1000 m^2) the slopes were richer, probably reflecting higher habitat diversity than on plains (Jürgens 1986). Measures of rainfall evenness were the strongest predictors of total species numbers per site (Cowling *et al.* 1994).

Richness of leaf-succulent species is negatively correlated with an index of climatic heterogeneity (Cowling *et al.* 1994). Highest values are found in the strongly winter rainfall area along the Atlantic coastline. Warm climates with non-seasonal rainfall provide the widest variety of opportunities for structural niche differentiation (Cody 1989), and Cowling *et al.* (1994) found that cover was more equitably distributed among growth forms in the non-seasonal-rainfall southern karoo than in the winter-rainfall region (Fig. 7.18). Species richness in the Nama-karoo was correlated with growth-form diversity, but in the winter rainfall area, species richness was high relative to growth-form diversity, implying a considerable functional redundancy among dwarf leaf succulents. This need not mean that the species-rich communities of the western succulent karoo are resilient to species loss. Similar growth forms may vary in phenology (Van Rooyen *et al.* 1979a; Struck 1992), germination requirements (Esler 1993; Milton 1995) and their responses to such disturbances as drought, herbivory, hail and the rare event of fire. The differential responses of leaf succulents to catastrophic events may be one of the mechanisms that maintains exceptionally high species richness at the community level (Cowling *et al.* 1994).

7.4.2   Dynamics

The dynamics of some succulent karoo plant communities have been studied with respect to competition, disturbance and herbivory. The scales at which the dynamics of these succulent karoo plant communities function are reasonably well known, but the rate of change and the frequencies of the major events that activate these changes are unknown.

7.4.2.1   *Competition, facilitation and cyclic succession*

Interspecific competition, as the mechanism leading to species replacements along spatial gradients of resource availability, was inferred from nearest-neighbour data at Tierberg (Esler & Cowling 1993) and verified experimentally by P. Drennan (unpubl. data), who found that competitive interactions between *Pteronia* species, grown under nursery conditions, varied with soil pH and calcium concentration. At Worcester most of the competitive interactions between establishing plant species were not influenced by changes in moisture supply (Stokes 1994). An exception was the interaction between non-succulent *Pteronia pallens* and succulent *Ruschia spinosa* (L. Dehn) shrub seedlings in which the former dominated under moist condition, but was outperformed by the succulent at low moisture levels. Thus the proximate factor in the distribution of species is edaphic or climatic, but the ultimate factor, in the absence of disturbance, is interspecific competition.

Some species gain a competitive advantage through addition of salt to the soil and others through rapid extraction of water from the soil. Despite their short life spans, salt-accumulating succulents *Mesembryanthemum crystallinum* and *M. guerichianum* can dominate communities for decades by depositing saline detritus on the soil surface, thereby inhibiting germination of other species (Vivrette & Muller 1977; Wentzel *et al.* 1994). Economical

use of water when the soil is moist may be disadvantageous, particularly when plants are competing for water with neighbours that use this resource extravagantly (Midgley & Moll 1993). This notion, based on laboratory studies of stomatal conductance by *Eriocephalus ericoides* and *Pentzia incana* from the Karroid Broken Veld, suggest a situation where individuals that over-exploit a common resource gain relative to conservative users of the resource.

Competition for water within the shrub guild may determine spacing of perennial plants and hence the abundance of the ephemeral component. Near Springbok, for example, the arid climate, infrequent establishment of the dominant succulent shrub (*Leipoldtia constricta*) and strong intraspecific competition resulted in wide spacing of shrubs and consequently, an abundance of the annual forb *Gorteria diffusa* (Cunliffe *et al.* 1990). In higher rainfall regions of the succulent karoo, where perennials are more densely packed, the annual component of undisturbed vegetation is minor (Milton 1990a). Perennial species coexist through spatial and temporal partitioning of resources (Van der Heyden 1992) and microsites (Milton 1990a; Esler 1993), and through small-scale disturbances that provide refuges for less competitive species (Wiegand *et al.* 1995).

Although deep-rooted perennials out-compete colonizers of open ground, they facilitate the establishment of members of the shallow-rooted, shade-loving succulent guild. At Anysberg in the Karroid Broken Veld, Beukman (1991) found that *Tylecodon wallichii* survived and flowered better beneath *Pteronia pallens* than beneath *Lycium* or *Ruschia*. Temperatures were better buffered and nutrient levels higher beneath *P. pallens* than beneath other nurse plants, and there was minimal overlap in the root distribution of the succulent and shrub. Some stem succulents that need shade for establishment later outgrow and suppress their host plants, and thus occupy exposed positions as adults (Gibson & Yeaton 1994).

A conceptual model for cyclical succession in the succulent karoo (Yeaton & Esler 1990), where the vegetation consists of a series of plant clusters at different stages of development, emphasizes the role of biotic interactions in determining recruitment patterns and community composition. Facilitation and site modification by colonizing plants may be a major determinant of composition in plains communities comprising succulent Mesembryanthemaceae and non-succulent dwarf shrubs, usually Asteraceae.

Research by Yeaton & Esler (1990) into cluster phase dynamics at Tierberg indicates that several species of small-seeded Mesembryanthemaceae (e.g. *Ruschia spinosa* (L. Dehn), *Brownanthus ciliatus*) germinate and establish in open areas between established plants. These low succulents accumulate detrital materials and fine soils

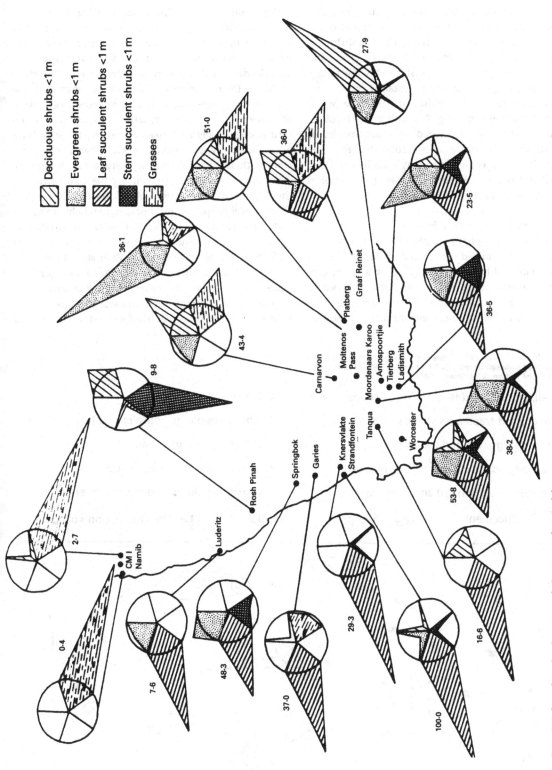

Figure 7.18 Star plots showing mean relative cover of perennial growth forms. Mean total cover of each site is indicated next to the star plots. Reproduced, with permission, from Cowling *et al.* 1994.

under their canopies and their roots stabilize the soil in their immediate vicinity. They also trap tumble-seeds and serve as nurse plants, facilitating the establishment of non-succulent shrubs that later out-compete and replace them (Fig. 7.19). Sheet wash erodes the unvegetated areas, giving the cluster a pedestalled appearance. The non-succulent shrubs persist until they die or are removed through overgrazing. The nutrient-rich soil of the pedestals is either colonized by other species of woody shrubs or erodes and is transported to other mounds. The open, eroded sites may then be colonized by small-seeded Mesembryanthemaceae and a cyclical succession results (Yeaton & Esler 1990).

7.4.2.2 *Disturbance and drought*

Yeaton & Esler (1990) have suggested that dynamics of succulent karoo mixed shrubland communities is directly influenced by *heuweltjies*. Differences in the conductivity, nutrient status and permeability of *heuweltjie* soils relative to the surrounding matrix (see 7.2.3) are further accentuated by the concentration of animal

activities on *heuweltjies*. Disturbances such as frass dumping by termites (*Microhodotermes viator*), burrowing and browsing by rodents (*Cryptomys, Parotomys, Rhabdomys*), excavation for termites by aardvarks (*Orycteropus afer*) and dung burial by antelopes are more frequent on *heuweltjies* (Milton & Dean 1990a) than in surrounding areas where soils are compacted (Dean 1992a). Moreover, vegetation of *heuweltjies* is preferentially browsed by sheep (Armstrong & Siegfried 1990). Together, these disturbances cause plant turnover on *heuweltjies* and further elevate their nutrient status. Disturbance intensities differ among *heuweltjies*, influencing the composition of their plant communities (Helme 1990; Stokes 1994). Many *heuweltjie* species are short-lived and able to regenerate from dormant seed (Esler & Cowling 1995), whereas those in the relatively stable matrix are perennial and often lack soil-stored seed banks (Esler *et al.* 1996).

Once animal activity on a *heuweltjie* ceases, organic material is no longer incorporated into the mound but is moved downslope by wind and water where *Pteronia pallens* and other calcicoles are favoured. As the soil of

Figure 7.19 **Cyclic succession in plains shrublands of the succulent karoo, based on conceptual models by Yeaton & Esler (1990), Esler (1993) and Milton & Hoffman (1994).**

0 Small, smooth and large, winged seeds	4 Shrubs
1A Bare ground traps small seeds ●	5 Shrub seedlings complete with succulents
1B Pits trap large, winged seeds ✳	6 Shrubs grow and seed
2 Small-seeded succulents grow and seed	7 Shrubs die, litter traps seed
3 Succulents trap winged shrub seeds	8 Shrub seedlings establish in litter

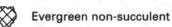 Succulent  Evergreen non-succulent | Deciduous non-succulent

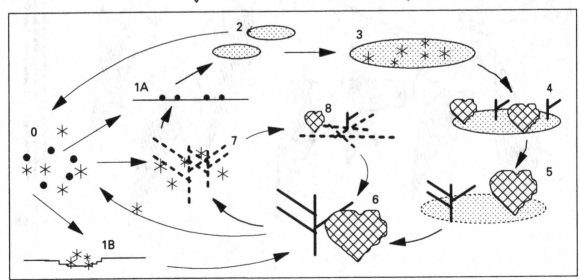

an inactive *heuweltjie* gradually becomes more acidic through leaching, it is colonized by shrubs from the surrounding plains. Thus a major determinant of the dynamics of these succulent karoo communities may be the erosion and movement from *heuweltjies* of base-rich soils to adjacent, more acidic, zonal soils (Yeaton & Esler 1990). This process may be accelerated by overgrazing, which denudes *heuweltjies*, promoting erosion (Steinschen *et al.* 1996).

Animals influence the dynamics of plant communities in other ways. Harvester ants, *Messor capensis*, construct mounds of organic matter and soil over their brood chambers and granaries. These nest-mounds, which occur at densities of 2–23 ha^{-1} (Dean & Yeaton 1993b), locally increase soil fertility and plant productivity and influence vegetation dynamics (Dean & Yeaton 1993a,b). The granaries contain viable seeds (Milton & Dean 1993), some of which are dispersed into sites suitable for seedling establishment when the nest mounds are excavated by mammals foraging for ants (Dean & Yeaton 1992).

Disturbance reduces density of perennials to the advantage of weaker competitors. At Tierberg, in 25-m^2 plots from which shrubs had been cleared in the previous year, annuals increased relative to their densities in surrounding undisturbed vegetation (Milton 1995). Ploughed lands and livestock concentration points, where both vegetation and soil have been disturbed, may take more than 80 years to be recolonized by long-lived plants, and are generally dominated for decades by pauciennial succulents (Beukes, Cowling & Ellis 1994; Wentzel *et al.* 1994; Dean & Milton 1995).

Seasonal changes in vegetation composition are most marked in Namaqualand, where annual forbs and geophytes emerge in the winter and spring. The abundance of annuals is influenced by the quantity and distribution of rainfall (Rösch 1977) as well as by interactions with perennial shrubs and grasses (Cunliffe *et al.* 1990). Perennial vegetation cover at Goegap Nature Reserve remained between 26% and 37% over 15 years, despite deviations of up to 80% above and 60% below mean annual rainfall (150 mm) during this period (Fairall & Le Roux 1991). Cover variation probably reflected canopy growth or die-back in stem-deciduous and pauciennial species rather than turnover in long-lived shrubs.

Responses of succulent karoo plant communities to drought appear to vary with the intensity and duration of the drought. Drought in the Richtersveld in 1979 killed about 80% of a predominantly leaf-succulent plant community, whereas exceptionally high rainfall from 1981 to 1983 initiated a recovery in the vegetation (Von Willert *et al.* 1985). During 1990/91, the Steytlerville karoo received 30% of its mean annual rainfall. This

drought killed 65% of the non-succulent shrubs, mainly *Pentzia* and *Eriocephalus* and 42% of the succulents. Annual forbs, grasses and pauciennial succulents regenerated from soil-stored seed and replaced dead perennials within one year, but non-succulent shrub populations, lacking a durable seed bank, recovered very slowly (Milton *et al.* 1995). Moisture stress affects the competitive interactions between seedlings of certain species, and hence determines their recruitment success during dry years (Stokes 1994), which may lead, in turn, to changes in community composition.

Hail, which is uncommon in the succulent karoo, damaged succulents (38%) more than non-succulents (11%) at Tierberg in September 1987 when plants were growing and flowering (Milton & Collins 1989). Similarly Powrie (1993) found that 14 months after hail at Williston in the Nama-karoo, the only succulent species (*Eberlanzia horrida*) in the vegetation had suffered more damage and subsequent recruitment than non-succulent shrubs. Succulent karoo vegetation seldom burns, but fire in January 1988 near Worcester killed succulent and non-succulent dwarf shrubs. Arborescent succulents (*Tylecodon paniculatus*, *Aloe*) and a tree (*Euclea undulata*) survived the fire. There was little regeneration after two months (Forrester 1988). However, smoke-induced germination has been reported for some Mesembryanthemaceae from fire-free habitats (Pierce, Esler & Cowling 1995).

Despite the longevity of some perennials in succulent karoo communities, continual disturbance by animals and drought maintains a niche for ephemeral and pauciennial species. Differences in scale of disturbances and in intensity and duration of drought may partly account for the mixture of survival strategies and longevities found in superficially homogeneous dwarf shrublands.

7.4.2.3 *Grazing and browsing*

Carbon isotopes in the soil indicate that, although C_4 grasses have decreased in recent times in the eastern Karoo, little change in the $C_3 : C_4$ ratio has occurred in the southwestern Karoo, where C_3 shrubs were dominant before European settlement (Bond *et al.* 1994). This does not imply that the composition of succulent karoo vegetation was unaffected by changes in land use. Herbivory has altered plant community composition at the patch and ranch scale (Dean *et al.* 1995; Chap. 17, this volume). Increases in toxic perennials and reductions in forage plants (Milton 1994b; Beukes *et al.* 1994) necessitated a reduction of 75% in the density of domestic livestock in this biome between 1850 and 1980 (Dean & Macdonald 1994).

The mechanisms for grazing-induced changes in succulent karoo plant communities are:

1 Increased mortality of grazed/browsed plants (Stokes 1994);

2 Reduced competitive ability of grazed/browsed plants, allowing them to be out-competed by species less palatable to livestock (Yeaton & Esler 1990; Stokes 1994);

3 Reduced establishment of grazed plants through direct and indirect impacts on seed production (Milton 1992a, 1994b; Milton & Hoffman 1994; Stokes 1994);

4 Changes in cover, radiation, runoff, soil texture and salinity, which prevent the establishment of plants that require sheltered microsites (Milton *et al.* 1994a) but favour the establishment of self-perpetuating halophyte assemblages (Beukes *et al.* 1994).

Exclusion of domestic livestock does not result in a rapid increase in palatable species. After 15 years without domestic livestock the perennial vegetation cover of Goegap Nature Reserve increased from 26% to 37%, but unpalatable species, which had been dominant before protection, contributed more to the cover increase than palatable plant species (Fairall & Le Roux 1991). At Tierberg, there were no measurable changes in cover, composition or seedling abundance six years after the exclusion of sheep, although highly palatable species set more seed on protected than on grazed plots (Milton & Dean 1990b, 1993; Milton 1992a). Restoration of overgrazed and eroded sites at Worcester to diverse perennial communities was incomplete after 45 years (Smitheman & Perry 1990).

Established plants of species that are superior competitors limit rates of population growth in less competitive species. This may partly explain why the return of palatable perennials to overgrazed rangeland is so slow (Milton & Hoffman 1994; Stokes 1994). Clearing of shrubs in the succulent karoo generally leads to high-density recruitment of a variety of small-seeded leaf-succulents and Aizoaceae within three years (Stokes 1994; Milton 1995).

The dynamics of a succulent karoo plant community has recently been simulated by means of spatial models that incorporate available plant life-history information and climatic data. Results suggest that decades of quasi-stability may be interrupted by sudden changes in species composition (Wiegand *et al.* 1995). The reason for this episodic behaviour is that both mortality and establishment of plants depend on particular, rare sequences of events that differ between species. Abundant recruitment of less competitive species occurs only when unusually favourable rainfall coincides with below-average density of established plants. Such models are helpful in conceptualizing community dynamics on temporal and spatial scales that cannot be monitored in the field, and can be used for testing long-term effects of management interventions including grazing, clearing and reseeding.

7.5 Ecosystem function

7.5.1 Biomass, productivity and energy flow

Above-ground biomass in the succulent karoo ranges from 400 to 7000 kg (dry mass) ha^{-1} (Werger 1985; Rutherford & Westfall 1986; Milton 1990b), which is similar to the Nama-karoo but low relative to fynbos and savanna. Fresh mass comprises about 60–80% water compared with 25–35% in the Nama-karoo (Milton 1990b; Werger & Morris 1991). Above-ground phytomass (dead and live) sampled in the southern Mojave and Sonoran Deserts of north America (annual rainfall 138–279 mm), was 2100 to 7400 kg ha^{-1} dry weight (MacMahon & Wagner 1985; Hunter 1994). Therefore, although the upper biomass limits were comparable for the American and southern African winter-rainfall deserts, the lower limit for the Mojave was five times higher than for the succulent karoo. More recently, Esler *et al.* (1994) have reported that biomass values in winter-rainfall deserts of Baja California and Chile are higher than in the succulent karoo.

Productivity of arid and semi-arid communities of drought-tolerant succulents and shrubs is expected to be 10 to 20% of the biomass (Noy-Meir 1973). The productivity of both the succulent karoo (200 to 1000 dry kg ha^{-1} yr^{-1}) (Rutherford & Westfall 1986) and the Mojave Desert International Biological Programme sites, where annuals produce 0 to 700, and perennials 200 to 600 dry kg ha^{-1} yr^{-1} (MacMahon & Wagner 1985), falls within predicted limits.

In the western part of the succulent karoo, ephemerals contribute a major portion (122–548 kg ha^{-1}) of the grazing available to domestic livestock during the winter and spring (Barnard *et al.* 1993). Many of the perennials are unpalatable to domestic livestock (Louw 1969; Milton *et al.* 1994b; Table 7.3). The concentrated salts, oxalic acid crystals (Milton *et al.* 1994b), tannins (Van der Heyden 1992) and cardiac glycosides found in some succulents reduce the nutritional value of these plants, or are toxic to ruminants (Kellerman *et al.* 1988). Non-succulent evergreen shrubs, particularly Asteraceae, may contain resins, phenolic compounds or toxic alkaloids (Louw 1969; Milton *et al.* 1994b) which likewise make them unacceptable as food for most ruminants (Chap. 17, this volume).

Insects appear to be more important than indigenous vertebrates as consumers of primary productivity in this biome. It is probably the parasite (sapsuckers, endophages) and detritivore guilds, rather than the folivore guild, that convert most carbohydrate to animal protein. Species-specific insect herbivores may also help to maintain the diversity of succulent karoo plant communities, because their impacts are patchy in space and time (Milton 1993). Below-ground biomass of root-feeding cicadas (2.5 kg ha^{-1}) at Tierberg (Dean & Milton 1991a) was equal to 25% of the recommended biomass for stocking sheep and was estimated to exceed the biomass of indigenous above-ground mammals (1.4 kg ha^{-1}).

7.5.2 Decomposition and redistribution of nutrients

The most prominent detritivore in the succulent Karoo is the termite *Microhodotermes viator* (Coaton & Sheasby 1974), but other important groups are the harvester ant *Messor capensis* (Milton & Dean 1993; Dean & Yeaton 1993a), tenebrionid beetles and little-known collembola and mites. In common with parasites, detritivores are seldom seen and their role in the functioning of succulent karoo ecosystems is largely unexplored. Harvester ants and harvester termites are a major food of the aardvark and bat-eared fox (Kuntzsch & Nel 1992), mammals which apparently maintain diversity in succulent-karoo vegetation by dispersing seed and generating small-scale disturbances (Dean & Milton 1991b; Dean 1992a; Dean & Yeaton 1992).

There is little published information on decomposition processes or nutrient cycling in the succulent karoo, despite the apparent importance of detritivores in ecosystem functioning and community dynamics. The activities of harvester ants and termites, which collect large quantities of seeds and dead organic matter and store it beneath the ground for direct consumption, increase the organic carbon content of the soil, locally improving infiltration and water-holding capacity (Dean 1992a). A microbiological study of soils at Worcester indicated that bacterial and actinomycete densities were higher on *M. viator* mounds (*heuweltjies*), but that fungal spores were more numerous in matrix vegetation. Vegetation clearing reduced the microbial population of both sites (Trusler & Yeaton 1991).

It is possible that decay of surface litter is water limited in summer and temperature limited in winter. Under arid conditions, dung and urine deposition and soil compaction cause salinization around water points (Beukes *et al.* 1994) and in livestock holding pens (Wentzel *et al.* 1994). Altered plant assemblages presumably reflect changes in water and nutrient assimilation and ecosystem function.

The significance of parasitic arthropods and detritivores in the succulent karoo is thus four-fold. Firstly, they are keystone resources for the vertebrates that disperse plants or alter microtopography; secondly, arthropod herbivores maintain plant communities in a dynamic state by debilitating individual plants; thirdly, soil dwelling arthropods and detritivores increase the permeability and water-holding capacity of soils; and finally, detritivores concentrate nutrients and salts patchily in the landscape, increasing spatial heterogeneity.

7.6 Conservation and management

Agriculture, mining and ecotourism represent both threats and opportunities for the conservation of succulent karoo ecosystems. On one hand, they generate public awareness of the region and place an economic value on the land. On the other hand, the future of many plant and animal species may be endangered by these activities.

7.6.1 Agriculture

Nomadic pastoralists first brought sheep into the succulent karoo about 2000 years ago, and cattle some 1500 years later (Smith 1986; Chap. 21, this volume). Indigenous antelope (Skead 1980) and nomadic hunting and pastoral people used the succulent karoo in the winter and spring growing seasons, but moved northwards to the Nama-karoo in summer (Penn 1986; Archer, Hoffman & Danckwerts 1989). The European pastoralists (*trekboers*), who moved northwards from the Cape Peninsula in the 18th century, were also nomadic, moving with their flocks to suitable grazing (Penn 1986). In the 19th century, the succulent karoo became the first biome to be used for settled European pastoralism (Christopher 1982). However, the extremely arid summers make much of the region unsuitable for settled pastoralism, even now when boreholes provide perennial water and forage can be imported from other biomes. Initially, boreholes gave domestic livestock access to vegetation that was unavailable to nomadic pastoralists because of the lack of surface water. Today, stocking rates in the succulent karoo are inversely related to water-point density, suggesting that previously untapped forage reserves, far from natural surface water, have now been depleted (Dean & Macdonald 1994).

Between 1800 and 1980, stocking rates for domestic livestock declined further in the succulent karoo than in any other arid and semi-arid region in South Africa. Stock reduction exceeded 50% in seven of the biome's eight magisterial districts (Dean & Macdonald 1994). Assuming that stocking rates are an index of carrying capacity, it may be inferred that domestic herbivores have reduced palatable plant populations in the succulent karoo more than they have in other biomes. This may be the result of either a relatively long history of

European pastoralism, or a higher susceptibility of these plant species to herbivore-induced elimination.

The tendency for herbivory by domestic livestock to alter the composition of vegetation to a less useful 'state', which cannot easily be reversed by resting, further complicates ranching in this biome. Perennial species lost through grazing mismanagement do not re-establish within a few decades, because they lack dormant seeds, have short dispersal distances, or are excluded by competition from other perennials that have filled the spaces left by their demise (Milton & Hoffman 1994). There appears to be no rapid, reliable or economically feasible method for restoring function and species diversity to such rangeland.

Most soils in the succulent karoo are not deep enough for ploughing, but few alluvial terraces have escaped cultivation. In southern Namaqualand, 2449 km² have been ploughed (Macdonald 1989). Although the biome is too arid for regular dry-land agriculture, less than 10% of ploughed lands are under irrigation and most are currently unused. Wind erosion is a problem in ploughed and overgrazed parts of Namaqualand and the Richtersveld (Pickett & Hoffman 1992).

Annual grasses, including *Stipa capensis* and European species of *Avena*, *Bromus* and *Hordeum*, increase on deep loamy soils in the Vanrhynsdorp and Nieuwoudville area where perennial cover has been reduced by ploughing and grazing (Steinschen *et al.* 1996). These species could pose a threat to indigenous annuals and geophytes in areas that are renowned for the diversity and floral display of these groups (Snijman & Perry 1987; Vlok 1988, 1994).

7.6.2 Mining

The arid winter rainfall region is rich in minerals deposited by evaporation and igneous activity. In the north of the region, intrusive granites and overlying metamorphic rocks contain copper, silver, zinc and uranium. Surface deposits of diamondiferous gravel occur along the northwestern coast (Fig. 7.20). Further south near Vanrhynsdorp there are surface deposits of kaolin, gypsum and rare earths. The Tanqua Karoo is rich in manganese, and phosphates and limestone are mined on the southwestern coastal plains (Geological Survey 1984). Surface mineral deposits are usually strip-mined with consequent destruction of the overlying vegetation. Subterranean mining has a relatively small impact on vegetation, most of the damage being caused by the establishment of mining settlements. Apart from its unquantified impact on succulent plant populations, past surface mining in Namaqualand has left large heaps of unvegetated spoil within sight of major tourist routes. Changes in microtopography and salinity of strip-mined surfaces and paucity of information on establishment

Figure 7.20 **The exploitation of surface deposits of diamondiferous gravel has resulted in large-scale habitat transformation of a narrow coastal strip in the northwestern part of the succulent karoo** (Photo: P.G. Desmet).

requirements of indigenous plants are among the problems that face restoration ecologists on these sites (De Villiers 1993).

7.6.3 Ecotourism

Major assets of the succulent karoo are the spring flower displays and succulent flora, which, together with spectacular scenery, habitat diversity, proximity to major cities and well-developed infrastructure, provide potential for the development of a tourist industry. A successful ecotourist industry should satisfy the desire of its clients to see both spectacular flowering and diversity. Prediction of the best time to visit various sites depends on the understanding of vegetation responses to rainfall and management. The flowering-season model (Steyn *et al.* 1994) demonstrates that successful predictions can be generated if adequate information is available.

Ecotourism is an option that could bring the biome to the attention of world conservation bodies as well as generating employment opportunities in transport, accommodation and interpretation. However, natural resources will need careful management. Small populations of endemic succulents are threatened by unscrupulous collecting (Hilton-Taylor & Le Roux 1989). Disturbances, including heavy grazing, ploughing and ripping are sometimes carried out in Namaqualand to increase the extent of the spring flower displays (Milton & Dean 1991), but they also reduce diversity (Vlok 1994; Steinschen *et al.* 1996). An economic comparison of the financial (income) and social (employment) benefits of ranching and tourism in the most arid and overgrazed parts of the biome is needed to guide its future development.

7.6.4 Conservation implications

The formal conservation status of the succulent karoo is poor (Chap. 23, this volume). Land use has changed and continues to change the vegetation of the karoo, but prospects for acquisition of additional land for conservation zoning appear bleak given the public apathy towards this arid region. In a recent opinion survey conducted among government and private conservation bodies, karoo conservation received the lowest priority rating (Macdonald, Van Wijk & Boyd 1993). Conservation of much of this unique and exceptionally rich succulent flora therefore rests in the hands of landowners.

7.7   Conclusions

Our understanding of vegetation diversity, physiology and dynamics in the succulent karoo biome has improved substantially over the past decade. However, vegetation research in this biome is in its infancy and many large gaps in our knowledge need to be filled if its management and conservation status are to be improved. We conclude with a list of research priorities.

- The region is generally under-collected, distribution ranges are poorly documented and

the taxonomy of some of its most diverse groups, particularly Mesembryanthemaceae and Liliaceae (*sensu lato*), require major revision.

- There is very little information on ecosystem function, particularly decay rates, nutrient cycling and effects of various types of land use on these processes.
- Our understanding of the role of vertebrates and invertebrates in maintaining the diversity of succulent karoo plant communities is inadequate.
- Information on factors limiting recolonization of denuded and altered soils in this biome is urgently needed, to restore lost productivity and maintain plant species diversity in this unique arid biome.
- Comparative studies, which highlight differences in diversity, structure and dynamics between the succulent karoo and other winter-rainfall deserts, are needed, to gain an understanding of the physical and evolutionary forces that have conspired to produce this unique flora.

.8   References

cocks, J.P.H. (1953). Veld Types of South Africa. *Memoirs of the Botanical Survey of South Africa*, **28**, 1–192.

cocks, J.P.H. (1976). Riverine vegetation of the semi-arid and arid regions of South Africa. *Journal of South African Biological Society*, **17**, 21–35.

cocks, J.P.H. (undated). *Veld Types of the Western Half of the Republic of South Africa*. Unpublished typescript. Cape Town: National Botanical Institute.

damson, R.S. (1938). Notes on the vegetation of the Kamiesberg. *Memoirs of the Botanical Survey of South Africa*, **18**, 1–25.

rcher, F., Hoffman, M.T. & Danckwerts, J.E. (1989). How economic are the farming units of Leliefontein, Namaqualand. *Journal of the Grassland Society of Southern Africa*, **6**, 211–15.

rmstrong, A.J. & Siegfried, W.R. (1990). Selective use of *heuweltjie* earth mounds by sheep in the Karoo. *South African Journal of Ecology*, **1**, 77–80.

rnold, T.H. & de Wet, B.C. (1993). Plants of southern Africa: names and distribution. *Botanical Survey of South Africa Memoirs*, 62. Pretoria: Botanical Research.

Barnard, S.A., Gerber, H.S., van Heerden, J.M. & Frey, D.J. (1993). 'Opslag' production in arid regions of the winter rainfall region of the RSA. *Bulletin of the Grassland Society of southern Africa*, **4**, 33.

Bayer, M.B. (1978). Pollination in Asclepiads. *Veld & Flora*, **64**, 21–3.

Beneke, K., van Rooyen, M.W., Theron, G.K. & van der Venter, H.A. (1993). Fruit polymorphism in ephemeral species of Namaqualand: III. Germination differences between the polymorphic diaspores. *Journal of Arid Environments*, **24**, 333–44.

Beukes, P.C., Cowling, R.M. & Ellis, F. (1994). Vegetation and soil changes across a succulent karoo grazing gradient. *Arid Zone Ecology Forum Abstracts*, p. 23. Pretoria: Foundation for Research Development.

Beukman, R.P. (1991). *The Role of Nurse Plants in the Vegetation Dynamics of the Succulent Karoo*. MSc Thesis. Cape Town: University of Cape Town.

Bond, W.J. (1983). On alpha-diversity and the richness of the Cape flora: a study in southern Cape fynbos. In *Mediterranean-Type Ecosystems: the Role of Nutrients*, ed. F.J. Kruger, D.T. Mitchell & J.U.M. Jarvis, pp. 337–56. Berlin: Springer.

Bond, W.J. & Slingsby, P. (1983). Seed dispersal by ants in shrublands of the Cape Province and its evolutionary implications. *South African Journal of Science*, **79**, 231–3.

Bond, W.J., Stock, W.D. & Hoffman M.T. (1994). Has the Karoo spread? A test for desertification using carbon isotopes from soils. *South African Journal of Science*, **90**, 391–7.

Boobyer, M.G. (1989). *The Eco-ethology of the Karoo Korhaan Eupodotis vigorsii*. MSc Thesis. Cape Town: University of Cape Town.

Boshoff, C.R. (1989). *Structure, Distribution and Phenology of Perennial Plant Species in the Worcester Veld Reserve, in the Arid Winter Rainfall Region of the Southwestern Cape*. MSc Thesis. Cape Town: University of Cape Town.

Boucher, C. & Jarman, M.L. (1977). The vegetation of the Langebaan area, South Africa. *Transactions of the Royal Society of South Africa*, **42**, 241–72.

Breytenbach, G.J. (1988). Why are myrmecochorous plants limited to fynbos (Macchia) vegetation types? *South African Journal of Forestry*, **144**, 3–5.

Campbell, B.M. (1985). A Classification of the Mountain Vegetation of the Fynbos Biome. *Memoirs of the Botanical Survey of South Africa*, 50, 1–119.

Christopher, A.J. (1982). Towards a definition of the nineteenth century South African frontier. *South African Geographic Journal*, 64, 97–113.

Coaton, W.G.H & Sheasby, J.L. (1974). National survey of the Isoptera of southern Africa. 6. The genus *Microhodotermes* Sjostedt (Hodotermitidae). *Cimbebasia*, A3, 47–59.

Cody, M.L. (1989). Growth-form diversity and community structure in desert plants. *Journal of Arid Environments*, 17, 199–209.

Cody, M.L. (1993). Do cholla cacti (*Opuntia* spp., subgenus Cylindropuntia) use or need nurse plants in the Mojave Desert? *Journal of Arid Environments*, 24, 139–54.

Cole, D.T. (1988). *Lithops – Flowering Stones*. Johannesburg: Acorn Books.

Compton, R.H. (1929). The vegetation of the Karoo. In *The Botanical Features of the South Western Cape Province*, pp. 40–67. Cape Town: Speciality Press.

Compton, R.H. (1931). The flora of the Whitehill District. *Transactions of the Royal Society of South Africa*, 19, 269–329.

Cowling, R.M., Esler, K.J., Midgley, G.F. & Honig, M.A. (1994). Plant functional diversity, species diversity and climate in arid and semi-arid southern Africa. *Journal of Arid Environments*, 27, 141–58.

Cowling, R.M., Gibbs Russell, G.E. Hoffman, M.T. & Hilton-Taylor C. (1989). Patterns of plant species diversity in southern Africa. In *Biotic Diversity in Southern Africa. Concepts and Conservation*, ed. B.J. Huntley, pp. 19–50. Cape Town: Oxford University Press.

Cunliffe, R.N., Jarman, M.L., Moll, E.J. & Yeaton, R.I. (1990). Competitive interactions between the perennial shrub *Leipoldtia constricta* and an annual forb, *Gorteria diffusa*. *South African Journal of Botany*, 56, 34–8.

Dean, W.R.J. (1992a). Effects of animal activity on the absorption rate of soils in the southern Karoo, South Africa. *Journal of the Grassland Society of southern Africa*, 9, 63–74.

Dean, W.R.J. (1992b). Temperatures determining activity patterns of some ant species in the southern Karoo, South Africa. *Journal of the Entomological Society of southern Africa*, 55, 149–56.

Dean, W.R.J., Hoffman, M.T., Meadows, M.E. & Milton, S.J. (1995). Desertification in the semi-arid Karoo, South Africa: review and reassessment. *Journal of Arid Environments*, 30, 247–64.

Dean, W.R.J. & Macdonald, I.A.W. (1994). Historical changes in stocking rates of domestic livestock as a measure of semi-arid and arid rangeland degradation in the Cape Province, South Africa. *Journal of Arid Environments*, 26, 281–98.

Dean, W.R.J., Midgley, J.J. & Stock, W.D. (1994). The distribution of mistletoes in South Africa: patterns of species richness and host choice. *Journal of Biogeography*, 21, 503–10.

Dean, W.R.J. & Milton, S.J. (1991a). Emergence and oviposition of *Quintillia* cf. *conspersa* Karsch (Homoptera: Cicadidae) in the southern Karoo, South Africa. *Journal of the Entomological Society of Southern Africa*, 54, 111–19.

Dean, W.R.J. & Milton, S.J. (1991b). Disturbances in semi-arid shrubland and arid grassland in the Karoo, South Africa: mammal diggings as germination sites. *African Journal of Ecology*, 29, 11–16.

Dean, W.R.J. & Milton, S.J. (1993). The use of *Galium tomentosum* (Rubiaceae) as nest material by birds in the southern Karoo. *Ostrich*, 64, 187–9.

Dean, W.R.J. & Milton, S.J. (1995). Plant and invertebrate assemblages on old fields in the arid southern Karoo, South Africa. *African Journal of Ecology*, 33, 1–13.

Dean, W.R.J., Milton, S.J. & Siegfried, W.R. (1990). Dispersal of seeds as nest material by birds in semi-arid Karoo shrubland. *Ecology*, 71, 1299–306.

Dean, W.R.J., Williams, J.B. & Milton, S.J. (1993). Breeding of the White-backed Mousebird *Colius colius* in relation to rainfall and the phenology of fruiting plants in the southern Karoo, South Africa. *African Journal of Zoology*, 107, 105–11.

Dean, W.R.J. & Yeaton, R.I. (1992). The importance of harvester ant *Messor capensis* nest mounds as germination sites in the southern Karoo, South Africa. *African Journal of Ecology*, 30, 335–45.

Dean, W.R.J. & Yeaton, R.I. (1993a). The effects of harvester ant *Messor capensis* nest mounds on physical and chemical properties of soils in the southern Karoo, South Africa. *Journal of Arid Environments*, 25, 249–60.

Dean, W.R.J. & Yeaton, R.I. (1993b). The influence of harvester ant *Messor capensis* nest mounds on the productivity and distribution of some plant species in the southern Karoo, South Africa. *Vegetatio*, 106, 21–35.

de Villiers, A.J. (1993). *Ecophysiological Studies on Several Namaqualand Pioneer Species, With Special Reference to the Revegetation of Saline Mined Soil*. MSc Thesis. Pretoria: University of Pretoria.

Eliovson, S. (1972). *Namaqualand in Flower*. Johannesburg: Macmillan.

Ellery, W.N., Scholes, R.J. & Mentis, M.T. (1991). An initial approach to predicting the sensitivity of the South African grassland biome to climatic change. *South African Journal of Science*, 87, 499–503.

Ellis, F. & Lambrechts, J.J.N. (1986). Soils. In *The Karoo Biome: a Preliminary Synthesis. Part I – Physical Environment*, ed. R.M. Cowling, P.W. Roux & A.J.H. Pieterse, pp. 18–38. *South African National Scientific Programmes Report* 124. Pretoria: CSIR.

Esler, K.J. (1993). *Vegetation Patterns and Plant Reproductive Processes in the Succulent Karoo*. PhD Thesis. Cape Town: University of Cape Town.

Esler, K.J. & Cowling, R.M. (1993). Edaphic factors and competition as determinants of pattern in South African karoo vegetation. *South African Journal of Botany*, 59, 287–95.

Esler, K.J. & Cowling, R.M. (1995). The comparison of selected life history characteristics of *Mesembryanthema* species occurring on and off Mima-like mounds (*heuweltjies*) in semi-arid southern Africa. *Vegetatio*, 116, 41–50.

...ler, K.J., Cowling, R.M. & Ivey, P. (1992). Seed biology of three species of Mesembryanthema in the southern succulent Karoo. *South African Journal of Botany*, **58**, 343–7.

...ler, K.J., Cowling, R.M. & Juritz, J.M. (1996). Dynamics of soil-stored seed banks in semi-arid, succulent Karoo vegetation. *Journal of Vegetation Science* (in press).

...ler, K.J. & Phillips, N. (1994). Experimental effects of water stress on semi-arid Karoo seedlings: implications for field seedling survivorship. *Journal of Arid Environments*, **26**, 325–37.

...ler, K.J., Rundel, P.W. & Cowling, R.M. (1994). Intercontinental comparisons along mediterranean/desert transition zones: climate and vegetation structure. *Noticiero de Biologia*, **2**, 22.

...enari, M., Noy-Meir, I. & Goodall, D.W. (1985). *Hot Deserts and Arid Shrublands*. Amsterdam: Elsevier.

...airall, N. & Le Roux, A. (1991). Game management in arid areas: the non-equilibrium alternative. In *Wildlife Production: Conservation and Sustainable Development*, ed. L.A. Renecker & R.J. Hudson, pp. 251–4. Fairbanks: University of Alaska.

...orrester, J. (1988). Fire in the Karoo N.B.G. Floral Reserve. *Veld & Flora*, **74**, 5.

...eological Survey (1984). Geological Series Maps. Pretoria: Department of Surveys and Mapping.

...ess, S.K. & Gess, F.W. (1989). Flower visiting by masarid wasps in southern Africa (Hymenoptera: Vespoidea: Masaridae). *Annals of the Cape Provincial Museums* (Natural History), **18**, 95–134.

...ibbs Russel, G.E. (1987). Preliminary floristic analysis of the major biomes in southern Africa. *Bothalia*, **17**, 213–27.

...ibson, J.A. & Yeaton, R.I. (1994). A preliminary study on *Euphorbia stellaespina* on the farm Trakaskuilen in the Great Karoo. *Arid Zone Ecology Forum Abstracts*, pp. 47–8. Pretoria: Foundation for Research Development.

...oldblatt, P. (1978). An analysis of the flora of southern Africa: its characteristics, relationships and origins. *Annals of the Missouri Botanical Gardens*, **65**, 369–436.

Grubb, P.J. (1988). The uncoupling of disturbance and recruitment, two kinds of seed bank, and persistence of plant populations at the regional and local scales. *Annali Zoologici Fennici*, **25**, 23–36.

Hammer, S.A. (1991). Scents and sensibility. *Veld & Fora*, **77**, 70–1.

Hammer, S.A. (1993). *The Genus Conophytum – A Conograph*. Pretoria: Succulent Plant Publications.

Hartmann, H.E.K. (1983). Monographien der Subtribus Leipoldtiinae. IV. Monographie der Gattung *Vanzijlia* (Mesembryanthemaceae). *Botanische Jahrbücher für Systematik*, **103**, 499–538.

Hartmann, H.E.K. & Gölling H. (1993). A monograph of the genus *Glottiphyllum* (Mesembryanthema, Aizoaceae). *Bradleya*, **11**, 1–49.

Hartmann, H.E.K. & Stüber, D. (1993). On the spiny Mesembryanthema and the genus *Eberlanzia* (Aizoaceae). *Contributions from the Bolus Herbarium*, **15**, 1–75.

Helme, N. (1990). Disturbance and community dynamics on *heuweltjies*. Unpublished Honours project, Botany Department, University of Cape Town, Rondebosch.

Hilton-Taylor, C. (1986). Growth form distribution along a rainfall seasonality gradient in the southern Karoo. *Karoo Biome Project 2nd. Annual Research Meeting Abstracts*, pp. 8–9. Pretoria: Foundation for Research Development.

Hilton-Taylor, C. & Le Roux, A. (1989). Conservation status of the Fynbos and Karoo Biomes. In *Biotic Diversity in Southern Africa. Concepts and Conservation*, ed. B.J. Huntley, pp. 202–23. Cape Town: Oxford University Press.

Hoffman, M.T. (1989a). *Vegetation Studies and the Impact of Grazing in the Semiarid Eastern Cape*. PhD Thesis. Cape Town: University of Cape Town.

Hoffman, M.T. (1989b). A preliminary investigation of the phenology of subtropical thicket and karroid shrubland in the lower Sundays River Valley, SE Cape. *South African Journal of Botany*, **55**, 586–98.

Hoffman, M.T., Barr, G.D. & Cowling, R.M. (1990). Vegetation dynamics in the semi-arid eastern Karoo, South Africa: the effect of seasonal rainfall and competition on grass and shrub basal cover. *South African Journal of Science*, **86**, 462–3.

Hoffman, M.T. & Cowling, R.M. (1987). Plant physiognomy, phenology and demography. In *The Karoo Biome: a Preliminary Synthesis. Part 2 – Vegetation and History*, ed. R.M. Cowling & P.W. Roux, pp. 1–34. *South African National Scientific Programmes Report 142*. Pretoria: CSIR.

Hunter, R.B. (1994). *Status of the Flora and Fauna of the Nevada Test Site, 1989–91*. Las Vegas, Nevada: United States Department of Energy.

Ihlenfeldt, H.D. (1971). Some aspects of the biology of dissemination of the Mesembryanthemaceae. In *The Genera of the Mesembryanthemaceae*, ed. H. Herre, pp. 28–34. Cape Town: Tafelberg Publications.

Jacobsen, H. (1960). *A Handbook of Succulent Plants*, vol. 1. Dorset, UK: Blandford Press.

Jacobsen, H. & Volk, O.H. (1960). The form and mode of life of the Mesembryanthemums. In *A Handbook of Succulent Plants, vol. 3: Mesembryanthemums*, ed. H. Jacobsen, pp. 875–9. Dorset, UK: Blandford Press.

Joubert, J.G.V. (1968). *Die Ekologie Van die Weiveld van die Robertson-Karoo*. DSc Thesis. Stellenbosch: Stellenbosch University.

Jump, J.A. (1988). Phytogeographic and evolutionary trends in *Lithops*. *Aloe*, **25**, 33.

Jürgens, N. (1986). Untersuchungen zur Ökologie sukkulenter Pflanzen des südlichen Afrika [Contributions to the ecology of southern African succulent plants], *Mitteilungen aus dem Institut für Allgemeine Botanik Hamburg*, **21**, 129–365.

Kellerman, T.S., Coetzer, L.A.W & Naudé, T.W. (1988). *Plant Poisoning and Mycotoxicoses of Livestock in Southern Africa*. Cape Town: Oxford University Press.

Knight, R.S. (1988). *Aspects of Plant Dispersal in the Southwestern Cape with Particular Reference to the Roles of Birds as Dispersal Agents*. PhD Thesis. Cape Town: University of Cape Town.

Knight, R.S., Rebelo, A.G. & Siegfried, W.R. (1989). Plant assemblages on mima-like earth mounds in the Clanwilliam district, South Africa. *South African Journal of Botany*, **55**, 465–72.

Kuntzsch, V. & Nel, J.A.J. (1992). Diet of bat-eared foxes *Otocyon megalotis* in the Karoo. *Koedoe*, 35, 37–48.

Lane, S.B. (1978). An ecological study of the Tanqua/Doorn Karoo based on a structural–physiognomic survey. *South African Archaeological Bulletin*, 33, 128–33.

Leistner, O.A. (1994). Annuals of the arid transition zone between winter and summer rainfall regions in southern Africa. In *Proceedings of the 13th Plenary Meeting of AETFAT*, ed. J.H. Seyani & A.C. Chikuni, pp. 1181–93. Zomba, Malawi: Association for the Taxonomic Study of the Flora of Tropical Africa.

Le Roux, A. (1984). *'n Fitososiologiese Studie van die Hester Malan-Natuurreservaat.* MSc Thesis. Pretoria: University of Pretoria.

Le Roux, A., Perry, P. & Kyriacou, X. (1989). South Africa. In *Plant Phenomorphological Studies in Mediterranean Type Ecosystems*, ed. G. Orshan, pp. 159–346. Dordrecht: Kluwer.

Levyns, M.R. (1950). The relations of the Cape and the Karoo floras near Ladismith, Cape. *Transactions of the Royal Society of South Africa*, 32, 235–46.

Lloyd, J.W. (1985). *A Plant Ecological Study of the Farm Vaalputs, Bushmanland, with Special Reference to Edaphic Factors.* MSc Thesis. Cape Town: University of Cape Town.

Lloyd, J.W. (1989). Discriminate analysis and ordination of vegetation and soils on the Vaalputs radioactive waste disposal site, Bushmanland, South Africa. *South African Journal of Botany*, 55, 127–36.

Louw, G.N. (1969). The nutritive value of natural grazings in South Africa. *Proceedings of the South African Society for Animal Production*, 8, 57–61.

Lovegrove, B.G. & Siegfried, W.R. (1989). Spacing and origin(s) of Mima-like earth mounds in the Cape Province of South Africa. *South African Journal of Science*, 85, 108–12.

Macdonald, I.A.W. (1989). Man's role in changing the face of southern Africa. In *Biotic Diversity in Southern Africa. Concepts and Conservation*, ed. B.J. Huntley, pp. 51–77. Cape Town: Oxford University Press.

Macdonald, I.A.W., van Wijk, K. & Boyd, L. (1993). *Conservation Priorities in Southern Africa.* Stellenbosch: Southern African Nature Foundation.

MacMahon, J.A. & Wagner, F.H. (1985). The Mojave, Sonoran and Chihuahuan Deserts of North America. In *Hot Deserts and Arid Shrublands*, ed. M. Evenari, I. Noy-Meir & D.W. Goodall, pp. 105–202. Amsterdam: Elsevier.

Marais, A. (1989). Die nuttige rol wat termiete in die fynbos speel. *Veld & Flora*, 75, 62–3.

Midgley, G.F. & Bösenberg, J. de W. (1990). Seasonal and diurnal plant water potential changes in relation to water availability in Mediterranean climate western Karoo. *South African Journal of Ecology*, 1, 45–59.

Midgley, G.F. & Moll, E.J. (1993). Gas exchange in arid-adapted shrubs: when is efficient water use a disadvantage? *South African Journal of Botany*, 59, 491–5.

Midgley, G.F. & Musil, C.F. (1990). Substrate effects of zoogenic mounds on vegetation composition in the Worcester-Robertson valley, Cape Province. *South African Journal of Botany*, 56, 158–66.

Midgley, J.J. (1991). Beetle daisies and daisy beetles. *African Wildlife*, 45, 318–19.

Midgley, J.J. & Cowling, R.M. (1993). Regeneration patterns in Cape subtropical transitional thickets: where are all the seedlings? *South African Journal of Botany*, 59, 496–9.

Milton, S.J. (1978). Plant communities of Andriesgrond, Clanwilliam District. Habitat Working Group Pamphlet. Bolus Herbarium, University of Cape Town.

Milton, S.J. (1990a). Life-styles of plants in four habitats in an arid Karoo shrubland. *South African Journal of Ecology*, 1, 63–72.

Milton, S.J. (1990b). Above-ground biomass and plant cover in succulent shrubland in the southern Karoo, South Africa. *South African Journal of Botany*, 56, 587–9.

Milton, S.J. (1991). Plant spinescence in arid southern Africa: does moisture mediate selection by mammals? *Oecologia*, 87, 279–87.

Milton, S.J. (1992a). Effects of rainfall, competition and grazing on flowering of *Osteospermum sinuatum* (Asteraceae) in arid Karoo rangeland. *Journal of the Grassland Society of southern Africa*, 9, 158–64.

Milton, S.J. (1992b). Plants eaten and dispersed by *Geochelone pardalis* (Reptilia: Chelonii) in the southern Karoo. *South African Journal of Zoology*, 27, 45–9.

Milton, S.J. (1993). Insects from the shrubs *Osteospermum sinuatum* and *Pteronia pallens* (Asteraceae) in the southern Karoo. *African Entomology*, 1, 257–61.

Milton, S.J. (1994a). Small-scale reseeding trials in arid rangeland: effects of rainfall, clearing and grazing on seedling survival. *African Journal of Range and Forage Science*, 11, 54–8.

Milton, S.J. (1994b). Growth, flowering and recruitment of shrubs in grazed and protected rangeland in the arid Karoo. *Vegetatio*, 111, 17–27.

Milton, S.J. (1995). Spatial and temporal patterns in the emergence and survival of seedlings in arid Karoo shrubland. *Journal of Applied Ecology*, 32, 145–56.

Milton, S.J. & Collins, H. (1989). Hail in the southern Karoo. *Veld & Flora*, 75, 69–73.

Milton, S.J. & Dean, W.R.J. (1990a). Mima-like mounds in the southern and western Cape: are the origins so mysterious? *South African Journal of Science*, 86, 207–8.

Milton, S.J. & Dean, W.R.J. (1990b). Seed production in rangelands of the southern Karoo. *South African Journal of Science*, 86, 231–3.

Milton, S.J. & Dean, W.R.J. (1991). Disturbances in dune grassland: colourful consequences of clearing. *African Wildlife*, 45, 199–203.

Milton, S.J. & Dean, W.R.J. (1993). Selection of seed by harvester ants (*Messor capensis*) in relation to condition of arid rangeland. *Journal of Arid Environments*, 24, 63–74.

Milton, S.J. & Dean, W.R.J. (1995). Factors influencing recruitment of forage plants in arid Karoo shrublands, South Africa. In *Proceedings: Wildland Shrub and Arid Land Restoration Symposium*, ed. B.A. Roundy, D.E. McArthur, J.S. Haley & D.K. Mann, pp. 216–22. Ogden, Utah: US Department of Agriculture.

ilton, S.J., Dean, W.R.J., du Plessis, M.A. & Siegfried, W.R. (1994a). A conceptual model of arid rangeland degradation: the escalating cost of declining productivity. *BioScience*, **44**, 70–6.

ilton, S.J., Dean, W.R.J. & Kerley, G.I.H. (1992a). Tierberg Karoo Research Centre: history, physical environment, flora and fauna. *Transactions of the Royal Society of South Africa*, **48**, 15–46.

ilton, S.J., Dean, W.R.J. & Marincowitz, C.P. (1992b). Preferential utilization of pans by springbok (*Antidorcas marsupialis*). *Journal of the Grassland Society of southern Africa*, **9**, 114–18.

ilton, S.J., Dean, W.R.J., Marincowitz, C.P. & Kerley, G.I.H. (1995). Effects of the 1990/91 drought on rangeland in the Steytlerville Karoo. *South African Journal of Science*, **91**, 78–84.

ilton, S.J., Dean, W.R.J. & Siegfried, W.R. (1994b). Food selection by ostrich in southern Africa. *Journal of Wildlife Management*, **58**, 234–48.

ilton, S.J. & Hoffman, M.T. (1994). The application of state-and-transition models to rangeland research and management in arid succulent and semi-arid grassy Karoo, South Africa. *African Journal of Range and Forage Science*, **11**, 18–26.

oore, J.M. & Picker, M.D. (1991). *Heuweltjies* (earth mounds) in the Clanwilliam district, Cape Province, South Africa: 4000 year old termite nests. *Oecologia*, **86**, 424–32.

oy-Meir, I. (1973). Desert ecosystems: environment and producers. *Annual Review of Ecology and Systematics*, **4**, 25–51.

livier, M.C. (1966). *Die Plantegroei en Flora van die Worcester Veldreservaat*. DSc Thesis. Stellenbosch: Stellenbosch University.

livier, M.C. (1979). An annotated systematic checklist of Angiospermae of the Worcester Veld Reserve. *Journal of South African Botany*, **45**, 49–62.

enn, N.G. (1986). Pastoralists and pastoralism in the northern Cape Frontier zone during the eighteenth century. *South African Archaeological Society Goodwin Series*, **5**, 62–8.

ickett, G.A. & Hoffman, M.T. (1992). Karosion: soil erosion in the Karoo. *Veld & Flora*, **78**, 8–9.

Pierce, S.M., Esler, K.J. & Cowling, R.M. (1995). Smoke-induced germination of succulents (Mesembryanthmaceae) from fire-prone and fire-free habitats in South Africa. *Oecologia*, **102**, 520–2.

Powrie, L.W. (1993). Responses of Karoo plants to hail damage near Williston, Cape Province. *South African Journal of Botany*, **59**, 65–8.

Rees, M. (1994). Delayed germination of seed.: a look at the effects of adult longevity, the timing of reproduction, and population age/stage structure. *American Naturalist*, **144**, 43–64.

Rösch, M.W. (1977). *Enkele Plantekologiese Aspekte van die Hester Malannatuurreservaat*. MSc Thesis. Pretoria: University of Pretoria.

Roux, P.W. & Vorster, M. (1983). Vegetation change in the Karoo. *Proceedings of the Grassland Society of southern Africa*, **18**, 25–9.

Rundel, P.W., Cowling, R.M., Esler, K.J., Mustart, P.J., van Jaarsveld, E. & Bezuidenhout, H. (1995). Winter growth phenology and leaf orientation in *Pachypodium namaquanum* (Apocynaceae) in the Succulent Karoo of the Richtersveld, South Africa. *Oecologia*, **101**, 472–7.

Rutherford, M.C. (1978). Karoo-fynbos biomass along an elevational gradient in the western Cape. *Bothalia*, **12**, 555–60.

Rutherford, M.C. & Westfall, R.H. (1986). Biomes of Southern Africa – an objective categorization. *Memoirs of the Botanical Survey of South Africa*, **54**, 1–98.

Schulze, R.E. & McGee, O.S. (1978). Climatic indices and classifications in relation to the biogeography of southern Africa. In *Biogeography and Ecology of Southern Africa*, ed. M.J.A. Werger, pp. 19–52. The Hague: Junk.

Scott, J.D. & van Breda, N.G. (1937a). Preliminary studies of the root system of the renosterbos *Elytropappus rhinocerotis* on the Worcester Veld Reserve. *South African Journal of Science*, **23**, 560–9.

Scott, J.D. & van Breda, N.G. (1937b). Preliminary studies of the root system of *Galenia africana* on the Worcester Veld Reserve. *South African Journal of Science*, **24**, 268–74.

Scott, J.D. & van Breda, N.G. (1939). Preliminary studies of the root systems of *Euphorbia mauritanica*, *E. burmanni* and *Ruschia multiflora* on the Worcester Veld Reserve. *South African Journal of Science*, **26**, 227–35.

Skead, C.J. (1980). *Historical Mammal Incidence in the Cape Province*. Cape Town: Cape Nature Conservation.

Smith, A.B. (1986). Competition, conflict and clientship: Khoi and San relationships in the western Cape. In *Prehistoric Pastoralism in Southern Africa*, Goodwin Series 5, ed. M. Hall & A.B. Smith, pp. 36–41. Vlaeberg: South African Archaeological Society.

Smitheman, J. & Perry, P. (1990). A vegetation survey of the Karoo National Botanic Garden Reserve, Worcester. *South African Journal of Botany*, **56**, 525–41.

Snijman, D. & Perry, P. (1987). A floristic analysis of the Nieuwoudtville Wild Flower Reserve. *South African Journal of Botany*, **53**, 445–54.

Steinschen, A.K., Görne, A. & Milton, S.J. (1996). Threats to the Namaqualand flowers: out-competed by grass or exterminated by grazing? *South African Journal of Science*, **92** (in press).

Steyn, H.M., van Rooyen, M.W., van Rooyen, N. & Theron, G.K. (1994). Namaqualand ephemerals: flowering time prediction. *South African Association of Botanists Congress Abstracts*, **29**, 93.

Stokes, C.J. (1994). *Degradation and Dynamics of Succulent Karoo Vegetation*. MSc Thesis. Pietermaritzburg: University of Natal.

Struck, M. (1989). The biology of the fruit capsules in *Dorotheanthus* Schwantes (Mesemb.) *Veld & Flora*, **75**, 41–3.

Struck, M. (1992). Pollination ecology in the arid winter rainfall region of southern Africa: a case study. *Mitteilungen aus dem Institut für Allgemeine Botanik Hamburg*, **24**, 61–90.

Struck, M. (1994). Flowering phenology in the arid winter rainfall regions of southern Africa. *Bothalia*, **24**, 77–90.

Stuart, C.T. (1981). Notes on the mammalian carnivores of the Cape Province, South Africa. *Bontebok*, **1**, 1–58.

Theron, C.K. (1964). *'n Outelologiese Studie van Plinthus karooicus Verdoorn*. MSc Thesis. Pretora: University of Pretoria.

Tölken, H.R. (1977). A revision of the genus *Crassula* in southern Africa. *Contributions from the Bolus Herbarium*, 8, 1–560.

Tölken, H.R. (1985). Crassulaceae. *The flora of Southern Africa*, 14, 1–244.

Trusler, A.E. & Yeaton, R.I. (1991). A preliminary study of the soil microbiology of the Worcester Veld Reserve. *Arid Zone Ecology Forum Abstracts*, pp. 23–4. Pretoria: Foundation for Research Development.

Turner, J.S. & Picker, M.D. (1993). Thermal ecology of an embedded dwarf succulent from southern Africa (*Lithops* spp: Mesembryanthemaceae). *Journal of Arid Environments*, 24, 361–86.

van Breda, N.G. (1939). An improved method of sowing grass and Karoo shrub seed. *South African Journal of Science*, 36, 328–35.

van Breda, P.A.B. & Barnard, S.A. (1991). *100 Veld Plants of the Winter-Rainfall Region*. Bulletin 422. Pretoria: Department of Agriculture.

van der Heyden, F. (1992). *Effects of Defoliation on Regrowth and Carbon Budgets of Three Semi-arid Karoo Shrubs*. PhD Thesis. Cape Town: University of Cape Town.

van der Merwe C.R. (1962). *Soil groups and subgroups of South Africa*. Department of Agriculture, Chemistry Series 165. Pretoria: Government Printer.

van Jaarsveld, E. 1994. *Gasterias of South Africa – A New Revision of a Major Succulent Group*. Cape Town: Fernwood Press.

van Rooyen, M.W. & Grobbelaar, N. (1982). Seed populations in the soil of the Hestermalan Nature Reserve in the Namaqualand Broken Veld. *South African Journal of Botany*, 1, 41–50.

van Rooyen, M.W., Grobbelaar, N., Theron, G.K. & van Rooyen, N. (1991). The ephemerals of Namaqualand: effects of photoperiod, temperature and moisture stress on development and flowering of three species. *Journal of Arid Environments*, 20, 15–29.

van Rooyen, M.W., Grobbelaar, N., Theron, G.K. & van Rooyen, N. (1992). The ephemerals of Namaqualand: effect of germination date on development of three species. *Journal of Arid Environments*, 22, 51–66.

van Rooyen, M.W., Theron, G.K. & Grobbelaar, N. (1979a). Phenology of the vegetation in the Hester Malan Nature Reserve in the Namaqualand Broken Veld: 1 General observations. *Journal of South African Botany*, 45, 279–93.

van Rooyen, M.W., Theron, G.K. & Grobbelaar, N. (1979b). Phenology of the vegetation in the Hester Malan Nature Reserve in the Namaqualand Broken Veld: 2 The therophyte population. *Journal of South African Botany*, 45, 433–52.

van Rooyen, M.W., Theron, G.K. & Grobbelaar, N. (1990). Life form and dispersal spectra of the flora of Namaqualand, South Africa. *Journal of Arid Environments*, 19, 133–45.

van Wyk, B.E. 1989. The taxonomic significance of cyanogenesis in *Lotononis* and related genera. *Biochemical Systematics and Ecology*, 17, 297–307.

Visser, J. (1981). *Parasitic Flowering Plants*. Cape Town: Juta & Co.

Visser, J.N.J. (1986). Geology. In *The Karoo Biome: A Preliminary Synthesis. Part 1 – Physical Environment*, ed. R.M. Cowling, P.W. Roux & A.J.H. Pieterse, pp. 1–17. *South African National Scientific Programmes Report* 124. Pretoria: CSIR.

Vivrette, N.J. & Muller, C.H. (1977). Mechanism of invasion and dominance of coastal grassland by *Mesembryanthemum crystallinum*. *Ecological Monographs*, 47, 301–18.

Vlok, J. (1988). Alpha diversity of lowland fynbos herbs at various levels of infestation by alien annuals. *South African Journal of Botany*, 54, 623–7.

Vlok, J. (1994). How is your valley green? *South African Association of Botanists 20th Annual Congress Abstracts*, p. 108. Johannesburg: University of Witwatersrand.

Vogel, J., Fuls, A. & Ellis, R. (1978). The geographic distribution of Kranz grasses in South Africa. *South African Journal of Science*, 74, 209–15.

Volk, O.H. (1960). Flowers and fruits of the Mesembryanthemums. In *A Handbook of Succulent Plants, vol. 3: Mesembryanthemums*, ed. H. Jacobsen, pp. 896–901. Dorset, UK.: Blandford Press.

von Willert, D.J., Brinckmann, E., Scheitler, B. & Eller, B.M. (1985). Availability of water controls Crassulacean acid metabolism in succulents of the Richtersveld (Namib desert, South Africa). *Planta*, 164, 44–55.

von Willert, D.J., Eller, B.M., Werger, M.J.A. & Brinckmann, E. (1990). Desert succulents and their life strategies. *Vegetatio*, 90, 133–43.

von Willert, D.J., Eller, B.M., Werger, M.J.A., Brinckmann, E. & Ihlenfeldt, H.D. (1992). *Life Strategies of Succulents in Deserts*. Cambridge: Cambridge University Press.

Wentzel, H.E., van Rooyen, M.W., Theron, G.K. & de Villiers, A.J. (1994). *Mesembryanthemum guerichianum*: dominance on old kraals. *Arid Zone Ecology Forum Abstracts*, p. 45. Pretoria: Foundation for Research Development.

Werger, M.J.A. (1978). The Karoo-Namib region. In *Biogeography and Ecology of Southern Africa*, ed. M.J.A. Werger, pp. 233–99. The Hague: Junk.

Werger, M.J.A. (1985). The Karoo and southern Kalahari. In *Hot Deserts and Arid Shrublands*, ed. M. Evenari, I. Noy-Meir & D.W. Goodall, pp. 282–358. Amsterdam: Elsevier.

Werger, M.J.A. & Morris, J.W. (1991). Climatic control of vegetation structure and leaf characteristics along an aridity gradient. *Annali di Botanica*, 49, 203–15.

Whitehead, V.B. (1984). Distribution, biology and flower relationships of fideliid bees of southern Africa (Hymenoptera, Apoidea, Apoidea, Fideliidae). *South African Journal of Zoology*, 19, 87–90.

Whitehead, V.B. & Steiner, K. (1985). Oil-collecting bees in South Africa. *African Wildlife*, 39, 144–7.

Wiegand, T., Milton, S.J. & Wissel, C. (1994). A simulation model for a shrub-ecosystem in the semi-arid Karoo, South Africa. *Ecology*, 76, 2205–21.

Wiens, D. (1982). Mimicry in plants. *Evolutionary Biology*, 7, 365–403.

Willis, C. (1992). Richtersveld: land of contrasts. *Veld & Flora*, 78, 14–17.

Wood, J. (1991). Threats facing the Robertson Karoo. *Veld & Flora*, 77, 16–17.

Yeaton, R.I. & Esler, K.J. (1990). The dynamics of a succulent Karoo vegetation. *Vegetatio*, 88, 103–13.

Nama-karoo

8

A.R. Palmer and M.T. Hoffman

8.1 Introduction

The Nama-karoo, succulent karoo and desert biomes comprise the Karoo-Namib Region (Werger 1978; White 1978; Jürgens 1991), the largest phytochorion in southern Africa (Rutherford & Westfall 1986; Chap. 3, this volume). The Nama-karoo biome is the largest of the three and occupies 607 235 km² or 22.7% of the southern African region (Chap. 5, this volume). Its vegetation is characterized as a dwarf open shrubland (Campbell *et al.* 1981) or open dwarf-shrub steppe (Werger 1980), dominated by the Asteraceae, Poaceae, Aizoaceae, Mesembryanthemaceae, Liliaceae (*sensu lato*) and Scrophulariaceae (Werger 1978). Although sparsely populated, the Nama-karoo sustains an important meat- and wool-based, small-stock industry and much research has focused on understanding the agricultural resource base of the region. In terms of its ecology, however, it is one of the least studied of southern Africa's biomes.

Early descriptions of the biome (Marloth 1908; Bews 1916; Cannon 1924) emphasized the diversity of growth forms, which include large and dwarf shrubs, leaf-succulents, stem-succulents, bulbous monocotyledons, grasses and annuals. This structural approach continues to provide a basis for understanding patterns in the biome. Variability in structure is evident in the range of terms used to describe the biome's vegetation: 'desert shrub' (Pole-Evans 1936), 'karroid scrub' (Dyer 1937), 'arid bush' (Adamson 1938), 'dwarf and succulent shrub' (Edwards 1970) and 'dwarf shrub steppe formation' (Martin & Noel 1960).

Since no single nomenclature encompasses the diversity of floristic, growth-form and structural classes in the biome, we have used a landscape approach to describe the vegetation. We focus almost exclusively on the South African part of the Nama-karoo, since no data are available on the vegetation of the biome in Namibia and the outliers in Botswana and Lesotho (see Chap. 5, this volume). First, we describe the important environmental gradients within the biome and within landscapes of the biome. Next, we discuss the climatic and geological determinants of the biome in relation to the other biomes that surround it. Using a topo-moisture gradient analysis we sub-divide the Nama-karoo into three sub-biomes. We describe selected landscapes, vegetation units and communities in terms of their floristic, structural and environmental characteristics within each of these sub-biomes. We then present details of the growth-form composition of the vegetation of the biome and describe the vegetation dynamics in terms of a state-and-transition model. We also highlight a number of autecological studies conducted on key taxa within the region. We conclude with an account of the important issues that have dominated the land-use and management history of the biome.

8.2 Environmental gradients

8.2.1 Topography

Almost all of the biome lies at an elevation of 550 to 1500 m and is divided by the Great Escarpment into two broad classes: 550–900 m and 900–1300 m. The Orange River valley runs east-west across the northern part of the biome. The landscape is unique in southern Africa, with characteristic buttes and mesas (King 1942; Chap. 1, this volume), and large interior basins drained by ephemeral rivers (e.g. the Great Fish, Sundays and Seacow Rivers). The Nama-karoo biome abuts on all the other biomes of southern Africa, except the forest biome.

8.2.2 Precipitation

Annual rainfall varies from 60 to 400 mm and decreases from east to west with areas of very low precipitation in the rain shadow of major mountain ranges. These effects are pronounced in the east (e.g. near Hofmeyer and Tarkastad), where the Winterberg prevents the rain from reaching the interior. Similar effects are obvious in the rain shadows of the Baviaans-Kouga mountains, the Swartberg and the Nuweveldberge.

Rainfall is highly seasonal with unimodal peaks occurring from December to March at all stations within the biome (Fig. 8.1). Approaching the Atlantic Ocean in the west, there is a trend towards autumn rain with a concomitant decline in rainfall variability. These western sites are transitional to the succulent karoo, having an increasing proportion of succulent taxa (Chap. 7, this volume). Rainfall variability, as described by the coefficient of variation (CV) of annual rainfall, decreases with increasing mean annual rainfall (Hoffman & Cowling 1987; Chap. 2, this volume). Analysis of the relationship between annual rainfall and CV reveals several stations with higher than expected variability; these are mainly situated on the eastern boundary of the biome in areas reported to be undergoing change from grassland to dwarf-shrubland, e.g. Phillipolis, Graaff-Reinet, Petrusville. The contemporary vegetation at these sites is ascribed by Acocks (1953) to recent overgrazing (see also Chap. 21, this volume), but there is a strong link between grass–shrub cycles and high rainfall variability (Bousman & Scott 1994).

The east–west aridity gradient outlined above has important ecophysiological implications for growth-form mix and species distributions across the biome. Montane grasslands and shrublands predominate in the east, whereas drought-tolerant grasses and succulent dwarf shrubs are more conspicuous in the west.

There is also a moisture gradient at the landscape scale. Conditions of higher soil moisture are associated with rocky dykes and sills, which are refugia for mesic grasses and woody shrubs with strong grassland and savanna biome affinities, respectively. Drainage lines cross the landscape support a variety of phreatophytic woody shrubs. In the east, a grassy and largely evergreen dwarf shrubland occurs on the dry pediments and bottomlands, whereas this is replaced by drought-deciduous and succulent shrubland in the west.

8.2.3 Temperature

The biome experiences high January mean maximum temperatures (>30 °C) and relatively low July mean minimum temperatures (<0 °C) (Fig. 8.1). A strong temperature gradient is evident across the biome, with temperature range increasing with elevation. Frost occurs throughout the Nama-karoo (Hoffman & Cowling 1987; Chap. 2, this volume), and the length of the growing season declines with increasing elevation.

At a landscape scale there are distinct slope- and aspect-related differences in radiation levels and temperature regimes (Schulze 1976). This is reflected in species composition and structure. Tall shrubs are more abundant on cooler, wetter, southern slopes, whereas on warmer, drier, northern slopes plants are shorter and cover is sparse, with dwarf shrubs dominating at lower elevations and C_3 grasses at higher elevations.

8.2.4 Geology and soils

The lithology of the biome comprises the Karoo Supergroup, of Permian age, which consists of the Dwyka Formation, Ecca Group and Beaufort Group; and the Namaqualand Metamorphic Province (NMP), comprising meta-sedimentary, meta-volcanic and intrusive rocks (Visser 1989). The NMP is exposed in the north, with the Ecca and Beaufort Groups occupying the central region (Visser 1986). The Beaufort Group overlies the Ecca Group and consists of alternating layers of mudstone and sandstone. The sandstones represent river-channel deposits and the mudstone, floodplain deposits. The Beaufort Group has been extensively intruded by Jurassic-age dolerite, occurring as dykes and sills. The inclined sills tend to form crescent-shaped or concentric intrusions, which range from a few metres to >100 m in thickness. Dolerite has metamorphosed adjacent host rocks, with mudstone altered to hornfels and the sandstones to quartzites. Quaternary deposits include river terrace gravel, calcrete, alluvium and colluvial debris. The terrace gravel consists of rounded cobbles and boulders composed largely of dolerite. These gravels are partly calcrete-cemented and occur on terrace remnants that lie some 4–30 m above the general landsurface. Calcrete occurs directly on bedrock as well as forming extensive deposits within some of the larger areas of alluvium, and may attain a maximum thickness of a few metres. The alluvium embraces both alluvial slope (sheetwash) and alluvial valley (channel-related) deposits, with the former predominating (Johnson & Keyser 1979).

The region is dominated by two broad soil patterns: (1) shallow soils of pedologically young landscapes, and (2) red and yellow, apedal to weakly structured, freely-drained soils of the Dwyka formation and the NMP (Ellis & Lambrechts 1986). Soils are mainly shallow (<0.3 m), except along drainage lines. A calcareous layer (depth: 0.2–0.3 m) is a feature of the profile, restricting

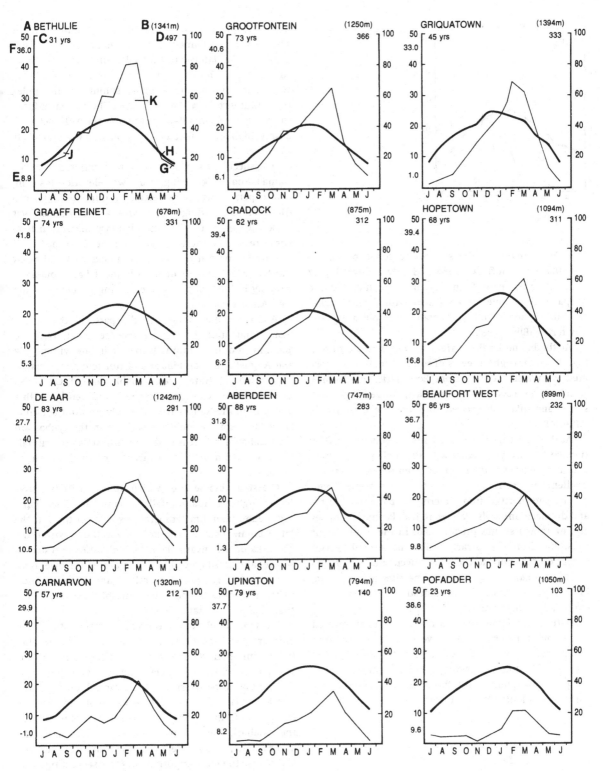

Figure 8.1 **Walter-Lieth climate diagrams for selected sites in the Nama-karoo biome. A, climate station; B, height above sea level; C, duration of observation; D, mean annual precipitation (mm); E, mean daily minimum temperature (°C) of the coldest month (July); F, mean daily maximum temperature (°C) of the warmest month (January); G, curve of mean monthly precipitation; H, curve of mean monthly temperature; J, relative period of drought; K, relative humid season. Fig. continued on p. 170.**

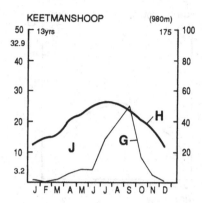

Figure 8.1 contd.

root development to the shallow topsoil (Hoffman & Cowling 1987; Hoffman 1989a). Highly erosive duplex soils occur adjacent to major drainage lines, and a Mispah-rock complex is associated with dolerite dykes and sills. By southern African standards, soils are generally fertile (Ellis 1988).

At the biome level, nutrient gradients relate to geological strata. Highest levels of soil nutrients are associated with soils derived from Ecca series mudstones, which are exposed in the southeast, whereas the sandstone- and quartzite-derived soils are the least fertile (Ellis 1988).

At the landscape scale, nutrients are concentrated by transport down the catena, with the highest phosphate and nitrate levels being recorded from the pediments. A gradient of declining phosphate levels occurs from upland sites derived from dolerite, to bottomland sites and drainage line soils (Palmer 1991a). Perennial grasses and evergreen shrubs predominate in the low-nutrient sites, and deciduous dwarf shrubs and annual grasses occurr on the more fertile sites. Succulents (e.g. *Brownanthus, Ruschia*) occupy sandy, saline sites (Lloyd 1989a; Chap. 7, this volume), whereas non-succulent shrubs generally occur on the mudstone substrata.

Within the landscape, nutrients are concentrated into enriched patches by biotic vectors (small and large herbivores, termites, ants). Zoogenic disturbance is widespread and frequent, providing an unstable environment for perennial plants, and favouring pauciennial and annual succulents (Palmer 1987).

8.3   Boundaries

8.3.1   Biome-scale
The Nama-karoo is differentiated from the other biomes by a range of environmental factors including the amount, seasonality and variability of precipitation, elev-

ation, temperature and substratum. A summary of the important environmental determinants of the distribution of the Nama-karoo and adjacent biomes is shown in Table 8.1. The desert biome is separated from the Nama-karoo by having a higher summer-aridity index (>5.0) (Rutherford & Westfall 1986). The annual rainfall in deserts is <60 mm (but see Chap. 9, this volume), with a very high coefficient of variation (132% at Gobabeb, Namibia).

The succulent karoo biome is differentiated from the Nama-karoo by occurring at a lower elevation (<550 m) and experiencing more reliable rainfall (CV 30–40%) (Hoffman & Cowling 1987) with a July–August wet season (Chap. 2, this volume). January maximum temperatures in the succulent karoo are lower (<30 °C) and July minimum temperatures are higher (>4 °C) than in the Nama-karoo. Most of the succulent karoo biome is strongly influenced by the moderating influence of the cold Atlantic Ocean.

Annual rainfall totals for the parts of the fynbos biome that abut the Nama-karoo are generally above 300 mm, and much rain falls mainly in late winter (July and August). The combination of nutrient-poor, quartzitic-derived substrata, a less variable rainfall (CV <30%), and moderate elevation (<550 m) differentiates this biome from the Nama-karoo. The only sites in the Nama-karoo with soils as infertile as those in the fynbos are the Kalahari sand areas of the northwest, where rainfall is low (<150 mm yr^{-1}) and concentrated in the summer months.

Grassland biome sites which abut the Nama-karoo have a higher annual rainfall (>400 mm) and lower January mean maximum temperatures (<30 °C) than in the latter biome. Patches of grassland vegetation within the Nama-karoo are invariably at high elevations (>1500 m), where rainfall is higher and temperatures are lower than the surrounding areas. Generally, annual precipitation in the grassland biome is less variable (CV <30%) than that in the Nama-karoo biome

Sites in the subtropical thicket vegetation of the Eastern Cape (see Chap. 12, this volume) have higher rainfall (>300 mm) than Nama-karoo sites (Hoffman 1989a,b). There is, however, not a clear climatic or topographic differentiation between the Nama-karoo biome and the arid Kalahari vegetation of the savanna biome, with the exception of July minimum temperatures being <1 °C in the Kalahari. The dominance of Kalahari sands in the arid savanna, with greater volumes for moisture storage, is probably the most important variable separating these biomes.

8.3.2   Regional- and landscape-scale
At the landscape level, we divided the biome using a topo-moisture class analysis (Whittaker 1975). A cross-

Table 8.1 Environmental variables which define the boundaries of the Nama-karoo and other biomes that surround it

Biome	Elevation (m)	Mean annual rainfall (mm)	Coefficient of variation in mean annual rainfall (%)	Mean January maximum temperature (°C)	Mean July minimum temperature (°C)
Nama-karoo	550–1500	60–400	30–60	>30	<0
Desert	<550	<60	30–40	<30	>4.0
Succulent karoo	<550	60–400	30–40	<30	>4.0
Subtropical thicket	<550	>400	<30	<30	>4.0
Savanna (Kalahari)	>550	60–400	30–60	>30	<1.0
Grassland	>1500	>400	<30	<30	<0
Fynbos	<550	>400	<30	<30	>4.0

Table 8.2 Environmental variables characterizing the subdivisions of the Nama-karoo biome

Biome	Elevation (m)	Mean annual rainfall (mm)	Coefficient of variation in mean annual rainfall (%)	Mean July minimum temperature (°C)	Mean January maximum temperature (°C)
Griqualand West and Bushmanland	550–1300	60–200	>50	>2.0	>35
Great Karoo and Central Lower Karoo	550–1500	60–200	40–50	<2.0	30–35
Upper Karoo and Eastern Cape Midlands	900–1300	200–400	40–50	<2.0	30–35

tabulation of elevation and annual rainfall provided three geographically distinct regions, namely Griqualand West and Bushmanland; the Great Karoo and Central Lower Karoo; and the Upper Karoo and Eastern Cape Midlands (Table 8.2). Within these categories we recognized structurally defined vegetation types based on subdivisions of these topo-moisture classes (Fig. 8.2).

8.3.2.1 Griqualand West and Bushmanland

This is the largest subdivision within the biome (Fig. 8.2). Elevation within this region varies from 550 to 1300 m, annual rainfall is low (60–200 mm) and the CV of annual rainfall is high (>50%). July minimum temperatures are moderate (>2 °C) with very high January maxima (>35 °C) (Table 8.2).

We describe three main units within this sub-biome and list some of the important vegetation formations and communities which have been discussed in the literature.

Succulent dwarf shrubland This unit occurs where the Griqualand West and Bushmanland sub-biome abuts the succulent karoo (Fig. 8.2) and has been described by Lloyd (1989b). With an elevation of 900–1000 m and an annual rainfall of 70–100 mm, this is one of the more arid vegetation types in the biome. Rainfall variability is low (CV 36–38%). Generally abundant taxa include *Lycium* sp., *Salsola zeyheri* and *Sphalmanthus tetragonus*.

Arid shrubland The arid shrublands (Fig. 8.2) are synonymous with Acocks' (1953) Orange River Broken Veld (ORBV) and parts of the Namaqualand Broken Veld

along the Orange River. The ORBV is the second largest Veld Type in the Nama-karoo (Scheepers 1983), covering some 37 000 km², and occurs at moderate (800–1100 m) elevation in the Orange River valley. The substratum consists of highly reflective stony surfaces (quartzites and calc-silicates) derived from the Namaqualand complex, and the vegetation is dominated by *Acacia mellifera, Rhigozum trichotomum* and *Zygophyllum suffruticosum*. Where the terrain becomes hilly, and elevation exceeds 1100 m, *Aloe dichotoma* and *Euphorbia avasmontana* occur. Annual rainfall varies from 150 to 200 mm. The CV of annual rainfall is high (50–60%) and January maximum temperatures exceed 34 °C.

Arid grassland Covering much of this subdivision (Fig. 8.2) is a matrix of arid grassland and dwarf shrubland, synonymous with Werger's (1980) Stipagrostion and parts of Acocks' (1953) Arid Karoo. It is associated with high elevations (900–1300 m), annual rainfalls of 60–200 mm, rainfall variability of 43–45% and a mean January maximum temperature of 31 °C. Subdivisions include a grassy dwarf shrubland dominated by *Stipagrostis ciliata* and *S. obtusa* on soils derived from dolerites; and a succulent dwarf shrubland dominated by *Ruschia spinosa* (L. Dehn), occurring mainly on soils derived from mudstone and sandstone. Dominant shrubs include *Eriocephalus* spp., *Felicia* spp., *Lycium prunus-spinosa, Pentzia incana* and *Pteronia* spp. These grassy dwarf shrubland communities are transitional to the dwarf shrublands of the Upper Karoo.

The extensive, spiny, caespitose *Stipagrostis* spp. grass-

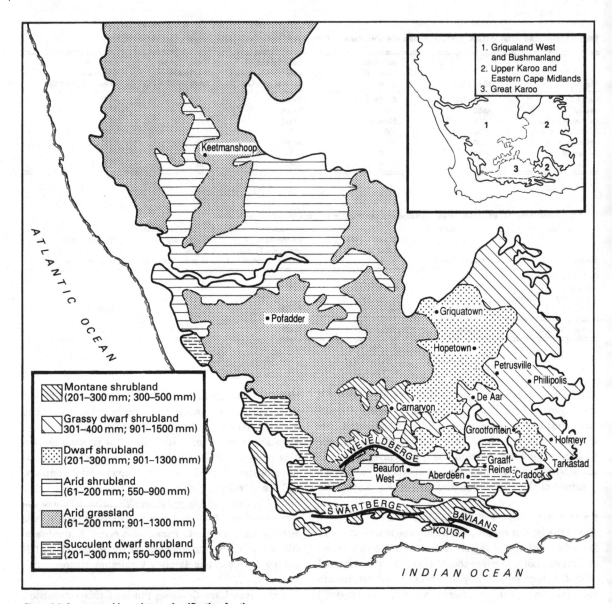

Figure 8.2 **A topographic–moisture classification for the Nama-karoo biome, showing the expected vegetation types for each class. Prepared using elevation and median annual rainfall data provided by Dent et al. (1987).**

lands are unique to Bushmanland (Fig. 8.3). They have strong affinity with the desert grasslands of the southern Namib (Yeaton 1988). Where dolerite or sandstone are overlain by Kalahari sand, a community dominated by *Stipagrostis obtusa*, *S. ciliata* and *Enneapogon desvauxii* occurs. The arid grasslands are described more fully by Werger (1980).

8.3.2.2 *Great Karoo and Central Lower Karoo*

This sub-biome, which is sandwiched between the Great Escarpment and the Cape Folded Belt (Fig. 8.1), extends

over a considerable elevation range from 550 to 1500 m (Table 8.2). Generally, winters are cold with regular frost and summers are moderately hot. Rainfall is low to extremely low, but more predictable than sites north of the Great Escarpment. The important rain-bearing southerly winds are influenced first by the Swartberg range to the south of the Great Karoo and the Nuweveldberge to the north. In the rainshadow of the Swartberg, the rainfall is extremely low and uncertain (Dent *et al.* 1987). However, precipitation increases with increasing altitude along the Great Escarpment. The region comprises

Figure 8.3 **Heavily grazed arid grassland of** *Stipagrostis brevifolia* **on deep sands near Pofadder in Bushmanland** *(Photo: M.J.A. Werger).*

two major topo-moisture classes: the western plain, or Great Karoo, and the eastern plain or Central Lower Karoo (Acocks 1953). Both are dominated by dwarf shrublands with grass increasing in abundance towards the east. Although community-level descriptions of the sub-biome are rare (Acocks 1953; Palmer & Van Heerden 1994), we describe three common formations within this region.

Grassy scrub Situated in a narrow belt from the plains to the edge of the escarpment (and not mapped in Fig. 8.2), grassy scrub occurs on rocky surfaces (Fig. 8.4). Differential species are *Digitaria eriantha* and *Rhus burchellii*. The grass component consists of high cover-abundance of *Aristida diffusa*, with a lower abundance of *Digitaria eriantha*, *Heteropogon contortus* and *Themeda triandra*. Clumps of subtropical thicket, dominated by *Rhus burchellii*, are common; other common species are *Acacia karroo*, *Carissa haematocarpa*, *Grewia robusta* and *Maytenus polyacantha*. The substratum consists of dolerite boulders overlying sandstone. Erosion is minimal here, and soil between the rocks is fertile with high aluminium, magnesium and organic carbon levels. Elevation varies from 900 to 1250 m. There is a steep rainfall gradient from the

edge of the escarpment to the pediments below. Rainfall varies from 180 to 300 mm (Dent *et al.* 1987).

Arid shrubland The arid and largely dwarf shrublands of the Great Karoo and Central Lower Karoo (Fig. 8.2) are inadequately described, with the exception of Acocks' (1953) general account and recent treatment by Palmer & Van Heerden (1994). Dominant dwarf shrubs include species of *Drosanthemum*, *Eriocephalus*, *Pentzia*, *Pteronia* and *Ruschia* (Fig. 8.5). Principal perennial grasses include species of *Aristida*, *Enneapogon*, *Digitaria* and *Stipagrostis*. Taller shrubs and trees that are scattered throughout this type include *Acacia karroo*, *Euclea undulata* and *Rhigozum obovatum*.

Riparian woodlands The major drainage lines of the Nama-karoo support low woodlands. They are, however, particularly common in the basins of the Great and Central Lower Karoo. Unpredictable flooding events suggest that drainage lines possess high disturbance regimes and relatively high levels of soil movement. Taxa from many different communities are represented, with the taller woody species being most successful and obvious. These are dominated by *Acacia karroo*, but may include *Diospyros lycioides*, *Grewia robusta*, *Lycium oxycar-*

Figure 8.4 **Grassy shrubland at the foot of the Great Escarpment in the Central Lower Karoo. Grasses include *Aristida diffusa* and *Digitaria eriantha*. Shrubs in background are *Rhus burchellii*** (Photo: F. van Heerden).

pum, *Maytenus polyacantha*, *Rhus lancea* and *Tamarix usneo-ides*. The dominant grass is *Cynodon incompletus*, with other perennial taxa such as *Cenchrus ciliaris*, *Fingerhuthia africana*, *Hyparrhenia hirta* and *Stipagrostis namaquensis* forming scattered clumps. Grass cover varies both spatially and temporally as moisture levels fluctuate, and the deep, sandy alluvium provides an ideal germination environment for many annual taxa, e.g. *Atriplex lindleyi*. This community dissects the landscape and appears uncoupled from the controlling influence of the topo-moisture gradients of the region.

8.3.2.3 Upper Karoo and Eastern Cape Midlands
The boundaries of this sub-biome are defined by elevation (900–1300 m), annual rainfall (200–400 mm), and CV of annual rainfall (40–50%) (Table 8.2). July minimum temperatures are <2 °C and January maximum temperatures range from 30 to 35 °C.

The structure and floristic composition of the vegetation of this sub-biome have been surveyed at a regional (Acocks 1953; Werger 1973, 1980; Vorster 1985;

Palmer 1991a) and a local scale (Werger 1973; Jooste 1980; Van der Walt 1980; Palmer 1988, 1989). We discuss four formations within the sub-biome that have been recognized on the basis of broad structural and floristic attributes.

Montane shrubland These shrublands, which often grade into xeric grasslands (Chap. 10, this volume), are associated with high-altitude (>1300 m; Fig. 8.2), steeply sloping landforms. On average annual rainfall varies from 200 to 300 mm, but may be as high as 400 mm in the almost-pure grasslands of the Nuweveld escarpment. Rainfall shows a unimodal peak in late summer. Geology varies from sandstone to a mixture of blue and grey mudstone, with occasional dolerite outcrops. The grassy part of this formation is synonymous with Acocks' (1953) Karroid *Merxmuellera* Mountain Veld, the C_3 grass, *Merxmuellera disticha*, being the dominant species (Fig. 8.6). Other differential species include *Cymbopogon plurinodis*, *Elytropappus rhinocerotis*, *Felicia filifolia*, *Nenax microphylla* and *Pentzia globosa*. Species found towards the mesic end of the gradient are *Melica decumbens*, *Passerina montana*,

Figure 8.5 **Arid shrubland on sandstone-derived soils near Beaufort West in the Central Lower Karoo. Dominant species are *Rhigozum obovatum* (Bignoniaceae), *Pentzia incana* (Asteraceae), *Ruschia spinosa* (L. Dehn; Mesembryanthemaceae) and *Aristida* spp. (Poaceae)** *(Photo: F. van Heerden).*

Tetrachne dregei and *Themeda triandra*. These grasslands have floristic affinities with drier communities of the fynbos biome (see Chap. 6, this volume). Drier sites are dominated by dwarf shrubs including species of *Eriocephalus, Osteospermum, Pentzia, Pteronia, Rosenia* and *Sutera*. Principal grasses include species of *Aristida, Eragrostis* and *Stipagrostis*.

Dwarf shrubland This type occupies the largely featureless country of the Upper Karoo (Fig. 8.7) and was probably much more grassy in the past (Acocks 1953; Chap. 21, this volume). These shrublands occur on the gently sloping, arid pediments, and soils are very weakly structured. Soils have generally developed *in situ* from colluvium, with lime present in the entire landscape, and skeletal forms predominate (Ellis & Lambrechts 1986). Dominant shrub species include *Eriocephalus spinescens, Monechma desertorum, Pentzia spinescens* and *Ruschia spinosa* (L. Dehn). The main grasses are *Eragrostis lehmanniania* and *Stipagrostis obtusa*.

Grassy dwarf shrubland According to Acocks (1953) the 'development of this Veld Type constitutes the most

spectacular of all the changes in the vegetation of South Africa'. Essentially, this type represents the conversion of grassland to karoo (Chaps 10 and 21, this volume). Rainfall is generally the highest of all Nama-karoo vegetation types, ranging from 300 to 400 mm yr^{-1} (Fig. 8.2). Dominant grasses include *Eragrostis chloromelas, E. obtusa Sporobolus* spp., while *Themeda triandra* is scattered throughout, depending on grazing history. The principal shrub species is *Chrysocoma ciliata*.

The rocky hills and low mountains are grassier than the adjacent plains and have a low cover of tall shrubs, particularly *Rhus erosa*. Vegetation in these habitats is synonymous with Werger's (1980) Rhoetea erosae and Acocks' (1953) False Karroid Broken Veld.

Succulent dwarf shrubland This type occurs in the eastern sector of the Great Karoo around Graaff-Reinet at altitudes of 600–1000 m and an annual rainfall of c. 250 mm falling mainly in spring and autumn (Fig. 8.2). Dominant species are *Drosanthemum lique, Enneapogon desvauxii, Pentzia incana* and *Ruschia spinosa* (L. Dehn; Acocks 1953). Vegetation in this area shows complex

Figure 8.6 **Montane grassland on dolerite-derived soils on the rim of the Great Escarpment above Beaufort West. The dominant grass is the C₃ species *Merxmuellera disticha*. Scattered individuals of *Elytropappus rhinocerotis* (Asteraceae), a shrub of fynbos biome affinity, are present in the background** (Photo: F. van Heerden).

transitions to subtropical thicket (Fig. 8.8; Chap. 12, this volume). These transitional types are dominated by relatively tall (1–2 m) succulent shrubs, e.g. *Portulacaria afra* and *Euphorbia coerulescens* (Hoffman 1989b; see also Chap. 7, this volume).

8.3.3 Community-scale

To describe some of the major community and higher-order vegetation types within the sub-biomes, we synthesized available synecological data for the biome (Werger 1973, 1980; Jooste 1980; Van der Walt 1980; Palmer 1988, 1989, 1991a; Lloyd 1989b; Palmer & Van Heerden 1994). Within the tradition of vegetation survey in southern Africa, the methods of the Zürich–Montpellier school were employed and a synoptic table comprising 93 associations and 1341 species was constructed. Although not reproduced here, this table forms the basis for our community-level descriptions.

8.3.3.1 *Griqualand West and Bushmanland*
Lloyd (1989b) has described four communities within the transitional succulent dwarf shrublands, including (1) *Aptosimum procumbens* var. *procumbens* dwarf shrubland on calcrete, calcareous and saline soils (2) *Eberlanzia stylosa* succulent dwarf shrubland on shallow soils overlying granite; (3) *Stipagrostis brevifolia* open grassy shrubland on soils mixed with aeolian sand; and (4) *Stipagrostis ciliata* var. *capensis* grassland on deeper, strongly acid aeolian sand.

8.3.3.2 *Great Karoo and Central Lower Karoo*
Pteronia tricephala arid shrublands Differential species are *Eriocephalus ericoides*, *Euryops nodosus*, *Melica racemosa* and *Pteronia tricephala*. The community occurs in open, level terrain at an elevation of 1600–1700 m. Associated species include the C₃ grasses, *Karroochloa purpurea* and *Merxmuellera disticha*. The duplex soils are derived from mudstone, siltstone and sandstone. The

Figure 8.7 **Vegetation on a doleritic koppie in the Upper Karoo of the southwestern Free State.** *Cussonia paniculata* and *Rhus* spp. (Anacardiaceae) grow in a matrix of grasses (*Heteropogon contortus*, *Themeda triandra*, etc.). On the plains below, sweet grasses coexist with karroid shrubs, principally *Chrysocoma ciliata* (Asteraceae) *(Photo: J.P.H. Acocks).*

community is associated with the rainfall gradient away from the edge of the escarpment with precipitation declining from >300 mm to approximately 260 mm. There is a concomitant decline in rockiness and canopy cover. This is reflected in the floristics and structure, with dwarf shrubs replacing the larger shrubs and grasses as aridity increases.

Pentzia incana arid shrublands Differential taxa of these communities (Table 8.3) include the dwarf shrubs *Astroloba herrei*, *Eriocephalus ericoides*, *Felicia filifolia*, *Monechma divaricatum*, *Pentzia incana*, *Sericocoma avolans* and with high cover-abundance of *Chrysocoma ciliata* and *Walafrida geniculata*. These communities are associated with the arid part of the moisture gradient, and occur on both sandstone-derived and calcareous substrata. On rocky soils, i.e. where dolerite boulders overlay the sandstone, woody shrubs such as *Grewia robusta* and *Maytenus polyacantha* are found. On the flat, rocky plateaux, *Ruschia spinosa* (L. Dehn) becomes diagnostic, with *Rhigozum obovatum* remaining common throughout. On the calcareous

Table 8.3 **Examples of plant associations from the Great Karoo and Central Lower Karoo, with associated substratum, elevation and rainfall (after Palmer & Van Heerden 1994)**

Community[a]	Differential species	Common taxa	Substratum	Elevation (m)	Rainfall (mm yr^{-1})
1	Garuleum bipinnatum Blepharis villosa Cenchrus ciliaris	Rhigozum obovatum Pentzia incana Fingerhuthia africana	Mudstone Sandstone	800–1100	200–260
2	Pentzia globosa Felicia hirsuta Astroloba herrei Trichodiadema setuliferum	Rhigozum obovatum Pentzia incana	Sandstone	890–960	160–260
3	Rosenia humilis Pteronia glomerata Helichrysum lucilioides	Pentzia incana	Sandstone	900–1100	200–250
4	Trichodiadema barbatum Delosperma pruinosum Crassula muscosa Ruschia spinosa (L. Dehn) Enneapogon desvauxii Aptosimum procumbens	Rhigozum obovatum Pentzia incana	Sandstone	900–1200	180–200
5	Malephora sp. Hermannia cuneifolia Lepidium divaricatum Blepharis capensis	Rhigozum obovatum Aristida congesta Eragrostis lehmanniana Stipagrostis obtusa	Calcareous sandstone	900–1000	<200
6	Kleinia longiflora Galenia frutescens Trichodiadema setuliferum Pteronia glauca	Rhigozum obovatum Stipagrostis obtusa Pentzia incana Eriocephalus ericoides	Sandstone	800–900	
7	Lycium prunus-spinosa Pteronia sordida	Pentzia incana Stipagrostis obtusa S. ciliata Aristida congesta Eragrostis lehmanniana Eriocephalus ericoides	Sandy alluvium	900–1060	180–250
8	Monechma incanum Zygophyllum gilfillanii Cadaba aphylla Psilocaulon absimile	Stipagrostis obtusa Pentzia incana Stipagrostis ciliata Aristida congesta Eragrostis lehmanniana Fingerhuthia africana Cenchrus ciliaris	Alluvium with mudstone		
9	Eriocephalus spinescens Galenia africana Lycium schizocalyx	Pentzia incana Stipagrostis obtusa S. ciliata	Mudstone and sandstone	800–900	170
10	Salsola tuberculata	Stipagrostis obtusa Pentzia incana	Mudstone, siltstone and sandstone with surface lime	800–900	160–170
11	Salsola aphylla Kochia salsoloides Malephora crocea Delosperma pubescens Psilocaulon absimile	Lycium sp. Stipagrostis obtusa	Calcareous silt		160–180
12	Acacia karroo Grewia robusta Lycium oxycarpum Maytenus polyacantha Diospyros lycioides Rhus lancea Cenchrus ciliaris Stipagrostis namaquensis Hyparrhenia hirta	Atriplex lindleyi Delosperma sp. Atriplex spp.	Sandy alluvium		

[a]Given on opposite page.

Figure 8.8 **In the Sundays River Basin south of Graaff-Reinet, there is a transition from Nama-karoo shrublands to subtropical thicket via a band of Noorsveld. This vegetation type is dominated by the stem-succulent 'noors' (***Euphorbia coerulescens***) and many shrubs of tropical affinity such as** *Euclea undulata* **(Ebenaceae),** *Pappea capensis* **(Sapindaceae) and** *Schotia afra* **(Fabaceae)**
(Photo: R.M. Cowling).

substrata where large dolerite bedrock is exposed on the surface, elements of the Stipagrostion (Werger 1980) are represented by *Stipagrostis ciliata* and *S. obtusa*.

Stipagrostis obtusa arid grasslands *Stipagrostis ciliata* and *S. obtusa* are the differential species for this community, which has been provisionally described elsewhere in the Nama-karoo as the Stipagrostion (Werger 1980). The class occurs on shallow stony substrata with surface lime. Lithology is predominantly sandstone, with rarer occurrences of mudstone and alluvium.

8.3.3.3 *Upper Karoo and Eastern Cape Midlands*
Some 68 associations and sub-associations have been published for this region. These include four by Jooste (1980), 44 by Werger (1980), 17 by Palmer (1988, 1989) and three by Van der Walt (1980). Analysis of the relationship between some of these concepts and environmental variables (Palmer 1991b) suggests that topo-moisture and substratum gradients are the major determinants of floristic change (Palmer & Cowling 1994).

8.4 Plant structure and function

8.4.1 Plant types
Growth forms occurring in the biome include large and dwarf shrubs, leaf-succulents, stem-succulents, bulbous monocotyledons, grasses and annuals (Marloth 1908; Bews 1916; Cannon 1924). This list has been further expanded by Cowling *et al.* (1994), who showed that growth-form diversity in 25-m² sites is positively related to species richness. Gibbs Russell (1987) reports that Nama-karoo is similar to grassland in the proportion of life forms (Raunkiaer 1934), with a predominance of hemicryptophytes (50%) followed by equal proportions of chamaephytes and cryptophytes (25%). An analysis of the deviations from random expectation by Palmer & Cowling (1994) shows how life forms are replaced along the east–west aridity gradient in the Orange River valley. On dolerite, nanophanerophytes are rarer than expected in the xeric west, becoming dominant in the east, whereas the reverse is true on sandstone. This may be due to greater availability of phreatic water in the east. Chamaephytes are less common on sandstone in the west, and become dominant in the mesic conditions in the east.

Leaf and root anatomical studies of southern African xerophytic plants have been neglected (Jordaan & Kruger 1992). Despite the early descriptive work of Cannon (1924) and Scott & Van Breda (1938) on the root systems of a few important Nama-karoo species, very little structural–functional research on different plant types has been done. Theron, Schweickert & Van der Schyff (1968) have shown how stem and root anatomical adaptations may relate to the success of an important forage species, *Plinthus karooicus*. In secondary stem growth, periderm bridges may be formed, leading eventually to the longitudinal splitting of the main stem axis into a number of separate plants. Theron *et al.* (1968) suggest that this facilitates vegetative propagation. However, it may also be important in the survival of the individual, since Jones (1984) has measured significant differences in

---

Notes to Table 8.3
Community names:
1 *Garuleum bipinnatum–Rhigozum obovatum* shrubland on mudstone slopes.
2 *Pentzia globosa* shrubland on sandstone footslopes.
3 *Rosenia humilis–Ruschia spinosa* shrubland on lower pediments.
4 *Trichodiadema barbatum–Ruschia spinosa* (L. Dehn) shrubland on sandstone pediments.
5 *Malephora* sp. grassy shrubland on valley bottoms.
6 *Kleinia longiflora* shrubland on rocky pediments.
7 *Lycium prunus–spinosa* grassy shrubland on sandstone.
8 *Monechma incanum* grassy shrubland adjacent to drainage lines.
9 *Eriocephalus spinescens* grassy shrubland.
10 *Salsola tuberculata* arid shrubland.
11 *Salsola aphylla* shrubland on Quaternary alluvium.
12 *Acacia karroo–Stipagrostis namaquensis* riparian shrubland.

xylem pressure potentials of different branches of other species with split axes.

8.4.2 Reproductive ecology

In attempting to lay the foundation for an understanding of pollination systems within the Nama-karoo, Hoffman (1989b) characterized the phenological profiles of a number of Nama-karoo species growing in succulent dwarf shrubland in the Central Lower Karoo. He argued that species responded opportunistically to rainfall, thus supporting Noy-Meir's (1973) 'pulse and reverse' model, widely advocated for other deserts of the world. This contrasts with Roux's (1968a) 'average growth cycle' model, in which he suggested that different growth forms in the Nama-karoo each grow, flower and set seed predictably at different times of the year.

The phenology of the flora should be reflected in the activity patterns of floral herbivores. The role of aculeate wasps and bees as flower visitors and pollinators have been studied throughout the biome. The majority of these insects have a short adult phase synchronized with the flowering of the flora (Gess 1981; Gess & Gess 1994). Visitors to the flowers of 35 families of plants have been sampled. Apiaceae, Celastraceae and subfamily Mimosoideae are, as a rule, generalists attracting generalist feeders. Asteraceae attract both generalist and specialist feeders, which, though more dependable visitors, may have a narrower geographical distribution than their forage plants. In general, Papilionoideae are visited principally by Megachilidae, Anthophoridae and the honeybee *Apis melifera* (Apidae). Gess & Gess (1994) found that, although the megachilids and anthophorids successfully strip the flowers, they should be considered potential pollinators. Honeybees are frequently able to steal nectar from the flowers without tripping them and should therefore be considered thieves. Examples of genera with specialist flowers are *Aptosimum* and *Peliostomum* (Scrophulariaceae). Their pollinators need to be large enough to receive a pollen load dorsally but must be able to reach the nectar at the base of a very narrow tube. Such specialized requirements are met by the long-tongued *Celonites* (Masarinae), which are almost the sole attendants throughout their range in the western, northern, southern and eastern karoo (Gess & Gess 1989,1991; Gess 1992).

In a study of the pollination ecology of an important arborescent succulent, *Aloe ferox*, Hoffman (1988b) showed how both birds and Cape honeybees are involved. Unlike the majority of karroid plants, which flower after rain during the warm to hot months, this plant flowers in the winter, when the solitary wasps and bees are not active.

Hoffman & Cowling (1987) summarize the autecological work on the reproductive ecology of Nama-karoo species and highlight the paucity of studies in this field. Henrici's investigations (1935b, 1939) into the germination ecology of a number of Nama-karoo species showed that germination success of all test species was higher at 20 °C than 30 °C. Little else is known of the reproductive ecology of important plants in the region, other than a considerable body of anecdotal information (e.g. Roux 1968b; Hobson *et al.* 1975; Hoffman, Barr & Cowling 1990). Although considerable work has been undertaken in communities of the succulent karoo biome (Chap. 7, this volume), the Nama-karoo has not received the same degree of attention.

8.4.3 Physiological traits and processes

Although there are a number of studies on the water relations of Nama-karoo species, more than half were conducted over 30 years ago. The early pioneering experiments of Henrici (1937, 1940, 1941) compared the transpiration rates of a range of karroid shrubs and grasses. She showed that transpiration rates of both grasses and shrubs were high in soils with high soil water contents but that shrubs, unlike grasses, were able to restrict their water loss with declining water supply. Midgley & Moll (1993) have further suggested that for the shrubs *Eriocephalus ericoides* and *Pentzia incana*, this combination of opportunistic and parsimonious water use may be advantageous. Conservative water use has obvious advantages in arid and semi-arid environments. However, optimal stomatal opening, which occurs when soil water content is high, may increase growth rates and, hence, competitive success, by diverting water from neighbours. Moore, *et al.* (1988) and Moore (1989) have shown that competition for water between *Rhigozum trichotomum*, an indigenous invasive shrub, and a range of key forage grass species, is important in structuring these invaded communities. *Rhigozum trichotomum* is of little grazing value and an increase in its density depresses grass production exponentially. In shrub densities of more than about 1500 tree equivalents ha^{-1} (where one tree equivalent = a single 1.5-m high shrub), grass recruitment is precluded.

The small stock industry in the Nama-karoo is directly dependent on production from natural pastures, and defoliation of key forage species has formed a major focus for autecological studies in the region. Hoffman & Cowling's (1987) review of this field (e.g. Du Preez 1972; Hobson & Sykes 1980) indicates that these early studies focused on dry matter yields only. They have usually taken the form of extensive field trials aimed at developing, testing and refining management policy. More detailed ecophysiological responses to defoliation such as the translocation, storage and use of photosynthate following grazing have not been carried out on Nama-karoo species. These details are important, since some of the key principles underlying management strategies for the region use untested physiological arguments to support their rationale (see Hoffman 1988a).

Nearly all of the studies on the palatability and toxicity of Nama-karoo species were conducted before 1970. They also focused on applied management objectives. Henrici (1932, 1935a, 1945, 1952) pioneered much of this early work, relating the chemical content of a number of key Nama-karoo forage species to one or more measures of digestibility. She concluded from these experiments that a mixture of grasses and shrubs in any Nama-karoo community provided the most nutritious grazing. Louw, Steenkamp & Steenkamp (1967, 1968) analysed the chemical composition of a large number of Nama-karoo species. On the basis of these data, they advised on the most appropriate supplementary feedings in each region that they studied. The dwarf shrub *Chrysocoma ciliata* is a common invasive species, especially in overgrazed grasslands. General observations have indicated that it may possess allelopathic properties and two studies have attempted to quantify this (Hewitt & Nel 1969; Squires & Trollope 1979).

This brief survey of the autecological work conducted on Nama-karoo species confirms Hoffman & Cowling's (1987) conclusion that the biota of this region have largely been ignored by southern African ecologists.

8.5 Community structure and dynamics

8.5.1 Growth-form mix

Our description of vegetation structure in the Nama-karoo biome is derived from unpublished data collected by the noted South African botanist John Acocks. After the publication of his classic work *The Veld Types of South Africa* (Acocks 1953), he spent the next 26 years refining his data set for a smaller part of South Africa. The region he chose encompasses the more arid and semi-arid vegetation units roughly south of 28° S and west of 27° E and includes the Nama-karoo, succulent karoo and parts of the fynbos biome. We present data for Acocks' (1953) Veld Types of the Nama-karoo biome only and refer to M.T. Hoffman, G.F. Midgley & R.M. Cowling (unpubl. data) for a broader context of the data set. Unfortunately we could not link this analysis directly to the regional vegetation classification described above, but have instead used Acocks' (1953) Veld Type definitions.

At each site Acocks listed all plant species in a 1–2-km radius, compiling a summary of total plant species richness for each Veld Type. Acocks also assigned each species to a growth form: herbaceous annual or biennial (excluding grasses); geophyte; perennial grass; herbaceous perennial; succulent shrub (usually <0.5 m);

woody shrub <1 m; woody shrub, tree or liana >1 m. The distribution of the different growth forms within the 19 Veld Types and their variations as delimited by J.P.H. Acocks (unpubl. data) are shown in Fig. 8.9 (M.T. Hoffman *et al.* unpubl. data).

On average, annuals comprise a quarter of all species in Nama-karoo vegetation types. The largest numbers of species are consistently found in this growth form. The proportion of annual species in a Veld Type or its variation ranges from 19 to 34%: lowest where subtropical thicket trees are abundant as an overstorey such as in the Noorsveld (24) and False Karroid Broken Veld (37) in the southeast; and highest in the open and arid vegetation types in the northwest. Hoffman *et al.* (unpubl. data) suggest that for the arid and semi-arid parts of the subcontinent as a whole, the proportion of annual species in a Veld Type or its variation is highest in environments characterized by a low and unpredictable rainfall such as the Arid Karoo (29).

Geophytes typically comprise 10–26% of a Veld Type. They are most poorly represented in the Veld Types of the Great Karoo situated between the escarpment and the Cape Fold Mountains and are most common in those Veld Types that abut the grassland biome in the northeast and the Western Mountain Karoo (Acocks 1953). The latter has close affinity with the renoster shrublands of the fynbos biome to the southwest (see Chap. 6, this volume).

The proportion of grasses in the Veld Types of the Nama-karoo ranges typically from 3 to 11%. It is highest for Veld Types that are adjacent to the grassland biome in the northeast and lowest for those adjacent to the winter rainfall succulent karoo and fynbos biomes in the west and southwest. Hoffman *et al.* (unpubl. data) found a highly significant negative relationship for both the number and the proportion of perennial grass species in a Veld Type and an index of summer aridity.

The proportion of perennial herbs ranges from 11 to 25% and is lowest in the arid western and southern Veld Types and highest in the northeast in Veld Types that have numerous grassland biome elements. Hoffman *et al.* (unpubl. data) have shown a significant positive relationship between the number and proportion of perennial herbs in a Veld Type and the mean annual rainfall of the region.

Succulents are well represented in the vegetation of the Nama-karoo and comprise 11–26% of taxa in a Veld Type. Succulents are most poorly represented in Veld Types that abut the grassland biome in the northeast and have their greatest representation in the Great Karoo, especially in the southeast, where many of the vegetation types comprise species with strong succulent karoo and subtropical thicket affinities (Acocks 1953; see also Cowling *et al.* 1994).

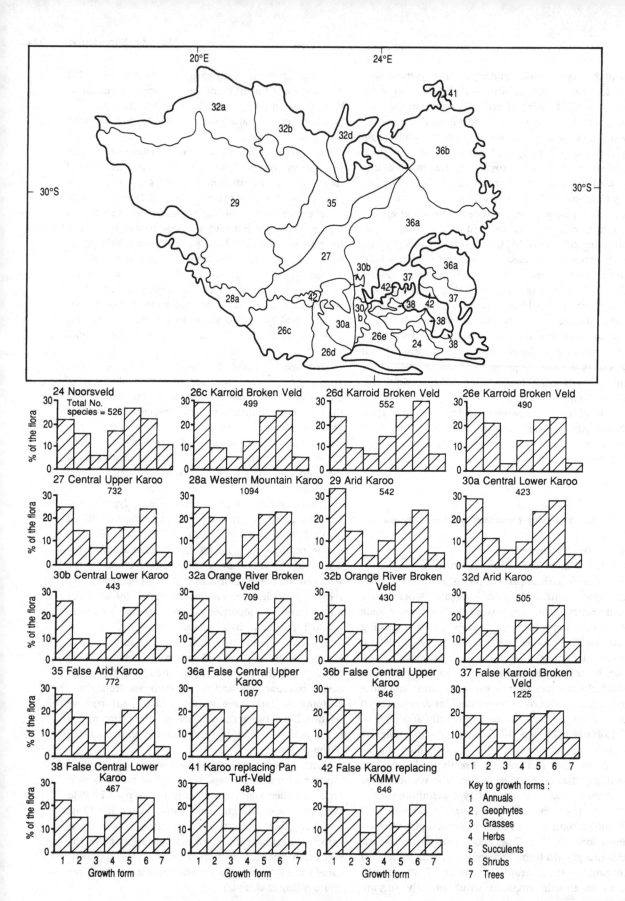

182

Next to annuals, shrubs <1 m in height usually comprise the largest proportion (14–30%) of the flora of a Veld Type. Their proportion is highest in the Karroid Broken Veld of the Great Karoo and the Orange River and is lowest in vegetation types that have a high proportion of grasses, such as those in the northeast.

Of all the growth forms investigated here, trees, shrubs >1 m and lianes are the most poorly represented and never comprise more than 10% of a regional flora in the Nama-karoo biome. For vegetation of the upper plateau region (e.g. Western Mountain Karoo; False Arid Karoo; Acocks 1953), where frosts are common, trees may make up as little as 3% of the total flora. The proportion of trees is highest for vegetation types adjacent to the arid savanna biome in the north and the subtropical thicket in the southeast.

8.5.2 Succession and dynamics

Studies on the dynamics of the Nama-karoo biome have emphasized the responses of grasses and shrubs to rainfall and grazing regimes (Roux 1966; Novellie & Strydom 1987; Hoffman *et al.* 1990). Many of these debates are summarized in Milton & Hoffman's (1994) state-and-transition model (Westoby, Walker & Noy-Meir 1989) for the eastern karoo (Fig. 8.10). Although the model refers to the dynamics of Acocks' (1953) False Upper Karoo vegetation (Veld Type 36), it may also have relevance for surrounding vegetation types.

Milton & Hoffman (1994) present six vegetation states and discuss the transition processes between states. In State 1 (Fig. 8.10) perennial grasses such as *Aristida diffusa, Digitaria eriantha, Eragrostis curvula* and *Themeda triandra* coexist with shrubs such as *Pentzia incana, Plinthus karooicus, Pteronia* spp., *Rhigozum obovatum* and *Rosenia humilis.* The stability of this matrix is poorly known, but it is suggested that it can be transformed by both grazing and climatic factors to states that contain either more grasses (State 2) or more shrubs and fewer perennial grasses (State 3) (see also O'Connor & Roux 1995; Chap. 10, this volume).

The possible mechanisms for these transitions are poorly known and have not been tested experimentally. Some suggestions, however, have come from Roux (1966)

and Novellie & Strydom (1987), who measured the response of grasses and shrubs to different rainfall and grazing regimes over several years. From his surveys, Roux (1966) suggests that 'grass cycles' and 'shrub cycles' occur in the karoo, driven largely by seasons of above-average summer rainfall and above-average winter rainfall, respectively. A detailed partitioning of grazing and climatic influences on grass and shrub recruitment, however, was not undertaken. Following a re-analysis of Roux's (1966) data, Hoffman *et al.* (1990) find evidence for a negative effect of a dense grass sward on shrub survival and suggest that a more general 'cyclical' succession may occur in the eastern part of the Nama-karoo, driven largely by the amount of summer rainfall received. If this is high for several seasons then productive C_4 grasses recruit and dominate the landscape. Such dominance, through competition for both light and moisture, may kill many of the dwarf shrubs that appear to be poorer competitors for these resources. However, during the frequently long periods of extensive summer drought in the eastern Karoo, grasses may themselves die, leaving patches in the landscape into which shrubs may recruit, following favourable autumn rains (Hobson *et al.* 1975; Hoffman *et al.* 1990).

Although there is little evidence for the precise mechanisms, Milton & Hoffman (1994) suggest that continuous selective grazing at high stocking rates, if practised for long enough, will result in a degradation gradient from State 1 to State 4 and then finally to an unproductive State 5 (Fig. 8.10). Once this state is reached it may be more economical to convert the degraded land to a forage bank of alien shrubs such as *Atriplex nummularia* or *Opuntia* spp.

The state-and-transition model proposed by Milton & Hoffman (1994) provides the first synthesis of our understanding of the dynamics of Nama-karoo vegetation. However, O'Connor & Roux's (1995) recent multivariate analysis of vegetation changes over 23 years in the eastern karoo suggests that a number of refinements to the model are necessary. Firstly, they find little evidence for the influence of winter rainfall on shrub recruitment (i.e. no 'shrub rains' or 'shrub cycles') and suggest that as a growth form, 'shrubs' are too variable to possess identical functional responses to climatic influences. Secondly, O'Connor & Roux (1995) find no evidence for a competitive effect between grasses and shrubs, except in one plot protected from grazing. Their work clearly underpins the need for an experimental approach to clarify some long-held perceptions about the dynamics of eastern karoo vegetation.

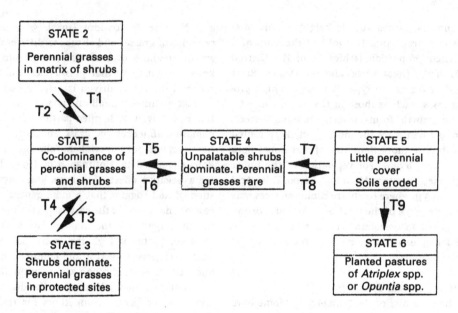

Synopsis of states and transitions

Catalogue of states	Catalogue of transitions
1 Co-dominance of shrubs, perennial and short-lived grasses.	T1 Consecutive years of above-average summer rain with very light grazing in summer provide an opportunity for C_4 tussock grasses to recruit.
2 Perennial grasses dominant in a matrix of palatable and unpalatable shrubs. Ephemerals and short-lived grasses are confined to local disturbance sites.	T2 Average winter rainfall seasons with summer drought and rotational grazing, including grazing in summer, at recommended stocking rates.
3 Shrubs dominate. These may be palatable or unpalatable, depending on management. Perennial grasses present but confined to protected microsites.	T3 Consecutive years of above-average winter rain and extended summer drought combined with grazing in summer at recommended stocking rates provides an opportunity for shrub recruitment.
4 Unpalatable, toxic and spiny shrubs and geophytes dominant. Perennial grasses rare to absent. Ephemerals common after seasonal rains.	T4 Average summer rainfall season with winter drought and rotational grazing at recommended stocking rates.
5 Little perennial cover, eroded. Ephemerals may be locally common after seasonal rains.	T5 Rotational grazing at recommended stocking rates, rests at seed set and following key rainfall events.
6 Planted forage banks or pastures.	T6 Continuous selective grazing at recommended stocking rates probably exacerbated by drought.
	T7 Active intervention including soil reclamation and reseeding.
	T8 Heavy continuous selective grazing above recommended stocking rates, exacerbated by extended drought. Little recruitment of perennial plants as a result of trampling and altered soil conditions.
	T9 Cultivation and, in some cases, irrigation.

Figure 8.10 **State-and-transition model (Milton & Hoffman 1994) of the False Upper Karoo (Acocks 1953). Reproduced, with permission, from Milton & Hoffman (1994).**

8.6   Land use and management

Excellent reviews of the land-use history of the Nama-karoo have been provided by Talbot (1961) and Christopher (1982). Hoffman (Chap. 21, this volume) outlines a general chronology for the colonization of the Nama-karoo biome and discusses the nature of pre-colonial environments with their attendant large herds of indigenous ungulates (Cronwright-Schreiner 1925). Hoffman (Chap. 21) also describes, in some detail, the long-standing desertification debate that has influenced the management practices of the region and which has also driven much of the ecological research in the Nama-karoo. It is this perception of environmental degradation that has spawned a number of management systems aimed largely at minimizing the impact of sheep and goats on this arid and semi-arid ecosystem. The Department of Agriculture has played a pivotal role in this regard, although it has not always been without criticism (Acocks 1966; Hoffman 1988a).

To develop a more scientific approach to the management of these arid environments, the Department of Agriculture began, as long ago as 1934, to develop a set of management guidelines for the region. This culminated in their Group Camp Approach, which incorporated a deferred rotational grazing system, at relatively light stocking rates coupled with a carefully constructed fencing network aimed at minimizing the negative effects of patch- and area-selective grazing (Roux 1968a). This remains the advocated and most widely used approach in the region today.

Probably the most important challenge to the Group Camp system was initiated in the mid-1960s following the publication of the Acocks–Howell or Non-Selective Grazing (NSG) system of livestock management (Acocks 1966). Acocks (1966) proposed that the dominance of unpalatable shrubs in the Nama-karoo was the direct result of selective grazing. He suggested that one way to rehabilitate these degraded rangelands was to simulate the pre-colonial herding patterns of wild ungulates and to use heavy concentrations of domestic animals in small camps for short periods, thus forcing the animals to graze non-selectively. This, he suggested, would not only lead to increased carrying capacity of the vegetation of the Nama-karoo but also reduce the number of unpalatable shrubs and increase grass cover. Although there arose a brief and positive flurry of interest in the approach, experimental tests of the system proved controversial (see Hoffman 1988a). Today, NSG enjoys only limited support and it has rather found expression in a number of related approaches to semi-arid land management such as short-duration grazing (SDG) and Savory's (1988) Holistic Resource Management concept. Hoffman

(1988a) reviewed the empirical and statistical basis for all currently advocated management approaches in the Nama-karoo and suggested that experimental tests were sorely needed to support the rationale and purported successes of the different systems.

8.7   Conclusions

The Nama-karoo biome has not attracted the same level of ecological interest as have the other biomes of southern Africa. Efforts to focus on the biome by research agencies (notably the Foundation for Research Development) were short-lived. Although the contributions from the agricultural institutes in the region have been considerable, their research has seldom been published in the primary literature. Where they have made important ecological contributions is in describing the composition and structure of the vegetation of the region. Despite this work, however, and despite more than a century of scientific investigation, the functioning of Nama-karoo communities and ecosystems is still poorly understood.

This paucity of ecological knowledge of the region is regrettable for a number of reasons. The Nama-karoo biome covers an extensive area on the subcontinent and, largely as a result of its size and central location, it abuts on all the other major biomes of the subcontinent. It also exists at the arid end of a moisture gradient which extends into adjacent, more productive vegetation types, especially on its southern and eastern margins. Degradation of these more mesic regions is thought to facilitate the expansion of typical Nama-karoo species (Hoffman & Cowling 1990a) and a general 'expanding karoo' hypothesis has had a profound influence on agricultural policy and financial commitment in the past and continues to do so today (see Chap. 21, this volume). Despite recent attempts to discount the severity and extent of land degradation in the Karoo (Hoffman & Cowling 1990b) the matter has not been laid to rest (Bond, Stock & Hoffman 1994; Dean & Macdonald 1994). It is crucial that there is renewed ecological interest in the region, not only to develop a deeper ecological understanding of the Nama-karoo biome itself but also to understand its dynamics in relation to the vegetation of the rest of the subcontinent.

Extensive pastoralism and wildlife ranching continue to be the major economic activities in the biome. Future research priorities should focus on these activities and include a re-assessment of herbivore preference tables (Hobson *et al.* 1975), which have been presented for the dominant taxa in the biome. Preliminary results (A.R.

Palmer & F. Van der Heyden, unpubl. data) suggest that taxa vary both in photosynthetic rate and palatability, depending on substratum. Patchiness in the distribution of resources in the landscape appears to impact on the productivity and dynamics of the ecosystem, with both wild and domestic herbivores showing seasonal and spatial preferences for patches. Elucidation of these trends will impact significantly on the rangeland condition assessment techniques (Vorster 1982) currently being applied by pastoralists.

Within the framework of global climate-change scenarios, testable predictions should be made on the direction that vegetation changes will take in the biome. Long-term climatic records in the biome are vital to our continued understanding of climatic processes, and with the reduction in the number of manned recording stations, due to de-population, the time length and spatial extent of records is declining.

8.8 References

Acocks, J.P.H. (1953). Veld Types of South Africa. *Memoirs of the botanical survey of South Africa*, **28**, 1–192.

Acocks, J.P.H. (1966). Non-selective grazing as a means of veld reclamation. *Proceedings of the Grassland Society of South Africa*, **1**, 33–9.

Adamson, R.S. (1938). *The Vegetation of South Africa*. London: British Empire Vegetation Committee.

Bews, J.W. (1916). An account of the chief types of vegetation in southern Africa, with notes on plant succession. *Journal of Ecology*, **4**, 129–59.

Bond, W.J., Stock, W.D. & Hoffman, M.T. (1994). Has the Karoo spread? A test for desertification using carbon isotopes from soils. *South African Journal of Science*, **90**, 391–7.

Bousman, B. & Scott, L. (1994). Climate or overgrazing?: the palynological evidence for vegetation change in the eastern Karoo. *South African Journal of Science*, **90**, 575–8.

Campbell, B.M., Cowling, R.M., Bond, W. & Kruger, F.J. (1981). *Structural Characterization of Vegetation in the Fynbos Biome. South African National Scientific Programme Report 52*. Pretoria: CSIR.

Cannon, W.A. (1924). General and physiological features on the vegetation of the more arid portions of South Africa with notes on the climatic environment. *Yearbook of the Carnegie Institution of Washington*, **8** (No. 354), 1–159.

Christopher, A.J. (1982). Towards a definition of the nineteenth century South African frontier. *South African Geographical Journal*, **64**(2), 97–113.

Cowling, R.M., Esler, K.J., Midgley, G.F. & Honig, M.A. (1994). Plant functional diversity, species diversity and climate in arid and semi-arid southern Africa. *Journal of Arid Environments*, **27**, 141–58.

Cronwright-Schreiner, S.C. (1925). *The Migratory Springbucks of South Africa*. London: Unwin.

Dean, W.R.J. & Macdonald, I.A.W. (1994). Historical changes in stocking rates of domestic livestock as a measure of semi-arid and arid rangeland degradation in the Cape Province, South Africa. *Journal of Arid Environments*, **26**, 281–98.

Dent, M.C., Lynch, S.D. & Schulze, R.E. (1987). *Mapping Mean Annual and Other Rainfall Statistics Over Southern Africa. Agricultural Catchments Research Unit Report No. 27*. Pietermaritzburg: University of Natal.

Du Preez, C. M. R. (1972). The production potential of karoo veld: growth and yield of karoo bush and reaction to pruning. Unpublished report K-Gf 109/1. Middelburg: Department of Agriculture.

Dyer, R.A. (1937). The vegetation of the divisions of Albany and Bathurst. *Memoirs of the Botanical Survey of South Africa*, **17**, 1–138.

Edwards, D. (1970). Vegetation map of South Africa, unpublished. Botanical Research Institute: Pretoria.

Ellis, F. (1988). *Die Gronde van die Karoo*. PhD Thesis. Stellenbosch: Stellenbosch University.

Ellis, F. & Lambrechts, J.J.N. (1986). Soils. In *The Karoo Biome: A Preliminary Synthesis. Part I – Physical Environment*, ed. R.M. Cowling, P.W. Roux & A.J.H. Pieterse, pp. 18–38. *South African National Scientific Programme Report 124*. Pretoria: CSIR.

Gess, F.W. (1981). Some aspects of an ethological study of the aculeate wasps and bees of a karroid area in the vicinity of Grahamstown, South Africa. *Annals of the Cape Provincial Museums*, **14**(1), 1–80.

Gess, S.K. (1992). *Ecology and Natural History of the Masarid Wasps of the World with an Assessment of their Role as Pollinators in Southern Africa (Hymenoptera: Vespoidea: Masaridae)*. PhD Thesis. Grahamstown: Rhodes University.

Gess, S.K. & Gess, F.W. (1989). Flower visiting by masarid wasps in southern Africa (Hymenoptera: Vespoidea: Masaridae). *Annals of the Cape Provincial Museums*, **18**, 95–134.

Gess, S.K. & Gess, F.W. (1994). Potential pollinators of the Cape Group of Crotalarieae (*sensu* Polhill) (Fabales: Papilionaceae), with implications for seed production in cultivated rooibos tea. *African Entomology*, **2**, 97–106.

Gibbs Russell, G.E. (1987). Preliminary floristic analysis of the major biomes in southern Africa. *Bothalia*, **17**, 213–27.

enrici, M. (1932). Cystine and sulphur content of bushes and grasses in a Karroid area (Fauresmith). *Report on Veterinary Research, Union of South Africa*, 18, 479–595. Pretoria: Department of Agriculture.

enrici, M. (1935a). Fodder plants of the Broken Veld. Their chemical composition, palatability and carrying capacity. *Science Bulletin No. 142*. Pretoria: Department of Agriculture.

enrici, M. (1935b). Germination of Karoo bush seeds. Part 1. *South African Journal of Science*, 32, 223–34.

enrici, M. (1937). Transpiration and water supply of South African plants. *South African Journal of Science*, 34, 61–72.

enrici, M. (1939). Germination of Karoo bush seeds. Part 2. *South African Journal of Science*, 36, 212–19.

enrici, M. (1940). Transpiration of different plant associations in South Africa. Part I: Transpiration of Karoo bushes. *Science Bulletin No 185*. Department of Agriculture and Forestry. Pretoria: Government Printer.

enrici, M. (1941). Transpiration of large Karoo bushes. *South African Journal of Science*, 37, 156–63.

enrici, M. (1945). Digestion experiments with fresh Karoo plants. *South African Journal of Science*, 41, 213–17.

enrici, M. (1952). Comparative study of the content of starch and sugars of *Tribulis terrestris*, lucerne, some Gramineae and *Pentzia incana* under different meteorological, edaphic and physiological conditions. *Onderstepoort Journal of Veterinary Research*, 25, 45–92.

ewitt, P.H. & Nel, J.J.C. (1969). Toxicity and repellency of *Chrysocoma tenuifolia* (Berg) (Compositae) to the harvester termite *Hodotermes mossambicus* (Hagen) (Hodotermitidae). *Journal of the Entomological Society of Southern Africa*, 32, 113–36.

obson, F.O. & Sykes, E. (1980). Defoliation frequency with respect to three karoo bush species. *Karoo Agric*, 1(5), 9–11.

obson, N.K., Jessop, J.P., van Der Ginn, M.C. & Kelly, J. (1975). *Veld Plants of Southern Africa*. Johannesburg: MacMillan.

Hoffman, M.T. (1988a). Rationale for karoo grazing systems: criticisms and research implications. *South African Journal of Science*, 84, 556–9.

Hoffman, M.T. (1988b). The pollination of *Aloe ferox* Mill. *South African Journal of Botany*, 54, 345–50.

Hoffman, M.T. (1989a). A preliminary investigation of the phenology of subtropical thicket and karroid shrubland in the lower Sundays River Valley, S.E. Cape. *South African Journal of Botany*, 55, 586–97.

Hoffman, M.T. (1989b). *Vegetation Studies and the Impact of Grazing in the Semiarid Eastern Cape*. PhD Thesis. Cape Town: University of Cape Town.

Hoffman, M.T., Barr, G.D.& Cowling, R.M. (1990). Vegetation dynamics in the semi-arid, eastern Karoo, South Africa: The effect of seasonal rainfall and competition on grass and shrub basal cover. *South African Journal of Science*, 86, 462–3.

Hoffman, M.T. & Cowling, R.M. (1987). Plant physiognomy, phenology and demography. In *The Karoo Biome: A Preliminary Synthesis. Part 2. Vegetation and History*, ed. R.M. Cowling & P.W. Roux, pp. 1–34. *South African National Scientific Programmes Report 142*. Pretoria: CSIR.

Hoffman M.T. & Cowling, R.M. (1990a). Desertification in the lower Sundays River Valley, South Africa. *Journal of Arid Environments*, 19, 105–17.

Hoffman M.T. & Cowling, R.M. (1990b). Vegetation change in the semi-arid eastern Karoo over the last 200 years: an expanding Karoo – fact or fiction? *South African Journal of Science*, 86, 286–94.

Johnson, M.R. & Keyser, A.W. (1979). *Explanatory notes on the 1 : 250 000 Geological Series 3226 King William's Town*. Pretoria: Government Printer.

Jones, S. (1984). The effect of axis splitting on xylem pressure potentials and water movement in the desert shrub *Ambrosia dumosa* (Gray) Payne (Asteraceae). *Botanical Gazette*, 145, 125–31.

Jooste, J.F. (1980). *A Study of the Phytosociology and Small Mammals of the Rolfontein Nature Reserve, Cape Province*. MSc Thesis. Stellenbosch: Stellenbosch University.

Jordaan, A. & Kruger, H. (1992). Leaf surface and anatomy of two xerophytic plants from southern Africa. *South African Journal of Botany*, 58, 133–8.

Jürgens, N. (1991). A new approach to the Namib Region I: phytogeographic subdivision. *Vegetatio*, 97, 21–38.

King, L.C. (1942). *South African Scenery*. London: Oliver & Boyd.

Lloyd, J.W. (1989a). Discriminant analysis and ordination of vegetation and soils on the Vaalputs radioactive waste disposal site, Bushmanland, South Africa. *South African Journal of Botany*, 55, 127–36.

Lloyd, J.W. (1989b). Phytosociology of the Vaalputs radioactive waste disposal site, Bushmanland, South Africa. *South African Journal of Botany*, 55, 372–82.

Louw, G.N., Steenkamp, C.W.P. & Steenkamp, E.L. (1967). Die verwantskap tussen die eterekstrakinhoud van karoobsossies en hul smaaklikheid vir skape. *South African Journal of Agricultural Science*, 10, 867–73.

Louw, G.N., Steenkamp, C.W.P. & Steenkamp, E.L. (1968). Chemiese samestelling van die vernaamste plantspesies in die dorre, skyn-sukkulente en Sentrale Bo-Karoo. *Tegniese mededeling No. 79*. Pretoria: Department of Agricultural Technical Services.

Marloth, R. (1908). *Das Kapland*. Jena: Gustav Fischer.

Martin, A.R.H. & Noel, A.R.A. (1960). *The Flora of Albany and Bathurst*. Grahamstown: Rhodes University.

Midgley, G.F. & Moll, E.J. (1993). Gas exchange in arid-adapted shrubs: when is efficient water use a disadvantage? *South African Journal of Botany*, 59, 491–5.

Milton, S.J. & Hoffman, M.T. (1994). The application of state-and-transition models to rangeland research and management in arid succulent and semi-arid grassy Karoo, South Africa. *African Journal of Range and Forage Science*, 11, 18–26.

Moore, A. (1989). *Die Ekologie en Ekofisiologie van Rhigozum trichotomum (Driedoring)*. PhD Thesis. Port Elizabeth: University of Port Elizabeth.

Moore, A., van Eck, J.A.J., van Niekerk, J.P. & Robertson, B.L. (1988). Evapotranspirasie in drie plantgemeenskappe van 'n *Rhigozum trichotomum* habitat te Upington. *Journal of the Grassland Society of Southern Africa*, 5, 80–4.

Novellie, P. & Strydom, G. (1987). Monitoring the response of vegetation to use by large herbivores: an assessment of some techniques. *South African Journal of Wildlife Research*, 17, 109–17.

Noy-Meir, I. (1973). Desert ecosystems: environment and producers. *Annual Review of Ecology and Systematics*, 4, 25–51.

O'Connor, T.G. & Roux, P.W. (1995). Vegetation changes (1949–1971) in a semi-arid, grassy dwarf shrubland in the Karoo, South Africa: influence of rainfall variability and grazing by sheep. *Journal of Applied Ecology*, 32, 612–26.

Palmer, A.R. (1987). The vegetation associated with whistling rat colonies in the karoo. *The Naturalist*, 31, 33–6.

Palmer, A.R. (1988). *Vegetation Ecology of the Camdebo and Sneeuberg Regions of the Karoo Biome, South Africa.* PhD thesis. Grahamstown: Rhodes University.

Palmer, A.R. (1989). The vegetation of the Karoo Nature Reserve, Cape Province. 1. A phytosociological reconnaissance. *South African Journal of Botany*, 55, 215–30.

Palmer, A.R. (1991a). A syntaxonomic and synecological account of the vegetation of the eastern Cape midlands. *South African Journal of Botany*, 57, 76–94.

Palmer, A.R. (1991b). The potential vegetation of the upper Orange River, South Africa: concentration analysis and its application to rangeland assessment. *Coenoses*, 6, 131–8.

Palmer, A.R. & Cowling, R.M. (1994). An investigation of topo-moisture gradients in the eastern Karoo, South Africa, and the identification of factors responsible for species turnover. *Journal of Arid Environments*, 26, 135–47.

Palmer, A.R. & van Heerden, F. (1994). The vegetation of the Karoo National Park. I. A phytosociological reconnaissance. Unpublished report to the National Parks Board. Grahamstown: Roodeplaat Grassland Institute.

Pole-Evans, I.B. (1936). A vegetation map of South Africa. *Memoirs of the Botanical Survey of South Africa*, 15, 1–23.

Raunkiaer, C. (1934). *The Life Forms of Plants and Statistical Plant Geography.* Oxford: Clarendon.

Roux, P.W. (1966). Die uitwerking van seisoenreenval en beweiding op gemengde Karooveld. *Proceedings of the Grassland Society of southern Africa*, 1, 103–10.

Roux, P.W. (1968a). Principles of veld management in the Karoo and adjacent sweet grassveld regions. In *The Small Stock Industry*, ed. W.J. Hugo, pp. 318–40. Pretoria: Government Printer.

Roux, P.W. (1968b). Ken die Karoo bossie. *Landbouweekblad*, 8 June, 4–5.

Rutherford, M.C. & Westfall, R.H. (1986). Biomes of Southern Africa – an objective categorization. *Memoirs of the Botanical Survey of South Africa*, 54, 1–98.

Savory, A. (1988). *Holistic Resource Management.* Washington DC: Island Press.

Scheepers, J.C. (1983). The present status of vegetation conservation in South Africa. *Bothalia*, 14, 991–5.

Schulze, R.E. (1976). Incoming radiation on sloping surfaces: a general model for use in southern Africa. *Agrochemophysika*, 8, 55–9.

Scott, J.D. & van Breda, N.G. (1938). Preliminary studies on the root systems of *Pentzia incana*-forma on the Worcester Veld Reserve. *South African Journal of Science*, 35, 280–7.

Squires, V.R. & Trollope, W.S.W. (1979). Allelopathy in the Karoo shrub, *Chrysocoma tenuifolia. South African Journal of Science*, 75, 88–9.

Talbot, W.J. (1961). Land utilization in the arid regions of southern Africa. Part I: South Africa. In *A History of Land Use in Arid Regions. Arid Zone Research Vol. 17*, ed. L. D. Stamp, pp. 299–338. Paris: UNESCO.

Theron, G.K., Schweickert, H.G. & van der Schyff, H.P. (1968). 'n Anatomiese studie van *Plinthus karooicus* Verdoorn. *Tydskrif vir Natuurwetenskap*, 8, 69–104.

van der Walt, P.T. (1980). A phytosociological reconnaissance of the Mountain Zebra National park. *Koedoe*, 23, 1–32.

Visser, D.J.L. (1989). *The Geology of the Republics of South Africa, Transkei, Bophuthatswana, Venda and Ciskei and the Kingdoms of Lesotho and Swaziland.* Pretoria: Government Printer.

Visser, J.N.J. (1986). Geology. In *The Karoo Biome: A Preliminary Synthesis. Part 1 – Physical Environment*, ed. R.M. Cowling, P.W. Roux & A.J.H. Pieterse, pp. 1–17. *South African National Scientific Programmes Report 124.* Pretoria: CSIR.

Vorster, M. (1982). The development of the Ecological Index for assessing veld condition in the karoo. *Proceedings of the Grasslands Society of Southern Africa*, 17, 84–9.

Vorster, M. (1985). *Die Ordenering van die Landtipes in die Karoostreek in Redelik Homogene Boerderygebiede deur Middel van Plantegroei – en Omgewingsfaktore.* DSc Thesis. Potchefstroom: Potchefstroom University.

Werger, M.J.A. (1973). An account of the plant communities of the Tussen die Riviere Game Farm, Orange Free State. *Bothalia*, 11, 165–76.

Werger, M.J.A. (1978). The Karoo–Namib Region. In *Biogeography and Ecology of Southern Africa*, ed. M.J.A. Werger, pp. 231–99. The Hague: Junk.

Werger, M.J.A. (1980). A phytosociological study of the Upper Orange River Valley. *Memoirs of the Botanical Survey of South Africa*, 46, 1–98.

Westoby, M., Walker, B.H. & Noy-Meir, I. (1989). Opportunistic management for rangelands not at equilibrium. *Journal of Range Management*, 42, 266–74.

White, F. (1978). The Afro-montane Region. In *Biogeography and Ecology of Southern Africa*, ed, M.J.A. Werger, pp. 463–513. The Hague: Junk.

Whittaker, R.H. (1975). *Communities and Ecosystems.* New York: MacMillan.

Yeaton, R. I. (1988). Structure and function of the Namib dune grasslands: characteristics of the environmental gradients and species distributions. *Journal of Ecology*, 76, 744–58.

Desert

N. Jürgens, A. Burke, M.K. Seely and K.M. Jacobson

9.1   Introduction

The desert biome of southern Africa (defined by Rutherford & Westfall 1986, but extended in this contribution) is represented almost exclusively by the Namib Desert. It covers 111 147 km² or 4.15% of the region discussed in this volume. The Namib has been the focus of much interest (Koch 1962; Werger 1978; Ward, Seely & Lancaster 1983; Walter & Breckle 1984; Walter 1986; Jürgens 1991; Seely 1991), as it is regarded as a very old desert (Ward & Corbett 1990) and possesses unique environmental conditions, unique adaptations and highly interesting biota including the monotypic gymnosperm *Welwitschia mirabilis* (Bornman 1972, 1977; Von Willert 1985). Nevertheless, no complete treatment of the vegetation of the Namib has been published hitherto. Only certain parts of the Namib, notably the vegetation of the dunes and plains of the central Namib in the vicinity of the desert research station at Gobabeb have been studied intensively, while other parts of this vast strip of land, nearly 2000 km long in a north–south direction (Ward *et al.* 1983), have been neglected. In this chapter we review both published data and unpublished studies throughout the Namib.

From many perspectives, there is not one Namib Desert, but two: one comprising the most arid portions of the succulent karoo biome, and the other the Nama-karoo biome. Because of the continuous transition between extreme desert and the more humid parts of the two adjacent biomes, some vegetation units are discussed in this desert chapter, which could just as well be included in the chapters on the succulent karoo biome (Chap. 7, this volume) or the Nama-karoo biome (Chap. 8, this volume) (for details see 9.3.1).

Landscapes in the Namib Desert range from areas devoid of any vegetation, to many areas where there is a relatively dense ground layer of herbs and dwarf shrubs with taller shrubs and even small trees – at least at sites with a slightly better water regime. Such a wide spectrum of habitats, ranging from 'extreme desert' to 'the edge of the savanna', is found in many deserts in the world. However, the Namib is special with respect to the rapid change of vegetation over short distances, owing to extremely steep ecological gradients. Over intervals of sometimes less than 100 km, the complete spectrum can be observed from succulent karoo biome shrubland through absolute desert to semi-desert with patches of savanna. Consequently the analysis of these steep ecological gradients and the related vegetation is essential for the understanding of desert vegetation.

9.2   Environmental gradients

Numerous environmental parameters of the Namib Desert co-vary in space and can be understood as parts of more general gradients (complex gradients; Walter 1971). Such general gradients are: tropical to temperate conditions along a north–south axis; and coastal to inland conditions along an east–west axis. Both gradients have a major influence on the flora and vegetation. The temperate–tropical gradient is gentle, whereas the coast–inland gradient is very steep. The southern extremity of the temperate–tropical gradient is represented by the temperate winter-rainfall region of the most arid parts of the succulent karoo biome, e.g. Namaqualand, Richtersveld, Ceres Karoo and the western parts

(a)

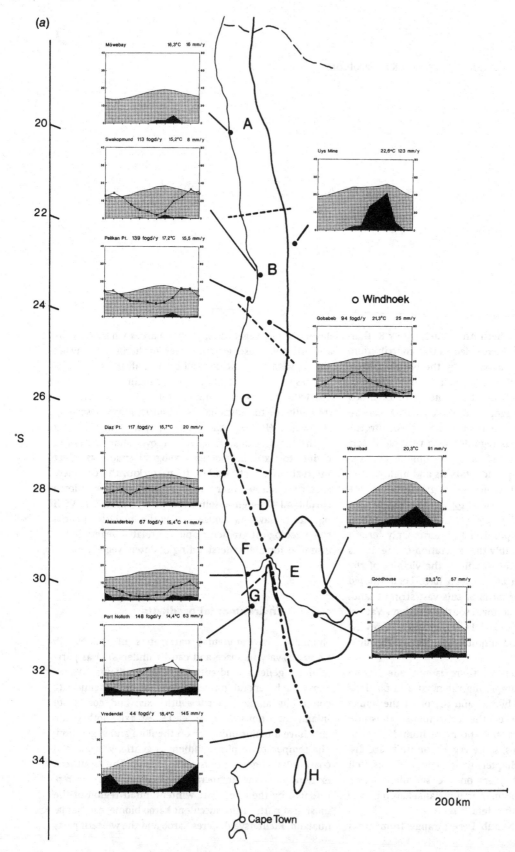

(b)

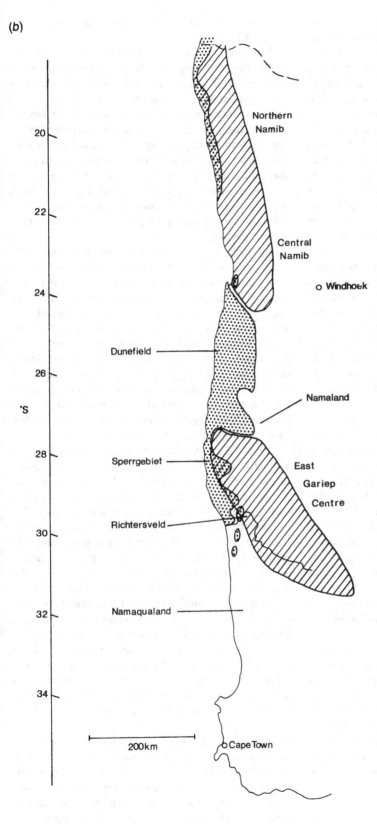

Figure 9.1 (a) **Arid regions below 100 mm mean annual precipitation (solid black line) with subdivisions of the Namib Desert and selected Walter–Lieth climate diagrams (x-axis = July to December; black shading, mean monthly precipitation; grey shading, mean monthly temperature; in some of the diagrams an additional line shows mean number of fog days month^{-1}) and the eastern limit of the winter-rainfall climate (dotted line): A, Northern Namib; B, Central Namib plains; C, Central Namib dunefield; D, Southern Namib (Nama-karoo part); E, East-Gariep Namib; F, Southern Namib (succulent karoo part); G, Northwestern Namaqualand Sandveld; H, Ceres Karoo. (b) Dunefields of the Namib Desert (dotted). Two phytogeographical subdivisions (hatched) inside the region with less than 100 mm mean annual precipitation are separated by the large Central Namib dunefield: East Gariep Centre in the south, Central and Northern Namib Centre in the north.**

of the Little Karoo. Here mild temperatures and higher relative humidity lead to relatively low potential evapotranspiration (Chap. 2, this volume). The northern extremity is represented by the tropical summer-rainfall region in arid northern Namibia and southwestern Angola, characterized by high temperatures, low humidity and high potential evapotranspiration.

The coast–inland gradient shows some similarities to the temperate–tropical gradient. Throughout the extent of the Namib coast, from Namaqualand to northern Namibia, almost identical climatic conditions prevail, characterized by mild temperatures, very high humidity and high fog incidence (Fig. 9.1a; Olivier 1995). In contrast, the inland conditions are characterized by much higher temperatures and lower humidity, but (in Namibia) higher rainfall.

Owing to these similarities, the two gradients amplify each other with respect to temperature, humidity and evapotranspiration. As the coast–inland gradient is much steeper than the temperate–tropical gradient, the dominating ecological gradient of the Namib Desert runs in a WSW–ENE direction. In general, the ecology, flora and vegetation are most conveniently interpreted as a coast–inland zonation with certain variations due to geomorphological, edaphic or historical features.

9.2.1 Coast–inland gradients

The sharp contrast between the coast with its high humidity and mild temperatures and the hot and often extremely dry hinterland has already been described in earlier publications (e.g. Walter 1986). Recently, more detailed information on the steepness and shape of this gradient has contributed additional insights.

Many parameters do not increase or decrease uniformly along the gradient. This is true for temperature, with maxima at the foot of the Great Escarpment, but milder temperatures near the coast and at higher altitudes on the escarpment (Fig. 9.2). Similarly, there is a zone of minimum humidity located between the humid oceanic air and (in summer) the humidity of the summer-rainfall zone further inland.

The position of the zone of lowest humidity varies almost hourly. In the late morning, humidity can be evenly distributed across the gradient or, as a result of hot east or 'bergwinds', be lowest close to the coast. During the day the sea breeze penetrates progressively further inland and in early afternoon this zone of lowest humidity can move some 50 or 100 km inland. The additive effect of the temperature and humidity curve results in conditions least favourable for plant growth being located somewhere between the coast and the escarpment.

The gradients show extreme steepness along the first 20 km near the coast, but become more gentle further inland (Fig. 9.3).

The coast–inland gradients show latitudinal variations; they are steeper in the north and shallower in the south. This latitudinal pattern is caused mainly by the larger extension of the winter-rainfall zone in the south and by the more extreme differences between the cool and foggy Benguela coast and the hot hinterland in the northern subtropical region. The maximum of aridity is close to the escarpment in the south, because of the strong influence of the winter rainfall along the coast which reaches relatively far inland. In the north the maximum aridity is closer to the sea, because the influence of the summer rainfall is stronger and reaches further westwards.

The highest incidence of fog is along the coast (Fig. 9.1a), particularly in the central Namib area between Cape Cross (north of Swakopmund) and Walvis Bay, where >100 fog days were recorded during 1984 (Olivier 1995). Fog day frequency decreases northwards and southwards along the coast from this zone; fog penetrates further inland most frequently in this central Namib zone. There are considerable seasonal variations in the frequency of fog. This shows a peak in the winter months at Swakopmund, Alexander Bay and Vredendal, but is highest in summer at Gobabeb and Port Nolloth (but see Olivier 1995). This indicates several types of fog, each possessing different dynamics.

Three types of fog have been reported hitherto, and a fourth is added here. Besides the normal sea fog, driven inland by slow air movements at low altitude, ground fog layers can be produced by condensation after rapid cooling of the soil surface at night. A third type of fog involves the seaward flow of cold air, which mixes with humid and warmer coastal air (K. Loris, pers. comm.). In addition to these three, there are also low cloud layers, which are driven inland by air movements. Near the coast these can be observed as stratified cloud layers, but they touch the ground at 30 to 80 km inland of the coast, e.g. at Gobabeb or east of the Rössingberge, resulting in fog precipitation without salt content. This fog is similar to the 'Garua fog' in the Lomas in Peru and Chile (Walter & Breckle 1984). The seasonality pattern of Gobabeb (Fig. 9.1a) suggests that this fog is strongest in summer. Further investigation of the spatial and temporal variation of these fog types is urgently needed.

9.2.2 Geomorphological and edaphic features

The most important latitudinal variation in geomorphology and soils is due to the existence of vast dunefields, especially in the region of the southern Namib sand sea between Lüderitz and Walvis Bay, and to a lesser degree in the Sperrgebiet (Diamond Area) south of Lüderitz and

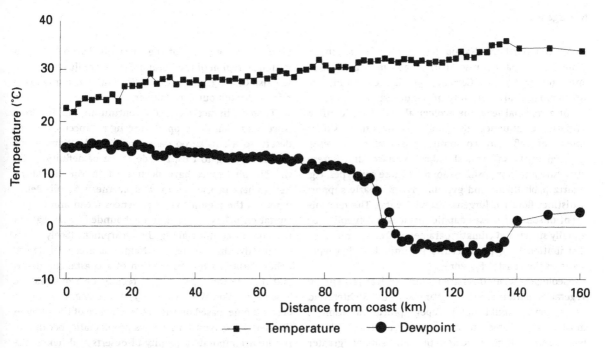

Figure 9.2 **Coast–inland transect (Swakopmund to Karibib) measurement of air temperature and humidity (dewpoint) around midday, showing the extension of moist oceanic air masses (sea breezes) to nearly 100 km inland and the position of the minimum in humidity lying between the eastern limit of oceanic air (c. 100 km) and the western limit of moist air originating from summer rains further inland (140 km).**

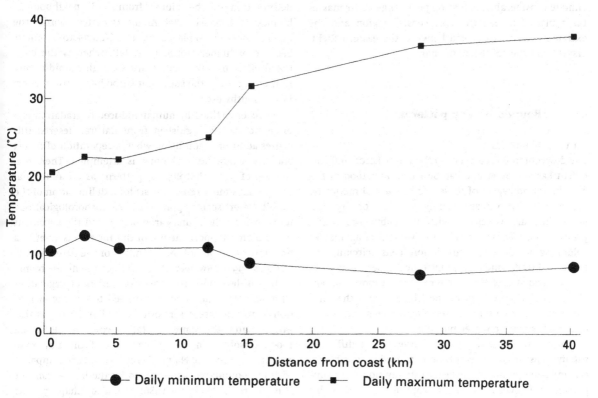

Figure 9.3 **Measurement of daily maximum and minimum temperature along a transect running inland from Lüderitz (S. Fanroth, unpubl. data). Gradients are steep close to the coast and shallow further inland.**

along the northern Namib (Fig. 9.1b). The southern Namib sand sea, because of its great age of at least 15 million years (Ward & Corbett 1990), is the most important biological discontinuity of the desert.

At a regional level the ecological and floristic differentiation is influenced by edaphic features such as duricrusts of different composition (silcretes, calcretes, gypsum crusts, salt crusts) and soil surfaces of different structure and composition. Some of these soil types, e.g. quartz pebblefields and gypsum crusts, clearly support a distinct flora (N. Jürgens, unpubl. data). The development of some of these edaphic surfaces is controlled by zonally structured climatic parameters. For example, the distribution of gypsum crusts is limited to low-lying areas in the coastal fog zone.

Geological formations are important determinants of vegetation patterns, e.g. the granites in the Richtersveld (N. Jürgens, unpubl. data). Topographical variation is most clearly visible on outcrops in the Namib plains, which receive higher precipitation and support greater plant biomass. A more complex pattern associated with topography is caused by the interruptions of the Great Escarpment, such as in the central Namib and in the Richtersveld. In the vicinity of the lower Orange River, the lack of the Great Escarpment influences the regional climate considerably, as a strong exchange of air masses takes place between the cool coastal region and the extremely hot inland basin between the eastern Richtersveld and the southern Kalahari.

9.3   Boundaries and patterns

9.3.1   Biome-scale

The desert biome is adjacent to the Nama-karoo and succulent karoo biomes and can best be understood as the extremely arid region of these two biomes. Climatically, the arid conditions are mainly a result of dry and descending air associated with the subtropical high-pressure zone. Because of the involvement of air masses belonging to different circulation cells (circumpolar fronts and tradewind cells), subtropical deserts generally can be divided into two different parts, a more temperate one and a more tropical one. The Namib is the zone between these two rain-transporting systems and consequently has very low precipitation.

These basic climatic patterns define the two different subdivisions of the Namib Desert, a tropical northeastern and a temperate southwestern subdivision, both supporting their own flora. In the tropical subdivision the flora is part of the Nama-karoo phytogeographical region of the Palaeotropics, while the temperate subdivision is

occupied by the flora of the succulent karoo phytogeographical region of the Greater Cape Flora (Jürgens 1991, see also Chap. 3, this volume). This subdivision is maintained throughout this chapter.

There is, in fact, a climatic continuum, ranging from 'succulent karoo proper' through 'succulent-karoo desert' to 'Nama-karoo desert' and finally 'Nama-karoo proper'. Owing to this, a wide range of definitions for the Namib Desert have been used in the literature. Researchers in the central and northern Namib define desert as the region that experiences mean annual precipitation below 100 mm and is bounded in the east by the 1000-m contour along the escarpment (Seely 1987). Essentially, this defines the Namib as a coastal desert. Other authors stress the notion of a coastal fog desert and include the foggy coastal regions of Namaqualand (Koch 1962; Von Willert *et al.* 1992). The definition of the desert biome, based on the predominance of therophytes (Rutherford & Westfall 1986), is problematic, because of the high temporal variability of deserts and hence, the episodic occurrence of this life form.

Furthermore, because of the strong climatic differences between the Nama-karoo and succulent karoo portion of the desert, no definition for the desert biome can be proposed that allows an entirely satisfactory delimitation of the biome from both neighbouring biomes (cf. Evenari 1985). Along its eastern border the desert biome is replaced by the Nama-karoo biome, because of higher summer rainfall, whereas the combined effect of lower temperatures, high humidity and rainfall separates the succulent karoo biome from desert in the southwest.

Areas desertified by human-induced degradation can often not be distinguished from natural deserts: this causes additional confusion when a separation of desert and the neighbouring biomes is attempted. Therefore, the use of phytochorological patterns as an additional and perhaps more reliable basis for a delimitation of the Namib Desert seems appropriate. Phytochorological patterns in both the Nama-karoo biome and the succulent karoo biome support the use of the 100-mm isohyet as a first approximation of the boundary of the desert biome (Jürgens 1991). However, if one defines the desert biome as having less than 10% cover of perennial vegetation (Shmida 1985), then one would need to use the 60-mm isohyet to separate desert from succulent karoo, as this would match the same vegetation cover as along the 100-mm isohyet in the Nama-karoo. If the phytogeographical patterns of Namib Desert species are compared with the boundaries as proposed by the biome concept (Rutherford & Westfall 1986; see also Chap. 5, this volume), two implications emerge.

Firstly, there is a large extension of the eastern

border. The distribution of 107 taxa occurs exclusively in the Nama-karoo phytogeographical region (Jürgens 1991) below 100 mm mean annual precipitation (Fig. 9.1*b*). There are clear discontinuities at nearly all margins, although the dunefield of the central Namib divides the Nama-karoo phytogeographical region into two subunits. The inland region between the Richtersveld, Hunsberge and the Pofadder–Upington region has recently been recognized as a part of the Namib, namely the East-Gariep Centre (Jürgens 1991).

Secondly, there is a vague southern border. There is no clear discontinuity between the flora of the winter-rainfall Namib and the succulent-karoo phytogeographical region proper. Nevertheless, the change in composition does not form a continuum but rather a cascade of steps. One of the more important steps confirms a Namib border close to a 60-mm isohyet in the northern Richtersveld, as proposed by Werger (1978). Another more important step close to the 100-mm isohyet includes the coastal lowlands and intermontane plains from the Richtersveld to near Kleinzee. On the other hand there is a large number of species linking the whole strip of the succulent karoo phytogeographical region from the Sperrgebiet to the Knersvlakte, even including the Little Karoo. Considering that large parts of the Ceres Karoo with very sparse vegetation also receive less than 100 mm annual rainfall, the inclusion of these regions into a larger Namib concept is appropriate.

9.3.2 Regional- and landscape-scale

Within the boundaries of the desert biome, regional subunits can be recognized. These have been described as very pronounced centres of endemism (Jürgens 1992).

East-Gariep Centre The most important centre of endemism in the Nama-karoo portion of the Namib lies in the triangle bounded by Lüderitz, Steinkopf and Upington. Here a wide belt of inland landscapes has to be included in the Namib, including the inland areas east of the Hunsberge–Richtersveld mountain ridges (Fig. 9.1*b*).

Central Namib Centre The second important centre in the Nama-karoo portion of the Namib is found in the plains north of the dunefield.

Southern Namib Centre This centre of endemism belongs predominantly to the succulent karoo phytogeographical region and is situated in the triangle between the Richtersveld, Alexander Bay and Lüderitz.

Obviously the major climatic differences, especially seasonality of rainfall, are the major determinants that separate the Southern Namib Centre (winter-rainfall zone) from the two other phytogeographical centres. The separation of the East-Gariep Centre and the Central

Namib Centre (interrupted by the great dunefield of the Namib) may be due to historical reasons. These two centres are held together by the occurrence of a large number of species with disjunct distributions and a high number of closely related species, e.g. *Euphorbia gregaria* and *Euphorbia damarana*, which occur north and south of the dunefield, respectively (N. Jürgens, unpubl. data).

9.3.3 Community-scale

The ecological gradients, described in 9.2, are clearly correlated with changes in floristic composition and vegetation structure. The majority of species have an elongated distribution along a NNW–SSE axis and a narrow west–east distribution, owing to the steepness of the coast-inland ecological gradient. The distribution of community types shows a similar pattern.

A community-level analysis of the Namib vegetation raises a number of problems. Firstly, plant cover is often too low for meaningful community delineation. Secondly, the frequent occurrence of monotypic vegetation units, the high importance of local endemism in the southern Namib and the steep ecological gradients result in a considerable number of communities. Thirdly, the high temporal variability, especially of short-lived species, makes it difficult to document the potential vegetation of many desert habitats.

A comprehensive treatment of the plant communities is beyond the scope of this chapter. Therefore, we present a broad description of the vegetation of the most important major habitats of the desert, i.e. plains, mountains, inselbergs, dunes and rivers (Table 9.1). Tables 9.2 and 9.3 provide brief descriptions of broad community types. These types have been delineated by multivariate analysis of 948 relevés (N. Jürgens, unpubl. data).

9.3.3.1 *Plains*

Vast plains comprise most of the surface area of the Namib Desert. Vegetation of the plains represents the zonal vegetation in the sense that water balance is based on the regional climate without additional benefits such as runoff or groundwater in river beds or sand dunes. Composition, structure and functional aspects of the vegetation of the Namib plains have been treated by numerous authors qualitatively (Giess 1981; Walter 1986), and Robinson (1976) and Moisel & Moll (1981) provide quantitative phytosociological treatments of parts of the plains of the central Namib. Vegetation analyses of the plains of the southern Namib and parts of the East-Gariep Namib are based on our unpublished data. The vegetation of the plains shows large differences across the desert. The eastern summer-rainfall parts of the central Namib, characterized by the dominant stem succulent, *Euphorbia damarana*, shares almost no species

Table 9.1 **Major vegetation zones in the desert biome of southern Africa**

Vegetation zone	Structural characterization	Dominant species	Mean species richness/ 100 m^{-2}	Environment
I Coastal Zone (succulent karoo phytogeographical region)	Dwarf open succulent shrubland and lichen fields	*Zygophyllum clavatum, Drosanthemum paxianum, Ramalina capensis, Teloschistes capensis, Cladoraphis cyperoides*	8–11	Very high humidity, fog frequent, mild temperatures
II Temperate Zone of Namib plains (succulent karoo phytogeographical region)	Dwarf open succulent shrubland	*Zygophyllum stapffii, Z. prismatocarpum, Z. cordifolium, Arthraerua leubnitziae, Othonna furcata, Brownanthus schlichtianus, B. arenosus, Stipagrostis sabulicola, S. lutescens*	3–12	High humidity, rel. mild to warm temperatures
III Temperate Namib mountains (succulent karoo phytogeographical region)	Dwarf to low open succulent shrubland	*Eberlanzia* spp., *Tetragonia* spp., *Ruschia* spp., *Sphalmanthus* spp., *Stoeberia* spp., *Pteronia* spp., *Tylecodon paniculatus, Galenia dregeana, Ceraria fruticulosa*	6–12	High humidity, mild temperatures
IV Zone of minimum vegetation	Extremely sparse dwarf to small succulent shrubland	*Salsola tuberculata, Sarcocaulon* spp., *Zygophyllum simplex*	0–3	Mild to high temperatures, low humidity
V Eastern grassland and stem succulent shrubland zone (Nama-karoo phytogeographical region)	Grasslands (on fine textured soils) and tall stem succulent shrubland (on rocky ground)	*Calicorema capitata, Stipagrostis* spp., *Euphorbia gregaria, E. damarana, E. virosa, Acacia reficiens, Aloe dichotoma*	3–7	Warm to very high temperatures, low humidity
VI Transition to Nama-karoo shrubland (Nama-karoo phytogeographical region)	Tall stem succulent shrubland or open savanna	*Rhigozum trichotomum, Parkinsonia africana, Boscia* spp., *Euphorbia guerichiana, Cyphostemma* spp., *Commiphora* spp., *Moringa ovalifolia, Myrothamnus flabellifolius*	5–12	Warm to very high temperatures, low humidity

Table 9.2 **Major plant communities in the succulent karoo phytogeographical region of the desert biome**

Zone	Community	Other diagnostic species	Environment	Geography
I	*Zygophyllum clavatum*	*Brownanthus marlothii, Othonna furcata, Didelta carnosa var. tomentosa*	Coastal dunes, saline fine-grained soils	Coast from Richtersveld to Northern Namib
I	Subcommunity with *Ramalina capensis*	*Limonium dyeri, Hypertelis angrae-pequenae*	Fog dependent, sand storm resistant	Port Nolloth to Lüderitz
I	Subcommunity with *Teloschistes capensis*		Fog dependent, on gypsum soils	Alexander Bay to Northern Namib
I	Subcommunity with *Lebeckia multiflora, Cladoraphis cyperoides*	*Lebeckia cinerea, Othonna cylindrica, Cladoraphis spinosa, Brownanthus arenosus, Lycium cinereum, Eberlanzia sedoides*	Coastal dunes and sandy valleys	Namaqualand to Northern Namib
1	Subcommunity with *Salsola zeyheri, Cephalophyllum ebracteatum*	*Psilocaulon dinteri, Pteronia glabrata, Sarcocaulon patersonii, Drosanthemum paxianum, Tetragonia decumbens, Juttadinteria deserticola*	Coastal, saline and rocky soils	Richtersveld to Lüderitz
I	Subcommunity with *Arthraerua leubnitziae*		Coastal, saline and loamy or sandy soils	Central Namib to Northern Namib
I	Subcommunity with *Dracophilus dealbatus*	*Eberlanzia* sp., *Ruschianthemum gigas*	Stony hills and plains	Lower Orange River
I/II	*Stipagrostis sabulicola*		Coastal to inland dunes	Sperrgebiet to Namib dunefield
II	*Brownanthus schlichtianus*	*Trachyandra muricata, Oncosiphon piluliferum, Senecio cardaminifolius, Aridaria noctiflora, Hypertelis salsoloides, Mesembryanthemum squamulosum*	Silty or loamy soils (often desert loess) over calcrete	Richtersveld to Lüderitz
II	*Zygophyllum prismatocarpum*	*Mesembryanthemum pellitum, Cheiridopsis robusta, Trianthema parvifolia, Euphorbia chersina, Psammophora modesta, P. nissenii*	Eroded calcretes	Richtersveld to Lüderitz

Table 9.2 **contd.**

Zone	Community	Other diagnostic species	Environment	Geography
II	*Arthraerua leubnitziae*	*Zygophyllum stapffii*	Mainly gypsum crusts	Central Namib
II	*Zygophyllum stapffii*	*Salsola tuberculata, Sesuvium sesuvioides*	Mainly calcretes	Central Namib
II/III	*Brownanthus pubescens*	*Delosperma pergamentaceum, Portulacaria pygmaea*	Shallow soils with gypsum crusts or calcrete and high salinity	Lower Orange River region
III	*Galenia dregeana*	*Ruschia* sp., *Euphorbia chersina, Crassula grisea, Othonna opima*	Neutral to acid loamy soils	Richtersveld and Sperrgebiet
IV	Ephemerals	*Synaptophyllum juttae, Opophytum aquosum, Stipagrostis obtusa, Zygophyllum simplex,* etc.	Plains and valleys	Richtersveld to Northern Namib

Table 9.3 **Major plant communities in the Nama-karoo phytogeographical region of the desert biome**

Zone	Community	Other diagnostic species	Environment	Geography
V	*Sisyndite spartea*	*Zygophyllum microcarpum, Rogeria longiflora*	Sheetwash plains and rivers	East Gariep Namib
V	*Sphalmanthus tetragonus*		Sandy and loamy plains	East Gariep Namib
V	*Euphorbia gregaria*		Rocky plateaux or crusts	East Gariep Namib
V	*Euphorbia virosa*	*Ceraria namaquensis, Pachypodium namaquanum*	Rocky mountains	East Gariep Namib
V	*Euphorbia damarana*		Rocky plateaux or crusts	Central Namib to Angola
V	*Stipagrostis ciliata (Euphorbia gummifera)*	*Calicorema capitata, Stipagrostis obtusa, S. geminifolia, Eragrostis nindensis*	Sandy plains, mostly with calcrete	East Gariep Namib to Central Namib
V	*Stipagrostis seelyae*		Eastern dunes	Central Namib dunefield
VI	*Rhigozum trichotomum*	*Parkinsonia africana, Boscia foetida*	Silty, loamy to gravelly plains	East Gariep Namib to Namaland
VI	*Commiphora virgata*	*Commiphora tenuipetiolata, C. glaucescens*	Rocky ground	Eastern Central and Northern Namib

with the plains covered by *Brownanthus schlichtianus* and other leaf-succulent chamaephytes in the winter-rainfall part of the Namib.

Lichen Fields Lichen fields form a characteristic constituent of the Namib vegetation near the coast. In the Central Namib, on fine-grained soils, *Teloschistes capensis* is the most important lichen species, whereas on stones and rocks, lichen communities are species-rich (Schieferstein & Loris 1992). High fog precipitation is a necessary prerequisite of the lichen fields, while the sand-blasting effects of strong easterly winds might explain the absence of lichen fields in other fog-rich coastal areas (Schieferstein & Loris 1992). The vast lichen fields of the central Namib have their southern limit near the great dunefield, where the incidence of sand blasting is very high. However, two very well-developed, but isolated, lichen fields occur near Alexander Bay and Lüderitz (Jürgens & Niebel 1991) and at least two more lichen fields exist between Lüderitz and Oranjemund, both situated a few kilometres inland of the coast near Elizabeth Bay and Chameis (G. Williamson, pers. comm.).

Extensive fruticose lichen vegetation on fine soil is limited to sites where environmental conditions are disadvantageous for angiosperms (e.g. high salinity, extremely low rainfall). Generally, taller angiosperms harvest the fog before it touches the ground and therefore epiphytic lichens predominate (Jürgens & Niebel-Lohmann 1995). Lichen fields dominated by *Teloschistes capensis* are limited to soils containing gypsum (Fig. 9.4), but only *Ramalina capensis* is able to withstand frequent sand storms and stabilizes dunefields.

Central and northern Namib The western part of the central Namib, from the coast to some 40 to 60 km inland, is dominated by two low (0.6–0.7 m tall), halophytic nanophanerophytes, namely *Arthraerua leubnitziae* and *Zygophyllum stapffii*. *Arthraerua leubnitziae* occurs frequently in saline habitats in almost any topographic position in the western half of this belt, but further inland it is concentrated along large drainage channels. This species extends about 80 km inland in the southern cen-

Figure 9.4 **Lichen field on eroded gypsum soils at Alexander Bay, South Africa, dominated by *Teloschistes capensis* with occasional *Ramalina capensis* and individuals of *Sarcocaulon* sp. (Geraniaceae) and various Mesembryanthemaceae** *(Photo: D.M. Richardson).*

Figure 9.5 **Zone of minimum vegetation on the central Namib plains, about 80 km east of Walvis Bay, Namibia. Plant life comprises a few individuals of the annual form of *Stipagrostis ciliata*** *(Photo: R.M. Cowling).*

Figure 9.6 ***Acacia erioloba* on the eastern fringes of the central Namib** *(Photo: R.M. Cowling).*

tral Namib. *Zygophyllum stapffii* is commonly found in this temperate coastal zone, but, in most regions, reaches 10 to 20 km further inland than *A. leubnitziae*, often following drainage lines with deeper soils. Both species occur on the coast and are clearly tolerant of high soil salinity. The habitat spectrum of *Z. stapffii* is wider than that of *A. leubnitziae* and includes rocky outcrops (Robinson 1976). Both species may be accompanied by *Salsola tuberculata* over their entire distribution ranges. These species exist at the extreme end of the spectrum of alkalinity and salinity in the zonal habitats of the central Namib (N. Jürgens, unpubl. data).

Inland of the *Zygophyllum–Arthraerua* belt, there is a zone of extreme aridity almost devoid of vegetation (Besler 1972; Fig. 9.5). Shallow rocky soils are more favourable sites, supporting a sparse vegetation, whereas homogeneous fine-grained soils in this zone are often totally devoid of perennial plant life. Common species include *Salsola tuberculata*, which also occurs in the coastal region and further inland. Phreatophytes such as *Welwitschia mirabilis* or *Acacia reficiens* may represent the only perennial plants along the shallow drainage lines. However, these species are more frequent further inland and are more appropriately described as extrazonal outliers of the transition zone to the Nama-karoo biome. As the amount and frequency of summer rainfall increases east of the zone of minimum vegetation, there is an increase in the vegetation cover and species richness (Cowling *et al.* 1994). Many species in this region are drainage-line specialists and never or rarely occupy zonal habitats of the plains. Examples are *Petalidium setosum* (Günster 1993c), *Pechuel-Loeschea leubnitziae*, *Asclepias buchenaviana* and *Adenolobus pechuelii*. Many other species, e.g. *Calicorema capitata*, are concentrated along drainage lines in the most arid parts of the Namib, but form diffuse vegetation on the plains further eastward, owing to increasing summer rainfall.

Another unit east of the zone of minimum vegetation is dominated by the stem-succulent *Euphorbia gregaria*. These tall plants (up to 3 m high) are associated with a number of other taxa belonging to the Nama-karoo phytogeographical region such as *Boscia foetida*, *Calicorema capitata*, *Acacia reficiens* and *Maerua schinzii*. On sandy soils, this community is replaced by grasslands with an admixture of *Kissenia capensis*, *Petalidium* spp. and *Geigeria* spp. Dominant grasses that form dense swards on sandy soils, and also on the calcrete plains at the eastern margin of the Namib, are *Eragrostis nindensis* and *Stipagrostis ciliata*. These areas are dissected by drainage channels which, depending on their size, depth and water storage, support the above-mentioned chamaephytes or tall trees such as *Acacia erioloba* (Fig. 9.6).

The absence or extreme rarity of young plants of *Wel-*

Figure 9.7 **Succulent vegetation dominated by *Cheiridopsis robusta* (Mesembryanthemaceae) in the southern Namib near Alexander Bay** *(Photo: P.G. Desmet).*

Figure 9.8 ***Fenestraria rhopalophylla* ssp. *aurantiaca* (Mesembryanthemaceae), a 'window plant' endemic to the foggy southern Namib** *(Photo: P.G. Desmet).*

witschia mirabilis in the coastal zone and in the zone of minimum vegetation is puzzling in view of the ability of adult plants of this long-lived species to cope with these habitats. In the case of *W. mirabilis*, only populations in the sparse savannas of Damaraland, Kaokoland and southwestern Angola show a normal demographic structure. This suggests that old plants may have persisted in desert conditions after recruiting in savanna-like environments many thousands of years ago (Jürgens 1992). Interestingly, the transpiration rate of 1 l m^{-2} leaf surface day^{-1} (Von Willert *et al.* 1982) is high compared to other evergreen plants of the Namib (Von Willert *et al.* 1992).

Southern Namib The plains of the southern Namib are covered by species of the succulent karoo phytogeographical region (Jürgens 1991; Fig. 9.7) which is best described as a more sparse extension of the vegetation of northwestern Namaqualand, especially the Richtersveld and the lower Orange River valley. However, many species are endemic to the region between about 26 and 28° S.

Fog incidence as a measure of distance from the cool foggy coast is a major determinant of community structure (Fig. 9.8). A very important factor is the impact of storms, which are of extraordinary force and frequency in the coastal part of the southern Namib. These winds are ecologically extremely important, because of their sand-blasting effect.

Along the coast, *Zygophyllum clavatum*, *Brownanthus marlothii*, *Didelta carnosa*, *Drosanthemum paxianum* and *Tetragonia decumbens* comprise a community that can be subdivided into several subunits. The immediate coastal zone, where fog incidence is highest, supports numerous lichen species, including *Ramalina capensis*, and a high frequency of the angiosperms *Limonium dyeri* and *Hypertelis angrae-pequenae*. Further inland on sandy soils in warm sheltered valleys, this community is replaced by one dominated by *Lebeckia multiflora*, *Eberlanzia sedoides*, *Lycium cinereum*, *Othonna arbuscula* and *Helichrysum obtusum*.

A community, almost exclusively composed of perennial species including *Salsola zeyheri*, *Cephalophyllum ebracteatum*, *Juttadinteria deserticola*, *Sarcocaulon patersonii*, *Pteronia glabrata* and *Euphorbia verruculosa*, is found on very shallow soils on rocky ground from the coast to some 20 km inland. In the Lüderitz region, east of the coastal zone, very little vegetation occurs on the wide plains to the east with mass germination of annual grasses and succulents following good rains.

South of Grillental and inland of the above-mentioned units, the plains are covered by several species-rich communities. On silty loamy soils on calcrete, *Brownanthus schlichtianus*, *Aridaria noctiflora*, *Mesembryanthemum squamulosum*, *Foveolina albida*, and *Hypertelis salsoloides* form a distinctive unit that can include more than 10 species 100 m^{-2}. Dolomite rock is typically occupied by a community with *Euphorbia chersina*, *Zygophyllum prismatocarpum*, *Psammophora modesta*, and *P. nissenii*, whereas eroded calcretes often support a community dominated by *Z. prismatocarpum*.

A community characterized by the tall stem-succulent *Euphorbia gummifera* covers large areas of calcrete. These areas are sometimes covered by sands, in which case grasses such as *Stipagrostis obtusa* and *S. geminifolia* commonly occur, as well as *Brownanthus arenosus*. *Augea capensis* is frequently found on more saline plains.

The northeastern part of the southern Namib is covered by species of the Nama-karoo phytogeographical region (Jürgens 1991). In the hot valleys of the Hunsberge and the northern Richtersveld (succulent karoo phytogeographical region), fine-grained soils support communities with *Rhigozum trichotomum* and *Parkinsonia africana* (both Nama-karoo species). Similarly, rocky ground is covered by vast *Euphorbia gregaria* stands, and sometimes *Aloe dichotoma* and *Kissenia capensis* are found. Grasses such as *Stipagrostis geminifolia*, *S. ciliata* and *S. obtusa* are found on sandy soils. Aspects similar to the eastern margin of the central Namib are found, where drainage channels form tree lines with *Acacia erioloba*.

Inland of the Hunsberge, from Namaland to the Upington region, vast rocky plains are covered by stands of *Euphorbia gregaria*. In the East-Gariep Namib east of the Hunsberge and the Richtersveld, sheetwash valleys form an important part of the landscape, and support vegetation similar to that of river beds. A characteristic species is *Sisyndite spartea* (Fig. 9.9).

9.3.3.2 Mountains

True montane deserts occur in the high elevation areas of the Diamond Area, especially the Hunsberge and the Richtersveld. This mountainous area supports an exceptionally rich flora, clustered in numerous species-rich communities. These communities include species of both the karoo and the Nama-karoo floras.

Composition and community structure are influenced by altitude, topography, geology and aspect. Aspect can be very important: moist air and fog from the coast enter the mountains along regular routes, bringing additional moisture to certain slopes. Nocturnal dew-fall evaporates in the morning more rapidly on northeastern slopes than on southwestern slopes. This also strongly influences vegetation patterns.

Owing to the cooler and more humid conditions, mountains can house enclaves of succulent karoo flora in a Nama-karoo matrix. e.g. Tsaukheib, Hunsberge around Witpüts (Jürgens 1991). Close to the central mountain ridge of the Richtersveld and the Hunsberge, many mountains and valleys show a rapid transition from vegetation of succulent karoo affinity on southwestern slopes to taxa of Nama-karoo affinity on northeastern slopes.

9.3.3.3 Inselbergs

In deserts, the vegetation on inselbergs is often more diverse than in the surrounding plains and comprises taxa that usually occur in areas of higher rainfall (Le Houerou 1986; Monod 1986). This is also the case with inselberg vegetation in the Namib. Taxa such as *Xerophyta viscosa*, *Aloe dichotoma* and *Commiphora* spp., which are normally associated with the transitional savanna zone of the escarpment, commonly occur on inselbergs in the desert zone (Giess 1971). In a study in the southern Namib, A. Günster (unpubl. data) found 84 species on an inselberg and fewer than ten species in the surrounding sand sea. Similar figures can be obtained for the central Namib, where inselbergs are surrounded by gravel plains.

The composition of inselberg floras is largely dependent on their geographical position. In the southern Namib, they harbour mainly winter-rainfall taxa such as a variety of Mesembryanthemaceae (*Conophytum saxetanum*, *C. gratum*, *Namibia pomonae*) and Crassulaceae (*Crassula sladenii*). Similarly, inselberg vegetation in the central and northern Namib comprises mainly summer-rainfall taxa such as *Sterculia africana*, *Euphorbia virosa* and numerous *Commiphora* spp. Coastal inselbergs often harbour a variety of lichens as well as dwarf and low succulents such as species of *Lithops*, *Lycium*, *Othonna* and *Pelargonium*, whereas inselbergs further inland are dominated by larger shrubs and trees (*Aloe dichotoma*, *Sterculia africana* and *Commiphora* spp.). The life-form spectra differ accordingly. Leaf-succulents dominate in the south and along the coast (Jürgens 1986), whereas evergreen and deciduous shrubs as well as stem-succulents dominate on inselbergs in the central and northern Namib (Figs. 9.10 and 9.11).

Figure 9.9 *Sisyndite spartea* (Zygophyllaceae), a stem-photosynthetic shrub, forms monospecific stands in the sheetwash plains of the East-Gariep Namib *(Photo: N. Jürgens)*.

Figure 9.10 *Euphorbia virosa* (foreground) and *Maerua schinzii* on an inselberg west of the Hartmannberge in the Kaokoland region of the northern Namib *(Photo: R.M. Cowling)*.

Figure 9.11 The lower slopes of the Brandberg, a massive inselberg on the fringes of the central Namib that reaches an elevation of 2574 m, supports a diverse tree flora. Shown here is *Boscia albitrunca* (foreground), *Commiphora* sp. and *Acacia montis-usti* *(Photo: R.M. Cowling)*.

9.3.3.4 Dunes

The active Namib dune sea is not devoid of vegetation, despite statements to this effect in the earlier literature (Giess 1962; Walter 1971, 1986; Leistner 1979). Numerous authors have recently described various aspects of the composition and functioning of the central Namib dune sea vegetation (Robinson 1976; Seely & Louw 1980; Yeaton 1988, 1990; Boyer 1989; Seely 1990, 1991; Jacobson 1992). Species diversity in the dunes is low relative to that of other habitats of the Namib. For example, Robinson (1976) recorded 20 species from the central Namib dune sea and 250 from the nearby gravel plains. Of the ten plant species found most commonly in the central Namib dunes, seven are endemic (Robinson & Seely 1980). Generally, the level of endemism in the Namib dunes is exceptional relative to similar areas elsewhere in the world (Bowers 1982; Walter & Box 1983).

The dominant plants of the central Namib dunes are perennial grasses of the genera *Cladoraphis*, *Centropodia* and *Stipagrostis* (De Winter 1990). Succulents are generally uncommon, but the woody-stemmed leaf-succulent *Trianthema hereroensis* (Aizoaceae) is co-dominant with *Stipagrostis sabulicola* in the coastal part of the main dunefield (Seely 1991; Fig. 9.12). *Acanthosicyos horridus* (Cucurbitaceae) occurs from the coast to about 80 km inland, in sandy interdune valleys from near Port Nolloth to at least the Curoca River in southwestern Angola (M.K. Seely, pers. obs.). Annual grasses and herbs form the dominant component on the lower dune slopes and inter-dune valleys, following average rain events. Perennial plants always dominate the upper slopes and, even after unusually high rains, germinants of perennial species are more frequent here (Seely 1990, 1991).

The favourable water relations of the deep dune sands facilitate the development of perennial grassland communities that are not found on the adjacent gravel plains, despite similar rainfall regimes. Grass community structure and richness vary across the dunefield in response to the rainfall gradient and concomitant levels of sand stability (Yeaton 1988).

In the eastern half of the dunefield where rainfall is higher, dune slopes support distinct vegetation zones.

Figure 9.12 **The vegetation of the central Namib dunes is dominated by species of *Stipagrostis* (*S. sabulicola* in the foreground) and other Poaceae** *(Photo: N. Jürgens).*

Perennial plant species richness and cover are highest on the dune base and plinth zones; lowest richness and cover are on the dune crest and inter-dune plains (Boyer 1989), but annual cover on the inter-dune plains is highly variable from year to year, depending on the amount of rainfall (Seely & Louw 1980; K.M. Jacobson, pers. obs.). Vegetation zonation on the dunes has been attributed to differential tolerance of plant species to the abiotic stresses imposed by the desert climate, namely, moisture availability and sand stability (Yeaton 1988, 1990; Seely 1990, 1991) as well as inter- and intra-species competition (Yeaton 1990).

9.3.3.5 *Rivers*

The 12 major ephemeral rivers crossing the Namib bring additional moisture to the desert. Variation in hydrology and hence composition of the woodlands among riverine systems is largely a function of catchment size and rainfall regimes within the catchments. Small catchments with irregular flows, e.g. the Tsondab and the Tsauchab Rivers, support woodlands composed primarily of *Acacia erioloba, A. tortilis* and *Parkinsonia africana*. Larger systems, with regular annual flow, support larger and more tree species. In the central Namib, ephemeral rivers such as the Kuiseb and Swakop support woodlands dominated by *Acacia erioloba, Faidherbia albida, Ficus cordata, F. sycomorus, Euclea pseudebenus, Salvadora persica* and *Tamarix usneoides*. In addition to these species, the Omaruru and other northern Namibian rivers support *Combretum imberbe* and *Colophospermum mopane*. Species richness of the riparian forests is generally low, but increases from south to north. Forest composition and plant cover varies along the length of the rivers: for example, *Ficus* sp., *Faidherbia albida* and *Tamarix usneoides* occur in more mesic habitats than *Acacia erioloba* and *Euclea pseudebenus* (Seely *et al.* 1980; K.E.A. Leggett unpubl. data; Fig. 9.13). Some river mouths form meandering deltas supporting saline or freshwater marsh vegetation; however, the form of the river mouths varies greatly depending on their annual flood volume.

9.3.3.6 *Special features, occurring in several community types*

Fairy rings A curious phenomenon in the eastern Namib, from Angola to Namaqualand, are circular patches, devoid of vegetation, often referred to as fairy rings (Fig. 9.14). Several hypotheses have been proposed to explain their origin. One proposal is that fungi and/or allelopathic plants (e.g. *Euphorbia* spp.) may exude toxins and thus make the soil unsuitable for establishment of seedlings (Eicker, Theron & Grobbelaar 1982; R.I. Yeaton, pers. comm.). In addition, the decomposition of dead euphorbias, which are abundant in many parts of

Figure 9.13 **The Kuiseb River forms the boundary between the Namib sand sea to the south and the gravel plains of the central Namib (foreground). The riverine vegetation, shown here near Gobabeb, is a woodland dominated by *Acacia erioloba* and *Faidherbia albida*** *(Photo: R.M. Cowling).*

Figure 9.14 **Fairy rings in the sandy plains near Sesfontein in the northern Namib** *(Photo: R.M. Cowling).*

the eastern Namib, may be responsible for the release of toxins into the soil (Brückner 1977). Another hypothesis invokes the activities of ants and termites (Moll 1994). However, these hypotheses remain to be tested. There are some similarities to the *heuweltjies* (mima-like mounds) found in the succulent karoo biome (Chap. 7, this volume). Both fairy rings and *heuweltjies* are circular in form, and show a regular distribution pattern with 'rings' of different ages in close proximity. However, while *heuweltjies* form mounds, fairy rings form slight depressions. Furthermore, *heuweltjies* often have a biogeneous calcrete formation, whereas fairy rings are associated with gaps in the calcrete crust (Cox 1987). The southernmost fairy rings occur on a plain in the northern Richtersveld, some tens of metres from a *heuweltjie* at the base of a slope (N. Jürgens, pers. obs.).

Microphytic crusts Other important components of desert vegetation are microphytic crusts, consisting of bluegreen and green algae and sometimes bryophytes and lichens (West 1990; Danin 1991; Lange *et al.* 1992). These crusts contribute to nutrient turnover and the redistribution of surface runoff (Gillis 1992). Although microphytic crusts have been observed in the Namib at various locations, nothing is known about their distribution, composition or role in Namib Desert ecosystems. In the coastal fog belt, where gypsum crusts predominate and there is a low but frequent moisture supply, microphytic crusts play an important role. Crusts of lichens (*Heppia* spp.) may be important for silty, loamy and sandy soils in resisting erosion. Similarly, the 'window algae', located several centimetres below the soil surface level and receiving light through quartz stones (Vogel 1955), contribute to biomass production in the Namib Desert.

9.4 Plant types

Very little autecological research has been undertaken on the Namib flora. Therefore, it is not possible to provide an extensive review on the characteristics of all the different plant types in the biome. Rather we highlight a few interesting features in this section.

9.4.1 Annuals and ephemerals

Ephemerals and annuals comprise by far the largest proportion of species in most desert regions (Inouye 1991). Annuals are seasonal and germinate only when rain falls at a specific time of the year, whereas ephemerals germinate whenever there is sufficient moisture, irrespective of season (Von Willert *et al.* 1990). Both plant types have very short life spans, some as short as a few weeks, and may produce seeds within weeks after establishment. Since conditions after sufficient rainfall are almost mesic and ephemerals complete their life cycle before arid conditions prevail again, adaptations to prevent water loss are usually not well developed. Nevertheless, there is a remarkable number of succulent annuals in the temperate part of the Namib Desert, e.g. *Mesembryanthemum* spp., *Opophytum* spp., *Zygophyllum simplex* and *Crassula dichotoma*.

The ephemeral and annual flora in the Namib differs greatly between regions, mainly due to the differences in prevailing form and pattern of moisture supply. Annual plants add to the species spectrum in nearly all different habitat types, but some habitats such as the zone of minimum vegetation harbour only annual or ephemeral species, e.g. *Stipagrostis* spp. on the gravel plains of the central Namib (Nel 1983; see also 9.3.3.1).

9.4.2 Facultative perennials

To cope with the unpredictability and variability of the Namib Desert environment, many desert plants show a high plasticity in life history parameters, such as life span, growth and reproductive output (Günster 1994a). An example is the large number of species that can either be annual or perennial, depending on the prevailing moisture conditions. They usually grow in drainage lines where plants have access to underground water stores and die when this water source is exhausted, usually after two to several years. The facultative perennials in the Namib flower and produce seeds continuously. Common examples are *Codon royenii* (Günster 1995a), *Rogeria longiflora* and *Kissenia capensis*.

Many species of *Stipagrostis* respond opportunistically, in terms of vegetative growth and reproductive output, to variable precipitation (Walter 1986; Seely 1990). For example, 12 mm of rainfall is sufficient for *S. ciliata* to set seed in a sandy substratum. However, with 25 mm the same plants can build up sufficient reserves to set seed during the first season and then sprout again in response to rains during a second season (Jacobson 1992). Walter (1986) describes the morphology of the root and shoot structure of *Stipagrostis* which enables survival in the dry period. These facultative perennial *Stipagrostis* spp. can grow into large bunchgrasses which, once established, have many advantages over individuals newly established from seed; namely, well-developed root systems and associated mycorrhizae for water and nutrient acquisition (Jacobson 1992, 1995); more rapid development of photosynthetic material in response to rainfall; and more economic use of moisture for growth and seed set (P. Jacobson, pers. obs.).

9.4.3 Geophytes

As in other desert regions, geophytes are a common sight after rains in the Namib. In the succulent karoo portion of the Namib they are more prominent in species number and cover than in the Nama-karoo (Chap.7, this volume). After favourable rains some species (e.g. *Ornithogalum stapffii*, *Dipcadi longifolium* and *Trachyandra laxa*) form extensive stands of one or two species on plains in the central and southern Namib. Apart from casual observations (e.g. Olivier 1984), there are no data on geophyte life cycles and species interactions in the Namib.

9.4.4 Phreatophytes

Phreatophytes, species that are specialized in the exploitation of groundwater, are a common feature of the Namib Desert. A good example is *Acanthosicyos horridus* (Cucurbitaceae), the !nara plant, which is endemic to the Namib dunefields. It is a leafless evergreen (stem photosynthetic) and is active all year round. Its long taproot always reaches water stores deep in the ground, sometimes down to 15 m below the surface.

Welwitschia mirabilis is the most studied plant in the central Namib. However, many unanswered questions still puzzle botanists today. Many consider it to be a tree with its stem half-buried. From the top of the cone-shaped stem, two large leaves grow continually, dying off at the leaf tips. Thus, leaves may reach more than 10 m in length (Bornman 1977). Some individuals are considered to be more than 1000 years old. Recent ^{14}C dating of a small plant of 180 × 550 mm stem diameter indicated an age of 550 years (N. Jürgens, unpubl. data).

A number of physiological studies on *Welwitschia mirabilis* have been concerned with its photosynthetic pathway and its ability to take up fog precipitation. It is now established that *W. mirabilis* is not a primitive CAM plant, as was previously assumed (Bornman 1977), but possesses an ordinary C_3 pathway (Eller *et al.* 1983). It is able to take up fog water, which condenses on the leaf surface and is channelled to the base of the stem, but its ability to take up water through stomata as demonstrated by Bornman (1977) has not been confirmed by Von Willert (1985). Comparisons of water relations, CO_2 exchange, water-use efficiency and growth of *W. mirabilis* plants in three contrasting habitats, including the *Welwitschia* plain approximately 45 km east of the coast and the foot-hills of the Brandberg, showed highly contrasting results (Von Willert & Wagner-Douglas 1994) which allow no generalizations to be made about adaptations.

9.4.5 Succulents
Succulents, defined by their ability to store water (Von Willert *et al.* 1990), form a very important part of the flora of the Namib (Van Jaarsveld 1987), and include a wide spectrum of life forms (Jürgens 1986, 1990). Namib Desert succulents include annuals (e.g. *Zygophyllum simplex*, summer-rainfall annual), ephemerals (e.g. *Mesembryanthemum guerichianum*), pauciennials (*Mesembryanthemum squamulosum*, *Sphalmanthus* spp.) and perennials (*Ruschia* spp.). Tall stem-succulents are the dominant succulent growth-form in the Nama-karoo portion of the desert biome. However, leaf succulent chamaephytes with a minor admixture of stem succulents that are either small (e.g. *Euphorbia stapelioides*) or thin-stemmed (e.g. *E. chersina*) predominate in the succulent karoo portion of the biome (Jürgens 1986; Chap 7, this volume). Ecologically, this pattern of separation can be explained by the more opportunistic behaviour of the small succulents of the succulent karoo, which are able to take up water from fog or even humid air, but require regular water availability for long-time survival (Von Willert *et al.* 1992) compared to the larger and more drought-resistant stem-succulents with their larger water-storage

volumes. Therefore, tall stem-succulents are better adapted to the less predictable rainfall pattern of the Nama-karoo (Hoffman & Cowling 1987). However, the concentration of tall stem-succulents in regions with a mean daily maximum temperature in the hottest month of more than 32.5 °C suggests that hot climates may favour this growth-form (Jürgens 1986).

Highest growth-form diversity of succulents is found in the succulent karoo portion of the desert biome (see also Chap.7, this volume), Here, many species belonging to the same growth-form (e.g. dwarf leaf-succulent shrubs) may coexist in small sites (Cowling *et al.* 1994). Despite their similar appearance, many of these leaf-succulents are not functional analogues, but respond differently to drought conditions (Cowling *et al.* 1994). For example, leaf-succulent Mesembryanthemaceae and Crassulaceae have very different leaf anatomies and phenological rhythms (Jürgens 1985,1986,1990), resulting in different responses to extreme droughts (Von Willert *et al.* 1985,1992). These differences should be incorporated into any functional group classification of desert plants (Jürgens 1986, 1990).

9.4.6 Psammophorous plants
There are many Namib Desert species, termed psammophorous plants, which accumulate a cover of sand and dust, thereby forming a protective layer against the destructive force of sand storms (sand blasting) (Jürgens 1995). In some cases sand grains are fixed by a sticky substance covering most of the leaf surface (e.g. *Psammophora nissenii*), in others, sand grains adhere to sticky hairs or glands borne on only certain parts of the plant surface (e.g. *Monechma* spp.). A preliminary analysis of the phytogeographic distribution of psammophorous plants shows a clear concentration in the southern Namib Desert (N. Jürgens, unpubl. data), a region with the strongest winds and the highest sand transport rate in southern Africa (Lancaster 1985). Psammophorous species such as *Psammophora nissenii*, *P. modesta* and *Chlorophytum viscosum* are important constituents of the vegetation in the Diamond Area No. 1 (Sperrgebiet). Many other structural features, e.g. the subterranean growth of *Fenestraria rhopalophylla*, also provide protection against the strong sand blasting. It is a remarkable fact that psammophorous plants are extremely rare outside the Namib Desert. Furthermore, in these rare examples (e.g. *Silene viscosa* and *Ifloga spicata* in the Mediterranean Basin) the ability to fix sand to their surface is not as strongly developed as in the Namib species. The rarity of this adaptation at a worldwide scale can be explained by the fact that, usually, sand blasting takes place only in very extreme deserts, which are devoid of vegetation. In the southern Namib, however, fog, high humidity and

low temperatures support rich vegetation in a virtually rainless environment where sand blasting is a frequent occurrence.

9.5 Ecophysiology

9.5.1 Modes of carbon dioxide fixation

All three photosynthetic pathways (C_3, C_4 and CAM) are present in Namib Desert plants (Von Willert *et al.* 1992). Among succulents the most common photosynthetic pathways are the CAM and C_3 types whereas the C_4 type is restricted to some halophytic succulents and a few annuals such as *Zygophyllum simplex*, which occur in dry areas with summer rainfall (Von Willert *et al.* 1992; see also Chap. 16, this volume). Some desert plants exhibit two photosynthetic pathways, following each other in an ontogenetically controlled sequence: for example, the leaf-succulent *Mesembryanthemum crystallinum* switches from C_3 to CAM (Herppich, Herppich & Von Willert 1992). There is a direct relationship between rainfall regime and photosynthetic pathway in the grasses of the Namib Desert and in southern Africa in general: C_3 grasses occur predominantly in the winter-rainfall region and C_4 grasses in the summer-rainfall region (Ellis, Vogel & Fuls 1980).

9.5.2 Water uptake

Many desert plants have developed additional means of water uptake besides uptake by roots. Water can be taken up directly via leaf or stem surfaces. For example, several species of *Crassula* take up water through their leaves (Barthlott & Capesius 1974) via specialized exposed structures, referred to as hydathodes (Tölken 1985; Von Willert *et al.* 1992) or adsorption fissures (Jürgens 1985, 1986). Von Willert *et al.* (1992) showed that many Mesembryanthemaceae are able to regain nearly 50% of their daily water losses by transpiration during the following night, provided a clear sky allows sufficient cooling of the leaf surfaces. Under these circumstances and with sufficient humidity, 'reverse transpiration' seems to be possible for plants under drought stress or with very negative osmotic potentials. Thus, water uptake is driven by a reversed water vapour gradient that can be directed from the atmosphere into the substomatal cavities of leaves. This may be one advantage of the accumulation of salts observed in most Mesembryanthemaceae (Von Willert *et al.* 1992). Uptake of fog water by the stems of the halophilic *Arthraerua leubnitziae* has been confirmed by ecophysiological observations (K. Loris, pers. comm.) and water uptake from fog by *Trianthema hereroensis* has been observed in the central

Namib dunes (Seely, De Vos & Louw 1977; Nott & Savage 1985).

9.6 Community structure and dynamics

9.6.1 Phenology

The phenology of most desert plants is closely related to rainfall, temperature and soil moisture. However, some species flower at any time of the year, provided there is sufficient rainfall (Von Willert *et al.* 1992; Günster 1994b). Other species show a photoperiodic response and only flower at certain times of the year (Ackerman & Bamberg 1974). Both strategies are present in the Namib, sometimes within the same species, and depending on climatic conditions at a particular site. Inland populations of the evergreen desert shrub *Calicorema capitata*, for example, have a distinct flowering period independent of the rainfall pattern in a particular season, whereas coastal populations of the same species flower all year round, apparently because of the continuous moisture supply (M.K. Seely, unpubl. data).

9.6.2 Seed biology and germination

Little work has been done on seed biology, germination and dispersal of Namib desert plants. Life histories and reproductive strategies differ in the summer and winter rainfall portions of the Namib. In the northern and central Namib with predominantly summer rains, plants respond directly to the amount of moisture provided and germinate readily as soon as a certain soil moisture level is reached (Günster 1994b). In the southern Namib, besides rain, temperature and photoperiod play an important role in germination (Van Rooyen & Grobbelaar 1982). Another important factor determining the dynamics of desert plant communities is the minimum amount of precipitation required to trigger germination. Grasses on the central Namib plains germinate after 17 mm of precipitation (Seely 1978a), whereas trees and shrubs require more moisture. *Welwitschia mirabilis*, for example, germinates after 25 mm of rain (Bornman 1972). Besides the minimum amount of moisture, seed dispersal mode can also influence germination patterns. In serotinous species in the central Namib, seeds that are ballistically dispersed germinate earlier than those that are dispersed by raindrops (Günster 1994b).

The frequent but low moisture supplied by fog along the coast also affects germination patterns. Populations of the same annual species in the southern Namib showed different dormancy periods, depending on whether they grew inland or near the coast (Hartmann 1983). The coastal populations showed a longer dormancy period, thus preventing germination after heavy fog, which might provide sufficient moisture to trigger germination, but not enough for seedling establishment.

Dormancy and germination behaviour are usually variable within and between species, sometimes even between seeds positioned on different parts of the same plant. These are strategies to spread the risk of unsuccessful seedling establishment in a particular season. There are several annuals and perennials in the northern and central Namib that show this behaviour (Stopp 1958; Marsh 1987, Günster 1995b) and which warrant further investigation. In an annual *Mesembryanthemum* sp. from the southern Namib, not more than 27% of seeds germinated after 80 days (Ihlenfeldt 1985). In contrast, a perennial succulent, *Delosperma pergamentaceum*, germinated rapidly after rains and most seedlings emerged at the same time, thus indicating a relationship between life-history strategy and germination pattern (see also Chap. 7, this volume).

Some species are adapted to the environmental conditions that prevail after germination. During germination in these groups, e.g. the Mesembryanthemaceae, a seed lid (operculum) covers the radicle, which remains protected by the inner integument for some time (Bittrich & Ihlenfeldt 1984). If dry conditions continue for a number of days, the seed suspends germination until favourable conditions return.

After germination, numerous environmental factors including microtopography, soil moisture and nutrient conditions, as well as competition with other seedlings determine levels of establishment. Survival rates are highly variable within and between species. On the eastern Namib plains, survival rates fluctuate between zero and 100% in different seasons. Rainfall patterns, determined by the timing, spacing and intensity of rain events, are important factors influencing seedling establishment. In the eastern Namib, early rains in summer result in lower seedling survival than later rains in autumn, when temperatures are lower and evaporation rates are reduced (Günster 1994a,c).

As for perennial plants, even-aged stands of shrubs indicate that establishment of shrub seedlings occurs only in particularly good seasons which, in most arid parts of the Namib, might be once in decades. The same is true for the perennial dune grasses (Seely 1991) and trees in ephemeral water courses (M.K. Seely, pers. obs.).

Seed dispersal has a crucial impact on community structure and dynamics. As in many other deserts, harvester ants in the Namib gather seeds in their nests and leave most behind when they move to another site (Marsh 1987). Plants that retain seeds on the parent plant for several seasons are common in many parts of the eastern Namib (Günster 1994c). It has been proposed that these seed-retaining or serotinous plants are abundant in the eastern, central Namib, because of unpredictable and intense rains as well as seed predation by insects and small mammals (Günster 1992). To increase the chances of distributing seeds to favourable microhabitats, several plant species show dimorphic dispersal strategies. *Geigeria alata*, for example, is a facultative perennial that produces diaspores adapted for short-range dispersal and diaspores adapted for long-range dispersal (Günster 1995b). The Namib is similar to other deserts (Ellner & Shmida 1981) in having a high incidence of species with special adaptations for seed dispersal (Ihlenfeldt 1983; Günster 1992).

9.6.3 Population dynamics

There are no long-term studies of populations of Namib plants. However, short-term fluctuations in composition of annual species were observed in the eastern Namib in response to variation in the timing and amount of rainfall: these were associated with different dormancy and germination behaviour of the species involved (Günster 1993c, 1994a).

Climatic oscillations at a scale of several years can have a clear effect on the composition of the vegetation, as was demonstrated in the northern Richtersveld. Monitoring of a 100-m^2 permanent plot at the end of each winter growing season showed that species composition did not change significantly over a ten-year period (1980–1990) (N. Jürgens unpubl. data). However, massive changes in biomass and number of individuals of different species and of life-form types were observed (Fig. 9.15). These changes were determined by rainfall patterns: lower than average rainfall was recorded from 1977 to 1980, followed by wet years from 1981 to 1983 (Fig. 9.16). Over the monitoring period there was a steady increase in individuals of leaf-succulent chamaephytes which have a xeromorphic epidermis. These evergreen plants provide safe sites for germination for many other species (Jürgens 1986). This increase is interpreted either as a period of recovery after the dry period or as the response of vegetation to a climatic oscillation. The number of individuals of leaf-succulents, which possess a mesomorphic epidermis (i.e. an epidermis with thin-walled bladder-cell idioblasts) and which shed their leaves during the dry season (Jürgens 1986) also increased in 1981. However, the number of individuals remained fairly constant thereafter. This group behaved more opportunistically and established numerous individuals in the years with the highest winter rains (1983 and 1989). Oscillations in the density of annuals and geophytes were extreme and these two groups showed different responses to the rains of the different years. Both groups increased in the good rain years of 1983 and 1989.

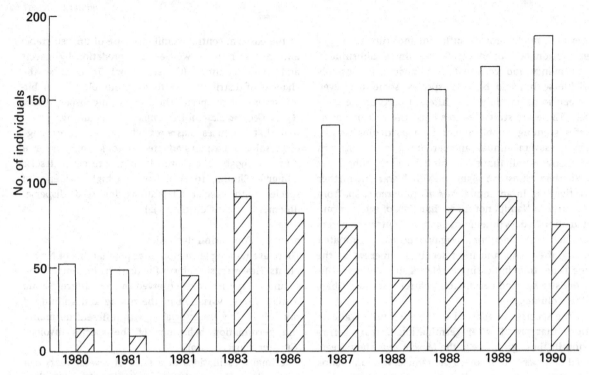

Figure 9.15 **Increase in the number of individuals in a permanent observation plot in the Richtersveld over a period of ten years after the drought of 1978 and 1979. Two life-form groups are shown: white bars, chamaephytes with evergreen succulent leaves with a xeromorphic epidermis, showing a continuous recovery; hatched bars, chamaephytes with annually shed succulent leaves with mesomorphic epidermis (with bladder cell idioblasts), showing a more opportunistic response to rainfall.**

Figure 9.16 **Precipitation in the Richtersveld (Lekkersing) 1976–1990, showing drought situation in 1978 and 1979. Precipitation shown for year (white bars) and winter (hatched bars).**

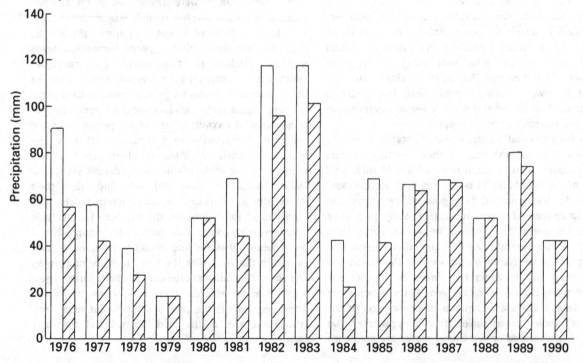

9.6.4 Species interactions

The role of competition in determining the structure of Namib Desert communities ranges from weak to strong, depending on the habitat and the climatic region. Competition for water and nutrients is important in structuring the hyper-arid communities on dunes (Yeaton 1990), and is probably of increasing importance towards the eastern Namib, where higher rainfall results in denser vegetation (Polis 1991). Nothing is known about other communities in this respect.

Namib desert plants support a variety of consumers, ranging from seed-eating insects, birds and small mammals to large browsers and grazers such as ostrich, antelopes, elephants and black rhinoceros (Stuart 1975; Seely & Louw 1980; Loutit, Louw & Seely 1987; Viljoen 1989). Although large-scale movements of game species in response to rainfall and consequently primary production have been recognized (Tarr & Tarr 1989), very few quantitative data are available on their impacts on plant communities. Elephants in the northern Namib, for example, have had no detrimental effect on the tree populations in the rivers over the last 20 years, although their food preferences for species such as *Faidherbia albida* might affect the composition of the riverine vegetation in the long term (Viljoen & Bothma 1990).

An association between a leaf- and a stem-succulent species has been reported from the Namib gravel plains (Dean *et al.* 1992). This study showed that seedling establishment of the stem-succulent *Euphorbia damarana* was confined to sites in the shelter of the leaf rosette of *Aloe asperifolia*. Many more associations of this type are expected between different species and in different parts of the Namib, e.g. between *Salvadora persica* and *Boscia foetida* in Damaraland (A. Günster, pers. obs.). In the Richtersveld, beneath adult individuals of low evergreen leaf-succulent chamaephytes, up to five species are regularly found. Furthermore, about 80% of the newly established individuals (less than one year old) of perennial Mesembryanthemaceae had established beneath these nurse plants (N. Jürgens, pers. obs.).

9.7 Ecosystem function

Although a direct relationship between rainfall and annual grass production has been recognized in the Namib (Seely 1978a,b; Günster 1995c), there have been no studies on the productivity of the perennial vegetation. In general, productivity increases with increased moisture availability. Yeaton (1988) showed that the standing crop of *Stipagrostis sabulicola* increases in an easterly direction with increasing rainfall across the Namib dunefield, as does the average standing crop of the dunegrass communities. The average standing crop of individuals in dune slope communities is also highest on the mid- to lower dune slope, where moisture conditions are optimal (Yeaton 1988).

Namib dune sand is deficient in all three macronutrients: nitrogen, phosphorus and potassium (Seely & Louw 1980; Jacobson 1992). Mycorrhizal fungi are well known for improving plant growth under low-nutrient conditions (see Chap. 16, this volume) and have well-developed associations with grasses in the Namib dunefield (Jacobson 1992). The role that these fungi play in nutrient cycling in the Namib Desert has not yet been investigated. However, it is generally known that in addition to improving nutrient and moisture relations of the plants, the activities of fungi support other organisms in the plant rhizosphere (Allen 1991; Chap. 16, this volume).

Detritus from dune plants represents a largely aseasonal, relatively persistent form of organic matter in mobile dunes, on which many organisms depend (Louw & Seely 1982; Crawford & Seely 1993). Detritivores are generally more abundant than carnivores in the Namib dunes, even though similar numbers of species exist (Crawford & Seely 1987). Arthropods probably decompose the most plant detritus in the Namib Desert, but the role that micro-organisms play in the decomposition and mineralization of detritus and litter has not yet been investigated. Fungi are ubiquitous following rain events and, as with plant species, their activity patterns are synchronized with optimal utilization of available moisture (P. Jacobson, unpubl. data). Information concerning interactions between micro-organisms, the material that they decompose, and other organisms involved in decomposition, are lacking for the Namib Desert. Some studies in other deserts suggest, however, that termites may preferentially consume plant material colonized by fungi, because of the higher carbon : nitrogen ratios resulting from fungal decomposition of the plant substrate (West 1991). *Psammotermes mossambicus* has been observed eating this high-quality equivalent in the Namib dunes (Crawford & Seely 1994; K. Jacobson, pers. obs.), but how important this is to nutrient acquisition by the termites is not known. As termites are thought to be major consumers in the Namib Desert (Crawford & Seely 1994), such interactions may have important implications for nutrient cycling and deserve further study.

9.8   Management and human use

9.8.1   Desertification

Hunter-gathering and nomadic pastoralism were the primary land uses for the Namib Desert up to the last century (Kinahan 1991). There is ample evidence that much of the plains landscape was used on an irregular basis, presumably after good rains. More recently, parts of the eastern fringe of the Namib have been used for emergency grazing during drought by commercial farmers. Today, communal farmers on the desert borders use the perennial vegetation during drought periods and permanent settlements are being established on the desert fringe. These communal farmers are particularly dependent on the vegetation of the ephemeral river courses (Jacobson *et al.* 1995). As the populations in the communal areas increase, so their activities extend further into the Namib. In contrast, many of the private farmers owning land on the desert edge are abandoning livestock farming and focusing on tourism and other alternative development strategies. Nevertheless, overgrazing appears to be of growing concern as farmers under all land-tenure systems fail to adjust stocking rates to the natural wet and dry cycles (Seely & Jacobson 1994). High proportions of unpalatable and poisonous plant species in the current vegetation are a clear indicator of incipient degradation (Günster 1993a).

9.8.2   Tourism

Tourism in Namibia is rated as one of the three top earners of foreign exchange before the turn of the century (Hoff & Overgaard 1993). The Namib Desert, along with wildlife throughout the country, is a major tourist attraction and the sand dunes such as those at Sossusvlei are popular destinations. Part of this attraction is the dune vegetation, seen to be thriving in an extremely hostile environment. This attraction could be enhanced by information for tourists on the natural history of the endemic plants and animals (Seely 1987).

Further tourist attractions in the desert are the linear oases provided by the large trees growing along the ephemeral river courses (Jacobson, Jacobsen & Seely 1995). These trees provide browse and pods for wildlife as well as shade for tourists. Their conservation status is, however, affected by activities in the higher catchment areas. The mountains of the Namib, for example the Brandberg and to a lesser extent the Naukluft Mountains, are also foci for tourism and recreation. The unusual succulent flora of the high Brandberg is particularly interesting to visitors. Less well-protected coastal areas of the Namib are also used by tourists and fishermen, although the coastal vegetation is subjected to heavy vehicular traffic (Seely 1987).

Tourism in the Namib is supported by the Government of Namibia in that the entire core desert, lying between the Kunene River in the north and Lüderitz in the south, is incorporated into protected areas: the Skeleton Coast Park in the north, the West Coast Recreation area in the centre and the Namib–Naukluft Park to the south. With the exception of a narrow coastal strip, the entire Sperrgebiet (Diamond Area) south to Lüderitz is also a protected area. Within these protected areas, some land is managed as private tourist concessions with long-term government contracts. Adjoining the protected areas, many private farming concerns are developing guest farms where tourists can enjoy the local environment or travel to state-controlled tourist areas. Similarly, community-based tourist camps are also situated to take advantage of the desert landscape and state-controlled land. In the state-controlled communal lands inland from the coast, community-based game guards at present play a major role in maintaining the integrity of the landscape and its wildlife. Tourism, present and future, is enhanced by their efforts and those of their supporting non-government organizations. Nevertheless, conflicts over land and water use in this area are escalating.

9.9   Conclusions

Although the Namib has been the focus of much botanical interest, scientific research on the vegetation has been limited. More research is required on all aspects of the biome's flora and vegetation. More taxonomic research is still required in a number of taxa, e.g. Mesembryanthemaceae and various petaloid monocots (see also Chap. 7, this volume). Distribution data for species and vegetation units are also inadequate. Basic knowledge on important environmental parameters, e.g. fog precipitation, is missing. Very few sites for the analysis of vegetation dynamics have been established in this highly variable environment. Few data are available on species interactions, life-history strategies or soil–plant interactions, and comprehensive ecosystem-level studies have yet to be initiated. Data are lacking on the history of the desert environment, especially with respect to changing patterns of human influence. Studies on desertification processes and the potential for restoring degraded environments are in their infancy. It would be a great tragedy if the unique flora and vegetation of the Namib Desert were to remain so poorly known and understood.

9.10   Acknowledgements

We gratefully acknowledge the input of various people whose comments helped to complete the picture of the current knowledge of the vegetation of the Namib Desert. Our thanks go especially to S. Brown, P. Craven, E. Erb, S. Fanroth, M. Hessing, C. Hines, E. van Jaarsveld, U. Jähnig, P. Jacobson, R. Kubirske, H. Kolberg, K. Loris, B. Loutit, G. Maggs, A. Niebel-Lohmann, B. Schieferstein, B. and M. Strohbach, D.J. von Willert and G. Williamson. We also thank the editors for their constructive comments on several versions of the manuscript.

.11   References

Ackerman, T.L. & Bamberg, S.A. (1974). Phenological studies in the Mojave Desert at Rock Valley. In *Phenology and Seasonality Modeling*, pp. 215–26. Berlin: Springer-Verlag.

Allen, M.F. (1991). *The Ecology of Mycorrhiza*. Cambridge: Cambridge University Press.

Barthlott, W. & Capesius, I. (1974). Wasserabsorption durch Blatt-und Sproßorgane einiger Xerophyten. *Zeitschrift für Pflanzenphysiologie*, 72, 443–55.

Besler, H. (1972). *Klimaverhältnisse und klima-geomorphologische Zonierung der zentralen Namib (Südwestafrika)*. Stuttgarter Geographische Studien 83.

Dittrich, V. & Ihlenfeldt, H.-D. (1984). Morphologie früher Keimungsstadien bei Mesembryanthemaceae: eine Anpassung an aride Umweltbedingungen. *Mitteilungen des Instituts für Allgemeine Botanik Hamburg*, 19, 123–39.

Bornman, C.H. (1972). *Welwitschia mirabilis*: paradox of the Namib Desert. *Endeavour*, 31(113), 95–9.

Bornman, C.H. (1977). *Welwitschia mirabilis*: structural and functional anomalies. *Madoqua*, 10, 21–31.

Bowers, J.E. (1982). The plant ecology of inland dunes in western North America. *Journal of Arid Environments*, 5, 199–220.

Boyer, D.C. (1989). Some characteristics of the plant communities of three dunes situated across a climatic gradient in the Namib Desert. *Madoqua*, 16, 141–8.

Brückner, V. (1977). Über einige Inhaltsstoffe der Wolfsmilchgewächse unter besonderer Berücksichtigung der Krebsverstärker. *Dinteria*, 13, 1–24.

Cowling, R.M., Esler, K.J., Midgley, G.F. & Honig, M.A. (1994). Plant functional diversity, species diversity and climate in arid and semi-arid southern Africa. *Journal of Arid Environments*, 27, 141–58.

Cox, G.W. (1987). The origin of vegetation circles on stony soils of the Namib Desert near Gobabeb, South West Africa/Namibia. *Journal of Arid Environments*, 13, 237–44.

Crawford, C.S. & Seely, M.K. (1987). Assemblages of surface-active arthropods in the Namib dunefield and associated habitats. *Journal of African Zoology*, 101, 397–421.

Crawford, C.S. & Seely, M.K. (1993). Dunefield detritus: its potential for limiting population size in *Lepidochora discoidalis* (Gebein) (Coleoptera: Tenebrionidae) in the Namib Desert. *Journal of African Zoology*, 107, 527–34.

Crawford, C.S. & Seely, M.K. (1994). Detritus loss in the Namib Desert dunefield: influence of termites, gerbils and exposure to surface conditions. *Journal of African Zoology*, 108, 49–54.

Danin, A. (1991). Plant adaptations in desert dunes. *Journal of Arid Environments*, 21, 193–212.

Dean, W.R.J., du Plessis, M.A., Milton, S.J., Adams, N.J. & Siegfried, W.R. (1992). An association between a stem succulent (*Euphorbia damarana*) and a leaf succulent (*Aloe asperifolia*) in the Namib Desert. *Journal of Arid Environments*, 22, 67–72.

De Winter, B. (1990). A new species of *Stipagrostis* (Aristideae) from the Dune–Namib Desert, Namibia. *Bothalia*, 20, 82–7.

Eicker, A., Theron, G.K. & Grobbelaar, N. (1982). A microbiological study of 'bare patches' in the Giribes plains of Kaokoland, South Africa. *South African Journal of Botany*, 1, 69–74.

Eller, B.M., von Willert, D.J., Brinckmann, E. & Baasch, R. (1983). Ecophysiological studies on *Welwitschia mirabilis* in the Namib Desert. *South African Journal of Botany*, 2, 209–23.

Ellis, R.P., Vogel, J.C. & Fuls, A. (1980). Photosynthetic pathways and the geographic distribution of grasses in South West Africa/Namibia. *South African Journal of Science*, 76, 307–14.

Ellner, S. & Shmida, A. (1981). Why are adaptations for long-range seed dispersal rare in desert plants? *Oecologia*, 51, 133–44.

Evenari, M. (1985). The desert environment. In *Ecosystems of the World*, vol. 12A, ed. M. Evenari, I. Noy-Meir & D.W. Goodall, pp. 1–22. Amsterdam: Elsevier.

Giess, W. (1962). Some notes on the vegetation of the Namib Desert with a list of plants collected in the areas visited by the Cape-Transvaal Museum Expedition during May, 1959. *Cimbebasia*, 2, 3–35.

Giess, W. (1971). A preliminary vegetation map of South West Africa. *Dinteria*, 4, 5–114.

Giess, W. (1981). Die in der Zentralen Namib von Südwestafrika/Namibia festgestellten Pflanzenarten und ihre Biotope. *Dinteria*, 15, 13–71.

Gillis, A.M. (1992). Israeli researchers planning for global climate change on the local level. *BioScience*, 42, 587–9.

Günster, A. (1992). Aerial seed banks of the central Namib: distribution of serotinous species in relation to climate and habitat. *Journal of Biogeography*, 19, 563–72.

Günster, A. (1993a). *Ökologische Untersuchungen zur Zeitverzögerten Samenverbreitung in der Zentralen Namib*. PhD Thesis. Münster: University of Münster.

Günster, A. (1993b). Does the timing of rainfall events affect resource allocation in serotinous desert plants? *Acta Oecologica*, 8, 153–9.

Günster, A. (1993c). Microhabitat differentiation among serotinous plants in the Namib Desert. *Journal of Vegetation Science*, 4, 585–90.

Günster, A. (1994a). Variability in life history parameters of four serotinous plants in the Namib Desert. *Vegetatio*, 114, 149–60.

Günster, A. (1994b). Phenological niches of serotinous plants in the central Namib Desert. *Journal of Tropical Ecology*, 10, 531–44.

Günster, A. (1994c). Seed bank dynamics – longevity, viability and predation of seeds of serotinous plants in the central namib. *Journal of Arid Environments*, 28, 195–205.

Günster, A. (1995a). The multi seasonal desert plant *Codon royenii*. *Cimbebasia*, 14, 23–30.

Günster, A. (1995b). *Geigeria alata* in the Namib Desert – seed heteromorphism in an extremely arid environment. *Journal of Vegetation Science*, 6, 473–8.

Günster, A. (1995c). Grass cover distribution in the central Namib – a rapid method to assess regional and local rainfall patterns of arid regions. *Journal of Arid Environments*, 29, 107–14.

Hartmann (1983). Monographie der Gattung *Vanzijlia* (Mesembryanthemaceae). *Botanische Jahrbücher für Systematik*, 103, 499–538.

Herppich, W., Herppich, M. & von Willert, D.J. (1992). The irreversible C₃ to CAM shift in well-watered and salt-stressed plants of *Mesembryanthemum crystallinum* is under strict ontogenetic control. *Botanica Acta*, 105, 34–40.

Hoff & Overgaard (Pty) Ltd. (1993). *Namibia Tourism Development Study Summary*. Windhoek: Ministry of Wildlife, Conservation and Tourism/Commission of the European Communities.

Hoffman, M.T. & Cowling, R.M. (1987). Plant physiognomy, phenology and demography. In *The Karoo Biome: A Preliminary Synthesis. Part 2 – Vegetation and History*, ed. R.M. Cowling & P.W. Roux, pp. 1–34, South African National Scientific Programmes Report 124. Pretoria: CSIR.

Ihlenfeldt, H.-D. (1983). Dispersal of Mesembryanthemaceae in arid habitats. *Sonderband des naturwissenschaftlichen Vereins Hamburg*, 7, 381–90.

Ihlenfeldt, H.-D. (1985). Lebensformen und Überlebensstrategien bei Sukkulenten. *Berichte der Deutschen Botanischen Gesellschaft*, 98, 409–23.

Inouye, R.S. (1991). Population biology of desert annual plants. In *The Ecology of Desert Organisms*, ed. G.A. Polis, pp. 25–54. Tucson: University of Arizona Press.

Jacobson, K.M. (1992). Factors affecting VA-Mycorrhizal Community Structure in the Namib Dune Field; and the Population Biology of an ectomycorrhizal Basidiomycete: *Suillus granulatus*. PhD Thesis. Blacksburg, USA: Virginia Polytechnic Institute and State University.

Jacobson, K.M. (1996). Fungal ecology of the Etosha National Park. *Madoqua*, (in press).

Jacobson, P.J., Jacobson, K.M. & Seely, M.K. (1995). *Ephemeral Rivers and Their Catchments: Sustaining People and Development in Western Namibia*. Windhoek: Desert Research Foundation of Namibia and Department of Water Affairs, 160 pp.

Jürgens, N. (1985). Konvergente Evolution von Blatt- und Epidermismerkmalen bei blattsukkulenten Familien. *Berichte der Deutschen Botanischen Gesellschaft*, 98, 425–46.

Jürgens, N. (1986). Untersuchungen zur Ökologie sukkulenter Pflanzen des südlichen Afrika. *Mitteilungen aus dem Institut für Allgemeine Botanik Hamburg*, 21, 139–365.

Jürgens, N. (1990). A life form concept including anatomical characters, adapted for the description of succulent plants. *Mitteilungen aus dem Institut für Allgemeine Botanik Hamburg*, 23A, 321–42.

Jürgens, N. (1991). A new approach to the Namib region. I. Phytogeographic subdivision. *Vegetatio*, 97, 21–38.

Jürgens, N. (1992). Die Wüste der lebenden Wasserspeicher. *Uni HH Forschung*, 27, 68–76.

Jürgens, N. (1996). Psammophorous plants and other adaptations to desert ecosystems with high incidence of sand storms. *Feddes Repertorium*, 107(3/4).

Jürgens, N. & Niebel, A. (1991). The unknown lichen hill. *Veld & Flora*, 77, 24–6.

Jürgens, N. & Niebel-Lohmann, A. (1995). Geobotanical observations on lichen fields in the Southern Namib Desert. *Mitteilungen des Institut für Allgemeine Botanik Hamburg*, 25, 135–56.

Kinahan, J. (1991). *Pastoral Nomads of the Central Namib. The People That History Forgot*. Windhoek: New Namibia Books.

Koch, C. (1962). The Tenebrionidae of Southern Africa XXXI. Comprehensive notes on the Tenebrionid fauna of the Namib Desert. *Scientific Papers Namib of the Desert Research Station*, 5, 61–106.

Lancaster, N. (1985). Winds and sand movements in the Namib sand sea. *Earth Surface Processes and Landforms*, 10, 607–19.

Lange, O.L., Kidron, G.J., Büdel, B., Meyer, A., Kilian, E. & Abeliovich, A. (1992). Taxonomic composition and photosynthetic characteristics of the 'biological soil crust' covering sand dunes in the western Negev Desert. *Functional Ecology*, 6, 519–27.

Le Houerou, H. N. (1986). The desert and arid zones of northern Africa. In *Ecosystems of the World, vol. 12B, Hot Deserts and Arid Shrublands*, ed. M. Evenari, I. Noy-Meir & D.W. Goodall, pp. 101–47. Amsterdam: Elsevier.

Leistner, O. A. (1979). Southern Africa. In *Arid-land Ecosystems: Structure, Functioning and Management, vol. 2*, ed. D.W. Goodall & R.A. Perry, pp. 109–44. Cambridge: Cambridge University Press.

utit, B.D., Louw, G.N. & Seely, M.K. (1987). First approximation of food preferences and the chemical composition of the diet of the desert-dwelling black rhinoceros, *Diceros bicornis* L. *Madoqua*, 15, 35–41.

uw, G.N. & Seely, M.K. (1982). *Ecology of Desert Organisms*. New York: Longmans.

arsh, A.C. (1987). The foraging ecology of two Namib harvester ant species. *South African Journal of Zoology*, 22, 130–6.

oisel, A. & Moll, E.J. (1981). A Braun-Blanquet survey of the vegetation of the *Welwitschia* plain. *Dinteria*, 15, 3–12.

oll, E. (1994). The origin and distribution of fairy rings in Namibia. In *Proceedings of the 13th Plenary Meeting of AETFAT*, ed. J.H. Seyani & A.C. Chikuni, pp. 1203–9. Zomba, Malawi: Association for the Taxonomic Study of the Flora of Tropical Africa.

onod, T. (1986). The Sahel zone north of the equator. In *Ecosystems of the World, vol.12B, Hot Deserts and Arid Shrublands*, ed. M. Evenari, I. Noy-Meir & D.W. Goodall, pp. 203–44. Amsterdam: Elsevier.

el, P. S. (1983). *Monitering van die Beskikbaarheid, Gehalte en Benutting van Voer op Deigruisvlaktes van die Kuisebstudiegebied*. MSc Thesis. Bloemfontein: University of the Orange Free State.

ott, K. & Savage, M.J. (1985). Variation in seasonal and diurnal leaf water potential of a Namib dune succulent. *Madoqua*, 14, 177–9.

livier, J. (1995). Spatial distribution of fog in the Namib. *Journal of Arid Environments*, 29, 129–38.

livier, W. (1984). *Pancratium tenuifolium*. *Veld & Flora*, 70, 77–8.

olis, G.A. (1991). Desert communities: an overview of patterns and processes. In *The Ecology of Desert Communities*, ed. G.A. Polis, pp. 1–26. Tuscon: University of Arizona Press.

obinson, E.R. (1976). *Phytosociology of the Namib Desert Park, South West Africa*. MSc Thesis. Pietermaritzburg: University of Natal.

Robinson, M.D. & Seely, M.K. (1980). Physical and biotic environments of the southern Namib dune ecosystem. *Journal of Arid Environments*, 3, 183–203.

Rutherford, M.C. & Westfall, R.H. (1986). Biomes of Southern Africa – an objective categorization. *Memoirs of the Botanical Survey of South Africa*, 54, 1–98.

Schieferstein, B. & Loris, K. (1992). Ecological investigations on lichen fields of the Central Namib. *Vegetatio*, 98, 113–28.

Seely, M.K. (1978a). Grassland productivity: the desert end of the curve. *South African Journal of Science*, 74, 295–7.

Seely, M. K. (1978b). The Namib dune desert: an unusual ecosystem. *Journal of Arid Environments*, 1, 117–28.

Seely, M.K. (1987). *The Namib*. Windhoek: Shell Oil SWA Ltd.

Seely, M.K. (1990). Patterns of plant establishment on a linear desert dune. *Israel Journal of Botany*, 39, 443–51.

Seely, M.K. (1991). Sand dune communities. In *The Ecology of Desert Communities*, ed. G.A. Polis, pp. 348–82. Tucson: University of Arizona Press.

Seely, M.K., Buskirk, W.H., Hamilton, W.J. III & Dixon, J.E.W. (1980). Lower Kuiseb River perennial vegetation survey. *Journal of the South West African Society*, 35, 57–86.

Seely, M.K., de Vos, M.P. & Louw, G.N. (1977). Fog imbibition, satellite fauna and unusual leaf structure of a Namib Desert dune plant, *Trianthema hereroensis*. *South African Journal of Science*, 73, 169–72.

Seely, M.K. & Jacobson, K.M. (1994). Desertification and Namibia: a perspective. *Journal of African Zoology*, 108, 21–36.

Seely, M.K. & Louw, G.L. (1980). First approximation of the effects of rainfall on the ecology and energetics of a Namib Desert dune ecosystem. *Journal of Arid Environments*, 3, 25–54.

Shmida, A. (1985). Biogeography of the desert flora. In *Ecosystems of the World*, vol. 12A, ed. M. Evenari, E. Noy-Meir & D.W. Goodall, pp. 23–77. Amsterdam: Elsevier.

Stopp, K. (1958). Die verbreitungshemmenden Einrichtungen der südafrikanischen Flora. *Botanische Studien*, 8.

Stuart, C.T. (1975). Preliminary notes on the mammals of the Namib Desert Park. *Madoqua*, 11, 5–68.

Tarr, P.W. & Tarr, J.G. (1989). Veld dynamics and utilisation of vegetation by herbivores on the Ganias flats, Skeleton Coast Park, South West Africa/Namibia. *Madoqua*, 16, 15–22.

Tölken, H. R. (1985). Crassulaceae. *Flora of Southern Africa*, 14, 1–244.

van Jaarsveld, E. (1987). The succulent riches of South Africa and Namibia and their adaptive strategies. *Aloe*, 24, 46–92.

van Rooyen, M.W. & Grobbelaar, N. (1982). Saadbevolkings in die grond van die Hester Malan-Natuurreservaat in die Namakwalandse Gebroke Veld. *South African Journal of Botany*, 1, 41–50.

Viljoen, P.J. (1989). Habitat selection and preferred food plants of a desert-dwelling elephant population in the northern Namib Desert, South West Africa/Namibia. *African Journal of Ecology*, 27, 227–40.

Viljoen, P.J. & Bothma, J.D.P. (1990). The influence of desert dwelling elephants on vegetation in the northern Namib Desert, South West Africa/Namibia. *Journal of Arid Environments*, 18, 85–96.

Vogel, S. (1955). Niedere 'Fensterpflanzen' in der südafrikanischen Wüste. *Beiträge zur Biologie der Pflanzen*, 31, 45–133.

von Willert, D.J. (1985). *Welwitschia mirabilis* – new aspects of the biology of an old plant. *Advances in Botanical Research*, 11, 157–91.

von Willert, D.J., Brinckmann, E., Scheitler, B. & Eller, B.M. (1985). Availability of water controls crassulacean acid metabolism in succulents of the Richtersveld (Namib Desert, South Africa). *Planta*, 164, 45–55.

von Willert, D.J., Eller, B.M., Brinckmann, E. & Baasch, R. (1982). CO_2 gas exchange and transpiration of *Welwitschia mirabilis* Hook. fil. in the central Namib Desert. *Oecologia*, 55, 21–9.

von Willert, D.J., Eller, B.M., Werger, M.J.A. & Brinkmann, E. (1990). Desert succulents and their life strategies. *Vegetatio*, 30, 133–43.

von Willert, D.J., Eller, B.M., Werger, M.J.A., Brinkmann, E. & Ihlenfeldt, H.-D.(1992). *Life Strategies of Succulents in Deserts*. Cambridge Studies in Ecology. Cambridge: Cambridge University Press.

von Willert, D.J. & Wagner-Douglas, U. (1994). Water relations, CO2 exchange, water-use efficiency and growth of *Welwitschia mirabilis* Hook. fil. in three contrasting habitats of the Namib Desert. *Botanica Acta*, **107**, 291–9.

Walter, H. (1971). *Ecology of Tropical and Subtropical Vegetation*. Edinburgh: Oliver & Boyd.

Walter, H. (1986). The Namib Desert. In *Ecosystems of the World. Hot Deserts and Arid Shrublands*, vol. 12B, ed. M. Evenari, I. Noy-Meir & D.W. Goodall, pp. 245–82. Amsterdam: Elsevier.

Walter, H. & Box, E.O. (1983). The Karakum desert, an example of a well-studied eubiome. In *Ecosystems of the World*, vol. 5, *Temperate Deserts and Semideserts*, ed. N.E. West. Amsterdam: Elsevier.

Walter, H. & Breckle, S.-W. (1984). *Ökologie der Erde. Band 2: Spezielle Ökologie der tropischen und subtropischen Zonen*. Stuttgart: UTB-Gustav Fischer Verlag.

Ward, J.D. & Corbett, I. (1990). Towards an age of the Namib. In *Namib Ecology. 25 Years of Namib Research*, ed. M.K. Seeley, pp. 17–26, *Transvaal Museum Monograph 7*. Pretoria: Transvaal Museum.

Ward, J.D., Seely, M.K. & Lancaster, N. (1983). On the antiquity of the Namib. *South African Journal of Science*, **79**, 175–83.

Werger, M.J.A. (1978). Biogeographical division of Southern Africa. In *Biogeography and Ecology of Southern Africa*, ed. M.J.A. Werger, pp. 147–70. The Hague: Junk.

West, N.E. (1990). Structure and function of microphytic soil crusts in wildland ecosystems of arid and semi-arid regions. *Advances in Ecological Research*, **20**, 179–223.

West, N.E. (1991). Nutrient cycling in soils of semiarid and arid regions. In *Semiarid Lands & Desert. Soil Resource & Reclamation*, ed. J. Skujins, pp. 295–332. New York: Marcel Dekker Inc.

Yeaton, R.I. (1988). Structure and function of the Namib dune grasslands: characteristics of the environmental gradients and species distributions. *Journal of Ecology*, **76**, 744–58.

Yeaton, R.I. (1990). The structure and function of the Namib Dune grasslands: species interactions. *Journal of Arid Environments*, **18**, 343–9.

Grassland

10

T.G. O'Connor and G.J. Bredenkamp

10.1 Introduction

The grassland biome, occupying 349 174 km², is centrally located in southern Africa, and adjoins all except the desert, fynbos and succulent karoo biomes (Chap. 5, this volume). It spans a rainfall gradient from ca. 400 to >1200 mm yr⁻¹, a temperature gradient from frost-free to snow-bound in winter, ranges in altitude from sea level to >3300 m, and occurs on a spectrum of soil types from humic clays to poorly structured sands. The main geological substrata are Karoo sediments and volcanics, but geomorphological history is diverse (Chap. 1, this volume). Although grassland has a general uniformity of structure, there is considerable variation in the floristic composition, functional attributes of species, vegetation dynamics and ecosystem functioning across such wide environmental variation. Summaries of the grassland biome include those by Killick (1978), Werger & Coetzee (1978), White (1978), Mentis & Huntley (1982) and Tainton & Mentis (1984).

This chapter aims at providing a novel synthesis and explanation of the reason that grasslands occur where they do, the relation between grassland and surrounding biomes, the variation in floristic composition and functional attributes of species along environmental gradients; it also aims at understanding the vegetation dynamics of communities and populations and of patterns of primary production. Our purpose is to show that most variation across the biome can be related mainly to rainfall but also to soil properties and temperature. On occasion, reference is made to the grassland layer of the savanna biome.

10.2 Boundaries

Four lines of evidence help to identify the factors that determine the extent and boundaries of grassland. These are: variables that discriminate between biome types; vegetation changes on the boundaries between biome types; the occurrence of satellite grasslands within other biomes and the converse; and the factors that preclude woody individuals from the grassland biome.

The extent of the grassland biome is limited by climatic factors (Rutherford & Westfall 1986; Chap. 5, this volume), and is distinguished from neighbouring biomes by the number of days with sufficient soil moisture for plant growth, the mean temperature of such days and the mean temperature of days too dry for growth to occur (Ellery, Scholes & Mentis 1991). The boundaries between the grassland and other biomes can be determined by a water balance approach (Ellery, Mentis & Scholes 1992). Savannas, grassland and forest are distinguished from the Nama-karoo by longer growing seasons and higher temperatures during the non-growing season. Forest receives higher seasonal rainfall than grassland or savanna. Savanna experiences higher temperatures during the non-growing season than grassland. Forest patches can occur within grassland provided there are sufficient growing days, appropriate soils and protection from fire (Chaps 12 and 18, this volume).

10.2.1 Dynamics of biome boundaries
Much of the drier southwestern boundary of the grassland biome interfaces with the Nama-karoo (Fig. 5.2, this volume). This boundary is related to altitude and hence to annual rainfall (Palmer 1991a), and relates specifically to a zone of high rainfall uncertainty. It is further influ-

enced by substratum texture (sands versus clays), and supports a generalist flora capable of tolerating such uncertainty (Palmer & Cowling 1994; Chap. 8, this volume). Following the advent of settled pastoralism, overgrazing by domestic livestock, particularly sheep, is considered to be the major factor responsible for the eastward expansion of the Nama-karoo into the grassland biome (Acocks 1953; Roux & Vorster 1983; Roux & Theron 1987; Chaps 8 and 21, this volume). This process was accompanied by the near elimination of perennial grasses. An alternative proposal is that these putative, grazing-induced changes reflect cyclic shifts in the seasonality of rainfall (Hoffman & Cowling 1990), with years of higher spring–summer rainfall promoting perennial grasses at the expense of karoo bushes; and years of relatively higher autumn–winter rainfall promoting the converse (Roux 1966). Long-term grazing experiments confirm the overriding effect of year-to-year rainfall variation on vegetation dynamics. They further illustrate the potential for almost pure perennial grassland in the absence of grazing, and the presence of mainly dwarf shrubs and short-lived grasses in the face of sustained grazing during the growth period of grasses (O'Connor & Roux 1995).

The largest portion of the grassland boundary interfaces with savanna (Fig. 5.2, this volume), which often consists of an almost mono-specific woody component of *Acacia karroo*, a tall shrub or tree (Chap. 11, this volume). This species has invaded grassland throughout the biome in recorded time (Bews 1917; Dyer 1937; Acocks 1953; Comins 1962; Morris 1976; Bredenkamp & Bezuidenhout 1990) as a result of reduced grass biomass, owing to sustained heavy grazing by livestock (Friedel 1987). Reduced grass biomass increases the seedling survival and growth of *A. karroo* (Story 1951; Du Toit 1966) and reduces the frequency and intensity of fires, thereby reducing the mortality of *A. karroo* seedlings (Du Toit 1972) and top-kill of small individuals (Trollope 1974; Trollope & Tainton 1986; Chap. 18, this volume). With invasion of moister grassland, individuals of *A. karroo* often serve as nuclei for the establishment of mostly bird-dispersed woody species (Bews 1917; Dyer 1937; Story 1951; Comins 1962; Morris 1976).

Although not classified as part of the grassland biome, there are fairly extensive grasslands in the mountains of the Eastern Cape (within the fynbos biome; Chap. 6, this volume). Small fynbos satellites occur throughout the grassland of the eastern escarpment, especially on nutrient-poor soils derived from aeolian sands of the Clarens Formation, and are restricted to localized areas protected from fire by virtue of topography (Killick 1963; Du Preez & Bredenkamp 1991; Du Preez 1992). Transition from grassland to fynbos with

protection from fire occurs over an unknown period and is facilitated by heavy grazing (Story 1951; Trollope 1971). However, the transformation of fynbos to grassland by burning can occur within four years (Trollope 1973; Downing et al. 1978; Chap. 6, this volume). For example, a single fire killed 33% and 97% of established *Cliffortia linearifolia* and *Erica brownleeae* plants, respectively (Trollope & Booysen 1971). Fire almost eliminated the seed-reproducing *Erica* (= *Phillipia*) *evansii* in the Drakensberg (Smith & Tainton 1985), markedly reduced seedling densities of fynbos species (Trollope & Booysen 1971) and established a dense grass sward dominated by fire-adapted perennials (Downing et al. 1978). However, studies showed that subsequent exclusion of fire allowed fynbos to re-establish within five years (Trollope 1973). The potential of some fynbos species to invade grassland in the eastern Cape is also constrained by the vulnerability of their seedlings to drought (Martin 1966). This is possibly because this region is transitional between winter and summer rainfall.

The ericoid *Stoebe vulgaris*, a shrub of fynbos affinity, has become prominent in the central inland and northern plateau grasslands. It was thought that heavy grazing promoted this species (Gillman 1934). However, heavy grazing over more than 20 years resulted in the elimination of *S. vulgaris* (Krupko & Davidson 1961). This is because this period exceeded the longevity of individuals (Hattingh 1953), germination was depressed by increased irradiance (Cohen 1937), seedling establishment was enhanced under shade (Lecatsas 1962), and seedling growth and adult survival was depressed by the increased nitrogen derived from urine and dung deposition (Roux 1969). Fire resulted in a marked increase in seedling and adult mortality (Lecatsas 1961). In contrast, protection from fire and grazing promoted ten-fold increases of *S. vulgaris*, mainly because seedlings can establish within a dense sward.

Grassland interfaces with forest patches in the eastern sector of the Great Escarpment (Chap. 12, this volume). Most forest patches occur on mesic, steep, southern slopes (Killick 1963; Moll 1967; Van Zinderen Bakker 1973; Du Preez & Bredenkamp 1991; Smit, Bredenkamp & Van Rooyen 1993c), because the reduced radiation load of southern aspects greatly increases their soil moisture regime in the dry season (Granger & Schulze 1977). In contrast, the water stress resulting from 28% higher radiation on northern than on southern aspects precludes the shallow-rooted forest precursor *Erica* (= *Phillipia*) *evansii* from invading northern aspects in the Drakensberg (Everson & Breen 1983). Forest margins may be particularly vulnerable to fire, because, relative to grassland, these open scrub formations often have a higher fuel availability (the quantity of fuel consumed in a fire), which is a primary determinant of fire intensity (Everson, Smith & Everson 1985b; Chap. 18, this volume). However, some common forest precursor species (*Buddleja salviifolia, Leucosidea sericea, Widdringtonia nodiflora*) can resprout (Smith & Tainton 1985). Prolonged protection (c. 40 years) from fire can lead to grassland adjacent to forest converting to scrub forest (Westfall, Everson & Everson 1983). Similarly, secondary grassland derived from the clearing of dune forest may almost disappear within 37 years with protection from fire (Weisser 1978).

10.2.2 Satellite grasslands

Local areas of grassland within the Nama-karoo and savanna biomes exhibit distinctive environmental profiles relative to surrounding plant communities. Grassland outliers in the karoo, characterized by *Merxmuellera disticha, Ehrharta calycina* (both C₃), *Themeda triandra, Cymbopogon plurinodis* and *Eustachys paspaloides* (C₄ species), occur on higher escarpments of variable geology. They intercept considerably more precipitation than their surroundings (Palmer 1989; A.R. Palmer & F. Van Heerden unpubl. data; Chap. 8, this volume). However, the perennial *Stipagrostis* grassland of Bushmanland, characterized by *Stipagrostis brevifolia, S. obtusa, S. ciliata* and *Centropodia glauca*, covers extensive areas within the Nama-karoo biome (Acocks 1953; Chap. 8, this volume), and seems to occur on tableland relief where soils are covered with a sand veneer (Tinley 1982). For example, a grassland dominated by *S. ciliata* and *C. glauca* was not topographically distinct from surrounding dwarf shrub communities, but occurred on a locally unique soil of relatively deep (0.45–1.50 m), well-drained aeolian sands, which would ameliorate soil moisture conditions in a region receiving c. 150 mm rainfall yr⁻¹ (Lloyd 1989a,b).

Satellite grassland within the savanna biome seems restricted to distinctive topographic–geological complexes and specific soil conditions. Grasslands occur on plateau summits of hills with poorly drained, sandy soils overlying impermeable quartzitic or sandstone rocks in areas of relatively high rainfall (>700 mm yr⁻¹) in Kwa-Zulu-Natal, and are characterized by *Aristida junciformis, Cymbopogon validus, Diheteropogon (= Andropogon) amplectens* and *Eragrostis* spp. (Killick 1958; Morris 1969). A large satellite grassland occurs on the sandy soils derived from granite of the frost-prone Pietersberg plateau in the Northern Province (Van der Schijff 1971).

10.2.3 Absence of a woody component

The following hypotheses, not necessarily mutually exclusive, have been put forward to account for the lack of woody elements in grassland.

1 Afromontane grassland along the eastern rim of southern Africa has been anthropogenically derived in historical time and maintained by fire (Story 1951; Acocks 1953; White 1978; Tainton 1981; Smit 1992), and is often termed 'false grassland'.

2 Within the Transkei, soils with a shallow impervious layer, or a well-developed illuvial B-horizon (including duplex, vertic, or waterlogged soils), but not soils with a poorly developed illuvial B-horizon (lithocutanic soils), limit the occurrence of woody elements (Feely 1987).

3 Soil-moisture balance determines the distribution of grassland, savanna and forest as it overrides all other properties or influences their effects (Tinley 1982). Specifically, grasslands are maintained as a result of impeded drainage, owing to a shallow (0.4 m) impermeable layer that results in waterlogging during the wet season but also excessive drying out of the pan horizon during the dry season. Woody plants occur only at sites with appropriate water balance and aeration (e.g. rock outcrops) or where the water balance is perturbed by erosion.

4 Frost excludes forest at high altitudes in the Drakensberg (Acocks 1953), and may exclude woody elements in other frost-prone areas.

5 The climate of the grassland biome promotes frequent fires, which determine the distribution of the biome (Chap. 18, this volume). Topography and soils affect the fire-proneness of a site, and thereby promote patches of woody elements, but soil characteristics are not responsible for grassland (Ellery 1992 – adapted from postulates for the central plains of the USA; e.g. Wells 1965, 1970; Axelrod 1985).

The hypothesis that grasslands were anthropogenically derived and maintained by fire is now generally discounted (Meadows & Meadows 1988; Ellery & Mentis 1992; Meadows & Linder 1993; Coetzee, Bredenkamp & Van Rooyen 1994). Evidence for this includes palynological data that clearly show the presence of Afromontane grassland since the late Pleistocene (>10 000 years BP) (Meadows & Meadows 1988; Scott 1989; Meadows & Linder 1993; Chap. 4, this volume), well before any period of cultivation (Feely 1987). In addition, the biogeographical distribution of locally endemic animals and plants (especially geophytes) is closely centred on these grasslands (Chap. 3, this volume). Pedological evidence provides further support, especially the occurrence of deep humic A-horizons (Feely 1987; Ellery & Mentis 1992; Meadows & Linder 1993). However, some areas of grassland are derived from localized forest destruction (Meadows & Linder 1993), and these grasslands are currently maintained by fire (Ellery 1992; Meadows & Linder 1993). Certainly, the great demand for wood by Iron Age societies for smelting (Feely 1980, 1987) could have had a local effect on the position of grassland boundaries.

The proposals that woody elements are excluded from grassland by the degree of illuviation (Feely 1987) or by the waterlogging desiccation effect of shallow pan horizons (Tinley 1982) is incompletely supported across the grassland biome (Ellery 1992). Although poorly developed illuvial B-horizons invariably have a woody cover of >2%, some soils with a shallow, impermeable layer also support >2% woody cover, provided these sites are located in fire-protected sites or experience reduced fire frequency owing to heavy grazing. For example, grassland with a well-developed illuvial B-horizon developed a scrub forest when protected from fire for 50 years (Westfall et al. 1983). Also, grasslands occur extensively on well-drained, loose, deep sandy soils of the Hutton and Clovelly forms (Scheepers 1975; Morris 1976; Werger & Coetzee 1978).

Grassland associated with well-developed cutanic B-horizon occurs on gentle slopes without rocks, whereas shallow lithocutanic soils occur on steeper slopes with rocks (Ellery 1992). Increased slope or rock exposure reduces fuel loads and therefore fire frequency and intensity (Everson, Van Wilgen & Everson 1988). Across the biome, tree cover is related to the number of growing days and slope (Ellery 1992), and woodland patches within the grassland biome appear to be restricted to areas of topographic complexity (Potts & Tidmarsh 1937; Mogg 1955; Moll 1967; Bredenkamp & Theron 1980). From this evidence, it has been proposed that climate contributes indirectly to the maintenance of grassland by supporting a fire regime that excludes woody plants not tolerant of fire (Ellery 1992). Specifically, the relatively high rainfall of the grassland biome results in high productivity, while seasonal growth and frost curing results in high fuel loads. Because of the high incidence of lightning strikes (Manry & Knight 1986), and anthropoeic activities, the high fuel loads lead to frequent fires. Less fire-prone environments convert to wooded patches (e.g. Van Daalen & Geldenhuys 1988). In contradiction, many southern African savannas are burned regularly. Indeed, the fire-regime north and south of Pretoria, which are areas of savanna and grassland, respectively, does not differ (Coetzee et al. 1994). The high incidence of lightning strikes is associated with a high incidence of hail-storms, which can also cause considerable damage to trees (Moll 1967).

Frost does not seem an adequate explanation for the absence of woody elements from the grassland biome, because there are exceptions, such as Acacia karroo, which tolerate frost (Story 1951; Ellery 1992). It may be important in the high-altitude areas of the eastern Great Escarpment (Acocks 1953). However, frost does limit the pool of species available for colonization of grassland, because most of these woody species are of tropical affinity. For example, recurrent frosts during a particularly cold winter in a savanna area killed 4% of all woody individuals outright, and top-killed >40% of the individuals of four species (Smit 1990). Inselbergs within grassland often support scattered trees (e.g. Rhus erosa shrubland; Table 10.1), which may in part be due to frost-free mountainsides maintained by adiabatic cooling and temperature inversion in regions of frosty winter conditions (Roberts 1975).

Invasion of grassland by alien species, notably Australian Acacia and Eucalyptus spp. of more temperate origin (Henderson & Musil 1984; Henderson 1989), suggest a strong role for climate rather than simply fire in determining the distribution of the grassland biome (see also Chap. 22, this volume).

10.2.4 Summary: determinants of grassland

The distribution of the grassland biome is the result of the subtle interplay of climate, topography, fire and grazing. Although the overall extent of grassland seems to be strongly determined by climatic variables, fire and grazing exert considerable influence over the boundaries of this biome (Fig. 10.1). Fire is a secondary determinant. Its pattern is dependent on climate, and its effects are not independent of topography and climate. Furthermore, the role of fire in maintaining grassland is greater in humid than in semi-arid regions. The notion that climate determines the grassland biome indirectly through its effect on the fire regime is possibly inadequate given the occurrence of satellite grasslands not dependent on fire, and some savannas well adapted to high fire frequencies. The patterns described above suggest that the importance of fire, relative to climate, depends inter alia on climate (rainfall and temperature), topography and soil type.

10.3 Community types

The grassland vegetation of the southern African inland plateau largely represents the Zambesian Domain of the Sudano-Zambesian Region (Werger & Coetzee 1978; Chap. 3, this volume). However, more restricted grasslands of the high-altitude eastern mountains and escarpment areas show affinity to the Afroalpine or Afromontane regions (White 1978).

Table 10.1 **Floristic and habitat features of the major vegetation types of the grassland biome**

Dominant and diagnostic grass species	Conspicuous or diagnostic non-grassy forbs	Mean no. spp 100 m^{-2}	Habitat	References
A Central inland plateau				
A1 *Themeda triandra–Eragrostis curvula* grassland				
Themeda triandra	*Helichrysum rugulosum*	23	Geology: Ecca sandstone, shale, lava	Bezuidenhout & Bredenkamp (1991)
Eragrostis curvula	*Crabbea acaulis*		Land type: B and C	Bezuidenhout *et al.* (1994b,c)
Cymbopogon plurinodis	*Hermannia depressa*		Soil: deep red, yellow eutrophic, often ploughed	Bredenkamp & Theron (1980)
Setaria sphacelata	*Rhynchosia totta*		Altitude: 1400–1600 m	Bredenkamp *et al.* (1989)
Heteropogon contortus	*Tephrosia semiglabra*		Rainfall: 600–700 mm	Coetzee (1993)
Brachiaria serrata	*Ipomoea obscura*			Fuls (1993)
Elionurus muticus	*Deverra burchellii*			Fuls *et al.* (1992)
				Kooij *et al.* (1990b, 1992)
A2 *Panicum coloratum–Eragrostis curvula* grassland				
Themeda triandra	*Ophioglossum polyphyllum*	20	Geology: Ecca mudstone	Kooij *et al.* (1990a)
Panicum coloratum	*Pentzia globosa*		Land type: D	
Eragrostis curvula	*Felicia filifolia*		Soil: dry clayey duplex	
Eragrostis plana	*Lycium cinereum*		Altitude: 1400–1500 m	
Setaria nigrirostris	*Bulbine narcissifolia*		Rainfall: 600 mm	
B Dry western areas				
B3 *Eragrostis obtusa–Eragrostis lehmanniana* grassland				
Themeda triandra	*Chamaecrista mimosoides*	15	Geology: Ecca mudstone, shale	Du Preez & Bredenkamp (1991)
Eragrostis lehmanniana	*Solanum panduriforme*		Land type: A	Kooij *et al.* (1990c)
Anthephora pubescens	*Dicoma anomala*		Soil: yellow apedal sand	Malan (1992)
Eragrostis obtusa	*Anthospermum hispidulum*		Altitude: 1400 m	Werger (1973a)
Eragrostis trichophora	*Chrysocoma ciliata*		Rainfall: 500–600 mm	
Aristida stipitata	*Felicia muricata*			
B4 *Stipagrostis uniplumis–Fingerhuthia africana* grassland				
Stipagrostis uniplumis	*Helichrysum callicomum*	48	Geology: dolomite	Bezuidenhout *et al.* (1993, 1994d)
Aristida diffusa	*H. cerastioides*		Land type: Bd, Fa	Morris (1976)
Fingerhuthia africana			Soil: dry aeolian sand	
			Altitude: 1370–1450 m	
			Rainfall: 450 mm	
B5 *Rhus erosa* shrubland				
Melinis nerviglumis	*Olea europaea* ssp. *africana*	18	Rocky hills at Winburg, Bloemfontein, Thaba'Nchu, Aliwal North	Bosch *et al.* (1986)
Enneapogon scoparius	*Rhus erosa*		Geology: dolerite	Du Preez (1991)
Setaria lindenbergiana	*R. burchellii*		Soil: Shallow lithosols	Du Preez & Venter (1992)
Aristida diffusa	*R. ciliata*		Altitude: 1200–1300 m	Muller (1986)
Eustachys paspaloides	*Buddleja saligna*		Rainfall: 550 mm	Roberts (1966b)
Schismus barbatus	*Printzia auriculata*			Rossouw (1983)
	Phymaspermum aciculare			Werger (1973a,b)

Table 10.1 (cont.)

Dominant and diagnostic grass species	Conspicuous or diagnostic non-grassy forbs	Mean no. spp 100 m^{-2}	Habitat	References
C Northern areas				
C6 Rhus leptodictya–Acacia caffra Mountain Bushveld				
Trachypogon spicatus	Acacia caffra	32	Sheltered sites on hills of Magaliesberg, Witwatersrand, Vredefort Dome, Suikerbosrand	Behr & Bredenkamp (1988)
Diheteropogon amplectens	Protea caffra			Bezuidenhout et al. (1994a)
Loudetia simplex	Rhus magalismontana		Geology: quartzite, shale, andesitic lava, dolomite	Bredenkamp & Theron (1978, 1980)
Tristachya leucothrix	R. leptodictya		Land type: Fb, Ib	Coetzee et al. (1994)
Alloteropsis semialata	Dombeya rotundifolia		Soil: shallow, rocky	Du Preez & Venter (1990)
Panicum natalense	Tapiphyllum parvifolium		Altitude: 1500–1600 m	
Aristida transvaalensis	Senecio venosus		Rainfall: 650–750 mm	
	Helichrysum setosum			
C7 Loudetia simplex–Trachypogon spicatus grassland				
Trachypogon spicatus	Sphenostylis angustifolia	37	Exposed sites on hills and ridges of Magaliesberg, Gatsrand, Witwatersrand, Vredefort Dome, Suikerbosrand	Bezuidenhout et al. (1994a)
Schizachyrium sanguineum	Justicia anagalloides			Bredenkamp et al. (1994)
Digitaria tricholaenoides	Senecio coronatus			Coetzee (1974, 1975)
D. diagonalis	Pearsonia cajanifolia		Geology: quartzite, shale, andesitic, lava, dolomite	Myburgh (1993)
D. monodactyla	Haplocarpha scaposa		Soil: shallow, rocky	Van Wyk & Bredenkamp (1986)
Sporobolus pectinatus	Vernonia natalensis		Altitude: 1500–1600 m	
			Rainfall: 650–750 mm	
D Eastern inland plateau				
D8 Aristida junciformis–Eragrostis plana grassland				
Aristida junciformis	Anthospermum rigidum	18	Geology: Beaufort sandstones, shales	Bloem et al. (1993)
Eragrostis plana	Hermannia geniculata		Soil: deep, yellow, grey sandy loam	Bosch et al. (1986)
Microchloa caffra	Senecio erubescens		Altitude: 1600–1800 m	Du Preez (1991)
Helictotrichon turgidulum	Falckia oblonga		Rainfall: 700–950 mm	Eckhardt et al. (1993b)
Andropogon appendiculatus	Berkheya onopordifolia		Frost: severe	Fuls et al. (1993a,b)
				Jarman (1977)
				Scheepers (1975)
				Turner (1989)
D9 Themeda triandra–Aristida bipartita grassland				
Themeda triandra	Berkheya pinnatifida	19	Geology: Ecca shales	Breytenbach et al. (1993a,b,c)
Aristida bipartita	Chaetacanthus costatus		Soil: black vertic clay	
Digitaria ternata	Salvia repens		Altitude: 1500–1700 m	
Setaria nigrirostris	Abildgaardia ovata		Rainfall: 650–750 mm	
S. incrassata	Bulbostylis contexta			
Panicum coloratum	Pseudognaphalium luteo-album			
E Eastern mountains and escarpment				
E10 Rhus dentata–Leucosidea sericea shrubland				
Poa annua	Leucosidea sericea	29	Lower slopes of the Drakensberg	Eckhardt et al. (1993b,c)
Hyparrhenia hirta	Rhus dentata		Altitude: >1750 m	
Aristida diffusa	Myrsine africana		Rainfall: >900 mm	
Trachypogon spicatus	Sutera polensis			
	Stachys kuntzei			

E11 Monocymbium ceresiiforme–Tristachya leucothrix grassland

Monocymbium ceresiiforme
Diheteropogon filifolius
Sporobolus centrifugus
Harpochloa falx
Rendlia altera
Cymbopogon dieterlenii
Eulalia villosa
Helichrysum cerastioides
H. oreophilum
H. spiralepis
Rhus discolor
Selago galpinii
Clutia monticola
Sebaea sedoides

Drakensberg slopes and plateaux
Altitude: >1700 m
Rainfall: >1000 mm

34

Du Preez (1991)
Du Preez & Venter (1992)
Eckhardt et al. (1995)
Kay et al. (1993)
C.M. Shackleton (1989)
Smit et al. (1993b,c)

E12 Loudetia simplex–Diheteropogon filifolius grassland

Ctenium concinnum
Koeleria capensis
Rendlia altera
Alloteropsis semialata
Loudetia simplex
Eriosema angustifolium
Vernonia centaureoides
Helichrysum platypterum
H. mariepscopicum
Rabdosiella calycina
Craterocapse tarsodes
Wahlenbergia squamifolia

Eastern Transvaal Escarpment
Altitude: 1650–1850 m
Rainfall: 1100–1500 mm

30

Deall et al. (1989)
Matthews et al. (1991, 1992a,b, 1994)

E13 Merxmuellera drakensbergensis–Festuca caprina Afro-alpine grassland

Merxmuellera disticha
M. drakensbergensis
Festuca caprina
Eragrostis caesia
Poa binata
Pentaschistis galpinii
Carex clavata
Scirpus falsus
Helichrysum flanaganii
H. trilineatum
H. witbergense
Erica frigida

Alpine Drakensberg plateaux
Altitude: 2500–3480 m
Rainfall: >1000 mm

19

Herbst & Roberts (1974)
Jacot-Guillarmod (1971)
Van Zinderen-Bakker & Werger (1974)
Killick (1978)
Wieland (1982)
Martin (1986)
Mokuku (1991)
Morris (1994)

F Eastern lowlands

F14 Hyparrhenia hirta tall grassland

Hyparrhenia hirta
Sporobolus pyramidalis
Acacia sieberiana
Rhus rehmanniana
Walafrida densiflora
Spermacoce natalensis
Kohautia cynanchica
Phyllanthus glaucophyllus

Eastern Drakensberg foothills, KwaZulu-Natal and Eastern Cape
Altitude: 1200–1400 m
Rainfall: 850 mm

25

Edwards (1967)
Smit (1992)

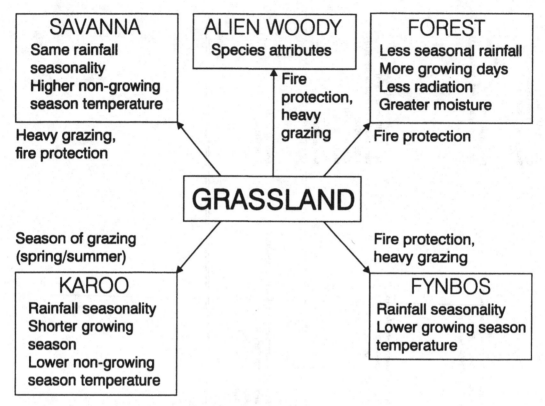

Figure 10.1 **Schematic representation of the determinants of the boundaries between grassland and adjacent biomes. Boxes denote the climatic differences between each biome and the grassland biome; arrows denote the biotic effects responsible for the change from grassland to an alternative vegetation structure. See Figure 5.2 (Chap. 5, this volume) for the geographical distribution of these biomes in southern Africa.**

Prominent grass species encountered throughout most of the grassland biome include *Themeda triandra*, *Eragrostis curvula*, *Cymbopogon plurinodis*, *Setaria sphacelata*, *Digitaria eriantha*, *Hyparrhenia hirta* and *Cynodon dactylon*. *Themeda triandra* is by far the most dominant species in the greater part of the biome. Grassland with *T. triandra* is widespread over Africa, but the phytosociological diagnostic species of the Themedetalia triandrae (order) from east Africa (Lebrun 1947) are absent from southern Africa. Some mountain slopes are characterized by woody species, where grasses such as *Trachypogon spicatus*, *Tristachya leucothrix*, *Panicum natalense*, *Schizachyrium sanguineum*, *Loudetia simplex*, *Monocymbium ceresiiforme*, *Alloteropsis semialata* and *Eulalia villosa* predominate (Smit, Bredenkamp & Van Rooyen 1993b; Bezuidenhout *et al.* 1994a). Karroid shrubs and forbs may become dominant at overgrazed sites (Kooij *et al.* 1990a; Fuls 1992), whereas wetland areas are dominated by sedges (Eckhardt, Van Rooyen & Bredenkamp 1993a; Chap. 14, this volume).

Major plant community differentiation is generally related to climate (rainfall and temperature) associated with altitude (Fuls 1993). In the dry northwestern grasslands, differences in geology, topography and associated land-type influence plant community distribution (Bezuidenhout 1988, 1993). Land-type is defined as an area mapped at 1 : 250 000 scale that displays a marked degree of uniformity with respect to terrain form, soil pattern and climate (Land Type Survey 1984). However, in the eastern areas, higher rainfall dampens the influence of physiography and soil type, so that gradual transitions in the plant communities become the rule, and abrupt discontinuities the exception (Du Preez & Bredenkamp 1991). Here, position in the landscape (plateaux, slopes, plains) within land-types rather than geology or land-type determine the distribution of plant communities (Fuls *et al.* 1993c).

Six major regions, comprising 14 vegetation types (each numbered when named below), were identified (Fig. 10.2, Table 10.1).

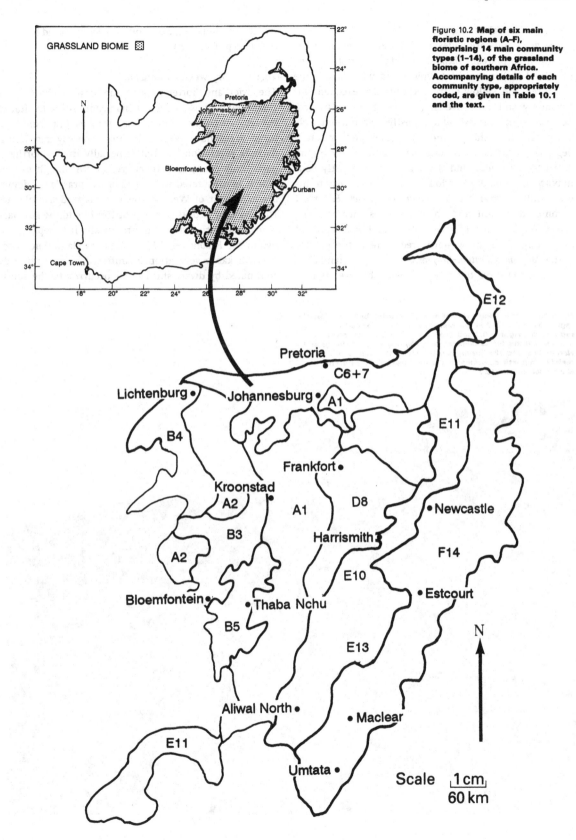

Figure 10.2 **Map of six main floristic regions (A–F), comprising 14 main community types (1–14), of the grassland biome of southern Africa. Accompanying details of each community type, appropriately coded, are given in Table 10.1 and the text.**

10.3.1 Central inland plateau (A)

These grasslands are widespread, covering much of the central high plateau. *Themeda triandra–Eragrostis curvula* grassland (A1) is the most typical and widespread. The deep, red and yellow eutrophic soils are excellent for agronomy and extensive areas have been cultivated. Relict patches of grassland are mostly grazed intensively by cattle and sheep, and are consequently often degraded. Grasslands in pristine condition are dominated by *T. triandra*, and the vegetation is mostly poor in both grass and dicotyledonous forb species. When this grassland is overgrazed, *T. triandra* decreases, *E. curvula* becomes dominant and karroid shrubs increase. The clayey duplex soils of vast bottomland Pan Turfveld (Acocks 1953) in the western Free State is the habitat of the *Panicum coloratum–Eragrostis curvula* grassland (A2). Owing to the droughty clay soil type, this area is nor-

mally not ploughed, but is often heavily grazed by cattle and sheep (Fig. 10.3).

10.3.2 Dry western areas (B)

The drier and hotter area to the west of the Central Inland Plateau, with deep aeolian sand, is the habitat of the *Eragrostis obtusa–Eragrostis lehmanniana* grassland (B3). Large areas have been cultivated, thereby restricting natural vegetation, so that, especially in the northern parts, only overgrazed relicts remain. Although at present it is considered to be part of the grassland biome, the area west of Wesselsbron was mapped as Kalahari Thornveld by Acocks (1953), probably due to the Kalahari sands, but floristically and structurally it is representative of the *E. obtusa–E. lehmanniana* grassland. Encroachment of karoo vegetation, manifested by communities dominated by dwarf shrubs, has occurred to the south

Figure 10.3 **A fence-line contrast at the transition between floristic region types A and D in the eastern Free State (for code to regions, see Figure 10.2 and Table 10.1). The well-managed grassland to the left is dominated by *Heteropogon contortus* and *Harpochloa falx*. The heavily grazed veld to the right comprises mainly *Eragrostis chloromelas* and the shrub, *Felicia filifolia* (Asteraceae)** *(Photo: J.C. Scheepers).*

in an area of the Ae land-type (Land Type Survey 1984) west of Bloemfontein. The dry *Stipagrostis uniplumis–Fingerhuthia africana* grasslands (B4) (Bezuidenhout *et al.* 1994c) are found west of Lichtenburg and of 26° E, occurring on dry sandy aeolian soils overlying dolomite and limestone. This grassland merges with the bordering Kalahari Thornveld to the west. Rainfall is erratic, making this a high-risk area for agronomy, but despite its suitability for cattle and sheep farming the best veld has been ploughed, and natural vegetation is restricted to shallow or aeolian soils and pans.

Hills and ridges situated in the central Free State south to Aliwal North, and the low valleys of the Senqu River in Lesotho, are covered with *Rhus erosa* shrubland (B5). This area is physiographically complex, representing both the grassland and Nama-karoo biomes (Du Preez 1991; Fig. 10.4), with some Afromontane species intruding into the lowland phytochoria (White 1978), and with floristic affinity to ravine forest and grassland (Du Preez & Bredenkamp 1991).

10.3.3 Northern areas (C)
The vegetation of this area is transitional between grasslands of the high inland plateau and savanna of the low inland plateau. This area is characterized by rocky hills and ridges, with shallow rocky soils. The southern slopes of the Magaliesberg, Timeball Hill and Daspoort ridges, and the ridges of the Witwatersrand, Suikerbosrand and mountains of the Vredefort Dome are the habitat of the *Rhus leptodictya–Acacia caffra* Mountain Bushveld (C6), which forms part of the Bankenveld 'false grassland' of Acocks (1953). This vegetation, which occurs as islands of temperate mountain bushveld within the grassland biome, is characterized by the presence of many woody species typical of Sour Bushveld (Acocks 1953; Coetzee *et al.* 1994). The grass species, on the other hand, have a definite Drakensberg affinity. The *Loudetia simplex–Trachypogon spicatus* grassland (C7) occurs in the same distribution range as the *Rhus leptodictya–Acacia caffra* Mountain Bushveld, but is restricted to exposed sites, on rocky soils in the irregular undulating, high-altitude landscape, especially on the crests of quartzite hills. This vegetation is at present considered as typical Bankenveld (Acocks 1953). The great variation in topography has resulted in a great diversity of plant communities. Floristic relationships between this vegetation type and the Sour Bushveld are emphasized by Bredenkamp & Theron (1978), B.J. Coetzee (1974), J.P. Coetzee (1993) and Werger & Coetzee (1978).

In the vast area underlain by dolomite in the western Transvaal, Bankenveld (Acocks 1953) is called 'klipveld' (Louw 1951), owing to the abundance of surface rock.

Figure 10.4 **The arid fringes of the grassland biome extend deep into the Nama-karoo on the uplands of the Eastern Cape Midlands. Shown here is a dry form of B5 in the Sneeuberg Mountains north of Graaff-Reinet (for code to regions, see Figure 10.2 and Table 10.1)** *(Photo: R.M. Cowling).*

Because of the variety of microhabitats created by the rocky habitat, with mosaics of extremely shallow and deep soils occurring within a few metres, no single species attains dominance. Instead, the vegetation is characterized by mosaics of many co-dominants with high species diversity.

10.3.4 Eastern inland plateau (D)
This vegetation type occurs east of the Central Inland Plateau, at higher altitude and at higher rainfall, on the plains immediately west of the eastern escarpment. *Aristida junciformis–Eragrostis plana* grassland (D8) is found on sandy loam soils in the eastern Free State. Extensive areas have been cultivated. Land-type and soil type have a lesser influence on vegetation distribution than terrain form and associated soil depth, soil moisture (clay content), rockiness and grazing. Variation in vegetation is determined mostly by position in the landscape

Figure 10.5 **The Natal Drakensberg, a grass-covered massif with summits exceeding 3000 m in height, supports types E11 (in foreground) and E13 (in background at altitudes of >2500 m) (for code to regions, see Figure 10.2 and Table 10.1). Note the patch of Afromontane forest on a fire-protected scree slope**
(Photo: J.C. Paterson-Jones).

Figure 10.6 **The foothills of the Natal Drakensberg showing E11 (for code to regions, see Figure 10.2 and Table 10.1). The midsummer sward is dominated by *Themeda triandra* and *Monocymbium ceresiiforme*. The tree fern is *Cyathea dregei***
(Photo: J.C. Paterson-Jones).

(plateaux, slopes, plains) within land-types. This is probably due to the higher rainfall, which moderates variation in the physical environment, so that plant communities show gradual transitions rather than abrupt discontinuities (Du Preez & Bredenkamp 1991; Fuls *et al.* 1993c). The vegetation of the wet, poorly drained floodplains and water courses within the drier western grassland types represent outliers of this grassland. It is a typical example of seasonally moister habitats within drier western grasslands. This grassland is typically poor in species, with *E. plana* usually dominant, but *Hyparrhenia hirta* is often prominent.

In the southeastern Transvaal, a landscape of slightly undulating plains of black vertic or near vertic montmorillonitic clay soils is the habitat of the *Themeda triandra–Aristida bipartita* grassland (C9), corresponding with the Turf Highveld of Acocks (1953). As in the *T. triandra–E. curvula* grassland, *T. triandra* dominates wherever the vegetation is not severely degraded.

10.3.5 Eastern mountains and escarpment (E)

The high-altitude slopes and plateaux of the Drakensberg and associated mountains constitute the habitat of this vegetation. These areas are generally cooler and have higher rainfall than the other grassland types in southern Africa. Rocky, lower slopes are the habitat of the *Rhus dentata–Leucosidea sericea* shrubland (E10). The small tree *Leucosidea sericea* is the most conspicuous diagnostic species, and tends to encroach where the herbaceous layer is degraded.

The *Monocymbium ceresiiforme–Tristachya leucothrix* grassland (E11) occurs on leached, shallow, rocky soils of the moist, cool, steep Drakensberg mountain slopes and plateaux, from Swaziland through to the Eastern Cape (Fig. 10.5). It occurs on a variety of land-types, but seems to be restricted to Karoo sediments in the Newcastle area (Smit, Bredenkamp & Van Rooyen 1992). The rugged topography creates a great variety of habitats, and consequently many different plant communities can be distinguished (Fig. 10.6). The area may be climatically suited for forest or woodland, but grassland is maintained by fire and grazing, and this grassland is therefore considered a fire climax (Tainton 1981), although frost may be responsible for the persistence of grassland on exposed ridges (Edwards 1967). In some places, forest precursors, typical of forest fringes, encroach into grassland (Smit, Bredenkamp & Van Rooyen 1993a).

A narrow grassland peninsula, the *Loudetia simplex–Diheteropogon filifolius* grassland (E12) is wedged between the northern Transvaal bushveld on the west and eastern Transvaal lowveld on the east. Acocks (1953) classified this grassland as North-eastern Mountain Sourveld under Inland Tropical Forest Types, because of the patches of forest occurring in sheltered valleys of the escarpment. This grassland contains many endemic plant species (Matthews, Van Wyk & Bredenkamp 1993), and is therefore considered to represent a separate vegetation unit. A total of 78 endemic or near endemic grassland species, mostly of the Liliaceae (*sensu lato*), Iridaceae and Asteraceae, Lamiaceae and Orchidaceae, have been identified on the Black Reef quartzites situated at the edge of the escarpment, with a further 31 endemic species identified from the drier dolomites of the Chuniespoort Formation. All endemics are grassland-associated species, and no forest species are considered to be endemic to the area (see also Chap. 3, this volume).

Patches of Afromontane fynbos occur scattered in this vegetation type, especially on plateau areas of the Drakensberg, Korannaberg, Thaba'Nchu and Lesotho Mountains (Killick 1963; Roberts 1966a; Edwards 1967; Jacot-Guillarmod 1971; Martin 1986; Du Preez & Bredenkamp 1991).

The *Merxmuellera drakensbergensis–Festuca caprina* Afroalpine grassland (E13) of the Drakensberg plateaux occurs in the treeless alpine regions of Lesotho and adjacent Drakensberg, which is composed of the massive Drakensberg Basalt. The terrain includes plateaux and steep slopes with terracettes. The vegetation consists mainly of tussock grasses, ericoid dwarf shrubs and creeping or mat-forming plants (Van Zinderen-Bakker & Werger 1974; Fig. 10.7).

10.3.6 Eastern lowlands (F)

This vegetation type comprises the *Hyparrhenia hirta* tall grassland (F14), which occurs east of the eastern escarpment on the undulating, drier landscapes of KwaZulu-Natal, southeastern Transvaal and northeastern Cape (Fig. 10.8). It is often associated with shallow, rocky soils, often derived from dolerite.

10.4 Organisms

This section describes the considerable variation in the functional attributes of grass species across the grassland biome and of coexisting species within communities, and relates these patterns to environmental factors.

10.4.1 Biome patterns

The C_4 grasses are the most abundant over most of the country, with the NADP-me (malate) form of the Andro-

Figure 10.7 **Mont aux Sources plateau on the top of the Natal Drakensberg (c. 3200 m) supports fynbos-like vegetation (E13) with strong alpine characteristics (for code to regions, see Figure 10.2 and Table 10.1). Dominant species here include** *Erica dominans, Helichrysum trilineatum* **and** *Festuca caprina* (Photo: D. Edwards).

Figure 10.8 **A dense sward of 'climax' grasses – the green gold of the South African veld – is shown here in the KwaZulu-Natal midlands. This form of type F14 is dominated by *Themeda triandra, Tristachya leucothrix* and *Andropogon appendiculatus* (for code to regions, see Figure 10.2 and Table 10.1)**
(Photo: D. Edwards).

pogoneae dominating in the moister, and the NAD-me (aspartate) form of mainly the Chlorideae dominating in the more arid, and the PCK (aspartate) form dominating in the intermediate regions (Vogel, Fuls & Ellis 1978; Ellery 1992). These subgroups of C_4 plants differ in their mean net assimilation rates in the eastern Cape, where they occur together (Botha & Russell 1988). Both non-Kranz and Kranz forms of *Alloteropsis semialata* exist (Ellis 1974), although the two forms are usually geographically separated (Vogel *et al.* 1978), and are morphologically and physiologically distinct (Frean, Barrett & Cresswell 1980; Frean & Cresswell 1981). There are no C_3 grasses in the warmer parts of the grassland biome, and C_4 grasses are absent where mean daily maximum temperatures are below 25 °C (Vogel *et al.* 1978), corresponding at one locality to altitudes above 2950 m on the northern and 2750 m on the southern aspect in the Drakensberg (Morris, Tainton & Boleme 1993). The distribution of C_3 and C_4 grasses on some hills reflects this pattern (Roberts 1966b).

Tannin-like polyphenols occur in the epidermal cells of the leaf blades of 101 grass species of 39 genera (Ellis 1990), belonging mainly to the Andropogoneae and Arundineae (Chesselet, Wolfson & Ellis 1992). They are found most commonly in C_4 grasses of the NADP-me photosynthetic type, particularly within moist areas (Ellery 1992). These substances have been confirmed as leucoanthocyanins, which have the properties of condensed tannins, for *Hyparrhenia filipendula, H. hirta, Imperata cylindrica, Themeda triandra* and *Bothriochloa insculpta* (Bate-Smith & Swain 1967) and as condensed tannin for *Eulalia villosa* (Du Toit, Wolfson & Ellis 1991). Tannin-like substances may interfere with protein digestion by animals, and may inhibit nitrification in the soil by reducing the number of micro-organisms present. This would favour grasses sensitive to high concentrations of soil nitrogen (Ellis 1990). The unpalatable *Elionurus muticus* possesses tannin-like substances (Ellis 1990) but does, however, have crude protein and fibre values comparable to those of the palatable *T. triandra* (Opperman, Roberts & Nel 1974). Aromatic compounds, especially terpenes, are characteristic of the genera *Cymbopogon* and *Bothriochloa*, and occur at sites intermediate along the moisture gradient of the biome (Ellery 1992).

A moisture gradient across the grassland biome influences variation in many functional attributes of species (Ellery 1992). Grass species of high-rainfall grassland are unpalatable (sour), tufted, seed-reproducing species that produce lower numbers of larger-sized seeds and show pronounced underground development, but only slight woodiness of the base. Grass species of arid grassland exhibit the converse characteristics, with a greater proportion of tufted species that reproduce vegetatively by rhizomatous growth. Grasses of arid grassland also tend to flower late in the growing season, whereas species of moist grassland flower early in or throughout the season. However, *Themeda triandra* and *Heteropogon contortus* can flower twice a year in semi-arid areas, depending on soil moisture conditions (Opperman & Roberts 1978). Many other species flower regularly either early (e.g. *Tristachya leucothrix, Eustachys paspaloides, Elionurus muticus, Microchloa caffra*) or late (e.g. *Cymbopogon excavatus, Eragrostis plana, Diheteropogon amplectens, Sporobolus pyramidalis*) in the season (Steinke 1968; Gibbs Russell 1983), with flowering or phenology of at least some species (e.g. *Aristida junciformis, Hyparrhenia hirta*) being dependent on daylength (Mes 1952; Edwards, Jones & Tainton 1979). In other cases, the flowering response to photoperiod in tropical and subtropical grasses is particularly variable (Tainton 1969). Commencement of growth is determined by available moisture in semi-arid grassland (Danckwerts 1984), but by soil temperatures in montane (Everson & Everson 1987) and more temperate grassland (Leigh 1960; Steinke & Nel 1967; Rethman & Booysen 1969). Very close grazing of temperate grassland can, because of absorption of radiation by the soil, result in raised minimum temperatures at ground level just above 0 °C, thereby stimulating midwinter growth (Du Toit, Ingpen & Nel 1974). Growth during the growing season is generally influenced by moisture availability rather than by temperature – temperature had little effect on the growth, allocation or time taken for flowering to commence of four populations of *T. triandra* from sites with very different temperature profiles (frost-free to prolonged frost season) (Downing & Groves 1985). In semi-arid grassland, active leaf growth is restricted to short sporadic spells following a rainfall event (e.g. one week for a 50-mm event) (Danckwerts 1988), whereas, at higher rainfall sites, growth is almost continuous, but at a variable rate (Everson 1985).

10.4.2 Within-community patterns
Within a community, there is considerable variation in almost any functional attribute, including water relations (Weinbrenn 1938, 1939; Henrici 1943), production and nutrient content of foliage (Henrici 1930a,b;

Du Toit *et al.* 1932, 1935a,b; Louw 1938; Roberts, Anderson & Fourie 1975; Fourie & Roberts 1976, 1977; Van der Westhuizen, Van Den Berg & Opperman 1978), root distribution (Murray & Glover 1935), anthesis (Davidson 1941), and flowering and phenology (Bews 1918; Glover 1937; Potts & Tidmarsh 1937; Gilliland 1955; Jones 1968). In a semi-arid grassland, recovery of species from wilting generally follows the inverse pattern in which they wilt with diminishing moisture availability (Table 10.2). These species do not differ much in evapotranspiration, but vary in production because of differences in water-use efficiency (Table 10.2). The water-use efficiency and, as a result, production, of *Cymbopogon plurinodis, Digitaria eriantha* and *Themeda triandra* is higher than that of the other species, although crude protein content tends to be lower. However, differences between species in their production per unit area of tuft depend on soil type – *Eragrostis lehmanniana* is superior to *T. triandra* on sandy soils, but the reverse holds for soils derived from dolomite, and *E. lehmanniana* is more capable than *T. triandra* of responding to small rainfall events (Fourie & Roberts 1976). A notable feature of some southern African grass species belonging to the genera *Eragrostis, Tripogon, Sporobolus, Oropetium, Microchloa* and *Brachyachne* (Eragrostoideae) is the ability of their foliage to tolerate dehydration to the point of air-dryness (Gaff 1971; Gaff & Ellis 1974).

Individual species may show considerable geographic variation in functional attributes. Forms of *Themeda triandra* have different leaf types, which differ in their transpiration rate (Mes & Aymer-Ainslie 1935) and occupy different habitats within a landscape (Mostert 1958). Seven widely distributed species, namely *T. triandra, Heteropogon contortus, Digitaria eriantha, Setaria sphacelata, Cenchrus ciliaris, Eragrostis curvula* and *Cynodon dactylon*, have a range of morphological variants, many ecotypes and occupy a range of habitats (Gibbs Russell & Spies 1988). Their success is attributed to their ability to form polyploid complexes, reproduce both sexually and either apomictically or vegetatively, and hybridize or cross-fertilize (Spies & Gibbs Russell 1988).

Seeds of grass species of the seasonal grassland and savanna climates generally show innate dormancy for 6–12 months, an attribute usually interpreted as a drought-avoidance mechanism (West 1951; Veenendaal & Ernst 1991; O'Connor & Pickett 1992). For *Themeda triandra*, innate dormancy is longer for cooler, montane sites than for lower-altitude savanna sites, with after-ripening of seed having the effect of widening the range of temperature over which germination can occur (Baxter, Van Staden & Granger 1993). Dormancy can be broken by exposure of seeds to fire smoke (Baxter *et al.* 1994). Mechanisms responsible for innate dormancy in southern

Table 10.2 Rain use and production attributes of species from a drier variant of the *Themeda triandra-Eragrostis curvula* grassland

Species	Leaf water potential (Pa) at wilting[a]	Daily evapotranspiration[b] (mm)	Rain use efficiency[b] (kg ha⁻¹ mm⁻¹)	Tuft production[c] (g cm⁻²)	Crude protein[c] (%)		Heat of combustion[c] (± S.E.) (kJ kg⁻¹)	
					Mean of months	Range	Vegetative	Mature
Digitaria argyrograpta	-1450	—	—	0.25[d]	—	—	—	—
Digitaria eriantha	-1800	2.0	6.39	0.30	4.66	3.00-5.94	16722±140	17538±96
Tragus koelerioides	-3400	—	—	—	6.96[f]	—	—	—
Themeda triandra	-2450	2.39	6.01	0.51	3.61	2.36-4.82	17170±17	17727±44
Eragrostis lehmanniana	-2050	1.72	4.73	0.55[d]	—	—	—	—
E. obtusa	-2100	—	—	0.22	4.91	2.57-6.57	—	—
Sporobolus fimbriatus	-2750	1.66	4.86	0.31	5.33	3.75-6.84	17543±157	17212±46
Panicum stapfianum	-2300	1.77	4.72	0.35	4.70	2.72-7.71	—	—
Cymbopogon plurinodis	—	2.1	7.83	0.6	4.27	2.34-5.79	17643±48	18133±46
Eragrostis chloromelas	—	1.78	5.17	0.24	4.35	3.17-5.24	—	—

[a] Snyman et al. (1987).
[b] Snyman (1989).
[c] Van der Westhuizen et al. (1978).
[d] Roberts et al. (1975).
[e] Trollope (1983).
[f] Van Schalkwyk, Lombard & Vorster (1968a).

African grasses include complete embryonic dormancy, a partial dormancy of the coleoptile and physical constraints including the presence of glumes or the fixation of the caryopsis to the dispersal unit (Opperman *et al.* 1974; Ernst, Kuiters & Tolsma 1991; Veenendaal & Ernst 1991). *Cynodon dactylon* lacks innate dormancy (Veenendaal & Ernst 1991), but its germination is influenced by temperature (Weinbrenn 1946). Secondary dormancy has not been recorded for southern African grasses, except for light-enforced and phytochrome-regulated secondary dormancy of the forest gap species *Setaria megaphylla* (Erasmus & Van Staden 1984).

Patterns of tiller development, particularly elevation of shoot apices, have special implications for grazing and fire management (Tainton 1981). For most species, the inflorescence can develop without marked internodal elongation raising the shoot apex above ground level. However, for *Themeda triandra* and *Cymbopogon excavatus*, the shoot apex is raised above ground before floral initiation (Rabie 1964). The resultant difference between species in the time of exposure of the growing point (Fig. 10.9) determines their vulnerability, especially of flower and seed production, to defoliation (Tainton & Booysen 1963; Rethman, Beukes & Malherbe 1971). For example, spring burning should kill all the apices of *T. triandra*

and thereby prevent flowering, but should not affect the flowering of *Tristachya leucothrix*. An increased frequency of grazing during inflorescence development therefore diminishes inflorescence production (Daines 1976), although this depends on further differences between species, i.e. in the proportion of tillers that flower. For example, 77%, 44% and 30% of the tillers of *Eragrostis lehmanniana*, *T. triandra* and *Cymbopogon plurinodis*, respectively, flowered in a season (Viljoen 1966). However, the pattern of shoot elevation is not a fixed characteristic of a species, but can differ between ecotypes and seasons, and in response to treatment in different seasons (Booysen, Tainton & Scott 1963; Tainton & Booysen 1965a,b; Rethman & Booysen 1967, 1968; Rethman 1971; Drewes & Tainton 1981). Moisture availability further influences the elevation of shoot apices (Rethman 1964), even overriding the effects of all except the highest frequencies of defoliation (Danckwerts & Nel 1989). In semi-arid grassland, moisture availability also markedly influences tiller growth (Danckwerts, Aucamp & Du Toit 1984, 1986), leaf production, expansion, senescence and longevity (Danckwerts & Aucamp 1985; Danckwerts 1988), mortality of the shoot growing point or tiller and the production of secondary tillers (Danckwerts 1988). However, defoliation of primary tillers invariably stimu-

Figure 10.9 **The seasonal pattern of elevation of the shoot apex of *Themeda triandra* (x——x), *Cymbopogon excavatus* (. . . .), *Heteropogon contortus* (– – –), *Tristachya leucothrix* (•——•), *Hyparrhenia hirta* (——), *Eragrostis curvula* (■——■) and *Aristida junciformis* (+——+). Only the first season is shown for *E. curvula* and *A. junciformis* to avoid congestion. Adapted from Bridgens (1968).**

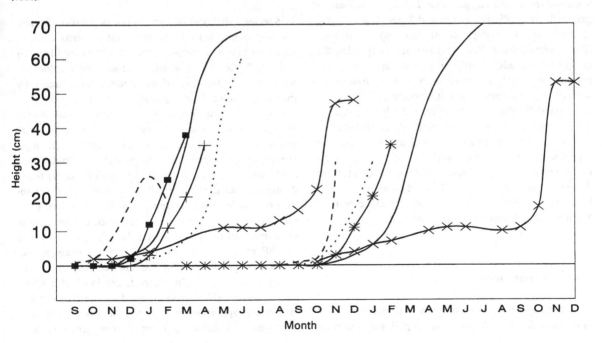

232 / T.G. O'Connor & G.J. Bredenkamp

Table 10.3 **The percentage mortality of individual grass tufts in reponse to drought for a number of species from localities throughout southern Africa**

Species	Sandveld savanna[a]	Savanna-clays[b]	Acacia savanna[c]	Semi-arid grassland[d]	Savanna[e]	Kalahari grassland[f]
Mean annual rainfall (mm)	650	480	460	560	480	450
Pogonarthria squarrosa	65					
Digitaria eriantha	87	8	70	47	74	
Aristida stipitata	25					
Perotis patens	36					
Aristida bipartita		17				
Bothriochloa insculpta		29				
Heteropogon contortus		23				
Themeda triandra		18	57	55		
Cymbopogon plurinodis			0			
Eragrostis chloromelas			70	100		
Eragrostis obtusa			100	100		
Microchloa caffra			97			
Panicum stapfianum			43	100		
Sporobolus fimbriatus			53	17		
Eragrostis lehmanniana				71		90
Panicum maximum					86	
Panicum coloratum					84	
Schmidtia pappophoroides					97	
Bot...iochloa radicans					7	

[a] O'Connor (1991a). Mortality over three years of successively less rainfall, under light grazing.
[b] O'Connor (1994). Annual tuft mortality in 1989 under light grazing. Highest of two values given.
[c] Danckwerts & Stuart-Hill (1988). Tuft mortality in response to the 1982/83 drought. Highest of two values given.
[d] Snyman & Van Rensburg (1990). Tuft mortality after two years (1982/83 and 1983/84) of drought.
[e] Scholes (1985). Tuft mortality over the 1982/83 drought.
[f] Fourie & Roberts (1977). Mortality during one season of summer drought.

lates secondary tillering, although parent tillers only regrow after lenient defoliation (Tainton & Booysen 1965a,b; Drewes & Tainton 1981).

Differences in the patterns of tiller and leaf demography and growth in response to defoliation between *Themeda triandra* and *Sporobolus fimbriatus* growing in a semi-arid grassland suggest that *T. triandra* is adapted primarily to defoliation by fire and *S. fimbriatus* adapted mainly to year-long defoliation by herbivores (Danckwerts & Stuart-Hill 1987). Features supporting this conclusion include the following: shoot apices of *S. fimbriatus* elevate only at flowering, whereas *T. triandra* tillers elevate their apices while still vegetative; defoliation is necessary for the lateral tillering of *T. triandra*, whereas *S. fimbriatus* can reproduce vegetatively at all times of the year over a range of utilization intensities (Danckwerts 1984; Danckwerts *et al.* 1984, 1986; Danckwerts & Nel 1989); and *T. triandra* has greater production at the tiller, tuft and sward level than *S. fimbriatus* and is a better competitor for light (Danckwerts & Aucamp 1985; Danckwerts 1987, 1988).

10.5 **Populations**

Population dynamics of grasses can be considered at the tiller and tuft levels. Almost catastrophic levels of com-

plete tuft mortality commonly occur in response to drought, although species with terpenoid compounds (*Cymbopogon* or *Bothriochloa* spp.) seem less affected (Table 10.3). Tiller mortality can be as severe – 94% of drought-stressed in comparison to only 40% of irrigated tillers of *S. fimbriatus* succumbed to a severe drought (Danckwerts *et al.* 1986).

Species' attributes influence their population dynamics, and hence spatial and temporal patterns of community change in response to rainfall variability and grazing (Table 10.4). *Digitaria eriantha* rapidly colonized accessible patches created by drought and grazing by means of stoloniferous growth. For seed-reproducing species, the considerably greater recruitment under grazing of the unpalatable species *Aristida bipartita*, relative to the palatable species *Themeda triandra* or *Heteropogon contortus*, was due in part to greater seed production, seed bank size and emergence, leading ultimately to its domination of the community. The slow clonal expansion of *Setaria incrassata* appears to be limited by high rainfall years (O'Connor 1994). Although change in this grassland community was driven largely by inter-annual rainfall variability, the direction of these changes was reversed, depending on the sequence of years from wet to dry or dry to wet. In contrast, grazing had a small annual but consistent and cumulative effect on change over years, operating most noticeably on the regeneration phase. Palatable, seed-reproducing perennials such

Table 10.4 Attributes and population dynamics of a suite of coexisting species in a savanna grassland

Attribute/dynamic	Aristida bipartita	Heteropogon contortus	Themeda triandra	Bothriochloa insculpta	Digitaria eriantha	Setaria incrassata	Source
Growth form, mode of reproduction	Tufted, seed	Tufted, seed	Tufted, seed	Tufted, mainly seed, also stolon	Tufted, mainly stolon, also seed	Clonal rhizomatous, occasionally seed	Carter & O'Connor (1991)
Palatability to livestock	Unpalatable	Intermediate	Palatable	Unpalatable	Intermediate	Palatable	Carter & Robinson (1993)
Diaspore weight (mg) (without awns)	0.33	2.81	4.05	0.85	—	2.90	O'Connor (1992)
Viable seed mm^{-2} of tuft (maximum recorded)	591 990	40 109	51 956	387 857	3069	47 858	Capon & O'Connor (1990)
Seed bank densities (m^{-2}) (maximum recorded)							O'Connor & Pickett (1992)
heavy grazing	1465	118	81	68	36	448	
light grazing	2672	158	109	262	81	561	
Seedling recruits (mm^{-2}) (mean of four seasons)							O'Connor (1994)
heavy grazing	14.4	3.1	6.6	1.0	1.6	0.0	
light grazing	13.3	2.6	2.5	0.8	1.2	0.3	
Population turnover (%)[a] (over four years)							O'Connor (1994)
heavy grazing	47	48	13	49	11	<2	
light grazing	36	18	8	13	5	<2	
Population growth rate (individuals yr^{-1} m^{-1}) (mean of four seasons)							O'Connor (1993)
heavy grazing	0.260	−0.090	0.002	0.015	0.143	0.008	
light grazing	0.123	0.018	0.108	0.064	0.162	0.059	
Colonizing ability[b]	22.5	4.3	0.2	5.0	4.1	0.1	O'Connor (1991c)

[a] Fraction of the basal area of a species after four years constituted by individuals extant at the beginning, i.e. that fraction not comprising seedling or stolon recruits.
[b] Percentage contribution to experimental patches after three years/percentage contribution to surrounding community.

as *T. triandra*, which experience extensive mortality as a result of drought (Table 10.3) and curtailment of seedling recruitment by grazing, may be prone to local extinction because of these attributes (O'Connor 1991b). Anecdotal accounts support this (Bews 1918; Potts 1923; Acocks 1953; Killick 1958). In contrast, the attributes of an unpalatable perennial such as *A. bipartita* (Table 10.4) can result in a burgeoning population (O'Connor 1994). A consideration of species' attributes would therefore seem partly to explain the grazing 'succession' of this community. Selective defoliation can also shift the competitive relationship between palatable and unpalatable species in favour of the non-defoliated, unpalatable species (Jones 1967; Jones & Cross 1971), a process that has been partly responsible for the replacement of *T. triandra* by *Aristida junciformis* in mesic grassland (Morris & Tainton 1993).

Tiller, rather than tuft, dynamics account for the response of grasses of moister grassland to defoliation by fire or grazing. Tiller populations of *Cymbopogon excavatus*, whose tillers live a maximum of two years, are eventually eliminated by complete protection from grazing, because tiller mortality is marginally increased and tiller recruitment is suppressed (S.E. Shackleton 1989). Tillers are also eliminated by severe and frequent defoliation, because primary tillers are killed and secondary tillers are prevented from developing (Shackleton & Mentis 1991). However, they increase under annual or biennial burning, or annual harvesting for thatch, because the increased production of secondary tillers compensates for the increased mortality of harvested or burnt tillers (S.E. Shackleton 1989). Tiller populations of *Themeda triandra* (with tiller life-spans a maximum of two years) declined rapidly with summer burning, because of catastrophic mortality (<6% survival) (Everson, Everson & Tainton 1985a). Tillers increased with annual winter burning, because of increased recruitment of daughter tillers and no marked mortality of primary tillers, but remained stable under biennial spring burning, because mortality of primary tillers was not markedly heightened and tiller recruitment was not increased. These responses to fire correspond to the effect of fire on irradiance regime (Everson, Everson & Tainton 1988). Decreased irradiance (<30% full sunlight) eliminated *Heteropogon contortus* and *Trachypogon spicatus*, depressed tiller initiation of *Tristachya leucothrix*, *Alloteropsis semialata* and *T. triandra*, but had only a slight effect on *Harpochloa falx*.

The dynamics of many herbaceous dicotyledons in grasslands seem entrained by the state of the grass sward, because they may increase quite dramatically in abundance when the sward is heavily grazed or burned. Examples are the perennial legume *Chamaecrista (Cassia) mimosoides* (Steinke & Nel 1990), the poisonous *Senecio*

retrorsus (Brynard 1953; Hildyard 1967; Du Toit & Aucamp 1985), *Helichrysum* spp. (Story 1951; Noel 1961) and the legume *Elephantorrhiza elephantina* (Ellery & Walker 1986). The success of these species under heavy grazing is due to different attributes, although all are relatively unpalatable to livestock. *Chamaecrista mimosoides* can increase 15-fold when swards are grazed very short, partly because of a persistent seed bank (>2.5 years) with a variable release of seeds from dormancy, depending on temperature and moisture. Optimal temperature restricts germination to the summer and seedling survival is far higher in the open (52%) than within a sward (6%) (Steinke & Nel 1990). Similarly, the success of *S. retrorsus* under heavy grazing would seem to be due to the reduction of competition from grass (Brynard 1953; Hildyard 1967; Hildyard & Booysen 1971). This species can produce 1400 wind-dispersed achenes per plant that do not form a persistent seed bank. Plants germinating from seed in early spring can flower within six months, but competition from a dense sward can depress germination, reproductive output and growth. In contrast, seedlings of *E. elephantina* establish best within the protection of a closed sward, then grow rhizomatously at a rate of 0.4 m yr^{-1} for 15 years, during which time one of the roots differentiates to function as a storage organ. Their roots reach deeper than do those of the grassland species (Ellery & Walker 1986).

10.6 Community dynamics

10.6.1 Secondary succession
Early influential workers embraced the Clementsian paradigm of succession, emphasizing that succession converged towards a climatic monoclimax, could link different vegetation types and comprised discrete stages of integrated biotic communities behaving as complex organisms (Bews 1916; Phillips 1934, 1935a,b). This viewpoint is still largely retained (e.g. Tainton 1981).

In the 1930s Phillips initiated a long-term study of succession on abandoned fields at Frankenwald, which has already been summarized (Roux 1969) and is therefore only briefly recounted. A sequence of discrete stages, each with distinctive edaphic and life-history correlates, was recognized for the succession at Frankenwald, an area of sandy soil receiving rainfall of 790 mm yr^{-1}. A short-lived ruderal stage comprising mostly exotic annual weeds gives way to a grassland of *Cynodon dactylon* and *Eragrostis* spp. and some *Hyparrhenia hirta*. Within about eight years, *H. hirta* forms almost pure stands, but its replacement by a 'sub-climax' community does not take place within 30 years and has not been directly demonstrated (Table 10.5 lists the main species of each

Table 10.5 **Reproductive characteristics of species for different successional stages at Frankenwald (adapted from Jones 1964, 1968).**
Sub-climax vegetation type is a *Loudetia simplex–Trachypogon spicatus* grassland (C7). NA, not applicable; ND, no data available

Species	Average seed output plant⁻¹	Seed weight (mg)[a]	Disseminule weight (mg)	Sound seed (%)
Ruderals				
Bidens schimperi	166		ND	92
Tagetes minuta	306		0.82	81
Conyza ambigua	73 500		0.03	67
Pseudognaphalium luteo-album	24 800		0.01	ND
Oenothera biennis	4560		0.70	NA
Primary grasses				
Eragrostis curvula	6050	0.19	NA	24
E. gummiflua	3060	0.03	ND	ND
Aristida congesta	1110	0.32	0.40	78
Pogonarthria squarrosa	14 600	0.08	0.04	ND
Eragrostis racemosa	ND	0.05	NA	ND
Secondary grasses				
Hyparrhenia hirta	194	0.55	1.06	10
Cymbopogon excavatus	350	0.61	0.65	23
Eragrostis capensis	ND	0.30	NA	ND
Sub-climax grasses				
Dihetoropogon amplectens	83	1.39	2.44	15
Themeda triandra	17	2.00	ND	23
Panicum natalense	52	0.49	0.18	5
Urelytrum agropyroides	7	5.62	9.50	4
Trachypogon spicatus	5	1.82	3.51	4
Tristachya leucothrix	0	ND	ND	0
Monocymbium ceresiiforme	4	ND	0.70	1

[a] For grasses this is caryopsis weight.

Table 10.6 **Relative seed bank densities of different successional stages at Frankenwald, as determined by germination of soil cores.**
Sub-climax vegetation type is a *Loudetia simplex–Trachypogon spicatus* grassland (C7) (adapted from Jones 1964)

Species	Normal stage	Stage of succession				
		Ruderal	Primary	Secondary	Sub-climax	Fallow
Herbaceous dicotyledons						
Acanthospermum glabratum	Ruderal	59	0	0	0	0
Corrigiola litoralis	Ruderal	10	20	17	0	3
Tagetes minuta	Ruderal	75	4	1	0	9
Pseudognaphalium luteo-album	Ruderal	9	2	32	4	4
Conyza ambigua	Ruderal	9	2	6	0	16
Portulaca oleracea	Ruderal	0	1	0	0	9
Oldenlandia sp.	Ruderal	17	2	6	0	1
Other dicotyledons		66	9	44	3	3
Monocotyledons						
Cyperus esculentus	Ruderal	72	59	8	12	5
Fimbristylis sp.	Ruderal	93	62	55	0	1
Eleusine indica	Ruderal	13	52	0	1	24
Eragrostis curvula	Primary grass	2	12	4	0	0
Cynodon dactylon	Primary grass	3	5	17	6	14
Hyparrhenia hirta	Secondary grass	0	0	9	0	0
Trachypogon spicatus	Sub-climax	0	0	0	3	0
Other monocotyledons		28	26	51	37	0

stage). However, the floristic composition of long-abandoned Iron Age sites suggests that the *Hyparrhenia* stage eventually changes to a *Trachypogon spicatus*–other species grassland (Roux 1970).

As succession proceeds, there is an increase in soil organic matter, improved moisture relations, increased soil colloids and total nitrogen, but lower plant-available nitrogen and phosphorus, a decrease in irradiance, an increased depth from which water is extracted (Coetzee,

Page & Meredith 1946), an increase in seed parasitism, seed and disseminule weights and a decrease in fertile seed set (Table 10.5) and in dispersibility of diaspores. The far greater seed output of early-successional sub-climax grassland (20954 seeds m⁻² to 115 seeds m⁻²), with 62% of the seeds produced in sub-climax grassland originating from 'secondary' grasses (Jones 1968), has a marked influence on seed banks (Table 10.6).

Most workers believed this succession was driven by

edaphic factors (principally nitrogen) (Davidson 1962, 1964) corresponding to the facilitation model of Connell & Slatyer (1977). A gradual decrease in nitrates, nitrites and mineralization rates occurs as succession progresses, with associated changes in the composition of the microbial community (Warren 1965). Supporting evidence indicates that fertilizing grassland with nitrogenous compounds rapidly transforms grasslands of *Hyparrhenia* or *Trachypogon* to those of *Cynodon* and *Eragrostis* spp. (Davidson 1964). Similar transformations resulting from the application of nitrogenous, but not phosphatic, compounds have been widely reported (Grunow, Pienaar & Breytenbach 1970; Barnes, Bransby & Tainton 1987; Janse van Rensburg *et al.* 1990), although the detrimental effects of nitrogen fertilization for certain species can be mitigated by severe defoliation (Edwards 1975). Species at Frankenwald exhibit a hierarchy in growth response to low nitrogen concentration – poor in early successional species and relatively good in late successional species – that is reversed for growth in high nitrogen concentration (Roux 1954; Jong & Roux 1955; Warren 1965). Climax perennial grasses do, however, perform better when supplied with ammonium nitrogen than with nitrate nitrogen (Wiltshire 1972, 1973). The sensitivity of *Hyparrhenia hirta* and *Themeda triandra*, but not of *Eragrostis curvula*, to increased nitrogen availability is due to the differential effect of nitrogen on the photosynthetic activity, CO_2 compensation point and photorespiratory activity and enzymes of these species (Grossman & Cresswell 1974; Tew *et al.* 1974; Amory & Cresswell 1981; Wolfson, Amory & Cresswell 1982).

A second school proposed that the availability of seeds determined the initial occupation, and thereafter competition for water determined the direction of succession (Rose Innes 1939; Jones 1964). Support for this hypothesis comes from the fact that a ruderal stage is not required for establishment of later successional grassland (Jones 1964). The 'high-successional' species, *Themeda triandra*, can be rapidly re-established on fallow land by reseeding (Dyer 1937; Scott 1937; Kruger & Smit 1973), and *Cynodon dactylon* is the first grass to appear, because its seeds are transported by livestock (Jones 1964). Unfortunately, research at Frankenwald had been discontinued by the time this polemic had crystallized.

10.6.2 Effects of rainfall, grazing and fire
Changes in community structure in response to grazing have usually been interpreted as uni-directional in accordance with a Clementsian concept of succession. More recently, ecologists have recognized the possibility of alternate states (Westoby, Walker & Noy-Meir 1989) or of communities within a 'domain of attraction' with a

high inertia to change (Bosch 1989; Bosch & Gauch 1991). For grasslands with high year-to-year variability in rainfall, there is now increasing evidence that grazing-induced changes are contingent upon rainfall patterns (O'Connor 1985).

The early Frankenwald work provided support for a unidirectional sequence of change in terms of both patterns and processes. Heavy grazing resulted in the same floristic changes as those brought about by nitrogenous fertilizers (Glover & Van Rensburg 1938; Van Rensburg 1939). This is because grazing led to a more rapid cycling of nitrogen and increased the proportion of plant-available forms (Roux 1969). The bulk (83%) of this nitrogen was contained in dung and not urine, and the rapid burial of fresh dung by dung beetles minimized nitrogen loss by volatilization of ammonia from 80% to 20% (Gillard 1963). The fertilization effect of grazing on the nutrient-poor soils of Frankenwald increased herbage quality and animal production (Gillard 1965), challenging concepts of 'overgrazing' that held that 'sub-climax' communities were the most productive and desirable. A wealth of observation of apparently similar change firmly ensconced Clementsian succession as the paradigm for the dynamics of grazed grassland in southern Africa (e.g. Acocks 1953; Tainton 1981).

However, long-term studies suggest the subtle interplay of a number of factors in determining community change. For the *Hyparrhenia hirta* tall grassland (F14) of KwaZulu-Natal, floristic changes over 30 years in response to grazing treatments were determined (Fig. 10.10). Although a number of treatments resulted in an increase of *Aristida junciformis* at the expense of *Themeda triandra* and other palatable perennial grasses, only rotational rest at a high stocking rate showed no subsequent re-establishment of *T. triandra* once grazing was removed. This provides evidence of a threshold and partly supports the concept of alternate states as proposed for *H. hirta* tall grassland (F14) by Westoby *et al.* (1989). Nevertheless, direct gradient analysis identified a linear ordering of *Diheteropogon amplectens*, *Tristachya leucothrix*, *T. triandra*, *A. junciformis* and *Eragrostis curvula* from least to most tolerant of grazing intensity (Mentis 1982), thus corroborating the decreaser–increaser classification of species (Foran, Tainton & Booysen 1978).

High-rainfall grassland experiences low variation in inter-annual rainfall (Chap. 2, this volume) and seems less susceptible to fundamental change in community structure in response to grazing (e.g. continuous grazing of Dohne Sourveld; Du Toit & Aucamp 1985). Indeed, the sour grasslands of the communally grazed regions of the Transkei, considered by commercial standards to be 2.5 times overstocked for 30 years, have maintained populations of *Themeda triandra*, although *Aristida junciformis*

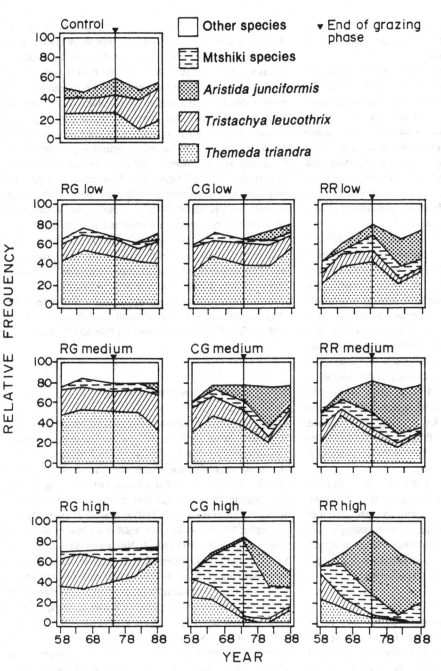

Figure 10.10 **Changes over 30 years in the relative frequency of dominant species or species guilds ('Mtshiki' = *Eragrostis curvula*, *E. plana* and *Sporobolus africanus*) for rotational grazing (RG), continuous grazing (CG) and rotational rest (RR) at three stocking rates (low: 0.31–0.75 AU ha⁻¹; medium: 0.62–1.11 AU ha⁻¹; high: 0.93–2.22 AU ha⁻¹) in the *Hyparrhenia hirta* Tall Grassland (F14), Ukulinga, KwaZulu-Natal. From Morris, Tainton & Hardy (1992). Grazing treatments were imposed between 1958 and 1974. Thereafter the area was burnt periodically in spring for the next 14 years.**

has become locally dominant in some areas (McKenzie 1982).

In contrast, marked changes in floristic composition can take place in areas of lower rainfall in response to grazing pressure. In a semi-arid district, *Themeda triandra* and *Eragrostis lehmanniana* decreased, whereas *Aristida congesta* and *Tragus koelerioides* increased with more intensive grazing during the summer. These changes persisted for nine years after the withdrawal of grazing (Potts 1923; Mostert 1958; Van den Berg, Roberts & Vorster 1975). Similar responses were shown in a *Cymbopogon plurinodis–T. triandra* grassland (Coetsee & Benadie 1975; Coetsee & Van Rensburg 1975a,b; Edwards & Nel 1973). However, experimental manipulation of moisture availability has shown that drought seasons, even without grazing, are responsible for much community change (Snyman & Van Rensburg 1990). Furthermore, the pattern of grazing following a drought season can determine the recovery of species (e.g. *T. triandra*) that suffer extensive mortality during a drought (Table 10.3) (Danckwerts & Stuart-Hill 1988).

Indirect gradient analysis, after appropriate stratification for landscape features, has revealed that the decreaser–increaser response of species to grazing severity (Bosch & Janse van Rensburg 1987; Bosch, Janse van Rensburg & Truter 1987; Bosch & Gauch 1991) depends on both climate – *Themeda triandra* reacts as an increaser in higher rainfall areas but as a decreaser in lower rainfall areas (Bosch 1989) – and topo-edaphic conditions (Truter 1988; Strohbach 1989; Janse van Rensburg & Bosch 1990). The increaser–decreaser classification of a species is therefore not absolute.

Protection of a community from fire and grazing can result in almost complete transformation, even without the invasion of woody elements (Glover & Van Rensburg 1938; Edwards 1968; Everson & Tainton 1984). For example, a *Themeda triandra–Elionurus muticus* vegetation-type changed to a *Cymbopogon plurinodis–Setaria sphacelata* type within 20 years of protection (Coetsee & Van Rensburg 1975b). However, various burning regimes do not appear fundamentally to change the composition of grassland, although both the frequency of defoliation by fire and the season of burn can cause shifts in the relative abundance of constituent species (Roux 1969; Scotcher & Clarke 1981; Le Roux 1989). For example, increasing intervals between fires in the Drakensberg favours *Tristachya leucothrix* and *Harpochloa falx* but not *T. triandra* (Everson & Tainton 1984).

These empirical studies suggest that change in community composition in response to grazing will depend on year-to-year rainfall variability and the interaction of grazing, rainfall and fire. Grazing-induced changes may not necessarily be reversible, but may result in apparently persistent alternate states maintained by, as yet incompletely identified, factors or processes.

10.6.3 Patch effects

Patch-selective grazing by livestock (Hatch 1991) can initiate patches of a degraded nature characterized by a low basal cover and a community of mostly shorter-lived perennial grasses (e.g. *Aristida congesta*) with few palatable perennial species (e.g. *Themeda triandra* and *Digitaria eriantha*) (Fuls & Bosch 1991), and a lower water-use efficiency than the surrounding grassland (Snyman & Opperman 1983). Large inputs of water are required for the re-establishment of a good cover of large, tufted perennials (Fuls 1991).

Piosphere effects – the compositional gradient extending from waterpoints – occur in grasslands, with *Aristida congesta, Eragrostis plana, E. curvula* and *Trichoneura grandiglumis* characteristic of heavy livestock densities associated with proximity to watering points (Janse van Rensburg & Bosch 1990). Notably, piosphere effects have not developed in the semi-arid grassland of the Kalahari Gemsbok National Park (Van Rooyen *et al.* 1994).

Termites are the most conspicuous of other animals responsible for patch effects. Intact mounds of *Trinervitermes* spp. (0.8 m mean diameter, density of 53.3 ha⁻¹) are not colonized by vegetation as are mounds of the *Macrotermitinae* in the savanna regions (e.g. Griffioen & O'Connor 1991). However, the area around the mound that receives run-off water has three times the growth response and a greater reproductive output than the intervening grassland (Steinke & Nel 1989). Mounds of *Trinervitermes trinervoides* are enriched in clay, silt, bases, organic carbon and nitrogen relative to the surrounding soil, with a C : N ratio that has higher plant-available nitrogen and cation ratios that are more favourable for plant growth (Laker *et al.* 1982a,b). When mounds collapse, often because of digging by aardvark (*Orycteropus*), they are colonized by the nitrophilous *Cynodon dactylon* and encircled by *Eragrostis* species (Steinke & Nel 1989).

10.7 Ecosystem functioning

Primary production across the grassland biome is related to mean annual rainfall (Fig. 10.11). However, fertilization experiments indicate that production is limited by the availability of nitrogen and phosphorus (yield can be doubled by nitrogen fertilization), but only slightly by the availability of phosphorus alone (Hall *et al.* 1937, 1941; Weinmann 1943a,b; Hall & Meredith 1945; Davidson 1962; Vorster & Mostert 1968; Grunow *et al.* 1970; Van Ginkel 1971; Edwards & Nel 1973; Van Zyl 1973, 1975). Nonetheless, water-balance models are adequate for predicting phytomass production in semi-arid grassland (De Jager, Opperman & Booysen 1980). Variation in grassland production of a locality is also strongly dependent on rainfall (Fig. 10.12). Production is similarly influenced by the availability of nitrogen and phosphorus (Tainton *et al.* 1970; Le Roux 1989; Snyman & Fouché 1991, 1993), and by the seasonal mean daily maximum temperature for a higher latitude, semi-arid grassland (Trollope 1983). Local variation in herbage production is also a consequence of spatial variation in moisture and nitrogen availability (Barnes *et al.* 1991).

Production at any one locality depends on sward composition. In a semi-arid region, grassland with mostly longer-lived perennial species and high basal cover had a pronounced linear relationship between production and rainfall, whereas a degraded grassland with mostly shorter-lived perennial and annual species and a poor

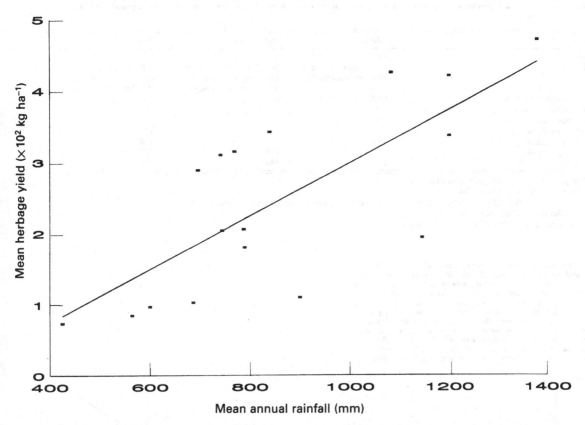

Figure 10.11 **Relationship across the grassland biome between mean annual rainfall and mean herbage yield at or near the end of the growing season. Relationship described by a linear regression: $y = 3.72x - 742$ ($R^2 = 0.53$; $F = 19.0$; $P < 0.0006$), although note that the serial dependence of observations (because of perenniality) violates the assumptions of a model 1 regression. Only one value was entered for each locality. Combined values for different estimates of the same locality were weighted by their length of record (range 1–15 years). Variability weakening the relationship is possibly derived from different times and methods of sampling, soil type and mean annual temperature. Sources: Hall, Meredith &** Murray 1937, 1941; Weinmann 1938, 1943a, b, 1948; Davidson 1962; Van Schalkwyk, Lombard & Vorster 1968c; Vorster & Mostert 1968; Tainton, Booysen & Scott 1970; Rethman *et al.* 1971; Van Ginkel 1971; Smit 1972; Du Toit 1972; Edwards & Nel 1973; Rethman & Beukes 1973; Van Zyl 1973, 1975; Coetsee & Van Rensburg 1975c; Vorster 1975; Drewes 1979; Scotcher & Clarke 1981; Downing & Marshall 1983; Nel 1983; Everson 1985; Le Roux 1989; Barnes *et al.* 1991; Shackleton 1991; Snyman & Fouche 1991.

basal cover showed a far weaker relationship (Fig. 10.12). Different sward composition can also dramatically influence the fire behaviour of a grassland. This is not only because the fuel load for fire intensity is obviously important (Trollope 1983), but also because the heat of foliage combustion differs among species (Table 10.2).

Differences in production among compositional states depend on processes influencing available soil moisture and on the rain-use efficiency of the species of each community (Table 10.7). Infiltration capacity is linearly related to the basal cover of perennial grasses (Van den Berg, Roberts & Vorster 1976). Therefore, loss of runoff from 'pioneer' swards is three times greater than from perennial grass swards (Haylett 1960; Snyman & Opperman 1984; Snyman, Van Rensburg & Opperman 1985; Snyman & Fouché 1991). Therefore, long-term management practices can influence runoff markedly (Table 10.8; Edwards 1961; Du Plessis & Mostert 1965; Van den Berg *et al.* 1976). Evapotranspiration losses from a sward of *Themeda triandra* are about 10 mm day^{-1}, with nearly uniform removal of water by roots from throughout the profile (Opperman, Human & Viljoen 1977). These losses are not appreciably different from swards with a greater proportion of shorter-lived species, but are higher than the losses from a pioneer sward (Table 10.7). However, the loss of water per unit of dry matter produced is considerably lower for a *T. triandra* sward than for a sub-climax or pioneer sward. Moderate levels of defoliation can further increase the rain-use efficiency of *T. triandra* swards (Snyman & Opperman 1983).

Long-term above-ground production is not affected by fire (Tainton *et al.* 1978) or season of fire (Everson 1985), despite the destruction of a large proportion of tillers (Tainton, Groves & Nash 1977). However, it is inversely related to the frequency of physical defoliation, and is

Table 10.7 **Production, hydrological characteristics and soil loss of three compositional states (climax, sub-climax and pioneer in successional terms) of a semi-arid grassland, specifically a drier variant of the *Themeda triandra–Eragrostis curvula* grassland (A1). The soil is a fine sandy loam**

	Climax	Sub-climax	Pioneer
Basal cover (%)[a]	8.47	6.56	2.86
Composition[b]	70% *Themeda triandra*	50% *Eragrostis lehmanniana*	60% *Aristida congesta*
	30% *Digitaria eriantha*	25% *Eragrostis chloromelas*	20% *Cynodon hirsutus*
		25% *Sporobolus fimbriatus*	20% *Tragus koelerioides*
Above-ground production (kg ha^{-1})[a]	1022	598	281
Evapotranspiration (Et) mm yr^{-1})[c]	615	603	557
Et/soil evaporation (Eo)[c]	1.32	1.29	1.19
Rain use efficiency (kg ha^{-1} mm^{-1})[a]	2.0	1.2	0.6
Expected soil loss (tons ha^{-1} yr^{-1})[d]	0.61	1.42	3.86
Relative loss organic C (kg ha^{-1})[e]	0	2659	5225
Relative loss of N (kg ha^{-1})[e]	0	180	331

[a] Mean of 12 years (1977–1989) (Snyman & Fouché 1991).
[b] Snyman, Opperman & Van den Berg (1980).
[c] Mean of four years (1979/80–1982/83) of a lysimeter study, where Eo is evaporation from bare soil (Snyman 1988).
[d] Snyman, Van Rensburg & Opperman (1986).
[e] Calculated loss for 15 years (Du Preez & Snyman 1993).

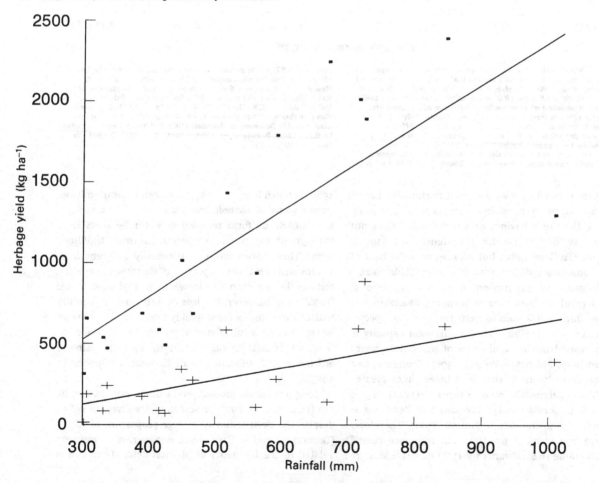

Figure 10.12 **Relationship between annual rainfall (July–June inclusive) and herbage yield (harvested July) for 'good' (squares) and 'poor' (+) condition grassland (details in Table 10.7) of a drier variant of *Themeda triandra–Eragrostis curvula* Grassland (A1) over 17 successive years. After Snyman & Fouche (1991) and H.A. Snyman (unpubl. data). Both relationships are described by a linear model 1 regression (note that an assumption of a model 1 regression is violated by the serial dependence of observations because of perenniality): $y = 2.66x - 272$ ($R^2 = 0.58$; F = 23.5; $P < 0.0002$) for grassland in good condition; $y = 0.78 x - 105$ ($R^2 = 0.40$; F = 11.7; $P < 0.0038$) for grassland in poor condition.**

Table 10.8 **Effect of different defoliation treatments on runoff and soil loss from** *Hyparrhenia hirta* **Tall Grassland (F14), Estcourt, KwaZulu-Natal; adapted from Le Roux (1989). Superscripts within columns denote significant (*P* < 0.05) differences between treatments**

	Annual runoff (kl ha⁻¹) (*n* = 24)		Annual soil loss (kg ha⁻¹) (*n* = 21)	
	Mean	Max	Mean	Max
Not mowed or burned	3555[A]	12 660	332[B]	977
Burned annually – August	2477[A]	6885	1134[A]	4851
Burned annually – spring	1394[B]	5764	481[B]	1782
Burned biennially – autumn	1370[B]	4662	356[B]	943
Burned biennially – spring	3226[A]	7676	452[B]	1184
Burned triennially – spring	1467[B]	6635	732[A]	2448
Mowed for hay – mid-summer/August	1358[B]	6032	411[B]	1519

Figure 10.13 **Cumulative differences over 15 years between treatments in herbage yield for Tall Grassveld, Ukulinga, KwaZulu-Natal (from Tainton *et al.* 1970). Key to treatments: cut in early summer (solid); cut in late summer (dotted); cut in early and late summer (dashed); cut in mid-winter (×).**

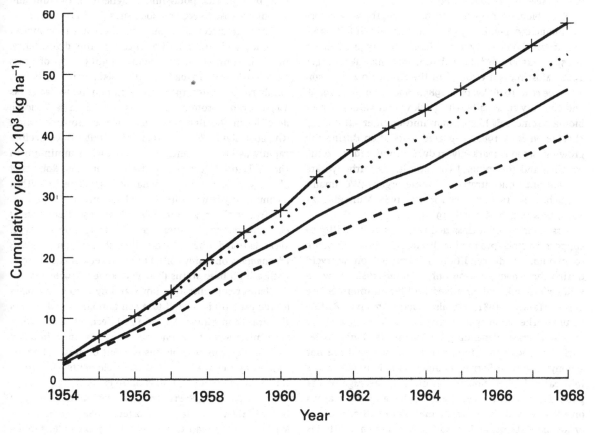

particularly depressed by physical defoliation during the growing rather than the dormant season (Fig. 10.13). In semi-arid grassland, production is affected more by sward composition than by defoliation, although defoliation had a greater effect on the production of grassland with 'good' than with 'poor' species composition (Danckwerts & Barnard 1981). However, compensatory above-ground growth occurs at intermediate levels of defoliation for many species and environments, but almost always at the expense of root biomass (Weinmann 1943a,b, 1944, 1948, 1949; Opperman, Roberts & Van Rensburg 1969, 1970; Burger, Grunow & Rabie 1975; Danckwerts & Nel 1989). Increased frequency of defoliation invariably reduces herbage yields, but results in herbage with higher crude protein, mineral content and digestibility and with roots with reduced concentrations and amounts of carbohydrates and minerals (Weinmann 1944, 1949; Barnes 1960a,b, 1961; Downing & Marshall 1983). This is partly because of reduced seasonal translocation of nitrogen, phosphorus and sugars from shoots to roots at the end of the growing season (Weinmann 1940, 1942). Regrowth of perennial grasses following defoliation is dependent on the remobilization of stored carbon from roots and stem bases (Steinke & Booysen 1968; Bartholomew & Booysen 1969; Steinke 1969; Danckwerts & Gordon 1990; Danckwerts 1993), unless sufficient leaf material remains (Steinke 1975). Moisture stress can also increase the root mass and carbohydrate content of the roots and stem bases of *Themeda triandra* (Nursey 1971; Opperman, Human & Roberts 1976), which may explain why defoliation of *T. triandra* at wilting point in the growing season has a carry-over effect on above-ground production in the following season (Snyman 1993).

Root biomass densities (to 0.5 m depth) vary across the biome independently of mean annual rainfall, ranging from about 900 g m^{-2} (Weinmann 1944; Everson 1985) to >4500 g m^{-2} (Shackleton, McKenzie & Granger 1988). Most roots are found in the surface 0.2 m of soil (Coetzee *et al.* 1946; Everson 1985; Shackleton *et al.* 1988), and can vary seasonally with peak winter values of root biomass some 20% higher than mid-summer values (e.g. Shackleton *et al.* 1988). Repeated defoliation during the growing season markedly reduces root and rhizome weights, and the mineral concentrations, carbohydrate content and concentration of these organs (Weinmann 1943a,b, 1944, 1948; Barnes 1960a,b, 1961; Mufandaedza 1976; Downing & Marshall 1983).

Grassland is often classified in terms of its ability to support livestock production. Palatable forage, adequate to maintain body condition, is provided on sourveld during the growing season only (c. 6 months), on sweetveld year-round, and on mixed veld for an intermediate period (Tainton 1981; see also Chaps 17 and 18, this volume). The same species may occur as both a sweet or a sour grass, depending on locality (Kirkman 1988; Zacharias 1990); therefore, sweet- and sourveld are not synonymous with floristic composition. Sourveld occurs in areas of higher moisture availability (average of 134 potential growth days yr^{-1}) and cooler growing seasons on soils derived from parent material of intermediate- or low base status, and on soils derived from quartzites or sandstones at low moisture availability. Sweetveld occurs in more arid (average of 52 potential growth days yr^{-1}), warmer areas on soils derived from intermediate- or high base status material, or in mesic areas on soils derived from basic igneous rocks. Mixed veld is intermediate (Ellery 1992). Correlates of sourveld are a decreasing availability of phosphorus, an increasing pool of readily mineralizable nitrogen but a decreasing ratio of the number of growth days to this pool. Seasonal vari-

ation in forage quality may be determined by the strength of nutrient-supplying processes in the soil-plant system relative to the strength of carbon-assimilation metabolism of the plants. Sweetveld occurs where the environment favours nutrient assimilation relative to carbon assimilation; sourveld occurs where carbon assimilation predominates over nutrient assimilation (Ellery 1992).

Plant quality attributes corresponding with sour- and sweetveld have not been clearly identified, although some time ago Henrici (1930a,b) recognized that sourveld grasses had lower crude protein, phosphorus and calcium, but higher fibre and nitrogen-free extractions than sweetveld grasses. For *Themeda triandra*, leaf digestibility is closely connected with concentrations of nitrogen, phosphorus, potassium, magnesium, calcium and sulphur across sweet and sour sites (Zacharias 1990).

For both sweet- and sourveld, there are pronounced seasonal and annual differences in the digestibility, nutrient content and nutrient standing crop of grass swards (Henrici 1930a,b; Preller 1950; Van Schalkwyk, Lombard & Vorster 1968a,b,c; Kirkman 1988; Zacharias 1990). Crude protein can be as high as 23% (*Cynodon dactylon*) in the first five weeks of the growing season (Grunow, Rabie & Grattorola 1977), but values decline rapidly as growth continues, and reach a minimum in the mid- to late dry season (Oellerman 1965; Roberts *et al.* 1975; Kirkman 1988; Zacharias 1990). Species within a community usually differ in foliage quality (Table 10.2; Du Toit *et al.* 1932; Fourie & Roberts 1976; Edwards *et al.* 1979), but these differences can depend on soil type. *Themeda triandra* has a higher digestibility than *Eragrostis lehmanniana* on sandy soils, but the reverse holds for soils with a high base status (Fourie & Roberts 1976, 1977).

Neither concentration nor standing crop of nitrogen (crude protein) is related to mean annual rainfall across the grassland biome (Fig. 10.14). Effects of fertilization with nitrogenous or phosphatic compounds indicate that nitrogen or phosphorus content of grass foliage is limited by the availability of these elements (Hall 1932; Hall & Meredith 1937; Hall *et al.* 1937, 1941, 1949; Weinmann 1943a,b, 1948; Preller 1950; Scott & Booysen 1956; Gillard 1965; Vorster & Mostert 1968; Rethman & Malherbe 1970; Van Ginkel 1971). For example, supplying 22 kg ha^{-1} of nitrogen can increase crude protein 1.35 times the content of unfertilized foliage. However, peak amounts of nitrogen in five grasslands were closely related to the amount of mineralizable nitrogen in the soil (Wiltshire 1978), which limited productivity (Wiltshire 1980). In turn, nitrogen mineralization, immobilization and turnover is restricted both spatially and temporally to the layer wetted by rain, and its amplitude in the heterotrophic cycle may be determined largely by

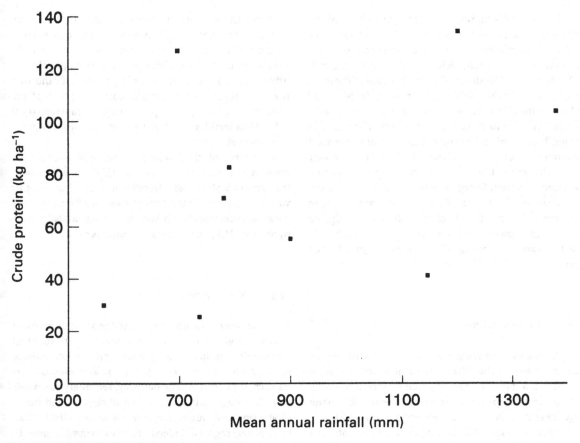

Figure 10.14 **Relationship (*P* > 0.2) across the grassland biome between mean annual rainfall and peak standing crop of crude protein. The relationship between rainfall and concentration of crude protein was also non-significant. Only one value was entered for each locality. Combined values for different estimates of the same locality were weighted by their length of record (range 1–13 years). Variability obscuring the relation possibly derives from different times of sampling, different methods of analysis and soil type. Sources: Hall *et al.* 1941; Weinmann 1943a,b, 1948; Hall, Meredith & Altona 1949; Van Schalkwyk *et al.* 1968c; Vorster & Mostert 1968; Rethman & Malherbe 1970; Van Ginkel 1971; Van Zyl 1973, 1975; Tainton *et al.* 1977; Scotcher & Clarke 1981; Everson 1985; Barnes *et al.* 1991; Shackleton & Mentis 1992.**

the availability of water to the soil microflora (Wiltshire 1990). Physical defoliation or fire invariably leads to increased leaf nutrient concentrations (Du Toit *et al.* 1935a,b; Louw 1938; Weinmann 1943a; Mes 1958; Opperman *et al.* 1969, 1970; Mufandaedza 1976; Tainton *et al.* 1977; Everson 1985; Shackleton & Mentis 1992) but not necessarily leaf nutrient standing crop (Shackleton & Mentis 1992; but see Everson 1985) or root nutrient concentrations (Opperman *et al.* 1969, 1970).

Nonetheless, most aspects of nutrient cycling have been all but ignored in southern African grassland. In a warm, high-rainfall area, breakdown of grass litter is rapid (e.g. 50% of mass in three months), but the rate depends on moisture availability and species

(Shackleton, McKenzie & Granger 1989). In a semi-arid region, the decomposition of swards of poor composition relative to those of good composition, on the same soil type, had lower input of nitrogen, resulting in markedly lower organic carbon, nitrogen and carbon : nitrogen ratios in the surface soil (Table 10.7). Erosion was responsible for only 3.4% of the loss of soil organic matter (5.1% of nitrogen), with an increased rate of decomposition resulting from higher soil temperatures possibly accounting for the remainder. Notably, increased yields and therefore organic inputs, as effected over 13 years of nitrogen fertilization, did not increase the soil carbon content and only slightly increased the soil nitrogen content in a mesic region (Theron 1961, 1965). Loss of nitrogen through leaching was low, although 83 ppm of cations were lost (Theron 1964). Soil loss, and therefore nutrient loss, can vary markedly on the same soil type in response to different defoliation treatments (Table 10.8; Du Plessis & Mostert 1965), and because of different species composition (Table 10.7). For example, predicted sediment losses in the Drakensberg are greater with biennial burning in spring than with annual burning in winter, because of the timing of exposure of the ground to summer storms, despite the greater average canopy and mulch cover found on a biennial spring burning treatment (Everson, George & Schulze 1989). Despite the soil losses resulting from fire, burning over the long term seems to have the effect of depleting soil organic matter and nitrogen only slightly, although exchangeable bases are more greatly affected (Cook 1939; Edwards 1961; White & Grossman 1972).

10.8 Conservation

The grassland biome is considered rich in plant species diversity – even richer than fynbos at a 1000-m² scale (Cowling et al. 1991). Floristically, the grassland biome is distinct, and comprises a centre of diversity for many large genera, with an estimated 3788 plant species in the core region (Gibbs Russell 1987). On the basis of quarter-degree grid congruency between dispersions of plants and nature reserves, about 78% of these plant species are currently conserved in 142 publicly-owned nature reserves (Siegfried 1991), although only 2% of the grassland biome area is preserved (see also Chap. 23, this volume).

The grassland biome is considered to be seriously threatened by agriculture, industrialization, urbanization and forestry (Chap. 21, this volume). Gauteng, the major urban complex of South Africa, and many other cities (Bloemfontein, Welkom, Klerksdorp, Witbank, Newcastle) are situated within the biome. Associated urbanization and industrialization have already destroyed large areas of grassland, especially of vegetation types A1, B3, C6, C7 and F14 (see Fig. 10.2). Many of these cities developed as a result of the mining industry: gold mining on the Witwatersrand and northern Free State, coal mining on the eastern Transvaal Highveld and northern KwaZulu-Natal, and large-scale diamond diggings in the Northwest Province. The mining industry occupies large tracts of land, where virtually all natural vegetation has been destroyed.

In the Highveld Agricultural Region, representing most of the Transvaal grasslands, 49% of the land has been ploughed for the production of maize, sunflower and other crops (Schoeman & Scotney 1987).

A relatively new but serious threat to grassland, especially of vegetation types E10–13 (Fig. 10.2), is afforestation in the higher rainfall areas of the eastern plateau with mainly pine (Pinus) and eucalypts (Eucalyptus). Afforestation is occurring over large areas of land in a region of high biodiversity and containing 30% of the endemic and rare plant species in the Transvaal (Raal 1986; Macdonald 1991; Matthews et al. 1993; Chap. 21, this volume).

A number of alien woody species have successfully invaded the grassland biome, particularly near towns (Henderson & Musil 1984; Henderson 1989; Chap. 22, this volume). The most important of these are Rubus spp., Solanum mauritianum, Acacia mearnsii, Acacia dealbata, Eucalyptus spp., Melia azederach and Prunus persica.

10.9 Conclusion

We have sought to illustrate that the rainfall gradient across the grassland biome is a main determinant of community composition, primary production, foliage nutrient content, nutrient cycling and attributes of species such as photosynthetic pathway, secondary chemicals and phenology. Rainfall in semi-arid regions, and hence production and nutrient cycling, is more variable than in moister regions. Indeed, rainfall regime seems to determine the distribution of the biome both directly (i.e. water balance) and indirectly through fire regime, although biotic effects of grazing can influence biome boundaries. A temperature gradient is also undoubtedly important, and is partly independent of rainfall, although this relationship has not been well investigated. Soil type is a critical modifier of the influence of rainfall regime at a local or regional scale. Although all grasslands of the biome comprise mainly tufted perennials, it is tentatively suggested that semi-arid grassland has faster turnover of individual tufts, because of the increased frequency of drought-related mortality, and therefore has the potential for rapid compositional change. In contrast, tuft turnover and change in high-rainfall regions is slow, because of the stable rainfal regime. It would appear that as a result of these different rainfall patterns, grazing has a more immediate effect on community change in semi-arid than moist grassland. Changes in community composition can dramatically influence water balance, production, nutrient cycling, foliage quality, soil loss and fire behaviour. Community change depends on the influence of communities on the abiotic environment and on species attributes, but the response of species to environment is contextual rather than absolute.

10.10 References

cocks, J.P.H. (1953). Veld Types of South Africa. *Memoirs of the Botanical Survey of South Africa*, 28, 1–192.

mory, A.M. & Cresswell, C.F. (1981). A stimulation of an alternative photorespiratory CO_2 pathway by nitrogen in *Themeda triandra* – a possible explanation for its sensitivity to nitrogen. *Proceedings of the Grassland Society of southern Africa*, 16, 145–9.

xelrod, D.I. (1985). The rise of the grassland biome, central North America. *Botanical Review*, 51, 163–201.

arnes, D.L. (1960a). Growth and management studies on Sabi panicum and Star grass. Part I. *Rhodesia Agricultural Journal*, 57, 399–411.

arnes, D.L. (1960b). Growth and management studies on Sabi panicum and Star grass. Part II. *Rhodesia Agricultural Journal*, 57, 451–7.

arnes, D.L. (1961). Residual effects of cutting-frequency and fertilizing with nitrogen on root and shoot growth, and the available carbohydrate and nitrogen content of the roots of Sabi panicum. *Rhodesia Agricultural Journal*, 58, 365–9.

arnes, D.L., Smith, M.F., Swart, M. & Wiltshire, G.H. (1991). Relations between soil factors and herbage yields of natural grassland on sandy soils in the south-eastern Transvaal. *Journal of the Grassland Society of southern Africa*, 8, 92–8.

arnes, G.R., Bransby, D.I. & Tainton, N.M. (1987). Fertilization of Southern Tall Grassveld of Natal: effects on botanical composition and utilization under grazing. *Journal of the Grassland Society of southern Africa*, 4, 63–7.

artholomew, P.E. & Booysen, P. de V. (1969). The influence of clipping frequency on reserve carbohydrates and regrowth of *Eragrostis curvula*. *Proceedings of the Grassland Society of southern Africa*, 4, 35–43.

ate-Smith, E.C. & Swain, T. (1967). New leuco-anthocynanins in grasses. *Nature*, 213, 1033–4.

Baxter, B.J.M, van Staden, J. & Granger, J.E. (1993). Seed germination response to temperature, in two altitudinally separate populations of the perennial grass *Themeda triandra*. *South African Journal of Science*, 89, 141–4.

Baxter, B.J.M., van Staden, J., Granger, J.E. & Brown, N.A.C. (1994). Plant-derived smoke and smoke extracts stimulate seed germination of the fire-climax grass *Themeda triandra*. *Environmental and Experimental Botany*, 34, 217–23.

Behr, C.M. & Bredenkamp, G.J. (1988). A phytosociological classification of the vegetation of the Witwatersrand National Botanic Garden. *South African Journal of Botany*, 54, 525–33.

Bews, J.W. (1916). An account of the chief types of vegetation in South Africa, with notes on the plant succession. *Journal of Ecology*, 4, 129–59.

Bews, J.W. (1917). Plant succession in the thornveld. *South African Journal of Science*, 14, 153–73.

Bews, J.W. (1918). *The Grasses and Grasslands of South Africa*. Pietermaritzburg: Dawies & Sons.

Bezuidenhout, H. (1988). *'n Plantsosiologiese Studie van die Mooirivieropvanggebied*. MSc Thesis. Potchefstroom: University of Potchefstroom.

Bezuidenhout, H. (1993). *Syntaxonomy and Synecology of Western Transvaal Grasslands*. PhD Thesis. Pretoria: University of Pretoria.

Bezuidenhout, H. & Bredenkamp, G.J. (1991). The vegetation of the Bc land type in the western Transvaal grassland. *Phytocoenologia*, 19, 497–518.

Bezuidenhout, H., Bredenkamp, G.J. & Theron, G.K. (1993). The vegetation of the Bd and Ea land types in the grassland of the western Tranvaal, South Africa. *South African Journal of Botany*, 59, 319–31.

Bezuidenhout, H., Bredenkamp, G.J. & Theron, G.K. (1994a). Syntaxonomy of the vegetation of the Fb land type in western Tranvaal grasslands. *South African Journal of Botany*, 60, 72–81.

Bezuidenhout, H., Bredenkamp, G.J. & Theron, G.K. (1994b). The vegetation syntaxa of the Ba land type in western Transvaal grassland, South Africa. *South African Journal of Botany*, 60, 214–24.

Bezuidenhout, H., Bredenkamp, G.J. & Theron, G.K. (1994c). Phytosociological classes of the western Transvaal grassland, South Africa. *Koedoe*, 37, 1–18.

Bezuidenhout, H., Bredenkamp, G.J., Theron, G.K. & Morris, J.W. (1994d). Braun–Blanquet reclassification of the *Cymbopogon–Themeda* grassland in the Lichtenburg area, south-western Transvaal. *South African Journal of Botany*, 60, 306–14.

Bloem, K.J., Theron, G.K. & van Rooyen, N. (1993). Grassland communities of the Verlorenvalei Nature Reserve in the north-eastern sandy Highveld, Transvaal. *South African Journal of Botany*, 59, 273–80.

Booysen, P. de V., Tainton, N.M. & Scott, J.D. (1963). Shoot-apex development in grasses and its importance in grassland management. *Herbage Abstracts*, 33, 209–13.

Bosch, O.J.H. (1989). Degradation of the semi-arid grasslands of southern Africa. *Journal of Arid Environments*, 16, 165–75.

Bosch, O.J.H. & Gauch, H.G. (1991). The use of degradation gradients for the assessment and ecological interpretation of range condition. *Journal of the Grassland Society of southern Africa*, 8, 138–46.

Bosch, O.J.H. & Janse van Rensburg, F.P. (1987). Ecological status of species on grazing gradients on the shallow soils of the western grassland biome in South Africa. *Journal of the Grassland Society of southern Africa*, 4, 143–7.

Bosch, O.J.H., Janse van Rensburg, F.P. & Truter, S. Du T. (1987). Identification and selection of benchmark sites on litholitic soils of the western grassland biome of South Africa. *Journal of the Grassland Society of southern Africa*, 4, 59–62.

Bosch, O.J.H., Jordaan, E.E. & Bredenkamp, G.J. (1986). *A reconnaissance survey of the vegetation types and grazing capacity of the Thaba'nchu area, Bophuthatswana*, unpublished manuscript, Potchefstroom: Department of Botany, University of Potchefstroom.

Botha, C.E.J. & Russell, S. (1988). Comparison of photosynthetic temperature and light optima in selected C_3 and C_4 grasses from the E. Cape/Ciskei region. *South African Journal of Botany*, **54**, 133–6.

Bredenkamp, G.J. & Bezuidenhout, H. (1990). The vegetation of the Faan Meintjes Nature Reserve, Transvaal. *South African Journal of Botany*, **56**, 54–64.

Bredenkamp, G.J., Bezuidenhout, H., Naude, C. & Joubert, H. (1994). The vegetation of the Boskop Dam Nature Reserve. *Koedoe*, **37**, 19–33.

Bredenkamp, G.J., Joubert, A.F. & Bezuidenhout, H. (1989). A reconnaissance survey of the vegetation of the Potchefstroom-Fochville–Parys area. *South African Journal of Botany*, **55**, 199–206.

Bredenkamp, G.J. & Theron, G.K. (1978). A synecological account of the Suikerbosrand Nature Reserve I. The phytosociology of the Witwatersrand geological system. *Bothalia*, **12**, 513–29.

Bredenkamp, G.J. & Theron, G.K. (1980). A synecological account of the Suikerbosrand Nature Reserve. II. The phytosociology of the Ventersdorp Geological System. *Bothalia*, **13**, 199–216.

Breytenbach, P.J.J., Myburgh, W.J., Theron, G.K. & Bredenkamp, G.J. (1993a). The phytosociology of the Villiers-Grootvlei area, South Africa. 2. Plant communities of the Ba land type. *South African Journal of Botany*, **59**, 218–26.

Breytenbach, P.J.J., Myburgh, W.J., Theron, G.K. & Bredenkamp, G.J. (1993b). The phytosociology of the Villiers-Grootvlei area, South Africa. 3. Plant communities of the Ib land type. *South African Journal of Botany*, **59**, 227–34.

Breytenbach, P.J.J., Myburgh, W.J., Theron, G.K. & Bredenkamp, G.J. (1993c). The phytosociology of the Villiers-Grootvlei area, South Africa. 4. Plant communities of the Ea land type. *South African Journal of Botany*, **59**, 235–46.

Bridgens, A.B. (1968). Aspects of shoot apex morpho-genesis, development and behaviour in grasses, with reference to the utilization and management of natural grassland. *Technical Communication 67*. Pretoria: Department of Agricultural and Technical Services.

Brynard, A.M. (1953). *'n Fisiologies-Ekologiese Studie van Senecio retrorsus D.C.* MSc Thesis. Potchefstroom: University of Potchefstroom.

Burger, S.J., Grunow, J.O. & Rabie, J.W. (1975). Die reaksie van *Anthephora pubescens* Nees op verskillende intensiteite en frekwensies van ontblaring. *Proceedings of the Grassland Society of southern Africa*, **10**, 29–34.

Capon, M.H. & O'Connor, T.G. (1990). The predation of perennial grass seeds in Transvaal savanna grasslands. *South African Journal of Botany*, **56**, 11–15.

Carter, A.J. & O'Connor, T.G. (1991). A two-phase mosaic in a savanna grassland. *Journal of Vegetation Science*, **2**, 231–6.

Carter, A.J. & Robinson, E.R. (1993). Genetic structure of a population of the clonal grass *Setaria incrassata*. *Biological Journal of the Linnean Society*, **48**, 55–62.

Chesselet, P., Wolfson, M.M. & Ellis, R.P. (1992). A comparative histochemical study of plant polyphenols in southern African grasses. *Journal of the Grassland Society of southern Africa*, **9**, 119–25.

Coetsee, G. & Benadie, C.C. (1975). Seasonal grazing of *Themeda–Cymbopogon* veld: grazing of summer rested veld during autumn, winter and early summer. 1968-1974. Final Report, H-Ps. 60/2. Pretoria: Department of Agricultural and Technical Services.

Coetsee, G. & van Rensburg, P.H.J.J. (1975a). Summer veld grazing systems. Final Report, H-Ps. 58. Pretoria: Department of Agricultural and Technical Services.

Coetsee, G. & van Rensburg, P.H.J.J. (1975b). Year long rotational grazing 1948-1969. Final Report, H-Ps. 57. Pretoria: Department of Agricultural and Technical Services.

Coetsee, G. & van Rensburg, P.H.J.J. (1975c). Veld fertilizer experiment: cutting of hay and winter grazing plus hay. Final Report, H-Ps. 59. Pretoria: Department of Agricultural and Technical Services.

Coetzee, B.J. (1974). A phytosociological classification of the Jack Scott Nature Reserve. *Bothalia*, **11**, 329–47.

Coetzee, B.J. (1975). A phytosociological classification of the Rustenburg Nature Reserve. *Bothalia*, **11**, 561–80.

Coetzee, J.A., Page, M.I. & Meredith D. (1946). Root studies in highveld grassland communities. *South African Journal of Science*, **42**, 105–18.

Coetzee, J.P. (1993). *Phytosociology of the Ba and Ib Land Types in the Pretoria-Witbank-Heidelberg Area*. MSc Thesis. Pretoria: University of Pretoria.

Coetzee, J.P., Bredenkamp, G.J. & van Rooyen, N. (1994). An overview of the physical environment and vegetation units of the Ba and Ib land types of the Pretoria-Witbank-Heidelberg area. *South African Journal of Botany*, **60**, 49–61.

Cohen, C. (1937). Some aspects of the reproduction of *Stoebe vulgaris*, and certain local grass and weeds seeds. *South African Journal of Science*, **33**, 351–4.

Comins, D.M. (1962). The vegetation of the districts of East London and King William's Town, Cape Province. *Memoirs of the Botanical Survey of South Africa*, **33**, 1–184.

Connell, J.H. & Slatyer, R.O. (1977). Mechanisms of succession in natural communities and their role in community stability and organization. *American Naturalist*, **111**, 1119–44.

Cook, L. (1939). A contribution to our information on grass burning. *South African Journal of Science*, **36**, 270–82.

Cowling, R.M., Gibbs Russell, G.E., Hoffman, M.T. & Hilton-Taylor, C. (1991). Patterns of plant species diversity in southern Africa. In *Biotic Diversity in Southern Africa. Concepts and Conservation*, ed. B.J. Huntley, pp. 19–50. Cape Town: Oxford University Press.

ines, T. (1976). *Grazing Sequence Pattern and Species Selection by Cattle in the Dohne Sourveld.* MSc Thesis. Grahamstown: Rhodes University.

nckwerts, J.E. (1984). *Towards Improved Livestock Production of Sweet Grassveld.* PhD Thesis. Pietermaritzburg: University of Natal.

nckwerts, J.E. (1987). Growth analysis of *Themeda triandra* and *Sporobolus fimbriatus* tillers in semi-arid grassveld. *Journal of the Grassland Society of southern Africa,* 4, 7–12.

nckwerts, J.E. (1988). The effect of leaf age on the photosynthetic rate of *Themeda triandra* Forsk. *Journal of the Grassland Society of southern Africa,* 5, 227–8.

nckwerts, J.E. (1993). Reserve carbon and photosynthesis: their role in regrowth of *Themeda triandra*, a widely distributed subtropical graminaceous species. *Functional Ecology,* 7, 634–41.

nckwerts, J.E. & Aucamp, A.J. (1985). The rate of leaf emergence and decay as criteria for optimising the grazing rotation in semi-arid grassveld. *Journal of the Grassland Society of southern Africa,* 2, 28–34.

nckwerts, J.E., Aucamp, A.J. & Du Toit, L.P. (1984). Ontogeny of *Themeda triandra* tillers in the False Thornveld of the eastern Cape. *Journal of the Grassland Society of southern Africa,* 1, 9–14.

nckwerts, J.E., Aucamp, A.J. & Du Toit, L.P. (1986). Ontogeny of *Sporobolus fimbriatus* tillers in the False Thornveld of the eastern Cape. *Journal of the Grassland Society of southern Africa,* 3, 96–102.

nckwerts, J.E. & Barnard, H.J. (1981). Short-term effects of defoliation on herbage yield at three veld condition sites in the False Thornveld of the Eastern Province. *Proceedings of the Grassland Society of southern Africa,* 16, 79–84.

nckwerts, J.E. & Gordon, A.J. (1990). Partitioning, storage and remobilization of ^{14}C assimilated by *Themeda triandra* Forsk. *Journal of the Grassland Society of southern Africa,* 7, 97–105.

Danckwerts, J.E. & Nel, L.O. (1989). The effect of frequency of defoliation on *Themeda triandra* in the False Thornveld of the eastern Cape. *Journal of the Grassland Society of southern Africa,* 6, 32–6.

Danckwerts, J.E. & Stuart-Hill, G.C. (1987). Adaptation of a decreaser and an increaser grass species to defoliation in semi-arid grassveld. *Journal of the Grassland Society of southern Africa,* 4, 68–73.

Danckwerts, J.E. & Stuart-Hill, G.C. (1988). The effect of severe drought and management after drought on the mortality and recovery of semi-arid grassveld. *Journal of the Grassland Society of southern Africa,* 5, 218–22.

Davidson, R.L. (1941). A note on anthesis in some common grasses near Johannesburg, and the relation of anthesis to collection of pollen for medical purposes. *Journal of South African Botany,* 7, 145–52.

Davidson, R.L. (1962). The influence of edaphic factors on the species composition of early stages of the subsere. *Journal of Ecology,* 50, 401–10.

Davidson, R.L. (1964). An experimental study of succession in the Transvaal Highveld. In *Ecological Studies in Southern Africa,* ed. D.N.S. Davis, pp. 113–25. The Hague: Junk.

Deall, G.B., Theron, G.K. & Westfall, R.H. (1989). The vegetation ecology of the Eastern Transvaal Escarpment in the Sabie area. 2. Floristic classification. *Bothalia,* 19, 53–67.

De Jager, J.M., Opperman, D.P.J. & Booysen, J. (1980). Produksie van natuurlike grasveld in die sentrale Oranje-Vrystaat in verhouding tot klimaat. *Proceedings of the Grassland Society of southern Africa,* 15, 65–8.

Downing, B.H. & Groves, R.H. (1985). Growth and development of four *Themeda triandra* populations from southern Africa in response to temperature. *South African Journal of Botany,* 51, 350–4.

Downing, D.H. & Marshall, D.J. (1983). Burning and grazing of a *Themeda* grassland: estimates of phytomass and root element concentrations. *Proceedings of the Grassland Society of southern Africa,* 18, 155–8.

Downing, B.H., Robinson, E.R., Trollope, W.S.W. & Morris, J.W. (1978). Influence of macchia eradication techniques on botanical composition of grasses in the Döhne Sourveld of the Amatole mountains. *Proceedings of the Grassland Society of southern Africa,* 13, 111–15.

Drewes, R.H. (1979). *The Response of Veld to Different Winter Removal Treatments.* MSc Thesis. Pietermaritzburg: University of Natal.

Drewes, R.H. & Tainton, N.M. (1981). The effect of different winter and early spring removal treatments on *Themeda triandra* in the tall grassveld of Natal. *Proceedings of the Grassland Society of southern Africa,* 16, 139–43.

du Plessis, M.C.F. & Mostert, J.W.C. (1965). Afloop en grondverliese by die landbouavorsingsinstituut Glen. *South African Journal of Agricultural Research,* 8, 1051–60.

du Preez, C.C. & Snyman, H.A. (1993). Organic matter content of a soil in a semi-arid climate with three long-standing veld conditions. *African Journal of Range & Forage Science,* 10, 108–10.

du Preez, P.J. (1991). *A Syntaxonomical and Synecological Study of the Vegetation of the South-eastern Orange Free State and Related Areas with Special Reference to Korannaberg.* PhD Thesis. Bloemfontein: University of the Orange Free State.

du Preez, P.J. (1992). The classification of the vegetation of Korannaberg, eastern Orange Free State, South Africa. I. Afromontane fynbos communities. *South African Journal of Botany,* 58, 165–72.

du Preez, P.J. & Bredenkamp, G.J. (1991). Vegetation classes of the southern and eastern Orange Free State and the highlands of Lesotho. *Navorsinge van die nasionale Museum Bloemfontein,* 7, 478–526.

du Preez, P.J. & Venter, J.H.T. (1990). The phytosociology of woody vegetation in the southern part of the Vredefort Dome area. *South African Journal of Botany,* 56, 637–44.

du Preez, P.J. & Venter, J.H.T. (1992). The classification of the vegetation of Korannaberg, eastern Orange Free State, South Africa, II: Grassland communities. *South African Journal of Botany*, 58, 173–81.

du Toit, E.W., Wolfson, M.M. & Ellis, R.P. (1991). The presence of condensed tannin in the leaves of *Eulalia villosa*. *Journal of the Grassland Society of southern Africa*, 8, 74–6.

du Toit, P.F. (1966). 'n Outekologiese Studie van die Acacia karroo-saailing. MSc Thesis. Pretoria: University of Pretoria.

du Toit, P.F. (1972). *Acacia karroo* intrusion: the effect of burning and sparing. *Proceedings of the Grassland Society of southern Africa*, 7, 23–7.

du Toit, P.F. & Aucamp, A.J. (1985). Effect of continuous grazing in the Döhne Sourveld on species composition and basal cover. *Journal of the Grassland Society of southern Africa*, 2, 41–5.

du Toit, P.F., Ingpen, R.A. & Nel, L.O. (1974). Effect of winter defoliation of Dohne sourveld on spring growth. *Agroplantae*, 6, 67–72.

du Toit, P.J., Louw, J.G. & Malan, A.E. (1935a). A study of the mineral content and feeding value of natural pastures in the Union of South Africa. IV. The influence of season and frequency of cutting on the yield, persistency, and chemical composition of grass species. *Onderstepoort Journal of Veterinary Science and Animal Industry*, 5, 215–70.

du Toit, P.J., Malan, A.I., Louw, J.G., Holzapfel, C.R. & Roets, G.C.S. (1932). A study of the mineral content and feeding value of natural pastures in the Union of South Africa – (First Report). *18th Report of the Director of Veterinary Service and Animal Industry, Union of South Africa*, August, 1932, pp. 525–77.

du Toit, P.J., Malan, A.I., Louw, J.G., Holzapfel, C.R. & Roets, G. (1935b). A study of the mineral content and feeding value of natural pastures in the Union of South Africa – (Third Report). *Onderstepoort Journal of Veterinary Science and Animal Industry*, 5, 201–14.

Dyer, R.A. (1937). The vegetation of the divisions of Albany and Bathurst. *Memoirs of the Botanical Survey of South Africa*, 17, 1–138.

Eckhardt, H.C., van Rooyen, N. & Bredenkamp, G.J. (1993a). Wetland plant communities of the Vrede–Memel–Warden area, north-eastern Orange Free State. *Navorsinge van die nasionale Museum Bloemfontein*, 9, 245–62.

Eckhardt, H.C., van Rooyen, N. & Bredenkamp, G.J. (1993b). An overview of the vegetation of the Vrede–Memel–Warden area, north-eastern Orange Free State. *South African Journal of Botany*, 59, 391–400.

Eckhardt, H.C., van Rooyen, N. & Bredenkamp, G.J. (1993c). The phytosociology of thicket and woodland vegetation of the north-eastern Orange Free State. *South African Journal of Botany*, 59, 401–9.

Eckhardt, H.C., van Rooyen, N. & Bredenkamp, G.J. (1995). The grassland communities of the slopes and plains of the north-eastern Orange Free State. *Phytocoenologia*, 25, 1–21.

Edwards, D. (1967). A plant ecological survey of the Tugela basin. *Memoirs of the Botanical Survey of South Africa*, 35, 1–285.

Edwards, P.J. (1961). *Studies on Veld Burning and Mowing in the Tall Grassveld of Natal*. MSc Thesis. Pietermaritzburg: University of Natal.

Edwards, P.J. (1968). The long-term effects of burning and mowing on the basal cover of two Veld Types in Natal. *South African Journal of Agricultural Science*, 11, 131–40.

Edwards, P.J. (1975). The effect of selective defoliation and fertilization on *Cymbopogon–Themeda* veld. *Proceedings of the Grassland Society of southern Africa*, 10, 141–6.

Edwards, P.J., Jones, R.I. & Tainton, N.M. (1979). *Aristida junciformis* Trin. et Rupr.: a weed of the veld. *Proceedings of the 3rd National Weeds Conference of South Africa*, Pretoria: pp. 25–32.

Edwards, P.J. & Nel, S.P.J. (1973). Short-term effects of fertilizer and stocking rates on the Bankenveld: 1. Vegetational changes. *Proceedings of the Grassland Society of southern Africa*, 8, 83–8.

Ellery, W.N. (1992). *Classification of Vegetation of the South African Grassland Biome*. PhD Thesis. Johannesburg: University of the Witwatersrand.

Ellery, W.N. & Mentis, M.T. (1992). How old are South Africa's grasslands? In *Forest–Savanna Boundaries*, ed. P.A. Furley, J. Proctor & J.A. Ratter, pp. 283–92. London: Chapman & Hall.

Ellery, W.N., Mentis, M.T. & Scholes, R.J. (1992). Modelling the location of woody-grassland boundaries. In *Forest–Savanna Boundaries*, ed. P.A. Furley, J. Proctor & J.A. Ratter, pp. 659–82. London: Chapman & Hall.

Ellery, W.N., Scholes, R.J. & Mentis, M.T. (1991). An initial approach to predicting the sensitivity of South African grassland biome to climate change. *South African Journal of Science*, 87, 499–503.

Ellery, W.N. & Walker, B.H. (1986). The distribution and dynamics of *Elephantorrhiza elephantina* on the farm Maccauvlei. *South African Journal of Botany*, 52, 100–4.

Ellis, R.P. (1974). The significance of the occurrence of both Kranz and non-Kranz leaf anatomy in the grass species *Alloteropsis semialata*. *South African Journal of Science*, 70, 169–73.

Ellis, R.P. (1990). Tannin-like substances in grass leaves. *Memoirs of the Botanical Survey of South Africa*, 59, 1–80.

Erasmus, D.J. & van Staden, J. (1984). Germination of *Setaria chevalieri* caryopses. *Weed Research*, 23, 225–9.

Ernst, W.H.O., Kuiters, A.T. & Tolsma, D.J. (1991). Dormancy of annual and perennial grasses from a savanna of south-eastern Botswana. *Acta Oecologica*, 12, 727–39.

Everson, C.S. (1985). *Ecological Effects of Fire in the Montane Grasslands of Natal*. PhD Thesis. Pietermaritzburg: University of Natal.

Everson, C.S. & Breen, C.M. (1983). Water stress as a factor influencing the distribution of the ericoid shrub *Philippia evansii* in the Natal Drakensberg mountains, South Africa. *South African Journal of Botany*, 2, 290–6.

Everson, C.S. & Everson, T.M. (1987). Factors affecting the timing of grassland regrowth after fire in the montane grasslands of Natal. *South African Forestry Journal*, 142, 47–52.

Everson, C.S., Everson, T.M. & Tainton, N.M. (1985a). The dynamics of *Themeda triandra* tillers in relation to burning in the Natal Drakensberg. *Journal of the Grassland Society of southern Africa*, 2, 18–25.

Everson, C.S., Everson, T.M. & Tainton, N.M. (1988). Effects of intensity and height of shading on the tiller initiation of six grass species from the highland sourveld of Natal. *South African Journal of Botany*, 54, 315–18.

Everson, C.S., George, W.J. & Schulze, R.E. (1989). Fire regime effects on canopy cover and sediment yield in the montane grasslands of Natal. *South African Journal of Science*, 85, 113–16.

Everson, T.M., Smith F.R. & Everson C.S. (1985b). Characteristics of fire behaviour in the montane grasslands of Natal. *Journal of the Grassland Society of southern Africa*, 2, 13–21.

Everson, C.S. & Tainton, N.M. (1984). The effect of thirty years of burning on the highland sourveld of Natal. *Journal of the Grassland Society of southern Africa*, 1, 15–20.

Everson, T.M., van Wilgen, B.W. & Everson, C.S. (1988). Adaptation of a model for rating fire danger in the Natal Drakensberg. *South African Journal of Science*, 84, 44–9.

Feely, J.M. (1980). Did Iron Age man have a role in the history of Zululand's wilderness landscapes? *South African Journal of Science*, 76, 150–2.

Feely, J.M. (1987). The early farmers of Transkei, southern Africa, before A.D. 1870. *Cambridge Monographs in African Archaeology 24 BAR International Series 378.* Oxford: BAR Series.

Foran, B.D., Tainton, N.M. & Booysen, P. de V. (1978). The development of a method for assessing veld condition in three grassveld types in Natal. *Proceedings of the Grassland Society of southern Africa*, 13, 27–33.

Gourie, J.H. & Roberts, B.R. (1976). A comparative study of three Veld Types of the northern Cape: species evaluation and yield. *Proceedings of the Grassland Society of southern Africa*, 11, 79–85.

Gourie, J.H. & Roberts, B.R. (1977). Seasonal dry matter production and digestiblity of *Themeda triandra* and *Eragrostis lehmanniana. Agroplantae*, 9, 129–33.

Frean, M., Barrett, D. & Cresswell, C.F. (1980). Variability in leaf surface features and water efficiency utilisation in C₃ and C₄ forms of *Alloteropsis semialata* (R. Br.) Hitchc. *Proceedings of the Grassland Society of southern Africa*, 15, 99–103.

Frean, M. & Cresswell, C.F. (1981). An ontogenetic study with special reference to leaf development in C₃ and C₄ forms of *Alloteropsis semialata. Proceedings of the Grassland Society of southern Africa*, 16, 155–60.

Friedel, M.H. (1987). A preliminary investigation of woody plant increase in the western Transvaal and implications for veld assessment. *Journal of the Grassland Society of southern Africa*, 4, 25–30.

Fuls, E.R. (1991). The effect of nutrient enriched sediment deposits on the vegetational traits of a patch-grazed semi-arid grassland. *Vegetatio*, 96, 177–83.

Fuls, E.R. (1992). Ecosystem modification created by patch-overgrazing in semi-arid grasslands. *Journal of Arid Environments*, 23, 59–69.

Fuls, E.R. (1993). *Vegetation Ecology of the Northern Orange Free State.* PhD Thesis. Pretoria: University of Pretoria.

Fuls, E.R. & Bosch, O.J.H. (1991). The influence of below-average rainfall on the vegetational traits of a patch-grazed semi-arid grassland. *Journal of Arid Environments*, 21, 13–20.

Fuls, E.R., Bredenkamp, G.J. & van Rooyen, N. (1992). The plant communities of the undulating grassland of the Vredefort–Kroonstad–Lindley–Heilbron area, northern Orange Free State. *South African Journal of Botany*, 58, 224–30.

Fuls, E.R., Bredenkamp, G.J. & van Rooyen, N. (1993a). Grassland communities of the midslopes in the northern Orange Free State. *South African Journal of Botany*, 59, 478–84.

Fuls, E.R., Bredenkamp, G.J. & van Rooyen, N. (1993b). Grassland communities of the footslopes in the northern Orange Free State. *South African Journal of Botany*, 59, 485–90.

Fuls, E.R., Bredenkamp, G.J., van Rooyen, N. & Theron, G.K. (1993c). The physical environment and major plant communities of the Heilbron–Lindley–Warden–Villiers area, northern Orange Free State. *South African Journal of Botany*, 59, 345–59.

Gaff, D.F. (1971). Desiccation-tolerant flowering plants in southern Africa. *Science*, 174, 1033–4.

Gaff, D.F. & Ellis, R.P. (1974). Southern African grasses with foliage that revives after dehydration. *Bothalia*, 11, 305–8.

Gibbs Russell, G.E. (1987). Preliminary floristic analysis of the major biomes of southern Africa. *Bothalia*, 17, 213–27.

Gibbs Russell, G.E. & Spies, J.J. (1988). Variation in important pasture grasses: I. Morphological and geographical variation. *Journal of the Grassland Society of southern Africa*, 5, 15–21.

Gillard, P. (1963). The effect of grazing intensity on cattle gains and on the vegetation of the Bankenveld. *South African Journal of Science*, 59, 64–5.

Gillard, P. (1965). Responses to grazing intensity on the Transvaal highveld. *Experimental Agriculture*, 2, 217–24.

Gilliland, H.B. (1955). On the phenology of the veld around Johannesburg. I. Monocotyledons. *Journal of South African Botany*, 21, 77–82.

Gillman, H. (1934). *The Indication Significance of 'slangbos', Stoebe cinerea Thunb.* MSc Thesis. Johannesburg: University of the Witwatersrand.

Glover, P.E. (1937). A contribution to the ecology of the highveld flora. *South African Journal of Science*, 34, 224–59.

Glover, P.E. & van Rensburg, H. (1938). A contribution to the ecology of the highveld grassland at Frankenwald, in relation to grazing and burning. *South African Journal of Science*, 35, 274–9.

Granger, J.E. & Schulze, R.E. (1977). Incoming solar radiation patterns and vegetation response: examples from the Natal Drakensberg. *Vegetatio*, 35, 47–54.

Griffioen, C. & O'Connor, T.G. (1991). The influence of termite mounds on the soils and herbaceous composition of a savanna grassland. *South African Journal of Ecology*, 1, 18–26.

Grossman, D. & Cresswell, C.F. (1974). The influence of nitrate and ammonia nitrogen on the photosynthetic and photorespiratory activity of selected highveld grasses exhibiting C-4 photosynthesis. *Proceedings of the Grassland Society of southern Africa*, 9, 89–93.

Grunow, J.O., Pienaar A.J. & Breytenbach C. (1970). Long-term nitrogen application of veld in South Africa. *Proceedings of the Grassland Society of southern Africa*, 5, 75–90.

Grunow, J.O., Rabie, J.W. & Grattorola, L. (1977). Standing crop dry matter accumulation and quality patterns of certain subtropical species. *Proceedings of the Grassland Society of southern Africa*, 12, 37–44.

Hall, T.D. (1932). Intensive grazing on veld – II. *South African Journal of Science*, 29, 389–413.

Hall, T.D. & Meredith D. (1937). Intensive grazing on veld – IV: the effect of rotational grazing and fertilising on veld over a six year period. *South African Journal of Science*, 33, 404–30.

Hall, T.D. & Meredith, D. (1945). Residual effects of fertilizers on veld hay. *South African Journal of Science*, 41, 194–203.

Hall, T.D., Meredith, D. & Altona, R.E. (1949). A comparison of four nitrogenous fertilisers on veld. *South African Journal of Science*, 45, 100–5.

Hall, T.D., Meredith, D. & Murray, S.M. (1937). The productivity of fertilised natural highveld pastures. *South African Journal of Science*, 34, 275–85.

Hall, T.D., Meredith, D. & Murray, S.M. (1941). Fertilising natural veld and its effect on sward, chemical composition, carrying capacity and beef production. *South African Journal of Science*, 37, 111–29.

Hatch, G.P. (1991). *Species and Area Selection Patterns in the Southern Tall Grassveld of Natal*. MSc Thesis. Pietermaritzburg: University of Natal.

Hattingh, E.R. (1953). Observations on the ecology of *Stoebe vulgaris* Levyns. *Empire Journal of Experimental Agriculture*, 21, 309–13.

Haylett, D.G. (1960). Run-off and soil erosion studies at Pretoria. *South African Journal of Agricultural Science*, 3, 379–92.

Henderson, L. (1989). Invasive alien woody plants of Natal and the north-eastern Orange Free State. *Bothalia*, 19, 237–61.

Henderson, L. & Musil, K.J. (1984). Exotic woody plant invaders of the Transvaal. *Bothalia*, 15, 298–313.

Henrici, M. (1930a). Mineral and feeding stuff analyses of grasses of the eastern Transvaal highveld. *16th Report of the Director of Veterinary Services and Animal Industry, Union of South Africa*, August, 1930, pp. 421–33.

Henrici, M. (1930b). The phosphorus content of grasses in the eastern Transvaal. *16th Report of the Director of Veterinary Services and Animal Industry, Union of South Africa*, August, 1930, pp. 435–99.

Henrici, M. (1943). Transpiration of grasses in the sour mountain grassveld of the Drakensberg in comparison with the water loss of indigenous forests. *South African Journal of Science*, 34, 155–63.

Herbst, S.N. & Roberts, B.R. (1974). The alpine vegetation of the Lesotho Drakensberg: a study in quantitative floristics at Oxbow. *Journal of South African Botany*, 40, 257–67.

Hildyard, P. (1967). *A Study of Shoot Growth and Development in Senecio retrorsus D.C.* PhD Thesis. Pietermaritzburg: University of Natal.

Hildyard, P. & Booysen, P. de V. (1971). The influence of sward competition on shoot growth and development in *Senecio retrorsus*, D.C. *Proceedings of the Grassland Society of southern Africa*, 6, 39–43.

Hoffman, M.T. & Cowling, R.M. (1990). Vegetation change in the semi-arid eastern Karoo over the last 200 years: an expanding Karoo – fact or fiction? *South African Journal of Science*, 86, 286–94.

Jacot-Guillarmod, A. (1971). Flora of Lesotho. *Flora et vegetation Mundi*, 3, 1–474.

Janse van Rensburg, F.P.J. & Bosch, O.J.H. (1990). Influence of habitat differences on the ecological grouping of grass species on a grazing gradient. *Journal of the Grassland Society of southern Africa*, 7, 11–15.

Janse van Rensburg, F.P., Drewes, R.H., Cilliers, J.W. & Beckerling, A.C. (1990). Die invloed van bemesting op die spesiesamestelling en basale bedekking van beweide veld in die Potchefstroom-omgewing. *Journal of the Grassland Society of southern Africa*, 7, 232–7.

Jarman, N.G. (1977). *An Evaluation of Different Types of Aerial Photographs for Surveying and Mapping Grass and Dwarf Shrub Vegetation*. MSc Thesis. Johannesburg: University of the Witwatersrand.

Jones, R.I. (1967). Comparative effects of differential defoliation of grass plants in pure and mixed stands of two species. *South African Journal of Agricultural Science*, 10, 429–44.

Jones, R.I. & Cross, G.W. (1971). Effect of selective defoliation on species in mixed swards. *Proceedings of the Grassland Society of southern Africa*, 6, 123–8.

Jones, R.M. (1964). *A Further Study of Secondary Succession on the Highveld*. MSc Thesis. Johannesburg: University of the Witwatersrand.

Jones, R.M. (1968). Seed production of species in the highveld secondary succession. *Journal of Ecology*, 56, 661–6.

Jong, K. & Roux, E.R. (1955). A further investigation of the nitrogen sensitivity of veld grasses. *South African Journal of Science*, 52, 27–8.

Kay, C., Bredenkamp, G.J. & Theron, G.K. (1993). The plant communities in the Golden Gate Highlands National Park in the north-eastern Orange Free State. *South African Journal of Botany*, 59, 442–9.

Killick, D.J.B. (1958). An account of the plant ecology of the Table Mountain area of Pietermaritzburg, Natal. *Memoirs of the Botanical Survey of South Africa*, 32, 1–133.

Killick, D.J.B. (1963). An account of the plant ecology of the Cathedral Peak area of the KwaZulu-Natal Drakensberg. *Memoirs of the Botanical Survey of South Africa*, 34, 1–178.

Killick, D.J.B. (1978). The Afro-Alpine Region. In *Biogeography and Ecology of Southern Africa*, ed. M.J.A. Werger, pp. 515–60. The Hague: Junk.

Kirkman, K.P. (1988). *Factors Affecting the Seasonal Variation of Veld Quality in South Africa*. MSc Thesis. Pietermaritzburg: University of Natal.

ooij, M.S., Bredenkamp, G.J., Scheepers, J.C. & Theron, G.K. (1992). The vegetation of the Kroonstad area: a description of the grassland communities. *South African Journal of Botany*, **58**, 155–64.

ooij, M.S., Bredenkamp, G.J. & Theron, G.K. (1990a). The vegetation of the north-western Orange Free state, South Africa 2. The D land type. *Bothalia*, **20**, 241–8.

ooij, M.S., Bredenkamp, G.J. & Theron, G.K. (1990b). Classification of the vegetation of the B land type in the north-western Orange Free State. *South African Journal of Botany*, **56**, 309–18.

ooij, M.S., Bredenkamp, G.J. & Theron, G.K. (1990c). The vegetation of the deep sandy soils of the A land type in the north-western Orange Free state, South Africa. *Botanica Bulletin Academia Sinica*, **31**, 235–43.

ruger, J.A. & Smit, I.B.J. (1973). Herwinning van oulande in die Oos-Vrystaat deur die insaai van *Digitaria smutsii*, *Eragrostis curvula* en *Themeda triandra*. *Agroplantae*, **5**, 101–6.

rupko, I. & Davidson, R.L. (1961). An experimental study of *Stoebe vulgaris* in relation to grazing and burning. *Empire Journal of Experimental Agriculture*, **29**, 176–80.

ker, M.C., Hewitt, P.H., Nel, A. & Hunt, R.P. (1982a). Effects of the termite *Trinervitermes trinervoides* Sjöstedt on the pH, electrical conductivities, cation exchange capacities and extractable base contents of soils. *Fort Hare Papers*, **7**, 275–86.

ker, M.C., Hewitt, P.H., Nel, A. & Hunt, R.P. (1982b). Effects of the termite *Trinervitermes trinervoides* Sjöstedt on the organic carbon and nitrogen contents and particle size distribution of soils. *Revue d'ecologie et de biologie du sol*, **19**, 27–39.

nd Type Survey (1984). Land types of the maps 2626 West Rand, 2726 Kroonstad. *Memoirs of the Agricultural Natural Resources of South Africa*, **4**, 1–441.

brun, J. (1947). La vegetation de la plaine alluviale au sud du lac Eduard. *Explor. Parc. National Albert*, **1**, 1–800.

Lecatsas, G. (1961). *Some Aspects of the Ecology of Stoebe vulgaris*. MSc Thesis. Johannesburg: University of the Witwatersrand.

Lecatsas, G. (1962). Solar radiation as a factor in the establishment of *Stoebe vulgaris*. *South African Journal of Science*, **58**, 304–6.

Leigh, J.H. (1960). Temperature, moisture and day-length effects in love-grass, *Eragrostis curvula* (Schrad.) Nees. *South African Journal of Science*, **56**, 268.

Le Roux, C.J.G. (1989). Die invloed van brand en maai op onbeweide langgrasveld in Natal. *Journal of the Grassland Society of southern Africa*, **6**, 59–64.

Lloyd, J.W. (1989a). Discriminant analysis and ordination of vegetation and soils on the Vaalputs radioactive waste disposal site, Bushmanland, South Africa. *South African Journal of Botany*, **55**, 127–36.

Lloyd, J.W. (1989b). Phytosociology of the Vaalputs radioactive waste disposal site, Bushmanland, South Africa. *South African Journal of Botany*, **55**, 372–82.

Louw, J.G. (1938). The influence of frequency of cutting on the yield, chemical composition, digestibility and nutritive value of some grass species. *Onderstepoort Journal of Veterinary Science and Animal Industry*, **11**, 163–244.

Louw, W.J. (1951). An ecological account of the vegetation of the Potchefstroom area. *Memoirs of the Botanical Survey of South Africa*, **24**, 1–105.

Macdonald, I.A.W. (1991). Man's role in changing the face of southern Africa. In *Biotic Diversity in Southern Africa. Concepts and Conservation*, ed. B.J. Huntley, pp. 51–77. Cape Town: Oxford University Press.

Malan, P.W. (1992). *Plantsosiologie in die Bloemfontein-wes Distrik*. MSc Thesis. Bloemfontein: University of the Orange Free State.

Manry, D.E. & Knight, R.S. (1986). Lightning density and burning frequency in South African vegetation. *Vegetatio*, **66**, 67–76.

Martin, A.R.H. (1966). The plant ecology of the Grahamstown Nature Reserve: II. Some effects of burning. *Journal of South African Botany*, **32**, 1–39.

Martin, N.L. (1986). *Classification and Ordination of the Plant Communities of Lesotho*. PhD Thesis. Boise: University of Idaho.

Matthews, W.S., Bredenkamp, G.J. & van Rooyen, N. (1991). The grassland-associated vegetation of the Black Reef Quartzite and associated large rocky outcrops in the north-eastern mountain sourveld of the Transvaal escarpment. *South African Journal of Botany*, **57**, 143–50.

Matthews, W.S., Bredenkamp, G.J. & van Rooyen, N. (1992a). The phytosociology of the high altitude hygrophilous vegetation regions of the north-eastern mountain sourveld in the Transvaal, South Africa. *Phytocoenologia*, **20**, 559–74.

Matthews, W.S., Bredenkamp, G.J. & van Rooyen, N. (1992b). The vegetation of the dry dolomitic regions of the north-eastern mountain sourveld in the Transvaal, South Africa. *Phytocoenologia*, **20**, 467–88.

Matthews, W.S., Bredenkamp, G.J. & van Rooyen, N. (1994). The phytosociology and syntaxonomy of relatively low altitude areas in the north-eastern mountain sourveld in the Eastern Transvaal Escarpment region. *Koedoe*, **37**. 33–9.

Matthews, W.S., van Wyk, A.E. & Bredenkamp, G.J. (1993). Endemic flora of the north-eastern Transvaal escarpment, South Africa. *Biological Conservation*, **63**, 83–94.

McKenzie, B. (1982). Resilience and stability of the grasslands of Transkei. *Proceedings of the Grassland Society of southern Africa*, **17**, 21–4.

Meadows, M.E. & Linder, H.P. (1993). A palaeoecological perspective on the origin of Afromontane grasslands. *Journal of Biogeography*, **20**, 345–55.

Meadows, M.E. & Meadows, K.F. (1988). Late, Quaternary vegetation history of the Winterberg mountains, eastern Cape, South Africa. *South African Journal of Science*, **84**, 253–9.

Mentis, M.T. (1982). *A Simulation of the Grazing of Sour Grassveld*. PhD Thesis. Pietermaritzburg: University of Natal.

Mentis, M.T. & Huntley, B.J. (1982). *A Description of the Grassland Biome Project. South African National Scientific Programmes Report 62*. Pretoria: CSIR.

Mes, M.G. (1952). The influence of some climatic factors on the growth and seed production of grasses. In *Veld Gold. A South African Book of Grassland Farming*, pp. 35–51. Johannesburg: National Veld Trust.

Mes, M.G. (1958). The influence of veld burning or mowing on the water, nitrogen and ash content of grasses. *South African Journal of Science*, 54, 83–6.

Mes, M.G. & Aymer-Ainslie, K.M. (1935). Studies on the water relations of grasses. 1: *Themeda triandra*, Forsk. *South African Journal of Science*, 32, 280–304.

Mogg, A.O.D. (1955). A preliminary note on the woody plants of the Witwatersrand region in relation to the geology and physiography. *South African Journal of Science*, 51, 301–7.

Mokuku, C. (1991). *Classification of the Alpine Plant Communities of Mafika-Lisiu*. MSc Thesis. Harare: University of Zimbabwe.

Moll, E.J. (1967). A plant ecological reconnaissance of the upper Mgeni catchment. *Journal of South African Botany*, 34, 401–20.

Morris, C.D. (1994). *The Influence of Environment and Livestock Grazing on the Mountain Vegetation of Lesotho*. MSc Thesis. Pietermaritzburg: University of Natal.

Morris, C.D. & Tainton, N.M. (1993). The effect of defoliation and competition on the regrowth of *Themeda triandra* and *Aristida junciformis* subsp. *junciformis*. *African Journal of Range & Forage Science*, 10, 124–8.

Morris, C.D., Tainton, N.M. & Boleme, S. (1993). Classification of the eastern alpine vegetation of Lesotho. *African Journal of Range & Forage Science*, 10, 47–53.

Morris, C.D., Tainton, N.M. & Hardy, M.B. (1992). Plant species dynamics in the southern Tall Grassveld under grazing, resting and fire. *Journal of the Grassland Society of southern Africa*, 9, 90–5.

Morris, J.W. (1969). An ordination of the vegetation of Ntshongweni, Natal. *Bothalia*, 10, 89–120.

Morris, J.W. (1976). Automatic classification of the highveld grassland of Lichtenburg, south-western Transvaal. *Bothalia*, 12, 267–92.

Mostert, J.W.C. (1958). Studies of the vegetation of parts of the Bloemfontein and Brandfort districts. *Memoirs of the Botanical Survey of South Africa*, 31, 1–226.

Mufandaedza, O.T. (1976). Effects of frequency and height of cutting on some tropical grasses and legumes. I. *Hyparrhenia filipendula* (Hochst.) Stapf and *Heteropogon contortus* (L.) Beauv. ex Roem. & Schult. *Rhodesian Journal of Agricultural Research*, 14, 21–38.

Muller, D.B. (1986). *Plantekologie van die Willem Pretorius Wildtuin*. PhD Thesis. Bloemfontein: University of the Orange Free State.

Murray, S.M. & Glover, P. (1935). A preliminary study of the root development of certain South African highveld grasses. *Journal of South African Botany*, 6, 65–70.

Myburgh, W. (1993). *Die Fitososiologie van die Suurgrasveld in die Suidoos-Transvaalse Hoëveld*. MSc Thesis. Pretoria: University of Pretoria.

Nel, L.O. (1983). *Elionurus muticus* in the Dohne sourveld. *Proceedings of the Grassland Society of southern Africa*, 18, 84–8.

Noel, A.R.A. (1961). A preliminary account of the effect of grazing upon species of *Helichrysum* in the Amatole mountains. *Journal of South African Botany*, 27, 81–5.

Nursey, W.R.E. (1971). Starch deposits in *Themeda triandra* Forsk. *Proceedings of the Grassland Society of southern Africa*, 6, 157–60.

O'Connor, T.G. (1985). *A Synthesis of Field Experiments Concerning the Grass Layer in the Savanna Regions of Southern Africa. South African National Scientific Programmes Report 114*. Pretoria: CSIR.

O'Connor, T.G. (1991a). Influence of rainfall and grazing on the compositional change of the herbaceous layer of a sandveld savanna. *Journal of the Grassland Society of southern Africa*, 8, 103–9.

O'Connor, T.G. (1991b). Local extinction in perennial grasslands: a life-history approach. *American Naturalist*, 137, 753–73.

O'Connor, T.G. (1991c). Patch colonisation in a savanna grassland. *Journal of Vegetation Science*, 2, 245–54.

O'Connor, T.G. (1992). Patterns of plant selection by grazing cattle in two savanna grasslands: a plant's eye view. *Journal of the Grassland Society of southern Africa*, 9, 97–104.

O'Connor, T.G. (1993). The influence of rainfall and grazing on the demography of some African savanna grasses: a matrix modelling approach. *Journal of Applied Ecology*, 30, 119–32.

O'Connor, T.G. (1994). Composition and population responses of an African savanna grassland to rainfall and grazing. *Journal of Applied Ecology*, 31, 155–71.

O'Connor, T.G. & Pickett, G.A. (1992). The influence of grazing on seed production and seed banks of some African savanna grasslands. *Journal of Applied Ecology*, 29, 247–60.

O'Connor, T.G. & Roux, P.W. (1995). Vegetation changes (1949–1971) in a semi-arid, grassy dwarf shrubland in the Karoo, South Africa: influence of rainfall variability and grazing by sheep. *Journal of Applied Ecology*, 32, 612–26.

Oellerman, R.A. (1965). The nutritive value of *Themeda triandra*. III. Seasonal variations in forage quality as determined by different criteria. *South African Journal of Agricultural Science*, 8, 607–20.

Opperman, D.P.J., Human, J.J. & Roberts, B.R. (1976). Die invloed van ontblaring en vogstremming op die groeikragtigheid van *Themeda triandra* Forsk. onder gekontroleerde toestande. *Proceedings of the Grassland Society of southern Africa*, 12, 65–9.

Opperman, D.P.J., Human, J.J. & Viljoen, M.F. (1977). Evapotranspirasie – studies op *Themeda triandra* Forsk. onder veldtoestande. *Proceedings of the Grassland Society of southern Africa*, 12, 71–6.

Opperman, D.P.J. & Roberts, B.R. (1978). Die fenologiese ontwikkeling van *Themeda triandra, Elionurus argenteus* en *Heteropogon contortus* onder veldtoestande in die sentrale Oranje-Vrystaat. *Proceedings of the Grassland Society of southern Africa*, 13, 135–40.

perman, D.P.J., Roberts, B.R. & Nel, L.O. (1974). *Elionurus argenteus* Nees – a review. *Proceedings of the Grassland Society of southern Africa*, 9, 123–31.

perman, D.P.J., Roberts, B.R. & van Rensburg, W.L.J. (1969). The influence of defoliation on dry matter production and nutritive value of perennial veld grasses. *Agroplantae*, 1, 133–8.

perman, D.P.J., Roberts, B.R. & van Rensburg, W.L.J. (1970). Die invloed van ontblaring op die wortelgewigte, stoppelge-gewigte en reserwestatus van drie meerjarige veldgrasse. *Agroplantae*, 2, 51–60.

mer, A.R. (1989). The vegetation of the Karoo Nature Reserve, Cape Province. I. A phytosociological reconnaissance. *South African Journal of Botany*, 55, 215–30.

mer, A.R. (1991a). The potential vegetation of the upper Orange river, South Africa: concentration analysis and its application to rangeland assessment. *Coenoses*, 6, 131—8.

mer, A.R. (1991b). A syntaxonomic and synecological account of the vegetation of the eastern Cape midlands. *South African Journal of Botany*, 57, 76–94.

mer A.R. & Cowling, R.M. (1994). An investigation of topo-moisture gradients in the eastern Karoo, South Africa, and the identification of factors responsible for species turnover. *Journal of Arid Environments*, 26, 135–47.

illips, J. (1934). Succession, development, the climax, and the complex organism: an analysis of concepts. Part I. *Journal of Ecology*, 22, 554–71.

illips, J. (1935a). Succession, development, the climax, and the complex organism: an analysis of concepts. Part II. Development and climax. *Journal of Ecology*, 23, 210–46.

illips, J. (1935b). Succession, development, the climax, and the complex organism: an analysis of concepts. Part III. The complex organism: conclusions. *Journal of Ecology*, 23, 488–508.

tts, G. (1923). The plant succession in the Orange Free State, and the need for maintaining a covering of vegetation. *South African Journal of Science*, 20, 196–201.

Potts, G. & Tidmarsh, C.E. (1937). An ecological study of a piece of Karoo-like vegetation near Bloemfontein. *Journal of South African Botany*, 2, 51–92.

Preller, J.H. (1950). *Pasture and Pasture Crops*. Science Bulletin 280, Department of Agriculture. Pretoria: Government Printer.

Raal, P.A. (1986). The Transvaal threatened plants programme. *Fauna and Flora*, 44, 17–21.

Rabie, J.W. (1964). Developmental studies on veld grasses. *South African Journal of Agricultural Science*, 7, 583–8.

Rethman, N.F.G. (1964). *Studies on Grass Growth in a Tall Grassveld Sward*. MSc Thesis. Pietermaritzburg: University of Natal.

Rethman, N.F.G. (1971). Elevation of shoot-apices of two ecotypes of *Themeda triandra* on the Transvaal highveld. *Proceedings of the Grassland Society of southern Africa*, 6, 86–92.

Rethman, N.F.G. & Beukes, B.H. (1973). Overseeding of *Eragrostis curvula* on north-eastern sandy highveld. *Proceedings of the Grassland Society of southern Africa*, 8, 57–9.

Rethman, N.F.G., Beukes, B.H. & Malherbe, C.E. (1971). Influence on a north-eastern sandy highveld sward of winter utilization by sheep. *Proceedings of the Grassland Society of southern Africa*, 6, 55–62.

Rethman, N.F.G. & Booysen, P. de V. (1967). Growth and development in *Cymbopogon excavatus* tillers. *South African Journal of Agricultural Science*, 10, 811–22.

Rethman, N.F.G. & Booysen, P. de V. (1968). Growth and development in *Heteropogon contortus* tillers. *South African Journal of Agricultural Science*, 11, 259–72.

Rethman, N.F.G. & Booysen, P. de V. (1969). The seasonal growth patterns of a tall grassveld sward. *Proceedings of the Grassland Society of southern Africa*, 4, 56–60.

Rethman, N.F.G. & Malherbe, C.E. (1970). The influence of fertilization on the production and digestibility of natural veld. *Agroplantae*, 2, 43–4.

Roberts, B.R. (1966a). *The Ecology of Thaba'nchu. A Statistical Study of Vegetation/Habitat Relationships*. PhD Thesis. Pietermaritzburg: University of Natal.

Roberts, B.R. (1966b). Observations on the temperate affinities of the vegetation of Hangklip mountain near Queenstown, C.P. *Journal of South African Botany*, 32, 243–60.

Roberts, B.R. (1975). Microclimatic differences in habitat on a high mountain in the Orange Free State. *Proceedings of the Grassland Society of southern Africa*, 10, 49–55.

Roberts, B.R., Anderson, E.R. & Fourie J.H. (1975). Evaluation of natural pastures: quantitative criteria for assessing condition in the *Themeda* veld of the Orange Free State. *Proceedings of the Grassland Society of southern Africa*, 10, 133–40.

Rose Innes, R. (1939). *Some Observations on the Nature and Causes of Secondary Succession on Certain Areas at Frankenwald and in the Highveld of the Transvaal*. MSc Thesis. Johannesburg: University of the Witwatersrand.

Rossouw, L.F. (1983). *'n Ekologiese Studie van die Boomgemeenskappe van die Bloemfontein-omgewing, Oranje Vrystaat*. MSc Thesis. Bloemfontein: University of the Orange Free State.

Roux, E.R. (1954). The nitrogen sensitivity of *Eragrostis curvula* and *Trachypogon plumosus* in relation to grassland succession. *South African Journal of Science*, 50, 173–6.

Roux, E. (1969). *Grass. A Story of Frankenwald*. Cape Town: Oxford University Press.

Roux, E.R. (1970). Plant succession on Iron Age I sites at Melville Koppies (Johannesburg). *South African Journal of Science*, 66, 48–50.

Roux, P.W. & Theron, G.K. (1987). Vegetation change in the Karoo Biome. In *The Karoo Biome: A Preliminary Synthesis. Part 2 – Vegetation and History*, ed. R.M. Cowling & P.W. Roux. South African National Scientific Programmes Report 142. Pretoria: CSIR.

Roux, P.W. & Vorster, M. (1983). Vegetation change in the Karoo. *Proceedings of the Grassland Society of southern Africa*, 18, 25–9.

Rutherford, M.C. & Westfall, R.H. (1986). Biomes of southern Africa – an objective categorization. *Memoirs of the Botanical Survey of South Africa*, 54, 1–98.

Scheepers, J.C. (1975). *The Plant Ecology of the Kroonstad and Bethlehem Areas of the Highveld Agricultural Region*. DSc Thesis. Pretoria: University of Pretoria.

Schoeman, J.L. & Scotney, D.M. (1987). Agricultural potential as determined by soil, terrain and climate. *South African Journal of Science*, 83, 260–8.

Scholes, R.J. (1985). Drought related grass, tree and herbivore mortality in a southern African savanna. In *Ecology and Management of the World's Savannas*, ed. J.C. Tothill & J.J. Mott, pp. 350–3. Canberra: Australian Academy of Science.

Scotcher, J.S.B. & Clarke, J.C. (1981). Effects of certain burning treatments on veld condition in Giant's Castle Game Reserve. *Proceedings of the Grassland Society of southern Africa*, 16, 121–7.

Scott, J.D. (1937). The possibilities of reseeding damaged veld and old lands. *South African Journal of Science*, 33, 617–24.

Scott, J.D. & Booysen, P. de V. (1956). Effects of certain fertilizers on veld at Ukulinga. *South African Journal of Science*, 52, 240–3.

Scott, L. (1989). Late Quaternary vegetation history and climatic change in the eastern Orange Free State, South Africa. *South African Journal of Botany*, 55, 107–16.

Shackleton, C.M. (1989). *An Ecological Survey of a Selected Area of Pondoland Sourveld With Emphasis on its Response to the Management Practices of Burning and Grazing*. MSc Thesis. Umtata: University of Transkei.

Shackleton, C.M. (1991). Seasonal changes in above-ground standing crop in three coastal grassland communities in Transkei. *Journal of the Grassland Society of southern Africa*, 8, 22–8.

Shackleton, C.M., McKenzie, B. & Granger, J.E. (1988). Seasonal changes in root biomass, root/shoot ratios and turnover in two coastal grassland communities in Transkei. *South African Journal of Botany*, 54, 465–71.

Shackleton, C.M., McKenzie, B. & Granger, J.E. (1989). Breakdown and decomposition in three coastal grassland communities in Transkei. *South African Journal of Botany*, 55, 551–9.

Shackleton, C.M. & Mentis, M.T. (1992). Seasonal changes in nutrient content under three defoliation treatments in two coastal grassland communities of Transkei. *Journal of the Grassland Society of southern Africa*, 9, 30–7.

Shackleton, S.E. (1989). *Autecology of Cymbopogon validus (Stapf) Stapf ex Burtt Davy in Mkambati Game Reserve, Transkei*. MSc Thesis. Johannesburg: University of the Witwatersrand.

Shackleton, S.E. & Mentis, M.T. (1991). Response of *Cymbopogon validus* tillers to three clipping frequencies. *Journal of the Grassland Society of southern Africa*, 8, 35–6.

Siegfried, W.R. (1991). Preservation of species in southern African nature reserves. In *Biotic Diversity in Southern Africa. Concepts and Conservation*, ed. B.J. Huntley, pp. 186–201. Cape Town: Oxford University Press.

Smit, C.M. (1992). *Phytosociology of the Newcastle–Memel–Chelmsford Dam area*. M.Sc. Thesis. Pretoria: University of Pretoria.

Smit, C.M., Bredenkamp, G.J. & van Rooyen, N. (1992). Phytosociology of the B land type in the Newcastle–Memel–Chelmsford Dam area. *South African Journal of Botany*, 58, 363–73.

Smit, C.M., Bredenkamp, G.J. & van Rooyen, N. (1993a). Woodland plant communities of the Fa land type in the Newcastle–Memel–Chelmsford Dam area. *South African Journal of Botany*, 59, 14–20.

Smit, C.M., Bredenkamp, G.J. & van Rooyen, N. (1993b). Plant communities in the Ad land type in the Newcastle–Memel–Chelmsford Dam area. *South African Journal of Botany*, 59, 116–22.

Smit, C.M., Bredenkamp, G.J. & van Rooyen, N. (1993c). Phytosociology of the Ac land type in the foothills in the low Drakensberg in north-western KwaZulu-Natal. *South African Journal of Botany*, 59, 203–14.

Smit, G.N. (1990). Kouebeskadiging van houtagtige plante in die Suuragtige-Gemengde Bosveld. *Journal of the Grassland Society of southern Africa*, 7, 196–200.

Smit, I.B.J. (1972). Veld reclamation studies 1954–1971. Final Report, H-Bl 26. Pretoria: Department of Agricultural and Technical Services.

Smith, F.R. & Tainton, N.M. (1985). Effects of season of burn on shrub survival, regeneratiion and structure in the Natal Drakensburg. *Journal of the Grassland Society of southern Africa*, 2, 4–10.

Snyman, H.A. (1988). Bepaling van waterverbruiksdoeltreffendheid van veld in die sentrale Oranje-Vrystaat vanaf evapotranspirasiemetings. *Water SA*, 14, 153–8.

Snyman, H.A. (1989). Evapotranspirasie en waterverbruiksdoeltreffendheid van verskillende grasspesies in die sentrale Oranje-Vrystaat. *Journal of the Grassland Society of southern Africa*, 6, 146–51.

Snyman, H.A. (1993). The effect of defoliation during wilting on the production of *Themeda triandra* and *Eragrostis lehmanniana* in semi-arid grassland. *African Journal of Range & Forage Science*, 10, 113–17.

Snyman, H.A. & Fouché, H.J. (1991). Production and water-use efficiency of semi-arid grasslands of South Africa as affected by veld condition and rainfall. *Water SA*, 17, 263–7.

Snyman, H.A. & Fouché, H.J. (1993). Estimating seasonal herbage production of a semi-arid grassland based on veld condition, rainfall, and evapotranspiration. *African Journal of Range & Forage Science*, 10, 21–4.

Snyman, H.A. & Opperman, D.P.J. (1983). Die invloed van vog- en ontblarings-behandelings in hidrologiese eenhede op natuurlike veld van die sentrale Oranje-Vrystaat. *Proceedings of the Grassland Society of southern Africa*, 18, 124–30.

Snyman, H.A. & Opperman, D.P.J. (1984). Afloopstudies vanaf natuurlike veld in verskillende suksessiestada van die sentrale Oranje-Vrystaat. *Journal of the Grassland Society of southern Africa*, 1, 11–15.

ayman, H.A., Opperman, D.P.J. & van den Berg, J.A. (1980). Hidrologiese siklus en waterverbruiksdoeltreffendheid van veld in verskillende suksessiestadia. *Proceedings of the Grassland Society of southern Africa*, 15, 69–72.

ayman, H.A. & van Rensburg, W.L.J. (1990). Korttermyn invloed van strawwe droogte op veldtoestand en waterverbruiksdoeltreffendheid van grasveld in die sentrale Oranje-Vrystaat. *Journal of the Grassland Society of southern Africa*, 7, 249–56.

ayman, H.A., van Rensburg, W.L.J. & Opperman, D.P.J. (1985). Grond- en afloopverlies bepalings vanaf natuurlike veld, met behulp van 'n reënvalnabootser. *Journal of the Grassland Society of southern Africa*, 2, 35–40.

ayman, H.A., van Rensburg, W.L.J. & Opperman, D.P.J. (1986). Toepassing van 'n gronderosievergelyking op natuurlike veld van die sentrale Oranje-Vrystaat. *Journal of the Grassland Society of southern Africa*, 3, 4–9.

ayman, H.A., Venter, W.D., van Rensburg, W.L.J. & Opperman, D.P.J. (1987). Ranking of grass species according to visible wilting order and rate of recovery in the central Orange Free State. *Journal of the Grassland Society of southern Africa*, 4, 78–81.

pies, J.J. & Gibbs Russell, G.E. (1988). Variation in important pasture grasses. II. Cytogenetic and reproductive variation. *Journal of the Grassland Society of southern Africa*, 5, 22–5.

teinke, T.D. (1969). The translocation of ^{14}C-assimilates in *Eragrostis curvula*: an autoradiographic survey. *Proceedings of the Grassland Society of southern Africa*, 4, 19–34.

teinke, T.D. (1975). Effect of height of cut on translocation of ^{14}C-labelled assimilates in *Eragrostis curvula* (Schrad.) Nees. *Proceedings of the Grassland Society of southern Africa*, 10, 41–17.

teinke, T.D. & Booysen, P. de V. (1968). The regrowth and utilization of carbohydrate reserves of *Eragrostis curvula* after different frequencies of defoliation. *Proceedings of the Grassland Society of southern Africa*, 3, 105–10.

Steinke, T.D. & Nel, L.O. (1967). The growth of veld in response to defoliation by various means in late winter and spring. *Proceedings of the Grassland Society of southern Africa*, 2, 113–17.

Steinke, T.D. & Nel, L.O. (1989). Some effects of termitaria on veld in the eastern Cape. *Journal of the Grassland Society of southern Africa*, 6, 152–5.

Steinke, T.D. & Nel, L.O. (1990). A preliminary account of growth characteristics and seed germination of *Cassia mimosoides* L. in Dohne Sourveld. *Journal of the Grassland Society of southern Africa*, 7, 166–73.

Story, R. (1951). A botanical survey of the Keiskammahoek district. *Memoirs of the Botanical Survey of South Africa*, 27, 1–184.

Strohbach, B.J. (1989). *Die Verfyning van Populasieveranderingsmodelle met Behulp van Outekologiese Studies van die Belangrike Grasspesies in die Westelike Grasveldbioom*. MSc Thesis. Potchefstroom: University of Potchefstroom.

Tainton, N.M. (1969). Environmental control of flowering in tropical-subtropical grasses. *Proceedings of the Grassland Society of southern Africa*, 4, 49–55.

Tainton, N.M. (1981). *Veld and pasture management in South Africa*. Pietermaritzburg: Shuter & Shooter.

Tainton, N.M. & Booysen, P. de V. (1963). The effects of management on apical bud development and seeding in *Themeda triandra* and *Tristachya hispida*. *South African Journal of Agricultural Science*, 6, 21–30.

Tainton, N.M. & Booysen, P. de V. (1965a). Growth and development in perennial veld grasses. I. *Themeda triandra* tillers under various systems of defoliation. *South African Journal of Agricultural Science*, 8, 93–110.

Tainton, N.M. & Booysen, P. de V. (1965b). Growth and development in perennial veld grasses. II. *Hyparrhenia hirta* tillers under various systems of defoliation. *South African Journal of Agricultural Science*, 8, 745–60.

Tainton, N.M., Booysen, P. de V., Bransby, D.I. & Nash, R.C. (1978). Long-term effects of burning and mowing on tall grassveld in Natal: dry matter production. *Proceedings of the Grassland Society of southern Africa*, 13, 41–4.

Tainton, N.M., Booysen, P. de V. & Scott, J.D. (1970). Response of tall grassveld to different intensities, seasons and frequencies of clipping. *Proceedings of the Grassland Society of southern Africa*, 5, 32–41.

Tainton, N.M., Groves, R.H. & Nash, R. (1977). Time of mowing and burning veld: short-term effects on production and tiller development. *Proceedings of the Grassland Society of southern Africa*, 12, 59–64.

Tainton, N.M. & Mentis, M.T. (1984). Fire in grassland. In *Ecological Effects of Fire in South African Ecosystems*, ed. P. de V. Booysen & N.M. Tainton, pp. 115–47. Berlin: Springer-Verlag.

Tew, J., Grossman, E.S., Fair, P. & Cresswell, C.F. (1974). A study of the fine structure, enzyme activities and pattern of $^{14}CO_2$ incorporation of highveld grasses from different successional stages. *Proceedings of the Grassland Society of southern Africa*, 9, 95–103.

Theron, J.J. (1961). Die herstel van grond-humus deur middel van bemeste grasroeste. *South African Journal of Agricultural Science*, 4, 415–30.

Theron, J.J. (1964). Lisimeterproewe. II: 1945–1961. *South African Journal of Agricultural Science*, 7, 109–22.

Theron, J.J. (1965). The influence of fertilizers on the organic matter content of the soil under natural veld. *South African Journal of Agricultural Science*, 8, 525–34.

Tinley, K.L. (1982). The influence of soil moisture balance on ecosystem patterns in southern Africa. In *Ecology of Tropical Savannas*, ed. B.J. Huntley & B.H. Walker, pp. 175–92. Berlin: Springer Verlag.

Trollope, W.S.W. (1971). Fire as a method of eradicating macchia vegetation in the Amatole mountains of South Africa – experimental and field scale results. *Proceedings Tall Timbers Fire Ecology Conference – Fire in Africa*, 11, 99–120.

Trollope, W.S.W. (1973). Fire as a method of controlling macchia (fynbos) vegetation on the Amatole mountains of the eastern Cape. *Proceedings of the Grassland Society of southern Africa*, 8, 35–41.

Trollope, W.S.W. (1974). Role of fire in preventing bush encroahment in the eastern Cape. *Proceedings of the Grassland Society of southern Africa*, **9**, 67–72.

Trollope, W.S.W. (1983). *Control of Bush Encroachment with Fire in the Arid Savannas of Southeastern Africa*. PhD Thesis. Pietermaritzburg: University of Natal.

Trollope, W.S.W. & Booysen, P. de V. (1971). The eradication of macchia (fynbos) vegetation on the Amatole mountains of the eastern Cape. *Proceedings of the Grassland Society of southern Africa*, **6**, 28–38.

Trollope, W.S.W. & Tainton, N.M. (1986). Effect of fire intensity on the grass and bush components of the Eastern Cape Thornveld. *Journal of the Grassland Society of southern Africa*, **3**, 37–42.

Truter, S. du T. (1988). *Die Kwantitatiewe Karakterisering van Meervoudige Veldverwysingspersele vir Vier Redelik Homogene Boerderygebiede in die Grasveldbioom*. MSc Thesis. Potchefstroom: University of Potchefstroom.

Turner, B.J. (1989). *A Phytosociological Study of the South-eastern Transvaal Highveld Grasslands*. MSc Thesis. Pretoria: University of Pretoria.

van Daalen, J.C. & Geldenhuys, C.J. (1988). Southern African forests. In *Long-term Data Series Relating to Southern Africa's Renewable Natural Resources*, ed. I.A.W. Macdonald & R.J.M. Crawford, pp. 237–49. *South African National Scientific Programmes Report 157*. Pretoria: CSIR.

van den Berg, J.A., Roberts, B.R. & Vorster, L.F. (1975). The effect of seasonal grazing on the cover and composition of *Cymbopogon–Themeda* veld. *Proceedings of the Grassland Society of southern Africa*, **10**, 111–17.

van den Berg, J.A., Roberts, B.R. & Vorster, L.F. (1976). Die uitwerking van seisoenbeweiding op die infiltrasievermoë van gronde in 'n *Cymbopogon–Themeda* veld. *Proceedings of the Grassland Society of southern Africa*, **11**, 91–5.

van der Schijff, H.P. (1971). Die plantegroei van die distrikte Potgietersrus, Pietersburg en Soutpansberg in die noordelike Transvaal. *Tydskrif Natuurwetenskap*, **11**, 108–44.

van der Westhuizen, F.G.J., van den Berg, J.A. & Opperman, D.P.J. (1978). Die benutting van grasveld in die sentrale Oranje-Vrystaat met skape. *Proceedings of the Grassland Society of southern Africa*, **13**, 83–9.

van Ginkel, B. (1971). Plantkundige en drakragveranderinge deur bemesting van 'n *Cymbopogon–Themeda*-veldtipe. *Technical Communication 98*. Pretoria: Department of Agricultural and Technical Services.

van Rensburg, H.J. (1939). A further contribution to the ecology of the highveld grassland at Frankenwald. *South African Journal of Science*, **36**, 238–45.

van Rooyen, N., Bredenkamp, G.J., Theron G.K., Bothma, J. du P. & Le Riche, E.A.N. (1994). Vegetation gradients around artifical watering points in the Kalahari Gemsbok National Park. *Journal of Arid Environments*, **26**, 349–61.

van Schalkwyk, A., Lombard, P.E. & Vorster, L.F. (1968a). Evaluation of the nutritive value of a *Themeda triandra* pasture in the central Orange Free State. I. Botanical composition and nutritive value. *South African Journal of Agricultural Science*, **11**, 113–22.

van Schalkwyk, A., Lombard, P.E. & Vorster, L.F. (1968b). Evaluation of the nutritive value of a *Themeda triandra* pasture in the central Orange Free State. II. Comparison between quadrat and simulated grazing samples. *South African Journal of Agricultural Science*, **11**, 249–58.

van Schalkwyk, A., Lombard, P.E. & Vorster, L.F. (1968c). Evaluation of the nutritive value of *Themeda triandra* pasture in the central Orange Free State. III. Grazing samples and selecting ability of cattle. *South African Journal of Agricultural Science*, **11**, 483–92.

van Wyk, S. & Bredenkamp, G.J. (1986). 'n Braun–Blanquet klassifikasie van die plantegroei van die Abe Bailey Natuurreservaat. *South African Journal of Botany*, **52**, 321–31.

van Zinderen Bakker, E.M. (1973). Ecological investigations of forest communities in the eastern Orange Free State and the adjacent Natal Drakensberg. *Vegetatio*, **28**, 299–334.

van Zinderen-Bakker, E.M. Snr & Werger, M.J.A. (1974). Environment, vegetation and phytogeography of the high altitude bogs of Lesotho. *Vegetatio*, **29**, 37–49.

van Zyl, L.G. (1973). Fertilizer experiment with nitrogen, phosphate and potash on veld. 1949–1963. Final Report, T-Al 9. Pretoria: Department of Agricultural and Technical Services.

van Zyl, L.G. (1975). Veld fertilizer experiment with nitrogen, phosphate and lime, 1950–1963. Final Report, T-Al 10. Pretoria: Department of Agricultural and Technical Services.

Veenendaal, E.M. & Ernst, W.H.O. (1991). Dormancy patterns in accessions of caryopses from savanna grass species in south-eastern Botswana. *Acta Botanica Neerlandica*, **40**, 297–309.

Viljoen, L. (1966). *Fenologiese Studie van Meerjarige Grasse*. MSc Thesis. Bloemfontein: University of the Orange Free State.

Vogel, J.C., Fuls, A. & Ellis, R.P. (1978). The geographical distribution of Kranz grasses in South Africa. *South African Journal of Science*, **74**, 209–14.

Vorster, L.F. (1975). The influence of prolonged seasonal defoliation on veld yields. *Proceedings of the Grassland Society of southern Africa*, **10**, 119–22.

Vorster, L.F. & Mostert, J.W.C. (1968). Veld fertilization trends over a decade in the central Orange Free State. *Proceedings of the Grassland Society of southern Africa*, **3**, 111–19.

Warren, M. (1965). *A Study of Soil Nutritional and Other Factors Operating in Secondary Succession in Highveld Grassland in the Neighbourhood of Johannesburg*. PhD Thesis. Johannesburg: University of the Witwatersrand.

Weinbrenn, C. (1938). A comparative study of the osmotic values of the leaf saps of certain South African highveld grasses. *South African Journal of Science*, **35**, 317–18.

Weinbrenn, C. (1939). Further investigation on the osmotic values of the leaf saps of certain South Africa highveld grasses. *South African Journal of Science*, **36**, 265–9.

Weinbrenn, C. (1946). Preliminary account of investigation into the germination of seeds of *Cynodon dactylon*. *South African Journal of Science*, 42, 142–3.

Weinmann, H. (1938). Effect of fertiliser treatment on Transvaal Highveld. *South African Journal of Science*, 35, 246–9.

Weinmann, H. (1940). Seasonal chemical changes in the roots of some South African highveld grasses. *Journal of South African Botany*, 6, 131–45.

Weinmann, H. (1942). On the autumnal remigration of nitrogen and phosphorus in *Trachypogon plumosus*. *Journal of South African Botany*, 8, 179–96.

Weinmann, H. (1943a). Yields and chemical composition of pasture herbage as influenced by fertilizing and frequent clipping. *South African Journal of Science*, 40, 127–34.

Weinmann, H. (1943b). Effects of defoliation intensity and fertilizer treatment on Transvaal highveld. *Empire Journal of Experimental Agriculture*, 11, 113–24.

Weinmann, H. (1944). Root reserves of South African highveld grasses in relation to fertilizing and frequency of clipping. *Journal of South African Botany*, 10, 37–54.

Weinmann, H. (1948). Effects of grazing intensity and fertilizer treatment on Transvaal highveld. *Empire Journal of Experimental Agriculture*, 16, 111–18.

Weinmann, H. (1949). Productivity of Marandellas sandveld pasture in relation to frequency of cutting. *Rhodesia Agricultural Journal*, 46, 175–89.

Weisser, P.J. (1978). Changes in area of grasslands on the dunes between Richards Bay and the Mfolozi river. *Proceedings of the Grassland Society of southern Africa*, 13, 95–7.

Wells, P.V. (1965). Scarp woodlands, transported grassland soils, and concept of grassland climate in the Great Plains region. *Science*, 148, 246–9.

Wells, P.V. (1970). Postglacial vegetational history of the Great Plains. *Science*, 167, 1574–82.

Werger, M.J.A. (1973a). An account of the plant communities of the Tussen die Riviere Game Farm, Orange Free State. *Bothalia*, 11, 165–76.

Werger, M.J.A. (1973b). *Phytosociology of the Upper Orange River Valley*. PhD Thesis. Nijmegen: University of Nijmegen.

Werger, M.J.A. & Coetzee, B.J. (1978). The Sudano-Zambezian Region. In *Biogeography and Ecology of southern Africa*, ed. M.J.A. Werger, pp. 303–462. The Hague: Junk.

West, O. (1951). The vegetation of Weenen County, Natal. *Memoirs of the Botanical Survey of South Africa*, 23, 1–183.

Westfall, R.H., Everson, C.S. & Everson, T.M. (1983). The vegetation of the protected plots of Thabamhlope Research Station. *South African Journal of Botany*, 2, 15–25.

Westoby, M., Walker, B. & Noy Meir, I. (1989). Opportunistic management for rangelands not at equilibrium. *Journal of Range Management*, 42, 266–74.

White, F. (1978). The Afro-montane region. In *Biogeography and Ecology of Southern Africa*, ed. M.J.A. Werger, pp. 463–513. The Hague: Junk.

White, R.E. & Grossman, D. (1972). The effect of prolonged seasonal burning on soil fertility under *Trachypogon* – other species grasslands at Frankenwald. *South African Journal of Science*, 68, 234–9.

Wieland, R.G. (1982). *Vegetation and Ecosystem Classification and Range Inventory of Molumong Prototype, Lesotho, southern Africa*. MSc Thesis. Seattle: Washington State University.

Wiltshire, G.H. (1972). Response of highveld grass species to ammonium and nitrate nitrogen. *Proceedings of the Grassland Society of southern Africa*, 7, 67–75.

Wiltshire, G.H. (1973). Response of grasses to nitrogen source. *Journal of Applied Ecology*, 10, 429–35.

Wiltshire, G.H. (1978). Nitrogen uptake into unfertilized pasture of the Willem Pretorius Game Reserve, central Orange Free State. *Proceedings of the Grassland Society of southern Africa*, 13, 99–102.

Wiltshire, G.H. (1980). The responses of *Eragrostis chloromelas* Steud. to levels of water, nitrogen and phosphorus applied to soil from the Willem Pretorius Game Reserve. *Proceedings of the Grassland Society of southern Africa*, 15, 117–21.

Wiltshire, G.H. (1990). Nitrogen mineralization, nitrification and nitrogen balance in laboratory incubation of soil from natural grassland in the central Orange Free State. *South African Journal of Plant and Soil*, 7, 45–9.

Wolfson, M.M., Amory, A.M. & Cresswell, C.F. (1982). The effect of night temperature and leaf inorganic nitrogen status on the C_4 pathway enzymes in selected C_4 photosynthetic grasses. *Proceedings of the Grassland Society of southern Africa*, 17, 106–11.

Zacharias, P.J.K. (1980). *The Seasonal Patterns in Plant Quality in Various Ecological Zones in Natal*. MSc Thesis. Pietermaritzburg: University of Natal.

Savanna

R.J. Scholes

11.1 Introduction

The spirit of the savanna concept is that it is a tropical vegetation type co-dominated by woody plants and grasses. Savannas have at least a two-layered above ground structure: a tree layer with a discontinuous crown cover 2–10 m tall overlies a grassy layer 0.5–2 m tall. The grass layer is often said to be continuous, but this is an illusion caused by its finer scale of organization. In general it is highly clumped at the scale of individual tufts. An intermediate layer of small trees or shrubs is sometimes present, and the grass layer may be temporarily absent, or replaced by dicotyledonous herbs, during periods of drought or disturbance. Since there is no reliable floristic, structural or functional distinction between trees and shrubs, this chapter will use the shorthand 'trees' for all woody plants, except where a specific growth form is intended.

Savannas are part of a continuum that includes arid shrublands, lightly wooded grasslands, deciduous woodlands and dry forests. Any definition of the limits of savannas on this continuum is unavoidably arbitrary. For the purposes of this chapter, tree canopy cover can range from 5 to 90%. It is possible to subdivide savannas further on the basis of the height and degree of canopy cover of the tree layer; for instance into shrublands, parklands, woodlands, etc. (Edwards 1983; Cole 1986). This is useful for descriptive purposes, but is also arbitrary. The canopy cover, in particular, may change substantially at a given location over a period of one or two decades (Van Vegten 1984). Savannas are commonly referred to as 'bushveld' or 'bosveld' in South Africa and Namibia, and there are many other local vernacular names, some of which are mentioned in Table 11.3.

The key feature common to savannas all around the world is a climate that has a hot wet season of four- to eight-month duration and a warm dry season for the rest of the year (Nix 1983). In southern African savannas the wet season is uni-modal, and falls in the summertime, between October and April (Table 11.1). The strongly seasonal water availability leads to the accumulation of fine, dry, easily-ignited fuels that have the potential to burn virtually every year, although the actual fire frequency ranges from every year in moist savannas to once every ten or more years in arid savannas (Chap. 18, this volume). The high frequency of grass-layer fires is therefore also a unifying factor in savannas.

Savannas are one of the world's major biomes, and are the dominant vegetation of Africa. Savannas occupy 54% of southern Africa (Chap. 5, this volume), 60% of sub-saharan Africa (Scholes & Walker 1993) and 12% of the global land surface (Scholes & Hall 1996). Their importance lies in the large contribution that they make to the informal and subsistence economies through the supply of grazing, fuelwood, timber and other resources; their contribution to the formal economy as the main location of the livestock and ecotourism industries; their global impact through the emissions of trace gases from fires, soils, vegetation and animals (Justice, Scholes & Frost 1994); the sequestration of carbon in their soils and biomass (Scholes & Hall 1996); and their biological diversity.

Southern African savannas are part of the Sudano-Zambezian phytochorion (White 1983), and therefore share many genera and species with the savannas of Central and East Africa. They share fewer species, but many families with the savannas of West Africa. There are some floristic links with the savannas of the Indian Pen-

Table 11.1 **The climate at a moist (Harare) and arid (Windhoek) savanna site in southern Africa. Data: International Station Meteorological Climate Summary (1992)**

Month	Harare 17°55′ S 31°06′ E 1495 m 1940–1990			Windhoek 22°34′ S 17°06′ E 1725 m 1891–1984		
	Rainfall (mm)	Mean minimum temperature (°C)	Mean maximum temperature (°C)	Rainfall (mm)	Mean minimum temperature (°C)	Mean maximum temperature (°C)
Jan	196	15.3	20.8	78	17.2	30.0
Feb	177	15.3	20.5	77	16.5	28.6
Mar	117	15.0	20.8	79	15.4	27.2
Apr	28	14.0	20.2	38	12.8	25.6
May	13	12.2	16.2	7	9.2	22.7
Jun	3	10.9	18.4	1	6.7	20.2
Jul	1	10.9	18.4	1	6.3	20.5
Aug	3	11.7	19.5	1	8.6	23.4
Sep	5	13.2	21.3	3	11.9	26.5
Oct	27	14.2	22.3	11	14.6	29.1
Nov	97	15.0	22.0	27	15.6	29.6
Dec	163	15.3	21.3	42	16.9	30.7
Annual	828	13.5	20.5	365	12.7	26.1

insula, but very few with the savannas of America, Australia or South-East Asia (Johnson & Tothill 1985). Despite their floristic differences, the savannas of the world are believed to share the same basic patterns of structure and function (Frost *et al.* 1986).

Southern African scientists have been among the leaders in savanna ecological research. Within South Africa, ecosystem-level studies were first formally undertaken in savannas. The large body of work produced by the South African Savanna Ecosystem Programme, with its main research site at Nylsvley, was synthesized by Huntley & Walker (1982) and Scholes & Walker (1993). The southern African savanna research literature is biased towards studies on large mammals, the tree–grass interaction, fire and production ecology.

11.2 Boundaries

11.2.1 Biome-level

The southern African savannas extend from about 34° S in the Eastern Cape, northward along the east side of the subcontinent (below altitudes of about 1000 m; Fig. 11.1). At about 26° S they spread westwards, skirting the northern edge of the Highveld plateau at Pretoria, across the Kalahari to Namibia, and northward until they meet the tropical forests of Central Africa. In the southeastern region they form a fairly well-defined boundary with the grassland biome, which is strongly correlated with the dry-season temperature (Ellery, Scholes & Mentis 1991; Chap. 10, this volume). Savannas in southern Africa experience mild frosts on a few occasions per year and severe frosts every few decades, whereas grasslands are exposed to frequent, severe frosts. Since temperature

decreases with altitude, the savanna–grassland boundary often coincides with a change in altitude between two erosional surfaces. Where no such abrupt change occurs, such as in the western Free State, the grassland–savanna boundary is more diffuse.

Savannas grade continuously into the arid shrublands of the Nama-karoo in the eastern and northern Cape and Namibia (Chap. 8, this volume). As water availability decreases, the trees become sparser and lower (and confined to drainage lines) until they are more appropriately called shrubs. The mapped boundary occurs at about 350 mm mean annual rainfall, which is equivalent to about 40 days of potential evaporation (Ellery *et al.* 1991). Many elements of savanna ecological theory apply equally to the Nama-karoo shrublands, and the two biomes share many species and genera (Gibbs Russel 1987). In western Namibia, as the annual rainfall decreases below about 200 mm, the savannas grade into desert grasslands without necessarily passing through an intervening shrubland (Chap. 9, this volume).

In the north, moist savannas grade into woodlands and open semi-deciduous forests. Where savannas abut directly onto evergreen tropical forests, the savannas are usually secondary, having been derived from forest by clearing and agriculture, and are maintained as savannas by fire or repeated disturbance. Within the core of the savanna region, forests are restricted to areas of enhanced moisture status, such as river banks.

11.2.2 Landscape-level

The main functional distinction within southern African savannas is between the broad- and fine-leaved savannas (Huntley 1982; Scholes 1990). The underlying ecological difference is between savannas in nutrient-rich, arid environments (fine-leaved; Fig. 11.2) and savannas in

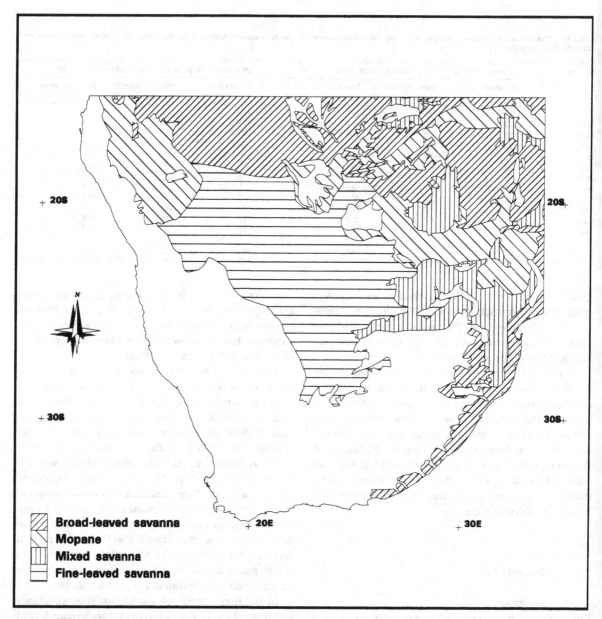

Figure 11.1 **The distribution of broad classes of savanna in southern Africa. The types of communities that are included in these classes are noted in Table 11.3.**

Legend:
- Broad-leaved savanna
- Mopane
- Mixed savanna
- Fine-leaved savanna

nutrient-poor, moister environments (broad-leaved; Fig. 11.3). There are many other factors associated with this distinction (Table 11.2). Leaf size is used as a short-hand label for the entire cluster of attributes; it is based on the observation that each leaf (or leaflet, in the case of compound leaves) of the dominant trees in broad-leaved savannas has a surface area of at least 5 cm^2 (mesophyllous or larger; Mueller-Dombois & Ellenberg 1974), whereas those of the fine-leaved savannas are usually less than 1 cm^2 (nanophyllous). There are some

exceptions: for instance *Colophospermum mopane* savannas (Fig. 11.4) have broad leaves, but are functionally arid/fertile savannas.

The main spatial organizing process within savanna landscapes is geomorphological. At the scale of hundreds of metres to kilometres, the migration of fine soil particles and ions from ridge crests to the valleys under the influence of water movement and gravity establishes a toposequence of soils and associated vegetation known as a catena (Milne 1936; Dye & Walker 1980; Yeaton,

Table 11.2 **A comparison of features of the two broad categories of African savannas. These are general trends, to which there are specific exceptions. Source: Justice *et al.* (1994)**

Feature	Broad-leaved	Fine-leaved
Age of erosional surface	Ancient	Recent
Parent material	Acid crystalline igneous rocks	Basic igneous lavas
	Aeolian sands	Mud- and silt-stones
	Sandstones	Limestones
Phosphorus availability	Low	Moderate
Dominant clay minerals and CEC per unit clay	Kaolinite, iron oxides (low CEC)	Montmorillonite (high CEC)
Mean annual rainfall	600–1500 mm	400–800 mm
Dominant tree family or subfamily	Caesalpinoideae (wet)	Mimosoideae (dry)
	Combretaceae (dry)	Burseraceae (v. dry)
Dominant grass subfamily (and tribe)	Panicoideae (Paniceae, Arundinelleae, Andropogoneae)	Arundinoideae (Chloridoideae, Panicoideae)
Mean tree leaf nitrogen content at maturity	<2.5%	>2.5%
Grass nitrogen content at senescence	<1%	>1%
Mycorrhizal associations and biological *N*-fixation	Predominantly ECM	Predominantly VAM
	Low *N*-fixation	Moderate *N*-fixation
Main tree anti-herbivore defence mechanism	Chemical (mainly polyphenols especially condensed tannins)	Structural (mainly thorns)
Tree leaf size	2–10 cm	0.1–1 cm
Grass growth form	Bunch (caespitose)	Creeping (stoloniferous)
Large mammal herbivory	Low (5–10%)	High (10–50%)
Insect herbivory	Episodic, mostly of woody plants by lepidoptera larvae	Seasonally recurrent, mostly of grass by grasshoppers and harvester termites, also episodic by locusts
Fire fuel load	High	Low
Fire frequency	Annual–triennial	Quintennial or longer

CEC, cation exchange capacity; ECM, ectomycorrhiza; VAM, vesicular–arbuscular mycorrhiza.

Figure 11.2 Fine-leaved, arid savanna in the southern Kalahari. Although the Kalahari sands are infertile in an absolute sense, under these conditions of aridity they have sufficient fertility to satisfy the strongly water-limited growth requirements of a vegetation that is more typical of nutrient-rich soils. Springbok (*Antidorcas marsupialis*) seek the shade of an *Acacia erioloba* growing along the Nossob River in the Kalahari Gemsbok Park, South Africa *(Photo: J.C. Paterson-Jones)*.

Frost & Frost 1986; Chappel & Brown 1993). Catenas are most prominent on long, gentle slopes in semi-arid climates, especially on soils derived from granites (Fig. 11.5). The interfluves support broad-leaved savannas, while the valley bottoms support fine-leaved savannas, unless they are seasonally saturated, in which case they support a *dambo* grassland (Fig. 11.6). The interface between the sandy upland and clayey bottomland is often marked by a seep-line (an intermittent perched water table, creating a strip of grassland with a band of *Terminalia sericea* immediately above it; Tinley 1982). The transition from broad-leaved savanna to dambo is often marked by a row of evenly spaced bush-clumps on termitaria.

Distinctive striped clumping of the woody vegetation in savannas is sometimes visible from the air. Whorled, finger-print like patterns ('*gilgai*') are a result of the micro-relief variation found on swelling clays. Bush stripes parallel or perpendicular to the contour on gentle slopes with loamy soils are due to an alternation of runoff–runon bands (Van der Meulen & Morris 1979).

11.2.3 Community-level

There is no widely accepted and over-arching classification system for southern African savannas. The AETFAT/UNESCO vegetation map of Africa (White 1983) is based on national maps such as those for South Africa, Lesotho and Swaziland (Acocks 1953), Zimbabawe, Botswana and Mozambique (Wild & Granvaux-Barbosa 1967) and Namibia (Giess 1970). The units in these maps are somewhere between the level of formation (biome) and community. Classification at the community-level has only been performed on small portions of the savanna biome, mostly in nature reserves (e.g. Coetzee *et al.* 1976; Van Rooyen, Theron & Grobbelaar 1981; Coetzee 1983; Bredenkamp & Theron 1985; Le Roux *et al.* 1988). A sum-

Figure 11.3 **Broad-leaved, moist savanna on deep, infertile Kalahari sands in Kavango, northern Namibia. The dominant tree is *Baikiaea plurijuga* (Fabaceae–Caesalpinioideae)** *(Photo: B. de Winter).*

mary of some of the most widespread communities is given in Table 11.3.

The broad-scale distribution of the main structural and floristic types in southern African savannas is highly predictable from a knowledge of the water and nutrient availability in the environment (Carter 1993; Fig. 11.7). In general, there are close relationships between species-level vegetation composition and soil at the landscape scale (Werger, Morris & Louppen 1979; Frazer, Von Rooyen & Verster 1987; O'Connor 1992), and strong genus- and family-level patterns at the subcontinental level.

The major spatial organizing process within the community is the effect of trees on the herbaceous layer (Whittaker, Morris & Goodman 1984; Yeaton *et al.* 1986). The shading, rainfall- and nutrient-redistributing influence of the tree canopy imposes a pattern on the understorey at a scale of a few metres to a few tens of metres (Fig. 11.8).

11.3 Organisms

11.3.1 Functional types

The plant functional types found in southern Africa are discussed in detail elsewhere in this volume (Chap. 16). This section expands on some of the dominant and unusual types associated with savannas (Scholes *et al.* 1996).

Sarmiento (1984) classifies the plant functional types in South American savannas according to the mechanism that they use to survive the seasonal drought. Trees in southern African savannas mostly avoid desiccation by dropping their leaves. Evergreen species typically comprise less than 5% of the leaf biomass. The evergreen species are very strongly sclerophyllous.

Most deciduous trees in southern African savannas are also sclerophyllous according to the normal definitions (Chap. 16, this volume), but to a lesser degree. The leaf area : mass ratio is typically $6-7 \, m^2 \, kg^{-1}$

Figure 11.4 *Colophospermum mopane* savanna at Klaserie, South Africa. Mopane is unusual in several respects. It is a broad-leaved member of the Caesalpinioideae but occurs in arid areas on relatively fertile, clay-rich soils. It usually grows in mono-specific stands of uneven size classes and several growth forms, ranging from low, multi-stemmed shrubs to tall, few-stemmed trees. This plasticity reflects growth conditions and history of disturbance, especially by elephants *(Photo: R.J. Scholes)*.

(Rutherford 1979). The sparse and sunken stomata contribute to a conservative water use pattern, but also place a limit on the maximum photosynthetic rate.

Savanna trees can be further classified on the basis of their adaptations to herbivory: the broad-leaved species are rendered unpalatable by high leaf concentrations of secondary chemicals, particularly tannins (Owen-Smith & Cooper 1987); whereas the fine-leaved palatable species typically have thorns (Cooper & Owen-Smith 1986; Chap. 17, this volume). The evergreen species are typically the least palatable. The few that are palatable (e.g. *Boscia albitrunca*) are a key browsing resource, and are heavily utilized.

Perennial grasses survive the dry season by dying back to ground level. There are three broad morphologi-

cal groups of savanna grasses in southern Africa (Blackmore 1992). The broad-leaved savannas are dominated by tall, caespitose (tuft-forming) species. The fine-leaved savannas have a higher proportion of stoloniferous (lawn-forming) species, better adapted to high levels of mammalian herbivory. The arid savannas are dominated by wiry grasses, with narrow or rolled leaves resistant to desiccation. Another way of classifying savanna grasses functionally is according to their photosynthetic pathways. Virtually all have the C_4 pathway, but the arid areas are dominated by the NAD-me variant and the mesic areas by NADP-me, with PCK somewhere in between (Vogel, Fuls & Ellis 1978; Ellis, Vogel & Fuls 1980). The grasses and some shrubs generally have an opportunistic pattern of water use (Scholes 1988; Baines 1989), which combines a high maximum stomatal conductance with a highly negative wilting-point xylem pressure potential.

Savanna trees and grasses generally have a high proportion of their biomass underground relative to forests or temperate grasslands, but the habit of growing mostly below-ground is displayed in its extreme form by the geoxylic suffrutices ('under-ground trees'). These plants typically occur on deep, infertile sands and are usually contra-seasonal, producing their new leaves and flowers at the height of the dry season. As a consequence, they tend to be highly unpalatable (e.g. *Parinari capensis*) or extremely toxic (e.g. *Dichapetalum cymosum*; Fig. 11.9) to avoid excessive herbivory. This growth-form is interpreted as an adaptation to fire and low fertility (White 1976).

Annual grasses and forbs survive the dry season as seeds. Most savannas have a small proportion of dicotyledonous plants in the grass layer, which may be annuals or facultative perennials. The forbs can become dominant following a severe disturbance such as a drought or sustained heavy grazing (O'Connor 1991), but disappear gradually once the disturbance ceases. Most of the symbiotic nitrogen fixation in southern African savannas is performed by leguminous forbs from the subfamily Papilionoidae (Grobbelaar & Rosch 1981; Zietsman, Grobbelaar & Van Rooyen 1988).

11.3.2 Reproductive biology

Vegetative reproduction is an important mode of propagation for both trees and grasses in savannas. Grasses such as *Setaria incrassata* and trees such as *Ochna pulchra* (Rutherford 1983) form large, almost mono-dominant clonal colonies. Most savanna grass seeds have adaptations for dispersal by wind or adherence to the fur of large mammals, but even with these adaptations, the vast majority are transported only a few metres from the

Table 11.3 Some widespread savanna communities. The dominant species listed do not occur in all the communities in the given functional class, but usually occur in many of them, and are dominant in the communities indicated with a number and letter combination. Where there is some uncertainty regarding the species that is present, or where several species of the same genus are present, the notations sp. and spp. are used respectively. The numbers in parentheses refer to the sources in which the community is described, as listed in the footnote. The letters in parentheses link the dominant species to the communities in which they occur. Where a source describes several communities within one functional class, they are given sequential letters

Functional class	Community names	Dominant woody species	Herbaceous species
Moist, infertile, mesophyllous	Brachystegia–Julbernadia moist savanna (1a) Dry forest thickets and woodlands (1b) Savanna woodland (3a) Miombo (5a) Mixed broadleaf savanna (1b) South Zambezian undifferentiated woodland and wooded grassland (2) Savanna parkland (3b) Lowveld deciduous tree savanna (4a) Lowveld dry deciduous forest (4b) Bushveld small tree savanna (4c) Broad orthophyll plains bushveld (5b and c) Terminalia sandveld (5b) Combretum sandveld (5c) Lowveld sour bushveld (6a) Lowveld (6b) Mixed bushveld (6c) Sour bushveld (6d)	Acacia spp. (5c), Afzelia quanzensis (1b), Albizia versicolor (4a), Brachystegia spiciformis (1a, 3a, 5a), Brachystegia spp. (5a), Baikiaea plurijuga (1a, 1b, 3a), Burkea africana (1a,b, 2, 3a, 4, 5b), Combretum apiculatum (2, 3b, 5b,c, 6c), Combretum molle (6d), Combretum zeyheri (2, 3b, 5b, 6c), Combretum spp. (4a,c, 5b,c, 6a,b,c,d), Diplorhynchus condylocarpon (5), Faurea saligna (6d), Guibourtia coleosperma (1b), Isoberlinia spp. (5a), Julbernadia globiflora (1a, 3a, 5a), Marquesia macroura (5a), Ochna pulchra (5b), Peltophorum africanum (4b), Pseudolachnostylus maprouneifolia (5), Pterocarpus rotundifolius (1b, 4b, 6a), Sclerocarya birrea (5b, 6a,b), Securidaca longepedunculata (4b), Terminalia sericea (1b, 2, 3b, 4b, 5b, 6a and c)	Tribe Andropogoneae (1), Aristida sp. (6b,c,d), Cymbopogon plurinodis (4b), Eragrostis sp. (6b,c,d), Hyparrhenia sp. (4a), Hyperthelia dissoluta (6a,d), Schizachyrium jefferysii (6a,d), Setaria sp. (4c), Themeda triandra (4c), Trichopteryx sp. (4a)
Arid, fertile, mesophyllous	Colophospermum mopane arid shrublands, savannas, woodlands (1) South Zambezian undifferentiated woodland and wooded grassland (2a) Colophospermum woodland (2b) Savanna woodland (3) Mopane (4, 5) Mopane (6)	Acacia spp. (1, 2b), Adansonia digitata (4, 6), Colophospermum mopane (1, 2a,b, 3, 4, 5, 6), Commiphora africana (1, 6), Kirkia acuminata (1, 2b, 6)	Anthephora pubescens (6), Aristida spp. (1), Brachiaria nigropedata (6), Enneapogon cenchroides (1), Eragrostis spp. (1), Schmidtia pappophoroides (1)
Arid, fertile, spinescent, nanophyllous	Arid savanna – microphyllous Acacia (1) Kalahari deciduous bushland and wooded grassland (2) Savanna parkland (3a) Low tree and shrub savanna (3b) Temperate savanna – Acacia grassland (4a) – Acacia savanna (4b) Bushveld – scattered bush savanna (4c) Bush savanna – Acacia (giraffae) gerrardii (4d) – Acacia bush (4e) – Tarchonanthus bush (4f) Microphyllous thorny plains bushveld (5) Springbok flats thornveld (6a) Kalahari thornveld (6b) Arid sweet bushveld (6c)	Acacia caffra (4a), A. erioloba (4d,f, 5, 6b), A. hereroensis (3b), A. karroo (4a, 5, 6a), A. mellifera (1, 2, 3b, 5, 6b), A. nigrescens (2, 6a), A. robusta (4d), A. tortilis (2, 5, 6a,c), Acacia spp. (3a, 4b,c,e, 5, 6a,c), Commiphora pyracanthoides (1, 3b), Dichrostachys cinerea (5, 6a,c), Euclea divinorum (4c, 5), Grewia flava (3b), Piliostigma thonningii (3a), Sclerocarya birrea (2, 6a), Tarchonanthus camphoratus (4f)	Aristida spp.(2, 3b, 6), Bothriochloa insculpta (5), Dactyloctenium sp. (5), Digitaria (pentzii) nuda (2), Eragrostis spp.(5), Chrysopogon sp.(3b), Panicum kalaharense (2), Heteropogon contortus (4a), Panicum maximum (5), Sporobolus spp.(5), Stipagrostis uniplumis (3b), Themeda triandra (5)

Moist, fertile, herbaceous	Savanna grassland (3)		*Hyparrhenia hirta–Themeda triandra* (3)

Transitional between moist/infertile and arid fertile, with mixed nano- and mesophyllous leaves	Transitional undifferentiated Zambesian woodland (2)
	Intermediate types (3a)
	Bushveld
	  – Small tree savanna (4a)
	  – Bush clump savanna (4b)
	Bush savanna
	  – Low bush (4c)
	  – *Olea* community (4d)
	Spiny arid bushveld (5)
	Arid lowveld (6a)
	Sourish mixed bushveld (6b)
	Thicket and scrub (3b)
	Temperate savanna
	Valley bushveld (6)
		Acacia caffra (4b, 6a,b), *Acacia karroo* (6b), *Acacia tortilis* ssp. *heteracantha* (4a), *Acacia* spp. (3b), *Burkea africana* (3a, 4a), *Combretum imberbe* (3a, 6a), *Combretum zeyheri* (4a,b, 6a,b), *Commiphora* spp.(3b), *Cussonia paniculata* (4e, 6), *Dombeya rotundifolia* (6), *Euphorbia ingens* (6), *Lycium* spp. (5), *Olea europaea* subsp. *africana* (4d), *Rhigozum obovatum* (5),*Diospyros (Royena) pallens* (4c), *Terminalia sericea* (3a), *Terminalia prunioides* (3a)	*Digitaria eriantha* (2), *Themeda triandra* (2, 4a,c), *Panicum* spp. (4b, 6a), *Setaria* spp. (4a)

(1) Wild & Fernandes (1968)
(2) White (1983)
(3) Cole (1986)
(4) Adamson (1939)
(5) Werger (1978)
(6) Acocks (1953)

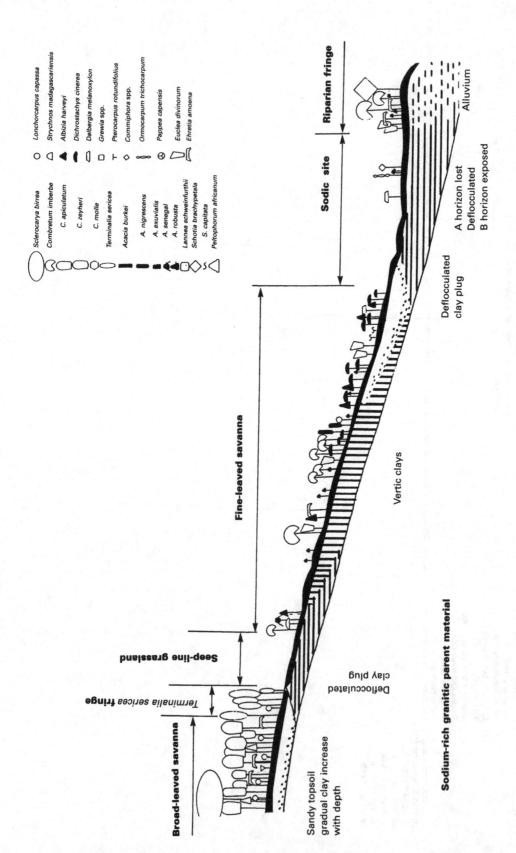

Figure 11.5 **Profile of a typical catena in the Eastern Transvaal Lowveld, on soils derived from Archaen granites (After Chappel 1993).**

Legend (tree/shrub species):

Sclerocarya birrea
Combretum imberbe
C. apiculatum
C. zeyheri
C. molle
Terminalia sericea
Acacia burkei
A. nigrescens
A. exuvialis
A. senegal
A. robusta
Lannea schweinfurthii
Schotia brachypetala
S. capitata
Peltophorum africanum

Lonchorcarpus capassa
Strychnos madagascariensis
Albizia harveyi
Dichrostachys cinerea
Dalbergia melanoxylon
Grewia spp.
Pterocarpus rotundifolius
Commiphora spp.
Ormocarpum trichocarpum
Pappea capensis
Euclea divinorum
Ehretia amoena

Broad-leaved savanna

Terminalia sericea fringe

Seep-line grassland

Fine-leaved savanna

Sodic site

Riparian fringe

Sandy topsoil gradual clay increase with depth

Deflocculated clay plug

Vertic clays

Deflocculated clay plug

A horizon lost
Deflocculated
B horizon exposed

Alluvium

Sodium-rich granitic parent material

266

Figure 11.6 **Nylsvley, in the Northern Transvaal, represents the southernmost extent of a landscape very widespread in south-central Africa: broad-leaved savanna on the interfluves and hydromorphic grassland (*vlei* or *dambo*) in the valley bottoms of a gently undulating surface. The South African Savanna Ecosystem study site is in the foreground** *(Photo: R.J. Scholes).*

There is no general survey of the pollination spectra of southern African savanna trees; informal observation suggests pollination is predominantly by insects, and especially by several species of bees. Along with a small proportion of bird-, moth- and wind-pollinated species, some other interesting specializations occur. For instance *Kigelia africana* is pollinated by nectar-feeding bats, and it has been suggested that *Acacia nigrescens* is pollinated by giraffes (Du Toit 1990). Species such as *Terminalia prunioides* and *Parinari curatellifolia* produce a foul odour attractive to flies. The main dispersal mechanisms for tree seeds are ballistic (often by explosive splitting of the pod; Strang 1966), wind, or ingestion by mammals. Passage through the mammal gut improves the germination success of many species with a hard testa (Bainbridge 1965). Dispersal by birds is common but not dominant.

parent plant (Veenendal 1991). The disseminule mass in eleven savanna grass species from Botswana varied from 0.13 to 5.62 mg, of which 0.08 to 1.54 mg was contributed by the caryopsis; the calculated distance over which they would be dispersed by wind was up to 13 m. On soils where surface water movement occurs during rainfall, many seeds are transported by water and are deposited in debris drifts.

11.3.3 Ecophysiological adaptations

The predominance of flat-crowned canopies of nanophyllous, compound leaves in the arid, nutrient-poor savannas may be an adaptation to water stress. The leaflets are relatively thick, therefore the small size minimizes the surface area : volume ratio and may assist in tolerating low water potentials without structural collapse. The small dimensions also increase the rate at which the leaf

Figure 11.7 **The association between environmental factors and floristic and structural characteristics in southern African savannas. DT, dominant tree genus; TF, dominant tree family; GF, dominant grass tribe; V, vernacular name.**

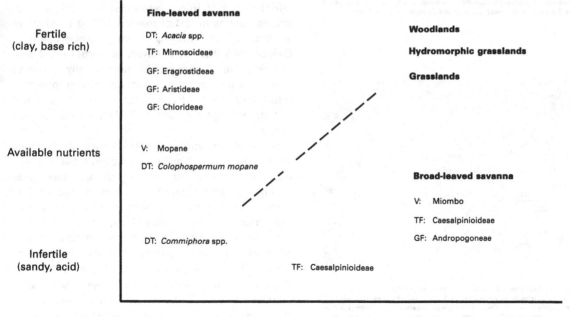

Figure 11.8 **Fine-leaved savanna at northen KwaZulu-Natal. Note the typical flat-crowned ('umbrella tree') form of *Acacia tortilis*, and the difference in the grass layer beneath the tree canopy (dominance by *Panicum maximum*) and in the open (mixed sward with much *Themeda triandra*)** *(Photo: R.J. Scholes).*

Figure 11.9 **Geoxylic suffrutices or 'underground trees' are especially common in broad-leaved savannas on deep, infertile sands. Shown here is *Dichapetalum cymosum*, an evergreen that produces new green leaves and flowers in the late dry season, especially after fire. During this time it is extremely toxic. Although it is avoided by indigenous herbivores, it frequently kills cattle** *(Photo: R.J. Scholes).*

can dissipate excess heat, without transpiring (Bate, Furniss & Pendle 1982). The flat form and compound leaves may assist by 'coupling' the leaves and thus reducing their water loss (Jarvis & McNaughton 1986). Each leaflet is in the slipstream of the upwind leaflet, and therefore benefits from the reduced vapour pressure deficit caused by upwind transpiration.

The subject of plant–herbivore relations is covered in Chap. 17 (this volume). The degree of herbivory in savannas is high relative to other biomes, and shows marked differences between the fine-leaved and broad-leaved savannas (East 1984; Scholes 1993). The current dominant theory of plant defence in savannas is based on the carbon to nutrient availability balance within the plant (Bryant *et al.* 1989). These adaptations can therefore be viewed as being driven by both herbivory and plant nutrition.

The fine-leaved savanna trees are generally also thorny. Spines retard the rate of leaf ingestion to a level tolerable by the plant, rather than preventing it entirely (Cooper & Owen-Smith 1986). The broad-leaved savannas, on the other hand, are almost thornless, yet are hardly browsed. This is due to the very high leaf concentrations of digestion-retarding substances, particularly tannins. Toxins such as alkaloids are relatively rare, and mostly confined to the 'underground trees' and some short-lived forbs (Cooper 1985). There are two main forms of tannin: the condensed tannins are located in the cell wall, and are therefore principally effective against mammals (Cooper & Owen-Smith 1985); whereas the hydrolysable tannins are found in the cell contents, and are mainly effective against insect herbivores. The ratio of condensed tannins to leaf nitrogen is a good predictor of acceptability to browsing antelope (Owen-Smith & Cooper 1987). The anti-herbivore properties could be a beneficial consequence of substances initially evolved to protect the leaf against microbial attack. Their presence, coupled with the low leaf nutrient content at senescence and the sclerophylly, causes the leaves to decompose slowly.

Some savanna grasses contain aromatic oils that may deter herbivores, and many have significant amounts of tannin (Ellis 1990; Chap. 10, this volume). Grass leaves are high in silica bodies, which restricts the mammalian grazers to those with hypsodont dentition. Many savanna grasses are very hairy, and have serrated margins; adaptations thought to deter invertebrate grazers. The principal adaptation in grasses with respect to herbivory is the ability to regrow leaf tissue after defoliation, by virtue of the basal location of the bud. Since grasses came to dominate the African flora several tens of millions of years before ruminants evolved (Clayton 1981; Stebbins 1981), this adaptation was probably initially selected by fire (Chap. 18, this volume). The grasses of the heavily grazed, fertile savannas respond to defoliation by adopting a prostrate, stoloniferous growth-form, which remains highly productive (Chap. 17, this volume). When grazing is removed, they revert to a more upright growth-form.

Most savanna grasses and trees have mycorrhizae-infected roots (predominantly VAM), which may assist in the acquisition of both water and nutrients (Hogberg 1989; Chap. 16, this volume). Although the dominant trees in many savannas are legumes, surprisingly few form root nodules containing symbiotic nitrogen-fixing bacteria. Those with the capacity to do so, which are mainly in the subfamilies Mimosoidae and Papilionoidae, apparently seldom do under field conditions. Almost all of the nitrogen fixation is performed by nodulated leguminous forbs or by free-living bacteria and soil surface crusts of blue-green algae (Zietsman *et al.* 1988).

Both trees and grasses in savannas have a remarkable ability to recover following fire and drought (see also Chap. 18, this volume). In trees, the resprouting can take place from the cambium of the stem if the terminal buds have been damaged or, if the stem has also been killed, from the root collar (Rutherford 1981). Physically damaged stems or exposed roots, for instance resulting from

feeding by elephants, will also produce sprouts. Suckering is a major mechanism of propagation in the broad-leaved trees of *miombo* woodlands (Strang 1966) but is less prevalent in the fine-leaved species.

The first leaves of many savanna species are red in colour, due to a high anthocyanin content (Ludlow 1987). The traditional explanation is that the young leaves require protection from damage by ultraviolet radiation, but other explanations are equally plausible. Anthocyanins are the precursors to condensed tannins, which are one of the main compounds used for defence against herbivory. The anthocyanin content does not decline with age; it simply becomes masked by chlorophyll and diluted by leaf growth.

11.4 Populations

11.4.1 Population dynamics of grasses

Savanna grasses are predominantly perennial, with the proportion of annuals increasing with aridity, and on mesic sites in arid years (Fig. 11.10; see also Chap. 10, this volume). Individual tillers of the perennial grasses

Figure 11.10 **Late dry season aspect of heavily grazed savanna on the Maputaland coastal plain in northern KwaZulu-Natal. Trees are *Acacia nigrescens* and *Sclerocarya birrea*** (Photo: E.J. Moll).

have a life span of about two months. Grasses begin to produce seed in December, about two months after the onset of the rains, but the peak period of seed set is in March (Veenendal 1991). The production of grass seed in southern African savannas is in the range 5000–100 000 m² yr⁻¹ (Veenendal 1991; O'Connor & Pickett 1992); viability ranges from 0 to 70% (Janse van Rensburg 1982; O'Connor & Pickett 1992). The steep decrease from the number of viable seeds produced to the viable seed store (100 to 2000 propagules m⁻²), and then to the number of individual grass tufts (10–100 per m²) implies high seed predation, a lifetime in the soil of one to a few years only (although most species remain viable for several years if stored in a dry, cool place) and a low rate of establishment success. Seed predation is mainly by ants, rodents and birds (Capon & O'Connor 1990). Almost all savanna grasses have innate dormancy mechanisms that prevent the seeds from germinating in the year in which they were produced (O'Connor & Pickett 1992). After a delay of several months, exposure of the seeds to a wetting event of about 20 mm will cause germination within six days, but not all seeds of one species will germinate at the first wetting. This protects the species against false starts (Veenendal 1991). Drying and rewetting will trigger further germination.

Seedling establishment success varies greatly with microsite. Survival beneath tree canopies is much higher than between canopies (Kennard & Walker 1973; Veenendal 1991). Mortality of mature perennial tufts can result from severe or frequently repeated defoliation, or from drought (Scholes 1985; O'Connor 1991).

11.4.2 Population dynamics of trees

About 2% of the primary production by trees in a broad-leaved savanna at Nylsvley is invested in reproductive structures every year. Little is known about the inter-annual variation in seed production in savanna trees, except anecdotal evidence that it is highly variable. Seeds are typically large, up to a few grams each. Few are viable, since seed parasitism by insect larvae is high, and mostly occurs while the seed is still on the tree. Unparasitized seeds can remain viable for many years when stored. There is usually no dormancy mechanism other than the hard testa; therefore, most of the viable seeds germinate in the first year. The soil store of viable tree seeds is about 30 m⁻² (Janse van Rensburg 1982).

The main controls on recruitment of woody plants are imposed by the interactive effects of fire, herbivory and competition with grasses while the tree is still within the grass layer (Frost 1985; Chap. 18, this volume). There are usually many small trees in the grass layer, with strongly developed root systems. Some could be many years old. These saplings-in-waiting escape from the grass layer following a simultaneous release from competition, fire and herbivory; for instance, by a drought that kills the perennial grasses and results in a reduction in the numbers of herbivores. Once clear of the grass canopy and flame zone, trees have a competitive advantage over grasses (see section 11.5.1 below). Palatable trees may continue to be held in a sub-mature state even once they have escaped the herbaceous layer, due to intense browsing by tall species such as kudu, giraffe and elephant.

The life span of savanna trees varies from a few decades in pioneer species such as *Dichrostachys cinerea*, to several centuries in *Adansonia digitata* (Guy 1970). Tree mortality can be associated with physical damage (such as wind-throw, ring-barking or felling by elephants), drought (Scholes 1985) or fire. Mortality is frequently a consequence of a combination of several factors. For instance, at Nylsvley a key mechanism of tree mortality involves a porcupine-created scar at the base of the trunk, which over a period of many years is enlarged by successive fires until the tree is hollow and susceptible to high winds (Yeaton 1988).

11.4.3 Population dynamics of forbs

The proportion of forbs in the herbaceous layer is normally less than 5% on a biomass basis, but following prolonged drought, heavy grazing or soil disturbance, can increase and dominate for several months or years (Von Maltitz 1990; O'Connor 1991). By inference, their seeds persist for long periods in the soil (but see Chap. 10, this volume). Seed size is intermediate between that of grasses and trees.

11.5 Community structure and dynamics

11.5.1 Tree–grass interactions

The major challenge of savanna community ecology has been to understand the mechanism of co-dominance between trees and grasses, and the factors that determine their proportions (see also Chap. 16, this volume). From a theoretical point of view, there are two main mechanisms that allow competitors to coexist: equilibrium theories based on niche separation; and disequilibrium theories, in which disturbance prevents competitive exclusion from proceeding to its logical conclusion. Both have been applied to the savanna condition.

The classical conceptual model of tree–grass interaction in savannas, the 'Walter Hypothesis' (Walter 1971), is based on rooting-depth separation with respect to competition for water. This hypothesis proposes that trees have roots in both the surface and deeper soil layers, while grasses root only in the surface layer. For this combination to result in coexistence, grasses must have a superior water-use efficiency. The model has been invoked to explain the generally greater tree biomass on soils with a low water-holding capacity (i.e. sands) and in wetter climates (Walker 1985), and suggests that the result of grazing is to allow more water through to the deeper layers, increasing tree growth (Walker *et al.* 1981).

Empirical observations of root distributions (Scholes 1988; Scholes & Walker 1993) show that the rooting depth separation between trees and grass is real but slight. The grass-root density exceeds tree-root density to a depth of nearly 1 m. Trees and grasses both have most of their roots in the top 40 cm of the soil, which is not surprising, since in dry climates water seldom penetrates below this depth in significant quantities. The popular notion of savanna trees having massive tap roots is generally incorrect. In fact, savanna trees have a few, small-diameter roots that can penetrate to great depths. These probably confer a survival advantage during droughts, but are insufficient to supply the transpiration needs of a complete tree canopy. This explains why the majority of savanna trees are deciduous.

The Walter Hypothesis requires preferential, rather than exclusive, access to subsoil water by trees (Walker & Noy Meir 1982). However, the observed degree of root separation and realistic water-use efficiency values are unable to account for the observed patterns of tree and grass primary production. The major axis of niche separation with respect to water acquisition is not depth, but season of use (Scholes & Walker 1993). Woody plants are able to store water and carbohydrates from the previous season, and therefore expand their leaves before or immediately after the first rains (the few deep roots may assist here). This allows trees several weeks of preferential resource access before grasses are able to grow enough leaf area to be serious competitors. Most tree growth and shoot extension takes place in this first part of the season, and sometimes again at the end of the wet season (Rutherford 1984). An early start also gives trees first access to the early-season nutrient flush, which may be the fundamental limitation, rather than water.

There is much evidence that trees have a marked competitive effect on grasses and on other trees, but little to suggest that grasses have much effect on trees once they are taller than the grass layer (Knoop & Walker 1984). Typically, the grass production decreases steeply with the first increments in tree biomass, basal area or density above zero, and then levels off at some small value (e.g. Donaldson & Kelk 1970). An experiment from the eastern Cape indicates that *Acacia karroo* density has no suppressive effect on grass production until it exceeds 1000 trees ha^{-1} (Aucamp *et al.* 1983), but this is an unusual finding. Trees have both competitive and facilitative effects on grasses in their vicinity (Stuart-Hill, Tainton & Barnard 1987; Belsky *et al.* 1989).

The water-use niche of grasses, both in depth and time, is completely included by the tree niche. The water-use efficiency of grasses is not substantially or consistently higher than that of trees, hence mature trees should always out-compete grass. This is experimentally confirmed (Knoop & Walker 1984). The implication is that, in general, the savanna tree : grass ratio is not a deterministic equilibrium, but a stochastic, unstable combination that persists as a result of disturbances which prevent dominance by trees (Chap. 18, this volume). The upper limit of tree biomass is probably determined by tree-on-tree competition, not grass-on-tree competition (Smith & Goodman 1986; Smith & Grant 1986).

Disequilibrium, or 'state-and-transition' models (Walker, Matthews & Dye 1986; Westoby, Walker & Noy-Meir 1989) have been used to describe the relatively rapid (and sometimes apparently irreversible) shifts in the structure of savannas. The instabilities and asymmetries of the tree–grass interaction allow communities to become trapped in a particular state. They can be shifted to another state only by some externally driven event or combination of events.

11.5.2 Disturbance and succession

The principal disturbance that keeps trees from dominating savannas is fire (Chap. 18, this volume). When fire is excluded, the tree biomass increases (Trapnell 1959). Fire acts predominantly by controlling the biomass of trees within the flame zone (those trees smaller than approximately 2 m), rather than as a cause of mortality. Fire and browsing together act as a powerful restriction on recruitment of trees to the mature, grass-dominating size classes. The short-term vegetation changes following a savanna fire do not comprise a succession, since they consist mainly of the regrowth of pre-existing vegetation (Rutherford 1981). Long-term fire exclusion does lead to a change in species in wetter savannas, as more fire-sensitive and shade-tolerant forest species appear (Trapnell 1959; see Chap. 18, this volume for further discussion).

At moderate levels, the trampling, soil disturbance and uprooting that occurs during grazing by large mammals helps to maintain diversity in the herbaceous layer by creating gaps. Burrowing animals, from dung-beetles to warthogs, turn over large amounts of soil and create patches suitable for the maintenance of early-succession species, even in an 'undisturbed' savanna. Deterministic succession patterns are not obvious in the grass layer.

Composition shifts due to species interactions and grazing treatments are overridden by the large inter-annual climate variations (O'Connor 1985; O'Connor 1991; Van Rooyen et al. 1993). Therefore, grass-layer condition indices based implicitly on orderly succession have little applicability in savannas.

Large-scale successional sequences in savannas are mostly associated with human disturbances, such as sustained heavy grazing or clearing for agriculture and settlement (Chap. 21, this volume). A significant part of the landscape structure and diversity in the 'undisturbed' savannas is as a consequence of human disturbances and fire-setting over the ages (Blackmore, Mentis & Scholes 1990). 'Megaherbivores' such as elephants can also alter the vegetation state over large areas (Chap. 17, this volume). *Dichrostachys cinerea* is a common pioneer tree species in arid areas, *Acacia karroo* in moister areas and *Terminalia sericea* on sandy soils.

11.5.3 Patterns of diversity

The plant species richness of the southern African savannas at a whole-biome scale is high, both in global terms and relative to other southern African biomes. Gibbs Russell (1987) reports 5788 species in the savanna areas of southern Africa (about 632 000 km²), which is second only to the fynbos biome. The non-plant diversity (mammals, reptiles, birds, amphibians and insects) in southern African savannas as a whole is similarly high (Siegfried 1989). The plant diversity per unit area is lower in savannas than in the fynbos, forest, grassland or succulent karoo biomes, although it remains high in global terms. This is because, although the species richness of southern African savannas at the small-plot scale ('alpha diversity') is high, the species turnover ('beta') and landscape diversity ('gamma') is generally low.

Southern African savanna contains 3–14 species m⁻², and 40–100 species 0.1 ha⁻¹ (Whittaker et al. 1984; Cowling et al. 1989), which is not significantly different from other southern African biomes. However, the level of local endemism is low and the environmental gradients are frequently gradual; therefore, the turnover of species with distance is relatively low. The community diversity within a landscape can be high or low, depending primarily on the underlying geomorphological variation. It is frequently high in the well-documented areas, such as nature reserves (Coetzee et al. 1976; Goodman 1990), since this is one of the reasons for their location, but on the extensive, stable erosional surfaces it is usually low. Goodman (1990) found the species richness of savanna communities in Mkuze Game Reserve to be positively correlated with above-ground productivity, and negatively correlated with standing biomass and degree of disturbance.

11.6 Ecosystem function

11.6.1 Primary production

In southern African savannas it is helpful to think of availability of water to plants as a switch that determines the period of time for which they can grow, and nutrient availability as a rheostat regulating the rate of production when the switch is on. Above-ground primary production of the herbaceous layer is linearly related to annual rainfall up to about 900 mm. The slope of this relationship is determined by soil fertility and the y-intercept is controlled by tree biomass and soil physical characteristics (Rutherford 1980; Dye & Spear 1982; Scholes 1993). Mean annual woody-plant litter-fall and stem diameter increment are also approximately linearly related to mean annual rainfall when sites are compared, but between years within a site they are related to a combination of the rainfall in the current year and the previous year. This is a reflection of the ability of trees to carry over production potential between years. Mean annual tree basal area increment in savannas is fairly consistent at about 4% per year (Scholes & Walker 1993). Litter-fall is also fairly consistent at 5% of the total tree standing biomass.

The fraction of primary production which appears below ground is approximately 50% of the total (Scholes & Walker 1993; Scholes & Hall 1996). This excludes root exudates and mycorrhizal respiration. Total net primary production in southern African savannas is in the range of 500 to 1500 g m⁻² yr⁻¹ (Rutherford 1978). Depending on savanna structure, 5 to 95% of this can be herbaceous layer production, of which 90% or more is typically produced by grasses.

11.6.2 Herbivory

The processes of herbivory in savannas are discussed in detail in Chap. 17 (this volume). Herbivory is relatively complex in savannas, because there are two main forage resources (grass and tree leaves) and many species of herbivores. Four points need to be stressed. First, large mammals are not the only herbivores in savannas, and in many cases may not be the main ones. Grasshoppers are significant grazers (Gandar 1982), and lepidopteran larvae are important, but episodic, browsers (Scholtz 1976; Scholes & Walker 1993). Second, the fraction of the primary production consumed by herbivores is strongly under the influence of the forage 'quality', in particular its nitrogen content and tannin content and their seasonal dynamics. In broad-leaved savannas, only a small proportion of the grass is grazed, even if the grass production is high, whereas in fine-leaved savannas, a large fraction may be consumed, resulting in a higher secondary production, even though the primary production is lower. The uneaten grass in broad-leaved savannas accumulates and is burned, with important consequences for nutrient cycling and the global atmosphere (Scholes, Justice & Ward 1996).

Third, as a generalization, savanna browsers are limited by the low quantity of available food, especially during the late dry season; whereas grazers are constrained by the low quality of what is available (Owen-Smith & Novellie 1982). Finally, the herbivores satisfy their quantity–quality constraints by selecting, at small and large scales, from different parts of the landscape. Many of the degradation problems now encountered in savannas result from the removal of this spatial flexibility (Walker *et al.* 1987; Behnke, Scoones & Kerven 1993).

11.6.3 Hydrology

Although the annual potential evaporation greatly exceeds the annual precipitation in savannas, there may be times of the year when there is more water in the soil profile than it can hold. Recharge of the deep aquifers occurs only sporadically, and constitutes 1–2% of mean annual precipitation (MAP), while streamflow is typically only 2–3% of MAP unless the vegetation cover has been reduced and the soil surface has become crusted. Most rainwater leaves savannas as evaporation (Chap. 2, this volume), a large fraction of which is from the soil surface and water intercepted by the plant canopy and litter layer (Scholes & Walker 1993). Mean ecosystem potential evaporation rates peak at about 4 mm day^{-1}, due to the low stomatal conductance (0.3–0.8 cm s^{-1}) and low total leaf area (typically less than 1).

11.6.4 Nutrient cycling

The structure, function and composition of southern African savannas are all highly influenced by nutrient availability, especially nitrogen and phosphorus (Scholes 1990, 1993). Since both of these elements are seldom present in available forms in significant quantities in the soil, it can be concluded that the rate-limiting step is their mineralization from the unavailable, largely organic form. Decomposition is slow in savannas for two reasons: the long, dry season and the high concentration of microbe-resistant compounds in the leaves. It can be speeded up by fire, or by the activities of soil fauna. Fire causes a major fraction of the nitrogen in the litter to volatilize ('pyrodenitrification'), which may be the main reason why savannas are frequently nitrogen-limited. It is also part of the reason for the low soil organic matter, which further reduces mineralization potential. Termites are conspicuous in many savannas, but millipedes, ants, earthworms and dung-beetles are also significant detritivores.

11.7 Management and use

Savannas supply a large proportion of the population of southern Africa with meat, edible and medicinal plants, edible insects, honey, timber, thatch, fuelwood and grazing. Much of this value is not reflected in national accounts, since it occurs in the informal and subsistence sectors.

11.7.1 Pastoralism

In the formal sector, the beef and goat industries are based on savanna lands. About two-thirds of the stock are ranched on large, commercial ranches, with the objective of meat production. The remainder are kept on communally grazed lands where meat production is a secondary objective to the accumulation of assets in the form of livestock, as well as the services that animals offer, such as milk and draught power. The stocking rates on the communal lands are typically substantially higher than on commercial lands, but in some cases these high stocking rates are apparently sustainable. There has been a lively debate in recent years around the issues of appropriate stocking policy in savannas, and what constitutes a valid measure of degradation (Behnke *et al.* 1993).

11.7.2 Conservation, wildlife and ecotourism

The largest and best-known national parks in southern Africa are all in savanna regions. There has been a trend in recent years to restock privately owned savanna areas with indigenous herbivores. In South Africa it is estimated that there is currently more private than state land under conservation management (in the broad sense). Large areas of communal land in Botswana, Zimbabwe and Namibia have wildlife as their main source of income. This trend was originally driven by the income to be made from licensed trophy hunting, but is increasingly orientated towards recreation and tourism.

The ecological promise of wildlife-based meat production systems in savannas (Dasmann 1964) has not yet been realized. In theory, multi-species herbivory should make more efficient use of the diverse plant resources in savannas than herbivory by cattle alone, and indigenous herbivores should require less veterinary care than introduced breeds. The main unresolved problems relate to harvesting the animals in a cost-effective manner and marketing the meat products.

11.7.3 Wood harvesting

Fuelwood, mainly harvested from savannas, is the principle energy source in most southern African households (Chap. 20, this volume). Urbanization and electrification is decreasing this dependence, but for the foreseeable future the demand will remain high. At a national scale, fuelwood production in southern African countries is close to the consumption rate (Kruger, Scholes & Geldenhuys 1993), but in the vicinity of habitations, the demand usually greatly exceeds the supply.

With the exception of the *Baikiea-* and *Pterocarpus-*woodlands of Zimbabwe, Botswana and Namibia, which yield high-value hardwoods, commercial exploitation of the timber resource in savannas is minimal. The shape and size of the trees limits the dimensions of the timber produced; therefore, most is used in the craft industry for carving and furniture.

A wide variety of savanna plants are traditionally used for food, medicinal and domestic purposes. The ethnobotanical literature is surveyed by Liengme (1983; see also Chap. 20, this volume). It is especially rich for the San hunter-gatherers of the Kalahari. The harvesting pressures are generally less intense in savannas than in the indigenous forests, owing to the more extensive nature of the former, but in some cases harvesting is locally unsustainable.

11.8 Conclusions

Savannas are important in the southern African context, owing to their extent, economic value and high non-plant diversity. The ecological features of particular interest in savannas are the strong coupling of the vegetation structure and function to sporadic driving variables such as rainfall, fire and herbivory, and the inherent instability of the competitive relationship between the two main vegetation components.

There is a wealth of information relating to the vegetation ecology of southern African savannas, with some conspicuous deficiencies. In particular, the fundamental biology of savanna plant species is poorly known: their reproductive biology, life-history characteristics and environmental limits. The pressures of human use in this biome will continue to focus research on production ecology in the coming decades, but an increasing emphasis on conservation, rehabilitation and possible climate change will require a more detailed understanding of processes at the organism level.

11.9 References

Acocks, J.P.H. (1953). Veld Types of South Africa. *Memoirs of the Botanical Survey of South Africa*, **28**, 1–192.

Adamson, R. (1939). *Vegetation of South Africa*. Monographs of British Empire Vegetation. London: British Empire Vegetation Committee.

Aucamp, A.J., Dankwerts, J.E., Teague, W.R. & Venter, J.J. (1983). The role of *Acacia karroo* in the false thornveld of the Eastern Cape. *Journal of the Grassland Society of southern Africa*, **8**, 151–4.

Bainbridge, W.R. (1965). Distribution of seed in elephant dung (*Acacia, Ricinodendron, Baikaea, Hyphaena*). *Puku*, **3**, 173–5.

Baines, K.A. (1989). *The Water Use, Growth and Phenology of Four Species of Savanna Grasses in Response to Changing Soil Water Availability*. MSc Thesis. Johannesburg: University of the Witwatersrand.

Bate, G.C., Furniss, P.R. & Pendle, P.G. (1982). Water relations of southern African savannas. In *Ecology of Tropical Savannas*, ed. B.J. Huntley & B.H. Walker, pp. 336–58. *Ecological Studies* 42. Berlin: Springer-Verlag.

Behnke, R.H., Scoones, I. & Kerven, C. (eds) (1993). *Range Ecology at Disequilibrium: New Models of Natural Variability and Pastoral Adaptation in African Savannas*. London: Overseas Development Institute.

Belsky, A.J., Amundson, R.G., Duxbury, J.M., Riha, S.J., Ali, A.R. & Mwonga, S.M. (1989). The effects of trees on their physical, chemical and biological environments in a semi-arid savanna in Kenya. *Ecology*, **26**, 1005–24.

Blackmore, A.C. (1992). *The Functional Classification of South African Savanna Plants Based on Their Ecophysiological Characteristics*. MSc Thesis. Johannesburg: University of the Witwatersrand.

Blackmore, A.C., Mentis, M.T. & Scholes, R.J. (1990). The origin and extent of nutrient enriched patches within a nutrient-poor savanna in South Africa. *Journal of Biogeography*, **17**, 463–70.

Bredenkamp, G.J. & Theron, G.K. (1985). A quantitative approach to the structural analysis and classification of the vegetation of the Manyeleti Game Reserve. *South African Journal of Botany*, **51**, 45–54.

Bryant, J. P., Kuropat, P. J., Cooper, S. M., Frisby, K. & Owen-Smith, N. (1989). Resource availability hypothesis of plant anti-herbivore defence tested in a South African savanna. *Nature*, **340**, 227–9.

Capon, M.H. & O'Connor, T.G. (1990). The predation of perennial grass seed. in Transvaal savanna grasslands. *South African Journal of Botany*, **56**, 11–15.

Carter, G. (1993). *A Functional Classification of a Range of Southern African Savanna Types*. MSc Thesis. Johannesburg: University of the Witwatersrand.

Chappel, C. (1993). *The Ecology of Sodic Sites in the Eastern Transvaal Lowveld*. MSc Thesis. Johannesburg: University of the Witwatersrand.

Chappel, C. & Brown, M.A. (1993). The use of remote sensing in quantifying rates of soil erosion. *Koedoe*, **36**, 1–14.

Clayton, W.D. (1981). Evolution and distribution of grasses. *Annals of the Missouri Botanical Gardens*, **68**, 5–14.

Coetzee, B.J. (1983). Phytosociology, vegetation structure and landscapes of the Central District, Kruger National Park, South Africa. *Dissertationes Botanicae*. Vaduz: J. Cramer.

etzee, B.J. & van der Muelen, F., Zwanziger, S., Gonsalves, P. & Weisser, P.J. (1976). A phytosociological classification of the Nylsvley Nature reserve. *Bothalia*, 12, 137–60.

le, M.M. (1986). *The Savannas: Biogeography and Geobotany*. London: Academic Press.

oper, S.M. (1985). *Factors Influencing the Utilization of Woody Plants and Forbs by Ungulates*. PhD Thesis. Johannesburg: University of the Witwatersrand.

oper, S.M. & Owen-Smith, R.N. (1985). Condensed tannins deter feeding by browsing ruminants in a South African savanna. *Oecologia*, 67, 142–6.

oper, S. M. & Owen-Smith, R. N. (1986). Effects of plant spinescence on large mammalian herbivores. *Oecologia*, 68, 446–55.

owling, R.M., Gibbs Russell, G.E., Hoffman, M.T. & Hilton Taylor, C. (1989). Patterns of plant species diversity in southern Africa. In *Biotic Diversity in Southern Africa. Concepts and Conservation*, ed. B.J. Huntley, pp. 19–50. Cape Town: Oxford University Press.

asmann, R.F. (1964). *African Game Ranching*. London: Pergamon.

onaldson, C.H. & Kelk, D.M. (1970). An investigation of the veld problems of the Molopo area. 1. Early findings. *Proceedings of the Grassland Society of southern Africa*, 5, 50–7.

1 Toit, J.T. (1990). Giraffe feeding on *Acacia* flowers: predation or pollination? *African Journal of Ecology*, 28, 63–8.

ye, P.J. & Spear, P.T. (1982). The effects of bush clearing and rainfall variability on grass yield and composition in south-west Zimbabwe. *Zimbabwe Journal of Agricultural Research*, 20, 103–18.

ye, P.J. & Walker, B.H. (1980). Vegetation–environment relations on sodic soils of Zimbabwe Rhodesia. *Journal of Ecology*, 68, 589–606.

ast, R. (1984). Rainfall, soil nutrient status and biomass of large African savanna mammals. *African Journal of Ecology*, 22, 245–70.

lwards, D. (1983). A broad-scale structural classification of vegetation for practical purposes. *Bothalia*, 14, 705–12.

Ellery, W.N., Scholes, R. J. & Mentis, M.T. (1991). An initial approach to predicting the sensitivity of the South African grassland biome to climate change. *South African Journal of Science*, 87, 499–503.

Ellis, R.P. (1990). Tannin-like substances in grass leaves. *Memoirs of the Botanical Survey of South Africa*, 59, 1–80.

Ellis, R.P., Vogel, J.C. & Fuls, A. (1980). Photosynthetic pathways and the geographical distribution of grasses in South West Africa/Namibia. *South African Journal of Science*, 76, 307–14.

Frazer, S.W., van Rooyen, T.H. & Verster, E. (1987). Soil-Plant relationships in the Central Kruger National Park. *Koedoe*, 30, 19–34.

Frost, P.G.H. (1985). The responses of savanna organisms to fire. In *Ecology and Management of the World's Savannas*, ed. J.C. Tothill & J.J. Mott, pp. 232–7. Canberra: Australian Academy of Science.

Frost, P.G.H., Menaut, J.C., Walker, B.H., Medina, E., Solbrig, O.T. & Swift, M. (1986). Responses of savannas to stress and disturbance. *Biology International Special Issue 10*. Paris: International Union of Biological Scientists.

Gandar, M. (1982). The dynamics and trophic ecology of grasshoppers (Acridoidea) in a South African savanna. *Oecologia*, 54, 370–8.

Gibbs Russell, G.E. (1987). Preliminary floristic analysis of the major biomes in southern Africa. *Bothalia*, 17, 213–27.

Giess, W. (1970). A preliminary vegetation map of South-West Africa. *Dinteria*, 4, 5–114.

Goodman, P.S. (1990). *Soil, Vegetation and Large Herbivore Relations in the Mkuzi Game Reserve, Natal*. PhD Thesis. Johannesburg: University of the Witwatersrand.

Grobbelaar, N. & Rosch, M. W. (1981). Biological nitrogen fixation in a Northern Transvaal savanna. *Journal of South African Botany*, 47, 493–506.

Guy, G.L. (1970). *Adansonia digitata* and its rate of growth in relation to rainfall in south central Africa. *Proceedings of the Transactions of the Rhodesian Scientific Association*, 54, 68–84.

Hogberg, P. (1989). Root symbioses of trees in savannas. In *Mineral Nutrients in Tropical Forest and Savanna Ecosystems*, ed. J. Proctor. pp. 121–36. Oxford: Blackwell.

Huntley, B.J. (1982). Southern African savannas. In *Ecology of Tropical Savannas*, ed. B.J. Huntley & B.H. Walker, pp. 101–19. Ecological Studies 42. Berlin: Springer-Verlag.

Huntley, B.J. & Walker, B.H. (ed.) (1982). *Ecology of Tropical Savannas*. Ecological Studies 42. Berlin: Springer-Verlag.

International Station Meteorological Climate Summary (1992). Database on CD-ROM. Asheville, North Carolina: National Climate Data Center.

Janse van Rensburg, J. (1982). *'n Outekologiese Studie van Enkele Plantsoorte op die Nylsvley-natuurreservaat*. MSc Thesis. Pretoria: University of Pretoria.

Jarvis, P.G. & McNaughton, K.G. (1986). Stomatal control of transpiration: scaling up from leaf to region. *Advances in Ecological Research*, 15, 1–45.

Johnson, & Tothill, J.C. (1985). Definition and broad geographic outline of savanna lands. In *Ecology and Management of the World's Savannas*, ed. J.C. Tothill & J.J. Mott, pp. 1–14. Canberra: Australian Academy of Science.

Justice, C.O., Scholes, R.J. & Frost, P.G.H. (1994). *African Savannas and the Global Atmosphere*. IGBP Report 31. Stockholm: International Geosphere-Biosphere Programme.

Kennard, D.G. & Walker, B.H. (1973). Relationships between tree canopy and *Panicum maximum* in the vicinity of Fort Victoria. *Rhodesian Journal of Agricultural Research*, 11, 145–53.

Knoop, W.T. & Walker, B.H. (1984). Interactions of woody and herbaceous vegetation in a southern African savanna. *Journal of Ecology*, 73, 235–53.

Kruger, F.J., Scholes, R.J. & Geldenhuys, C.J. (1993). Strategic assessment of land resources in southern Africa as a framework for sustainable development. In *National Veld Trust Jubilee Conference*, pp. 11–34. Pretoria: National Veld Trust.

Le Roux, C.J.G., Grunlow, J.O., Morris, J.W., Bredenkamp, G.J. & Scheepers, J.C. (1988). A classification of the vegetation of the Etosha National Park. *South African Journal of Botany*, **54**, 1–10.

Liengme, C.A. (1983). A survey of ethnobotanical research in southern Africa. *Bothalia*, **14**, 621–9.

Ludlow, A.E. (1987). *A Development Study of the Anatomy and Fine Structure of the Leaves of Ochna pulchra Hook*. PhD Thesis. Johannesburg: University of the Witwatersrand.

Milne, G. (1936). Some suggested units of classification and mapping, particularly for East African soils. *Soils Research*, **4**, 183–98.

Mueller-Dombois, D. & Ellenberg, H. (1974). *Aims and methods of vegetation ecology*. New York: John Wiley.

Nix, H.A. (1983). Climate of tropical savannas. In *Tropical Savannas*, ed. F. Bourliere, pp. 37–62. *Ecosystems of the World* 13. Amsterdam: Elsevier.

O'Connor, T.G. (1985). *A Synthesis of Field Experiments Concerning the Grass Layer in the Savanna Regions of Southern Africa*. South African National Scientific Programmes Report 114. Pretoria: CSIR.

O'Connor, T.G. (1991). Influence of rainfall and grazing on the compositional change of the herbaceous layer of a sandveld savanna. *Journal of the Grassland Society of Southern Africa*, **8**, 103–9.

O'Connor, T.G. (1992). Woody vegetation-environment relations in a semi-arid savanna in the northern Transvaal. *South African Journal of Botany*, **58**, 268–74.

O'Connor, T.G. & Pickett, G.A. (1992). The influence of grazing on seed production and seed banks of some African savanna grasslands. *Journal of Applied Ecology*, **29**, 247–60.

Owen-Smith, R.N. & Cooper, S.M. (1987). Palatability of woody plants to browsing ruminants in a South African savanna. *Ecology*, **68**, 319–31.

Owen-Smith, R. N. & Novellie, P. (1982). What should a clever ungulate eat? *The American Naturalist*, **119**, 151–78.

Rutherford, M.C. (1978). Primary production ecology in southern Africa. In *Biogeography and Ecology of Southern Africa*, ed. M.J.A. Werger, pp. 621–59. The Hague: Junk.

Rutherford, M.C. (1979). *Aboveground Biomass Subdivisions in Woody Species of the Savanna Ecosystem Project Study Area*. South African National Scientific Programmes Report 36. Pretoria: CSIR.

Rutherford, M.C. (1980). Annual plant-precipitation relations in arid and semi-arid regions. *South African Journal of Science*, **76**, 53–6.

Rutherford, M.C. (1981). Survival, regeneration and leaf biomass changes in woody plants following spring burns in *Burkea africana–Ochna pulchra* savanna. *Bothalia*, **13**, 531–52.

Rutherford, M.C. (1983). Growth rates, biomass and distribution of selected woody plant roots in *Burkea africana–Ochna pulchra* savanna. *Vegetatio*, **52**, 45–63.

Rutherford, M.C. (1984). Relative allocation and seasonal phasing of woody plant components in a South African savanna. *Progress in Biometeorology*, **3**, 200–21.

Sarmiento, G. (1984). *The Ecology of Neotropical Savannas*. Cambridge, Massachusetts: Harvard University Press.

Scholes, R.J. (1985). Drought-related grass, tree and herbivore mortality in a southern African savanna. In *Ecology and Management of the World's Savannas*, ed. J.H. Tothill & J.J. Mott, pp. 350–3. Canberra: Australian Academy of Science.

Scholes, R.J. (1988). *The Responses of Three Savannas on Contrasting Soils to Removal of the Woody Component*. PhD Thesis. Johannesburg: University of the Witwatersrand.

Scholes, R.J. (1990). The influence of soil fertility on the ecology of southern African savannas. *Journal of Biogeography*, **17**, 417–19.

Scholes, R.J. (1993). Nutrient cycling in semi-arid grasslands and savannas: its influence on pattern, productivity and stability. *Proceedings of the 17th International Grasslands Congress*. Palmerston North: International Grasslands Society.

Scholes, R.J., Ellery, W.N., Carter, G. & Blackmore, A.C. (1996). Plant functional types in African savannas and grasslands. In *Plant Functional Types*, ed. H. Schugart & I. Woodward. *Global Change Reports 1*. Cambridge: Cambridge University Press (in press).

Scholes, R.J & Hall, D.O. (1996) The carbon budget of tropical savannas, woodlands and grasslands. In *Global Change: Carbon Cycle in Coniferous Forests and Grasslands*, ed. J. M. Melillo & A. Breymeyer. New York: John Wiley & Sons (in press).

Scholes, R.J., Justice, C.O. & Ward, D. (1996). Trace gas emissions from biomass burning in Southern-Hemisphere Africa. *Journal of Geophysical Research* (in press).

Scholes, R.J. & Walker, B.H. (1993). *An African Savanna: Synthesis of the Nylsvley Study*. Cambridge: Cambridge University Press.

Scholtz, C.H. (1976). *Biology and Ecological Energetics of Lepidoptera Larvae Associated with Woody Vegetation in a Savanna Ecosystem*. MSc Thesis. Pretoria: University of Pretoria.

Siegfried, W.R. (1989). Conservation status of terrestrial ecosystems and their biota. In *Biotic Diversity in Southern Africa. Concepts and Conservation*, ed. B.J. Huntley, pp. 186–201. Cape Town: Oxford University Press.

Smith, T.M. & Goodman, P.S. (1986). The effect of competition on the structure and dynamics of *Acacia* savannas in southern Africa. *Journal of Ecology*, **74**, 1031–44.

Smith, T.M. & Grant, K. (1986). The role of competition in the spacing of trees in a *Burkea africana–Terminalia sericea* savanna. *Biotropica*, **18**, 219–23.

Stebbins, G.L. (1981). Coevolution of grasses and herbivores. *Annals of the Missouri Botanical Gardens*, **68**, 75–86.

Strang, R.M. (1966). The spread and establishment of *Brachystegia spiciformis* Benth. and *Julbernardia globiflora* Benth. Troupin in the Rhodesian highveld. *Commonwealth Forestry Review*, **48**, 26–40.

art-Hill, G.C., Tainton, N.N. & Barnard, H.J. (1987). The influence of an Acacia karroo on grass production in its vicinity. *Journal of the Grassland Society of Southern Africa*, **4**, 83–8.

ley, K.L. (1982). The influence of soil moisture balance on ecosystem patterns in southern Africa. In *Ecology of Tropical Savannas*, ed. B.J. Huntley & B.H. Walker, pp. 175–92, Berlin: Springer-Verlag.

pnell, C.G. (1959). Ecological results of woodland burning in Northern Rhodesia. *Journal of Ecology*, **47**, 161–72.

n der Meulen, F. & Morris, J.W. (1979). Striped vegetation patterns in a Transvaal savanna (South Africa). *Geo-Eco-Trop*, **3**, 523–66.

n Rooyen, N., Theron, G.K., Bezuidenhout, D., Bredenkamp, G.J. & van Rooyen, M.W. (1993). Patterns of change in the herbaceous layer of a mesic savanna, South Africa. *South African Journal of Botany*, **59**, 74–80.

n Rooyen, N., Theron, G.K. & Grobbelaar, N. (1981). A floristic description and structural analysis of the plant communities of the Punda Milia–Pafuri–Wambiya area in the Kruger National Park, Republic of South Africa: 3. The *Colophospermum mopane* communities. *Journal of South African Botany*, **47**, 585–626.

n Vegten, J. A. (1984). Thornveld invasion in a savanna ecosystem in eastern Botswana. *Vegetatio*, **56**, 3–7.

enendal, E. M. (1991). *Adaptive Strategies of Grasses in a Semi-arid Savanna in Botswana*. PhD thesis. Amsterdam: Amsterdam Vrije Universiteit.

gel, J.C., Fuls, A. & Ellis, R.M. (1978). The geographical distribution of Kranz grasses in South Africa. *South African Journal of Science*, **74**, 209–15.

n Maltitz, G.P. (1990). *The Effect of Spatial Scale on Patch Dynamics*. MSc Thesis. Johannesburg: University of the Witwatersrand.

alker, B.H. (1985). Structure and function of savannas: an overview. In *Ecology and Management of the World's Savannas*, ed. J.H. Tothill & J.J. Mott, pp. 350–3. Canberra: Australian Academy of Science.

Walker, B.H., Emslie, R.H., Owen-Smith, R.N. & Scholes, R.J. (1987). To cull or not to cull: lessons from a southern African drought. *Journal of Applied Ecology*, **24**, 381–401.

Walker, B.H., Ludwig, D., Holling, C.S. & Peterman, R.S. (1981). Stability of semi-arid savanna grazing systems. *Journal of Ecology*, **69**, 473–98.

Walker, B.H., Matthews, D.A. & Dye, P.J. (1986). Management of grazing systems – existing versus event-oriented approach. *South African Journal of Science*, **82**, 172–7.

Walker, B. H. & Noy-Meir, I. (1982). Aspects of the stability and resilience of savanna ecosystems. In *Ecology of Tropical Savannas*, ed. B.J. Huntley & B.H. Walker, pp. 556–609. Berlin: Springer-Verlag.

Walter, H. (1971). *Ecology of Tropical and Subtropical Vegetation*. London: Oliver & Boyd.

Werger, M.J.A. (1978). The Sudano-Zambesian region. In *Biogeography and Ecology of South Africa*, ed. M.J.A. Werger, pp. 301–462. The Hague: Junk.

Werger, M.J.A., Morris, J.W. & Louppen, J.M.W. (1979). Vegetation–soil relationships in the southern Kalahari. *Docums Phytosociolol*, **4**, 967–81.

Westoby, M., Walker, B.H. & Noy-Meir, I. (1989). Opportunistic management for rangelands not at equilibrium. *Journal of Range Management*, **42**, 266–74.

White, F. (1976) The underground forests of Africa: a preliminary review. *Gardens Bulletin (Singapore)*, **29**, 55–71.

White, F. (1983). *The Vegetation of Africa. Natural Resources Research XX*. Paris: UNESCO.

Whittaker, R.H., Morris, J.W. & Goodman, D. (1984). Pattern analysis in savanna woodlands at Nylsvley, South Africa. *Memoirs of the Botanical Survey of South Africa*, **49**, 1–51.

Wild, H. & Fernandes, A. (1968). *Flora Zambesiaca. Moçambique, Malawi, Zambia, Rhodesia, Botswana*. Salisbury, Rhodesia: M.O. Collins.

Wild, H. & Granvaux-Barbosa, L.A. (1967). *Vegetation map of the Flora Zambeziaca area*. Descriptive memoir (supplement to Flora Zambeziaca). Salisbury, Rhodesia: M.O. Collins.

Yeaton, R.I. (1988). Porcupines, fire and the dynamics of the tree layer in *Burkea africana* savanna. *Journal of Ecology*, **76**, 1017–29.

Yeaton, R. I., Frost, S. K. & Frost, P. G. H. (1986). A direct gradient analysis of grasses in a savanna. *South African Journal of Botany*, **82**, 482–6.

Zietsman, P. C., Grobbelaar, N. & van Rooyen, N. (1988). Soil nitrogenase activity of the Nylsvley Nature Reserve. *South African Journal of Botany*, **54**, 21–7.

Forest

J.J. Midgley, R.M. Cowling, A.H.W. Seydack and G.F. van Wyk

12.1   Introduction

Southern Africa is a largely semi-arid region (Chap. 2, this volume) not endowed with a large area of indigenous forests (probably less than 0.5% of the total land area, i.e. less than 5000 km²). The forest biome, as mapped by Rutherford & Westfall (1986; see also Chap. 5, this volume) is restricted to the largest block of continuous forest situated around Knysna on the southern coast of the Western Cape and covering an area of only 568 km² – only 0.02% of the area of southern Africa. The rest of the forested areas are very patchily distributed across the subcontinent (Fig. 12.1). Because of the limited area and economic value of forests, and also because most forests are situated far from the major universities, research in this biome has not moved much beyond a descriptive phase (Fig. 12.2). There is very little information on plant–plant or plant–animal interactions and energy or nutrient cycles, and very little experimental research has been undertaken.

White (1983), in his account of the vegetation of Africa, considered forests to be closed stands of trees taller than 10 m, and woodlands to be open stands of trees with canopies 8 to 20 m tall. His interpretation of bushland/thicket type is similar to woodland, except that trees are between 3 and 7 m tall, whereas his concept of scrub forest is intermediate between forest and thicket, with trees of 10 to 15 m in height. In this chapter we define forests as being closed-canopy plant communities comprising mainly woody plants more than 5 m tall. Patches of forest thus defined occur within almost all southern African biomes. For reasons given in 12.2.1 below, we also deal with aspects of the ecology of subtropical thicket, a closed shrubland, usually less than 5 m tall.

The patchiness of forests in southern Africa led

Acocks (1953) and others to hypothesize that the forests are the climatic climax over much of the eastern seaboard and southern coast of South Africa and that they are, therefore, relictual. Alternative hypotheses are that forest patchiness reflects recent expansion (e.g. Meadows & Linder 1993) or that forests have always been patchily distributed. Palaeoecological evidence for changes in the extent of forests during the Quaternary and in the last three centuries are reviewed in Chaps 4 and 21 (this volume), respectively.

The aims of this chapter are to describe the salient features of southern African forests and to review recent trends in forest research. For more detailed accounts of various topics we refer the reader to recent reviews and compilations such as those by Donald & Theron (1983) and White (1983; general description and phytochorology); Geldenhuys (1985; catalogue of forest research); Geldenhuys, Le Roux & Cooper (1986), Knight et al. (1987) and Chap. 22 (this volume; alien plants in forests); Geldenhuys & MacDevette (1989; diversity and conservation); Gordon (1989; KwaZulu-Natal forests); Geldenhuys (1990; biogeography); Everard (1987) and Zacharias, Stuart-Hill & Midgley (1991; subtropical thicket); Everard (1993; biogeography) and Midgley, Bond & Geldenhuys (1995a; conifer ecology).

12.2   Characterization

12.2.1   Floristics and phytochorology

White (1983) argued that southern African forests are associated with two phytochoria: an Afromontane archipelago-like regional centre of endemism (which extends

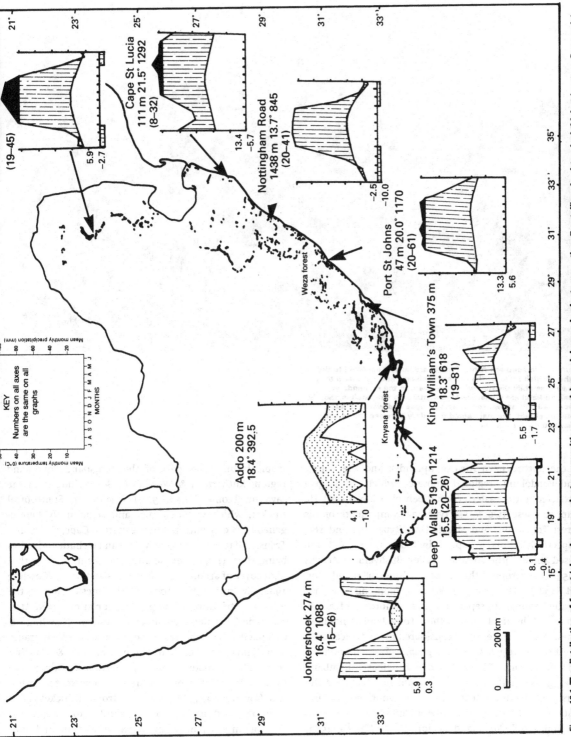

Figure 12.1 The distribution of forest (closed-canopy plant communities comprising mainly woody plants more than 5 m tall) and subtropical thicket (eastern Cape only) in southern Africa. Also shown are representative Walter–Lieth climate diagrams for Afromontane forests (Woodbush, Nottingham Road, Deep Walls, Jonkershoek), Indian Ocean forests (Cape St Lucia, Port St Johns) and subtropical thicket (King Williams Town, Addo).

Figure 12.2 **In southern Africa forests are largely confined to the higher rainfall areas of the eastern seaboard and even here they are very patchily distributed. Shown here in the midlands of KwaZulu-Natal are numerous Afromontane forest patches, mostly associated with steep, south facing slopes and water courses. Note the abrupt, fire-maintained boundary between forest and grassland** *(Photo: D. Edwards).*

to the northeastern African uplands); and the Tongaland–Pondoland regional mosaic (which is largely restricted to the eastern coastal belt of South Africa). He mapped these two forest types as components of his Forest Transitions and Mosaic Unit (types 19a and 16c, respectively). This grouping includes forest and the closed thicket [the so-called Valley Bushveld of Acocks (1953) or subtropical thicket of Cowling (1984) and Everard (1987)]. The scale of White's mapping was not detailed enough to separate the 'sand forests' of northern KwaZulu-Natal from other Tongaland–Pondoland types. Moll & White (1978) considered many typical sand forest species (e.g. *Balanites maughamii*, *Cleistanthus schlechteri* and *Newtonia hildebrandtii*) to be Zanzibar–Inhambane linking species (the Zanzibar–Inhambane Region is the coastal phytochorion extending from southern Mozambique to East Africa). This phytochorion, together with the Tongaland–Pondoland Region, form White's (1983) Indian Ocean Belt. White (1983) dealt with mangroves as 'azonal vegetation' and included swamp forest as a Tongaland–Pondoland type (16c). He also mapped sub-

tropical thicket as part of the Tongaland–Pondoland region; Rutherford & Westfall (1986) include it in the savanna biome (Chap. 5, this volume). Subtropical thicket, which is floristically and structurally heterogeneous, is concentrated in the eastern Cape, Ciskei and Transkei (Fig. 12.1). In general, it can be characterized as being short (1–5 m), dense and spinescent, often with a high cover of arborescent succulents and lianas (Cowling 1984; Everard 1987). Towards the mesic end of the thicket distribution it merges into forest of mixed Afromontane–Tongaland–Pondoland affinity (Cowling & Campbell 1983), whereas towards the xeric end it grades into Nama- or succulent karoo (Hoffman & Cowling 1990). Many woody plants in subtropical thicket are shrubby forms of trees that are well represented in forests (Cowling 1983). Moreover, subtropical thicket is not fire-prone and is functionally similar to forests, for example in nutrient-cycling processes and the high incidence of species with vertebrate-dispersed fruits (Cowling 1984). We, therefore, include it in this chapter (Table 12.1).

Table 12.1 **Dominant canopy species and environmental characteristics of selected Afromontane and Indian Ocean forest communities**

Site	Community number	Species	Canopy height (m)	Environment	Source
Afromontane					
Soutpansberg, Northern Province	1	*Acacia ataxacantha, Brachylaena discolor, Cussonia spicata, Nuxia floribunda, Rhus chirindensis, Vepris lanceolata, Zanthoxylum davyi*	10–15	Inselberg Geology: sandstone Alt: 1000–1500 m Rainfall: 800 mm	Geldenhuys & Murray (1993)
	2	*Celtis africana, Cussonia spicata, Drypetes gerrardii, Kiggelaria africana, Nuxia floribunda, Rhoicissus tomentosa, Xymalos monospora*			
	3	*Cassipourea gerrardii, Celtis africana, Diospyros whyteana, Ocotea kenyensis, Xymalos monospora*			
Wolkberg, Northern Province	1	*Olea capensis* ssp. *macrocarpa, Oxyanthus speciosus* ssp. *gerrardii, Pterocelastrus echinatus, Rinorea angustifolia, Syzygium gerrardii, Tricalysia lanceolata, Trichilia dregeana, Xymalos monospora*	10–15	Great Escarpment Geology: quartzites Alt: 1200–1700 m Rainfall: 500–1350 mm	Geldenhuys & Pieterse (1993)
	2	*Cryptocarya libertiana, Cussonia spicata, Oricia bachmannii, Trichilia dregeana*			
	3	*Bersama transvaalensis, Ekebergia pterophylla, Nuxia floribunda, Rhus chirindensis*			
Drakensberg, KwaZulu-Natal		*Maytenus peduncularis, Podocarpus latifolius, Pterocelastrus echinatus, Scolopia mundii*	10–15	Great Escarpment Geology: sandstone/shale/ dolerite Alt: 1300–1850 m Rainfall: 1400 mm	Everard *et al.* (1995)
Weza, KwaZulu-Natal		*Casearia gladiiformis, Ocotea bullata, Podocarpus henkelii, Vepris lanceolata*	15–25	Midlands Mistbelt Geology: sandstone/shale/ dolerite Alt: 1000–1500 m Rainfall: 1250 mm	Everard *et al.* (1995)
Ngome, KwaZulu-Natal		*Cassipourea* spp., *Drypetes natalensis, Podocarpus falcatus, Syzygium gerrardii*	20	Midlands Mistbelt Geology: sandstone/shale/ dolerite Alt: 900–1300 m Rainfall: 1530 mm	Everard *et al.* (1995)
Transkei Mountains, Eastern Cape	1	*Kiggelaria africana, Ocotea bullata, Podocarpus henkelii, Xymalos monospora*	10–20	Inland Mountains Geology: sandstone/dolerite/ shale Alt: 600–1100 m Rainfall: 900–1200 mm	Cawe (1990)
	2	*Celtis africana, Podocarpus falcatus, P. henkelii, P. latifolius, Scolopia mundii*			
Drakensberg, Free State	1	*Clausena anisata, Diospyros whyteana, Olinia emarginata, Podocarpus latifolius, Scolopia mundii*	10	Great Escarpment, inland margin Geology: dolerite Alt: 1200 m Rainfall: 500 mm+	Du Preez *et al.* (1991)
	2	*Celtis africana, Diospyros whyteana, Maytenus undata, Olea europaea* ssp. *africana, Pittosporum viridiflorum*	10		Du Preez & Bredenkamp (1991)
Knysna, Western Cape		*Ocotea bullata, Olea capensis* ssp. *macrocarpa, Podocarpus falcatus, P. latifolius*	25	Coastal Platform Geology: sandstone/shale Alt: 50–300 m Rainfall: 1000–1500 mm	Geldenhuys & MacDevette (1989)

Table continued

Table 12.1 (*cont.*)

Site	Community number	Species	Canopy height (m)	Environment	Source
Cape Peninsula		*Cassine peragua, Olea capensis* ssp. *capensis, O. europaea* ssp. *africana, Olinia ventosa, Podocarpus latifolius*	10–15	Cape Folded Belt Geology: granite Alt: 50–250 m Rainfall: 900–1100 mm	McKenzie, Moll & Campbell (1977)
Indian Ocean Maputaland, KwaZulu-Natal		*Cleistanthus schlechteri, Hymenocardia ulmoides, Newtonia hildebrandtii, Pteleopsis myrtifolia, Wrightia natalensis*	10–15	Coastal Plain Geology: recent sands Alt: 5–150 m Rainfall: 750 mm	Everard *et al.* (1995)
Sodwana, KwaZulu-Natal		*Cassine aethiopica, Cassipourea gerrardii, Catunaregam spinosa, Cavacoa aurea*		Coastal Plain Geology: recent dunes Alt: 25–100 m Rainfall: 1070 mm	Everard *et al.* (1995)
Yengele, KwaZulu-Natal		*Allophylus natalensis, Euclea natalensis, Mimusops caffra, Sideroxylon inerme*	15	Coastal Plain Geology: recent dunes Alt: 25–50 m Rainfall: 1170 mm	Everard *et al.* (1995)
Dukuduku, KwaZulu-Natal		*Celtis africana, Chaetacme aristata, Strychnos decussata, S. madagascariensis*	15	Coastal plain Geology: recent sand Alt: 15–65 m Rainfall: 1240 mm	Everard *et al.* (1995)
Ngoya, KwaZulu-Natal		*Drypetes gerrardii, Garcinia gerrardii, Millettia sutherlandii, Trichilia dregeana*	8	Coastal Scarp Geology: granite/schist Alt: 350–500 m Rainfall: 1100 mm	Everard *et al.* (1995)
Transkei Coast, Eastern Cape	1	*Cordia caffra, Maytenus peduncularis, Mimusops caffra, Sideroxylon inerme, Trichilia dregeana*	10–20	Coast Geology: recent dunes/shales Alt: 50–500 m Rainfall: 1000–1500 mm	Cawe (1990)
	2	*Buxus macowanii, Clerodendrum myricoides, Heywoodia lucens, Millettia grandis, Oricia bachmanii, Vepris lanceolata*			Cawe (1990)
	3	*Chaetacme aristata, Cola natalensis, Heywoodia lucens, Millettia sutherlandii, Trichilia dregeana, Vitellariopsis marginata*			Cawe (1990)
Albany, Eastern Cape	1	*Acacia karroo, Apodytes dimidiata, Brachylaena elliptica, Burchellia bubalina*	3–10	Coastal forelands Geology: shales Alt: 50–500 m Rainfall: 550–900 mm	Everard (1987)
	2	*Cassine aethiopica, Dovyalis rhamnoides, Euphorbia triangularis, Maytenus heterophylla, Scutia myrtina*	2–5		Everard (1987)
Humansdorp, Eastern Cape		*Cassine aethiopica, Euclea racemosa, Olea exasperata, Pterocelastrus tricuspidatus, Sideroxylon inerme*	3–5	Coastal forelands Geology: recent dunes Alt: 15 m Rainfall: 650 mm	Cowling (1984)

KwaZulu-Natal has the greatest diversity of forest types (Table 12.1). MacDevette *et al.* (1990) suggested that there were two basic types: the interior (Afromontane) and the coastal (Indian Ocean or Tongaland–Pondoland; Fig. 12.3), and these two basic types were further divided into two and five types, respectively. The two Afromontane types are the coast scarp and undifferentiated [comprising the midlands (mistbelt) and western (Drakensberg or montane)] types. The five coastal types are the tropical dry ('sand forests'), riverine, swamp,

(a)

(b)

(c)

(d)

Figure 12.3 **Some Indian Ocean forest types.** (*a*) **Dune forest at Banga Nek, northern Kwazulu-Natal** *(Photo: E.J. Moll)*; (*b*) **swamp forest (*Ficus trichopoda, Myrica serrata*) at Sodwana Bay, Northern KwaZulu-Natal** *(Photo: J. Vahrmeijer)*; (*c*) **species- and endemic-rich forest associated with ancient sandstones of the Cape Supergroup in the Umzimkhulu River valley, southern KwaZulu-Natal** *(Photo: E.J. Moll)*; (*d*) **subtropical thicket near East London in the Eastern Cape. The arborescent succulent, *Euphorbia triangularis*, is conspicuous. Evergreen trees include *Schotia latifolia* (Fabaceae) and *Harpephyllum caffrum* (Anacardiaceae)** *(Photo: D.M. Comins)*.

mangrove and undifferentiated coastal (comprising coastal plain and dune forests). Riverine, swamp and mangrove forests are fairly discrete types, owing to their specialized habitats (Chap. 13, this volume). Tinley (1967) also suggested the existence of only two major forest types in KwaZulu-Natal, namely a moist evergreen type and a tropical dry semi-deciduous ('sand forest') type. He preferred the term Afrotemperate to Afromontane, in part because elements of this type grow almost at sea-level in the western Cape where latitude compensates for altitude (Table 12.1; Fig. 12.4). However, many Afromontane species are also found on the oligotrophic coastal sands of KwaZulu-Natal, where the climate is distinctly subtropical (Chap. 2, this volume).

Cawe (1994) and Cawe, Moll & McKenzie (1994) also concluded that there are two basic types in the Transkei:

Figure 12.4 **Afromontane forest reaches sea level in the temperate southern Cape. Here *Podocarpus falcatus* emerges from the canopy of a forest on narrow alluvial plain, immediately below sandstone coastal cliffs, at Nature's Valley near Knysna** *(Photo: J.C. Paterson-Jones).*

an Afromontane type at higher altitudes and a subtropical type at lower altitudes towards the coast (Table 12.1). Rainfall is an important discriminatory variable in the subtropical type, distinguishing between moist, medium-moist, dry and dune subtypes. Among the Afromontane types, moisture (including soil water-holding capacity) and altitude appeared to discriminate among four types, namely: high-altitude, middle-altitude, moist and mistbelt types. The Swaziland forests show strong similarities with the essentially Afromontane forests of the Eastern Transvaal and KwaZulu-Natal (Masson 1994).

Whatever classification system is used, there is considerable overlap between types. For example, the coastal forests near Kosi Bay in northern KwaZulu-Natal include *Raphia australis* (coastal), *Podocarpus falcatus*, *Rapanea melanophloeos* (both Afromontane), *Hymenocardia ulmoides* (sand forest) and *Ficus trichopoda* (swamp forest) (J.J. Midgley pers. obs.). This emphasizes the fact that forest types exist as a continuum in southern Africa. There is, therefore, endless scope for classification and reclassification, depending on the aims of categorization. In this chapter we recognize, pragmatically, two major forest complexes: Afromontane types associated with a warm temperate environmental regime, and Indian Ocean types associated with subtropical climates (Table 12.1). We use as a term Indian Ocean in preference to Tongaland–Pondoland, since some of the forests of Maputaland in northeastern KwaZulu-Natal represent an intrusion of the Zanzibar–Inhambane region into southern Africa (Moll & White 1978).

One purpose of classification is to provide insight into aspects such as conservation (which types deserve higher conservation priority?) and explanations for distribution patterns (which types are where and why?). The distribution patterns of forest species are the result of many factors, including history (e.g. length of time for dispersal from centres of origin), dispersal (e.g. distribution patterns determined by dispersers), habitat requirements (e.g. bioclimate), biotic interactions (e.g. competition) and disturbance (e.g. incidence of fire). Little work has been done in determining the relative importance of these factors in different southern African forest types.

The Afromontane forests give the impression of having a southern, and thus temperate, origin, because of the presence of Gondwanan elements such as the Cunoniaceae, Cupressaceae, Podocarpaceae and Proteaceae (Levyns 1962; Table 12.1). The Indian Ocean forests, on the other hand, appear to have a tropical origin. Therefore, the distribution patterns of these two subdivisions of the forest floras could be interpreted as the result of historical patterns determined by limited dispersal from two geographically separated source areas.

These 'apparent' patterns are misleading for at least two reasons. Firstly, the temperate Gondwanan component is so small as to be unimportant; the Afromontane flora is actually dominated by 'tropical' taxa (White 1978). Africa has apparently never harboured characteristic Gondwanan forest taxa such as *Nothofagus* (Linder, Meadows & Cowling 1992). Unfortunately, there is no comprehensive phylogenetic study of important austral taxa such as the Podocarpaceae and Proteaceae to determine whether they have tropical or austral origins. The other point is that the southwards migration of the tropical flora and northwards migration of the putative southern flora has been so extensive across the continent that the floras have been well mixed in many areas, both now and in the past (Chap. 4, this volume). Many of the so-called austral species have extensive distribution ranges throughout Africa. Therefore, distribution patterns are unlikely to be much influenced by historical patterns such as limited dispersal from centres of origin.

Midgley *et al.* (1995a) investigated the distribution patterns of the Podocarpaceae (classic indicators of Afromontane forests) in southern African forests. They argued that biotic interactions (as opposed to history or site-specific requirements) were probably the most important factors in explaining distribution patterns of these species. Furthermore, they suggested that podocarp distribution patterns were determined by ecological interactions, such as competition with faster-growing angiosperms, and not history or lack of dispersal (podocarps have vertebrate-dispersed seeds), nor by problems with inefficient wind pollination in species-rich forest communities. As was discussed in Chap. 3 (this volume), most forest tree species, like the podocarps, have very wide distribution ranges. This, too, suggests that dispersal and highly specific habitat requirements have not been important in determining distribution patterns. Therefore, the differences between the Afromontane and Indian Ocean types, if the dichotomy is to be used, may, in part, be determined by biotic interactions rather than narrow habitat requirements.

Gentry (1988) argued that familial composition is fairly constant between forest types, such as between temperate and tropical floras on different continents. Therefore, species-rich tropical forests tend to be dominated by Fabaceae, Moraceae, Annonaceae, Lauraceae and Euphorbiaceae. These patterns are evident in the more strongly tropical Indian Ocean forests of southern Africa, where Fabaceae and Euphorbiaceae are dominant. Afromontane forests, especially in the south, are dominated by Podocarpaceae, Oleaceae, Celastraceae and Rubiaceae (Table 12.1).

12.2.2 Structure

Structural attributes and structural classifications of the forests of southern Africa have not been widely studied. Phillips (1931) produced a classification of the Afromontane forest in the Knysna area based on forest height. Rogers & Moll (1975) and Cawe (1986) used basal area, rather than stems ha^{-1}, for classifying forests. Huntley (1984) suggested that different forest types could be classified by basal area, with Afromontane forests in the range between 40 and 50 m^2 ha^{-1} and Indian Ocean forests with <40 m^2 ha^{-1}. Cawe (1986) found a large range (5–85 m^2 ha^{-1}) in basal area for his plots in the Afromontane forests of the Transkei. Geldenhuys & MacDevette (1989) indicated the rough proportion of various growth forms in a range of forest types. Few clear patterns emerged from this study, except that ferns are more abundant in Afromontane than in coastal forests.

12.3 Boundaries

The boundaries of forests are determined by numerous factors that operate at many scales; from climatic patterns at scales of kilometres to smaller-scale patterns of forest in fire-protected or mesic microsites. Here we discuss the role of climate and fire and review the processes that establish boundaries between forests and other vegetation types.

12.3.1 Climate

Rutherford & Westfall (1986) used a summer-aridity index (SAI) and a winter-rainfall index to discriminate among the biomes of southern Africa (see also Chap. 5, this volume). They suggest that an annual rainfall of 625 mm is an approximate lower limit for forest development, although in winter-rainfall areas, the minimum is less (525 mm) and in summer-rainfall areas, the minimum is greater (725 mm). Their SAI lower limit for forests varied from 2.5 in summer-rainfall areas to 4.5 in winter-rainfall areas. Furthermore, Rutherford & Westfall (1986) used climatic indices, based mainly on rainfall and surrogate variables for temperature, to map vegetation types. This led, in some instances, to the misclassification of forest areas owing to 'climatic anomalies'. For example, the presence of the tall mixed Afromontane–Indian Ocean forest in a low-rainfall area of the eastern Cape (Marker & Russell 1984) may in part be due to the fog that increases effective precipitation, or due to the fact that the water-holding capacity of the deep, sandy substratum is adequate to sustain forest growth.

Ellery, Scholes & Mentis (1991) identified climatic factors that were associated with various biomes in southern Africa. Their model was based on the duration of the growth season and provided a reasonable separation of biomes. They also invoked climatic controls on fire regime (Chap. 18, this volume) as a mechanism for the distribution of the biomes, including forest. Because their climatic indices match contemporary vegetation patterns, they argue that Acocks' (1953) notion that forests are relictual is incorrect. Many researchers have suggested that frost limits the establishment of forest vegetation in the high-altitude areas of southern Africa (Chap. 10, this volume). However, there are no data on the frost tolerance of forest species.

12.3.2 Fire and fire-responses

In many parts of southern Africa the boundary between forests and adjacent biomes is very abrupt, but does not follow environmental discontinuities. The abruptness of the boundary is due to fire, which is an important determinant of forest boundaries and forest composition (Fig. 12.5; Chap. 18, this volume). In the western Cape it is clear that fire determines the boundary between fynbos and forest over much of the area (see also Chaps 6 and 18, this volume). Forest species readily invade many fynbos sites when fire is excluded from fynbos (Luger & Moll 1992). Furthermore, Midgley & Bond (1990) suggest that 'islands' of fynbos in the Knysna forest region are the result of post-glacial forest expansion. At a finer scale, Geldenhuys (1994) suggested that distribution patterns of Afromontane forests in the Knysna area are determined by fires associated with warm, dry (or 'berg') winds. Also, charcoal is frequently found in soils in the Afromontane forests of the southern Cape (Scholtz 1983). This suggests that forests do burn, although there are no data on forest fire regimes.

There are few data on the responses of forest species

Figure 12.5 **Fire-free sites such as this rock outcrop at Giant's Castle in the KwaZulu-Natal Drakensberg provide the nucleus for forest establishment** *(Photo: D.M. Richardson).*

to components of the fire regime (Granger 1984). Everard (1985) noted that floristic differences in some Afromontane KwaZulu-Natal forests were probably a function of fire history. Many forest and subtropical thicket species are capable of sprouting, both in response to disturbance (e.g. fire or treefall) and as a normal process of ramet recruitment (Philips 1931; Midgley & Cowling 1993; Van Wyk, Everard & Midgley 1996). H. Homan & J.J. Midgley (unpubl. data) noted that three common Afromontane tree species (*Cunonia capensis*, *Kiggelaria africana* and *Rapanea melanophloeos*) acquired strong fire-resistance within five years of establishment. Saplings and adults, even of the apparently fire-sensitive species, *Podocarpus latifolius*, were able to survive a degree of canopy scorch.

The ability of a plant species to resprout has other life-history consequences (Le Maitre & Midgley 1992). In sprouters, resources allocated to thick bark and a bud bank could otherwise be allocated to growth or reproduction. Consequently, seedlings of resprouters should grow slower than those of reseeders, although resprouts

(ramets) may be able to grow more quickly than seedlings. Furthermore, the fecundity of resprouters should be lower than that of reseeders. Finally, resprouters, because of their multi-stemmed structure, seldom grow very tall, because each stem is poorly supported. Kruger, Midgley & Cowling (1996) have developed a model which suggests that short forests should be dominated by resprouting species and tall forests dominated by reseeding species. For example, short subtropical thicket is dominated by resprouting species and seedlings are notoriously scarce, whereas reseeding species and seedlings are more conspicuous in tall thicket and forest (Phillips 1931; Midgley & Cowling 1993).

12.3.3 Forest invasion into other biomes

Forest invasion of adjacent biomes, such as fynbos or grassland, generally occurs in the absence of fire and has two pathways. Firstly, many fire-tolerant forest species are components of these adjacent, fire-prone systems (e.g. species of *Diospyros*, *Euclea*, *Maytenus*, *Olea* and *Rhus*). In the absence of fire, they can grow taller, expand vegetatively and eventually suppress the surrounding fire-dependent vegetation (Manders & Richardson 1992; Cowling *et al.* 1997). Once this has happened, these forest nuclei will remain unburned in all but the most intense fires, because forest species are less flammable than the graminoids and shrubs of the surrounding matrix (Van Wilgen, Higgins & Bellstedt 1990). Thickening up of forest species may occur in areas where fires are excluded from an area actively (via management) or physically (via landform) (e.g. Luger & Moll 1992), or where heavy grazing, via fuel reduction, prevents fires or reduces their intensity. Forest succession may also occur as a result of genet recruitment. Many forest species have fleshy fruits that are eaten by birds and deposited under perches; seedlings which establish in these sites begin a nucleation process (Knight 1988; Manders & Richardson 1992; Cowling *et al.* 1997; Chap. 6, this volume). This pattern has been documented: in KwaZulu-Natal (Smith & Goodman 1987) where fleshy-fruited *Euclea divinorum* establishes under *Acacia nilotica*; in the fynbos/forest transition (Knight 1988; Manders & Richardson 1992; Manders, Richardson & Masson 1992); and in the fynbos/subtropical thicket transition (Cowling *et al.* 1997). Although the role of fire is probably critical in this invasion process, further work is necessary before it can be fully understood. For example, as was mentioned above, most of the fleshy-fruited forest species are vigorous sprouters and they are therefore, to some extent, fire-resistant. The role and relative importance of recruitment via seeds versus resprouts needs clarification.

In many parts of the country, depending on disturbance regimes, it is possible to have two alternative stable vegetation states (Chaps 10 and 11, this volume). This poses major problems for any models that try to correlate the distribution patterns of biomes with climatic data (e.g. Rutherford & Westfall 1986; Ellery *et al.* 1991). These states may differ in many characteristics, including structure (e.g. broad-leaved Afromontane forest versus fine-leaved fynbos shrublands) and dynamics (tree-by-tree replacement versus stand-destroying fires). Richardson *et al.* (1995) discuss the implications of the presence of either the forest or fynbos alternative states in terms of their impact on ecosystem processes and biodiversity. Unless the critical role of fire is considered, the enormous differences between the states that can occur in the same environment undermine much of convergence theory. In this context, Bond & Midgley (1995) have shown that adaptation to fire by way of increased inflammability could be a significant factor in determining which of these alternative stable states occurs in a particular area (see also Chap. 18, this volume).

12.4 Plant structure and function

12.4.1 Reproductive ecology

There has been almost no research on the reproductive biology of forest trees since Phillips' (1931) major contribution. Geldenhuys (1989) analysed the breeding systems of the Knysna forest flora and found that bisexuality predominated (311 out of 398 woody species). Dioecy, as expected, was more common in woody plants (51 out of 52 taxa). No correlation analysis was performed between breeding systems and dispersal, although it seems likely that dioecy and fleshy-fruitedness are highly correlated in the Knysna forest flora (cf. Bawa 1980). Small dry seeds (159 taxa) predominated over small fleshy fruits (111 taxa) and large fruits for specialized disperal are rare in the Knysna area.

12.4.2 Seed dispersal

Knight & Siegfried (1983) showed that species with fleshy (drupes and berries) and dry (pods, capsules and nuts) fruits are equally numerous in the southern African tree flora (including savanna and arid land species), although berry-producing species far out-number species with drupes. Fruits eaten by birds tend to be brightly coloured, whereas those dispersed by mammals are more cryptically coloured. Regional patterns in fruit types are complex (Knight 1986). Species with pods and arils occur mainly in arid areas (e.g. Kalahari), whereas mammal- and bird-dispersed species are best-represented along the humid east coast. At a finer scale, Midgley (1993) showed that sand forests are dominated by species with wind-dispersed seeds, whereas other Indian Ocean forests are dominated by fleshy-fruited species. However, a common sand forest species, *Balanites maughamii*, possesses large, date-like fruits that are dispersed by elephants (G.F. van Wyk, unpubl. data). Besides descriptive studies, very little work has been done on the actual dispersal process. Frost (1980) describes some features of fruit quality and dispersal patterns for selected species from a dune forest in KwaZulu-Natal. Knight (1987) studied the fruit display of eight tree and shrub species common in coastal thicket and Afromontane forest in the Western Cape. He found that the early colonizing species, *Chrysanthemoides monilifera* and *Myrica cordifolia*, exhibit intensive, strongly seasonal fruit displays, thereby ensuring efficient and effective dispersal (see also Knight 1988). *Sideroxylon inerme*, a dominant species of coastal thickets, maintains a large fruit crop (c. 50 000 seeds per tree) almost continuously. Thus, it exhibits a maximized strategy, acting as a keystone species for avian frugivores in these habitats. The remaining species, *Colpoon compressum, Euclea racemosa, Maytenus acuminata, Myrsine africana* and *Rapanea melanophloeos*, have small and prolonged fruit displays. With the exception of *C. compressum*, fruits of these species are infrequently consumed. Nonetheless, fruits of some of these species, especially *R. melanophloeos*, are available at times of overall fruit scarcity, suggesting a role as 'short-term' keystone species for frugivores.

Following Howe & Estabrook (1977; see also Howe 1993), Knight (1987) tested the relationship between avian frugivory and fruit quantity and quality in a western Cape coastal thicket dominated by *Cassine maritima, Olea exasperata, Rhus* spp. and *Sideroxylon inerme*. Contrary to predictions of coevolutionary models, he found that the overall use of fleshy fruits by birds was in direct proportion to the availability of fruits. A limited preference existed among frugivores for species bearing fleshy fruits that were continuously available (e.g. *Sideroxylon inerme*) or were seasonally predictable (e.g. *Chrysanthemoides monilifera*).

12.4.3 Seed germination

There are very few quantitative data on the germination ecology of southern African forest plants. Cowling *et al.* (1997) demonstrated faster rates and more successful germination in shade than in open sunlight for several dune forest and subtropical thicket species. Relatively high germination (50–80%) was shown for many species after 360 days, including *Sideroxylon inerme*, a dominant dune forest species that has a notoriously low seedling abundance (Midgley & Cowling 1993).

Almost no research has been done on the effects of bird ingestion on the germination of bird-dispersed forest fruits in southern Africa (cf. Barnea, Yom-Tov and

Friedman 1991). Seeds of *Euclea racemosa* and *Sideroxylon inerme* showed significantly faster and more complete germination after manual removal of pulp and ingestion by redwinged starlings, *Onychognathus morio*, than untreated controls (Cowling *et al.* 1997).

12.4.4 Seed banks

Seed banks of southern African forest species have not been studied in any detail. Pierce & Cowling (1991) demonstrated the absence of a persistent seed bank in a depauperate Afromontane forest in the Eastern Cape. Most forest species that are bird-dispersed, e.g. *Podocarpus* spp., lack dormant seeds (Geldenhuys 1993). An interesting variation is *Podocarpus falcatus*, a bat-dispersed species, which displays dormancy (J. Farrant, pers. comm.).

12.4.5 Growth forms

Bews (1925) pioneered the study of growth-forms in South Africa, but his descriptions and analyses were not at the biome level. Midgley, Van Wyk & Everard (1995d) investigated variation in leaf size and margin type and whether leaves were compound or simple in a subset of southern African forests. Leaf attributes were fairly similar in all forests. Leaves are generally evergreen; sclerophyllous (Chap. 16, this volume); micro- or mesophyllous; lacking in spines; with entire margins; and non-compound (Cowling & Campbell 1983; Midgley *et al.* 1995a). Only the sand forests of KwaZulu-Natal have a conspicuous component of deciduous species. The distribution and abundance of lianas and climbers, at least in southern Cape forests, is determined by the architecture of host trees, rather than directly by climatic or soil factors (Balfour & Bond 1993). Shorter forests, associated with drier, warmer sites (i.e. low forest and thicket of Indian Ocean affinity) have an abundance of fine branches at a height suitable for colonization by lianas; this is not the case for taller Afromontane forests at cooler, wetter sites.

12.5 Richness, diversity and endemism

12.5.1 Rarity and endemism

There have been few studies of distribution patterns of forest species that document levels of endemism. Nevertheless, it is safe to say that at local and regional scales, tree endemism is low (Cowling 1983; White 1983; Chap. 3, this volume). Similarly, few tree species are rare (MacDevette *et al.* 1989), and uncommon species are generally those in ecosystems such as sand forests that barely penetrate southern Africa from the northeast. Given the fairly large distribution of sand forest in adjacent countries such as Mozambique, genuine rarity of trees in the subcontinent is probably low. The one exception is the Pondoland Centre, where several rare endemics are associated with an outcrop of Cape Supergroup sandstone – many are palaeoendemics (Van Wyk 1990; Chap. 3, this volume). Van Wyk (1990) suggests that this relictual flora has persisted because of reduced competition on the unusual and infertile sandstone substratum. This view has been challenged by Cawe (1994), who noted that some of the endemics occur on non-sandstone substrata; he suggests that the high level of endemism is due to the unusually high rainfall of the Pondoland Centre. Levels of endemism and rarity are low in subtropical thicket, with the exception of geophytes and dwarf to low succulent shrubs in the Euphorbiaceae and Mesembryanthemaceae (Cowling 1983; Cowling & Holmes 1991; Moolman & Cowling 1994; Chap. 3, this volume). This component is adversely affected by heavy goat grazing on farmlands and moderately impacted by elephant grazing in reserves (Midgley & Joubert 1991; Stuart-Hill 1992; Moolman & Cowling 1994).

12.5.2 Regional richness

The correlates of regional richness in southern African forests are discussed in Chap. 19 (this volume; see also Geldenhuys 1992). In this section we focus on latitudinal patterns of the regional richness of the forest flora. It has long been known that both regional and local diversity of southern African tree species declines in a southwesterly direction from the northeastern Transvaal to the Cape Peninsula (e.g. Tinley 1985). Geldenhuys (1989) showed that 162 forest species drop out southwestwards from Humansdorp to Mossel Bay, whereas only 23 species drop out in the reverse direction. A necessary corollary to this is that the species in the more southerly areas tend to have large distribution ranges (since the basic pattern is one of species dropping out, rather than a decline in local endemism). Therefore, many of the species in the more temperate forests of the south (e.g. *Apodytes dimidiata*, *Podocarpus latifolius* and *Rapanea melanophloeos*) have wide distribution patterns in southern Africa (and indeed in Africa; White 1978). The pattern then is of a southwards decline in richness with the more southerly species having very wide distributions.

This pattern is consistent with a poleward decline in richness observed for many other groups of organisms, as predicted by Rapoport's Rule (Stevens 1989) and species-energy theory (Currie 1991). O'Brien (1993) showed that at the scale of 20 000 km² quadrats, species richness of southern African trees is strongly correlated with precipitation and energy (but see Chap. 19, this volume). She demonstrated a curvilinear (uni-modal) relationship between richness and minimum (winter) available energy: highest richness was recorded in the more productive northeastern areas with lowest values in the low-energy winter-rainfall and high-energy arid regions. However, Latham & Ricklefs (1993) have pointed out global anomalies in the relationship between tree richness and available energy. They contend that a historical (time) hypothesis is the best explanation for the latitudinal trends in richness. Most trees have a tropical origin and idiosyncratic effects on migration and speciation dominate over ecological controls on regional richness. Similarly, the low richness of trees observed by O'Brien (1993) in the winter-rainfall region could be a result of the anomalously low diversity of trees in the fynbos biome (Campbell, McKenzie & Moll 1979; Richardson & Cowling 1992; see also Chap. 19, this volume).

Stevens (1989) showed that Rapoport's Rule (species from higher latitudes have larger mean distribution ranges than more equatorial species) was upheld for trees. Unfortunately the distribution ranges of most southern African tree species are not sufficiently known for a rigorous analysis of Stevens' (1989, 1992) hypotheses. Nevertheless, it is safe to say that tree diversity declines with latitude, and that species from the higher latitudes tend to have wider distribution ranges. Stevens (1989) argued that the reason for this pattern is that only generalists can cope with the increasing variability of conditions at higher latitudes. A problem with this hypothesis is that many of the generalists are also present in the more equable tropical areas. Rohde (1992) has criticized Stevens (1992) on the basis that the pattern may be an artifact, owing to a few low-latitude species reaching high latitudes. Rohde postulated that the increase in species richness in the tropics was due to 'greater effective evolutionary time' and the greater 'evolutionary speed' in the tropics, the latter because of shorter generation times, faster mutation rates and faster selection. If species : genus ratios reflect rates of evolution then there is some support for Rohde's (1992) hypothesis for the southern African forest flora. Species : genus ratios for southern African forest floras are generally low (Geldenhuys 1990), and the genera that are rich in species (e.g. *Isoglossa*, *Plectranthus*) are either herbaceous or comprised mainly of small trees and shrubs (*Diospyros*, *Maytenus*, *Pavetta*, *Rhus*). This implies that patterns of speciation are linked to generation times, but further work is needed on this topic.

12.5.3 Diversity

Much remains to be studied regarding the diversity of southern African forests, and standards have yet to be set regarding plot size, minimum plant size and the guilds that are to be sampled. Nevertheless, despite the

Table 12.2 **Correlations with species richness in 400-m² plots in several KwaZulu-Natal forests (Everard *et al.* 1993)**

Forest	No. plots	Correlation (*r*) with richness		
		Evenness	No. stems	Basal area
Ngome	39	0.61	0.53	0.60
Dukuduku	20	0.49	0.61	0.17
Sodwana	34	0.81	0.80	0.52
Mapelane	88	0.77	0.63	0.04
Yengele	22	0.80	0.71	0.03

abundance of sampling problems, the available data indicate that alpha diversity declines southwards and with increasing altitude. Thus, Indian Ocean forests of KwaZulu-Natal have higher richness than the Afromontane forests at higher altitudes, and coastal and montane forests of the eastern and western Cape are poorer in species than forests of similar structure in KwaZulu-Natal (Geldenhuys & MacDevette 1989; Everard, Van Wyk & Midgley 1994).

At the 0.01-ha scale, southern African forests have richness values of 50–100 species (Geldenhuys & MacDevette 1989): levels that would be expected for their latitude on the basis of Gentry's (1988) global survey. Mean richness in 400-m² plots ranges from 28 in Indian Ocean forests of KwaZulu-Natal to 15 in the Afromontane forests of the Drakensberg foothills (Everard *et al.* 1994). Diversity at this scale drops to a mean of eight species in the montane forests on the Cape Peninsula (J.J. Midgley unpubl. data; see also Campbell & Moll 1977).

Two potentially important correlates of species richness are regional richness and evenness. The southwards decline in local richness corresponds with the parallel decline in regional richness (see 12.5.2), suggesting that local richness may be constrained by the size of the regional species pool (see Chap. 19, this volume for further details). Everard *et al.* (1994) found no strong relationship between mean evenness (Shannon–Wiener Index) and mean richness in 400-m² plots from different forest complexes when compared across all of these complexes. However, variation within forests is fairly strongly and positively correlated with evenness (Table 12.2). Variation in evenness in southern African forests is fairly low, as is evenness itself (using the number of stems as an importance value). Gentry (1988) found very high evenness values (>5) in some tropical forests and a very strong relationship between evenness and species richness. Richness in southern African forests does not appear to be linked to variation in basal area of a plot, but is often correlated with numbers of stems within a plot, at least for the forests investigated in KwaZulu-Natal.

Table 12.3. **Species–area relationships for some forest types in KwaZulu-Natal (double logarithmic regression analysis; based on the data by Everard *et al.* 1993)**

Forest type	No. forests	r^2	z
All	84	0.12	0.15
Scarp	16	0.17	0.17
Mistbelt	40	0.18	0.15
Lowland	9	0.17	0.65
Dune	5	0.98	0.40
Montane	9	0.04	0.09

12.5.4 Species–area relationships

Everard *et al.* (1994) investigated species–area relationships at the scale of 100–1000 ha for a number of forests in KwaZulu-Natal. They found lower intercepts and slopes for the Afromontane forests when compared to the Indian Ocean types (Table 12.3). Geldenhuys (1992), working at a larger scale, also found poor correlations between species number and area and a slope (z value) of only 0.14.

At a finer scale, Everard *et al.* (1994) investigated species–area relationships for smaller (mainly 1–50 ha) forest patches at Weza (a KwaZulu-Natal Afromontane forest) and at Dukuduku (KwaZulu-Natal Indian Ocean forest) (Table 12.1). Here, z values were again low (0.10 and 0.26, respectively). Bond & Euston-Brown (1993) reported a z value of 0.15 for Afromontane forest patches on the Cape Peninsula.

Area thus appears to have only a modest influence on species richness in southern African forests. In general, log–log models are weak, with low r^2 and z values. Midgley (1993) speculated that species–area curves are relatively flat for forest tree species, because edge effects augment richness in small patches. The low z values of forests contrast strongly with that ($z = 0.434$) found by Bond *et al.* (1988) for fynbos, a vegetation essentially without an edge effect. Another possible reason for the flat and weak species–area curves for forest trees is that beta diversity in southern African forests is low. There are very few studies of beta diversity in forests to test this idea. Everard *et al.* (1994) showed that cumulative species richness flattens off very quickly (after sampling between two and six 400-m² plots) in KwaZulu-Natal forests. Again this is in stark contrast to patterns found by Gentry (1988) in equatorial forests where high beta diversity prevents any flattening-off of cumulative species richness. Low beta diversity in forests would then be a reflection of a well-mixed and dispersed flora containing few habitat specialists (see also Chap. 19, this volume). Finally, extinction rates of forest trees in small patches may be low, because many tree species are good persisters (they are long-lived and can resprout almost indefinitely). Although demographic stochasticity is clearly an important determinant of extinction rates in small animal populations, thereby influencing species-area curves, its role in affecting species-area relationships for trees is not clear.

Another approach to investigating area effects is to use 'incidence functions' (*sensu* Diamond 1975) to determine whether some species are restricted to either small or large forest patches. Bond & Euston-Brown (1993) and Midgley (1993) investigated incidence functions for forest patches on the Cape Peninsula and on the western shores of Lake St. Lucia in KwaZulu-Natal, respectively. They found some evidence of area effects for trees and suggested that these patterns were due to fire effects rather than strict area effects. Small islands tend to be dominated by fire-resistant species and only as patches became relatively fire-proof did fire-sensitive species invade.

Everard *et al.* (1994) demonstrated a high degree of nestedness (*sensu* Wright & Reeves 1992) of tree species in Afromontane forest patches of various sizes in the Weza area of KwaZulu-Natal. Strong nestedness was also reported for Indian Ocean forest patches at Lake St. Lucia (G.F. van Wyk, J.J. Midgley & D.E. Everard, unpubl. data). These findings suggest that forest patches are not a random assemblage of species. Where the patches are relictual, high nestedness indicates an orderly process of extinction; where forest patches are the result of recent invasion, high nestedness indicates an orderly process of succession.

12.6 Dynamics

12.6.1 Populations

Most southern African tree species are difficult to age. However, data on size-class structure of many trees in different forests have provided some insights on the population structure of southern African forests. In the predominantly Afromontane forests of the eastern and western Cape, most species have a negative exponential distribution of stems in relation to size-class categories (Midgley *et al.* 1991; Midgley 1992; Midgley & Gobetz 1993; Geldenhuys 1994), whereas in the many Indian Ocean forests in KwaZulu-Natal, size-class distributions show a predominance of larger individuals, suggesting little *in situ* regeneration (Midgley, Everard & Van Wyk 1995c). The generalization is that proportionally more species in the Indian Ocean forests (Cawe 1990; Midgley *et al.* 1991; Everard *et al.* 1994) have flatter size-class distributions than do species in the Afromontane forests (Cawe 1986; Midgley *et al.* 1991; Midgley 1992). Furthermore, there are fewer individuals in the former forests (Midgley *et al.* 1991; Midgley 1992; Van Wyk *et al.* 1996).

These differences are probably due to the different life histories of the trees in the different forest types (Fig. 12.6). In particular, many Afromontane species are shade-tolerant (Midgley *et al.* 1991) and can accumulate large banks of saplings since the establishment of seedlings is not strongly linked to gap formation. In contrast, shade-intolerant species should accumulate fewer recruits, because usually less than 10% of a forest canopy is broken by gaps at any one time (Midgley *et al.* 1991). This suggests that densities of recruits of shade-tolerant and shade-intolerant species should be different. Additionally, many species in the tall Afromontane forests regenerate from saplings (genets), whereas many species in coastal forests regenerate from sprouts (ramets).

However, in some Indian Ocean forests there are

(a)

(b)

Figure 12.6 **Interior views of different forest types. (a) Afromontane forest in the KwaZulu-Natal midlands, showing dominance of the tree component by large specimens of *Podocarpus latifolius*; understorey stratum comprises *Carissa bispinosa* and *Diospyros whyteana*; sparse field layer, which probably includes a large seedling bank of podocarps and other canopy dominants** *(Photo: E.J. Moll);* **(b) Indian Ocean forest at Ngoye on the north coast of KwaZulu-Natal, showing a wide range of diameter classes amongst the canopy tree component (the large specimen on the left is *Milletia sutherlandii* and those on the right are *Syzygium gerrardii*); the presence of lianas and buttress roots; and the dense field layer** *(Photo: D. Edwards).*

fewer recruits of shade-intolerant species than there are adults (Midgley *et al.* 1995c). This may be because something has 'gone wrong' with recruitment processes (e.g. excessive predation of tree seeds or seedlings by small ungulates, or the absence of dispersal agents such as elephants) or disturbance processes (such as an absence of megaherbivores to create openings or to thin out understorey species). Whatever the cause, these observations suggest imminent successional change (Midgley *et al.* 1995c).

12.6.2 Succession and dynamics

The Afromontane forests around the Knysna region have been subject to long-term monitoring. Here, Van Daalen (1991) analysed a 35-year record of growth and mortality for a 44-ha stand which had previously experienced various degrees of exploitation. Over this period, rates of basal area increment declined to about $2 \, m^3 \, yr^{-1}$ and net growth was less than $0.5\% \, yr^{-1}$. Stand basal area appeared to stabilize at around $35 \, m^3$. Mortality of stems with d.b.h greater than 100 mm [dbh = diameter at 1.3 m (breast height)], measured over a 15-year period (1972–1987), amounted to 10.6% (i.e. less than 1% per year). Mortality was poorly correlated with size (dbh). These data were used to calculate the mean stand half-life (88.7 years) and the doubling time (100.12 years) (J.J. Midgley, unpubl. data). In a five-year study of an Afromontane forest on the Cape Peninsula, monitoring of stems with a dbh >100 mm gave a mean stand half-life of 262.2 years and the doubling time was 228 years (J.J. Midgley unpubl. data). This suggests that Afromontane forests have an extremely low turnover and a high degree of stability (*sensu* Korning & Balslev 1994). Data for the Indian Ocean forests are lacking.

Geldenhuys & Maliepaard (1983) and Midgley, Cameron & Bond (1995b) have studied patterns of disturbance and regeneration in the Knysna forest. Rates of disturbance are low (only 2–10% of the canopy occurred as measurable gaps) and most (70%) trees died standing. This implies that biotic factors rather than environmental factors (e.g. wind) dominate disturbance patterns in this forest. A Markov model of gap-maker versus gap-filler species (Midgley *et al.* 1995b) showed high compositional stability for most species, despite an absence of consistent species-level disturbance or replacement patterns. This is because most species are shade-tolerant and therefore recruit close to adults. Most gaps are extremely small with a mean diameter : height ratio of 0.24 and canopy gap : expanded gap ratio of 0.3. This system is probably functioning like a lottery, with species replacement patterns after tree death being largely determined by the frequency of individuals of different species in the sapling bank.

Figure 12.7 **Rapid regeneration of dune forest following the initial establishment of *Acacia karroo* (now c. 20 m tall) after agricultural abandonment at Mapelane, northern Kwazulu-Natal coast** *(Photo: E.J. Moll).*

This fine-scale pattern of replacement is not evident in Indian Ocean forests. Van Wyk *et al.* (1996) showed evidence of successional change at a coarse level in the Dukuduku forest in KwaZulu-Natal. Unlike the Knysna forest, seedling and canopy composition showed little similarity, thus suggesting little *in situ* recruitment by many species. Thus, many species have size-class distributions which change between successional stages (Fig. 12.7). For example, pioneer species in late-successional stages tend to be relatively rare and have a predominance of large stems. In contrast, climax species are more numerous and have inverse-J distribution curves

(Van Wyk *et al.* 1996). A less subtle form of succession is occurring on the Mlalazi dunes in KwaZulu-Natal where the deposition of river-borne silt is creating a prograding coastline (Van Daalen *et al.* 1986; see also Chap. 13, this volume).

Most woody species in subtropical thicket produce basal sprouts, and some species produce creeping above- or below-ground stems that can produce aerial shoots (Midgley & Cowling 1993). Mortality rates of these multi-stemmed species are low. The internal dynamics of sub-tropical thicket seem to depend on vegetative recruit-ment; there is an abundance of ramets and a notable absence of genets (seedlings) (Phillips 1931; Cowling & Pierce 1988; Midgley & Cowling 1993). In a survey of many sites, Midgley & Cowling (1993) found that seed-ling regeneration of subtropical thicket species was restricted mainly to arborescent succulents. Cowling *et al.* (1997) showed that seedling establishment is import-ant in the invasion of thicket by forest species in rela-tively mesic coastal dune sites where regular fire pre-vents forest invasion over most of the landscape. Generally, subtropical thicket appears to be a very stable vegetation type, and most species are tenacious persisters.

Holmes & Cowling (1993) investigated the effect of shade on leaf and photosynthetic attributes of seedlings of several species of subtropical thicket and coastal forest. They found a range of responses to shading. The forest species *Sideroxylon inerme* and *Cassine peragua* were tolerant of shade, but pioneer species (*Rhus glauca*) and species characteristic of dry thickets (*Pappea capensis* and *Schotia afra*) were less shade-tolerant, suggesting that they might require gaps for seedling recruitment.

12.7 Conservation and management

12.7.1 Fragmentation

Southern African forests are fairly well conserved in the sense that many exist in formal conservation areas and, given the wide ranges and commonness of most species (Geldenhuys & MacDevette 1989) the reserve network is likely to include most forest taxa and communities. For-ests have certainly become increasingly fragmented in recent times as a result of human activities (Chap. 21, this volume). Conservation of these protected, yet iso-lated, islands of forest now depends on the preservation of species and the maintenance of ecosystem processes within and between these islands. Southern African for-ests are, however, probably naturally patchily distrib-uted and further fragmentation may not necessarily lead to cascading extinctions.

12.7.2 Utilization

Only the Afromontane forests in the Knysna region are harvested in a systematic and controlled manner. It has been estimated that between 1.5 and 3 million m³ of timber was removed from these forests between 1776 and 1939 (whereafter harvesting was temporarily halted); harvesting was most intense at the beginning of the 19th century when 20 000 m³ of timber was removed annually (Von Breitenbach 1968; Donald & Theron 1983). The pressure on these forests declined as afforestation with *Eucalyptus* and *Pinus* species increased. The Amatole forests in the eastern Cape and Weza in southern Kwa-Zulu-Natal are utilized to a lesser degree.

The Knysna forests are managed for nature conser-vation, the sustained utilization of forest products and outdoor recreation. Various management systems have been applied to the forests during the past decades (Von Breitenbach 1968; Von Dem Bussche 1975; Donald & Theron 1983). Sustainable timber harvesting is currently practised according to the 'Senility Criteria Yield Regu-lation System' (SCYRS) (Seydack 1995; Seydack *et al.* 1995). Conventional yield-regulation systems attempt to utilize the net growth in a particular forest; they focus on har-vesting pre-mature trees and avoid very large or sen-escent individuals (Baur 1964; Poore 1989). The SCYRS that is currently used in the Knysna forests differs from conventional systems in that low-increment (i.e. dying or senescent) trees are harvested (Fig. 12.8).

A yield-regulation system must stipulate how many trees can be harvested per hectare and the spatiotempo-ral layout of these trees (spatial grain of harvesting oper-ations, felling cycle) and it must define which trees are harvestable (selection criteria). In the Knysna forests a ten-year felling cycle applies. It represents a compromise between the cost-effectiveness of management, which favours long felling cycles; and the prevailing disturb-ance regime (Geldenhuys & Maliepaard 1983; Midgley *et*

Figure 12.8 **Depot for wood selectively harvested from the Afromontane forests of the southern Cape** *(Photo: J.C. Paterson-Jones).*

Table 12.4 **Population turnover in comparison with actual harvest rates (1991–1993) (from Seydack *et al.* 1995).**
All data relate to trees with diameters of ≥300 mm at 1.3 m

| Species | Population turnover: ten-year mortality | | Harvest percentages over compartments (1991–1993)[c] | | Mean diameters of harvested trees at 1.3 m |
	% age[a]	Stems ha⁻¹[b]	Mean	Range	(mm)
Olea capensis macrocarpa	8.2	3.95	5.5	2.4–10.9	443
Podocarpus latifolius	3.2	1.11	2.9	1.5–7.1	443
Pterocelastrus tricuspidatus	5.0	0.95	5.8	0.9–13.6	441
Apodytes dimidiata	5.2	0.68	6.9	1.7–12.7	384
Curtisia dentata	7.1	0.80	4.6	1.5–9.8	393
Psydrax obovata	9.3	0.85	7.8	3.5–17.3	431
Platylophus trifoliatus	18.9	1.59	7.0	2.8–20.0	411
Ocotea bullata	8.1	0.58	4.6	0.9–15.3	394
Cunonia capensis	14.3	0.63	4.0	1.3–11.5	385
Podocarpus falcatus	1.9	0.06	0.5	0.0–2.9	591
Olinia ventosa	3.6	0.11	3.0	0.0–5.7	555
Diospyros whyteana	11.6	0.13	4.4	2.8–8.3	443
Mean	6.4	11.4	4.9		428

[a] Ten-year mortality percentages for the Deep Walls research compartments A11 + A22 (67.7 ha): 1972/1974–1987/1990.
[b] Mortality percentages applied to regional stem density data.
[c] Data from full-tally monitor compartments marked during 1991–1993 (542.2 ha).

al. 1991, 1995b), which favours short felling cycles. Most trees in this forest die standing, implying a protracted period of decline and/or stagnation. By assessing the canopy condition of each tree, the dying/stagnating trees can be identified and selected for harvesting (Seydack *et al.* 1995). These selection criteria are aimed at the pre-emption of mortality. The harvested trees will be replaced by 'waiting' recruits, because most species produce large banks of seedlings (Midgley *et al.* 1991), and there are no species-specific replacement patterns (Midgley *et al.* 1995b). Selection criteria have been established for each tree species so that the most senile trees (equivalent in number to the associated ten-year mortality rates) can be identified (Seydack *et al.* 1995).

The canopy species harvested under the SCYRS in the Knysna forests are listed in Table 12.4; of these, the most sought-after are *Ocotea bullata* (stinkwood), *Podocarpus latifolius* (real yellowwood) and *Olinia ventosa* (hard pear). Between eight and ten canopy trees ha⁻¹ are potentially harvestable every ten years. Table 12.4 also provides a comparison between mortality and harvesting rates (percentages of trees ≥300 mm dbh harvested), and an indication of how harvesting rates differ between compartments.

The SCYRS has the following advantages (Seydack *et al.* 1995): sustainability, compatibility with the aims of nature conservation, optimal productivity, managerial simplicity and flexibility. Its limitations are that: the existing species composition of the forest is likely to be maintained (this may not be desirable); timber quality is generally poorer (since the dying trees that are harvested are susceptible to fungal infection); and tree selection and harvesting are expensive.

Yield-regulation systems must fit in with prevailing circumstances, such as differences in forest dynamics or tree life-histories. This is clearly illustrated by the need for different yield-regulation systems for the Amatole (eastern Cape) and Weza (KwaZulu-Natal) forests. In the eastern Cape, only the two yellowwood species (*Podocarpus falcatus* and *P. latifolius*) are currently harvested. The mortality patterns of these species are clearly different from those in the southern Cape forests where individuals of these species usually die as a result of breakage of an apparently healthy crown. Mortality cannot be pre-empted in such circumstances, and trees are harvested after they have died or when most of the crown has died back. The harvest selection criteria are obvious (i.e. recent wind-throws), and the selection of harvest trees can be left to contractors. The harvesting operations are very dispersed, because with such an approach the felling cycle has to be very short (one year in this case). In contrast, the yield regulation differs for the the moister parts of the Weza forests, which are dominated by two species, *Ocotea bullata* and *Podocarpus henkelii*. The latter species has a very extended mature phase, with many trees becoming internally hollow and thus of limited use for timber. Under such circumstances, the SCYRS is of limited use, but a more appropriate system has yet to be developed.

The monitoring of forest dynamics in harvested and unharvested stands is an important part of yield regulation in the long term. Such monitoring is done in a number of permanent monitoring areas and sample-plot systems where trees of ≥100 mm dbh are marked (Seydack 1991). Full-count monitoring of all trees ≥300 mm dbh is also done before each harvesting event in selected harvesting compartments (Seydack *et al.* 1995).

Monitoring becomes even more important for safeguarding sustainability when harvesting deviates from the natural disturbance regime. An example here is the commercial harvesting from the forests of the seven-weeks fern (*Rumohra adiantiformis*) for florist greenery (Chap. 20, this volume). The income from fern harvesting is in the same order of magnitude as that presently realized from indigenous tree species. The harvesting interval is stipulated as 15 months and only 50% of the pickable fronds per plant may be removed. However, this fern species is typically adapted to a nutrient-poor environment and has associated life-history features (Milton 1987, 1991). This suggests that a high rate of artificial defoliation may have negative consequences in the long term. Monitoring the effects of long-term harvesting on frond size, reproductive output and yield is therefore essential.

12.8   Conclusions

Southern African forests may be ordinated along a continuum from relatively depauperate warm temperate (Afromontane) types to species-rich Indian Ocean types of subtropical affinity. There is also a gradient in the scale of dynamics, from the fine-grained Afromontane forests with many shade-tolerant species, to the coarse-grained Indian Ocean types with more shade-intolerant elements. Regeneration in Afromontane forest is dominated by recruitment from saplings, whereas in the subtropical types, recruitment is often achieved by resprouting.

Research in southern African forests has been dominated by descriptive studies. Only recently have attempts been made to understand the dynamics of different forest types. Clearly, much more research is required. Some priorities are given below.

1   There is an urgent need for research to provide guidelines for the conservation of forests while allowing sustainable harvesting of forest products (Chap. 20, this volume). Many forests far away from cities (and notably the sand forests of northern KwaZulu-Natal) are being threatened by non-sustainable utilization. In some instances,

forests are still being cleared for conventional agriculture.

2   Attention must be given to identifying critical ecological processes that need to be maintained to conserve the full species complement of forest patches. The fact that forests are fairly well represented in nature reserves (Chap. 23, this volume) does not imply that the ecological processes that maintain biodiversity (especially dispersal, pollination and disturbance) are fully operational.

3   More work is required to elucidate the role of persistence versus recruitment in tree demography. Many studies on the dynamics of trees (e.g. Midgley *et al.* 1991) have probably over-rated the role of seedlings (recruitment) versus resprouts (persistence). Researchers searched in vain (Midgley & Cowling 1993) before realizing that seedlings are probably redundant in subtropical thicket and low forest where almost every species recruits from basal sprouts or subaerial shoots. Interactions among resprouts are difficult to study and new techniques are needed for studying dynamics in these vegetation types. For example, where resprouters dominate, the ability of one species to penetrate the canopy and overtop another species may be more critical than the ability of seedlings to pre-empt space in gaps.

4   There are no data on ecosystem-level patterns and processes in southern African forests. There is an urgent need to investigate the role of forests in the maintenance of hydrological and pedological processes in water catchments and riverine habitats.

12.9   Acknowledgements

We thank Shirley Pierce and Dave Richardson for very useful comments on several drafts of this chapter.

12.10   References

Acocks, J.P.H. (1953) Veld Types of South Africa. *Memoirs of the Botanical Survey of South Africa*, **28**, 1–192.

Balfour, D.A. & Bond, W.J. (1993). Factors limiting climber distribution and abundance in a southern African forest. *Journal of Ecology*, **81**, 93–9.

Barnea, A., Yom-Tov, Y. & Friedman, J. (1991). Does ingestion by birds affect seed germination? *Functional Ecology*, **5**, 394–402.

Baur, G. N. (1964). *The Ecological Basis of Rainforest Management*. Sydney: Forestry Commission of New South Wales.

Bawa, K.S. (1980). Evolution of dioecy in flowering plants. *Annual Review of Ecology and Systematics*, **11**, 15–39.

Bews, J.W. (1925). *Plant Forms and their Evolution*. London: Longmans, Green and Co.

Bond, W.J. & Euston-Brown, D. (1993). Do forest species require minimum forest sizes? A study in Cape peninsula forests In *The Relevance of Island Biogeographic Theory in Commercial Forestry*, ed. D.A. Everard, pp. 72–80. Environmental Forum Report. Pretoria: Foundation for Research Development.

Bond, W.J. & Midgley, J.J. (1995). Kill thy neighbour: an individualistic argument for the evolution of flammability. *Oikos*, **73**, 79–85.

Bond, W.J., Midgley, J.J & Vlok, J. (1988). When is an island not an island; insular effects and their causes in fynbos shrublands. *Oecologia*, **77**, 512–21.

Campbell, B.M., McKenzie, B. & Moll, E.J. (1979). Should there be more tree vegetation in the mediterranean climate region of South Africa? *Journal of South African Botany*, **45**, 543–7.

Campbell, B.M & Moll, E.J. (1977). The forest communities of Table Mountain, South Africa. *Vegetatio*, **34**, 105–10.

Cawe, S.G. (1986). *A Quantitative and Qualitative Survey of the Inland Forests of Transkei*. MSc Thesis. Umtata: University of Transkei.

Cawe, S.G. (1990). *Coastal Forest Survey: a Classification of the Coastal Forests of Transkei and an Assessment of Their Timber Potential*. Umtata: University of Transkei.

Cawe, S.G. (1994). Rainfall and vegetation patterns in Transkei and environs. *South African Journal of Science*, **90**, 79–85.

Cawe, S.G., Moll, E.J. & McKenzie, B. (1994). An evaluation of the phytochorological classification of the forests of Transkei. In *Proceedings of the 13th Plenary Meeting of AETFAT*, ed. J.H. Seyani & A.C. Chikuni, pp. 1043–59. Zomba, Malawi: Association for the Taxonomic Study of the Flora of Tropical Africa.

Cowling, R.M. (1983). Phytochorology and vegetation history in the southeastern Cape, South Africa. *Journal of Biogeography*, **10**, 393–414.

Cowling, R.M. (1984). A syntaxonomic and synecological account of the vegetation of the Humansdorp region of the fynbos biome. *Bothalia*, **15**, 175–227.

Cowling, R.M. & Campbell, B.M. (1983). A comparison of fynbos and non-fynbos coenoclines in the lower Gamtoos River Valley, south east Cape. *Vegetatio*, **53**, 161–78.

Cowling, R.M. & Holmes, P.M. (1991). Subtropical thicket in the south eastern Cape: a biogeographical perspective. In *Proceedings of the First Valley Bushveld/Subtropical Thicket Symposium*, ed. P.J.K. Zacharias, G.C. Stuart-Hill & J. Midgley, pp. 3–4. Howick: Grassland Society of Southern Africa.

Cowling, R.M., Kirkwood, D., Midgley, J.J. & Pierce, S.M. (1997). Invasion and persistence of bird-dispersed, subtropical thicket and forest species in fire-prone fynbos. *Journal of Vegetation Science* (in press).

Cowling, R.M. & Pierce, S. M. (1988). Secondary succession in coastal dune fynbos: variation due to site and disturbance. *Vegetatio*, **76**, 131–9.

Currie, D.J. (1991). Energy and large-scale patterns of animal- and plant-species richness. *American Naturalist*, **137**, 27–49.

Diamond, J.M. (1975). Assembly of species communities. In *Ecology and Evolution of Communities*, ed. M.L. Cody & J.M. Diamond, pp. 342–444. Cambridge, Massachusetts: Belknap.

Donald, D.G.M. & Theron, J.M. (1983). Temperate broad-leaved evergreen forests of Africa south of the Sahara. In *Temperate Broad-Leaved Evergreen Forests*, ed. J.D. Ovington, pp. 135–68. Amsterdam: Elsevier.

du Preez, P.J. & Bredenkamp, G.J. (1991). The syntaxonomy and synecology of the forests in the eastern Orange Free State, South Africa. II. The Pittsporetalea viridiflorum. *South African Journal of Botany*, **57**, 207–12.

du Preez, P.J., Bredenkamp, G.J. & Venter, H.J. (1991). The syntaxonomy and synecology of the forests in the eastern Orange Free State, South Africa. I. The Podocarpetalia latifolia. *South African Journal of Botany*, **57**, 182–206.

Ellery, W.N., Scholes, R.J. & Mentis, M.T. (1991). An initial approach to predicting the sensitivity of the South African grassland biome to climate change. *South African Journal of Science*, **87**, 499–503.

Everard, D.A. (1985). The effects of fire on the *Podocarpus latifolius* forests of the Royal Natal National Park, Natal Drakensberg. *South African Journal of Botany*, **52**, 60–6.

Everard, D.A. (1987). A classification of subtropical transitional thicket in the eastern Cape, based on synecological and structural attributes. *South African Journal of Botany*, **53**, 320–40.

Everard, D.A. (ed.) (1993). *The Relevance of Island Biogeographic Theory in Commercial Forestry*. Environmental Forum Report. Pretoria: Foundation for Research Development.

Everard, D.A., Midgley, J.J. & van Wyk, G.F. (1993). Biogeography of forest patches in Natal: preliminary findings. In *The Relevance of Island Biogeographic Theory in Commercial Forestry*. Environmental Forum Report, ed. D.A. Everard. Pretoria: Foundation for Research Development.

Gerard, D.A., Midgley, J.J. & van Wyk, G.F. (1995). Dynamics of some forests in KwaZulu-Natal, South Africa, based on ordinations and size-class distributions. *South African Journal of Botany*, **61**, 283–92.

Gerard, D.A., van Wyk, G.F. & Midgley, J.J. (1994). Disturbance and the diversity of forests in Natal, South Africa: lessons for their utilisation. *Strelitzia*, **1**, 275–86.

Frost, P.G.H. (1980). Fruit–frugivore interactions in a South African coastal dune forest. In *Congressum Internationalem Ornithologiai*, ed. R. Nohring, pp. 1179–84.

Geldenhuys, C.J. (1985). *Annotated Bibliography of South African Indigenous Evergreen Forest Ecology. South African National Scientific Programmes Report* 107, pp. 1–125. Pretoria: CSIR.

Geldenhuys, C.J. (1989). *Environmental and Biogeographic Influences on the Distribution and Composition of Southern Cape Forests (Veld Type 4)*. PhD Thesis. Cape Town: University of Cape Town.

Geldenhuys, C.J. (ed.) (1990). *Biogeography of the Mixed Evergreen Forests of Southern Africa*. Occasional Report 45, Ecosystem Programmes. Pretoria: Foundation for Research Development.

Geldenhuys, C.J. (1992). Richness, composition and relationships of the floras of selected forests in southern Africa. *Bothalia*, **22**, 205–33.

Geldenhuys, C.J. (1993). Reproductive biology and population structures of *Podocarpus falcatus* and *P. latifolius* in southern Cape forests. *Botanical Journal of the Linnean Society*, **112**, 59–74.

Geldenhuys, C.J. (1994). Bergwind patches and the location of forest patches in the southern Cape landscape, South Africa. *Journal of Biogeography*, **21**, 49–62.

Geldenhuys, C.J., Le Roux, P.J. & Cooper, K.H. (1986). Alien invasions in indigenous evergreen forest. In *Ecology and Management of Biological Invasions in South Africa*, ed. I.A.W. Macdonald, F.J. Kruger & A.A. Ferrar, pp. 119–31. Cape Town: Oxford University Press.

Geldenhuys, C.J. & MacDevette, D.R. (1989). Conservation status of coastal and montane evergreen forests. In *Biotic Diversity in Southern Africa. Concepts and Conservation*, ed. B.J. Huntley, pp. 224–38. Cape Town: Oxford University Press.

Geldenhuys, C. J. & Maliepaard, W. (1983). The causes and sizes of canopy gaps in the southern Cape forests. *South African Forestry Journal*, **124**, 50–5.

Geldenhuys, C.J. & Murray, B. (1993). Floristic and structural composition of Hangklip Forest in Soutpansberg, Northern Transvaal. *South African Forestry Journal*, **165**, 9–20.

Geldenhuys, C.J. & Pieterse, F.J. (1993). Floristic and structural composition of Wonderwoud Forest in the Wolkberg, North-eastern Transvaal. *South African Forestry Journal*, **164**, 9–17.

Gentry, A.H. (1988). Changes in plant community diversity and floristic composition on environmental and geographic gradients. *Annals of the Missouri Botanical Gardens*, **75**, 1–34.

Gordon, I.G. (ed.) (1989). *Natal Indigenous Forests: a Preliminary Collection of Reports on Indigenous Forests in Natal*. Pietermaritzburg: Natal Parks Board.

Granger, J.E. (1984). Fire in forest. In *Ecological Effects of Fire in South African ecosystems*, ed. P. de V. Booysen & N.M. Tainton, pp. 177–97. Berlin: Springer-Verlag.

Hoffman, M.T. & Cowling, R.M. (1990). Desertification in the Sundays River Valley, South Africa. A multivariate approach. *Journal of Arid Environments*, **19**, 105–17.

Holmes, P.M. & Cowling, R.M. (1993). Effects of shade on seedling growth, morphology and leaf photosynthesis in six subtropical thicket species from the eastern Cape, South Africa. *Forest Ecology and Management*, **61**, 199–220.

Howe, H.F. (1993). Specialized and generalized dispersal systems: where does the paradigm stand? *Vegetatio*, **107**, 3–13.

Howe, H.F. & Estabrook, G.F. (1977). On intraspecific competition for dispersal agents in tropical trees. *American Naturalist*, **111**, 817–32.

Huntley, B.J. (1984). Characteristics of South African Biomes. In *Ecological Effects of Fire in South African Ecosystems*, ed. P. de V. Booysen & N.M. Tainton, pp. 2–17. Berlin: Springer-Verlag.

Knight, R.S. (1986). Interrelationships between fruit types in southern African trees and environmental variables. *Journal of Biogeography*, **13**, 99–108.

Knight, R.S. (1987). *Aspects of Plant Dispersal in the Southwestern Cape with Particular Reference to the Roles of Birds as Dispersal Agents*. PhD Thesis. Cape Town: University of Cape Town.

Knight, R.S. (1988). Fruit displays of indigenous and invasive alien plants in the southwestern Cape, South Africa. *South African Journal of Botany*, **52**, 249–55.

Knight, R.S., Geldenhuys, C.J., Masson, P.H., Jarman, M.L. & Cameron, M.J. (eds) (1987). *The Role of Aliens in Forest Edge Dynamics*. Occasional Report 22, Ecosystem Programmes. Pretoria: Foundation for Research Development.

Knight, R.S. & Siegfried, W.R. (1983). Distribution and evolution of aril-bearing trees in southern Africa. *South African Journal of Botany*, **1**, 117–23.

Korning, J. & Balslev, H. (1994). Growth and mortality of trees in Amazonian tropical rain forest in Ecuador. *Journal of Vegetation Science*, **4**, 77–86.

Kruger, L.M., Midgley, J.J. & Cowling, R.M. (1996). Recruitment versus persistence: a model of forest dynamics based on canopy height. *Functional Ecology*, (in press).

Latham, R.E. & Ricklefs, R.E. (1993). Global patterns of tree species richness in moist forests: energy-diversity theory does not account for variation in species richness. *Oikos*, **67**, 325–33.

Le Maitre, D.C. & Midgley, J.J. (1992). Plant reproductive biology. In *The Ecology of Fynbos. Nutrients, Fire and Diversity*, ed. R.M. Cowling, pp. 135–74. Cape Town: Oxford University Press.

Levyns, M. R. (1962). Possible antarctic elements in the South African flora. *South African Journal of Science*, **37**, 85–107.

Linder, H.P., Meadows, M.E. & Cowling, R.M. (1992). History of the Cape Flora. In *The Ecology of Fynbos. Nutrients, Fire and Diversity*, ed. R.M. Cowling, pp. 113–34. Cape Town: Oxford University Press.

Luger, A.D. & Moll, E.J. (1992). Fire protection and Afromontane forest expansion in Cape fynbos. *Biological Conservation*, **64**, 51–6.

MacDevette, D.R., MacDevette, D.K., Gordon, I.G. & Bartholomew, R.L.C. (1989). Floristics of the Natal Indigenous Forests. In *Biogeography of the Mixed Evergreen Forests of Southern Africa*, ed. C.J. Geldenhuys, pp. 124–44, Occasional Report 45, Ecosystem Programmes. Pretoria: Foundation for Research Development.

Manders, P.T. & Richardson, D.M. (1992). Colonization of Cape fynbos communities by forest species. *Forest Ecology and Management*, **48**, 277–93.

Manders, P.T., Richardson, D.M. & Masson, P.H. (1992). Is fynbos a stage to forest? Analysis of the perceived ecological distinction between two communities. In *Fire in South African Mountain Fynbos*, ed. B.W. van Wilgen, D.M. Richardson, F.J. Kruger & H.J. van Hensbergen, pp. 81–107. Berlin: Springer-Verlag.

Marker, M.E. & Russel, S. (1984). The application of biogeographic techniques to forest site-factor analysis. *South African Geographic Journal*, **66**, 65–78.

Masson, P. (1994). Forest composition and conservation status in the Swaziland highveld. In *Proceedings of the 13th Plenary Meeting of AETFAT*, ed. J.H. Seyani & A.C. Chikuni, pp. 993–1005. Zomba, Malawi: Association for the Taxonomic Study of the Flora of Tropical Africa.

McKenzie, B., Moll, E.J. & Campbell, B.M. (1977). A phytosociological study of Orange Kloof, Table Mountain, South Africa. *Vegetatio*, **34**, 41–53.

Meadows, M.E. & Linder, H.P. (1993). A palaeoecological perspective on the origin of Afromontane grasslands. *Journal of Biogeography*, **20**, 345–55.

Midgley, J.J. (1992). Aspects of the dynamics of the high forest at Langebos (Alexandria). *South African Forestry Journal*, **161**, 19–22.

Midgley, J.J. (1993). Biogeographic and other perspectives on the management and conservation of forest patches. In *The Relevance of Island Biogeographic Theory in Commercial Forestry*, ed. D. Everard, pp. 67–71. Pretoria: Foundation for Research Development.

Midgley, J.J. & Bond, W.J. (1990). Knysna fynbos 'islands'; origins and conservation. *South African Forestry Journal*, **153**, 18–21.

Midgley, J.J., Bond, W.J. & Geldenhuys, C.J. (1995a). The ecology of southern African Conifers. In *Southern Conifers*, ed. N.J. Enright & R. Hill, pp. 64–80. Melbourne: University of Melbourne Press.

Midgley, J.J., Cameron, M.J. & Bond, W.J. (1995b). Gap characteristics and replacement patterns in the Knysna forest, South Africa. *Journal of Vegetation Science*, **6**, 29–36.

Midgley, J.J. & Cowling, R.M. (1993). Regeneration patterns in Cape subtropical transitional thicket; where are all the seedlings? *South African Journal of Botany*, **59**, 496–9.

Midgley, J.J., Everard, D.A. & van Wyk, G. (1995c). Relative lack of regeneration of shade-intolerant canopy species in some South African forests. *South African Journal of Science*, **91**, 7–8.

Midgley, J.J. & Gobetz, P.N. (1993). Dynamics of the forest vegetation of the Umtiza Nature Reserve, East London. *Bothalia*, **23**, 111–16.

Midgley, J.J. & Joubert, D. (1991). Mistletoes, their host plants and the effects of browsing by large mammals in Addo Elephant National Park. *Koedoe*, **34**, 149–52.

Midgley, J.J., Seydack, A., Reynell, D. & McKelly, D. (1991). Fine-grain pattern of southern Cape plateau forests. *Journal of Vegetation Science*, **1**, 539–46.

Midgley, J.J., van Wyk, G.F. & Everard, D.A. (1995d). Leaf attributes of South African forest species. *African Journal of Ecology*, **33**, 160–8.

Milton, S.J. (1987). Growth of seven-week fern (*Rumohra adiantiformis*) in the southern Cape forests: implications for management. *South African Forestry Journal*, **143**, 1–4.

Milton, S.J. (1991). Slow recovery of defoliated seven-weeks *Rumohra adiantiformis* in Harkerville Forest. *South African Forestry Journal*, **158**, 23–8.

Moll, E.J. & White, F. (1978). The Indian Ocean coastal belt. In *Biogeography and Ecology of Southern Africa*, ed. M.J.A. Werger, pp. 561–98. The Hague: Junk.

Moolman, J.H. & Cowling, R.M. (1994). The impact of elephant and goat grazing on the endemic flora of South African succulent thicket. *Biological Conservation*, **68**, 53–61.

O'Brien, E.M. (1993). Climatic gradients in woody plant species richness: towards an explanation based on an analysis of southern Africa's woody flora. *Journal of Biogeography*, **20**, 181–98.

Phillips, J.F.V. (1931). Forest succession and ecology in the Knysna Region. *Memoirs of the Botanical Survey of South Africa*, **14**, 1–327.

Pierce, S.M. & Cowling, R.M. (1991). Disturbance regimes as determinants of seedbanks in coastal dune vegetation of the southeastern Cape. *Journal of Vegetation Science*, **2**, 403–12.

Poore, D. (ed.) (1989). *No Timber without Trees. Sustainability in the Tropical Forest*. London: Earthscan Publications.

Richardson, D.M. & Cowling, R.M. (1992). Why is mountain fynbos invasible and which species invade? In *Fire in South African Mountain Fynbos*, ed. B.W. van Wilgen, D.M. Richardson, F.J. Kruger & H.J. van Hensbergen, pp. 161–81. Berlin: Springer-Verlag.

Richardson, D.M., Cowling, R.M., Bond, W.J., Stock, W.D. and Davis, G.D. (1995). Links between biodiversity and ecosystem function in the Cape Floristic Region. In *Mediterranean-type Ecosystems. The Function of Biodiversity*, ed. G.W. Davis & D.M. Richardson, pp. 285–333. Berlin: Springer-Verlag.

Rogers, D.J. & Moll, E.J. (1975). A quantitative description of some coast forests of Natal. *Bothalia*, **11**, 523–37.

Rohde, K. (1992). Latitudinal gradients in species diversity: the search for the primary cause. *Oikos*, **65**, 514–27.

Rutherford, M.C. & Westfall, R.H. (1986). Biomes of Southern Africa: an objective categorization. *Memoirs of the Botanical Survey of South Africa*, **54**, 1–98.

choltz, A. (1983). Houtskool weerspieël die geskiedenis van inheemse woude. *Bosbounuus*, 3, 18–19.

eydack, A.W.H. (1991). Inventory of South African natural forests for management purposes. *South African Forestry Journal*, 158, 105–8.

eydack, A.W.H. (1995). An unconventional approach to timber yield regulation for multi-aged, multispecies forests. I. Fundamental considerations. *Forest Ecology & Management*, 77, 139–53.

eydack, A.W.H., Vermeulen, W.J., Heyns, H., Durrheim, G., Vermeulen, B., Willems, D., Ferguson, M., Huisamen, J. & Roth, J. (1995). An unconventional approach to timber yield regulation for multi-aged, multispecies forests. II. Application to a South African forest. *Forest Ecology & Management*, 77, 155–68.

mith, T.M. & Goodman, P.S. (1987). Successional dynamics in an *Acacia nilotica–Euclea divinorum* savannah in southern Africa. *Journal of Ecology*, 75, 603–10.

tevens, G.C. (1989). The latitudinal gradient in geographical range. How so many species co-exist in the tropics. *American Naturalist*, 133, 240–56.

tevens, G.C. (1992). The elevational gradient in altitudinal range: an extension of Rapoport's latitudinal range to altitude. *American Naturalist*, 140, 893–911.

tuart-Hill, G.C. (1992). Effects of elephants and goats on the Kaffrarian succulent thicket of the eastern Cape, South Africa. *Journal of Applied Ecology*, 29, 699–710.

inley, K.L. (1967). The moist evergreen forest dry-deciduous forest tension zone in north-eastern Zululand and hypotheses on past temperate/montane forest connections. *Palaeoecology of Africa*, 2, 82–5.

inley, K.L. (1985). *Coastal Dunes of South Africa*. South African National Scientific Programmes Report 109, pp. 1–300. Pretoria: CSIR.

van Daalen, J.C. (1991). Forest Growth: a 35-Year Southern Cape Case Study. *South African Forestry Journal*, 159, 1–10.

van Daalen, J.C., Geldenhuys, C.J., Frost, P.G.H. & Moll, E.J. (1986). *A Rapid Survey of Forest Succession at Mlalazi Nature Reserve*. Occasional Report 11, Ecosystem Programmes. Pretoria: Foundation for Research Development.

van Wilgen, B.W., Higgins, K.B. & Bellstedt, D.U. (1990). The role of *Vegetation* structure and fuel chemistry in excluding fire from forest patches in the fire-prone fynbos *Vegetation* of South Africa. *Journal of Ecology*, 78, 210–22.

van Wyk, A.E. (1990). Floristics of Natal/Pondoland Sandstone Forests. In *Biogeography of the Mixed Evergreen Forests of Southern Africa*, ed. C.J. Geldenhuys, pp. 145–57. Occasional Report 45, Ecosystem Programmes. Pretoria: Foundation for Research Development.

van Wyk, G.F., Everard, D.E. & Midgley, J.J. (1996). Floristics and dynamics of the Dukuduku lowland forest, South Africa. *South African Journal of Botany*, 62, 133–42.

Von Breitenbach, F. (1968). *Southern Cape Indigenous Forest Management Manual, II*. George: Department of Forestry.

Von dem Bussche, G.H. (1975). Indigenous forest conservation management. *South African Forestry Journal*, 93, 25–31.

White, F. (1978). The Afromontane Region. In *Biogeography and Ecology of Southern Africa*, ed. M.J.A. Werger, pp. 463–513. The Hague: Junk.

White, F. (1983). *The Vegetation of Africa*. Paris: UNESCO.

Wright, D.H. & Reeves, J.H. (1992). On the meaning and measurement of nestedness of species assemblages. *Oecologia*, 92, 416–28.

Zacharias, P.J.K., Stuart-Hill, G.C. & Midgley, J.J. (eds) (1991). *Proceedings of the First Valley Bushveld/Subtropical Thicket Symposium*. Howick: Grassland Society of Southern Africa.

Coastal vegetation 13

R.A. Lubke, A.M. Avis, T.D. Steinke and C. Boucher

13.1 Introduction

The coastal region includes vegetation that comes under the direct and strong influence of the ocean and its associated climate. It occurs mainly on substrata of marine origin, such as shells, sea sand and calcrete; but also on pebbles, cobbles, boulders and various types of bedrock. This chapter deals with the vegetation of mobile or partially fixed coastal dunes, exposed coastal cliffs and estuaries (including mangrove swamps). The vegetation of the marine intertidal and subtidal zones (principally algal communities and excluding estuaries) is described in Chap. 15 (this volume). Vegetation of fixed coastal dunes and other substrata immediately inland of this coastal zone is dealt with in the biome chapters of this volume. In some places, for example the coastal parts of the desert biome, it is very difficult to distinguish the boundary of the coastal zone.

Over half of the southern African coastline comprises sandy beaches backed by mobile or stable dunes. The nature of the coastline, its landform and geology, and the varied climate of the subcontinent results in a number of distinct coastal regions. In this chapter the different coastal regions and their component communities are described. The particular characteristics, form and function of the plants are discussed in the next section; followed by the characteristic community structure and ecosystem functions. The concluding section deals with the conservation and management of these ecologically important and fragile ecosystems, which are under extreme pressure from development.

13.2 Coastal regions

Various criteria have been used to classify the coast into regions. These include the climatic characteristics, the coastal trend (protection or exposure), the characteristic biomes along the coastline (Werger 1978a; Rutherford & Westfall 1986), the species distributions and even political boundaries. A combination of these is used in recognizing six regions that follow those of Tinley (1985), namely, West Coast, South West Coast, South Coast, South East Coast, Transkei and KwaZulu-Natal Coast (Fig. 13.1).

One of the major factors influencing plant communities is the type of shoreline, which may be fine- or coarse-grained, sand or pebble/shingle beaches, rocky headlands or wavecut rocky platforms (Lord 1984). The southern African coastline is not intricately dissected and has few islands. These types of shoreline occur along most of the coast, except pebble beaches, which are infrequent (Table 13.1). From the Transkei Coast northwards into KwaZulu-Natal, the sandy beaches consist of coarse-grained sands, whereas along the rest of the coast the sands are fine-grained. Pebble or shingle beaches occur only as pocket beaches in some regions, mainly on the South and Transkei Coasts.

The main environmental factors accounting for the varied vegetation types are: substratum stability (e.g. mobile sands driven by strong winds and concomitant sand blasting); salt spray (including occasional immersion under sea water); strong aridity gradients and intense radiation as a result of reflection from the sea and white sand.

An analysis of the distribution of estuaries around the southern African coast (Table 13.2) shows that there

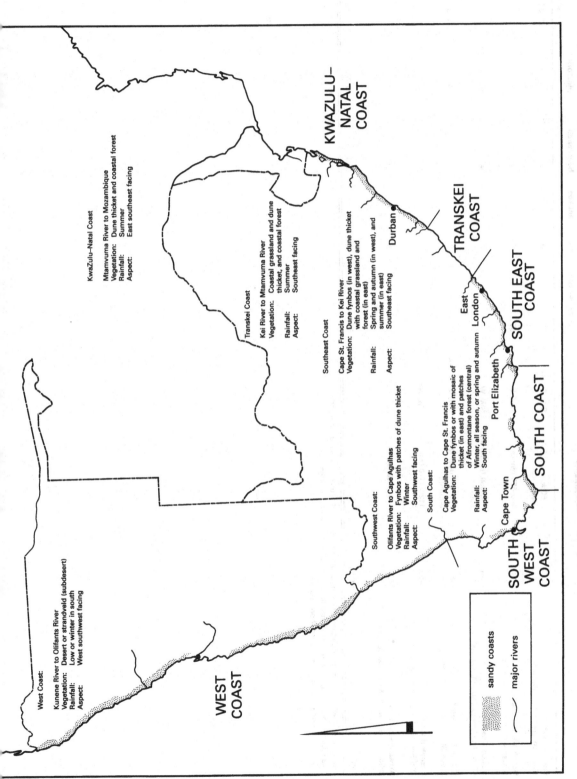

West Coast:

Kunene River to Olifants River
Vegetation: Desert or strandveld (subdesert)
Rainfall: Low or winter in south
Aspect: West southwest facing

WEST
COAST

KwaZulu–Natal Coast

Mtamvuma River to Mozambique
Vegetation: Dune thicket and coastal forest
Rainfall: Summer
Aspect: East southeast facing

Transkei Coast

Kei River to Mtamvuma River
Vegetation: Coastal grassland and dune
thicket, and coastal forest
Rainfall: Summer
Aspect: Southeast facing

Southeast Coast

Cape St. Francis to Kei River
Vegetation: Dune fynbos (in west), dune thicket
with coastal grassland and
forest (in east)
Rainfall: Spring and autumn (in west), and
summer (in east)
Aspect: Southeast facing

Southwest Coast:

Olifants River to Cape Agulhas
Vegetation: Fynbos with patches of dune thicket
Rainfall: Winter
Aspect: Southwest facing

South Coast:

Cape Agulhas to Cape St. Francis
Vegetation: Dune fynbos or with mosaic of
thicket (in east) and patches
of Afromontane forest (central)
Rainfall: Winter, all season, or spring and autumn
Aspect: South facing

KWAZULU–
NATAL
COAST

TRANSKEI
COAST

Durban

East
London

SOUTH EAST
COAST

Port Elizabeth

SOUTH COAST

Cape Town

SOUTH
WEST
COAST

sandy coasts

major rivers

Figure 13.1 **Coastal regions of southern Africa (adapted from Tinley 1985).**

Table 13.1 Environmental characteristics of southern African coastal regions. Data from Lord (1984) and Tinley (1985)

Region	Length of coastline (km)	Rainfall regime (mm yr^{-1}) and seasonality	Formative wind direction^{-1}	Habitats (percentage of coastline)				
				Fine-grained sandy beaches	Coarse-grained sandy beaches	Exposed rocky headlands	Wavecut rocky platforms	Pebble/ shingle beaches
West Coast (Angolan border to Olifants River)	1933	8–60 summer (N) winter (S)	SW, S	70.5	0	29[b]		0.5
South West Coast (Olifants River to Cape Agulhas)	746	400–550 (winter)	NW, S	55	0	42	3	0
South Coast (Cape Agulhas to Cape St. Francis)	473	400–950 (spring and autumn)	SW, W, E	34	0	25	39	2
South East Coast (Cape St. Francis to Kei River)	467	580–900 (spring and autumn)	SW, SE, E, NE	65	0	5	30	0
Transkei Coast (Kei River to Mtamvuma River)	260	1120–1170 (summer)	SW, NE	0	36	18	38	8
Kwazulu–Natal Coast (Mtamvuma River to Mozambique border)	580	750–1335 (summer)	S, SE, NE	0	79	14	7	0
Total	4459[c]			50.2	12.4	19.3	17.2	0.8

[a] Major winds, which have an important role in coastal conditions and cause a net sand transport.
[b] Namibian coast not divided into rocky headlands and wavecut rocky platforms.
[c] Excludes some peninsulas and all offshore islands.

Table 13.2 **Number of estuaries along the South African coast (data from Ward & Steinke 1982; Lord 1984). See Fig. 13.1 for coastal regions**

	Number of estuaries			Number of estuaries with mangrove communities
	Open	Closed	Total	
West Coast	9	13	22	0
South West Coast	4	20	24	0
South Coast	9	18	27	0
South East Coast	15	29	44	3
Transkei Coast	17	33	50	17
Natal Coast	2	30	32	20
Total	56	143	199	40

are fewer estuaries on the West and South East Coasts, but they become more frequent towards the northeast. The major determinants of the composition and structure of estuarine vegetation are the period over which estuaries are open to the sea, the size of the river, salinity gradients and the inundation period. The mangrove communities, which appear to be limited by the lower winter temperatures in the southern waters (MacNae 1968), become more frequent along the Transkei and KwaZulu-Natal Coasts.

In the following section the structure and composition of various coastal vegetation types and their distribution in the different coastal regions are described. The physical features and the structure (physiognomy and composition) of the communities in all the regions are discussed.

13.3 Dry coastal communities

Clear discontinuities in species distributions are found along the coast, with some species being confined to certain regions (Raal & Burns 1989a,b, 1991, 1992). These discontinuities form the basis for the subdivision of the South African and Namibian dry coastal areas into the zones described by Penrith (1993), Boucher & Le Roux (1993), Taylor & Boucher (1993) and Weisser & Cooper (1993). This distribution is related to the distribution of propagules, climatic differences (especially moisture, temperature and wind) and phytochorological gradients (Avis 1992). Owing to the low alpha diversity of pioneer communities, any changes in species distribution will be reflected in community composition. Furthermore, these communities may represent stages in a successional gradient towards the climax community of the stable dunes, the composition of which is also influenced by the above factors (see Fig. 13.2).

The influence of temperature and moisture is reflected in the size, density and structural complexity of dry coastal communities. Fig. 13.2 indicates the change in vegetation type, height and density, from the hyper-arid Namibian coast with short sparse succulents and spiky grasses (Robinson 1976; Boucher & Le Roux 1993) southwards to the temperate West and South East Cape Coast with denser, short herbs and taller small-leaved shrubs (Boucher & Jarman 1977; Lubke 1983; Boucher 1987; Raal & Burns 1991, 1992; Boucher & Le Roux 1993) to the Transkei and KwaZulu-Natal Coast where the littoral herb and shrub zones are narrow, with grasslands and forest occurring in close proximity to the sea (Tinley 1985; Raal & Burns 1989a,b; Avis 1992; Weisser & Cooper 1993; Chap. 12, this volume).

13.3.1 Pioneer communities

Because of the dynamic nature of coastal dunes, many of these communities are ephemeral, and can disappear in a single storm event or gradually change in response to changing environmental factors. Many of the pioneer species dominating these communities (e.g. *Ipomoea pes-caprae* and *Sporobolus virginicus*) are cosmopolitan, whereas others (e.g. *Ehrharta villosa* and *Cladoraphis cyperoides*) are adapted to a specific region of the coast, although endemism is generally low (Avis 1992; Taylor & Boucher 1993).

Pioneer species are able to colonize the bare, shifting sand above the intertidal beach by means of special adaptations and mechanisms. These include a low, creeping or rhizomatous growth form, and leaves with a thick, waxy cuticle tolerant of salt spray and sand abrasion. Many species can withstand burying for varying lengths of time, and have adaptations to reduce water loss. Once established, they are able to stabilize sand and grow in response to sand accretion. This leads to the formation of foredunes above the spring high tide mark unless winds are too strong and persistent, or rainfall too low. In such areas, foredunes are absent, and the dune landscape is dominated by mobile transverse or barchanoid dunes (Tinley 1985; Avis 1992).

The West Coast (Fig. 13.1) can be divided into two biogeographical zones. The flora of the northern region, from the Kunene River to the Holgat River (just south of the Orange River) shows certain affinities with Namib desert species. Boucher & Le Roux (1993) consider this vegetation to be a Namib intrusion into the Namaqualand strand vegetation. The boundary of this vegetation corresponds well with the southern boundary of the Namib domain (Werger 1978a,b). The coarse-grained sands of the Namib desert extend to the sea on the West Coast, and vegetation in this dry, windy environment is sparse. Small, isolated hummock dunes may be stabil-

Figure 13.2 **Profiles of the coastal regions of southern Africa showing the vegetation zonation (see Fig. 13.1 for coastal regions).**

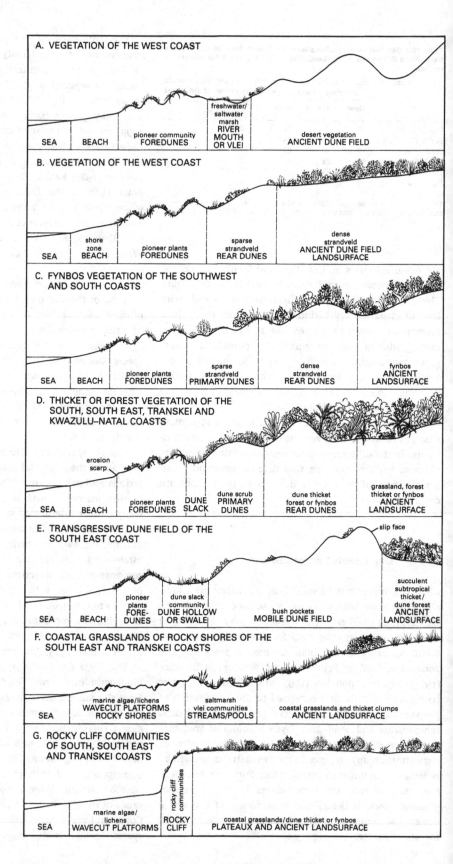

A. VEGETATION OF THE WEST COAST

SEA | BEACH | pioneer community FOREDUNES | freshwater/saltwater marsh RIVER MOUTH OR VLEI | desert vegetation ANCIENT DUNE FIELD

B. VEGETATION OF THE WEST COAST

SEA | shore zone BEACH | pioneer plants FOREDUNES | sparse strandveld REAR DUNES | dense strandveld ANCIENT DUNE FIELD LANDSURFACE

C. FYNBOS VEGETATION OF THE SOUTHWEST AND SOUTH COASTS

SEA | BEACH | pioneer plants FOREDUNES | sparse strandveld PRIMARY DUNES | dense strandveld REAR DUNES | fynbos ANCIENT LANDSURFACE

D. THICKET OR FOREST VEGETATION OF THE SOUTH, SOUTH EAST, TRANSKEI AND KWAZULU–NATAL COASTS

erosion scarp

SEA | BEACH | pioneer plants FOREDUNES | DUNE SLACK | dune scrub PRIMARY DUNES | dune thicket forest or fynbos REAR DUNES | grassland, forest thicket or fynbos ANCIENT LANDSURFACE

E. TRANSGRESSIVE DUNE FIELD OF THE SOUTH EAST COAST

slip face

SEA | BEACH | pioneer plants FORE-DUNES | dune slack community DUNE HOLLOW OR SWALE | bush pockets MOBILE DUNE FIELD | succulent subtropical thicket / dune forest ANCIENT LANDSURFACE

F. COASTAL GRASSLANDS OF ROCKY SHORES OF THE SOUTH EAST AND TRANSKEI COASTS

SEA | marine algae/lichens WAVECUT PLATFORMS ROCKY SHORES | saltmarsh vlei communities STREAMS/POOLS | coastal grasslands and thicket clumps ANCIENT LANDSURFACE

G. ROCKY CLIFF COMMUNITIES OF SOUTH, SOUTH EAST AND TRANSKEI COASTS

rocky cliff communities

SEA | marine algae/lichens WAVECUT PLATFORMS | ROCKY CLIFF | coastal grasslands/dune thicket or fynbos PLATEAUX AND ANCIENT LANDSURFACE

Figure 13.3 **Hummock dunes at the mouth of the Hoab River in the northern zone of the West Coast, colonized by *Zygophyllum stapffii*, *Brownanthus kuntzei* and *Salsola* sp.** *(Photo: C. Boucher).*

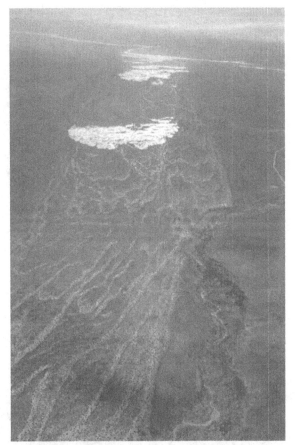

Figure 13.5 **A dune plume extending inland from its source along the southern zone of the West Coast, south of Alexander Bay. The history of these plumes is critical for understanding landscape-level vegetation patterns along this dynamic, sandy coastline** *(Photo: D.M. Richardson).*

Figure 13.4 **Dunes on the southern zone of the West Coast, south of Alexander Bay, dominated by the shrubby grass *Cladoraphis cyperoides*** *(Photo: D.M. Richardson).*

ized by the shrub *Salsola kali*, or the succulent *Tetragonia decumbens*. Robinson (1976) recognized five communities in the dune areas, three of which are found along the coast north of the Orange River. These are dominated almost exclusively by *Acanthosicyos horridus*, *Cladoraphis cyperoides* and *Stipagrostis sabulicola*. In places, dwarf shrubs such as *Arthraerua leubnitziae*, *Brownanthus kuntzei*, *Zygophyllum clavatum* and *Z. stapffii* may dominate the hummock dunes (Seely 1987; Fig. 13.3).

Cladoraphis cyperoides, a tough spiny grass, becomes more frequent and dominant south of the Orange River. Although this grass is an effective sand stabilizer, it is the accumulation of seaweed debris, especially of *Ecklonia maxima*, that initially stabilizes the sand and increases nutrient levels, thus enabling colonization by pioneer species (Boucher & Le Roux 1993).

The southern zone of the West Coast consists of eroding or retrograding beaches with cliffs, rocky platforms, boulder-strewn beaches and occasional estuaries. Extens-

ive dune plumes originate at river mouths and extend in a northeasterly direction, providing habitats for dune plant populations (Boucher & Le Roux 1993). The dunes along the coast consist of loose, coarse sand and are usually sharply angular, with the result that topography is undulating (Boucher & Le Roux 1993). The pioneer communities are dominated by dwarf shrubs, usually with succulent leaves and stems, and grasses up to 0.75 m tall. Common shrubs in this Namaqualand strand vegetation (*vide* Boucher & Taylor 1993) include *Euphorbia brachiata* and *Didelta carnosa* var. *tomentosa* and *Zygophyllum morgsana*. *Didelta carnosa* var. *tomentosa* is a common component of the strandveld further inland. Grasses include *Cladoraphis cyperoides* and *Eragrostis sabulosa* (Figs. 13.4 and 13.5).

The northern boundary of the South West Coast corresponds with the Olifants River, which also forms the northern boundary of the Cape Floristic Region (Werger 1978a, Chap. 3, this volume). Notable features

Figure 13.6 **Littoral vegetation at Bloubergstrand along the South West Coast (Table Mountain in the background). Conspicuous species are *Didelta carnosa* var. *tomentosa* (background), *Senecio elegans* (foreground) and *Arctotheca populifolia* (dominant sprawling herb)** *(Photo: C. Boucher).*

Figure 13.7 **Stabilization of mobile dune sands at Yzerfontein (South West Coast) using *Ammophila arenaria* (August 1980). This practice was subsequently discontinued** *(Photo: C. Boucher).*

Figure 13.8 **A natural 'blowout' at Blombos on the South Coast. The dominant shrub in the foreground is *Chrysanthemoides monilifera*** *(Photo: C. Boucher).*

of this region are the Saldanha–Langebaan lagoon complex, and a number of 'half-heart' or log-spiral bays – including False Bay (the largest), St. Helena and Table Bay. In the Cape Peninsula and False Bay areas, the shore is lined with spectacular cliffs of the 'concave' type (Heydorn & Tinley 1980) with massive talus slopes. These rocky coasts comprise just less than half the total length of this region, the remainder being predominantly sandy (Table 13.1). These sands, particularly in the northern portion, are fine, white and of marine origin.

The vegetation of the dry west coast between the Olifants River and Table Bay has been described by Boucher & Le Roux (1993) as Capensis Strand vegetation (Fig. 13.6). It is distinguished from the Namaqualand vegetation further north by the presence of *Chenolea diffusa*, *Exomis microphylla* and *Helichrysum micropoides*. Other widespread or dominant species include *Cladoraphis cyperoides*, *Didelta carnosa* var. *tomentosa*, *Arctotheca populifolia* and *Hebenstreitia cordata*. The last two mentioned species, together with *Tetragonia decumbens*, are common in False Bay and further east.

Tetragonia decumbens, a dominant species throughout the South West Coast often forms dense patches, especially in hollows or blowouts (Taylor & Boucher 1993), whereas *Arctotheca populifolia* forms characteristic hummock dunes. Towards the east the introduced grasses, *Thinopyrum distichum* (sea wheat) and *Ammophila arenaria* (marram grass) are locally dominant from the South West to the South East Coasts where they have been used in dune stabilization programmes (Walsh 1968; Fig. 13.7). The spread of these grass species into mobile sands (Avis 1989) is the subject of more intensive research studies by the two senior authors of this chapter. Marram grass in particular colonizes mobile dunes aggressively and, on the west coast of North America, is altering the dune landscape (Weidemann & Pickat 1996).

The South Coast lies within the Cape Fold Belt. Low limestone cliffs (under 150 m) of the Bredasdorp Formation on shales and sandstones of the Bokkeveld Group dominate the landscape in the western portion of this region (Heydorn & Tinley 1980). The area further east has long stretches of cliffs resulting from the remnant planation shelf falling abruptly to the sea (Heydorn & Tinley 1980). Sandy beaches are often restricted to the mouths of the larger rivers, but eastward towards Port Elizabeth (Fig. 13.1), east-facing log-spiral bays with large sandy beaches characterize the shoreline (Heydorn & Tinley 1980; Taylor & Boucher 1993; Fig. 13.8). The sands here are yellowish, contrasting with the pure white dune sands further west.

Although *Tetragonia decumbens* and *Hebenstreitia cordata* are still abundant, the graminoid element is enriched by *Ehrharta villosa* and *Sporobolus virginicus*. *Arctotheca populifolia* and *Dasispermum suffruticosum* still form a major component of the foredune vegetation, and the woody, succulent pioneer shrub *Scaevola plumieri* appears for the first time (Taylor & Boucher 1993). Where planted, *Ammophila arenaria* is locally important and *Gazania rigens*, which forms conspicuous, small hummock dunes is more common. Two other species, of tropical affinity, *Phylohydrax carnosa* and *Ipomoea pes-caprae* appear midway in this coastal region. The former is uncommon, but the latter increases in abundance further east.

A major feature of the South East Coast is the extensive, transgressive dunefield, with dunes that reach massive proportions on the Alexandria coast. The coastline towards the west of this region is embayed and aligned approximately east to west, with headlands on the western edges of half-heart bays. The northeastern sector is straighter in gross outline, and as one approaches Kei Mouth, spectacular dolerite headlands dominate the scenery. In the southwest, the calcareous sands have hardened by lime cementation to form aeolinite, which can form conspicuous headlands and promontories. Here, large crescentic and transverse dunes are common and, owing to the dominant southwesterly winds, show a net movement to the east (Avis 1992). The dunefields becomes narrower towards the northeast, and are finally replaced by dolerite headlands along the Transkei Coast.

The foredune communities of the South East Coast are similar in species composition to the South Coast, with the following exceptions. As one moves northeastward, *Ehrharta villosa*, *Thinopyrum distichum* and *Ammophila arenaria* become less abundant, reaching the eastern extent of their range. *Scaevola plumieri* and *Ipomoea pes-caprae* become more abundant northeastward, and the dominant dune-forming species are *S. plumieri*, *Arctotheca populifolia* and *Sporobolus virginicus* (Fig. 13.9).

The Transkei Coast is dominated by coastal sandstone cliffs of the KwaZulu-Natal Group, and rocky shores. Sandy beaches and dunes are uncommon, and in the northeastern portion cliffs of up to 160 m are formed (Weisser & Cooper 1993). The pioneer communities are similar to those of the South East Coast, and *Scaevola plumieri* and *Ipomoea pes-caprae* are abundant. They occupy the usually narrow zone from the high-tide level to about 10 m inland, which is characterized by a low foredune ridge. Landward of this zone is a narrow, rather ill-defined scrub zone dominated by *Passerina rigida*. Dune forests or cliffs occur behind this zone (Weisser & Cooper 1993; Fig. 13.10).

The KwaZulu-Natal Coast is characterized by beaches with low dunes of recent sands banked up against older rocks or red, almost pure quartz, sands. Further north

Figure 13.9 **A foredune, colonized by *Ipomoea pes-caprae*, at Christmas Rocks on the South East Coast** *(Photo: A.M. Avis)*.

Figure 13.10 ***Scaevola plumieri* on dunes at Mkambati on the Transkei Coast. Note the dune forest in the background** *(Photo: C. Boucher)*.

is the Zululand coastal plain dune cordon that stretches uninterrupted into Mozambique. Parallel beach ridges are common, and in localized areas where the coastline is prograding, the vegetation exhibits a clear successional gradient from pioneer communities to forest (Weisser & Backer 1983; Avis 1992; Fig. 13.11).

Two distinct zones of vegetation have been recognized by Weisser & Cooper (1993). The Dune Pioneer zone is dominated by *Scaevola plumieri*, and species abundant on the South and South East Coast (*Arctotheca populifolia* and *Gazania rigens* var. *uniflora*) occur sporadically. Tropical or subtropical (Tongaland-Pondoland) species such as *Canavalia maritima* become more abundant, and the graminoid component is represented by *Sporobolus virginicus* and *Imperata cylindrica*. The *Passerina* Scrub zone

is more clearly defined than along the Transkei Coast, and is the first woody community to become established behind the foredune ridge. Other species besides this dominant shrub include *Stipagrostis zeyheri* and *Helichrysum asperum* var. *comosum*, which are the first species to become established.

13.3.2 Dune slack communities

Dune slack communities are often found within large mobile dune systems on the South and South East Coasts. The West Coast, north of the Orange River, although endowed with large dune systems, has a very low rainfall. As a result, the interdune hollows are never moist or stable long enough for the establishment of slack vegetation. On the KwaZulu-Natal Coast, the inter-

Figure 13.11 **Dune vegetation at Mtunzini on the KwaZulu-Natal coast, showing the transition from *Scaevola plumieri* (foreground), through *Passerina rigida–Stipagrostis zeyheri* scrub, to thicket clumps and dune forest (background)** *(Photo: E.J. Moll).*

dune hollows are rapidly replaced by more permanent plant communities and these slack communities are therefore less frequent in this region.

The interdune hollows between the predominantly transverse dunes or within the transgressive dunefields have a shallow water table, so that the sand is both moist and stable. This allows significant quiescent time for more mesic species to colonize, and a more luxuriant vegetation develops. The dune slacks are also protected from salt spray, and damage from sand abrasion is reduced, owing to the moist soil.

Pioneer plants of dune slacks often share many characteristics of pioneer dune communities. In some cases the species may be the same as those of the dunes, e.g. *Sporobolus virginicus* on the South East Coast or *Cladoraphis cyperoides* on the West Coast. Dune slack species tolerate fluctuating wet and dry conditions, and occasional seawater inundation. Sedges, rushes and herbaceous species of saltmarshes are often found in the moister dune slacks (Lubke & Avis 1982a; McLachlan, Ascaray & Du Toit 1987).

13.3.3 Rocky substrata

Rocky shores and rocky cliff promontories differ from dune coasts in having stable substrata and shallow soils, and are often subject to high winds and salt spray. Consequently, the vegetation is usually very sparse or depauperate. In sites where sand has accumulated, pioneer species common to dune systems establish (Taylor & Boucher 1993).

Along rocky headlands with exposed cliff faces, wave splash may be quite common and the effect of salt spray above the cliffs is a limiting factor for the growth of plants. The moist, shallow pockets of soil are nutrient-poor and frequently saturated with salt. The soil dries out periodically and vegetation is subject to frequent changes in salinity concentrations (Raal & Burns 1989b). Herbs and grasses tolerant of these saline conditions constitute this halophytic community, often comprising species of saltmarshes or saline vleis above rocky shores (Fig. 13.12).

Under moderate rainfall regimes, specialized communities develop that are biogeographically linked to the inland flora, e.g. the *Coleonema album* short coastal fynbos on sandstones on the South West Coast (Boucher 1978), whereas along the high-rainfall Transkei and KwaZulu-Natal Coasts, rocky substrata support stunted and depauperate versions of inland grassland, thicket and forest communities. These communities do, however, support some strictly coastal species (mostly succulents) belonging to the genera *Carpobrotus*, *Drosanthemum*, *Euphorbia* and *Tetragonia* (Avis 1992; Weisser & Cooper 1993; D.B. Hoare unpubl. data).

Figure 13.12 **Coastal cliffs and rock shelf vegetation west of Port Elizabeth on the South East Coast. The coastal cliffs are dominated by salt-spray tolerant species (*Rhus crenata*, *Scutia myrtina*, *Sideroxylon inerme*) that are common components of inland thicket and forest vegetation. The rocky shelf community includes many species typical of salt marshes (e.g. *Chenolea diffusa*, *Disphyma crassifolium*, *Sarcocornia* sp. and *Triglochin striata*)** *(Photo: C. Boucher)*.

13.4 Wet coastal communities

13.4.1 Saltmarshes

There are a variety of different estuarine communities along the coast, depending on the nature of the estuary and its position along the coast (Heydorn & Tinley 1980). In some regions there are extensive saltmarshes, and along the Indian Ocean coast, mangrove communities occur.

The most important environmental factors determining the biotic structure of estuaries are salinity and tidal exposure. The vegetation of saltmarshes and of mangroves shows patterns of zonation correlated with tidal inundation (Fig. 13.13, Table 13.3). Submerged aquatic macrophytes grow attached to tidal mud banks. Commonly occurring in this zone from the Western Cape through to KwaZulu-Natal are eelgrass beds (*Zostera capensis*). In the warmer subtropical estuaries two other seagrasses are also found, viz. *Halophila ovalis* and *Halodule uninervis*. The latter is also found in pools on rocky intertidal shores along with *Thalassodendron ciliata*. The seagrass communities are transient: they may be washed out with the mud banks during floods and also show a cyclic dieback (T.D. Steinke, pers. obs.). In less saline pools and streams of the estuaries there are some submerged freshwater aquatics that can tolerate slightly saline conditions. *Potamogeton pectinatus* and *Zannichellia palustris* are two species that are often found as single species communities under these conditions.

Saltmarsh communities usually consist of stands dominated by single species of halophytic plants (Fig. 13.14, Table 13.3). In many of the Cape estuaries, an early colonizer is *Spartina maritima* (cord grass), putatively introduced from Europe (Pierce 1982). It is partially submerged at high tides. Samphires (*Sarcocornia* spp.) and sea lavender (*Limonium* spp.) form communities in some

Table 13.3 **Species contributing to the zonation patterns of different estuarine types in southern Africa (from O'Callaghan 1994). H, high; L, low; M, mean; N, neap; S, spring; W, water**

	MLWS	MLWN	MSL	MHWN	MHWS
Temperate estuaries					
Disphyma crassifolium					**
Sarcocornia pillansii					****
Suaeda inflata					***
Chenolea diffusa					***
Limonium linifolium					**
Cotula coronopifolia					***
Triglochin striata				****	
Sarcornia perennis				****************	
Spartina maritima			**************************		
Zostera capensis	************************************				
Seasonally open estuaries					
Conyza scabrida					***
Phragmites australis					***
Stenotaphrum secundatum					***
Juncus kraussii					***
Triglochin striata					****
Sarcocornia decumbens				**************************	
Sporobolus virginicus			*******************		
Ruppia maritima	*****************************				
Arid zone, seasonally closed estuaries					
Sarcocornia pillansii					****
Odyssea paucinervis					*****
Sporobolus virginicus				***************************	
Sarcocornia natalensis			****		

Figure 13.13 **Profiles of estuarine coasts of South Africa showing the vegetation zonation.**

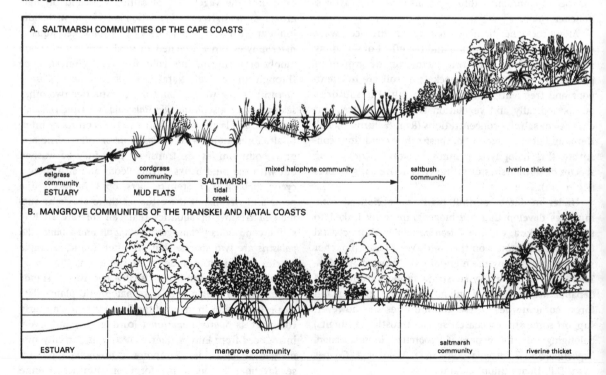

Fig. 13.14 **Saltmarsh vegetation growing in the estuary of the Duiwenhoks River on the South Coast.** *Sarcocornia perennis* **is prominent in the right foreground** *(Photo: C. Boucher).*

estuaries. In the drier higher regions of the estuary, grasses (*Stenotaphrum secundatum* and *Sporobolus virginicus*) and rushes (*Juncus* spp.) are the predominant species. Saltmarshes contain a variety of other species, many of which may become locally dominant, e.g. *Plantago carnosa* and *Disphyma crassifolium*. In less saline conditions in the upper estuarine reaches, freshwater hydrophytes may form large communities, e.g. *Phragmites australis*, the common reed, and *Bolboschoenus maritimus* is found in habitats subject to periodic dry conditions.

13.4.2 Mangroves

Mangroves are trees or shrubs which grow in bays and estuaries mainly along the subtropical Transkei and Kwa-Zulu-Natal Coasts, exhibiting their own pattern of zonation in response to tidal inundation (Fig. 13.13). Mangroves extend from the Kosi system (26° S) in the north to the Nahoon River (33° S) in the south (Moll & Werger 1978; Ward & Steinke 1982; see Fig. 13.15). The original stand of mangroves near the mouth of the Nahoon River was transplanted from Durban Bay, although it is possible that a stand further upstream arose naturally (Steinke 1972). At the northern Kosi system, five mangrove species are found: *Avicennia marina*, *Bruguiera gymnorrhiza*, *Ceriops tagal*, *Lumnitzera racemosa* and *Rhizophora mucronata* (Fig. 13.13). *Ceriops tagal* and *L. racemosa* do not occur further south, except at Beachwood (Mgeni River) where they were planted. In the estuaries of the southern Transkei Coast, *A. marina* is common, but *B. gymnorrhiza* is sparse south of the Mbashe River and its natural distribution ends at Wavecrest (Nxaxo-Nquisi Rivers). South of the Kobonqaba River, *A. marina* occurs only sporadically and appears to have a rather tenuous existence. *Rhizophora mucronata*, which is common as far south as the Mtata River, reaches its southern limit at the Bulungula River. *Acrostichum*

aureum, the mangrove fern, extends from the KwaZulu-Natal Coast into the northeastern Transkei. The so-called freshwater mangrove, *Hibiscus tiliaceus*, which is frequently associated with mangroves, occurs commonly southwards to the Mbashe River, whereafter it may be found sporadically as far as the Nahoon River.

A survey of mangroves was conducted in 1982 (Ward & Steinke 1982) and the distribution of mangroves in each estuary is given in Fig. 13.15. Since then, change has been negligible. There has been a general tendency for mangroves to increase in area during the past 20 years. The most significant increases have been in the sanctuary at Richards Bay where *Avicennia marina* and, to a lesser extent, *Bruguiera gymnorrhiza* have colonized the alluvial sediments from the Mhlatuze River. Mangroves now occupy more than 450 ha in this region, at present the largest area in the country. In contrast, recent harbour developments at Richards Bay and Durban, and flood damage on the Mtata and Mgeni Rivers, have destroyed relatively small areas of mangroves. In the 1960s and 1970s, however, there was a significant reduction in the area of mangroves in South Africa. The reductions were the result of a number of factors: harbour developments at Richards Bay and Durban, mass mortality following high water levels after a cyclone in the Kosi system, as well as high water levels caused by poorly planned bridge construction at Sodwana and Beachwood (Breen & Hill 1969; Moll *et al.* 1971; Bruton 1980). It is encouraging that in the Kosi system, mangroves are re-establishing (Ward, Steinke & Ward 1986).

Generally, mangroves are largest and reach their best development in the tropics. In South Africa, although there are local variations, trees in the northern stands are larger. In the southern estuaries, mangroves may be reduced to a narrow fringe of low-growing shrubs (1 to 2 m high) at their upper estuarine limits.

Algae also form a significant component of the mangrove community. They may be present as epiphytes on the above-ground parts of mangroves. They also occur as mats of bluegreen algae (cyanobacteria) on the mud substrate, either in the mangroves or in the adjacent saltmarshes. A survey of epiphytic algae on the mangroves revealed 12 reds (Rhodophyceae), 27 bluegreens (Cyanophyceae), eight greens (Chlorophyceae) and three browns (Phaeophyceae) from the Kosi system to the Nahoon River (Lambert, Steinke & Naidoo 1987, 1989). Most were collected from the aerial roots of *Avicennia marina*: this reflects the high intensity of sampling in the lowest zone where these pneumatophores are exposed for short periods only. Species-richness of bluegreen algal assemblages on pneumatophores are richer than those of red algae: individual bluegreen species, with a

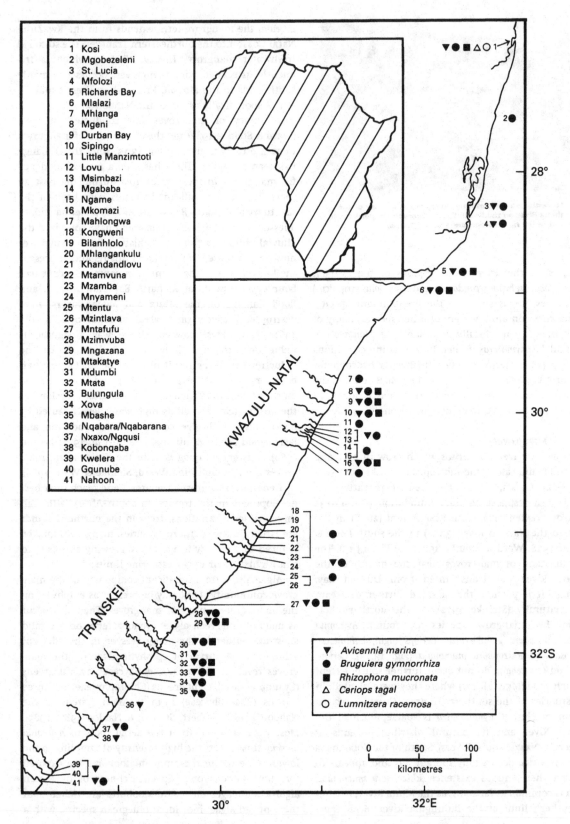

Distribution map of mangroves along the South African coast (KwaZulu-Natal and Transkei), with numbered locality list.

1	Kosi
2	Mgobezeleni
3	St. Lucia
4	Mfolozi
5	Richards Bay
6	Mlalazi
7	Mhlanga
8	Mgeni
9	Durban Bay
10	Sipingo
11	Little Manzimtoti
12	Lovu
13	Msimbazi
14	Mgababa
15	Ngame
16	Mkomazi
17	Mahlongwa
18	Kongweni
19	Bilanhlolo
20	Mhlangankulu
21	Khandandlovu
22	Mtamvuna
23	Mzamba
24	Mnyameni
25	Mtentu
26	Mzintlava
27	Mntafufu
28	Mzimvuba
29	Mngazana
30	Mtakatye
31	Mdumbi
32	Mtata
33	Bulungula
34	Xova
35	Mbashe
36	Nqabara/Nqabarana
37	Nxaxo/Ngqusi
38	Kobonqaba
39	Kwelera
40	Gqunube
41	Nahoon

KWAZULU-NATAL

TRANSKEI

▼ *Avicennia marina*
● *Bruguiera gymnorrhiza*
■ *Rhizophora mucronata*
△ *Ceriops tagal*
○ *Lumnitzera racemosa*

28°
30°
32°S

30°   32°E

0   50   100
kilometres

Figure 13.15 **Distribution of mangroves in South Africa.**
Reproduced, with permission, from Ward & Steinke (1982).

few exceptions, have a wider distribution than most red algal species. Most of the other mangroves occupy higher ground, which is inundated less frequently and consequently epiphytic algae are less abundant.

A survey of horizontal and vertical zonation of alga epiphytes on pneumatophores of *Avicennia marina* has been carried out at Beachwood. The red algae, chiefly *Bostrychia moritziana*, *B. radicans*, *Caloglossa leprieurii* and *Polysiphonia subtilissima*, occur predominantly near the mouth of the estuary. They decrease upstream, following a horizontal gradient from isosalinity to hyposalinity. The bluegreen algae, chiefly *Lyngbya confervoides* and *Microcoleus chthonoplastes* and green algae, dominated by *Rhizoclonium implexum* and *R. riparium*, were more obvious in the drier areas away from the creek. Vertical distribution along the pneumatophores occurred as an upper *Rhizoclonium-*, a middle *Bostrychia-* and a lower *Caloglossa-*zone, corresponding to tidal ebb and flow levels (Phillips *et al.* 1994).

13.5   Plant form and function

13.5.1   Dry coastal plants

Few ecophysiological and autecological studies have been made of plant species in pioneer dune communities. Germination studies showed 22% germination success in dune species left to germinate under natural conditions (A. Wrigley unpubl. data). Important shrub species, such as *Myrica cordifolia*, did not germinate unless subjected to extremes of low and high temperatures, such as those which occur naturally in the field (Lubke & Avis 1986). Soil seed banks of dune species were investigated at Mlalazi on the KwaZulu-Natal coast (Todd 1994). In the seed bank, seven dominant species accounted for 35.7% of the species composition, the bulk of which consisted of *Passerina rigida* seeds (94.8%). The soil seed-bank size distribution varied, with early and mid-successional species occurring in many series, but late successional species, such as *Mimusops caffra* and *Eugenia capensis*, being more restricted. Seeds of the latter were also found only under canopies, whereas the early successional species were found in the open and under canopies.

The role of mammals and birds in the maintenance and dynamics of vegetated bush pockets in the Alexandria dunefield (South East Coast) were investigated (Bruton 1992; Castley 1992). It was shown that there is a potential input of 323 million intact seeds into the dunefield, and that despite the localized distribution of bushpockets, 18 million of these seeds were found in these habitats. Large mammals and birds were responsible for this input of seed, with those animals that frequent the bushpockets acting as seed dispersers. These included bushpigs (*Potamochoerus porcus*), bushbuck (*Tragelaphus scriptus*), vervet monkey (*Cercopithecus aethiops*) and black-backed jackal (*Canis mesomelas*). Possible residents included Cape grysbok (*Raphicerus melanotis*) and large spotted genet (*Genetta tigrina*). About 30 species of birds were recorded in the bushpockets, but few species breed there. The majority fly between the thicket and pockets, thus dispersing seeds of thicket species. Mammals dispersed 29 plant species (Castley 1992) and birds 15 species (Bruton 1992), with considerable overlap between them. Seeds dispersed by both birds and mammals included the primary colonizing shrub, *Myrica cordifolia*, as well as secondary species such as *Euclea* spp., *Sideroxylon inerme* and *Olea exasperata* (Castley 1992). At the De Hoop Nature Reserve (South Coast), chacma baboons (*Papio ursinus*) are important dispersers of *Chrysanthemoides monilifera* (C. Boucher, pers. obs.). Bushpig were the most important mammalian dispersers, and birds dispersed large numbers of seeds of *Myrica cordifolia*, and the invasive alien *Acacia cyclops*. These studies concluded that the bushpockets are established by wind-dispersed species, and then maintained by birds and large mammals (Castley 1992).

Lubke & Avis (1982a,b) found that salt spray negatively affects the growth of *Scirpus nodosus*, particularly in the seedling stage. Establishment sites of *S. nodosus* seedlings may be controlled by the degree of salt spray (Lubke & Avis 1982b). Salt spray had a deleterious effect on plant growth only on the foredunes, whereas sand movement is important along the entire width of the dune system, at some sites being the only factor causing the death of plants (Avis & Lubke 1985). Sand movement, low percentage of soil moisture and high soil conductivity on the foredunes were the main factors preventing plants from establishing throughout the dune system.

In similar studies conducted in a greenhouse, *Scaevola plumieri* showed no significant response to salt spray, whereas *Arctotheca populifolia* and *Vellereophyton vellereum* showed a decrease in phytomass under heavy spray conditions (equivalent to foredune conditions) (J. Blood unpubl. data). The amount of salt spray on different regions of the dunes varies (Avis & Lubke 1985), explaining the distribution of species within the dunes and dune slack regions. Burns (1986) found that there was a marked difference between the spray received on the seaward side of the dune forests compared with the landward side and noted differences in species composition and structure. This factor is therefore also important for tree and shrub species, which are severely pruned by wind and salt spray. Donnelly & Pammenter (1983) showed that soil salinity and soil moisture correlated with the distribution of plants in a dune system near Durban, KwaZulu-Natal. They found that both burying and the abrasive action of the sand affected the distribution of plants in the dunefield.

Sand burial is an important limiting factor, since many of the dune species are inundated by drift sand. Experiments under field and laboratory conditions showed that different species responded differently to sand burial (J. Allsopp, unpubl. data). *Scaevola plumieri* seedlings managed to survive sand burial, as did *Vellereophyton vellereum*, whereas *Arctotheca populifolia* was erratic in its response to burial and showed a decrease in phytomass at higher rates of burial. All three species are successful pioneers of dunes or dune slacks. Studies indicate that *Ehrharta villosa* plants are fairly successful in sand stabilization and can withstand extremes of burying under high wind conditions (R.A. Lubke unpubl. data). The important aspect in their survival is not sand supply, as is the case with *Ammophila arenaria* (Ranwell 1972), but, more likely, the availability of nutrients washed up onto the driftline in the actively growing foredune zone community (D. B.Hoare unpubl. data; see below).

Ecophysiological studies on *Scaevola plumieri* indicated that this species grows well in accreting sand, shows high transpiration rates and seldom seems to suffer severe water stress (Pammenter 1983, 1985). This species is a C_3 plant, but its water-use efficiency is low when compared to other C_3 beach species. The plant responded directly to humidity, but transpiration was affected to a greater extent than net CO_2 assimilation, suggesting that internal conductances were lower than stomatal conductances. Leaves of *S. plumieri* from both young pioneer and senescing plants showed rates of photosynthesis that declined markedly with leaf age. It was concluded that this species did not show any noticeable photosynthetic characteristics particularly suited to its habitat. Its nitrogen acquisition ability, along with its physical characteristics to withstand high wind speeds, salt spray and sand abrasion in low-nutrient dune sands are given as the reasons for its success as a pioneer of coastal dunes (Pammenter 1983). It is limited in its distribution down the southern coast by low winter temperatures. Although seeds have been noted along the shores at Cape Point, individuals have only been found as far south as the Breede River Mouth (C. Boucher, pers. obs.).

13.5.2 Wet coastal plants

There have been few ecophysiological and autecological studies of wet coastal plants in South Africa. Germination of *Sporobolus virginicus*, an important pioneer on sandy beaches and saltmarshes, is generally very poor, often with less than 50% of the seeds germinating (Breen, Everson & Rogers 1977). Salinity above 10‰ adversely affected the germination of *Sporobolus* seeds. Other studies showed that *S. virginicus* was shown to be tolerant of salinity (NaCl up to 400 mol m^{-3}) and waterlogging (Naidoo & Mundree 1993). This may account for its successful colonization of a wide range of wet coastal environments.

Laboratory studies of growth under different salinity regimes show that saltmarsh plants such as *Chenolea diffusa*, *Sarcocornia natalensis* and *Sporobolus virginicus*, common to western Cape estuaries, are all more productive in fresher water (Blomerus 1992). However, these species differ from non-halophytes in their ability to continue production (albeit at slower rates) in saline water of up to 500 mol m^{-3} (sea water = 460 mol m^{-3}). Germination trials at saline solutions above 300 mol m^{-3} indicated that exposure to saline water promotes seedling establishment of these estuarine species. In the case of *Triglochin bulbosa* and *T. striata*, herbaceous perennials widely distributed in saltmarshes, seed germination was highest at salinities less than 250 mol m^{-3}, with less success in salinities up to 500 mol m^{-3} (Naidoo & Naicker 1992). It appears that dormancy at high salinities and rapid germination during periods of reduced salt stress, especially in *T. striata*, may be an efficient adaptive strategy in seedling establishment in saline areas (Naidoo & Naicker 1992).

Very little research has been carried out on the reproductive ecology and physiology of mangroves: readers are referred to general texts for an overview of the physiology (especially salinity tolerance mechanisms) of mangrove species (e.g. Scholander *et al.* 1962; Ball 1988). General information on the biology of selected mangrove species is given below.

Avicennia marina is generally regarded as the pioneer mangrove and is the most widespread and common mangrove along the coastline. It establishes rapidly, has a wide environmental tolerance and also prefers open, unshaded areas. This mangrove produces aerial pencil roots (or pneumatophores), which allow gaseous exchange in trees growing under waterlogged, anaerobic conditions. In addition, the relatively small, usually buoyant propagules facilitate a wide dispersal. Flowering is from September onwards and, in the southern mangrove stands, flowers may be found in trees during most of the year. Fruit drop usually occurs in March/April.

Bruguiera gymnorrhiza is also a very common mangrove. It is not generally regarded as a pioneer, although it has colonized some estuaries where the mouth closes occasionally. Under conditions of periodic closure, *B. gymnorrhiza* survives better than *Avicennia marina*, which is more sensitive to submersion. This mangrove has aerial roots for gaseous exchange. In its early growth stages, *B. gymnorrhiza* prefers shaded conditions. Flowering takes place in late summer and the trees are viviparous. Propagules of 120 mm long abscise in February/March almost one year after flowering. Its cigar shape allows the falling propagule to 'self-plant' in the soft mud below. Alternatively, its buoyancy allows wide dispersal by water.

In South Africa, *Rhizophora mucronata* is not as common as *Avicennia marina* and *Bruguiera gymnorrhiza*. The trees produce aerial prop roots. Flowering and propagule production are similar to those of *B. gymnorrhiza*. In the case of *R. mucronata*, however, the hypocotyl is approximately 250 mm in length. *Rhizophora mucronata* tends to form dense stands marginal to creeks and streams within the mangrove community.

13.6 Community structure and dynamics

13.6.1 Dry coastal communities

Community composition is related to successional changes in these dynamic regions. Along some coastlines there is a continuous supply of sand, and new habitats are available for colonization by pioneer species. Along eroding coastlines there may be a cyclical removal and deposition of sand so that the pioneer communities are continually being removed and replaced, and stable systems remain in a dynamic state of equilibrium.

Descriptions of the successional process are given by Moll (1969), Weisser, Garland & Drews (1982) and Lubke & Avis (1988). These accounts describe the pathways of succession, but our understanding of the processes and mechanisms is incomplete. The prograding dunes at the Mlalazi Nature Reserve, at Mtunzini on the KwaZulu-Natal Coast, provide an excellent example of a successional pathway. A detailed, quantitative study of the Mlalazi dune system was carried out by Avis (1992) and enabled a comparison of succession along the South East Coast. Changes at Mlalazi can be related to the chronology of dune development: aerial photographs were used to date the parallel dune ridges of the region, which advance some 5 m a year (Weisser *et al.* 1982). Avis (1992) found that seedlings of *Scaevola plumieri* establish on the drift line as a result of seeds rolling down from the first foredune covered with mature *S. plumieri*. Under high rainfall conditions, these seedlings establish rapidly and form hummock dunes. The hummocks coalesce to form parallel linear dune ridges. *Ipomoea pes-caprae* also occurs in the region, but the continuous stem elongation and root production of *S. plumieri* is the important factor

in initial sand stabilization and foredune development. On the second dune ridge there is an abundance of other herbaceous species, and as the *S. plumieri* dies due to the lack of sand accumulation, grasses, herbs and some shrubs establish.

Avis (1992) identified eight communities along a gradient of increasing distance from the sea, and increasing age of the dunes (from 5 to 120 years), as estimated from dating the parallel dune ridges. The pathway of change was as follows: pioneer, enriched pioneer, open dune scrub, closed dune scrub, bushclumps, bushclump/forest margin transition, forest margin and, finally, forest. These communities showed a general increase in species richness, cover, stature and biomass. Herbs were initially the dominant life form, but grasses became dominant in the intermediate communities. Shrubs, ferns and trees became more abundant in the later successional stages, but a more even distribution of life form was noted in forest. Changes in soil properties included a decrease in pH (8.5 to 7.3) and increase in percentage organic matter (0.5 to 6.1%), exchangeable bases and conductivity with increasing age. The successional pathway appears to support the facilitation model of Connell & Slatyer (1977), as changes were clearly autogenic and community driven. These results can be compared with the classical Clementsian succession on the Lake Michigan dunefields reported by Cowles (1899) and Olson (1958). However, the data are insufficient to elucidate the demographic and ecophysiological causes of turnover, or to apply multiple working hypotheses of community structure (Shipley & Keddy 1987). Studies on the population biology of key species were therefore undertaken (Todd 1994) in an attempt to elucidate the mechanisms of change (see section 13.5.1).

The phenology, germination and reproductive strategies of the major species of the Mtunzini dunefield help to explain the mechanisms affecting the community dynamics at the population level in the early stages of succession (Todd 1994). Flowering and seed production of the different key species is separated in time. Some species, such as *Passerina rigida*, produce high numbers of seeds with a low fruit : seed mass ratio within a short cycle. These seeds are mainly wind dispersed. Animal-dispersed fruits (e.g. *Mimusops caffra*) with large fleshy seeds and a high fruit : seed mass ratio are produced over longer cycles. *Chrysanthemoides monilifera* was found to be the most important animal-dispersed species, and provides seeds throughout the year. Thus the staging of fruiting and flowering ensures a year-round availability of fruit in the dunefield. Key species are important in succession in creating and modifying conditions through bush development and canopy closure, and creating suitable conditions for the growth of other species. The mechanism of change is therefore autogenic, following the facilitation pathway of Connell & Slatyer (1977).

In other regions of the coast there is no distinct chronological sequence, since the coastline is not prograding uniformly in this way and is often eroding. On the South East Coast, two approaches have been used to describe the vegetation dynamics. Avis (1992) sampled relevés ranging from pioneer communities to closed dune thicket in a number of localities at Kleinemond, near Port Alfred. Results were similar to those obtained at Mtunzini, except that some pioneer foredune communities were dominated by various grasses and *Scaevola plumieri*, whereas the dune thicket was overwhelmingly dominated by *Sideroxylon inerme*. At Mtunzini there is a variety of dune forest types that result from a single pioneer community. Although not spatially linked in the dunefield, the various communities sampled were positioned along a gradient of increasing community complexity (Avis 1992). Changes in soil properties in the different communities were similar to those observed at Mtunzini. These directional changes in the communities were also autogenic, and the dune slacks acted as centres of diversity within the mobile dunefield. At Kleinemond the foredunes provide indirect facilitation by protecting the more mesic dune slack habitats from factors such as salt spray and sand movement, whereas this facilitation is more direct on prograding systems such as that at Mtunzini.

Lubke & Avis (1988) monitored the temporal changes in dune slack vegetation at Kleinemond over ten years in fixed plots. These studies show how the dune slack changes progressively as the pioneer species of sedges (*Scirpus nodosus*) and rushes (*Juncus kraussii*) are replaced by woody shrubs (*Rhus crenata*, *Stoebe plumosa*, *Myrica cordifolia* and *Metalasia muricata*) (Fig. 13.16). The dynamic changes recorded in the dune slack community are the result of the movement of sands from the west to the east. This leads to the transgressive dunefield burying the western side of the slack, while other species are colonizing the eastern margins. Vegetation can establish in these slacks due to the high water table and the protection afforded by the foredunes. Here, the area was covered with a mass of *Scirpus nodosus* in 1978; other herbaceous species such as *Vellereophyton vellereum* and *Chironia decumbens*, abundant in the early stages (Lubke & Avis 1982a, 1988), were replaced by woody species after about four years.

McLachlan *et al.* (1987) carried out similar studies on the dune slacks that occur in successive interdune hollows from the Sundays River towards Woody Cape in the Alexandria dunefields (South East Coast). They surveyed the area and measured change in sand accumulation, plant species and other biota. The transverse dune ridges flanking the dune slacks showed a net easterly movement of 7 m yr^{-1}, and this succession therefore only covers a time span of six to seven years, since the slacks are less than 50 m wide. Therefore, a dune thicket and forest stage is never reached, since the climax *Scirpus nodosus* community is buried before it can develop further (McLachlan *et al.* 1987).

Studies of this type on the dune slacks provide interesting information on the dynamics of species change. However, it is unlikely that the dune thicket communities of the South and South East Coasts will progress in the same way as the parallel linear dunes of Mtunzini of the Kwazulu-Natal Coast. Lubke & Avis (1988) have postulated that this is a consequence of the rainfall in the eastern Cape, which is erratic, with long drought periods. Sufficient rainfall for an extended period is crucial in the successional development of patches of thicket vegetation from dune slack vegetation.

13.7 Ecosystem function

13.7.1 Dry coastal systems

The only study on food webs and energetics was carried out by McLachlan *et al.* (1987) on the Alexandria dune fields of the South East Coast. These food chains are supplied with energy from the primary production of the dune plants, as well as organic input from sea and land. Dune plants provide a source of food for the grazers such as herbivorous insects, mammals and seed- and fruit-eating birds. Important plants in the grazing food chain are *Chrysanthemoides monilifera*, *Gazania rigens* and *Ipomoea pes-caprae*. Gerbils (*Gerbillurus paeba*) are important seed-eaters, while bulbuls (Pycnonotidae) and monkeys (*Cercopithecus aethiops*) are fruit-eaters. Many insects feed on the detritus, fuelling the detrital food chain. Interstitial organisms, such as bacteria, fungi and meiofauna maintain this interstitial food chain.

McLachlan *et al.* (1987) point out that the interstitial food chain is more important in these moist dune soils than the grazing and production food chains. Buried organic matter in the form of decaying litter as well as marine input as wrack washed up on the shore may be an important supply for the interstitial fauna and flora. There have been few studies on the macrofauna food chains. The feeding and foraging behaviour of mammals and birds is important, and also leads to the dispersal of plants in the dune system (Ascaray 1986; Van der Merwe 1987), since they rely on these vectors for dispersal into

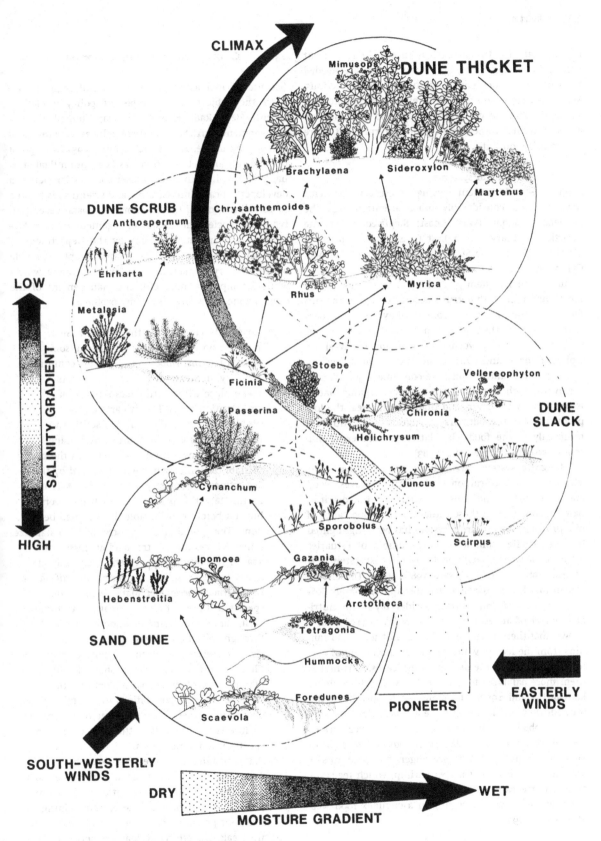

Figure 13.16 **Model of dune succession at Kleinemond. Reproduced,
with permission, from Lubke & Avis (1988).**

new habitats (Todd 1994). Little is known about nutrient cycling in coastal dunes, and more studies are needed to explain both succession and dune-beach interactions. McLachlan (1991) discusses the exchange of materials across the dune-beach interface, and also the gradients across a coastal dunefield in general terms, but more research is required.

13.7.2 Wet coastal systems

Food webs and nutrient cycling in estuarine systems have received considerably more attention, e.g. the Langebaan lagoon (West Coast; Siegfried 1977), the Swartkops estuary (South East Coast; Baird, Marais & Martin 1988) and the St. Lucia system (Whitfield 1980; Cyrus 1989) on the KwaZulu-Natal Coast.

In saltmarsh systems, the detrital food chain is the most important, with detritus accumulating on the surface of the mud, where it is consumed by benthic feeders that live in or on the substratum. These filter feeders (for example, polychaete worms, mussels and bivalves) are siphon feeders, and, with some crabs (which feed on larger particles and worms), are consumed by birds and carnivorous fish. In these systems, where organic matter is supplied by the river, and tidally by the sea, the low productivity of the saltmarsh cordgrass may contribute minimally to the foodweb relative to northern hemisphere systems (Pierce 1982, 1983).

Mangrove systems are extremely productive. They provide a source of carbon in the form of leaves, wood and other litter, and thus contribute to the detritus-based food chains in these estuaries. The importance of mangroves in the estuarine system has been highlighted by litter studies conducted in a number of estuaries along the KwaZulu-Natal and Transkei Coasts (Steinke & Charles 1986; Steinke & Ward 1988, 1990). These have shown that leaf material constitutes approximately 60% of the litter, and contributes considerably to the nutrient budget of an estuary. Steinke & Ward (1987) have shown that there is an initial loss of nutrients and tannins from the leaves, which results in a build-up of bacteria and fungi on the decomposing leaves. This makes them more palatable to crabs (Robertson 1988; Steinke, Rajh & Holland 1993). The crabs are important in the break-down of the leaves, and the leaf particles then support microbial growth, which is utilized by invertebrates such as worms, fiddler crabs, gastropods and young fish. There are many carnivores, scavengers and predators, a notable one being the snapper shrimp, which feeds on other invertebrates. Numerous birds and fish feed on these species as well, resulting in a complex food web (Berjak *et al.* 1977).

13.8 Conservation and management

Because of increasing pressures, coastal regions have been the subject of a number of policy guidelines recently (e.g. Raal & Burns 1989a; Council for the Environment 1991). Some of these policies were reviewed recently by Lubke, Avis & Hellstrom (1995), who point out that although legislation has been promulgated, it has yet to be implemented. Closer cooperation between developers, local administrators and scientists is needed to curtail the loss of these important coastal vegetation types and species. The Coastal Management Advisory Programme was initiated in 1992 by the Department of Environment Affairs and CSIR, in an attempt to provide better advisory management. In spite of many worthwhile attempts at more efficient management of the coastal zone, there are still many problems.

- Coastal development is the prime cause of pressure on coastal vegetation. Along some coastal regions, ribbon development comprising industry, urban sprawl or coastal resorts has taken place without the maintenance of conservation areas. This is particularly problematic along the coasts of southern KwaZulu-Natal and the southwestern Cape.
- Major problems were introduced with the policy of stabilization of dunefields, initiated by the Department of Forestry in the 1970s (Avis 1992). Fortunately, this policy has now been reversed, but not before considerable damage had been done. The spread of aliens introduced to stabilize dunes has resulted in tremendous losses to coastal vegetation types. Alien species include *Acacia cyclops*, *A. saligna*, *Casuarina equisetifolia* and *Leptospermam laevigatum*. Many of these alien species still present a real threat to our dunes.

  In some cases destabilization of littoral dunes through collection of washed-up seaweed (*Ecklonia maxima*) along the western seaboard (see Chap. 15, this volume) is a problem in the maintenance of these coastal communities. Increases in local populations in coastal regions may result in destabilization, due to the impact on forests and thickets of the need for firewood and traditional medicines.
- Diamond mining on the West Coast has devastated vast areas of the arid coastline (Chap. 7, this volume). Recent legislation will ensure attempts at rehabilitation of current mining sites, but prospects of restoration of older dumps are bleak. The effects of heavy mineral mining and their associated industries along the

KwaZulu-Natal and Namaqualand coasts will become apparent only in the future as our understanding of the coastal environment improves.

- Destruction of estuarine communities (especially mangrove communities in KwaZulu-Natal and Transkei) as a result of engineering works (harbours, bridge construction, land development) and agriculture is an important problem of management in these regions. Moreover, there is uncontrolled utilization of animals in the estuaries and the use of mangrove trees for wood and fishing traps in KwaZulu-Natal. Estuaries also come under pressure from silting by industrial pollution, unfavourable agricultural practices and domestic and industrial pollution.

Cape Nature Conservation in conjunction with the CSIR's Division of Earth, Marine and Atmospheric Sciences have recently carried out surveys on the coast, to identify sensitive areas. This approach is required if we are to maintain a sustainable use of the coastal region. Development must be limited to specific areas to curtail ribbon development. There must be more adequate control of stabilization programmes and the use of alien vegetation, and more importance must be given to protection of sensitive areas such as the dunes and estuarine systems.

3.9 References

Ascaray, C.M. (1986). *An Ecological Study of the Hairy-Footed Gerbil, Gerbillurus paeba, in an Eastern Cape Dunefield.* MSc Thesis. Port Elizabeth: University of Port Elizabeth.

Avis, A. M. (1989). A review of coastal dune stabilization in the Cape Province of South Africa. *Landscape and Urban Planning*, 18, 55–68.

Avis, A.M. (1992). *Coastal Dune Ecology and Management in the Eastern Cape.* PhD Thesis. Grahamstown: Rhodes University.

Avis, A.M. & Lubke, R.A. (1985). The effect of wind borne sand and salt spray on the growth of Scirpus nodosus in a mobile dune system. *South African Journal of Botany*, 1, 100–10.

Baird, D., Marais, J.F.K. & Martin, A.P. (eds) (1988). *The Swartskops Estuary.* South African Natural Scientific Programmes Report 156, 1–107.

Ball, M.C. (1988). Ecophysiology of mangroves. *Trees*, 2, 129–42.

Berjak, P., Campbell, G.K, Huckett, B. I. & Pammenter, N.W. (1977). *In the Mangroves of Southern Africa.* Durban: Wildlife Society of Southern Africa.

Blomerus, J.J. (1992). *'n Voorlopige Outekologiese Ondersoek na die Seleksie van Plante vir Hervestiging in Verbrakte Omgewings in Suid-Afrika.* MSc Thesis. Stellenbosch: Stellenbosch University.

Boucher, C. (1978). Cape Hangklip area. Part 2. The vegetation. *Bothalia*, 12, 455–97.

Boucher, C. (1987). *A Phytosociological Study of Transects Through the Western Cape Coastal Foreland, South Africa.* PhD Thesis. Stellenbosch: Stellenbosch University.

Boucher, C. & Jarman, M.L. (1977). The vegetation of the Langebaan area, South Africa. *Transactions of the Royal Society of South Africa*, 42, 214–72.

Boucher, C. & Le Roux, A. (1993). Dry coastal ecosystems of the South African West Coast. In *Dry Coastal Ecosystems – Africa, America, Asia and Oceania. Ecosystems of the World*, 2B, ed. E. Van der Maarel, pp. 75–88. Amsterdam: Elsevier.

Boucher, C. & Taylor, H.C. (1993). Are Sideroxylon inerme (milkwood) thickets relictual? Poster paper presented at the 19th Annual SAAB Congress, Department of Botany, University of the Western Cape.

Breen, C.M., Everson, C. & Rogers, K. (1977). Ecological studies on Sporobolus virginicus (L.) Kunth with particular reference to salinity and inundation. *Hydrobiologia*, 54, 135–40.

Breen, C.M. & Hill, B.J. (1969). A mass mortality of mangroves in the Kosi estuary. *Transactions of the Royal Society of South Africa*, 38, 285–303.

Bruton, M.N. (1980). An outline of the ecology of the Mgobezeleni Lake System at Sodwana, with emphasis on the mangrove community. In *Studies on the Ecology of Maputaland*, ed. M.N. Bruton & K.H. Cooper, pp. 408–26. Grahamstown: Rhodes University.

Bruton, J.S. (1992). *Seed Dispersal by Birds in the Alexandria Dunefield.* MSc Thesis. Port Elizabeth: University of Port Elizabeth.

Burns, M.E.R. (1986). *A Synecological Study of the East London Coast Dune Forests.* MSc Thesis. Grahamstown: Rhodes University.

Castley, J.G. (1992). *Role of Mammals in Seed Dispersal in the Alexandria Dunefield.* MSc Thesis. Port Elizabeth: University of Port Elizabeth.

Connell, J.H. & Slatyer, R.O. (1977). Mechanisms of succession in natural communities and their role in community stability and organization. *American Naturalist*, 111, 1119–44.

Council for the Environment (1991). *A Policy for Coastal Zone Management in the Republic of South Africa. Part 2. Guidelines for Coastal Land-use.* Pretoria: Academica Publishers.

Cowles, H.C. (1899). The ecological relations of the vegetation on the sand dunes of Lake Michigan. *Botanical Gazette*, 27, 95–117; 167–202; 281–308; 361–91.

Cyrus, D.P. (1989). The Lake St. Lucia system – a research assessment. *South African Journal of Aquatic Science*, 15, 13–25.

Donnelly, F.A. & Pammenter, N.W. (1983). Vegetation zonation on a Natal coastal sand dune in relation to salt spray and soil salinity. *South African Journal of Botany*, 2, 46–51.

Heydorn, A.E.F. & Tinley, K.L. (1980). *Estuaries of the Cape, Part I. Synopses of the Cape coast – Natural Features, Dynamics and Utilisation*. Stellenbosch: CSIR.

Lambert, G., Steinke, T.D. & Naidoo, Y. (1987). Algae associated with mangroves in southern African estuaries. I. Rhodophyceae. *South African Journal of Botany*, 53, 349–61.

Lambert, G, Steinke, T.D. & Naidoo, Y. (1989). Algae associated with mangroves in southern African estuaries. Cyanophyceae. *South African Journal of Botany*, 55, 476–91.

Lord, D.A. (1984). *Coastal Sensitivity Atlas of Southern Africa*. Pretoria: Department of Transport.

Lubke, R.A. (1983). A survey of the coastal vegetation near Port Alfred, Eastern Cape. *Bothalia*, 14, 725–38.

Lubke, R.A. & Avis, A.M. (1982a). Factors affecting the distribution of *Scirpus nodosus* plants in a dune slack community. *South African Journal of Botany*, 1, 97–103.

Lubke, R.A. & Avis, A.M. (1982b). The effect of salt spray on the growth of *Scirpus nodosus*. *South African Journal of Botany*, 1, 163–4.

Lubke, R.A. & Avis, A.M. (1986). Dune stabilisation in the eastern Cape. Attempts at replacement of exotics by indigenous species. In *Structure and Function of Sand Dune Ecosystems*, ed. D. van der Merwe, A. McLachlan & P.A. Hesp, pp. 51–4, Institute for Coastal Research, Report No.8. Port Elizabeth: University of Port Elizabeth.

Lubke, R.A. & Avis, A.M. (1988). Succession on the coastal dunes and dune slacks at Kleinemond, Eastern Cape, South Africa. *Monographs of Systematic Botany of the Missouri Botanical Gardens*, 25, 599–622.

Lubke, R.A., Avis, A.M. & Hellstrom, G.B. (1995). Current status of coastal zone management in the eastern Cape region, South Africa. In *Management and Habitat Conservation*, vol. 2., ed. A.H.P.M. Salman, H. Barends & M. Bonazountas, pp. 239–60. Leiden: EUCC.

MacNae, W. (1968). A general account of the fauna and flora of mangrove swamps and forests in the Indo-West-Pacific region. *Advances in Marine Biology*, 6, 73–270.

McLachlan, A. (1991). Ecology of coastal dune fauna. *Journal of Arid Environments*, 21, 229–43.

McLachlan, A., Ascaray, C. & du Toit, P. (1987). Sand movement, vegetation succession and biomass spectra in a coastal dune slack in Algoa Bay, South Africa. *Journal of Arid Environments*, 12, 9–25.

Moll, E.J. (1969). A preliminary account of the dune communities at Pennington Park, Mtunzini, Natal. *Bothalia*, 10, 615–26.

Moll, E.J., Ward, C.J., Steinke, T.D. & Cooper, K.H. (1971). Our mangroves threatened?. *African Wildlife*, 25, 103–7.

Moll, E.J. & Werger, M.J.A. (1978). Mangrove communities. In *Biogeography and Ecology of Southern Africa*, ed. M.J.A. Werger, pp. 1231–8. The Hague: Junk.

Naidoo, G. & Mundree, S.G. (1993). Relationship between morphological and physiological responses to waterlogging and salinity in *Sporobolus virginicus* (L.) Kunth. *Oecologia*, 93, 360–6.

Naidoo, G. & Naicker, K. (1992). Seed germination in the coastal halophytes *Triglochin bulbosa* and *Triglochin striata*. *Aquatic Botany*, 42, 217–29.

O'Callaghan, M. (1994). *Salt Marshes of the Cape (South Africa): Vegetation Dynamics and Interactions*. PhD Thesis. Stellenbosch: Stellenbosch University.

Olson, J.S. (1958). Rates of succession and soil changes on southern Lake Michigan sand dunes. *Botanical Gazette*, 119, 125–70.

Pammenter, N.W. (1983). Some aspects of the ecophysiology of *Scaevola thunbergii*, a subtropical coastal dune pioneer. In *Sandy Beaches and Ecosystems*, ed. A. McLachlan & T. Erasmus, pp. 675–85. The Hague: Junk.

Pammenter, N.W. (1985). Photosynthesis and transpiration of the subtropical coastal sand dune pioneer *Scaevola plumieri* under controlled conditions. *South African Journal of Botany*, 51, 421–4.

Penrith, M.-L. (1993). Dry Coastal Ecosystems of Namibia. In *Dry Coastal Ecosystems – Africa, America, Asia and Oceania*. Ecosystems of the World, 2B, ed. E. van der Maarel, pp. 71–4. Amsterdam: Elsevier.

Phillips, A., Lambert, G., Granger, J.E. & Steinke, T.D. (1994). Horizontal zonation of epiphytic algae associated with *Avicennia marina* (Forssk) Vierh. pneumatophores at Beachwood Mangroves Nature Reserve, Durban, South Africa. *Botanica Marina*, 3.7. 567–76.

Pierce, S. M. (1982). What is *Spartina maritima* doing in our estuaries? *South African Journal of Science*, 78, 229–30.

Pierce, S. M. (1983). Estimation of the non-seasonal production of *Spartina maritima* (Curtis) Fernald in a South African estuary. *Estuarine, Coastal and Shelf Science*, 16, 241–54.

Raal, P.A. & Burns, M.E.R. (1989a). *An Analysis of Vegetation Importance along the Natal Coastline*, Report EMA-C89136, 1–53. Stellenbosch: CSIR.

Raal, P.A. & Burns, M.E.R. (1989b). *The Mapping and Conservation Importance Rating of the Cape Coastal Vegetation as an Aid to Development Planning: Eastern Cape Coast*, Report EMA-C 89153, 1–51. Stellenbosch: CSIR.

Raal, P.A. & Burns, M.E.R. (1991). *The Mapping and Conservation Importance Rating of the Cape Coastal Vegetation as an Aid to Development Planning: Southern Cape Coast*, Report EMA-C 89153A, 1–47. Stellenbosch: CSIR.

Raal, P.A. & Burns, M.E.R. (1992). *The Mapping and Conservation Importance Rating of the Cape Coastal Vegetation as an Aid to Development Planning: South-western Cape Coast*, Report EMA-C 89153B, 1–67. Stellenbosch: CSIR.

Ranwell, D.S. (1972). *Ecology of Salt Marshes and Sand Dunes*. London: Chapman and Hall.

Robertson, A.I. (1988). Decomposition of mangrove leaf litter in tropical Australia. *Journal of Experimental Marine Biology and Ecology*, 116, 235–47.

obinson, E. R. (1976). *Phytosociology of the Namib Desert Park, South West Africa.* MSc Thesis. Pietermaritzburg: Natal University.

utherford, M.C. & Westfall, H. (1986). Biomes of southern Africa – an objective classification. *Memoirs of the Botanical Survey of South Africa,* **54,** 1–98.

cholander, P.F., Hammel, H.T., Hemmingsen, E. & Garey, W. (1962). Salt balance in mangroves. *Plant Physiology,* **37,** 723–9.

eely, M.K. (1987). *The Namib. The Shell Guide.* Windhoek: Shell Oil.

hipley, B. & Keddy, P.A. (1987). The individualistic and community-unit concepts as falsifiable hypotheses. *Vegetatio,* **69,** 47–55.

iegfried, W.R. (ed). (1977). A review of a symposium on research in the natural sciences at Saldanha Bay and Langebaan Lagoon. *Transactions of the Royal Society of South Africa,* **42,** 209–509.

teinke, T.D. (1972). Further observations on the distribution of mangroves in the Eastern Cape Province. *Journal of South African Botany,* **38,** 165–78.

teinke, T.D. & Charles, L.M. (1986). Litter production by mangroves. I. Mgeni Estuary. *South African Journal of Botany,* **52,** 552–8.

teinke, T.D, Rajh, A. & Holland, A.J. (1993). The feeding behaviour of the red mangrove crab *Sesarma meinerti* de Man, 1887 (Crustacea: Decapoda: Grapsidae) and its effect on the degradation of mangrove leaf litter. *South African Journal of Marine Science,* **13,** 151–60.

teinke, T.D. & Ward, C.J. (1987). Degradation of mangrove leaf litter in the St Lucia estuary as influenced by season and exposure. *South African Journal of Botany,* **53,** 323–8.

teinke, T.D. & Ward, C.J. (1988). Litter production by mangroves II. St. Lucia and Richards Bay. *South African Journal of Botany,* **54,** 445–54.

teinke, T.D. & Ward, C.J. (1990). Litter production by mangroves III. Wavecrest (Transkei) with predictions for other Transkei estuaries. *South African Journal of Botany,* **56,** 514–19.

Taylor, H.C. & Boucher, C. (1993). Dry coastal ecosystems of the South African South Coast. In *Dry Coastal Ecosystems – Africa, America, Asia and Oceania. Ecosystems of the World,* 2B, ed. E. van der Maarel, pp 89–107. Amsterdam: Elsevier.

Tinley, K.L. (1985). Coastal dunes of South Africa. *South African National Scientific Programmes Report,* **109,** 1–300.

Todd, C.B. (1994). *A Comparison of the Reproductive Strategies of Key Species of a Prograding Dune System in the Mlalazi Nature Reserve, Natal.* MSc Thesis. Grahamstown: Rhodes University.

van der Merwe, D. (1987). *The Ecology of the Whitefronted Plover (Charadrius marginatus).* MSc Thesis. Port Elizabeth: University of Port Elizabeth.

Walsh, B. N. (1968). *Some notes on the incidence and control of drift sands along the Caledon, Bredasdorp and Riversdale coastline of South Africa.* Department of Forestry Bulletin No. 44. Pretoria: Government Printer.

Ward, C.J. & Steinke, T.D. (1982). A note on the distribution and approximate areas of mangroves in South Africa. *South African Journal of Botany,* **1,** 51–3.

Ward, C.J., Steinke, T.D. & Ward, M.C. (1986). Mangroves of the Kosi system, South Africa: their re-establishment since a mass mortality in 1965/66. *South African Journal of Botany,* **52,** 501–12.

Weidemann, A.M. & Pickat, A. (1996). The *Ammophila* problem on the northwest coast of North America. *Landscape and Urban Planning* (in press).

Weisser, P. J. & Backer, A. P. (1983). Monitoring beach and dune advancement and vegetation changes 1937–1977 at the farm Twinstreams, Mtunzini, Natal, South Africa. In *Sandy Beaches as Ecosystems: Developments in Hydrobiology,,* ed. A. MacLachlan & T. Erasmus, pp. 727–32. The Hague: Junk.

Weisser, P.J. & Cooper, K. H. (1993). Dry coastal ecosystems of the South African east coast. In *Dry Coastal Ecosystems –Africa, America, Asia and Oceania. Ecosystems of the World,* 2B, ed. E.van der Maarel, pp. 109–28, Amsterdam: Elsevier.

Weisser, P.J., Garland, L.F. & Drews, B.K. (1982). Dune advancement 1937-1977 at the Mlalazi Nature Reserve, Natal, South Africa, and preliminary vegetation-succession chronology. *Bothalia,* **4,** 127–30.

Werger, M.J.A. (1978a). Biogeographical division of southern Africa. In *Biogeography and Ecology of Southern Africa,* ed. M.J.A. Werger, pp. 145–70. The Hague: Junk.

Werger, M.J.A. (1978b). The Karoo–Namib region. In *Biogeography and Ecology of Southern Africa,* ed. M.J.A. Werger, pp. 231–99. The Hague: Junk.

Whitfield, A.K. (1980). Food chains in Lake St. Lucia. In *Studies on The Ecology of Maputoland,* ed. M.N. Bruton & K.H. Cooper, pp. 427–31. Grahamstown: Rhodes University.

Freshwater wetlands

14

K.H. Rogers

14.1   Introduction

Wetlands are enigmatic ecosystems. They are neither fully aquatic nor terrestrial and are not easily classified as ecotonal between the two (Patten 1990). As such they have remained largely unstudied in many regions of the world, and southern Africa is no exception (Breen & Begg 1989).

The wide variability of landscapes and climates over the subcontinent manifests itself in an equally wide range of wetland types, distributed over the full range of other biomes. Because of the low rainfall and rugged topography over much of the region, most wetlands are small and ephemeral. The few exceptions are associated with specific geological phenomena such as the northern coastal plain of KwaZulu-Natal (Mkuze and Pongolo floodplains) and the tectonic subsidence in the northern Kalahari Basin (Okavango Delta; see Chap. 1, this volume).

The small size, scattered distribution and high degree of ephemerality of southern African wetlands belies the extremely important roles they play in the hydrological and chemical cycles of regional landscapes (Begg 1986). Their function as the 'kidneys of the landscape' (Mitsch & Gosselink 1986) is therefore seldom recognized by decision-makers. All of these circumstances are reflected in the database of the region's wetlands; data are scattered and highly unrepresentative of the variability both within and between wetland types.

Few wetland types have been thoroughly investigated, and of those that have been well studied, there is seldom more than one example. There is therefore little basis for the presentation of either consensus or debate on the nature of the vegetation of the wetland 'biome'.

Instead, this review integrates the disparate collection of studies to develop a broad view of what is particular to wetland vegetation science in southern Africa.

This review follows more or less the same sequence as the other biome chapters in this volume. However, where other chapters have delimited the study area in terms of a discussion of the biome boundary, this chapter begins with a general classification of wetland types and a description of their distribution across the region. Subsequent sections follow the ecological hierarchy from species autecology level to the ecosystem.

There are many definitions of what constitutes a wetland (e.g. Rogers, Rogers & Buzer 1985; Walmsley & Boomker 1988; Breen & Begg 1989). Almost any water body, from a mountain stream to an estuary or coral reef, has been called a wetland. The definition of wetland that is adopted for this chapter is: 'land where the water table is, at least periodically, at or above the land surface for long enough to promote the formation of hydric (waterlogged) soils and the growth of aquatic plants'. Even this definition can be widely interpreted, but this chapter deals specifically with inland lacustrine, palustrine and riverine wetlands. Estuarine wetlands are dealt with in Chap. 13 (this volume), and artificial water bodies are not considered, other than in the context of aquatic weeds (Chap. 22, this volume), because the extreme and varied water-level fluctuations they experience on the subcontinent generally prevents the growth of littoral vegetation.

14.2 Wetland classification and distribution

The above definition provides the basis for the physical, chemical and biotic description of the boundary of any particular wetland. This section describes the types of wetlands found in southern Africa, their distribution over the landscape and the factors that influence their distribution.

In most parts of the world the problem of classification is dealt with in two ways: firstly by delineating the large and 'priority' wetlands, and then by developing an objective classification – one that groups similar wetlands and so furnishes the units for a more inclusive inventory, which can be developed over time (Breen 1988). Such classifications are usually hierarchical and thus permit the consistent, systematic aggregation and disaggregation of large amounts of data and information. They also allow selection of the appropriate level of detail for decision making at a range of geographic or administrative scales. Wetland classification in southern Africa is in a state of flux and although important advances are being made, the products are not yet rigorous enough to serve the inventory and decision-making processes described above.

The first attempt at classifying southern African wetlands was made by Noble & Hemens (1978) in their review of research needs in freshwater ecosystems. Theirs was a very simple, non-hierarchical description of the general types of wetlands found in South Africa. No similar description has been published for Namibia or Botswana. Morant (1983) reviewed wetland classification and concluded that because of the diversity of wetlands in South Africa it would be appropriate to adopt the very comprehensive system used by the United States Fish and Wildlife Services. This system has, however, not been adopted and wetland diversity on a national scale remains undocumented. Several studies (Morant 1983; Breen & Begg 1989; Rogers 1995) have added to the Noble & Hemens model by expanding the range of wetlands it covers and developing a hierarchical structure. Unfortunately these studies have not had much impact on improving the potential of the model as a predictive tool; it remains more a general description of the range of wetland types than an objective classification based on multivariate criteria. Two recent publications (Allanson *et al.* 1990; Cowan 1995) do, however, provide the material for such an objective classification; one that can be based not only on wetland characteristics but also on the determinants of wetland structure and function.

It was beyond the scope of this chapter to generate a new objective classification, but a description of southern African wetland types and their distribution is provided (Table 14.1). Wetland types are presented as a hierarchical modification of the Noble & Hemens (1978) classification and distribution is described in terms of landscape classifications provided by Allanson *et al.* (1990) and Cowan (1995). The summaries of Allanson *et al.* and Cowan's proposals in sections 14.2.1 and 14.2.2 can be used to interpret Table 14.1.

14.2.1 Limnological regions

Allanson *et al.* (1990) used broad geomorphological, geochemical and climatological features of southern Africa to define five limnological regions (Fig. 14.1). Although this classification was not specifically designed for wetlands it is the only one that gives spatial definition to aquatic habitat types and their distribution over the whole subcontinent.

14.2.1.1 *Subtropical coastal peneplain*

Impermeable Cretaceous mudstone formations, which dip gently to the eastern seaboard, provide the physical basis for a high water table in overlying coastal sands. This area probably has the highest concentration of wetlands in the subcontinent. Examples are the Pongolo floodplain, the Muzi and Mkuzi swamps, the Kosi lakes, Lake Sibaya, Lake St. Lucia, swamp forests and the extensive pan systems of Maputaland (Fig. 14.2).

14.2.1.2 *Elevated plateau and southeastern coastal plain*

This summer-rainfall region is one of the largest and extends from the Great Fish River in the south across the highveld of South Africa into the savanna of the Limpopo valley. Wetland diversity is low (Fig. 14.3) and it is dominated by pans (highveld) and the riparian fringes of rivers that have cut through the overlying Karoo sediments and now flow over the harder, more ancient undermass (e.g. Sabie, Crocodile and other eastern Transvaal rivers). This region also contains the most spectacular of the subcontinent's wetlands – the 18 000-km^2 Okavango Delta in northeastern Botswana.

14.2.1.3 *Australomontane*

The montane massif of Lesotho lies at an altitude of 3000 m or more and is one of the highest rainfall regions in southern Africa (Chap. 2, this volume). The wetlands are typically alpine bogs of late- to post-glacial age (c. 8000 BP).

14.2.1.4 *Temperate acid waters of the Western Cape*

This winter-rainfall region is dominated by coastal rim mountains of erosion-resistant quartzites and hard sandstone, which produce oligotrophic, poorly buffered and acidic waters. The wetlands of this floristically rich area have received little attention and the aquatic vegetation

Table 14.1 Description of the wetland types of southern Africa and their distribution. Wetland types are presented as a hierarchical modification of the Noble & Hemens (1978) classification and distributions are described in terms of the regional landscape. Classifications of Allanson et al. (1990) (prefix A) and Cowan (1995) (prefix C). Examples of specific wetlands are provided where known

Wetland type	Description	Distribution	Example	Vegetation	References
VLEIS AND FLOODPLAINS					
River source sponges					
Bogs	Seasonally or perennially waterlogged areas; hydrologically static wetland found in volcanic soil at the source of rivers; mire with accumulated peat in mountain valleys, often raised above surrounding grassland	Mountains in Lesotho, Transkei and Drakensberg at altitudes between 1800 and 3500 m **A-3; C-Md**	Source of the Tugela river	Dominated by sedges and mosses; no woody vegetation; hydrophytes and algae in open pools in bogs; reeds at lower altitudes	Jacot-Guillarmod (1962) Breen & Begg (1989) Noble & Hemens (1978) Allanson et al. (1990) Cowan (1995)
Acid sponges	Seasonally or perennially waterlogged areas with acid (pH 3.5–6) and humic stained water found on slopes, some immature peat	Mountains of the Western Cape **A-3; C-MCF**		Fynbos, mostly Restionaceae and Bruniaceae	Noble & Hemens (1978) Breen & Begg (1989) Allanson et al. (1990) Cowan (1995)
Marshes and swamps (Vleis)	Generally flat terrain with stretches of land which are inundated and become waterlogged in the wet season. Found on soils of high clay content, sand or gravel. Marsh – water level not much above soil surface in wet season. Swamp – water level well above the soil surface in wet season			Marshes – emergent plants mostly less than 2 m in height. Swamps – less diverse vegetation with sedges, reeds or trees, many plants being 3 m or more in height	Noble & Hemens (1978) Breen & Begg (1989)
Sedge marshes	Perennially waterlogged and may have accumulated peat	Southern escarpment **C-SE**		Dominated by sedges and hydrophilous grasses	Noble & Hemens (1978) Cowan (1995)
Restioid marshes	Perennially waterlogged, acidic	Western Cape **A-4; C-MC, MD, SW, SE**		Dominated by Restionaceae and sedges	Noble & Hemens (1978) Allanson et al. (1990) Cowan (1995)
Reed marshes and swamps	Found in perennial or seasonal standing water on floodplains. Also fringe many coastal and estuarine lakes	Continental plateau of summer rainfall region **A-3; C-C, PE, PN, SE, SOR**	Chobe Marsh, Namibia; Olifantspruit, southern Transvaal; confluence of the Kavango and Cuito rivers, Namibia; Linyanti Swamp, Namibia	Generally dominated by Phragmites with Cyperus, Scirpus and Typha as subdominants	Allanson et al. (1990) Bethune (1991) Schlettwein et al. (1991) Venter (1991) Cowan (1995)
Papyrus swamps	Perennial, deep fresh-water situations	Northern KwaZulu-Natal, north eastern Botswana **A-1, 2; C-C**	Mkuze and Mfolosi Swamps, northern KwaZulu-Natal; Okavango Delta, Botswana	Dominated by Cyperus papyrus	Noble & Hemens (1978) Allanson et al. (1990) Cowan (1995)
Mediterranean-climate seasonal wetlands	Areas that become waterlogged in winter forming seasonal pools	Southwestern coastal flats **A-4**			Noble & Hemens (1978) Allanson et al. (1990)
Swamp forests	Swamps found in freshwater streams along the KwaZulu-Natal coast	KwaZulu-Natal coast, south of Kosi lakes **A-1; C-SE, C**	Streams on the eastern shores of Lake St. Lucia	Dominated by Ficus trichopoda and other trees; climbing fern Stenochlaena tenuifolia	Wessels (1991) Noble & Hemens (1978) Allanson et al. (1990) Cowan (1995)
Floodplains	Develop in the middle or lower reaches of southern African rivers, where the rivers attain grade and overflow their banks during floods. Most hydrologically dynamic wetland systems				Noble & Hemens (1978) Breen & Begg (1989)
Karoo salt flats	Connected to drainage systems, dry most of the time, highly saline soils	Southern plateau **A-5; C-PS**	Van Wyksvlei, Grootvloer and Verneukpan	Devoid of any macrophytic vegetation	Noble & Hemens (1978) Allanson et al. (1990) Cowan (1995)

324

	Description	Distribution	Location	Vegetation	References
	floodplain of varying width on either side. Riverine area normally seasonally inundated and the grassy floodplain is only inundated by occasional floods	A-2; C-PN, SL	Northern Transvaal	*Cymbopogon*. Few trees e.g. species of *Acacia* and *Rhus*	Allanson *et al.* (1990) Bethune (1991) Coetzee & Rogers (1991) Schlettwein *et al.* (1991) Rogers & Higgins (1993) Cowan (1995)
Storage floodplains	Comprise a riverine area and adjacent floodplain inundated by flooding; retain standing water on the floodplain for long periods between floods in small lakes and smaller water bodies (pans)	Northern Transvaal, coastal plain A-1, 2; C-SLV, C	Pongolo, KwaZulu-Natal	*Potamogeton crispus* and other aquatics, marginal *Phragmites*, *Cynodon* lawns and riparian forest	Heeg & Breen (1982) Noble & Hemens (1978) Allanson *et al.* (1990) Cowan (1995)
RIPARIAN FRINGE	Hydromorphic areas along a water course, periodically influenced by high water tables or flooding	Along all water courses **All A, Most C**	Sabie River fringe, eastern Transvaal; Kavango River, Namibia	Reeds, woody vegetation: trees and shrubs	Allanson *et al.* (1990) Bethune (1991) Van Coller (1993) Rogers (1995)
ENDORHEIC PANS	Circular to oval depressions; shallow, usually less than 3 m deep, even when fully inundated. Inundation is ephemeral. Have no outlet, semi-permanently or periodically filled with water. Occur in areas of less than 500 mm mean annual precipitation	Area south of the Zambezi, Kavango and Kunene Rivers. Highest concentrations found in the northern Cape, western and north central Free State and southern Transvaal A-1, 2, 5; C-PW	Makgadikgadi Pans, Botswana		Noble & Hemens (1978) Shaw (1988) Allanson *et al.* (1990) Allan *et al.* (1995) Cowan (1995)
Salt pans	Dry for most of the time, may contain perennial pools filled by springs, highly saline soils; shallow (<1 m), clayey substrate on calcrete	Northern Namibia, Ovambo and Etosha; Karoo, Kalahari, western Orange Free State and Transvaal A-2, 5; C-PW, SK	Etosha Pan and Kwarikarib, Namibia	Devoid of macrophytic vegetation	Noble & Hemens (1978) Allanson *et al.* (1990) Lindeque & Archibald (1991) Allan *et al.* (1995) Cowan (1995)
Temporary pans	Shallow, dry for long periods of time. Alkaline and moderately saline soils	Northern Cape, the western Orange Free State and Transvaal A-2, 5; C-PW		Salt-tolerant grasses	Noble & Hemens (1978) Allanson *et al.* (1990) Allan *et al.* (1995) Cowan (1995)
Grass pans	Seasonal, dry up in winter, inundated in summer, nutrient rich	Southern and eastern Transvaal A-1, 2		Hygrophilous grasses and other low vegetation (some salt tolerant), submerged hydrophytes in summer	Noble & Hemens (1978) Allanson *et al.* (1990) Allan *et al.* (1995)
Sedge pans	Seasonal, do not dry sufficiently in the middle for vegetation to establish	Northern escarpment A-1; C-SNE	Pans of the Lake Chrissie area	Thick marsh vegetation, mainly Cyperaceae	Noble & Hemens (1978) Allan *et al.* (1995) Cowan (1995)
Reed pans	Temporary or semi-permanent, clear water, sediments rich in organic matter			*Phragmites*	Noble & Hemens (1978) Allan *et al.* (1995)
Semi-permanent pans	Remain inundated for many consecutive years. May dry out occasionally	A-2, 5; C-SNE	Lake Chrissie, Barberspan, western Transvaal	*Potamogeton*, sparse grasses	Noble & Hemens (1978) Allanson *et al.* (1990) Allan *et al.* (1995)
LAKES					
Coastal lakes	Water body with relatively deep open waters adjacent to the coastline and distinct littoral fringe wetland	Northern Transvaal, Western Cape coast, west coast north of the Berg River mouth A-1, 4; C-C,SS	Kosi Lake system, St. Lucia, Wilderness lakes	*Potamogeton pectinatus*, *Phragmites australis*	Noble & Hemens (1978) Allanson *et al.* (1990) Howard-Williams (1980) Cowan (1995) Hart (1995)
Freshwater interior lakes	Water body with extensive and relatively deep open waters	Pietersberg plateau A-2B; C-PN	Fundudzi, Northern Transvaal		Noble & Hemens (1978) Shaw (1988) Allanson *et al.* (1990) Cowan (1995) Hart (1995)

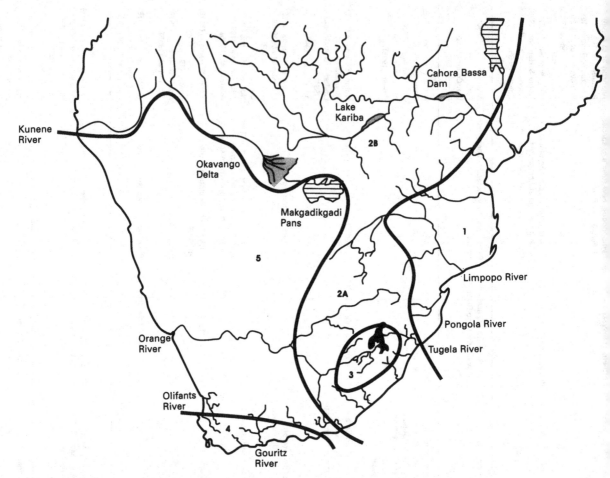

Figure 14.1 **Limnological regions of southern Africa (after Allanson et al. 1990). 1, Subtropical coastal peneplain; 2, Southeastern coastal plain (A) and Elevated plateau (B); 3, Australomontane; 4, Temperate acid waters of the Western Cape; 5, Arid west.**

of the mountainous areas has been described by Allanson et al. (1990) as 'relatively uncharacteristic'. However, the dominance of the emergent wetland flora in many marshes of this region by species of the Restionaceae contradicts this.

14.2.1.5 Arid west

This, the largest of the limnological regions of Allanson et al. (1990), is dominated by the Kalahari Basin. The principle limnological feature of this region is the Orange River with its extensive stands of Phragmites australis. There are also many ephemeral pans, vleis and rivers scattered across this arid to semi-arid area. They too have received little attention, with the notable exceptions of Sossus Vlei (Grobbelaar 1976) and the Kuiseb River in Namibia (see Chap. 9, this volume).

14.2.2 Classification of South African wetlands

Cowan (1995) used a Geographic Information System (GIS) to overlay topography, hydrology and 'nutrient regime' (a combination of stratigraphy, temperature and humidity) maps to devise a subjective classification of wetland regions of South Africa (Figs. 14.4 and 14.5). The four primary regions in this classification (Fig. 14.4) closely resemble the limnological regions of Allanson et al. (1990; Fig. 14.1). They are: (1) the subtropical coastal plain; (2) the coastal slope and rimland; (3) the mountain region; and (4) the plateau, which incorporates both the plateau and arid west of Allanson et al. (1990). These primary regions are further divided into secondary and tertiary regions (Fig. 14.5), but there are, as yet, no detailed descriptions of the types of wetlands characteristic of each region. The summary in Table 14.1 is a first attempt at such a description.

Figure. 14.2 **Swamp of *Phragmites australis* (foreground), *Cyperus papyrus* and *Typha capensis* on the inland margin of a dune barrier at Mapelane on the northern KwaZulu-Natal coast** *(Photo: H.C. Taylor).*

Although Walmsley & Boomker (1988) provided a detailed strategy for achieving a comprehensive inventory and functional classification of southern African wetlands, no such classification has been undertaken. This remains an urgent priority.

14.3 Autecology

Almost all the studies of aquatic macrophytes at this level of organization have been conducted in reservoirs and most have been on alien weeds (cf. Allanson *et al.* 1990). The exceptions are two detailed studies of submerged aquatic species, *Potamogeton crispus* on the Pongola River floodplain (Rogers & Breen 1980; Rogers 1984) and *P. pectinatus* in the coastal lake, Swartvlei (Howard-Williams & Allanson 1978; Howard-Williams 1980). The latter study focused on the role of *P. pectinatus* in primary productivity and nutrient cycling in Swartvlei and as such is more appropriately dealt with

in the section on ecosystem functioning. The study of *P. crispus* would therefore appear to be the only detailed study of the life history of an indigenous aquatic macrophyte in the region (Fig. 14.6).

Potamogeton crispus, a cosmopolitan species, is common in the freshwater systems of the summer-rainfall area. There are many ecotypic variations of this species (Mitchell & Rogers 1985) but, to generalize, it can be classified (*sensu* Grime 1979) as a 'ruderal with multiple regenerative strategies'. The regenerative strategies are 'seasonal regeneration' by means of asexual turions, a 'persistent seed (achene) bank' and 'persistent seedlings' which remain dormant over periods of poor light intensity.

On the Pongola River floodplain frequent floods and concomitant poor light climate during summer create an unfavourable period for plant growth, but winter conditions are more favourable. Droughts result in the lake habitats drying up with a recurrence interval of about 12 years. As such, this floodplain represents a typical habitat of the summer-rainfall plateau in which aquatic

Figure 14.3 **Vlei below Hlatikulu in the foothills of the KwaZulu-Natal Drakensberg, showing *Polygonum plebeium*, *Cyperus fastigiatus* and emergent stems of *Aponogeton ranunculiflorus*** (Photo: D. Edwards).

species experience pronounced seasonal fluctuations in habitat availability as well as drought that is less predictable and occurs on a longer time scale.

Since the unfavourable light conditions during summer vary in duration from 3 to 7 months (Rogers 1984), the autumnal decrease in temperature is a more reliable signal of the forthcoming favourable winter than is photoperiod. Thus, a drop in water temperature to 25 °C in autumn stimulates germination of turions (seasonal regeneration) which increases until the onset of the winter minimum temperature of 15 °C. Turion germination of *Potamogeton crispus* is thus staggered over a period of 3–4 months (April–July) when flooding and therefore unfavourable conditions are less likely. The chance of successful plant establishment is further increased by the ability of young plants to remain dormant for up to three months if light availability remains poor as a consequence of late floods (persistent seedling strategy). Individual plants have a life span of 4–5 months, but the staggered turion germination means that the population exploits the full 6–8-month period between floods. Furthermore, this germination pattern

ensures that reproduction by turions takes place over a 3–4-month period when the likelihood of floods increases. A single reproductive event would increase the risk of reproductive failure if floods should arrive early.

Achenes of *Potamogeton crispus* require drying and rewetting for germination. Thus, during wet periods the achene bank increases to several thousand per square metre, but rewetting after drought stimulates germination. This 'persistent seed bank strategy', therefore, ensures persistence through droughts.

There is much contradiction and argument in the literature about the life history of this species, particularly with reference to the relative importance of the various reproductive structures. A comparison of the life history of *Potamogeton crispus* on the subtropical Pongolo river with that of the same species in Canadian ponds that freeze over in winter (cf. Mitchell & Rogers 1985) indicates that, although entirely different environmental events are experienced in these habitats, ecotypic differentiation of the different regenerative strategies permits the species to survive over a very wide latitudinal range.

Although *Potamogeton crispus* is widespread and is rela-

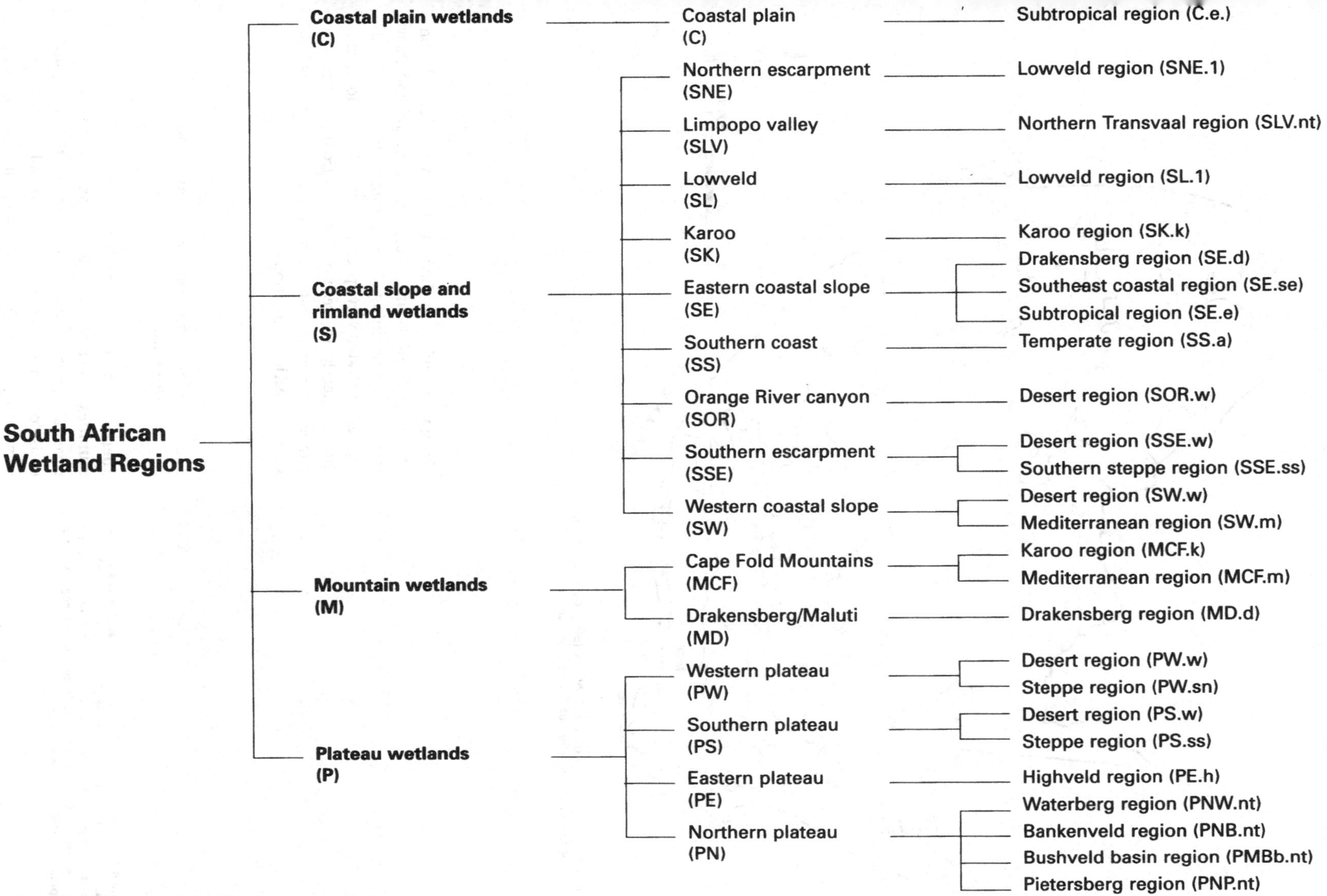

South African Wetland Regions

Coastal plain wetlands (C)
- Coastal plain (C)
  - Subtropical region (C.e.)
- Northern escarpment (SNE)
  - Lowveld region (SNE.1)
- Limpopo valley (SLV)
  - Northern Transvaal region (SLV.nt)
- Lowveld (SL)
  - Lowveld region (SL.1)
- Karoo (SK)
  - Karoo region (SK.k)

Coastal slope and rimland wetlands (S)
- Eastern coastal slope (SE)
  - Drakensberg region (SE.d)
  - Southeast coastal region (SE.se)
  - Subtropical region (SE.e)
- Southern coast (SS)
  - Temperate region (SS.a)
- Orange River canyon (SOR)
  - Desert region (SOR.w)
- Southern escarpment (SSE)
  - Desert region (SSE.w)
  - Southern steppe region (SSE.ss)
- Western coastal slope (SW)
  - Desert region (SW.w)
  - Mediterranean region (SW.m)

Mountain wetlands (M)
- Cape Fold Mountains (MCF)
  - Karoo region (MCF.k)
  - Mediterranean region (MCF.m)
- Drakensberg/Maluti (MD)
  - Drakensberg region (MD.d)

Plateau wetlands (P)
- Western plateau (PW)
  - Desert region (PW.w)
  - Steppe region (PW.sn)
- Southern plateau (PS)
  - Desert region (PS.w)
  - Steppe region (PS.ss)
- Eastern plateau (PE)
  - Highveld region (PE.h)
- Northern plateau (PN)
  - Waterberg region (PNW.nt)
  - Bankenveld region (PNB.nt)
  - Bushveld basin region (PMBb.nt)
  - Pietersberg region (PNP.nt)

Figure 14.4 A hierarchical classification of the wetland regions of South Africa. After Cowan 1995.

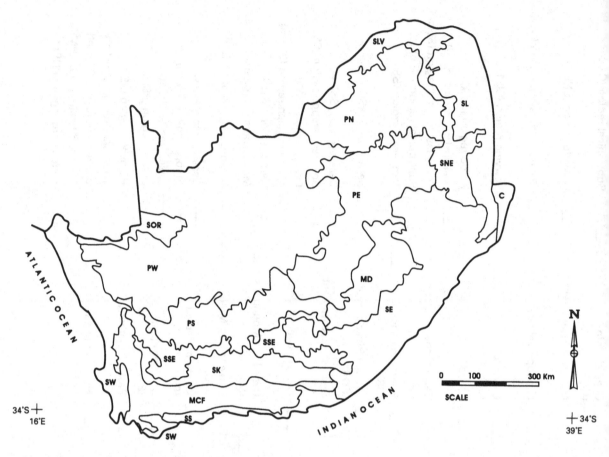

Figure 14.5 **Wetland regions of South Africa (after Cowan 1995).**
See Fig. 14.4 for an explanation of the abbreviations.

Figure 14.6 **The Pongola River floodplain pans and grasslands in KwaZulu-Natal. The Pongola River is shown by the line of scattered trees in the foreground. The riparian forest is sparse, owing to slash and burn agriculture in this fertile region of the subtropical coastal peneplain** (Photo: K.H. Rogers).

tively well studied in the rest of the world, elucidation of the details of its life-history characteristics and their response to the timing, frequency and duration of flooding demonstrated very specific relationships to local conditions. Clearly, the results of studies from the northern temperate regions cannot be reliably extrapolated to southern African situations.

14.4 Plant community structure and dynamics

There have been few quantitative studies of plant communities in southern African wetlands, and they do not cover all the wetland types. There is, therefore, little basis for generalization about regional wetland plant community structure and dynamics. The few site-specific studies do, however, provide the basis for comparing southern African wetlands with those in other parts of the world. The general conclusion reached by an analysis of these studies is that southern hemisphere aquatic systems differ in degree rather than kind from the northern temperate systems on which so much of the basic theory has been established. This view concurs with the findings of Davies & Walmsley (1985) and Mitchell & Rogers (1985).

The very wide range of southern African wetland types provides the potential for ecologists to extend the horizons of the theory of community structure in wetlands which has been based on a relatively narrow range of temperate systems. This section expands upon this northern/southern hemisphere contrast while presenting an account of wetland plant community structure and dynamics in southern Africa.

14.4.1 Community structure

Most studies of wetland vegetation have been qualitative (i.e. descriptive) or have arisen out of broad-scale regional vegetation studies. Descriptive studies include those of the coastal lakes Swartvlei (Howard-Williams & Allanson 1978) and Sibaya (Howard-Williams 1979); Barber's (Zimbatis 1975) and Rolfe's pans (Rogers *et al.* 1989); and the Kavango River wetlands (Bethune 1991). Many accounts of regional vegetation mention wetlands specifically. Some examples are: Phillips (1931; Knysna forests); Tinley (1976; Maputaland coastal plain); Coetzee, Bredenkamp & Van Rooyen (1994; highveld grasslands).

The first site-specific, quantitative study of wetland vegetation was conducted during the 1950s on Groenvlei, a fen on the southern Cape coast (Martin 1960). This very detailed study showed that the species composition and environmental conditions (physical and chemical) in Groenvlei were strikingly similar to those of northern temperate fens. Had Martin extended his study to address successional theory he would have had a major impact on the Gleasonian/Clementsian debate which had received much local impetus from the highly 'organismal' approach of Phillips (1931). Indeed, Martin's hypothesis of multiple successional pathways and the role of disturbance in their determination was conceptually very advanced for the time and was very similar to that which was so clearly elucidated by Walker (1970) for classic hydroseres in Britain. It was Walker who produced the first clear evidence that succession did not follow a predictable and linear sequence of events, but showed elements of stochasticity with multiple pathways that could lead to similar or different 'climax' communities.

Many recent studies of wetland plant community structure in southern Africa have been more quantitative, and have used various multivariate statistics to improve the objectivity of data analysis and interpretation. There are now detailed descriptions of a range of systems, including the Pongolo River floodplain (Musil, Grunow & Bornman 1973; Furness & Breen 1980), the Nyl River floodplain (Coetzee & Rogers 1991), the Okavango Delta (Ellery 1987; Ellery *et al.* 1991; Ellery, Ellery & McCarthy 1993a), Daggafontein reed-marsh (Venter 1991), the swamp forests of the eastern shores of Lake St. Lucia (Wessels 1991) and the Sabie River riparian communities (Bredenkamp & Van Rooyen 1993; Van Coller 1993). A review of these studies is beyond the scope of this section, where the emphasis is placed on vegetation dynamics rather than description.

Most of the above studies produced both community descriptions and correlations between species distribution and environmental conditions. In so doing they established that, in general, vegetation, hydrology and geomorphology can be linked along three hydrogeomorphic gradients. These are the lateral, vertical and longitudinal gradients along which there may be distinct changes in the frequency, duration and depth of flooding. The physical conditions along these gradients in turn lead to vegetation changes with distance away from the channel, elevation above the channel and distance downstream. Thus, the classical zonation of plant communities (cf. Denny 1985) is evident in many wetlands, but it may be spatially modified and manifest in the form of a mosaic, rather than linear bands of vegetation. Key determinants of such modification are hydrological disturbances in flowing systems (Okavango river channels; Ellery 1988; Ellery *et al.* 1995), differential peat accumulation (Okavango back-swamps; Ellery *et al.* 1991), geology and geomorphic landform (Sabie river; Van Coller 1993), the competitive growth of dominant clonal species such as *Phragmites australis* (highveld vleis; Venter 1991) and even soil chemistry (Okavango islands; Ellery *et al.* 1993a).

Studies of community structure in southern African wetlands have, therefore, generally corroborated the established view that species distribution in any wetland is primarily a function of the site-specific hydrological regime which, in turn, is a product of catchment runoff and local geomorphology.

14.4.2 Community dynamics

There have been very few studies on either the pattern or the process of vegetation change in southern African wetlands and most of our understanding is presented as hypotheses generated from the correlations of species distribution and environmental conditions discussed above (e.g. Martin 1960; Downing 1968; Van Coller 1993).

Studies of four very different systems (a highveld pan, a highveld vlei, the permanent swamps of the Okavango Delta, and the riparian zone of the rivers in the semi-arid Eastern Transvaal) do give important insights. They provide a good basis for developing a conceptual framework of wetland vegetation processes that is applicable to the diverse conditions of the subcontinent.

14.4.2.1 *Highveld pans*

The pans of the highveld experience large fluctuations in water level, both seasonally and in the longer term (Rogers et al. 1989). Although marked changes in species dominance can be expected as a result, only one study has documented such vegetation response. Rogers et al. (1989) described the change in species composition and spatial distribution in Rolfe's pan near Johannesburg during its recovery from the 1983 drought. In 1983 the pan was extremely shallow (mean depth 0.2 m) and was colonized by a mixed stand of the submerged aquatics *Lagarosiphon major*, *Potamogeton pectinatus* and *P. thunbergii*. The marginal zone showed a marked zonation with monospecific bands of *Panicum stapfianum*, *Paspalum distichum* and *Sporobolus fimbriatus* occurring in this order up the elevation gradient. By the end of winter of that year, only 7.7% of the 78 ha of pan area was inundated and all the submerged vegetation had died back.

In 1986, after three years of good rainfall, species composition and distribution had changed. The entire marginal zone was dominated by *Panicum stapfianum* and it was separated from the now uncolonized central regions of the pan by a zone of *Typha capensis* patches, interspersed with a mixed stand of the two *Potamogeton* species.

Such a change from submerged to emergent species following re-inundation suggests that highveld pans show vegetation 'cycles' similar to those described for prairie potholes (pans) in northern America (Kantrud, Millar & Van der Valk 1989). During periods of above-normal precipitation, the rise in water level in these potholes drowns out emergent vegetation, creating a 'lake' marsh dominated by submerged species. During the next drought, the drying out of the marsh bottom stimulates the germination of seeds of emergent species which then spread vegetatively on re-inundation. If water levels rise and remain too high for persistent growth of these emergents, the cycle is repeated.

These vegetation cycles have very important implications for breeding waterfowl in the Prairie Pothole Region (Weller 1981) and are carefully managed to ensure their continuance. Rogers et al. (1989) also noted marked changes in aquatic bird species composition and their use of Rolfe's pan during their study. Clearly more detailed studies of the patterns and mechanisms of veg-

etation change are needed for the management of the extensive pan systems which appear to play a similar role in avian ecology in southern Africa (Allan, Seaman & Kaletja 1995; Breen 1988).

14.4.2.2 *Riparian vegetation of semi-arid regions*

The riparian zone of rivers has very high conservation and resource value and has undergone much degradation as a result of human activities (Acocks 1976; Rogers 1995). However, there has been only one quantitative study of the vegetation dynamics in these systems. Carter & Rogers (1989) and Carter (1995) used a Markov analysis of the 50-year aerial photographic record to describe the change in vegetation cover of the rivers in the Kruger National Park (Fig. 14.7). They categorized the river bed into six states (water; sand; reed (*Phragmites mauritianus*); other vegetation states; and woody riparian vegetation), and described a complex set of state transitions in space and time. All the 36 possible interstate transitions for which these models catered, occurred in each of the rivers over the study period, but they did not occur simultaneously, nor with equal probability. Fur-

Figure 14.7 **View of a river in the Kruger National Park showing a very distinctive channel form created by bedrock control of the channel** *(Photo: K.H. Rogers).*

thermore, the probabilities of interstate transition, themselves, changed with time.

In general the models revealed that the most common sequence of change was water – sand – herbaceous species – reeds – woody vegetation, suggesting a successional sequence that occurs on areas of sediment deposition or exposure. It was hypothesized that decreasing flow in these rivers (as a result of catchment development) is leading to marked changes in the structure of the river beds and, since the state replacement sequence described above has become increasingly probable, the overall vegetative cover and composition of the rivers is becoming increasingly resistant to changes in the reverse direction.

There are many causes of state change in the rivers, but flood disturbance that removes vegetation, drought that kills it, and extended periods of intermediate water level that promote plant growth and establishment, appear to be the main agents. Increasing utilization of upstream water resources are therefore likely to have major implications for the conservation of riparian vegetation in the Kruger National Park.

Rivers are generally recognized as highly dynamic systems, and changing patterns of sediment deposition and erosion play a major role in riparian vegetation dynamics of alluvial rivers (Wissmar & Swanson 1990). Local studies present a unique perspective of bedrock-controlled rivers which receive high sediment loads in the semi-arid regions of southern Africa. Almost all other studies have been on alluvial, temperate rivers, which present entirely different hydrological and sediment regimes.

14.4.2.3 *Highveld vleis*

Whereas vegetation change in the hydrologically variable rivers of the Kruger National Park is primarily driven by allogenic disturbance, the marked changes that can occur in the more stable highveld vleis appear to be autogenically driven (Venter 1991; Fig. 14.8).

Most of the highveld vleis are permanently flooded marshes dominated by mono-specific stands of *Phragmites australis* growing on peat up to 6 m thick (Rogers 1990). Rapid urbanization and accompanying waste water disposal have, however, resulted in increased flow in many streams and, consequently, inundation of the associated floodplain (Smuts 1984). The vleis on these streams are characterized by more diverse vegetation growing primarily on inorganic soils. The nature of the soil and the history of recent inundation suggests that these communities represent an earlier successional stage than that represented by the *P. australis*-dominated systems. The seasonally inundated fringe is colonized by hygrophilous grasses and sedges (Noble & Hemens 1978)

Figure 14.8 **The Marievale vlei near Johannesburg on the Highveld. Wetland vegetation is dominated by reeds (*Phragmites australis* and *Typha capensis*). Agriculture and vlei grassland are evident at top right and bottom left** *(Photo: K.H. Rogers).*

and the permanent marsh by a complex mosaic of patches of open water colonized by patches of *P. australis*, *Potamogeton* spp. and stands of *Typha capensis* (Venter 1991).

Venter (1991) used aerial photographs to follow change in the distribution of the communities dominated by these species in Daggafontein vlei on the Blesbokspruit. Over the period 1968 to 1989, the area covered by *Phragmites australis* increased 132%, while that covered by *Typha capensis*, open water and grassland decreased by 14%, 50% and 69%, respectively, confirming a net change towards the hypothesized climax of *P. australis*-dominated communities. Studies of differential resource capture by species at the interfaces between communities led Venter to propose an autogenic process of succession dominated by the superior ability of *P. australis* to compete for light and root space. The process of change towards a community dominated by *P. australis* was, however, much more complex over smaller spatial and temporal scales. All possible combinations of change in state occurred over these smaller time scales and many appear to have been driven more by the allogenic forces of fire and waterfowl grazing. Waterfowl, especially the Spurwing Goose (*Plectropterus gambensis*), feed extensively on the rhizomes and young shoots of *T. capensis*, thereby converting areas to open water (Kemper 1984; A.P. Curwell, unpubl. data; see also Chap. 17, this volume). Fire, on the other hand, appears to reduce the competitive ability of *P. australis* and during the severe drought of the early 1980s both the grassland and the *T. capensis*-dominated communities showed a net encroachment on the *P. australis* community.

The vleis of the Blesbokspruit have been designated as a RAMSAR site, because of their importance as a waterfowl habitat, which is in turn dependent on the maintenance of the heterogeneity of the component wetland communities. An understanding of the dynamic interactions between waterfowl, fire and plant competition is therefore crucial to South Africa's commitment to the RAMSAR Convention on Wetlands of International Importance, but a great deal more research is required before the system can be managed with confidence.

14.4.2.4 *The Okavango Delta*

The Okavango River rises in southern Angola, but its course has been interrupted in northern Botswana by rifting, resulting in the collapse of a segment of the earth's crust along northeasterly striking faults. Sediment discharged by the Okavango River into this depression has accumulated to form a gently sloping (1 : 3600), convex alluvial fan. Water is dispersed through the fan by a complex system of distributary channels which intersects the delta to form a mosaic of rivers, lakes, swamps, floodplains and islands (McCarthy *et al.* 1986; Ellery *et al.* 1991). The delta, therefore, contains a very wide range of wetland types, the study of which has made many important contributions to wetland ecology in southern Africa. Indeed, the Okavango Delta is, from a vegetation point of view, probably the best studied wetland system in southern Africa.

Although actual vegetation change has not been researched as it has been in the Kruger National Park, these studies have attempted to test a number of hypotheses of community dynamics. There are four aspects of change in vegetation of the delta that are relevant to this discussion, each of which illustrates different wetland processes: (1) downstream vegetation change; (2) backswamp successional dynamics; (3) island vegetation; and (4) long-term landscape change (Fig. 14.9).

Downstream vegetation change The Okavango Delta is traditionally divided into the permanent swamp of the upper reaches, seasonal wetlands of the middle reaches and the floodplains of the lower reaches and swamp fringes (Smith 1976). This simple division, however, masks a complex vegetation response to the changes in hydrological regime which occur in a downstream direction as water spreads laterally over the fan and is lost by evapotranspiration and seepage to groundwater.

The uppermost reaches of the delta are characterized by a few deep, fast-flowing channels that are devoid of vegetation and flow through swamps dominated by *Cyperus papyrus* with an understorey of *Thelypteris interrupta*. The lower reaches of the permanent swamps are characterized by a larger number of shallower, slower-flowing channels colonized by a wide range of submerged species, such as *Eichhornia natans*, *Ottelia ulvifolia*, *Potamogeton octandrus* and *Rotala myriophylloides*. The back-swamps of

Figure 14.9 **A seasonal swamp in the Okavango Delta showing typical zonation of vegetation from deep water to riparian woodland. The water lily *Nymphaea nouchali* dominates the deep water. This gives way to a floating-leaved community of *Brasenia schreberi* and *Nymphoides indica*. *Oryza longistaminata* dominates the emergent vegetation in shallower water, giving way to *Miscanthus junceus*. Floodplain grasses such as *Sorghastrum friesii* dominate the middle floodplain, giving way to *Imperata cylindrica* and ultimately to the evergreen woodland and *Hyphaene petersiana*** (Photo: W.N. Ellery).

these regions consist of a complex mosaic of wetland communities, ranging from submerged (*Ceratophyllum demersum*, *Lagarosiphon ilicifolius*), to floating-leaved (*Brasenia schreberi*, *Nymphaea nouchali* var. *caerulea*) and a diverse array of emergent communities. Many of the emergent communities develop as floating mats, termed 'sudds' (*Cyperus pectinatus*, *Pycreus nitidus*, *Leersia friesii*).

As the permanent swamps intergrade with the seasonal swamps, the species composition of emergent communities changes first to grasses (*Leersia hexandra*, *Oryza longistaminata* and *Panicum repens*) and then, in the drier downstream reaches, back to large expanses of sedges (*Cyperus articulatus* and *Schoenoplectus corymbosus* var. *corymbosus*).

The floodplains of the most downstream reaches are also found bordering all island and mainland edges.

They are colonized by a complex set of communities that are flooded least frequently, for the shortest period and to the shallowest depth (Smith 1976). In general, the wettest of these floodplains are colonized by species of *Echinochloa, Leersia, Paspalum* and *Vossia*, while the drier parts support lawns of *Cynodon dactylon* or taller species such as *Cymbopogon excavatus* and *Hyparrhenia rufa*.

These complex downstream changes in plant community composition provide an excellent example of species response to the longitudinal hydrological gradient of a wetland. In northern temperate systems where the hydrological regime is less variable, such a gradient is only apparent in very long river systems. In semi-arid systems, however, the combination of highly seasonal inflows, very high evapotranspirational losses and low local rainfall, contracts this gradient and this is a major factor governing species distribution (Rogers & Higgins 1993).

Back-swamp successional dynamics The accumulation of organic sediments in the water bodies of the Okavango Delta leads to a classic successional sequence from submerged aquatics (*Najas horrida, Lagarosiphon verticillifolius, Websteria confervoides*) to floating-leaved species (*Brasenia schreberi, Nymphaea nouchali* var. *caerulea*) and, finally, emergents such as *Eleocharis acutangula, E. dulcis* and *Typha capensis* (Fig. 14.10 a–c; Ellery 1987). However, many hundreds of square kilometres of back-swamp are structurally and floristically distinct from the emergent communities which form the endpoint of this primary successional process. Furthermore, their distribution is unrelated to the environmental factors normally correlated with the change in water depth that follows sediment accumulation.

These communities arise through the formation of floating mats (sudds) of vegetation (14.10 d; Ellery et al. 1990), which develop regardless of water depth. These mats have two origins; in some instances gas-filled masses of organic debris, bound by the roots of *Nymphaea nouchali* var. *caerulea*, rise to the surface and provide exposed areas for colonization of the emergent species *Cyperus pectinatus, Fuirena pubescens* and *Ludwigia leptocarpa*. In other cases, rapid encroachment of shallow open water occurs as a result of the vegetative growth of the sudd-forming sedge *Pycreus nitidus*. This species grows over the canopy formed by dense growths of the submerged *Websteria confervoides*, and the accumulation of plant debris among its roots leads to the formation of a floating peat layer (Fig. 14.11).

Over time, peat accumulates under these mats to form a consolidated substratum for plant growth (Fig. 14.10 e). Concomitant autogenic succession leads to 'climax' communities dominated by *Cyperus papyrus* in the upper reaches of the delta and *Miscanthus junceus*

in the middle reaches (Fig. 14.10 f; Ellery et al. 1991).

These successional processes in the Okavango Delta provide an excellent example of the Mitsch & Gosselink (1986) model of community processes in wetlands. They propose that initially the major determinants of wetland succession are environmental, but as the ecosystem 'develops' there is a progressive, autogenically driven accumulation of nutrients and peat, which results in an 'insulation' of the community from environmental fluctuations. The floating mats in the Okavango Delta provide a substratum that insulates the community from both water-level fluctuations and the variety of consequent environmental changes (light climate, water chemistry and soil conditions) which are major determinants of species composition change earlier in the succession process.

Island vegetation Vegetation of the islands in the perennial swamps of the Okavango Delta exhibits a marked zonation pattern (Ellery et al. 1993a). Island fringes are often characterized by a broad-leaved, evergreen riparian community of *Diospyros mespiliformis, Ficus natalensis, F. sycomorus, F. verruculosa, Garcinia livingstonei, Phoenix reclinata* and *Syzygium cordatum*. This gives way, towards the island interior, to a community dominated by *Acacia nigrescens, Croton megalobotrys* and *Hyphaene petersiana*. The central regions are characterized either by short, sparse grassland dominated by *Sporobolus spicatus* or are completely devoid of vegetation, with sodium bicarbonate encrusted soil surrounding a central pan of extremely high conductivity.

As with the entire Okavango Basin, the basic substratum of the islands is aeolian Kalahari sands which have been re-worked by water flow to form mounds that are entirely surrounded by low-conductivity waters. McCarthy & Ellery (1994) have shown that the transpirational demand of the riparian trees results in a lateral movement of water from the swamp into the island. The differential transpiration of water over salts then results in increased salinity of the soil and groundwater in the centre of the island. Therefore, not only are plant species distributed along a gradient of increasing water-table depth and conductivity/salinity but their water-use actively increases the steepness of that gradient over space and time.

Ellery et al. (1993a) suggest that these sorts of interactions are likely to be more prevalent in wetlands of semi-arid regions than is suggested in the literature. By definition, the movement of surrounding water to the riparian groundwater will, because of transpirational demand, exceed the potential for rainwater to flush out the accumulated salts. The influence of soil and groundwater chemistry on wetland vegetation in southern Africa should, therefore, not be underestimated.

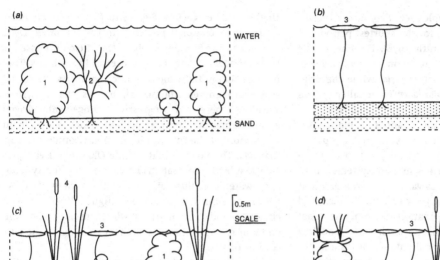

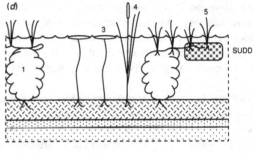

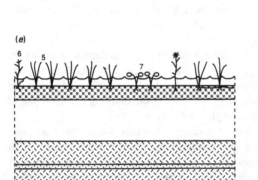

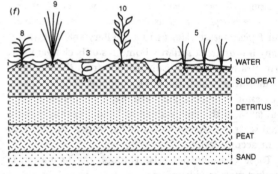

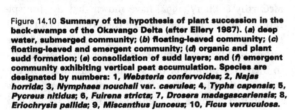

Figure 14.10 **Summary of the hypothesis of plant succession in the back-swamps of the Okavango Delta (after Ellery 1987). (a) deep water, submerged community; (b) floating-leaved community; (c) floating-leaved and emergent community; (d) organic and plant sudd formation; (e) consolidation of sudd layers; and (f) emergent community exhibiting vertical peat accumulation. Species are designated by numbers: 1, Websteria confervoides; 2, Najas horrida; 3, Nymphaea nouchali var. caerulea; 4, Typha capensis; 5, Pycreus nitidus; 6, Fuirena stricta; 7, Drosera madagascariensis; 8, Eriochrysis pallida; 9, Miscanthus junceus; 10, Ficus verruculosa.**

Long-term landscape change While the vegetation of islands, channels and back-swamps responds to changes in local conditions as described above, it is also subject to a set of unique changes at a broader landscape-scale (McCarthy et al. 1986; McCarthy, Stanistreet & Cairncross 1991; McCarthy, Ellery & Stanistreet 1992; McCarthy 1992; Ellery et al. 1993b). Studies at this scale are unparalleled elsewhere in Africa and have wide implications for the understanding of vegetation dynamics of the many other large wetlands on the continent.

The full process of channel avulsion and capture which leads to these landscape-scale processes is summarized in Fig. 14.12.

The main channels of the upstream, permanent swamps of the Okavango Delta are not incised into the Kalahari sands over which they flow as one would expect; rather, the water is confined laterally by dense vegetation rooted in a peat layer forming banks up to 5 m high (Fig. 14.12 a iii–c iii). Because the sediment is largely coarse-grained Kalahari sand it is transported as

Figure 14.11 **A patchy wetland community in the mid-delta region associated with the colonization of floating rafts of organic debris by *Pycreus nitidus* (initially) and *Miscanthus junceus* (ultimately)** *(Photo: W.N. Ellery).*

bedload and therefore also confined to the channels by the flanking vegetation and peat.

The channel banks are permeable to water, especially towards the surface, and this allows water to dissipate into the surrounding swamp matrix. The consequent reduction in flow rate translates into a decline in the competence of the channel to transport this bedload; this results in aggradation as a result of deposition on the channel bed. Aggradation is most rapid in areas close to the channels, since the lateral water-flow increases nutrient supply and promotes vegetation growth (Fig. 14.12 *a–c*).

This combination of channel bed and bank aggradation causes the whole channel to become raised relative to the surrounding swamps; this results in an increased hydraulic gradient at right angles to the channel axis. Eventually, the gradient becomes so great that the channel is subject to capture by headward erosion in a secondary channel that develops elsewhere on the delta surface (Fig. 14.12 *a* ii–*c* ii). Hippopotamus (*Hippopotamus amphibius*) paths often form the basis of these secondary channels, as they drain the lower elevation back-swamps and seasonal floodplains of water seeping from the main channel.

Once the process of capture is complete (Fig. 14.12 *c* iii) the abandoned channel and adjacent swamps dry up (Fig. 14.13). The peat then ignites, sometimes by spontaneous combustion (Ellery *et al.* 1989), reducing the level of the land surface to well below that of the original swamp (Fig. 14.12 *d* iii). In the short term these fires result in a release of plant nutrients, an improvement in soil water retention properties and the transformation of the area into a productive terrestrial habitat.

In the longer term, however, these low-lying areas are subject to reflooding when nearby channel systems are abandoned by the process explained above. If flooding is gradual and shallow, a series of pools interspersed with seasonal floodplains forms important buffalo (*Syncerus caffer*) and lechwe (*Kobus leche*) habitat. In contrast, if flooding is rapid, large and deep open water bodies may be formed, in which case the successional sequence described under *Back-swamp successional dynamics* above results in a typical back-swamp mosaic.

14.5 A conceptual framework of wetland vegetation processes

The preceding discussion of community structure and dynamics of southern African wetlands clearly illustrates the wide range of both allogenic and autogenic processes that determine vegetation processes. There is currently no single model that provides an adequate framework for either the study of wetland communities or the prediction of vegetation response to changing environmental conditions. Three, essentially complementary, models have, however, been proposed in recent years (Van der Valk 1981; Mitsch & Gosselink 1986; Breen, Rogers & Ashton 1988). The Van der Valk model has Gleasonian origins, whereas that of Mitsch & Gosselink is based on Clementsian principles. The Breen *et al.* model goes some way towards reconciling these differences. A contrast of these models forms a good basis from which to propose future research directions.

14.5.1 The Van der Valk model

This conceptual model of allogenic succession is based primarily on life-history features of wetland plants and their differential response to environmental change. The environment acts as a sieve, permitting the establishment of only certain species at a given time. To apply the model to a particular wetland, information is needed on: (1) the potential flora of the wetland; (2) the life-history type of each species; and (3) the nature of the environmental change.

In the model the potential flora is defined as all existing species in the wetland, plus any additional species present in the seed bank. Most of the information needed can therefore be obtained by examination of seed bank data. Species are categorized into one of 12 life-history types on the basis of their life span, propagule longevity and propagule establishment requirements. Environmental change is incorporated into the model in terms of the presence or absence of standing water, since establishment, growth and reproduction of wetland species are each influenced to some extent by these two

Figure 14.12 **A conceptual model of channel dynamics in the Okavango Delta. After Ellery *et al.* 1993b.**

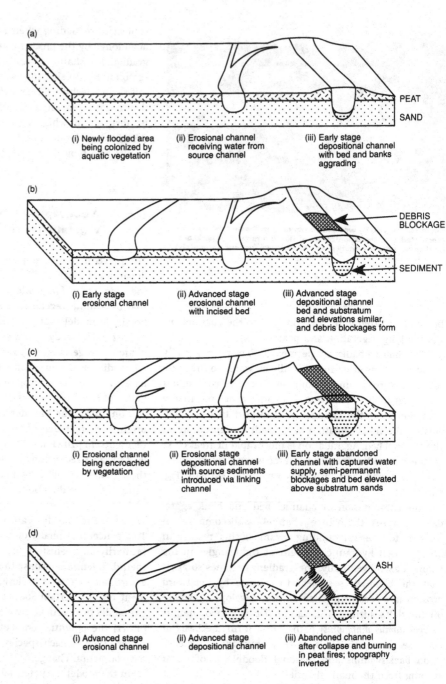

(a)

PEAT

SAND

(i) Newly flooded area being colonized by aquatic vegetation

(ii) Erosional channel receiving water from source channel

(iii) Early stage depositional channel with bed and banks aggrading

(b)

DEBRIS BLOCKAGE

SEDIMENT

(i) Early stage erosional channel

(ii) Advanced stage erosional channel with incised bed

(iii) Advanced stage depositional channel bed and substratum sand elevations similar, and debris blockages form

(c)

(i) Erosional channel being encroached by vegetation

(ii) Erosional stage depositional channel with source sediments introduced via linking channel

(iii) Early stage abandoned channel with captured water supply, semi-permanent blockages and bed elevated above substratum sands

(d)

ASH

(i) Advanced stage erosional channel

(ii) Advanced stage depositional channel

(iii) Abandoned channel after collapse and burning in peat fires; topography inverted

alternate states. The model is qualitative in that it only predicts which species will be present under a given water regime, but does not provide information about their relative abundance.

Although Van der Valk (1981) suggests that this model of succession should be applicable to any type of wetland, it is based on the assumption that periodic water-level fluctuations and reflooding are the primary environmental variables. The model therefore really only deals with littoral or periodically inundated wetlands. It may be applicable to pan and floodplain wetlands in southern Africa, but there are some basic shortcomings of the model that limit its widespread application. The most important of these are that many disturbances, such as those caused by fire, herbivory and high flow rates, are not considered, and that it totally ignores the influence of autogenic processes, such as competition.

(b)

Figure 14.13 (*a*) An aerial view of the Nqogha River in the Okavango Delta, flanked by a ribbon of luxuriant *Cyperus papyrus*, which gives way laterally to a more open community. It is expected that in the next 50–100 years this channel will become abandoned and will resemble (*b*) which shows the same river in a reach that was abandoned in the 1930s. Today, all that remains of the channel is a dry, sinuous, sandy tract, surrounded by a featureless plain (*Photo: W.N. Ellery*).

14.5.2   The Mitsch and Gosselink model

Besides its Clementsian overtones, this model invokes Odum's (1969) concept of ecosystem development. Mitsch & Gosselink (1986) hold that initially the major determinants of wetland succession are environmental (allogenic), but as the ecosystem 'develops' there is a progressive accumulation of nutrients and peat which results in an 'insulation' of communities from environmental fluctuations and perturbations. It is proposed that insulation results in a change from allogenic to autogenic processes (competition, allelopathy) during ecosystem development and that this in turn leads to increasingly stable communities. In support of their argument, they point out that most wetlands are very long-lived and do not necessarily fill in and develop to a terrestrial climax vegetation as classical theory suggests.

The model of succession in highveld vleis which leads to a *Phragmites australis* climax on deep peat supports this hypothesis and, as discussed above, so does that of backswamp vegetation in the Okavango Delta.

Ellery *et al.* (1990), however, also illustrated that peat accumulation in the delta can in fact increase the potential of other environmental factors (e.g. fire, differential sedimentation) to alter ecosystem processes and ultimately reset the successional sequence at the landscape scale. These processes would also ensure that succession in the Okavango Delta does not lead to a terrestrial climax. Thus, the very process that Mitsch & Gosselink propose as the driving force for insulation can generate a feedback mechanism that prevents continuous insulation.

Clearly it is too simplistic to consider that succession in permanently inundated wetlands consists only of an allogenic phase followed by the autogenic development of ecosystem insulation and perpetual stability, as is implied by the Mitsch & Gosselink model. Similarly, it is too simplistic to invoke autogenic causes as the determinants of vegetation processes in wetlands that experience wide fluctuations in water level (Van der Valk 1981).

14.5.3 The Breen *et al.* model

Breen *et al.* (1988) reviewed vegetation processes in swamps and floodplains as a contrast of wetlands at opposite ends of the inundation continuum. They proposed a model in which vegetation pattern is viewed as the consequence of how both allogenic and autogenic processes modify patterns of resource distribution in space and time.

A basic premise of the model is that since water is a dense medium and a very good solvent, it has a high potential to move resources about the wetland ecosystem. Differential hydraulic and hydrological conditions are, therefore, important determinants of resource distribution pattern in wetlands. However, differential patterns of plant architecture and density will also affect water flow characteristics and resource distribution, thereby generating feedback effects on the vegetation pattern.

The model pays particular attention to the manner in which both allogenic and autogenic processes modify the boundary conditions of vegetation patches ('communities') and the effect this has on the resource and vegetation distribution mosaic. This aspect of the model is expanded below.

It is generally accepted that changes in vegetation distribution in space primarily reflect the degree of discontinuity in the physical and chemical environment. The fact that aquatic vegetation usually exhibits marked zonation therefore implies environmental discontinuities as important determinants of community boundaries. For example, interactions between hydrology, principally water level, and geomorphology can create distinct boundaries between aerobic, dry habitats and anaerobic, wet habitats which are clearly reflected in plant species distribution along the elevation profile.

The community, however, reflects both these initial conditions and the extent to which they have been modified by the physical presence (e.g. shading) and metabolism (e.g. aeration of the soil) of the component species. Plant architecture (the way in which community components fill the available space) in particular has major influences on hydrodynamics. As plants increasingly fill space, they offer greater frictional drag, thereby reducing the flow rates and dispersing some water laterally. On the one hand, the reduced flow rate within the community increases residence time and hence opportunity to capture dissolved and suspended resources; however, on the other, deflection of water away from the community reduces the supply of resources across the boundary.

The influence of growth strategy is well illustrated by contrasting the effects that a sparse, creeping (guerilla) and a closely packed, upright (phalanx) growth form (Lovett-Doust 1981) would have on boundary integrity and resource distribution in a flowing system. Similarly, submerged, floating-leaved and tall emergent species would have different potential to capture light resources and thus modify the boundaries between the communities they dominate. Growth-form is clearly an important determinant of community boundary conditions and, thus, contributes to the definition of vegetation structure.

A general hypothesis of the model is that the more distinct the boundaries between communities, the stronger would be the mechanisms 'isolating' within-community processes, such as those generating Mistch & Gosselink (1986) type 'insulation'. In a well 'isolated' or 'insulated' community the differential patterns of resource partitioning of the component species (i.e. between maintenance, storage, growth and reproduction) would result in differential resource distribution in space and time, thereby promoting internal patchiness.

Clearly, though, the extent to which the community is isolated also depends on the existence and magnitude of mechanisms that can disrupt isolation and transport of resources across boundaries. Typical transporting mechanisms in wetlands are floods, fire and animal (e.g. hippopotamus) movements, the magnitude, frequency and timing of which will determine the distribution and nature of particular boundaries.

The Breen *et al.* (1988) model is therefore built on the premise that 'isolating' and 'transporting' mechanisms oppose one another in the development of patch boundaries and provide powerful forces which shape the temporal and spatial pattern of the wetland vegetation mosaic. As such, it invokes both allogenic and autogenic processes, but recognizes that they interact in a complex manner in space and time to modify resource and species distribution.

14.5.4 Towards a new predictive model

All three of the models described above provide valuable insight to wetland vegetation processes but, like most contemporary theories of community dynamics, they are too conceptual to be of predictive value in real situations. There is, therefore, an urgent need to develop them beyond their present heuristic character.

Keddy (1992) advocates a highly reductionist approach to developing predictive potential which builds on life-history models such as that of Van der Valk (1981). The basis of his approach is to establish a few simple 'assembly' and 'response' rules that will allow prediction of community change. The objective of assembly rules is to predict which subset of the total species pool for a given region will occur in a specified habitat. The habitat acts as a filter to eliminate those sets of life-history traits which are unsuitable to that environment. The species that comprise the community are those that survive the filter. Response rules specify how an initial vector of species composition will respond when an environmental factor is changed.

This approach has much potential, but the implicit assumption that species response to environmental change is direct and linear will limit it to making statements about short-term species response to small-scale environmental change. It has severe limitations for the longer-term large-scale prediction that would be needed to manage the Okavango Delta or the Sabie River. For example, it ignores the process of feedback of the vegetation on the environment and so cannot deal with the effects of insulation (Mitsch & Gosselink 1986) and isolation (Breen *et al.* 1988). It would also be very difficult to factor the many contingency effects that structure communities (McNaughton 1983; Pickett & McDonnell 1989) into the response-rule approach. Finally, Keddy's approach lacks the ability to make the spatially explicit predictions needed to deal effectively with change over a wetland landscape mosaic.

The Breen *et al.* (1988) model provides a large measure of the conceptual basis for dealing with the limitations outlined above. The partitioning of phenomena and processes into those that determine spatiotemporal characteristics of boundaries on the one hand and internal structure on the other, implies that it has strong roots in landscape ecology (cf. Forman & Godron 1986; Kotlier & Wiens 1990; Wiens, Crawford & Gosz 1985), the discipline which is underpinning the 'new paradigm in ecology' (termed patch dynamics by some) (Pickett, Parker & Fiedler 1991). The model, therefore, has considerable potential to provide a spatially explicit framework for wetland vegetation science that would be compatible with the spatially dynamic modelling techniques being developed with the aid of GIS (Sklar & Costanza 1991; Turner & Dale 1991).

There is considerable potential for linking the rule-based approach of predicting species response to environmental change (Keddy 1992) and landscape modelling which allows prediction of patch dynamics (Pickett *et al.* 1991). The rule-based approach would be appropriate for modelling small-scale, within-patch species response. GIS-based, landscape models, on the other hand, can be used for modelling between-patch (mosaic) dynamics over longer periods and over larger areas.

14.6 Ecosystem functioning

Detailed studies of ecosystem processes have only been conducted in two natural systems in southern Africa: the coastal lake, Swartvlei, in the southern Cape (Howard-Williams & Allanson 1978) and the Pongola River floodplain (Furness 1981; Rogers 1981; Rogers 1984). Some additional information is available for Lake Sibaya on the KwaZulu-Natal coastal peneplain (Allanson *et al.* 1990). This part of the review deals with the main ecosystem processes in three sections: (1) primary production and nutrient cycling, (2) decomposition, and (3) grazing.

14.6.1 Primary production and nutrient cycling

An important question in any ecosystem study is how much does each species or community contribute to each process? In this context, it is instructive to compare results from Swartvlei, where the aquatic macrophytes occur in the shallow littoral zone around the edge of an extensive pelagic zone, and Tete pan on the Pongola floodplain, where the entire system is littoral. These systems differ further in that Tete pan is subject to frequent and marked fluctuations in water level. Periodic inundation of floodplain communities result in transfers of both organic matter and nutrients to the aquatic phase. Water levels in Swartvlei, on the other hand, are relatively stable and the lake is typically surrounded by a zone of emergent species (mainly *Phragmites australis*) rooted in the shallowest region of the littoral.

These structural differences result in quite distinct carbon budgets in the two wetlands (Table 14.2). As can be expected, pelagic phytoplankton made an important contribution (14%) to the total carbon inputs to Swartvlei but was insignificant in Tete pan. Inputs from seasonally inundated *Cynodon dactylon* lawns of Tete pan, on the other hand, account for 28% of carbon input, which is twice that of the emergent species in Swartvlei (12%).

The net production of submerged species in Swartvlei (primarily *Potamogeton pectinatus*) was fourteen times higher per unit area than that of *P. crispus* in Tete pan. However, because submerged species cover a small area (28%) of Swartvlei, there was little difference in the total contribution of this community to carbon inputs in the two systems. The proportional contribution of the epiphyton was, however, about twice as high in Tete pan (12.3%) as in Swartvlei (7.6%).

In both these wetlands, a key issue was to determine the degree to which the nutrient cycles were closed. In Swartvlei the question was driven by the concern that the macrophytes acted as a nutrient pump, transferring nutrients from the sediments of the littoral zone to the

Table 14.2 **A comparison of the primary production carbon inputs to the coastal lake, Swartvlei, and Tete pan on the Pongolo River Floodplain. Data from Allanson *et al.* (1990) and Rogers (1984), respectively**

Source	Swartvlei			Tete Pan		
	Net production (P_n) (g cm^{-2} yr^{-1})	Whole lake P_n (g cm^{-2} yr^{-1})	% total P_n	Net input[a] (g cm^{-2} yr^{-1})	Whole pan net input (g cm^{-2} yr^{-1})	% total input
Phytoplankton	27	19	13.9	—	—	—
Submerged species	1007	90	66.1	68	68	59.6
Epiphyton[b]	60	10.5	7.7	14	14	12.3
Emergents	690	16.5	12.1	—	—	—
Marginal species	—	—	—	32	32	28.1

[a] Calculated as input, since only a proportion of the net production (P_n) of the marginal species enters the aquatic phase as detritus during flood events.
[b] Algal communities on both submerged species and the submerged portion of emergent species.

open water pelagic zone where they might stimulate excessive algal blooms. On the Pongolo floodplain, the concern was for the potential loss of nutrients during floods. In this system, the decomposition of the annual standing crop occurs at a time when floods are more likely and, if cycling were not closed, nutrients could be flushed from the system. This was of particular concern, as the construction of an upstream dam would hold back sediment that would normally replenish nutrient stocks.

In both cases, however, nutrient cycles were tightly closed with little loss of nutrients from the macrophyte stands. Howard-Williams & Allanson (1978) clearly demonstrated that the pathway of nutrient transfer was from the sediments to the water column via the aquatic plants. The lack of any concomitant increase in nutrient concentration in the pelagic zone, however, led them to hypothesize a closed system in which the epiphytic algae and filter-feeding invertebrates within the littoral zone trapped the nutrients released during decomposition.

Rogers (1984) tested this experimentally in Tete pan and showed how the consumption of senescing macrophyte tissue by molluscs during decomposition resulted in a rapid transfer of both nitrogen and phosphorus to the sediment as faeces. Thus, most of the internal nutrient loading was retained within the system and was not lost during floods.

14.6.2 Decomposition

Several studies in southern African wetlands have made important contributions to the understanding of the decomposition process. Before their publication it was generally accepted that plant decomposition in wetlands was primarily a consequence of microbial activity and that, in contrast to the process in stream ecosystems, invertebrates played an insignificant role. Furthermore, this general model of decomposition assumed that colonization of plant material by microbial decomposers occurred only after tissue death.

A series of studies on southern African wetlands (Howard-Williams, Davies & Cross 1978; Rogers & Breen 1981, 1982, 1983; Rogers & de Bruyn 1988) demonstrated

that the complex bacterial epiphyton on living plants was, in fact, necrotrophic and that it both promoted tissue senescence and acted as the primary decomposer. Furthermore, naturally senescent tissues that have been processed by these bacteria provide a nutritious food source for invertebrates, which are extremely important in the reduction of particle size (comminution) during decomposition. For example, in Tete pan on the Pongolo river floodplain, a small snail (*Bulinus natalensis*) with a biomass of 0.7 tonnes ash-free dry mass processed the entire annual vegetative production (54 tonnes ash-free dry mass) of *Potamogeton crispus* (Rogers 1984).

The original general model of decomposition was derived from studies that all adopted a similar experimental approach in which plant material, senescent, dead or living, was dried or lyophilized before immersion in water. The southern African studies showed that this practice killed off the epiphytic bacteria that normally promote senescence and altered the nutrient status of the plant material to such an extent that invertebrates found the plants unpalatable. The resulting interference with the processes of senescence and comminution reduced the decay rate in experiments by more than half.

Local studies have clearly demonstrated that the phenomena and processes of decomposition are inextricably linked to the life cycles of the plants, the epiphyton and the invertebrates and do not begin only after plant death. Therefore, care must be taken to ensure that life-history events (e.g. the onset of senescence or dormancy) and changes in the environment (e.g. the timing of inundation or drying events) are taken into account in the design of decomposition experiments. If they are not, erroneous conclusions will be drawn about both nutrient and energy transfers in the wetland ecosystem (Rogers & de Bruyn 1988).

14.6.3 Grazing

In general, only a small percentage of the annual production of aquatic macrophytes enters the grazing food web. For example, waterfowl grazing on the Pongolo floodplain utilized about 10% of the submerged plant production, whereas cattle grazing accounted for 55% of the production of the marginal *Cynodon dactylon* lawns. Waterfowl grazing stimulated net production of *Potamogeton crispus* reproductive organs by about 10%, but *C. dactylon* production was depressed by about 6%.

The consumption by waterfowl of the reproductive organs of aquatic plants has potentially important implications for the stability of these grazing systems (Rogers & Breen 1980), because this can reduce the 'residual growth potential' from which the plants regenerate (Noy-Meir 1975). Such 'unstable' systems are prone to over-grazing and local extinction of the resource. Kemper (1984), for example, describes how, in a highveld vlei, grazing of rhizomes by Egyptian geese (*Alopochen aegyptiacus*) and Spurwing geese can eliminate *Typha capensis* stands and create open water bodies that become colonized by submerged species.

Consumption of *Potamogeton crispus* fruits on the Pongolo floodplain had no effect on the size of the seed bank, thus not affecting the potential of this species to persist through periodic droughts. Grazing can, however, remove 90% of the standing crop of the asexual turions, which exclosure studies have shown represents up to 25% of the annual turion production (Rogers & Breen 1980). This anomaly is explained by the fact that the birds feed preferentially on large turions, but in doing so they uproot the plants, which then rapidly produce large numbers of small turions. These small turions are energetically unrewarding for the birds but still have a high viability. Thus, while grazing depletes the available food supply, it stimulates production of an unavailable reserve growth potential that maintains annual plant standing crop.

This study illustrated that waterfowl consumption of the 'reserve biomass' of food plants does not necessarily promote instability in the grazing system, because other life-history traits may control feeding activities. The effectiveness of these traits is, however, dependent on the environmental conditions. Rogers & Breen (1980) also showed that if water levels dropped below 0.2 m before the *Potamogeton crispus* plants had begun to reproduce, the waterfowl could up-end, feed on the parent turions and destroy the plant standing crop.

14.7 Future research and management

Several proposals for further research have been made in the individual sections of this review. The purpose of this section is to provide, in brief, a broader context in relation to important management issues.

The status of wetlands is declining across the subcontinent (Breen & Begg 1989; Simmons, Brown & Griffin 1991). The real extent of wetland degradation and loss cannot be estimated with the existing database, but studies in KwaZulu-Natal suggest that 58% of the original wetland area of the Mfolozi catchment (Begg 1988) and over 90% of that in parts of the Tugela basin and Sibayi catchments have been destroyed. It is not surprising, therefore, that concern has been expressed at the paucity of coherent management policies or strategies for wetlands in southern Africa (Breen & Begg 1989).

The causes of wetland degradation and loss are many, but in general they are still a consequence of the impression that wetlands are wastelands in need of 'reclamation'. Therefore, wetlands are drained, dyked or ploughed for their short-term agricultural potential (Begg 1986; Tarboton 1987), or used for disposal of urban, industrial and mine waste (Viljoen 1987). All forms of water abstraction for urban, industrial or agricultural use also result in wetland degradation of one sort or another, as does the poor management of their catchments.

The value of wetlands to society should be evaluated in terms of the resources and services that accrue from their function in the landscape, i.e. those functions which provide services as the 'kidneys of the landscape' or as prime recreation sites, and resources such as food, building materials, grazing and biotic diversity. These values have been discussed in detail elsewhere (Begg 1986; Rogers 1986; Rogers & Van der Zel 1989).

Fundamental to the resource and service values of wetlands is the status of the vegetation, but, once again, very few published studies have dealt directly with wetland vegetation management. In South Africa the current need for vegetation management arises from two main sources. Firstly, the policy of the Department of Water Affairs and Forestry that 'the environment' is a legitimate demand sector in water resources management requires that ecologists develop the potential to predict vegetation response to managed flow regimes; hence the emphasis placed on developing a new framework for prediction of vegetation change in section 14.5. Secondly, the Government's commitment to the RAMSAR Convention (Cowan 1995) is shifting some of the emphasis to the acquisition and conservation of specific wetland systems (Goodman 1987). The prime need in this context is to develop a system for the classification and inventory of wetlands (section 14.2) in terms of their biotic and functional diversity such that we may know which wetland types should receive priority attention. Without this basic information wetlands will continue to be the forgotten 'biome'.

Managers, therefore, need data of both a descriptive and a predictive nature if they are to meet their objectives. Detailed programmes for research and classification of South African wetlands have been compiled (Breen 1988; Walmsley & Boomker 1988), but these have not yet been implemented throughout the country. The techniques and strategy largely exist for meeting managers' needs.

It is crucial to recognize that, although wetlands may exist as satellite patches in all biomes, they cannot be managed in isolation of each other, or of the terrestrial biome matrix in which they are found. Management and prediction of change at the biome level must, therefore, take place in a landscape context (Rogers 1995).

14.8 Acknowledgements

Gina De Ornelas provided invaluable assistance in compiling the information used in the classification section. Both she and Wendy Midgely are also thanked for their assistance with manuscript preparation. Pat Morant provided useful comments on an early draft.

14.9 References

Acocks, J.P.H. (1976). Riverine vegetation of the semi-arid and arid regions of South Africa. *Journal of the South African Biological Society*, **17**, 21–35.

Allanson, J.P.R., Hart, R.C., O'Keefe, J.H. & Robarts, R.D. (1990). *Inland Waters of Southern Africa: An Ecological Perspective*. Monographiae Biologicae 64. The Hague: Kluwer Academic Publishers.

Allan, D. G., Seaman, M.T. & Kaletja, B. (1995). The endorheic pans of South Africa. In *Wetland Regions of South Africa*, ed. G. Cowan, pp. 75–101. Pretoria: Department of Environmental Affairs and Tourism.

Begg, G.W. (1986). *The Wetlands of Natal (Part 1). An Overview of their Extent, Role and Present Status*. Report 68. Pietermaritzburg: Natal Town and Regional Planning.

Begg, G.W. (1988). *The Wetlands of Natal (Part 2). The Distribution, Extent and Status of Wetlands in the Mfolozi Catchment*. Report 71. Pietermaritzburg: Natal Town and Regional Planning.

Bethune, S. (1991). Kavango River wetlands. *Madoqua*, **17**, 77–112.

Bredenkamp, G. & van Rooyen, N. (1993). *A Survey of the Riparian Vegetation of the Sabie River in the Kruger National Park*. Pretoria: ECOTRUST.

Breen, C. M. (1988). Wetland classification. In *Inventory and Classification of Wetlands in South Africa*, ed. R.D. Walmsley & E.A. Boomker, pp. 76–85. Ecosystem Programmes Occasional Report No. 34. Pretoria: Foundation for Research Development, CSIR.

Breen, C.M. & Begg, G.W. (1989). Conservation status of southern African wetlands. In *Biotic Diversity in Southern Africa. Concepts and Conservation*, ed. B.J. Huntley, pp. 254–64. Cape Town: Oxford University Press.

Breen, C.M., Rogers, K.H. & Ashton, P.J. (1988). Vegetation processes in swamps and flooded plains. In *Handbook of Vegetation Science*, Vol. 15, ed. J.J. Symoens & H. Lieth, pp. 223–48. The Hague: W. Junk.

Carter, A.J. (1995). *A Markovian Approach to the Investigation of Landscape Change in the Perennial Rivers of the Kruger National Park*. PhD Thesis. Johannesburg: University of the Witwatersrand.

Carter, A.J. & Rogers, K.H. (1989). Phragmites reedbeds in the Kruger National Park: the complexity of change in riverbed state. Proceedings of the Fourth South African National Hydrological Symposium, pp. 339–40. Pretoria: Water Research Commission.

Coetzee, J.P., Bredenkamp, G.J. & Van Rooyen, N. (1994). Phytosociology of the wetlands of the Ba and Ib land types in the Pretoria–Witbank–Heidelberg area of the Transvaal, South Africa. *South African Journal of Botany*, **60**, 61–7.

Coetzee, M.A.S. & Rogers, K.H. (1991). Environmental correlates of plant species distribution on the Nyl River floodplain. *South African Journal of Aquatic Sciences*, **17**, 44–50.

Cowan, G.I. (1995). Wetland Regions of South Africa. In *Wetlands of South Africa*, ed. G.I. Cowan, pp. 21–31. Pretoria: Department of Environmental Affairs and Tourism.

Curwell, A.P. (1986). *Waterfowl Induced Changes on the Growth and Distribution of Typha capensis on the Daggafontein Vlei*. BSc Hons. Dissertation. Johannesburg: University of the Witwatersrand.

Davies, B.R. & Walmsley, R.D. (eds) (1985). *Perspectives in Southern Hemisphere limnology. Development in Hydrobiology 28*. Dordrecht: Junk.

Denny, P. (1985). Submerged and floating leaved aquatic macrophytes (euhydrophytes). In *The Ecology and Management of African Wetland Vegetation*, ed. P. Denny, pp. 19–45. Dordrecht: Junk.

Downing, B.H. (1968). Notes on the Ecology of Natal Highland Sourveld Vleis. *Proceedings of the Grassland Society of Southern Africa*, **3**, 131–4.

Ellery, K. (1987). *Wetland Plant Community Composition and Successional Processes in the Maunachira River System of the Okavango Delta*. MSc Thesis. Johannesburg: University of the Witwatersrand.

Ellery, K., Ellery, W.N., Rogers, K. H. & Walker, B.H. (1990). Formation, colonisation and fate of floating sudds in the Maunachira river system of the Okavango Delta, Botswana. *Aquatic Botany*, **38**, 315–29.

ery, K., Ellery, W.N., Rogers, K.H. & Walker, B.H. (1991). Water depth and biotic insulation: major determinants of back-swamp plant community composition. *Wetlands Ecology and Management*, 1, 149–62.

ery, W.N. (1988). *Channel Blockage and Abandonment in the North-eastern Okavango Delta: The Role of the Cyperus papyrus*. MSc Thesis. Johannesburg: University of the Witwatersrand.

ery, W.N., Ellery, K. & McCarthy, T.S. (1993a). Plant distribution in islands of the Okavango Delta, Botswana: determinants and feedback interactions. *African Journal of Ecology*, 31, 118–34.

ery, W.N., Ellery, K., McCarthy, T.S., Cairncross, B. & Oelofse, R. (1989). A peat fire in the Okavango Delta, Botswana, and its importance as an ecosystem process. *African Journal of Ecology*, 27, 7–21.

ery, W.N., Ellery, K., Rogers, K.H. & McCarthy, T.S. (1995). The role of *Cyperus papyrus* L. in channel blockage and abandonment in the northeastern Okavango Delta, Botswana. *African Journal of Ecology*, 33, 25–49.

ery, W.N., Ellery, K., Rogers, K.H., Walker, B.H. & McCarthy, T.S. (1993b). Vegetation, hydrology and sedimentation processes as determinants of channel form and dynamics in the northeastern Okavango Delta, Botswana. *African Journal of Ecology*, 31, 10–25.

rman, R.T.T. & Godron, M. (1986). *Landscape Ecology*. New York: John Wiley and Sons.

arness, H.D. (1981). *The Plant Ecology of Seasonally Flooded areas of the Pongolo River Floodplain, with Particular Reference to Cynodon dactylon (L.) Pers.* PhD Thesis. Pietermaritzburg: University of Natal.

arness, H.D. & Breen, C.M. (1980). The vegetation of seasonally flooded areas of the Pongolo River floodplain. *Bothalia*, 13, 217–31.

oodman, P.S. (1987). Natal Parks Board. In *Proceedings of a Symposium on Ecology and Conservation of Wetlands in South Africa*, ed. R.D. Walmsley & M.L. Botten, pp. 30–7, Occasional Report Series No. 28, Ecosystem Programmes. Pretoria: CSIR.

Grime, J.P. (1979). *Plant Strategies and Vegetation Processes*. Chichester: J. Wiley & Sons.

Grobbelaar, J.U. (1976). Some limnological properties of an ephemeral waterbody at Sossus Vlei, Namib Desert South West Africa. *Journal of the Limnological Society of Southern Africa*, 2, 51–4.

Hart, R.C. (1995). Coastal Lakes. In *Wetlands of South Africa*, ed. G.I. Cowan. Pretoria: Department of Environment Affairs and Tourism.

Heeg, J. & Breen, C.M. (1982). *Man and the Pongolo Floodplain.South African National Scientific Programmes Report 56*. Pretoria: CSIR.

Howard-Williams, C. (1979). Distribution, biomass and the role of aquatic macrophytes in Lake Sibaya. In *Lake Sibaya*, ed. B.R. Allanson, pp. 88–107. Monographiae Biologicae 36. The Hague: Junk.

Howard-Williams, C. (1980). Aquatic macrophyte communities of the Wilderness Lakes: community structure and environmental conditions. *Journal of the southern African Limnological Society*, 6, 85–92.

Howard-Williams, C. & Allanson, B.R. (1978). *Swartvlei Project* Report Part ii. Pretoria: Water Research Commission.

Howard-Williams, C., Davies, B.R. & Cross, R.H.M. (1978). The influence of periphyton on the surface structure of a *Potamogeton pectinatus* leaf (an hypothesis). *Aquatic Botany*, 5, 87–91.

Jacot-Guillarmod, A. (1962). The bogs and sponges of the Basotho Mountains. *South African Journal of Science*, 58, 179–82.

Kantrud, H.A., Millar, J.B. & Van der Valk, A.G. (1989). Vegetation of the Prairie pothole region. In *Northern Prairie Wetlands*, ed. A.G. Van der Valk, pp. 132–85. Ames: Iowa State University Press.

Keddy, P.A. (1992). Assembly and response rules: two goals for predictive community ecology. *Journal of Vegetation Science*, 3, 157–64.

Kemper, N.P. (1984). *The Effects of Waterfowl on the Vegetation of the Daggafontein vlei*. BSc Hons. Thesis. Johannesburg: University of the Witwatersrand.

Kotlier, N.B. & Wiens, J.A. (1990). Multiple scales of patchiness and patch structure: a hierarchical framework for the study of heterogeneity. *Oikos*, 59, 253–60.

Lindeque, M. & Archibald, T.J. (1991). Seasonal wetlands in Owambo and the Etosha National Park. *Madoqua*, 17, 129–33.

Lovett-Doust, L. (1981). Population dynamics and local specialisation in a clonal perennial (*Ranunculus repens*). 1. The dynamics of ramets in contrasting habitats. *Journal of Ecology*, 69, 743–55.

Martin, A.R.H. (1960). The ecology of Groenvlei, a South African coastal fen. Part 1: The primary communities. *Journal of Ecology*, 48, 307–29.

McCarthy, T.S. (1992). Physical and biological processes controlling the Okavango Delta – a review of recent research. *Botswana Notes and Records*, 24, 57–86.

McCarthy, T.S. & Ellery, W.N. (1994). The effect of vegetation on soil and ground water chemistry and hydrology of islands in the seasonal swamps of the Okavango fan, Botswana. *Journal of Hydrology*, 154, 169–93.

McCarthy, T.S., Ellery, W.N. & Ellery, K. (1993). Vegetation induced subsurface precipitation of carbonate as an aggradational process in the permanent swamps of the Okavango (Delta) fan, Botswana. *Chemical Geology*, 107, 111–31.

McCarthy, T.S., Ellery, W.N., Rogers, K.H., Cairncross, B. & Ellery, K. (1986). The roles of sedimentation and plant growth in the changing flow patterns of the Okavango Delta, Botswana. *South African Journal of Science*, 82, 588–91.

McCarthy, T.S., Ellery, W.N. & Stanistreet, I.G. (1992). Avulsion mechanisms on the Okavango fan, Botswana: the control of a fluvial system by vegetation. *Sedimentology*, 39, 779–95.

McCarthy, T.S., Stanistreet, I.G. & Cairncross, B. (1991). The sedimentary dynamics of active fluvial channels on the Okavango fan, Botswana. *Sedimentology*, 38, 471–87.

McNaughton, S.J. (1983). Serengeti grassland ecology: the role of composite environmental factors and contingency in community organisation. *Ecological Monographs*, **53**, 291–320.

Mitchell, D.S. & Rogers, K.H. (1985). Seasonality/aseasonality of aquatic macrophytes in Southern Hemisphere inland waters. *Hydrobiologia*, **125**, 137–40.

Mitsch, W.J. & Gosselink, J.G. (1986). *Wetlands*. New York: Van Nostrand Reinhold.

Morant, P.D. (1983). Wetland classification: towards an approach for southern Africa. *Journal of the Limnological Society of southern Africa*, 9, 76–84.

Musil, C.F., Grunow, J.O. & Bornman, C.H. (1973). Classification and ordination of aquatic macrophytes in the Pongolo river pans Natal. *Bothalia*, **11**, 181–90.

Noble, R.G. & Hemens, J. (1978). *Inland Water Ecosystems in South Africa – A Review of Research Needs*. South African National Scientific Programmes Report 34. Pretoria: CSIR.

Noy-Meir, I. (1975). Stability of grazing systems: an application of predator-prey graphs. *Journal of Ecology*, **63**, 459–81.

Odum, E.P. (1969). The strategy of ecosystem development. *Science*, **164**, 262–70.

Patten, B.C. (1990). Introduction and overview. In *Wetlands and Shallow Continental Water Bodies*, ed. C. Patten. pp. 3–8. The Hague: SPB Academic Publishing.

Phillips, J.P. (1931). Forest Succession and Ecology in the Knysna Region. *Memoirs of the Botanical Survey of South Africa*, **14**, 1–327.

Pickett, S.T.A. & McDonnell, M.J. (1989). Changing perspectives in community dynamics: a theory of successional forces. *Trends in Ecology and Evolution*, **4**, 241–5.

Pickett, S.T.A., Parker, V.T. & Fiedler, P.L. (1991). The new paradigm in ecology: implications for conservation above the species level. In *Conservation Biology. The Theory and Practice of Nature Conservation, Preservation and Management*, ed. P.L. Fiedler & S.K. Jain, pp. 67–88. New York: Chapman & Hall.

Rogers, F.E.J. (1981). *Studies on the Epiphyton of Potamogeton crispus L*. MSc Thesis. Pietermaritzburg: University of Natal.

Rogers, F.E.J., Rogers, K.H. & Buzer, J.S. (1985). *Wetland for Wastewater Treatment with Special Reference to Municipal Wastewaters*. Olifants Working Group, Department of Botany, University of the Witwatersrand. Johannesburg: Witwatersrand University Press.

Rogers, K.H. (1984). *The Role of Potamogeton crispus L. in the Pongolo River Floodplain Ecosystem*. PhD Thesis. Pietermaritzburg: University of Natal.

Rogers, K.H. (1986). The value and function of wetlands in catchments. In *The Vaal River Ecosystem: Status and Problems. Occasional Report No. 5*. Pretoria: Foundation for Research Development, CSIR.

Rogers, K.H. (1990). *The Use of Olifantsvlei for Water Quality Improvement*. Final Research Report. Johannesburg: City Engineers Department.

Rogers, K.H. (1995) Riparian Wetlands. In *Wetlands of South Africa*, ed. G.I. Cowan. Pretoria: Government Printer.

Rogers, K.H. & Breen, C.M. (1980). Growth and reproduction of *Potamogeton crispus* L. in a South African lake. *Journal of Ecology*, **68**, 561–71.

Rogers, K.H. & Breen, C.M. (1981). Effects of epiphyton on *Potamogeton crispus* leaves. *Microbial Ecology*, **7**, 351–63.

Rogers, K.H. & Breen, C.M. (1982). Decomposition of *Potamogeton crispus* L.: the effects of drying on the pattern of mass and nutrient loss. *Aquatic Botany*, **12**, 1–12.

Rogers, K.H. & Breen, C.M. (1983). An investigation of macrophyte, epiphyte and grazer interactions. In *Periphyton of Freshwater Ecosystems*, ed. R.G. Wetzel, pp. 217–26. The Hague: Junk.

Rogers, K.H. & de Bruyn, J. (1988). Decomposition of *Paspalum disticum* L.: methodology in seasonally inundated systems. *Verhanderlungen der Internationale vereinigung für Theoretische und Angewandte Limnologie*, **232**, 1945–8.

Rogers, K.H., Ellery, W.N., Winternitz, N.L. & Domeier, R. (1989). Physical, chemical and biotic response to decreasing water depth in a highveld pan following wet and dry summers. *South African Journal of Aquatic Science*, **15**, 67–90.

Rogers, K.H. & Higgins, S.I. (1993). *The Nyl Floodplain as a Functional Unit in the Landscape – Preliminary Synthesis and Future Research*, Report 1/1993. Johannesburg: Centre for Water in the Environment.

Rogers, K.H. & Van der Zel, D.W. (1989). Water quantity requirements of riparian vegetation and floodplains. In *Ecological Flow Requirements for South African Rivers*, ed. A.A. Ferrar. pp. 94–108. South African National Scientific Programmes Report 162. Pretoria: CSIR.

Schlettwein, C.H.G., Simmons, R.E., McDonald, A. & Grobler, H.J.W. (1991). Flora, fauna and conservation of East Caprivi wetlands. *Madoqua*, **17**, 67–76.

Shaw, P.A. (1988). After the flood: the fluvio-lacustrine landform of northern Botswana. *Earth Science Review*, **25**, 449–56.

Simmons, R.E., Brown, C.J. & Griffin, M. (1991). Introduction: The status and conservation of wetlands in Namibia. *Madoqua*, **17**, 55–60.

Sklar, F.H. & Costanza, R. (1991). The development of dynamic spatial models for landscape ecology: a review and prognosis. In *Quantitative Methods in Landscape Ecology, The Analysis and Interpretation of Landscape Heterogeneity*, ed. M.G. Turner & R.H. Gardner, pp. 323–51. New York: Springer-Verlag.

Smith, P.A. (1976). An outline of the vegetation of the Okavango drainage system. In *Symposium on the Okavango Delta*. Gaborone: Botswana Society.

Smuts, G.L. (1984). *Harvesting and Managing Waterfowl on the Blesbokspruit*. Report to the Chairman's office. Johannesburg: Anglo American Corporation of South Africa Limited.

rboton, W.R. (1987). The Nyl floodplain: its significance, phenology and conservation status. In *Proceedings of a Symposium on Ecology and Conservation of Wetlands in South Africa*, ed. R.D. Walmsley & M.L. Botten, pp. 101–14. Occasional Report Series No. 28, Ecosystem Programmes. Pretoria: Foundation for Research Development, CSIR.

nley, K.L. (1976). *The Ecology of Tongaland*. Durban: Natal Branch of the Wildlife Society of S.A.

rner, M.G. & Dale, V.H. (1991). Modelling Landscape Disturbances. In *Quantitative Methods in Landscape Ecology, The Analysis and Interpretation of Landscape Heterogeneity*, ed. M.G. Turner & R.H. Gardner, pp. 323–51. New York: Springer-Verlag.

n Coller, A.L. (1993). *Riparian Vegetation of the Sabie River: Relating Spatial Distribution Patterns to the Physical Environment*. MSc Thesis. Johannesburg: University of the Witwatersrand.

n der Valk, A.G. (1981). Succession in wetlands: a Gleasonian approach. *Ecology*, **62**, 688–96.

nter, A.K. (1991). *Phragmites australis (Cav.) Stued. Encroachment in the Daggafontein Vlei*. BSc Hons. Thesis. Johannesburg: University of the Witwatersrand.

ljoen, V.C. (1987). A study of the impact of a waste disposal site on a section of the Elsbergspruit wetland system. In *Proceedings of a Symposium on Ecology and Conservation of Wetlands in South Africa*, ed. R.D. Walmsley & M.L. Botten, pp. 188–91. Occasional Report Series No. 28, Ecosystem Programmes. Pretoria: CSIR.

alker, D. (1970). Direction and rate of succession in some British post-glacial hydroseres. In *Studies in the Vegetational History of the British Isles*, ed. D. Walker & R.G. West, pp. 117–39. Cambridge: Cambridge University Press.

almsley, R.D. & Boomker, E.A. (1988). *Inventory and Classification of Wetlands in South Africa*. Occasional Report No. 34, Ecosystem Programmes. Pretoria: Foundation for Research Development, CSIR.

Weller, M.W. (1981). *Freshwater Marshes – Ecology and Wildlife Management*. Minneapolis: Minnesota University Press.

Wessels, N.G. (1991). *The Syntaxonomy and Synecology of Swamp Forest in the Lake St. Lucia Area, Natal*. Unpublished Report, Division of Forest Science and Technology, Pretoria: CSIR.

Wiens, J.A., Crawford, C.S. & Gosz, F.R. (1985). Boundary dynamics: a conceptual framework for studying landscape ecosystems. *Oikos*, **45**, 421–7.

Wissmar, R.C. & Swanson, F.J. (1990). Landscape disturbances and lotic ecotones. In *The Ecology and Management of Aquatic-Terrestrial Ecotones, Vol. 4*, ed. R.J. Naiman & H. Décamps, pp. 65–86. Paris: UNESCO.

Zimbatis, N. (1975). *A Floristic Survey of the Provincial Nature Reserves, Transvaal. Barbers Pan Nature Reserve*. Report on project No. T.N. 6.3.3. Pretoria: TPA Nature Conservation Division.

Marine vegetation

J.J. Bolton and R.J. Anderson

15

15.1 Introduction

The concept of vegetation is more difficult to apply in the marine than in the terrestrial environment, because, in many benthic marine environments, sedentary animals make up a large proportion of the visible biomass. This review describes the marine algal macrophyte vegetation of Namibia and South Africa, with comments on Angola and Mozambique. It confines itself to those algal communities extending from the deepest plants to the highest seaweed growth in the upper intertidal region on rocky coastlines. Estuarine vegetation and the terrestrial vegetation that grows above the high-tide mark on dunes and rocky substrata are dealt with in Chap. 13 (this volume). This synthesis serves to fill a gap in the southern African marine ecological literature, which is predominantly zoological (e.g. Field & Griffiths 1991). It will become clear that certain aspects of seaweed ecology in the region are poorly studied. The terminology of intertidal zones used in this chapter is that of Russell (1991) and, unless stated, species authorities are according to Seagrief (1984).

Most of the literature on southern African seaweeds has been, and continues to be, taxonomic (see Seagrief & Troughton 1973). In the *Ecological Survey of the South African Coasts*, which is still widely used for the description of intertidal communities (summarized by Stephenson 1939, 1944, 1948), seaweeds were included, but in a minor role. There was some activity in seaweed ecology by Isaac in the 1950s (e.g. Isaac 1951, 1956, 1957; Isaac & Hewitt 1953), stimulated primarily by the need for ecological studies on seaweeds of potential economic importance, but otherwise information was very limited until an upturn of interest after 1980.

A large marine research project on the west coast of

South Africa, the Benguela Ecology Programme, was initiated in the 1970s. Studies by its inshore section led to more detailed knowledge of the ecology of coastal systems on the west coast of South Africa (see review by Branch & Griffiths 1988) with most biological work on animals. However, information on the south and east coasts of South Africa, and Namibia, remains scanty. As will be seen, most detailed ecological studies on southern African seaweeds have been, and continue to be, carried out in the Western Cape. A detailed description of the seaweeds of the west coast of South Africa from Cape Agulhas to the Orange River has been prepared (Stegenga, Bolton & Anderson 1996). There is much less information on the other regions of the coastline.

15.2 Boundaries of marine provinces

Changes in the distribution of seaweed species along continuous coastlines are determined overwhelmingly by sea-water temperature regimes (Breeman 1988; Lüning 1990). Consequently, studies on large-scale biogeographic patterns of intertidal communities in South Africa reveal a close correlation with the temperature of inshore waters (Isaac 1938; Stephenson 1948; Bolton 1986; Bolton & Anderson 1990). The coastline of South Africa and Namibia has predominantly temperate sea-water conditions, bounded by tropical West Africa to the west and tropical East Africa to the east (tropical regions being those with monthly mean sea-water temperatures always above 20 °C; Lüning 1990). The region under discussion comprises two clearly delineated temperate marine provinces, grading into two tropical marine provinces (Fig. 15.1). This review concentrates on marine floristic data, although the patterns of marine provinces and their boundaries are closely paralleled by the marine fauna (Stephenson 1939, 1944, 1948; Brown & Jarman 1978; Emanuel *et al.* 1992).

15.2.1 Tropical West Africa Marine Province

The seaweeds of the northern section of this marine province are well documented (Lawson & John 1987; John & Lawson 1991), whereas those south of the equator are poorly known. Regions of southern Angola are affected by upwelling from the Benguela current (Shannon 1985), and their seaweed flora contains few species of tropical affinity (Lawson, John & Price 1975). The western limit of temperate southern Africa is therefore in the region of the Kunene River (the border between Namibia and Angola). Only 14 of 136 species recorded by Lawson, Simons & Isaacs (1990) in Namibia are known to occur in Angola, eight of these being widely distributed elsewhere. Our knowledge of the Angolan seaweed flora is,

however, based almost entirely on one brief expedition (Lawson *et al.* 1975).

15.2.2 Benguela Marine Province

The west coast of South Africa (north of Cape Point on the Cape Peninsula) and the entire coast of Namibia constitute the Benguela Marine Province. This is geographically equivalent to the 'Benguela upwelling region' (Shannon 1985), and the designation 'Benguela' is preferred to the name 'Southwestern Africa marine province' as coined by Lüning (1990). Most of the species in the northern Namibian marine flora occur in the southern regions of the South African west coast. There is, however, a continual northward reduction in species numbers, with the southern portion (the Southwestern Cape sub-province) having the most species, fewer in the Namaqua sub-province (which includes the coast of the Northern Cape and southern Namibia), and a species-poor northern Namibian region (the Namib sub-province) (Engledow, Bolton & Stegenga 1992). The two latter sub-provinces have very few endemic species,

Figure 15.1 **Southern Africa, showing positions of Marine Provinces recognized in this chapter.**

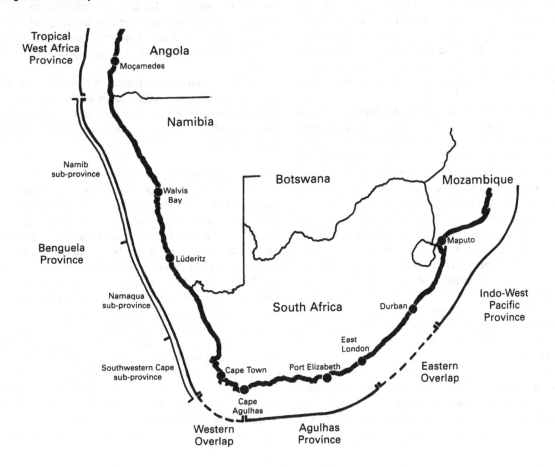

Figure 15.2 **Upwelling of cool, nutrient-rich water in the Benguela Province supports the highly productive beds of the kelp** *Ecklonia maxima* (Photo: R.J. Anderson).

and are distinguished by the absence of many species that occur further south. Oceanographically, the Benguela Marine Province is characterized by upwelling from the Benguela current, with inshore sea-water temperatures as low as 8 °C for short periods (Shannon 1985; Bolton 1986), although inshore monthly means never fall below 11 °C. The warmest months have average sea temperatures as low as 14 °C along some sections of the coast, up to around 20 °C in the Namib sub-province. The Benguela Marine Province is characterized by the dominance of large laminarian kelps in the subtidal (Fig. 15.2).

The South African portion of the Benguela Marine Province has commonly been treated as a biogeographic entity known as the West Coast (Stephenson 1948) or the West Cape (Hommersand 1986). The west coast of South Africa has been widely designated as 'cold temperate' since Stephenson's work in 1948. The classification of marine provinces into temperature categories has been widely used in the northern hemisphere, where a region with monthly means <10 °C in the winter and >10 °C in the summer is 'cold temperate', whereas one with >15 °C in the summer and <20 °C in the winter is 'warm temperate'. In the Benguela Marine Province the temperature conditions (reflected also in the seaweed temperature tolerances of the plants; Bolton 1986) are intermediate between these categories and it is best described as 'cool temperate' as used but not defined by Emanuel et al. (1992).

The Benguela and Agulhas Marine Provinces are separated by a clearly defined overlap region, with intermediate sea-water temperature conditions and consequent seaweed flora, which is commonly designated the 'Western Overlap' in the literature (Stephenson 1948;

Jackelman, Stegenga & Bolton 1991; Emanuel et al. 1992; Stegenga & Bolton 1992). The seaweed communities in this overlap region show affinities with the Benguela rather than the Agulhas region; *Ecklonia* beds occur throughout the overlap, almost to Cape Agulhas, and common west coast species generally dominate the intertidal.

15.2.3 Agulhas Marine Province

Annual and maximum monthly mean sea-water temperatures rise immediately east of Cape Agulhas (the southernmost tip of Africa), owing to the much lower incidence of cool water upwelling, and the influence of the warm Agulhas current (Shannon 1985; Bolton 1986). Along the 800 km from Cape Agulhas to East London, mean annual temperatures are around 17–18 °C, with typically warm temperate conditions. Upwelling of cool waters occurs sporadically and can cause fish kills (Hanekom et al. 1989). There have been few recent taxonomic studies in this region compared with the Benguela region and KwaZulu-Natal, and the section of coastline from De Hoop Nature Reserve (Bolton & Stegenga 1987) and Stilbaai (Stephenson et al. 1937) to Tsitsikamma (Seagrief 1967) is very poorly documented, floristically or ecologically. The most obvious distinguishing feature of the seaweed communities in this province is the generally low biomass of mid-eulittoral seaweeds, on many shores dominated by a single species, *Gelidium pristoides*.

The eastern limit of the Agulhas Province (the Eastern Overlap of Stephenson 1948) is much less clearly defined than the Western Overlap. Temperatures rise gradually from around East London to northern KwaZulu-Natal (Bolton 1986).

15.2.4 Indo-West Pacific Marine Province

The intertidal biota of the coast of KwaZulu-Natal was designated as the 'sub-tropical East Coast' by Stephenson (1948). A considerable number of KwaZulu-Natal seaweeds also occur in the Agulhas Province, but further northwards an increasing number of tropical elements are included in the flora, so that at Isipingo, on the central KwaZulu-Natal coast, 67% of the seaweed species have warm-water affinities *sensu* Stephenson (1948). From central KwaZulu-Natal northwards, the sea-water temperatures are tropical, with monthly means never below 20 °C. In this part of the region the southernmost reef-forming corals are found (Brown & Jarman 1978). The seaweed flora of Mozambique and northern KwaZulu-Natal (as well as the marine fauna) has close affinities with other regions in the enormous Indo-West Pacific Marine Province, which stretches to northern Australia (Isaac 1958; Pocock 1958; Hommersand 1986; Critchley *et al.* 1994; Farrell *et al.* 1994). Although this matter is not clearly resolved, the existence of a 'sub-tropical' marine province in KwaZulu-Natal is unlikely, as the flora seems to be composed of an eastwardly decreasing number of Agulhas Province species, replaced largely by Indo-West Pacific species as the water temperature rises. At this stage we consider an 'Eastern Overlap' between Agulhas and Indo-West Pacific Marine Provinces comprising the coastline from around East London to the Mozambique border.

15.2.5 Community gradients on a biogeographic scale

It has been shown that individual seaweed species distributions are overwhelmingly limited by sea water temperature regimes (Breeman 1988). In one southern African study, patterns of seaweed vegetation on a biogeographic scale have, similarly, been shown to correlate with temperature. The geographical changes in seaweed-dominated intertidal communities were documented in detail over a 600-km section of coastline, comprising the Western Overlap and portions of the adjoining Benguela and Agulhas Provinces (Bolton & Anderson 1990). The community gradient closely followed the temperature gradient around the coastline. For example, samples from the warmest region of False Bay were similar to those from geographically distant sites with comparable temperature regimes. One site which was sand-affected had a very different community from adjacent sites with similar temperature conditions. This study provides evidence that, on a biogeographic scale, seaweed-dominated communities can be correlated with sea-water temperature conditions, in addition to the more widely demonstrated links between distributions of individual species and temperature (Breeman 1988).

15.3 The seaweed flora of southern Africa

15.3.1 Diversity

The seaweed flora of temperate southern Africa has long been recognized as being rich in species with a high proportion of endemics, when compared with those of other world regions (see Van den Hoek 1984), although detailed data are not readily available. A rich seaweed flora (comprising Phaeophyta, Chlorophyta and Rhodophyta) is one with around 600–800 species, with the highest world figures being a little over 1000 (Lüning 1990; Bolton 1994). It is very difficult at this time to produce an accurate figure for the number of seaweeds in southern Africa, as much taxonomic work is in progress. A synthesis of available information gives a total of a

little under 800 species recorded for the area under review. The ratio of the three main algal groups is remarkably consistent among the different marine provinces, with a ratio of Phaeophyta (P) : Chlorophyta (C) : Rhodophyta (R) of 15%P : 19%C : 66%R in KwaZulu-Natal (Indo-West Pacific Province) (Farrell *et al.* 1993); 13% P : 16%C : 70%R in the Eastern Cape, west of the Fish River (Agulhas Province; Seagrief 1988); and 15%P : 13% C : 72%R on the South African coast west of Cape Agulhas (Benguela Province; Bolton 1986).

From the Cape Peninsula eastwards (i.e. in the Western Overlap, Agulhas Province and KwaZulu-Natal), species numbers along different sections of the coast appear to be fairly similar. Detailed collections along a few kilometres of coastline generally produce a little under 200 species, not including crustose coralline red algae (e.g. Bolton & Stegenga 1987; Jackelman *et al.* 1991; Farrell *et al.* 1993). False Bay, in the Western Overlap, appears to be particularly rich, with c. 270 spp. This is attributable to its biogeographical position (with interdigitation of species of the Benguela and Agulhas Provinces) and its high habitat variability (Bolton, Stegenga & Anderson 1991; Stegenga *et al.* 1996). However, moving westwards and northwards from the Cape Peninsula there is a gradual drop in species numbers (Engledow *et al.* 1992) which is difficult to correlate with any gradient in environmental factors. For example, the Lüderitz region has fewer species than southern regions of the Benguela Province, with fairly similar oceanographic conditions. The low species numbers in the Namib sub-province may be partially attributable to the limited areas of rocky substratum on this predominantly sandy stretch of coastline (Molloy 1990; Engledow *et al.* 1992; Chap. 13, this volume). However, it could be hypothesized that upwelling itself may be responsible for the general northward reduction in species diversity. A comparable pattern occurs in temperate Pacific South America, with a rich cold-water flora in Southern Chile, and a gradual drop in species numbers moving northwards (Santelices 1980). It could be that the seasonally irregular temperature/nutrient conditions in upwelling regions in the Benguela Province of southwestern Africa and Chile/Peru reduce species diversity. It could also be hypothesized that this factor is operative on an evolutionary time-scale, with the relatively young age of the Benguela upwelling (only 2–3 million years at its present intensity; Shannon 1985) being insufficient either for immigration or evolution of species adapted to these conditions. It is interesting to note that the Laminariales (kelps), a group that has spread relatively recently (Saunders & Druehl 1993) and is rapidly speciating, has produced *Ecklonia maxima*, which is presumed to have evolved since the origins of the Benguela upwelling and dominates

Table 15.1 **Biogeographic affinities of southern African marine red algae (adapted from Hommersand 1986), expressed as percentages**

Marine Province	Antiboreal Pacific Ocean	West and South Australia	Japan	Tropical Indian Ocean	Cosmopolitan	Endemic
Benguela Province ($n = 124$)	60	4	4	6	11	15
Agulhas Province ($n = 202$)	17	19	4	35	8	16
KwaZulu-Natal ($n = 169$)	2	14	3	65	5	10

subtidal ecosystems in much of the Benguela Province (Bolton & Anderson 1987, 1994).

15.3.2 Endemism
Accurate figures for endemism in the southern African marine algal flora are difficult to obtain. Generic-level endemism in the Rhodophyta, at about 7%, is fairly average on a world scale (Van den Hoek 1984). Of 64 genera of Phaeophyta, five are endemics: one west coast crust, *Zeacarpa* (Anderson, Simons & Bolton 1988) and, notably, four members of the Fucales, *Anthophycus*, *Bifurcariopsis*, *Oerstedtia* (see Jensen 1976) and *Axillariella* (see Clayton 1994).

Hommersand (1986) presents figures for endemism in the different sub-regions of southern Africa, which were based on his own system, in which 'species are listed as "endemic" only if their nearest relative outside of South Africa is a distantly related taxon, normally one belonging to a separate genus'. The figures for endemism determined in this manner (from 10 to 16%, Table 15.1) in the different southern African regions are thus more comparable to generic endemism.

Of the almost 400 species on the South African west coast, 58% of the red algae, 33% of the browns and 28% of the greens are endemic to southern Africa. These are high figures in a global context (Stegenga *et al.* 1996). For southern Africa as a whole, particular groups of brown and green algae have high species endemism, with 16 out of 19 species of *Codium* (Chlorophyta; Silva 1959), 11 out of 16 species of Fucales (Phaeophyta) and 14 out of 22 species of Dictyotales (Phaeophyta) confined to the region.

15.3.3 Affinities and evolutionary biogeography
In a study based on the red algal genera of all world regions, Van den Hoek (1984) showed that the southern African flora is unusual, with low similarities to other floras. In this analysis the local flora has closest links with the temperate floras of southern Australia and New Zealand.

Hommersand (1986) analysed the southern African Rhodophyta, setting up hypotheses on their evolutionary biogeography. His method compared the 'nearest neighbour' of south African species, and the resultant geographical affinities of the species are summarized in Table 15.1. Red seaweeds from the different marine provinces have different affinities, and thus different origins are hypothesized. As expected, the majority of the red seaweeds of KwaZulu-Natal have affinities with the tropical Indo-West Pacific. Most Benguela Province reds have affinities with an historical region designated the 'Antiboreal Pacific Ocean' by Hommersand. These species 'have their closest relationships with species occurring in the temperate floras of South Australia, Tasmania and New Zealand or South America' (Hommersand 1986). The species of the Agulhas Province have a large Indo-West Pacific component, but also significant affinities with Hommersand's 'Antiboreal Pacific Ocean' and Western and southern Australia. The first detailed test of these hypotheses was published with a re-analysis of the genera of the red algal family Gigartinaceae, first based on morphology and developmental criteria (Hommersand *et al.* 1993), followed by a comparative analysis of members of these genera based on molecular similarity of the rbcL gene (Hommersand, Fredericq & Freshwater 1994). These latter authors conclude that the origins of geographical affinities in this group are old, with ancestral taxa of the Gigartinaceae occupying the eastern edge of Gondwanaland in the Mesozoic, spreading around the coastline of Pangea and 'giving rise to present-day genera in more or less linear sequence'. The rbcL data include species-pairs distributed between New Zealand and South Africa, supporting the hypothesis that the Gigartinaceae of the Benguela Province 'originated on the Pacific Ocean side of Gondwanaland and were dispersed to south Africa (sic) through a passageway between western and eastern Antarctica . . . during warm periods in the Oligocene and Miocene when West Antarctica was little glaciated, and were subsequently rafted or dispersed by currents to southern Africa where they evolved into distinct species'. Other Benguela Province taxa with these affinities are likely to have originated in a similar manner, but this sort of detailed research needs to be more widely applied to provide conclusive tests of biogeographic hypotheses.

15.4   Community types

Seaweed communities are characterized by vertical zonation because of rapidly changing physical conditions and strong biological interactions along this gradient over very small distances (for a summary see Russell 1991). This section will describe the main physiognomic differences in the seaweed vegetation in the various southern African marine provinces and the major dominants in the different vertical zones of the intertidal and subtidal.

15.4.1   Intertidal

The intertidal vegetation of shores in the Benguela province is dominated by red algae (Fig. 15.3). The uppermost macroscopic vegetation occurs around the boundary of the littoral fringe and eulittoral zone, with usually a clearly distinct zone of *Porphyra capensis*, with *Bostrychia intricata* common in crevices on some shores (see Stephenson, Stephenson & Day 1940). Common and abundant mid- to lower shore species include *Gigartina radula, G. stiriata, Aeodes orbitosa, Ulva capensis* and *Splachnidium rugosum*. The sublittoral fringe often has a zone of *Champia lumbricalis*. Rock pools above mid-shore are often dominated by *Chordariopsis capensis* and *Grateloupia filicina*, and in lower regions of the intertidal by *Gigartina scutellata*.

Upper and mid-shore regions of the intertidal of the Western Overlap (Jackelman *et al.* 1991) have very similar dominants to Benguela Province intertidal regions with a few notable additions, such as the presence of *Gelidium pristoides* and *Iyengaria stellata*, and the replacement (in terms of dominance) of *Ulva capensis* by *U. rigida*. The subtidal fringe of shores in the Western Overlap is commonly dominated by large single-species patches of the fucoid *Bifurcaria brassicaeformis*, which is more or less confined to this small region (Manuel 1990).

The replacement in the overlap of Benguela Province dominants by Agulhas Province dominants is more marked in the subtidal fringe. Common warm-water species in the Western Overlap include *Anthophycus longifolius* (Turner) Kützing, *Sargassum heterophyllum, Laurencia* spp., *Dictyota dichotoma*, and the increased dominance of *Plocamium corallorhiza* and *Hypnea spicifera*.

In the Agulhas Province, *Porphyra* is generally present, although much reduced in both size and abundance (Isaac 1957; Graves 1969), and upper regions have small amounts of species such as *Bostrychia intricata, Gelidium reptans, Bachelotia antillarum* and *Gigartina acicularis* (Stephenson *et al.* 1937; Bolton & Stegenga 1990). By far the most abundant seaweed species in the mid- to lower shore is *Gelidium pristoides*, which often grows attached to the backs of the dominant limpets (Carter & Anderson

Figure 15.3 **The biological zones on a rocky shore in the Benguela Province (Paternoster). Kelp (*Ecklonia maxima*) dominates the sublittoral fringe and sublittoral, just below a dark zone (lower eulittoral) of fleshy reds such as *Champia lumbricalis* and *Gigartina radula*. Above this is a 'bare' mid-eulittoral zone dominated by the herbivorous limpet *Patella granularis*. A darker zone of *Porphyra capensis* marks the upper eulittoral, and above this is a broad dark zone of bluegreen algae in the supralittoral zone** *(Photo: J.J. Bolton).*

Figure 15.4 **In the mid-eulittoral zone of some Agulhas Province shores, compact algal turfs form mosaic patterns between the territories of limpets. The dark tufts are the economically important agarophyte *Gelidium pristoides* growing on limpet shells** *(Photo: R.J. Anderson).*

1991). Turfs in the eulittoral zone of this region (Fig. 15.4) commonly contain *Caulacanthus ustulatus, Chylocladia capensis, Bryopsis* spp., *Polysiphonia* spp. and a variety of other small red algae. Damp niches in the eulittoral zone often have stands of *Ulva rigida, Iyengaria stellata* and *Gigartina paxillata*. Rock pools and the subtidal fringe, where the grazing pressure is particularly high, are dominated by brown algae (e.g. *Anthophycus longifolius, Sargassum heterophyllum, Zonaria subarticulata*). Common and abundant red seaweeds in this zone include *Plocamium corallorhiza, P. cornutum, Gelidium capense* (Gmelin) Silva and *G. abbottiorum* Norris, *Hypnea spicifera* and a variety of articulated corallines (particularly species of *Arthrocardia* and *Amphiroa*, Fig. 15.5). Common

Figure 15.5 **Dense communities of articulated coralline red algae such as *Amphiroa ephedraea* characterize the lower eulittoral and shallow sublittoral of exposed shores in the Agulhas Province** *(Photo: J.J. Bolton).*

Figure 15.6 **The blades of *Ecklonia maxima* are suspended below the water surface by a gas-filled bladder at the top of the stipe, forming the canopy of kelp beds** *(Photo: R.J. Anderson).*

low shore turf-forming species are *Laurencia* spp., *Anotrichium tenue*, *Cheilosporum* spp., *Dictyota* spp., *Pterosiphonia cloiophylla*, etc. The intertidal at Hluleka in the Eastern Overlap (Bolton & Stegenga 1987) is similar to shores further west in the Agulhas Province, with one marked difference being the significant zone of *Caulerpa racemosa* in the lower eulittoral.

KwaZulu-Natal intertidal regions (such as that at Isipingo described by Farrell *et al.* 1993) have little seaweed biomass on open rock except in the lower eulittoral/subtidal fringe. Common species include *Bostryschia intricata* and *Rhizoclonium ambiguum* in the upper regions, *Caulacanthus ustulatus* and *Gelidium reptans* in the eulittoral, with pools having larger plants, of genera such as *Sargassum*, *Galaxaura* and *Codium*. The subtidal fringe has similar species to those on shores of the Agulhas Province, such as *Hypnea spicifera*, *Gelidium abbottiorum* and *Plocamium corallorhiza*. Further eastwards, the marine vegetation gradually takes on a more obvious Indo-West Pacific character (e.g. Inhaca Island, Mozambique; Isaac 1958), with many typically tropical taxa such as *Valonia*, *Microdictyon*, *Udotea*, *Hydroclathrus*, *Turbinaria* and a larger variety of *Caulerpa* species. Some pools and rocky slopes on Inhaca are covered by the marine angiosperm (sea grass) *Thalassodendron ciliata*, which is common on shores in tropical East Africa. It is notable that the uppermost significant seaweed vegetation on these more tropical shores occurs around the mid-tidal, at the lower limit of a belt of the oyster *Crassostrea*.

15.4.2 Subtidal

The west coast of southern Africa is characterized by kelp beds wherever there are rocky substrata in the shallow sublittoral (Fig. 15.6). In the Southwestern Cape subprovince the large hollow-stiped sea-bamboo, *Ecklonia maxima*, forms extensive beds to depths of around 8 m (Field *et al.* 1980a), with the solid-stiped, smaller kelp, *Laminaria pallida*, dominant from 8 to 14 m (occurring to 20 m) (Fig. 15.7). Further northwards, populations of *Laminaria pallida*, the form previously known as *Laminaria schinzii* (see Stegenga *et al.* 1996), become increasingly hollow-stiped. The hollow-stiped form dominates inshore as far south as Lamberts Bay. The Namaqua subprovince has both *Ecklonia maxima* and *Laminaria pallida* forming inshore beds, with *Ecklonia* becoming rarer in southern Namibia. In the Namib sub-province, to just north of Rocky Point (Penrith & Kensley 1970), the kelp beds comprise only *Laminaria*.

The main assemblages of species in kelp-bed communities of the southwest coast of the Western Cape Province are described by Velimirov *et al.* (1977) and Field *et al.* (1980a) who quantified the fauna and flora in terms

Figure 15.7 *Laminaria pallida* forms a subcanopy in deeper water in the kelp beds of the southern Benguela Province *(Photo: R.J. Anderson)*.

of energy units. Jackelman *et al.* (1991) provide a more detailed description of marine vegetation in a site at Cape Hangklip, in the Western Overlap. There are no published studies describing subtidal marine vegetation in Namibia. Several generalizations can be made from published studies on the southwestern Cape and our own observations. Inshore areas (down to 5–10 m depth) are characterized by a high biomass of kelps and other seaweeds. Understorey communities are usually dominated by red algae such as *Epymenia obtusa*, *Hymenena venosa*, *Botryoglossum platycarpum*, *Botryocarpa prolifera* and other smaller species. Brown algae tend to be more occasional (*Axillariella constricta*, small kelps) although the annual, *Desmarestia firma*, can occupy large areas of rock, particularly where disturbance has created space (Anderson & Hay 1986). Few green algae are found (e.g. several *Cladophora* species), except in open spaces, for example clearings in shallow water where kelp has been harvested, in which case *Ulva* may rapidly appear.

The offshore zone (deeper than about 10 m) of Western Cape kelp beds is characterized by a high biomass

of animals and a decrease in kelp, especially *Ecklonia maxima*, which reaches its depth limit. Algal communities in deeper water (below 10 m) are sparse, except for *Laminaria pallida*, which may form extensive beds. Encrusting corallines cover most of the rock that is not covered by sessile invertebrates down to depths of about 20 m, beyond which even these persistent algae disappear.

The kelps are themselves host to a distinctive biota of epiphytic plants and animals, including a variety of seaweeds and a canopy fauna of at least 27 invertebrate species, which are mainly associated with the most abundant seaweed epiphytes, *Polysiphonia virgata* (C. Agardh.) Sprengel, *Carpoblepharis flaccida* and *Suhria vittata*, and are an important source of food for fish (Allen & Griffiths 1981).

While the studies of Field *et al.* (1980a) and Velimirov *et al.* (1977) described the main communities in west coast kelp beds, they were aimed at quantifying important components of the biomass, and floristic studies in these communities are lacking. The subtidal benthic algae at Cape Hangklip, near Betty's Bay are being investigated in order to provide such information. Factors such as grazing, sand scour and burial, kelp canopy and depth are important in determining the floristic composition and structure of particular communities (J.J. Jackelman, pers. comm.).

The only study of subtidal algal communities in the Agulhas Province is that of Anderson & Stegenga (1989) on Bird Island, 10 km off the mainland in Algoa Bay in the Eastern Cape. Of 122 species recorded in this study, 40 were new to the Eastern Cape, including ten new records for southern Africa, illustrating the poor state of floristic knowledge of this coast.

Several of the algal community types described from Bird Island are typical of similar habitats along the south coast. Shallow reefs and gullies that are exposed to water movement but not influenced by sand are dominated by a mixture of articulated coralline algae (see Plate 15.4), especially *Amphiroa ephedraea*, *Arthrocardia duthiae* and *Corallina officinalis*. Occasional foliose algae include several species of *Plocamium* (especially *P. corallorhiza*), *Laurencia*, *Codium* and *Dictyota dichotoma*. Casual observations suggest that the Agulhas Province shallow subtidal algal communities are strongly influenced by herbivores, compared to the Benguela Province: the predominance of articulated and crustose coralline algae is striking, and most of the foliose algae present belong to genera, for example *Laurencia*, *Dictyota* and *Caulerpa*, that have been shown, in other parts of the world, to contain herbivore-deterrent chemicals. Important herbivores probably include sea urchins (Fig. 15.8), molluscs, crustaceans and fish. At Bird Island, *Plocamium corallorhiza* and *Pachychaeta*

Figure 15.8 **Intense urchin grazing in the shallow subtidal of the Western Overlap creates areas of encrusting coralline algae, and limits recruitment of kelps and other fleshy seaweeds** (Photo: R.J. Anderson).

brachyarthra were the most abundant foliose algae, and both were important in the diet of the abalone there.

On a reef at 10 m depth at Bird Island, *Gelidium pteridifolium* (Norris, Hommersand & Fredericq) formed a dense community (Anderson & Stegenga 1989). This species typically occurs on exposed rocks between the eulittoral zone and about 10 m. At 22 m depth the community was dominated by stunted articulated corallines, and the calcified *Peysonnelia capensis*, which also occurs in dark caves and overhangs in shallower water. This deep site produced most of the new algal records, illustrating the paucity of subtidal collections on the south coast. At these depths the substratum is increasingly covered by sessile filter-feeding animals such as sponges, soft corals and compound ascidians.

Shallow subtidal reefs that are affected by shifting sand have a different flora from the sand-free substrata at Bird Island. Species with well-developed creeping holdfasts (*Caulerpa bartoniae*, *C. filiformis* and *C. holmesiana*) often form conspicuous green stands where the rock is covered by thin sand. In other areas there are short turfs of small articulated corallines (e.g. *Jania* spp.), other algae (e.g. *Polysiphonia* spp.) and tubes of polychaete worms, that accumulate sand to form a compact mat.

Algal communities of the KwaZulu-Natal coast subtidal remain almost completely unknown. A few, mainly faunal, subtidal studies (e.g. Berry *et al.* 1979) mention algae only in passing, and provide no information on their floristics or community composition. Recent studies of crustose coralline algae at Sordwana Bay, the southernmost coral reef system in the western Indian Ocean (Chamberlain 1993; Keats & Chamberlain 1993, 1994a,b), describe six new species and several new records. These studies are particularly valuable, because they will help to provide a floristic basis for community studies of the structurally and functionally important

coralline crusts in southwestern Indian Ocean coral reef systems.

15.5 Population biology

Very little is known about the population biology of southern African seaweeds, with the exception of a handful of common species in the Benguela Province, and two species in the Agulhas Province. Species studied are generally those with some economic potential, but do allow for some generalizations.

Dominant species are generally present throughout the year, with a few species passing part of the year as a microscopic phase with an annually occurring macroscopic phase (Stephenson *et al.* 1937; Bolton & Anderson 1990). Possible exceptions are the furry, string-like brown algae of the family Chordariaceae, which are common in the Benguela and Agulhas Provinces in spring and early summer, and presumably overwinter as the microscopic phase (e.g. *Myriocladia capensis*; Bolton & Stegenga 1990).

The kelps (Laminariales), most of the fucoids and many other relatively large brown and red algae exist as individuals that survive for more than one year. Excepting the kelps, they have a reduced form in winter often consisting of little more than old main axes, but producing new branch growth in spring and summer. This has been observed in many genera such as *Sargassum*, *Anthophycus*, *Bifurcariopsis* (see Jensen 1976), *Dictyopteris*, *Zonaria*, *Botryoglossum*, *Neuroglossum*, etc., although not properly documented.

Many other species have perennating holdfasts from which new uprights are produced in spring. Many red seaweeds have perennating mechanisms such as holdfasts producing many uprights and/or basal crusts. Species of Gigartinaceae (e.g. *Iridaea* – Bolton & Joska 1993; *Gigartina* – Bolton & Levitt 1992) all have long-lived crusts, and species such as *G. radula* and *G. stiriata* in the Benguela Province, and *G. pistillata* and *G. paxillata* in the Agulhas Province, also have spreading long-lived, regenerating holdfasts. Despite the various perennating mechanisms, recruitment of these algae is very seasonal, with production of large numbers of juveniles in winter or early spring (Bolton & Levitt 1992). An interesting example of convergence in form is provided by one brown and two red algae which dominate intertidal rock in the lower eulittoral zone of different biogeographical regions. In the Agulhas Province and KwaZulu-Natal *Hypnea spicifera*, in the Western Overlap *Bifurcaria brassicaeformis*, and in the Benguela Province, *Champia lumbricalis*, are species that dominate space due to spreading, rhizome-like holdfasts, which give rise to many terete uprights a few millimetres in diameter. Two of these species show very little or no successful recruitment from propagules, despite the high numbers produced (1.7×10^9 spores m^{-2} of substratum in December in *Hypnea spicifera* – Van Zyl 1993; 1.2×10^5 eggs per upright per year in *Bifurcaria* – Manuel 1990). Similar studies have not been carried out on *Champia lumbricalis*.

The sublittoral understorey brown alga *Desmarestia* (Anderson & Hay 1986) and the intertidal and shallow subtidal red alga *Aeodes orbitosa* (Bolton & Levitt 1992), both very common in the Benguela Province, have blades that are functionally annual. In spring, large numbers of juveniles are recruited. Because cohorts from successive years have a slight overlap, the annual nature of the individual thalli is thus somewhat obscured by the year-round presence of the species. The blades of *Aeodes* become fertile in late summer/autumn. In *Desmarestia*, the seasonality of production of the macroscopic sporophyte generation is controlled by a low temperature/short-day requirement for reproduction of the microscopic gametophyte (Anderson & Bolton 1989).

All other west and south coast species investigated in detail are fertile throughout the year. In *Gigartina radula* and *G. stiriata*, the mass of reproductive relative to vegetative tissue is consistent over the course of the year, therefore a documented biomass peak in spring/summer (Bolton & Levitt 1992) would be expected to coincide with a peak spore production. Among the red algae, spore release has only been studied on a seasonal basis in *Gelidium pristoides* (Carter 1986) and shown to vary little throughout the year. Individual plants of the dominant west coast kelp *Ecklonia maxima* are fertile all year, although fertile area per plant, and number of spores produced per unit fertile area, vary seasonally (Joska & Bolton 1987). Fertile area was maximal in September, and spores released per unit area peaked in December.

15.6 Community structure and dynamics

In this section we discuss the physical and biological factors that determine the structure and dynamics of seaweed communities along the coast of South Africa. Most research has been done in the Benguela Province and Western Overlap.

15.6.1 Physical factors

In the intertidal zone of Cape Peninsula rocky shores, sea temperature and wave exposure were considered by McQuaid & Branch (1984) to be the main determinants of species composition and biomass, respectively. Whereas temperature determines the potential pool of species that may colonize a shore, the relative proportions of the main functional groups of organisms are determined by the degree of wave exposure. Algae dominate sheltered shores, but filter-feeders and omnivorous and carnivorous animals dominate exposed shores (McQuaid, Branch & Crowe 1985). However, in a geographically wider study of seaweeds on semi-exposed shores of the southern Benguela, Western Overlap and western Agulhas Provinces, Bolton & Anderson (1990) found that not only the absolute range of species distribution but also the structure and composition of communities on a biogeographical scale can be correlated with sea-water temperatures.

On sheltered shores, the habitat that algae provide for other species (particularly invertebrates) increases the overall species diversity (McQuaid *et al.* 1985). Competition for space between algae and filter feeders is important, particularly on exposed shores (McQuaid *et al.* 1985).

In general the upper limits of the vertical distribution of seaweeds on intertidal shores are set by physical factors associated with emersion during low tides, but biological factors such as competition for space become important in structuring communities lower on the shore (Underwood 1986; Peckol, Harlin & Krumscheid 1988). This was experimentally confirmed in the case of the economically important species *Gelidium pristoides*, a dominant species on shores in the Agulhas Province, by transplanting tufts that were growing on the shells of limpets to various positions on the shore (Carter & Anderson 1991). The upper limit was shown to be set by prolonged emersion, whereas below the normal range the species was displaced by articulated coralline red algae, and overgrown by epiphytic encrusting coralline red algae.

McQuaid *et al.* (1985) found that intertidal communities (including seaweeds) did not differ significantly with different rock types in their study of sites around the Cape Peninsula, but that there was a marked reduction in species richness (alpha diversity) on unstable substrata (i.e. boulders). However, the surface relief of different rock types (granite, sandstone and dolerite) at Smitswinkel Bay on the Cape Peninsula accounted for differences in species diversity and percentage cover of seaweed communities (Wells, Moll & Bolton 1989).

Desiccation controls the biomass of many False Bay seaweeds, on account of the significant correlation between the temporal biomass patterns and the predicted heights of the lowest low tides (McQuaid 1985). In the Benguela Province, especially north of Cape Columbine, thick fog is common in the mornings, when spring low tides occur. This may protect intertidal communities from severe desiccation (Stephenson & Stephenson 1972).

Although light levels are unlikely to be limiting for intertidal seaweeds, this physical factor undoubtedly controls the depth to which subtidal species can penetrate (Lüning 1990). The depth limits of southern African seaweeds have not been studied, but casual observations show that deeper offshore reefs, where the water is clearer, can support dense seaweed communities down to at least 50 m, which is about the limit for scuba diving. Southern African inshore waters are more turbid than offshore waters as a result of terrestrial runoff and greater bottom surge-related stirring up of bottom sediments, and large seaweeds are seldom found below about 20 m depth. Worldwide, as on our coasts, the deepest-growing algae are generally encrusting coralline rhodophytes. This is possibly due to their slow growth and persistence for long periods at very low light levels (see Mathieson, Penniman & Harris 1991 for 'extinction depths' of north Atlantic seaweeds), and their survival despite overgrowth by sponges and tunicates (Sebens 1986).

Intertidal studies have in the past concentrated on either purely rocky or purely sandy shores, but it has recently been recognized that many southern African shores are 'mixed' in that both types of substrata have a strong influence on the biota (Bally, McQuaid & Brown 1984). Seaweed and animal communities that live on rock are strongly affected by seasonal and episodic sand burial. This is mainly a result of competition for space between sand-tolerant and sand-intolerant species (Taylor & Littler 1982). On sand-inundated shores in False Bay, overall species diversity is relatively low, and certain sand-tolerant seaweeds increase their biomass when the sand excludes many of the grazers (Brown, Wynberg & Harris 1991). In contrast, a study by McQuaid & Dower (1990), comparing Agulhas and Benguela Province shores, found that sand inundation increased habit heterogeneity and thus species diversity. However, this study did not take into account biogeographical effects – eastern Cape shores may well be inherently more species-rich, having a larger species pool.

In a study of sand-affected shores in Namibia, Engledow & Bolton (1994) found that seaweed species diversity was reduced when more than 5.6 kg m^{-2} of sand was present. Below this level, the main factor controlling species number was mussel cover in wave-exposed areas. In quadrats without mussels, seaweed diversity was, however, positively correlated with wave action.

Several Benguela Province seaweeds have crustose holdfasts or crustose phases in their life histories that are particularly resistant to burial by sand. The crustose tetrasporophyte and gametophytic base of some *Gymnogongrus* species, and the crustose base of *Iridaea* can survive burial under a metre of sand for at least six months,

and once the sand is removed (e.g. by winter storms) rapidly produce uprights or tetraspores, so that the shore is dominated by these species within a few months (R.J. Anderson, unpubl. data). Different morphologies in the life history of a species endow it with the ability to survive different sources of mortality (Shannon, Crow & Mathieson 1988). Crusts are adapted to withstand physical factors such as grazing, shear stress, sand and low light (Taylor & Littler 1982).

15.6.2 Grazing

Grazers play a central role in structuring seaweed communities on southern African shores (Branch 1976, 1981; McQuaid & Branch 1985). In the Benguela and Agulhas Provinces a characteristic biological zone (the cochlear zone) has been named after the limpet *Patella cochlear*, which dominates the lower eulittoral of most exposed rocky shores. In the cochlear zone a pavement of encrusting coralline algae (*Spongites* spp.; Chamberlain 1993) covers the rock. This is grazed by the limpet, which also maintains a narrow belt of fine red algae around itself. These algae are eaten and defended from other individuals: a behaviour called 'gardening' (Branch 1976, 1981). The red algae are usually *Gelidium micropterum* or *Herposiphonia heringii* (Fig. 15.9).

The limpet *Patella granularis* is an important grazer in the upper eulittoral of Cape shores, where it keeps algal biomass low. In the Agulhas Province *P. oculus* is important in the mid-eulittoral. Within a few months of removal of the limpets, the rock is covered by a dense growth of *r*-selected ephemeral seaweeds such as *Enteromorpha* and *Ulva*, which are subsequently replaced by perennial seaweeds characteristic of the 'climax' state of that zone. For example, Carter & Anderson (1991) showed that removal of *P. oculus* from the mid-eulittoral of an

Figure 15.9 The limpet *Patella cochlear* 'gardens' a narrow band of red seaweed (here *Gelidium micropterum*) surrounded by species of the encrusting coralline red alga *Spongites* (Photo: R.J. Anderson).

Eastern Cape shore led to a four-fold increase in the cover of the dominant species, *Gelidium pristoides*, within 14 months. Conversely, the rate of production of intertidal algae was shown by Bosman & Hockey (1988) to affect directly the growth rate and fertility of *Patella granularis* on a western Cape shore.

Further evidence of the importance of grazers in structuring the whole intertidal community was obtained after the severe flooding of the Orange River in 1988 which led to the death of the main herbivores, and ephemeral algae rapidly dominated this Benguela Province shore (Branch, Eekhout & Bosman 1990). Here, even in the sublittoral fringe, the upper vertical limit of the kelps was raised, and foliose algae appeared, smothering and killing the normal cover of encrusting coralline species.

Although grazers such as limpets have a strong structuring effect by removing spores and sporelings, they may be unable to feed on large seaweeds. In southwestern Australia, limpets placed in clearings in a mature algal stand were displaced (starved out) as the algae grew and spread (Underwood & Jernakoff 1981).

The effect of grazers in structuring sublittoral communities is less well studied. Kelp beds support a number of grazing invertebrates, including sea urchins and the commercially important abalone (*Haliotis midae*, local name: perlemoen). The structuring effects of large numbers of sea urchins (*Parechinus angulosus*) are evident in most of the kelp beds between Cape Point and Cape Agulhas where, at depths down to about 5 m, they remove most of the foliose seaweeds, creating a rather barren understorey beneath the kelps, and leaving most of the rock surface covered by only a pink 'pavement' of encrusting coralline algae (R.J. Anderson, unpubl. data). In contrast, west of Cape Point, urchins are far less abundant in the shallower water, and there is a dense understorey community of red seaweeds (e.g. *Hymenena venosa*, *Botryocarpa prolifera*, *Neuroglossum binderianum*), sponges and other sessile organisms.

When urchins were experimentally removed from a kelp bed in False Bay, the density of kelp plants increased significantly (Fricke 1979), and at Oudekraal on the west coast of the Cape Peninsula these animals consumed about 14% of the kelp production in the form of kelp debris, per year (Buxton & Field 1983). Sea urchins may have a marked effect on the recovery of some kelp beds after commercial harvesting. Recent experiments show that recovery of kelp beds after harvesting is much slower at Danger Point, some 100 km to the southeast of Cape Point, than at Kommetjie, west of Cape Point (G.J. Levitt & R. J. Anderson, unpubl. data). The Danger Point kelp beds, like many in the Western Overlap, have dense populations of the urchin *Parechinus angulosus*, but

these are absent from west coast sites such as Kommetjie, where kelp beds regenerate fully within 2–3 years of cutting. Current experiments are testing the hypotheses that recruitment of juvenile kelps is severely limited by a high grazer biomass, and that holdfasts of adult kelp plants may provide important refugia for the young sporophytes (R.J. Anderson & G. J. Levitt, unpubl.). There may, therefore, be a fundamental difference in the ability to withstand disturbance, between the kelp beds with numerous grazers (east of Cape Point) and those with few (west of Cape Point).

In general, grazing of seaweeds by fish has not been studied in southern Africa, although certain species are known to be herbivores. For example the small sparid *Sarpa salpa* occurs around the South African coast, and has caused damage to beds of the commercial seaweed *Gracilaria* (Anderson et al. 1993). In the warmer waters of the east coast of South Africa the diversity of fish species is much higher and there are therefore more herbivorous species, particularly kyphosids and siganids (Rabbitfish). Here fish are likely to have an important effect in structuring seaweed and other benthic communities. For example, the main grazers of seaweeds at ORI Reef off Durban are fish (Berry et al. 1979).

15.6.3 Competition

The strong competition for space among sessile organisms on many rocky shores is illustrated when space is cleared in a community of a dominant species. There is a rapid succession of organisms, usually starting with ephemeral algae, followed by longer-living more persistent species. Space-clearing phenomena such as physical dislodgement (waves, moving rocks), sand burial and grazing cause much of the 'patchiness' that is characteristic of intertidal and subtidal communities (Fig. 15.10). Perennial intertidal seaweeds often have well-developed rhizomatous holdfasts that spread vegetatively to occupy space. Ephemerals such as *Ulva* and *Enteromorpha* were eventually replaced by species such as *Bifurcaria brassicaeformis* and *Champia lumbricalis* on certain exposed Benguela Province shores (Bokenham & Stephenson 1938), and by *Gelidium pristoides* on an Agulhas Province shore (Carter & Anderson 1991).

The distribution and productivity of sublittoral seaweeds is strongly affected by inter- and intraspecific competition for light and space (e.g. Denley & Dayton 1985). Storms remove an estimated 12.5% of the kelp from a typical west coast bed annually (Jarman & Carter 1981). In many cases the entire holdfast is torn off, and this mechanism is an important source of new primary space. Where most of the rock surface is otherwise covered by dense understorey communities, these cleared patches are often the main habitat for settle-

Figure 15.10 **Patchiness in an Benguela Province intertidal community: the red seaweed *Porphyra capensis* colonizes an area from which mussels were removed in a storm. The bare zone around the *Porphyra* results from grazing forays of small limpets, which shelter among the mussels** *(Photo: R.J. Anderson).*

ment of new kelp or species such as the annual *Desmarestia firma* (Anderson & Hay 1986).

The deeper-water kelp *Laminaria pallida* can compete for space by 'sweeping' the rock around it with its fronds (Velimirov & Griffiths 1979). Once a plant becomes established and reaches a certain size, other kelp sporophytes develop directly below it, and a *Laminaria* patch develops. This progressive spread is indicated by the structure of new patches, which have older, large plants in the middle and progressively smaller plants towards the periphery.

15.6.4 Life forms
A few generally unsatisfactory attempts have been made to place seaweeds in life-form categories, for comparison with regard to their distribution (Chapman & Chapman 1976; Garbary 1976). However, there is a clear pattern in southern Africa of a dominance of larger species in the cooler, more nutrient-rich waters of the west coast, to a predominance of small algae, particularly turf-forming species, in warmer waters to the east. This is well demonstrated in Fig. 15.11, which shows biomass of intertidal foliose seaweeds, articulated coralline red seaweeds and non-coralline turf algae along the entire South African coast. Benguela Province shores have a large biomass of foliose seaweeds, low coralline biomass, and an absence of turfs. The Agulhas Province has some sites with a few turfs and a moderate biomass of corallines, and some with very low seaweed biomass, which are animal dominated. The KwaZulu-Natal coast is characterized by coralline biomass and increasing turf biomass eastwards into tropical waters.

The high intertidal biomass of fleshy seaweeds in the Benguela Province is probably linked to the protective effects of fog banks, which greatly reduce insolation, and hence the desiccation effects of emersion, on this coastline (Stephenson & Stephenson 1972). The dominant vegetation of the intertidal is predominantly made up of rhodophytes, most of which are not red in colour but various shades from yellow-brown to blackish. For example, the ca. 3-km stretch of coast around Kommetjie on the west coast of the Cape Peninsula has been estimated to support over 112 tonnes of the carrageenophytes *Gigartina radula* (17 tonnes), *Iridaea stiriata* (30 tonnes) and *Aeodes orbitosa* (65 tonnes) in a spring/summer biomass estimate (Levitt, Bolton & Anderson 1995). The increase in articulated coralline red algae in warmer waters has been previously documented on a global scale (Garbary 1976).

On many intertidal regions of the Agulhas province, the mid-shore has almost no macroscopic seaweeds, with the exception of crispy tufts of the red seaweed *Gelidium pristoides*. The most obvious distinction on KwaZulu-Natal coasts is the increase in areas of turf algae. Advantages conferred by the turf life-form are varied, including resistance to desiccation, sand inundation and scouring, fish grazing and ability to recycle nutrients in nutrient-poor tropical waters (Hay 1981; Adey & Goertemiller 1987). Very few seaweeds with a turf habit grow in the Benguela Province, and these are upper shore species, their form presumably aiding in desiccation tolerance (e.g. *Caulacanthus ustulatus*, *Cladophora contexta*; Bolton, Stegenga & Anderson 1991). Many of the turfs in False Bay and regions of the Agulhas Province are associated with sand action. The increase in turfs in northern KwaZulu-Natal is presumably related to increases in fish grazing and low nutrient levels, both of which are major ecological factors in tropical waters.

15.7 Ecosystem function

15.7.1 Intertidal
There are substantial differences between species composition of intertidal communities on rocky shores in different geographic zones along the southern African coast, but the response of the communities to environmental factors, in terms of ecosystem functioning, is often very similar. Probably the most important such factor is wave exposure.

On the Cape Peninsula, sheltered rocky shores are characterized by a high algal biomass, and exposed shores are dominated by filter-feeding, omnivorous and carnivorous animals (McQuaid *et al.* 1985). On exposed shores there is a net inflow of energy in the form of phytoplankton and detritus, whereas sheltered shores

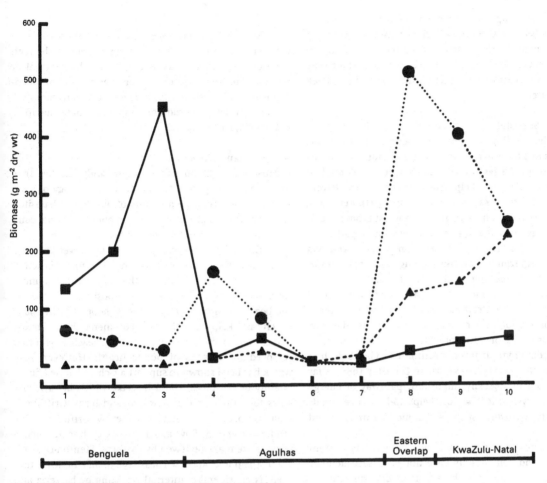

Figure 15.11 **Life forms of seaweeds (expressed as biomass) along the southern African intertidal zone (Bustamante *et al.* 1995).** Squares, foliose algae; circles, articulated corallines; triangles, turf algae. The sites are: 1, Kleinsee; 2, Groen River; 3, Kommetjie; 4, De Hoop; 5, Tsitsikamma; 6, Maitland; 7, Cape Recife; 8, Dwesa; 9, Pennington; 10, Umdoni. Marine Provinces are shown in Fig. 15.1.

are net exporters of energy, because much of the high algal production is transported offshore as detritus (McQuaid & Branch 1985).

On an exposed intertidal rocky shore ecosystem in the Benguela Province, inputs of carbon and nitrogen from seaweeds and micro-algae (including sporelings) are similar (Field 1983), while the input from phytoplankton comprises about 20% of the total. Zoospores of seaweeds can be considered a temporary component of the phytoplankton, and are not included in this figure. However, according to Joska & Bolton (1987), despite the huge numbers of spores produced by the dominant west coast kelp *Ecklonia* (estimated as 3.07×10^{10} per plant per year), these spores represent only 0.17% of the total annual production of the kelp. It is important to note that although micro-algae and sporelings are almost

invisible and thus negligible in terms of biomass, they are an important food source for grazers, and on some shores are estimated to produce as much as the macrophytic seaweeds (Field 1983). Furthermore, on a typical Cape Peninsula shore, sufficient nitrogen is released by all the animals to support the production by the small intertidal seaweeds.

On Agulhas Province rocky shores the degree of wave exposure produces patterns among the various trophic groups that are similar to those in the Benguela Province (McLachlan, Lombard & Louwrens 1981; Dobkins 1992). However, these authors also report differences in the proportions of different morphological groups of algae. Exposed shores have higher biomasses of foliose algae, and sheltered shores are dominated by encrusting algae (e.g. *Ralfsia* and non-geniculate corallines) and algal turfs (often comprising mainly geniculate coralline species), probably because these growth forms are more resistant to destruction by the relatively high biomass of grazers (e.g. limpets). The role of encrusting and turf-forming algae in ecosystem functioning has not been investigated.

15.7.2 Subtidal

An understanding of how an ecosystem functions depends to a large extent on an understanding of energy flow through the system, and this has been well studied in Benguela Province kelp beds (see reviews by Newell, Field & Griffiths 1982; Field & Griffiths 1991). Kelps and the other seaweeds have by far the highest biomass, and feed energy into the system via two main pathways: directly to herbivores (smaller component) and via detritus. Kelp fronds grow like moving belts of tissue and are constantly eroding at the tips, releasing dissolved and particulate organic matter that is rapidly colonized by bacteria. The bacteria-enriched particulate matter is an important food for mussels and other filter-feeders, which also feed on phytoplankton. The total input of energy from phytoplankton is estimated to be about half that from the macrophytes (Jarman & Carter 1981). A far smaller, but biologically important, pathway of energy flow in the system is that from kelps and other seaweeds directly to herbivores such as abalone, sea urchins and limpets.

The kelp bed ecosystem is not closed, and certain pathways and components have not been satisfactorily quantified. For example, the export of detritus and the import of phytoplankton are impossible to measure (Field & Griffiths 1991), particularly since the rate of water exchange in west coast kelp beds is weather-dependent and very variable: at Oudekraal on the Cape Peninsula, water turnover can occur more than seven times per day during active upwelling (Field et al. 1980b).

On the KwaZulu-Natal coast, energy flow has been studied in a shallow reef ecosystem (ORI Reef) at Durban (Schleyer 1980 in Field & Griffiths 1991; Berry et al. 1979; Berry 1982). The input from primary producers is mainly via detrital pathways, because phytoplankton primary production is low and the contribution from seaweeds, although it was not measured, appears small and sporadic. The main sources of detritus are terrestrial plant material washed into the sea by flooding rivers and, to a lesser extent, detached seaweeds. The reef biomass is dominated by filter-feeders and the top carnivores move offshore or migrate away from the reef, so that at the highest trophic level, energy is exported. This, and the terrestrial source of much of the primary production, makes the reef a very open system that depends to a large extent on environmental factors such as river flooding and water movement.

Under certain conditions seaweeds may become more apparent and presumably play a bigger role on the reef. For example, when space is created by the removal of any of the main filter feeders by waves or sand-deposition, algae such as *Ulva*, *Enteromorpha*, *Hypnea spicifera* and *Cheilosporum cultratum* soon dominate the affected areas (Jackson 1976).

15.7.3 Sandy beaches

Because seaweeds do not grow on sandy beaches they might not be expected to influence the biology of such areas, but where large amounts of wrack are deposited they can have a significant effect on both the faunal communities and energy flow in the beach ecosystem.

In the Benguela Province and Western Overlap, large quantities of kelp are washed ashore on many beaches. For example, at Kommetjie on the Cape Peninsula, more than 2000 kg m^{-2} of wet kelp is deposited annually on the beach from neighbouring beds (Koop & Field 1980). Beach-cast kelp is collected commercially in many places, and wrack on some urban beaches is often removed for reasons of public health. However, kelp wrack has been shown to support a rich and varied fauna (mainly of crustaceans and insects), which in turn provides food for larger crustaceans and birds (Griffiths & Stenton-Dozey 1981; Stenton-Dozey & Griffiths 1983). Studies of energy flow and nutrient cycling in such a beach system are reviewed by Griffiths, Stenton-Dozey & Koop (1983) and Koop & Lucas (1983), and indicate that there is considerable internal recycling by bacteria and meiofauna in the sand, making these systems very closed. It is clear that such beaches receive a considerable energy subsidy in the form of kelp, and have a distinctive fauna, indicating that large-scale kelp removal is likely to change such ecosystems (Griffiths & Stenton-Dozey 1981).

15.8 Economic use and resource management

Humans have used seaweeds for millennia, particularly in the Far East, where they form the basis of a food industry that is worth many times the monetary value of the entire southern African fishing industry. One Cape seaweed, *Suhria vittata*, is reported to have been eaten by Khoisan people (Rotmann 1985), and used by early inhabitants of the Cape colony for making jelly. It is still used for this purpose. It is not clear whether any other indigenous peoples used seaweeds: this is not the case at present.

In the Western World, commercial use of seaweeds

is based on the properties of their cell-wall polysaccharides. The colloids (substances which in solution maintain materials in suspension) extracted from certain genera find numerous uses in food, industrial and pharmaceutical products (see for example Lewis, Stanley & Guist 1988). In South Africa commercial use of seaweeds was first investigated during World War Two, when supplies of agar were unavailable from Japan (Isaac 1942). Since then a small industry has developed, based on harvesting these plants from the rocks or collection of beach-cast material (see review by Anderson, Simons & Jarman 1989).

Southern African commercial seaweeds are limited to the coast between Namibia and the Eastern Cape. Between four and six genera of red and brown seaweeds have been collected at various times, depending on demand from international markets. This industry is worth about R 15 million yr^{-1}, but is labour-intensive, and employs (on a part- and full-time basis) up to two thousand people.

15.8.1 Agarophytes, carrageenophytes and alginophytes

The cell walls of certain seaweeds contain commercially useful colloids (gelling and emulsifying agents) that fall into three main groups: agars and carrageenans (from certain red seaweeds) and alginates (from certain brown seaweeds).

Agar is used in the food and pharmaceutical industries but is most important as a neutral medium for the culture of bacteria and fungi in medical pathology. In southern Africa two genera of agarophytes, *Gelidium* and *Gracilaria*, are collected commercially.

About 400 tonnes (dry weight) of *Gelidium* are harvested annually in the Agulhas Province and Eastern Overlap. Half this crop comprises the strictly intertidal species *Gelidium pristoides*, obtained mostly between the Kei River and Cape St. Francis (Fig. 15.12). This species provides an excellent agar (Carter & Anderson 1986) and fetches a high price. It is collected directly off the rocks at low spring tides. Research on this species (see review by Anderson *et al.* 1991) indicates that, as a robust, fast-growing alga, capable of rapid vegetative spread from a persistent rhizoidal base, it is largely unaffected by harvesting.

The rest of the *Gelidium* harvest is made up of *G. abbottiorum*, *G. pteridifolium* and *G. capense*. These essentially subtidal species are collected from rock pools and gullies at low spring tides: the first two make up the bulk of material harvested east of the Kei River.

The other important agarophyte is *Gracilaria*, which occurs in economic quantities at two sites in the Benguela Province, namely in Saldanha Bay, South Africa and Lüderitz Bay, Namibia. Both populations are referred

Figure 15.12 **The commercially valuable red seaweed *Gelidium pristoides* is collected from intertidal shores of the Agulhas Province for the extraction of agar** *(Photo: R.J. Anderson).*

to as *G. verrucosa* (Hudson) Papenfuss, although the name is 'rarely correctly applied' (Santelices & Doty 1989; see Bird *et al.* 1994 for evidence that Namibian *Gracilaria* is best described as *G. verrucosa*). They are very similar in their ecology: both grow subtidally, usually anchored in the sediment. Material that washes ashore is collected. Propagation appears to be entirely (Lüderitz) or almost entirely (Saldanha) by vegetative growth from fragments of the string-like thallus that are distributed when material washes along the bottom. Pieces that become buried in the sediment are the basis for recovery of the beds (Anderson *et al.* 1993), as was first shown in certain Chilean *Gracilaria* populations (Santelices *et al.* 1984). The biology of *Gracilaria* at Lüderitz has been studied in detail by Molloy (1992).

At Lüderitz, *Gracilaria* has been collected since 1981, yielding about 1800–3000 tonnes dry weight yr^{-1} (Rotmann 1987; Molloy 1992; Molloy & Bolton 1992), although yields have decreased since 1993 (C.P. Dawes, pers. comm.) An agar extraction plant was commissioned in 1987, and some *Gracilaria* was processed locally for agar, although most continues to be exported. In 1991 the agar factory came under new ownership and was modernized (Critchley & Rotmann 1992) and experimental farming (mariculture) of *Gracilaria* started in the protected waters of the Lüderitz Lagoon. Currently, yields of natural (beach-cast) *Gracilaria* are still obtained at Lüderitz, and mariculture has reached a commercially successful scale, ensuring a stable supply of this valuable seaweed. Dawes (1992) describes some of the methods employed and yields obtained.

At Saldanha Bay, *Gracilaria* has been collected since at least 1951, when 20 tonnes of agar was produced locally and used mainly in the meat-canning industry. In the 1960s and early 1970s annual beach-casts often exceeded 1000 tonnes (dry weight), providing enough raw material for two agar factories (Rotmann 1990). In 1974–75 the *Gracilaria* resource collapsed, probably as a result of dredging of the bay and the construction of an ore jetty and breakwater, and the factories closed. Only in the 1980s was a modest recovery evident, with yields of a few hundred tonnes annually. At the end of 1988, beach-casts ceased again, and no material was obtained until 1992. The reasons for this collapse are still not clear, but may partly have been related to grazing by fish, keyhole limpets and sea urchins (Anderson *et al.* 1993).

This resource was plagued by further problems when a bloom of the green seaweed *Ulva lactuca* appeared in the summer of 1993–94, contaminating the *Gracilaria* wash-ups. The *Ulva* bloom resulted from ammonium pollution in fish-factory effluent (Anderson, Monteiro & Levitt 1996). This is the first local example of a harmful bloom of green seaweed resulting from marine eutrophication, a phenomenon that is becoming increasingly common in developed countries such as France, Italy and Japan.

The mariculture of *Gracilaria* is currently being investigated at Saldanha Bay (e.g. Anderson *et al.* 1992). Results indicate that yields of at least 250 tonnes fresh weight ha^{-1} yr^{-1} can be expected (R.J. Anderson unpubl. data). This would represent a gross annual income of about R 162 000 ha^{-1}, at current prices for this seaweed. Saldanha Bay is unquestionably the best mariculture site on the South African coast, and careful environmental management of this protected embayment will be required in future.

The endemic west coast red seaweed *Suhria vittata* has been used domestically for agar, but has limited commercial potential, because it grows epiphytically on *Ecklonia* and is not abundant. Its morphology and biology are described by Anderson (1994), and it may be suitable for mariculture (Anderson & Bolton 1985).

Carrageenan is used as a stabilizer and gelling agent in cosmetics, confectionery, sauces and foods: it is most useful in milk products, because it reacts with casein. No carrageenophytes are collected in southern Africa at present, although two species were harvested occasionally in the past (Anderson *et al.* 1989). Recently, research has been done on the biology of the Benguela Province species *Gigartina radula*, *G. stiriata*, *Aeodes orbitosa* and *Iridaea capensis* (see Bolton & Levitt 1992; Bolton & Joska 1993; Levitt *et al.* 1995), and *Hypnea spicifera*, which is most abundant in the Agulhas Province (van Zyl 1993). These studies are part of a planned investigation of current and potential seaweed resources of southern Africa (Anderson *et al.* 1989) and their findings will form the basis for management of these seaweeds should they be harvested in future.

Several South African large brown seaweeds (kelps) are collected for the extraction of alginates. Alginates are a group of colloids used as gelling, emulsifying and stabilizing agents in foods, pharmaceuticals, textile dyes, paper and other industries. They are obtained from certain brown seaweeds belonging to the orders Laminariales (kelps) and Fucales. Commercially used southern African kelps (*Ecklonia maxima* and *Laminaria pallida*) are found between northern Namibia and Cape Agulhas, and have been collected in various amounts since the mid 1950s (see Anderson *et al.* 1989). Beach-cast plants are collected, dried in the sun, milled to various sizes, and most of the material exported, although small quantities are used locally in fertilizers, soil stabilizers, etc. (Critchley & Rotmann 1992).

Far better prospects are assured for the horticultural and agricultural growth hormone/fertilizer that is extracted from *Ecklonia*. The kelp is harvested by divers and processed fresh, and requires relatively little material to produce a high-quality product that enjoys good local and international markets. The extract has been shown to increase growth, and drought and pest resistance in a number of crops (Mooney & Van Staden 1986). A new demand for fresh kelp as a feed has recently been created by the development of commercial abalone farming at several sites on the west coast.

15.8.2 Management

The scientific research for the management of South African seaweed resources is carried out or funded by the Sea Fisheries Research Institute. Exploitation of marine resources is regulated by the Sea Fisheries Act, which for commercial seaweed management divides the relevant coastline into Concession Areas. Rights to the seaweeds of each area are granted to one concession-holder by a five-year permit (see Anderson *et al.* 1989), subject to a fee and levy. Recent political changes mean that management is likely to be re-examined in the near future, but regulation is essential to avoid damage by over-exploitation.

15.8.3   Conservation

The conservation of seaweed vegetation is a difficult and little-studied topic. South African rocky shores are undoubtedly less affected by large-scale pollution than those in more developed countries and, except in a few more sheltered bays (e.g. Anderson *et al.* 1996), the sea is relatively unpolluted by human activities. Conserving seaweed diversity would be best carried out, as with coastal animals, by the maintenance of marine reserves covering the variety of habitats of marine organisms in the different marine provinces (Hockey & Buxton 1989; Emanuel *et al.* 1992).

In many parts of the world, large seaweeds have appeared in recent years in local floras; these can have been introduced only artificially, particularly as epiphytes on cultured oysters, on the hulls and in the ballast water of ships (Rueness 1989) or even as aquarium escapees (Meinesz *et al.* 1993). These may have drastic effects on the ecology of inshore systems. Perhaps because southern African waters are geographically distant from other regions with similar water temperature conditions, there is as yet no verified record of an introduced seaweed in the flora of southern Africa. It is extremely likely that the most successful and problematic introduced species in the North Atlantic, the large Japanese brown algae *Sargassum muticum* and *Undaria pinnatifida* (Rueness 1989) would thrive on South African coasts.

15.9   **Conclusions**

Southern Africa has an extremely rich and diverse seaweed flora, and a wide variety of marine habitats, from kelp beds to coral reefs. It is clear that there are enormous gaps in our understanding of southern African marine vegetation. Species identification remains a problem for marine ecologists. This has been largely remedied in the Benguela Province by the publication of a comprehensive seaweed Flora (Stegenga *et al.* 1996). In the rest of the region identification problems will continue to discourage studies of seaweed communities.

Historically, seaweed research has been concentrated in the southwestern Cape, with even basic descriptive studies lacking for many parts of the coast. We were unable to find a single published description of the subtidal vegetation of KwaZulu-Natal. There is obviously enormous scope for research on seaweed communities and populations.

The economic use of seaweeds in the region is likely to grow. It is possible to increase yields from natural populations, for example by harvesting more kelp, but as in other parts of the world the greatest developments are likely to involve cultivation.

15.10   **Acknowledgements**

We thank the University of Cape Town, the Sea Fisheries Research Institute and the Foundation for Research Development for support.

5.11   **References**

Adey, W.H. & Goertemiller, T. (1987). Coral reef algal turfs: master producers in nutrient-poor seas. *Phycologia*, **26**, 374–86.

Allen, J.C. & Griffiths, C.L. (1981). The fauna and flora of a kelp bed canopy. *South African Journal of Zoology*, **16**, 80–4.

Anderson, R.J. (1994). The genus *Suhria* (Gelidiaceae, Rhodophyta). In *The Biology of Economic Algae*, ed. I.A. Akatsuka, pp. 345–52. The Hague: SPB Academic Publishing.

Anderson, R.J. & Bolton, J.J. (1985). Suitability of the agarophyte *Suhria vittata* (L.) J. Ag. (Rhodophyta: Gelidiaceae) for mariculture: geographical distribution, reproductive phenology and growth of sporelings in culture in relation to light and temperature. *South African Journal of Marine Science*, **3**, 169–78.

Anderson, R.J. & Bolton, J.J. (1989). Growth and fertility in relation to temperature and photoperiod in South African *Desmarestia firma* (Phaeophyceae). *Botanica Marina*, **32**, 149–58.

Anderson, R. J. & Hay, C. H. (1986). Seasonal production of *Desmarestia firma* (C.Ag.) Skottsb. (Phaeophyceae, Desmarestiales) in kelp beds on the west coast of the Cape Peninsula, South Africa. *Botanica Marina*, **29**, 523–31.

Anderson, R. J., Levitt, G.J., Dawes, C.P. & Simons, R.H. (1992). Experimental growth of *Gracilaria* in Saldanha Bay, South Africa. In *Proceedings of the First International Workshop on Sustainable Seaweed Resource Development in Sub-Saharan Africa*, ed. K.E. Mshigeni, J.J. Bolton, A.T. Critchley & G. Kiangi, pp. 19–36. Windhoek: K.E. Mshigeni.

Anderson, R. J., Levitt, G. J., Keats, D. W. & Simons, R.H. (1993). The role of herbivores in the collapse of the *Gracilaria* resource at Saldanha Bay, South Africa. *Hydrobiologia*, **260/261**, 285–90.

Anderson, R.J., Monteiro, P. & Levitt, G.J. (1996). The effect of localized eutrophication on competition between *Ulva lactuca* L. and a commercial *Gracilaria* resource. *Hydrobiologia* (in press).

Anderson, R.J., Simons, R.H. & Bolton, J.J. (1988). *Zeacarpa leiomorpha* (Ralfsiaceae), a new genus and species of crustose marine Phaeophyceae from southern Africa. *Phycologia*, **27**, 319–26.

Anderson, R.J., Simons, R.H. & Jarman, N.G. (1989). Commercial seaweeds in southern Africa: a review of utilization and research. *South African Journal of Marine Science*, **8**, 277–99.

Anderson, R. J., Simons, R. H., Jarman., N. G. & Levitt, G.J. (1991). *Gelidium pristoides* in South Africa. *Hydrobiologia*, **221**, 55–66.

Anderson, R.J. & Stegenga, H. (1989). Subtidal algal communities at Bird Island, eastern Cape, South Africa. *Botanica Marina*, **32**, 299–311.

Bally, R., McQuaid, C.D. & Brown, A.C. (1984). Shores of mixed sand and rock: an unexplored ecosystem. *South African Journal of Science*, **80**, 600–3.

Berry, P.F. (1982). Biomass and density of detritivores on a littoral rocky reef on the Natal coast, with an estimate of population production for the ascidian *Pyura stolonifera*. *Investigational Reports of the Oceanographic Research Institute (Durban, South Africa)*, **53**, 1–15.

Berry, P.F., Hanekom, P., Joubert, C., Joubert, M., Schleyer, M., Smale, M. & van der Elst, R. (1979). Preliminary account of the biomass and major energy pathways through a Natal nearshore reef community. *South African Journal of Science*, **75**, 565.

Bird, C.J., Ragan, M.A., Critchley, A.T., Rice, E.L. & Gutell, R.R. (1994). Molecular relationships among the Gracilariaceae (Rhodophyta): further observations on some undetermined species. *European Journal of Phycology*, **29**, 195–202.

Bokenham, N.A.H. & Stephenson, T.A. (1938). The colonisation of denuded rock surfaces in the intertidal region of the Cape Peninsula. *Annals of the Natal Museum*, **9**, 47–81.

Bolton, J.J. (1986). Marine phytogeography of the Benguela upwelling region on the west coast of southern Africa: a temperature dependent approach. *Botanica Marina*, **29**, 251–6.

Bolton, J.J. (1994). Global seaweed diversity: patterns and anomalies. *Botanica Marina*, **37**, 241–5.

Bolton, J.J. & Anderson, R.J. (1987). Temperature tolerances of two South African *Ecklonia* species (Alariaceae: Laminariales) and of hybrids between them. *Marine Biology*, **96**, 293–7.

Bolton, J.J. & Anderson, R.J. (1990). Correlation between intertidal seaweed community composition and sea water temperature patterns on a geographical scale. *Botanica Marina*, **33**, 447–57.

Bolton, J.J. & Anderson, R.J. (1994). The genus *Ecklonia*. In *The Biology of Economic Algae*, ed. I.A. Akatsuka, pp. 385–406. The Hague: SPB Academic Publishing.

Bolton, J.J. & Joska, M.A.P. (1993). Population studies on a South African carrageenophyte: *Iridaea capensis* (Gigartinaceae, Rhodophyta). *Hydrobiologia*, **260/261**, 190–5.

Bolton, J.J. & Levitt, G.J. (1992). South African west coast carrageenophytes. In *Proceedings of the First International Workshop on Sustainable Seaweed Resource Development in Sub-Saharan Africa*, ed. K.E. Mshigeni, J.J. Bolton, A.T. Critchley & G. Kiangi, pp. 37–50. Windhoek: K.E. Mshigeni.

Bolton, J.J. & Stegenga, H. (1987). The marine algae of Hluleka (Transkei) and the warm temperate/sub-tropical transition on the east coast of southern Africa. *Helgoländer Meeresuntersuchungen*, **41**, 165–83.

Bolton, J.J. & Stegenga, H. (1990). The seaweeds of De Hoop Nature Reserve and their phytogeographical significance. *South African Journal of Botany*, **56**, 233–8.

Bolton, J.J., Stegenga, H. & Anderson, R.J. (1991). The seaweeds of False Bay. *Transactions of the Royal Society of South Africa*, **47**, 605–10.

Bosman, A.L. & Hockey, P.A.R. (1988). The influence of primary production rate on the population dynamics of *Patella granularis*, an intertidal limpet. *Marine Ecology*, **9**, 181–98.

Branch, G.M. (1976). Interspecific competition experienced by South African *Patella* species. *Journal of Animal Ecology*, **45**, 507–29.

Branch, G.M. (1981). The biology of limpets: physical factors, energy flow and ecological interactions. *Oceanography & Marine Biology Annual Review*, **19**, 235–380.

Branch, G.M., Eekhout, S. & Bosman, A.L. (1990). Short-term effects of the 1988 Orange River floods on the intertidal rocky shore communities of the open coast. *Transactions of the Royal Society of South Africa*, **47**, 331–54.

Branch, G.M. & Griffiths, C.L. (1988). The Benguela ecosystem Part V. The coastal zone. *Oceanography & Marine Biology Annual Review*, **26**, 395–486.

Breeman, A.M. (1988). Relative importance of temperature and other factors in determining goegraphic boundaries of seaweeds: experimental and phenological evidence. *Helgoländer Meeresuntersuchungen*, **42**, 199–241.

Brown, A.C. & Jarman, N.G. (1978). Coastal marine habitats. In *Biogeography and Ecology of Southern Africa, Vol. 2*, ed. M.J.A. Werger, pp. 1239–77. The Hague: Junk.

Brown, A.C., Wynberg, R.P. & Harris, S.A. (1991). Ecology of shores of mixed sand and rock in False Bay. *Transactions of the Royal Society of South Africa*, **47**, 563–73.

Bustamante, R.H., Branch, G.M., Eekhout, S., Robertson, B., Zoutendyk, P., Schleyer, M., Dye, A., Hanekom, N., Keats, D., Jurd, M. & McQuaid, C. (1995). Gradients of intertidal primary productivity around the coast of South Africa and their relationships with consumer biomass. *Oecologia*, **102**, 189–201.

ıxton, C.D. & Field, J.G. (1983). Feeding, defecation and absorption efficiency in the sea urchin, *Parechinus angulosus* Leske. *South African Journal of Zoology*, **18**, 11–14.

arter, A.R. (1986). *Studies on the Biology of the Economic Marine Red Alga Gelidium pristoides* (Turner) Kützing (Gelidiales: Rhodophyta). PhD Thesis. Grahamstown: Rhodes University.

arter, A. R. & Anderson, R.J. (1986). Seasonal growth and agar contents in *Gelidium pristoides* (Gelidiales, Rhodophyta) from Port Alfred, South Africa. *Botanica Marina*, **29**, 117–23.

arter, A. R. & Anderson, R. J. (1991). Biological and physical factors controlling the spatial distribution of the intertidal alga *Gelidium pristoides* in the eastern Cape, South Africa. *Journal of the Marine Biological Association of the United Kingdom*, **71**, 555–68.

hamberlain, Y.M. (1993). Observations on the crustose coralline red alga *Spongites yendoi* Foslie comb. nov. in South Africa and its relationship to *S. decipiens* Foslie comb. nov. and *Lithophyllum natalense* Foslie. *Phycologia*, **32**, 100–15.

hapman, V.J. & Chapman, D.J. (1976). Life forms in the algae. *Botanica Marina*, **19**, 65–74.

layton, M.N. (1994). Circumspection and phylogenetic relationships of the southern hemisphere family Seirococcaceae (Phaeophyceae). *Botanica Marina*, **37**, 213–20.

ritchley, A.T., Farrell, E.G., Aken, M.E. & Pienaar, R.N. (1994). A multivariate approach to the phycogeographical aspects of the seaweed flora of Inhaca Island, Moçambique. *Botanica Marina*, **37**, 261–5.

ritchley, A. T. & Rotmann, K.W.G. (1992). Industrial processing of seaweeds in Africa: the South African experience. In *Proceedings of the First International Workshop on Sustainable Seaweed Resource Development in Sub-Saharan Africa*, ed. K.E. Mshigeni, J.J. Bolton, A.T. Critchley & G. Kiangi, pp. 85–97. Windhoek: K.E. Mshigeni.

Dawes, C. P. (1992). Rope cultivation of *Gracilaria* in Namibia: Prospects? In *Proceedings of the First International Workshop on Sustainable Seaweed Resource Development in Sub-Saharan Africa*, ed. K.E. Mshigeni, J.J. Bolton, A.T. Critchley & G. Kiangi, pp. 99–109. Windhoek: K.E. Mshigeni.

Denley, E. J. & Dayton, P.K. (1985). Competition among macroalgae. In *Handbook of Phycological Methods, Ecological Field Methods: Macroalgae*, ed. M.M. Littler & D.S. Littler, pp. 511–31. Cambridge: Cambridge University Press.

Dobkins, G.S. (1992). *A Study of the Status of East Cape Rocky Shores with Special Reference to Anthropogenic Impacts*. MSc Thesis. Port Elizabeth: University of Port Elizabeth.

Emanuel, B.P., Bustamante, R.H., Branch, G.M., Eekhout, S. & Odendaal, F.J. (1992). A zoogeographic and functional approach to the selection of marine reserves on the west coast of South Africa. *South African Journal of Marine Science*, **12**, 341–54.

Engledow, H.R. & Bolton, J.J. (1994). Seaweed α-diversity within the lower eulittoral zone in Namibia: the effects of wave action, sand inundation, mussels and limpets. *Botanica Marina*, **37**, 267–76.

Engledow, H.R., Bolton, J.J. & Stegenga, H. (1992). The biogeography of the seaweed flora of Namibia. In *Proceedings of the First International Workshop on Sustainable Seaweed Resource Development in Sub-Saharan Africa*, ed. K.E. Mshigeni, J.J. Bolton, A.T. Critchley & G. Kiangi, pp. 117–30. Windhoek: K.E. Mshigeni.

Farrell, E.G., Critchley, A.T., Aken, M.A. & Pienaar, R.N. (1993). The intertidal algal flora of Isipingo Beach, Natal, South Africa and its phycogeographical affinities. *Helgoländer Meeresuntersuchungen*, **47**, 145–60.

Farrell, E.G., Critchley, A.T., Aken, M.A. & Pienaar, R.N. (1994). The biogeographical affinities of the seaweed flora of Natal, South Africa, as typified by Isipingo Beach: a multivariate approach. *Botanica Marina*, **37**, 277–85.

Field, J.G. (1983). Coastal ecosystems: flow patterns of energy and matter. In *Marine Ecology*, Vol. 5, ed. O. Kinne, pp. 758–94. Chichester: Wiley-Interscience.

Field, J.G. & Griffiths, C.L. (1991). Littoral and sublittoral ecosystems of southern Africa. In *Ecosystems of the World 24: Intertidal and Littoral Ecosystems*, ed. A.C. Mathieson & P.H. Nienhuis, pp 323–46. New York: Elsevier.

Field, J.G., Griffiths, C. L., Griffiths, R. J., Jarman, N.G., Zoutendyk, P., Velimirov, B. & Bowes, A. (1980a). Variation in structure and biomass of kelp communities along the west coast of South Africa. *Transactions of the Royal Society of South Africa*, **44**, 145–203.

Field, J.G., Griffiths, C.L., Linley, E.A., Carter, R.A. & Zoutendyk, P. (1980b). Upwelling in a nearshore marine ecosystem and its biological implications. *Estuarine and Coastal Marine Science*, **11**, 133–50.

Fricke, A. H. (1979). Kelp grazing by the common sea urchin *Parechinus angulosus* Leske in False Bay, Cape. *South African Journal of Zoology*, **14**, 143–8.

Garbary, D. (1976). Life-forms of algae and their distribution. *Botanica Marina*, **19**, 97–106.

Graves, J. (1969). The genus *Porphyra* on South African coasts. I. Observations on the autecology of *Porphyra capensis* sensu Isaac (1957), including a description of dwarf plants. *Journal of South African Botany*, **35**, 343–62.

Griffiths, C.L. & Stenton-Dozey, J.M.E. (1981). The fauna and rate of degradation of stranded kelp. *Estuarine and Coastal Shelf Science*, **12**, 645–53.

Griffiths, C.L., Stenton-Dozey, J.M.E. & Koop, K. (1983). Kelp wrack and the flow of energy through a sandy beach ecosystem. In *Sandy Beaches as Ecosystems*, ed. A. McLachlan & T. Erasmus, pp. 547–56. The Hague: Junk.

Hanekom, N., Hutchings, L., Joubert, P.A. & van der Byl, P.C.N. (1989). Sea temperature variations in the Tsitsikamma coastal national park, South Africa, with notes on the effect of cold conditions on some fish populations. *South African Journal of Marine Science*, **8**, 145–53.

Hay, M.E. (1981). The functional morphology of turf-forming seaweeds: persistence in stressful marine habitats. *Ecology*, **62**, 739–50.

Hockey, P.A.R. & Buxton, C.D. (1989). Conserving biotic diversity on southern Africa's coastline. In *Biotic diversity in southern Africa. Concepts and Conservation*, ed. B.J. Huntley, pp. 298–309. Cape Town: Oxford University Press.

Hommersand, M.H. (1986). The biogeography of the South African marine red algae: a model. *Botanica Marina*, 29, 257–70.

Hommersand, M.H., Fredericq, S. & Freshwater, D.W. (1994). Phylogenetic systematics and biogeography of the Gigartinaceae (Gigartinales, Rhodophyta) based on sequence analysis of rbcL. *Botanica Marina*, 36, 193–204.

Hommersand, M.H., Guiry, M.D., Fredericq, S. & Leister, G.L. (1993). New perspectives in the taxonomy of the Gigartinaceae (Gigartinales, Rhodophyta). *Hydrobiologia*, 260/261, 105–20.

Isaac, W.E. (1938). The geographical distribution of seaweed vegetation in relation to temperature and other factors, with special reference to South Africa. *International Geographaphical Congress*, 2, 12–28.

Isaac, W.E. (1942). Seaweeds of possible economic importance in the Union of South Africa. *Journal of South African Botany*, 8, 225–36.

Isaac, W.E. (1951). Observations on the ecology of *Bifurcaria brassicaeformis* (Kütz.) Barton. *Journal of Ecology*, 39, 94–104.

Isaac, W.E. (1956). The ecology of *Gracilaria confervoides* (L.) Grev. in South Africa with special reference to its ecology in the Saldanha–Langebaan lagoon. In *Second International Seaweed Symposium*, ed. T. Braarud & N.A. Soerenson, pp. 173–85. London: Pergamon Press.

Isaac, W.E. (1957). The distribution, ecology and taxonomy of *Porphyra* on South African coasts. *Proceedings of the Linnean Society, London*, 168, 61–5.

Isaac, W.E. (1958). Ecology of algae. In *A Natural History of Inhaca Island, Mozambique*, ed. W. Macnae & M. Kalk, pp. 18–22. Johannesburg: University of the Witwatersrand.

Isaac, W.E. & Hewitt, F.E. (1953). The morphology, geographical distribution and ecology of *Hypnea spicifera* (Suhr) Harv. *Journal of South African Botany*, 19, 73–84.

Jackelman, J.J., Stegenga, H. & Bolton, J.J. (1991). The marine benthic flora of the Cape Hangklip area and its phytogeographical affinities. *South African Journal of Botany*, 57, 295–304.

Jackson, L.F. (1976). Aspects of the intertidal ecology of the east coast of South Africa. *Investigational Reports of the Oceanographic Research Institute (Durban, South Africa)*, 46, 1–72.

Jarman, N.G. & Carter, R.A. (1981). The primary producers of the inshore regions of the Benguela. *Transactions of the Royal Society of South Africa*, 44, 321–6.

Jensen, J.B. (1976). Morphological studies in Cystoseiraceae and Sargassaceae (Phaeophyceae). *University of California Publications in Botany*, 68, 1–61.

John, D.M. & Lawson, G.W. (1991). Littoral ecosystems of tropical western Africa. In *Ecosystems of the World 24: Intertidal and Littoral Ecosystems*, ed. A.C. Mathieson & P.H. Nienhuis, pp. 297–322. New York: Elsevier.

Joska, M.A.P. & Bolton, J.J. (1987). *In situ* measurement of zoospore release and seasonality of reproduction in *Ecklonia maxima* (Alariaceae, Laminariales). *British Phycological Journal*, 22, 209–14.

Keats, D.W. & Chamberlain, Y.M. (1993). *Sporolithon ptychoides* Heydrich and *S. episporum* (Howe) Dawson: two crustose coralline red algae (Corallinales, Sporolithaceae) in South Africa. *South African Journal of Botany*, 59, 541–50.

Keats, D.W. & Chamberlain. Y.M. (1994a). Three species of *Hydrolithon* (Rhodophyta, Corallinales): *Hydrolithon onkodes* (Heydrich) Penrose and Woelkerling, *Hydrolithon superficiale* sp. nov. and *H. samoense* (Foslie) comb. nov. from South Africa. *South African Journal of Botany*, 60, 8–21.

Keats, D. W. & Chamberlain, Y.M. (1994b). Two melobesioid coralline algae (Rhodophyta, Corallinales), *Mesophyllum erubescens* (Foslie) Lemoine and *Mesophyllum funafutiense* (Foslie) Verheij from Sodwana Bay, South Africa. *South African Journal of Botany*, 60, 175–91.

Koop, K. & Field, J.G. (1980). The influence of food availability on the population dynamics of a supralittoral isopod *Ligia dilatata* (Brandt). *Journal of Experimental Marine Biology and Ecology*, 48, 61–72.

Koop, K. & Lucas, M.I. (1983). Carbon flow and nutrient regeneration from the decomposition of macrophyte debris in a sandy beach microcosm. In *Sandy Beaches as Ecosystems*, ed. A. McLachlan & T. Erasmus, pp. 249–62. The Hague: Junk.

Lawson, G.W. & John, D.M. (1987). The marine algae and coastal environment of tropical West Africa, 2nd edn. *Nova Hedwigia (Beihefte)*, 93, 1–415.

Lawson, G.W., John, D.M. & Price, J.H. (1975). The marine algal flora of Angola: its distribution and affinities. *Botanical Journal of the Linnean Society*, 70, 307–24.

Lawson, G.W., Simons, R.H. & Isaacs, W.E. (1990). The marine algal flora of Namibia: its distributions and affinities. *Bulletin of the British Museum of Natural History (Botany)*, 20, 153–68.

Levitt, G.J., Bolton, J.J. & Anderson, R.J. (1995). The potential harvestable biomass of four carrageenan-producing seaweeds of the Western Cape, Republic of South Africa. *South African Journal of Marine Science*, 15, 49–60.

Lewis, J.G., Stanley, N.F. & Guist, E.G. (1988). Commercial production and applications of hydrocolloids. In *Algae and Human Affairs*, ed. C.A. Lembi & J.R. Waaland, pp. 205–36. Cambridge: Cambridge University Press.

Lüning, K. (1990). *Seaweeds: Their Environment, Biogeography and Ecophysiology*. New York: John Wiley & Sons.

Manuel, T.L. (1990). *The Biology and Ecology of Bifurcaria brassicaeformis (Kütz.) Barton (Phaeophyta, Fucales)*. MSc Thesis. Cape Town: University of Cape Town.

Mathieson, A.C., Penniman, C.A. & Harris, L.G. (1991). Northwest Atlantic rocky shore ecology. In *Ecosystems of the World 24. Intertidal and Littoral Ecosystems*, ed. A.C. Mathieson & P.H. Nienhuis, pp. 109–91. Amsterdam: Elsevier.

cLachlan, A., Lombard, H.W. & Louwrens, S. (1981). Trophic structure and biomass distribution on two East Cape rocky shores. *South African Journal of Zoology*, **16**, 85–9.

cQuaid, C.D. (1985). Seasonal variation in biomass and zonation of nine intertidal algae in relation to changes in radiation, sea temperature and tidal regime. *Botanica Marina*, **28**, 539–44.

cQuaid, C.D. & Branch, G.M. (1984). Influence of sea temperature, substratum and wave exposure on rocky intertidal communities: an analysis of faunal and floral biomass. *Marine Ecology Progress Series*, **19**, 145–51.

cQuaid, C.D. & Branch, G.M. (1985). Trophic structure of rocky intertidal communities: response to wave action and implications for energy flow. *Marine Ecology Progress Series*, **22**, 153–61.

cQuaid, C.D., Branch, G.M. & Crowe, A.A. (1985). Biotic and abiotic influences on rocky intertidal biomass and richness in the southern Benguela region. *South African Journal of Zoology*, **20**, 115–22.

cQuaid, C.D. & Dower, K.M. (1990). Enhancement of habitat heterogeneity and species richness on rocky shores inundated by sand. *Oecologia*, **84**, 142–4.

einesz, A., de Vaugelas, J., Hesse, B. & Mari, X. (1993). Spread of the introduced tropical green alga *Caulerpa taxifolia* in northern Mediterranean waters. *Journal of Applied Phycology*, **5**, 141–7.

olloy, F.J. (1990). Utilized and potentially utilizable seaweeds of the Namibian coast: biogeography and accessibility. *Hydrobiologia*, **204/205**, 293–9.

olloy, F.J. (1992). *Studies on the Ecology and Production of Seaweeds of Economic and Potential Economic Importance on the Namibian Coast*. PhD Thesis. Cape Town: University of Cape Town.

olloy, F.J. & Bolton, J.J. (1992). The more important seaweed resources on the Namibian coast, with special reference to *Gracilaria*. In *Proceedings of the First International Workshop on Sustainable Seaweed Resource Development in Sub-Saharan Africa*, ed. K.E. Mshigeni, J.J. Bolton, A.T. Critchley & G. Kiangi, pp. 213–19. Windhoek: K.E. Mshigeni.

Mooney, P.A. & van Staden, J. (1986). Algae and cytokinins. *Journal of Plant Physiology*, **123**, 1–21.

Newell, R.C., Field, J.G. & Griffiths, C.L. (1982). Energy balance and significance of micro-organisms in a kelp bed community. *Marine Ecology Progress Series*, **8**, 103–13.

Peckol, P., Harlin, M.M. & Krumscheid, P. (1988). Physiological and population ecology of intertidal and subtidal *Ascophyllum nodosum* (Phaeophyta). *Journal of Phycology*, **24**, 192–8.

Penrith, M. & Kensley, B. (1970). The constitution of the fauna of rocky intertidal shores of South West Africa. Part II. Rocky Point. *Cimbebasia (A)*, **1**, 243–68.

Pocock, M.A. (1958). Preliminary list of marine algae collected at Inhaca and on the neighbouring mainland. In *A Natural History of Inhaca Island, Mozambique*, ed. W. Macnae & M. Kalk, pp. 23–7. Johannesburg: University of the Witwatersrand Press.

Rotmann, K.W.G. (1985). *A Strategic Plan for the Establishment of an Integrated Seaweed Industry in Southern Africa*. MComm Thesis. Johannesburg: University of the Witwatersrand.

Rotmann, K.W.G. (1987). The collection, utilization, and potential farming of red seaweeds in Namibia. *Hydrobiologia*, **151/152**, 301–5.

Rotmann, K.W.G. (1990). Saldanha Bay, South Africa: recovery of *Gracilaria verrucosa* (Gracilariales, Rhodophyta). *Hydrobiologia*, **204/205**, 325–30.

Rueness, J. (1989). *Sargassum muticum* and other introduced Japanese macroalgae: biological pollution of European coasts. *Marine Pollution Bulletin*, **20**, 173–6.

Russell, G. (1991). Vertical distribution. In *Ecosystems of the World 24. Intertidal and Littoral Ecosystems*, ed. A.C. Mathieson & P.H. Nienhuis, pp. 43–65. Amsterdam: Elsevier.

Santelices, B. (1980). Phytogeographic characterisation of the temperate coast of Pacific South America. *Phycologia*, **19**, 1–12.

Santelices, B. & Doty, M.S. (1989). A review of *Gracilaria* farming. *Aquaculture*, **78**, 95–133.

Santelices, B., Vasquez, J., Ohme, U. & Fonck, E. (1984). Managing wild crops of *Gracilaria* in central Chile. *Hydrobiologia*, **116/117**, 77–89.

Saunders, G.W. & Druehl, L.D. (1993). Revision of the kelp family Alariaceae and the taxonomic affinities of *Lessoniopsis* Reinke (Laminariales, Phaeophyta). *Hydrobiologia*, **260/261**, 689–97.

Seagrief, S.C. (1967). *The Seaweeds of the Tsitsikamma Coastal Park*. Pretoria: National Parks Board of Trustees.

Seagrief, S.C. (1984). A catalogue of South African green, brown and red marine algae. *Memoirs of the Botanical Survey of South Africa*, **47**, 1–72.

Seagrief, S.C. (1988). Marine algae. In *A Field Guide to the Eastern Cape Coast*, ed. R.A. Lubke, F.W. Gess & M.N. Bruton, pp. 35–72. Grahamstown: Wildlife Society of Southern Africa, Grahamstown Branch.

Seagrief, S.C. & Troughton, S.C. (1973). A guide to the literature on South African marine algae. *Journal of South African Botany*, **39**, 95–109.

Sebens, K.P. (1986). Spatial relationships among encrusting marine organisms in the New England subtidal zone. *Ecological Monographs*, **56**, 73–96.

Shannon, L.V. (1985). The Benguela ecosystem Part I. Evolution of the Benguela, physical features and processes. *Oceanography and Marine Biology Review*, **23**, 105–82.

Shannon, R.K., Crow, G.E. & Mathieson, A.C. (1988). Seasonal abundance and recruitment patterns of *Petalonia fascia* (O.F. Muller) Kuntze and *Scytosiphon lomentaria* (Lyngbye) Link var. *lomentaria*. *Botanica Marina*, **31**, 207–14.

Silva, P.C. (1959). The genus *Codium* (Chlorophyta) in South Africa. *Journal of South African Botany*, **25**, 103–65.

Stegenga, H. & Bolton, J.J. (1992). Ceramiaceae (Rhodophyta) of the Cape Province, South Africa: distribution in relation to concepts of marine provinces. *Botanica Marina*, **35**, 99–107.

Stegenga, H., Bolton, J.J. & Anderson, R.J. (1996). Seaweeds of the South African west coast. *Contributions from the Bolus Herbarium* (in press).

Stenton-Dozey, J.M.E. & Griffiths, C.L. (1983). The fauna associated with kelp stranded on a sandy beach. In *Sandy Beaches as Ecosystems*, ed. A. McLachlan & T. Erasmus, pp. 557–68. The Hague: Junk.

Stephenson, T.A. (1939). The constitution of the intertidal fauna and flora of South Africa, I. *Journal of the Linnean Society (Zoology)*, **40**, 487–536.

Stephenson, T.A. (1944). The constitution of the intertidal fauna and flora of South Africa, II. *Annals of the Natal Museum*, **10**, 261–358.

Stephenson, T.A. (1948). The constitution of the intertidal fauna and flora of South Africa, III. *Annals of the Natal Museum*, **11**, 207–324.

Stephenson, T.A. & Stephenson, A. (1972). *Life between Tidemarks on Rocky Shores*. San Francisco: W.H. Freeman.

Stephenson, T.A., Stephenson, A. & Day, J.H. (1940). The South African intertidal zone and its relation to ocean currents. 8. Lamberts Bay and the west coast. *Annals of the Natal Museum*, **9**, 345–80.

Stephenson, T.A., Stephenson, A. & du Toit, C.A. (1937). The South African intertidal zone and its relation to ocean currents. 1. A temperate Indian Ocean shore. *Transactions of the Royal Society of South Africa*, **24**, 341–82.

Taylor, P.R. & Littler, M.M. (1982). The roles of compensatory mortality, physical disturbance and substrate retention in the development and organisation of a sand influenced, rocky intertidal community. *Ecology*, **63**, 135–46.

Underwood, A.J. (1986). Physical factors and biological interactions: the necessity and nature of ecological experiments. In *The Ecology of Rocky Coasts*, ed. P.G. Moore & R. Seed, pp. 372–90. New York: Columbia University Press.

Underwood, A.J. & Jernakoff, P. (1981). Interactions between algae and grazing gastropods in the structure of a low-shore algal community. *Oecologia*, **48**, 221–33.

van den Hoek, C. (1984). Worldwide latitudinal and longitudinal seaweed distribution patterns and their possible causes, as illustrated by the distribution of Rhodophytan genera. *Helgoländer Meeresuntersuchungen*, **38**, 227–57.

van Zyl, P.F.F. (1993). *Ecology and Management of an Eastern Cape Carrageenophyte Hypnea spicifera (Suhr) Harv.* PhD Thesis. Port Elizabeth: University of Port Elizabeth.

Velimirov, B., Field, J.G., Griffiths, C.L. & Zoutendyk, P. (1977). The ecology of kelp bed communities in the Benguela upwelling system: analysis of biomass and spatial distribution. *Helgoländer wissenschaftliche Meeresuntersuchungen*, **30**, 495–518.

Velimirov, B. & Griffiths, C.L. (1979). Wave induced kelp movement and its importance for community structure. *Botanica Marina*, **22**, 169–72.

Wells, J., Moll, E. J. & Bolton, J.J. (1989). Substrate as a determinant of marine intertidal algal communities at Smitswinkel Bay, False Bay. *Botanica Marina*, **32**, 499–502.

The third and final part of this volume comprises eight chapters, each dealing with a particular theme in vegetation science. The chapter topics were chosen because they represent proven strengths in southern African vegetation science, or they describe unusual features of the region's plant life, or for both of these reasons. Some of these topics are touched on elsewhere in this volume, especially in Part 2. We have tried to minimize overlap and to integrate material through careful cross-referencing. Topics that appear in more than one chapter in this part as well as in Part 2 are grass–shrub/tree interactions, desertification, fire effects, grazing impacts and the ecology of succulents. Below we provide a brief introduction to the chapters by identifying their contributions to vegetation science in a southern African context.

In Chapter 16, Stock *et al.* explore several themes of plant form and function in southern African vegetation. Sclerophylly, a widespread feature of all biomes, has arisen in response to the different selective forces of water stress, nutrient poverty and herbivory. The authors document the anomalously high diversity and relative biomass of succulents in certain biomes, and attempt to explain this in terms of climatic factors. They provide an ecophysiological perspective of graminoid/woody plant interactions, a unifying theme across several biomes. The inherent instability of this growth-form mix is treated from different perspectives in Chaps 8, 10, 11, 17, 18 and 20 of this volume. The final section of the chapter deals with mycorrhizal and other plant–microbe interactions in an ecosystem context.

Whereas the previous chapter emphasizes the effects of climate and soil nutrients on southern African plant life, the next two chapters highlight the roles of herbivory (Owen-Smith & Danckwerts – Chap. 17) and fire (Bond – Chap. 18). A theme that runs through both chapters is that the primary factors – climate and soil nutrient content – determine grazing intensity and influence fire regime. Indeed, these four factors are inextricably linked – a powerful combination that determines the appearance of the vegetation of much of Africa. Many high-rainfall areas, especially those with sandy soils, support unpalatable graminoid foliage and, hence, low indigenous herbivore biomass. This results in the accumulation of large quantities of flammable fuel which, in turn, supports intense fires that hinder the recruitment of woody plants. On the other hand, many semi-arid regions, especially in clay-rich landscapes, have fertile soils, and 'sweet' or palatable grasses and nutritious browse that sustains large numbers of herbivores. It is only in exceptional years that sufficient fuel accumulates to support fires. In these environments, herbivores, especially the charismatic megaherbivores so evocative of the great African savannas, can eliminate the woody component entirely. The eradication of these herbivores has resulted in the densification of the tree component over vast tracts of semi-arid savanna. The unusual combination of these three forces in semi-arid

Part three

Ecological themes

ecosystems – rainfall, fire and herbivory – may convert a wooded land-scape to grassland or *vice versa*. These inherently unstable and 'event-driven' ecosystems require innovative management protocols.

Owen-Smith & Danckwerts (Chap. 17) stress the role of soil nutrients in determining plant responses to herbivory across southern African biomes. On fertile soils, plants produce palatable foliage; are protected mechanically against herbivory; frequently rely on herbivores for seed dispersal; and are vulnerable to herbivore-mediated local extinction. In nutrient-poor regions, plants are predominantly chemically defended and have large underground reserves that enable them to resist herbi-vore damage. The authors discuss the implications of these differences for plant persistence and community dynamics. They also describe the complementary partitioning of forage among different groups of herbiv-ores, namely invertebrates, indigenous ungulates and domestic live-stock. The replacement of diverse and dynamic assemblages of indigen-ous herbivores with relatively stable and simple communities of domestic livestock has resulted in spectacular changes in vegetation composition and structure, especially in semi-arid, 'event-driven' ecosys-tems. These grazing impacts are discussed further in Chap. 21.

Bond (Chap. 18) provides a thorough review of the very long history of fire-related research in southern Africa. He provides compelling evi-dence from all levels of organization that fire is a major determinant of plant form and function, community dynamics and biome bound-aries in the savanna, grassland and fynbos biomes. The widespread occurrence in southern Africa of species that produce flammable fuels is a relatively recent (post-Miocene) phenomenon. The establishment of regular fires, particularly in productive, grassy vegetation where annual fires are possible, must have had a profound impact on late Tertiary vegetation as fire-intolerant lineages were eliminated and fire-tolerant ones expanded their ranges and diversified. The importance of fire in determining contemporary biome boundaries in southern Africa is clearly evident in the results of fire exclusion experiments. However, there are insufficient data to resolve many questions regarding macro-scale fire effects. Bond makes a strong plea for experimental studies to disentangle the relative roles of fire, climate (i.e. productivity) and her-bivores in determining the structure and composition of fire-prone vegetation.

In Chap. 19, Cowling *et al.* depart from the review format by pre-senting an analysis of a previously unpublished data set on regional richness. Southern Africa is, overall, extremely rich in plant species, but nodes of species richness and impoverishment are neither ran-domly nor uniformly distributed across the subcontinent. Southern Africa is also characterized by great climatic, topographic and edaphic complexity and is home to several phylogenetically distinct floras. The area, therefore, offers great opportunity to test the predictions of the many hypotheses on the determinants of plant species richness at the regional scale. The detailed analyses of Cowling *et al.* show that environmental heterogeneity is the major predictor of richness in the 'southern' floristic zone of the fynbos and karroid biomes, whereas available energy is the most important determinant in the tropical-derived floristic zone of the grassland and savanna biomes. They interpret these results in terms of historical rather than contemporary phenomena.

The last four chapters in this part of the volume focus on the interface between humans and vegetation. Cunningham & Davis (Chap. 20) discuss the changing patterns of human use of indigenous plants over time. Before the relatively recent introduction of alien crops and the widespread availability of industrially manufactured foodstuffs, drugs and building materials, indigenous people depended entirely on locally available plants for these resources. This dependency on, and profound knowledge of, indigenous plant life persists to this day, especially in the infertile, sandy habitats that are unsuitable for the cultivation of crops. In the major urban centres, mining towns and fertile rural areas of the eastern seaboard, the thriving contemporary trade in indigenous plants is a testimony to their continuing importance as medicines and symbols. Research by southern African ethnobotanists is unusual in the impressive amount of quantitative data on plant use. Much of this research has been underpinned by theoretical considerations regarding the relationship between plant traits and sustainable yields. It is inevitable that for vulnerable and economically valuable species, as stocks become depleted and in response to market forces, there will be a shift from the exploitation of wild populations to the cultivation of superior genotypes. This process is already far advanced in the cut-flower and herbal tea industries in the fynbos biome.

In Chap. 21, Hoffman outlines the impacts of humans on the vegetation of the subcontinent from the earliest times to the present day. Hominids have been resident in the region for three million years and some of the earliest human fossils in the world have been discovered on the South African veld. Although early hunter-gatherer, nomadic pastoralist and agropastoralist societies did, with increasing impact, leave their mark on the landscape, it is only after European settlement that transformation occurred on a massive scale. This is most evident in the lowlands of the fynbos biome where vast tracts of species-rich vegetation have been replaced with a few crop plants of European origin. Hoffman goes on to describe the impacts of domestic herbivores, their numbers augmented by the provision of artificial water points, on the structure and composition of karroid, grassland and savanna vegetation. In particular, he highlights the plight of mesic grasslands, severely fragmented by agriculture and increasingly threatened by afforestation.

Richardson *et al.* (Chap. 22) continue with the theme of human impacts by focusing on perhaps the most problematic threat to the future persistence of indigenous vegetation in many parts of southern Africa, namely invasive alien plants. European colonists introduced to the subcontinent, both for pragmatic and aesthetic reasons, a formidable array of alien plants. Many of these, especially (and unusually so) trees and large shrubs, began to spread into the relatively treeless landscapes of the fynbos and grassland biomes. It was only in the second half of the twentieth century that the significance of this problem started to gain general recognition. The most devastating effects of alien plant invasions are evident in the fynbos biome and the extent and impacts of this phenomenon in this hot-spot of biodiversity stimulated the launch in the early 1980s of an international research programme on biological invasions. Recently, there have been attempts to quantify, in economic terms, the deleterious effects of invasions.

In Chap. 23, the final in this volume, Rebelo reviews the conservation status of southern African vegetation. The first reserves were established in economically marginal areas with the principal aim of preserving populations of large mammals. Over the past 50 years, and especially since the 1960s, there has been a steady increase in the area of land allocated for conservation. However, only Botswana and Namibia, both relatively depauperate floristically, have more than 10% of their land surface under conservation management. Rebelo discusses the thorny problems of scale, levels and representativeness in assessing conservation requirements. The much-quoted and strived-for figure of 10% conserved area for each nation is meaningless without due consideration to conserving most of the nation's species as well as adequate and representative tracts of each major vegetation category. Furthermore, the use of different units (e.g. species versus vegetation types) as the basis for determining an optimal reserve network may result in the identification of different priority areas. For example, the results of iterative procedures using species as the unit indicate that succulent karoo is the major conservation priority. However, when Veld Types are used as the unit and consideration is given to rates of transformation, then grassland areas emerge as conservation priorities. Rebelo concludes that adequate conservation of southern Africa's vegetational diversity will require a 1.5-fold increase in the conservation estate and most of the new areas will need to be located in South Africa. Given the socio-economic imperatives for equity in land distribution, the achievement of this goal will require imaginative solutions.

Plant form and function

W.D. Stock, N. Allsopp, F. van der Heyden and E.T.F. Witkowski

16.1 Introduction

The flora of southern Africa has held a fascination for botanists since the first plant collectors visited the region in the late seventeenth century. The structure of the vegetation, the large number of species, the distinctive growth-forms and the diversity of habitats are unparalleled in the northern hemisphere from whence these collectors originated; there was thus considerable excitement among scientists and horticulturalists when plants from this region reached Europe. The significance of the characteristic structures and distinctive growth-forms was, however, poorly understood until phytogeographers such as Kerner (1863), Grisebach (1872), Schimper (1903) and Warming (1909) attempted to relate the ecology of plant distribution to climate. This ecological approach attempted to explain the nature of adaptations of plants to their environment and was the philosophy underlying the first complete analysis of southern African vegetation in relation to climate, physiographic, edaphic and biotic factors undertaken by Adamson (1938).

Concern about the global effects of a continued increase in atmospheric concentrations of greenhouse gases (CO_2, CH_4, N_2O, and others) has re-awakened interest in the significance of climate in controlling the distributions of individual species and vegetation types (Woodward 1987). Tacitly, it is often assumed that present-day distribution limits of a species are determined by the ecological inability of the plant to survive outside areas characterized by the present-day climate (Chap. 2, this volume). In an unchanging environment this is possible when the evolution of adaptive characteristics reaches a stable equilibrium, with each species fully compatible with its physical environment. However, since abiotic and biotic conditions are constantly changing, individual species from a single geographical region might each have different limits to distribution because of their own specific evolutionary histories. Species might also contain traits unadapted to present-day climate which could pre-adapt them to specific climate changes (exaptations *sensu* Gould & Vrba 1982). Thus, in analyses of current or future vegetation–climate relationships, any adaptation resulting in a heritable, morphological, physiological or developmental character that improves the survival and reproductive efficiency of a species must be considered both in terms of the selective regime under which it evolved and its suitability to present or future conditions.

Southern Africa has long been recognized as an evolutionary centre of plant speciation, with sclerophyll leaves, leaf and stem succulence and the coexistence of similar and widely divergent growth-forms being particularly well represented (Bews 1925). The functional significance of many of these attributes was, however, poorly understood until recently. Some of the simple, untested explanations given to account for the predominance of sclerophylls (summer aridity; Adamson 1938) and succulents (water storage in arid environments; Bews 1925; Werger 1983) in certain regions now appear spurious as evidence accumulates from ecophysiological studies of plant performance. It is now evident that the selective forces responsible for shaping the southern African flora are diverse; they also vary immensely from region to region. Two selective factors emerge as being important across the subcontinent and a discussion of these will form the focus of this chapter. Water scarcity has had a major impact on the plant growth-form composition of the region which is not surprising considering that over two-thirds of the area is arid or semi-arid (Chap. 2, this volume). The very old geological landscapes with low-nutrient soils (Chap. 1, this volume) are also considered to be responsible for the evolution of certain plant traits and growth-forms.

This chapter attempts to explain the relationships between the geographical distributions of plants with traits of functional importance and present-day selective forces while appreciating the constraints imposed by the regimes under which these traits evolved. The chapter is also highly selective, since research results from only a few biomes are given. This uneven coverage is a consequence of both the bias of the authors and the deliberate selection of unique and physiologically interesting plant attributes ranging across a hierarchy of functional levels from the leaf (sclerophylls and succulents), whole-plant interactions (graminoid–woody plant interactions) through to below-ground plant–microbe interrelationships. The chapter is therefore not the definitive account of the structure and functioning of southern African plants, but rather a summary of the current state of knowledge of these selected topics.

16.2   Sclerophyll leaves

The remarkable convergence of leaf structure and morphology among unrelated taxa in the five mediterranean-type ecosystems of the world has provoked much interest in leaf sclerophylly. Studies initially focussed on sclerophylly as an adaptation to water stress, but also, more recently, to the significance of evergreen leaves as a physiological strategy to increase plant nutrient-use efficiency (NUE), and to reduce herbivory (Loveless 1962; Rundel 1988; Stock, Van der Heyden & Lewis 1992). The dominance of woody plants with evergreen sclerophyllous leaves in the overstorey of the mediterranean-climate regions led Schimper (1903) to suggest that summer droughts played a major role in the evolution of sclerophylly. The ground-level flora (geophytes, hemicryptophytes and therophytes) in these ecosystems are mostly seasonal, and non-sclerophyllous. On the limited areas of nutrient-poor soils in the Mediterranean Basin, sclerophyllous species are found in both the overstorey and understorey (Specht & Moll 1983) This pattern is observed throughout most of the fynbos biome and the sclerophyllous heaths of southern Australia where nutrient-poor soils are the norm (Beadle 1966; Specht & Moll 1983; Specht & Rundel 1990). In the dominant evergreen sclerophyllous shrubs of the mediterranean-type ecosystems of the world, leaf specific mass increases from

Chile, through California, South Africa and Western Australia, this representing a gradient of decreasing nutrient availability (Mooney 1983; Witkowski & Lamont 1991; Cowling & Witkowski 1994).

The term sclerophylly is poorly defined but is commonly used to describe several co-occurring leaf properties. In ecological studies leaves are often classified into 'leaf consistence' classes, such as orthophylls, sclerophylls, semi-succulent or succulent by the 'feel' (thickness and hardness) method which provides a rapid technique for broad-scale surveys (Werger & Ellenbroek 1978; Cowling & Campbell 1983b; Campbell & Werger 1988; Blackmore 1992; Cowling & Witkowski 1994). More precise measures of sclerophylly or leaf hardness have been obtained from various indices such as the ratio of crude fibre to nitrogen ['sclerophylly index' = (lignin + cellulose)/crude protein, where crude protein = $N \times 6.25$] (Rundel 1988), leaf thickness (Mooney *et al.* 1982), moisture content (Loveless 1962), resistance to puncture or tearing (tensile strength; Martens & Booysen 1968; Theron & Booysen 1968; Blackmore 1992), leaf specific mass (LSM; sometimes called specific leaf mass/weight) and specific leaf area (SLA, the leaf area/dry mass which is the inverse of LSM). Other indices such as the 'degree of sclerophylly' (leaf dry mass per unit bifacial leaf area) is merely LSM/2 (Camerik & Werger 1980). Both 'degree of sclerophylly' (and LSM) and succulence (wet mass/leaf area) show significant positive correlations with leaf size (Camerik & Werger 1980). Of all the indices, LSM has been used most often as an index of sclerophylly and it is usually highly negatively correlated with leaf nitrogen and phosphorus concentrations, which are particularly low in sclerophylls.

Sclerophylly is a common trait among xerophytes, but the possession of sclerophyllous leaves alone is not sufficient to predict the water relations of any plant species. The forces selecting for the evolution of sclerophylly and some of its correlates (e.g. leaf longevity and evergreenness) are poorly understood (Stock *et al.* 1992). Evidence that sclerophylly arose during the Cretaceous in tropical taxa (Specht 1963) and recent evidence on patterns of water use in Mediterranean Basin sclerophylls (Sallelo & Lo Gullo 1990) suggests that water scarcity has probably had little influence in the selection of sclerophylly *per se*. For instance, the incidence of sclerophylls shows little change along rainfall gradients in the fynbos biome (Campbell & Werger 1988; Cowling & Holmes 1992), being largely associated with the nutrient-poor soils derived from Table Mountain Group Sandstones (Taylor 1978, 1980). Furthermore, sclerophyllous shrubs are widespread in the tropical montane climate regimes of East Africa on nutrient-poor soils (White 1978).

Table 16.1 **Variation in mean specific mass (index of sclerophylly) and leaf area within representative species from five southern African biomes**

Biomes/species	Leaf specific mass (g dm^{-2})	Average leaf area (cm^2)
Fynbos[a,b]		
Fynbos		
Aulax umbellata	1.86	3.3
Erica phylicaefolia	0.66	0.3
Leucadendron cordifolium	3.03	6.2
L. elimense	2.27	0.9
L. meridianum	2.76	1.8
L. xanthoconus	1.52	0.9
Metalasia muricata	0.70	0.4
Protea compacta	3.39	17.0
P. obtusifolia	4.00	11.8
P. repens	2.37	3.5
Strandveld		
Chrysanthemoides monilifera	1.26	4.8
Eriocephalus africanus	0.60	0.2
Olea exasperata	2.82	2.9
Salvia africana	0.72	1.4
Sideroxylon inerme	2.10	12.3
Forest[c]		
Brabejum stellatifolium	2.51	13.8
Buxus macowanii	1.82	1.6
Celtis africana	0.89	13.0
Cola greenwayi	0.90	22.8
Cunonia capensis	1.82	13.8
Diospyros natalensis	1.28	2.1
Maytenus acuminata	1.56	4.8
Olea capensis subsp. *macrocarpa*	1.47	36.7
O. woodiana	1.58	14.8
Podocarpus elongatus	3.01	1.8
P. latifolius	3.08	6.6
Strychnos decussata	0.96	5.5
S. henningsii	0.97	13.5
Teclea natalensis	0.75	9.8
Vepris undulata	1.04	12.6
Wrightia natalensis	0.49	9.5
Xymalos monospora	0.80	24.9
Savanna[d]		
Combretum apiculatum	1.28	24.6
C. molle	1.49	26.3
Euclea undulata	1.54	1.8
Faurea saligna	1.23	18.9
Gardenia spatulifolia	1.54	6.4
Grewia bicolor	1.48	10.1
Maytenus tenuispina	1.43	3.7
Rhus pyroides	0.96	13.2
Nama karoo[b,e]		
Pentzia incana	0.86	ND
Portulacaria afra	1.78	1.2
Rhus longispina	1.75	ND
R. undulata	1.86	ND
Succulent karoo[b,e]		
Galenia africana	1.27	ND
Lampranthus multiradiatus	1.08	0.5
Osteospermum sinuatum	0.95	ND
Pteronia empetrifolia	2.73	ND
P. pallens	1.96	ND

ND, not determined.
[a] Cowling & Witkowski (1994).
[b] Cowling & Campbell (1983b).
[c] Midgley, van Wyk & Everard (1995).
[d] E.T.F. Witkowski (unpubl. data).
[e] Stock *et al.* (1992).

Plants with sclerophyllous leaves are found to a greater or lesser extent in most of the biomes of southern Africa (Table 16.1). In the Nama-karoo of the northwestern Cape, Werger & Ellenbroek (1978) investigated changes in sclerophylly and other leaf morphological characteristics along a 700-km east–west climatic gradient in 'riverine forests' of the Orange River. Deciduous, malacophyllous and microphyllous leaves predominate in the temperate eastern parts of the river valley which experience regular frosts. Further west, in the zone where frost events are infrequent and temperatures are high, small sclerophyllous nanophylls and leptophylls are common. In the extreme west between the Great Escarpment and the Atlantic coast (succulent karoo and desert biome), where temperatures are moderate, frosts absent or rare and radiation fluxes high, leaves have strong xeromorphic (even semi-succulent and succulent) features and are larger in size than in the more climatically extreme plateau areas. These changes in leaf characteristics were all attributed to differences in climate (Werger & Ellenbroek (1978). This was not the case in an analogous study at the interface of the fynbos with thicket/forest biomes in the Eastern Cape (Cowling & Campbell 1983a). Here the changes in leaf consistence (sclerophylly) along gradients in the Gamtoos River valley were interpreted largely as a response to changes in climate and soil moisture only in the 'non-fynbos' vegetation types that occur on fine-textured soils of relatively high-nutrient status. The predominance of sclerophyllous leaves throughout the fynbos gradient, irrespective of climate differences, was considered to be a response to the consistently low-nutrient status of the coarse-textured soils (Chap. 6, this volume).

The fynbos biome includes plants with exceptionally high degrees of sclerophylly (LSM, Table 16.1). Rundel (1988) used a 'sclerophyll index' to compare leaf properties of the three major growth forms to test whether patterns of sclerophylly differed between taxa in a single region. Ericoid shrubs are less sclerophyllous than restioids or proteoid shrubs. These differences are attributable to the higher nitrogen concentrations of ericoid leaves, which are used in the calculation of the 'sclerophyll index'. If an index that does not include nitrogen (such as LSM) is used, there are no significant differences between growth-forms.

Sclerophylly (LSM) and leaf longevity are inextricably linked traits that show a close correlation across a range of species which vary in leaf longevity more than 100-fold (Reich, Walters & Ellsworth 1992). Greater leaf duration is advantageous where water or nutrient resources are available in low amounts for a large part of the year, but is disadvantageous in situations where there are brief but predictable flushes in resource availability.

Although evergreen leaves are more 'expensive' to produce than deciduous leaves (Orians & Solbrig 1977; Miller & Stoner 1979), they can amortize their costs of production over a longer period. The relatively low photosynthetic capacities of evergreen sclerophylls are thus balanced by the potential ability to photosynthesize throughout the year and for more than one year, thereby increasing nutrient-use efficiency. Photosynthetic capacity is largely a function of leaf nitrogen content which, on a weight basis, declines with leaf age in almost all plants (Field & Mooney 1986). Older leaves often have much higher LSM than young fully expanded leaves, with reported increases in LSM with ageing of 79–270% in fynbos sclerophylls (Jongens-Roberts & Mitchell 1986; Rundel 1988). Fynbos and kwongan (Australian heath) shrubs appear to exhibit greater leaf longevities than those of the other mediterranean ecosystems (Kruger 1981; Bond & Midgley 1988; Witkowski *et al.* 1992), which is a function of their more nutrient-poor soils.

Various experiments have been undertaken to investigate the nutrient-use efficiency argument for the high incidence of sclerophylls in fynbos. Seedlings of the proteoid shrub (Chap. 6, this volume) *Protea repens* responded by decreasing LSM after the addition of phosphorus, nitrogen, or a mixture of other essential plant nutrients (Witkowski 1989b), and *Leucadendron laureolum*, another proteoid, showed decreases in LSM upon phosphorus addition (E.T.F. Witkowski, unpubl. data). Leaf morphological properties did not change in mature *Leucospermum parile* proteoid shrubs in the field when given low additions of phosphorus, nitrogen and other nutrients (Witkowski, Mitchell & Stock 1990); however, leaf turnover increased, resulting in shorter leaf life spans (Witkowski 1989a). Field-grown seedlings of the same species were more plastic in response to nutrient additions with small but significant decreases in LSM, the number of layers of mesophyll cells and leaf width being recorded (Stock *et al.* 1992). Thus, phenotypic plasticity in the degree of leaf sclerophylly (and the traits with which it is correlated) tends to be low in fynbos sclerophylls when compared with typical mesophytic shrubs (Witkowski & Lamont 1991).

In a study of the factors controlling the boundary between fynbos and a riparian forest, fynbos proteoid shrubs (*Protea repens* and *P. nitida*) were more sclerophyllous (higher LSM) and had features such as lower stomatal densities, more vertically inclined leaves and lower leaf conductances than the forest species (*Brabejum stellatifolium*, *Brachylaena neriifolia* and *Cunonia capensis*; Richardson & Kruger 1990). This shows that the fynbos species have a conservative pattern of water use. Seedlings of the forest species also exhibited much greater plasticity in LSM in response to variation in soil nutrients and

moisture than the fynbos species (Manders & Smith 1992a,b). These studies show that separating the ultimate and proximate factors responsible for the evolution of the sclerophyllous leaf is difficult, particularly since the leaves of fynbos species respond so conservatively to nutrient and water additions.

The leaves of trees from the forests of southern Africa appear to be comparatively more sclerophyllous than species from other subtropical and warm temperate forests (Table 16.1). Data for specific leaf area (SLA; inverse of LSM) and foliar nitrogen concentrations collected from 100 species of trees from nine forests representing a wide range of forest types (Midgley *et al.* 1995) show SLA to be positively related to nitrogen concentration on a species basis ($r^2 = 0.63$, $P < 0.01$). For forests as a whole, rainfall was negatively related to SLA on a 'mean species', or a 'mean community' basis (i.e. weighted by relative abundance) (spp. $r^2 = 0.59$, $P < 0.01$; community $r^2 = 0.43$, $P < 0.01$) and leaf n (community $r^2 = 0.35$, $P < 0.05$). Variation in leaf size between species was relatively small, despite a substantial rainfall gradient (800–2500 mm yr^{-1}). Members of the Podocarpaceae (evergreen), which tend to be slow growing, have the lowest SLA, whereas members of the Fabaceae (deciduous: e.g. *Erythrina*, *Acacia*, *Erythrophleum*), as well as the deciduous *Celtis* and *Wrightia* have the highest SLA. At the community- or whole-forest scale, subtropical Indian Ocean forests have relatively high SLA, and warm temperate Afromontane forests have low SLA (see also Chap. 12, this volume). The range in LSM for these forest species is 0.82–1.73 g dm^{-2}, which is within the range for tropical forest species (0.61–2.19 g dm^{-2}; Table 16.1, Turner & Tan 1991). South African forest species have leaves with relatively small surface area (range 9.5–26.9 cm^2) relative to tropical forests worldwide (range 15.9–202 cm^2; Turner & Tan 1991)). There are hardly any data on leaf longevity, but *Podocarpus latifolius* retains individual leaves for more than six years (J.J. Midgley, pers. comm.). Gymnosperms elsewhere have much greater leaf longevities, e.g. up to several decades in *Pinus longaeva* (Vasek & Thorne 1977). The sclerophyllous nature of southern African forests is a consequence of the leaf characteristics of many species of different phylogenies and is very different from the much more sclerophyllous but species-poor coniferous forests of the northern hemisphere.

In savannas, a broad distinction is observed between microphyllous woodlands on relatively nutrient-rich soils, and broad-leaved orthophyllous woodlands on nutrient-poor soils (Acocks 1953; Werger & Coetzee 1978; Witkowski & O'Connor 1996; Chap. 11, this volume). Most species in the former type have thinner and smaller (often microphyllous-compound) leaves than the largely broad-leaved trees of the latter, but both types are dominated by deciduous species. Blackmore (1992) determined the mean LSM for abundant species from 15 savanna sites on nutrient-rich and nutrient-poor soils in South Africa and found that they ranged from 1.0 to 1.7 g dm^{-2} for woody plants and 0.5 to 1.06 g dm^{-2} for grasses. Therefore, trees were (predictably) more sclerophyllous than grasses. Differences in LSM between tree species from nutrient-rich and nutrient-poor sites were not as great as expected. LSM of four species from nutrient-rich sites ranged from 1.00 to 1.81 g dm^{-2}, and 11 tree species in a broad-leaved nutrient-poor savanna at Nylsvley ranged from 0.96 to 1.65 g dm^{-2} [E.T.F. Witkowski unpubl. data: similar to values of 1.28–1.89 g dm^{-2} found by Scholes & Walker (1993) for eight broad-leaved species]. There were also small differences in leaf thickness between the different savanna types (0.16–0.41 mm and 0.17–0.34 mm on the nutrient-rich and nutrient-poor broad-leaved sites, respectively). Savanna tree leaves are thinner than those of fynbos sclerophylls.

Interest in the leaf properties of plant species has been centred mainly on the fynbos, forest and savanna biomes, but a few studies have reported leaf characteristics of other vegetation types. The degree of sclerophylly (LSM) has been determined for evergreen shrubs of strandveld vegetation of the fynbos biome which occurs on relatively nutrient-rich but sandy soils (Witkowski & Mitchell 1987; Chap. 6, this volume). Values for *Carissa bispinosa*, *Cassine peragua*, *Maytenus oleoides* and *Olea europaea* ssp. *europaea* range from 1.44 to 1.68 g dm^{-2}; these values are lower than for sclerophylls typical of nutrient-poor soils (Table 16.1). LSM values for eight karoo sclerophylls were highly variable, ranging from 0.86 to 2.73 g dm^{-2} (Stock *et al.* 1992). From these preliminary studies (Table 16.1) it appears that, for woody plants in southern Africa, the occurrence and degree of sclerophylly among biomes follows the general trend of fynbos > strandveld = karoo > savanna = forest.

The functional significance of the sclerophyll leaf in southern African biomes is poorly understood, because of the lack of comprehensive data. What is clear from the few studies that have been done is that both nutrient and water scarcity promote the incidence of sclerophylly. Nutrient impoverishment appears to be particularly important in the fynbos biome, whereas moisture shortage is the prime determinant in the Nama-karoo. Within biomes it appears that both selective forces are also responsible for the leaf characteristics of certain vegetation units (e.g. differences between broad-leaved and microphyllous savanna). At the subcontinental scale, the vegetation of southern Africa is highly sclerophyllous when compared to that of other regions of the world. It is only in Australia where aridity and low-nutrient soils coincide that more extreme examples of leaf sclerophylly are found (Cowling & Witkowski 1994).

16.3 Succulents

The southern African flora is extremely rich in species and growth forms at both the community and landscape levels (Cowling *et al.* 1989; see also Chap. 19, this volume). Of particular interest is the region's rich succulent flora; some 3700 species from 32 families which constitute over a third (36%) of the world's total of succulent plants occur here (Van Jaarsveld 1987; see also Chaps 7, 8 and 9, this volume). Some succulent families such as the Mesembryanthemaceae and Zygophyllaceae are found mainly in the succulent karoo biome (Fig. 16.1a and b; Jürgens 1991; Von Willert *et al.* 1992), where succulents constitute 32% of the flora (Table 16.2). Other succulent families are distributed more widely (e.g. Crassulaceae; Fig. 16.1c), but are still concentrated along the semi-arid west coast, and can comprise up to 18% of the floras in areas such as the southeastern Nama-karoo (Werger & Ellis 1981). Although succulents are conspicuous in other regions of the world, they do not have the same species richness nor do they constitute such a high proportion of the flora (Table 16.2). In southern Africa, succulents not only make a major contribution in terms of the total number of species, but also make the major contribution to the vegetation in terms of abundance and cover (up to 70% in the succulent karoo biome; Werger & Ellis 1981; Cowling *et al.* 1994a; Chap. 7, this volume).

Since there is no generally accepted definition of the term succulent (Von Willert *et al.* 1990), we use the term very loosely in this chapter to include plants that have tissues that can store water for later use. The degree of succulence (water at saturation/surface area of organ) may vary considerably. Different organs such as the leaf, stem and root, or combinations thereof, may be succulent. Succulents also range in growth-form from annuals to trees (>4 m), and root- (geophytes), stem- (Euphorbiaceae, Apocynaceae) and leaf-succulents (Asphodelaceae, Crassulaceae, Mesembryanthemaceae, Portulacaceae, Zygophyllaceae) are all well represented in southern Africa. In this chapter we restrict our discussion of the physiological properties of succulents in relation to climate to plants with above-ground succulence.

The predominance of succulents in semi-arid regions is most commonly attributed to their ability to absorb water during times of plenty for later use during droughts. This view was based on the strong global-scale correlation between succulent abundance and low rain-

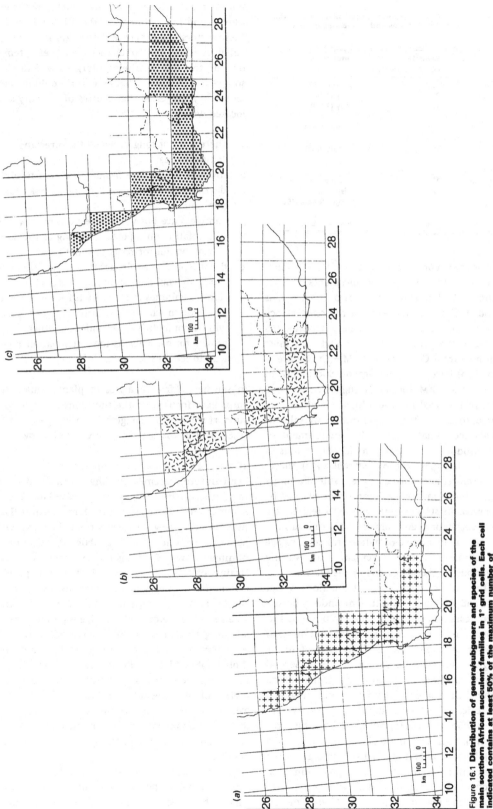

Figure 16.1 **Distribution of genera/subgenera and species of the main southern African succulent families in 1° grid cells. Each cell indicated contains at least 50% of the maximum number of genera/subgenera recorded for a single grid cell for (a) Mesembryanthemaceae, and 50% of the maximum number of species recorded for a single grid cell of the genera *Zygophyllum* (b) and *Crassula* (c).** *Adapted from Jürgens (1991).*

Table 16.2 **The proportion of succulents in various regions of the world, known for their high richness and abundance of this growth form**

	Succulents (% of the flora)	References
California	5.6	Shmida (1981)
Sonoran Desert	20.1	Cody (1989)
Mojave Desert	22.0	Cody (1989)
Israel	2.1	Shmida (1981)
Canary Islands	10.1	Shmida & Werger (1992)
Southern Africa		
Nama-karoo biome	18.5	Van Jaarsveld (1987)
Namaland	2.2	Leistner (1967)
Eastern karoo	18.0	Werger & Ellis (1981)
Succulent karoo biome	32.0	Werger & Ellis (1981)

fall (Ellenberg 1981; Von Willert *et al.* 1992). It does not, however, account for the lack of succulents in true deserts (Burgess & Shmida 1988; see also Chap. 9, this volume), and it also does not explain the close association between succulence and saline environments (Chaps 7 and 9, this volume), nor the fact that most plants with nocturnal CO_2 uptake (CAM – Crassulacean Acid Metabolism) have succulent organs. Succulence in saline-tolerant and CAM species is suggested to be a consequence of the need to have a high cell water content required to dilute the concentration of salts or organic acids accumulated from the environment or which are produced by plant metabolic processes (Jennings 1968; Ting 1989; Von Willert *et al.* 1992). Succulence has, therefore, not evolved as the result of a single selective agent but is rather the compromise reached by plants to survive in stressful environments (e.g. succulent leaves of epiphytic CAM bromeliads in tropical rainforests, the succulent nature of halophytes such as *Salicornia* in saltmarshes, and succulent water-storing species in various semi-arid regions of the world).

Despite the multiple physiological functions (water storage, salt and organic acid dilution) underlying the various hypotheses accounting for the evolution of succulence, there are specific ecological conditions that seem to favour the dominance and diversity of succulents in regions such as the succulent karoo. Several attempts have been made to account for succulent distribution patterns in relation to climate at both regional and global scales (Nobel 1980; Ellenberg 1981; Werger 1986; Burgess & Shmida 1988; Hoffman & Cowling 1987; Von Willert *et al.* 1992; Cowling *et al.* 1994a; Van Coller & Stock 1994). Two general themes have emerged from these studies, namely (1) that succulents predominate in regions of low but predictable rainfall; and (2) that they are absent or rare in regions with regular frosts. Unfortu-

nately most studies have not integrated these two aspects and few syntheses of the relative importance of these factors and others in determining patterns of succulent richness and abundance have been attempted. In one of the few syntheses, Burgess & Shmida (1988) sought to determine what climates should favour succulents; they suggested that centres of diversity should be located where:

1 there are not long periods of subfreezing temperatures;
2 drought periods are frequent enough to make daytime photosynthesis (C_3 and C_4) risky for perennials;
3 summers are hot with some rain, otherwise season of rainfall is unimportant as long as the soil is warm enough for roots to function;
4 despite prolonged droughts, temperatures are relatively cool and/or humidity is high;
5 rainfall volumes are small with some variance (with too much rain and little variance, succulents will be competitively excluded by other growth-forms, whereas too much variance could lead to local extinction).

In southern Africa analyses of plant–climate relationships have focused on the succulent karoo; evidence to support some of the suggestions of Burgess and Shmida (1988) will be discussed in more detail below.

16.3.1 Succulents and temperature

Temperature extremes limiting succulent distributions in southern Africa appear to include both high mean summer and low mean winter temperatures (Fig. 16.2; Jürgens 1991). Few succulents occur in areas with temperature summer maxima above 32 °C and winter minima below 0 °C. This is not surprising, since the constraints of being a CAM succulent include problems of cooling in summer, when the stomata are closed during the day and transpirational cooling is not operative. Succulents can therefore only survive high temperatures by adopting structural modifications such as self-shading, reduction of surface area orientated towards the sun (Nobel 1981), highly reflective cuticles, and/or cytoplasmic tolerance which is a relatively rare phenomenon in higher plants (Steponkus 1981). Leaf temperatures of over 50 °C have been recorded for *Aloe* species on days when the ambient temperature was only 28 °C (W.D. Stock, unpubl. data), suggesting that they have cytoplasmic heat tolerance.

A group of succulents expected to be rare in very hot areas because of the harshness of the microclimate in which they live are the dwarf succulents or 'stone

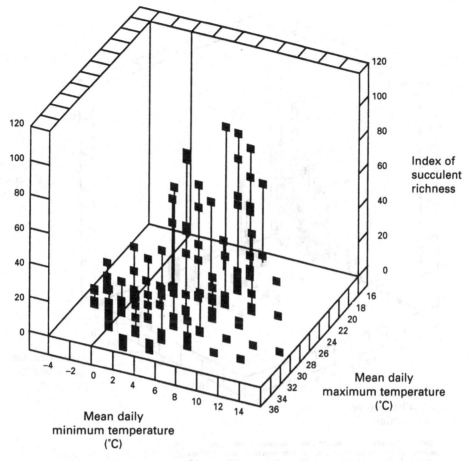

Figure 16.2 **Index of succulent plant richness (sum of number of genera/subgenera of the Mesembryanthemaceae and species of *Zygophyllum* and *Crassula* per 1° cell; from Jürgens 1991) plotted against mean daily minimum and maximum temperatures for each square degree grid of southern Africa (18° 00′–34° 30′ S and 12° 00′–32° 00′ E).**

plants', which include species of *Conophytum*, *Lithops* and *Gibbaeum*. These plants are unable to use convective cooling or other common mechanisms (see above) for keeping cool. Their leaf temperatures are maintained by the thermal coupling of the plant and soil, so that the potentially large daily temperature fluctuations of the plant are dampened by the heat sink capacity of the soil (Eller & Grobbelaar 1986; Turner & Picker 1993). With the chlorenchyma (photosynthetic tissue) in the coolest regions deep within the plant and with variable transmission of light through leaf windows, plants can maintain physiological functioning with leaf temperatures regularly reaching 40–50 °C. Turner & Picker (1993) also suggest that in cool and less sunny environments one would expect the clarity of the leaf window to be greater than in plants from hot, high-radiation environments. Circumstantial evidence supports this contention, since the species of *Lithops* identified by Cole (1988) as

having well-developed clear windows are found in regions with lower solar radiation loads (450–500 cal cm^{-2} day^{-1}) and the species with opaque or reduced windows are more abundant in the high-radiation areas (>500 cal cm^{-2} day^{-1}).

Studies on the low temperature limits on succulent distribution in the North American semi-arid regions suggest that succulents are highly sensitive to freezing temperatures because of their high cellular water content, or their inability to undergo cold hardening, or both (Nobel 1982). Succulent richness in southern Africa appears to be low in areas where winter mean temperatures are below 0 °C (Fig. 16.2; see also Werger & Ellis 1981). In a detailed study, Van Coller & Stock (1994) found that the lowest critical killing temperature of all subspecies of a widely distributed southern African member of the Crassulaceae (*Cotyledon orbiculata*) is about −9.8 °C. This is very similar to the physiological inhi-

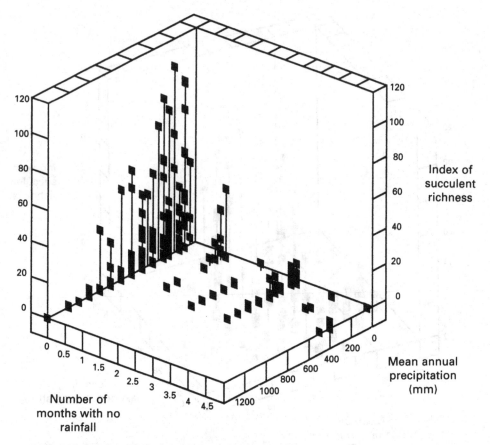

Figure 16.3 **Index of succulent plant richness (sum of number of genera/subgenera of the Mesembryanthemaceae and species of *Zygophyllum* and *Crassula* per degree grid: from Jürgens 1991) plotted against annual rainfall and number of months per annum with no measurable precipitation for each square degree grid of southern Africa (18° 00′–34° 30′ S and 12° 00′–32° 00′ E).**

bition range found for succulents such as Sonoran cacti and Macronesian Sempervivoideae (Nobel 1982; Lösch & Kappen 1981, respectively). *Cotyledon orbiculata* also showed a limited ability to acclimatize physiologically (harden) in relation to exposure to low temperatures, but this ability was insufficient to explain the geographical occurrence of *C. orbiculata* in areas with 60–90 frost nights per year. From the above studies it is evident that there is a paucity of information concerning the relationships between temperature extremes and succulent richness, distribution and abundance. Without further research we are unable to assess the impacts of past or future climate change on the survival and distribution of the succulent flora of southern Africa.

16.3.2 Succulents and rainfall
Generally a high abundance of succulents is associated with areas of low but predictable rainfall (Ellenberg 1981; Hoffman & Cowling 1987; Burgess & Shmida 1988;

Von Willert *et al.* 1992; Cowling *et al.* 1994a). Succulent richness in southern Africa appears to be greatest in regions with an annual rainfall of less than 500 mm and in regions where there are two months of the year or less which have no measurable precipitation (Fig. 16.3). Therefore, succulents are best represented in semi-arid regions where the drought conditions are regularly interrupted by climatic conditions that favour the uptake of small quantities of water (Von Willert *et al.* 1992).

In one of the first attempts to explain the occurrence of stem-succulents in the arid regions of the world, Ellenberg (1981) proposed that the presence of this growth-form coincided with an index of rainfall irregularity of less than five (Hellmann quotient – the annual rainfall divided by the minimum over a sufficiently long time series) and an annual rainfall of less than 500 mm. These measures explain the lack of succulents in Australia, where the rainfall of the various climate stations was either too abundant or too erratic (Westoby 1980). The strength of the relationship described by Ellenberg (1981) was questioned by Von Willert *et al.* (1992), since in many areas dominated by succulents the interruption of the drought period is not only accomplished through rainfall but might also occur as a result of moisture precipitated from fogs, dew-fall or high nocturnal air humidities (Chap. 9, this volume). The need for regular precipitation was emphasized by Von Willert *et al.* (1985) who noted the massive mortality of succulents during a severe drought (1979–1980) in the Richtersveld.

In a community-level analysis of growth-form diversity and species richness in semi-arid and arid southern Africa, Cowling *et al.* (1994a) found that seasonality of rainfall, particularly the occurrence of winter rainfall, was the strongest predictor of the total number of succulents per site (see also Werger 1978; Van Jaarsveld 1987, Hilton-Taylor & Le Roux 1989, Hartmann 1991). Regression models also showed a relationship between succulent richness and rainfall evenness that was sufficiently strong to account for the fact that certain centres of succulent diversity in southern Africa are located in areas receiving all-year-round rainfall (Cowling *et al.* 1994a). Growth-form diversity is weakly correlated with species richness in the succulent karoo, suggesting that among the leaf-succulent shrubs there are many species that could be considered to be functionally equivalent (Cowling *et al.* 1994a).

Despite some broad-scale correlative studies, very little is known about the determinants of the distribution of different succulent growth-forms in the southern African flora. In particular, almost no experimental research has been carried out on the physiological tolerances of this enormously diverse group.

16.4 Graminoid–woody plant interactions

Most southern Africa biomes are characterized by the co-existence of monocotyledonous graminoids and dicotyledonous trees and shrubs. In this section we examine the factors that determine the variation in the balance between graminoids and woody plants across and within biomes. In support of arguments developed we provide evidence to illustrate that (1) graminoids and woody plants respond differently to gradients of climatic and soil variables across the sub-continent; and (2) the balance between graminoids and woody plants and ultimately overall ecosystem functioning within biomes may be determined by relatively few, easily quantifiable environmental factors (soil fertility, soil texture, rainfall seasonality and temperature). Although factors such as

herbivory (Chap. 17, this volume), fire (Chap. 18, this volume), pollination syndromes and seed dispersal mechanisms play important roles in determining the structural appearance of some ecosystems, these influences are probably of secondary importance at the biome scale and we will not discuss them in detail in this chapter (see individual biome chapters).

16.4.1 Fynbos graminoid–woody plant interactions

The structural composition of plant communities generally varies spatially along distinct environmental gradients within most southern African biomes. In fynbos, however, growth-form diversity is low (Cowling *et al.* 1994b) and even along well-defined environmental gradients there is little change in the overall growth-form mix (Linder *et al.* 1993; Chap. 6, this volume). However, the relative importance of major fynbos growth-forms (proteoid shrubs, ericoid shrubs and restioids; Chap. 6, this volume) does show predictive change along environmental gradients; this enabled Campbell (1985b) to establish a biome-wide structural classification. Different elevations and aspects have characteristic structural compositions. Shallow, rocky, north-facing slopes that experience high solar radiation loads and high temperatures are dominated by low restioids (Bond 1981; Campbell 1983, 1985a, 1986; Euston-Brown 1995). Sites with other aspects and elevations, less hostile in terms of the thermal environment and water availability, are dominated by various combinations of ericoid and proteoid shrubs, with restioids being relatively less important (see Chap. 6, this volume, for details).

Restioids have anatomical, morphological and functional attributes that ensure survival in a range of habitats including dry mountain slopes, seasonally waterlogged soils and thermally stressed environments. In physiological terms, plant water deficits can be induced both by a deficit of water and by the breakdown of root function in anaerobic waterlogged soils. Therefore, many of the morphological features of these plants facilitate survival on shallow, dry soils. These include lignified photosynthetic culms, non-photosynthetic scale leaves, buried creeping rhizomes (Stock *et al.* 1992), the development of cappillaroid roots that increase absorptive area (Lamont 1982), thick cuticles and deeply sunken stomata on the culms (Moll & Sommerville 1985), and a thin cuticular layer covering the protective cells of the substomatal cavity (Botha, Van der Schyff & Van Tonder 1972). These attributes, as well as the development of aerenchyma tissue in the roots (Hardcastle & Schutte 1983), also contribute to survival under waterlogged conditions. Physiologically, the restioids show some interesting features such as an immediate elevation in stomatal conductance when water becomes available which maximizes photosynthetic carbon gain (Van der Heyden & Lewis 1990). Some species are also able to acclimatize photosynthetically to high temperatures (shift in photosynthetic temperature optimum; Van der Heyden & Lewis 1990) and the clonal structure of the plant allows for efficient nutrient use (Stock, Sommerville & Lewis 1987). These physiological properties all favour restioids in low-nutrient soils which experience regular water deficits because of either droughts or waterlogging.

In contrast, proteoid shrubs have their highest cover on pole-facing slopes, usually at lower elevations where the soils are deepest (Campbell 1986). Here, the proteoids can establish deep roots, which provide access to water throughout the year. This has been substantiated by physiological studies that show that these plants experience little summer water stress and can maintain high net photosynthetic rates throughout the year (Van der Heyden & Lewis 1989; Richardson & Kruger 1990; Richards, Stock & Cowling 1995). In fact, the seasonal pattern of water relations of proteoid shrubs and adjacent Afromontane forest trees are so similar that water availability is unlikely to be the determinant of fynbos/forest boundaries in the region (Manders 1990; Richardson & Kruger 1990). Ericoid shrubs are conservative water users and are distinctly different from the more opportunistic restioids (Van der Heyden & Lewis 1989).

The moisture hypothesis is inadequate to account for both the unique composition of the graminoid layer (Restionaceae as opposed to Poaceae, which occur in all other biomes) and the balance between graminoids and woody plants in fynbos. C_4 grasses have never been abundant in the region (Stock, Bond & Le Roux 1993); this is not surprising given that the climatic constraints associated with the low temperatures in the winter growing season select against this photosynthetic pathway (Vogel, Fuls & Ellis 1978). In the eastern part of the biome, with year-round rainfall and a growth season coincident with warmer temperatures, C_4 grasses are far more abundant (Cowling 1983; Euston-Brown 1995). Despite the widespread presence of C_3 grasses in fynbos, including many endemics (Vogel *et al.* 1978; Bond & Goldblatt 1984; Gibbs Russell 1987; Linder & Ellis 1990), they are rarely locally abundant and appear to be effectively suppressed by the Restionaceae. This may be the result of the low-nutrient soils, which favour the long-lived evergreen restioids. Grasses are only important on the more nutrient-rich soils derived from granite or shale, or in environments where the soils are temporarily enriched after fires (Hoffman, Moll & Boucher 1987). For example, the cover and biomass of grasses increased for the duration of enhanced nutrient availability associated with nitrogen fertilization of a nutrient-poor fynbos site (Witkowski & Mitchell 1987; Witkowski 1989c).

16.4.2 Nama-karoo graminoid–woody plant interactions

Nama-karoo vegetation is essentially a dwarf shrubland that experiences pulses of graminoid (C_4 grasses) abundance in the interspaces (Hoffman & Cowling 1990; see also Chap. 8, this volume). This growth-form mix has been the centre of much debate, since it is commonly believed that the grassy component was more conspicuous before the introduction of wide-scale commercial grazing by small livestock (Chap. 21, this volume). O'Connor (1991) has suggested that extensive overgrazing has diminished the potential of grasses to respond to favourable events by reducing the size of seed banks. Evidence from photographic records (Hoffman & Cowling 1990; Chap. 21, this volume) and soil isotopes (Bond, Stock & Hoffman 1994) indicate that pulses of grass abundance result when climate conditions are suitable, irrespective of grazing regime (see also Chap. 21, this volume). The amount of summer rain, which promotes grass growth, shows high inter-annual variation in the Nama-karoo, depending upon climatic circulation patterns (Tyson 1986). Grasses become conspicuous during cycles of high summer rainfall, whereas spring and autumn rains are thought to favour shrub recruitment and growth (Roux 1980; Novellie & Strydom 1987; Hoffman, Barr & Cowling 1990). This is supported by the findings of Bond *et al.* (1994) who have shown that the geographical boundary in the karoo for a 50:50 shrub : grass cover mix has remained remarkably constant through historical time (see also Chap. 21, this volume). They suggest that the relatively grassless shrublands of the arid southern and western regions of the Nama-karoo have remained stable for a considerable period (see also Chap. 8, this volume). It is only in the graminoid–shrub transition zone of the central and eastern Nama-karoo that land-use practices and not climate have reduced the amount of grass cover.

16.4.3 Savanna graminoid–woody plant interactions

Savannas are structurally simple ecosystems with seasonally (winter) water-stressed communities that are dominated by a continuous layer of graminoids with a discontinuous woody tree or shrub overstorey (Walker & Noy-Meir 1982). The mix of growth forms varies, however, across the biome, and savannas may even exist in several, apparently stable, states in uniform habitats (Scholes & Walker 1993; Chap. 11, this volume). Factors affecting the dynamics and composition of states may vary in time (season) or in space along environmental gradients (Skarpe 1992). The balance between trees, shrubs and grasses is maintained by fires (Chap. 18, this volume), herbivores (Chap. 17, this volume), physical stress, or by competition for limited resources (Smith & Goodman 1986; Stuart-Hill & Tainton 1989; Chap. 11, this volume). Belsky & Canham (1994) suggest that the dynamics of savannas can be understood in terms of patch dynamic models developed primarily for forest ecosystems. Instead of structural changes being initiated by gaps resulting from tree death or tree-fall, changes are induced by tree seedling establishment. Trees usually establish in savannas only upon disturbance of the graminoid layer (Archer 1990) since, in undisturbed grass swards, tree seed germination and seedling success is constrained by abiotic and biotic controls. Competition for light, water and nutrients by the grass sward is high, and regular burning and grazing of the grass results in high tree seedling mortalities. Once trees do establish, they alter local micro-climates and patterns of nutrient cycling (Walker 1985; Skarpe 1992) which can facilitate further invasion by trees and often lead to the development of closed woodlands (Chap. 11, this volume).

Zones of altered physical, chemical and biotic properties associated with savanna trees have been extensively researched, as it was believed that this patchiness in soil properties could explain the coexistence of the graminoid and woody layers (Huntley & Walker 1982). For example, light conditions below isolated *Acacia tortilis* trees can be reduced as much as 45% (Belsky *et al.* 1989), with the shading effect stretching some 40 m east and west of the tree. Soil temperatures are also reduced (by 8–11 °C in the mid-afternoon) and soil moisture conditions of tree-dominated patches have been reported to be higher than soils of adjacent communities (Kennard & Walker 1973; Joffre & Rambal 1988). Soils beneath savanna trees have consistently higher concentrations of organic matter, phosphorus, total and available nitrogen, potassium and other cations, and sulphur than adjacent soils supporting grasses (Kellman 1979; Bate & Gunton 1982; Belsky *et al.* 1989). These soils also have higher water-infiltration rates, higher microbial biomass, greater total cation-exchange capacities and lower bulk densities (Kennard & Walker 1973; Dunham 1991; Campbell *et al.* 1993; Smit & Swart 1994). These differences affect the functioning of the graminoid layer: some studies have shown that grass productivity is enhanced in the tree-dominated patches (Weltzin & Coughenour 1990), but in other cases lower productivity was observed (Monk & Gabrielson 1985; Stuart-Hill & Tainton 1989). Conflicting results such as these suggest that the balance between the graminoid and woody plant layers is dynamic and that savanna landscapes consist of a matrix of patches of different plant structural combinations (Whittaker, Morris & Goodman 1984). The frequency of such patches is probably the key determinant to understanding the species diversity, ecosystem

stability, productivity and nutrient fluxes of savanna ecosystems rather than any single equilibrium model of graminoid–woody plant interactions (Walter 1971). Scholes & Walker (1993) also present arguments against the 'Walter Hypothesis' which was the paradigm underlying much of the southern African savanna research initiative (Walker *et al.* 1981; Walker & Noy-Meir 1982; see Chap. 11, this volume). This model suggests that water is the major limiting factor in semi-arid savannas and that grasses are competitively superior to trees in competing for water in the surface layers of the soil. Trees, on the other hand, would preferentially use subsoil water. Experimental research in a moist savanna in South Africa did not support the model (Knoop & Walker 1985; Scholes & Walker 1993; Chap. 11, this volume) and disequilibrium models, such as the patch dynamic model described above (Belsky & Canham 1994) have been invoked to describe shifts in the inherently unstable mix of graminoids and woody plant species in savannas (Belsky 1990; Scholes & Walker 1993).

16.4.4 Grassland graminoid–woody plant interactions

Although there are large tracts of grassland in southern Africa, much of this biome can be regarded as the extreme state of a savanna ecosystem where woody species have been suppressed by regular human-induced disturbances (Frost *et al.* 1985; Chaps 10, 11 and 18, this volume). Many grasslands and sparse savannas were occupied by Iron-Age people for several thousand years (Inskeep 1978; Blackmore, Mentis & Scholes 1990). The heavy demand for wood used in iron-smelting and for domestic purposes, coupled with the firing of the veld to provide improved grazing for domestic livestock, may have been important in maintaining the graminoid–woody plant balance of grasslands and some open parkland landscapes, independently of climate (Chap. 21, this volume). The recent conversion of many of these areas, in the historical past, to commercial farming has been accompanied by a rapid increase in the woody component, probably because of active fire avoidance and extremely high levels of grass consumption (see sections on bush encroachment in Chaps 10, 11, 18 and 21, this volume).

Environmentally determined grasslands occur in the arid, summer-rainfall regions of southern Africa (Chaps 8 and 9, this volume). Skarpe (1992) has suggested that these pure grasslands may be transitory in that they exist in the years between the death of one cohort of trees and the establishment of the next, which might only occur after a few years of above-average rainfall. Other environmentally determined grasslands occur where poorly drained or shallow soils exclude woody

Table 16.3 **Root adaptations for improving nutrient acquisition by plants in southern Africa**

Type	Taxa involved	Function	Biome	Environmental determinant
Mycorrhizas				
Arbuscular mycorrhizas	Most angiosperms	Nutrient uptake, particularly of phosphorus; soil binding	All terrestrial ecosystems	Wide ranging, but absent under waterlogged conditions
Ericoid mycorrhizas	Ericaceae	Nutrient release from organic matter and nutrient uptake; detoxification of soil substratum	Fynbos or grassland at high altitude	Extremely nutrient-poor soils and waterlogged conditions
Ecto-mycorrhizas	Mainly angiosperm trees	Nutrient release and uptake	Absent except for occasional incidence in northern savannas	Soils with surface organic matter accumulation
Orchid mycorrhizas	Orchidaceae	Nutrient release and uptake	All terrestrial biomes (except deserts?)	Broad tolerances
Cluster roots	Proteaceae, Restionaceae, Fabaceae	Nutrient release and uptake	Fynbos, grassland and savanna	Nutrient-poor soils
Sand rhizosheaths	Poaceae, Restionaceae	Nutrient release and uptake	All terrestrial biomes	Sandy soils
Nitrogen-fixing nodules	Members of Papilionoideae and Mimosoideae (Fabaceae)	Atmospheric nitrogen fixation	All terrestrial biomes	Soils with low nitrogen content or high competition for nitrogen – often linked to early succession

species (Bezuidenhout, Bredenkamp & Theron 1994; Chap. 10, this volume). If some woody species establish as a result of drought- or human-induced impacts on the water table, the increased evapotranspiration is often sufficient to maintain a low water table and thus facilitate the establishment of other trees in the area (Bezuidenhout *et al.* 1994). Frost is another major climatic factor that could play a role in restricting the distribution of trees within biomes, particularly along altitudinal gradients (Chap. 10, this volume).

This analysis of the interaction of the graminoid and woody elements of the biomes of southern Africa has stressed the underlying role of climate and soil factors in determining the structural and functional characteristics of the region. Human-induced changes have been imposed upon this climatic and edaphic matrix to such an extent that some of the vegetation–climate relationships are almost totally obscured. The challenge for ecologists attempting to predict how southern African ecosystems will respond to climate change is going to be one of separating human-induced from abiotically driven changes.

16.5 **Below-ground acquisition of resources**

On a global scale, the distribution of different types of mycorrhizas is a result of interactions between soil and climate (Read 1991). Different environments in southern Africa are associated with distinctive groups of nutrient-capture mechanisms (Table 16.3). However, the possession of specialized structures for the acquisition from the soil of resources, such as water and nutrients, is seldom viewed at any level higher than that of the indi-

vidual plant. Thus, the significance of the below-ground component in ecosystem functioning is seldom assessed. This section addresses the role of various structural and functional attributes of roots of certain groups of plants, and assesses their influence upon ecosystem functioning.

16.5.1 **Mycorrhizal relationships**

The most common nutrient-acquiring modification found in plants is the mutualistic relationship between plant roots and various groups of fungi known generally as mycorrhizas. In common with other root modifications, mycorrhizas act to reduce interspecific competition by capturing resources not available to other plants, and by increasing the competitive advantages of the mutualistic plant species. In southern Africa, as in most terrestrial ecosystems (Read 1991), arbuscular mycorrhizas dominate (Allsopp & Stock 1993a; Dames 1991; Högberg & Piearce 1986). Besides their role in nutrient acquisition, a network exists in the soil of arbuscular mycorrhizal hyphae, which may often provide a linkage between plants in a community. At the community level, a hyphal network in conjunction with a continuous lateral root system can function to maintain tight nutrient cycling patterns (Pankow, Boller & Wienken 1991), and to improve soil structure (Miller & Jastrow 1990). Within biomes and landscapes, arbuscular mycorrhizal fungi are often patchily distributed; e.g. frequently burnt lowland fynbos has a much greater variability in infection levels among potentially mycorrhizal plants than seldom burnt renosterveld (Allsopp & Stock 1994). Such patchiness appears to be a consequence of disturbance and environmental conditions which reduce the abundance of arbuscular mycorrhizal propagules in the soil. Since the fungi are obligate mutualists, and cannot exist other than as dormant spores in the absence of their plant mutualist, factors that cause the death of plants may reduce mycorrhizal infectivity of the soil. Patchiness in arid regions may be exacerbated by the lack of a permanent mycorrhizal population. For example, in Botswana, where grass growth is interrupted by the long dry season, infection of the new season's grass occurs only from spores (Veenendal *et al.* 1992). Spore numbers decrease if drought or disturbance prevent plant growth as illustrated by the lower mycorrhizal propagule numbers on bare dunes compared to vegetated dunes in the southern Kalahari (N. Allsopp, unpubl. data). Such reduction in mycorrhizal infectivity of the soil can seriously restrict the establishment of seedlings (Allsopp & Stock 1995), with repercussions for future community composition.

Disruption or disorganization of the hyphal network is also associated with an increase in nutrient leakiness of the ecosystem. In moist environments of southern Africa, where the growing season is relatively long, arbuscular mycorrhizal hyphae may survive as a permanent hyphal network. This network could act as a source of infection for recruiting seedlings and to improve nutrient acquisition. In ecosystems with very different growth-form mixes, permanent and transient hyphal networks may exist simultaneously if the separate components react differently to fire, grazing or seasonal drought, for example in savannas where the graminoid and woody components respond differently to environmental stresses (Chap. 11, this volume). Preliminary investigations of a subtropical thicket/grassland mosaic in the Eastern Cape support this model, with arbuscular mycorrhizal spores being more numerous in the soils under bushclumps than in soils between clumps (N. Allsopp, unpubl. data).

Although ectomycorrhizas are common among northern hemisphere tree species (Read 1991), they are rare in African trees (Redhead 1968; Alexander & Högberg 1986; Högberg & Piearce 1986). They have not been reported as mutualists on indigenous trees south of Zambia (Högberg & Piearce 1986). The absence of ectomycorrhizas in southern Africa is probably due to ecological reasons, rather than taxonomic considerations (Högberg 1992). Ectomycorrhizas are favoured in soils with high levels of organic matter (Read 1991), which accumulates when environmental conditions (low temperatures or waterlogging) constrain decomposition. Organic-rich soils are scarce in southern Africa and the ability of ectomycorrhizal fungi to enhance decomposition of organic-rich substrata by the exudation of hydrolytic substances (Harley & Smith 1983) does not appear to be a significant requirement for local tree species. Rather, the patterns

of nutrient release in the mineral, low-organic soils of the region are temporally and spatially patchy (Ménaut *et al.* 1985); the finer morphology of the arbuscular mycorrhizal hyphal network provides a greater nutrient uptake benefit to the plant mutualist than ectomycorrhizas would under these conditions.

Unlike the ectomycorrhizas, the ericoid mycorrhizas are well represented in southern Africa, particularly among the species-rich Ericaceae (Allsopp & Stock 1993a). Ericoid mycorrhizas in fynbos appear to have a unique functional role. In most ecosystems where this symbiont is found it appears to enable host plants to tolerate high levels of organic acids found in waterlogged humic soils. The compounds often preclude colonization of these soils by other plant species (Leake, Shaw & Read 1989). In fynbos, however, the Ericaceae commonly grow in species-rich upland communities on leached sandy soils, which are possibly the most nutrient-poor conditions in the fynbos biome (Campbell 1985a,b; Chap. 6, this volume). The ericoid mycorrhizal symbiosis may provide an advantage under these conditions, since, unlike the arbuscular mycorrhizal symbiosis, the fungus can excrete substantial quantities of phosphatases (Straker & Mitchell 1986) and proteases (Stribley & Read 1980), which can accelerate the release of nutrients and mediate their subsequent uptake into the plant (Straker & Mitchell 1986). Rates of organic matter decomposition and nutrient mineralization in fynbos are extremely low, because of the poor litter quality (Mitchell *et al.* 1986). Root systems of the Ericaceae are much shallower and finer compared to other shrubby plants of similar above-ground morphology (N. Allsopp pers. obs.), placing the ericoid mycorrhizas in the zone of litter accumulation where the mycorrhizas facilitate decomposition. Ericoid mycorrhizas can, therefore, release the host plant from competition with other species for the very low levels of available nutrients in the soil. Furthermore, unlike the more substantial structure of ectomycorrhizas, the finely structured ericoid mycorrhizas are less of a drain on the carbon resources of the plant.

Mycorrhizas mediate plant/environment interactions at many stages of the life-cycle of the plant which can result in interesting evolutionary trade-offs. Most shrubby fynbos species (Proteaceae, Fabaceae, Rhamnaceae) provision their seeds so as to facilitate initial seedling establishment (Stock, Pate & Delfs 1990) independently of, or with restricted, mycorrhizal symbiont development (Allsopp & Stock 1995). The Ericaceae, on the other hand, have extremely small seeds (Allsopp & Stock 1992a) and are probably entirely dependent on ericoid mycorrhizas for nutrient acquisition. Typically, weedy, short-lived species, which proliferate after disturbance, are non-mycorrhizal (Trappe 1987), because, it is suggested, the carbon costs of the mutualism could deleteriously affect growth rates and subsequent competitive ability. Selection against the mycorrhizal state, as seen in the weedy species, also appears to have occurred in some dominant woody plant (proteoids) and graminoid groups (restioids) in fynbos and this loss of the mutualism is usually accompanied by morphological root alterations (Allsopp & Stock 1993a; Section 16.5.2). The Proteaceae, with abundant proteoid roots (cluster roots), lack mycorrhizal associations, whereas members of the Fabaceae often have cluster roots and mycorrhizal symbionts simultaneously (Allsopp & Stock 1993b). This pattern is not universal, since some Proteaceae outside fynbos have been reported to be ectomycorrhizal (*Faurea* in Zambia; Högberg & Piearce 1986) or have the capacity to form arbuscular mycorrhizas (Australian genera; Bellgard 1991). Two of seven Australian members of the Restionaceae were found to have arbuscular mycorrhizas for very short periods while environmental conditions were suitable; functionally important arbuscules appeared to be rare, however (Meney *et al.* 1993). Therefore, despite the presence of these mutualisms in non-fynbos taxa, they appear to be of limited functional significance which raises the possibility of phylogenetic constraints on the incidence of mycorrhizal association in these characteristic fynbos lineages. The Fabaceae (in particular *Aspalathus* spp.), which form mycorrhizas and cluster roots, are less dependent on the mutualism for seedling development, and phosphorus acquisition, than other mycorrhizal fynbos plant species (Allsopp & Stock 1992b, 1994).

The limited range of mycorrhizal types and the diversity of other modes of nutrition in fynbos (plant parasites, root modifications, etc.) suggests that there is no single solution to the problem of nutrient acquisition in the region.

16.5.2 Cluster roots and root sheaths

Cluster roots are dense aggregations of short, hairy roots along lateral roots (Purnell 1960) which capture nutrients as a result of their position in nutrient-rich pockets and their large surface area (Lamont 1982). Exudates from cluster roots may also increase the availability of some nutrients by secreting chelating agents and by increasing the hydrogen ion concentration in their rhizosphere (Grierson & Attiwell 1989; Marschner 1991). Cluster roots appear to be a response to very low nutrient availability, and the taxa in which they are prevalent (Proteaceae and some Fabaceae) are important in young and mature vegetation in these environments. Cluster roots have also been reported on Restionaceae (Lamont

1982), but they are not common in this group (N. Allsopp pers. obs.).

Despite the importance of grasses, especially C_4 grasses, in many southern African biomes little work has been done on their root sheaths. These form when water uptake is limited by the late development of open metaxylem vessels, but root exudation and microbial activity in the rhizosphere is high (Gochnauer, McCully & Labbe 1989). Similar sheaths develop on members of the Restionaceae in fynbos (N. Allsopp pers. obs.) and Australian kwongan (Dodd *et al.* 1984). A protective function against water loss has been proposed (Leistner 1967), but has still to be tested. Alternatively, they may play a role in plant nutrition (Bailey 1994). Although the presence of a grass sheath is genetically fixed, its thickness appears to be a phenotypic response to environmental factors induced by the sandiness of the soils, with sandier soils inducing thicker sheaths which, in turn, are responsible for greater phosphorus uptake (Bailey 1994). The role of sheaths in nitrogen nutrition, including associative N_2-fixation, has yet to be evaluated, although rhizosheaths of xeric grasses in North America have shown acetylene reduction, indicating nitrogen fixation (Wullstein, Bruening & Bollen 1979). Fixation of atmospheric N_2 in root sheaths may be important for replenishing nitrogen lost through fire or volatilization in several southern African ecosystems.

Research in southern Africa has shown that a wide diversity of nutrient-acquiring modifications are found, particularly in the fynbos biome. This is not surprising, considering the selective pressures that plants in this low-nutrient region have experienced. The implications of below-ground activity in determining the structure and functioning of southern African ecosystems are now being realized. More research is required, to obtain a better understanding of the ecological implications of different modes and abundances of nutrient-acquiring mechanisms on ecosystem functioning. Without a consideration of whole systems at a hierarchy of scales we will be unable to predict the effects of disturbances from local patches to landscapes, nor will we be able to predict changes in response to global climate change.

16.6 Conclusions

This chapter has highlighted some interesting, and sometimes unique, features of plant structure and functioning in the biomes of southern Africa. Detailed ecophysiological investigations of plant adaptive features are providing new insights into the functional significance of some of these adaptations. The possible role of soil-nutrient status, for instance, in explaining the incidence of sclerophyll leaves in fynbos contrasts with the climatic determinism implicit in hypotheses derived from correlative approaches. Without experimental testing of hypotheses many of the observed patterns of sclerophyll and/or succulent leaf distributions will remain unexplained.

The experimental approach has also been used to refine hypotheses accounting for the high incidence and geographical distribution of succulents in southern Africa. The understanding of the low but regular water requirement of succulents, the temperature tolerances and the photosynthetic mechanism underlying CAM plants has developed from ecophysiological experiments. The synthesis of these studies in this chapter has provided the basis for further studies to enhance the understanding of succulent performance and persistence in the biomes of southern Africa.

The analysis of graminoid interactions with woody plants undertaken in this chapter has shown that current explanations for the patterns of growth-form mixes found in the biomes of southern Africa are still largely descriptive. Water and soil factors are often invoked to explain the patterns, but little experimental evidence is available to support this in most of the biomes. The role of low temperatures in limiting the distribution of trees has, for instance, never been tested experimentally. There is, however, good experimental evidence for evaluating the relative roles of climate- and human-induced changes affecting the dynamics of the interface between C_4 grasses and C_3 shrubs in the region.

Without an experimental approach, the significance of nutrients and water (and their interaction) as selective forces in the evolution of plant structure and functioning in southern Africa would be incomplete. Knowledge concerning the linkages between the diversity of resource acquisition mechanisms, plant species richness and ecosystem functioning is still vague and poses a major challenge to southern African biologists. Without this knowledge, the accuracy of predictions of plant responses to human-induced climate change will remain poor.

5.7   **References**

ocks, J.P.H. (1953). Veld Types of South Africa. *Memoirs of the Botanical Society of Southern Africa*, **28**, 1–192.

amson, R.S. (1938). *The Vegetation of South Africa*. London: British Empire Vegetation Committee.

exander, I.J. & Högberg, P. (1986). Ectomycorrhizas of tropical angiospermous trees. *New Phytologist*, **102**, 541–9.

lsopp, N. & Stock, W.D. (1992a). Mycorrhizas, seed size and seedling establishment in a low nutrient environment. In *Mycorrhizas in Ecosystems*, ed. D.J. Read, D.H. Lewis, A.H. Fitter & I.J. Alexander, pp. 59–64. Wallingford: CAB International.

lsopp, N. & Stock, W.D. (1992b). Density dependent interactions between VA mycorrhizal fungi and even-aged seedlings of two perennial Fabaceae species. *Oecologia*, **91**, 281–7.

lsopp, N. & Stock, W.D. (1993a). Mycorrhizal status of plants growing in the Cape Floristic Region, South Africa. *Bothalia*, **23**, 91–104.

Allsopp, N. & Stock, W.D. (1993b). Mycorrhizas and seedling growth of slow-growing sclerophylls from nutrient-poor environments. *Acta Oecologia*, **14**, 577–87.

Allsopp, N. & Stock, W.D. (1994). VA mycorrhizal infection in relation to edaphic characteristics and disturbance regime in three lowland plant communities in the south-western Cape, South Africa. *Journal of Ecology*, **82**, 271–9.

Allsopp, N. & Stock, W.D. (1995). Relationships between seed reserves, seedling growth and mycorrhizal responses in fourteen related shrubs (Rosidae) from a low nutrient environment. *Functional Ecology*, **9**, 248–54.

Archer, S. (1990). Development and stability of grass/woody mosaics in a subtropical savanna parkland, Texas, U.S.A. *Journal of Biogeography*, **17**, 453–62.

Bailey, C.L. (1994). *The Distribution and Functioning of Rhizosheaths among South African Grass Species*. MSc Thesis. Johannesburg: University of the Witwatersrand.

Bate, G. C. & Gunton, C. (1982). Nitrogen in the Burkea savanna. In *Ecology of Tropical Savannas*, ed. B.J. Huntley & B.H. Walker, pp. 498–513. Berlin: Springer-Verlag.

Beadle, N.C.W. (1966). Soil phosphate and its role in molding segments of the Australian flora and vegetation with special reference to xeromorphy and sclerophylly. *Ecology*, **47**, 992–1007.

Bellgard, S.E. (1991). Mycorrhizal associations of plant species in Hawkesbury sandstone vegetation. *Australian Journal of Botany*, **39**, 357–64.

Belsky, A.J. (1990). Tree/grass ratios in East African savannas: a comparison of existing models. *Journal of Biogeography*, **17**, 483–9.

Belsky, A.J., Amundson, R.G., Duxbury, J.M., Riha, S.J., Ali, A.R. & Mwonga, S.M. (1989). The effects of trees on their physical, chemical, and biological environments in a semi-arid savanna in Kenya. *Journal of Applied Ecology*, **26**, 1005–24.

Belsky, A.J. & Canham, C.D. (1994). Forest gaps and isolated savanna trees: an application of patch dynamics in two ecosystems. *BioScience*, **44**, 77–84.

Bews, J.W. (1925). *Plant Forms and Their Evolution in South Africa*. London: Longmans.

Bezuidenhout, H., Bredenkamp, G.J. & Theron, G.K. (1994). A classification of the vegetation of the western Transvaal dolomite and chert grassland, South Africa. *South African Journal of Botany*, **60**, 152–60.

Blackmore, A.C. (1992). *The Variation of Ecophysiological Traits of Savanna Plants in Relation to Indices of Plant Available Moisture and Nutrients*. MSc Thesis. Johannesburg: University of the Witwatersrand.

Blackmore, A.C., Mentis, M.T. & Scholes, R.J. (1990). The origin and extent of nutrient-enriched patches within a nutrient-poor savanna in South Africa. *Journal of Biogeography*, **17**, 463–70.

Bond, P. & Goldblatt, P. (1984). Plants of the Cape Flora. A descriptive catalogue. *Journal of South African Botany*, Supp. Vol. 13, 1–455.

Bond, W.J. (1981). *Vegetation Gradients in Southern Cape Mountains*. MSc Thesis. Cape Town: University of Cape Town.

Bond, W.J. & Midgley, J. (1988). Allometry and sexual differences in leaf size. *American Naturalist*, **131**, 901–10.

Bond, W.J., Stock, W.D. & Hoffman, M.T. (1994). Has the karoo spread? A test for desertification using carbon isotopes from soils. *South African Journal of Science*, **90**, 391–7.

Botha, D.J., van der Schyff, H.P. & Van Tonder, E.M.A. (1972). The position, structure and ontogeny of the stomata in the stems of *Elegia vaginulata* Mast. *Tydskrif vir Natuurwetenskappe*, **12**, 193–9.

Burgess, T.L. & Shmida, A. (1988). Succulent growth forms in arid environments. In *Arid Lands Today and Tomorrow*, ed. E.E. Whitehead, C.F. Hutchinson, B.N. Timmermann & R.G. Varady, pp. 383–95. Boulder, Colorado: Westview Press.

Camerik, A.M. & Werger, M.J.A. (1980). Leaf characteristics of the flora of the high plateau of Itatiaia, Brasil. *Biotropica*, **3**, 39–48.

Campbell, B.M. (1983). Montane plant environments in the fynbos biome. *Bothalia*, **14**, 283–98.

Campbell, B.M. (1985a). *Montane Vegetation Structure in the Fynbos Biome. Structural Classification and Adaptive Significance of Structural Characters*. Utrecht: Drukkerij Elinkwijk BV.

Campbell, B.M. (1985b). A classification of the mountain vegetation of the fynbos biome. *Memoirs of the Botanical Survey of South Africa*, **50**, 1–119.

Campbell, B.M. (1986). Montane plant communities of the fynbos biome. *Vegetatio*, **66**, 3–16.

Campbell, B.M., Frost, P., King, J., Mawanza, M. & Mhlanga, L. (1993). The influence of trees on soil fertility and microbial activity, Matobo, Zimbabwe. In *The Ecology and Management of Indigenous Forest in Southern Africa*, ed. G.D. Piearce & D.J. Gumbo, pp. 197–208. Harare: Forest Commission.

Campbell, B.M. & Werger, M.J.A. (1988). Plant form in the mountains of the Cape, South Africa. *Journal of Ecology*, **76**, 637–53.

Cody, M.L. (1989). Growth-form diversity and community structure in desert plants. *Journal of Arid Environments*, **17**, 199–209.

Cole, D.T. (1988). *Lithops: Flowering Stones*. Randburg: Acorn/Russel Friedman Books.

Cowling, R.M. (1983). The occurrence of C_3 and C_4 grasses in fynbos and allied shrublands in the South Eastern Cape, South Africa. *Oecologia*, **58**, 121–7.

Cowling, R.M. & Campbell, B.M. (1983a). A comparison of fynbos and non-fynbos coenoclines in the lower Gamtoos river valley, south eastern Cape, South Africa. *Vegetatio*, **53**, 161–78.

Cowling, R.M. & Campbell, B.M. (1983b). The definition of leaf consistence in the fynbos biome and their distribution along an altitudinal gradient in the south eastern Cape. *Journal of South African Botany*, **49**, 87–101.

Cowling, R.M., Esler, K.J., Midgley, G.F. & Honig, M.A. (1994a). Plant functional diversity, species diversity and climate in arid and semi-arid southern Africa. *Journal of Arid Environments*, **27**, 141–58.

Cowling, R.M., Gibbs Russel, G.E., Hoffman, M.T. & Hilton-Taylor, C. (1989). Patterns of plant species diversity in southern Africa. In *Biotic Diversity in Southern Africa. Concepts and Conservation*, ed. B.J. Huntley, pp. 19–50. Cape Town: Oxford University Press.

Cowling, R.M. & Holmes, P.M. (1992). Flora and Vegetation. In *The Ecology of Fynbos. Nutrients, Fire and Diversity*, ed. R.M. Cowling, pp. 23–61. Cape Town: Oxford University Press.

Cowling, R.M., Mustart, P.J., Laurie, H. & Richards, M.B. (1994b). Species diversity: functional diversity and functional redundancy in fynbos communities. *South African Journal of Science*, **90**, 333–6.

Cowling, R.M. & Witkowski, E.T.F. (1994). Convergence and non-convergence of plant traits in climatically and edaphically matched sites in mediterranean Australia and South Africa. *Australian Journal of Ecology*, **19**, 220–32.

Dames, J.F. (1991). *The Distribution of Vesicular Arbuscular Mycorrhizal Fungi in the Savanna Region of Nylsvley Nature Reserve in Relation to Soil Fertility Factors*. MSc Thesis. Johannesburg: University of the Witwatersrand.

Dodd, J., Heddle, E.M., Pate, J.S. & Dixon, K.W. (1984). Rooting patterns of sandplain plants and their functional significance. In *Kwongan. Plant Life of the Sandplain*, ed. J.S. Pate & J.S. Beard, pp. 146–77. Nedlands: University of Western Australia Press.

Dunham, K.M. (1991). Comparative effects of *Acacia albida* and *Kigelia africana* trees on soil characteristics in Zambezi riverine woodlands. *Journal of Tropical Ecology*, **7**, 215–20.

Ellenberg, H. (1981). Reasons for stem succulents being present or absent in the arid regions of the world. *Flora*, **171**, 114–69.

...er, B.M. & Grobbelaar, N. (1986). Diurnal temperature variation in and around *Lithops lesliei* plant growing in its natural habitat on a clear day. *South African Journal of Botany*, 52, 467-71.

...ston-Brown, D. (1995). *Environmental and Dynamic Determinants of Vegetation Distribution in the Kouga and Baviaanskloof Mountains, Eastern Cape*. MSc Thesis. Cape Town: University of Cape Town.

...eld, C. & Mooney, H.A. (1986). The photosynthetic-nitrogen relationship in wild plants. In *On the Economy of Plant Form and Function*, ed. T.J. Givnish, pp. 25-55. Cambridge: Cambridge University Press.

...ost, P.G.H., Menaut, J.C., Walker, B.H., Medina, E., Solbrig, O.T. & Swift, M. (1985). Responses of savannas to stress and disturbance. *International Union of Biological Sciences News Magazine*, Special issue 10, Paris.

...obs Russel, G.E. (1987). Preliminary floristic analysis of the major biomes in southern Africa. *Bothalia*, 17, 213-27.

...ochnauer, M.B., McCully, M.E. & Labbe, H. (1989). Different populations of bacteria associated with sheathed and bare regions of field-grown maize. *Plant & Soil*, 114, 107-20.

...ould, S.J. & Vrba, E.S. (1982). Exaptation - a missing term in the science of form. *Paleobiology*, 8, 4-15.

...ierson, P.F. & Attiwell, P.M. (1989). Chemical characteristics of the proteoid root mat of *Banksia integrifolia* L. *Australian Journal of Botany*, 37, 137-43.

...risebach, A. (1872). *Die Vegetation der Erde*. 2 vols. Leipzig: Engelmann.

...ardcastle, J. & Schutte, K.H. (1983). Aspects of an experimental study on root aerenchyma development and the ecological implications thereof. *Bothalia*, 14, 791-4.

...arley, J.L. & Smith, S.E. (1983). *Mycorrhizal Symbiosis*. London: Academic Press.

...artmann, H. (1991). Mesembryanthema. *Contributions from the Bolus Herbarium*, 13, 75-157.

Hilton-Taylor, C. & Le Roux, A. (1989). Conservation status of the fynbos and karoo biomes. In *Biotic Diversity in Southern Africa. Concepts and Conservation*, ed. B.J. Huntley, pp. 202-23. Cape Town: Oxford University Press.

Hoffman, M.T., Barr, G.D. & Cowling, R.M. (1990). Vegetation dynamics in the semi-arid eastern karoo, South Africa: the effect of seasonal rainfall and competition on grass and shrub basal cover. *South African Journal of Science*, 86, 462-3.

Hoffman, M.T. & Cowling, R.M. (1987). Plant physiognomy, phenology and demography. In *The Karoo Biome: a Preliminary Synthesis. Part 2. Vegetation and History*, ed. R.M. Cowling & P.W. Roux, pp. 1-34, *South African National Scientific Programmes Report No. 124*. Pretoria: CSIR.

Hoffman, M.T. & Cowling, R.M. (1990). Vegetation change in the semi-arid eastern karoo over the last two hundred years: an expanding karoo - fact or fiction? *South African Journal of Science*, 86, 286-94.

Hoffman, M.T., Moll, E.J. & Boucher, C. (1987). Post-fire succession at Pella, a South African lowland fynbos site. *South African Journal of Botany*, 53, 370-4.

Högberg, P. (1992). Root symbioses of trees in African dry tropical forests. *Journal of Vegetation Science*, 3, 393-400.

Högberg, P. & Piearce, G.D. (1986). Mycorrhizas in Zambian trees in relation to host taxonomy, vegetation type and successional patterns. *Journal of Ecology*, 74, 775-85.

Huntley, B.J. & Walker, B.H. (1982). *Ecology of Tropical Savannas*. Ecological Studies 42. Berlin: Springer-Verlag.

Inskeep, R.R. (1978). *The Peopling of Southern Africa*. Cape Town: David Philip.

Jennings, D.H. (1968). Halophytes, succulence and sodium in plants - a unified theory. *New Phytologist*, 67, 899-911.

Joffre, R. & Rambal, S. (1988). Soil water improvement by trees in the rangelands of southern Spain. *Acta Oecologia Oecologia Plantarum*, 9, 405-22.

Jongens-Roberts, S. & Mitchell, D.T. (1986). The distribution of dry mass and phosphorus in an evergreen fynbos shrub species, *Leucospermum parile* (Salisb. ex J Knight) Sweet (Proteaceae), at different stages of development. *New Phytologist*, 103, 669-83.

Jürgens, N. (1991). A new approach to the Namib Region. I. Phytogeographic subdivision. *Vegetatio*, 97, 21-38

Kellman, M. (1979). Soil enrichment by neotropical savanna trees. *Journal of Ecology*, 67, 565-77.

Kennard, D.G. & Walker, B.H. (1973). Relationships between tree canopy cover and *Panicum maximum* in the vicinity of Fort Victoria. *Rhodesian Journal of Agricultural Research*, 11, 145-53.

Kerner, A. (1863). *Die Pflanzenleben der Donauländer*. Vienna.

Knoop, W.T. & Walker, B.H. (1985). Interactions of woody and herbaceous vegetation in a southern Africa savanna. *Journal of Ecology*, 73, 235-54.

Kruger, F.J. (1981). Seasonal growth and flowering rhythms. In *Heathlands and Related Shrublands. B. Analytical Studies*, ed. R.L. Specht, pp. 1-4. Amsterdam: Elsevier.

Lamont, B.B. (1982). Mechanisms for enhancing nutrient uptake in plants, with particular reference to mediterranean South Africa and Western Australia. *Botanical Review*, 48, 597-689.

Leake, J.R., Shaw, C. & Read, D.J. (1989). The role of ericoid mycorrhizas in the ecology of ericaceous plants. *Agriculture, Ecosystems & Environments*, 29, 237-50.

Leistner, O.A. (1967). The plant ecology of the southern Kalahari. *Memoirs of the Botanical Survey of South Africa*, 38, 1-172.

Linder, H.P. & Ellis, R. (1990). Vegetative morphology and interfire survival strategies in the Cape Fynbos grasses. *Bothalia*, 20, 91-103.

Linder, H.P., Vlok, J.H., McDonald, D.J., Oliver, E.G.H., Boucher, C., van Wyk, B.-E. & Schutte, A. (1993). The high altitude flora and vegetation of the Cape Floristic Region, South Africa. *Opera Botanica*, 121, 247-61.

Lösch, R. & Kappen, L. (1981). The cold resistance of Macronesian *Sempervivoideae*. *Oecologia*, 50, 98–102.

Loveless, A.R. (1962). Further evidence to support a nutritional interpretation of sclerophylly. *Annals of Botany*, 26, 551–61.

Manders, P.T. (1990). Fire and other variables as determinants of forest/fynbos boundaries in the Cape Province. *Journal of Vegetation Science*, 1, 483–90.

Manders, P.T. & Smith, R.E. (1992a). Effects of watering regime on growth and competitive ability of nursery-grown Cape fynbos and forest plants. *South African Journal of Botany*, 58, 188–94.

Manders, P.T. & Smith, R.E. (1992b). Effects of artificially established depth to water table gradients and soil type on the growth of Cape fynbos and forest plants. *South African Journal of Botany*, 58, 195–201.

Marschner, H. (1991). Mechanisms of adaptation of plants to acid soils. *Plant & Soil*, 134, 1–20.

Martens, P.O. & Booysen, P. de V. (1968). A tensilemeter for the measurement of the tensile strength of grass leaf blades. *Proceedings of the Grassland Society of Southern Africa*, 3, 51–6.

Meney, K.A., Dixon, K.W., Scheltema, M. & Pate, J.S. (1993). Occurrence of vesicular-arbuscular mycorrhizal fungi in dryland species of Restionaceae and Cyperaceae from south-western Australia. *Australian Journal of Botany*, 41, 733–7.

Ménaut, J.C., Barbault, R., Lauelle, P. & Lepage, M. (1985). African savannas: biological systems of humification and mineralization. In *Ecology and Management of the World's Savannas*, ed. J.C. Tothill & J. Mott, pp. 14–33. Canberra: Australian Academy of Science.

Midgley, J.J., van Wyk, G.R. & Everard, D.A. (1995). Leaf attributes of South African forest species. *African Journal of Ecology*, 33, 160–8.

Miller, P.C. & Stoner, W.A. (1979). Canopy structure and environmental interactions. In *Topics in Plant Population Biology*, ed. O.T. Solbrig, S. Jain, G.B. Johnson & P.H. Raven, pp. 428–60. New York: Columbia University Press.

Miller, R.M. & Jastrow, J.D. (1990). Hierarchy of root and mycorrhizal fungal interactions with soil aggregation. *Soil Biology & Biochemistry*, 22, 579–84.

Mitchell, D.T., Coley, P.G.F., Webb, S. & Allsopp, N. (1986). Litterfall and decomposition processes in the coastal fynbos vegetation, south-western Cape, South Africa. *Journal of Ecology*, 74, 977–93.

Moll, E.J. & Sommerville, J.E.M. (1985). Seasonal xylem pressure potentials of two South African coastal Fynbos species in three soil types. *South African Journal of Botany*, 51, 187–93.

Monk, C.D. & Gabrielson, F.C. (1985). Effects of shade, litter and root competition on old-growth vegetation in South Carolina. *Bulletin of the Torrey Botanical Club*, 112, 383–92.

Mooney, H.A. (1983). Carbon-gaining capacity and allocation patterns of mediterranean-climate plants. In *Mediterranean-type Ecosystems. The Role of Nutrients*, ed. F.J. Kruger, D.T. Mitchell & J.U.M. Jarvis, pp. 103–19. Berlin: Springer-Verlag.

Mooney, H.A., Kummerow, J., Moll, E.J., Orshan, G., Rutherford, M.C. & Sommerville, J.E.M. (1982). Plant form and function in relation to nutrient gradients. In *Mineral Nutrients in Mediterranean Ecosystems*, ed. J.A. Day, pp. 55–76. *South African National Scientific Programmes Report 71*. Pretoria: CSIR.

Nobel, P.S. (1980). Morphological, surface temperatures and northern limits of columnar cacti in the Sonoran Desert. *Ecology*, 61, 1–7.

Nobel, P.S. (1981). Influences of photosynthetically active radiation on cladode orientation, stem tilting and height of cacti. *Ecology*, 62, 982–90.

Nobel, P.S. (1982). Low-temperature tolerance and cold hardening of cacti. *Ecology*, 63, 1650–6.

Novellie, P. & Strydom, G. (1987). Monitoring the response of vegetation to use by large herbivores: an assessment of some techniques. *South African Journal of Wildlife Research*, 17, 109–17.

O'Connor, T.G. (1991). Local extinction in perennial grasslands: a life-history approach. *American Naturalist*, 137, 753–73.

Orians, G.H. & Solbrig, O.T. (1977). A cost–income model of leaves and roots with special reference to arid and semi-arid areas. *American Naturalist*, 111, 677–90.

Pankow, W., Boller, T. & Wienken, A. (1991). The significance of mycorrhizas for protective systems. *Experientia*, 47, 391–4.

Purnell, H.M. (1960). Studies of the family Proteaceae. I. Anatomy and morphology of the roots of some Victorian species. *Australian Journal of Botany*, 8, 38–50.

Read, D.J. (1991). Mycorrhizas in ecosystems. *Experientia*, 47, 376–91.

Redhead, J.F. (1968). Mycorrhizal associations in some Nigerian forest trees. *Transactions of the British Mycological Society*, 51, 377–87.

Reich, P.B., Walters, M.B. & Ellsworth, D.S. (1992). Leaf life-span in relation to leaf, plant and stand characteristics among diverse ecosystems. *Ecological Monographs*, 62, 364–92.

Richards, M.B., Stock, W.D. & Cowling, R.M. (1995). Water relations of seedlings and adults of two fynbos *Protea* species in relation to their distribution patterns. *Functional Ecology*, 9, 575–83.

Richardson, D.M. & Kruger, F.J. (1990). Water relations and photosynthetic characteristics of selected trees and shrubs of riparian and hillslope habitats in the south-western Cape Province, South Africa. *South African Journal of Botany*, 56, 214–25.

Roux, P. (1980). Vegetation change in the karoo region. *Karoo Agriculture*, 1, 15–16.

Rundel, P.W. (1988). Leaf structure and nutrition in mediterranean-climate sclerophylls. In *Mediterranean-type Ecosystems. A Data Source Book*, ed. R.L. Specht, pp. 157–67. Dordrecht: Kluwer.

Sallelo, S. & Lo Gullo, M.A. (1990). Sclerophylly and plant water relations in three mediterranean *Quercus* species. *Annals of Botany*, 65, 259–70.

Schimper, A.F.W. (1903). *Plant Geography on a Physiological Basis*. Oxford: Clarendon Press.

:holes, R.J. & Walker, B.H. (1993). *An African Savanna. Synthesis of the Nylsvley Study.* Cambridge: Cambridge University Press.

amida, A. (1981). Mediterranean vegetation of Israel and California: similarities and differences. *Israel Journal of Botany*, 30, 105–23.

amida, A. & Werger, M.J.A. (1992). Growth form diversity on the Canary Islands. *Vegetatio*, 102, 183–99.

:arpe, C. (1992). Dynamics of savanna ecosystems. *Journal of Vegetation Science*, 3, 293–300.

nit, G.N. & Swart, J.S. (1994). Influence of leguminous and non-leguminous woody plants on the herbaceous layer and soil under varying competition regimes in mixed bushveld. *African Journal of Range and Forage Science*, 11, 27–33.

nith, T.M. & Goodman, P.S. (1986). The effect of competition on the structure and dynamics of *Acacia* savannas in southern Africa. *Journal of Ecology*, 74, 1031–44.

echt, R.L. (1963). Dark Island heath (Ninety-Mile Plain, South Australia). VII. The effects of fertilizers on composition and growth 1950–1960. *Australian Journal of Botany*, 11, 67–94.

echt, R.L. & Moll, E.J. (1983). Mediterranean-type heathlands and sclerophyllous shrublands of the world: an overview. In *Mediterranean-type Ecosystems. The Role of Nutrients*, ed.. F.J. Kruger, D.T. Mitchell & J.U.M. Jarvis, pp. 41–65. Berlin: Springer-Verlag.

echt, R.L. & Rundel, P.W. (1990). Sclerophylly and foliar nutrient status of mediterranean-climate plant communities in southern Australia. *Australian Journal of Botany*, 38, 459–74.

ponkus, P.L. (1981). Responses to extreme temperatures. Cellular and sub-cellular bases. In *Physiological Plant Ecology I. Responses to the Physical Environment*, ed. O.L. Lange, P. Nobel, C.B. Osmond & H. Ziegler, pp. 372–437. New York: Springer-Verlag.

ck, W.D., Bond, W.J. & Le Roux, D. (1993). Isotope evidence from soil carbon to reconstruct vegetation history in the south-western Cape Province. *South African Journal of Science*, 89, 153–4.

Stock, W.D., Pate, J.S. & Delfs, J. (1990). Influence of seed size and quality on seedling development under low nutrient conditions in five Australian and South African members of the Proteaceae. *Journal of Ecology*, 78, 1005–20.

Stock, W.D., Sommerville, J.E.M. & Lewis, O.A.M. (1987). Seasonal allocation of dry mass and nitrogen in a fynbos endemic Restionaceae species *Thamnochortus punctatus* Pill. *Oecologia*, 72, 315–20.

Stock, W.D., van der Heyden, F. & Lewis, O.A.M. (1992). Plant structure and function. In *The Ecology of Fynbos. Nutrients, Fire and Diversity*, ed. R.M. Cowling, pp. 225–40. Cape Town: Oxford University Press.

Straker, C.J. & Mitchell, D.T. (1986). The activity and characterization of acid phosphatases in endomycorrhizal fungi of the Ericaceae. *New Phytologist*, 104, 243–56.

Stribley, D.P. & Read, D.J. (1980). The biology of mycorrhiza in the Ericaceae VII. The relationship between mycorrhizal infection and the capacity to utilize simple and complex organic nitrogen sources. *New Phytologist*, 86, 365–71.

Stuart-Hill, G.C. & Tainton, N.M. (1989). The competitive interaction between *Acacia karroo* and the herbaceous layer and how this is influenced by defoliation. *Journal of Applied Ecology*, 26, 285–98.

Taylor, H.C. (1978). Capensis. In *Biogeography and Ecology of Southern Africa*, ed. W.J.A. Werger, pp. 171–229. The Hague: Junk.

Taylor, H.C. (1980). Phytogeography of fynbos. *Bothalia*, 13, 231–5.

Theron, E.P. & Booysen, P. de V. (1968). The tensile properties of ten indigenous grasses. *Proceedings of the Grassland Society of South Africa*, 3, 57–61.

Ting, I.P. (1989). Photosynthesis of arid and subtropical succulent plants. *Aliso*, 12, 387–406.

Trappe, J.M. (1987). Phylogenetic and ecologic aspects of mycotrophy in the angiosperms from an evolutionary standpoint. In *Ecophysiology of VA Mycorrhizal Plants*, ed. G. Safir, pp. 5–25. Boca Raton, Florida: CRC Press.

Turner, I.M. & Tan, H.T.W. (1991). Habitat-related variation in tree leaf form in four tropical forest types on Palau Ubin, Singapore. *Journal of Vegetation Science*, 2, 691–8.

Turner, J.S. & Picker, M.D. (1993). Thermal ecology of an embedded dwarf succulent from southern Africa (*Lithops* spp: Mesembryanthemaceae). *Journal of Arid Environments*, 24, 361–85.

Tyson, P.D. (1986). *Climatic Change and Variability in Southern Africa*. Cape Town, Oxford University Press.

van Coller, A. & Stock, W.D. (1994). Cold tolerance of the southern African succulent, *Cotyledon orbiculata* L. across its geographical range. *Flora*, 189, 89–94.

van der Heyden, F. & Lewis, O.A.M. (1989). Seasonal variation in photosynthetic capacity with respect to plant water status of five species of the mediterranean climate region of South Africa. *South African Journal of Botany*, 55, 509–15.

van der Heyden, F. & Lewis, O.A.M. (1990). Environmental control of photosynthetic gas exchange characteristics of fynbos species representing three growth forms. *South African Journal of Botany*, 56, 654–8.

van Jaarsveld, E. (1987). The succulent riches of South Africa and Namibia. *Aloe*, 24, 45–92.

Vasek, F.C. & Thorne, R.F. (1977). Transmontane coniferous vegetation. In *Terrestrial Vegetation in California*, ed. M.G. Barbour & J. Major, pp. 797–832, Special Publication 9. New York: California Native Plant Society.

Veenendal, E.M., Monnaapula, S.C., Gilika, T. & Magole, I. (1992). Vesicular-arbuscular mycorrhizal infection of grass seedlings in a degraded semi-arid savanna in Botswana. *New Phytologist*, 121, 477–85.

Vogel, J.C., Fuls, A. & Ellis, R.P. (1978). The geographical distribution of krantz grasses in South Africa. *South African Journal of Science*, 74, 209–15.

von Willert, D.J., Brinckmann, E., Scheitler, B. & Eller, B.M. (1985). Availability of water controls Crassulacean acid metabolism in succulents of the Richtersveld (Namib desert, South Africa). *Planta*, 164, 44–55.

von Willert, D.J., Eller, B.M., Werger, M.J.A. & Brinckmann, E. (1990). Desert succulents and their life strategies. *Vegetatio*, **90**, 133–43.

von Willert, D.J., Eller, B.M., Werger, M.J.A., Brinckmann, E. & Ihlenfeldt, H.-D. (1992). *Life Strategies of Succulents in Deserts with Special Reference to the Namib Desert.* Cambridge: Cambridge University Press.

Walker, B.H. (1985). Structure and function of savannas: an overview. In *Ecology and Management of the World's Savannas*, ed. J.C. Tothill & J.J. Mott. pp. 91–83. Canberra: Australian Academy of Science.

Walker, B.H., Ludwig, D., Holling, C.S. & Peterman, R.M. (1981). Stability of semi-arid savanna grazing systems. *Journal of Ecology*, **69**, 473–98.

Walker, B.H. & Noy-Meir, I. (1982). Aspects of the stability and resilience of savanna ecosystems. In *Ecology of Tropical Savannas*, ed. B.J. Huntley & B.H. Walker, pp. 556–609. Berlin: Springer-Verlag.

Walter, H. (1971). *Ecology of Tropical and Subtropical Vegetation.* Edinburgh: Oliver & Boyd.

Warming, E. (1909). *Oecology of Plants.* Oxford: Clarendon Press.

Weltzin, J.F. & Coughenour, M.B. (1990). Savanna tree influence on understorey vegetation and soil nutrients in northwestern Kenya. *Journal of Vegetation Science*, **1**, 325–34.

Werger, M.J.A. (1978). The Karoo–Namib Region. In *Biogeography and Ecology of Southern Africa*, ed. M.J.A. Werger, pp. 231–99. The Hague: Junk.

Werger, M.J.A. (1983). Vegetation geographical patterns as a key to the past, with emphasis on the dry vegetation types of South Africa. *Bothalia*, **14**, 405–10.

Werger, M.J.A. (1986). The karoo and southern Kalahari. In *Ecosystems of the World 12B. Hot Deserts and Arid Shrublands*, ed.. M. Evenari, I. Noy-Meir & D.W. Goodall, pp. 283–359. Amsterdam: Elsevier.

Werger, M.J.A. & Coetzee, B.J. (1978). The Sudano–Zambezian Region. In *Biogeography and Ecology of Southern Africa*, ed. M.J.A. Werger, pp. 301–462. The Hague: Junk.

Werger, M.J.A. & Ellenbroek, G.A. (1978). Leaf size and leaf consistence of a riverine forest formation along a climatic gradient. *Oecologia*, **34**, 297–308.

Werger, M.J.A. & Ellis, R.P. (1981). Photosynthetic pathways in the arid regions of South Africa. *Flora*, **171**, 64–75.

Westoby, M. (1980). Elements of a theory of vegetation dynamics in arid rangelands. *Israel Journal of Botany*, **28**, 169–94.

White, F. (1978). The Afromontane region. In *Biogeography and Ecology of Southern Africa*, ed. M.J.A. Werger, pp. 463–513. The Hague: Junk.

Whittaker, R.H., Morris, J.W. & Goodman, D. (1984). Pattern analysis in savanna woodlands at Nylsvley, South Africa. *Memoirs of the Botanical Survey of South Africa*, **49**, 1–51.

Witkowski, E.T.F. (1989a). Effects of nutrients on the distribution of dry mass, nitrogen and phosphorus in seedlings of Protea repens (L.) L. (Proteaceae). *New Phytologist*, **112**, 481–7.

Witkowski, E.T.F. (1989b). Effects of nutrient additions on litter production and nutrient return in a nutrient-poor Cape fynbos ecosystem. *Plant & Soil*, **117**, 227–35.

Witkowski, E.T.F. (1989c). Response to nutrient additions by the plant growth forms of sandplain lowland fynbos, South Africa. *Vegetatio*, **79**, 89–97.

Witkowski, E.T.F. & Lamont, B.B. (1991). Leaf specific mass confounds leaf density and thickness. *Oecologia*, **88**, 486–93.

Witkowski, E.T.F., Lamont, B.B., Walton, C.S. & Radford, S. (1992). Leaf demography, sclerophylly and ecophysiology of two banksias with contrasting leaf life spans. *Australian Journal of Botany*, **40**, 849–62.

Witkowski, E.T.F. & Mitchell, D.T. (1987). Variations in soil phosphorus in the fynbos biome, South Africa. *Journal of Ecology*, **75**, 1159–71.

Witkowski, E.T.F., Mitchell, D.T. & Stock, W.D. (1990). Response of a Cape fynbos ecosystem to nutrient additions: shoot growth and nutrient contents of a proteoid (*Leucospermum parile*) and an ericoid (*Phylica cephalantha*) evergreen shrub. *Acta Oecologica*, **11**, 311–26.

Witkowski, E.T.F. & O'Connor, T.G. (1996). Topo-edaphic, floristic and structural variation in a southern African savanna. *Vegetatio*, (in press).

Woodward, F. I. (1987). *Climate and Plant Distribution.* Cambridge: Cambridge University Press.

Wullstein, L. H., Bruening, M. L. & Bollen, W. B. (1979). Nitrogen fixation associated with sand grain root sheaths (rhizosheaths) of certain xeric grasses. *Physiologia Plantarum*, **46**, 1–4.

Herbivory

17

N. Owen-Smith and J.E. Danckwerts

17.1 Introduction

The early European settlers and explorers entering southern Africa during the seventeenth and eighteenth centuries encountered a rich abundance and diversity of large mammalian herbivores. They also found substantial herds of cattle, sheep and goats, kept by the Khoikhoi and Bantu-speaking residents of the region. Archaeological evidence indicates that pastoralists had already reached the southernmost tip of the continent 2000 years ago (Voigt 1983; Hall 1984).

Before the end of the nineteenth century, African elephant (*Loxodonta africana*), white rhinoceros (*Ceratotherium simum*), black rhinoceros (*Diceros bicornis*), quagga (*Equus burchelli quagga*) and black wildebeest (*Connochaetes gnou*) had been almost exterminated by hunting. Populations of other wild ungulates were decimated by rinderpest, a viral disease, which spread as far as the Cape by 1896. By the early twentieth century, indigenous large herbivores had become confined mostly to national parks and other conserved areas, which cover about 10% of southern Africa. Domestic livestock are today the principal grazers and browsers over most of the region.

We use the term herbivory in the broadest sense to encompass consumption of not only leaves and supporting stems (folivory), but also seeds (granivory), fruits (frugivory), roots (rhizophagy), bark and twigs. The southern African region offers an opportunity to compare, for a range of biomes, the impacts on vegetation of intact large mammal communities, domestic ungulates, small mammals, birds and insects. Herbivores affect vegetation in other ways by trampling or breaking plants, disturbing soils and changing fire regimes (Chap. 18, this volume). Consequences of these impacts will be considered in Chap. 21 (this volume). We are concerned primarily with long-term consequences of herbivory for vegetation structure and composition, and plant phenotypes. Less attention will be paid to short-term influences on plant growth, survival and reproductive output. We do not address the very important roles of herbivores in affecting the community structure and dynamics of marine vegetation, as this is dealt with in Chap. 15 (this volume).

A theme running through this chapter is the pervasive influence of soil nutrient levels on plant–herbivore interactions. Rainfall governs primary production, and therefore sets a basic constraint on the abundance of mammalian herbivores. However, food quality is strongly affected by soil fertility, and greatly modifies overall herbivore biomass. The distinction between 'sweetveld', capable of supporting livestock year-round, and 'sourveld', where cattle must be given supplementary forage during the dry season, has long been recognized by South African pastoralists (see Chaps 10 and 18, this volume). Soil fertility also influences levels of herbivory on woody plants, the kinds of anti-herbivore defences that these plants exhibit, and dispersal syndromes. In nutrient-rich savannas herbivores consume a large fraction of primary production, but in nutrient-poor savannas most of it passes to decomposers, or is incinerated by fire (Chap. 11, this volume).

17.2 Herbivore abundance and distribution

17.2.1 Indigenous large mammals

In African savannas, the regional biomass of large mammalian herbivores is proportionally related to the annual rainfall total (Coe, Cumming & Phillipson 1976; McNaughton & Georgiadis 1986). However, total herbivore biomass is 2–3 times greater on average in regions of high soil nutrient status, than in areas with infertile soils receiving similar rainfall (Bell 1982; Fritz & Duncan 1994; Fig. 17.1). This contrast operates even at small scales within landscapes. At Nylsvley in the northern Transvaal, the disturbed, nutrient-enriched sites associated with former human settlements supported nearly four times the biomass of herbivores as the surrounding nutrient-poor savanna, although the geological substratum was identical (Scholes & Walker 1993). In nutrient-deficient regions, local concentrations of animals in 'hot-spots' such as drainage depressions (vleis or dambos) may be high, but elsewhere herbivores occur transiently, perhaps lingering after fires. The influence of soil fer-

tility weakens as rainfall decreases, because less leaching of nutrients occurs (Chap. 11, this volume).

Megaherbivores (species exceeding 1000 kg in adult body weight) typically form 50% or more of the total biomass of large herbivores in savanna communities (Owen-Smith 1988). Included in this category are African elephant, white rhinoceros, black rhinoceros and hippopotamus (*Hippopotamus amphibius*). African buffalo (*Syncerus caffer*) are generally the predominant grazer (15–35% of grazing biomass), except where white rhinoceros or hippopotamus are abundant (Owen-Smith & Cumming 1994). In drier regions of southern Botswana and northern Namibia, blue wildebeest (*Connochaetes taurinus*) were until recently the most abundant grazer. In Highveld grassland and grassy parts of the Nama-karoo biome, black wildebeest, blesbok (*Damaliscus dorcas* ssp.

Figure 17.1 Relationship between biomass of large mammalian herbivores (both wild and domestic) and annual rainfall for African savanna ecosystems distinguished by soil nutrient status. Regression lines: high nutrients (top); log (biomass) = 1.58 log (precipitation) – 0.67 ($r^2 = 0.82$); medium nutrients (middle); log (biomass) = 1.78 log (precipitation) – 1.32 ($r^2 = 0.85$); low nutrients (bottom); log (biomass) = 1.96 log (precipitation) – 2.04 ($r^2 = 0.85$). *Redrawn, with permission, from Fritz & Duncan (1994).*

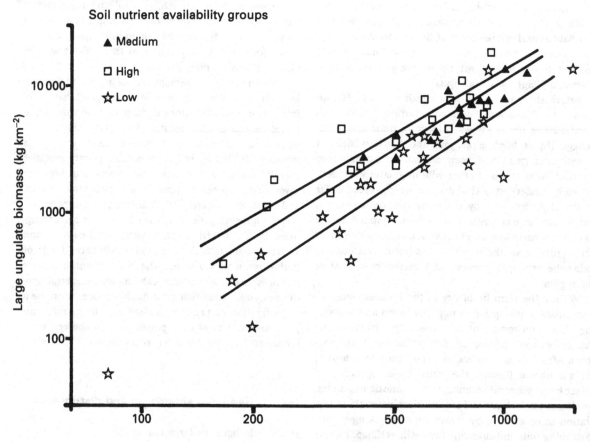

phillipsi), quagga, springbok (*Antidorcas marsupialis*) and eland (*Taurotragus oryx*) were the predominant grazing ungulates.

Although trees typically contribute nearly half the primary production in savannas, browsers (excluding elephant) generally form only 5–20% of large herbivore biomass (Owen-Smith 1993a). Browsers including elephant, black rhinoceros, eland and kudu (*Tragelaphus strepsiceros*) are or were especially abundant in the succulent subtropical thickets of the Eastern Cape (Stuart-Hill 1992). These species formerly occurred in the fynbos biome, possibly associated with renosterveld growing on shale-derived soils. Elephant, bushbuck (*Tragelaphus scriptus*), duiker (*Cephalophus* spp.) and bushpig (*Potamochoerus porcus*) penetrate the forest biome. Arboreal mammalian folivores in forests are restricted to tree hyrax (*Dendrohyrax arboreus*) and the mainly frugivorous samango monkey (*Cercopithecus mitis*) (see Skinner & Smithers 1990 for general distributions of these species).

In the karoo biomes, Namib desert and savannas on Kalahari sand, surface water availability becomes the main factor governing the abundance of large herbivores, especially grazers. Even in the savanna biome, there is a typical pattern of concentration by animals near waterpoints in the dry season (McNaughton & Georgiadis 1986). Elephant penetrate the Namib desert, but depend on the vegetation fringing dry rivers traversing the region (Viljoen 1989; Chap. 9, this volume).

Ungulate migrations may involve seasonal movements along rainfall gradients, as with wildebeest and zebra in the Serengeti region of east Africa (Maddock 1979). Similar migrations by black wildebeest, quagga and blesbok may formerly have occurred in regions of the grassland biome adjoining the Drakensberg mountains. Other so-called migrations involve a dry-season concentration near rivers or other permanent water sources, and a wet-season dispersal away from water. Seasonal movements of the latter type, over distances of up to 100 km, are exhibited by blue wildebeest and zebra in parts of the Kruger National Park (Whyte & Joubert 1988) and in northern Botswana (Joos-Vandewalle 1988). Similar migrations occurred until recently in the southern Kalahari (Williamson, Williamson & Ngwamotsoko 1988) and northern Namibia (Berry 1981). Springbok formerly moved in large aggregations between the Bushmanland region of the Northern Cape, with predominantly summer rainfall, and the Koue Bokkeveld region of Namaqualand, with more reliable winter rainfall, although it is unclear how regular these movements were (Skinner 1993). In other parts of the Nama- and succulent karoo biomes, so-called springbok treks appeared more erratic in direction, and were probably governed by local food availability, or lack thereof. Movements of black wildebeest, blesbok, quagga and eland also occurred in past times in the Nama-karoo and Kalahari regions in response to drought.

17.2.2 Domestic livestock

Domestic ungulates, especially cattle, are today the major consumers of vegetation over most of southern Africa. Over the region as a whole, livestock biomass averages 30 kg (livemass) ha^{-1}. Highest livestock concentrations, exceeding 100 kg ha^{-1}, occur in the grassland biome and humid regions of savanna (Table 17.1). In the Transvaal savanna region, the overall stocking level of domestic livestock (47 kg ha^{-1}) is 1.5 times the biomass density of wild herbivores in the Kruger National Park (32 kg ha^{-1}). However, in the Umfolozi Wildlife Park in KwaZulu-Natal, the biomass of wild ungulates totalled 89 kg ha^{-1} during the 1970s (Table 17.2), with white rhinoceros comprising almost half of this (Owen-Smith 1988). Hence, biomass levels of communities of indigenous large herbivores that include megaherbivores can match those attained by domestic ungulates (see Fritz & Duncan 1994). Nevertheless, livestock densities are maintained at a high level over a vast region, largely through augmentation of water supplies.

There is much local variation in stocking levels within biomes (expressed in 'large stock units', LSU; one LSU represents the metabolic equivalent of a 454-kg cow). In the grassland biome, stocking rate ranges between 35 and 15 LSU km^{-2} for areas of commercial pastoralism, compared with a mean stocking rate of 47 LSU km^{-2} in communally managed grasslands in the Eastern Cape. In the Nama-karoo biome, commercial stocking rates range from around 1.5 LSU km^{-2} in the northwest to 10 LSU km^{-2} or higher in the east (Anon. 1993, 1994). The stocking rate for Botswana is considerably lower than that of the Kalahari thornveld (arid savanna) of South Africa, which is ecologically similar to southwestern Botswana. In the Nama- and succulent karoo biomes and in the fynbos biome, the LSU equivalent of sheep and goats exceeds that of cattle. Goats are prevalent in savanna and succulent thicket in the Eastern Cape, and in adjacent parts of the Nama-karoo.

In the past, the low country of the eastern and northern parts of southern Africa was infested with tsetse flies (*Glossina* spp.), vectors of the protozoans (*Trypanosoma* spp.) that cause nagana in cattle, and sleeping sickness in humans. Following the rinderpest epizootic, the flies, deprived of their hosts, disappeared from eastern Transvaal, although they persisted in the coastal region of northern KwaZulu-Natal (Ford 1971). In the latter area tsetse flies were eliminated by aerial spraying with insecticides during the 1940s (Vincent 1970). Tsetse flies per-

Table 17.1 Total biomass and estimated consumption levels for domestic livestock in different regions of southern Africa. LSU, large stock unit defined as metabolic equivalent of 454 kg bovine (Meissner et al. 1983). Ratio = total LSU (cattle + equines)/total LSU (sheep + goats). Consumption derived from feeding standards (Meissner et al. 1983)

Country	Biome	Mean rainfall (mm)	Estimated primary production (kg ha^{-1} yr^{-1})	Mean stocking (kg ha^{-1})	Mean stocking (ha LSU^{-1})	Large/small stock ratio (LSU/LSU)	Estimated consumption (kg ha^{-1} yr^{-1}) Total	Grass	Browse	Consumption/production (%) Total	Grass	Browse
South Africa	Semi-arid savanna (Transvaal)	600	4500	47	9.0	7.6	360	325	35	8	11	2
	Humid savanna (coastal region)	1000	8000	114	3.6	4.6	900	840	60	12.5	14	3
	Arid savanna (Kalahari)	350	2700	26	15.4	2.7	213	185	28	8	13	2
	Grassland	750	6000	101	4.0	2.6	834	834	0	14	14	0
	Nama-karoo	250	1250	12	25.0	0.2	133	70	63	10.5	14	8
Lesotho	Grassland	1500	11000	96	3.9	1.9	864	777	87	8	8	8
Swaziland	Savanna	900	7500	128	3.4	11.6	966	900	66	13	15	5
Mozambique	Savanna	700	5000	8	55	4.3	59	50	9	1	1.5	0.5
Botswana	Savanna	400	2500	12	34.5	8.0	95	85	10	4	5.5	1
Namibia	Savanna and Karoo	400	2500	13.5	28.6	2.7	115	90	25	5	6	2.5

Table 17.2 Total biomass and estimated consumption levels for indigenous large herbivore communities in selected conservation areas in southern Africa. Data on population levels (with aerial census data corrected for undercount bias) from Owen-Smith (1988), Owen-Smith (1993b) and Stuart-Hill (1992), species weights from Owen-Smith (1988), consumption rates relative to body mass from Owen-Smith (1988) and Owen-Smith (1993b). Assumed primary production equals above-ground production of leaves and shoots, from rainfall–primary production relations presented by Rutherford (1980) and Scholes & Walker (1993), modified for general soil nutrient status

Area	Biome	Mean rainfall (mm)	Assumed primary production (kg ha^{-1} yr^{-1}) Total	Grass	Browse	Large herbivore biomass (kg ha^{-1})	Estimated consumption (kg ha^{-1} yr^{-1}) Total	Grass	Browse	Elephants	Consumption/production (%) Total	Grass	Browse
Umfolozi Wildlife Park	Nutrient-rich savanna	700	6000	4000	2000	89.5	634	544	90	0	10.6	13.6	4.5
Kruger National Park	Intermediate savanna	575	4000	2600	1400	31.9	235	155	45	35	5.9	6.5	4.6
Chobe and Hwange National Parks	Nutrient-poor savanna	550	3000	1500	1500	34.5	205	58	21	126	6.8	6.7	7.0
Addo Elephant National Park	Succulent thicket	440	3500	500	3000	46.0	278	20	77	182	8.0	8.0	8.0

sist in northern Botswana and over much of Mozambique. The presence of tsetse flies and associated cattle disease limits the abundance and distribution of people and their livestock, and restricted availability of surface water and poor forage quality are additional limitations. The major expansion in the distribution of livestock followed industrialization towards the end of the nineteenth century, and was facilitated by the exploitation and reticulation of subterranean water, feed supplementation and immunization against disease. However, although stocking rates of livestock increased in northern Cape savanna between the 1911–30 and 1971–81 periods, a large (c. 45%) decline occurred in the Nama-karoo and succulent karoo regions of this province (Dean & Macdonald 1994). The lowered stocking in the karoo biomes, despite additional waterpoints, suggests that the productive potential of the land had been reduced as a result of vegetation change (see section 21.4.2 in Chap. 21, this volume, for further discussion).

17.2.3 Small vertebrates

The abundance of small mammals may rival or exceed that of ungulates, even in the savanna biome, if transformed into metabolic biomass equivalents (based on raising body mass to the power 0.75; Korn 1987). However, the distribution of small rodents is patchy, and their abundance can vary widely over time (Nel 1978), as well as being depressed by heavy grazing pressure from ungulates (Bowland & Perrin 1988). Small rodents appear to be somewhat more common in the southern than in the southwestern parts of the fynbos biome (Van Hensbergen *et al.* 1992). Notable also are fossorial molerats (Bathyergidae), which are especially common in the fynbos biome (Bigalke 1979). Tree squirrel (*Paraxerus cepapi*) and springhare (*Pedetes capensis*) may figure prominently in rodent biomass in parts of the savanna biome (Temby 1977; Viljoen & Du Toit 1985; Korn 1987). Porcupine (*Hystrix africaeaustralis*) are widespread through grassland and savanna. Several species of hare are common and distributed through the savanna, grassland, Nama-karoo and succulent karoo biomes. The biomass density of scrub hare (*Lepus saxatilis*) in northern Transvaal savanna was low relative to that of rodents (Temby 1977).

In the avifauna of a northern Transvaal savanna, mixed-diet species predominated in terms of biomass, notably guineafowl (Numididae) and francolins (Phasianidae) (Tarboton 1980). Specialist granivores or frugivores formed 8% of avian biomass in fertile Acacia savanna, but only 1.5% in infertile *Burkea* savanna. Overall, the avian biomass was twice as great in *Acacia* as in *Burkea* savanna.

Tortoises are the only reptilian herbivores in the region. Most species are restricted to the southwestern region of South Africa, but the mountain or leopard tortoise (*Geochelone pardalis*) and hinged tortoises (*Kinixys* spp.) are distributed through the savanna and grassland biomes.

17.2.4 Insects and other invertebrates

Grasshoppers are generally the most prominent insect herbivores, although few estimates of their abundance are available. In savanna the biomass of grasshoppers may be such that their consumption of grass rivals or exceeds that of large mammals (Gandar 1982a; Chap. 11, this volume). Although strictly detritivores rather than herbivores, harvester termites can reach a high abundance and cause severe denudation of the grass layer during drought periods (Coaton 1958; Nel 1968).

Outbreaks of brown locust (*Locustana pardalina*) are a notable feature of the Nama-karoo (Lea 1964). Before the widespread application of insecticides, locust plagues reportedly persisted for about 13 years, with intervals of about 11 years between outbreaks (Lounsbury 1915, cited by De Villiers 1988). Chemical control measures have shortened the duration of outbreaks to about seven years, but the intervals between them have been reduced correspondingly to about seven years (Botha & Lea 1970, cited by De Villiers 1988).

Irregular outbreaks of defoliating caterpillars (especially Saturnidae) are a feature particularly of savannas dominated by species of *Brachystegia*, *Burkea* and *Colophospermum mopane* (Reeler, Campbell & Price 1991). Outbreaks do not seem to occur on the *Acacia* species which predominate in most nutrient-rich savanna areas. In the Nama-karoo, the karoo caterpillar (*Loxostege frustalis*, Pyralidae), a polyphagous species feeding primarily upon shrubby Asteraceae, can reach abundance levels so great as to cause severe defoliation of these shrubs over huge areas (Annecke & Moran 1977). In the fynbos biome, Lepidoptera are notably depauperate in both species richness and abundance (Cottrell 1985), with sapsucking Hemiptera far more important than leaf chewers (Theron 1984).

17.3 Consumption of leaves and plant structural tissues

In this section we cover the impacts of herbivores that primarily consume leaves, supporting stems, bark and roots. Among large mammalian herbivores, there is a fairly distinct division between grazers, feeding largely or entirely on leaves and stems of grasses and other graminoids, and browsers, consuming mainly the leaves of trees, shrubs and forbs (Owen-Smith 1992). A similar distinction between grass-feeders and dicot-feeders is evident among other groups of herbivores. Small mammals, birds and other small vertebrates feeding on leaves, stems or roots will be termed nibblers and gnawers.

17.3.1 Grazing

For estimating levels of consumption relative to production, we assume that primary production of edible and accessible vegetation (leaves, fruits and supporting stems) amounts to 0.5–1 tonne (dry mass) km^{-2} mm^{-1} of rainfall, and that variation within this range is influenced primarily by soil fertility (Rutherford 1980; Scholes & Walker 1993). In the savanna biome, typically from half to three-quarters of the production takes place in the herbaceous layer. The partitioning of production between grasses, forbs and the woody component varies more widely in other biomes.

From allometric relations between daily food intake and body mass (Owen-Smith 1988), it is possible to estimate the fraction of available production consumed by large mammals. In savanna communities of indigenous large herbivores, about 75–90% of consumption is concentrated on the grass component (Owen-Smith 1993a). In the fertile savannas of the Umfolozi Wildlife Park, grass consumption amounted to about 13% of production when based on the 1970 estimates of ungulate populations. For the Kruger National Park as a whole, which is intermediate in soil fertility, estimated grass consumption amounts to about 6% of production. For the Chobe–Hwange region extending from Botswana into Zimbabwe, where elephants predominate and soils are largely infertile Kalahari sands, estimated grass consumption amounts to about 7% of production (Table 17.2). These figures seem to be typical of infertile and fertile savannas elsewhere in Africa that retain intact communities of wild herbivores (Owen-Smith 1993a).

Estimated consumption levels of grass by domestic livestock amount to 8–15% of production in different regions of South Africa, Lesotho and Swaziland (Table 17.1). Livestock population levels are somewhat lower in Mozambique, Botswana and Namibia, where surface water, and in some places the presence of tsetse fly, impose contraints.

The above figures represent regional averages. In sections of Umfolozi Wildlife Park favoured by white rhinoceros and other grazing ungulates, estimated grass consumption exceeds 25% of production. Favoured grass species incurred somewhat higher levels of defoliation. In communal grazing areas, over 50% of grass production may be consumed annually by cattle and other livestock.

Severe grazing pressure owing to high stocking levels

of cattle and sheep is commonly believed to have been the prime factor responsible for the perceived degradation of the grass component in the grassland, savanna and Nama-karoo biomes (Anon. 1923; Acocks 1953; Chap. 21, this volume). In the humid grasslands of KwaZulu-Natal, the fibrous and hence unpalatable ngongoni grass (*Aristida junciformis*) has increased in abundance at the expense of the highly valued red grass (*Themeda triandra*) (Morris, Tainton & Hardy 1992). In the Eastern Cape, grazing by cattle in the season following an intense drought, which had caused severe tuft mortality, suppressed recovery of *Themeda*. However, in an area where grazing livestock were withdrawn for six months after the drought had broken, palatable grasses quickly recovered their former abundance (Danckwerts & Stuart-Hill 1988).

Heavy grazing pressure may have a strong negative effect on seed production by grasses, both through consumption of inflorescences and by suppressing the production of flowering culms (O'Connor 1991, 1994; O'Connor & Pickett 1992). Sensitive species, such as *Themeda triandra*, are perennial grasses that are palatable to grazers, are obligate seed-reproducers and produce few large diaspores that are poorly dispersed (Chap. 10, this volume). Seeds persist no more than 2–3 years in a state of enforced or induced dormancy, and are subject to heavy predation by rodents and other granivores. Such grasses are vulnerable to local extinction in regions that experience both variable rainfall and sustained heavy grazing. Following drought-related mortality of tufts, extinction-prone species become replaced by unpalatable or annual grasses that produce small seeds prolifically, such as *Aristida* spp., or species that regenerate vegetatively through stolons, such as *Digitaria eriantha* and *Urochloa mosambicensis* (Chap. 10, this volume).

This mechanism could have been responsible for the putative disappearance of the perennial grass component over much of the Nama-karoo biome, associated with heavy stocking with sheep during the second half of the nineteenth century (Acocks 1953; but also see Chaps 8, 16 and 21, this volume). There is evidence that a slow but perceptible increase of dwarf shrubs at the expense of the grass component has continued in areas grazed in summer by sheep over the past 40 years (O'Connor & Roux 1995). Sheep favour higher-quality plant parts, select lateral tillers and concentrate their feeding in particular localities. Therefore, their grazing effects may be more damaging to plants than those caused by equivalent stocking levels of cattle (Tainton 1972; Danckwerts 1989).

These and other observations on the dynamics of South African rangelands in regions with high variability in rainfall led to the conceptualization of these as

Figure 17.2 **Hippopotamus (*Hippopotamus amphibius*) can exert considerable pressure on the vegetation bordering rivers and pans. Ndumu Wildlife Park, northern KwaZulu-Natal** *(Photo: E.J. Moll).*

Figure 17.3 **White or square-lipped rhinoceros (*Ceratotherium simum*) in heavily grazed veld at Hluhluwe Wildlife Park, northern KwaZulu-Natal** *(Photo: E.J. Moll).*

Figure 17.4 **Grazing pathway of a white rhinoceros etched in dew in a short grass lawn promoted by rhino grazing pressure. Dominant lawn grasses are *Digitaria argyrograpta*, *Urochloa mosambicensis* and the decumbent form of *Panicum coloratum*** (Photo: N. Owen-Smith).

migration (e.g. wildebeest, zebra) or by moving over large home ranges (buffalo). However, many wild ungulate species are sedentary. The denuded state of the grass cover bordering rivers in the Kruger National Park is due largely to persistent grazing by hippopotamus (N. Owen-Smith and I.R. McDonald, unpubl. data; Fig. 17.2). In Umfolozi Wildlife Park, the expanding white rhinoceros population transformed large sections of *Themeda*-dominated grassland into short grass lawns where stoloniferous or very short grass species (e.g. *Panicum coloratum, Urochloa mosambicensis, Sporobolus iocladus*) become predominant (Downing 1972; Melton 1987; Owen-Smith 1988). Except in regions susceptible to erosion, the grazing lawns created by white rhinoceros (Figs. 17.3 and 17.4), hippopotamus and herds of wildebeest and other antelope appear to maintain high cover and production despite severe defoliation pressure (Olivier & Laurie 1974; McNaughton 1984; Owen-Smith 1988; Novellie 1990). Warthog (*Phacochoerus aethiopicus*) have a major impact on floodplain grasslands through their digging activities for the rhizomes of grasses (e.g. *Oryza* spp.) and sedges (G.C. Marneweck, unpubl. data).

In the communal grazing lands, stocking levels of livestock are commonly 3–4 times those recommended for commercial cattle production (Shackleton 1993). Nevertheless, many of these areas have remained in a relatively stable, although apparently degraded, state for decades. The unpalatable *Aristida junciformis* has become predominant in sour grasslands in northern Transkei, where nutrient-poor quartzitic soils prevail. However, palatable species such as *Themeda triandra* and *Tristachya leucothrix* remain abundant in heavily stocked southern parts of Transkei where relatively fertile Karoo sediments predominate, although associated with them are highly fibrous species such as *Eragrostis plana* (McKenzie 1982, 1987). Grassland composition in parts of the heavily grazed communal area appears to be more favourable for livestock than that in adjacent areas rested from grazing. In a fertile savanna region of KwaZulu-Natal, measurements made across the fenceline between Umfolozi Wildlife Park and the adjacent area, which is communally grazed by livestock, showed only small differences in grass species richness and biomass accumulation (Venter et al. 1989). Nevertheless, *Themeda triandra* occurred only in the lower slope in the communal area, while in the wildlife area it was abundant on the upper and middle slopes as well. In fertile savanna in the eastern Transvaal, droughts coupled with heavy communal grazing led to reduced basal cover of perennial grasses and a prevalence of annual grasses and forbs (O'Connor 1994).

In areas of communal grassland which had previously been heavily grazed, but where livestock had since been

event-driven systems, with corresponding implications for their management (Walker, Matthews & Dye 1986; Mentis et al. 1989; Westoby, Walker & Noy-Meir 1989; Chaps 8 and 11, this volume). Because rainfall exerts an overriding influence on grass dynamics, management actions, such as destocking, may yield disappointing results if timed inappropriately in relation to rainfall variability. Actions taken in certain critical periods, such as during the post-drought recovery phase, can leave a long-lasting impact on the state of the vegetation. This viewpoint is an alternative to the Clementsian paradigm of successional changes in plant species composition in relation to gradients in grazing pressure (Foran, Tainton & Booysen 1978; Tainton 1986; Bosch, Kellner & Scheepers 1989; Chap. 10, this volume).

The impact of wild grazers is commonly believed to be less detrimental to grassland condition, because these free-ranging animals graze rotationally through

excluded for varying periods, grass species composition improved relative to adjacent areas still heavily grazed (Harrison 1993). This change towards recovery took place over four to nine years for fertile sites, and six to nine years for infertile sites. Some sites showed a predominance of palatable *Themeda triandra* in the enclosed area, while unpalatable *Aristida junciformis* prevailed in the adjacent grazed area. However, no consistent difference in basal cover between the enclosed and the adjoining grassland was found.

Sustained heavy grazing commonly leads to bush thickening in the savanna biome by reducing fire frequency or intensity (Watson & Macdonald 1983; C.R. Hurt & R.I. Yeaton unpubl. data; Chaps 11 and 18, this volume). Spinescent trees and shrubs (*Acacia* spp., *Dichrostachys cinerea*) and other shallow-rooted species (e.g. *Grewia* spp.) are particularly promoted (Van Vegten 1983; Sweet & Mphinyane 1986; Skarpe 1990).

17.3.2 Browsing

Relative to production, levels of consumption of tree and shrub foliage by indigenous large herbivores are lower than those of grasses, even when elephants are present (Table 17.2). For savanna regions, the estimated consumption of browse was equivalent to about 4% of leaf and shoot production in both Umfolozi Wildlife Park, where elephant were absent, and the Kruger National Park, where elephant are maintained at low abundance by culling, but was about 7% in the Chobe–Hwange region, where the elephant population is large and growing (Table 17.2). In succulent thicket vegetation in Addo Elephant National Park, where elephant occur at high density, browse consumption amounted to about 8% of estimated production (Stuart-Hill 1992).

In northern Transvaal savanna, palatable deciduous woody plant species lost less than 10% of their foliage to browsing ruminants during the growing season (Owen-Smith & Cooper 1987). Leaf losses were more severe for the more palatable species among evergreens, and some of these had virtually all of their foliage below 2 m height consumed by kudus and impalas by the late dry season. Certain unpalatable tree and shrub species were browsed for a brief period when they grew a flush of new leaves before the onset of the rains. In East Africa, small antelope have an important impact on woodland regeneration, by suppressing seedling recruitment (Belsky 1984; Prins & Van der Jeugd 1993), and the same is likely to be true in southern Africa (Trollope 1980).

Regional levels of browsing by domestic ungulates tend to be somewhat lower than those of wild ungulates, because cattle are primarily grazers (Table 17.1). Nevertheless, goats are locally abundant in parts of the savanna and Nama-karoo biomes, especially in the Eastern

Cape, and in association with fire can have an important effect on the woody canopy cover (Trollope 1980; W.S.W. Trollope, unpubl. data). Experimental browsing of the leaves and shoots of *Acacia karroo* by goats (25–75% removal) stimulated regrowth so that the total production of these parts greatly exceeded that by non-defoliated plants (Teague & Walker 1988). This stimulation persisted through to the next growing season. However, little or no increase in production occurred when severe defoliation was repeated at intervals as short as two weeks (Teague 1989a). Goats stocked at high density following a burn in Botswana had more impact on the abundance of spineless than of spinescent shrubs, and *Acacia tortilis* in particular increased in abundance (Sweet & Mphinyane 1986).

In succulent thicket vegetation, goats, by suppressing vegetative regeneration, have a far more devastating effect on the dominant spekboom (*Portulacaria afra*) than do elephants and other indigenous browsers (Stuart-Hill 1992). With heavy stocking by goats, dense thickets are transformed to open shrubland characterized by small unpalatable shrubs and widely scattered umbrella-shaped trees (Fig. 17.5). Therefore, long rest periods between browsing are needed for maximum production by spekboom (Aucamp 1979).

Browsing ungulates such as kudu, impala and goats, as well as mainly grazing sheep, include a high proportion of forbs and dwarf shrubs in their diets (Owen-Smith & Cooper 1985, 1989). In the succulent karoo biome, severe browsing by sheep greatly reduced the canopy size of the dwarf shrub *Osteospermum sinuatum*, with a consequent reduction in flower and hence seed production (Milton & Dean 1990; Milton 1994; Chap. 7, this volume). This favoured the competitive

Figure 17.5 **Destruction of subtropical succulent thicket by goats, northwest of Uitenhage in the Eastern Cape. The once-dense thicket has been reduced by continuous browsing to an open community of umbrella shaped trees of mainly *Pappea capensis* and *Schotia afra*. Palatable species such as *Portulacaria afra* have been eliminated** *(Photo: M.T. Hoffman)*.

Table 17.3 **Stages of degradation of semi-arid or arid grasslands or shrublands as a result of severe pressure from large herbivores (modified from Milton et al. 1994a)**

Stage number	State of vegetation
1	Increased abundance of unpalatable, spinescent or decumbent plant species relative to palatable undefended species
2	Plant species sensitive to foliage and seed losses become locally extinct; resistant plants predominate
3	Ephemeral plant species predominate at the expense of perennials so that ground layer cover fluctuates widely in response to rainfall variability; in savanna woody plants may increase greatly at the expense of the grass component
4	Soil changes resulting from denudation lead to extensive bare areas devoid of herbaceous vegetation except temporarily after heavy rain; surviving shrubs are mostly spinescent or chemically defended

replacement of this palatable shrublet by unpalatable or spinescent dwarf shrubs and ephemeral herbs. This is explained by greater seedling recruitment of the latter species under conditions of heavy stocking coupled with droughts. Springbok, observed during drought in the central Nama-karoo, favoured shrubs such as *Pteronia* spp., unpalatable to sheep, while avoiding species of *Osteospermum* and *Walafrida*, which are palatable to sheep (Davies, Botha & Skinner 1986).

In the succulent karoo, mima-like mounds termed *heuweltjies* are sites of soil and nutrient accumulation created through the activities of termites and molerats. They attract intense herbivory by both sheep and rodents which in turn promotes short-lived forb and succulent species capable of regeneration from dormant seed banks (Chap. 7, this volume).

The exclusion of domestic ungulates from areas of succulent karoo did not result in much improvement in plant species composition or cover over periods of 5–15 years (Chap. 7, this volume). These observations led to the formulation of a conceptual model of degradation in arid shrublands, emphasizing interactions between herbivory, seed production, seedling recruitment and episodic droughts, similar in outline to the event-driven concepts developed for semi-arid savannas (Milton et al. 1994a; Table 17.3).

Black rhinoceros have a severe pruning effect on regenerating woody plants less than 1 m in height, biting off stems up to 10 mm in diameter (Goddard 1970; Joubert & Eloff 1971; Mukinya 1977; Hall-Martin, Erasmus & Botha 1982; Loutit, Louw & Seely 1987; Kotze & Zacharias 1993). They favour species of Mimosaceae and Euphorbiaceae, including stem-succulents. Black rhinos may use their horns to break stems up to 170 mm in diameter to bring foliage within reach. Eland and kudu bulls also use their horns to break branches of trees to reach inaccessible foliage (N. Owen-Smith pers. obs.). Black rhinos consume a high proportion of forbs, eating entire above-ground parts of plants as large as *Ammocharis* spp. (Amaryllidaceae; N. Owen-Smith, pers. obs.).

African elephant exert a major impact on woody vegetation, by felling, debarking, breaking leader shoots, or otherwise damaging trees and shrubs (Owen-Smith 1988; Fig. 17.6). Species of *Acacia* seem particularly vulnerable, through being shallow-rooted and having bark that is easily stripped. Baobab (*Adansonia digitata*) and other trees with soft, pithy trunks may be destroyed completely. Regenerating saplings may be broken level to the ground level or uprooted.

Along the Chobe river front in Botswana, the standing dead trunks of large trees of *Acacia* and other genera testify to the destruction of the former riparian woodland by the high local concentration of elephant. In compensation, *Croton megalobotrys* trees and the scrambler *Capparis tomentosa*, not favoured by elephants, have increased in abundance (Simpson 1974, 1975; Addy 1993). Shrubs including *Terminalia*, *Baphia*, *Bauhinia* and *Combretum* spp. seem able to maintain their populations despite severe pruning of their annual shoot growth by elephant (J. Chafota, unpubl. data). The transformation of woodlands to a stable shrub coppice state seems to be a feature of savannas growing on Kalahari sands or other sandy soils, where woody shrubs have a large fraction of their biomass underground (Rushworth 1975; Guy 1989).

Figure 17.6 **Elephant (Loxodonta africana) browsing in arid, thorny savanna at Etosha, northern Namibia** (Photo: D.M. Richardson).

Severe elephant damage to canopy trees may occur episodically, in association with events such as severe droughts, fires and severe frost (J. Chafota, unpubl. data). Mopane woodlands seem to be somewhat resistant to elephant impacts through the ability of the dominant *Colophospermum mopane* to coppice or persist in a shrub stage (Guy 1981; Fig. 17.7). In fertile savanna where trees are relatively shallow-rooted, the destruction of trees by elephant, coupled with their suppressant effect (in conjunction with fire) on regeneration, can lead to a progressive opening of the woodland component. This trend has been documented for *Acacia–Sclerocarya* parkland in the eastern basaltic region of the Kruger National Park, despite the management policy of restricting elephant abundance by annual culling (Viljoen 1988).

In succulent thicket of the Addo Elephant National Park, elephant confined at high density (2–3.8 km⁻²) reduced woody plant biomass by as much as 50% in places, over a 12-year period (Penzhorn, Robbertse & Olivier 1974). Nevertheless, there was only a 12% decline in shrub canopy volume. The predominant shrub species, including *Portulacaria afra*, seem well-adapted in their regrowth patterns to tolerate regular and severe pruning by elephant (Barratt & Hall-Martin 1991; Stuart-Hill 1992;

Fig. 17.8). However, the succulent rosette-shrub *Aloe africana* and other succulent species, as well as certain geophytes, have suffered declines in abundance in the area occupied by elephant (Moolman & Cowling 1994).

The progressive thickening of woody vegetation in the Hluhluwe Wildlife Park, resulting in some places in the replacement of grassland by thicket, could have been due largely to the elimination of elephant by hunting during the nineteenth century (Owen-Smith 1989). Heavy grazing by ungulates and resultant fire suppression promotes an expansion in the woody plant component (Chap. 18, this volume). It has proved impossible to reverse this trend by management actions conducted in the absence of elephant (Watson & Macdonald 1983). The recent reintroduction of elephant into this area will in time reveal the effectiveness of these animals in opening the thickets. The general perception that savannas throughout southern Africa are more densely wooded today than they were at the turn of the century could be due as much to the extermination of elephant as to the suppression of fires.

Before the elimination of megaherbivores by hunters, the vegetation over much of southern Africa may have existed in what would have been perceived today

Figure 17.7 **Mopane trees (*Colophospermum mopane*) felled by elephants – near the Linyati River, northern Botswana** *(Photo: N. Owen-Smith).*

Figure 17.8 **Top-down browsing by elephants results in severe pruning, but long-term persistence, of *Portulacaria afra* (spekboom) individuals in the Addo Elephant National Park, Eastern Cape** *(Photo: N. Owen-Smith).*

17.3.3 Nibbling and gnawing

Judging from the stomach contents of animals collected in the Nama-karoo biome and arid savanna of the Kalahari region, green plant material is the predominant dietary constituent of most of the common species of murid rodents in southern Africa (Smithers 1983; Kerley 1989, 1992; Kerley, Knight & Erasmus 1990). This is true even for murine field mice (e.g. *Mastomys* and *Rhabdomys* spp.) and gerbils (*Gerbillurus* and *Tatera* spp.). Vlei rats (*Otomys* spp.) and other species in the Otomyinae are strictly herbivorous, consuming the leaves and stems of reeds, grasses and dicots (Smithers 1983; Du Plessis, Kerley & Winter 1991). Succulent forbs constituted over half the diet of whistling rats (*Parotomys brantsii*) in the succulent karoo (Du Plessis 1989, cited by Milton 1992).

Gerbils can have a major below-ground and above-ground impact on grass biomass around their colonies (Korn & Korn 1989). Their feeding activities favour palatable species of *Digitaria* at the expense of unpalatable *Eragrostis* and *Aristida* species. Field mice generally concentrate their feeding under shrub or dense grass canopies, although gerbils tend to forage more in open patches than do other murid species (Kerley *et al.* 1990). Cane rats (Thryonomyidae) occupy reedbeds and other areas of tall cane-like grass, where they feed on the shoots and rhizomes of reeds and other grasses (Skinner & Smithers 1990).

In the fynbos biome, rodents are important predators of shrub seedlings, consuming the cotyledons shortly after emergence (Bond 1984). Small mammals may be equally important as seedling predators in other biomes, judging from the green material in their diets, but their specific impacts have not been documented.

Porcupines gnaw sections of bark from the base of trunks of certain tree species (Thomson 1974; Yeaton 1988). Together with fire, this damage may lead to tree mortality. Porcupines also dig for bulbs, tubers and roots (Smithers 1983). The rock hyrax (*Procavia capensis*) is both a browser and a grazer. Hyraxes can exert a severe pruning effect on trees, and compete with sheep for grazing near rocky outcrops. Springhares, usually associated with regions of Kalahari sand, consume green grass plus the bark and soft stems of woody plants, rhizomes and corms (Curtis & Perrin 1979; Smithers 1983; Kerley 1989). Ground squirrels (*Xerus inauris*), which are common in the Kalahari, feed on the leaves and stems of grasses, forbs and dwarf shrubs as well as on rhizomes and bulbs (Herzig-Straschil 1978). Hares (*Lepus* spp.) are primarily grazers (Smithers 1983), although the Cape hare (*Lepus capensis*) is largely a browser in the Nama-karoo (Kerley 1990).

In the fynbos biome, molerats are important herbiv-

as a chronically overgrazed and overbrowsed state (Owen-Smith 1988). The tendency towards an increase in woody plants in savanna as a result of heavy grazing by white rhinos and hippos would have been suppressed by the impact of elephants and other browsers on regeneration of woody plants. This would almost certainly have been the state near permanent water, although there would have been vast areas away from permanent water where fire would have been the main consumer of vegetation. Today there is no region in Africa where both a grazing megaherbivore and a browsing megaherbivore coexist at population levels limited by vegetation capacity. Such a situation may eventually develop in the Kruger National Park, if the current population trend of white rhinos continues, and policies of elephant population management are changed. It remains to be seen whether the effects thereof on the diversity of other species are positive or negative.

ores feeding on underground plant parts, including bulbs, tubers and rhizomes. They also pull whole forbs underground for feeding (Smithers 1983; Lovegrove & Jarvis 1986). In the forest biome, bushpig feed largely on rhizomes, especially those of ferns (*Pteridium aquilinum*) and various climbers (e.g. *Rhoicissus* spp.; Breytenbach & Skinner 1982; Seydack 1990).

Among birds, ostrich (*Struthio camelus*) consume primarily green annual grasses and forbs when these are available, switching to leaves, flowers and fruits from succulents at other times. Small plants may be uprooted and eaten whole (Robinson & Seely 1975; Kok 1980; Milton, Dean & Siegfried 1994b). Spurwing and Egyptian geese (Anatidae) are largely grazers on short grass in the vicinity of water bodies (Chap. 14, this volume). Guineafowl, francolins and weaver-finches (Ploceidae) include green leaves as well as seeds and insects in their diets. Several francolin species also commonly dig for corms and tubers (Milstein 1989). Frugivores such as mousebirds (Coliidae) and louries (Musophagidae) supplement their diet with flowers and green buds at times of the year when fruits are unavailable.

Tortoises consume a variety of plant parts including leaves, stems, flowers, fruits and seeds (Branch 1988). The widespread mountain or leopard tortoise favours grasses and forbs, whereas the serrated tortoise (*Psammobates oculifer*), largely restricted to the two karoo biomes and the fynbos biome, shows a strong preference for succulents and leguminous herbs when these are available (Milton 1992; Rall & Fairall 1993).

17.3.4 Invertebrates

In northern Transvaal savanna, grasshoppers were responsible for the consumption plus wastage of 15–20% of estimated primary production in the herbaceous layer, exceeding estimated consumption levels by large mammalian herbivores (Gandar 1982a; Chap. 11, this volume). However, these measurements were made during a peak in grasshopper numbers, and population levels were less than half as great in subsequent years. Notably, two of the grass species heavily used by grasshoppers produced more above-ground biomass in the presence of grasshoppers than in an experimental area from which grasshoppers were excluded (Gandar 1980). In the Nama-karoo, brown locusts are notorious for denuding the herbaceous layer along the paths traversed by hopper bands and at sites where swarms settle temporarily. However, lasting impacts have not been reported.

Grasshoppers feed primarily on leaf laminae, with much wastage occurring due to terminal parts of blades being excised (Gandar 1982b). Different species of grasshopper favour different kinds of grasses. Members of the Acridinae eat mainly soft grasses such as *Digitara eriantha*. Species in the Truxalinae tend to select tough grasses such as *Eragrostis pallens*, which are largely avoided by ungulates. Missing leaf area varied between 2% and 25% among individual grass species in northern Transvaal savanna. Interestingly, the least damaged species, *Brachiaria serrata*, is favoured by cattle, whereas the most damaged species, *Aristida aequiglumis*, is little used by ungulates. Grasshoppers in the Pyrgomorphidae, as well as certain of the Pamphagidae, Cantatopinae and Tettigoniidae, feed primarily on forbs and small shrubs. Evidently nibbling by the grasshopper guild is more evenly spread over the herbaceous layer than is grazing by ungulates.

Consumption of woody plant foliage by insects, mainly lepidopteran larvae, amounted to only 2–3% of leaf production in infertile broad-leaf savanna in northern Transvaal (Gandar 1982a; Scholtz 1982), although missing leaf area indicated higher losses of about 5% (Owen-Smith & Cooper 1987). Sporadic outbreaks can result in the removal of 20–80% of the foliage of certain of the predominant tree species in broad-leaf savanna over fairly extensive areas (Scholes & Walker 1993). Dwarf shrubs in the Nama-karoo may be subjected to similar levels of defoliation from outbreaks of the karoo caterpillar (Annecke & Moran 1977).

Lepidopteran larvae feeding on leaves of trees are active mainly in the early growing season while leaves are young. In northern Transvaal savanna, peak levels of consumption occurred during October and November (Scholtz 1982). For some tree and shrub species there was a second surge in defoliation around March. The pre-rain leaf flush of many savanna trees, which commences during September, may be a way of reducing exposure to insect herbivores (Aide 1993; see also Chap. 11, this volume). In Zimbabwe, early-flushing trees of *Brachystegia spiciformis* escaped being defoliated during two successive outbreaks of a chrysomelid beetle (Reeler *et al.* 1991). For *Acacia karroo*, leaf area loss to insect herbivores did not stimulate regrowth, so that total leaf production was reduced as a result of insect damage (Teague & Walker 1988). Insect herbivores tend to be concentrated in the upper canopies of trees, and on early phenophases, where the most negative impacts on leaf production were found (Teague 1989b).

In northern Transvaal savanna, the tree species susceptible to caterpillar outbreaks were generally those that were unpalatable to browsing ungulates, whereas several of the palatable species showed relatively little insect damage (Owen-Smith & Cooper 1987). In the succulent-karoo, the shrub species palatable to sheep showed a higher species richness of leaf-chewing insects than unpalatable shrubs, and some of the palatable plants were subject to defoliation by larvae of a chrysomelid beetle (Milton 1993).

The role of underground invertebrates that are herbivores, such as nematode worms, remains largely unknown in southern Africa.

17.4 Granivory and frugivory

17.4.1 Granivory and seed predation

In the fynbos biome, small rodents have an important influence on the recruitment patterns of shrubs, especially among the Proteaceae (Bond 1984; Mustart, Cowling & Wright 1995). Many proteoid shrubs bear serotinous cones that protect seeds from granivorous rodents and birds (Chap. 6, this volume). Few seeds are shed between fire events, and emergent seedlings in the interfire period are subject to heavy predation by rodents. The infructescences open and shed their seeds following fire, making available more seeds than preexisting rodent populations can consume. Moreover, the abundance of rodents declines in the post-fire period, probably due to lack of cover (Fraser 1990; Van Hensbergen et al. 1992). Accordingly, seedling establishment is restricted almost entirely to the post-fire stage, leading to even-aged stands (Chap. 6, this volume). Models considering plant vital attributes alone (e.g. Van Wilgen & Forsyth 1992) are inadequate for explaining the post-fire succession (Chap. 18, this volume).

Other members of the Proteaceae disperse their seeds annually (Chap. 6, this volume). For these species, the impact of rodent granivores is strongly controlled by the activities of ants, which are the main seed dispersal agents (Bond & Slingsby 1983). Ants of several genera carry the elaiosome-bearing seeds back to their nests, where they consume the elaiosomes and discard the seeds in their underground nests. Thus, buried seeds are hidden from rodents (Bond & Breytenbach 1985). The ants forage primarily during the afternoon, and remove seeds rapidly, before most rodents become active. Seeds buried by ants germinate rapidly after fire, and rates of predation on seedlings are generally low in post-burn sites (Bond 1984; Van Hensbergen et al. 1992). This mutualism between fynbos shrubs and ants is threatened by the spread of the alien Argentine ant (Linpithema humile), which consumes the elaiosome but does not bury the seeds (Bond & Slingsby 1984).

Seed-harvesting ants may facilitate the invasion of alien Acacia spp. into fynbos by burying the aril-bearing seeds, thereby removing seeds from the attention of rodents (Holmes 1990). Once these acacias attain high canopy cover, rodents decline drastically in numbers and become ineffective as seed predators.

Seeds stored in the closed infructescences of Protea spp. are subject to predation by insects such as weevils. In P. repens only 16% of seeds survived undamaged for two years post-flowering, so that continual flowering was necessary to maintain a seed store (Coetzee & Giliomee 1987). Bark-feeding beetles also cause the annual shedding of seeds stored in the cones of Clanwilliam cedar (Widdringtonia cedarbergensis), a relict tree species in the fynbos biome (Botha 1990). Those seeds released between fires are mostly eaten by rodents and do not contribute to regeneration. Baboons are also major predators on the cones of this tree species, although some seeds that are spilled may germinate.

Seed predation by birds appears to be negligible in the fynbos biome (Fraser 1990). In the Nama- and succulent karoo biomes, birds and harvester ants (Messor spp.), rather than rodents, are the main consumers of seeds (Kerley 1991; Milton & Dean 1993). Seed collection by harvester ants at a southern karoo site amounted to 10% of the annual seed crop produced by common plants. By disturbing ant-mounds during their feeding, aardvark (Orycteropus afer) promote the germination of these seeds (Dean & Yeaton 1992).

In savanna vegetation in the eastern Transvaal, ants and rodents were the major consumers of grass seeds, with birds appearing to be unimportant (Capon & O'Connor 1990). Coupled with the consumption or suppression of inflorescences by heavy grazing pressure, granivory could contribute to the eventual elimination of perennial grasses producing relatively few, large seeds, such as Themeda and Heteropogon (O'Connor 1991; O'Connor & Pickett 1992).

Tree squirrel, widespread in savanna, favour feeding on large seeds, such as those of Acacia spp. and the kernels of marula (Sclerocarya birrea) fruits, destroying them in the process (Viljoen 1977). They may collect these seeds from ungulate dung or from accumulations where antelope have ruminated. Pouched mouse (Saccostomus campestris) select medium to large seeds such as those of Acacia and Grewia spp. Other small rodents consume a variety of dicot seeds, especially those of Acacia spp., as well as grass seeds (Smithers 1983; Hoffman et al. 1989; Miller 1994b). Ground squirrels in the Kalahari consume grass seeds and various berries (Smithers 1983).

In the forest biome, red squirrel (Paraxerus palliatus) consume a variety of fruits and seeds, preferring those with a large endosperm, which they extract and consume (Viljoen 1983). Surplus seeds are buried individually, after removal of the exocarp and mesocarp, usually against a tree trunk or rock. Germination success was considerably higher for nuts of *Garcinia gerrardii* that had been dehusked by squirrels than for intact seeds. Certain fruits with very hard or possibly toxic seeds are discarded by squirrels after consumption of the mesocarp, e.g. monkey oranges (*Strychnos* spp.). In the Knysna forest, mice (*Myomyscus verreauxii*) consume a large fraction of the seeds of yellowwood trees (*Podocarpus falcatus*) beneath tree canopies within ten months of seed production (Koen 1991). Giant rats (*Cricetomys gambianus*), restricted to forests or well-wooded savanna along the east coast, eat a wide range of soft fruits, which they carry back to be consumed or store temporarily in nest chambers (Skinner & Smithers 1990).

Widely distributed avian granivores include guineafowl, francolins, doves (Columbidae) and various subfamilies of weaver-finches, canaries and other seedeaters (Fringillidae). Individual helmeted guineafowl (*Numida meleagris*) may strip thousands of grass seeds from inflorescences (Milstein in Ginn, McIlleron & Milstein 1989). Redbilled quelea (*Quelea quelea*) are notorious for their huge swarms, which can devastate cultivated grain fields as well as the seeds of indigenous grasses in arid savannas (C.J. Vernon in Ginn *et al.* 1989). Waxbills (Estrildinae) generally favour grass seeds, whereas canaries and related seedeaters tend to take daisy (Asteraceae) seeds, in both cases directly from the plant rather than off the ground. In fynbos vegetation, the Protea seedeater (*Serinus leucopterus*) specializes in consuming the large seeds of certain *Protea* spp. from the serotinous cones (Milewski 1978).

Bruchid beetles are notable predators on the seeds of *Acacia* species and other mimosaceous trees and shrubs (Coe & Coe 1987; Ernst, Decille & Tolsma 1990). Species with indehiscent pods seem to suffer the highest levels of seed predation. However, following consumption by ungulates, *Acacia* seeds infested by bruchid beetles show increased germination success (Hoffman *et al.* 1989; Miller & Coe 1993).

17.4.2 Frugivory and seed dispersal

Species of *Acacia* and other Mimosoideae prevalent in highly fertile savannas, and along riverlines in the Nama- and succulent karoo biomes, produce dry indehiscent pods that are shed when ripe (Milton, Siegfried & Dean 1990; Miller & Coe 1993; Miller 1994a). The pods are rapidly consumed from the ground by large mammalian herbivores ranging in size from steenbok to elephants. In contrast, dehiscent pods are characteristic of members of the Caesalpinioideae that predominate in infertile savanna, and also of those *Acacia* spp. that occur on less fertile soils. The fraction of *Acacia* seeds passing intact through the gut of ungulates is variable (Jarman

1976; Coe & Coe 1987; Miller & Coe 1993). Elephants pass a large fraction of seeds undamaged, while small ruminants such as impala eject most seeds while ruminating (Coe & Coe 1987). Medium-large ruminants such as kudu and cattle fragment and digest a significant fraction of the seeds consumed, especially those weakened by bruchid beetle infestation (Coe & Coe 1987; Miller & Coe 1993). Giraffe pluck immature pods of both indehiscent and dehiscent species from the tree canopy, and hence probably destroy most of the seeds (Miller 1994a).

Other savanna tree species produce large fleshy fruits that are sought out by browsing antelope as well as by baboons and monkeys. Notable examples include the marula tree and various monkey oranges. Kudu and impala spit out the kernels of marula fruits when they ruminate (N. Owen-Smith, pers. obs.). Black rhino and giraffe feed on the enormous fibrous fruits of the sausage tree (*Kigelia africana*; Owen-Smith 1988).

The benefit to the plants of fruit consumption by elephants or ungulates probably lies primarily with the translocation of seeds away from parent tree canopies, where seed predators concentrate their foraging. The deposition of *Acacia* seeds in dung does not enhance germination success directly (Miller & Coe 1993). However, burial of seeds contained in dung by dung beetles may remove seeds from the attention of seed predators and facilitate successful germination (Coe & Coe 1987; Stuart-Hill 1992).

Fruits form a major component of the diets of baboons (*Papio ursinus*), monkeys (*Cercopithecus aethiops* and *C. mitis*) and thick-tailed bushbabies (*Otolemur crassicaudatus*; Stoltz & Saayman 1970; Smithers 1983; Lawes, Henzi & Perrin 1990; Lawes 1991). Baboons in particular occupy a wide range of habitats, from high-altitude grassland through savanna to fynbos. Fruits of shrubs such as stamvrug (*Bequaertiodendron magalismontanum*), sourplum (*Ximenia caffra*) and wild apricot (*Ancyclobotris capensis*), common in rocky hills in the Transvaal, have a slippery coating over the seeds. This may facilitate the fruit being swallowed intact by baboons and other primates.

Bushpigs consume many types of fruits in the forest biome, and seedlings emerge from their faeces (Seydack 1990). Elephants persisting in the Knysna forest, in contrast to elephants elsewhere in Africa, do not play an important role in dispersing the seeds of tree and shrub species (Koen 1983).

Frugivorous birds such as mousebirds, louries, starlings (Sturnidae), barbets (Capitonidae) and green pigeons (*Treron calva*) disperse seeds of many tree, shrub, vine and epiphyte species. In forest patches in the fynbos biome, a quarter or more of the trees are dependent primarily on birds for seed dispersal (Knight & Siegfried

1983; see also Koen 1988; Manders & Richardson 1992; Chap. 12, this volume). In savanna vegetation, the bush-clumps associated with termite mounds may be constituted largely of species with bird-dispersed fruits (Bews 1917). Tortoises disperse viable seeds of grasses and forbs in their dung (Milton 1992).

17.4.3 Consumption of flowers and inflorescences

Heavy grazing by cattle and sheep can result in the near total consumption of grass inflorescences (Tainton & Booysen 1964). Zebra actively seek out the inflorescences of certain grass species (Winkler 1992; M.E. Joos-Vandewalle, unpubl. data), while oribi (*Ourebia ourebi*), the smallest grazing antelope, selectively consume the inflorescences of both grasses and forbs (Reilly, Theron & Du Plessis 1990). A small fraction of the small seeds of certain grasses (e.g. *Cynodon* spp.) and other plants ingested by cattle, and possibly other ungulates, pass through the digestive tract intact and germinate in the dung (Wilson & Hennessy 1977; Janzen 1984).

Browsing ruminants consume flowers of trees or shrubs when these are abundantly available, particularly those produced early in the season before leaf flush (Owen-Smith & Cooper 1985: du Toit 1990). The flowers of *Acacia nigrescens* form much of the diet of giraffe in the pre-leaf flush period (du Toit 1990). Since the inflorescences are readily accessible to feeding giraffe, it is possible that the trees benefit by achieving long-distance pollen transfer in exchange for losing some of their flower crop. The inflorescences of certain Proteaceae are designed to attract rodents as pollination agents (Wiens *et al.* 1983; Chap. 6, this volume).

Ostrich feed selectively on the flowers of many dwarf shrub species (Milton & Dean 1990). Flowers and flower buds can form a significant proportion of the diets of seed- or fruit-eating birds such as mousebirds, louries and canaries (Milewski 1978; Ginn *et al.* 1989).

17.5 Anti-herbivore defences

17.5.1 Plant morphology

A number of common grass species in southern African have a prostrate growth habit and regenerate by means of stolons (e.g. *Cynodon dactylon*, *Urochloa mosambicensis*). These are classified by rangeland scientists as 'increaser' grasses that show increased abundance in areas heavily grazed by livestock (Tainton 1981; Danckwerts 1989). Some grass species exhibit both erect and decumbent growth forms, depending on grazing pressure (e.g. *Panicum coloratum*). Other grasses have tough, fibrous leaves (e.g. *Eragrostis plana*; O'Reagain 1993).

The shrub *Portulacaria afra* shows a growth pattern that seems designed to tolerate browsing by elephants (Stuart-Hill 1992). Branches bend downwards and root from the nodes that touch the ground, producing an apron of rooted side branches. The multi-stemmed shrub spreads laterally, with separate individuals forming where connecting branches become detached – a counterpart of the stoloniferous growth pattern of grasses! Other potential defences exhibited by trees or shrubs include thick terminal stems, which inhibit browsing of shoot tips by large ungulates such as kudus (e.g. *Ehretia rigida*, *Gardenia volkensii*), and brittle stems, which dilute the nutritional value of associated leaves (e.g. *Strychnos pungens*; N. Owen-Smith pers. obs.). The small, scale-like leaves and twiggy growth forms of many dwarf shrub species may also serve to reduce their nutritional value to mammalian herbivores.

Some 10% of the southern African woody flora exhibits spinescence of various forms, including sharp-tipped branches, peduncles or leaves, or sharp outgrowths from the leaves or stems (Milton 1991). Thorns and similar structures do not prevent browsing, because ungulates have counter-adaptations to cope with these structures. However, thorns and spines restrict bite sizes and slow down biting rates, thereby reducing leaf losses to herbivores (Cooper & Owen-Smith 1986). This effect is enhanced by having small leaves distributed along stems, as is typical of *Acacia* spp. Certain tree species occurring locally within the grassland biome, where giraffe and other large browsers are absent, lack thorns when mature, although these species consistently bear thorns in savanna (e.g. *Acacia caffra*, *Ziziphus mucronata*).

Spinescent trees and shrubs predominate in fertile savanna (Huntley 1982; Owen-Smith & Cooper 1987). Within the savanna biome, the highest levels of spinescence (>30% of plant species) are associated with endoreic drainage basins or river floodplains (Milton 1991). The proportion of spinescent species is lower on sandy soils than on shale- or silt-derived soils. In the succulent karoo, trees and shrubs associated with pans and drainage lines show a high level of spinescence (Milton *et al.* 1990; Milton 1991). Prickles are typical of many leaf- and stem-succulents such as *Aloe* and *Euphorbia* spp. Relatively few plants are spinescent in the fynbos biome, where, owing to low soil fertility, ungulates were probably never abundant, and the exceptions have leaf prickles rather than thorns or stem spines (Campbell 1986).

Leaf hairs of various forms may inhibit feeding by insect herbivores. Several of the grass species that are palatable to large herbivores, but are eaten relatively little by grasshoppers, have pubescent leaves (e.g. *Brachiaria serrata*, *Panicum maximum*; Gandar 1982b). Among the savanna shrubs showing little damage by insects, *Grewia flavescens* has stellate hairs appressed against the leaf surface. The tree *Combretum molle*, with very pubescent leaves, had no Lepidoptera larvae recorded on it at Nylsvley in the Northern Province, but had its leaves nibbled by beetle larvae (Scholtz 1982).

17.5.2 Secondary chemical contents

In broad-leaf savanna growing on infertile sandy soil, 75% of the biomass of woody plant foliage was made up of species unpalatable to large herbivores, compared with only 6% in *Acacia*-dominated savanna on fertile sites (Owen-Smith & Cooper 1987). The main chemical factor discriminating palatable from unpalatable woody plants was the condensed tannin content in leaves, relative to protein levels (Cooper & Owen-Smith 1985; Cooper, Owen-Smith & Bryant 1988; Owen-Smith 1993a). High levels of total polyphenols, including hydrolysable tannins, had little inhibiting effect on feeding by the browsing ruminants studied (kudu, impala and goats).

In the same area, trees with high tannin levels were subject to irregular outbreaks of defoliating insects, as well as chronic leaf area losses to caterpillars (Owen-Smith & Cooper 1987). This suggests that tannins may be less effective as defences against insect herbivores than against ruminants. There may also be differences among mammalian herbivores in their tolerance for different forms of secondary chemical. Elephant discard the leaves of *Combretum* shrubs (J. Chafota, unpubl. data), and black rhinoceros do not favour *Combretum* species (R. H. Emslie pers. comm.), although trees within this genus are readily browsed by kudu and giraffe (Hall-Martin 1974; Owen-Smith 1979; Owen-Smith & Cooper 1987). The preferences of forest duikers among species of fallen leaves were unrelated to condensed tannin levels (Perrin, Bowland & Fauries 1992). Plant species with aromatic leaves, indicating the presence of monoterpenoid oils, are readily eaten by kudu (Cooper *et al.* 1988). However, both sheep and ostrich avoided dwarf shrub species with high concentrations of phenolics or ether-extractable oils in the succulent karoo (Milton *et al.* 1994b).

Typical fynbos shrubs such as *Protea* spp. do not show high levels of phenolics in their leaves (Glyphis & Puttick 1988). Their unpalatability may be largely a consequence of low levels of nutrients, coupled with sclerophylly (Chap. 16, this volume).

Aromatic grasses such as *Bothriochloa* and *Cymbopogon* spp. are generally avoided by cattle (Danckwerts 1989), although wild ungulates such as zebras are less deterred (Winkler 1992). Condensed tannins are present in the epidermal cells of certain grasses in the Andropogonae (Ellis 1990; Du Toit, Wolfson & Ellis 1991; Chesselet, Wolfson & Ellis 1992; see also Chap. 10, this volume).

Locusts were less inclined to feed on an ecotype of *Eragrostis chloromelas* containing condensed tannins than on plants of the same species lacking this compound (Dini & Owen-Smith 1995). The deterrent effects of grass tannins on ungulates remain to be demonstrated.

Certain herbaceous plants are rejected by ungulates as food, and some of these species have been shown to contain toxic chemicals. The low, lignotuberous shrub gifblaar (*Dichapetalum cymosum*) contains the highly potent toxin monofluoroacetate. Domestic cattle and goats are poisoned and die if they eat more than very small amounts of its leaves (Vickery & Vickery 1973). Wild ungulates such as eland are somewhat more tolerant of the poison, allowing them to learn to avoid the species (Basson *et al.* 1982). Pyrrolizidine alkaloids, which are potent liver toxins, occur in certain *Senecio* and *Helichrysum* spp. (Asteraceaé). Many succulent plants are toxic to livestock, including members of the Mesembryanthemaceae, Crassulaceae, Euphorbiaceae and Asclepiaceae (Kellerman, Coetzer & Naude 1988), owing to the presence of oxalic acid, milky latex or other factors.

17.6 Summary and conclusions

The southern African region offers examples of a wide range of interactions between plants and herbivores, in an exceptional variety of environmental contexts. Herbivores of various forms change the grass–woody plant balance in savanna, alter post-fire succession in fynbos and modify plant species composition in all biomes. Morphological and chemical attributes of many plant species testify to the strong impact of herbivory on plant design.

The impacts of herbivores on vegetation differ markedly between regions of nutrient-rich and nutrient-poor soils. Where soils are inherently fertile, plants produce mostly palatable foliage, are commonly spinescent or show other growth forms restricting losses to herbivores, frequently rely on herbivores for seed dispersal and are vulnerable to extinction in the localities heavily used by herbivores. In nutrient-deficient regions, plants are predominantly chemically defended, commonly have large underground reserves enabling them to resist both fire and herbivore damage and are subject to shifts in competitive balance mediated by herbivory. This distinction is most evident in the savanna biome, and was recognized by cattle pastoralists in South Africa long before it was given scientific form (Huntley 1982; Scholes & Walker 1993). Similar contrasts are apparent in semi-arid shrublands, where nutrients accumulate in certain localities in the landscape. The distinction does not feature in the fynbos biome, perhaps only because the renosterveld that formerly prevailed on the more fertile sub-strata has largely been cultivated. The effects of geological substratum on soil nutrient status, and hence on herbivore impacts, appear more strikingly evident in southern Africa and other regions of the African conti-nent, than elsewhere in the world. This is perhaps because much of Africa, and southern Africa in particu-lar, is 'high and dry', relative to other continents. This associates land surfaces with geology (Cole 1986), and reduces the effects of leaching on soils.

Different kinds of herbivore may be complementary to some extent in their impacts on different plant spec-ies. The grasses favoured by cattle may not be those sought out by wild ungulates, small mammals or grass-hoppers. There is suggestive evidence that insects and large mammals differ in their sensitivity to specific sec-ondary chemicals, and that differences may even exist in chemical tolerance between ruminants and non-ruminants among large herbivores. Physical deterrents against large mammalian herbivores are ineffective against insects, or small mammals. No plant defence is effective against all kinds of herbivores, nor against any one herbivore under all conditions. Apparent deterrents may be exaptative consequences of adaptations for other functions. For example, tannins may be designed to pro-tect leaves against microbial pathogens (Cooper & Owen-Smith 1985). In the international literature, generaliza-tions about plant anti-herbivore defences tend to be based largely on insect–plant interactions. The southern African region, with its variety of herbivores from very large to small, highlights the need to be far more cau-tious in postulating general patterns.

In semi-arid to arid, hence highly variable, environ-ments the major impacts of herbivores on vegetation composition occur during restricted periods following droughts or other disturbances. Accordingly the concept of event-driven ecosystems was developed largely in sou-thern Africa, although this model may be widely appli-cable to semi-arid regions elsewhere in the world (Walker *et al.* 1986; Ellis & Swift 1988; Mentis *et al.* 1989; Westoby *et al.* 1989). Processes operate more through the effects of herbivores on seed production, seed bank per-sistence and seedling survival, than via mortality among established plants, which is primarily drought-related. For both grasses and dwarf shrubs, there is a strong interaction between the effects of grazing ungulates on inflorescence production, and those of granivorous rod-ents, birds and ants on seed survival. In combination with droughts, these effects can lead to the extinction of susceptible plant species, at least on a local scale.

Dispersal of seeds from the vicinity of parent plants where seed predators concentrate, through the agency of frugivorous mammals and birds, may be crucially important for successful reproduction, as well as for re-establishment following local extinction. Among shrubs in the fynbos biome, there is a fascinating interaction between granivorous rodents and ants that act as seed dispersal agents. Herbivores feeding on flowers may facilitate cross-pollination over long distances.

High stocking-levels of domestic livestock have been blamed for widespread degradation of the grass cover, exacerbation of the problem of 'bush encroachment' in savanna, and for the spread of karroid shrubland into the eastern section of the grassland biome (Acocks 1953; cf. Chap. 21, this volume). High concentrations of indigenous large herbivores in favoured localities can also change vegetation cover and composition. The main difference between the impacts of domestic versus wild herbivores lies in the spatial scale of vegetation trans-formation. Widespread provision of surface water, nutri-ent supplementation through licks, elimination or sup-pression of diseases and ectoparasites, and eradication of predators, have promoted the expansion of domestic livestock at high densities through most of the southern African region. In commercial farming areas, drought-related mortality of animals is averted by fodder banks or supplementary forage, so that grazing pressure is maintained during the post-drought recovery phase. When similar support is extended to wildlife ranches, corresponding patterns of vegetation deterioration occur (Walker *et al.* 1987). In the past, grazing and browsing pressure was far more heterogeneous both spatially and temporally. The extent to which the perceived over-stocking of domestic livestock in communal grazing lands has altered vegetation cover and production remains unclear (Chap. 21, this volume). Species compo-sition may have changed, but the grass and shrub species that persist are those that have adaptations allowing them to survive despite high herbivore pressure. These herbivore-tolerant plants are an inherent component of the natural flora.

Elephants can drastically alter vegetation physiog-nomy through their destructive impact on trees. In sou-thern Africa, the corresponding impact of a grazing megaherbivore, the white rhinoceros, on the structure of the grass layer is strikingly evident. In combination, these keystone herbivores may in the past have exerted the prime controlling influence on the tree–shrub–grass balance over much of the savanna biome (Owen-Smith 1988, 1989). Fire may have had a somewhat more restric-ted role, in areas far from surface water, or where infer-tile soils predominated (Chap. 18, this volume). In the Nama-karoo, succulent karoo, and fynbos biomes, where indigenous large mammals were less abundant or highly vagile, it is small mammals, and insects such as the karoo caterpillar and perhaps locusts, that are the keystone herbivores. However, today it is domestic livestock that have largely taken over this role through most of southern Africa.

17.7 Acknowledgments

We are grateful to Sue Milton, Richard Dean, Tim O'Connor, Tony Milewski, Dave Richardson and Richard Cowling for their many helpful comments and suggestions for improvements to this chapter.

7.8 References

:ocks, J.P.H. (1953). Veld Types of South Africa. *Memoirs of the Botanical Survey of South Africa,* 28, 1–192.

ldy, J.E. (1993). *Impact of Elephant Induced Vegetation Change on the Status of the Chobe Bushbuck Along the Chobe River, Northern Botswana.* MSc Thesis. Johannesburg: University of the Witwatersrand.

de, T.M. (1993). Patterns of leaf development and herbivory in a tropical understorey community. *Ecology,* 74, 455–66.

nnecke, D. & Moran, V.C. (1977). Critical reviews of biological test control in South Africa. 1. The Karoo caterpillar *Loxostege frustalis* Zeller (Lepidoptera: Pyralidae). *Journal of the Entomology Society of Southern Africa,* 40, 127–45.

non. (1923). *Drought Investigation Commission Final Report.* Pretoria: Government Printer.

non. (1993). *Livestock Census.* Pretoria: Department of Agriculture.

non. (1994). *Livestock Numbers in TBVC Countries.* Pretoria: Development Bank of South Africa.

ucamp, A.J. (1979). *Die Produksiepotensiaal van die Vallei Bosveld as Weiding vir Boer en Angora bokke.* PhD Thesis. Pretoria: University of Pretoria.

arratt, G.G. & Hall-Martin, A. (1991). The effect of indigenous browsers on valley bushveld in the Addo Elephant National Park. In *Proceedings of the First Valley Bushveld/Subtropical Thicket Symposium,* ed. P.J.K. Zacharias, G.G. Stuart-Hill & J. Midgley, pp. 14–16. Howick: Grassland Society of Southern Africa.

Basson, P.A., Norval, A.G., Hofmeyr, J.M., Ebedes, H. & Schultz, R.A. (1982). Antelope and poisonous plants: I. Gifblaar containing monofluoroacetates. *Madoqua,* 13, 59–70.

Bell, R.H.V. (1982). The effect of soil nutrient availability on community structure in African ecosystems. In *Ecology of Tropical Savannas,* ed. B.J. Huntley & B.H. Walker, pp. 193–216. Berlin: Springer-Verlag.

Belsky, A.J. (1984). Role of small browsing mammals in preventing woodland regeneration in the Serengeti National Park, Tanzania. *African Journal of Ecology,* 22, 271–80.

Berry, H.H. (1981). Population structure, mortality patterns and a predicted model for estimating future trends in wildebeest numbers in the Etosha National Park. *Madoqua,* 12, 255–66.

Bews, J.W. (1917). Plant succession in the thornveld. *South African Journal of Science,* 14, 153–73.

Bigalke, R.C. (1979). Aspects of vertebrate life in fynbos, South Africa. In *Ecosystems of the World. 9A. Heathlands and Related Shrublands,* ed. R.L. Specht, pp. 81–95. Amsterdam: Elsevier.

Bond, W.J. (1984). Fire survival of Cape Proteaceae – influence of fire season and seed predators. *Vegetatio,* 56, 65–74.

Bond, W.J. & Breytenbach, G.J. (1985). Ants, rodents and seed predation in Proteaceae. *South African Journal of Zoology,* 20, 150–4.

Bond, W.J. & Slingsby, P. (1983). Seed dispersal by ants in shrublands of the Cape Province and its evolutionary implications. *South African Journal of Science,* 79, 231–4.

Bond, W.J. & Slingsby, P. (1984). Collapse of an ant plant mutualism: the Argentine ant and myrmechochorous Proteaceae. *Ecology,* 65, 1031–7.

Bosch, O.J.H., Kellner, K. & Scheepers, S.H.E. (1989). Degradation models and their use in determining the condition of southern African grasslands. *Proceedings of the 16th International Grassland Congress,* pp. 1643–4.

Botha, S.A. (1990). Seed-bank dynamics in the Clanwilliam cedar and the implications of beetle-triggered seed release for the survival of species. *South African Journal of Ecology,* 1, 53–9.

Bowland, A.E. & Perrin, M.R. (1988). The effect of overgrazing on the small mammals in Umfolozi Game Reserve. *Zeitschrift für Saugetierkunde,* 54, 251–60.

Branch, W. (1988). *Field Guide to the Snakes and other Reptiles of Southern Africa.* Cape Town: Struik.

Breytenbach, G.J. & Skinner, J.D. (1982). Diet, feeding and habitat utilization by bushpigs. *South African Journal of Wildlife Research,* 12, 1–7.

Campbell, B.M. (1986). Plant spinescence and herbivory in a nutrient poor ecosystem. *Oikos,* 47, 168–72.

Capon, M.H. & O'Connor, T.G. (1990). The predation of perennial grass seed. in Transvaal savanna grasslands. *South African Journal of Botany,* 56, 11–15.

Chesselet, P., Wolfson, M.M. & Ellis, R.P. (1992). A comparative biochemical study of plant polyphenols in southern African grasses. *Journal of the Grassland Society of Southern Africa,* 9, 119–25.

Coaton, W.G.H. (1958). *The Hodotermitid Harvester Termites of South Africa*. Science Bulletin No 375. Pretoria: Department of Agriculture.

Coe, M. & Coe, C. (1987). Large herbivores, acacia trees and bruchid beetles. *South African Journal of Science*, **83**, 624-35.

Coe, M.J., Cumming, D.H. & Phillipson, J. (1976). Biomass and production of large African herbivores in relation to rainfall and primary production. *Oecologia*, **22**, 341-54.

Coetzee, J.H. & Giliomee, J.H. (1987). Seed predation and survival in infructescences of *Protea repens* (Proteaceae). *South African Journal of Botany*, **53**, 61-4.

Cole, M.M. (1986). *The Savannas. Biogeography and Geobotany*. New York: Academic Press.

Cooper, S.M. & Owen-Smith, N. (1985). Condensed tannins deter feeding by browsing ungulates in a South African savanna. *Oecologia*, **67**, 142-6.

Cooper, S.M. & Owen-Smith, N. (1986). Effects of plant spinescence on large mammalian herbivores. *Oecologia*, **68**, 446-55.

Cooper, S.M., Owen-Smith, N. and Bryant, J.P. (1988). Foliage acceptability to browsing ruminants in relation to seasonal changes in the leaf chemistry of woody plants in South African savanna. *Oecologia*, **75**, 336-42.

Cottrell, C.B. (1985). The absence of coevolutionary associations with Capensis floral element plants in the larval/plant relationships of southwestern Cape butterflies. In *Species and Speciation*, ed. E.S. Vrba, pp. 115-24. *Transvaal Museum Monographs 4*. Pretoria: Transvaal Museum.

Curtis, B.R. & Perrin, M.R. (1979). Food preferences of the vlei rat and the four-striped mouse. *South African Journal of Zoology*, **14**, 224-9.

Danckwerts, J.E. (1989). The animal/plant interaction. In *Veld Management in the Eastern Cape*, ed. J.E. Danckwerts & W.R. Teague, pp. 37-46. Stutterheim: Pasture Research Section, Eastern Cape Region.

Danckwerts, J.E. & Stuart-Hill, G.C. (1988). The effect of severe drought and management after drought on mortality and recovery of semi-arid grassveld. *Journal of the Grassland Society of Southern Africa*, **5**, 218-22.

Davies, R.A.G., Botha, P. & Skinner, J.D. (1986). Diet selection by springbok and merino sheep during a karoo drought. *Transactions of the Royal Society of South Africa*, **46**, 165-76.

Dean, W.R.J. & Macdonald, I.A.W. (1994). Historical changes in stocking rates of domestic livestock as a measure of semi-arid and arid rangeland degradation in the Cape Province, South Africa. *Journal of Arid Environments*, **26**, 281-98.

Dean, W.R.J. & Yeaton, R.J. (1992). The importance of harvester ant nest mounds on germination sites in the southern Karoo, South Africa. *African Journal of Ecology*, **30**, 335-45.

de Villiers, W.M. (1988). On the plague dynamics of the brown locust. In *Proceedings of the Locust Symposium*, ed. B. McKenzie & M. Longridge, pp. 41-9. South African Institute of Ecologists Bulletin Special Issue. Cape Town: South African Institute of Ecologists.

Dini, J. & Owen-Smith. N. (1995). Condensed tannin in *Eragrostis chloromelas* leaves deters feeding by a generalist grasshopper. *African Journal of Range and Forage Science*, **12**, 49-52.

Downing, B.H. (1972). *A Plant Ecological Survey of the Umfolozi Game Reserve, Zululand*. PhD Thesis. Pietermaritzburg: University of Natal.

du Plessis, A., Kerley, G.J.H. & Winter, P.E.D. (1991). Dietary patterns of two herbivorous rodents: *Otomys unisulcatus* and *Paratomys brantsi* in the Karoo. *South African Journal of Zoology*, **26**, 51-4.

du Toit, E.W., Wolfson, M.M. & Ellis, R.P. (1991). The presence of condensed tannin in the leaves of *Eulalia villosa*. *Journal of the Grassland Society of Southern Africa*, **8**, 74-6.

du Toit, J.T. (1990). Giraffe feeding on *Acacia* flowers: predation or pollination? *African Journal of Ecology*, **218**, 63-8.

Ellis, J.E. & Swift, D.M. (1988). Stability of African pastoral ecosystems: alternate paradigms and implications for development. *Journal of Range Managment*, **41**, 450-9.

Ellis. R.P. (1990). Tannin-like substances in epidermal cells in the South African Poaceae: a taxonomic survey of a previously unrecorded attribute of tropical grasses. *Memoirs of the Botanical Survey of South Africa*, **59**, 1-80.

Ernst, W.H.O., Decille, J.E. & Tolsma, D.J. (1990). Predispersal seed predation in native leguminous shrubs and trees in savannas of southern Botswana. *African Journal of Ecology*, **28**, 45-54.

Foran, B.D., Tainton, N.M. & Booysen, P. de V. (1978). The development of a method for assessing veld condition in three grassveld types in Natal. *Proceedings of the Grassland Society of Southern Africa*, **13**, 27-33.

Ford, J. (1971). *The Role of Trypanosomiasis in African Ecology*. Oxford: Clarendon Press.

Fraser, M.W. (1990). Small mammals, birds and ants as seed predators in post-fire mountain fynbos. *South African Journal of Wildlife Research*, **20**, 52-6.

Fritz, H. & Duncan, P. (1994). On the carrying capacity for large ungulates of African savanna ecosystems. *Proceedings of the Royal Society of London*, **256**, 77-82.

Gandar, M.V. (1980). The short-term effects of the exclusion of large mammals and of insects in a broad-leafed savanna. *South African Journal of Science*, **76**, 29-31.

Gandar, M.V. (1982a). Trophic ecology and plant herbivore energetics. In *The Ecology of Tropical Savannas*, ed. B.J. Huntley & B.H. Walker, pp. 514-34. Berlin: Springer-Verlag.

Gandar, M.V. (1982b). The dynamics and trophic ecology of grasshoppers in a South African savanna. *Oecologia*, **54**, 370-8.

Ginn, P.J., McIlleron, W.G. & Milstein, P.D.S. (1989). *The Complete Book of Southern African Birds*. Cape Town: Struik.

yphis, J.P. & Puttick, G.M. (1988). Phenolics in some southern African mediterranean shrubland plants. *Phytochemistry*, 27, 743–51.

ddard, J. (1970). Food preferences of black rhinoceros in the Tsavo National Park. *East African Wildlife Journal*, 8, 145–61.

ay, P.R. (1981). Changes in the biomass and productivity of woodlands in the Sengwa Wildlife Research Area, Zimbabwe. *Journal of Applied Ecology*, 18, 507–19.

ay, P.R. (1989). The influence of elephants and fire on a *Brachystegia-Julbernardia* woodland in Zimbabwe. *Journal of Tropical Ecology*, 5, 215–25.

all, M. (1984). Man's historical and traditional use of fire in southern Africa. In *Ecological Effects of Fire in South African Ecosystems*, ed. P. de V. Booysen & N.M. Tainton. Berlin: Springer-Verlag.

all-Martin, A.J. (1974). Food selection by Transvaal lowveld giraffe as determined by analysis of stomach contents. *Journal of Southern African Wildlife Management Association*, 4, 191–202.

all-Martin, A.J., Erasmus, T. & Botha, B.P. (1982). Seasonal variation of diet and faeces composition of black rhinocerous in the Addo Elephant National Park. *Koedoe*, 25, 63–82.

arrison, Y. (1993). *Herbaceous Recovery of Communal Grasslands After Removal of High Continuous Grazing Pressure*. MSc Thesis. Johannesburg: University of the Witwatersrand.

erzig-Straschil, B. (1978). On the biology of *Xerus inauris. Zeitschrift für Saugetierkunde*, 43, 262–78.

offman, M.T., Cowling, R.M., Dowie, C. & Pierce, S.M. (1989). Seed predation and germination of *Acacia erioloba* in the Kuiseb River Valley, Namib Desert. *South African Journal of Botany*, 55, 103–6.

olmes, P.M. (1990). Dispersal and predation in alien *Acacia. Oecologia*, 83, 288–90.

untley, B.J. (1982). Southern African savannas. In *Ecology of Tropical Savannas*, ed. B.J. Huntley & B.H. Walker, pp. 101–19. Berlin: Springer-Verlag.

Janzen, D.H. (1984). Dispersal of small seeds by big herbivores: foliage is the fruit. *American Naturalist*, 123, 338–53.

Jarman, P.J. (1976). Damage to *Acacia tortilis* seeds eaten by impala. *East African Wildlife Journal*, 14, 223–5.

Joos-Vandewalle, M.E. (1988). *Abundance and Distribution of Large Herbivores in Relation to Environmental Factors in Savuti, Chobe National Park, Botswana*. MSc Thesis. Johannesburg: University of the Witwatersrand.

Joubert, E. & Eloff, F.C. (1971). Notes on the ecology and behaviour of the black rhinoceros in South West Africa. *Madoqua*, 3, 5–54.

Kellerman, T.S., Coetzer, J.A.W. & Naude, T.W. (1988). *Plant Poisonings and Mycotoxicoses of Livestock in Southern Africa*. Cape Town: Oxford University Press.

Kerley, G.J.H. (1989). Diet of small mammals from the Karoo, South Africa. *South African Journal of Wildlife Research*, 19, 67–72.

Kerley, G.J.H. (1990). Browsing by *Lepus capensis* in the Karoo. *South African Journal of Zoology*, 25, 199–200.

Kerley, G.J.H. (1991). Seed removal by rodents, birds and ants in the semi-arid Karoo, South Africa. *Journal of Arid Environments*, 29, 63–139.

Kerley, G.J.H. (1992). Trophic status of small mammals in the semi-arid Karoo, South Africa. *Journal of Zoology, London*, 226, 563–72.

Kerley, G.J.H., Knight, M.H. & Erasmus, T. (1990). Small mammal microhabitat use and diet in the southern Kalahari, South Africa. *South African Journal of Wildlife Research*, 20, 123–6.

Knight, R.S. & Siegfried, W.R. (1983). Inter-relationships between type, size and colour of fruits and dispersal in southern African trees. *Oecologia*, 56, 405–12.

Koen, J.H. (1983) Seed dispersal by the Knysna elephants. *South African Forestry Journal*, 124, 56–8.

Koen, J.H. (1988). Stratal distribution and resource partitioning of birds in the Knysna Forest, South Africa. *African Journal of Ecology*, 26, 229–38.

Koen, J.H. (1991). The effect of rodent granivory on recruitment of the irregularly fruiting *Podocarpus falcatus* in the southern Cape. *South African Forestry Journal*, 159, 25–8.

Kok, O.B. (1980). Voedselinname van volstruise in die Namib-Naukluftpark, Suidwes-Afrika. *Madoqua*, 12, 155–61.

Korn, H. (1987). Densities and biomasses of non-fossorial southern African savanna rodents during the dry season. *Oecologia*, 72, 410–13.

Korn, H. & Korn, V. (1989). The effects of gerbils on primary production and plant species composition in a South African savanna. *Oecologia*, 79, 271–8.

Kotze, D.C. & Zacharias, P.J.K. (1993). Utilisation of woody browse and habitat by the black rhino in western Itala Game reserve. *African Journal of Range and Forage Science*, 10, 36–40.

Lawes, M.J. (1991). Diet of samango monkeys in the Cape Vidal dune forest, South Africa. *Journal of Zoology, London*, 224, 149–73.

Lawes, M.J., Henzi, S.P. & Perrin, M.R. (1990). Diet and feeding behaviour of samango monkeys in Ngoye Forest, South Africa. *Folia Primatologica*, 54, 57–69.

Lea, A. (1964). Some major factors in the population dynamics of the brown locust. In *Ecological Studies in Southern Africa*, ed. D.H.S. Davis, pp. 269–83. The Hague: W. Junk.

Loutit, B.D., Louw, G.N. & Seely, M.K. (1987). First approximation of food preferences and the chemical compaction of the diet of the desert-dwelling black rhinoceros. *Madoqua*, 15, 35–54.

Lovegrove, B.G. & Jarvis, J.U.M. (1986). Coevolution between mole-rats (Bathyergidae) and a geophyte *Micranthus* (Iridaceae). *Cimbebasia (Series A)*, 8, 79–85.

Maddock, L. (1979). The 'migration' and the grazing succession. In *Serengeti. Dynamics of an Ecosystem*, ed. A.R.E. Sinclair & M. Norton-Griffiths, pp. 104–29. Chicago: University of Chicago Press.

Manders, P.T. & Richardson, D.M. (1992). Colonization of Cape fynbos communities by forest species. *Forest Ecology and Management*, **48**, 277–93.

McKenzie, B. (1982). Resistance and stability of the grasslands of the Transkei. *Proceedings of the Grassland Society of Southern Africa*, **17**, 21–4.

McKenzie, B. (1987). Composition, pattern and diversity of some Transkeian grasslands. *Journal of the Grassland Society of Southern Africa*, **4**, 135–8.

McNaughton, S.J. (1984). Grazing lawns: animals in herds, plant form and coevolution. *American Naturalist*, **124**, 863–86.

McNaughton, S.J. & Georgiadis, N.J. (1986). Ecology of African grazing and browsing mammals. *Annual Review of Ecology and Systematics*, **17**, 39–65.

Meissner, H.H., Hofmeyr, H.F., van Rensburg, W.J.J.J. & Pienaar, J.P. (1983). *Classification of Livestock for Realistic Prediction of Substitution Values in Terms of a Biologically Defined Large Stock Unit*. Technical Communication No. 175. Pretoria: Department of Agriculture.

Melton, D.A. (1987). Habitat selection and resource scarcity. *South African Journal of Science*, **83**, 646–51.

Mentis, M.T., Grossman, D., Hardy, M.B., O'Connor, T.G. & O'Reagain, P.J. (1989). Paradigm shifts in South African range science, management and administration. *South African Journal of Science*, **85**, 684–7.

Milewski, A.V. (1978). Diet of *Serinus* species in the southern Cape. *Ostrich*, **49**, 174–84.

Miller, M.F. (1994a). The fate of mature African *Acacia* pods and seeds during their passage from the tree to the soil. *Journal of Tropical Ecology*, **10**, 183–96.

Miller, M.F. (1994b). Seed predation by nocturnal rodents in an African savanna ecosystem. *South African Journal of Zoology*, **29**, 262–6.

Miller, M.F. & Coe, M. (1993). Is it advantageous for *Acacia* seeds to be eaten by ungulates? *Oikos*, **66**, 364–8.

Milstein, P. le S. (1989). Francolins, partridges and quails. In *The Complete Book of Southern African Birds*, ed. P.J. Ginn, W.G. McIlleron & P. le S. Miltein, pp. 176–89. Cape Town: Struik Winchester.

Milton, S.J. (1991). Plant spinescence in arid southern Africa: does moisture mediate selection by mammals? *Oecologia*, **87**, 279–87.

Milton, S.J. (1992). Plants eaten and dispersed by adult leopard tortoises in the southern Karoo. *South African Journal of Zoology*, **27**, 45–9.

Milton, S.J. (1993). Insects from the shrub *Osteospermum sinuatum* and *Pteronia pallens* (Asteraceae) in the southern Karoo. *African Entomology*, **1**, 257–61.

Milton, S.J. (1994). Growth, flowering and recruitment of shrubs in grazed and in protected rangeland in the arid Karoo, South Africa. *Vegetatio*, **111**, 17–28.

Milton, S.J. & Dean, W.R.J. (1990). Seed production in rangelands of the southern Karoo. *South African Journal of Science*, **86**, 231–3.

Milton, S.J. & Dean, W.R.J. (1993). Selection of seeds by harvester ants in relation to condition of arid rangeland. *Journal of Arid Environments*, **24**, 63–74.

Milton, S.J., Dean, W.R.J., du Plessis, M.A. & Siegfried, W.R. (1994a). A conceptual model of arid rangeland degradation. *Bioscience*, **44**, 70–6.

Milton, S.J., Dean, W.R.J. & Siegfried, W.R. (1994b). Food selection by ostrich in southern Africa. *Journal of Wildlife Management*, **58**, 234–48.

Milton, S.J., Siegfried, W.R. & Dean, W.R.J. (1990). The distribution of epizochoric plant species: a clue to the prehistoric use of arid Karoo rangelands by large herbivores. *Journal of Biogeography*, **17**, 25–34.

Moolman, H.J. & Cowling, R.M. (1994). The impact of elephant and goat grazing on the endemic flora of South African succulent thicket. *Biological Conservation*, **68**, 53–61.

Morris, C.D., Tainton, N.M. & Hardy, M.B. (1992) Plant species dynamics in the Southern Tall Grassveld under grazing, resting and fire. *Journal of the Grassland Society of Southern Africa*, **9**, 90–5.

Mukinya, J.G. (1977). Feeding and drinking habits of the black rhinoceros in Masai Mara Game Reserve. *East African Wildlife Journal*, **15**, 125–38.

Mustart, P.J., Cowling, R.M. & Wright, M.G. (1995). Clustering of fertile seed in infructescences of serotinous *Protea* species: an anti-predation mechanism. *African Journal of Ecology*, **23**, 224–9.

Nel, J.A.J. (1978). Habitat heterogeneity and changes in small mammal community structures and resource utilisation in the southern Kalahari. *Bulletin of the Carnegie Museum of Natural History*, **6**, 118–32.

Nel, J.J.C. (1968). Die grasdraertermiet as plaag van natuurlike weiveld. *Journal of the Entomological Society of South Africa*, **31**, 309–21.

Novellie, P. (1990). Habitat use by indigenous grazing ungulates in relation to sward structure and veld condition. *Journal of the Grassland Society of Southern Africa*, **7**, 16–23.

O'Connor, T.G. (1991). Local extinction in perennial grasslands: a life-history approach. *American Naturalist*, **137**, 753–73.

O'Connor, T.G. (1994). Composition and population responses of an African savanna grassland to rainfall and grazing. *Journal of Applied Ecology*, **31**, 155–71.

O'Connor, T.G. & Pickett, G.A. (1992). The influence of grazing on seed production and seed banks of some African savanna grasslands. *Journal of Applied Ecology*, **29**, 247–60.

O'Connor, T.G. & Roux, P.W. (1995). Vegetation changes (1949–1971) in a semi-arid grassland in the Karoo, South Africa: influence of rainfall variability and grazing by sheep. *Journal of Applied Ecology*, **32**, 612–29.

Olivier, R.C.D. & Laurie, W.A. (1974). Habitat utilization by hippopotamus in the Mara River. *East African Wildlife Journal*, **12**, 249–71.

Reagain, P.J. (1993). Plant structure and the acceptability of different grasses to sheep. *Journal of Rangeland Management*, **46**, 232–6.

wen-Smith, N. (1979). Assessing the foraging efficiency of a large herbivore, the kudu. *South African Journal of Wildlife Research*, **9**, 102–10.

wen-Smith, N. (1988). *Megaherbivores: The Influence of Very Large Body Size on Ecology*. Cambridge: Cambridge University Press.

wen-Smith, N. (1989). Megafaunal extinctions: the conservation message from 11 000 years BP. *Conservation Biology*, **3**, 405–12.

wen-Smith, N. (1992). Grazers and browsers: ecological and social contrasts among African ruminants. In *Ongules/Ungulates 91*, ed. F. Spitz, G. Janeau, G. Gonzalez & S. Aulagnier, pp. 175–82. Paris: SFEPM and Toulouse: IRGM.

wen-Smith, N. (1993a). Woody plants, browsers and tannins in southern African savannas. *South African Journal of Science*, **89**, 505–10.

wen-Smith, N. (1993b). Consumption and metabolism of vegetation by large mammalian herbivores. Workshop on *African Savannas, Land Use and Global Climatic Change*, Victoria Falls, Zimbabwe, 2–5 June.

Jwen-Smith, N. & Cooper, S.M. (1985). Comparative consumption of vegetation components by kudus, impalas and goats in relation to their commercial potential as browsers in savanna regions. *South African Journal of Science*, **81**, 72–6.

Jwen-Smith, N. & Cooper, S.M. (1987). Palatability of woody plants to browsing ungulates in a South African savanna. *Ecology*, **68**, 319–31.

Jwen-Smith, N. & Cooper, S.M. (1989). Nutritional ecology of a browsing ruminant, the kudu, through the seasonal cycle. *Journal of Zoology, London*, **219**, 29–43.

Jwen-Smith, N. & Cumming, D.H.M. (1994). Comparative foraging strategies of grazing ungulates in African savanna grasslands. *Proceedings of the XVII International Grassland Congress 1993*, 691–8.

Penzhorn, B.L., Robbertse, P.J. & Olivier, M.C. (1974). The influence of the African elephant on the vegetation of the Addo Elephant National Park. *Koedoe*, **17**, 137–58.

Perrin, M.R., Bowland, A.E. & Fauries, A.S. (1992). Niche segregation between the blue duiker and the red duiker. In *Ongules/Ungulates 91*, ed. F. Spitz, G. Jareau, G. Gonzalez & S. Aulagnier, pp. 201–4. Paris and Toulouse: SFEPM-IRGM.

Prins, H.H.T. & Van der Jeugd, H.P. (1993). Herbivore population crashes and woodland structure in East Africa. *Journal of Ecology*, **81**, 305–14.

Rall, M. & Fairall, N. (1993). Diets and food preferences of two South African tortoises. *South African Journal of Wildlife Research*, **23**, 63–70.

Reeler, B., Campbell, B. & Price, L. (1991). Defoliation of *Brachystegia spiciformis* by a species-specific insect *Melasoma quadrilineata*, over two growing seasons. *African Journal of Ecology*, **29**, 271–74.

Reilly, B.K., Theron, G.K. & du Plessis, J. (1990). Food preferences of oribi in the Golden Gate Highlands National Park. *Koedoe*, **33**, 55–61.

Robinson, E.R. & Seely, M.K. (1975). Some food plants of ostriches in the Namib Desert Park, South West Africa. *Madoqua*, **4**, 99–100.

Rushworth, J.E. (1975). *The Floristic, Physiognomic and Biomass Structure of Kalahari Sand Scrub Vegetation in Relation to Fire and Frost in Wankie National Park, Rhodesia*. MSc Thesis. Salisbury: University of Rhodesia.

Rutherford, M.C. (1980). Annual plant production–precipitation relations in arid and semi-arid regions. *South African Journal of Science*, **76**, 53–6.

Scholes, R.J. & Walker, B.H. (1993). *An African Savanna. Synthesis of the Nylsvley Study*. Cambridge: Cambridge University Press.

Scholtz, C.H. (1982). *Trophic Ecology of Lepidoptera Larvae Associated with Woody Vegetation in a Savanna Ecosystem. South African National Scientific Programmes Report 55*. Pretoria: CSIR.

Seydack, A.H.W. (1990). *Ecology of Bushpig (Potamochoerus porcus) Linn 1758 in the Cape Province, South Africa*. PhD Thesis. Stellenbosch: University of Stellenbosch.

Shackleton, C.M. (1993). Are the communal grazing lands in need of saving? *Development Southern Africa*, **10**, 67–78.

Simpson, C.D. (1974). *Ecology of the Zambezi bushbuck Tragelaphus scriptus ornatus Pocock*. PhD Thesis. Galveston: Texas A&M University.

Simpson, C.D. (1975). A detailed vegetation study on the Chobe River in North-East Botswana. *Kirkia*, **10**, 185–227.

Skarpe, C. (1990). Shrub layer dynamics under different herbivore densities in arid savanna, Botswana. *Journal of Applied Ecology*, **27**, 873–85.

Skinner, J.D. (1993). Springbok treks. *Transactions of the Royal Society of South Africa*, **48**, 291–305.

Skinner, J.D. & Smithers, R.H.N. (1990). *The Mammals of the Southern African Subregion*. 2nd Edition. Pretoria: University of Pretoria.

Smithers, R.H.N. (1983). *The Mammals of the Southern African Subregion*. Pretoria: University of Pretoria.

Stoltz, L.P. & Saayman, G.A. (1970). Ecology and social organization of chacma baboon troops in the northern Transvaal. *Annals of the Transvaal Museum*, **26**, 99–143.

Stuart-Hill, G.C. (1992). Effects of elephants and goats on the Kaffrarian succulent thicket of the eastern Cape, South Africa. *Journal of Applied Ecology*, **29**, 699–710.

Sweet, R.J. & Mphinyane, W. (1986). Preliminary observations on the ability of goats to control post-burning regrowth in *Acacia nigrescens/Combretum apiculatum* savanna in Botswana. *Journal of the Grassland Society of Southern Africa*, **3**, 79–84.

Tainton, N.M. (1972). The relative contribution of overstocking and selective grazing to the degeneration of tall grassveld in Natal. *Proceedings of the Grassland Society of Southern Africa*, **7**, 39–43.

Tainton, N.M. (1981). The ecology of the main grazing lands of South Africa. In *Veld and Pasture Management in South Africa*, ed. N.M. Tainton, pp. 27–56. Pietermaritzburg: Shuter & Shooter.

Tainton, N.M. (1986). A system of assessing range condition in South Africa. In *Rangelands: a Resource Under Siege*, ed. P.J. Joss, P.W. Lynch & O.B. Williams. Canberra: Australian Academy of Sciences.

Tainton, N.M. & Booysen, P. de V. (1964). Growth and development in perennial veld grasses: *Themeda triandra* tillers under various systems of defoliation. *South African Journal of Agricultural Science*, 8, 93–116.

Tarboton, W.R. (1980). Avian populations in a Transvaal savanna. *Proceedings of the 4th Pan African Ornithological Conference*, pp. 113–24.

Teague, W.R. (1989a). Effect of intensity and frequency of defoliation on aerial growth and carbohydrate reserve levels in *Acacia karroo* plants. *Journal of Grassland Society of Southern Africa*, 6, 132–8.

Teague, W.R. (1989b). The response of *Acacia karroo* plants to defoliation of the upper or lower canopy. *Journal of the Grassland Society of Southern Africa*, 6, 225–9.

Teague, W.R. & Walker, B.H. (1988). Effect of intensity of defoliation by goats at different phenophases on leaf and shoot growth of *Acacia karroo*. *Journal of the Grassland Society of Southern Africa*, 5, 197–206.

Temby, I. (1977). *The Non-Ungulate Mammals at Nylsvley*. Unpublished Report. Pretoria: CSIR.

Theron, J.G. (1984). Leafhoppers (Hemiptera: Cicadellidae) associated with the renosterbos, *Elytropappus rhinocerotis* less. 1. The genus *Renosteria* Theron. *Journal of the Entomological Society of Southern Africa*, 47, 83–97.

Thomson, W.R. (1974). Tree damage by porcupine in south east Rhodesia. *Journal of the Southern African Wildlife Management Association*, 4, 123–7.

Trollope, W.S.W. (1980). Controlling bush encroachment with fire in the savanna areas of South Africa. *Proceedings of the Grassland Society of Southern Africa*, 15, 173–7.

Van Hensbergen, H.J., Botha, S.A., Forsyth, G.G. & Le Maitre, D.C. (1992). Do small mammals govern vegetation recovery after fire in fynbos? In *Fire in South African Mountain Fynbos*, ed. B.W. van Wilgen, D.M. Richardson, F.J. Kruger & H.J. van Hensbergen, pp. 182–202. Berlin: Springer-Verlag.

Van Vegten, J.A. (1983). Thornbush invasion in a savanna ecosystem in eastern Botswana. *Vegetatio*, 56, 3–7.

van Wilgen, B.W. & Forsyth, G.G. (1992). Regeneration strategies in fynbos plants and their influence on the stability of community boundaries after fire. In *Fire in South African Mountain Fynbos*, ed. B.W. van Wilgen, D.M. Richardson, F.J. Kruger & H.J. van Hensbergen, pp. 54–80. Berlin: Springer-Verlag.

Venter, J., Liggitt, B., Tainton, N.M. & Clarke, G.P.Y. (1989). The influence of different land-use practices on soil erosion, herbage production and on grass species, richness and diversity. *Journal of the Grassland Society of Southern Africa*, 6, 89–98.

Vickery, B. & Vickery, M.L. (1973). Toxicity for livestock of organofluorine compounds present in *Dichapetalum* plant species. *Veterinary Bulletin*, 43, 537–42.

Viljoen, A. (1988). Long-term changes in the tree component of the vegetation in the Kruger National Park. In *Long-term Data Series Relating to Southern Africa's Renewable Natural Resources*, ed. I.A.W. Macdonald & R.J.M. Crawford, pp. 310–15. *South African National Science Progress Report 157*. Pretoria: CSIR.

Viljoen, P.J. (1989). Habitat selection and preferred food plants of a desert-dwelling elephant population in the northern Namib Desert, South West Africa/Namibia. *African Journal of Ecology*, 27, 227–40.

Viljoen, S. (1977). Feeding habits of the bush squirrel. *Zoologica Africana*, 12, 459–68.

Viljoen, S. (1983). Feeding habits and comparative feeding rates of three southern African arboreal squirrels. *South African Journal of Zoology*, 18, 378–87.

Viljoen, S. & du Toit, S.H.C. (1985). Postnatal development and growth of southern African tree squirrels in the genera *Funisciurus* and *Paraxerus*. *Journal of Mammalogy*, 66, 119–27.

Vincent, J. (1970). The history of Umfolozi Game Reserve, Zululand. *Lammergeyer (Natal)*, 11, 7–48.

Voigt, E.A. (1983). Mapungubwe. An archaeological interpretation of an iron age community. *Transvaal Museum Monographs*, 1, 1–204.

Walker, B.H., Emslie, R.H. Owen-Smith, R.N. & Scholes, R.J. (1987). To cull or not to cull: lessons from a southern African drought. *Journal of Applied Ecology*, 24, 381–402.

Walker, B.H., Matthews, D.A. & Dye, P.J. (1986). Management of grazing systems: existing versus an event orientated approach. *South African Journal of Science*, 82, 172.

Watson, H.K. & Macdonald, I.A.W. (1983). Vegetation changes in the Hluhluwe-Umfolozi Game Reserve complex from 1937–1975. *Bothalia*, 14, 265–9.

Westoby, M., Walker, B. & Noy-Meir, I. (1989). Opportunistic management for rangelands not at equilibrium. *Journal of Rangeland Management*, 42, 266–74.

Whyte, I.J. & Joubert, S.C.J. (1988). Blue wildebeest population trends in the Kruger National Park and the effects of fencing. *South African Journal of Wildlife Research*, 18, 78–87.

Wiens, D., Rourke, J.P., Casper, B.B., Rickart, E.A., Lapine, T.R., Peterson, C.J. & Channing, A. (1983). Non-flying mammal pollination of South African Proteas: a non-coevolved system. *Annals of the Missouri Botanical Garden*, 70, 1–31.

Williamson, D., Williamson, J. & Ngwamotsoko, K. T. (1988). Wildebeeste migration in the Kalahari. *African Journal of Ecology*, 26, 269–80.

Wilson, G.P.M. & Hennessy, D. (1977). The germination of excreted kikuyu grass seed in cattle dung pats. *Journal of Agricultural Science, Cambridge*, 88, 247–9.

Winkler, A. (1992). *The Feeding Ecology of the Cape Mountain Zebra in the Mountain Zebra National Park, South Africa*. MSc Thesis. Johannesburg: University of the Witwatersrand.

Yeaton, R.I. (1988). Porcupines, fire and the dynamics of the tree layer of the *Burkea africana* savanna. *Journal of Ecology*, 76, 1017–29.

Fire

18

W.J. Bond

18.1 Introduction

The trio of fire, climate and the great herds of mammalian herbivores have long been considered the primeval sculptors of southern African landscapes. Although climate may set the limits to plant growth, more often fire and herbivores determine vegetation patterns in the region. Fire can be viewed as a large generalist herbivore, sometimes competing, sometimes replacing and sometimes facilitating vertebrate herbivores. Fire is only possible where there is sufficient fuel to burn. It rarely occurs in the arid regions of the west and the interior. Fire is also rare in palatable grasslands or arid savannas when the offtake by herbivores leaves nothing to burn. In these 'sweetveld' areas (see Chap. 10, this volume), fire may occur only after exceptional rains and grass growth exceeds grazing capacity. Fire is most frequent in the more humid regions and in the least palatable vegetation types of southern Africa where fuel is continuous and where herbivore impact is minor. Fire is a regular feature of fynbos and related shrublands in the southwest and in the 'sour' grasslands and savannas of the more mesic summer-rainfall regions. Since these are among the most productive regions of the country, there has been a long history of research on fire in South Africa with several major reviews in the last 12 years (Booysen & Tainton 1984; Cowling 1992; Van Wilgen *et al.* 1992b). A single chapter cannot possibly do justice to the large, diverse literature on fire in southern Africa and the reader is referred to these reviews and to other chapters in this volume (especially Chaps 6, 10, 11 and 21).

The aim of this chapter is to provide a guide to the significance of fire in southern African vegetation for the non-specialist reader. The chapter describes how fire influences plants at different scales:

- Individual plants, and those features that determine their fire responses, and whether they reflect an evolutionary history of fire in different components of the southern African flora;
- plant populations and communities, and how fire influences their ecology;
- vegetation patterns, and how these are influenced at the landscape and subcontinental scale.

The use of fire in vegetation management is discussed briefly.

Most research has been aimed at the population and community level. Studies at the level of the individual plant have been comparatively neglected, but are beginning to yield interesting surprises. At the landscape scale, there has been plenty of speculation but little rigorous work, and longstanding questions are still unanswered. Finally, in common with the field of fire ecology generally, the role of fire in the evolution of plants or of floras has been astonishingly neglected.

18.2 Distribution and abundance of fire in southern African landscapes

Fire is a regular phenomenon in all parts of southern Africa except in the arid regions of the west and the

421

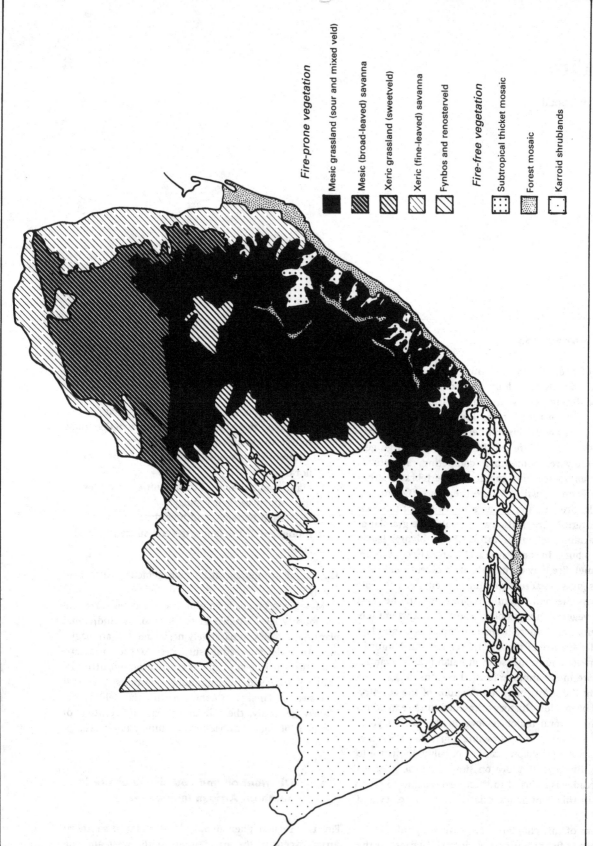

Fire-prone vegetation

- ■ Mesic grassland (sour and mixed veld)
- ▨ Mesic (broad-leaved) savanna
- ▨ Xeric grassland (sweetveld)
- ▨ Xeric (fine-leaved) savanna
- ▨ Fynbos and renosterveld

Fire-free vegetation

- ⬚ Subtropical thicket mosaic
- ⬚ Forest mosaic
- ☐ Karroid shrublands

Figure 18.1 **South African vegetation – broad fire regimes based on plant cover and rainfall. Fire-prone vegetation formations (black and hatched shading) include grasslands, savanna, fynbos and renosterveld. Formations that seldom burn, except under extreme conditions (areas with dotted shading), include forests, thickets**

southwestern interior (e.g. Edwards 1984; Manry & Knight 1986). Fires supported by grassy fuels predominate in the eastern half of the region and in the arid savannas of the northern Cape, Namibia and Botswana. Fires supported by woody fuels predominate in the fynbos of the southern and southwestern regions. In the karoo regions, which very seldom burn, islands of flammable C_3-dominated grasslands are common in the higher mountains (Chap. 8, this volume). In winter-rainfall regions, renosterveld, a shrubland dominated by Asteraceae (see Chap. 6, this volume) tends to replace these grassy islands on karoo mountains.

Islands of non-flammable forests or thicket vegetation, which seldom burn, occur as fire refugia in fynbos, grassland and savanna biomes. They are most common in dissected landscapes, occurring in ravines, rocky sites and along water courses. In mesic areas, Afromontane forests make up the non-flammable vegetation (Chap. 12, this volume). In more arid areas, subtropical thicket (Acocks 1953) predominates, especially in the steeply dissected valleys along the east coast. Non-flammable evergreen thickets are common as a band of varying width (up to 1 km) along the coast from the southwestern coast to northeastern KwaZulu-Natal (Chap. 13, this volume). In the savanna biome, patches of non-flammable thickets often occur on deep, well-drained sandy soils (e.g. the 'sand forests' of the Mozambique coastal plain; Moll & White 1978; Chap. 12, this volume) and on termitaria and, more rarely, on clay-rich sites.

A map indicating generalized fire regimes from diverse sources (e.g. Trollope 1980; Edwards 1984; Van Wilgen & Van Hensbergen 1991) is shown in Fig. 18.1. The map units do not conform to biome boundaries. Within both the savanna and grassland biomes, the fire regime differs in arid and mesic variants. In arid areas (<650 mm yr^{-1}) rainfall is erratic and fires are limited to high-rainfall years when there is sufficient grass fuel (e.g. Trollope 1984a; Tainton & Mentis 1984). In mesic areas, fire frequencies are limited more by ignition events than by fuel (see below).

18.3 Determinants of fire

18.3.1 Fuel characteristics

The importance of fire as a defoliator depends, like herbivory, on the availability of suitable 'forage' (the quantity and quality of fuel) but also on climate and ignition events (Trollope 1984b). Just as the chemical and morphological properties of plants make them more or less susceptible to herbivory (Chap. 17, this volume), so too

are some kinds of plants more susceptible to fire than others. However, fire differs from herbivory in being a very unselective consumer of biomass and in depending not only on the properties of individual species but also on the spatial arrangement of mixtures of species (Rundel 1981; Pyne 1984). The moisture content of plants is one of the chief determinants of flammability. Dead fuels have the lowest moisture content, and live leaves will burn more easily if their moisture content is low. Therefore, leaves that are well defended against herbivores with high fibre content and high leaf specific weight will tend to burn more easily, because of their lower moisture content and because the litter they produce decomposes more slowly. The shape, size and arrangement of plant parts also has a major influence on flammability. Plants made up of small particle sizes (with a large surface area to volume ratio), such as narrow leaves and thin branches, are more flammable, because they lose water more rapidly and have a high ratio of fuel to air. Leaves with high levels of oils, fats, waxes and terpenes also burn more readily, either because they release flammable gases when heated or because they have a higher heat of combustion.

In terms of southern African vegetation, the tussock grasses that cover much of the region make excellent fuels, because of their high surface-area to volume ratio and low moisture content when cured by frost or drought in winter. Flammability is enhanced by the fine leaves and low moisture content of fynbos plants as well as the accumulation of litter and dead material in the canopy. In contrast, forests seldom burn except under extreme drought conditions or if the forest structure has been greatly modified. Van Wilgen, Higgins & Bellstedt (1990b) studied the fuel properties of adjoining fynbos and forest. Besides the obvious differences in the vertical distribution of biomass, they found that foliar moisture contents of forest leaves were 50–100% higher than in fynbos plants. This difference alone is likely to account for the lack of canopy fires in forests, although there were also slight differences in heat yields and crude fat contents (Table 18.1). The lack of bridging fuels that carry fire from the understorey to the canopy is also a major reason for the scarcity of fire in forests. The low, dense structure of subtropical thicket would seem more conducive to canopy fires, but this and related vegetation types seldom burn, probably because of high foliar moisture content.

There have been a number of studies of fuel properties of South African vegetation aimed at developing fire behaviour models for prescribed burning (e.g. Van Wilgen, Le Maitre & Kruger 1985 for fynbos; Van Wilgen & Wills 1988, Trollope & Tainton 1986 for savannas; Everson, Van Wilgen & Everson 1988 for Drak-

Table 18.1 **Some characteristics influencing flammability of fynbos versus forest fuels. Moisture content, energy content and fat content were measured for six common fynbos, and six forest species in different seasons. Flame length was predicted using a three-strata Rothermel model. Data from Van Wilgen et al. 1990b**

Attribute	Fynbos	Forest
Amount of fuel (<6 mm diameter) (g m^{-2})	1696	3685
Seasonal range in moisture content (%)	98–126	167–216*
Mean energy content (J g^{-1})	21860	20703*
Range of fat content (% of dry weight)	0.4–10.3	1.0–5.7
Predicted flame length (m) under extreme fire hazard	4.9	0

* $P < 0.05$.

ensberg grasslands). There have been no studies of comparative flammability of different grass or shrub species. These would be interesting from an evolutionary perspective and would give a better understanding of how change in veld composition causes changes in the fire regime.

18.3.2 Climate
Weather conditions are important in determining how and when fires will burn. Rainfall, relative humidity and temperature all affect fire through their effect on the moisture content of the fuels. Fire can occur under all southern African climate conditions, since all experience dry conditions at some time of the year. Even the humid southern Cape, a centre of non-flammable evergreen forests, experiences fires in appropriate fuels especially during hot, dry, bergwind conditions (Van Wilgen 1984). Hot, dry periods are associated with extreme fire danger and largely determine the seasonal occurrence of fires in fynbos (Van Wilgen 1984) and the risk of run-away wildfires in grasslands (Barclay, Jury & Washington 1993). Large fires have occurred in recent years associated with El Niño events, but there appears to have been no analysis of a cyclical change in fire hazard associated with climate cycling at this inter-annual temporal scale.

18.3.3 Ignition
It has often been assumed that fire frequency is limited by ignition events. For example, Manry & Knight (1986) argued that a correlation between lightning flash density and fire frequency in South African biomes was a causal relationship. However, lightning flash density is low in both fynbos, which burns, and the karoo, which does not (Chap. 2, this volume). An alternative explanation is that lighting flash density is correlated with annual rainfall in summer-rainfall regions and that rainfall determines the availability of fuel, and therefore fire frequency.

Similar assumptions regarding the importance of ignition are implicit in the view that human use of fire precipitated major vegetation change by increasing fire

frequency. Hominids have used fire for >1.5 million years in southern Africa (Brain & Sillen 1988). Human pressures on vegetation must have increased markedly with the arrival of Iron Age cultures about a thousand years ago (Feely 1980; Hall 1984; Chap. 21, this volume). However, it is far from clear how humans influenced fire regimes and whether changes in seasonality or frequency had the greatest effect.

Today, most fires, particularly in the grassland and savanna biomes, are ignited by humans. Lightning fires are still common in the more remote areas of fynbos where they may account for about 20% of the area burnt (Horne 1981; Van Wilgen 1981). In the Kruger National Park, lightning fires account for about 10% of the area burnt. Lightning accounts for the larger proportion of fires recorded, but many of these are doused by subsequent rain or are controlled by fire crews. Rockfalls also cause fire and resulted in many fynbos fires after the Ceres earthquake in 1969 (reviewed by Edwards 1984). The frequency of fynbos fires ignited by rockfalls was estimated as 5% in the Cederberg, rising to 23% in the period including the Ceres earthquake. However, these estimates should be treated with some scepticism, since observations of rockfall-induced fires are extremely rare.

Fires started by humans have a different seasonal distribution from those induced by lightning. Fires started by humans often occur outside the thunderstorm season. For example, the lightning-fire season in southern and western Cape fynbos extends from spring to autumn, peaking in late summer (February) (Horne 1981; Brown et al. 1991). Lightning fires are very rare in late autumn, winter and early spring. However, fires ignited by humans occur throughout the year and can burn very large areas under winter bergwind conditions. Similar differences in the seasonal pattern of burning have occurred in grasslands and savannas. Lightning fires are most common early in the rainy season, whereas fires of human origin peak in the dry season (Fig. 18.2). The biological effects of these 'out of season' burns are discussed below (sections 18.6 and 18.9).

18.4 Fire regimes

18.4.1 Fire frequency
The biological effects of fire are determined by the fire regime. This is the frequency, type, intensity and season of burn. The frequency of fires is determined by the availability of fuel, suitable climate, and an ignition event. Fuel accumulation in fynbos depends largely on time since the last burn. Generally, fuel has been assumed to

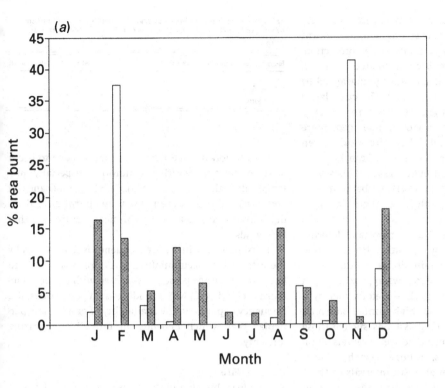

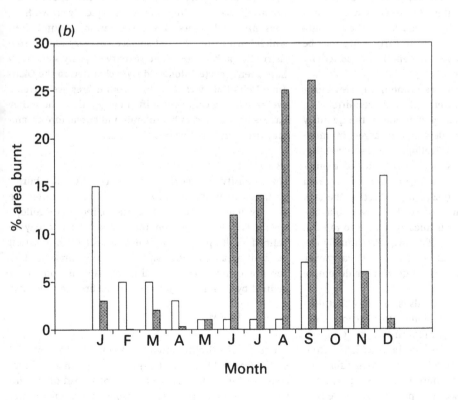

Figure 18.2 **Seasonality of lightning fires (open bars) versus fires of 'other', usually human, origin (shaded bars) as a percentage of the total area burnt. (a) Data for the Cederberg (fynbos) are for the period 1956–1986 (from Brown *et al.* 1991). The 'other' category includes prescribed burns (28% of the area) between the months of May and October. (b) Data for the Kruger National Park (savanna) are for the period 1980–1992 (from Trollope 1993).**

be too sparse to support fires before about four years after a burn (Kruger 1983). Few stands remain unburnt longer than 40 to 50 years without special protection. Mean fire frequencies have been estimated as 11–15 years in the Cederberg, an area relatively unaffected by prescribed burning (Brown et al. 1991). The cumulative probability of a fire in this area is shown in Fig. 18.3, where it is contrasted with a closely managed, more mesic area in the Jonkershoek valley. Fire return intervals greater than 20 years were significantly more common at low (<800 m) and high (>1400 m) elevations in the Cederberg (Brown et al. 1991), probably due to lower fuel accumulation rates at these climatic extremes.

The availability of fuel is also an important determinant of fire frequency in grassy vegetation. However, fuel accumulation is limited by rainfall rather than time since last burn. Fire frequency follows approximately a productivity gradient, which itself is determined by a rainfall gradient (West 1965). In high rainfall sourveld, fires occur annually. More xeric grasslands, such as those in the southern Free Sate, burn at longer intervals. Similar patterns occur in savannas. Trollope (1993) has shown that sourveld savannas burn at mean intervals of three years in the Kruger National Park, whereas sweetveld burns at mean frequencies of eight years (Table 18.2). Sweetveld forage remains relatively palatable even when the grass is dry during winter. Herbivory of sweetveld is therefore greater. In heavily grazed areas, the accumulation of adequate grass biomass to carry a fire may be limited to higher rainfall years when grass productivity exceeds herbivore offtake.

Whereas in fynbos annual variation in the area burnt may be determined by extreme climatic conditions, in grass fuels it is determined by variation in the quantity of grass, which is, in turn, determined largely by interannual variation in rainfall. Trollope (1993) showed that the area burnt in Kruger National Park reached a minimum in the drought years of 1983 (3.1% of the area burnt over 12 years) and 1992 (1.5%), reflecting the lack of fuel. In the arid savannas of Etosha in Namibia, the number of fires in any year is linearly related to rainfall over the preceding two years (Fig. 18.4). Presumably the cumulative effect of rainfall is due to fuel carry-over from one year to the next before grazing or decomposition takes its course.

Finally, fire in karoo shrublands is very rare, but fires did occur in the Nama-karoo biome after exceptionally good rains in the early 1970s (Edwards 1984, Chap. 8, this volume). Fire is also occasional in succulent karoo, especially where the biome adjoins fynbos shrublands. In most years fuel is too discontinuous to support fires, but dense ephemeral growth can provide adequate fuel at intervals of decades.

Table 18.2 Fire frequencies in sourveld versus sweetveld savannas of the Kruger National Park between 1981 and 1992 (Trollope 1993)

Number of years between fires	Sourveld	Sweetveld
Mean	3	8
Minimum	1	2
Maximum	6	11

Fires in forests, like fynbos, are most likely to occur after exceptional weather conditions, especially prolonged drought (Granger 1984). In 1869, thousands of hectares of forests in the Knysna area burnt after three dry months and six weeks of near continuous hot bergwinds.

Thus, in woody fuels, fire frequency is determined by the rate of fuel accumulation. Fires are more likely in hot, dry conditions possibly associated with periodic southern oscillation (El Niño) conditions. In grassy fuels, fire frequency depends on the availability of grass in any particular year, which depends on rainfall and grazing intensity.

18.4.2 Fire type
Surface fires, burning in the grass sward, are the rule in savanna and, of course, in grasslands. Crown fires, which burn the aerial portions of trees and shrubs, are exceptional in savannas. Fynbos fires are typically crown fires, consuming all layers of the vegetation. Ground fires, which burn below the soil surface, are very rare in southern Africa, because of the scarcity of peaty soils. They have been reported along old river channels in the Okavango Delta (Ellery et al. 1989). Ground fires are extreme events, causing complete shifts in vegetation where they have been reported in heathlands and boreal forests, and may occur in peaty or swampy sites.

18.4.3 Fire intensity
Fire intensity is a measure of how fiercely a fire burns. Variation in intensity is important in determining the severity of a fire in terms of vegetative recovery (Trollope 1984b; Trollope & Tainton 1986). Fire intensity may be defined as the product of available heat of combustion per unit area of ground and the rate of spread of fire. For fire control purposes, fire intensity is commonly defined by Byram's (1959) measure of fireline intensity:

$$I = Hwr$$

where I is the fireline intensity (kW m^{-1} or kJ s^{-1} m^{-1}), H is the heat yield of the fuel (J g^{-1}), w is the mass of fuel consumed (g m^{-2}), and r is the rate of spread of the fire (m s^{-1}). Fire intensity is very variable, depending on fuel properties, wind and weather conditions at the time of

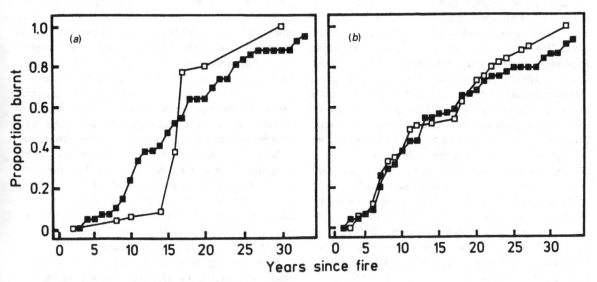

Figure 18.3 **The cumulative probability of fire (open squares) with increasing age since last burn in: (a) intensively managed fynbos at Swartboskloof, Western Cape; and (b) extensively managed fynbos in the Cederberg. Lines with closed squares show the predicted probabilities of fire from a stochastic model of fire frequency (after Van Wilgen & Van Hensbergen 1991). The low initial probability of fires in Swartboskloof is the result of restricting prescribed burns largely to older stands.**

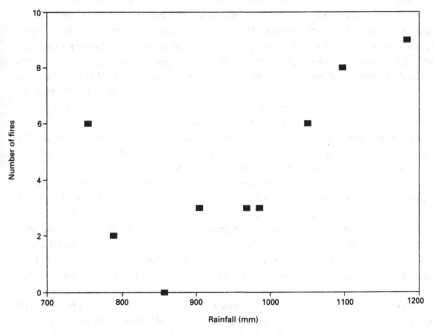

Figure 18.4 **Relationship between cumulative rainfall over the preceding two wet seasons and the number of fires in Etosha National Park, Namibia. After Siegfried (1981).**

fire, and on topography. Values of 100–2000 kW m^{-1} have been reported for grasslands, 200–6000 kW m^{-1} for savannas and 500–20 000 kW m^{-1} for fynbos (Trollope 1984b; Van Wilgen et al. 1985; Van Wilgen & Wills 1988). Fires under extreme conditions will exceed these values.

Byram's fire intensity measure is useful for designing safe burning prescriptions, but is not necessarily a good predictor of biological effects (Alexander 1982). Nevertheless, fireline intensity is correlated with flame length, which is a good predictor of top-kill in savanna trees. Trollope & Tainton (1986) concluded that intensities of 2500 kW m^{-1} were necessary for significant (>40%) top-kill of savanna trees up to a height of 2 m. In the Kruger National Park, Trollope (1993) noted that intense fires were possible only with grass fuel loads exceeding 4000 kg ha^{-1}. With the use of information on fire behaviour, burning can be prescribed under appropriate fuel and weather conditions to achieve the desired top-kill of trees in savannas (Van Wilgen & Wills 1988).

18.5 The natural history of plant responses to burning

The way plants persist through fires and the timing of regeneration in relation to fire determine their fire-response. Studies on modes of persistence and regeneration are an essential prerequisite for understanding population responses to burning and have been the focus of much research in South Africa and elsewhere. In southern African fynbos, the diversity of plant responses to burning is greater, and the phenomenon more intensively studied (reviewed by Le Maitre & Midgley 1992) than in grass-dominated fuels (reviewed by Frost 1984).

18.5.1 Fire and persistence

Individual persistence through a fire is determined by the degree to which living buds are protected from heat. The effects of fire depend on the extent to which the live crown is scorched and on survival of stem and root cambial tissue. Bark thickness, more than its type, protects the cambium from heat injury (e.g. Hare 1965; Uhl & Kauffman 1990). There are interesting differences in bark attributes in closely related taxa in southern African trees, but their significance in terms of bark insulation is unstudied. For example, Acacia species show considerable diversity in bark characteristics. Thin-barked species, such as A. mellifera, typically sprout from the base of the tree after burning, whereas species with thicker bark, such as A. erioloba or A. sieberiana, sprout from undamaged buds in the canopy, except when plants are very young. Bark attributes may differ between forest and savanna members of a genus, reflecting an adaptive response to burning. Pterocarpus angolensis, a savanna tree, has thick ridged bark, whereas P. lucens ssp. antunessii, a species of dry forest that rarely burns, has thin, smooth bark. A comparative study of bark attributes would be interesting in indicating alternative strategies of persistence within a fire-prone community or evolutionary changes associated with adaptive radiation of trees into fire-prone environments.

Many fynbos species are thin-barked, have no capacity to sprout at all and are routinely killed by fire. Non-sprouters make up as much as 50% of the species in fynbos communities (Le Maitre & Midgley 1992; Chap. 6, this volume). The dominant woody shrubs in fynbos are very often non-sprouters. Many locally abundant species of Restionaceae are also non-sprouters.

Some capacity for sprouting seems the norm in savanna and forest trees, although a few forest species cannot sprout (e.g. all Podocarpus species except the fynbos endemic, P. elongatus). However, there is considerable variation in the location of sprouting tissues on the plant and in the extent to which individuals sprout among species, size classes and in different fires. This variation is very important in determining how a plant population will respond to burning. The location of sprouting tissue on the plant determines the extent of architectural changes precipitated by burning. Some plants have little or no capacity for sprouting, but survive fires if shoot apices are not damaged by fire heat (e.g. Leucospermum conocarpodendron and Mimetes fimbriifolius in fynbos). Others have persistent bud banks, which are protected by bark and from which the canopy regenerates (e.g. Protea nitida). These species are better able to maintain their canopy position regardless of the intensity of a burn. Many woody plants sprout from the base of the stem if the canopy is burnt. The result is a multi-stemmed shrub which, in savannas, brings canopy foliage within reach of herbivores and shading by grasses. In some woody plants, buds proliferate at the bases of stems and form a woody swelling known as a lignotuber. Lignotubers are a developmental feature and not simply a response to wounding (James 1984). A few species can sprout from root suckers. These are generally extremely resilient to burning, because the buds are well insulated by soil layers. In fynbos, examples include several Cliffortia species. In savannas, species include Ochna pulchra (Rutherford 1983), Spirostachys africana, and Euclea divinorum. Thicket examples are Euclea racemosa and Olea exasperata (Cowling & Pierce 1988). White (1976) has referred to certain dwarf shrubs with this capacity as 'geoxylic suffrutices'. Many are poisonous and are very difficult to eradicate.

Sprouting behaviour is dependent on tree size and the fire event (Fig. 18.5). The effects of a burn on vegetation dynamics thus depend both on population structure immediately before the burn and on the much less predictable conditions at the time of the event, especially fire intensity. Information on sprouting behaviour is important in predicting fire effects on vegetation dynamics and has been used in prescribing fires for bush control (e.g. Van Wilgen, Everson & Trollope 1990a; Starfield *et al.* 1993).

Grasses are among the most fire-resistant components of plant communities. In tussock grasses, tiller initials are insulated by persistent leaf sheaths in a basal tuft. Some grasses also resprout from subterranean rhizomes where they are well insulated from fire heat. The geophytic growth form is common in monocotyledons and affords effective protection because of insulation by the soil. Some geophytes have contractile roots, which pull the storage organ of young plants below the ground, reducing their vulnerability to fire and other herbivores. Arborescent monocotyledons, especially the conspicuous 'tree' aloes, are insulated by persistent dead leaves (Bond 1983).

18.5.2   Fire and reproduction

For many species, the combination of open space, increased availability of resources and temporary reduction in seed predators ('enemy-free space') is highly favourable for seedling establishment in the post-fire environment. Species exploit these conditions in diverse ways. Some survivors are stimulated by fire and flower profusely after a burn. Others flower regularly between fires but accumulate seeds in seed banks. Some plants delay dispersal of seed until they are released *en masse* when the canopy burns. Many have soil-stored seed banks and use a variety of specialized cues to synchronize germination with the post-burn environment.

18.5.2.1   *Fire-stimulated flowering*

Fire-stimulated flowering is very common in monocotyledons. The fire lilies, *Cyrtanthus* spp., are geophytes with brilliant crimson flowers. They persist vegetatively between fires but only flower in the first week or two after a burn (Le Maitre & Brown 1992). Fire-stimulated flowering has been reported in Orchidaceae, Iridaceae, Liliaceae (*sensu lato*), Amaryllidaceae, Poaceae and Cyperaceae (Hall 1959; Kruger 1983; Le Maitre & Brown 1992; Le Maitre & Midgley 1992). Fire also stimulates flowering in some sprouting shrubs such as *Leucadendron salignum* and *Colpoon speciosum* in fynbos (Le Maitre & Midgley 1992). These are all growth forms that suffer little material loss in fires (e.g. geophytes, graminoids, lignotuberous shrubs) and can rapidly marshal existing resources to reproduction.

The association between fire and flowering varies from near obligate in small herbaceous geophytes to

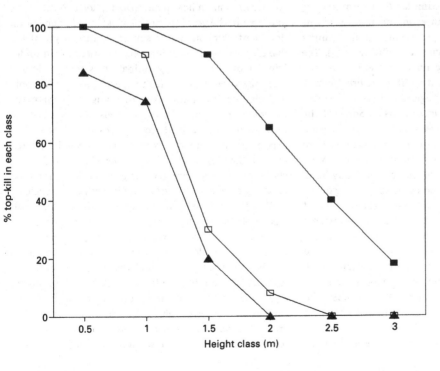

Figure 18.5 **The relationship between plant height and top-kill in *Acacia karroo* at three fire intensities. Top-kill causes the plant to coppice from the base. Data from Trollope (1984a). Triangles, 300 kJ s⁻¹ m⁻¹; open squares, 745 kJ s⁻¹ m⁻¹; closed squares, 2063 kJ s⁻¹ m⁻¹.**

weakly facultative in resprouting shrubs. In most examples of fire-stimulated flowering, fire is not obligatory as a flowering stimulus. The reduction in competition and increased availability of light, water and nutrients is probably a sufficient stimulus in many species (Le Maitre & Brown 1992). Keeley (1993) recently reported that flowering in *Cyrtanthus ventricosus* was stimulated by smoke, possibly an obligatory fire cue. The seeds of fire-flowering species typically germinate readily. However, seeds only become available after a burn, so that population growth is episodic and stimulated by fire.

18.5.2.2 *Fire-stimulated seed release*
In fynbos, some species accumulate seeds in a seed bank stored in the canopy of a plant, insulated from fire by cones, woody capsules, or persistent woody inflorescences. The protective structures open and release seeds *en masse* after fire. This condition is known as serotiny. Serotinous species are often dominant in the communities in which they occur. Serotinous species occur in a variety of unrelated families and genera (e.g. Bruniaceae, Cupressaceae, Proteaceae, Asteraceae, *Erica sessiliflora*; Kruger 1977, 1983; Lamont et al. 1991). Serotinous members of the Proteaceae have been more intensively studied than any other growth form in fynbos communities.

18.5.2.3 *Fire-stimulated seed germination*
Species with soil-stored seed banks occur in a much wider range of vegetation types than species with serotinous seed banks. Many are stimulated to germinate by fire. There is a large literature on the nature of fire-stimulated germination cues (e.g. Went, Juhren & Juhren 1952; Keeley 1991; Bell, Plummer & Taylor 1993). The nature of the cues can have important effects on the ecology of populations exposed to different fire regimes. For example heat-stimulated germination of legume seeds can fail if fires are of low intensity. Some of the most well-studied cases of fire-stimulated germination involve 'hard-seededness', with dormancy being due to an impermeable seed coat (e.g. Quinlivan 1971; Manning & Van Staden 1987). In the characteristically hard-seeded legumes, dormancy is broken by rupturing of the lens, a specialized part of the seed coat. The opening of the lens in *Acacia* is induced by heat and effected through the splitting of thin-walled cells within the lens (Cavanagh 1980). After the lens is lifted, water enters the seed via the lens cavity and germination can begin. Hard-seededness occurs in other families, such as the Rhamnaceae and Rutaceae (Jeffery, Holmes & Rebelo 1988). The mechanism for breaking dormancy is less well known in these groups and may be quite different. For example, in a fynbos *Leucospermum* species, heat causes desiccation

of the seed coat, which then cracks on moistening, and germination is initiated (Brits et al. 1993). Dormancy of hard seeds may be broken not only by fire but also by scarification in stream beds, passage through the gut of a vertebrate or the cracking of the seed coat caused by trampling of the soil. The trait can, therefore, not be considered a unique adaptation to fire.

As with most forms of seed dormancy, hard-seededness is quite variable within a species or population and, in some, seeds germinate without a heat stimulus. Nevertheless the heat associated with fire is important in the breaking of seed dormancy in many fire-prone systems. Other effects of fire may be indirect. For example, the removal by fire of the insulating effect of the vegetation layer results in a wider diurnal range in soil surface temperatures during autumn: wide diurnal temperature fluctuations have been found to enhance germination in some fynbos species (Brits 1986; Pierce & Moll 1994).

Fire alters the chemistry of the environment and these chemical changes can cue fire-stimulated germination. For example, seed germination is stimulated by charred wood in Californian chaparral (e.g. Keeley 1991) and also some fynbos species (J.E. Keeley & W.J. Bond, unpubl. data). Recent studies have also shown that seed germination can be directly stimulated by smoke. Smoke stimulation was first reported for *Audouinia* (Bruniaceae), a rare shrub species in the fynbos (De Lange & Boucher 1990). It has since been reported in many other fynbos species (Brown 1993), in ecologically similar Australian heathlands (Dixon, Roche & Pate 1995) and even in the grass species *Themeda triandra*, one of the dominant components of southern African grasslands (Baxter et al. 1994). The physiological mechanism underlying smoke-stimulated germination is not yet known and may vary from species to species. Although smoke and charate stimulation suggest obligate reproductive dependence on fire, not all seeds, or populations, respond to these cues. Karoo species that very seldom experience fire are also stimulated by smoke (Pierce, Esler & Cowling 1995). It is not yet clear whether the chemicals involved are general germination stimulants or whether some are active only in fire-prone species.

The distribution of fire-stimulated reproductive traits among biomes is rather poorly known in southern Africa except for fynbos (reviewed by Le Maitre & Midgley 1992). Fire-stimulated flowering seems to occur among a number of grasses and geophytes in grasslands. Fire-stimulated seed release (a form of serotiny) seems absent in savanna trees. Serotinous proteas occur in fynbos patches in the Drakensberg, but serotiny is absent in protea species that form savannas (Rourke 1980). Fire-stimulated seed germination has not been studied in

Table 18.3 **Fire life histories based on vegetative survival and reproductive timing**

	Vegetative response		
	Survive fire (sprouters, fire-resistant bark, etc.)	Killed by fire (non-sprouting)
Stimulated by fire		
Not stimulated by fire	Fire-recruiting sprouters	Fire-recruiting non-sprouters
	Non-fire-recruiting sprouters	Non-fire-recruiting non-sprouters

grasslands and savannas in the region. Perennial grasses do not accumulate long-lived (>1 yr) seed banks (Chap. 10, this volume). *Acacia karroo* seeds do not require heating to germinate (Story 1952); neither does *Pterocarpus angolensis* (Van Daalen 1991). Tolsma (1989) reported better germination success in unburned seeds, than in burned seeds of *Acacia erioloba*, *A. fleckii*, *A. tortilis*, *Combretum apiculatum*, *Peltophorum africanum* and *Terminalia sericea* in arid savannas of Botswana. However, fire may indirectly enhance germination by killing bruchid larvae in seeds or by creating a mineral seed bed (Sabiiti & Wein 1987; Van Daalen 1991). Therefore, from the limited evidence available, fire-stimulated recruitment seems rare in grasslands and savannas, in striking contrast to the findings for fynbos.

18.5.3 Fire life histories

Functional classifications are useful for organizing information on the population responses of plants to fire. A number of such classifications have been suggested (e.g. Gill 1981; Noble & Slatyer 1980) serving a variety of purposes. A simple but general approach is to consider, independently, survival and reproductive responses to burning. This produces four major fire life histories based on the dichotomy between reproductive timing (fire-recruiting and non-fire-recruiting) and survival modes (sprouting and non-sprouting) (Table 18.3). The four groups have distinctly different population responses to fire, have evolved under differing selective pressures, comprise different proportions of plant communities and appear at different stages in post-fire succession. The prevalence of the different life histories changes from one geographic region to another and is one reason for the fact that it is so difficult to make generalizations about the role of fire in ecosystems. Long-lived non-sprouting species are most common in fynbos with fire intervals greater than ten years or in forests which do not usually burn. Sprouting species occur in all vegetation types. Species that are not cued to fire for reproduction seem to predominate in grasslands and savannas with relatively short fire frequencies,

whereas fire-recruiting species are characteristic of fynbos with longer fire cycles. However, studies on fire-responses are very unevenly distributed and it is difficult to gain a general view of the distribution of different life histories in the region from present knowledge.

Further refinement is possible within each of the major categories. Because sprouting is so common among forest and savanna species, it would be useful to develop functional groups within this broad category. One possibility is illustrated in Fig. 18.6. This emphasizes size-specific variation in sprouting. Information of this kind is valuable in determining when and whether fire can be used to control the densities of plant populations. The quantitative shape of age-specific sprouting curves is known to vary among fires of different intensity (Fig. 18.5; see Trollope 1984a), but the behaviour of species relative to each other may stay more constant. Among sprouting species, there appears to be a trade-off between the certainty of persistence by sprouting and regeneration by seedling establishment. The most vigorously sprouting species seldom produce seedlings. One example from a comparative study of fynbos species is shown in Fig. 18.7. The trade-off appears to be quite general and may be a useful indicator of alternative recruitment strategies in co-existing species which also provides guidelines for manipulating plant populations.

18.6 Fire and vegetation dynamics

Fire has long been used, along with animal stocking rates, as a tool for manipulating rangeland composition in southern Africa. Experimental studies on the most appropriate fire-regimes for veld management began early in the century (Phillips 1919) and have continued to the present. Some of the experiments have been maintained for decades. There have been far fewer experiments in fynbos, because the vegetation is less conducive to burning in small replicated blocks. Here, population studies of key species, especially serotinous proteas, have been used to develop an understanding of the impact of fire on vegetation dynamics. Because fynbos has less economic value as a grazing resource, studies of fire ecology were far fewer until the impetus of the Fynbos Biome Project in the 1980s. The role of fire in each of the major South African biomes is discussed in the relevant biome chapters in this volume and was reviewed in Booysen & Tainton (1984). Below, only a brief summary of key findings is made for fynbos, savanna and grasslands.

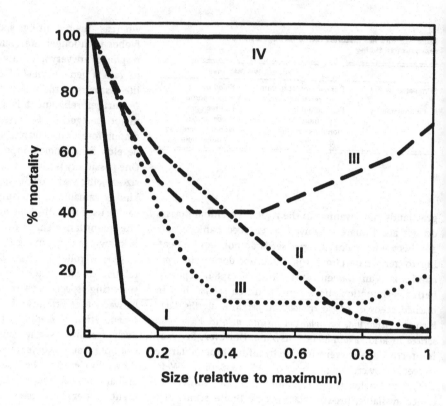

Figure 18.6 **Fire-induced tree and shrub mortality patterns in relation to plant size. Four broad patterns can be recognized: I, very low post-seedling mortality; II, decreasing mortality with increasing size; III, mortality highest in smallest and largest size classes; IV, non-sprouters killed across all size classes. The level of mortality can change greatly from fire to fire, especially in type III plants, shown here by two curves representing high- (upper curve) and low-intensity (lower curve) burns.** *Reproduced, with permission, from Bond & Van Wilgen (1996).*

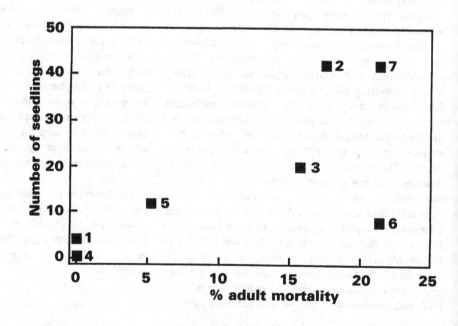

Figure 18.7 **The relationship between fire-induced mortality (%) and seedling recruitment in sprouting fynbos shrubs (after Le Maitre et al. 1992). Seedlings were counted two years after an autumn fire. Species are: 1, *Widdringtonia nodiflora*; 2, *Penaea mucronata*; 3, *Mimetes cucullatus*; 4, *Aulax pallasia*; 5, *Nebelia paleacea*; 6, *Erica coccinea*; 7, *Leucadendron salignum*.** *Reproduced, with permission, from Bond & Van Wilgen (1996).*

18.6.1   Fynbos

Fynbos has the most diverse range of fire life histories relative to the other southern African biomes, producing complex responses to variation in the fire regime. Our understanding of how fire influences community dynamics is still far from complete. Most recent research on fynbos dynamics has been based on an analysis of species traits and their interaction with fire-regime. Changes in species composition after a fire are attributed to the interaction between the fire and the regenerative properties of the plant. This approach has been widely used in other woody communities which burn at relatively long intervals. The effects of fire frequency on species composition have been most studied and the key or 'vital' attributes that influence the fate of a species are easily identified (Noble & Slatyer 1980).

Fynbos composition, unlike other biomes, can be drastically altered after a single fire. This is because the proteoid shrubs that dominate many communities are killed by fire and recruit only from seeds stored in canopy seed banks (Chap. 6, this volume). The effect of fire on protea demography has been extensively studied. Non-sprouting proteas can be eliminated from an area by burning at short fire frequencies before plants have accumulated seed reserves (Kruger 1977; Van Wilgen 1981). Densities are reduced by fires at long intervals (>40 yr), burning after the shrubs, and their seed banks, begin to die (Bond 1980). Non-sprouting proteas are also very sensitive to season of burn, with heavy recruitment after autumn burns and poor recruitment after spring burns (Jordaan 1965; Bond, Vlok & Viviers 1984; Van Wilgen & Viviers 1985; Midgley 1989). High temperatures can destroy seeds in the laboratory, but there have been no field studies of the effects of variation in fire intensity on recruitment of serotinous proteas (Midgley & Viviers 1990). Conditions governing recruitment of proteas with soil-stored seed banks is less well studied. Seeds are thought to be able to persist in the soil for at least 15 years: a population may 'disappear' from an area after an unfavourable burn then 're-appear' after the next burn (e.g. Boucher 1981). Species with soil-stored seeds are sensitive to fire intensity, showing higher recruitment after high-intensity burns (Bond, Le Roux & Erntzen 1990).

Recent studies on protea demography have begun to explore the effects of interactions among species, or density-dependent interactions within species, in addition to the vital attributes approach (Cowling 1987; Bond, Cowling & Richards 1992). For example, some of the variation in post-burn recruitment has been attributed to density-dependent effects on fecundity of serotinous proteas (Bond, Maze & Desmet 1995). At high densities, some protea species become so crowded that cone production

in a stand declines. This could result in a density-induced population shrinkage after the next fire. A confounding implication is that fires burnt to maximize protea seedling density may result in population collapses due to dense crowding.

Strong interactions have also been shown between a serotinous and an ant-dispersed proteoid shrub (Yeaton & Bond 1991). The ant-dispersed species was strongly suppressed by the serotinous species: plants growing next to serotinous neighbours had slower growth rates, greatly reduced reproduction and increased port-fire mortality relative to plants without near neighbours. A Markov simulation of seedling recruitment indicated that the two species could co-exist only where variable fire regimes favoured the weaker competitor at the expense of the stronger competitor. This is one of the very few population studies testing the argument that variable fire regimes are needed to maintain high stand diversity.

Another important perspective on interactions has been the finding that, given sufficient time between fires, the proteoid overstorey suppresses understorey species (Cowling & Gxaba 1990). Dense proteoid stands also act as nucleation sites for bird-dispersed species, thus initiating succession towards non-flammable bush clumps or forests (Manders & Richardson 1992; Manders, Richardson & Masson 1992; Chap. 6, this volume). In this way, the effects of fire on the proteoid layer have important repercussions, because proteas are strong interactors with other components of the community. Although the vital attributes approach has been very useful in indicating the impact of fire on dynamics, these new studies on interactions between species and how they are mediated by fire are beginning to give us a more complete view of fire and fynbos dynamics.

Although protea demography has dominated studies of fynbos responses to burning, there is growing interest in other fynbos elements. Using a vital attributes approach, Van Wilgen & Forsyth (1992) showed that short fire frequencies would have little effect on species other than proteoids. This is because, after fire, most non-sprouting species appear to flower and set seed earlier than the proteas. Long intervals without fire may cause loss of species with short-lived seed banks that do not survive long after the parents have stopped producing seeds, either through death or cessation of flowering as they become shaded. In one study, a fire in 50-year-old fynbos resulted in considerably fewer species and lower cover than in 18-year-old fynbos (Bond 1980).

Non-proteoids are also sensitive to season of burn. Experimental burns at different seasons have shown that the density and diversity of seedlings is greatest after autumn burns and least after spring burns (Le Maitre

1988). The flowering of geophytes is also sensitive to fire season: in the case of *Watsonia borbonica*, there was very poor flowering after spring burns compared to abundant flowering after autumn burns (Le Maitre & Brown 1992). The seasonal effect of fire on flowering can therefore have substantial effects on the number of seeds available for recruitment.

Fire intensity can influence regeneration from seed, through heat stimulation of germination of some species or lethal exposure of others. Bond & Van Wilgen (1996) have suggested that the differing effects of fire intensity on recruitment can be predicted from seed size. The depth from which seedlings can successfully emerge is limited by their stored resources. Hence, small-seeded species will only recruit well after low-intensity burns, and large-seeded species are able to emerge from deeper levels and survive more intense fires. Many of the larger-seeded species are myrmecochores, and may be too deeply buried to receive the appropriate heat cue from low-intensity burns, and therefore recruit successfully only after more intense fires. Thus, low-intensity burns will tend to favour small-seeded species and high-intensity burns will selectively favour large-seeded species. Shifts of species composition consistent with these predictions have been observed after high-intensity burns in cleared stands of invasive alien trees and shrubs (Richardson & Van Wilgen 1986) and, in a limited way, after low-intensity burns in proteoids. However, more field studies are needed.

18.6.2 Grasslands

Fire influences grassland dynamics in the following ways (see for example, Tainton & Mentis 1984; Chap. 10, this volume):

1 Unlike fynbos, fire does not initiate a distinct successional sequence, because the plants recover so rapidly by resprouting;
2 Changes in fire frequency cause marked shifts in grass composition;
3 Changes in fire season cause marked changes in grass composition;
4 Vertebrate herbivory acting in conjunction with fire can cause marked changes in species composition of the sward;
5 Changes in biomass are often less than changes in species composition;
6 The effects of fire on grass dynamics varies along a moisture gradient with the greatest effects in high-rainfall regions.

Some of the most common grass species, such as *Themeda triandra* and *Heteropogon contortus*, become moribund in the absence of fire, and decline in abundance. Other species are more tolerant of self-shading and their abundance increases in the absence of fire. Fire-dependent grass species also occur elsewhere in the world (Silva *et al.* 1991; Tilman & Wedin 1991). These grasses typically produce abundant tillers and are light-demanding. In the absence of defoliation, self-shading by litter causes a reduction in tiller production (Everson, Everson & Tainton 1988). Without fire, these species undergo endogenous oscillations in year-to-year biomass production which, in productive environments, can generate chaotic dynamics (Tilman & Wedin 1991). The demography of grasses has been studied with the use of transition matrix models that specify tiller production under different burning techniques. For example, a two-year study of tiller dynamics in a Drakensberg grassland predicted the relative ranking of grass species under different burning regimes. The ranking was similar to that generated by a 30-year burning experiment (Everson, Everson & Tainton 1985). The possibility of chaotic dynamics was not detected in these studies, because transition matrix models assume exponential growth of tillers.

The point of this example is that in some environments and for some species, defoliation by fire is essential for the continued maintenance in the sward. *Themeda triandra* appears to be one such 'fire weed'. In the eastern Cape, however, this species occurs alongside fynbos in extensive stands that burn at intervals of decades and it shows no sign of becoming moribund in the absence of fire (W.J. Bond, pers. obs.). The form of *Themeda* in these environments differs from the 'typical' form in bearing tillers from aerial shoots. This shrub-like growth form, therefore, does not suffer from self-shading (pers. obs). Intra-specific variation in grass behaviour complicates interpretation of grass ecology and reduces the generality of results based on field experiments (Gibbs Russell & Spies 1988). More study is needed on the biological causes of fire responses to improve our predictions.

There have been relatively few studies of the population biology of grasses in the more arid regions. Although semi-arid grasslands share many common species with mesic grasslands, growth patterns and population processes seem to differ. In mesic grasslands, the death of grass tufts is rare and compositional changes are mostly due to the dynamics of tiller production as influenced by competitors, and defoliation by both fire and grazers (Tainton & Mentis 1984; Everson *et al.* 1985). In arid grasslands, whole tufts may die during severe drought, and seedling recruitment occurs in the resulting gaps (O'Connor 1994). Rainfall variability seems to be of greater importance as a determinant of changes in grass composition than fire, competition or grazing (see Chap. 10, this volume). Shrub invasion tends to be initiated by heavy grazing rather than, as in mesic grasslands, fire exclusion.

18.6.3 Savannas

Savannas have not been studied as intensively as the grasslands in South Africa. A classic study by Trapnell (1959) in Zambian savannas was the first to show the power of fire in shaping savanna vegetation. Using three treatments (annual burns at two different seasons and protection from fire), Trapnell showed that savanna could be modified to a wooded grassland, closed woodland or forest, depending on fire frequency and season. Similar experimental studies elsewhere have also shown major changes in tree densities through manipulations of fire (e.g. Kennan 1971; Van Rensburg 1971; Van Wyk 1971). In some instances, similar experimental burning treatments had negligible effects on the tree layer (Geldenhuys 1977). In general, however, the occurrence of fires soon after trees have flushed their first leaves in spring favours grass at the expense of trees, whereas fires during or towards the end of the rainy season favour trees at the expense of grasses.

One of the most intractable problems in savanna management is the thickening up or invasion by trees, which then suppress productivity of the grass layer (e.g. Du Toit 1967; Donaldson 1969; Trollope 1980, 1984a; Scholes 1988; Smit & Swart 1994; Chap. 11, this volume). The use of fire in the management of bush encroachment has been studied for many years, but the problem is still far from resolved. Fire has the following effects on tree populations (Trollope 1984a; Frost & Robertson 1987):

1 Tree seedlings are killed by fire, but many species acquire resistance to burning within a year or two of germination.
2 With the exception of very young seedlings, very few woody plants are killed by fire (e.g. Rutherford 1980; Trollope 1980).
3 Burning of the canopy causes top-kill and coppicing of the plant from the base or from surviving buds in the canopy. The amount of top-kill and subsequent basal sprouting increases with flame height and fire intensity (Trollope 1984a).
4 The architecture of large trees (>2–3 m) is unaffected by burning. However, some species lose the capacity to sprout in larger size classes. These can be killed by fire, especially if the bark has been weakened by previous fires or damaged by animal activity such as from termites or porcupine foraging (Yeaton 1988).

Thus, fire may help to maintain an open physiognomy by: (a) inhibiting seedling establishment, by killing seedlings before they have acquired the capacity to sprout; and (b) by occasionally killing large trees that have lost the insulating properties of their bark. The most important effect of fire, however, is probably in bringing about architectural changes – caused by death of the canopy in small plants and their reversion to multi-stemmed shrubby forms. This reversion in height lowers the canopy down to the range of small- and medium-sized vertebrate browsers, holds the stems at a height where they can be shaded by grasses and keeps the trees at a height where they will still be vulnerable to the next fire (Trollope 1980, 1984a). If fires are withheld for a sufficient period, the woody plants escape to a height above the influence of the flames and thus become relatively immune to burning.

Fire in savannas is limited by the availability of grass fuel. Grass production is linearly dependent on moisture availability (Rutherford 1981; Scholes & Walker 1993). Therefore, the frequency of fires is highest in mesic savannas. Moisture availability also varies with soil type and slope position. Grass fuels are generally most abundant in bottomlands or on more clay-rich soils. Intense fires may spread from such sites to vegetation growing on soils at higher slope positions, which are less favourable for grass growth. Biologically important determinants of grass fuels are herbivores and the density of the tree layer. The escape of suppressed tree juveniles to the canopy occurs when the fire-return interval is prolonged through lack of ignition, active fire suppression, sparse fuels due to low rainfall or heavy grazing, or the absence of browsers. Once they have reached the canopy, trees suppress the grass fuels and fire is no longer an effective tool for reducing tree densities. Savannas appear to be an intrinsically unstable system and it is remarkable that this structural type is so extensive in Africa and other continents. Equilibrial models of grass/tree coexistence (e.g. Walker & Noy-Meir 1982) have failed to take into account size-dependent asymmetries in the relationship between trees and grasses. However, attempts to model the grass/tree balance from a population perspective, based on the above generalizations, usually predict dominance either by grasses or by trees rather than the usual mixture of the two growth forms (Menaut *et al.* 1990; Hochberg, Menaut & Gignoux 1994). The reason for grasses failing to eliminate tree seedlings through competition or fire, and for trees failing to exclude grasslands and form non-flammable thickets, except in local parts of the landscape, remains an intriguing and economically important puzzle for savanna ecologists.

Table 18.4 **Examples of observed transitions from flammable to less-flammable communities as a result of fire exclusion**

Flammable vegetation	Less-flammable vegetation	Time period for transition (yr)	Source
Fynbos	Forest	>50	Masson & Moll (1987) Luger & Moll (1992)
Fynbos	Forest	>25	Manders et al. (1992)
Grassland (eastern Cape)	Grassy fynbos	>5	Trollope (1973)
Grassland (Transvaal)	Asteraceous shrubs	>5	Roux (1969)
Grassland (Drakensberg)	Scrub forest	>35	Westfall et al. (1983)
Grassland (Zululand)	Dune forest	>35	Weisser (1978)
Savanna (Kruger National Park)	Scrub thicket	?	Van Wyk (1971)
Savanna (Hluhluwe)	Scrub thicket	?	Smith & Goodman (1987)
Savanna (Zambia)	Scrub forest	10	Trapnell (1959)

18.7 Fire and biome distribution

From a pyrocentric view, biome distribution in southern Africa is largely determined by growing conditions that promote flammable grass fuels. Wherever the quantity and seasonality of rainfall, as modified by local soils and topography, is adequate for dense growth of perennial grasses, fire becomes a major factor in structuring vegetation. Grassland fires, alone or in association with vertebrate browsers, restrict non-flammable woody vegetation to fire-proof refugia in all regions with adequate grass growth. In winter-rainfall regions, deep-rooted woody plants generally dominate the vegetation, as is found elsewhere in mediterranean climate regions of the world (Cody & Mooney 1978). However, here too, a flammable vegetation has evolved that has also displaced non-flammable alternative formations to fire refugia or geological refugia where fynbos cannot grow (Specht & Moll 1983).

The evidence that continental vegetation patterns are dominated by grasses comes from both landscape patterns as well as fire and herbivore exclusion experiments. Many landscapes contain mosaics of grasslands and non-flammable thickets or forests on steep scree slopes, ravines or river margins, where they gain some protection from fire (e.g. Geldenhuys 1994). In savanna regions, such as the coastal plains of northeastern KwaZulu-Natal and southern Mozambique, forest patches often occur on deep, well-drained sands or, more rarely, as dense thickets on clay-rich soils in a matrix of vegetation with flammable grassy layers (Moll & White 1978). Savanna vegetation often occurs on the same soils, presumably after they have been cleared by humans, intense fires, or concentrations of large mammal browsers. The co-occurrence of different vegetation types under the same macro-climate, and often on similar soils, has long impressed on South African ecologists the importance of disturbance in determining vegetation patterns (e.g. Phillips 1930).

The evidence for grass domination reflected in vegetation pattern is supported by experimental and observational study of processes. Successional changes from grass-dominated fuels to non-flammable woody vegetation have been observed in many vegetation types after the experimental or accidental exclusion of fire. These studies confirm the existence of alternative stable states determined by disturbance history, rather than directly by climate. Some examples are listed in Table 18.4 and are discussed in the relevant chapters in this volume. Although many of these changes occur as a result of exclusion of fire alone, some depend on interactions with vertebrate herbivores. The role of herbivores in determining biome structure and biome boundaries seems most important in the low-rainfall regions of the Karoo and in arid savannas (<600 mm yr^{-1}). Here the productivity of grasses is strongly dependent on the amount and seasonality of rainfall (Roux & Vorster 1983; Hoffman and Cowling 1990; Bond, Stock & Hoffman 1994). The pre-settlement distribution of grasses in the Central Karoo appears to have been determined by rainfall (Bond et al. 1994) with periodic advancement of grasses in high-rainfall years and retreats in drought years. With the appearance of settled pastoralists, grasslands have been invaded by karoo shrubs over quite extensive areas (Acocks 1953; Bond et al. 1994). This invasion has been attributed to reduction of grass cover by grazing animals and not to the exclusion of fire (see also Chap. 21, this volume).

As in the case of grasslands, fynbos seems to have displaced alternative non-flammable woody vegetation over much of its range through the agency of fire (Chap. 6, this volume). Fynbos protected from fire for long periods has been colonized by forest trees in mesic areas of the southern and southwestern Cape (Van Daalen 1981; Luger & Moll 1992; Manders et al. 1992). On a landscape scale, the mosaic of forest and fynbos in the southern Cape has been attributed to fires driven by hot, dry bergwinds (Geldenhuys 1994). Strandveld, a non-flammable broad-leaved shrubby vegetation, also invades coastal fynbos after long periods of fire suppression (Cowling & Pierce 1988). At its eastern limits, fynbos adjoins spiny and succulent thicket and here too, vegetation boundaries have been attributed to fire. Thicket species can survive and grow when transplanted to fynbos soils, but fynbos cannot survive on the clay-rich thicket soils (Euston-Brown 1995). Thus, like grasslands, fynbos appears to have spread far beyond those soils and climates to which it would be confined by taller vegetation, because it is flammable and can displace fire-intolerant species through the agency of fire.

The determinants of vegetation boundaries have come under much closer scrutiny with threats of global warming. Many attempts to predict future vegetation distribution are based on correlations with contemporary climates. Ellery *et al.* (1989) have used such an analysis for South African biomes and found a good fit between most biomes and rainfall and temperature components of climate. From the above discussion, this correlation cannot be viewed as indicating that climate directly influences vegetation distribution. The indirect effects of climate on fire regimes have been little studied in South Africa and the relative importance of climate, soils and defoliators (fire and herbivores) in determining biome patterns is far from resolved. Pleistocene and Holocene changes in vegetation have almost universally been attributed to climate change (e.g. Chap. 4, this volume) with little cognizance of the importance of fire and herbivory as agents of change.

18.8   Fire and evolution of flora

Given its apparent importance in shaping southern African vegetation, there has been remarkably little work on the importance of fire in the evolution of the flora. The evolution of flammable grasses in summer-rainfall regions and flammable shrubs in winter-rainfall regions would have introduced regular fire and had a profound effect on the pre-existing vegetation and flora of the region. There is growing evidence that the spread of grasses and grass-dominated vegetation was comparatively recent. C_4 grasses, which now dominate southern African grassy vegetation, first evolved in the Miocene. Isotopic evidence from both paleosols and tooth enamel indicates that C_4 grasses first spread, and rapidly became dominant, only some 6–8 Myr ago (MacFadden & Cerling 1994). Modern types of grasslands may not have existed until the Middle Pleistocene. The spread of C_4 grasslands was followed rapidly by increasing predominance of grazing species that form large herds, such as bovids,

from 7 to 5 Myr in Africa (Janis 1993). A similar late Miocene origin for C_4 grasslands and savannas is now apparent over most parts of the world (Janis 1993; MacFadden & Cerling 1994). Evidence from the evolution of equid teeth suggests that C_3 grasslands may have existed for a few million years prior to the explosion of C_4 grasses (MacFadden & Cerling 1994), but the emerging picture is that grass-dominated vegetation is of comparatively recent origin.

The evolution of fire-prone grasses must have had an extraordinary impact on global vegetation patterns. Grass fuels create a distinctive fire regime that must have acted as a powerful selective filter, permitting access to only a small subset of woody plants from ancestral forests that seldom burnt. Many African savanna species have close relatives in forests. Examples include savanna and forest species in genera such as *Faurea*, *Hymenocardia*, *Ochna*, *Olea*, *Pterocarpus*, *Strychnos* and *Terminalia*. There are unfortunately no phylogenetic studies on the evolutionary history of these genera, which would throw light on the characteristics that allowed species to colonize fire-prone environments. Phylogenetic studies would also be useful in testing whether ancestral taxa occur in the less flammable vegetation types, with more derived taxa occurring in fire-prone vegetation. Some genera, such as *Acacia*, *Brachystegia* and *Combretum*, are more or less confined to fire-prone environments, whereas some genera (e.g. *Colophospermum*) can be considered 'endemic' to savanna. Presumably more taxa will transgress forest and grassy formations where savannas are of recent origin than in fire-derived systems of greater antiquity. The *Acacia karroo* and bushclump savannas of the eastern Cape would appear to be of recent origin by this criterion. There appear to be no species confined to the fire-prone vegetation other than this species. Bushclump species appear to be transgressors from Valley Bushveld.

The appearance and diversification of fynbos is as recent as grasslands and savanna. The embryonic flora is thought to have been in place by the late Miocene (although elements had a more ancient origin) with major diversification occurring only in the Pliocene (Linder, Meadows & Cowling *et al.* 1992). As in the case of savannas, some species of subtropical and temperate forest ancestry have penetrated this fire-prone vegetation, although often only in relatively fire-protected habitats (Moll, McKenzie & McLachlan 1980; Linder *et al.* 1992). Here too, the evolution of a new fire regime appears to have acted as a powerful selective filter, limiting access to the pre-existing forest taxa.

Adaptations that enable plants to survive, and even promote, fires have been discussed by a number of authors. Studies are summarized by Frost (1984) for grasslands and savanna and Le Maitre & Midgley (1992) for fynbos. It will be clear from section 18.3 that many plants, especially in fynbos, have a near obligate dependence on fire for reproduction and/or recruitment. In some cases, such as fire-stimulated flowering, traits are not fixed and the evolutionary significance of the phenomenon is far from clear. The evolutionary importance of fire in determining plant attributes of savanna trees and grasses has received little attention.

18.9   Fire in management

18.9.1   The development of fire research

Fire has long been used in the management of vegetation in southern Africa. Scientific studies on the use of fire for vegetation management began early in the twentieth century. Approaches to burning have changed from hostility in the 1920s by leading botanists of the day, to widespread acceptance in rangeland management by the 1950s and in fynbos management by the 1970s (Scott 1984). The value of fire as a tool in rangeland management was recognized very early by southern African scientists in comparison to their Australian and North American colleagues (Scott 1984). Experimental studies on the effects of varying fire frequency, fire season and grazing intensity were initiated before 1920 (Phillips 1919). By the 1950s, numerous studies on fire in agricultural practice were in progress in southern Africa, stimulated by the leadership of J.F.V. Phillips and his students. In the late 1970s, Trollope initiated studies on fire behaviour, until then a neglected area of research in comparison to Australia and North America where there had been much greater emphasis on fire suppression (Trollope 1984b). Fire research lagged in the fynbos biome until it was linked to water catchment studies at Jonkershoek in the western Cape. The hydrological significance of burning led to a rapid increase in fire ecology research in the 1970s (Wicht & Kruger 1973). The Fynbos Biome Programme, a large cooperative scientific programme, led to many new studies on fynbos fire ecology with increasing emphasis on burning to achieve conservation aims (Cowling 1992; Van Wilgen et al. 1992b, Chap. 6, this volume).

Principles used in prescribed burning in different systems in southern Africa have been extensively reviewed (Trollope 1980; Booysen & Tainton 1984; Van Wilgen et al. 1990a, 1992b). Some of these are discussed briefly below.

18.9.2   Management objectives

Fire is extensively used in the management of southern African vegetation for a variety of objectives. There are many, and diverse, management objectives that can influence the type of fire applied in ecosystems (Van Wilgen et al. 1990a). For example:

1   Livestock production is dependent on the quality and quantity of fodder plants, which can be influenced by manipulating the fire regime to favour such plants, or to reduce the abundance of unpalatable species. Fire is an important tool for reducing woody plant cover in tree/grass mixtures.

2   Forestry operations use fire for site preparation and fuel reduction on the borders and within plantation forests.

3   Conservation aims are often met by applying fire, particularly in ecosystems where fire-dependent species and communities are involved. Habitat structure can be manipulated by burning to favour certain plant species, or to improve habitat for animals.

4   Fire is used in controlling invasion by undesirable plants. This is of particular importance in fynbos where fire can either promote, or help suppress, alien invasion.

5   Fire is often applied to influence the hydrological cycle, for example to enhance water yield from catchments, or to prevent erosion from unplanned wildfires (e.g. Scott & Van Wyk 1992). Such applications depend on the manipulation of vegetation structure in catchments, often reducing catchment cover and biomass to low levels (e.g. Bosch & Hewlett 1982). Invasive trees, often escaping from plantations, cause the greatest reduction in streamflow.

6   Reduction of fire hazard is a common aim of prescribed burning. Fires are applied frequently, and under mild weather conditions, to reduce the build-up of fuel loads and prevent intense fires under adverse conditions.

Quite often, conflicting objectives lead to controversy or compromise solutions. For example, reduction in streamflow is proportional to above-ground biomass so that short fire frequencies may favour maximum water yield in fynbos catchments (Bosch & Hewlett 1982). However, if fire-return intervals are too short, slow-maturing species can be eliminated from communities, and this conflicts with conservation aims (Kruger 1977; Van Wilgen 1981). There are also legal and safety constraints on burning that limit flexibility of managed fire regimes (e.g. Van Wilgen & Richardson 1985). For example, seasons of burning prescribed for safety reasons may cause loss of desirable species from a grazing or conservation perspective.

18.9.3 Fynbos

Fynbos composition is sensitive to variation in frequency, season and intensity of fire (Van Wilgen, Bond & Richardson 1992a). Largely for economic reasons, vast areas of fynbos are currently managed under a 'natural' burning regime in which lighting fires are allowed to burn but fires of known human origin are suppressed (Seydack 1992). Prescribed burning is used mostly where fires might endanger adjoining property (such as urban areas). Fire regimes are applied on the basis of responses by key species, especially those of serotinous Proteaceae, and economic and safety constraints. Both short (<5–8 yr) and very long (>40 yr) fire-return intervals can eliminate the proteoid layer. Fire season can also have a large impact on fynbos recovery after a burn. Autumn burns favour protea regeneration and that of many other species, whereas winter or spring burns result in poorest recovery. Fire intensity also influences regeneration with better recovery of myrmecochorous species after high-intensity burns and of small-seeded species of shrubs and graminoids after low-intensity burns. Managers can utilize these principles to achieve different objectives. For example, to reduce shrubs in grassy fynbos, and promote grass for grazing, winter or early spring burns would be preferred. Autumn burns would be favoured for most recruitment of non-sprouters and intense autumn burns for rare myrmecochorous species such as *Mimetes splendidus*.

The impact of fire on invasion by alien plants has been well researched, and management regimes have been devised for effective control (see reviews by Richardson & Cowling 1992; Richardson *et al.* 1992; Van Wilgen *et al.* 1992b; Chap. 22, this volume). Invasive plants alter fuel properties, thereby changing the fire regime (Van Wilgen & Richardson 1985). Mechanical felling of invasives can cause unusually heavy fuel loads, very intense fires and extreme conditions for post-burn vegetative recovery (Richardson & Van Wilgen 1986). Severe erosion can also follow such burns (Scott & Van Wyk 1992).

18.9.4 Grasslands

The role of fire (and grazing by large mammals) in managing vegetation with grassy fuels varies under different climate and soil conditions (e.g. Trollope 1980). A great deal of grassland research has been done in the humid grasslands of KwaZulu-Natal. Principles of range management, developed there and based on classical successional principles, have been applied somewhat uncritically to more arid regions. Recently, much more emphasis has been placed on event-driven dynamics and the contingencies that influence plant recruitment or mortality during an event (Frost *et al.* 1986; Mentis *et al.* 1989; Westoby, Walker & Noy-Meir 1989). The shift in

thinking on rangeland management principles has been too recent to have had a wide influence in management practice. The relationship between vegetation dynamics and rainfall which underlies the debate on management principles is indicated as a tentative set of generalizations (properties) in Table 18.5.

Fire is widely used for management purposes in the more humid 'sour' grasslands to maintain the vigour and palatability of the grass sward. Fires are burnt at annual or biennial frequencies, since fuel accumulates rapidly in these systems. Season of burn is determined by effects on sward productivity and composition, the condition of plant cover in relation to wind and water erosion and the hazards of runaway fires. In general, fires in the dormant season or early in the growing season are most favoured for maintaining grass productivity and composition. Fires in the summer growing season have detrimental effects on grass cover and composition. Many grasslands are also being colonized by invasive alien species including Australian *Acacia* species and pines (especially *Pinus patula*). Frequent fires can be used to suppress the invasion process (Chap. 22, this volume).

Fire is seldom used in arid grasslands (rainfall <650 mm yr^{-1}), except in high-rainfall years when it can help suppress invasion of shrubby species (Tainton & Mentis 1984). In general, fire is seen as an agent removing important forage in these grasslands. Unlike the sourveld, grasses remain palatable in the dry season. Furthermore, self-shading, which leads to reduced productivity of the sward, is not a factor. The exception is found in the C$_3$ grasslands on the higher karoo mountains. These behave more like sourveld and are burnt to increase grass production and to make grasses more accessible to livestock.

18.9.5 Savannas

Fire in savannas, as in grasslands, is used to manage the composition and productivity of the grass sward. However, a major objective of fire management is usually the control of the woody layer. The key principle driving the use of fire for controlling bush encroachment involves burning at frequencies and intensities that will prevent juvenile trees from escaping to a height where the canopy is no longer killed by fire (Trollope 1980). In mesic savannas (rainfall >650 mm yr^{-1}), fire alone can control woody plant densities, since the grass fuel load can support annual or biennial fires. Fires burnt at the end of the dry season, after trees have leafed out but before rains have initiated grass growth, are most effective at reducing tree biomass. Fires burnt during, or towards the end of the rainy season, favour woody plants at the expense of grass (Trapnell 1959; Trollope 1984a).

Table 18.5a **Grassland and savanna formations in relation to rainfall**

	Rainfall >650 mm yr⁻¹ (or sandy soils at lower rainfall)	Rainfall <650 mm yr⁻¹ (or clay soils at higher rainfall)
No trees	Sour grassveld	Sweet grassveld
With trees	Moist savanna	Arid savanna

Table 18.5b **Properties attributed to grassy systems in relation to rainfall**

	Rainfall >650 mm yr⁻¹ (sour grassveld and moist savanna)	Rainfall <650 mm yr⁻¹ (sweet grassveld and arid savanna)
System characteristics		
Growing conditions	Relatively invariant	Highly variable
Herbivory	Seasonally unpalatable, offtake usually small fraction of grass production	Palatable year-round, offtake large fraction of production, except in high-rainfall years
Fire frequency (yr)	1–3	3–10+
Dynamics		
Community dynamics driven by:	Competitive interactions and fire	Rainfall events, contingencies
Population dynamics of grasses	Density-dependent productivity, biomass oscillations due to self-shading	Weak density dependence, biomass oscillations follow rainfall
Grass regeneration	Seedlings rare, mostly vegetative (tillers)	Seedlings common, and vegetative
Persistence of individual grass tufts	Mostly very persistent, mortality very rare	Variable, some common species killed by drought
Grassland dynamics		
Effect of fire exclusion	Grass composition changes. Trees, shrubs invade	Grass composition unchanged. No woody plant invasion
Savanna dynamics		
Effect of fire on tree densities	Fire sufficiently frequent to control trees on its own	Fire and browsers needed to control trees
Effect of fire exclusion	Directional succession to non-flammable thicket or forest	Often cyclical succession when cohort of woody invasive senesces and dies

In arid savannas (rainfall <650 mm yr⁻¹), grass production is more variable and fuels are often too sparse to support fires. Fires that do occur in dry periods can cause more damage to the grass layer than to the trees (Gertenbach & Potgieter 1979). Fire frequencies in arid savannas vary from five to 30 or more years (in Kalahari thornveld). Trollope (1980) has argued that under these conditions fire alone cannot control tree densities. However, burning causes top-kill in juvenile trees and the basal coppice can be controlled by browsers. This principle has been tested effectively with cattle and goats in eastern Cape thornveld (Trollope 1980) and the principle has been extended to other savanna regions (Van Wilgen et al. 1990a). Since top-kill increases with increasing fire intensity, large grass fuel loads and dry weather conditions are prerequisites for management burns designed to control tree densities.

18.9.6   Fire and conservation

Fire prescriptions developed for livestock farming have often been applied to nature reserves to promote the large mammal component. Recently this 'farming' approach to fire has come under criticism, since it might promote some components of the biota at the expense of others. Both in the fynbos (e.g. Cowling 1987) and

grass and savanna biomes (e.g. Mentis & Bailey 1990), calls have been made for more variable fire-regimes (varied frequencies, varied seasons, varied burn intensities) in conservation areas in line with the objective of conserving biotic diversity. Where feasible, 'natural' fire-regimes have been proposed, meaning, in practice, that lightning fires be allowed to burn and other fires suppressed (e.g. for savannas: Stander, Nott & Mentis 1993; for fynbos: Seydack 1992; Van Wilgen, Richardson & Seydack 1994). The rationale is that the fire regime under which the biota evolved is most likely to ensure survival.

As noted by Van Wilgen et al. (1990a), one problem with this approach is that it has no predictive value regarding the components of the biota that will be influenced by the developing fire-regime. A 'natural' fire-regime may produce undesirable changes in terms of other management objectives, but the causes of these changes require understanding of fire/plant interactions. Another problem is that the properties of 'natural' fire-regimes are unknown and probably unknowable. Humans have ignited fire in the region for hundreds of thousands of years (Hall 1984; Brain & Sillen 1988) at unknown seasons, frequencies and intensities. The spread of fires from an ignition site has been curtailed by extensive fragmentation of nearly all fire-prone vegetation by roads, buildings, croplands and afforestation. Fires ignited by lightning in most contemporary landscapes will not be able to travel as far as they did in the past, thus confounding attempts to recreate 'natural' fire-regimes. Nevertheless, for economic reasons alone, fire management is likely to become more extensive and fire-regimes less rigid in conservation areas in the future.

In some instances, fire-regimes are managed to promote particular species of special conservation interest. *Widdringtonia cedarbergensis*, a conifer endemic to the Cederberg mountains in the fynbos biome, is the most intensively researched example (see Manders 1986 for a review). A past history of exploitation and fire has reduced populations of the species to a critical state. Manders (1987) developed a transition matrix model to explore the effects of different fire-regimes on population growth, drawing on a wealth of research on the biology of the species. He suggested that autumn burns at intervals of 15 or more years would sustain the species. However, additional data, and a re-analysis of the sensitivity of the matrix model, suggests that such a fire-regime would cause populations to decline (Privett 1994). A cedar reserve has been set aside in the Cederberg where low-intensity fires are burnt every 4–8 years to reduce mortality of large trees and to prevent the spread of large wildfires. Natural recruitment has been supplemented by manual planting of seedlings. This level of management intensity for species preservation is very rare in South Africa.

18.10 Conclusions

The study of fire and fire effects on vegetation has been a central feature of southern African ecology for most of this century and a large body of information has accumulated. Much of this research has been aimed at the level of populations and communities. Future progress in this area is likely to come from synthesis of existing information and an evaluation of important gaps in our understanding. Sensitivity analyses of vegetation dynamics, using population and community modelling approaches, hold much promise for highlighting the most critical processes for further study (e.g. Hochberg *et al.* 1994; Starfield *et al.* 1993 for savannas).

Because of global warming, there is growing interest in the landscape-level consequences of fire for vegetation distribution. South Africa's complex topography and diverse vegetation provide a model system for the study of the interrelationships between climate, vegetation and its consumers. The view taken in this chapter is that fire is a major determinant of the distribution of most southern African biomes and of the dynamics of communities within biomes. Fire is important because of the existence of flammable components of the biota: grasses, especially tufted perennial grasses, and fynbos shrubs and graminoids. The current dominance of fire-prone vegetation in the region is relatively recent in evolutionary terms (late Miocene/Pliocene). Fluctuations of vegetation in the Pleistocene, from this perspective, probably owed as much to climatically induced changes in the fire-regime, and possibly the intensity of vertebrate herbivory, as to direct influences of climate on plant growth and survival.

The view that fire is a key determinant of the structure and composition of southern African vegetation is an old one (e.g. Phillips 1930; Acocks 1953). However, there are many unanswered questions and untested hypotheses underlying this view. For example, the relative importance of vertebrate herbivores or fire as major consumers of plants, and therefore shapers of vegetation, is still obscure. Experimental studies of fire have often excluded vertebrate herbivores. Where herbivores have access, small experimental burns provide oases and have complicated the interpretation of results (e.g. Van Wyk 1971). The relative impact of fire and herbivory across gradients of primary productivity and 'sour'/ 'sweet' or nutrient availability axes would be interesting to explore further.

Archaeologists have championed the effect of humans, especially Iron Age cultivators, as agents of vegetation change (e.g. Feely 1980; Hall 1984; Chap. 21, this volume), but their arguments have not been critically examined by ecologists.

Phytosociological studies have been popular in South Africa since the pioneering studies of Acocks (1953). Though Acocks was well aware of the dynamic nature of vegetation, the Veld Type concept has produced a static view of plant distribution. The pattern of communities is commonly explained in terms of correlations with site variables such as climate and soils. The inference that these correlations are causal, and that site determines vegetation type, is not justified. Exclusion or addition of fire, and herbivores, can result in major compositional changes. The extent to which major vegetation patterns are constrained by physical limits on production, as opposed to consumption by fire and herbivores, and how these patterns will shift under changing climates are important questions for the future in a warming world.

18.11 References

Acocks, J.P.H. (1953). Veld types of South Africa. *Memoirs of the Botanical Survey of South Africa*, **28**, 1–128.

Alexander, M.E. (1982). Calculating and interpreting forest fire intensities. *Canadian Journal of Botany*, **60**, 349–57.

Barclay, J.J., Jury, M.R. & Washington, R. (1993). Meteorology of fire danger in the Natal Drakensberg. *South African Journal of Science*, **89**, 341–9.

Baxter, B.J.M., Van Staden, J., Granger, J.E. & Brown, N.A.C. (1994). Plant-derived smoke and smoke extracts stimulate seed germination of the fire-climax grass *Themeda triandra*. *Environmental and Experimental Botany*, **34**, 217–23.

Bell, D.T., Plummer, J.A. & Taylor, S.K. (1993). Seed germination ecology in Southwestern Western Australia. *Botanical Review*, **59**, 24–73.

Bond, W.J. (1980). Fire and senescent fynbos in the Swartberg, Southern Cape. *South African Forestry Journal*, **114**, 68–71.

Bond, W.J. (1983). Dead leaves and fire survival in Southern African tree aloes. *Oecologia*, **58**, 110–14.

Bond, W.J., Cowling, R.M. & Richards, M.B. (1992). Competition and coexistence in proteoid shrubs. In *The Ecology of Fynbos. Nutrients, Fire and Diversity*, ed. R.M. Cowling, pp. 206–25. Cape Town: Oxford University Press.

Bond, W.J., Le Roux, D. & Erntzen, R. (1990). Fire intensity and regeneration of myrmecochorous Proteaceae. *South African Journal of Botany*, **56**, 326–31.

Bond, W.J., Maze, K. & Desmet, P. (1995). Fire life histories and the seeds of chaos. *Ecoscience*, **2**, 252–60.

Bond, W.J., Stock, W.D. & Hoffman, M.T. (1994). Has the karoo spread? a test for desertification using carbon isotopes from soils. *South African Journal of Science*, **90**, 391–7.

Bond, W.J. & van Wilgen, B.W. (1996). *Fire and Plants*. London: Chapman & Hall.

Bond, W.J., Vlok, J. & Viviers, M. (1984). Variation in seedling recruitment of Cape Proteaceae after fire. *Journal of Ecology*, **72**, 209–21.

Booysen, P. de V. & Tainton, N.M. (eds) (1984). *Ecological Effects of Fire in South African Ecosystems*. Berlin: Springer-Verlag.

Bosch, J.M. & Hewlett, J.D. (1982). A review of catchment experiments to determine the effect of vegetation changes on water yield and evapotranspiration. *Journal of Hydrology*, **55**, 3–23.

Boucher, C. (1981). Autecological and population studies of *Orothamnus zeyheri* in the Cape of South Africa. In *The Biological Aspects of Rare Plant Conservation*, ed. H. Synge, pp. 343–53. New York: Wiley.

Brain, C.K. & Sillen, A. (1988). Evidence from the Swartkrans cave for the earliest use of fire. *Nature*, **336**, 464–6.

Brits, G.J. (1986). Influence of fluctuating temperatures and H_2O_2 treatments on germinaiton of *Leucospermum cordifolium* and *Serruria florida* (Proteaceae) seeds. *South African Journal of Botany*, **52**, 286–90.

Brits, G.J., Calitz, F.J., Brown, N.A.C. & Manning, J.C. (1993). Desiccation as the active principle in heat-stimulated seed germination of *Leucospermum* R.Br. (Proteaceae) in fynbos. *New Phytologist*, **125**, 397–403.

Brown, N.A.C. (1993). Promotion of germination of fynbos seeds by plant-derived smoke. *New Phytologist*, **122**, 1–9.

Brown, P.J., Manders, P.T., Bands, D.P., Kruger, F.J. & Andrag, R.H. (1991). Prescribed burning as a conservation management practice: a case history from the Cederberg mountains, Cape Province, South Africa. *Biological Conservation*, **56**, 133–50.

Byram, G.M. (1959). Combustion of forest fuels. In *Forest Fire. Control and Use*, ed. K.P. Davis, pp. 155–82, New York: McGraw-Hill.

Cavanagh, A.K. (1980). A review of some aspects of the germination of acacias. *Proceedings of the Royal Society of Victoria*, **91**, 161–80.

Cody, M.L. & Mooney, H.A. (1978). Convergence versus nonconvergence in mediterranean-climate ecosystems. *Annual Review of Ecology and Systematics*, **9**, 265–321.

Cowling, R.M. (1987). Fire and its role in coexistence and speciation in Gondwana shrublands. *South African Journal of Science*, **83**, 106–12.

Cowling, R.M. (ed.) (1992). *The Ecology of Fynbos. Nutrients, Fire and Diversity*. Cape Town: Oxford University Press.

Cowling, R.M. & Gxaba, T. (1990). Effects of a fynbos overstorey shrub on understorey community structure: implications for the maintenance of community-wide species richness. *South African Journal of Ecology*, **1**, 1–7.

Cowling, R.M. & Pierce, S.M. (1988). Secondary succession in coastal dune fynbos: variation due to site and disturbance. *Vegetatio*, **76**, 131–9.

De Lange, J.H. & Boucher, C. (1990). Autecological studies on *Audouinia capitata* (Bruniaceae). I. Plant-derived smoke as a germination cue. *South African Journal of Botany*, **56**, 700–3.

Dixon, K.W., Roche, S. & Pate, J.S. (1995). Promotive effect of smoke derived from burnt native vegetation on seed germination of Western Australian plants. *Oecologia*, **101**, 185–92.

Donaldson, C.H. (1969). *Bush Encroachment with Special Reference to the Blackthorn Problem of the Molopo Area*. Pretoria: Government Printer.

du Toit, P.F. (1967). Bosindringing met spesifieke verwysing na *Acacia karroo* indringing. *Proceedings of the Grassland Society of southern Africa*, **2**, 119–26.

Edwards, D. (1984). Fire regimes in the biomes of South Africa. In *Ecological Effects of Fire in South African Ecosystems*, ed. P. de V. Booysen & N.M. Tainton, pp. 19–38. Berlin: Springer-Verlag.

Ellery, W.N., Ellery, K., McCarthy, T.S., Cairncross, B. & Oelofse, R. (1989). A peat fire in the Okavango Delta, Botswana and its importance as an ecosystem process. *African Journal of Ecology*, **27**, 7–21.

...ston-Brown, D. (1995). *Environmental and Dynamic Determinants of Vegetation Distribution in the Kouga and Baviaanskloof Mountains, Eastern Cape.* MSc Thesis. Cape Town: University of Cape Town.

...erson, C.S., Everson, T.M. & Tainton, N.M. (1985). The dynamics of *Themeda triandra* tillers in relation to burning in the Natal Drakensberg. *Journal of the Grassland Society of southern Africa,* **2,** 18–25.

...erson, C.S., Everson, T.M. & Tainton, N.M. (1988). Effects of intensity and height of shading on the tiller initiation of six grass species from the Highland sourveld of Natal. *South African Journal of Botany,* **54,** 315–18.

...erson, T.M., Van Wilgen, B.W. & Everson, C.S. (1988). Adaptation of a model for rating fire danger in the Natal Drakensberg. *South African Journal of Science,* **84,** 44–9.

...ely, J.M. (1980). Did Iron Age Man have a role in the history of Zululand's wilderness landscapes? *South African Journal of Science,* **76,** 150–2.

...ost, P.G.H. (1984). The responses and survival of organisms in fire-prone environments. In *Ecological Effects of Fire in South African Ecosystems,* ed. P. de V. Booysen & N.M. Tainton, pp. 273–309. Berlin: Springer-Verlag.

...ost, P.G., Medina, E., Menaut, J.C., Solbrig, O., Swift, M. & Walker, B. (1986). Responses of savannas to stress and disturbance. *Biology International,* Special Issue 10.

...ost, P.G.H. & Robertson, F. (1987). The ecological effects of fire in savannas. In *Determinants of Tropical Savannas,* ed. B.H. Walker, pp. 93–140. Miami: ICSU Press.

...ldenhuys, C.J. (1977). The effect of different regimes of annual burning on two woodland communities in Kavango. *South African Forestry Journal,* **103,** 32–42.

...ldenhuys, C.J. (1994). Bergwind fires and the location of forest patches in the southern Cape landscape, South Africa. *Journal of Biogeography,* **21,** 49–62.

...rtenbach, W.P.D. & Potgieter, A.L.F. (1979). Veldbrandnavorsing in die struikmopanieveldvan die Krugerwildtuin. *Koedoe,* **22,** 1–28.

Gibbs Russell, G.E. & Spies, J.J. (1988). Variation in important pasture grasses: I. Morphological and geographical variation. *Journal of the Grassland Society of southern Africa,* **5,** 15–21.

Gill, A.M. (1981). Adaptive responses of Australian vascular plant species to fires. In *Fire and The Australian Biota,* ed. A.M. Gill, R.H. Groves & I.R. Noble, pp. 243–72. Canberra: Australian Academy of Sciences.

Granger, J.E. (1984). Fire in forest. In *Ecological Effects of Fire in South African Ecosystems,* ed. P. de V. Booysen & N.M. Tainton, pp. 177–98. Berlin: Springer-Verlag.

Hall, A.V. (1959). Observations on the distribution and ecology of Orchidaceae in the Muizenberg mountains, Cape Peninsula. *Journal of South African Botany,* **25,** 265–78.

Hall, M. (1984). Man's historical and traditional use of fire in southern Africa. In *Ecological Effects of Fire in South African Ecosystems,* ed. P. de V. Booysen & N.M. Tainton, pp. 39–52. Berlin: Springer-Verlag.

Hare, R.C. (1965). Contribution of bark to fire resistance in southern trees. *Journal of Forestry,* **63,** 248–51.

Hochberg, M.E., Menaut, J.C. & Gignoux, J. (1994). The influence of tree biology and fire in the spatial structure of the West African savannah. *Journal of Ecology,* **82,** 217–26.

Hoffman, M.T. & Cowling, R.M. (1990). Vegetation change in the semi-arid eastern karoo over the last two hundred years: an expanding karoo – fact or fiction? *South African Journal of Science,* **86,** 286–94.

Horne, I.P. (1981). The frequency of veld fires in the Groot Swartberg Mountain Catchment Area, Cape Province. *South African Forestry Journal,* **118,** 56–60.

James, S. (1984). Lignotubers and burls – their structure, function and ecological significance in mediterranean ecosystems. *Botanical Review,* **50,** 225–66.

Janis, C.M. (1993). Tertiary mammal evolution in the context of changing climates, vegetation, and tectonic events. *Annual Review of Ecology and Systematics,* **24,** 467–500.

Jeffery, D.J., Holmes, P.M. & Rebelo, A.G. (1988). Effects of dry heat on seed germination in selected indigenous and alien legume species in South Africa. *South African Journal of Botany,* **54,** 28–34.

Jordaan, P.G. (1965). Die invloed van 'n winterbrand op die voortplanting van vier soorte van die Proteaceae. *Tydskrif vir Natuurwetenskappe,* **5,** 27–31.

Keeley, J.E. (1991). Seed germination and life history syndromes in the California chaparral. *Botanical Review,* **57,** 81–116.

Keeley, J.E. (1993). Smoke-induced flowering in the fire lily *Cyrtanthus ventricosus. South African Journal of Botany,* **59,** 638.

Kennan, T.C.D. (1971). The effects of fire on two vegetation types of Matopos. *Proceedings of the Tall Timbers Fire Ecology Conference,* **11,** 53–98.

Kruger, F.J. (1977). Ecology of Cape fynbos in relation to fire. In *Proceedings of the Symposium on Environmental Consequences of Fire and Fuel Management in Mediterranean Ecosystems,* ed. H.A. Mooney & C.E. Conrad, pp. 230–44. USDA Forest Service, General Technical Report WO-3.

Kruger, F.J. (1983). Plant community diversity and dynamics in relation to fire. In *Mediterranean-type Ecosystems. The Role of Nutrients,* ed. F.J. Kruger, D.T. Mitchell & J.U.M. Jarvis, pp. 446–72. Berlin: Springer-Verlag.

Lamont, B.B., Le Maitre, D.C., Cowling, R.M. & Enright, N.J. (1991). Canopy seed storage in woody plants. *Botanical Review,* **57,** 277–317.

Le Maitre, D.C. (1988). Effects of season of burn on the regeneration of two Proteaceae with soil-stored seed. *South African Journal of Botany,* **54,** 575–80.

Le Maitre, D.C. & Brown, P.J. (1992). Life cycles and fire-stimulated flowering in geophytes. In *Fire in South African Mountain Fynbos,* ed. B.W. van Wilgen, D.M. Richardson, F.J. Kruger & H.J. van Hensbergen, pp. 145–60. Berlin: Springer-Verlag.

Le Maitre, D.C., Jones, C.A. & Forsyth, G.G. (1992). Survival of eight woody sprouting species following an autumn fire in Swartboskloof, Cape Province, South Africa. *South African Journal of Botany*, **58**, 405–13.

Le Maitre, D.C. & Midgley, J.J. (1992). Plant reproductive ecology. In *The Ecology of Fynbos. Nutrients, Fire and Diversity*, ed. R.M. Cowling, pp. 135–74. Cape Town: Oxford University Press.

Linder, H.P., Meadows, M.E. & Cowling, R.M. (1992). History of the Cape Flora. In *The Ecology of Fynbos. Nutrients, Fire and Diversity*, ed R.M. Cowling, pp. 113–34. Cape Town: Oxford University Press.

Luger, A.D. & Moll, E.J. (1992). Fire protection and Afromontane forest expansion in Cape fynbos. *Biological Conservation*, **64**, 51–6.

MacFadden, B.J. & Cerling, T.E. (1994). Fossil horses, carbon isotopes and global change. *Trends in Ecology and Evolution*, **9**, 481–6.

Manders, P.T. (1986). An assessment of the current status of the Clanwilliam cedar (*Widdringtonia cedarbergensis*) and the reasons for its decline. *South African Forestry Journal*, **139**, 48–53.

Manders, P.T. (1987). A transition matrix model of the population dynamics of the Clanwilliam cedar (*Widdringtonia cedarbergensis*) in natural stands subjected to fire. *Forest Ecology and Management*, **20**, 171–86.

Manders, P.T. & Richardson, D.M. (1992). Colonization of Cape fynbos communities by forest species. *Forest Ecology and Management*, **48**, 277–93.

Manders, P.T., Richardson, D.M. & Masson, P.H. (1992). Is fynbos a stage in succession to forest? Analysis of the perceived ecological distinction between two communities. In *Fire in South African Mountain Fynbos*, ed. B.W. van Wilgen, D.M. Richardson, F.J. Kruger & H.J. van Hensbergen, pp. 81–107. Berlin: Springer-Verlag.

Manning, J.C. & van Staden, J. (1987). The role of the lens in seed imbibition and seedling vigour of *Sesbania punicea* (Cav.) Benth. (Leguminosae: Papilionoideae). *Annals of Botany*, **56**, 705–13.

Manry, D.E. & Knight, R.S. (1986). Lightning density and burning frequency in South African vegetation. *Vegetatio*, **66**, 67–76.

Masson, P.H. & Moll, E.J. (1987). The factors affecting forest colonisation of fynbos in the absence of fire at Orange Kloof, Cape Province, South Africa. *South African Forestry Journal*, **143**, 5–10.

Menaut, J.C., Gignoux, J., Prado, C. & Clobert, J. (1990). Tree community dynamics in a humid savanna of the Cote-d'Ivoire: modelling the effects of fire and competition with grass and neighbours. *Journal of Biogeography*, **17**, 471–81.

Mentis, M.T. & Bailey, A.W. (1990). Changing perceptions of fire management in savanna parks. *Journal of the Grassland Society of southern Africa*, **7**, 81–5.

Mentis, M.T., Grossman, D., Hardy, M.B., O'Connor, T.G. & O'Reagain, P.J. (1989). Paradigm shifts in South African range science, management and administration. *South African Journal of Science*, **85**, 684–7.

Midgley, J.J. (1989). Season of burn of serotinous fynbos Proteaceae: a critical review and further data. *South African Journal of Botany*, **55**, 165–70.

Midgley, J.J. & Viviers, M. (1990). The germination of seeds from heated serotinous cones of eight shrubland species. *South African Forestry Journal*, **155**, 5–9.

Moll, E.J., McKenzie, B. & McLachlan, D. (1980). A possible explanation for the lack of trees in the fynbos, Cape Province, South Africa. *Biological Conservation*, **17**, 221–8.

Moll, E.J. & White, F. (1978). The Indian Ocean coastal belt. In *Biogeography and Ecology of Southern Africa*, ed. M.J.A. Werger, pp. 561–98. The Hague: Junk.

Noble, I.R. & Slatyer, R.O. (1980). The use of vital attributes to predict successional changes in plant communities subject to recurrent disturbances. *Vegetatio*, **43**, 5–21.

O'Connor, T.G. (1994). Composition and population responses of an African savanna grassland to rainfall and grazing. *Journal of Applied Ecology*, **31**, 155–71.

Phillips, E.P. (1919). A preliminary report on the veld burning experiments at Groenkloof, Pretoria. *South African Journal of Science*, **16**, 286–99.

Phillips, J.F.V. (1930). Fire: its influence on biotic communities and physical factors in South and East Africa. *South African Journal of Science*, **27**, 352–67.

Pierce, S.M., Esler, K.J. & Cowling, R.M. (1995). Smoke-induced germination of succulents (Mesembryanthemaceae) from fire-prone and fire-free habitats in South Africa. *Oecologia*, **102**, 520–2.

Pierce, S.M. & Moll, E.J. (1994). Germination ecology of six shrubs in fire-prone Cape fynbos. *Vegetatio*, **110**, 25–41.

Privett, S. (1994). *Restoration of the Clanwilliam Cedar*. BSc Hons. Thesis. Cape Town: University of Cape Town.

Pyne, S.J. (1984). *Introduction to Wildland Fire. Fire Management in the United States*. New York: Wiley.

Quinlivan, B.J. (1971). Seed coat impermeability in legumes. *Journal of the Australian Institute of Agricultural Science*, **37**, 283–95.

Richardson, D.M. & Cowling, R.M. (1992). Why is mountain fynbos invasible and which species invade? In *Fire in South African Mountain Fynbos*, ed. B.W. Van Wilgen, D.M. Richardson, F.J. Kruger & H.J. Van Hensbergen, pp. 161–81. Berlin: Springer-Verlag.

Richardson, D.M., Macdonald, I.A.W., Holmes, P.M. & Cowling R.M. (1992). Plant and animal invasions. In *The Ecology of Fynbos. Nutrients, Fire and Diversity*, ed. R.M. Cowling, pp. 271–308. Cape Town: Oxford University Press.

Richardson, D.M. & van Wilgen, B.W. (1986). The effects of fire in felled *Hakea sericea* and natural fynbos and implications for weed control in mountain catchments. *South African Forestry Journal*, **131**, 63–8.

Rourke, J.P. (1980). *The Proteas of Southern Africa*. Cape Town: Purnell.

Roux, E. (1969). *Grass. A Story of Frankenwald*. Cape Town: Oxford University Press.

ux, P.W. & Vorster, M. (1983). Vegetation change in the karoo areas. *Proceedings of the Grasslands Society of South Africa*, 18, 25–9.

ndel, P.W. (1981). Structural and chemical components of flammability. In *Proceedings of the Conference on Fire Regimes and Ecosystem Properties*, ed. H.A. Mooney, T.M. Bonnicksen, N.L. Christensen, J.E. Lotan & W.A. Reiners, pp. 183–207. General Technical Report WO-86. Amsterdam: USDA Forest Service.

therford, M.C. (1980). Annual plant-production–precipitation relations in arid and semi-arid regions. *South African Journal of Science*, 76, 53–6.

therford, M.C. (1981). Survival, regeneration and leaf biomass changes in woody plants following spring burns in *Burkea africana–Ochna pulchra* savanna. *Bothalia*, 13, 531–52.

therford, M.C. (1983). Growth rates, biomass and distribution of selected woody plant roots in *Burkea africana–Ochna pulchra* savanna. *Vegetatio*, 52, 45–63.

biiti, E.N. & Wein, R.W. (1987). Fire and *Acacia* seeds: a hypothesis of colonization success. *Journal of Ecology*, 74, 937–46.

holes, R.J. (1988). The regrowth of *Colophospermum mopane* following clearing. *Journal of the Grassland Society of southern Africa*, 7, 147–51.

holes, R.J. & Walker, B.H. (1993). *An African Savanna. Synthesis of the Nylsvley Study*. Cambridge: Cambridge University Press.

ott, D.F. & Van Wyk, D.B. (1992). The effects of fire on soil water repellency, catchment sediment and yields and streamflow. In *Fire in South African Mountain Fynbos*, ed. B.W. Van Wilgen, D.M. Richardson, F.J. Kruger & H.J. Van Hensbergen, pp. 216–39. Berlin: Springer-Verlag.

ott, J.D (1984). An historical review of research on fire in South Africa. In *Ecological Effects of Fire in South African Ecosystems*, ed. P. de V. Booysen & N.M. Tainton, pp. 53–66. Berlin: Springer-Verlag.

ydack, A.H.W. (1992). Fire management options in fynbos mountain catchment areas. *South African Forestry Journal*, 161, 53–8.

Siegfried, W.R. (1981). The incidence of veld fire in the Etosha National Park, 1970–1979. *Madoqua*, 12, 225–30.

Silva, J.F., Raventos, J., Caswell, H. & Trevisan, M.C. (1991). Population responses to fire in a tropical savanna grass, *Andropogon semiberbis*: a matrix model approach. *Journal of Ecology*, 79, 345–56.

Smit, G.N. & Swart, J.S. (1994). Influence of leguminous and non-leguminous woody plants on the herbaceous layer and soil under varying competition regimes in Mixed Bushveld. *African Journal of Range and Forage Science*, 11, 27–33.

Smith, T.M. & Goodman, P.S. (1987). Successional dynamics in an *Acacia nilotica–Euclea divinorum* savannah in southern Africa. *Journal of Ecology*, 75, 603–10.

Specht, R.L. & Moll, E.J. (1983). Heathlands and sclerophyllous shrublands – an overview. In *Mediterranean-type Ecosystems. The Role of Nutrients*, ed. F.J. Kruger, D.T. Mitchell & J.U.M. Jarvis, pp. 41–65. Berlin: Springer-Verlag.

Stander, P.E., Nott, T.B. & Mentis, M.T. (1993). Proposed burning strategy for a semi-arid African savanna. *African Journal of Ecology*, 31, 282–9.

Starfield, A.M., Cumming, D.H.M., Taylor, R.D. & Quadling, M.S. (1993). A frame-based paradigm for dynamic ecosystem models. *AI Applications*, 7, 1–13.

Story, R. (1952). A botanical survey of the Keiskammahoek district. *Memoirs of the Botanical Survey of South Africa*, 27, 1–184.

Tainton, N.M. & Mentis, M.T. (1984). Fire in grassland. In *Ecological Effects of Fire in South African Ecosystems*, ed. P. de V. Booysen & N.M. Tainton, pp. 115–48. Berlin: Springer-Verlag.

Tilman, D. & Wedin, D. (1991). Oscillations and chaos in the dynamics of a perennial grass. *Nature*, 353, 653–5.

Tolsma, D.J. (1989). *On the Ecology of Savanna Ecosystems in South-Eastern Botswana*. PhD Thesis. Amsterdam: Free University of Amsterdam.

Trapnell, C.G. (1959). Ecological results of woodland burning experiments in northern Rhodesia. *Journal of Ecology*, 47, 129–68.

Trollope, W.S.W. (1973). Fire as a method of controlling Macchia (Fynbos) vegetation on the Amatole mountains of the eastern Cape. *Proceedings of the Grasslands Society of South Africa*, 8, 35–41.

Trollope, W.S.W. (1980). Controlling bush encroachment with fire in the savanna areas of South Africa. *Proceedings of the Grasslands Society of South Africa*, 15, 173–7.

Trollope, W.S.W. (1984a). Fire in savanna. In *Ecological Effects of Fire in South African Ecosystems*, ed. P. de V. Booysen & N.M. Tainton, pp. 149–76. Berlin: Springer-Verlag.

Trollope, W.S.W. (1984b). Fire behaviour. In *Ecological Effects of Fire in South African Ecosystems*, ed. P. de V. Booysen & N.M. Tainton, pp. 199–218. Berlin: Springer-Verlag.

Trollope, W.S.W. (1993). Fire regime of the Kruger National Park for the period 1980–1992. *Koedoe*, 36, 45–52.

Trollope, W.S.W. & Tainton, N.M. (1986). Effect of fire intensity on the grass and bush components of the Eastern Cape Thornveld. *Journal of the Grassland Society of southern Africa*, 2, 27–42.

Uhl, C. & Kauffman, J.B. (1990). Deforestation, fire susceptibility, and potential tree responses to fire in the eastern Amazon. *Ecology*, 71, 437–49.

van Daalen, J.C. (1981). The dynamics of the indigenous forest–fynbos ecotone on the southern Cape. *South African Forestry Journal*, 119, 14–23.

van Daalen, J.C. (1991). Germination of *Pterocarpus angolensis* seed. *South African Forestry Journal*, 158, 33–6.

van Rensburg, H.J. (1971). Fire: its effects on grasslands, including swamps – southern, central and eastern Africa. *Proceedings of the Tall Timbers Fire Ecology Conference*, 11, 147–73.

van Wilgen, B.W. (1981). Some effects of fire frequency on fynbos plant community composition and structure at Jonkershoek, Stellenbosch. *South African Forestry Journal*, 118, 42–55.

van Wilgen, B.W. (1984). Fire climates in the Southern and Western Cape Province and their potential use in fire control and management. *South African Journal of Science*, 80, 358–62.

van Wilgen, B.W., Bond, W.J. & Richardson, D.M. (1992a). Ecosystem management, In *The Ecology of Fynbos. Nutrients, Fire and Diversity*, ed. R.M. Cowling, pp. 345–71. Cape Town: Oxford University Press.

van Wilgen, B.W., Everson, C.S. & Trollope, W.S.W. (1990a). Fire management in southern Africa: some examples of current objectives, practices, and problems. In *Fire in the Tropical Biota*, ed. J.G. Goldammer, pp. 179–215. Berlin: Springer-Verlag.

van Wilgen, B.W. & Forsyth, G.G. (1992). Regeneration strategies in fynbos plants and their influence on the stability of community boundaries after fire. In *Fire in South African Mountain Fynbos*, ed. B.W. van Wilgen, D.M. Richardson, F.J. Kruger & H.J. van Hensbergen, pp. 81–107. Berlin: Springer-Verlag.

van Wilgen, B.W., Higgins, K.B. & Bellstedt, D.U. (1990b). The role of vegetation structure and fuel chemistry in excluding fire from forest patches in the fire-prone fynbos shrublands of South Africa. *Journal of Ecology*, 78, 210–22.

van Wilgen, B.W., Le Maitre, D.C. & Kruger, F.J. (1985). Fire behaviour in South African fynbos (macchia) vegetation and predictions from Rothermel's fire model. *Journal of Applied Ecology*, 22, 207–16.

van Wilgen, B.W. & Richardson, D.M. (1985). Factors influencing burning by prescription in mountain fynbos catchment areas. *South African Forestry Journal*, 134, 22–32.

van Wilgen, B.W., Richardson, D.M., Kruger, F.J. & van Hensbergen, H.J. (eds) (1992b). *Fire in South African Mountain Fynbos*. Berlin: Springer-Verlag.

van Wilgen, B.W., Richardson, D.M. & Seydack, A.H.W. (1994). Managing fynbos for biodiversity: constraints and options in a fire-prone environment. *South African Journal of Science*, 90, 322–9.

van Wilgen, B.W. & van Hensbergen, H.J. (1991). Fuel properties of vegetation in Swartboskloof. In *Fire in South African Mountain Fynbos*, ed. B.W. van Wilgen, D.M. Richardson, F.J. Kruger & H.J van Hensbergen, pp. 37–53. Berlin: Springer-Verlag.

van Wilgen, B.W. & Viviers, M. (1985). The effect of season of fire on serotinous Proteaceae in the Western Cape and the implications for fynbos management. *South African Journal of Forestry*, 133, 49–53.

van Wilgen, B.W. & Wills, A.J. (1988). Fire behaviour prediction in savanna vegetation. *South African Journal of Wildland Research*, 18, 41–6.

van Wyk, P. (1971). Veld burning in the Kruger National Park, an interim report of some aspects of research. *Proceedings of the Tall Timbers Fire Ecology Conference*, 11, 9–31.

Walker, B.H. & Noy-Meir, I. (1982). Aspects of the stability and resilience of savanna ecosystems. In *Ecology of Tropical Savannas*, ed. B.J. Huntley & B.H. Walker, pp. 556–609. Berlin: Springer-Verlag.

Weisser, P.J. (1978). Changes in area of grasslands on the dunes between Richards Bay and the Mfolozi river, 1937 to 1974. *Proceedings of the Grasslands Society of South Africa*, 13, 95–7.

Went, F.W., Juhren, G. & Juhren, M.C. (1952). Fire and biotic factors effecting germination. *Ecology*, 33, 351–63.

West, O. (1965). *Fire in Vegetation and its Use in Pasture Management with Special Reference to Tropical and Subtropical Africa*. Mimeographed Publication No. 1. Hurley Berkshire: Commonwealth Bureau of Pastures and Field Crops.

Westfall, R.H., Everson, C.S. & Everson, T.M. (1983). The vegetation of the protected plots at Thabamhlope Research Station. *South African Journal of Botany*, 2, 15–25.

Westoby, M., Walker, B. & Noy-Meir, I. (1989). Opportunistic management for rangelands not at equilibrium. *Journal of Range Management*, 42, 266–74.

White, F. (1976). The underground forests of Africa: a preliminary review. *Gardens Bulletin, Singapore*, 29, 57–71.

Wicht, C.L. & Kruger, F.J. (1973). Die ontwikkeling van bergveld bestuur in Suid Afrika. *South African Forestry Journal*, 86, 1–17.

Yeaton, R.I. (1988). Porcupines, fires and the dynamics of the tree layer of the *Burkea africana* savanna. *Journal of Ecology*, 76, 1017–29.

Yeaton, R.I. & Bond, W.J. (1991). Competition between two shrub species: dispersal differences and fire promote coexistence. *American Naturalist*, 138, 328–41.

Species diversity at the regional scale 19

R.M. Cowling, D.M. Richardson, R.E. Schulze, M.T. Hoffman, J.J. Midgley and C. Hilton-Taylor

The study of diversity has become part of the study of evolution   (Whittaker 1977)

19.1   Introduction

Early in the twentieth century, patterns of species diversity were studied by biogeographers who were concerned with the richness of faunas and floras in different parts of the globe. They developed bold and imaginative hypotheses to explain these patterns, stressing historical processes such as migration, isolation and speciation (Shmida & Wilson 1985). During the late 1950s and especially in the 1960s, biologists studied local diversity as an ecological problem. They ignored the role of history in an attempt to develop deterministic models based solely on local processes (Kingsland 1985; Schluter & Ricklefs 1993). The predictions of these models fared rather badly, especially for plant communities (Whittaker 1977), and ecologists became somewhat disillusioned with diversity theory. Today, there are many indications of a new synthesis of diversity theory which is mindful of spatial scale and acknowledges the role of ecology, geography and evolution in determining diversity patterns at the local, meso- and regional scales (Ricklefs 1987; Ricklefs & Schluter 1993). It is in the spirit of this emerging new synthesis that we attempt to analyse patterns of plant regional richness in southern Africa.

Southern Africa has an extraordinarily rich flora (Chap. 3, this volume). It includes some of the most species-rich regions in the world (southwestern fynbos biome); the world's richest semi-arid vegetation (succulent karoo); and also some areas of extreme floristic impoverishment (Cowling et al. 1989; Cowling, Holmes & Rebelo 1992; Cowling & Hilton-Taylor 1994; Chap. 3, this volume). What determines these patterns of plant diversity at a regional scale, and is it possible to develop simple models to predict them? Such models would not only provide insights on the evolution and maintenance of diversity in a species-rich area, but also enable the identification of speciose areas for conservation (Chap. 23, this volume) without resorting to tedious and time-consuming survey.

In this chapter we review and develop models on the determinants of regional plant richness in southern Africa. First, we review the theory on the controls of diversity at this scale; next we introduce new data and describe our approach in analysing them; then we evaluate the published models as well as the new models in terms of these theories; and finally, we discuss the generalizations that have emerged from this study.

19.2   Theory

In a discussion of the theory on the determinants of species diversity, spatial scale is of crucial importance (Whittaker 1972; Auerbach & Shmida 1987). Theory that has strong explanatory power at the local or community scale may have little relevance at the regional scale (Rosenzweig & Abramsky 1993). Much confusion arises when theories that are mechanistically appropriate for a particular scale are invoked to explain patterns at another scale, even though, under certain circumstances, their predictive power may be strong at both scales. This analysis focuses explicitly on species number at the regional scale, i.e. in areas within southern Africa which encompass more than one habitat or community but are largely nested within biomes (see Chap. 5, this volume).

At the regional scale, diversity is the product of the number of species within communities (alpha diversity), the compositional change along environmental gradients (beta diversity) and the compositional change between equivalent environments along geographical gradients (gamma diversity) (Whittaker 1972; Cody 1975; 1986; Shmida & Wilson 1985; Westoby 1988; 1993; Cowling 1990; Cowling *et al.* 1992; Ricklefs & Schluter 1993). These different components of diversity are theoretically independent, although in practice correlations often exist among them (Cody 1986; Ricklefs & Schluter 1993). Any conditions that enhance the values of these diversity components will increase regional richness. These include ecological factors that determine alpha diversity (Whittaker 1977; Brown 1988; Diamond 1988; Tilman & Pacala 1993); the size and number of habitats (or communities) in a region (Shmida & Wilson 1985; Williamson 1988; Rosenzweig 1992); historical factors associated with the evolution of habitat specialist and ecologically equivalent species (Shmida & Wilson 1985; Cody 1986; Cowling *et al.* 1992; Ricklefs & Schluter 1993; Westoby 1993); the inherent biological properties of lineages that determine rates of diversification (Vrba 1980; Fiedler 1986; Cowling & Holmes 1992a; Cornell 1993; Lawton 1993); and the interaction between these traits and the physical environment in which they occur (Brown 1988; Cowling *et al.* 1992; Ricklefs & Schluter 1993). Therefore, the development of a predictive understanding of the determinants of regional richness is a complex problem, and explanations are likely to vary in different ecological circumstances and with phylogenetically different assemblages.

Nonetheless, considerable advances have been made in explaining patterns of plant regional richness in terms of primary environmental factors such as habitat heterogeneity, rainfall and available energy (Johnson, Mason & Raven 1968; Richerson & Lum 1980; Wright 1983; Currie & Paquin 1987; Adams & Woodward 1989; O'Brien 1993; Hoffman, Midgley & Cowling 1994). Although the generality of some of these models has been challenged recently (e.g. Latham & Ricklefs 1993), there is much to be gained from attempts to model patterns of diversity in terms of first-order or primary environmental factors (Pianka 1966; Brown 1988; Rohde 1992; Wright, Currie & Maurer 1993). Below, we describe the hypotheses commonly invoked to explain patterns of plant diversity at a regional scale. Most of these relate to patterns of primary resource quantity and quality (Brown 1988; Diamond 1988) although we do discuss secondary- (dispersal) and tertiary-level (speciation history) hypotheses. We also discuss theory regarding the relationships between local and regional diversity patterns and processes, and the predictions deduced from convergence theory, for global comparisons of regional richness. Many of these hypotheses are related and make similar predictions; wherever possible, we attempt to expose and disentangle these similarities.

19.2.1 Area

Ecologists and biogeographers have known for more than 70 years that species number increases in samples of larger area, and that the rate of increase decreases with progressively larger area (Williamson 1988). For most groups of organisms and over most area sizes, the double logarithmic form of the species–area curve, championed by Arrhenius (1921), provides the best linear fit, but there are many exceptions (see Connor & McCoy 1979). What theories have been developed to explain this species–area effect?

Preston (1962) suggested that the species–area relationship could be a sampling effect: if species in an area are randomly and log normally distributed, then increasing numbers of species will be found with increasing area in a manner consistent with the log–log plot. Connor & McCoy (1979) suggest a different argument to explain the species–area slope as an artefact of random sampling. Both explanations have been convincingly rejected by Williamson (1988).

Many have argued that area size is merely a surrogate for environmental heterogeneity – as the area of a sample increases, so does the number of habitats, each supporting a relatively distinct biota (McGuinness 1984; Williamson 1988; Rosenzweig 1992). In very large areas that include environmentally similar but geographically disjunct habitats, the accumulation of ecologically similar species contributes to a positive species–area relationship (Shmida & Wilson 1985; Cowling *et al.* 1992).

The quantity of resources in an area is a direct function of its size: larger samples support larger total resources and, hence, larger population sizes, thus facilitating the persistence of more species, especially naturally rare ones (MacArthur 1972; Wright 1983; Diamond 1988). Research has shown that models which employ total resources (i.e. area × resource per unit area) as explanatory variables are more powerful predictors of regional richness than models which use area alone (Wright 1983; Wylie & Currie 1993).

In conclusion, area is best viewed as a surrogate for more proximate factors that explain patterns of regional richness (Shmida & Wilson 1985). When both area size and resource availability are variables, the former should be incorporated into models to provide an estimate of total resource availability.

19.2.2 Heterogeneity

There is strong support, both theoretical and empirical, for the hypothesis that environmental heterogeneity promotes species richness at a regional scale: larger samples support a greater variety of habitats and species (Johnson et al. 1968; Whittaker 1972; Richerson & Lum 1980; Shmida & Wilson 1985; Brown 1988; Diamond 1988; Williamson 1988; Cornell 1993). This is especially the case where particular biological properties of lineages interact over long periods with the ecological and geographical characteristics of a region in such a way as to promote the diversification of habitat specialist and ecologically equivalent species (Cowling & Holmes 1992a; Ricklefs & Schluter 1993; McDonald et al. 1995).

19.2.3 Favourableness

The notion that climatically benign or favourable conditions promote regional richness has a long history in ecology (e.g. Dobzhansky 1950; Fischer 1960). Researchers often treat favourableness as the effect of the mean of climatic variables on diversity and model the effect of the variance of these variables within and between years as seasonality and irregularity, respectively (Richerson & Lum 1980; Currie 1991).

A problem with favourableness is that conditions are defined relative to species preferences within the regional pool, rendering the hypothesis tautologous (Terborgh 1973; Brown 1988; Rohde 1992). A way out of this circularity is to judge favourableness by the ability of alien species, not the native and presumably well-adapted biota, to tolerate the environmental conditions; and to view unfavourable or harsh conditions as those which promote high extinction and low colonization rates of even the native, well-adapted species (Brown 1988). However, it is difficult to derive explanatory variables to model these determinants of regional richness patterns.

Brown (1988) discusses an important link between history and favourableness. Most lineages are constrained to some extent by the environments in which they have evolved. These acquired constraints tend to limit a lineage to a restricted range of environments, although some do colonize different environments and even diversify there. Brown suggests that maximum diversity is often associated with the 'favourable' environment where the lineage evolved.

19.2.4 Energy

Simply stated, 'energy-diversity' or 'species-energy' theory argues that the number of organisms and, hence, the number of species a region can support, increases monotonically with environmentally available energy (Hutchinson 1959; MacArthur 1972; Richerson & Lum 1980; Brown 1981; 1988; Wright 1983; Currie & Paquin 1987; Adams & Woodward 1989; Currie 1991; Wright et al. 1993). In a sense, species-energy theory has strong similarities to theories of favourableness (see Wright et al. 1993) but most researchers separate the two. Energy

is measured with the use of a wide range of variables including precipitation (in the arid zone), solar radiation, actual evapotranspiration, potential evapotranspiration and primary production.

There are a number of problems with species-energy theory. Firstly, the local scale is arguably the appropriate scale for investigating the relationship between energy and diversity (Latham & Ricklefs 1993). Studies at this scale show that richness peaks in sites of intermediate energy availability for a wide range of organisms (see references in Tilman & Pacala 1993; Rosenzweig & Abramsky 1993). Wright et al. (1993), however, provide a theoretical basis for a positive interactive effect of energy supply and heterogeneity (which would promote turnover, and therefore, richness) at the regional scale.

Secondly, the energy hypothesis fails to explain why energy should be partitioned among increasing numbers of species rather than increasing numbers of individuals of a few species within a regional biota (Currie 1991). Brown (1988) has suggested that under conditions of high energy availability, which favour large numbers of individuals, the benefits of specialization outweigh the costs of commonness. Thus, areas with high total productivity will be able to support populations of rare specialists which have escaped the predators and pathogens of the commoner species.

19.2.5 Seasonality and irregularity

The hypothesis that large variations in resource abundance within and between years (seasonality and irregularity, respectively) leads to reduced regional richness is linked to the notion that stability augments diversity by permitting specialization (Sanders 1968; Stevens 1989). On the other hand, there is evidence that predictable climatic seasonality may enhance plant diversity through trophic niche differentiation, at least at the local scale (Whittaker 1977; Cody 1989; Cowling et al. 1994a). Grassle & Maciolek's (1992) finding that sporadic input of ephemeral resources is important in maintaining regional richness in the deep sea is further evidence that irregular resource pulses, induced by climate, disturbance or other means, may promote diversity in species-rich regions (see also Connell 1978).

19.2.6 Dispersal

Generally, dispersal has a positive effect on regional richness. In areas of transition between two or more phylogenetically distinct biotas, regional richness may be considerably elevated, owing to macro-scale mass effects whereby species from adjacent biotas establish outside their normal range in weakly persistent or transitory populations (Shmida & Wilson 1985). Regions located at the junction of major dispersal routes for different biotas (river systems, mountain ranges, etc.) are likely to be species-rich. These patterns should be strongest for organisms that are both habitat generalists and well dispersed (Holt 1993; Westoby 1993).

19.2.7 Speciation history

Speciation augments regional richness, whereas extinction reduces it (Whittaker 1977; Diamond 1988; Cowling *et al.* 1992; Ricklefs & Schluter 1993). In areas where ecological factors that promote diversification of resident lineages have persisted for a long time, species will accumulate without apparent limit (Stebbins 1974; Whittaker 1977; Walker & Valentine 1984; Ricklefs & Latham 1993; Ricklefs & Schluter 1993). This 'time' hypothesis, which is by no means new to biology (see Willis 1922), is supported by three lines of evidence (Latham & Ricklefs 1993): lack of convergence in regional richness among environmentally similar areas (e.g. Westoby 1988; 1993; Cowling *et al.* 1992; Ricklefs & Latham 1993); long-term evolutionary conservatism of ecologically important traits (Ricklefs & Latham 1992); and the steady increase in angiosperm diversity through the Cretaceous and Tertiary (Knoll 1986).

19.2.8 Effect of local processes

It is now well established that local richness of certain assemblages is profoundly influenced by the size of the regional species pool (Ricklefs 1987), especially when the organisms involved are well-dispersed generalists (Holt 1993; Westoby 1993). On the other hand, ecological and evolutionary circumstances that promote high local richness of habitat specialists will enhance the richness of heterogeneous regions (Westoby 1988; Cowling 1990). In these cases, habitat specialization produces a ceiling on local richness so that it is lower than expected from the size of the regional pool (Cornell 1993; Richardson *et al.* 1995).

19.2.9 Convergence of regional richness

Physiographically matched regions with biotas that are phylogenetically distinct, but possess similar arrays of ecological traits, should support similar numbers of species. Convergence at the regional scale would result from similar local controls on alpha diversity (Cody & Mooney 1978) and similar rates of diversification of habitat specialists and ecological equivalents (Cowling *et al.* 1992; 1994b). Lack of convergence of the richness of analogous regions suggests the influence of different geographical circumstances and speciation histories (Ricklefs & Latham 1993; Westoby 1993).

19.3 Approach and methods

The approach adopted in this study was to review published data on the determinants of plant regional richness in southern Africa in terms of the theories described above, as well as to present new models based on unpublished data. The remainder of this section describes the approach and methods used to develop the latter.

19.3.1 Data

We compiled data, from published and unpublished sources, on species number (native, higher plants only) for 63 sites located throughout South Africa (Fig. 19.1). All sites were independent samples (i.e. no nested samples); nesting and overlap evident in two instances in Fig. 19.1 result from the need to identify square or rectangular areas for deriving estimates, for the samples, of explanatory variables that are stored in their primary form at a scale of $1' \times 1'$ lat./long. (see Table 19.1). As far as was possible, we developed models for areas with phylogenetically homogeneous floras, in addition to models for the entire country. Each region was allocated to a biome or biome grouping (fynbos biome, Nama- and succulent karoo biomes, savanna and grassland biomes) and identical analyses were subsequently performed on the total data set and each of the three subsets. Each of these subsets include floras that are relatively distinct phylogenetically (Gibbs Russell 1987; Chap. 3, this volume). However, phylogenetic constraints could have been avoided more effectively were it possible to group samples according to phytochoria (Goldblatt 1978; Chap. 3, this volume), but insufficient samples were available for this. Details on the physical and plant biological features of each of these biomes are given in Chaps 5–12 of this volume.

Data for explanatory variables were derived from a computerized database described originally by Dent, Lynch & Schulze (1988) and Dent *et al.* (1989), and subsequently refined (see also Chap. 2, this volume). The area of each site was constrained to cover a square or rectangle (Fig. 19.1) that maximized the range of values of the explanatory variables for the site and facilitated data retrieval from the gridded database (Table 19.1). The variables, which were selected to test hypotheses on regional richness (cf. Richerson & Lum 1980; Currie 1991; O'Brien 1993), are shown in Table 19.1, where they are grouped according to the appropriate hypothesis, and the methods of derivation are described. We acknowledge that some of the variables are appropriate for more than one hypothesis. For example, mean annual rainfall, used here as a descriptor of favourableness, is a good surrogate for energy in arid and semi-arid regions (Rosenzweig & Abramsky 1993). Table 19.2 provides some

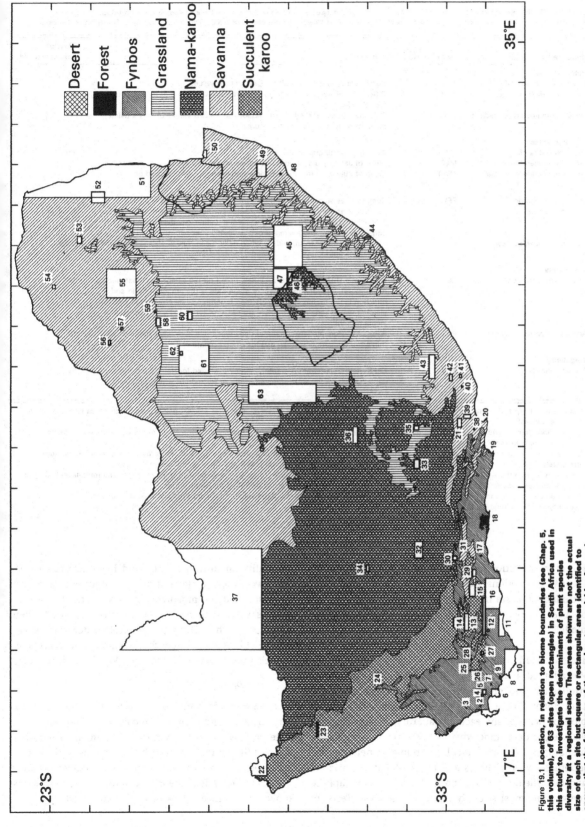

Figure 19.1 Location, in relation to biome boundaries (see Chap. 5, this volume), of 63 sites (open rectangles) in South Africa used in this study to investigate the determinants of plant species diversity at a regional scale. The areas shown are not the actual size of each site but square or rectangular areas identified to ensure that the full range of the explanatory variables for each site would be derived from the database where information was gridded at the 1' × 1' lat./long. scale (see text and Table 19.1 for details). Site data are given in Appendix 19.1.

Table 19.1 **Explanatory variables used to model regional plant species richness in South African biomes. The variables are grouped according to hypotheses widely invoked to explain patterns of regional richness (see text). Abbreviations used throughout the text**

Hypothesis variable	Abbreviation	Derivation	Unit of measurement
Heterogeneity			
Topographic diversity[a]	TD1	Coefficient of variation of all the grid altitude values	%
Length of rainfall gradient[a]	RAR	Difference between the highest and lowest gridded value of mean annual rainfall (MAR)[b]	mm
Length of temperature gradient[a]	RWT	Difference between the highest and lowest gridded values of July's (winter) mean of daily minimum temperatures	°C
Favourableness			
Mean annual rainfall[a]	MAR	Averaged for all grid points	mm
Summer temperature regime[a]	MST	Mean of January daily maximum temperatures averaged for all grid points	°C
Winter temperature regime[a]	MWT	Mean of July daily minimum temperatures averaged for all grid points	°C
Energy			
Potential evaporation[a]	PEV	Computed as summations of the 12 monthly A-pan equivalent values[c]	mm
Primary production[d]	PRO	Computed as the generic (i.e. non-biome specific) net above-ground primary production based on Rosenzweig's (1968) equation and determined for a July–June season and averaged over a 30–100-year period[e]	t ha^{-1} yr^{-1}
Duration of growing season[a]	GRS	The number of consecutive days when daily mean rainfall exceeds daily mean potential evaporation	days
Seasonality			
Rainfall concentration[a]	RCO	Markham's (1970) concentration index, based on the vector representation of mean monthly rainfall totals, where magnitude is the amount of rain and direction is the month of year. Summed monthly values are divided by MAR, yielding values ranging from 0% (zero seasonality) to 100% (all rainfall in a single month)[f]	%
Annual temperature range[a]	RAS	Range between January's mean of daily maximum and July's mean of daily minimum temperatures averaged for all grid points	°C
Irregularity			
Rainfall reliability[d]	RCV	Coefficient of variation of monthly rainfall averaged for the wettest three consecutive months and computed in the *ACRU* model[d] when run for the 712 zones[e]	%

[a] Data derived from a 1′ × 1′ lat./long. Digital Elevation Model containing gridded information at over 440 000 points over South Africa (see Chap. 2, this volume for details).
[b] Where more than 50 grid points constitute an area, highest and lowest MAR values were computed as the mean of the upper 2% of gridded values in order to eliminate extreme outliers.
[c] Estimated by multiple regression on a month-by-month basis from 13 defined evaporation regions in South Africa using maximum temperature, extraterrestrial radiation, altitude and median monthly rainfall. See Schulze & Maharaj (1991) and Chap. 2, this volume for details.
[d] Data derived from the *ACRU* agrohydrological model (Dent, Schulze & Angus 1989; Chap. 2, this volume), a daily time step and multi-layer soil water budget simulator, used in conjunction with 712 relatively homogeneous climatic zones delimited for South Africa.
[e] Where an area covered more than one of the 712 zones (see above), an area-weighted value of primary production was computed.
[f] See Chap. 2, this volume for details.

summary statistics for the variables in each data set. Clearly, the regions were not randomly selected, thus introducing biases in the explanatory variables for each data set. We refer to these biases in the discussion of the models.

19.3.2 Model development

We adopted a more-or-less standard approach used in studies that model plant regional richness (Richerson & Lum 1980; Currie 1991). We first examined bivariate plots of richness and each explanatory variable for each data set, to determine whether these relationships were monotonic. We then calculated non-parametric correlations between richness and the explanatory variables and identified, for each hypothesis within each data set, the variable most strongly related to richness. Tests for

normality on untransformed and log-transformed variables were then carried out using chi-square goodness of fit tests and Kolmogorov–Smirnov tests. Regression models specific to each of the hypotheses were then developed in the following manner. In order to incorporate the effects of area in each model, multiple regression models were developed as follows (Wylie & Currie 1993):

$$\log S = a + b \log A + bX$$

where S = species number, A = area, X = other explanatory variable, and a and b are constants. Step-wise multiple regression with forward selection, incorporating the best-fit variable for each hypothesis, as well as area, was used in an attempt to maximize explained variance in terms of independent predictors. No regressions showed a significant lack of fit or heteroscedacity.

Table 19.2 **Mean, mode and range for species number, area and explanatory variables listed in Table 19.1, where explanations for abbreviations are given**

Variable	Fynbos biome	Nama- and succulent karoo biomes	Savanna and grassland biomes	All biomes
No. species	829;585	372;358	690;559	660.6;481
	173–2256	128–768	248–1800	128–2256
Area (km²)	383;28	178;65	6580;133	2990;77
	0.7–2860	0.8–1620	1.3–122880	0.7–122880
TD1 (%)	42;41	17;17	19;20	26;21
	9–121	1–43	1–45	1–121
RAR (mm)	541;394	170;121	311;218	354;216
	15–2581	20–417	23–1192	15–2581
RWT (°C)	2.6;3.0	2.1;2.0	4.8;3.2	3.4;2.7
	0.2–5.5	0.1–5.7	0.4–19.8	0.1–19.8
MAR (mm)	674;607	290;321	687;615	588;554
	301–1460	48–474	189–1561	48–1561
MST (°C)	26.1;26.0	30.2;30.3	28.2;28.4	28.0;27.9
	24.0–29.2	26.8–33.2	19.6–33.8	19.6–33.8
MWT (°C)	6.9;7.0	3.3;3.2	4.5;4.1	5.0;5.6
	3.9–8.4	0.8–5.6	−1.8–10.8	−1.8–10.8
PEV (mm)	1768;1795	2316;2335	2034;2018	2012;1978
	1488–2099	1899–2611	1435–3013	1435–3013
PRO (t ha⁻¹ season⁻¹)	6.2;5.9	2.7;2.6	8.1;8.3	6.2;5.8
	2.6–9.4	0.4–4.8	1.8–14.8	0.4–14.8
GRS (days)	191;185	53;25	137;146	135;145
	46–346	0–138	0–301	0–346
RCO (%)	26;22	38;40	48;56	38;40
	5–53	3–56	11–68	3–68
RAS (°C)	19.1;18.1	26.8;27.3	23.7;22.9	22.9;22.9
	15.8–24.7	21.3–30.2	14.2–31.5	14.2–31.5
RCV (%)	66;63	96;94	69;67	75;74
	49–92	68–154	46–110	46–154

19.4 Evaluation of theories

19.4.1 Area

Area is a reasonably good predictor of regional plant richness in some South African biomes (Tables 19.3 and 19.4). However, models in previous studies (Kruger & Taylor 1979; Cowling *et al.* 1989; 1992) may have overestimated the contribution of area, owing to the use of nested samples (Boecklen & Gotelli 1984). In our data (independent samples) the strongest relationship with area was observed for the fynbos biome; area was a relatively poor predictor of richness in the two karroid biomes and the savanna and grassland biome (Table 19.3). Geldenhuys (1992) found that for South African forest patches, area explained only about 30 and 40% of the variance in herbaceous and woody species number, respectively (Table 19.4). In arid and semi-arid Veld Types of South Africa, log area is not significantly related to log species number and explains none of the variance in the latter variable (Hoffman *et al.* 1994). This may result in part from a sample bias: larger regions (or Veld Types) are topographically more uniform and have the most severe climatic regimes (Acocks 1953; Hoffman *et al.* 1994), both of which would serve to limit richness.

Area showed a significant relationship with species number for our entire data set and explained 35% of the variance in a simple regression model (Tables 19.3 and 19.4). Samples with high positive (> 1.5) standardized residuals in this model were either from the southwestern fynbos biome (Cape Peninsula, Cape Hangklip, Agulhas Plain) or the subtropical east coast (Umtamvuna, Ubisalia) (Fig. 19.1; see Appendix 19.1); samples with high negative standardized residuals were located in the semi-arid interior (Carnarvon, Southern Kalahari) (see also Cowling *et al.* 1989).

We have already stated that area is, in most instances, a surrogate for other explanatory variables, especially environmental heterogeneity (Williamson 1988). In our data, log area showed strong positive correlations with the heterogeneity variables that were the best predictors of species richness in each subset (Table 19.3) (fynbos: log TD1, $r = 0.72$, $P < 0.001$; savanna and grassland: log RWT, $r = 0.55$, $P < 0.01$; all biomes: log RAR, $r = 0.42$, $P < 0.001$). Only for the Nama- and succulent karoo data set was there no significant relationship between area and heterogeneity.

Table 19.3 Spearman rank correlation coefficients[a] and adjusted R^2 values[b] for relationships between regional plant species richness in South African biomes and explanatory variables grouped according to hypotheses invoked to explain patterns. Abbreviations for variables given in Table 19.1. Significance levels for correlations: NS, not significant; * $P < 0.05$; ** $P < 0.01$; *** $P < 0.001$. Unless otherwise indicated, all regressions significant at $P < 0.01$

Hypothesis variable	Fynbos biome (n = 21)		Nama- and succulent karoo biomes (n = 15)		Savanna and grassland biomes (n = 27)		All biomes (n = 63)	
	r_s	R^2 (%)	r_s	R^2 (%)	r_s	R^2 (%)	r_s	R^2 (%)
Area	0.76***	60[c]	0.65*	36[c]	0.49*	24[c]	0.63***	35[c]
Heterogeneity								
TD1	0.88***	76[c]	0.58*	—	NS	—	0.51***	—
RAR	0.54*	—	0.68*	58[c]	NS	—	0.55***	50[c]
RWT	0.69**	—	0.64*	—	0.53**	36[c]	0.61***	—
Favourableness								
MAR	NS	—	NS	—	NS	47[c]	0.39**	51
MST	NS	60[c]	NS	52[c]	NS	—	NS	—
MWT	NS	—	NS	—	NS	—	NS	—
Energy								
PEV	NS	—	NS	—	NS	—	-0.33**	—
PRO	NS	70	NS	—	0.48*	50[c]	0.45***	49
GRS	NS	—	NS	48	0.39*	—	0.38**	—
Seasonality								
RCO	NS	79[c]	NS	—	NS	22[c,d]	NS	—
RAS	NS	—	NS	38[d]	NS	—	-0.25*	54
Irregularity								
RCV	NS	84[c]	NS	26[c,e]	NS	22[c,d]	-0.33**	44[c]

[a] Correlation analysis using untransformed variables.
[b] Derived from multiple regressions of the following form: $\log S = a + b\log A + bX$ where A = area and X = the explanatory variable within each hypothesis group for each data set that contributed most to the coefficient of multiple determination (R^2).
[c] Variable log-transformed.
[d] $P < 0.05$.
[e] NS.

19.4.2 Heterogeneity

In bivariate analyses, measures of heterogeneity emerged as the strongest predictors of regional richness for all of our data sets, explaining between 74% (fynbos biome) and 38% (savanna and grassland biomes) of the variance in simple regressions (Table 19.3; Fig. 19.2). Interestingly, the best-fit explanatory variables for each data set were different (topographic diversity in fynbos; rainfall gradient in the karroid biomes, and temperature gradient in savanna and grassland), suggesting biome-specific controls. With the exception of RWT, the fynbos data set had the highest values of the heterogeneity variables (Table 19.2), suggesting that its extraordinary richness may simply be a function of this biome having the longest environmental gradients in the subcontinent. The relatively species-poor regions of the karroid biomes had the lowest values for all heterogeneity variables. At a subcontinental level, richness was most strongly related to the length of the rainfall gradient. Heterogeneity variables were those first selected in step-wise regressions of all data sets (Table 19.4).

Measures of heterogeneity combined with area explained the most variance in species number, relative to the other variables, in the fynbos and two karroid biomes; contributions of heterogeneity to explained variance in these models ranged from 12% (savanna and grassland) to 22% (karroid biomes) (Table 19.3). For the whole data set, area and length of the rainfall gradient explained 50% of the variance; regions with high positive standardized residuals (>2.0) in this model were the Cape Peninsula and Agulhas Plain in the fynbos biome, and Ubisalia along the subtropical east coast. The unusually high richness of the Cape Peninsula may be a result of this region's extremely high topographic diversity (TD1 = 121%), whereas edaphic rather than climatic heterogeneity explains the high richness of the Agulhas Plain (Cowling 1990). Ubisalia, located on a low (c. 400–500 m), granitic scarp a short distance from the coast, includes a wide array of habitats and vegetation types, including savanna and several phytochorologically mixed forest and grassland communities (Moll & White 1978). The only site with a standardized residual lower than 2.0 was the Southern Kalahari, an area of low heterogeneity where richness is negatively influenced by edaphic uniformity resulting from a deep mantle of Quaternary sand (Werger 1978).

Few other studies have explored the role of heterogeneity in determining regional plant richness in southern Africa. Linder (1991) reported strong correlations between richness of selected taxa and both rainfall and altitudinal range in quadrats of 170 km² in the southwestern fynbos biome. However, in a step-wise

Table 19.4 **Regression models explaining plant regional richness in southern Africa. Other than area, only those models with the highest** R^2 **values for each data set (Table 19.3) are shown. Explanations for variable abbreviations are given in Table 19.1, except for** A **= area, COR = number of dispersal corridors (1–4), MAL = mean altitude (m), MXR = mean maximum rainfall (mm), PEM = minimum monthly potential evapotranspiration (mm), PET = mean potential evapotranspiration (mm), PRX = proximity to other large forests (1 = close, 2 = intermediate, 3 = distant), RAA = altitudinal range (m),** S **= species number, TD2 = number of landscape types (min = 1, max = 5).**

Model	n	Area range (km^{-2})	R^2 (%)[a]	Type	Source
Fynbos biome					
$\text{Log}S = 2.69 + 0.25\text{log}A$	8 Southwest	0.27–471[b]	93	Simple	Kruger & Taylor (1979)
$\text{Log}S = 2.66 + 0.21\text{log}A$	13	0.27–36 600[b]	79	Simple	Cowling et al. (1989)
$S^c = \text{MXR} + \text{RAA} + \text{RAR}^d$	55 Southwest	Fixed 170	72	Step-wise	Linder (1991)
$\text{Log}S = 2.58 + 0.253\text{log}A$	19 Southwest	0.20–1609.3[b]	90	Simple	Cowling et al. (1992)
$\text{Log}S = 2.21 + 0.27\text{log}A$	15 Southeast	1.41–4800[b]	90	Simple	Cowling et al. (1992)
$\text{Log}S = 2.47 + 0.21\text{log}A$	21	0.68–2860	60	Simple	This study
$\text{Log}S = 5.91 + 0.27\text{log}A - 1.94\text{log}RCV$	21	0.68–2860	84	Multiple	This study
$\text{Log}S = 1.65 + 0.77\text{log}TD1$	21	0.68–2860	74	Step-wise	This study
Succulent karoo biome					
$\text{Log}S = 2.29 + 0.20\text{log}A$	6	1.15–50 600[b]	85	Simple	Cowling et al. (1989)
Nama- and succulent karoo biomes					
$\text{Log}S = 2.38 + 0.11\text{log}A$	15	0.75–1620	36	Simple	This study
$\text{Log}S = 1.88 + 0.07\text{log}A + 0.27\text{log}RAR$	15	0.75–1620	58	Multiple	This study
$\text{Log}S = 1.51 + 0.35\text{log}RAR + 0.003\text{log}RCV$	15	0.75–1620	59	Step-wise	This study
Grassland biome					
$\text{Log}S = 2.22 + 0.25\text{log}A$	6	0.62–111 900[b]	80	Simple	Cowling et al. (1989)
Savanna biome					
$\text{Log}S = 2.29 + 0.20\text{log}A$	10	2–632 000[b]	76	Simple	Cowling et al. (1989)
Savanna and grassland biomes					
$\text{Log}S = 2.55 + 0.09\text{log}A$	27	1.3–122 800	24	Simple	This study
$\text{Log}S = 0.25 + 0.12\text{log}A + 0.79\text{log}MAR$	27	1.3–122 800	47	Multiple	This study
$\text{Log}S = 2.63 + 0.32\text{log}RWT$	27	1.3–122 800	35	Step-wise	This study
Forest biome					
$\text{Log}S^e = 1.71 + 0.14\text{log}A$	14	1.5–605	30	Simple	Geldenhuys (1992)
$\text{Log}S^f = 1.48 + 0.18\text{log}A$	13	1.5–605	38	Simple	Geldenhuys (1992)
$\text{Log}S^e = 2.29 + 0.17\text{COR} - 0.06\text{PRX} + 0.018\text{log}MAL$	14	1.5–605	82	Step-wise	Geldenhuys (1992)
$\text{Log}S^f = 1.36 + 0.49\text{TD2} + 0.12\text{PRX}$	13	1.5–5605	75	Step-wise	Geldenhuys (1992)
Arid and semi-arid South Africa					
$\text{Log}S = 6.10 + 0.05\text{log}A$	40	1105–69 197	0	Simple	Hoffman et al. (1994)
$S = 2234.0 - 0.686\text{PET}$	40	1105–69 197	41	Simple	Hoffman et al. (1994)
$S = 449.1 + 1.108\text{MAR}$	40	1105–69 197	19	Simple	Hoffman et al. (1994)
$\text{Log}S = 16.5 - 1.74\text{log}PET + 0.18\text{log}A + 0.33\text{log}MAR$	40	1105–69 197	57	Step-wise	Hoffman et al. (1994)
Southern and South Africa					
$\text{Log}S = 2.48 + 0.18\text{log}A$	55	0.68–632 000[b]	56	Simple	Cowling et al. (1989)
$S^g = -1.55 + 0.436\text{MAR}$	65	Fixed 20 000	60	Simple	O'Brien (1993)
$S^g = -2.71 + 41.426\text{PEM} - 0.483\text{PEM}$	65	Fixed 20 000	59	Simple	O'Brien (1993)
$S^g = -2.69 + 0.29\text{MAR} + 30.045\text{PEM} - 0.371\text{PEM}$	65	Fixed 20 000	78	Multiple	O'Brien (1993)
$\text{Log}S = 2.48 + 0.13\text{log}A$	63	0.68–122 800	35	Simple	This study
$\text{Log}S = 3.13 + 0.16\text{log}A - 0.31\text{RAS}$	63	0.68–122 800	54	Multiple	This study
$\text{Log}S = 2.42 + 0.15\text{log}RAR + 0.12\text{log}A - 0.017\text{RAS} + 0.020\text{PRO}$	63	0.68–122 800	63	Step-wise	This study

[a] Adjusted for d.f.
[b] Not all data independent (nested samples).
[c] Species of *Aspalathus* (Fabaceae), *Pentaschistis* (Poaceae), Ericaceae, Proteaceae and Restionaceae.
[d] Regression statistics not given in source.
[e] Woody plants only.
[f] Herbaceous plants only.
[g] 'Trees' only (see text).

Figure 19.2 **Strongest relationships (simple regressions) between regional richness and explanatory variables shown in Table 19.1 (see also Tables 19.3 and 19.4). (a) Fynbos biome: (filled squares), southwestern and (open triangles), southeastern areas. (b) Succulent (filled triangles) and Nama-karoo (open triangles) biomes. (c) Savanna (filled circles) and grassland (open squares) biomes. (d) All biomes (symbols for biomes as above).**

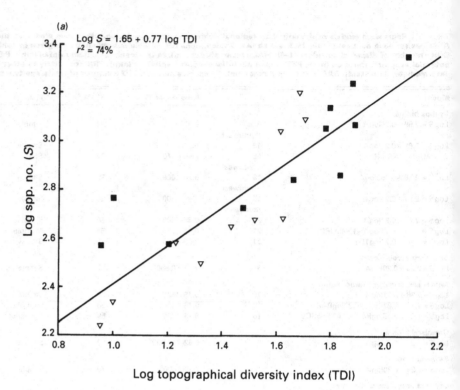

(a)

$\text{Log } S = 1.65 + 0.77 \log \text{TDI}$
$r^2 = 74\%$

Log spp. no. (*S*)

Log topographical diversity index (TDI)

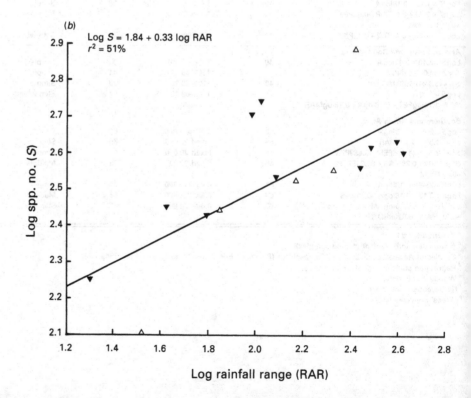

(b)

$\text{Log } S = 1.84 + 0.33 \log \text{RAR}$
$r^2 = 51\%$

Log spp. no. (*S*)

Log rainfall range (RAR)

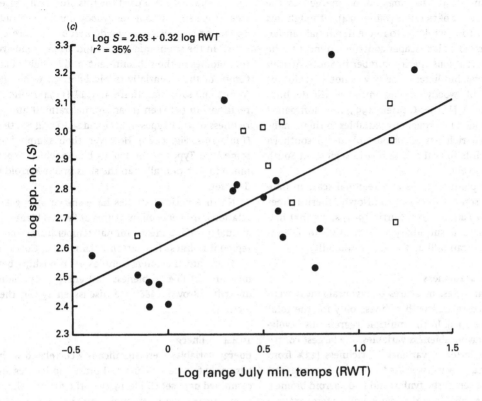

(c)

Log S = 2.63 + 0.32 log RWT
r² = 35%

Log spp. no. (S)

Log range July min. temps (RWT)

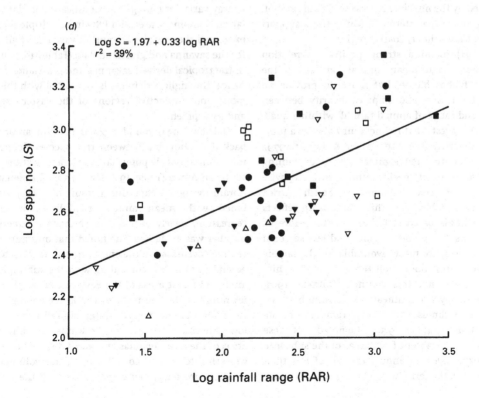

(d)

Log S = 1.97 + 0.33 log RAR
r² = 39%

Log spp. no. (S)

Log rainfall range (RAR)

regression, annual rainfall maximum emerged as the strongest predictor of richness (Table 19.4), although this variable was strongly related to range in annual rainfall ($r = 0.99$). Number of landscapes contributed most to the variance in herbaceous species number in South African forested regions, but heterogeneity was not a significant determinant of woody species number (Geldenhuys 1992; Table 19.4). Neither O'Brien (1993) nor Hoffman *et al.* (1994) included heterogeneity variables in their analyses of regional richness of 'woody' plants in southern Africa and plants in arid and semi-arid regions of South Africa, respectively.

Studies of plant richness at a regional scale in California (Richerson & Lum 1980) and North America (trees only; Currie & Paquin 1987; Currie 1991) show that heterogeneity plays a subsidiary role relative to favourableness (mean rainfall) and energy availability.

19.4.3 Favourableness

In bivariate analyses, measures of favourableness were significantly correlated with richness only for the total data set (Table 19.3). In the multiple regressions involving area and favourableness variables, the largest contribution of the latter to variance in richness (23% from MAR in this case) was associated with the savanna and grassland data set. In the fynbos and two karroo biomes, a negative relationship with mean summer temperature (MST) contributed zero and 16%, respectively to the variance explained by the multiple regressions. Summer-hot, semi-arid regions in the interior of South Africa support fewer species than milder, coastal regions.

Linder (1991) found a strong positive correlation between richness and mean annual rainfall in the southwestern fynbos biome, but failed to provide an explanation for this relationship; colinearity between this variable and range of annual rainfall within a quadrat (see above) suggests that richness may also be a function of beta diversity (see also Kruger & Taylor 1979). Mean altitude contributed significantly to explaining variance in woody plant species number in a step-wise regression for forested regions in South Africa (Geldenhuys 1992; Table 19.4). This relationship reflects the higher richness of coastal than montane forests, where conditions, being both warmer and less seasonal than montane sites, are most favourable for the largely tropical-derived forest flora (Moll & White 1978; White 1978; Chaps 3 and 12, this volume). O'Brien (1993) observed a relatively strong linear relationship between 'woody' species richness and annual rainfall for southern Africa (Table 19.4; Fig. 19.3a). She noted that these species are essentially absent from areas of the subcontinent which experience an annual rainfall of less than 140 mm. However, the term 'woody' is misleading, since

the floristic database used for this study comprises only trees (1372 spp.) among the woody flora; it precludes the 15 000 or so species of woody shrubs that are concentrated in the subhumid fynbos and arid to semi-arid karroid biomes of the subcontinent, and is highly biased in favour of the essentially tropical-derived woody flora of forests and savannas. Hoffman *et al.* (1994) found a weak relationship between mean annual rainfall and species richness of Veld Types in arid and semi-arid South Africa (Table 19.4; Fig. 19.4a). However, their sample included some Veld Types of the fynbos biome which experience much higher rainfall than the sites in our karroid biome data set.

North American studies have shown strong positive relationships between measures of favourableness (mean annual rainfall, mean annual temperature) and plant regional richness (Richerson & Lum 1980; Currie 1991). There are the often complex relationships between measures of favourableness and energy, especially productivity (Brown 1988): we discuss energy in the next section.

19.4.4 Energy

Energy variables were significantly correlated with richness only for the savanna and grassland biomes and the combined data set (Table 19.3). In the former, significant relationships were observed for primary production (PRO) and duration of growing season (GRS), whereas all energy variables emerged as significant correlates in the latter. Combining area and PRO in a multiple regression explained the greatest amount of variance of all models for the savanna and grassland data (Tables 19.3 and 19.4). In the tropical-derived savannas and grasslands of South Africa, the highest richness is associated with the warm, moist and productive regions of the eastern seaboard and escarpment.

O'Brien (1993) found a good curvilinear or hump-backed relationship between tree species richness and minimum monthly potential evapotranspiration (PET) in southern Africa (Table 19.4; Fig. 19.3b; cf. Rosenzweig & Abramsky 1993). When this measure of energy was combined with mean annual rainfall in a multiple regression, nearly 78% of the variance in tree species number was explained. She noted that optimum species richness occurs when the energy minimum falls between 36 and 50 mm PET, coinciding with the subtropical savanna and forest areas of the eastern coastal plain. Species numbers decline towards the lower energy areas of the fynbos biome (where winter rainfall results in very low minimum values), and towards the high-energy areas of the arid savannas and karroid shrublands of the western interior and coast (Chap. 2, this volume).

There are a number of problems with the interpret-

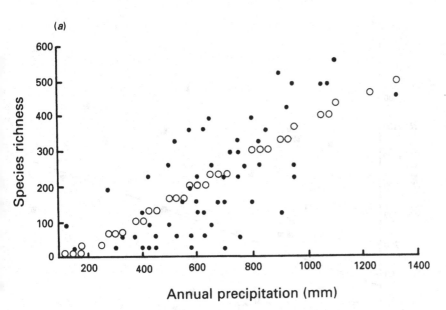

(a)

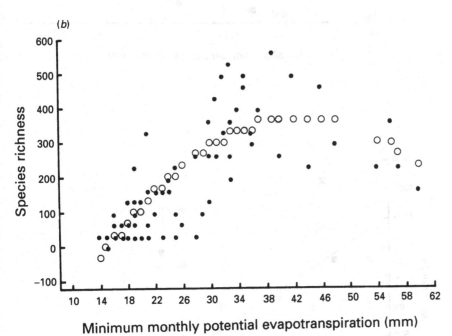

(b)

Figure 19.3 **Relationship between (a) mean annual rainfall (MAR) and (b) minimum monthly potential evapotranspiration (PEM) and tree species richness in equal area units (ca. 20 000 km²) in southern Africa; (filled circles), actual values; (open circles), predicted values. See Table 19.4 for regression equations. Redrawn, with permission, from O'Brien (1993).**

ation of this model. Firstly, as indicated in the previous section, the response variable represents a biased sample of the woody flora of southern Africa. Secondly, the low number of tree species in the fynbos biome cannot be explained by climatic factors alone: ecologically similar environments in other parts of the world support relatively diverse tree floras (Richardson & Cowling 1992). Thirdly, the low richness of trees in the high-energy karroid areas is largely a consequence of their restriction to azonal habitats with underground water such as drain-

age lines and rocky hills (Chaps 7 and 8, this volume); zonal habitats are too dry for trees. An analysis of total woody plant richness in southern Africa would probably show a pattern consistent with ours of a weak negative relationship with potential evaporation (Table 19.3).

Hoffman *et al.* (1994) found a reasonably good negative and linear relationship between species number and PET in South African arid and semi-arid vegetation types (Fig. 19.4b; Table 19.4). They interpreted this pattern as the negative arm of a more general hump-backed

Figure 19.4 **Relationship between (*a*) mean annual rainfall (MAR) and (*b*) potential evapotranspiration (PET) and plant species richness in 40 physiographically homogeneous units (Veld Types) ranging in size from 1104 km² to 69 196 km² in arid to semi-arid South Africa. See Table 19.4 for regression equations. From Hoffman *et al.* (1994).**

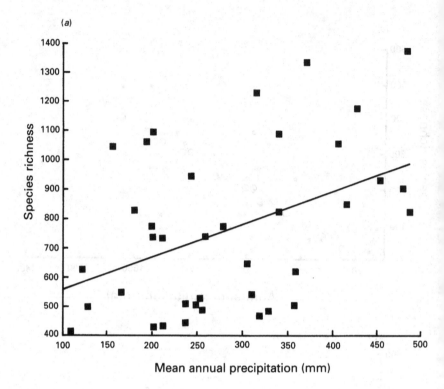

(*a*)

Mean annual precipitation (mm)

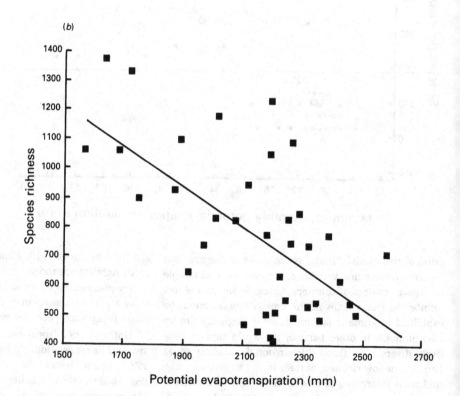

(*b*)

Potential evapotranspiration (mm)

relationship between energy availability and species richness, albeit that this relationship generally holds at the local rather than the regional scale (Rosenzweig 1992; Rosenzweig & Abramsky 1993; Tilman & Pacala 1993; Wright *et al.* 1993). We found no evidence of a curvilinear relationship between richness and potential evaporation (PEV) in any of our data sets; the weak negative correlation for the combined data mirrors the results of Hoffmann *et al.* for western South Africa. They argue that in arid areas, high energy availability, as estimated by PET (which is mainly a measure of the integrated, crude, ambient energy available to evaporate water and is independent of water availability) has an antagonistic effect on plant performance, resulting in reduced diversity. Thus, energy-rich, dry areas support fewer species than wetter areas with less available energy, such as the Veld Types of the fynbos biome in their sample. However, as pointed out previously, these former areas also have lower environmental heterogeneity than the fynbos types.

For plants, actual evapotranspiration (AET), which represents the joint availability of energy and water and which correlates closely with primary productivity (Rosenzweig 1968), is generally a better predictor of diversity than PET (Currie 1991; Rosenzweig & Abramsky 1993). In all our data sets, potential evaporation (PEV) showed a highly significant, linear and negative relationship with annual rainfall. Both Richerson & Lum (1980) and Currie & Paquin (1987) demonstrated strong positive relationships between AET (or its main component variables of temperature and precipitation) and plant regional richness in North America. Primary productivity emerged as the strongest predictor, among the energy variables, of richness in our data (Table 19.3). However, Latham & Ricklefs (1993) argue that the predictions of species–energy theory are more appropriately tested at the local scale and that the relationship observed by Currie & Paquin between tree species richness and AET at a regional scale 'arose primarily from the fact that more species of trees occur in the mesic forest habitats of east-cental North America than in the tundra, boreal and montane needle-leaf forests, grasslands and deserts of the north and west'. They argue that as moisture (and thus, AET) increases, so does the proportional area and variety of forested habitats in a sample, implying that heterogeneity may be an important determinant of richness at this scale. In arid areas with low values of AET, trees are restricted to azonal sites. The low richness of trees in high-latitude, energy-poor environments is explained as a consequence of the evolutionary history of broad-leaved trees and the relative newness of continental arctic climates (Latham &

Ricklefs 1993), a history-favourableness argument (see Brown 1988, p. 78).

We do not believe that the species–energy theory is generally useful in explaining regional plant richness in southern Africa. In the phylogenetically distinct southern floras of the fynbos and karroid biomes (Gibbs Russell 1987), relatively high richness occurs in environments where productivity is limited by nutrient and moisture availability, respectively (Chaps 6, 7 and 8, this volume). Certain lineages which are tolerant of these ecological conditions have undergone spectacular diversification in these biomes (Chap. 3, this volume). In the savanna and grassland biomes, richness is positively related to productivity, as would be expected in a tropical-derived flora. Thus, patterns of richness are largely biome-specific with phylogenetic history, or clade-specific tolerances of 'favourable' ecological conditions, explaining certain relationships (Brown 1988; Latham & Ricklefs 1993).

19.4.5 Seasonality and irregularity

In the bivariate analyses, measures of seasonality and irregularity were significantly correlated with richness for the total data set only (Table 19.3). Here richness was negatively and weakly related to annual temperature range (RAS), reflecting the low diversity of the arid, western interior zone, where continentality is most pronounced in southern Africa (Chap. 2, this volume). Similarly, richness was also negatively related to rainfall reliability (RCV), which is lowest in the low-rainfall interior of the subcontinent (Chap. 2, this volume). Therefore, thermal seasonality and rainfall irregularity have a negative influence on species number, possibly because lineages in both the southern and tropical-derived floras of the subcontinent are not generally tolerant of these extreme and uncertain environments. Additionally, the strongest heterogeneity predictor of richness for the total data set (log length of rainfall gradient (RAR)) was negatively correlated with both log RAS ($r = 0.55$, $P < 0.001$) and log RCV ($r = 0.69$, $P < 0.01$), indicating that regions with harsh climatic regimes also have low environmental heterogeneity.

In the fynbos biome, multiple regressions incorporating area and a measure of seasonality (RCO – rainfall concentration) and irregularity (RCV – rainfall reliability) explained 79% and 84% of the variance in species number, respectively (Table 19.3). These represent the highest values of all the regression models. This pattern is largely associated with differences in richness between regions in the southwestern (winter rainfall) and southeastern (non-seasonal rainfall) zones of the fynbos biome (Cowling *et al.* 1992; Fig. 19.5). We investigated

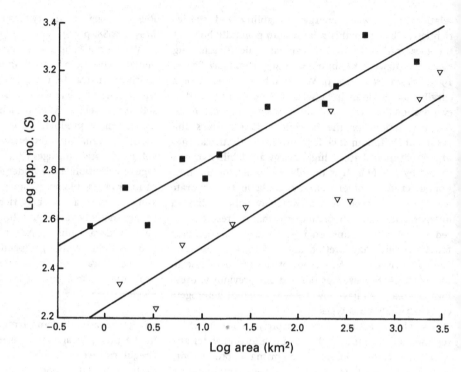

Figure 19.5 **Species–area relationships for floras from the southwestern (black square)** ($\log S = 2.61 + 0.227 \log A$; $R^2 = 0.90$) **and southeastern (open triangle)** ($\log S = 2.23 + 0.257 \log A$; $R^2 = 0.84$) **fynbos biome.**

this further by performing an ANCOVA on log species richness, where geographical locality (southwest and southeast) was the factor and log area was the covariate. This model predicted a 2.12 times higher species richness for equivalent-sized areas in the southwest than the southeast (overall ANCOVA, $F_{1,17} = 104.2$, $P < 0.0001$, $R^2 = 0.85$) (see also Cowling *et al.* 1992). The fit of the residuals of the model were satisfactory, showing no systematic pattern.

Why do southwestern fynbos landscapes support more than double the number of species than equivalent-sized areas in the southeast? Both RCO and RCV show significant variation between the two zones of the biome ($\text{RCO}_{\text{SW}} = 39.1 \pm (\text{SD})10.2\%$, $\text{RCO}_{\text{SE}} = 11.2 \pm 5.7\%$, $P < 0.001$ (Mann–Whitney U test); $\text{RCV}_{\text{SW}} = 56.5 \pm 8.4\%$, $\text{RCV}_{\text{SE}} = 77.2 \pm 8.0\%$, $P < 0.001$), thus explaining their contribution, in addition to area, to variance in species number in the multiple regressions (Table 19.3). Whereas rainfall in the southwestern zone is almost entirely associated with predictable frontal systems during winter, the situation is much more complex in the southeast (Deacon, Jurie & Ellis 1992). Here, the eastwards penetration of frontal rains is restricted by the north–south trending axes of the Cape Folded Belt and most rain is associated with post-frontal events, especially in the spring and autumn. Rain from these sources is much less predictable than the frontal rains of the southwest (Cowling & Holmes 1992b).

This lack of seasonality and high irregularity in rainfall patterns in the southeast may negatively influence richness by limiting the post-fire germination success of non-sprouting shrubs (Mustart & Cowling 1993). However, both classical niche theory and lottery theory predict higher local richness in environments of high temporal heterogeneity (Warner & Chesson 1985; Cody 1989; Bond, Cowling & Richards 1992) which, in turn, should positively influence regional richness. There is no evidence for alpha diversity to increase along a west–east gradient in the fynbos biome (Cowling *et al.* 1992).

A further possible explanation for this pattern is that the ecological heterogeneity of southwestern landscapes is higher than that in the southeast (Campbell 1983). However, we found no significant differences between the two regions in measures of area, TD1, RAR (Mann–Whitney U tests) and RWT (t-test).

We suggest that differences in the contemporary environment cannot satisfactorily explain variation in regional richness across the fynbos biome. Instead, as we will argue later in this chapter, the answer may lie in the different speciation histories of the two zones.

19.4.6 Dispersal

We would expect that floristic transition zones, where phylogenetically distinct floras are juxtaposed, should support higher than average numbers of species, owing to high phytochorological mixing (Cowling 1983) and macro-scale mass effects (Shmida & Wilson 1985). In South Africa, such transition zones occur in the southeastern Cape, where five phytochoria converge (Gibbs Russell & Robinson 1981; Cowling 1983; Chap. 3, this volume), and along the subtropical east coast, where steep rainfall and temperature gradients result in the compression, into relatively small areas, of vegetation types characteristic of four phytochoria (Moll & White 1978). We inspected the residuals of the significant regression models for each data set (Table 19.3) to see whether high positive values were associated with these transition zones.

For the total data set, regions with high positive standardized residuals were in the southwestern zone of the fynbos biome (Cape Peninsula – five models; Agulhas Plain – two models) and the subtropical east coast (Ubisalia – three models). The higher-than-average richness of the Ubisalia area, largely associated with the juxtaposition of many distinct but phytochorologically mixed vegetation types, is consistent with the dispersal model, but this is not the case for the fynbos sites that are located in the heart of the Cape phytochorion. Two regions had high residuals in the models for the savanna and grassland data set, namely Umtamvuna on the subtropical east coast (four models) and the Kruger National Park (four models). Only in the former site, which includes phytochorologically mixed forest and grassland vegetation, as well as a number of endemics (Van Wyk 1990), could transition zone patterns and dynamics play a role in determining richness. For the fynbos data set, sites in the southwest rather than the southeast (which overlaps with the phytochorologically complex western zone of the Eastern Cape) had high positive standardized residuals. For the karroid biomes, Niewoudtville in the southwestern succulent karoo and the Karoo National Park in the central Nama-karoo emerged as positive outliers. The latter site spans a broad gradient from semi-succulent karroid shrubland to grassland typical of the Kalahari–Highveld Transition Zone, whereas the former occurs in the transition between the fynbos and succulent karoo biomes.

Generally, however, it appears that dispersal-mediated transition zone and mass effects are most pronounced in forested areas with their well-dispersed and well-mixed floras (Geldenhuys 1989; Chap. 12, this volume). This is consistent with Geldenhuys' (1992) analysis of the determinants of woody species richness in South African forests. He found that the number of dispersal corridors in a region emerged as the strongest predictor of species number (Table 19.4). Interestingly, despite great phytochorological complexity (Chap. 3, this volume) and hence, access to propagules from many different floras, regions in the southeastern Cape do not have unusually high richness. Indeed, in many of the models, sites in this area had high negative residuals.

19.4.7 Speciation history

With some 80% of its plant species restricted to southern Africa, the region has an exceptionally high level of endemism for a continental area (Goldblatt 1978; Cowling & Hilton-Taylor 1994; Chap. 3, this volume). Most of these endemics are recently evolved species (neoendemics) associated with a limited number of genera; exhibit little intra-lineage ecological and morphological variation; and are concentrated mainly in the fynbos and succulent karoo biomes, with outlying 'hotspots' along the eastern escarpment and subtropical coast (Chap. 3, this volume). Clearly, *in situ* diversification has played an important role in the development of regional richness on the subcontinent. However, this topic has been studied in some detail only in the fynbos biome.

Cowling *et al.* (1992) argued that the two-fold difference in regional richness between the southwestern and southeastern zones of the fynbos biome (see 19.4.5 and Fig. 19.5) is a result of differences in the speciation histories of the two areas. Given that beta and gamma diversities are about twice as high in the southwest than the southeast (but alpha diversity is the same; Cowling *et al.* 1992), physiographically similar regions in the former zone should support twice the number of species, a prediction that has good empirical support (Fig. 19.5). In both areas local endemics are mainly habitat specialists associated with the same limited number of lineages and biologies (Chap. 3, this volume), but levels of endemism are lower overall in the southeast (Cowling *et al.* 1992). Therefore, lower richness in that area is the result of lower rates of diversification of habitat specialists and ecological equivalents.

The reason for these different speciation histories lies in the difference in the historical (and contemporary) climates of the two zones. In the southeast, colder glacial conditions substantially reduced the precipitation from post-frontal events, resulting in drier climates than those of interglacials, including the current one (Deacon, Hendey & Lamprechts 1983; Deacon *et al.* 1992). Under these conditions, the area covered by fynbos in the southeast would have been extensively reduced, contracting to the upper mountain peaks (Cowling 1983; Cowling & Holmes 1992b). Climates sufficiently moist to support fynbos over large areas of the southeast are an ephemeral phenomenon in evolutionary time; this would limit diversification of fynbos lineages. On the other hand, frontal rains would have increased in intensity and frequency during glacials in the southwest, as was the case for the west coasts of mediterranean California (Raven & Axelrod 1978) and Chile (Arroyo *et al.* 1988), providing for extensive and uninterrupted fynbos cover throughout the Plio-Pleistocene. While there is some palaeoecological evidence for this hypothesis (Meadows & Sugden 1991; Linder, Meadows & Cowling 1992; J. Parkington unpubl. data; Chap. 4, this volume), cladistic analyses of lineages common to the two zones would provide additional tests.

Given the antiquity and edaphic diversity of the South African landscape (Chap. 1, this volume), the steep climatic gradients over much of its area (Chap. 2, this volume), the absence of catastrophic change associated with Plio-Pleistocene climatic cycles (Chap. 4, this volume) and the high numbers of local and habitat specialist neoendemics (Chap. 3, this volume), it is likely that relatively recent diversification of certain lineages, resulting in the development of 'species flocks' (Brooks 1950), is a major determinant of regional richness in the subcontinent. The high beta diversity of species-rich regions such as the fynbos biome (Cowling *et al.* 1992), the succulent karoo (Cowling *et al.* 1989) and subtropical east coast (Goodman 1990), as well as the importance of environmental heterogeneity as a predictor of regional richness, support this contention. However, more research is required to test specific predictions of this model.

19.4.8 Local and regional patterns and processes

There is a weak positive relationship for South African biomes between local richness and regional richness, as measured by the size of the biome's flora (Fig 19.6). This is not unexpected, given the discrepancy in scale (0.1-ha versus biome-scale) and the fact that habitat specialization dilutes the effect of regional processes on local communities (Cornell 1993; Holt 1993; Richardson *et al.* 1995). The latter is better shown in the fynbos biome where local richness is invariant across the biome (Cowling *et al.* 1992) despite an eastwards decline in regional richness (Fig. 19.5). However, in the more generalist and well-dispersed forest flora, local richness is strongly influenced by the size of the species pool in a forest patch (Fig. 19.7).

In the other biomes local richness may have a positive influence on regional richness in terms of the contribution of richness at this scale to the alpha.beta.gamma equation. When resources in landscapes are at levels that maximize local richness (e.g. intermediate productivity in fynbos (Bond 1983) and savanna (Goodman 1990), and low thermal and rainfall seasonality in arid and semi-arid biomes (Cowling *et al.* 1994a), they should be richer than landscapes with similar ecological heterogeneity but less optimal resource levels. This hypothesis remains to be tested.

Table 19.5. **Observed plant species richness in different sized areas of three mediterranean-climate regions compared with values predicted from the species–area relationship (log$S$ = 2.61+0.227log$A$) for the winter-rainfall (southwestern) part of the fynbos biome (see Fig. 19.5). Data from sources in Cowling *et al.* (1992) and Arroyo *et al.* (1995)**

Region	Area (km^2)	Species number Observed	Species number Predicted	Predicted/ observed
Chile				
La Plata valley	10	249	687	2.76
Marga Marge valley	450	457	1627	3.56
Conception area	2100	672	2313	3.44
Valparaiso area	3300	799	2564	3.21
Santiago valley	4000	654	2674	4.09
California				
San Bruno Mts	12	389	716	1.78
Tiburon Peninsula	15	370	753	2.04
San Francisco Co.	117	640	1200	1.98
Mt Diablo Co.	146	525	1262	2.40
Santa Monica Mts	828	642	1872	2.92
Marin Co.	1369	1004	2098	2.09
Orange Co.	2024	568	2293	4.04
Santa Cruz Mts	2590	1246	2421	1.94
S.W. Australia				
Mt Leseur	0.5	286	348	1.22
Mt Adams	13	290	729	2.51
South Eneabba	20	429	804	1.87
Tutanning	22.5	628	826	1.32
Two Peoples Bay	46	614	972	1.58
Stirling Range	1156	874	2020	2.31
Barrens	3230	1422	2550	1.79

19.4.9 Global comparisons

Convergence theory predicts that physiographically and climatically matched regions with phylogenetically distinct biotas should support similar numbers of species (Orians & Paine 1983; Ricklefs 1987; Latham & Ricklefs 1993). Mediterranean-climate regions are ideal for testing aspects of convergence theory, since they are well isolated from each other and all have diverse, distinctive and endemic-rich biotas (Cody & Mooney 1978; Cowling *et al.* 1994b; Hobbs, Richardson & Davis 1995b). We compared patterns of regional richness in the mediterranean-climate regions of California, Chile, southwestern Australia and South Africa (southwestern fynbos biome). Specifically, we compared observed richness in the non-African regions with the value predicted from the species–area curve for the (winter rainfall) southwestern fynbos biome (Fig. 19.5). Area size spanned the same range on all continents. A similar analysis (but excluding Chile) was carried out by Cowling *et al.* (1992) using a species–area curve with nested data (Table 19.4). Although the results were broadly similar to those reported here, we discuss only the latter below.

Regions in the southwestern fynbos biome supported on average 3.4 times as many species as equal-sized areas of Chile, 2.4 times the number in California and 1.8 times the number in Australia (Table 19.5). Only in some

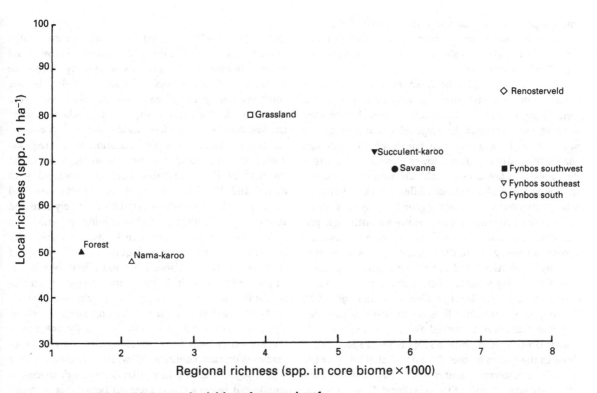

Figure 19.6 **Relationship between local richness [mean number of species in 0.1-ha plots, data from Cowling *et al.* (1992)] and regional richness [number of species in core biomes, data from Gibbs Russell (1987), Geldenhuys (1992) and Chap. 3 (this volume) for the biomes of South Africa].**

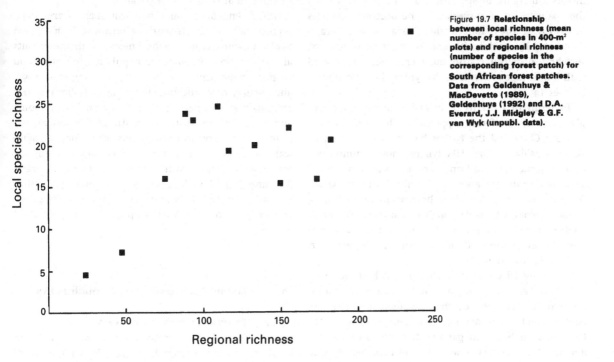

Figure 19.7 **Relationship between local richness (mean number of species in 400-m² plots) and regional richness (number of species in the corresponding forest patch) for South African forest patches. Data from Geldenhuys & MacDevette (1989), Geldenhuys (1992) and D.A. Everard, J.J. Midgley & G.F. van Wyk (unpubl. data).**

small regions of Australia was there evidence for convergence in regional richness. Why is the fynbos so much richer at the regional scale than the other mediterranean-climate ecosystems?

The answer includes both an ecological (environmental heterogeneity) and a phylogenetic (speciation history) component. In the mediterranean-climate zone of Chile, environmental heterogeneity is at least as pronounced as the southwestern fynbos biome, although edaphic diversity is probably lower (Rundel 1981; Arroyo *et al.* 1995). The few data on alpha diversity of indigenous species at the 0.1-ha scale in Chilean matorral suggest values comparable to, if not higher than, fynbos biome communities (Westman 1988). However, although precise data are lacking, both beta and gamma diversity appear to be very low in Chile: many species show high phenotypic plasticity and are widely distributed along environmental gradients; little compositional change occurs along geographical gradients (Rundel 1981; R.M. Cowling, pers. obs.). Therefore, on the basis of considerably lower turnover, we would expect physiographically similar landscapes in Chile to support fewer species than those in the fynbos biome. The historical absence of fire in Chilean matorral (Fuentes *et al.* 1995) and the comparatively high fertility of its soils (Rundel 1981; Lamont 1995) may explain the lower diversification of habitat-specialist and ecologically equivalent species in this, compared to fynbos and other mediterranean-climate ecosystems (but see Arroyo *et al.* 1995). The much lower beta and gamma diversities may also be an effect of life-history constraints on speciation rates in woody taxa and the paucity of fast-speciating herbaceous lineages (Arroyo *et al.* 1995). Indeed, the Chilean woody flora is dominated by species with exactly the traits (sprouters, vertebrate-dispersal, continuous recruitment) associated with low lineage turnover in the fynbos flora (Cowling & Holmes 1992a).

In terms of patterns of diversity and incidence of plant traits, California appears to be intermediate between Chile and the fynbos biome (Keeley 1992; Le Maitre & Midgley 1992). Like fynbos, most communities in mediterranean California are fire-prone, although median fire intervals are several times shorter in fynbos (Kruger 1983). Environmental heterogeneity, including edaphic complexity, is very high in California (Raven & Axelrod 1978; Richerson & Lum 1980). Generally, soil nutrient levels, especially phosphorus, are higher than in fynbos (Lamont 1995).

Alpha diversity of mature chaparral is half the value of fynbos (Westman 1988) although post-fire communities may be as rich as the richest fynbos sites (J.E. Keeley, pers. comm.). There are no community-scale data for California on beta and gamma diversity. Patterns for individual taxa suggest lower levels of turnover than in

fynbos (Cody 1986). In order to explain the more than two-fold lower regional richness in California, we would expect the turnover components of diversity there to be at least half the values of those recorded in the southwestern fynbos biome. This implies lower speciation rates in the California flora which is borne out by the low number of speciose woody genera with ecological traits conducive to diversification in a fire-prone environment (Cowling *et al.* 1992). However, much of the diversity of the Californian flora is associated with annual and short-lived perennial herbs (Raven & Axelrod 1978), growth forms which are both relatively rare and species-poor in fynbos (Le Maitre & Midgley 1992).

The southwestern Australian landscape shares with the fynbos biome great antiquity, a wide array of nutritionally impoverished soils and relatively short (*c.* 15 yr) fire intervals; it differs in its extreme topographical uniformity and, consequently, its shallow climatic gradients (Lamont, Collins & Cowling 1985). Levels of alpha diversity are similar (Rice & Westoby 1983) as is the extremely high turnover along edaphic and other gradients on each continent (Cowling *et al.* 1994b; Hobbs *et al.* 1995a), resulting in similar numbers of species in small-sized regions (Table 19.5). In both areas, richness is maximized in regions of high environmental heterogeneity (Hopper 1979; Kruger & Taylor 1979; Lamont, Hopkins & Hnatiuk 1984; Cowling 1990; Fig. 19.2a). However, the more subdued topography of Australia results in shallower environmental gradients and, hence, lower richness than fynbos in larger areas.

Local endemics on both continents are over-represented among fire-sensitive shrubs with small seed banks (Cowling *et al.* 1994b): lineages with these traits are vulnerable to fire-induced population reduction and fragmentation (Lamont *et al.* 1985; Cowling *et al.* 1992) and, hence, diversification (Cowling & Holmes 1992a; McDonald *et al.* 1995). The southwestern fynbos biome and southwestern Australia constitute the best known example for terrestrial ecosystems where nearly identical contemporary and historical ecologies, acting on unrelated biotas with similar ecological traits (Cowling & Witkowski 1994), have resulted in similar speciation histories (Cowling *et al.* 1994b), producing convergent patterns of richness in physiologically matched regions.

19.5 General discussion and conclusions

19.5.1 Patterns of regional richness
Species are neither uniformly nor randomly distributed across the South African landscape. When the entire

country is considered, regions of especially high richness occur in the southwestern fynbos biome, and some areas along the subtropical east coast and eastern escarpment where there is a complex mosaic of forest, savanna and grassland communities; regions of relatively low diversity occur mainly in the arid, central areas, but also in parts of the southern and southeastern coastal belt (see also Cowling *et al.* 1989). However, the situation is quite complex, with species-rich regions juxtaposed with species-poor ones, probably as a result of rapidly changing ecological conditions over short distances. For example, Ubisalia, a moist and partially forested site on the subtropical east coast with 662 species in 1.3 km^2, is only 45 km away from the Umfolozi Game Reserve, a site with only 425 species in 500 km^2. There are many other examples. These patterns are clearly evident in Fig. 19.8 where sites are classified on the basis of their standardized residuals in the step-wise regression model of the full data set (Table 19.4).

19.5.2 Models and explanations
Plant species richness at the regional scale in South Africa is determined largely by environmental heterogeneity. This is especially true of the fynbos and karroid regions, where explosive diversification within certain lineages has resulted in the evolution of numerous habitat specialists and ecological equivalents, resulting in very high beta and gamma diversities (Cowling *et al.* 1989). It is noteworthy that the very high regional richness in the fynbos biome is largely a function of extraordinarily high ecological heterogeneity, although the unparalleled diversification of certain lineages is of undoubted importance. In the tropical-derived savanna and grassland flora, total productivity of a region is also an important predictor of richness. Rather than being consistent with species–energy theory, we interpret this pattern as reflecting the environmental conditions (warm, moist and aseasonal) where these tropical lineages originated. The thermally seasonal environments with high rainfall irregularity, which are areas of low productivity and diversity in these biomes, are relatively young (Chap. 4, this volume); few tropical lineages have diversified there.

None of the explanations invoked in this study for the variation in regional richness in South Africa are deterministic in that they specify some resource-controlled asymptote on regional richness. Indeed, other than the pervasive influence of human-induced disturbance leading inexorably to extinctions (Chap. 21, this volume), there is no reason to believe that plant species will not continue to accumulate in certain South African landscapes.

The amount of variance in species richness explained by the models was not particularly high, except for those from the forest and fynbos biomes with their relatively biologically and phylogenetically homogeneous floras. We expect that when sites are nested within phytochoria, rather than across biomes as was the case for this study, patterns will be more fully explained. Therefore, future studies should attempt to locate sufficient numbers of sites within the Zambezian, Tongaland–Pondoland, Kalahari–Highveld, Afromontane grass and heath phytochoria and analyse these separately. The Karoo–Namib area should be subdivided into Nama- and succulent karoo zones (Chap. 3, this volume). We lacked sufficient samples for this level of analysis. The major advantage of this approach is that it will reduce unexplained variance associated with phylogenetic constraints.

The global comparisons showed that patterns of regional diversity in the speciose fynbos biome are not unique. Given the similar environmental history and mix of traits in the southwestern Australian flora, similar patterns to the southwestern fynbos biome have developed. If Australia had similar topographical diversity, it is likely that these patterns would have converged over all spatial scales (cf. Westoby 1993).

19.5.3 Ecology versus history
A new ecological synthesis is emerging where ecological and historical explanations are not seen as contending assertions, but as integrated parts of realistic and general explanations of ecological diversity. There is little doubt that patterns of regional richness are produced to a large extent by differential rates of diversification as determined by the biological traits of lineages and the ecological and geographical characteristics of the regions where they evolved and where they now occur (Ricklefs & Schluter 1993). The challenge is to produce a predictive science of ecological diversity that does not recognize a dualism between ecology and history (Kingsland 1985).

19.6   Acknowledgements

Many thanks to Jeanne Hurford for help in preparing the figures. RMC and DMR acknowledge the financial support of the University of Cape Town and the Foundation for Research Development.

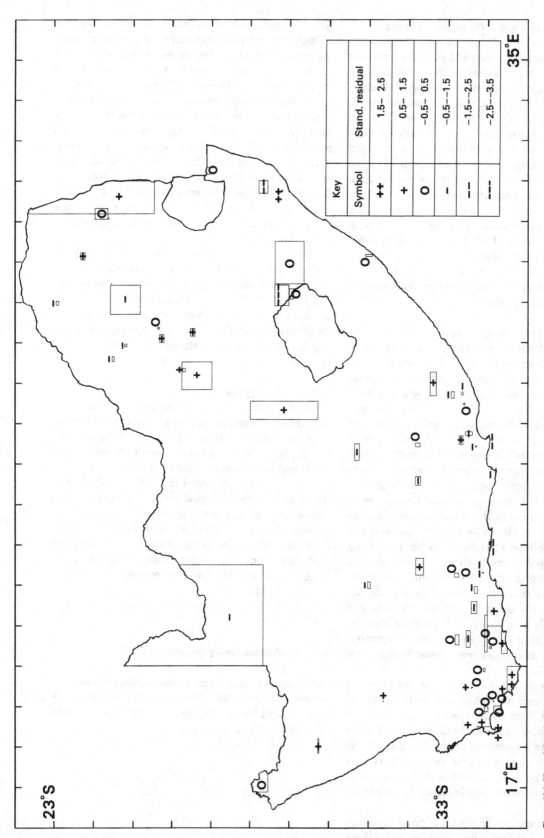

Figure 19.8 **Map of South Africa showing standardized residuals for different samples (regions) from the best step-wise regression model of total data set (Table 19.4). The model explained 63% of the variance in species number, and variables entered were as follows (contribution to R^2 given in brackets): log RAR (39%), log area (13%), RAS (10%) and PRO (5%).**

Appendix 19.1 **Data and sources for sites used in this study. See Fig. 19.1 for location of sites. N.R., nature reserve; N.P., National Park; NPB, National Parks Board; CNC, Western Cape Nature Conservation; N.B.G., National Botanical Garden**

No.	Site	Area (km²)	No. spp.	Source
Fynbos biome				
1	Cape Peninsula	471	2256	Cowling *et al.* (1992)
2	Tygerberg N.R.	0.68	373	Cowling *et al.* (1992)
3	Pella	2.7	379	Cowling *et al.* (1992)
4	Stellenbosch Flats	10.4	585	Cowling *et al.* (1992)
5	Jonkershoek N.R.	45.3	1142	Cowling *et al.* (1992)
6	Cape Hangklip	240	1383	Cowling *et al.* (1992)
7	Jakkalsrivier	1.58	533	Cowling *et al.* (1992)
8	Fernkloof N.R.	14.46	733	Cowling *et al.* (1992)
9	Vogelgat N.R.	6.02	697	Cowling *et al.* (1992)
10	Agulhas Plain	1609.3	1751	Cowling *et al.* (1992)
11	De Hoop N.R.	180	1179	Cowling *et al.* (1992)
12	Bontebok N.P.	27.86	466	Cowling *et al.* (1992)
13	Langeberg	1748	1228	McDonald *et al.* (1995)
14	Anysberg N.R.	340	473	Cowling *et al.* (1992)
15	Rooiberg N.R.	250	481	Cowling *et al.* (1992)
16	Riversdale Plain	2860	1580	Cowling *et al.* (1992)
17	Moordkuils	6.17	313	Cowling *et al.* (1992)
18	Goukamma N.R.	20.55	380	Cowling *et al.* (1992)
19	Seekoei N.R.	1.41	217	Cowling *et al.* (1992)
20	Cape Recife N.R.	3.36	173	Cowling *et al.* (1992)
21	Zuurberg N.P.	207.8	1100	Cowling *et al.* (1992)
Nama- and succulent karoo biomes				
22	Richtersveld N.P.	1620	506	NPB (unpubl. data)
23	Goegab N.R.	148.64	550	CNC (unpubl. data)
24	Nieuwoudtville N.R.	1.15	281	Cowling *et al.* (1989)
25	Karoo N.B.G.	1.54	362	Smitheman & Perry (1990)
26	Worcester Veld Reserve	0.75	267	Cowling *et al.* (1989)
27	Vrolijkheid N.R.	20.39	340	Cowling *et al.* (1989)
28	Whitehill	80	410	Compton (1930/31)
29	Gamka Mountain N.R.	94.28	426	Cowling *et al.* (1989)
30	Tierberg	1	178	Milton, Dean & Kerley (1992)
31	Boomplaas Cave	4.5	397	Moffett & Deacon (1977)
32	Karoo N.P.	330	768	A.R. Palmer (unpubl. data)
33	Karoo N.R.	162	336	A.R. Palmer (unpubl. data)
34	Carnarvon State Farm	10	128	A.R. Palmer (unpubl. data)
35	Mountain Zebra N.P.	65.2	358	Penzhorn (1970)
36	Oviston N.R.	130	278	Cowling *et al.* (1989)
37	Southern Kalahari	122880	454	Cowling *et al.* (1989)
Savanna and grassland biomes				
38	Springs N.R.	4.1	318	Cowling *et al.* (1989)
39	Addo N.P.	77.35	301	A.R. Palmer (unpubl. data)
40	Thomas Baines N.R.	2.6	374	Cowling *et al.* (1989)
41	Blaaukranz	2	253	Cowling *et al.* (1989)
42	Andries Vosloo N.R.	64.9	295	Cowling *et al.* (1989)
43	Amatole Mountains	900	1215	Phillipson (1987)
44	Umtamvuna N.R.	33	1260	A.T.D. Abbott (unpubl. data)
45	Estcourt and Weenen District	5969	1584	Cowling *et al.* (1989)
46	Cathedral Peak	300	907	Killick (1963)
47	Upper Tugela Location	1000	474	C.S. Everson (unpubl. data)
48	Ubisalia	1.3	662	A. Abbott (unpubl. data)
49	Umfolozi N.R.	500	425	Cowling *et al.* (1989)
50	Ndumu N.R.	101	644	Cowling *et al.* (1989)
51	Kruger N.P.	18833	1800	Cowling *et al.* (1989)
52	Klaserie N.R.	628.18	618	Zimbatis (1994)
53	Westfalia Estate	62.5	1061	Cowling *et al.* (1989)
54	Blouberg	400	559	Cowling *et al.* (1989)
55	Springbok Flats	17316	584	Cowling *et al.* (1989)
56	Groothoek	40	332	Westfall, van Rooyen & Theron (1986)
57	Lake Kyle	15.23	248	Van Schalkwyk (1993)
58	Magaliesberg	31	524	Cowling *et al.* (1989)
59	Roodeplaat Dam N.R.	1300	556	Van Rooyen (1983)
60	Suikerbosrand N.R.	133.4	742	Cowling *et al.* (1989)
61	Potchefstroom District	4480	1016	Cowling *et al.* (1989)
62	Abe Bailey N.R.	18.9	433	S. van Wyk & G.J. Bredenkamp (unpubl. data)
63	Bloemfontein and Brandfort	2560	989	Cowling *et al.* (1989)

19.7 References

Acocks, J.P.H. (1953). Veld Types of South Africa. *Memoirs of the Botanical Survey of South Africa*, 28, 1–192.

Adams, J.M. & Woodward, F.I. (1989). Patterns in tree species richness as a test of the glacial extinction hypothesis. *Nature*, 339, 699–701.

Arrhenius, O. (1921). Species and area. *Journal of Ecology*, 9, 95–9.

Arroyo, M.T.K., Cavieres, L., Marticorena, C. & Munoz-Schlick, M. (1995). Convergence in the mediterranean floras in central Chile and California: insight from comparative biogeography. In *Ecology and Biogeography of Mediterranean Ecosystems in Chile, California and Australia*, ed. M.T.K. Arroyo, M. Fox & P. Zedler, pp. 43–88. Berlin: Springer-Verlag.

Arroyo, M.T.K., Squeo, F.A., Armesto, J.J. & Villagran, C. (1988). Effects of aridity on plant diversity in the northern Chilean Andes: results of a natural experiment. *Annals of the Missouri Botanical Garden*, 75, 55–78.

Auerbach, M. & Shmida, A. (1987). Spatial scale and the determinants of plant species richness. *Trends in Ecology and Evolution*, 2, 238–42.

Boecklen, W.J. & Gotelli, N.J. (1984). Island biogeographic theory and conservation practice: species–area or specious area relationships. *Biological Conservation*, 29, 63–80.

Bond, W.J. (1983). On alpha diversity and the richness of the Cape flora: a study in southern Cape fynbos. In *Mediterranean-type Ecosystems. The Role of Nutrients*, ed. F.J. Kruger, D.T. Mitchell & J.U.M. Jarvis, pp. 225–43. Berlin: Springer-Verlag.

Bond, W.J., Cowling, R.M. & Richards, M.B. (1992). Competition and coexistence. In *The Ecology of Fynbos. Nutrients, Fire and Diversity*, ed. R.M. Cowling, pp. 206–25. Cape Town: Oxford University Press.

Brooks, J.L. (1950). Speciation in ancient lakes. *Quarterly Review of Biology*, 60, 131–76.

Brown, J.H. (1981). Two decades of a homage to Santa Rosalia: toward a general theory of diversity. *American Zoologist*, 21, 877–88.

Brown, J.H. (1988). Species diversity. In *Analytical Biogeography. An Integrated Approach to the Study of Animal and Plant Distributions*, ed. A.A. Myers & P.S. Giller, pp. 57–89. London: Chapman & Hall.

Campbell, B.M. (1983). Montane plant environments in the fynbos biome. *Bothalia*, 14, 283–98.

Cody, M.L. (1975). Towards a theory of continental diversities: bird distribution over mediterranean habitat gradients. In *Ecology and Evolution of Communities*, ed. M.L. Cody & J. Diamond, pp. 214–57. Cambridge, Massachusetts: Harvard University Press.

Cody, M.L. (1986). Diversity, rarity and conservation in mediterranean-climate regions. In *Conservation Biology. The Science of Scarcity and Diversity*, ed. M.E. Soulé, pp. 122–52. Sunderland, Massachusetts: Sinauer Associates.

Cody, M.L. (1989). Growth form diversity and community structure in desert plants. *Journal of Arid Environments*, 17, 199–209.

Cody, M.L. & Mooney, H.A. (1978). Convergence versus non-convergence in mediterranean-climate ecosystems. *Annual Review of Ecology and Systematics*, 9, 265–321.

Compton, R.H. (1930/31). The flora of the Whitehill district. *Transactions of the Royal Society of South Africa*, 19, 269–326.

Connell, J.H. (1978). Diversity in tropical rain forests and coral reefs. *Science*, 199, 1302–9.

Connor, E.F. & McCoy, E.O. (1979). The statistics and biology of the species-area relationship. *American Naturalist*, 113, 791–833.

Cornell, H.V. (1993). Unsaturated patterns in species assemblages: the role of regional processes in setting local species richness. In *Species Diversity in Ecological Communities: Historical and Geographical Perspectives*, ed. R.E. Ricklefs & D. Schluter, pp. 243–52. Chicago: University of Chicago Press.

Cowling, R.M. (1983). Phytochorology and vegetation history in the south eastern Cape, South Africa. *Journal of Biogeography*, 10, 393–419.

Cowling, R.M. (1990). Diversity components in a species-rich area of the Cape Floristic Region. *Journal of Vegetation Science*, 1, 699–710.

Cowling, R.M., Esler, K.J., Midgley, G.F. & Honig, M.A. (1994a). Plant functional diversity, species diversity and climate in arid and semi-arid southern Africa. *Journal of Arid Environments*, 27, 141–58.

Cowling, R.M., Gibbs Russell, G.E., Hoffman, M.T. & Hilton-Taylor, C. (1989). Patterns of plant species diversity in southern Africa. In *Biotic Diversity in southern Africa. Concepts and Conservation*, ed. B.J. Huntley, pp. 19–50. Cape Town: Oxford University Press.

Cowling, R.M. & Hilton-Taylor, C. (1994). Patterns of plant diversity and endemism in southern Africa: an overview. *Strelitzia*, 1, 31–52.

Cowling, R.M. & Holmes, P.M. (1992a). Endemism and speciation in a lowland flora from the Cape Floristic Region. *Biological Journal of the Linnean Society*, 47, 367–83.

Cowling, R.M. & Holmes, P.M. (1992b). Flora and vegetation. In *The Ecology of Fynbos*, ed. R.M. Cowling, pp. 23–61. Cape Town: Oxford University Press.

Cowling, R.M., Holmes, P.M. & Rebelo, A.G. (1992). Plant diversity and endemism. In *The Ecology of Fynbos. Nutrients, Fire and Diversity*, ed. R.M. Cowling, pp. 62–112. Cape Town: Oxford University Press.

Cowling, R.M. & Witkowski, E.T.F. (1994). Convergence and non-convergence of plant traits in climatically and edaphically matched sites in Mediterranean Australia and South Africa. *Australian Journal of Ecology*, 19, 220–32.

Cowling, R.M., Witkowski, E.T.F., Milewski, A.V. & Newbey, K.R. (1994b). Taxonomic, edaphic and biological aspects of endemism on matched sites in mediterranean Australia and South Africa. *Journal of Biogeography*, 21, 651–64.

Currie, D.J. (1991). Energy and large-scale patterns of animal- and plant-species richness. *American Naturalist*, 137, 27–49.

rrie, D.J. & Paquin, V. (1987). Large-scale biogeographical patterns of species richness of trees. *Nature*, 329, 326–7.

acon, H.J., Hendey, Q.B. & Lamprechts, J.J.N. (eds) (1983). *Fynbos Palaeoecology: a Preliminary Synthesis. South African National Scientific Programmes Report 75.* Pretoria: CSIR.

acon, H.J., Jury, M.R. & Ellis, F. (1992). Selective regime and time. In *The Ecology of Fynbos. Nutrients, Fire and Diversity*, ed. R.M. Cowling, pp. 6–22. Cape Town: Oxford University Press.

ent, M.C., Lynch, S.D. & Schulze, R.E. (1988). *Mapping Mean Annual and Other Rainfall Statistics Over Southern Africa. WRC Report 109/1/89.* Pretoria: Water Research Commission.

ent, M.C., Schulze, R.E. & Angus, G.R. (1989). *Crop Water Requirements, Deficit and Water Yield for Irrigation Planning in Southern Africa. WRC Report 118/1/88.* Pretoria: Water Research Commission.

iamond, J.M. (1988). Factors controlling species diversity: overview and synthesis. *Annals of the Missouri Botanical Garden*, 75, 117–29.

obzhansky, T. (1950). Evolution in the tropics. *American Scientist*, 38, 209–21.

edler, P.L. (1986). Concepts of rarity in vascular plants with special reference to the genus *Calochortus* Pursh (Liliaceae). *Taxon*, 35, 502–18.

scher, A.G. (1960). Latitudinal variations in organic diversity. *Evolution*, 14, 64–81.

uentes, E.R., Montenegro, G., Rundel, P.W., Arroyo, M.T.K., Ginnochio, R. & Jaksic, F.M. (1995). Functional approaches to biodiversity in the mediterranean-type ecosystems of central Chile. In *Mediterranean-type Ecosystems. The Function of Biodiversity*, ed. G.W. Davis & D.M. Richardson, pp. 185–232. Berlin: Springer-Verlag.

eldenhuys, C.J. (1989). *Environmental and Biogeographic Influences on the Distribution and Composition of the Southern Cape Forests (Veld Type 4).* PhD Thesis. Cape Town: University of Cape Town.

eldenhuys, C.J. (1992). Richness, composition and relationships of the floras of selected forests in southern Africa. *Bothalia*, 22, 205–33.

Geldenhuys, C.J. & MacDevette, D.R. (1989). Conservation status of coastal and montane evergreen forests. In *Biotic Diversity in Southern Africa. Concepts and Conservation*, ed. B.J. Huntley, pp. 224–38. Cape Town: Oxford University Press.

Gibbs Russell, G.E. (1987). Preliminary floristic analysis of the major biomes in southern Africa. *Bothalia*, 17, 213–27.

Gibbs Russell, G.E. & Robinson, E.R. (1981). Phytogeography and speciation in the vegetation of the eastern Cape. *Bothalia*, 13, 467–72.

Goldblatt, P. (1978). An analysis of the flora of southern Africa: its characteristics, relationships and origins. *Annals of the Missouri Botanical Garden*, 65, 369–436.

Goodman, P.S. (1990). *Soil, Vegetation and Large Herbivore Relations in Mkuzi Game Reserve, Natal.* PhD Thesis. Johannesburg: University of the Witwatersrand.

Grassle, J.F. & Maciolek, N.J. (1992). Deep-sea species richness: regional and local diversity estimates from quantitative bottom samples. *American Naturalist*, 139, 313–41.

Hobbs, R.J., Groves, R.H., Hopper, S.D. Lambeck, R.J., Lamont, B.B., Lavorel, S., Main, A.R., Majer, J.D. & Saunders, D.A. (1995a). Function of biodiversity in the mediterranean-type ecosystems of southwestern Australia. In *Mediterranean-type Ecosystems. The Function of Biodiversity*, ed. G.W. Davis & D.M. Richardson, pp. 233–84. Berlin: Springer-Verlag.

Hobbs, R.J., Richardson, D.M. & Davis G.W. (1995b). Mediterranean-type ecosystems: opportunities and constraints for studying biodiversity. In *Mediterranean-type Ecosystems. The Function of Biodiversity*, ed. G.W. Davis & D.M. Richardson, pp. 1–42. Berlin: Springer-Verlag.

Hoffman, M.T. Midgley, G.F. & Cowling, R.M. (1994). Plant richness is negatively related to energy availability in semi-arid southern Africa. *Biodiversity Letters*, 2, 35–8.

Holt, R.D. (1993). Ecology at the mesoscale: the influence of regional processes on local communities. In *Species Diversity in Ecological Communities: Historical and Geographical Perspectives*, ed. R.E. Ricklefs & D. Schluter, pp. 77–88. Chicago: University of Chicago Press.

Hopper, S.D. (1979). Biogeographical aspects of speciation in the southwest Australian flora. *Annual Review of Ecology and Systematics*, 10, 399–422.

Hutchinson, G.E. (1959). Homage to Santa Rosalia, or why are there so many kinds of animals? *American Naturalist*, 93, 145–59.

Johnson, M.P., Mason, L.G. & Raven, P.H. (1968). Ecological parameters and plant species diversity. *American Naturalist*, 102, 297–306.

Keeley, J.E. (1992). A Californian's view of fynbos. In *The Ecology of Fynbos. Nutrients, Fire and Diversity*, ed. R.M. Cowling, pp. 372–88. Cape Town: Oxford University Press.

Killick, D.J.B. (1963). An account of the plant ecology of the Cathedral Peak area of the KwaZulu-Natal Drakensberg. *Memoirs of the Botanical Survey of South Africa*, 34, 1–178.

Kingsland, S.E. (1985). *Modeling Nature: Episodes in the History of Population Ecology.* Chicago: Chicago University Press.

Knoll, A.H. (1986). Patterns of change in plant communities through geological time. In *Community Ecology*, ed. J. Diamond & T. Case, pp. 126–44. New York: Harper & Row.

Kruger, F.J. (1983). Plant community diversity and dynamics in relation to fire. In *Mediterranean-type Ecosystems: the Role of Nutrients*, ed. F.J. Kruger, D.T. Mitchell & J.U.M. Jarvis, pp. 446–72. Berlin: Springer-Verlag.

Kruger, F.J. & Taylor, H.C. (1979). Plant species diversity in Cape fynbos: gamma and delta diversity. *Vegetatio*, 47, 85–93.

Lamont, B.B. (1995). Mineral nutrient relations in mediterranean regions of California, Chile and Australia. In *Ecology and Biogeography of Mediterranean Ecosystems in Chile, California and Australia*, ed. M.T.K. Arroyo, M. Fox & P. Zedler, pp. 211–35. Berlin: Springer-Verlag.

Lamont, B.B., Collins, B.G. & Cowling, R.M. (1985). Reproductive biology of the Proteaceae in Australia and South Africa. *Proceedings of the Ecological Society of Australia*, **14**, 213–24.

Lamont, B.B., Hopkins, A.J.M. & Hnatiuk, R.J. (1984). The flora – composition, diversity and origins. In *Kwongan. Plant Life of the Sandplain*, ed. J.S. Pate & J.S. Beard, pp. 27–50. Nedlands: University of Western Australia Press.

Latham, R.E. & Ricklefs, R.E. (1993). Global patterns of tree species richness in moist forests: energy-diversity theory does not account for variation in species richness. *Oikos*, **67**, 325–33.

Lawton, J.H. (1993). Range, population abundance and conservation. *Trends in Ecology and Evolution*, **8**, 409–13.

le Maitre, D.C. & Midgley, J.J. (1992). Plant reproductive ecology. In *The Ecology of Fynbos*, ed. R.M. Cowling, pp. 133–74. Cape Town: Oxford University Press.

Linder, H.P. (1991). Environmental correlates of patterns of richness in the south-western Cape Province of South Africa. *Journal of Biogeography*, **18**, 509–18.

Linder, H.P., Meadows, M.E. & Cowling, R.M. (1992). History of the Cape flora. In *The Ecology of Fynbos. Nutrients, Fire and Diversity*, ed. R.M. Cowling, pp. 113–34. Cape Town: Oxford University Press.

MacArthur, R.H. (1972). *Geographical Ecology*. New York: Harper & Row.

Markham, C.G. (1970). Seasonality of precipitation in the United States. *Annals of the Association of American Geographers*, **60**, 593–7.

McDonald, D.J., Juritz, J.M., Cowling, R.M. & Knottenbelt, W.J. (1995). Modelling the biological aspects of local endemism in South African fynbos. *Plant Systematics and Evolution*, **195**, 137–47.

McGuinness, K.A. (1984). Equations and explanations in the study of species-area curves. *Biological Review*, **59**, 423–40.

Meadows, M.E. & Sugden, J.M. (1991). A vegetation history of the last 14 500 years on the Cederberg, SW Cape. *South African Journal of Science*, **87**, 34–43.

Milton, S.J., Dean, W.R.J. & Kerley, G.I.H. (1992). Tierberg Karoo Research Centre: history, physical environment, flora and fauna. *Transactions of the Royal Society of South Africa*, **48**, 15–46.

Moffett, R.O. & Deacon, H.J. (1977). The flora and vegetation of the surrounds of Boomplaas Cave: Cango Valley. *South African Archaeological Bulletin*, **32**, 127–45.

Moll, E.J. & White, F. (1978). The Indian Ocean coastal belt. In *Biogeography and Ecology of Southern Africa*, ed. M.J.A. Werger, pp. 561–98. The Hague: Junk.

Mustart, P.J. & Cowling, R.M. (1993). The role of regeneration stages in the distribution of edaphically restricted fynbos Proteaceae. *Ecology*, **74**, 1490–3.

O'Brien, E. (1993). Climatic gradients in woody plant species richness: towards an explanation based on an analysis of southern Africa's woody flora. *Journal of Biogeography*, **20**, 181–98.

Orians, G.H. & Paine, R.T. (1983). Convergent evolution at the community level. In *Coevolution*, ed. D.J. Futuyuma & M. Slatkin, pp. 431–58. Sunderland, Massachusetts: Sinauer Associates.

Penzhorn, B.L. (1970). A checklist of the flowering plants in the herbarium of the Mountain Zebra National Park. *Koedoe*, **13**, 131–46.

Phillipson, P.B. (1987). A checklist of the vascular plants of the Amatole Mountains, Eastern Cape Province. *Bothalia*, **17**, 237–56.

Pianka, E.R. (1966). Latitudinal gradients in species diversity: a review of concepts. *American Naturalist*, **100**, 33–46.

Preston, F.W. (1962). The canonical distribution of commonness and rarity. *Ecology*, **43**, 185–215, 410–32.

Raven, P.H. & Axelrod, D.I. (1978). Origin and relationships of the California flora. *University of Californian Publications in Botany*, **72**, 1–134.

Rice, B. & Westoby, M. (1983). Plant species richness at the 0.1 ha scale in Australian vegetation compared to other continents. *Vegetatio*, **52**, 129–40.

Richardson, D.M. & Cowling, R.M. (1992). Why is mountain fynbos invasible and which species invade? In *Fire in South African Mountain Fynbos*, ed. B.W. van Wilgen, D.M. Richardson, F.J. Kruger & H.J. van Hensbergen, pp. 161–81. Berlin: Springer-Verlag.

Richardson, D.M., Cowling, R.M., Lamont, B.B. & van Hensbergen, H.J. (1995). Coexistence of *Banksia* species in southwestern Australia: the role of regional and local processes. *Journal of Vegetation Science*, **6**, 329–42.

Richerson, P. & Lum, K.L. (1980). Patterns of plant species diversity in California: relation to weather and topography. *American Naturalist*, **116**, 504–36.

Ricklefs, R.E. (1987). Community diversity: relative roles of local and regional processes. *Science*, **235**, 167–71.

Ricklefs, R.E. & Latham, R.E. (1992). Intercontinental correlation of geographic ranges suggests stasis in ecological traits of relict genera of temperate perennial herbs. *American Naturalist*, **139**, 1305–21.

Ricklefs, R.E. & Latham, R.E. (1993). Global patterns of diversity in mangrove floras. In *Species Diversity in Ecological Communities: Historical and Geographical Perspectives*, ed. R.E. Ricklefs & D. Schluter, pp. 215–29. Chicago: University of Chicago Press.

Ricklefs, R.E. & Schluter, D. (1993). Species diversity: regional and historical influences. In *Species Diversity in Ecological Communities: Historical and Geographical Perspectives*, ed. R.E. Ricklefs & D. Schluter, pp. 350–63. Chicago: University of Chicago Press.

Rohde, K. (1992). Latitudinal gradients in species richness: the search for a primary cause. *Oikos*, **65**, 514–27.

Rosenzweig, M.L. (1968). Net primary productivity of terrestrial communities: prediction from climatological data. *American Naturalist*, **102**, 67–74.

Rosenzweig, M.L. (1992). Species diversity gradients: we know more and less than we thought. *Journal of Mammalogy*, **73**, 715–30.

osenzweig, M.L. & Abramsky, Z. (1993). How are diversity and productivity related? In *Species Diversity in Ecological Communities: Historical and Geographical Perspectives*, ed. R.E. Ricklefs & D. Schluter, pp. 52–65. Chicago: University of Chicago Press.

ndel, P.W. (1981). The matorral zone of central Chile. In *Mediterranean-type Shrublands*, ed. F. Di Castri, D.W. Goodall & R.L. Specht, pp. 175–201. Amsterdam: Elsevier.

nders, H.L. (1968). Benthic marine diversity: a comparative study. *American Naturalist*, 102, 243–82.

hluter, D. & Ricklefs, R.E. (1993). Species diversity: an introduction to the problem. In *Species Diversity in Ecological Communities: Historical and Geographical Perspectives*, ed. R.E. Ricklefs & D. Schluter, pp. 1–10. Chicago: University of Chicago Press.

hulze, R.E. & Maharaj, M. (1991). Mapping A-pan equivalent potential evaporation over southern Africa. *Proceedings, 5th South African National Hydrological Symposium*, Stellenbosch, 4B-4-1 to 4B-4-9.

mida, A. & Wilson, M.V. (1985). Biological determinants of species diversity. *Journal of Biogeography*, 12, 1–20.

nitheman, J. & Perry, P. (1990). A vegetation survey of the Karoo National Botanic Garden Reserve, Worcester. *South African Journal of Botany*, 56, 525–41.

ebbins, G.L. (1974). *Flowering Plants: Evolution Above the Species Level*. Cambridge, Massachusetts: Harvard University Press.

evens, G.C. (1989). The latitudinal gradient in geographical range: how so many species coexist in the tropics. *American Naturalist*, 133, 240–56.

rborgh, J. (1973). On the notion of favorableness in plant ecology. *American Naturalist*, 107, 481–501.

Iman, D. & Pacala, S. (1993). The maintenance of species richness in plant communities. In *Species Diversity in Ecological Communities: Historical and Geographical Perspectives*, ed. R.E. Ricklefs & D. Schluter, pp. 13–25. Chicago: University of Chicago Press.

van Rooyen, N. (1983). Die plantegroei van die Roodeplaatdam-natuurreservaat. I. 'n Voorlopige plantspesielys. *South African Journal of Botany*, 2, 105–14.

van Schalkwyk, A.G. (1993). *Plategroeiklassifikasie, Grondklassifikasie en Drakragbepaling van Lake Kyle – Gedeelte van die Groter Mabula Natuurreservaat*. Unpublished Report. Pretoria: Centre for Wildlife Management, University of Pretoria.

van Wyk, A.E. (1990). The sandstone regions of Natal and Pondoland: remarkable centres of endemism. *Palaeoecology of Africa*, 21, 243–57.

Vrba, E.S. (1980). Evolution, species and fossils: how does life evolve. *South African Journal of Science*, 76, 61–84.

Walker, T.D. & Valentine, J.W. (1984). Equilibrium models of evolutionary species diversity and the number of empty niches. *American Naturalist*, 124, 887–9.

Warner, R.R. & Chesson, P. (1985). Coexistence mediated by recruitment fluctuations: a field guide to the storage effect. *American Naturalist*, 125, 769–87.

Werger, M.J.A. (1978). The Karoo-Namib Region. In *Biogeography and Ecology of Southern Africa*, ed. M.J.A. Werger, pp. 147–70. The Hague: Junk.

Westfall, R.H., Van Rooyen, N. & Theron, G.K. (1986). The plant ecology of the farm Groothoek, Thabazimbe district. III. An annotated checklist. *Bothalia*, 16, 77–82.

Westman, W.E. (1988). Vegetation, nutrition and climate – data-tables (3). Species richness. In *Mediterranean-type Ecosystems: A Data Source Book*, ed. R.L. Specht, pp. 81–92. Dordrecht: Kluwer Academic Publishers.

Westoby, M. (1988). Comparing Australian ecosystems to those elsewhere. *Bioscience*, 38, 549–56.

Westoby, M. (1993). Biodiversity in Australia compared with other continents. In *Species Diversity in Ecological Communities: Historical and Geographical Perspectives*, ed. R.E. Ricklefs & D. Schluter, pp. 170–77. Chicago: University of Chicago Press.

White, F. (1978). The Afromontane region. In *Biogeography and Ecology of Southern Africa*, ed. M.J.A. Werger, pp. 465–510. The Hague: Junk.

Whittaker, R.H. (1972). Evolution and measurement of species diversity. *Taxon*, 21, 213–51.

Whittaker, R.H. (1977). Evolution of species diversity in land communities. *Evolutionary Biology*, 10, 1–67.

Williamson, M. (1988). Relationship of species number to area, distance and other variables. In *Analytical Biogeography. An Integrated Approach to the Study of Animal and Plant Distributions*, ed. A.A. Myers & P.S. Giller, pp. 91–115. London: Chapman & Hall.

Willis, J.C. (1922). *Age and Area: a Study in Geographical Distribution and Origin in Species*. Cambridge: Cambridge University Press.

Wright, D.H. (1983). Species-energy theory an extension of species-area theory. *Oikos*, 41, 495–506.

Wright, D.H., Currie, D.J. & Maurer, B.A. (1993). Energy supply and patterns of species richness on local and regional scales. In *Species Diversity in Ecological Communities: Historical and Geographical Perspectives*, ed. R.E. Ricklefs & D. Schluter, pp. 66–74. Chicago: University of Chicago Press.

Wylie, J.L. & Currie, D.J. (1993). Species-energy theory and patterns of species richness: I. Patterns of bird, angiosperm, and mammal species richness on islands. *Biological Conservation*, 63, 137–44.

Zimbatis, N. (1994). Ferns and flowering plants of Klaserie Private Nature Reserve, eastern Transvaal: an annotated checklist. *Bothalia*, 24, 37–53.

Human use of plants

A.B. Cunningham and G.W. Davis

20

20.1 Introduction

The study of human use of plants has been a multidisciplinary activity in southern Africa, with significant contributions made to this field by anthropologists, archaeologists, architects, chemists, linguists and naturalists as well as botanists. Six features characterize this research. Each of these is introduced here, then discussed in more detail in this chapter.

The first feature of this review of human use of plants is that its context is the African continent, where our species, *Homo sapiens*, originated; and more specifically the southern African region, with its great diversity of human lifestyles. A second feature is the late arrival of agriculture in southern Africa about 2000 years ago. This occurred after about 200 000 yr of low-impact hunter-gatherer interaction with vegetation and its arrival resulted in a period of increasingly rapid cultural and environmental change. Further change followed with the establishment of European settlement some 400 years ago, and rapid urbanization over the past 40 years (Chap. 21, this volume). Thirdly, there is remarkable continuity in the use of key plant species through time, from harvest by hunter-gatherer peoples to present-day commercial use. Fruits and nuts of marula (*Sclerocarya birrea*, Anacardiaceae), seeds of the !nara melon (*Acanthosicyos horrida*, Cucurbitaceae), herbal tea from rooibos (*Aspalathus linearis*, Fabaceae) and aromatic leaves of buchu (*Agathosma betulina*, Rutaceae) for example, were highly significant resources for hunter-gatherer peoples, and continue to be valued today. Several of these species are in a process of transition from wild species to crop plants. Conversely, since the late 1800s, new ornamental and industrial uses have been found for many indigenous plant species. As a result, indigenous plant species such as rooibos tea, protea flowers, ferns and marula fruits have become the focus of 'formal sector' national or international trade directly worth at least US$30–40 million yr^{-1} – and considerably more if the value-adding benefits to local business are taken into account.

Part of this continuity of use by hunter-gatherers, pastoralists, subsistence and commercial farmers is governed by a fourth feature of human use of plants in the region: the influence of soil type. A large area of southern Africa is covered by nutrient-poor, drought-susceptible sands (Chap. 1, this volume). This has had a strong influence on the use of wild plants across the region; on the high-rainfall Mozambique coastal plain, the arid Namib coast, the Kalahari sands and over much of the fynbos biome. Wild plant foods continue to supplement the lifestyles of subsistence agriculturalists in sandy areas on land of low arable value. Even commercial farmers in areas of low arable potential have looked to wild plant species such as thatch reed (Restionaceae), protea cut-flowers, or the sale of marula fruits or jellies as exploitable resources.

A fifth aspect of people–plant interaction has been the rapid loss of unrecorded cultural knowledge, particularly the ecological insights and in some cases entire languages, of hunter-gather peoples. Before the arrival of the Dutch in the Cape in 1652, there were probably 150 000 to 300 000 San people in southern Africa (Biesele *et al.* 1989). Two hundred years later, San populations south of the Orange River had declined dramatically as a result of conflict and disease. The loss of knowledge of

474

plant species and culture of the indigenous San people has been a tragic consequence of this, and has been most complete in the western region: the home of the San hunter-gatherers known as the /Xam. They undoubtedly had a deep knowledge of ecological issues. They also had a deep symbolic and religious attachment to the landscape. What was saved of this cultural legacy can be attributed to a few individuals. Starting in the 1870s, for example, philologists Wilhelm Bleek and Lucy Lloyd worked with /Xam people who were serving prison sentences at Breakwater Convict Station in Cape Town. The result was nearly 13 000 pages of interviews and a permanent record of the /Xam language. However, along with the /Xam people, most of the knowledge of this rich flora has gone.

Finally and fortunately, there has been a far longer period of formal study of human use of wild plants in southern Africa than most other parts of Africa. Starting with the study of traditional uses of Cape medicinal plants in the 1600s, this research has resulted in a remarkably detailed record of plant use in a botanically and culturally diverse region. Quantitative ethnobotanical studies also developed in southern Africa at an early stage compared to other parts of the world. The works of Quin (1959) and Scudder (1962) on edible wild plant use in southern Africa are two fine examples of pioneering quantitative ethnobotanical research. For a long period, however, the major focus of ethnobotanical research was on recording vernacular names and plant uses. Valuable contributions were certainly made through records of the plant names of the !Kung San (Story 1958; W. Geiss & J.W. Snyman, unpubl. data), Shona (Wild 1952), Sotho (Jacot Guillarmod 1971), Swazi (Compton 1966), Zulu (Bryant 1909; Gerstner 1938, 1939, 1941; Doke & Vilakazi 1964), Venda (Van Warmelo 1937) and KwaNyama Owambo (Rodin 1985). Many nutritional or chemical analyses of edible and medicinal wild plant species were also undertaken over a 50-year period. These were compiled in Watt & Breyer-Brandwijk's (1962) classic book on medicinal plants and by Fox & Norwood-Young (1982) and Wehmeyer (1986) on edible plants. More recently, since the 1980s, quantitative ethnobotanical studies have focused on the value of plants to people, the human impact on harvested plants and participatory approaches to ecologically sustainable use. These studies first concentrated on the measurement of wood use for fuel and building purposes (Best 1979; Whitlow 1979; Gandar 1983; Liengme 1983; Erkiila & Siiskonen 1992) and on the use of introduced *Acacia* species in the fynbos biome (Azorin 1992). Quantitative studies were done on the ecological effects of traditional indigenous tree conservation, the values and social importance of trees (Campbell 1986; Campbell & Du Toit 1988; Campbell,

Vermeulen & Lynam 1991), the marketing and impact of palm sap tapping (Cunningham 1990a,b), craft materials (Cunningham & Milton 1987; Cunningham 1987, 1988b), traditional medicines (Cunningham 1991, 1993a), *Phragmites australis* reeds (Cunningham 1985), *Cymbopogon* thatching grass (Shackleton 1990) and the indirect effects of aloe leaf removal for collection of sap (Bond 1983).

20.2 People and plants in the cradle of humankind

To appreciate the present and future use of plants by people in southern Africa, it is useful to look back into the past. Southern African researchers have played a significant role in this regard. Raymond Dart, for example, discovered the remains of the ape-like hominid *Australopithecus* as long ago as 1924. Subsequently, the Pliocene hominid localities of Sterkfontein and Makapansgat in South Africa have become as well known as those of Olduvai Gorge in Kenya or Hadar in Ethiopia. Although the period when hominids first made deliberate use of fire is still debated, there is no doubt that this would have provided a tool to alter vegetation (Chap. 18, this volume). It would also have expanded the number of plant species used for food, as detoxification of otherwise inedible foodstuffs would have been possible. Fagan (1990) suggests that *Homo erectus* may have used fire as early as 1.5 Myr for this purpose at Swartkrans, South Africa and in East Africa. Meticulous archaeological work in southern Africa has also provided a valuable perspective of early human history (Beaumont, De Villiers & Vogel 1978; Deacon 1989). Fossil bones provide direct evidence of hominids and the animals they hunted, but plant use by people today provides the major insight into plant use in the past. Detailed records of hunter-gatherer plant use, for example, have facilitated our understanding of early hominid diet (Peters & Maguire 1981; Peters & O'Brien 1981). Anthropological and ethnobotanical studies with the !Ko (!Xo) San (Heinz & Maguire 1974) and the !Kung San (Story 1958; Lee 1965, 1973) have also provided valuable insights into the resource value and cultural importance of plants, not only to the Kalahari San, but also to the ancestors of all humanity (/Xam, !Kung and !Ko are societies or linguistic groups of San people). It was the persistence of societies with this ancient hunter-gatherer lifestyle that stimulated Lee (1965) and subsequent researchers to undertake world-renowned anthropological studies of foraging societies in the Kalahari savanna.

20.3 Plants, culture and change

For probably 200 000 years, the effects of people on vegetation were limited to the patchy effects of fire and dispersal of favoured edible plants by low numbers of mobile hunter-gatherers. The major effect on vegetation over this period was climate change, not people (Chap. 21, this volume). A uniquely detailed example of this for southern Africa developed from interpretations of climate and vegetation change from archaeological work at Boomplaas Cave in the Cango Valley (Deacon 1988). Eighteen thousand years ago the valley was covered by semi-karroid scrubland. With warmer, moister conditions 15 000 years ago, it become woodland dominated by *Olea*. About 10 000 years ago this changed again, to thicket. In the later Holocene, with the southwards spread of *Acacia karroo*, the vegetation changed to a woodland dominated by *A. karroo* (Chap. 4, this volume). These changes altered the plant and mammal component of this valley, and also the diet and movement of hunter-gatherer people living there. Humphreys (1987) suggests that staple food plants may have been so reduced by some periods of environmental change that people were forced to withdraw from marginal to more resource-rich areas, an explanation he suggests for the periodic occupation of the Karoo between 12 000 and 9000 years ago and from about 4500 to the present time.

The late arrival of agriculture in a region with such a long human involvement makes the people–plant interface in southern Africa particularly interesting. Although Africa has the longest history of human occupation, agricultural production only developed 3000 years ago, spreading into southern Africa a thousand years later. By comparison, agriculture developed between 6000 and 8000 years ago in the Indus valley, South-East Asia, northern China and Europe and 7000 years ago in Central America (Fagan 1990). Although plant production was manipulated with fire as a tool to 'domesticate' landscapes, there was no domestication of plant species. One possible reason for the emphasis on gathering rather than plant domestication was the rich source of edible plants, game, fish and birds available to hunter-gatherers in the African savannas and adjacent wetlands. 'Why should we plant, when there are so many mongongos (*Schinziophyton rautanenii*) in the world?' asked Xashe of anthropologist Richard Lee (1973) during his research with !Kung San people in the Kalahari savanna. Xashe had a good point. Peters (1987), for example, estimates yields of 250–300 kg of nuts ha⁻¹ in savanna dominated by *Schinziophyton* (= *Ricinodendron*) *rautanenii* in Angola and 800 kg ha⁻¹ in Namibia. Marula (*Sclerocarya birrea*) trees have similarly high yields. Quin (1959) recorded fruit yields of 21 600–91 272 fruits per tree or an average yield of 644 kg fruit per tree from *S. birrea* trees. In the fynbos biome, plant remains from archaeological sites have led Deacon (1983) to speculate that Holocene hunter-gatherers, aware of the high below-ground production of geophytes, used fire as a tool to increase the production of *Watsonia borbonica* bulbs. This 'fire stick' farming is still practised by herders on the Mozambique coastal plain today, with the deliberate burning of *Hyphaene* palm savanna and coastal grasslands to stimulate fruit production of *Eugenia albanensis*, *Parinari curatellifolia* and *Salacia kraussii* (Cunningham 1988a).

Changes in lifestyle over the past 2000 years have come about as the result of more intensive pressure on the landscape by pastoralists and agriculturalists. One consequence of a sedentary agricultural lifestyle was a higher impact on surrounding vegetation, not only for fields, fuel and building materials, but also for localized iron smelting (Feeley 1980, 1985). These impacts are covered in detail in Chap. 21, this volume.

20.3.1 Wild plants: nature and culture

Wild species that were outstandingly important food sources developed a special place at the culture–nature interface across southern Africa. This is expressed in territorial rights, protection in customary law and the symbolic and religious significance of these key food sources today and in the past. Inherited rights by extended family groups (!hao-!nas) are attached to !nara melon patches (*Acanthosicyos horridus*) in the Kuiseb delta of the desert biome, for example (Dentlinger 1977; Budack 1983). Similarly, in the Kalahari savanna, mongongo nut (*Schinziophyton rautanenii*) groves are associated with the !Kung San family unit who have lived nearby for a long time: others require permission to collect from the grove (Lee 1973). Amongst farming communities throughout southern Africa, private rights are also accorded to marula (*Sclerocarya birrea*) and other wild fruit trees in cleared fields or near to homesteads, whereas anyone can collect fruits from uncleared woodlands. Private rights are also given to individual palm-wine tappers in *Hyphaene coriacea* savanna on the sandy coastal plain of southeastern Africa (Cunningham 1990b).

All four of the above-mentioned species are key resources, not only for their flavour but also owing to their year-round availability. The kernels of !nara melons, marula and mongongo fruits are all edible and can be stored as a year-round food source. *Hyphaene* palms are a year-round source of edible palm heart and sap for palm wine. It is no wonder, therefore, that the most favoured wild food sources are celebrated in song and ceremony, and frequently protected by customary tenure. Marula fruits or beer are a central feature of first fruit ceremonies, through much of southeastern Africa, with individual trees often selected as places where offerings are made to the ancestors. Councillors among the Topnaar peoples of the Namib desert are called 'fathers rich in !naras' (!naraaxa ||gun) for their role in settling boundary disputes about !nara 'fields' whose importance is recorded in song (Budack 1983):

!Gubu ǂūse	You round food
||khuxa |khase	with many thorns
ǂgui samese	you many-breasted
=Aoni-|gõan di	foster-mother of the
gai-gai aose	Topnaar children
!nuse ta ga hã	even if I am far away
xawe ta nǐ ǂeisi	I will think of you
ti ||naon ǂūse	you food of my ancestors
|urusi ta tite	I will never forget you
sas khemi ge deisi	there is no wet-nurse
khois a |khai	like you

20.4 Patterns of plant use across southern Africa

Archaeologist Martin Hall (1987) reviews the evidence for the two 'moving frontiers' of pastoralism and agriculture into a region occupied by hunter-gatherer communities, outlining how climate strongly influenced the spread of these new systems of food production in the region. Agriculture and agro-pastoralism developed in the summer-rainfall and relatively moist central, eastern and southeastern parts of southern Africa, whereas pastoralism spread into the dry west and southwestern parts where low or winter rainfall precluded crop production (Chap. 21, this volume).

Although distorted by rapid urbanization and the socially engineered crowding in the 'homelands' in South Africa, this basic pattern of high-density, sedentary populations along the moist east coast is still evident today (Chap. 21, this volume). Two major changes have been key factors in human impact on plants in the twentieth century. First, a rapid growth in human needs and numbers and second, rapid urbanization. In 1904, for example, the total population of South Africa, the most populous country in the region, was 5.1 million (Grobbelaar 1985). By 2000, the South African population will have increased over eight times to at least 44.7 million, out of a total population of 62.8 million in the region (South Africa, Lesotho, Swaziland, Zimbabwe and Namibia) (Bulatao et al. 1990).

20.4.1 African crops and introduced staples

The spread of subsistence and large-scale commercial agriculture has had a profound effect on the southern African landscape, yet most of the crops in the region today were introduced from Asia and Central or South America less than 400 years ago (Chap. 21, this volume). Finger millet (Eleusine coracana, Poaceae), the Livingstone potato (Plectranthus esculentus, Lamiaceae), sorghum (Sorghum bicolor, Poaceae) pearl millet Pennisetum glaucum, Poaceae and the melon Citrullus lanatus (Cucurbitaceae) were the indigenous crops grown by Iron Age agriculturalists. Each of these was selectively bred by African farmers from wild species into crop plants. Although the first two are seldom grown as staple foods today, Eleusine, Plectranthus, Pennisetum and Sorghum all retain special ritual significance. This ritual value is rarely attached to introduced staple crops from South or Central America (sweet potatoes, tomatoes, cassava, pineapples, chili peppers, peanuts, potatoes), south-east Asia (bananas, sugarcane) or the Mediterranean (cabbages), which are now important food crops on the subcontinent.

The acute observation of these early farmers in their selection of crop cultivars is shown in the folk taxonomy of seed types, soils and crop varieties today. Kwanyama Owambo farmers, for example, distinguish six Citrullus lanatus varieties (Fig. 20.1d) on the basis of fruit size, taste, skin or seed colour and whether seed margins are ridged or not (Rodin 1985). Van Oosterhout (1990), through her work with peasant farmers in Zimbabwe, provides insight into the selection processes and priorities of farmers that must have played a role in plant domestication. The most important selection factors were resistance to seed predation, good storage qualities, large seeds, large Sorghum heads, drought resistance and loss of the seed-shattering habit. Scudder (1962), also working in the Zambezi valley, documented the careful timing and selection of soil type when these seeds were sown in fields. Many of these practices illustrate the strategies that early farmers used in spreading the risk of a sedentary lifestyle where they placed faith in a few staple crops. Grivetti (1979) considered that the principal factor contributing to Tswana nutritional success at the peak of drought in the Kalahari was a diversified food base with an emphasis on wild food plants. Vegetation change and reduced biotic diversity have the effect of removing this buffer against rural poverty.

20.4.2 Wild plants for food

Starch-based staple diets are frequently deficient in nicotinic acid, vitamin C, calcium, riboflavin and protein. Bush foods are a valuable source of these nutrients, particularly nicotinic acid from wild spinaches (Hennessy & Lewis 1971; Lewis, Shanley & Hennessey 1971), vitamin

Figure 20.1 **Plants for people:** (*a*) *Landolphia petersiana*, a tasty fruit with potential as a new crop; (*b*) Beekeeping, an important activity in miombo woodland systems; (*c*) *Cyperus* rhizomes collected from wetlands of the Okavango Delta, Botswana – one of the many aromatic plants traditionally used for perfume; (*d*) *Citrullus lanatus*, ancestor of the watermelon and a crop plant developed in Africa; (*e*) Brewing marula (*Sclerocarya birrea*) beer, an important activity of nutritional, social and religious importance; (*f*) Fruits of *Catunaregam spinosa* and bark of *Warburgia salutaris*, widely used and sold for medicinal purposes; (*g*) A traditional medicine and a potential new pharmaceutical – the bulbous grassland plant *Hypoxis hemerocallidea*.

(Photos: A. B. Cunningham).

Figure 20.2. **The distribution of continental and coastal sands in southern Africa, a major determinant of the marginal environment, where people most value wild plants for housing materials, basketry and food, and to generate cash income. Redrawn, with permission, from Cooke (1964).**

C from wild fruits (Quin 1959; Wehmeyer 1966) and protein from *Sclerocarya birrea*, *Schinziophyton rautanenii* and *Tylosema esculentum* seeds and edible insects (Quin 1959). Bush foods are of great importance to the rural poor living in the vast area of southern Africa covered by nutrient-poor, drought-susceptible sands of the Mozambique coastal plain to the east, the Namib coast to the west and the Kalahari sands region in the centre (Fig. 20.2). Despite the change from a hunter-gatherer lifestyle to pastoralism or agriculture, these activities remain important to a high proportion of rural households in southern Africa (Ogle & Grivetti 1985; Cunningham 1988a; Wilson 1990, Campbell *et al.* 1991; Figs. 20.2 and 20.3).

Differences in climate, soil and vegetation type are reflected in significant differences in the availability and use of edible plants across southern Africa. Some bush foods are widely eaten, whereas other types of gathering characterize a particular biome. *Berchemia discolor*, *Vangueria infausta* and *Sclerocarya birrea* fruits (Fig. 20.1e), for example, are popular throughout the savanna of southern Africa. In contrast, gathering seed stores of grass and *Monsonia* collected by harvester ants (Fig. 20.4) is unique to the desert biome (Malan & Owen-Smith 1974; Steyn & Du Pisani 1984). O'Brien (1988), documents the decline in diversity of edible woody plants across the region from east to west (Fig. 20.5). Her data show that species richness of woody edible plants was lowest in the

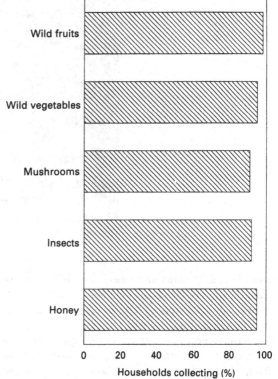

Figure 20.3 **The frequency of small-scale farming households collecting various wild foods in a Zimbabwean savanna (from a sample population of 359 households), after Campbell *et al.* (1991).**

Figure 20.4 **Winnowing grass seeds gathered from ant seed stores in Damaraland, Namibia, using a ≠goub, or winnowing bowl. After collection, the seeds are pounded, winnowed, ground and made into beer or porridge. This animal–plant-people interaction is characteristic of life in the desert biome along the arid western seaboard**
(Photo: A. B. Cunningham).

desert, succulent karoo and fynbos, and in the central Kalahari region of the savanna biome. The highest diversity of woody edible species is in the eastern sector of the Great Escarpment and eastern seaboard of the savanna biome. Differences in the extent of wild spinach use are also apparent in different lifestyles across southern Africa, with a greater diversity of species used by agricultural and agropastoral communities than by hunter-gatherers; disturbed habitats are created for these 'weedy' species at cattle posts or in fallow fields.

The diversity in use of underground plant parts and seeds (as opposed to fleshy fruits) across southern Africa shows the opposite trend to that of woody edible plants. A low number of edible species with root, tuber, bulb or corms are gathered on the coastal plain of the moist east coast (Cunningham 1985), but there is a high level of use of these plants in the Kalahari savanna region. O'Brien's analysis was based on the distribution maps for 264 woody edible species from Coates Palgrave's (1977) book *Trees of Southern Africa*. This analysis should therefore not

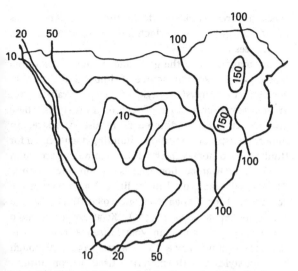

Figure 20.5 **An isometric map showing the east–west decline in diversity of woody plant species (trees) bearing edible wild fruits across southern Africa. Redrawn, with permission, from O'Brien (1988).**

Table 20.1 **Relishes (side-dishes) used to supplement starch-based staple foods are more often supplied by the gathering of wild plants (primarily the edible leaves of *Ceratotheca sesamoides*, *Cleome gynandra*, *Corchorus olitorius*, *Commelina diffusa* and *Triplochiton zambesiacus* than they are from agriculture, animal husbandry, hunting and fishing combined. Data are shown for three different sites in the middle Zambezi valley (at Miyaka, Chilola and Ntuni) during the period 1956–1957 (Scudder 1962)**

	Number of times relish provided		
Relish source	Miyaka	Chilola	Ntuni
Gathering of wild plants	547	48	47
Agriculture	106	75	33
Animal husbandry	35	31	1
Hunting	12	4	2
Fishing	12	0	0
Totals	712	158	83

be seen in isolation, as the dietary importance of non-woody edible species also needs to be considered. Underground plant parts provide an important source of food in those biomes (such as the succulent karoo and Kalahari savanna) which O'Brien (1988) recorded as having a low diversity of woody edible species. The Asclepiadaceae (*Brachystelma*, *Ceropegia*, *Duvalia*, *Fockea*, *Orbeopsis*, *Raphionacme* and *Stapelia*), Cucurbitaceae (*Acanthosicyos*, *Coccinia*, *Corallocarpus*, *Cucumis*, *Momordica*, *Trochomeria*) and Iridaceae (*Babiana*, *Lapeirousia*) are particularly important in this regard (Story 1958; Archer 1990; W. Geiss & J.W. Snyman, unpubl. data). In the Kalahari sand savanna of the Kavango area, for example, O'Brien's (1988) analysis records 31 woody edible species. In contrast, W. Geiss & J.W. Snyman (unpubl. data) record 101 edible plant species in the same area. Use of underground parts of 43 species in 15 families used by !Kung San people in the northwestern Kalahari is of particular botanical significance, as the 'underground forests of Africa' described by White (1976) form part of this food resource. The Kalahari savanna is a centre of diversity worldwide for plants with geoxylic suffrutices, large underground woody structures, which White (1976) records as having evolved independently in 31 families. Several of these 'underground trees' are a source of fruits (*Diospyros*, *Eugenia*, *Landolphia*, *Lannea*, *Parinari*, *Salacia*). To a lesser extent this also applies to the Mozambique coastal plain (White 1976).

Quantitative studies of bush foods (*veldkos*) have been carried out at two main levels. Firstly, the nutritional content of more than 300 bush food species was analysed over a period of 20 years by A.S. Wehmeyer and col-

leagues (Wehmeyer 1966, 1986; Van der Merwe, Burger & Wehmeyer 1967; Wehmeyer, Lee & Whiting 1969). In isolation, data on nutrient composition give limited insight into the dietary importance of bush foods. Therefore, a second level of study, the quantitative analyses of the amounts of foods collected and eaten, has been extremely useful. Two seminal studies quantified the dietary importance of bush foods when most ethnobotanical work was at a descriptive stage: Quin's (1959) work with Pedi people in the Northern Province of South Africa, and Scudder's (1962) work on the ecology of Gwembe Tonga people in the Zambezi valley. Quin's (1959) study recorded the cultural importance as well as the identity and nutrient composition of edible insects, the cultivated and wild foods plants in the diet of Pedi people and also the quantities consumed in a meal ration. Thus, through the meticulous records of recipes and quantities of foods that comprise the meal ration, nutrient composition could be interpreted in context. Quin (1959) and Scudder (1962) also obtained records of the frequency with which different foods, including bush foods, were consumed. Both studies illustrate the great importance of wild spinaches in the diet of rural people (Table 20.1). Popular wild spinaches with wide distributions in southern Africa are *Amaranthus hybridus*, and *A. spinosus* (Amaranthaceae), *Pentarrhinum insipidum* (Asclepiadaceae), *Cleome gynandra* (Capparaceae), *Corchorus tridens* and *C. trilocularis*) (Tiliaceae) and the introduced *Bidens pilosa* (Asteraceae).

In the savanna of the Northern Province of South Africa, Quin (1959) found that wild spinach featured in 31% (757) of recorded meal-days, primarily as a side dish to maize porridge. Forty years later, the Pedi proverb '*nama e a etela, morogo ke wa ka mehla*' ('meat is a visitor whereas *morogo* (spinaches) are a daily food') is as valid as ever to millions of people in southern Africa. Wild spinaches were similarly reported as the main side-dish to porridge in 39% of the 133 meals surveyed in Swaziland (Ogle & Grivetti 1985). A similar situation applies on the Maputaland coastal plain (Cunningham 1988a) and in *Colophospermum mopane* savanna in northern Namibia, where members of all homesteads gathered *Amaranthus* spp; *Cleome gynandra* and *Sesuvium sesuvioides* (Aizoaceae) as potherbs. Ironically, most of the plant species providing this nutritionally important food resource are considered to be useless weeds by commercial farmers. Commercial farmers also tend to fell all trees in fields. To subsistence farmers, however, trees bearing edible fruit form a crucial part of the food production system. Many other wild fruit trees are also valued as food sources, and conservation of favoured fruit-, fodder- or shade-producing trees has been an important factor in maintaining woody plant cover in agricultural lands of communal areas in southern Africa (Cunningham 1985; Campbell 1986; Wilson 1990).

20.4.3 Plants for fencing, housing and fuel

The influence of climate on lifestyle is also reflected in the diversity of architectural styles and settlement patterns in southern Africa. In the desert biome, homes of ≠Aonin (Topnaar) and Dama Nama communities are located along major river valleys such as the Kuiseb and Ugab ephemeral river courses. Houses are still made of slabs of *Faidherbia albida* bark over a wooden framework, but corrugated iron sheets are increasingly replacing stone or bark as construction materials. Further south in the succulent karoo, where rainfall is higher and sedges are more readily available along river valleys, uniquely portable *matjieshuise* comprising woven mats of sedges (*Scirpus inanis* and *S. dioecus*) are built over a framework of curved branches, usually cut from *Ziziphus mucronata* (Archer 1989, 1990).

'Mobile' *matjieshuise* are similar to Khoikhoi pastoralist houses encountered in the fynbos and karoo biomes by colonial explorers such as William Burchell. This architectural style has shown little change over 2000 years. Today, however, though their form and framework are often the same, *Scirpus* sedge mats are being replaced by plastic, canvas or sacking. The need for mobility in the seasonal movement in search of good pasture was one reason for the low volume of timber material used in *matjieshuise*. Low levels of wood use were also a feature of traditional architecture of the grassland biome. Possibly because of the problems posed by limited fuel or construction timber, the grassland biome was not settled at all by pioneer agriculturalists of the first millennium and it was only in the second millennium when these farmers had built up large herds of livestock that they moved into this biome, constructing houses of stone in villages of up to 1500 people (Hall 1987). As an alternative to stone, both Zulu and Swazi people developed a

specialist architectural style in these grassland areas (Knuffel 1973; Fig. 20.6) which minimizes wood use and maximizes use of grass.

In comparison to the grassland biome, the savanna biome was an excellent source of timber and thatch. In addition, the settled lifestyle of agropastoralists gave greater scope for a range of architectural styles. These also reflect cultural differences in use of space, an important aspect of traditional building documented for Himba-Herero homes in the desert biome by Jacobsohn (1988) and Owambo homesteads in mopane savanna (Mills 1984). Plants used in traditional architecture are documented for Tsonga (Liengme 1983), Tembe-Thonga (Cunningham & Gwala 1986), KwaNyama Owambo (Rodin 1985), Tswana (Van Voorthuisen & Odell 1976) and Cape Nguni (Shaw & Van Warmelo 1972). Although building styles and the materials used mirror cultural diversity, they also reflect variation in vegetation type. For example, vegetation patterns across the Maputaland coastal plain in KwaZulu-Natal are reflected in building style: in the sand forest zone (Chap. 12, this volume), 56% of homes were constructed with woven hardwood laths (2000–3000 laths per home), whereas *Phragmites australis* reeds were used for wall construction in 59–90% of homes in the palm savanna, coastal lake and coastal grassland zones, where only 5–35% had lath-woven walls (Cunningham 1985).

Fuelwood, fencing and building materials account for the highest volume of plant material used annually (Liengme 1983; Gandar 1983; Fig. 20.7). Fuelwood gathered from forest, woodland and plantations accounts for 51% of the domestic energy use in South Africa (Basson 1987) and is the biggest consumption of plant biomass, with rates varying between 0.27 and 1.12 tons per capita yr^{-1} (Gandar 1983). Wood consumption for building varies considerably with traditional building style and the availability of materials. The most spectacular of these is the traditional Owambo building style, which requires more wood than any other form of traditional construction in southern or central Africa (Fig. 20.8). A recent study of wood use in a homestead in northern Namibia showed that a single palisade fence, 302 m long surrounding the main homestead, was made of 7700 poles. The entire homestead required the removal of more than 100 m^3 of construction wood from surrounding woodland. Most of the wood (43 m^3) was used for palisade fencing, primarily from *Colophospermum mopane* and *Combretum* trees (Erkiila & Siiskonen 1992). By comparison, the mean timber volume of Tsonga huts in mopane savanna in the Northern Province of South Africa was 1.22 m^3 for round huts and 1.86 m^3 for square huts (Liengme 1983). Erkiila & Siiskonen (1992) calculated that 15 m^3 per household yr^{-1} of wood is used in

Figure 20.6 **A traditional Ngwane home, showing the specialized grass architecture developed as a result of wood scarcity in the grassland biome.** *(Photo: A. B. Cunningham).*

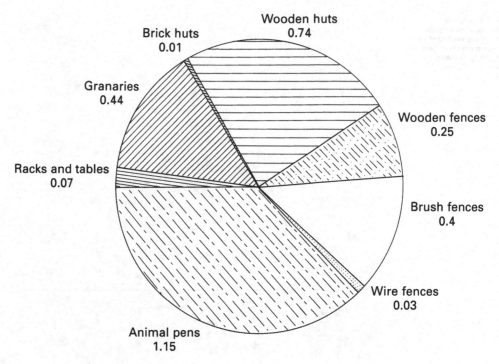

Figure 20.7 **Volumes of wood (m³ per household yr⁻¹) used for construction purposes by households in the woodland savanna of the Mutanda Resettlement Area in Zimbabwe. After Campbell *et al.* 1991.**

constructing an Owambo homestead, which is much smaller than that shown in Fig.20.8. This is five times more wood than the 3.02 m³ per household yr⁻¹ used for construction purposes in woodland in Zimbabwe (Grundy *et al.* 1991).

20.4.4 Plants in domestic use

Wild plants continue to provide a wide range of items for home use in many rural communities in southern Africa, ranging from drying racks for pots and pans, axe- and hoe-handles and grain-stamping mortars to general-purpose rope. In Zimbabwe, for example, woodland trees were a source of 96% of domestic utensils, 98% of agricultural tools and 94% of rope and cord (Campbell *et al.* 1991). Just as wood-carving skills are best developed in woodland savanna, so basketry skills are most highly developed among agricultural communities in the large portion of southern Africa covered by sandy soils (Fig. 20.1), where pottery clay is scarce and weaving fibre abundant from sedges and palms growing in high water-table sites. Although over 100 indigenous plant species are used as traditional dyes and fibres for basketry in southern Africa (Cunningham 1987; Shaw 1992), three plant genera (six species) form the major resource base of most basket production in the region. These are: the long and strong palm leaf blades of *Hyphaene petersiana*

(Arecaceae) (alkaline, clay-rich sands in Namibia, Botswana and western Zimbabwe) and *H. coriacea* (acid, high water-table sands of the eastern seaboard) as the main fibre for weaving baskets; the bark and roots of *Berchemia discolor* (Rhamnaceae), *Euclea divinorum* and *E. natalensis* (Ebenaceae), which are a favoured dye source, owing to their strong colour and colourfast nature.

As industrially produced goods become easily available at lower cost and as lifestyles change, so plant use for household items such as baskets and grinding mortars also changes. This is particularly evident as baskets give way to plastic and enamel bowls. Furthermore, baskets do not last long in normal household use, leaving little evidence of changing basket styles. For this reason it is fortunate that detailed studies and well documented, representative collections have been made of domestic use of wood (Hooper 1981), basketry (Shaw 1992) and material culture in general (Shaw & Van Warmelo 1972, 1988; Davison 1984). In a professional career spanning over 50 years, ethnologist Margaret Shaw has made an outstanding contribution in this field, ranging from publications on Owambo knives (Shaw 1938) to her detailed study of basketry techniques, as well as a collection of over 2500 specimens in the South African Museum, Cape Town (Shaw 1992).

Although basket production for local use is in

Figure 20.8 **A photograph taken in 1937 of complex palisade fences in a traditional homestead at Omedi, in Owambo, northern Namibia, representing massive consumption of** *Colophospermum, Combretum* **and** *Terminalia* **timber. Note that edible fruit-bearing trees (***Diospyros mespiliformis*** and** *Sclerocarya birrea***) in the background have been conserved. Since that time, changes from a polygamous to a monogamous lifestyle, coupled with wood scarcity, have resulted in smaller homesteads being built.** *(Photo: Duggan-Cronin print DC 4769 used with permission of the McGregor Museum, Kimberley.)*

decline, it has been promoted as a rural development option for rural people in several drought-prone, semi-arid areas of Africa, including Botswana, Namibia, South Africa and Zimbabwe. The largest volume of baskets produced for export are made in Botswana, Zimbabwe and South Africa, with the use of the same wild plant species or genera described above. Similarly, basketmakers are generally from subsistence farming communities at the interface between savanna woodlands and the seasonally flooded or permanent wetlands. The number of people involved in basketry sales from western Zimbabwe is a good example of increased commercial demand. In 1984, 20 women were selling baskets in the Binga area. By 1989, 200 women were involved, earning US$17 000 or an average of US$85 yr^{-1}. One result has been an increase in frequency and intensity of harvesting, sometimes with negative consequences for the natural resource base of the basketry industry.

20.4.5 Plants for medicine

Western and traditional African medicine are based on very different views of health and disease. Traditional medicine takes a broad approach, where disease or misfortune results from an imbalance between the individual and the social environment (Berglund 1976; Ngubane 1977), whereas western biomedicine takes a narrower technical and analytical approach. These different approaches to the cause of disease are one of the reasons that demand for traditional medicine continues in the urban environment even when western biomedicine is available. Another reason is that western medical care is limited in remoter rural areas of southern Africa. In

Table 20.2 Four symbolic mixes (*ubulawu, imbiza, ikhubalo* and *intelezi*) and 20 of the 400 traditional medicinal plant species sold in the largest volume in KwaZulu-Natal, showing quantity sold and the main vegetation types where they are collected. Volumes are those bought by 54 herb traders yr⁻¹ (in 50-kg maize bags) and probably reflect 25% of the quantity sold by the 100 herb traders in KwaZulu-Natal (after Cunningham 1989)

Species sold	Zulu name	Part used	Medicinal uses	Quantity (bags yr⁻¹)	Coastal forest	Afro-montane forest	Savanna	Grass-land
Mix, high in saponins[a]	Ubulawu		Love charm emetic, dream communication with ancestors	1966	•	•	•	•
Mix, mainly succulent spp.[b]	Intelezi		Protective charm	1942	•	•		•
Mix, pink/red barked trees[c]	Ikhubalo		Ward off disease; cleanse after death of relative or associate	1883	•	•	•	
Mix with *Scilla, Eucomis*[d]	Imbiza		Enema to cleanse blood	1211	•			•
Scilla natalensis	Inguduza	Bulb	Enema (imbiza)	774				•
Eucomis autumnalis	Umathunga	Bulb	Enema (imbiza)	581				•
Alepidea amatymbica	Ikhathazo	Root	Coughs and colds	519				•
Adenia gummifera	Imfulwa	Stem	To rebound evil spells (intelezi)	459	•			
Albizia adianthifolia	Usolo	Bark	Love charm emetic	424			•	
Clivia species[e]	Umayime	Bulb	Used in intelezi	397	•			
Pentanisia prunelloides	Icimamlilo	Tuber	Respiratory aid	343				•
Senecio serratuloides	Insukumbili	Stem/leaves	Purify blood	340				•
Gunnera perpensa	Ughobo	Rhizome	Eases childbirth	340				•
Rapanea melanophloeos	Umaphipha-ikhubalo	Bark	Ritual cleansing (ikhubalo)	327		•		
Dioscorea sylvatica	Ingwevu	Tuber	Chest complaints	326		•	•	
Warburgia salutaris	Isibhaha	Bark	Coughs; colds; upset stomach	315	•	•	•	
Bersama spp.[f]	Undiyaza	Bark	Menstrual complaints; to confuse opponents in battle or court	295	•	•		
Unidentified species	Ubhadlangu	Bark		288				
Kalanchoe crenata	Umahogwe	Stem/leaves	Enema; love charm	284			•	
Bowiea volubilis	Igibisila	Bulb	Protective charm (intelezi)	257				•
Trichilia emetica (and *T. dregeana*)	Umkuhlu	Bark	Enema	252	•			
Turbina oblongata	Uboqo	Tuber	Protection against lightning (intelezi); love charm (ubulawu)	249				•
Rhoicissus tridentata	Isinwasi	Tuber	Assist in childbirth	244	•		•	
Bulbine latifolia	Ibhucu	Bulb	Protect home against misfortune	240				•
Ocotea bullata	Unukane	Bark	To make competitors unpopular (bad smell)	234		•		
Stangeria eriopus	Imfingo	Stem/tuber	Protective charm (intelezi)	233				
Cryptocarya species[g]	Umkhondweni	Bark	Substitute for *Ocotea*	228	•	•		
Anemone fanninii	Amanzemnyama	Rhizome	Causes others to be disliked	227				

[a] *Maesa lanceolata, Hippobromus pauciflorus, Helinus integrifolius, Secamone* species, *Bulbine latifolia, Synaptolepis kirkii.*

[b] *Adenia gummifera, Foeniculum vulgare, Berkheya* species, *Clivia miniata/C. nobilis, Drimia* and *Urginea* species, *Dioscorea sylvatica, Dioscorea dregeana, Eriospermum cooperi, Cephalaria natalensis, Kalanchoe* species, *Stangeria eriopus.*

[c] *Harpephyllum caffrum, Ekebergia capensis, Rapanea melanophloeos, Protorhus longifolia, Sclerocarya birrea, Curtisia dentata, Cassine papillosa, Pterocelastrus* species.

[d] *Scilla natalensis, Eucomis autumnalis, Gnidia kraussiana, Euclea divinorum, E. natalensis, Combretum kraussii, Zanthoxylum davyi, Sideroxylon inerme.*

[e] *Clivia miniata* and *C. nobilis.*

[f] *Bersama lucens, B. tysoniana, B. stayneri* and *B. swinnyi.*

[g] *Cryptocarya latifolia* and *C. myrtifolia.*

Swaziland, for example, the ratio of medical doctor : total population was 1 : 10 000 compared to a traditional healer : total population ratio of 1 : 100 (Green 1985). The overall medical doctor : total population ratio in 'homeland' areas of South Africa was 1 : 17 400 in 1976, but with ratios as low as 1 : 119 000 people in QwaQwa and 1 : 30 000 people in Lebowa in 1982 (Savage 1985). In contrast, traditional practitioner : total population ratios are high. In Venda, Arnold & Gulumian (1984) estimated a traditional practitioner : total population ratio of between 1 : 720 and 1 : 1200 people. In Zimbabwe, the urban demand for traditional remedies is reflected in a higher traditional healer : total population ratio (1 : 234) than in rural areas (1 : 956) (Gelfand *et al.* 1985). Holdstock (1978) estimated that 80–85% of people in Soweto consult traditional practitioners. In Umlazi, one of the largest 'townships' in the Durban area, 30% of a random sample of residents had used the highly toxic medicinal plant *Callilepis laureola* (impila) (Wainwright, Schonland & Candy 1977), which is not even the most popular species sold in KwaZulu-Natal (Table 20.2). Use of medicinal plants is even higher in rural areas in KwaZulu-Natal, as all respondents interviewed in a random sample in the Estcourt area acknowledged using medicinal plants (Ellis 1986).

Well-trained traditional healers and traditional mid-wives provide an important health care service whose value has been underestimated (Holdstock 1979; Buhrm-ann 1984; Anderson & Staugard 1986). Although species differ, herbal remedies are widely used in southern Africa. Use also reflects distinct cultural preferences. In the fynbos biome, Ferreira (1987) found that 88% of elderly Coloured people used *boererate* ('home remedies'), primarily *bossiesmiddels* ('bush drugs') from aromatic fynbos plants such as *Salvia africana-caerulea* (Lamiaceae) or *Pelargonium antidysentericum* tubers. In contrast, most migrants to the fynbos today are Cape Nguni people from the summer-rainfall region. This is reflected in the fact that most species in the herbal remedy trade in Cape Town are derived from the summer-rainfall region or are widespread trees such as assegaai (*Curtisia dentata*, Cornaceae) and boekenhout (*Rapanea melanophloeos*, Myrsinaceae) growing in Afromontane forest remnants on the Cape Peninsula. The high volumes traded also reflect the value placed on traditional medicines. Trade

in traditional medicines is a multi-million rand 'hidden economy' in southern Africa, where a high level of urbanization generates high demand, particularly in mining towns or large urban centres (Fig. 20.9). Over 400 indigenous plant species and 20 alien species are commercially sold for this purpose in KwaZulu-Natal (Cunningham 1991) out of a total of about 1030 species used in this province (Hutchings *et al.* 1996). This trade poses a significant threat to many plant species. One example is the collection from the wild of more than 3000 *Stangeria eriopus* cycads per month, sold as an *intelezi* in traditional medicine markets in Durban alone, posing 'an enigma of the South African situation to which it is difficult to find a solution' (Osborne *et al.* 1994).

20.4.6 Plants as symbols

Southern African plants are widely used as symbols both nationally and internationally. The king protea, *Protea cynaroides*, for example, is South Africa's national flower, and the symbol of Los Angeles is the eastern Cape endemic, the crane flower (*Strelitzia reginae*). Other spec-

Figure 20.9 Distribution of traditional medicines in southern Africa determined from daily sales records of a major traditional medicine wholesaler based in KwaZulu-Natal. Although these data show sales of animal parts used in traditional medicine, herbal medicines would show the same pattern. Large urban areas and mining towns are major centres of demand. CT, Cape Town; D, Durban; EL, East London; F, Francistown; G, Gaberone; JHB, Johannesburg; PE, Port Elizabeth; W, Welkom; WH, Windhoek; V, Vryheid (from Cunningham 1993b). 1 Rand = 0.275 US$ (August 1995).

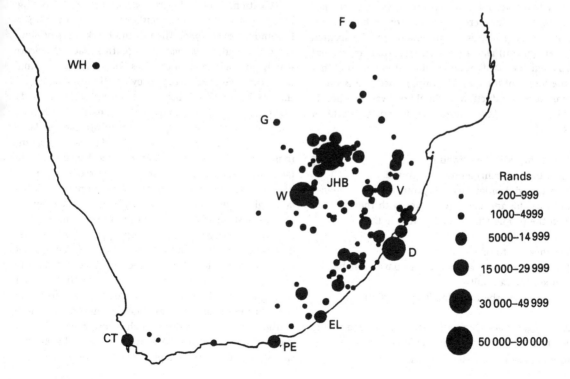

Rands

•	200–999
•	1000–4999
●	5000–14 999
●	15 000–29 999
●	30 000–49 999
●	50 000–90 000

ies have a much deeper symbolic, religious or magical meaning to the majority of southern Africa's people. The belief that nothing happens by chance, but is subject to influence by other living people or ancestral spirits, is central to this aspect of plant use (Berglund 1976; Hammond-Tooke 1989). For perfectly logical reasons, it is important to determine the cause of a particular illness or misfortune. Determining the root cause of such problems and guarding against them is the role of the diviner (*izangoma* (Zulu); *iqhira* (Xhosa), *mungome* (Venda, Tsonga)) rather than the herbalist (*inyanga* (Zulu, Tsonga, Nbebele), *ixhwele* (Xhosa), *nganga* (Venda, Shona)). In the past and today, plants with symbolic value are used by diviners and shamans to control events by supernatural means. Lewis-Williams & Dowson (1989) describe their importance to hunter-gatherer people: 'There are several areas of life in which the Bushmen (San) feel they cannot cope by themselves: they must reach out for supernatural power. Curing the sick, going on out-of-body journeys, and controlling the movements of animal herds – all tasks of shamans – demand this extra power.' Animals, because they are visibly powerful and at hand are seen as a source of the desired power. In Biesele's words, animals are 'metaphors with the strength to bridge worlds ...'.

Many plant species also symbolize this bridge to supernatural power described above, and their species-specific use, frequently reflected in their names, is often widespread. To farmers and pastoralists it is important not only to attract rain, but also to drive away storms, which could destroy crops, or to protect homesteads against lightning strikes. To the /Xam-San, the scented *Agathosma* (buchu) leaves were an important ingredient of this medicine, used to soothe the mythic rain-bull that was the source of the rain. The importance of this supernatural 'food production technology' was eloquently described by the /Xam hunter Dia!kwain to Wilhelm Bleek in 1874:

Where they kill the rainbull,
There the rain runs along the ground.
Then the wild onion* leaves sprout for the people,
And they dig and feed themselves with them.
Then the people who are at home see the
    rainclouds
And they say to one another,
'The medicine men really seem to have their hands
    upon the rainbull,
For you to see the rain clouds come gliding'.

Although *Agathosma* species are still widespread in the fynbos biome and rain-making ceremonies feature in

* Probably edible bulbs of *Cyanella hyacinthoides*

cave paintings, the /Xam people are gone. However, the intertwined symbolic, religious and magical uses of plants continue to be topical today. *Agathosma betulina* leaves are still sold by herbalists in Cape Town's parade market. In the same way that buchu was used by /Xam shamans over millennia in the fynbos biome, so the aromatic *Helichrysum odoratissimum* is used to draw the attention of the ancestral spirits in the Eastern Cape and Kwa-Zulu-Natal. Known as *imphepho* in Zulu (a name also used for incense), plaited stems of *H. odoratissimum* are the single biggest selling medicinal plant by volume in urban herbalist shops in South Africa today.

The cultural context of this use is very important. Diviners, who communicate with ancestral spirits, play a crucial role in society. This has been well described in detailed anthropological work (Berglund 1976; Ngubane 1977; Hammond-Tooke 1981) and will not be dealt with here. Misunderstanding of divining, and legislation against it, has not diminished the role of diviners in rural and urban society, particularly in time of conflict, social upheaval or uncertainty. Plants are used symbolically in love or war, social and physical health and success or failure, whether by hunter-gatherers, pastoralists, farmers or in business and political affairs.

Characteristics such as colour, aromatic scent or physical traits such as milky latex, touch-sensitive leaves, fruit, flower or root shape are all important symbolic characteristics of plants. The important symbolic triad of the colours red (intensification, consolidation), white (purification) and black (strengthening, cooling, protecting) applies across southern Africa (Turner 1967; Hammond-Tooke 1989). The red and black seeds of *Abrus precatorius* and *Afzelia quanzensis* (both Fabaceae) feature widely in traditional medicines for this reason. Similarly, the tree species that provide the ingredients for the *ikhubalo* mix all have pink or red colour, as do many of the plant parts used in battle medicines. The remarkable white lignotubers of *Synaptolepis kirkii* (Thymelaeaceae) are also powerful, and are an important ingredient of a special *ubulawu* mix used by diviners to facilitate interpretation of dreams. The red-brown roots of the coastal forest shrub *Acridocarpus natalitius* (Malpighiaceae), known as *umabophe*, are used throughout the eastern seaboard of southern Africa to improve crop yields (Krige 1937), to stop approaching danger (Gerstner 1941) or to stop evidence being given against one in a court of law. The resurrection plant *Myrothamnus flabellifolius* (Myrothamnaceae) is commonly used throughout southern and central Africa, from arid west coast to the moist east coast, both as an aromatic herb and for its symbolic value, which is also reflected in its vernacular names. Milky latex from stems of the climber *Sarcostemma viminale* (Asclepiadaceae) is used throughout

Botswana, Zimbabwe and South Africa to promote lactation in women and cattle (Gelfand *et al.* 1985; Hutchings 1989; Cunningham & Zondi 1991).

There are great psychological and physical pressures associated with the crowded and competitive environment of townships in southern African cities. Where does one find a job – and keep it when unemployment is so high? How does one survive dangerous mining work? How does one maintain a relationship with one's family when working away from home as a migrant labourer? The protective and cleansing function of medicinal plants is one way of dealing with the conflict-ridden and competitive urbanized environment, where individuals become 'polluted' through proximity to undesirable people, discarded medicinal charms or the activity of witches (Berglund 1976; Ngubane 1977). It is for this reason that the four major plant preparations are sold in such high volume by urban herbal traders in KwaZulu-Natal (Table 20.2). Three of these mixes (*ubulawu, imbiza, ikhubalo*) are used to rid the body of 'pollution'. The fourth, the *intelezi* mixture, protects the body against this pollution after it has been cleansed, or the household against misfortune. Again, a different pattern of plant use between moist eastern and drier western parts of southern Africa is apparent. On the eastern seaboard, bodily pollution is traditionally conceptualized as dirt (*umlaza*) by Zulu and Venda people, and as heat (*fisa*) by Sotho people living on the central plateau parts of the grassland biome. This malady is counteracted symbolically in different ways by the two groups, probably owing to the lower rainfall of the central plateau (Hammond-Tooke 1989).

Intelezi mixtures are also used in warfare. The advice and protective-battle medicines provided by diviners (*izangoma*) to participants in violent political conflicts are an important form of plant use in South Africa, where 13 000 people have died in the past six years. Tragically, this knowledge of traditional plant use has led to the death of *izangoma* caught between warring groups trying to undermine the 'supernatural power-base' of their opposition (Rickard & Mthembu 1992). Diviners also played a crucial role in the Zimbabwean war of independence (Lan 1985). The ingredients of battle medicines used in the KwaZulu-Natal conflict in 1994 are unchanged from those used in the Anglo-Zulu war of 1879. They are equally common ingredients in magical medicines used by soccer players in urban South Africa, in urban court cases, in love charms or to boost agricultural production. The magical uses of these plants are alluded to in their names. A few examples from some of the most commonly used species are *igibisila* ('to defeat bad luck', *Boweia volubilis*), *impikayiboni* ('the army doesn't see', *Cephalaria natalensis*), *uboqo* ('to give in', *Turbina*

oblongata) and *undiyaza* or *ndiyandiya*, both names for *Bersama* spp. that refer to the perplexed or confused effect this will have on opponents. It is precisely this symbolism that describes anthropologist Victor Turner's (1967) classic study of Ndembu divination in Zambia: 'The forest of symbols'.

20.5 Back to the future: the neglected harvest

20.5.1 The colonial interest in plant resources

During the nineteenth and early twentieth centuries, British colonial botanists and agriculturalists screened the southern African flora for its potential for commercial exploitation. Indigenous knowledge and observed patterns of use helped to identify potentially important plants, and likely candidates were sent to the Imperial Institute in London for testing. Such material included: the fibre of *Hibiscus cannabinus* from Zimbabwe; the roots of *Mondia whitei*; a traditional Zulu medicine proposed as a flavouring for soft drinks; wax from fruits of the coastal fynbos shrub *Myrica cordifolia* for polish; gum from *Acacia karroo* for gum arabic; and latex as a source of rubber from *Landolphia kirkii*, as well as from *Ficus verruculosa, Maytenus acuminata, Tabernaemontana elegans* and *Voacanga thouarsii*. Some species exported during this period continue to be exported today. Two examples are the almost 400 tonnes of bitter aloe (*Aloe ferox*) resin exported annually to Europe in 1929–1932 and the 71 tonnes of buchu (*Agathosma betulina*) exported to Japan for medicinal purposes as early as 1930 (Brown & Brown 1935). Other business schemes were short-lived. The Tongaland Rubber Corporation Ltd, Amatongaland Rubber Corporation and Pongola Rubber Estates Ltd were all formed by 1906 to exploit *Landolphia kirkii* latex in the Ingwavuma district in KwaZulu-Natal. Between 1894 and 1908, about 735 tonnes of *L. kirkii* latex valued at that time at £23 000 yr^{-1} were exported from Mozambique (Cunningham 1985). In the years that followed, the apparently good prospects offered by latex from *Euphorbia tirucalli* led to the formation of large syndicates (Sim 1920). Latex was also exported from Namaqualand and the Eastern Cape during this period for the manufacture of American chewing gum (Brown & Brown 1935). By 1916, the *L. kirkii* and *E. tirucalli* companies in Natal had gone into voluntary liquidation or compulsory closure. The Pongolo Rubber Estates Ltd, formed to harvest *Landolphia* latex, was floated with £125 000 of shareholders money – but its entire rubber output was only worth £101! (Von Wissel undated).

More recently the formal sector has established mar-

Table 20.3 **The type and source of some commercially harvested plant products in southern Africa. The best available data describing their economic value, and the quantities harvested for local and foreign markets is included. It should be noted that the figures cited are derived from a wide variety of sources, and that each should be used only in the context of the original reference. The purpose here is to give an approximate and comparative economic overview of indigenous plant resources in the region**

Biome	Product	Species (family)	Value (US$ millions)	Quantity sold (tonnes yr⁻¹)	Reference
Fynbos	Wildflower industry	*Leucospermum* spp., *Protea* spp. (Proteaceae); *Brunia* spp. and others (Bruniaceae)	13.8	1000	Department of Agriculture (unpubl. data)
	Buchu oil (diuretic and blackcurrent flavouring)	*Agathosma betulina* (Rutaceae)	1.5		E. Godfrey (pers. comm.)
	Herbal teas: honey-bush tea,	*Cyclospia* spp. (Fabaceae)	—	20	Welgemoed (1993)
	rooibos tea	*Aspalathus linearis* (Fabaceae)	2.0	4000	Anon. (1994)
	Fuelwood and charcoal	*Acacia cyclops*ᵃ (Fabaceae)	8.7 (fuelwood)	44 000	Azorin (1992)
		*A. saligna*ᵃ (Fabaceae)	1.2 (charcoal)	2400	
	Thatching reed	*Thamnochortus insignis* (Restionaceae)	2–4	—	Linder (1990)
Nama-karoo	Prickly pear (fruit and drought fodder)	*Opuntia ficus-indica*ᵃ (Cactaceae)	0.15		Brutsch & Zimmerman (1993)
Savanna	Bitter aloe (medicinal)	*Aloe ferox* (Asphodelaceae)	0.9	165–280 (dry matter)	Newton (1993)
	Amarula liqueur	*Sclerocarya birrea* (Anacardiaceae)	—	2000–3000	
	Harpago tea (medicinal and arthritis remedy)	*Harpagophytum procumbens, H. zeyheri* (Bignoniaceae)	0.1–0.7	90–190	Nott (1986) N. Marshall (pers. comm.)
	Vegetable fat, also for soap	*Trichilia dregeana, T. emetica* (Meliaceae)	—	—	
Desert	Butter-pits (almond flavouring)	*Acanthosicyos horridus* (Curcurbitaceae)	—		Dentlinger (1977)
Grassland	Medicinal (putative prostate remedy)	*Hypoxis* spp. (Hypoxidaceae)	—	—	
Forest	Florist greens	*Rumorha adianthiformis, Blechnum attenuatum, B. punctulatum* (Aspidiaceae), *Polystichum pungens* (Blechnaceae)	2.6	300 (= 14 million fronds)	Milton (1987); Milton & Moll (1988)

ᵃ Denotes invasive alien species (see Chap. 22, this volume)

kets both regionally and abroad, and ideally seeks to operate sustainably along conventional business lines, boosting profits by adding value to plant products. A typical example is that of the wildflower industry of South Africa. This industry utilizes, for the most part, the floristic diversity of the fynbos region as both cut and dried flowers. The European flower markets provide the outlet for 90% of production (Middelmann, Gibson & Bell 1989; Davis 1990). In 1993 this industry provided jobs for as many as 20 000 people, and produced material estimated to be worth US$14 million on export (Department of Agriculture, unpubl. data). In the absence of formal consolidated records, however, the total value of the industry, which includes dried flower production and a substantial local market, could be nearly twice as much (see Table 20.3). Many species utilized by the trade are taken directly from the veld (Chap. 6, this volume), usually from privately owned land. These include up to 300 species in about 30 families, of which the Proteaceae, Ericaceae, Asteraceae, Restionaceae and Bruniaceae are the most common (Davis 1990). The cutflower industry has demonstrated that strategic manage-

ment of these veld resources can help to avoid the dramatic 'boom and bust' syndrome experienced by the short-lived *Euphorbia* and *Landolphia* latex industry. The sustainability of fynbos flora is partly a product of fire-related life-histories resulting in relatively fast growth rates and short juvenile periods (Le Maitre & Midgley 1992; see also Chap. 6, this volume). The most valuable decorative fynbos plant products are species and cultivars of the family Proteaceae (primarily within the genera *Protea* and *Leucospermum*). The development of new cultivars has added value to the original stock, and cultivation has ensured a high degree of product conformity (Jacobs 1989). Formal agricultural research has assisted this process.

Another successful commercial venture based on the fynbos flora, is that of the 'rooibos' tea industry. Used by early farming communities in the Cederberg region of the Western Cape as a substitute for tea and coffee, leaf material of *Aspalathus linearis* (Fabaceae) was first marketed in 1902. It has now become an important product, and strategic marketing has gained it a place in the international health food market as caffeine-free tea. An estimated 4000 tonnes (Anon. 1994), worth approximately US$8 million, is harvested annually. About 20% of this is exported.

A third successful commercial resource from the fynbos biome is *Thamnochortus insignis*, a member of the Restionaceae. This reed-like plant is endemic to the coastal fynbos of the Western Cape, and is used as a roof-thatching material. Less desirable, but also used for roofing, are the shorter and less durable members of the family, *Thamnochortus erectus* and *Chondropetalum tectorum*. Unlike the grass thatches used in the summer-rainfall areas of southern Africa (such as *Cymbopogon validus*), the sclerophyllous culms of these species form a thatch able to withstand the persistent and penetrating winter rain of the mediterranean-climate region in the fynbos biome. This type of roofing epitomizes both the labourer cottages in the Western Cape and the grand colonial houses built during the Dutch occupation of the Cape. This use was probably styled on moveable shelters built by the nomadic pastoralists who occupied the area before European settlement. As a roofing material it is currently highly fashionable, in spite of its flammability, expense and limited lifespan in comparison to modern roofing tiles. In the region of its natural distribution it is regarded as a supplementary (though potentially lucrative) crop, and along with some harvesting of decorative wildflowers, boosts the income of livestock farmers in this marginal agricultural area (Linder 1990; Davis 1993). Unlike the wildflower and rooibos tea industries, thatch production is not yet formalized, and estimates of total production and crop value are not readily available, although initial estimates place the total annual value at upward of US$2 million (Linder 1990).

The savanna biome, home of the livestock industry, also has plants that are potentially important as commercial crops. The marula (*Sclerocarya birrea*), for instance, is valuable as a possible source of substitutes for products generally derived from other plants. The kernel of the seed is rich in protein and oil, the latter being ten times more stable than olive or sunflower oils (Burger, De Villiers & Du Plessis 1987). This makes it highly suitable as a replacement for oleic oil from safflowers (*Carthamus tinctorius*), which is imported for use in baby food formulas. Marula oil is suitable as a substitute for the oil used to coat dried fruit – as much as 250 tonnes of this oil is used annually in South Africa (Du Plessis 1988). Marula fruits are also used in the manufacture of an export-quality liqueur. Another indigenous plant-derived product, which can also substitute for current imports, is the gum tapped from *Acacia karroo*. This can be used in place of true gum arabic (from *Acacia senegal*). In 1987 a private company in Zimbabwe paid out about US$150 500 to local people in Matabeleland

for collecting gum from *A. karroo* (R. Barnes, pers. comm.) for use as an emulsifier in industry. This provides a useful substitute for gum arabic imported from Sudan, which currently sells at US$3250 tonne⁻¹, thereby providing an important saving on the unfavourable rate of foreign exchange against the Zimbabwean currency.

20.5.2   The hidden economy
Far less obvious than the sale of rooibos tea or marula liqueur is the 'informal sector' trade in plant products. The economic value of this trade is also far more difficult to assess, yet it is important at a household level (Fig. 20.10) and frequently on a regional scale. Trade or barter in plant products occurs in rural areas throughout Africa, where wild plant resources provide a wide range of products that are bartered or sold: dietary supplements, thatch, fuel, craftwork materials, honey, edible insects such as mopane 'worms'(caterpillars of the moth *Gonimbrasia belina*), traditional dyes, perfumes and medicines (Fig. 20.1). In addition to barter or trade at a homestead level, sales are made from roadside stalls, at cattle auctions, bus stops and taxi ranks. Naturalized species are also a source of lucrative trade. 'Dagga' (*Cannabis sativa*), originally from Asia, is an important

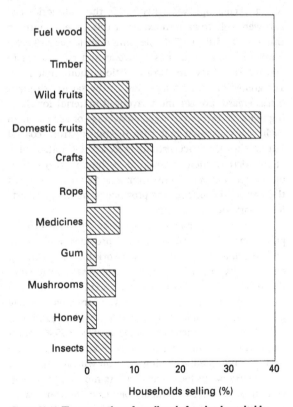

Figure 20.10 **The proportion of small-scale farming households selling woodland products in Zimbabwe (from Campbell *et al.* 1991).**

(but illegal) cash crop of small-scale farmers in the Eastern Cape and KwaZulu-Natal. Prickly pears (*Opuntia ficus-indica*), introduced from South America and particularly abundant in the savanna and Nama-karoo biomes in the Eastern Cape (Chap. 22, this volume), are traded informally and at urban fresh produce markets (Brutsch & Zimmerman 1993). Invasive Australian *Acacia* species (*A. cyclops* and *A. saligna*) form the basis of a charcoal industry valued at US$1.2 million and fuelwood sales of US$7.6 million yr^{-1} (Azorin 1992). Harvesting any of these plant products is usually labour intensive with low financial returns; however, it provides a source of income to probably hundreds of thousands of rural families in southern Africa. Informal trade networks are as complex as they are unobtrusive. Trade in crafts, palm wine and medicinal plants (see 20.4.5.) are good examples of these intricate marketing chains of sale and resale and of the value they can have in rural areas with high levels of unemployment.

In 1984, 14 years after commercial basket production first started in Botswana, about half of the female population of the villages Etsha (c. 1500 women) and Gomare/Tubu (400 women) in Ngamiland were making baskets commercially (Terry 1984, 1987). Preston-Whyte (1987) estimated that, in the KwaZulu-Natal region of South Africa, 7500 people were involved in the craftwork trade and were able to earn in excess of US$44 per month. In this area, traditional Zulu sleeping mats (*amacansi*) are a symbolic gift at weddings in urban and rural areas in South Africa. They are also a traditional household item. The golden brown, strong culms of the saltmarsh plant *Juncus kraussii* are the most favoured material for sleeping mats. From 1970 to 1990, an average of 2823 women each year harvested mat rush (*J. kraussii*) from coastal saltmarshes at St Lucia estuary in KwaZulu-Natal. In 1990 alone, 3831 women removed an estimated 50.9 tonnes of selected mat rush culms from a 20-ha area which was then resold throughout the province or made into crafts for export (Heinsohn 1991).

A study of the palm-wine trade in *Hyphaene coriacea* palm savanna in South Africa provides a detailed example of an informal trade network linking palm sap tappers with the sale, transport and resale away from the origin of the resource (Fig. 20.11). Although the volume of palm wine transported was large, with nearly 980 000 litres of undiluted palm-wine sold in a 12-month period, individual incomes were low (US$28–US$65 mo^{-1}) and required labour-intensive work. A single tapper constantly maintained a tapping rotation within a set area, tapping 712 palms (902 stems) in a 12-month period (Fig. 20.11a). Sales varied markedly between the four palm-wine sale points in the study area (Cunningham 1990a). They also fluctuated in response to factors ranging from competition with marula (*Sclerocarya birrea*) beer-brewing season (Fig. 20.11b) through to vehicle breakdowns. Despite these difficulties, the palm-wine trade provided a means of self-employment to 460–480 people in the study area and together with cattle-grazing, basket weaving and gathering of bush foods, was an important component in the multiple-use of palm savanna.

20.5.3 Developing new crops from wild plants

Selected southern African wild plants could follow the same path of commercial development as seen in species in other parts of the world. The African oil-palm (*Elaeis guineensis*), for instance, has been domesticated and bred for rapid maturation (Comte 1991), and now forms an important part of the Malaysian economy. The banana, too, has been adopted as an economically very important crop in many tropical countries, and the kiwi fruit, indigenous to China, is now synonymous with the fruit industry of New Zealand.

The only systematic survey of wild plant resources that we are aware of in southern Africa was carried out in Botswana. It identified several wild plant resources with commercial potential (Taylor 1982; Taylor & Moss 1982). Within the southern African region, however, there are several categories of plants that could generate income and employment, whether from wild harvest or through development as new crops. Some of these are:

- Edible wild plants and oilseeds. According to Wehmeyer (1986) there are about 1400 edible plant species in southern Africa. Nutrient analyses have only been carried out on 300 of these, and horticultural research on less than 2% of these. Many of these wild species grow in leached, nutrient-poor sands with marginal arable potential (Fig. 20.1).
- Aromatic plants as potential sources of essential oils. These occur particularly frequently in the fynbos, Nama-karoo, succulent karoo and desert biomes, and are members of the Asteraceae (e.g. *Pteronia*, 74 species; *Eriocephalus*, 26 species) and Rutaceae (*Agathosma*, 139 species; *Diosma*, 28 species; *Coleonema*, 8 species).
- Genetic resources. There are many wild relatives of crop, forage and pasture species that could be valuable as sources of genetic diversity in the future (e.g. melons (Curcurbitaceae) in the desert and semi-arid savanna of Namibia and Botswana).
- Horticultural plants. Exploitable species include several spectacular *Crinum* species (Amaryllidaceae) which grow on the floodplains of northern Namibia and Zimbabwe in the savanna biome; *Gazania* and *Gerbera* species

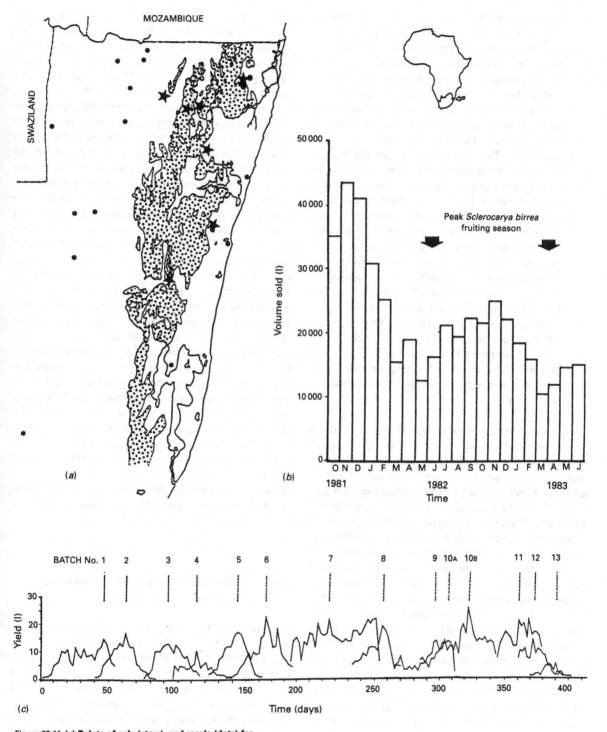

Figure 20.11 (a) **Points of sale (stars), and resale (dots) for** *Hyphaene coriacea* **and** *Phoenix reclinata* **palm-wine from palm savanna; with** (b) **seasonal variation in sales, showing a sharp drop coinciding with marula (***Sclerocarya birrea***) beer production.** (c) **Fluctuating yields to individual tappers from different batches of palms. A total of 712 palms (902 stems) were tapped by a single representative tapper during a 12-month period (from Cunningham 1990a,b).**

(Asteraceae) of the grassland and karroid biomes; and the spectacular bulbous species in the Iridaceae e.g. *Babiana* (36 species in the Cape flora), *Freesia* (ten species in the Cape flora), *Gladiolus* (69 of c. 150 species in the Cape flora) and the Hyacinthaceae *Lachenalia* (80 species, mainly in the Cape flora, but some in Namibia), *Ornithogalum* (31 species in the Cape flora, and others widespread in southern Africa).

- Herbal teas. These are primarily from the fynbos biome, and are mainly members of the Fabaceae. *Aspalathus* is a genus with 255 species, but only one (*A. linearis*) is a significant commercial crop plant with export value. *Cyclopia*, another legume endemic to the Cape flora, comprises 20 species, of which eight have been identified as tea sources by Kies (1951). Three species are currently showing commercial promise (Viljoen 1994): *C. intermedia, C. subternata* and *C. sessiliflora*.

- Medicinal plants. New technologies such as automated screening programmes, new cancer cell lines and HIV screens for new natural products are recent developments in the pharmaceutical industry, boosting work on medicinal plants such as *Hypoxis hemerocallidea* (Fig. 20.1g).

Commercial development of plant resources in southern Africa has been slow, and the bulk of the development has been driven primarily from agencies outside the region. For instance, research is being carried out in Israel on the edible fruit-bearing trees *Schinziophyton rautanenii, Sclerocarya birrea* and *Strychnos spinosa* (Cherfas 1989), and plantations of *Euphorbia tirucalli* have been developed on Okinawa island by a Japanese plastics company (Calvin 1979). The major commercial production of southern African bulbs for floriculture is in Holland, and cultivation of the Kalahari marama bean *Tylosema esculentum*, internationally recognized as an important crop plant (NAS 1979), has been implemented in Texas (Bousquet 1982). The value of lost opportunities in generating income and employment within southern Africa can only be guessed. What we do know is that *Freesia* flowers sold at two Dutch auctions were valued at US$92 million annually. More cultivars of *Agapanthus, Gladiolus, Kniphofia* and *Lachenalia* are available to gardeners in the United Kingdom than to gardeners in southern Africa (Ivey 1993). Noristan Ltd, a local pharmaceutical company that has been screening medicinal plants since 1974 (Fourie, Swart & Snykers 1992), was recently sold to a German multinational chemical company.

There are, however, some encouraging signs that regional capacity is being built for the development of natural plant resources. Some of these are: the selection of marula (*Sclerocarya birrea*) cultivars (Von Teichmann 1983; Goosen 1985); cultivation of plants for essential oils (Piprek, Graven & Whitfield 1982; Graven, Gardiner & Webber 1988); development of the rooibos tea industry (*Aspalathus linearis*) (Morton 1982) and preliminary work on *Cyclopia* honey-bush tea production (Viljoen 1994); growth of the cut-flower industry in the fynbos biome (Davis 1990; Van Wilgen, Bond & Richardson 1992); and management of ferns for the florist industry (Milton 1991). Far more applied research and resource development could be done, however. Although tree crops are slow-growing, several species with commercial potential for fruits, such as *Sclerocarya birrea* and *Schinziophyton rautanenii* grow easily from truncheons. This greatly speeds up the selection and commercial cultivation process. Wild stocks can also supplement commercial production until plantations can be harvested. In South Africa, production and marketing of marula (*Sclerocarya birrea*) liqueurs and fruit juice were developed from collected wild fruits bought from rural communities before superior commercial cultivars were produced. By 1987, 2000 tonnes of fruit were being processed into liqueur, 500 tonnes into fruit juice and 40 000 bottles of marula jelly were being made annually (Van Wyk 1987). The selection of marula fruits with commercially desirable qualities was made from wild genotypes over a relatively short period, and registered varieties are now being mass produced and planted out in the savanna of the Northern Province of South Africa. Graven *et al.* (1988) have worked on the selection of *Artemisia afra* genotypes that have high yields of selected essential oils.

20.6 Plant use, land use and harvesting impacts

The impact of harvesting has to be seen against the impact of agriculture, afforestation, grazing and fire in reducing supplies of some species and increasing populations of others. In the Eastern Cape, populations of the toxic medicinal plant *Ornithogalum thyrsoides* increased up to 60 times under selective grazing (Story 1952). Heavy grazing also results in a marked increase in *Helichrysum odoratissimum* populations (Noel 1961) as it does for the grapple plant, *Harpagophytum procumbens* in semi-arid savanna, possibly compensating for the effects of harvesting. The effects of reduction of forest area are particularly significant, as forests only cover 0.5% of the area of South Africa (Cooper 1985: Chap. 12, this volume) and are a source of over 130 plant species used for traditional medicines (Cunningham 1988d). The reduction in area of forest, grassland and woodland, which were formerly actual or potential sources of medicinal plants, has led to increased use of remaining areas.

If the role of wild plants as a buffer against poverty and as providing a range of basic needs or in generating income is to be maintained, then resource management rather than over-exploitation should take place. The degree of disturbance to the species population and vulnerability to over-exploitation depends on demand, supply, the part used, life-form and plant life-history. Two categories of plants are of concern: first, those slow-growing species with a limited distribution, which are the focus of commercial gathering; second, those species thatmay be popular, and not endangered, and which undergo a habitat change through commercial harvesting. Demand for fast-growing species with a wide distribution, high population density and high reproductive rate is often easily met. Sustained harvests are also more likely when leaves or fruits are used, for example, in the gathering of leaves and fruits for medicinal purposes; dietary supplements, thatch, weaving materials and reeds. Examples of easily managed vegetation types are *Phragmites australis* wetlands and *Cymbopogon validus* stands, which have a wide distribution, low species diversity and high biomass production of annual stems that are resilient to harvesting (Cunningham 1985; Shackleton 1990). Yields of *Cymbopogon* thatch grass, for example, are highest two years after burning, with yield of thatch bundles increasing from 97 bundles ha^{-1} (1.08 tonnes ha^{-1}) one year after burning to 200 bundles ha^{-1} (2.24 tonnes ha^{-1}) two years after the burn, declining to 1.8 tonnes ha^{-1} three years after burning (Shackleton 1990).

Similarly, at the species level in savannas, resilience to harvesting applies to common, fast-growing medicinal plants or encroaching species such as *Terminalia sericea* or *Acacia nilotica*, *Dichrostachys cinerea* (Fabaceae) and *Euclea divinorum* (Ebenaceae). These species may be harvested as an aid to management objectives for the savanna parks. In Hluhluwe Game Reserve, for example, an estimated 480 tonnes of wood were cleared from a 58-ha area by 15–20 local people, providing a useful source of fuelwood to the local community (Grobler 1987).

20.6.1 Leaf, flower and fruit harvesting

In theory, sustainable harvesting of fruits, leaves or flowers from wild populations is possible. It can also provide an incentive for conservation. In communal areas of southern Africa, trees or palms with edible fruits are most valued and conserved for food or fodder, particularly in areas of low arable potential (Fig. 20.12a). Savanna woodlands with few edible fruit-bearing trees (Fig. 20.12b) are completely cleared, apart from stumps (Fig. 20.12c). In Zimbabwe, Campbell (1986) found that the three most frequently used fruit trees (*Azanza garckeana*, *Diospyros mespiliformis* and *Strychnos cocculoides*) remained relatively constant in cover between untransformed woodland and cleared agricultural lands. As a result, their proportional contribution to total woody cover increased from 0.5% in uncleared woodland to 5% in cultivated fields. In northern Namibia this has created open, anthropogenic savannas of edible fruit-bearing trees (*Diospyros mespiliformis*, *Hyphaene petersiana* and *Sclerocarya birrea*) in place of formerly closed woodland dominated by *Terminalia sericea*, which was completely removed for hut-building. However, populations of valuable trees in transformed savannas will not be maintained unless recruitment balances mortality, a complex process (Shackleton 1993).

Several studies show the vulnerability of some species to over-exploitation of flowers, fruits and leaves. Commercially harvested species with slow leaf-production rates, long-lived leaves and slow reproductive rates are particularly vulnerable to over-harvesting. Many of these are plants (palms, proteas, aloes, ferns) growing on sandy nutrient-poor soils. Rebelo & Holmes (1988) quantify the unsustainably high level of damage to *Brunia albiflora* (Bruniaceae) populations as a result of harvesting for the fynbos cut-flower trade. Repeated harvests of 70% of the current year's inflorescences or fruit of three commercially harvested Proteaceae (*Leucadendron coniferum*, *L. meridianum* and *Protea obtusifolia*) in controlled harvesting experiments showed a decline in infructescence or 'cone' number (Mustart & Cowling 1992).

What is often over-estimated is the potential yield from wild stocks. This is a common problem worldwide, as Homma (1992) has described for wild-plant products from the Brazilian Amazon in terms of a 'boom and bust' model – a model equally applicable to the situation in southern Africa. An early expansion phase, during which profits are high, is followed by a stable period with no growth, and then finally the profit margin declines as the resource base is degraded by the impact of harvesting. The model goes on to suggest that sustainable growth of an industry based on natural plant resources is characterized by a fourth phase, which relies on domestication and cultivation of the target crop species. Several commercial harvesting ventures based on wild populations, including harvesting of *Landolphia kirkii* for rubber in the 1900s, are characterized by a "boom and bust" situation in which initial harvests are followed by

declining resource availability. Yields of marketable seven-weeks fern (*Rumohra adiantiformis*) leaves from a virgin stand were 46.7 kg ha^{-1} yr^{-1}, for example, compared to a sustainable yield of less than 16 kg ha^{-1} yr^{-1} (Milton & Moll 1988). Essential oil production from *Artemisia afra* was initially based on wild populations until it was realized that regrowth was lower than anticipated due to high-intensity harvesting (Graven *et al.* 1988). Milton (1988) has also documented a sharp decline in basal area increment with increased frequency of pruning of *Acacia tortilis*. Intense harvesting for the dried flower trade of the fynbos everlasting *Syncarpha vestita*, in response to a strong export demand, resulted in virtual extinction over large areas of its range. This was followed by a crash in the market when demand exceeded supply to a massive degree. Today, attempts are being made to cultivate this species on a wide scale (R.M. Cowling, pers. comm).

In contrast to the easily sustainable harvest of leaves of fast-growing 'weedy' species such as wild spinaches, is the harvest of long-lived fern, palm and aloe leaves. Leaf-production rates are low, increasing the potential for over-exploitation. Under natural conditions, seven-weeks ferns only produce 1–2 leaves yr^{-1} (Milton & Moll 1988) and *Hyphaene coriacea* and *H. petersiana* palms produce 3–4 leaves yr^{-1} in the size class selected for basketry fibre (Cunningham 1988b,c). In 1985, for example, 300 tonnes or 14 million leaves from seven-weeks ferns (*Rumohra adiantiformis*) were picked for florist materials from Afromontane forest (Milton & Moll 1988). In Ngamiland, Botswana, 1900 basket makers used an estimated 85 500 leaves yr^{-1} (Cunningham 1988b). *Aloe ferox* leaves are cut and stacked to collect the sap. About 600 tonnes of this sap is exported each year to Europe and North America for medicinal purposes (Bond 1983). What levels of harvest are likely to be sustained – particularly where

money and personnel for monitoring and resource management are generally scarce?

Evidence from pruning trials on ferns has shown that productivity can be sustained with a leaf harvest of up to 50% of annual leaf production, but more frequent or intense defoliation leads to declining production and shortened life span of plants (Milton 1991). Furthermore, monitoring of fern harvest has shown that commercial picking of fern leaves resulted in a decrease in mean leaf length between 1983 and 1988 and, even after five years, ferns had still not recovered from total or partial defoliation. Total defoliation for two consecutive years killed 89% of *Rumohra adiantiformis* rhizomes (Milton & Moll 1988). Timely implementation of frond harvesting at a lower intensity and frequency has led to an increase in frond number and stabilization of frond size (Milton 1991).

The effects of *Aloe ferox* leaf harvesting are even more subtle. Regular leaf harvesting results in the removal of the persistent dead leaves that form a fire-proof skirt around these fire-sensitive plants. An evaluation of *Aloe ferox* survival after three fires showed that survival after fire was strongly related to the presence of an intact skirt (Bond 1983). Where leaf harvesting had removed this skirt, 96% of all aloes were killed. By comparison, 97% of aloes with stems either completely covered by dead leaves or bare for less than 0.25 m survived (Bond 1983).

In Ngamiland, Botswana, commercial harvesting is concentrated on small areas of *Hyphaene petersiana* palms close to the villages that support 1900 basketmakers. As a result, 42–47% of annual leaf production was either cut or damaged by non-selective cutting (Cunningham & Milton 1987). By comparison, levels of *H. coriacea* palm leaf harvesting were low (up to 8% of annual leaf production) in the extensive palm savanna of Maputaland (Cunningham 1988c). In this case, there was potential for increased harvest from the 10 000–11 000 leaves yr^{-1} used in local craftwork production compared to an average of 140 leaves ha^{-1} yr^{-1} or an estimated 2.5 million leaves yr^{-1} in the same area (Cunningham 1987, 1988b).

There is also potential in southern Africa for the cultivation phase that Homma (1992) describes. Some examples are the cultivation of additional fynbos, cut and dried flower species, marula (*Sclerocarya birrea*) fruit trees and honey-bush tea (*Cyclopia* species), as well as several traditional medicinal plants such as the pepper-bark tree, *Warburgia salutaris*. Cultivation of slow-growing species subject to destructive harvesting is unlikely to be economically viable. It is this category of plants that are of greatest concern for conservation and resource management.

Figure 20.12 **People and vegetation change: (a) A woman sows millet seed in an anthropogenic *Hyphaene petersiana* woodland created through conservation of these palms, which produce fibre, fuel and edible fruits (Owambo, Namibia); (b) *Pterocarpus* woodland with a low proportion of large trees bearing edible fruits (although with many *Grewia* shrubs) (Eastern Caprivi, Namibia); (c) Millet fields cleared of *Pterocarpus* woodland, where a low frequency of trees with edible fruits has resulted in no woody cover (Etsha, Botswana); (d) Coppicing mopane trees in what was previously tall *Colophospermum mopane* woodland around Gweta village, Botswana; (e) Selective removal of *Spirostachys africana* (umthombothi) trees from woodland (Tshongwe, South Africa); (f) At least 300 000 *Scilla natalensis* (inguduza) bulbs are selectively dug up from montane grasslands each year to supply the herbal medicine trade in KwaZulu-Natal; (g) Siyabonga Zondi next to a stripped *Curtisia dentata* tree, its bark used to supply the urban demand for symbolic medicines (eMalowe State Forest, Eastern Cape); (h) Julio Rivero next to a debarked and uprooted *Berchemia discolor* tree, 20 years after the start of commercial basketry (Qoroga Island, Botswana)**
(Photos: A.B. Cunningham).

20.6.2 Harvesting bulbs, bark, roots, timber or whole plants

Popular species that are habitat-specific, slow-growing and slow-reproducing plants, and which are debarked or dug out for dyes, medicinal or horticultural purposes, are particularly vulnerable to over-harvesting. Well-publicized examples of plants illegally exploited from the wild for the horticultural trade are slow-growing, long-lived plants such as endemic cycads (*Encephalartos* species and *Stangeria eriopus*) from the Eastern Cape, Kwa-Zulu-Natal and Transvaal; the endemic *Aloe polyphylla* (Asphodelaceae) in Lesotho (Talukdar 1983); and the half-mens, *Pachypodium namaquanum* (Apocynaceae) from the Richtersveld. The effects of harvesting hardwood trees for building material or fuelwood are apparent, through complete removal of woody vegetation, low levels of recruitment, or formation of coppices. Demand for building timber and fuelwood around densely settled sites in the mopane savannas of Namibia and Botswana has led to the transformation of the tall savanna to short, multi-stemmed coppices (Fig. 20.12d). Palm-wine tapping has similarly altered palm savanna of northeastern KwaZulu-Natal from tall *Hyphaene coriacea* savanna to

short coppice growth (Cunningham 1990a,b). The effect of the selective removal of poles and laths on forest is less obvious. Muir (1991), in an innovative project where he worked with local woodcutters, modelled the effect of different levels of pole cutting on forest canopy composition of the Hlatikulu forest in KwaZulu-Natal. Patterns of use varied between different forest communities. Mesic forest dominated by *Celtis africana*, *Drypetes gerrardii* and *Olea capensis*, carried 73% of pole and 77% of lath utilization. The harvesting of thin laths has had the greatest impact of this forest type, but impact was lessened by the high proportion (54%) of laths coppicing (see Chap. 12, this volume). At current levels of utilization, estimated to be 33 600 laths and 2350 poles yr^{-1}, Muir (1991) predicted a 6–7% reduction in understorey stems (10–100 mm diameter) and a 6–12% reduction in canopy stems (100–800 mm and greater in diameter). Selective exploitation of forests for timber in the late nineteenth and early twentieth centuries in South Africa (Chap. 12, this volume) had a large impact on forest structure (King 1941). It also affected supplies of traditional medicinal plants. By 1901, 52% of forests in the Transkei were exhausted of exploitable timber, includ-

Figure 20.13 **The demise, between 1982 and 1992, of a population of *Berchemia discolor* trees scattered on termitaria over a 10-ha area, owing to bark removal for commercial basketry (Qoroga, Botswana).**

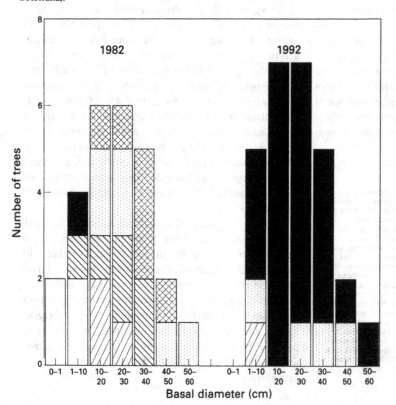

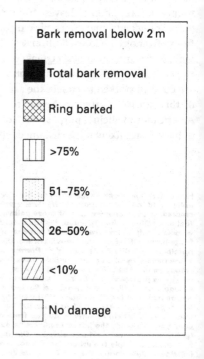

Bark removal below 2 m

- Total bark removal
- Ring barked
- \>75%
- 51–75%
- 26–50%
- <10%
- No damage

ing species such as *Ocotea bullata* and *Curtisia dentata* (King 1941). Both are also important medicinal plants. At present the basal area of *O. bullata* in Afromontane forest in the Transkei is 27.6 m³ ha⁻¹ (Cawe 1986) compared with King's (1941) estimate of 60 m³ ha⁻¹, a difference that Cawe (1986) attributes to exploitation. Harvesting of poles and laths uses much smaller size classes than the timber industry, but nevertheless can impact on tree recruitment and forest structure.

In the past, under subsistence demand, plants used as sources of medicine or basketry dye were rarely ring-barked or uprooted. This has changed drastically in response to commercial demand, with ring-barking and uprooting taking place in an opportunistic, species-specific scramble for roots, bulbs or bark. These plant parts represent the main ingredients of herbal medicine in the southern African region (Cunningham 1991; 1993b). Bark and roots also represent the main source of traditional dye for commercial craftwork production. As a result of commercial demand for bark, *Berchemia discolor* populations near basket-making villages in northern Botswana have been decimated (Fig. 20.13). This has occurred despite the widespread practice of conserving *Berchemia discolor* trees valued for their edible fruits. Coastal and Afromontane forest trees, exploited for their medicinal bark, such as *Curtisia dentata* (Cornaceae), *Rapanea melanophloeos* (Myrsinaceae), *Ocotea bullata* (Lauraceae) and *Cassipourea gerrardii* (Rhizophoraceae), suffer excessive debarking and resultant die-off. What is of concern, however, is ring-barking in 'conserved' forests, with the consequent development of canopy gaps that change forest structure and lead to an influx of invasive alien species. Extensive damage has taken place in some state forests in South Africa, which were theoretically set aside for the maintenance of habitat and species diversity (Cunningham 1988b, 1991). In the eMalowe State Forest, Transkei, if coppice stems less than 20 mm diameter are excluded, then the level of damage to *Curtisia dentata* (Cornaceae) and *Ocotea bullata* (Lauraceae) trees encountered would represent 51% and 57% of trees, respectively, with more than half the trunk bark removed. This also represents a loss of *Ocotea bullata* timber worth millions of US$ from these forests (see Chap. 12, this volume).

20.7 What are wild plants worth?

The expansion of agriculture in southern Africa has cleared most natural vegetation on arable soils. Alien pine or eucalyptus plantations now cover large areas of coastal or upland grasslands in the moist, eastern part of the subcontinent (Chap. 21, this volume). These plantations are now expanding into areas of marginal potential. To some urban-based (or biased) land-use planners these marginal, often nutrient-poor sites are regarded as 'useless' and need to be converted to more productive monocropping. To many rural people on communal or tribal lands in southern Africa, these same areas are valued multiple-use areas for grazing livestock or gathering a diversity of plant resources. Similarly in the fynbos biome, vegetation on nutrient-poor soils or mountain lands with little arable potential is generating income from cut-flowers or thatch. One thing is clear, whether documented through careful research, or experienced by resource users who walk further or pay more for increasingly scarce resources: harvested wild plants in many cases are not a 'free-good' harvested from a bountiful, overflowing cornucopia. Harvest is labour intensive with a high cost in time and effort that increases as more accessible resources are depleted. Why should local individuals, communities or policy makers at a local or national level show any concern about over-exploitation? What would we lose if these resources were lost?

Although difficult to assess, the value of indigenous or alien plants in providing income in impoverished areas should not be underestimated. On the sandy Maputaland coastal plain, *Hyphaene coriacea* palm savanna was valued at US$13 ha⁻¹ yr⁻¹ from sale and resale of palm-wine, leaf harvest for basketry and cattle sales (Cunningham 1985). In comparison, Milton & Bond (1986) estimated that *Acacia tortilis* thornveld savanna is worth at least US$187 family⁻¹ yr⁻¹ or US$21 ha⁻¹ yr⁻¹ to inhabitants of the Msinga area, KwaZulu-Natal. The main annual values per family were for fuelwood (US$100), animal feed (US$74) and construction timber (US$11). Campbell *et al.* (1991) calculated that miombo (*Brachystegia*) woodland in Zimbabwe had a similar value to *Acacia* woodland studied by Milton & Bond (1986) (Fig. 20.14). Detailed valuation studies by Campbell *et al.* (1991) in Zimbabwe indicate that the value of trees fell into three main groups: first, the direct value to households for fuel, farm and household materials; second, their value in terms of production as crop inputs such as leaf litter, animal feed and cash income; third, trees provide a range of services, ecologically, socially and for shade. Although fuelwood and construction timber made up a high proportion of the total value in miombo woodland, the value of wild fruits, browse and litter inputs for fields were as high.

These values decline with reduction in tree cover or with species-specific over-exploitation. *Colophospermum mopane*, *Combretum hereroense* and *C. imberbe* are all favoured sources of building timber in the savanna biome. All are becoming scarce, owing to the competing uses of

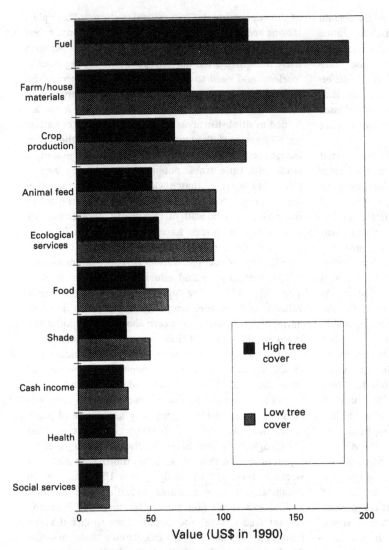

Figure 20.14 **Mean estimated value for commodities and services provided by trees in Zimbabwe, and the market response to diminished supply (from Campbell et al. 1991).**

timber for fencing, fuelwood and grain-stamping mortars and pestles. As a result, people are walking further or paying more for building timber. The commercial trade in building poles provides employment for commercial gatherers and the few people who can afford vehicles to transport the poles for sale. This is at the expense, however, of the rural communities in the outlying areas where the poles are gathered. Local resources are depleted and local self-sufficiency reduced as building poles are taken for use elsewhere. In Zimbabwe, dense woodland cover in Runde Communal Land decreased by 69% from 1968 to 1985, with a corresponding increase in grassland of 16%, a factor that buffered the effects of a 50% increase in cattle numbers on grass-

lands in this area (Meltzer & Hastings 1992). Agricultural clearing and the demand for building timber have similarly contributed to a decline in woodland area and structure in Owambo, northern Namibia, over at least a century. In 1866, missionary Hugo Hahn recorded a 60-km belt of *Colophospermum mopane* woodland between the Uukwanyama and Ondonga kingdoms. Fifty years later this woodland was 40-km wide, and, in the 1950s, 10-km wide. Today, this woodland no longer exists (Erkkila & Siiskonen 1992).

There is no doubt that resource users, most frequently the rural poor, *are* concerned. What is required is similar concern at a policy level with the aim of halting or reversing such changes. Determining the value of wild plants as 'natural resource capital' is a useful way of better illustrating land-use values and options. We know that plants provide a 'green social security' in the form of low-cost housing, fuel, subsistence income, food supplements and herbal medicines, on a continent where social security is rarely provided. Loss of this green safety net through agricultural clearing, intensive grazing or over-exploitation has serious consequences for both rural and urban areas, as the rural poor become refugees in urban shanty towns, at great social and economic cost.

20.8 Conclusions

The combination of high biological and cultural diversity in a region with a well-developed research capability has resulted in a unique record of people–plant interaction in southern Africa. Neither ecological nor social processes are static and we need to continue to be able to study – and predict – future changes. Economic factors are a major driving force of this change. One of the future challenges will be to bring together conservation biology, economics and social science in order to develop appropriate protocols for the study of the human use of plants. Recent studies have taken steps towards this (Bishop & Scoones 1994; Campbell *et al.* 1995). It is essential, however, that this blend of ecological and social science does not result in a less rigorous approach. It should instead enhance hypothesis formation and testing through new insights, leading to a predictive capability. This would strengthen the means to influence policy on conservation and management of regional plant resources.

20.9 Acknowledgements

Support for A.B.C. from the WWF-International while writing up this chapter is gratefully acknowledged, as is that of the National Botanical Institute of South Africa for G.W.D. The views expressed in this paper are a reflection of enjoyable field trips and discussions with many people, particularly Mlingo Gwala, Bekazitha Gwala, Chris Hines, Michelle Cunningham and Siyabonga Zondi. The patience, help and constructive criticism of the editors are acknowledged in production of the final draft. Joanne Turner is also thanked for her help.

.10 References

derson, S. & Staugard, F. (1986). *Traditional Midwives. Traditional Medicine in Botswana.* Gaborone: Ipelegeng Publishers.

on. (1994). Heuningbostee kan sterk bedryf word in Wes- en Suid-Kaap. *Die Burger*, 1 July 1994.

:her, F. (1989). The construction of a traditional Nama matjieshuis. *Sagittarius*, 4, 20–2.

:her, F. (1990). Planning with people – ethnobotany and African uses of plants in Namaqualand (South Africa). *Mitteilungen aus dem Institut für Allgemeine Botanik Hamburg*, 23b, 959–72.

nold, H.-J. & Gulumian, M. (1984). Pharmacopoeia of traditional medicine in Venda. *Journal of Ethnopharmacology*, 12, 35–74.

orin, E.J. (1992). *The Potential of Alien Acacias as a Woodfuel Resource in the South Western Cape.* Cape Town: University of Cape Town/National Energy Council.

Basson, J.A. (1987). Energy implications of accelerated urbanisation. *South African Journal of Science*, 83, 284–90.

Beaumont, P.B., De Villiers, H. & Vogel, J.C. (1978). Modern man in sub-Saharan Africa prior to 49 000 years BP: a review and evaluation with particular reference to Border Cave. *South African Journal of Science*, 74, 409–19.

Berglund, A.-I. (1976). *Zulu Thought Patterns and Symbolism.* Cape Town: David Philip.

Best, M. (1979). *The Consumption of Fuel for Domestic Purposes in Three African Villages.* MSc Thesis. Cape Town: University of Cape Town.

Biesele, M., Guenther, M., Hitchcock, R., Lee, R. & MacGregor, J. (1989). Hunters, clients and squatters: the contemporary socioeconomic status of Botswana Basarwa. *African Study Monographs*, 6, 109–51.

Bishop, J. & Scoones, I. (1994). *Beer and Baskets. The Economics of Women" Livelihoods in Ngamiland, Botswana.* London: International Institute for Environment and Development.

Bond, W. (1983). Dead leaves and fire survival in southern African tree aloes. *Oecologia*, 58, 110–14.

Bousquet, J. (1982). The morama bean of the Kalahari desert as a potential food crop, with a summary of current research in Texas. *Desert Plants*, 3, 213–15.

Brown, A.S. & Brown, G.G. (1935). *The South African Year Book and Guide.* London: Sampson, Low, Marston & Co, Ltd.

Brutsch, M.O. & Zimmerman, H.G. (1993). The prickly pear (*Opuntia ficus-indica* [Cactaceae]) in southern Africa: utilization of the naturalised weed and of cultivated plants. *Economic Botany*, 47, 154–62.

Bryant, A.T. (1909). Zulu medicine and medicine men. *Annals of the Natal Museum*, **2**, 1–123.

Budack, K.F.R. (1983). A harvesting people on the South Atlantic coast. *South African Journal of Ethnology*, **6**, 1–7.

Buhrmann, M.V. (1984). *Living Between Two Worlds*. Cape Town: Human & Rousseau.

Bulatao, R.A., Bos, E., Stephens, P.W. & Vu, M.T. (1990). *World Population Projections 1989–1990 Edition. Short and Long-term Estimates*. London: Johns Hopkins University Press.

Burger, A.E.C., De Villiers, J.B.M. & Du Plessis, L.M. (1987). Composition of the kernel oil and protein of the marula seed. *South African Journal of Science*, **83**, 733–5.

Calvin, M. (1979). Petroleum plantations and synthetic chloroplasts. *Energy*, **4**, 851–69.

Campbell, B.M. (1986). The importance of wild fruits for peasant households in Zimbabwe. *Food and Nutrition*, **12**, 38–44.

Campbell, B.M., Clarke, J., Luckert, M., Matose, F., Musvoto, C. & Scoones, I. (1995). *Local Level Economic Valuation of Savanna Woodland Resources. Village Cases from Zimbabwe*. London: International Institute for Environment and Development.

Campbell, B.M. & du Toit, R.F. (1988). Relationships between wood resources and use of species for construction and fuel in communal lands of Zimbabwe. *Monographs in Systematic Botany Missouri Botanical Gardens*, **25**, 331–41.

Campbell, B.M., Vermeulen, S.J. & Lynam, T. (1991). *Value of Trees in the Small-scale Farming Sector of Zimbabwe*. Ottawa: IDRC.

Cawe, S.G. (1986). *A Quantitative and Qualitative Survey of the Inland Forests of Transkei*. MSc Thesis. Umtata: University of Transkei.

Cherfas, J. (1989). Nuts to the desert. *New Scientist*, **19 August**, 44–7.

Coates Palgrave, K. (1977). *Trees of Southern Africa*. Cape Town: Struik.

Compton, R.H. (1966). An annotated checklist of the flora of Swaziland. *Journal of South African Botany*, **6** (Supplement), 1–191.

Comte, M.-S. (1991). A palm with promise. *Ceres*, **127**, 26–9.

Cooke, H.B.S. (1964). The Pleistocene environment in southern Africa. In *Ecological Studies in Southern Africa*, ed D.H.S. Davis, pp. 1–23. The Hague: Junk.

Cooper, K.H. (1985). *The Conservation Status of Forests in Transvaal, Natal and OFS, South Africa*. Durban: Wildlife Society of South Africa.

Cunningham, A.B. (1985). *The Resource Value of Indigenous Plants to Rural People in a Low Agricultural Potential Area*. PhD Thesis. Cape Town: University of Cape Town.

Cunningham, A.B. (1987). Commercial craftwork: balancing out human needs and resources. *South African Journal of Botany*, **53**, 259–66.

Cunningham, A.B. (1988a). *Botswana Basketry Resources. Research and Management of Plant Resources Supporting the Ngamiland Basketry Trade, 1982–1988*. Gaborone: Botswanacraft Marketing Company (Pty) Ltd.

Cunningham, A.B. (1988b). Collection of wild plant foods in Tembe Thonga society: a guide to Iron Age gathering activities? *Annals of the Natal Museum*, **29**, 433–46.

Cunningham, A.B. (1988c). *An Investigation of the Herbal Medicine Trade in Natal/KwaZulu*. Investigational Report no. 29, Institute of Natural Resources. Pietermaritzburg: University of Natal.

Cunningham, A.B. (1988d). Leaf production and utilization in *Hyphaene coriacea*: management guidelines for commercial harvesting. *South African Journal of Botany*, **54**, 189–95.

Cunningham, A.B. (1989). Indigenous plant use: balancing human needs and resources. In *Biotic Diversity in Southern Africa. Concepts and Conservation*, ed. B.J. Huntley, pp. 93–106. Cape Town: Oxford University Press.

Cunningham, A.B. (1990a). The regional distribution, marketing and economic value of the palm wine trade in the Ingwavuma district, Natal, South Africa. *South African Journal of Botany*, **56**, 191–8.

Cunningham, A.B. (1990b). Income, sap yields and effects of palm wine tapping on palms in south-eastern Africa. *South African Journal of Botany*, **56**, 137–44.

Cunningham, A.B. (1991). Development of a conservation policy on commercially exploited medicinal plants: a case study from southern Africa. In *Conservation of Medicinal Plants*, ed. O. Akerele, V. Heywood & D.H. Synge, pp. 337–58. Cambridge: Cambridge University Press.

Cunningham, A.B. (1993a). *Imithi isiZulu. The Traditional Medicine Trade in Natal/KwaZulu*. M Soc Sci Thesis. Pietermaritzburg: University of Natal.

Cunningham, A.B. (1993b). African medicinal plants: setting priorities at the interface between conservation and primary health care. *People and Plants Working Paper 1*, pp. 1–50. Paris: UNESCO.

Cunningham, A.B. & Gwala, B.R. (1986). Plant species and building methods used in Tembe Thonga hut construction. *Annals of the Natal Museum*, **27**, 491–511.

Cunningham, A.B. & Milton, S.J. (1987). Effects of basket weaving industry on Mokola palm and dye plants in northwestern Botswana. *Economic Botany*, **41**, 386–402.

Cunningham, A.B. & Zondi, A.S. (1991). *Use of Animal Parts for the Commercial Trade in Traditional Medicines*. Investigational Report, Institute of Natural Resources. Pietermaritzburg: University of Natal.

Davison, P. (1984). Lobedu material culture: a comparative study of the 1930's and the 1970's. *Annals of the South African Museum*, **94**, 41–201.

Davis, G.W. (1990). *Commercial Wildflower Production in the Fynbos Biome and its Role in the Management of Land-use*. PhD Thesis. Cape Town: University of Cape Town.

Davis, G.W. (1993). Don't underestimate dekriet. *Farmer's Weekly*, **12 November**, 52–3.

Deacon, H.J. (1983). The peopling of the fynbos. In *Fynbos Palaeoecology. A preliminary synthesis*, ed. H.J. Deacon, Q.B. Hendey & J.J.N. Lamprechts, pp. 183–204. *South African National Programmes Report 75*. Pretoria: CSIR.

acon, H.J. (1988). Palaeoecological perspectives on recent environmental change in southern Africa. In *Long-term Data Series Relating to Southern Africa's Renewable Resources*, South African National Scientific Programmes report no. 157, ed. I.A.W. Macdonald & R.J. Crawford, pp. 379–83. Pretoria: CSIR.

acon, H.J. (1989). Late Pleistocene palaeoecology and archaeology in the southern Cape, South Africa. In *The Human Revolution*, ed. P. Mellars & C. Stringer. pp. 185–215. Edinburgh: Edinburgh University Press.

ntlinger, U. (1977). The Nara plant in the Topnaar Hottentot culture of Namibia. *Munger Africana Library Notes*, 38. Pasadena: California Institute of Technology.

ke, C.M. & Vilakazi, B.W. (1964). *Zulu-English Dictionary*. Johannesberg: Witwatersrand University Press.

Plessis, L. (1988). Fruitful future for marula oil. *Scientiae*, 4, 15–16.

is, C.G. (1986). Medicinal plant use – a survey. *Veld & Flora*. 72, 72–3.

kiila, A. & Siiskonen, H. (1992). Forestry in Namibia 1850–1990. *Silva Carelica* 20. Joensuu: University of Joensuu.

gan, B.M. (1990). *The Journey from Eden. The Peopling of Our World*. London: Thames & Hudson.

eley, J.M. (1980). Did Iron Age man have a role in the history of Zululand's wilderness landscapes? *South African Journal of Science*, 76, 150–2.

eley, J.M. (1985). Smelting in the Iron Age of Transkei. *South African Journal of Science*, 81, 10–11.

rreira, M. (1987). Medicinal use of indigenous plants by elderly Coloured.: a sociological study of folk medicine. *South African Journal of Sociology*, 18, 139–43.

urie, T.G., Swart, I. & Snykers, F.O. (1992). Folk medicine: a viable starting point for pharmaceutical research. *South African Journal of Science*, 88, 190–3.

x, F.W. & Norwood-Young, M.E. (1982). *Food from the Veld*. Johannesburg: Delta Books.

Gandar, M.V. (1983). *Wood as a Source of Fuel in South Africa*. Monograph No 4, Institute of Natural Resources. Pietermaritzburg: University of Natal.

Gelfand, M., Mavi, S., Drummond, R.B. & Ndemera, B. (1985). *The Traditional Medical Practitioner in Zimbabwe*. Harare: Mambo Press.

Gerstner, J. (1938). A preliminary checklist of Zulu names of plants. *Bantu Studies*, 12, 215–36, 321–42.

Gerstner, J. (1939). A preliminary checklist of Zulu names of plants. *Bantu Studies*, 13, 49–64, 131–49, 307–26.

Gerstner, J. (1941). A preliminary checklist of Zulu names of plants. *Bantu Studies*, 15, 277–301, 369–83.

Goosen, H. (1985). Die maroela word getem. *South African Panorama*, 30, 21–5.

Graven, E.H., Gardiner, J.B. & Webber, L. (1988). *Ciskeian Aromatic Plants. Appropriate New Crops for Rural Development*. Agricultural & Rural Development Research Institute. Alice: University of Fort Hare.

Green, E.C. (1985). Traditional healers, mothers and childhood diarrhoeal disease in Swaziland: the interface of anthropology and health education. *Social Science and Medicine*, 20, 277–85.

Grivetti, L.E. (1979). Kalahari agro-pastoral hunter-gatherers: the Tswana example. *Ecology of Food and Nutrition*, 7, 235–56.

Grobelaar, J.A. (1985). The population of Natal/KwaZulu 1904–2010. *Natal Town and regional Planning Report 65*, Pietermaritzburg: Natal Town and Regional Planning Commission.

Grobler, H. (1987). *Aid for NPB Neighbours. Natal: News from NPB 1*. Pietermartizburg: Natal Parks Board.

Grundy, I.M., Campbell, B.M., Balenereho, S., Cunliffe, R., Tafangeyasha, C., Ferguson, R. & Parry, D. (1991). Availability and use of trees in Mutanda Resettlement area, Zimbabwe. *Forest Ecology and Management*, 56, 243–66.

Hall, M. (1987). *The Changing Past. Farmers, Kings and Traders in Southern Africa 200–1860*. Cape Town: Creda Press.

Hammond-Tooke, W.D. (1981). *Boundaries and Belief. The Structure of a Sotho Worldview*. Johannesberg: Witwatersrand University Press.

Hammond-Tooke, W.D. (1989). Rituals and Medicines. Indigenous Healing in South Africa. Johannesburg: Ad Donker.

Heinsohn, D. (1991). *The Potential for Cultivation of Juncus kraussii and other Wetland Species Used for Craftwork in Natal/KwaZulu*. Investigational Report no. 63, Institute of Natural Resources. Pietermaritzburg: University of Natal.

Heinz, H.J. & McGuire, B. (1974). *The Ethnobiology of the !Ko Bushmen – Their Botanical Knowledge and Plant Lore*. Occasional Paper no. 1. Gaborone: Botswana Society.

Hennessy, E.F. & Lewis, O.A.M. (1971). Anti-pellagragenic properties of wild plants used as dietary supplements in Natal (South Africa). *Plant Foods in Human Nutrition*, 2, 75–8.

Holdstock, T.L. (1978). Panel discussion. Proceedings of a conference on traditional medicine. *The Leech*, 48, 22.

Holdstock, T.L. (1979). Indigenous healing in South Africa: a neglected potential. *South African Journal of Psychology*, 9, 118–24.

Homma, A.K.O. (1992). The dynamics of extraction in Amazonia: a historical perspective. *Advances in Economic Botany*, 9, 23–31.

Hooper, L. (1981). Some Nguni crafts. Part 3. Woodcarving. *Annals of the South African Museum*, 70, 157–312.

Humphreys, A.J.B. (1987). Archaeology. In *The Karoo Biome. A Preliminary Synthesis. Part 2 – Vegetation and History*,. ed. R.M. Cowling & P.W. Roux, pp. 117–32. *South African National Scientific Programmes Report 142*. Pretoria: CSIR.

Hutchings, A. (1989). Observations on plant usage in Zulu and Xhosa medicine. *Bothalia*, 19, 225–35.

Hutchings, A., Scott, A., Lewis, G. & Cunningham, A.B. (1996). *Zulu Medicinal Plants. An inventory*. Pietermaritzburg: University of Natal Press. (in press).

Ivey, P. (1993). *South Africa's Petaloid Geophytes. At Home and Abroad*. MSc Thesis. Birmingham: University of Birmingham.

Jacobs, G. (1989). Recommendations for the future. In *The Wildflower Resource: Commerce, Conservation and Research*, ed. T. Greyling & G.W. Davis. Pretoria: Foundation for Research Development, Occasional Report **40**, 25–9.

Jacobsohn, M. (1988). Preliminary notes on the symbolic role of space and material culture among semi-nomadic Himba and Herero herders in western Kaokoland, Namibia. *Cimbebasia*, **10**, 75–99.

Jacot Guillarmod, A. (1971). The flora of Lesotho. *Flora et Vegetatio Mundi*, **3**, 459–65.

Kies, P. (1951). Revision of the genus *Cyclopia* and notes on some other sources of bush tea. *Bothalia*, **6**, 161–72.

King, N.L. (1941). The exploitation of the indigenous forests of South Africa. *Journal of the South African Forestry Association*, **6**, 26–48.

Knuffel, W.E. (1973). *The Construction of the Bantu Grass Hut*. Graz: Akademishe Druk.

Krige, E. (1937). *The Social System of the Zulus*. Pietermaritzburg: Shuter & Shooter.

Lan, D. (1985). *Guns and Rain. Guerillas and Spirit Mediums in Zimbabwe*. Harare: Zimbabwe Publishing House.

Lee, R.B. (1965). *The Subsistence Ecology of the Kung Bushmen*. PhD Thesis. Berkeley: University of California.

Lee, R.B. (1973). Mongongo: the ethnography of a major wild food resource. *Ecology of Food and Nutrition*, **2**, 307–21.

Le Maitre, D.C. & Midgley, J.J. (1992) Plant reproductive ecology. In *The Ecology of Fynbos. Nutrients, Fire and Diversity*, ed. R.M. Cowling, pp. 135–74. Cape Town: Oxford University Press.

Lewis, O.A.M., Shanley, B.M.G. & Hennessey, E.F. (1971). The leaf protein nutritional value of four wild plants used as dietary supplements by the Zulu. In *Protein and Food Supply in the Republic of South Africa*. Cape Town: Balkema.

Lewis-Williams, D. & Dowson, T. (1989). *Images of Power. Understanding Bushman Rock Art*. Cape Town: Southern Book Publishers.

Liengme, C.A. (1983). A study of wood use for fuel and building in an area of Gazankulu. *Bothalia*, **13**, 501–18.

Linder, H.P. (1990). The thatching reed of Albertinia. *Veld & Flora*, **76**, 86–9.

Malan, J.S. & Owen-Smith, G.L. (1974). The ethnobotany of the Kaokoland. *Cimbebasia Series*, **B2**, 131–78.

Meltzer, M.I. & Hastings, H.M. (1992). The use of fractals to assess the ecological impact of increased cattle population: case study from the Runde Communal Land, Zimbabwe. *Journal of Applied Ecology*, **29**, 635–46.

Middelmann, W.J., Gibson, B. & Bell, I. (1989). The commercial perspective. In *The Wildflower Resource. Commerce, Conservation and Research*, ed. T. Greyling & G.W. Davis, pp. 10–13. Occasional Report 40. Pretoria: Foundation for Research Development.

Mills, G.T. (1984). *An Inquiry into the Structure and Function of Space in Indigenous Settlement in Ovamboland*. M Arch Thesis. Cape Town: University of Cape Town.

Milton, S.J. (1987). Effects of harvesting on four species of forest ferns in South Africa. *Biological Conservation*, **1**, 133–46.

Milton, S.J. (1988). Effects of pruning on shoot production and basal increment of *Acacia tortilis*. *South African Journal of Botany*, **54**, 109–17.

Milton, S.J. (1991). Slow recovery of defoliated seven-weeks Fern *Rumhora adiantiformis* in Harkerville forest. *South African Journal of Forestry*, **158**, 23–7.

Milton, S.J. & Bond, C. (1986). Thorn trees and the quality of life in Msinga. *Social Dynamics*, **12**, 64–76.

Milton, S.J. & Moll, E.J. (1988). Effects of harvesting on frond production of *Rumohra adiantiformis* (Pteridophyta: Aspidiaceae) in South Africa. *Journal of Applied Ecology*, **25**, 725–43.

Morton, J.F. (1982). Rooibos tea, *Aspalathus linearis*, a caffeinless, low-tannin beverage. *Economic Botany*, **37**, 164–73.

Muir, D.P. (1991). *Indigenous Forest Utilization in KwaZulu. A Case Study of the Hlatikulu forest, Maputaland*. MSc Thesis. Pietermaritzburg: University of Natal.

Mustart, P.J. & Cowling, R.M. (1992). Impact of flower and cone harvesting on seed banks and seed set of serotinous Agulhas Proteaceae. *South African Journal of Botany*, **58**, 337–42.

NAS (1979). *Tropical Legumes. Resources for the Future*. Washington DC: National Academy of Sciences.

Newton, D. (1993). Is export industry endangering the bitter aloe? *Our Living World*, **September 1993**, 10.

Ngubane, H. (1977). *Body and Mind in Zulu Medicine*. New York: Academic Press.

Noel, A.R.A. (1961). A preliminary account of the effect of grazing upon species of *Helichrysum* in the Amatola mountains. *Journal of South African Botany*, **27**, 81–6.

Nott, K. (1986). *A Survey of the Harvesting and Export of Harpagophytum procumbens and H. zeyheri in SWA/Namibia*. Okaukuejo: Etosha Ecological Institute.

O'Brien, E.M. (1988). Climatic woody correlates of species richness for woody 'edible' plants across southern Africa. *Monographs in Systematic Botany from the Missouri Botanical Garden*, **25**, 385–401.

Ogle, B.R. & Grivetti, L.E. (1985). Legacy of the chameleon: edible wild plants in the Kingdom of Swaziland, southern Africa. A cultural, ecological, nutritional study. Part 4 – nutritional analysis and conclusions. *Ecology of Food and Nutrition*, **17**, 41–64.

Osborne, R., Grove, A., Oh, P., Mabry, T., Ng, J. & Seawright, A. (1994). The magical and medicinal usage of *Stangeria eriopus* in South Africa. *Ethnopharmacology*, **43**, 67–72.

Peters, C.R. (1987). *Ricinodendron rautanenii* (Euphorbiaceae): Zambezian wild food plant for all seasons. *Economic Botany*, **41**, 494–502.

Peters, C.R. & Maguire, B. (1981). Wild plant foods of the Makapansgat area: a modern ecosystems analogue for *Australopithecus africanus* adaptations. *Journal of Human Evolution*, **10**, 565–83.

ers, C.R. & O'Brien, E.M. (1981). The early hominid plant-food niche: insights from an analysis of plant exploitation by *Homo, Pan,* and *Papio* in eastern and southern Africa. *Current Anthropology*, **22**, 127–40.

rek, S.R.K., Graven, E.H. & Whitfield, P. (1982). *Some Potentially Important Indigenous Aromatic Plants for the Eastern Aspects*. The Hague: Martinus Nijhoff Publishers.

ston-Whyte, E.M. (1987). *Black Women and the Craft and Curio Trade in KwaZulu and Natal*. Pretoria: Human Sciences Research Council.

in, P.J. (1959). *Food and Feeding Habits of the Pedi*. Johannesburg: Witwatersrand University Press.

belo, A.G. & Holmes, P.M. (1988). Commercial exploitation of Brunia albiflora (Bruniaceae) in South Africa. *Biological Conservation*, **45**, 195–207.

kard, C. & Mthembu, E. (1992). Natal's sangomas need the muti. *Weekly Mail*, **December 11**, 7.

din, R.J. (1985). The ethnobotany of the KwaNyama Ovambos. *Monographs in Systematic Botany from the Missouri Botanical Garden*, **9**, 1–165.

age, M.T.D. (1985). Health. In *Basic Need. in Rural Areas. South African National Scientific Programmes Report*, **116**, 49–66.

udder, T. (1962). *The Ecology of the Gwembe Tonga. Kariba Studies 2*. Manchester: Manchester University Press.

ackleton, C. (1993). Demography and dynamics of the dominant and woody species in a communal land and protected area of the eastern Transvaal Lowveld. *South African Journal of Botany*, **59**, 569–74.

ackleton, S.E. (1990). Socio-economic importance of Cymbopogon validus in Mkambati Game Reserve, Transkei. *South African Journal of Botany*, **56**, 675–82.

aw, E.M. (1938). Owambo knives. *Annals of the South African Museum*, **24**, 253–75.

aw, E.M. (1992). The basketwork of southern Africa. Part 1. Technology. *Annals of the South African Museum*, **10**, 53–248.

Shaw, E.M. & Van Warmelo, N.J. (1972). The material culture of the Cape Nguni. Part 1. Settlement. *Annals of the South African Museum*, **58**, 1–101.

Shaw, E.M & Van Warmelo, N.J. (1988). The material culture of the Cape Nguni. Part 4. Personal and general. *Annals of the South African Museum*, **58**, 447–949.

Sim, T.R. (1920). South African Rubber. *The South African Journal of Industries Bulletin*, **53**, 1–28.

Steyn, H.P. & du Pisani, E. (1984). Grassseed., game and goats: an overview of Dama subsistence. *Journal of the South West Africa Scientific Society*, **39**, 37–52.

Story, R. (1952). A botanical survey of the Keiskammahoek district. *Memoirs of the Botanical Survey of South Africa*, **27**, 1–227.

Story, R. (1958). Some plants used by the Bushmen in obtaining food and water. *Memoirs of the Botanical Survey of South Africa*, **30**, 1–115.

Talukdar, S. (1983). The conservation of *Aloe polyphylla* endemic to Lesotho. *Bothalia*, **14**, 985–9.

Taylor, F.W. (1982). *Final Report on the Potential for Commercial Utilization of Veld products. Volume 2. The Resource and its Commercial Potential*. Gaborone: Government Printer.

Taylor, F.W. & Moss, H. (1982). *Final Report on the Potential for Commercial Utilization of Veld Products. Volume 1. The Resource and its Management*. Gaborone: Government Printer.

Terry, M.E. (1984). *Handcraft Survey Ngamiland CFDA. A Survey of Basketmakers, Etsha, Ngamiland, Botswana*. Gaborone: Botswanacraft Marketing Company.

Terry, M.E. (1987). *Handcraft Survey Ngamiland CFDA. The Basket Industry of Gomare and Tubu*. Gaborone: Botswanacraft Marketing Company.

Turner, V. (1967). *The Forest of Symbols*. Ithaca: Cornell University Press.

van der Merwe, A. le R., Burger, I.M. & Wehmeyer, A.S. (1967). *Suid-Afrikaanse Veld-kosse: 1. Makhatini Vlakte, Noord Natal*. Pretoria: National Food Research Institute, CSIR.

van Oosterhout, S.A.M. (1990). A question of cultural context: formal taxonomy versus peasant classification of *Sorghum bicolor* in Zimbabwe. *Mitteilungen aus dem Institut für Allgemeine Botanik Hamburg*, **23b**, 953–8.

van Voorthuisen, E.G. & Odell, M. (1976). Thatching in Botswana: the socioecology of traditional construction. *Botswana Notes and Records*, **8**, 165–74.

van Warmelo, N.J. (1937). *Tshivenda–English Dictionary. Ethnological Publications 6*. Pretoria: Government Printer.

van Wilgen, B.W., Bond, W.J. & Richardson, D.M. (1992). Ecosystem management. In *The Ecology of Fynbos. Nutrients, Fire and Diversity*, ed R.M. Cowling. pp. 345–71. Cape Town: Oxford University Press.

van Wyk, M.A. (1987). *'n Kritiese Evaluering van die Verdeeling van die Maroela*. BSc (Hons) Project. Pretoria: University of Pretoria.

Viljoen, B. (1994). Honeybush tea for extra income. *Farmers Weekly*, **4 March**, 24–5.

von Teichmann, I. (1983). Notes on the distribution, morphology, importance and uses of the Anarcardiaceae. 2. The importance and uses of *Sclerocarya birrea* (the marula). *Trees in South Africa*, **35**, 1–7.

von Wissel, L.C. (undated). *Dad's Story*. Unpublished manuscript compiled from the notes of L.C. von Wissel. Durban: Killie-Campbell Library.

Wainwright, J., Schonland, M.M. & Candy, H.J. (1977). The toxicity of *Callilepis laureola*. *South African Medical Journal*, **52**, 313–15.

Watt, J.M. & Breyer-Brandwijk, M.G. (1962). *Medicinal and Poisonous Plants of Southern and Eastern Africa*, 2nd Edition. Edinburgh: E & S Livingstone.

Wehmeyer, A.S. (1966). The nutrient composition of some edible wild fruit found in the Transvaal. *South African Medical Journal*, **40**, 1102–4.

Wehmeyer, A.S. (1986). *Edible Wild Plants of Southern Africa: Data on the Nutrient Contents of over 300 species*. Unpublished report. Pretoria: CSIR.

Wehmeyer, A.S., Lee, R.B. & Whiting, M. (1969). Nutrient composition and dietary importance of some vegetable foods eaten by the !Kung Bushmen. *South African Medical Journal*, **43**, 1529–30.

Welgemoed, Z. (1993). Heuningtee straks kommersioel verbou. *Landbou Weekblad*, **18 June**, 52–4.

White, F. (1976). The underground forests of Africa: a preliminary review. *Gardens Bulletin (Singapore)*, **29**, 55–71.

Whitlow, J.R. (1979). *The Household Use of Woodland Resources in Rural Areas*. Salisbury: Natural Resources Board.

Wild, H. (1952). *A Rhodesian Botanical Dictionary of African and English Plant Names*. Salisbury: Government Printers.

Wilson, K.B. (1990). *Ecological Dynamics and Human Welfare. A Case Study of Population, Health and Nutrition in Southern Zimbabwe*. PhD Thesis. London: University College.

Human impacts on vegetation 21

M.T. Hoffman

21.1 Introduction

This account deals with the influence of humans on the vegetation of southern Africa since hominids first emerged in this region at least 3 Myr BP. Its main objective is to show where and when humans have lived, for how long, and what their major activity and impact has been on the vegetation of the biomes of the region. Details on the impacts of harvesting of individual plant species by humans for food, fibre and shelter are given in Chap. 20 (this volume). An important theme throughout this account is a comparison of the extent of different human impacts over time. In doing this I have tried also to examine the 'classical' view of land degradation in southern Africa. This view holds that because of the ability of modern technologies to transform extensive areas with relative ease, and the demands made on the land by an ever-increasing human population, it has been during the modern industrial era of the last several decades that the most devastating influences on the vegetation of the subcontinent have taken place (Scholes & Walker 1993).

Why is it important to know this, or indeed any region's human disturbance history? Knowing the extent of transformation enables one to partition climatic and anthropogenic effects and to seek important interactions where they might exist. This facilitates prediction of the dynamics and successional trajectories, if present, of the vegetation. It also leads to a greater understanding of the patterns and mosaics that occur in many southern African landscapes at different levels of scale, since many human activities leave their imprint on the landscape long after they occurred. But, more important, a knowledge of southern Africa's disturbance history also helps in the construction of baseline vegetation patterns

that may have existed before the arrival of humans in significant numbers in the region. A comparison of today's conditions with these hypothesized pre-human constructs enables us to measure the extent of change that has taken place in the region (Forman & Russell 1983) and to devise appropriate management strategies for the sustainable use of our plant resources (Campbell & Child 1971; Macdonald 1989; Thomas & Shaw 1991; Seely & Jacobson 1994).

21.2 Humans in southern Africa: a general chronology

Hominids have been in southern Africa for at least the last 3 million years (Volman 1984; Thackeray et al. 1990) and some of the earliest human fossils in the world have been found at comtemporary savanna and savanna-grassland biome ecotonal sites in South Africa (Fig. 21.1a, Table 21.1). Although the Australopithecines and other forms ancestral to modern humans used primitive stone tools and undoubtedly used fire (Thackeray et al. 1990), no significant trace of their impact on the landscape remains today.

The Stone Age in southern Africa is presented as a single technological stage in Fig. 21.1b and Table 21.1. This is a simplification of an interesting and complex period characterized by both long spells of technological stability and shorter periods of innovation with considerable spatial variation across the subcontinent (Deacon 1986). Although archaeological sites are scarce in some areas (e.g. parts of the Nama-karoo biome and arid sav-

Table 21.1 General chronology of hominids in southern Africa indicating the broad time frame for each technological stage, the economy, distribution, tools and main impacts on the vegetation and landscapes of the region (from Volman 1984; Deacon 1986; Thackeray et al. 1990; Thomas & Shaw 1991)

Time	Technological stage	Distribution	Economy	Tools	Main impacts on the vegetation and landscape
3–1 Myr BP	Australopithecines	Sites rare: savanna biome and savanna–grassland biome ecotone	Hunter-gatherer-scavenger	Simple, large flaked stone tools, such as hand axes	Selected harvesting of plants and animals. Burning of vegetation. Populations small, probably little significant impact
1.5 Myr BP to historic times	Stone Age	Throughout southern Africa	Hunter-gatherer-fisher	Mostly stone tools: large hand axes and cleavers; microliths and hafted bladelets appear later	Selected harvesting of plants and animals. Burning of vegetation. Low impact of seasonal encampments
2000 BP to historic times	Khoikhoi pastoralists	West of 400 mm isohyet: desert (in part), Nama-karoo (in part), fynbos, succulent karoo biomes	Hunter-gatherer-fisher; pastoralism	Stone tools, microliths, bone and wooden tools	Selected harvesting of plants and animals. Burning of vegetation. Grazing by domesticated animals. Low impact of nomadic settlements
AD 300 to AD 1000	Early Iron Age	Savanna biome	Mixed subsistence farming	Stone, wooden and iron tools	Plant foods harvested. Trees for iron smelter fuel and construction. Hunting and trade in animal products. Burning. Grazing. Woodlands cleared for cropping especially in savanna lowlands
AD 1000 to present	Late Iron Age	East of 400 mm isohyet: savanna, grassland biomes	Mixed subsistence farming	Stone, wooden and iron tools	As above, but human populations now greater and impact extended to highland grasslands. Local impact of settlements
AD 1652 to AD 1800	Early period of European settlement	Fynbos, succulent karoo, Nama-karoo biomes	Cash crops; mixed subsistence farming	Wooden and iron tools; firearms	Trees harvested for construction and fuel. Burning, grazing and cropping. Increase in hunting activities. Local and regional landscape transformation
19th century	Middle period of European settlement	Throughout southern Africa	Cash crops; mixed subsistence farming; mining	Wooden and iron tools; firearms; machines make their appearance	As above, but impact extended to wider southern African region. Widespread reduction in wild ungulate and mega-herbivore biomass. Increase in domestic stock numbers. Mining impacts. Widespread introduction of alien plants
20th century	Industrialized southern Africa	Throughout southern Africa	Modern agricultural, mining and industrial economy	Industrialized agriculture	Regional landscape transformation as a result of the impact of settlements, agriculture practices and transportation networks. Air, water and soil pollution

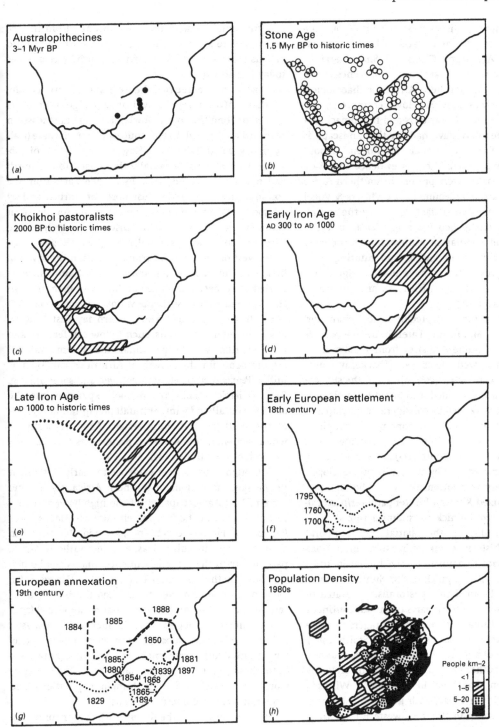

Figure 21.1 **General chronology of hominids in southern Africa.** (*a*)
**Locations where Australopithecine fossils have been found (Wood
1993).** (*b*) **Earlier, Middle and Later Stone Age sites in southern
Africa (Deacon 1986; Thomas & Shaw 1991).** (*c*) **Generalized
distribution of Khoikhoi pastoralists in southern Africa (Van Der
Merwe 1983; Bergh & Visagie 1985; Klein 1986).** (*d*) **Generalized
distribution of Early Iron Age agropastoralists in southern Africa
(Maggs 1984; Thomas & Shaw 1991).** (*e*) **Distribution of Late Iron
Age agropastoralists in southern Africa. Solid line indicates**
**confirmed record; dotted line is unconfirmed distribution (Van Der
Merwe 1983; Maggs 1984; Ajayi & Crowder 1985).** (*f*) **Boundaries
of the Cape Colony during the 18th century (Reader's Digest
1984).** (*g*) **The annexation of southern Africa by European
countries in the 19th century (Fullard 1967; Reader's Digest 1984).**
(*h*) **Population density of southern Africa in the late 1980s (ENPAT
1994).**

anna of the Kalahari), this may be the result of inadequate research in the area rather than a general absence of Stone Age people (Thomas & Shaw 1991).

Evidence of stone tool-making and subsistence activities are present in the southern African archaeological record from as early as 2 Myr BP. However, more extensive knowledge of human behaviours and the impact that these communities may have had on the landscape is very limited before the Upper Pleistocene (about 128 000 BP) (Volman 1984). There is evidence that San-like *Homo sapiens sapiens* were present in southern Africa about 115 000 years ago (Beaumont, De Villiers & Vogel 1978) and Later Stone Age artifacts appear in the archaeological record from 40 000 years ago (Deacon 1984). However, hominid population numbers were probably too small and their technological and hunting techniques too unsophisticated to have made a significant impact on the biota of Africa before 40 000 BP (Siegfried & Brooke 1995).

Details of the cultural and technological characteristics of the Earlier, Middle and Later Stone Age periods and the location of Stone Age sites from 1.5 Myr BP to historic times are well described (Thackeray 1981; Volman 1984; Deacon 1984, 1986). Although the subcontinent's history may be divided into broad technological stages, each is characterized by considerable overlap and interaction with other dominant economies of the time, particularly during the last 2000 years (Denbow 1984; Elphick 1985; Hall 1987; Smith 1992a). For example, where contact occurred, San hunter-gatherers developed complex client–servant relationships with Khoikhoi pastoralists and a mixed Khoisan hunter-gatherer/pastoralist economy resulted (Elphick 1985; Smith 1992a).

Despite much debate, the spatial and temporal sequence of prehistoric Khoikhoi pastoral migrations into southern Africa remains unresolved (Deacon 1984; Klein 1986; Hall 1987; Elphick 1985; Smith 1992a,b). However, it is undisputed that pastoralists migrated to the western and southwestern parts of the subcontinent (Fig. 21.1c), often displacing aboriginal foragers to more marginal environments (Parkington 1984; Smith 1987). The first evidence of a pastoral economy (sheep and pottery remains) in southern Africa emerges in the archaeological record from about 2000 BP (Denbow & Wilmsen 1986; Hall 1987; Smith 1992a) with indications of a significant sheep presence in the southwestern Cape from around 1600 BP (Smith 1992a). Although the origin of cattle pastoralism among the Khoikhoi is unknown, cattle appear in the archaeological record only later, around 1300 BP in the western Cape (Smith 1992b) and then in low numbers. However, an analysis of bones in archaeological remains of sites in the western Cape (Smith 1992b) suggests that by 900 BP the ratio of sheep

to cattle remains had changed significantly from an earlier ratio of 10 : 1 to 4 : 1. This is similar to the 3 : 1 ratio of small to large stock in African herding economies today (Smith 1992b).

Most authors agree that the earliest occupation of the southern African region by Iron Age agropastoralists occurred along the coastal lowlands of the northeastern seaboard and spread further southwards from there (Hall 1987) (Fig. 21.1d, Table 21.1). However, the details of subsequent movements, especially those back to the north and northwest, are disputed. By AD 200 Early Iron Age communities were settled just north of Durban and by AD 500 the colonization of the eastern and southeastern savanna regions of southern Africa by Early Iron Age agropastoralists was essentially complete (Maggs 1984). Areas west of the Great Escarpment, within the savanna biome of South Africa, Botswana and Namibia, were settled later (between AD 400 and AD 800) than areas in the east and southeast (Denbow 1984; Maggs 1984; Hall 1987). In the southeast, near modern-day East London, environmental constraints, such as low soil fertility and poor grazing and in particular the summer-rainfall requirements for the growth of African cereals such as millet (*Pennisetum* sp., *Eleusine coracana*) and sorghum (e.g. *Sorghum bicolor*), hampered further expansion into the all-year rainfall and winter-rainfall precipitation regimes of the southern and southwestern Cape, respectively, where semi-nomadic Khoikhoi pastoralists were already well established (Hall 1987).

Although there is clear evidence for early settlement in the savanna regions of southern Africa, there is no trace of Iron Age occupation of the high-lying and treeless grasslands in the first millennium AD (Maggs 1984; Fig. 21.1e). From around AD 900–1000, however, an abrupt change in cultural aspects in southern Africa, particularly in the ceramic sequence, marks the boundary between the Early and Late Iron Age (Maggs 1984; Table 21.1). From AD 1300 to AD 1400 the first signs of settlement in the highveld grasslands appear (Maggs 1984). Hall (1987) suggests that the increase in domestic animal numbers and subsequent resource depletion may have facilitated the movement of Late Iron Age farmers onto these excellent grazing lands that had not previously supported sedentary human populations and large numbers of domestic animals.

From the middle of the seventeenth century onwards, southern Africa was host to yet another human migration. However, this time it did not come from the interior regions of the subcontinent but from Europe. Initially, these tradesmen and farmers, who had come to establish a victualling station to service travellers on the important East Indian trade route, were content to remain close to their point of entry at Cape Town. By

Table 21.2 **The population of southern Africa in the 20th century and population density in 1995. Values are calculated from exponential models of population growth which used the actual returns of population censuses for the five southern African countries from 1904 to 1991 (Mitchell 1982; Domschke & Goyer 1986; Thomas & Shaw 1991; Famighetti 1993; Cooper et al. 1994). The models are of the form $y = \exp^{A(X-B)}$ where y is the number of people, A is the annual rate of population increase, X is the year and B is a constant. The different models have between seven and nine degrees of freedom and all were significant at $p < 0.0001$.**

Country	Area (km^2)	No. of people km^{-2}	Population				Model parameters		
			1900	1950	1980	1995	A	B	R^2
Botswana	581 730	2.35	68 771	331 858	853 259	1 368 185	0.0315	1546.15	0.9835
Lesotho	30 355	61.12	209 332	660 026	1 314 607	1 855 297	0.0230	1366.56	0.9429
Namibia	823 144	2.09	83 485	410 880	1 069 001	1 724 287	0.0319	1544.45	0.9957
South Africa	1 220 927	36.39	3 263 041	12 896 547	29 416 026	44 426 210	0.0275	1354.33	0.9946
Swaziland	17 363	54.87	44 373	222 900	587 086	952 792	0.0323	1568.53	0.9883
Total	2 673 519	18.82[a]	3 669 002	14 522 211	33 239 979	50 326 771			

[a] Mean density (No. of people km^{-2}) for southern Africa in 1995.

1700, nearly 50 years after their arrival, the colonists had expanded little further than the Berg River in the north and the Hottentots Holland Mountains in the east (Penn 1987) (Fig. 21.1*f*). However, by the end of 1739, after a number of devastating conflicts between the KhoiSan on the one hand and the *trekboere* (semi-nomadic European pastoralists) and the authorities at the Cape on the other, all suitable pastoral land south of Namaqualand and west of the Bokkeveld mountain range near modern-day Ceres was controlled by the colonists (Penn 1987).

For the next century, a steady diffusion of hunters, pioneer farmers and semi-nomadic, *trekboer* pastoralists filtered across the Cape landscape (Table 21.1; Fig. 21.1*f*; Christopher 1982). By the end of the eighteenth century, the European authorities at the Cape laid claim to much of the territory south of the Orange River (excluding the Upper Karoo and Bushmanland) and west of the Bamboesberg/Fish River axis (Bergh & Visagie 1985). From 1835 onwards, the floodgates to the interior of the rest of the subcontinent were opened. The discovery of enormous mineral and agricultural wealth in these regions meant that they were quickly annexed by European powers, all vying to extend and support their growing colonial empires (Fig. 21.1*g*).

21.3 Population growth and land degradation

It is difficult to estimate the number of people living in southern Africa prior to and during the early period of European colonization. Lee & De Vore (1976) speculate that there were between 150 000 and 300 000 San hunter-gatherers in the broader southern African region in AD 1650 but that this number dropped precipitously following assimilation into other cultures and the coordinated extermination campaigns of the Dutch and English settlers in the seventeenth, eighteenth and nine-

teenth centuries. Only a few speculative and patchy data for Khoikhoi pastoralists and Bantu-speaking farmers exist (Elphick & Malherbe 1989), but Christopher (1982) has documented the growth of the European population in the seventeenth and eighteenth centuries. By the beginning of the twentieth century, there were between three and four million people in the southern African region (Table 21.2). Today, there are more than 50 million people in southern Africa, and most are concentrated in the savanna and grassland biomes of the northeastern and southeastern parts of the subcontinent (Fig. 21.1*h*). In fact, just 17 of the more populous magisterial districts of the Gauteng urban complex in South Africa, covering less than 0.5% of the surface area of southern Africa, contain more people today (6 018 702) (Anon. 1993) than were recorded in the combined 1904 census figures of all five of the southern African countries put together (5 837 226) (Namibian values derived from exponential model – see Table 21.2 for data sources).

Despite the roughly 3% increase in the numbers of humans added to the southern African population per annum since AD 1900, there is a danger of being too simplistic in linking increased human population directly with environmental deterioration. More people on the land does not automatically translate to increased degradation, as the example from the Machakos district in southwestern Kenya has shown (Tiffen, Mortimore & Gichuki 1994). In this savanna landscape, documented evidence suggests that despite a five-fold increase in human numbers between 1930 and 1989, there has been a ten-fold increase in the value of output per hectare and an apparent general 'improvement' in the overall condition of the environment. Unfortunately, no equivalent analysis of the environmental history of a region exists for any area in southern Africa.

Understanding the ecological limits to population growth is vital if we are to develop accurate predictions of the future impacts of humans on southern African landscapes. However, as is the case in the rest of the

Table 21.3 **The impact of cultivation and urbanization on six Veld Types (Acocks 1953) within the fynbos biome. Data are from Moll & Bossi's (1984) analysis of 1981 Landsat imagery, in which areas of remaining vegetation**[a] **of 100 ha or more were mapped, and from Rebelo (1992)**

Veld Type (and synonymy)[b]	Original Area[c] (km²)	Remaining area (km²)	Natural vegetation remaining (%)	Main impact on the vegetation
Strandveld (dune thicket)[d]	4 453	2 072	47	Alien plants, cultivation, urbanization
West coast[e]	—	(1 920)	(47)	
South coast	—	(1 306)	(95)	
Mountain renosterveld	4 754	3 448	73	Cultivation, grazing
Coastal renosterveld	15 285	2 256	15	Cultivation grazing
West coast	—	(273)	(4)	
Southwest coast	—	(200)	(4)	
South coast	—	(3 570)	(46)	
Coastal macchia (fynbos)	8 770	4 627	53	Cultivation, aliens, urbanization
Macchia (fynbos)	18 345	16 305	89	Alien plants
False macchia (fynbos, grassy fynbos)	18 965	18 347	97	Alien plants

[a] Moll & Bossi's (1984) assessments do not take into account the impact of alien plant species on natural vegetation as it was not always possible from the satellite imagery to separate, in a consistent manner, indigenous vegetation from alien vegetation. Therefore, the percentage of natural vegetation is an overestimate.
[b] Synonymy according to Cowling & Holmes (1992).
[c] As mapped from Acocks' (1953) Veld Type delimitation.
[d] Only includes that part in the Cape Floral Kingdom.
[e] Data for lower divisions of dune thicket and renosterveld are from Rebelo (1992).

world (Pulliam & Haddad 1994), southern African ecologists have not forged links with social scientists, economists, educationists, health workers and politicians to develop appropriate protocols that deal effectively with this issue.

21.4 Human impacts on the biomes of southern Africa

In selecting specific human impacts on the biomes of southern Africa, I have focused on broad issues such as overgrazing, cultivation, mining, pollution, deforestation and afforestation. It is these issues that have formed the basis for concern over the sustainable land use of southern African environments and the livelihoods of its people. I have tried to place these themes in a historical and chronological context, suggesting when, where and at whose hands the disturbance took place and the severity of the impact on the flora as well as on the human populations of the region and their livelihoods.

21.4.1 Displacement, new crops and aliens in the fynbos biome

Humans have influenced the structure and composition of the vegetation of the fynbos biome in a number of ways including the use of fire and the harvesting of food items; grazing by their domestic animals; cultivation and alien plant introductions. More recently, localized and poorly quantified impacts include the harvesting of the flowers of attractive fynbos species for an overseas export market (Mustart & Cowling 1992) and culm har-

vesting of restioids such as *Thamnochortus insignis* for thatching purposes (Linder 1990).

The fynbos biome is topographically and geologically heterogeneous, and not all landscapes and vegetation types within the biome have been used by humans to the same extent. It has been mainly the easily accessible and more productive coastal lowlands, comprising renosterveld and, to a lesser extent, fynbos and coastal thicket (strandveld), that have supported the many and varied human activities of the region from antiquity to the present (Talbot 1971). Although significantly transformed by alien plant species in places (Chap. 22, this volume) the mountainous areas of the fynbos biome have not been subjected to the same level of human impact as the coastal lowlands (Moll & Bossi 1984; Rebelo 1992) (Table 21.3). This theme is expanded upon below.

The subsistence economy of the San, the first human inhabitants of the region, rested on the collection of underground plant organs (Deacon 1984; Parkington 1984; Kaplan 1987). The fibrous corm casings of genera within the Iridaceae (*Gladiolus*, *Ixia*, *Moraea*, *Watsonia*) usually dominate the food debris found in archaeological excavations throughout the fynbos biome. However, there is a range of other species, without easily preserved plant parts, that could also have been important to the hunter-forager economy (Metelerkamp & Sealy 1983). Although this is speculative, it is possible that populations of rare, yet valuable food items, such as *Dioscorea elephantipes*, *Fockea* spp., *Caralluma* spp. and *Hoodia* spp. could well have been affected by harvesting (see also Chap. 20, this volume). Story's (1958) contemporary study of the use of plants by Kalahari San communities supports the suggestion that food collection may decimate populations of rare species.

Both geophytes and grasses of the fynbos biome coastal plain communities are most abundant in immediate post-fire environments (Hoffman, Moll & Boucher 1987; Chap. 6, this volume). To create a dependable food supply, as well as to promote forage for prey such as zebra, eland and hartebeest, there is evidence that fire-stick farming has been practised for at least the last 100 000 years in this region (Deacon 1992). However, these human-imposed fire regimes, with a shorter fire interval, were probably not widespread across all landscapes in the biome. Archaeological evidence suggests that the activities and impacts of San hunter-gatherers would have focused initially on the coastal plains. The mountains would have been subject to significant human disturbance only after the arrival of Khoikhoi pastoralists with their domestic animals during the last two millennia. This occupation and utilization of the lowlands by the pastoralists, and the reduction of large wild ungulates as a result of the influx of sheep and later cattle, meant that the hunter-gatherer economy had to shift both in space and content as pastoralism took hold in the region. There is archaeological evidence for a displacement of hunter-gatherers to the more mountainous and marginal environments of the fynbos biome around 1800 BP. Furthermore, there was a shift in tool assemblages and a greater emphasis on smaller food parcels, dominated, even more than before, by the collection of plants (Metelerkamp & Sealy 1983; Parkington 1984; Kaplan 1987; Manhire 1987; Smith 1992b). It follows then that the mountain environments of the fynbos biome would have been subjected to significant human-imposed fire regimes and impacts only after the relatively recent arrival of nomadic pastoralists in the area. They may thus represent environments with some of the shortest human disturbance histories on the subcontinent.

The relatively rapid influx of Khoikhoi pastoralists into the fynbos biome coastal lowlands brought an entirely new disturbance regime for the region, caused mainly by the increase in animal and human population numbers. Unfortunately the vegetation that existed before the domination of a pastoralist economy is not well described for either the west or the south coast lowlands. Nevertheless, there is little archaeological and historical evidence to suggest that the 1500 or so years of Khoikhoi pastoralism in the region resulted in anything that could be described as environmental degradation or that the colonists, who very rapidly dispossessed and displaced the Khoikhoi, had inherited a wasteland. On the contrary, the regular seasonal movements of Khoikhoi and their livestock between vegetation types appeared to be well-suited for the maintenance of livestock and veld condition (Smith 1987).

Initially, these wide-ranging transhumance patterns were practised by the colonial stock-farmers. However, these movements were not adopted because they were perceived as the best way to exploit these environments on a sustainable basis. Rather, groups of *trekboere* shared grazing and water resources and moved their animals in large herds as protection against their Khoikhoi and San protagonists (Penn 1987). However, as this threat declined in the early eighteenth century as a result of disease and dispossession and under the exploitation by the colonists of natural resources upon which the Khoikhoi and San economies were based (Cruz-Uribe & Schrire 1991), so rights to individual tenure were implemented and more sedentary herding systems emerged. The first signs of degradation in the west coast lowlands were recorded as early as 1728. Vleis were drained and pasturage depleted (Penn 1986; 1987) under the pressure of significantly higher human and especially domestic animal numbers.

One of the greatest impacts that humans have had on the fynbos biome landscape is undoubtedly that of cultivation. Cultivation was possible only following the introduction by European colonists of food crops that could thrive in the mediterranean-type climates of the region, such as wheat, oats and vines. This impact was concentrated on the relatively fertile shale- and granite-derived soils of the lowlands. Initially, crop cultivation was profitable only if conducted close to the markets and if croppers could make use of established transportation networks (Talbot 1971). Consequently, farmlands around Stellenbosch and Paarl formed the nucleus for this early industry, with extensive livestock farming associated with the outlying areas on the edge of the colony (Penn 1986). However, as the frontier expanded in the late eighteenth century and as markets increased and transportation networks were improved, so too did the focus of the grain-cropping industry shift to the relatively clay-rich renosterveld soils of the west, southwest and south coast lowlands. Despite the current beleaguered status of all coastal renosterveld (Table 21.3; Chap. 6, this volume), different regions appear to have had distinctly different cultivation histories. Data from agricultural censuses suggest that the west coast lowlands were cleared earlier than those of the southwest and south coast (Fig. 21.2). In the latter areas, extensive clearing for wheat, oats, barley and rye increased significantly between 1921 and 1963, probably influenced by the incentives fostered by the promulgation of legislation in 1930 that protected South African producers (Talbot 1971). Islands of natural vegetation within croplands of the southwest and south coast lowlands are thus probably younger than those of the west coast; an important historical consideration in studies on the effects of landscape fragmentation on plant and animal extinctions.

Whatever the temporal sequence of transformation,

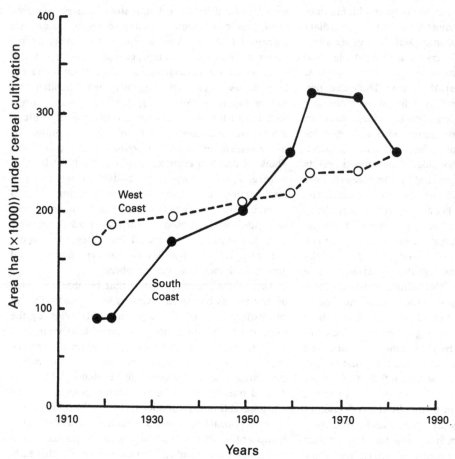

Figure 21.2 **Area (ha) under cereal cultivation (barley, oats, rye and wheat) between 1918 and 1981 in four magisterial districts on the west coast of the fynbos biome (Hopefield, Malmesbury, Piketberg, Vredenburg) (open circles and dashed lines) and seven magisterial districts on the southwest and south coast (Bredasdorp, Caledon, Heidelberg, Hermanus, Mossel Bay, Riversdale, Swellendam) (closed circles and solid lines). Data are from Agricultural Censuses for 1918, 1921, 1934, 1949, 1959, 1963, 1973 and 1981.**

the statistics are staggering. Nearly 96% of the original vegetation of both the west and southwest coast renosterveld has been transformed, mainly by agriculture (McDowell 1988; Rebelo 1992). Similarly, just more than half of the south coast renosterveld has been transformed (Table 21.3). What makes this loss of habitat even more alarming is the knowledge that these shrublands once contained an extraordinary wealth of plants, many of which are now listed as rare and endangered species (Parker 1982; Rebelo 1992; Fig. 21.3).

One other very significant and relatively recent impact that humans have had on the vegetation of the fynbos biome arises from alien plant infestations. All important introductions postdate the permanent settlement of European colonists in the Cape in 1652 and most originate from intentional introductions between 1830 and 1900 (Macdonald & Richardson 1986; Macdonald, Kruger & Ferrar 1986; Shaughnessy 1986). The motivation for the introductions are varied: aesthetic (*Hakea* spp., Pinus spp.); supply of construction materials (*Pinus* spp., *Eucalyptus* spp.); and stabilization of mobile sand dunes (*Acacia* spp.). In general, fynbos shrublands of both

the mountains and coastal lowlands have been most affected (Table 21.3). The history of alien plant introductions and the extent of transformation are dealt with elsewhere (Chap. 22, this volume).

Figure 21.3 **West coast renosterveld remnants near Cape Town. Threats from both cultivation and creeping urbanization confine pockets of this vegetation type to isolated fragments on low hillocks or koppies** *(Photo: D.M. Richardson).*

21.4.2 Desertification in the arid zone

The Nama-karoo, succulent karoo and desert biomes comprise the most arid portions of the subcontinent with much of the region experiencing rainfall of less than 250 mm yr^{-1} (Chaps 7, 8 and 9, this volume). Changes in land-use practices over time reveal a shift from hunter-gatherer economies to localized, and in some cases relatively short-lived, nomadic pastoralism, and finally to the present widespread small stock industry, which is based largely on the sedentary agricultural practices of commercial farmers. The timing and impacts on the vegetation of these different land-use practices are discussed below.

The archaeology of arid and semi-arid southern Africa is poorly known and has been described mostly for areas on the periphery of the region (Humphreys 1987). Nevertheless, there is evidence for hunter-gatherer occupation of this region from well before 200 000 BP with a series of distinct breaks during the last 30 000 to 40 000 years. The central interior of the Nama-karoo biome, for example, may have been devoid of humans for 5000 years between 9500 and 4500 BP (Humphreys 1987). Hunter-gatherers coexisted with Khoikhoi pastoralists (Elphick 1985; Sampson *et al.* 1989) when the latter arrived in the region after 2000 BP. By the middle of the nineteenth century, however, and much earlier in some cases, hunting and gathering as a way of life had ceased to exist in these biomes.

What impact did hunter-gatherers have on the vegetation of the arid zone? In the Seacow River valley of the eastern Karoo, San camps were responsible for a significant degree of disturbance covering more than 5% of the area of the doleritic habitats around spring sources (Sampson 1986). Many surface and gully erosion scars as well as the dominance of disturbance-related plant species such as *Lycium* spp. and *Tribulis terrestris*, can be traced to hunter-gatherer occupation of the region (Sampson 1986).

From about 2000 BP Khoikhoi nomadic pastoralists began to use the drier portions of the subcontinent (Klein 1986; Smith 1992b). However, their occupation was neither widespread throughout the region nor was it continuous. Also, in some areas, unequivocal evidence for the presence of pastoralists in the archaeological record is present only from the tenth or eleventh centuries AD (Klein 1986). Current indications are that Khoikhoi were present during the last two millennia in the high-rainfall parts of Namaqualand and the Little Karoo of the succulent karoo biome (Parkington 1984; Smith 1992b; see Fig. 21.1c) and in parts of southern Namibia and in isolated pockets within the desert biome particularly along rivers and at the Brandberg (Elphick 1985; Kinahan 1986; Klein 1986; Smith 1992b). They appeared a little later in places along the Orange River and in the Northern Cape (Butzer, *et al.* 1979; Klein 1986) and have a discontinuous record of occupation from the tenth century in the southern part of the Seacow River valley, on the eastern periphery of the Nama-karoo biome (Sampson *et al.* 1989).

Little is known of the possible impacts that these early pastoralist communities may have had on the vegetation of the region. However, reconstructed transhumance patterns for the Khoikhoi groups in Namaqualand (Webley 1986) suggest that frequent movements within a 100-km transhumance orbit would have prevented over-exploitation. Adjoining winter-rainfall (succulent karoo) and summer-rainfall (Nama-karoo) regions were also exploited seasonally (Webley 1986). The most arid parts of the succulent karoo biome, however, would probably have been avoided. Simon Van Der Stel's 1668 expedition, for example, noted that the arid region north of the bend in the Olifants River, in what is today known as the Knersvlakte, supported only San hunter-gatherers (Parkington 1984).

The arrival of European farmers on the subcontinent and their slow but steady diffusion into the arid zones during the seventeenth to nineteenth centuries has been well described (Botha 1923, 1926; Thom 1936; Van der Merwe 1938; Talbot 1961; Christopher 1976, 1982; Penn 1986, 1987). These settlers had two profound impacts. Firstly, populations of indigenous animals such as zebra, wildebeest and springbok were either substantially reduced or eliminated and some animals, such as the quagga, were hunted to extinction. In a relatively short time a diverse array of indigenous ungulates were replaced by three or four domestic species. Secondly, much of southern Africa's arid zone that had never before supported a pastoralist economy, was settled. Although farming was initially confined to transhumance orbits centred on perennial springs (Van der Merwe 1938), modern technologies, and in particular the establishment of boreholes in the late nineteenth century, enabled farmers to settle in previously inhospitable environments (Talbot 1961). The first written records that warn of the degradation of the karoo environment date from this time (Brown, 1875; Shaw, 1875).

The southern African desertification debate has been well documented (Hoffman & Cowling 1990; Dean *et al.* 1995; Hoffman, Bond & Stock 1995). The issue is not only of academic interest, as the South African, Namibian and Botswanan governments have all acted at one time or another to 'combat the threat of desertification' either through stock reduction schemes or some other form of state control such as drought subsidies. While the debate has benefited enormously from the contributions of a wide range of disciplines, there is some disagreement as to the severity and cause of the problem (Table 21.4).

The debate is centred on two key issues: (a) the nature

Table 21.4 palaeoecological, historical, ecological and stable isotopic perspectives on rainfall conditions and the dominance of grass and shrub cover in the eastern karoo environment from AD 700 to the present

Period AD	Palaeoecological evidence	Historical evidence	Ecological evidence	Stable isotope evidence (Bond et al. 1994)
700–1700	Alternating grassland/shrubland physiognomy depending on rainfall conditions (Avery 1991; Sampson 1994; Bousman & Scott 1994)	No data	Alternating grassland/shrubland physiognomy depending on rainfall conditions (Roux 1966; Hoffman et al. 1990; Milton & Hoffman 1994) Perennial grassland (Acocks 1953)	Broad but relatively stable transition zone with a decrease in grass cover from northeast to southwest
1700–1800	High rainfall, good grass cover (Avery 1991) Low rainfall, poor grass cover (Sampson 1994; Bousman & Scott 1994)	High rainfall towards end of 18th century, good grass cover (Raper & Boucher 1988)	Good grass cover towards end of 18th century (Hoffman & Cowling 1990)	
1800–1900	Poor grass cover at beginning but increasing towards end of 19th century (Avery 1991) Poor grass cover at beginning and end of 19th century (Sampson 1994)	Large fluctuations in rainfall and general environmental conditions 1800–1840 (Van der Merwe 1938; Hoffman & Cowling 1990) High rainfall 1887–1902 (Hoffman et al. 1995)	Poor grass cover at beginning but increasing towards end of 19th century (Hoffman & Cowling 1990; Hoffman et al. 1995) Decrease in grass cover towards end of 19th century (Roux & Vorster 1983)	
1900–1950	Decrease in grass cover (Bousman & Scott 1994)	Poor grass cover (Hoffman & Cowling 1990)	Poor grass cover (Roux & Vorster 1983; Hoffman & Cowling 1990; Hoffman et al. 1995)	
1950–present	Increase in grass cover (Bousman & Scott 1994)	Increase in grass cover in 1950s, decrease in 1960s and general increase since early 1970s (Hoffman & Cowling 1990; O'Connor & Roux 1995)	Increase in grass cover in 1950s, decrease in 1960s and general increase since early 1970s (Hoffman et al. 1995) Increase in short-lived grasses, increase in shrub cover (Roux & Vorster 1983)	Decrease in grass cover especially in more recent samples and an increase in shrub cover

of pre-colonial environments; and (b) the rate of expansion of karroid shrublands into more productive grasslands to the north and east of the Nama-karoo biome. Initial suggestions were that the pre-colonial eastern karoo was dominated by perennial and highly palatable 'climax' grasses, and that this sward has been replaced in historical times by unpalatable dwarf shrubs (De Klerk 1947; Tidmarsh 1948; Acocks 1953; Roux & Vorster 1983). These changes were ascribed either to overgrazing or to incorrect management methods. Traveller's records between 1773 and 1836, however, suggest no consistent differences in vegetation appearance between early colonial times and the present, suggesting that claims of a perennial grassland in the eastern karoo may have been exaggerated (Hoffman & Cowling 1990). These authors suggest an environment characterized by large fluctuations in grass and shrub components, largely dependent on summer rainfall conditions over short- (one year) and medium-term (2–5 years) rainfall cycles (Milton & Hoffman 1994; O'Connor & Roux 1995; Chap. 8, this volume). Recent palaeoecological studies find good evidence for an alternating grassland/shrubland physiognomy dependent on rainfall conditions during the last millennium (Avery 1991; Bousman & Scott 1994; Sampson 1994; Fig. 21.4; Table 21.4).

An analysis of stable carbon isotopes in the soil organic matter (SOM) of eastern karoo soils has recently contributed significantly to the debate (Bond, Stock & Hoffman 1994). Since C_4 grass and C_3 shrub SOM contains different $\delta^{13}C$ signatures it is possible to distinguish between grass- and shrub-dominated environments within different soil layers and thereby construct past vegetation histories. This study has shown that the eastern karoo has always represented a broad but relatively stable transition zone with a decrease in grass cover from northeast to southwest (Table 21.4). Signatures from soil samples at the bottom of profiles at all sites indicate a greater C_4 component, suggesting more widespread grassiness at some time in the past, supporting the early ideas of Acocks (1953). However, because of dating problems it is not possible to time this reduction in grassiness accurately. This is important because the onset of shrubby conditions in the eastern karoo predated the arrival of European stock farmers in the region (Scott & Bousman 1990; Bousman & Scott 1994; Sampson 1994) and must, therefore, have been driven by climate change. There are other suggestions that stock-herding had little effect, if any, on vegetation structure up to the beginning of the present century (Avery 1991).

The debate concerning the recent degradation of semi-arid rangelands and the rate of expansion of shrub vegetation in the eastern Karoo has been every bit as lively as that concerning pre-colonial vegetation composition. Many earlier reports suggest a karoo 'desertification front' advancing at a predictable rate of between 1 and 5 km yr^{-1} (see Hoffman & Cowling 1990). These alarming statistics have been quoted uncritically in many popular and scientific reviews (e.g. Huntley, Siegfried & Sunter 1989; Moll & Gubb 1989). Using both resurvey and matched photographic techniques, Hoffman & Cowling (1990) demonstrated little evidence for this expansion and questioned the entire notion of a degraded eastern karoo. Recently, however, a study of historical stocking rates for all magisterial districts within semi-arid South Africa for the period 1911–1981 concluded that the dramatic drop in stock numbers in nearly all districts in recent years provides incontrovertible evidence for the degradation of these environments (Dean & Macdonald 1994). In support of this, the stable carbon isotope analysis of SOM also shows that the most dramatic changes in eastern karoo environments have occurred during the past few decades, when the greatest shift from grass to shrub cover occurred (Bond *et al.* 1994; Fig. 21.5; Table 21.4).

It is possible, however, that the reduction in stock numbers in the eastern karoo is related to the expensive and effective stock reduction and subsidy schemes implemented by the South African government over the last two decades (Du Toit, Aucamp & Bruwer 1991) rather than rangeland degradation (Hoffman *et al.* 1995). The above-average rainfall in the eastern Karoo over the last two decades as well as survey data indicate an increase in overall cover and palatable species composition since the early 1960s (Hoffman *et al.* 1995). Clearly, these divergent theories on desertification in the eastern Karoo must be underpinned by a solid temporal framework before any consensus can be reached (Fig. 21.6).

21.4.3 Agricultural history and impacts in the grassland and savanna biomes

21.4.3.1 *The impact of the Iron Age*
Crop farmers first entered southern Africa along the northeastern coastal margins in or before the third century AD (Maggs 1984). Initially they survived on a mixed agricultural base of 'slash-and-burn' agriculture, hunting and marine mollusc collection augmented possibly by domestic sheep and goats (Maggs 1984; Hall 1987). At first only the vegetation around the coastal forest margins was cleared, but within a few hundred years descendants of these early farmers had moved westwards along river valleys and further southwards along the coast. The clearing of parts of the original forests and woodlands and subsequent cropland abandonment led, in the space of a few hundred years, to an increase in the extent of scrub and grassland habitats on the eastern coastal forelands and river valleys (Feely 1980). This may

(a)

(b)

Figure 21.4 **The eastern Karoo is subject to dramatic fluctuations in grass cover influenced largely by summer rainfall events as are shown in the three matched photographs of Koffiebus and Teebus near Hofmeyr in the eastern Karoo.** (*a*) was taken by I.B. Pole Evans in September 1925 and shows a badly degraded site with almost no grass cover at all. (*b*), taken by M.T. Hoffman in August 1989 after a number of seasons of above-average summer rainfall, indicates increased grassiness of short-lived perennial *Aristida* spp. and *Eragrostis* spp. By September 1993 (*c*), however, the grass cover had virtually disappeared, leaving the landscape open, with a distinct shrubby appearance.

Figure 21.5 **The timing of grazing has an important influence on species composition and structure in the vegetation of the eastern karoo, as shown in this demonstration experiment conducted over many years at the Middelburg Agricultural Development Institute.** If a paddock is grazed only in summer (right) then grass cover is dramatically reduced and dwarf shrub species dominate. However, if only grazed during the winter months (left) then grass cover dominates the landscape with shrubby species assuming a less prominent role.

have facilitated the range expansion of a number of indigenous plants (e.g. many early-successional *Acacia* species) and animals (e.g. white rhinoceros) (Feely 1980). It also brought with it a shift in domestic livestock composition. The reliance of the Early Iron Age farmers on browsing animals such as goats changed fairly rapidly to an increasing dependence on a cattle-based economy that now thrived on the abundant grass- and scrubland mosaic created by this early slash-and-burn agriculture (Maggs 1984; Hall 1987). Recently, however, McKenzie (1989) has challenged this general model of increased grassiness following Iron Age occupation of the eastern seaboard. He argues that in the Transkei, Iron Age population densities would have been too low for their activities to have resulted in significant increases in the extent of grasslands.

Archaeological remains of Early Iron Age settlements are found almost exclusively within the savanna biome.

(a)

(b)

Figure 21.6 **Some parts of the succulent karoo biome have shown
enormous stasis through time, as is shown in this pair of matched
photographs taken south of the Touwsberg near Ladismith.
Although there has been some mortality, the majority of** *Pappea
capensis* **and** *Euclea undulata* **trees and tall shrubs, evident in I.B.
Pole Evans' May 1919 photograph (***a***), are still present in M.T.
Hoffman's May 1993 image (***b***). One important difference is that
many of the trees and tall shrubs appear to have increased in size
and volume in the intervening 84 years. Boundaries demarcating
low shrubland vegetation dominated by** *Pteronia pallens* **also
appear not to have changed.**

These farmers chose the valley bottoms to build their villages (Maggs 1984), preferring alluvial soils for their crops of sorghum, millet and cucurbits such as pumpkins, gourds and melons. Other ecological prerequisites affecting site location were an abundant supply of wood and adjacent pasturage for cattle (Maggs 1984). Although Early Iron Age agropastoralists owned livestock, they also relied on hunting to supplement their diets. Remains of hippopotamus, crocodile and especially fish are present at these early sites. Villages which generally contained a few hundred people and varied in size from 8 to 20 ha, enjoyed a high level of self-sufficiency. Village density was surprisingly high, with one located every few kilometers (Maggs 1984).

The success of Early Iron Age farmers and their impact on the savanna landscape of the time was mainly due to their use of iron for a variety of agricultural and domestic purposes. Iron axes, for example, were essential for woodland clearing, and iron hoes also increased the range of tillable soils (Hall 1987). Iron tools, such as adzes and hoes increased the efficiency of tilling the soils and tending and harvesting the millet, sorghum, cow pea and cucurbit crops. In fact, these Early Iron Age farmers were so successful that Huffman (1979, 1982) has suggested, albeit in an unresolved and controversial hypothesis (Hall 1987), that it was population pressure and competition for resources that precipitated the north and westward migration of farmers across the Drakensberg escarpment into savanna lowland environments during the sixth, seventh and eighth centuries.

The manufacture of iron tools requires not only a good supply of iron ore but also an abundant supply of fuelwood to fire the furnaces and iron smelters. Van der Merwe & Killick (1979) have calculated that nearly 7000 trees (mostly hardwoods such as *Colophospermum mopane*, *Combretem imberbe* and *Terminalia sericea*) would have been required to produce the 180 metric tons of slag produced from six furnaces at a site near Phalaborwa over 'an arbitrary lifetime of 30 years'.

This extensive clearing of the bottomlands by Iron Age people for fuelwood, iron production and cultivation has left its mark on many savanna landscapes of today. In northern KwaZulu-Natal, for example, as much as 70% of the area which currently forms part of the lowland nature reserve network may be derived directly from Iron Age land-use patterns (Feely 1980). The 'wilderness model' concept for these reserves has been questioned (Feely 1980; Granger *et al.* 1985), since a set of secondary successional pathways adequately explains the structure and composition of the comtemporary vegetation. Clearing of the original closed deciduous woodland on the interfluves in the Eastern Transvaal lowveld by Late Iron Age farmers for iron smelting, construction

materials and fuelwood purposes probably increased runoff and erosion rates (Feely 1980). This would have led to significant changes in the vegetation of vleis and marshes in drainage lines. The draining and subsequent cultivation of these wetlands, coupled with the changed runoff regimes from the interfluves, would have led to an invasion of these drainage lines by woody plants following their later abandonment (Feely 1980).

The transition from Early to Late Iron Age towards the end of the first millennium AD is marked by dramatic cultural, agricultural and economic developments with concomitant changes to the disturbance regime of the savanna and grassland biomes at both landscape and regional scales. First, settlement location shifted from bottomland sites to hilltops with a greater reliance on stone materials for hut and perimeter wall construction (Maggs 1984; Hall 1987). Second, the interior, treeless grasslands, including those west of the escarpment, were colonized for the first time in the Late Iron Age. However, this spread was not uniform across the grassland biome. There were clear initial preferences for savanna/ grassland biome ecotonal sites where transhumance patterns presented a range of ecological bet-hedging strategies well suited to the agricultural economies of the time (Maggs 1984). The 'sourveld' grasslands on the acid soils at the base of the Drakensberg in the Free State, for example, were avoided (Maggs 1984). Finally, during the Late Iron Age the earlier emphasis on self-sufficiency at the village level changed to an economy based on the association of political power and wealth with cattle and the development of regional population centres with long-distance trade links (Hall 1987).

The increasing importance of cattle in the agricultural economy of the Late Iron Age led to a range of ecological problems. The rise and fall, in the ninth and fourteenth centuries, respectively, of a number of particularly well-developed economic centres in the Limpopo Basin (Hall 1987), and eastern Kalahari (Denbow 1984), provides evidence for the potentially devastating impact that these early farmers could have had on southern African savanna and grassland biome landscapes. The collapse in the fourteenth century of the Limpopo Basin state centred around Mapungubwe may have been as closely related to the deterioration of the grazing lands in the eastern Kalahari as to the shift in trade networks to more northerly centres in Great Zimbabwe (Denbow 1984; Maggs 1984; Hall 1987). These authors propose that the large cattle holdings of the settlements in the eastern Kalahari were crucial for supplying and maintaining the regional political centres in the Limpopo Basin. The deterioration of the grazing lands as a result of excessive grazing pressure resulted in a reduction of cattle numbers below that which was needed to maintain the regional trade and political networks.

The impact of Iron Age settlements, kraals and iron smelters appears widespread in savanna and grassland biome landscapes (Maggs 1984; Granger et al. 1985). Evidence of these early settlements remains and their impacts have contributed significantly to the level of patchiness and productivity in modern savanna landscapes (Scholes & Walker 1993). In the eastern Kalahari, for example, the productive and palatable blue buffalo grass, *Cenchrus ciliaris*, is consistently associated with vitrified dung deposits of Iron Age, nineteenth century and modern kraal sites (Denbow 1979). The ability of *C. ciliaris* to tolerate high nitrate and phosphate levels as well as its dense, mat-forming growth habit preclude the establishment of surrounding arid savanna trees on old kraal sites. These grassy sites are easily discernible as 'bald spots' on aerial photographs and are common, especially on hilltops, where they have not been destroyed by recent cultivation. The kraals are generally 50–150 m in diameter with vitrified and semi-vitrified dung deposits up to a metre deep, providing some indication of the extent of utilization of these arid savanna landscapes by Iron Age and recent agropastoralists (Denbow 1979).

One final example of the impact of Iron Age farmers on pre-colonial landscapes concerns the Difequane (the 'scattering'; Hall 1987). The military conquests of Shaka Zulu and others during the early part of the nineteenth century resulted in mass regional displacement and political restructuring, especially within the eastern and northern parts of the subcontinent. Although, it is rejected by Hall (1987), there remains a popular perception (Barker et al. 1988) that the region underwent a rapid increase in human and cattle populations under the favourable climatic conditions of the late eighteenth century. This led ultimately to an ecological collapse during the series of severe droughts that occurred early in the nineteenth century. In the ensuing territorial conflict, particularly over the lowlands of KwaZulu-Natal (Maggs 1984), smaller clans and tribes were amalgamated within larger groups to form more effective armies and ultimately consolidated into a broader northern Nguni society under Shaka Zulu, who died in 1828.

In the aftermath of the Difequane the European colonists entered the largely depopulated grassland and savanna biomes from the 1830s onwards, adding further to the territorial displacement of some Iron Age farming communities and to the restructuring of land-use practices in the two biomes. Within a few decades much of the subregion had been either annexed or colonized, but not necessarily controlled by Europeans, and some entirely new impacts on the vegetation of the savanna and grassland biomes were introduced.

21.4.3.2 *Megaherbivores and the 'bush encroachment' problem*

One of the first and most significant, albeit indirect, impacts that the early colonists had on the grassland and savanna biomes was the decimation of indigenous herbivore populations and their replacement with a few species of domestic animals (Chap. 17, this volume). Megaherbivores, such as elephant, rhinoceros and hippopotamus, and large grazing animals, such as wildebeest, hartebeest and zebra play key roles in a number of important population and ecosystem processes within the savanna and grassland biomes (Tinley 1977; Milchunas, Sala & Lauenroth 1988; Owen-Smith 1988; La Cock 1992; Chap. 17, this volume). Their elimination is thought to have had catastrophic implications for the normal functioning of ecosystems within these two biomes (Grossman & Gandar 1989). Although Iron Age people had traded in ivory, rhinoceros horn and animal skins for centuries before the arrival of Europeans (Hall 1987), there are no indications that they had a major impact on the populations of southern African megaherbivores, since they lacked the weaponry suitable for mass slaughter. The colonists, however, had firearms and could draw on the hunting and tracking skills of local hunter-gatherers, pastoralists and agropastoralists (Gordon 1984) to supply the huge demand of international markets for animal products, especially ivory. Estimates suggest that there were more than 100 000 elephants in South Africa alone prior to the big-game hunter era in the late eighteenth and nineteenth centuries, but that by the end of 1920 there were fewer than 120 individuals left (Hall-Martin 1992). These were confined to just four small populations mostly in remote parts of the savanna biome. This historical removal of megaherbivores from the savanna landscape, the alteration of fire regimes, the reduced use of trees, and overgrazing, has been blamed for the general 'bush encroachment' problem in the savanna biome today (Grossman & Gandar 1989; Chaps 10, 11, 17 and 18, this volume). Of the approximately 43 million ha comprising the savanna biome in South Africa, bush encroachment has rendered 1.1 million ha unusable, threatens 27 million additional ha and has reduced the carrying capacity of much of the rest of the region by up to 50% (Grossman & Gandar 1989; Figs. 21.7 and 21.8).

21.4.3.3 *Overgrazing*

Extensive livestock ranching is the most common agricultural practice in southern Africa; 84% of land in the savanna biome of South Africa is used for this purpose (Grossman & Gandar 1989). Despite the fact that, in South Africa, cattle, sheep and goat numbers during the last decade have been at their lowest level in 60 years

Figure 21.7 **The influence of land tenure on savanna–grassland dynamics near Kei Road in the Eastern Province. Communal farmers on the right cultivate the landscape and burn the grassland to provide grazing for mixed cattle, sheep and goat herds. Harvesting of wood for construction and fuel requirements also maintains an open landscape. The commercial farm on the left excludes fire and maintains lower stocking rates of mainly cattle herds. A shift to *Acacia karroo*-dominated thornveld is evident on the left, as is the development of dense bush clumps, comprising a wide variety of subtropical thicket elements** *(Photo: D.M. Comins).*

(Fig. 21.9), overgrazing by domestic animals is seen as the main cause for vegetation degradation in the region (e.g. Bosch & Theunissen 1992). For other southern African countries, which lack accurate stock census figures, overgrazing is also cited as one of the most important causes of landscape degradation (Cook 1983; Ringrose *et al.* 1990; Wolters 1994). This debate is most developed in Botswana where controversy exists as to the cause and severity of degradation of this country's savanna biome landscapes. Some studies (e.g. De Queiroz 1993) blame overgrazing by domestic stock for the region's range degradation. Others (e.g. Abel 1993; Dahlberg 1993; White 1993) point to the lack of reputable scientific evidence linking the livestock industry to widespread rangeland degradation. For Botswana's communal farmers the debate is of vital importance, since it impinges directly on their country's livestock development policy.

21.4.3.4 *Cultivation*

Of all modern agricultural practices, crop cultivation probably has the greatest impact on the terrestrial biota of a region. Not only is the relatively diverse cover and composition of natural vegetation replaced by one or a few alien species, but soil destruction and water and nutrient additions further transform the environment. The total area under cultivation in South Africa in 1988 was around 130 000 km² or about 10.6% of the land surface (Anon. 1994; Fig. 21.10). This is very close to the 12 to 15% estimate of total potential arable land area in South Africa (Schoeman & Scotney 1987). Data from agricultural censuses show that there has been a steady increase in the area cultivated between 1911 and 1965, but that this has levelled off in the last two decades. This suggests that most of the productive lands have already been cultivated. Thus, any agricultural expansion of croplands in the future will encroach increasingly on economically and ecologically marginal environments, where yields are lower and environmental impacts, such as wind and water erosion, probably greater. The impli-

(a)

(b)

Figure 21.8 **The general trend towards increased cover of tree and shrub species in savanna environments is shown in this matched photograph pair taken near Peelton in the Eastern Cape. The top photograph (a), taken by D.M. Comins in March 1958, shows a relatively open landscape with a grass-dominated foreground, scattered *Acacia karroo* trees on the hillslopes and subtropical thicket species (e.g. *Euphorbia triangularis, Harpephyllum caffrum, Schotia latifolia*) confined to dense bush clumps along drainage lines. (b), taken by M.T. Hoffman in January 1992, indicates a substantial increase in woody species, especially of *Acacia karroo* in the foreground and on the hillslopes. There is also a significant increase in the size of the bush clumps, which are now beginning to overflow the drainage lines.**

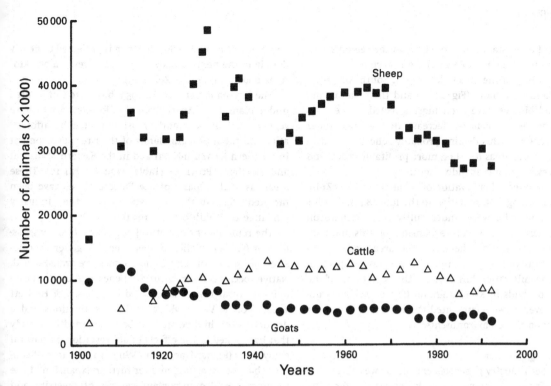

Figure 21.9 **Total number of sheep (filled squares), cattle (open triangles) and goats (closed circles) in South Africa for the period 1904–1992. Data are from Agricultural Censuses (Anon. 1961, 1977, 1994).**

Figure 21.10 **Total area cultivated (ha; filled squares) and the area under maize (closed circles) and sugarcane (open triangles) in South Africa (excluding 'homelands') for the period 1911–1992. Data are from Agricultural Censuses (Anon. 1961, 1977, 1994).**

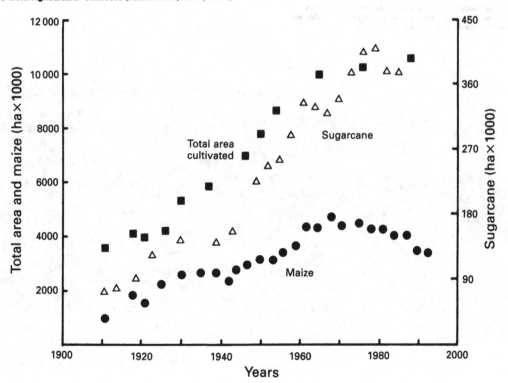

cations of these statistics in the light of the region's 3.0% population increase (Table 21.2) are sobering.

Nearly half of the area cultivated in South Africa has been planted to maize (Fig. 21.10) and the savanna and grassland biomes have been most affected. Since 1965 there has been a general decrease in the area under maize. However, this decline probably reflects a shift in maize-growing areas to other, more profitable crops, and not necessarily land abandonment.

The commercial cultivation of sugarcane in KwaZulu-Natal has a long history dating to the late 1840s. By 1866 just over 5000 ha were under cultivation (Richardson 1985). There was a steady expansion of this industry until the late 1970s, whereafter the area under cultivation decreased (Fig. 21.10). The greatest impact of sugarcane cultivation has been on the vegetation of the coastal lowlands of KwaZulu-Natal (Fig. 21.11). No studies, however, have documented the impact of the sugar industry on these environments.

21.4.3.5 Afforestation

The timber industry generates enormous wealth for the South African economy from both domestic and export sales. It employs more than 150 000 people and is one of the fastest growing sectors of the economy (Endangered Wildlife Trust 1994). Furthermore, the total area under plantations of wattle (*Acacia* spp.), gum (*Eucalyptus* spp.)

and pine (*Pinus* spp.) in South Africa is projected to nearly double in the next 30 years from the current 14 000 km² to 24 000 km² (Van der Zel 1989).

The Knysna forest (Acocks 1953) has 21.30% of its area under plantation. Apart from this forested area, those regions that are currently most impacted by afforestation are the moist grasslands of the Great Escarpment in northern KwaZulu-Natal and in the Eastern Transvaal and Northern Province (Table 21.5; Fig. 21.12). These areas, as well as parts of the Eastern Cape, have been targeted for further expansion of the industry (Endangered Wildlife Trust 1994).

The replacement of natural vegetation (mostly grassland or fynbos) by alien timber species is generally catastrophic, since soil and light regimes are transformed. Native species can rarely survive beneath plantation canopies. Natural runoff is also reduced (Bosch & Hewlett 1982; Scott & Van Wyk 1990). Local extinctions and a dramatic drop in diversity has been recorded in a study that has looked at the effects of afforestation on natural ecosystems (Richardson & Van Wilgen 1986). It is of concern that the areas targeted for further expansion of the industry are also important centres of diversity and endemism (Endangered Wildlife Trust 1994) and in the Eastern and Northern Transvaal provinces, 30% of threatened plants occur in forestry areas (Raal 1986). Afforestation also has severe impacts on catchment runoff and

Figure 21.11 **Fragmentation of coastal lowlands as shown by this relic patch of swamp forest surrounded by sugarcane fields near Stanger, KwaZulu-Natal** *(Photo: E.J. Moll).*

Table 21.5 **Total area (km²) of 29 Veld Types (Acocks 1953) and the area (km²) and proportion of the Veld Type (%) that currently comprises either indigenous forest or plantation timber species (chiefly** *Eucalyptus* **spp. and** *Pinus* **spp.). Veld Types not listed contain neither indigenous forests nor alien timber plantations. Data layers of forest boundaries (Van der Zel 1989) and Veld Types (Acocks 1953) were superimposed using a GIS (ARC/INFO)**

Veld Type No.	Veld Type	Total area (km²)	Indigenous forest (km²)	Plantation (km²)	Indigenous forest (%)	Plantation (%)
45	Natal mist belt 'Ngongoni veld	3 963	70	1030	1.77	25.99
4	Knysna forest	4 051	576	863	14.22	21.30
8	Northeastern mountain sourveld	9 834	371	2030	3.77	20.64
62	Bankenveld to sour sandveld transition	1 500	0	279	0.00	18.60
63	Piet Retief sourveld	9 970	0	1852	0.00	18.58
9	Lowveld sour bushveld	15 658	64	2706	0.41	17.28
57	Northeastern sandy highveld	14 575	1	976	0.01	6.70
64	Northern tall grassveld	6 746	3	451	0.04	6.69
5	'Ngongoni veld	11 619	103	730	0.89	6.28
1	Coastal forest and thornveld	20 574	233	1131	1.13	5.50
44	Highland sourveld and Döhne sourveld	47 960	434	1226	0.90	2.56
3	Pondoland coastal plateau sourveld	792	0	18	0.00	2.27
69	Macchia (fynbos)	18 703	0	302	0.00	1.61
70	False macchia (fynbos)	18 881	15	235	0.08	1.24
65	Southern tall grassveld	15 751	53	180	0.34	1.14
10	Lowveld	29 778	4	331	0.01	1.11
6	Zululand thornveld	5 259	9	51	0.17	0.97
21	False thornveld of eastern Cape	2 622	2	16	0.08	0.61
46	Coastal renosterveld	14 994	3	64	0.02	0.43
2	Alexandria forest	2 002	4	7	0.20	0.35
7	Eastern province thornveld	6 136	14	20	0.23	0.33
23	Valley bushveld	26 730	29	84	0.11	0.31
66	Natal sour sandveld	5 984	0	17	0.00	0.28
61	Bankenveld	28 405	0	54	0.00	0.19
19	Sourish mixed bushveld	32 776	3	23	0.01	0.07
47	Coastal macchia (fynbos)	8 810	0	3	0.00	0.03
56	Highland sourveld to *Cymbopogon–Themeda* veld transition	11 824	0	2	0.00	0.02
20	Sour bushveld	12 417	2	0	0.02	0.00
11	Arid lowveld	17 335	1	0.4	0.01	0.00

Figure 21.12 **Plantations of the alien** *Pinus patula* **covering the Northeastern Mountain Sourveld (Acocks' Veld Type 8) near Kaapsche Hoop, Mpumulanga. A few small remnant indigenous forest patches remain in the foreground** *(Photo: D.M. Richardson).*

(a)

Figure 21.13 **Since the turn of the century there has been a steady shift away from the use of natural products for household construction to imported, synthetic materials. However, this does not always translate into a general decrease in the impact of humans on surrounding landscapes, as these three matched photographs from Taung in the Northern Cape Province show. Photographs (a) and (b) were taken by H.L. Shantz in September 1919 and October 1956, respectively. They suggest a foreground dominated by *Acacia robusta*, with the 1956 image indicating a grass sward as well. (c), taken by M.T. Hoffman in August 1989, indicates a complete transformation of the foreground to an open grassland. Axe-marks, evident on old stumps in the foreground, suggest that this region was cleared for firewood. In recent years there has been a build-up of soil around the previously exposed rocks.**

hence the maintenance of riverine biodiversity and eco-system functioning (Chaps 14 and 22, this volume).

21.4.3.6 *Industrialization and urbanization*

Macdonald (1989) provides a comprehensive analysis of the impact of mining, urbanization and transportation networks on southern African landscapes and these subjects will not be discussed further here. Regarding atmospheric pollution, there is no clear evidence to show that it has had a significant impact on either natural ecosystems or forest plantations (Tyson, Kruger and Louw 1988). However, 'the null hypothesis that (it) has had no measurable impact, needs to be tested for each of the kinds of resources potentially at risk in the (region)' (Tyson *et al.* 1988).

21.4.4 Deforestation and the forest biome

Acocks (1953) proposed that the southern and eastern seaboard of the subcontinent was, until relatively recently (AD 1400?), covered in forest and scrub forest and that this vegetation has 'largely disappeared' as a direct result of both Iron Age and European impacts on the region. However, there is no evidence to indicate that the forest patches on the Winterberg mountains in the Eastern Cape were ever more extensive during the last 12 500 years than they are at present (Meadows & Meadows 1988; see also Chap. 12, this volume). Macroclimatic fluctuations have probably had a greater impact on the vegetation of this region than either Iron Age or European settlers (Meadows & Meadows 1988). Similarly, despite the increased impact of humans on coastal landscapes in the Transkei this century, woody vegetation

(b)

(c)

Table 21.6 Tree products extracted from indigenous forests and alien plantations (chiefly *Pinus* spp. and *Eucalyptus* spp.) in the Union of South Africa between 1918 and 1950 and used for wagon wood, mine props and railway sleepers. (Data from Anon. 1961)

Date	Wagon wood (m³)		Mine props (tonnes)		Railway sleepers (number)	
	Indigenous forest	Plantation	Indigenous forest	Plantation	Indigenous forest	Plantation
1918	—	—	13 287	6 833	—	—
1921	6 954	286	7 851	171 548	65 800	6 100
1926	9 327	498	7 168	264 966	91 654	16 497
1937	4 837	788	914	544 337	13 953	14 599
1946	1 110	1 089	405	717 013	1 900	7 047
1950	—	—	246	685 920	—	—

continues to increase in many areas (McKenzie 1989). Lower population densities of Early and Late Iron Age communities would have been unable to clear the coastal forests in the manner hypothesized by Acocks (1953), although large compositional changes have undoubtedly occurred as a result of Iron Age, and later, human activities (Feely 1980; Granger *et al.* 1985; McKenzie 1989).

Besides the impact of slash-and-burn agriculture on forest landscapes, Iron Age people would also have selected a range of forest products for domestic and agricultural use (Chap. 20, this volume). However, small- to medium-sized poles and saplings would have been favoured and only rarely would large specimens be felled. European colonists, however, not only possessed a different set of tools, but also had very different needs from the forests. Selection focused on large, mature individuals for building timber and for the manufacture of furniture, wagons, ploughs and even boats (Brink & Van Der Zel 1980). Exploitation was swift and numerous government proclamations in the seventeenth, eighteenth and nineteenth centuries attempted to control timber harvesting in the Cape colony (Laughton 1937; Grut 1977; Brink & Van der Zel 1980).

Although the exploitation of the Knysna forests probably began about 1750 (Laughton 1937), the greatest impact of humans during the last 100 years, especially on the indigenous forests of the southern and south eastern coastal regions, was undoubtedly between about 1890 and 1940 (Grut 1965); Table 21.6). Since then, the rise of motorized transport and the increasing dependence on alien plantation species for construction timber has seen a dramatic decline in the use of indigenous forest products.

While much of the industrialized economy of southern African today depends directly on the forestry industry for its wood and paper supply, many rural communities still use indigenous trees and forest products directly for their shelter, tool, fuel, food and spiritual needs; Chap. 20, this volume). Although the full extent of this impact on indigenous forests is not well known, there are clear indications that over-exploitation has

occurred in many areas of southern Africa. Indeed, in a number of regions such as in the Tugela Basin (Edwards 1967), the coastal lowlands of KwaZulu-Natal (Laughton 1937) and in northern Namibia (Shanyengana 1994), Acocks' (1953) assertion that many forests have 'disappeared' as a result of human impacts is supported by available evidence (Fig. 21.13).

21.5 Conclusions

The 'classical' view of land degradation in southern Africa, which holds that the most devastating influences on the vegetation of the region have occurred in the past several decades, is rejected as being too general. The extent and severity of human impacts cannot be viewed simply as an exponential increase with time. Each region has its unique history and even within-biome generalizations are problematic. Certainly many areas in southern Africa fit this general model and remain under threat from increasingly excessive exploitation. Also, the demands made on the land by a burgeoning population are cause for deep concern within both governmental and environmental agencies. However, there are cases (e.g. the Knysna forest, the eastern Karoo, the Limpopo Basin and eastern Kalahari) where severe and nonsustainable exploitation of the resource base occurred earlier this century or even earlier this millennium. In many instances, current legislation, coupled with an increasing environmental awareness amongst land users and managers, has changed exploitation practices to those that are more sustainable in the long term.

It seems that one of the best ways to prepare for the future is to understand the trends and patterns of the past. Although extrapolation of these historical trends into the future is problematic, southern African policy makers and indeed the community at large should take note of the political and cultural devastation created by past environmental degradation such as has been postulated for the Limpopo Basin state in and around Mapungubwe more than 600 years ago (Hall 1987). Perhaps more than anything else we need to understand the long-term implications of the region's rapid population growth rate on our environment, since the health of our primary resource base affects people's livelihoods directly. Indeed, the future of all southern Africa's people rests ultimately on its renewable, natural resources.

21.6 **Acknowledgments**

The Department of Environment Affairs are thanked for financial assistance. Many colleagues have contributed ideas, data and expertise in the development of this manuscript. I thank Richard Cowling, Janette Deacon, Jeanne Hurford, Tim Maggs, Guy Midgley, Jeremy Midgley, Shirley Pierce, Dave Richardson, Garth Sampson, Louis Scott, Andrew Smith and Dan Sonnenberg for their help.

.7 **References**

el, N. (1993). Carrying capacity, rangeland degradation and livestock development policy for the communal rangelands of Botswana. *Overseas Development Institute, Pastoral Development Network Paper,* 35c, 1–9.

ocks, J.P.H. (1953). Veld Types of South Africa. *Memoirs of the Botanical Survey of South Africa,* 28, 1–192.

ayi, J.F.A. & Crowder, M. (1985). *Historical Atlas of Africa.* Essex: Longman.

non. (1961). *Handbook of Agricultural Statistics.* Pretoria: Department of Agriculture.

non. (1977). *Abstract of Agricultural Statistics.* Pretoria: Department of Agriculture.

non. (1993). *Demographic Statistics.* Pretoria: Central Statistical Services.

non. (1994). *Abstract of Agricultural Statistics.* Pretoria: Department of Agriculture.

very, D.M. (1991). Micromammals, owls and vegetation change in the eastern Cape Midlands, South Africa, during the last millennium. *Journal of Arid Environments,* 20, 357–69.

arker, B.J., Bell, P., Duggan, A., Horler, V., Leroux, V., Maurice, P., Reynierse, C. & Schafer, P. (1988). *Reader's Digest Illustrated History of South Africa.* Cape Town: The Reader's Digest Association of South Africa.

aumont, P.B., de Villiers, H. & Vogel, J.C. (1978). Modern man in sub-Saharan Africa prior to 49 000 years B.P.: a review and evaluation with particular reference to Border Cave. *South African Journal of Science,* 74, 409–19.

Bergh, J.S. & Visagie, J.C. (1985). *The Eastern Cape Frontier Zone: 1660–1980.* Durban: Butterworths.

Bond, W.J., Stock, W.D. & Hoffman, M.T. (1994). Has the Karoo spread? A test for desertification using carbon isotopes from soils. *South African Journal of Science,* 90, 391–7.

Bosch, J.M. & Hewlett, J.D. (1982). A review of catchment experiments to determine the effect of vegetation changes on water yield and evapotranspiration. *Journal of Hydrology,* 55, 3–23.

Bosch, O.J. & Theunissen, J.D. (1992). Differences in the response of species on the degradation gradient in the semi-arid grasslands of southern Africa and the role of ecotypic variation. In *Desertified Grasslands: Their Biology and Management,* ed. G.P. Chapman, pp. 95–109. London: Academic Press.

Botha, C.G. (1923). The dispersion of the stock farmer in Cape Colony in the eighteenth century. *South African Journal of Science,* 20, 574–80.

Botha, C.G. (1926). *Place Names in the Cape Province.* Johannesburg: Juta.

Bousman, B. L. & Scott, L. (1994). Climate or overgrazing: the palynological evidence for vegetation change in the eastern Karoo. *South African Journal of Science,* 90, 575–8.

Brink, A.J. & Van Der Zel, D.W. (1980). Die geskiedenis van bosbou in Suider-Afrika. Deel I: Die inheemse bosse. *South African Forestry Journal,* 114, 13–18.

Brown, J. C. (1875). *Hydrology of South Africa.* London: King.

Butzer, K.W., Fock, G.J., Scott, L. & Stuckenrath, R. (1979). Dating and context of rock engravings in southern Africa. *Science,* 203, 1202–14.

Campbell, A.C. & Child, G.F. (1971). The impact of man on the environment of Botswana. *Botswana Notes and Records,* 3, 91–109.

Christopher, A.J. (1976). Emergence of livestock regions in the Cape Colony 1855–1911. *South African Geographer,* 5, 310–20.

Christopher, A.J. (1982). Towards a definition of the nineteenth century South African frontier. *South African Geographer,* 64, 97–113.

Cook, H. J. (1983). The struggle against environmental degradation – Botswana's experience. *Desertification Control Bulletin,* 8, 9–15.

Cooper, C., Hamilton, R., Mashabela, H., Mackay, S., Sidiropoulos, E., Gordon-Brown, C., Murphy, S. & Frielinghaus, F. (1994). *Race relations survey – 1993/94.* Johannesburg: South African Institute of Race Relations.

Cowling. R.M. & Holmes, P.M. (1992). Flora and vegetation. In *The Ecology of Fynbos. Nutients, Fire and Diversity,* ed. R.M. Cowling, pp. 23–61. Cape Town: Oxford University Press.

Cruz-Uribe, K. & Schrire, C. (1991). Analysis of faunal remains from Oudepost I, an early outpost of the Dutch East India Company, Cape Province. *South African Archaeological Bulletin,* 46, 92–106.

Dahlberg, A. (1993). The degradation debate: is clarification possible? *Overseas Development Institute, Pastoral Development Network Paper,* 35c, 10–14.

Deacon, H.J. (1992). Human settlement. In *The Ecology of Fynbos. Nutrients, Fire and Diversity*, ed. R.M. Cowling, pp. 260–70. Cape Town: Oxford University Press.

Deacon, J.C.G. (1984). Later Stone Age people and their descendants in southern Africa. In *Southern African Prehistory and Palaeoenvironments*, ed. R.G. Klein, pp. 221–328. Rotterdam: A. A. Balkema.

Deacon, J. C. G. (1986). Human settlement in South Africa and archaeological evidence for alien plants and animals. In *The Ecology and Management of Biological Invasions in Southern Africa*, ed.. I.A.W. Macdonald, F.J. Kruger & A.A. Ferrar, pp. 3–19. Cape Town: Oxford University Press.

Dean, W.R.J. & Macdonald, I.A.W. (1994). Historical changes in stocking rates of domestic livestock as a measure of semi-arid and arid rangeland degradation in the Cape Province, South Africa. *Journal of Arid Environments*, 26, 281–98.

Dean, W.R.J., Hoffman, M.T., Meadows, M. & Milton, S.J. (1995). Desertification in the semi-arid Karoo, South Africa: review and reassessment. *Journal of Arid Environments*, 30, 247–64.

De Klerk, J.C. (1947). Pastures of the southern Orange Free State a century ago and today. *Farming in South Africa*, April, 347–54.

Denbow, J. R. (1979). *Cenchrus ciliaris*: an ecological indicator of Iron Age middens using aerial photography in eastern Botswana. *South African Journal of Science*, 75, 405–8.

Denbow, J.R. (1984). Prehistoric herders and foragers of the Kalahari: the evidence for 1500 years of interaction. In *Past and Present in Hunter Gatherer Studies*, ed. C. Schrire, pp. 175–93. London: Academic Press.

Denbow, J.R. & Wilmsen, E.N. (1986). Event and course of pastoralism in the Kalahari. *Science*, 234, 1509–15.

De Queiroz, J.S. (1993). Range degradation in Botswana: myth or reality? *Overseas Development Institute, Pastoral Development Network Paper*, 35b, 1–17.

Domschke, E. & Goyer, D.S. (1986). *The Handbook of National Population Censuses*. New York: Greenwood Press.

du Toit, P.F., Aucamp, A.J. & Bruwer, J.J. (1991). The national grazing strategy of the Republic of South Africa: Objectives, achievements and future challenges. *Journal of the Grassland Society of southern Africa*, 8, 126–30.

Edwards, D. (1967). A plant ecological survey of the Tugela River Basin. *Memoir of the Botanical Survey of South Africa*, 36, 1–285.

Elphick, R. (1985). *KhoiKhoi And the Founding of White South Africa*. Johannesburg: Raven Press.

Elphick, R. & Malherbe, V.C. (1989). The Khoisan to 1828. In *The Shaping of South African Society, 1652–1840*, ed. R. Elphick & H. Giliomee, pp. 3–65. Cape Town: Maskew Miller Longman.

Endangered Wildlife Trust (1994). *Biodiversity and Afforestation: a Conservation Strategy*. Unpublished draft report No WO87, May 1994. Johannesburg: Endangered Wildlife Trust.

ENPAT (1994). *Environmental Potential Atlas*, 1st edn. Pretoria: Department of Environment Affairs.

Famighetti, R. (ed) (1993). *The World Almanac and Book of Facts*. New Jersey: World Almanac.

Feely, J.M. (1980). Did Iron Age man have a role in the history of Zululand's wilderness landscapes? *South African Journal of Science*, 76, 150–2.

Forman, R.T.T. & Russell, E.W.B. (1983). Evaluation of historical data in ecology. *Bulletin of the Ecological Society of America*, 64, 5–7.

Fullard, H. (1967). *Philips' College Atlas for Southern Africa*. London: George Philip & Son.

Gordon, R.J. (1984). The !Kung in the Kalahari exchange: an ethnohistorical perspective. In *Past and Present in Hunter Gatherer Studies*, ed. C. Schrire, pp. 195–224. London: Academic Press.

Granger, J.E., Hall, M. ,McKenzie, B. & Feely, J.M. (1985). Archaeological research on plant and animal husbandry in Transkei. *South African Journal of Science*, 81, 12–15.

Grossman, D. & Gandar, M.V. (1989). Land transformation in South African savanna regions. *South African Geographical Journal*, 71, 38–45.

Grut, M. (1965). *Forestry and Forest Industry in South Africa*. Cape Town: Balkema.

Grut, M. (1977). Notes on the history of forestry in the western Cape 1652–1872. *South African Forestry Journal*, 100, 32–7.

Hall, M. (1987). *The Changing Past: Farmers, Kings and Traders in southern Africa*. Cape Town: David Philip.

Hall-Martin, A.J. (1992). Distribution and status of the African elephant *Loxodonta africana* in South Africa, 1652–1992. *Koedoe*, 35, 65–88.

Hoffman, M.T., Barr, G.D. & Cowling, R.M. (1990). Vegetation dynamics in the semi-arid eastern Karoo, South Africa: the effect of seasonal rainfall and competition on grass and shrub basal cover. *South African Journal of Science*, 86, 462–3.

Hoffman, M.T., Bond, W.J. & Stock, W.D. (1995). Desertification of the eastern Karoo, South Africa: conflicting palaeoecological, historical and soil isotopic evidence. *Journal of Environmental Monitoring and Assessment*, 37, 159–77.

Hoffman, M.T. & Cowling, R.M. (1990). Vegetation change in the semi-arid eastern Karoo over the last two hundred years: an expanding Karoo – fact or fiction? *South African Journal of Science*, 86, 286–94.

Hoffman, M.T., Moll, E.J. & Boucher, C. (1987). Post-fire succession at Pella, a South African lowland fynbos site. *South African Journal of Botany*, 53, 370–4.

Huffman, T.N. (1979). African origins. *South African Journal of Science*, 75, 233–7.

Huffman, T. N. (1982). Archaeology and ethnohistory of the African Iron Age. *Annual Review of Anthropology*, 11, 133–50.

Humphreys, A.J.B. (1987). Archaeology. In *The Karoo Biome: a Preliminary Synthesis. Part 2 –Vegetation and History*, ed. R.M. Cowling & P.W. Roux, pp. 117–31. *South African National Scientific Programmes Report* 142. Pretoria: CSIR.

Huntley, B., Siegfried, R. & Sunter, C. (1989). *South African Environments in the 21st Century*. Cape Town: Human & Rousseau Tafelberg.

.plan, J. (1987). Settlement and subsistence at Renbaan Cave. *BAR International Series*, **332**(ii), 350–72.

nahan, J. (1986). The archaeological structure of pastoral production in the central Namib desert. *South African Archaeological Society Goodwin Series*, **5**, 69–82.

ein, R.G. (1986). The prehistory of Stone Age herders in the Cape Province of South Africa. *South African Archaeological Society, Goodwin Series*, **5**, 5–12.

Cock, G.D. (1992). *The Conservation Status of Subtropical Transitional Thicket, and Regeneration Through Seeding of Shrubs in the Xeric Succulent Thicket of the Eastern Cape*. Port Elizabeth: Cape Department of Nature Conservation.

ughton, F.S. (1937). The sylviculture of the indigenous forests of the Union of South Africa with special reference to the forests of the Knysna region. *Science Bulletin*, **157**, 1–168.

e, R.B. & De Vore, I. (1976). *Kalahari Hunter-Gatherers*. Cambridge, Massachusetts: Harvard University Press.

nder, H. P. (1990). Thatching reed. of Albertina. *Veld & Flora*, **76**, 86–9.

acdonald, I.A.W. (1989). Man's role in changing the face of southern Africa. In *Biotic Diversity in Southern Africa. Concepts and Conservation*, ed. B.J. Huntley, pp. 51–77. Cape Town: Oxford University Press.

acdonald, I.A.W., Kruger, F.J. & Ferrar, A.A. (eds) (1986). *The Ecology and Management of Biological Invasions of Southern Africa*. Cape Town: Oxford University Press.

acdonald, I.A.W. & Richardson, D.M. (1986). Alien species in terrestrial ecosystems of the fynbos biome. In *The Ecology and Management of Biological Invasions of Southern Africa*, ed. I.A.W. Macdonald, F.J. Kruger & A.A. Ferrar, pp. 77–91. Cape Town: Oxford University Press.

aggs, T. (1984). The Iron Age south of the Zambezi. In *Southern African Prehistory and Palaeoenvironments*, ed. R.G. Klein, pp. 329–60. Rotterdam: A.A. Balkema.

anhire, A. H. (1987). Sandveld deflation hollows: a study of open site assemblages in the south-western Cape. *BAR International Series*, **332**(ii), 326–49.

McDowell, C. R. (1988). *Factors Affecting the Conservation of Renosterveld by Private Landowners*. PhD Thesis. Cape Town: University of Cape Town.

McKenzie, B. (1989). Medium-term changes of vegetation pattern in Transkei. *South African Forestry Journal*, **150**, 1–9.

Meadows, M. E. & Meadows, K. F. (1988). Late Quaternary vegetation history of the Winterberg Mountains, eastern Cape, South Africa. *South African Journal of Science*, **84**, 253–9.

Metelerkamp, W. & Sealy, J. (1983). Some edible and medicinal plants of the Doorn Karoo. *Veld & Flora*, **69**, 4–8.

Milchunas, D.G., Sala, O.E. & Lauenroth, W.K. (1988). A generalized model of the effects of grazing by large herbivores on grassland community structure. *American Naturalist*, **132**, 87–106.

Milton, S.J. & Hoffman, M. T. (1994). The application of state-and-transition models to rangeland research and management in arid succulent and semi-arid grassy Karoo, South Africa. *African Journal of Range & Forage Science*, **11**, 18–26.

Mitchell, B.R. (1982). *International Historical Statistics – Africa and Asia*. London: MacMillan.

Moll, E.J. & Bossi, L. (1984). Assessment of the extent of the natural vegetation of the fynbos biome of South Africa. *South African Journal of Science*, **80**, 355–8.

Moll, E.J. & Gubb, A.A. (1989). Southern African shrublands. In *The Biology and Utilization of Shrubs*, ed. C.M. Mackell, pp. 145–75. New York: Academic Press.

Mustart, P.J. & Cowling, R.M. (1992). Impact of flower and cone harvesting on seed banks and seed set of serotinous Agulhas Proteaceae. *South African Journal of Botany*, **58**, 337–42.

O'Connor, T.G. & Roux, P.W. (1995). Vegetation changes (1949–1971) in a semi-arid, grassy dwarf shrubland in the Karoo, South Africa: influence of rainfall variability and grazing by sheep. *Journal of Applied Ecology*, **32**, 612–26.

Owen-Smith, R.N. (1988). *Megaherbivores. The Influence of Very Large Body Size on Ecology*. Cambridge: Cambridge University Press.

Parker, D. (1982). The western Cape lowland Fynbos: What is there left to conserve?! *Veld & Flora*, **68**(4), 98–101.

Parkington, J.E. (1984). Soaqua and Bushmen: hunters and robbers. In *Past and Present in Hunter Gatherer Studies*, ed. C. Schrire, pp. 151–74. London: Academic Press.

Penn, N.G. (1986). Pastoralists and pastoralism in the northern Cape frontier zone during the eighteenth century. *South African Archaeological Society Goodwin Series*, **5**, 62–8.

Penn, N.G. (1987). The frontier in the western Cape, 1700–1740. *BAR International Series*, **332**(ii), 462–503.

Pulliam, H.R. & Haddad, N.M. (1994). Human population growth and the carrying capacity concept. *Bulletin of the Ecological Society of America*, **75**, 141–57.

Raal, P.A. (1986). Opname van Transvaalse bedreigde plante. *Fauna and Flora*, **44**, 17–21.

Raper, P.E. & Boucher, M. (eds) (1988). *Robert Jacob Gordon. Cape Travels, 1777 to 1786*, vol. 1. Johannesburg: Brenthurst Press.

Reader's Digest, (1984). *Atlas of Southern Africa*. Cape Town: Reader's Digest Association South Africa.

Rebelo, A.G. (1992). Preservation of biotic diversity. In *The Ecology of Fynbos. Nutrients, Fire and Diversity*, ed. R.M. Cowling, pp. 309–44. Cape Town: Oxford University Press.

Richardson, D.M. & van Wilgen, B.W. (1986). Effects of thirty-five years of afforestation with *Pinus radiata* on the composition of mesic mountain fynbos near Stellenbosch. *South African Journal of Botany*, **52**, 309–15.

Richardson, P. (1985). The Natal sugar industry, 1849–1905: an interpretative essay. In *Enterprise and Exploitation in a Victorian Colony*, ed. B. Guest & J.M. Sellers, pp. 180–97. Pietermaritzburg: University of Natal Press.

Ringrose, S., Matheson, W., Tempest, F. & Boyle, T. (1990). The development and causes of range degradation features in southeast Botswana using multi-temporal Landsat MSS imagery. *Photogrammetric Engineering and Remote Sensing*, **56**, 1253–62.

534 / M.T. Hoffman

Roux, P.W. (1966). Die uitwerking van seisoensreënval en beweiding op gemengde Karooveld. *Proceedings of the Grassland Society of South Africa*, 1, 103–10.

Roux, P.W. & Vorster, M. (1983). Vegetation change in the Karoo areas. *Proceedings of the Grassland Society of South Africa*, 18, 25–9.

Sampson, C.G. (1986). Veld damage in the Karoo caused by its pre-trekboer inhabitants: preliminary observations in the Seacow Valley. *The Naturalist*, 30, 37–40.

Sampson, C.G. (1994). Ostrich eggs and Bushman survival on the north-east frontier of the Cape Colony, South Africa. *Journal of Arid Environments*, 26, 383–99.

Sampson, C.G., Hart, T.J.G., Wallsmith, D.L. & Blagg, J.D. (1989). The ceramic sequence in the upper Seacow valley: problems and implications. *South African Archaeological Bulletin*, 44, 3–16.

Schoeman, J.L. & Scotney, D.M. (1987). Agricultural potential as determined by soil, terrain and climate. *South African Journal of Science*, 83, 260–8.

Scholes, R.J. & Walker, B.H. (1993). *An African Savanna – Synthesis of the Nylsvley study*. Cambridge: Cambridge University Press.

Scott, D.F. & Van Wyk, D.B. (1990). The effects of wildfire on soil wettability and hydrological behaviour of an afforested catchment. *Journal of Hydrology*, 121, 239–56.

Scott, L. & Bousman, C.B. (1990). Palynological analysis of hyrax middens from southern Africa. *Palaeogeography, Palaeoclimatology, Palaeoecology*, 76, 367–79.

Seely, M.K. & Jacobson, K.M. (1994). Desertification and Namibia: a perspective. *Journal of African Zoology*, 108, 21–36.

Shanyengana, E. (1994). Deforestation in northern Namibia. In *Proceedings of Namibia's National Workshop to Combat Desertification*, ed. S. Wolters, pp. 89–94. Windhoek: Desert Research Foundation of Namibia.

Shaughnessy, G. (1986). A case study of some woody plant introductions to the Cape Town area. In *The Ecology and Management of Biological Invasions of Southern Africa*, ed. I.A.W. Macdonald, F.J. Kruger & A.A. Ferrar, pp. 37–43. Cape Town: Oxford University Press.

Shaw, J. (1875). On the changes going on in the vegetation of South Africa through the introduction of the Merino sheep. *Botanical Journal of the Linnean Society*, 14, 202–8.

Siegfried, W.R. & Brooke, R.K. (1995). Anthropogenic extinctions in the terrestrial biota of the Afrotropical Region in the last 500,000 years. *Journal of African Zoology*, 109, 5–14.

Smith, A.B. (1987). Seasonal exploitation of resources on the Vredenburg Peninsula after 2000 BP. *BAR International Series*, 332(ii), 393–403.

Smith, A.B. (1992a). Origins and spread of pastoralism in Africa. *Annual Review of Anthropology*, 21, 125–41.

Smith, A.B. (1992b). *Pastoralism in Africa. Origins and Development Ecology*. Johannesburg: Witwatersrand University Press.

Story, R. (1958). Some plants used by the Bushmen in obtaining food and water. *Memoir of the Botanical Survey of South Africa*, 30, 1–115.

Talbot, W.J. (1961). Land utilization in the arid regions of southern Africa. Part I. South Africa. In *A History of Land Use in Arid Regions*, Arid Zones Research, Volume 17, ed. L. D. Stamp, 299–338. Paris: UNESCO.

Talbot, W.J. (1971). *South Western Cape Province*. Cape Town: South African Geographical Society.

Thackeray, A.I. (1981) *The Holocene Cultural Sequence in the Northern Cape Province, South Africa*. PhD Thesis. New Haven: Yale University.

Thackeray, A.I., Deacon, J., Hall, S., Humphreys, A.J.B., Morris, A.G., Malherbe, V.C. & Catchpole, R.M. (1990). *The Early History of Southern Africa to AD1500*. Cape Town: The South African Archaeological Society.

Thom, H.B. (1936). *Die Geskiedenis van Skaapboerdery in Suid-Afrika*. Amsterdam: Swetz en Zeitlinger.

Thomas, D.S.G. & Shaw, P.A. (1991). *The Kalahari Environment*. Cambridge: Cambridge University Press.

Tidmarsh, C.E.M. (1948). Conservation problems of the Karoo. *Farming in South Africa*, 23, 519–30.

Tiffen, M., Mortimore, M. & Gichuki, F. (1994). *More People, Less Erosion. Environmental Recovery in Kenya*. New York: John Wiley & Sons.

Tinley, K.L. (1977). *Framework of the Gorongoza Ecosystem*. DSc Thesis. Pretoria: University of Pretoria.

Tyson, P.D., Kruger, F.J. & Louw, C.W. (1988). *Atmospheric Pollution and its Implications in the Eastern Transvaal Highveld*. South African National Scientific Programmes Report 150. Pretoria: CSIR.

van der Merwe, J.H. (1983). *National Atlas of South West Africa*. Cape Town: National Book Printers.

van der Merwe, N.J. & Killick, D.J. (1979). Square: an iron smelting site near Phalaborwa. *South African Archaeological Society, Goodwin Series*, 3, 86–93.

van der Merwe, P.J. (1938). *Die Noordwaartse Beweging van Boere Voor die Groot Trek (1770–1842)*. The Hague: Van Stockum & Zoon.

van der Zel, D.W. (1989). *Strategic Forestry Development Plan for South Africa*. Unpublished Report. Pretoria: Department of Environment Affairs.

Volman, T.P. (1984). Early prehistory of southern Africa. In *Southern African Prehistory and Palaeoenvironments*, ed. R.G. Klein, pp. 169–220. Rotterdam: A.A. Balkema.

Webley, L. (1986). Pastoralist ethnoarchaeology in Namaqualand. *South African Archaeological Society, Goodwin Series*, 5, 57–61.

White, R. (1993). Comments. *Overseas Development Institute, Pastoral Development Network Paper*, 35c, 15–18.

Wolters, S. (ed.) (1994). *Proceedings of Namibia's National Workshop to Combat Desertification*. Windhoek: Desert Ecological Research Unit of Namibia.

Wood, B.A. (1993). *Patterns of Hominid Evolution in Africa*. Johannesburg: Witwatersrand University Press.

Alien plant invasions

D.M. Richardson, I.A.W. Macdonald, J.H. Hoffmann and L. Henderson

22.1   Introduction

The importance of invasive alien organisms as major agents of land transformation, disrupters of ecosystem functioning and a threat to biodiversity has increased rapidly over the last 200 years throughout the world. Widespread and damaging invasions of alien organisms are, however, mostly a twentieth century phenomenon (Elton 1958). In many regions, biological invasions currently rank alongside cultivation, deforestation, urbanization and chemical pollution as the major agents of land transformation (Cronk & Fuller 1995).

Organisms that are transported to distant areas by humans usually arrive there without important natural enemies that control population growth in their native ranges. Many of these organisms persist and proliferate only in disturbed areas, for example where resource levels, altered through tillage, fertilization and irrigation, provide environments suitable for invading species. Alien plants that dominate these habitats are termed agrestal or ruderal weeds (Holzner & Humata 1982). There is considerable overlap between the alien floras of such areas in different parts of the world (e.g. Holm et al. 1977, 1979). We do not consider these invaders in this chapter, since they have no marked impact on the natural vegetation.

A small proportion of introduced species is able to invade natural or semi-natural habitats; such plants may be termed 'environmental weeds' (Holzner & Humata 1982). In this chapter we concentrate on species in this category that establish self-perpetuating populations and that spread without the deliberate assistance of people into natural or semi-natural habitats (i.e. more or less intact ecosystems) *and* which produce a marked change in the composition and structure of communi-

ties as well as ecosystem processes. This subset of environmental weeds corresponds roughly with the 'transformer species' as defined by Wells *et al.* (1986b; see later) and, in many cases, with Swarbrick's (1991) category of 'canopy dominant weeds'. The invasive spread of such species poses one of the major threats to the conservation of the remaining natural and semi-natural vegetation of southern Africa. Although alien organisms from almost all taxonomic groups have invaded southern African ecosystems to some extent, alien vascular plants have had the most severe impact on natural vegetation (Macdonald, Kruger & Ferrar 1986). We therefore confine our discussion to alien plants. We also do not deal with the invasive spread of indigenous plants, such as the widespread phenomenon of 'bush encroachment' (see Chap. 21, this volume).

Biological invasions have been relatively well studied in some parts of southern Africa, but poorly studied in others. By far the most work has been done in the fynbos biome where invasions have caused the greatest problems (Macdonald & Richardson 1986). The invasion of unmodified fynbos by alien trees and shrubs and the marked alteration of ecosystem functioning caused by these invaders captured the attention of ecologists from many parts of the world during the *Third International Conference on Mediterranean-Type Ecosystems* held in Stellenbosch in 1982. At that time the prevailing view, greatly influenced by Elton's (1958) classic review of plant and animal invasions, was that disturbance was a prerequisite for the invasion of natural systems by alien organisms. The situation in fynbos appeared to refute this notion. This provided the incentive for launching an international programme, under the auspices of SCOPE

(Scientific Committee on Problems of the Environment of the International Council of Scientific Unions), to reassess key issues regarding biological invasions worldwide (Drake *et al.* 1989).

The southern African part of the SCOPE programme (outlined by Ferrar & Kruger 1983) led to the publication of a series of reviews of the status of alien organisms and of methods for dealing with the problem (references in Drake *et al.* 1989, p. 507). These publications fed into a regional synthesis volume (Macdonald *et al.* 1986) which, with those from other regions, paved the way for global syntheses (Usher *et al.* 1988; Drake *et al.* 1989) in which southern African case studies featured prominently. The SCOPE programme stimulated a considerable amount of research on alien plant invasions in southern Africa and resulted in a steady stream of dissertations and publications through the late 1980s and early 1990s. Most of the studies focused on impacts, aspects of the autecology of prominent invaders, patterns and processes of invasion, and control options and management implications. Several recent papers dealt with the issue of predicting invasiveness. More than two-thirds of published studies dealt with invasions in the fynbos biome. The SCOPE programme also stimulated several intercontinental comparisons that helped to place southern African invasions in perspective and shed light on the processes that drive these invasions (Macdonald *et al.* 1988; Kruger *et al.* 1989; Richardson & Bond 1991; Richardson, Williams & Hobbs 1994b).

Considerable progress has been made in documenting the extent of invasion in South African biomes (Henderson 1992 and references therein). A questionnaire survey of invasions in protected areas was conducted in the mid 1980s (Macdonald 1986, 1991) and this was supported by rapid field surveys of selected reserves (e.g. Macdonald & Nott 1987; Macdonald, Clark & Taylor 1987; Macdonald *et al.* 1988; Macdonald 1988, 1991). The Southern African Plant Invaders Atlas project (SAPIA) was launched in 1994 with the aim of producing maps for all important invaders for the entire region at a resolution of quarter-degree cells.

The control of invasive alien organisms has received considerable attention (see e.g. Macdonald, Jarman & Beeston 1985 for a review of control efforts in fynbos). Southern Africa has an outstanding record of successes with biological control of invasive plants. If one considers the number of target weeds and the number of biological control agents released, then South Africa is the third most active country in the world in this field; only in the USA and Australia have more weeds been subjected to biological control. The southern African biological control programmes were recently reviewed in a special issue of *Agriculture Ecosystems & Environment* (Vol.

37 (1–3); 1991). Another milestone publication was the *Catalogue of Problem Plants in Southern Africa* (Wells *et al.* 1986a), which listed 789 alien species ('a species of foreign origin, but established and reproducing itself as though a native'), of which 47 were labelled 'transformer species' that 'change the character, condition, form or nature of a natural ecosystem over a substantial area'.

The recent reviews and the considerable advances made in understanding processes of invasion allow for improved management of the problem. This chapter draws on this literature to review the problem and to assess the current level of understanding of the processes involved in alien plant invasions in southern Africa. We first summarize the history of the introduction of alien plants to the region and of the appreciation of the problem. Next, we determine which species have invaded, where they came from, why they were introduced and whether they share any features that can explain their success in particular biomes. We then review the extent of invasions by the most important species in each biome. Patterns and processes of invasion are described for some prominent categories of invaders. The range of impacts of plant invasions are then assessed. We then survey the various methods and strategies that have been used to combat invasive alien plants. We also attempt to define trends in the invasion of southern African biomes by alien plants and consider the prospects for dealing with the current invaders and for screening potential invaders. Finally, we discuss whether recent research on plant invasions in southern Africa has shed light on any of the major questions that emerged from the recent SCOPE synthesis. The review is, inevitably, biased in favour of examples from South Africa, and the fynbos biome in particular, where most research has been done and where most progress has been made towards gaining a predictive understanding of invasions. Readers seeking more detailed information for particular biomes or species should consult the reviews cited above.

22.2 The history of plant invasions in southern Africa

22.2.1 Pre-colonial introductions

Deacon (1986) reviewed patterns of human settlement in southern Africa and the archaeological evidence for alien plants and animals. Cereals and other food crops indigenous to central and northern Africa were introduced approximately 1735 years ago, about 265 years after the arrival of sheep. Several other plant species, including *Medicago polymorpha* and *Ricinus communis*, were possibly introduced in sheep's wool at about this time. There is no evidence that any plant species introduced then, or any of the crops such as *Eleusine*, *Pennisetum* and *Sorghum* spp. introduced by Early and Late Iron Age communities which were concentrated in the eastern and southern Transvaal and KwaZulu-Natal, had any significant impact on the vegetation away from cultivation sites. Indeed, no plant species introduced by humans prior to the arrival of European settlers in 1652 seem to have had any significant impact on the natural vegetation (Chap. 21, this volume).

22.2.2 The urge to alter, tame or improve: the impact of colonialism

Plant introductions from Europe and Asia between 1650 and 1806 reflected the need of Dutch colonists to cultivate a wide variety of agricultural and horticultural crops, mainly from their homeland in Europe and from the Dutch possessions in the East. Fifty or more crop plants were introduced within the first few years of European settlement, including many present-day agrestal weeds (Wells *et al.* 1986b). A few species were also introduced from South America for ornament or food. *Pinus pinaster*, now an important transformer species in the fynbos biome, probably arrived in about 1685 (under the name *P. sylvestris*; Shaughnessy 1986) and has the longest history of all transformer species in the region (Fig. 22.1).

Many groups of settlers arrived in southern Africa, especially in the Eastern Cape and KwaZulu-Natal, between 1800 and 1870. Large groups arrived from Britain, Mauritius, Germany, Java and India. Thousands of migrants of many nationalities flocked to the diamond fields in the northern Cape between 1867 and 1870. Hundreds of species of useful plants were introduced (or reintroduced as different strains of species that had arrived much earlier) during this period, and many more with no obvious use arrived as accidental imports. Many of these species are now weeds. Most of the species that now cause the greatest problems as invaders in fynbos arrived between 1825 and 1860 (11 of the 12 most widespread transformers in that biome; Richardson *et al.* 1992). About 65% of the approximately 150 invasive alien plant species in the grassland and savanna biomes of South Africa arrived between 1800 and 1900, and 90% of aliens (and 89% of transformer species) arrived between 1820 and 1950 (Henderson & Wells 1986).

Early attempts at forestry in the fynbos and grassland biomes, and the 'rehabilitation' of indigenous forests (by planting faster-growing aliens in gaps) led to the introduction of many tree species (Poynton 1979a,b; Geldenhuys, le Roux & Cooper 1986). The second half of the nineteenth century saw a concerted effort by the settlers to achieve a total cover of alien trees in the fynbos biome where indigenous trees are scarce. Besides the programme of conventional plantings, very large quantities of seeds of *Pinus pinaster* were scattered in the mountains by foresters and hikers. Many tree species were also planted in the southern Cape (Phillips 1963) and in coastal parts of KwaZulu-Natal (Sim 1905). Afforestation on the northern KwaZulu-Natal coast increased rapidly after 1910, about the same time that the first plantations were

Figure 22.1 **Early European settlers introduced hundreds of species of plants and animals from their countries of origin. The Dutch and Huguenot settlers introduced, amongst others, species of *Quercus*, *Pinus* and *Vitis*, all of which form part of the 'europeanized' landscape which is now regarded as part of the region's cultural legacy. Some of the species introduced at this time, e.g. *Pinus pinaster*, are now major environmental weeds.** *(Photo: D.M. Richardson).*

(a)

(b)

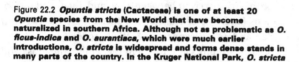

Figure 22.2 *Opuntia stricta* (Cactaceae) is one of at least 20 *Opuntia* species from the New World that have become naturalized in southern Africa. Although not as problematic as *O. ficus-indica* and *O. aurantiaca*, which were much earlier introductions, *O. stricta* is widespread and forms dense stands in many parts of the country. In the Kruger National Park, *O. stricta* is effectively controlled by a phycitid moth, *Cactoblastis cactorum*, in conjunction with herbicide applications. (*a*) shows a thicket of *O. stricta* near Skukuza, Kruger National Park; (*b*) shows severe damage caused by *C. cactorum* on *O. stricta* near Skukuza, Kruger National Park *(Photos: J.H. Hoffmann)*.

established on the Transvaal escarpment (Geldenhuys *et al.* 1986). The rate of afforestation increased rapidly until the early 1980s, when plantations in South Africa covered about 1.1 million ha (see Chap. 21, this volume for further discussion). More than 400 species of alien trees and shrubs (a number equivalent to about 40% of the native tree) have been successfully cultivated in South Africa (Poynton 1984), including at least 134 species of *Eucalyptus* (Poynton 1979b) and more than 80 species of *Pinus* (Poynton 1979a). Large plantations of these species provided major sources of propagules for invasions. A number of the major invasive species were originally introduced as 'barrier plants', including *Caesalpinia decapetala*, the *Hakea* species and *Pereskia aculeata*. In recent years, several woody species introduced for horticultural purposes have become invaders, e.g. *Jacaranda mimosifolia*, *Lantana camara*, *Melia azedarach*, *Nerium oleander* and *Sesbania punicea*. Very few of the woody invaders were introduced unintentionally.

At least 20 *Opuntia* species (Cactaceae) were introduced to the region where they were widely planted throughout the Cape Province, initially for ornamental purposes but later, in the case of several species, as barrier plants, for fodder and for their edible fruits. The two most invasive species, *O. ficus-indica* and *O. aurantiaca*, were already well established in South Africa by 1750 and 1860, respectively (Pettey 1947). By 1772 *Opuntia vulgaris* occurred in the Western Cape, where it became a major weed problem in the coastal regions until it was successfully brought under biological control through the introduction of a cochineal insect in 1913. The other *Opuntia* species (Fig. 22.2) were more recent introductions; *O. rosea* by 1913 and *O. imbricata* by 1930. Other invasive species in different genera (e.g. *Harrisia martinii*)

were introduced in about 1930 and now have a scattered distribution throughout South Africa (Moran & Zimmermann 1991).

In contrast to the situation for the alien woody and perennial succulent plants, most of the non-woody invasive alien species that are now widespread in southern Africa were introduced and dispersed widely within the region unintentionally. Several of the important species in this category, e.g. *Chromolaena odorata*, *Datura innoxia* and *Tagetes minuta*, probably reached southern Africa in fodder or building material imported from South America during wartime [during the Second World War (1939–1945), the Herero Rebellion (1904–1907) and the Boer War (1899–1902), respectively in the case of the three examples cited above].

22.2.3 Appreciation of the problem

Several obscure references from about 1850 allude to the start of widespread invasion of the region's natural vegetation by alien plants. The first campaign against an alien plant, *Xanthium spinosum*, was initiated in 1860, and in 1863 farmers around Bathurst in the Eastern Cape met to discuss the rapid spread of *Hakea sericea* (Phillips 1938). The first attempts to control *Opuntia ficus-indica* and *O. aurantiaca* using various mechanical and chemical methods started in 1883 and 1892, respectively (Moran & Annecke 1979). Despite these developments, it appears that the invasion of *natural* vegetation by alien plants was still not considered a major problem at the time.

The changing views on invasive alien plants over the last 100 years are most clearly chronicled for the fynbos biome. Bolus (1886), in probably the earliest reference to the potential hazard of alien plant invasions in southern African botanical literature (Taylor 1969), wrote that 'it is remarkable how small upon the whole is the influence exerted upon the aspect of the vegetation and how weak ... is their aggressive power against the indigenous flora'. This view was echoed by several botanists (e.g. Bews 1916) over the ensuing 30 years. One of the first references by a botanist to the fact that invasive alien plants were having any serious impact on the indigenous vegetation was made by Sim (1927), who observed that 'the extent to which *Pinus pinaster* can take possession indicates that, if given a long enough period without check, it would probably kill out some of the endemic monotypes'. Eleven years later, Adamson (1938) mentioned that several trees and shrubs, 'spreading near towns and villages', '. . . have either altered or completely changed the character of the vegetation'. He referred to the Australian *Acacia* species which had 'spread over quite large areas and become completely dominant' on parts of the Cape Flats, and the *Hakea* species which were behaving in a similar way on mountain slopes. By 1945 the appreciation of the problem had increased to such an extent that Wicht (1945) wrote that 'One of the greatest, if not the greatest, threats to which the Cape vegetation is exposed, is suppression through the spread of various exotic plant species'. The next three decades saw a rapid increase in awareness of the problem (e.g. Woods 1950; Taylor 1969; Stirton 1978).

The early 1930s marked the start of the rapid spread of *Opuntia aurantiaca*, mainly in the succulent karoo (Moran & Zimmermann 1991), which led to an increase in the awareness of the problem in the more arid parts of southern Africa (*O. vulgaris* had already been brought under control; see 22.2.2). Realization of the problem of invasive alien plants in other biomes increased several decades later [e.g. Macdonald (1983) for Hluhluwe–Umfolozi; Henderson & Musil (1984) for grassland and savannas of the Transvaal; Macdonald (1988) for the Kruger National Park].

In general, awareness of the threats that alien plant invasions pose to the natural vegetation grew more rapidly in southern Africa (particularly South Africa) than in most other continental areas of the world. This rapid increase in awareness was probably due to the fact that many of the important invaders were conspicuous, large woody plants. Another possible reason was the relatively short period of large-scale human alteration of the natural vegetation in southern Africa. In places such as the Mediterranean Basin, this alteration had occurred prehistorically, and many aliens had been established for so long that the distinction between aliens and natives was not as obvious as in South Africa. In some other areas, such as Western Australia, human-induced changes to the environment were much more drastic

and extensive than in southern Africa (i.e. total transformation over larger areas), and there was therefore less interface between human-altered and natural vegetation.

Most of the early management programmes in the fynbos biome were prompted by the threat that tree and shrub invasions were considered to pose to water supplies in the region (Fenn 1980). More recently, regional and local surveys have highlighted the major threat to the endemic flora (Richardson *et al.* 1992). In other parts of southern Africa direct economic implications of invasions were the major reason for increasing awareness of the problem. For example, in the grassland biome most concern was generated by species that threatened grazing capacity [e.g. *Nassella trichotoma* (= *Stipa trichotoma*)]. In the Karoo, *Opuntia aurantiaca* injured the feet and mouths of livestock, and in aquatic habitats *Eichhornia crassipes* reduced the value of water resources (see 22.6). In coastal KwaZulu-Natal and the Eastern Transvaal Lowveld the aesthetic impact of dense stands of climbing and scrambling shrubs/creepers, notably *Cardiospermum grandiflorum*, *Chromolaena odorata*, *Lantana camara* and *Pereskia aculeata*, prompted the concern of authorities regarding alien plant invasions. In several cases (e.g. *C. odorata*, *Solanum mauritianum*), economic impacts of invasive plants in commercial forestry plantations gave further weight to calls by conservationists for effective control programmes to be initiated.

22.3 Which species have invaded?

It is sometimes difficult to decide which alien plants have invaded native vegetation and, in some cases, which plant species are alien to a region (Wells *et al.* 1986a). Given these uncertainties, between 780 and 970 invasive alien vascular plant species had been recorded in South Africa by the mid-1980s (Wells *et al.* 1986a,b). Of these, some 281 to 368 species were recorded as invading native vegetation in nature reserves in a southern African survey (Macdonald 1991).

For this chapter we selected a subset of environmental weeds (invaders of natural or semi-natural habitats), namely the 47 transformer species listed by Wells *et al.* (1986b) and the 67 taxa recorded from at least 10% of reserves in any one biome and which were classified as 'widespread' in a survey of southern African nature reserves (Macdonald 1991). Combining these two lists gives 84 species of environmental weeds (Appendix 22.1) which are discussed according to the biomes in which they were most frequently recorded during the nature reserves study. The percentage of reserves in which the

Figure 22.3 **At least eight *Pinus* species (out of more than 80 species that have been introduced) have invaded natural vegetation in southern Africa. Invasive pines are particularly widespread and troublesome in the fynbos biome, where dense stands of *P. halepensis*, *P. pinaster* and *P. radiata* cover thousands of hectares. The photograph shows a mixed stand of all three species near Stellenbosch. Reproduced, with permission, from Richardson *et al.* (1994b)** *(Photo: D.M. Richardson).*

species was recorded invading (if this was > 10%) and whether the species was categorized as a 'transformer' (T) are listed for the biome in which the species most frequently invades. Many species invade more than one biome (Appendix 22.1).

22.3.1 Major invaders in southern African biomes
22.3.1.1 *The fynbos biome*
The most important category of invaders in this biome is trees and shrubs, mainly from other mediterranean-type climate regions. Species from Australia are *Acacia cyclops* (52T), *A. longifolia* (26T), *A. melanoxylon* (27T), *A. pycnantha* (T), *A. saligna* (51T), *Eucalyptus cladocalyx* (T), *E. diversicolor* (11), *E. lehmannii* (T), *Hakea gibbosa* (T), *H. sericea* (37T), *H. drupacea* (= *H. suaveolens*) (T), *Leptospermum laevigatum* (22T) and *Paraserianthes lophantha* (12). From the Mediterranean Basin come *Pinus pinaster* (45T), *P. pinea* (T), *Populus alba* and *P. canescens* (combined as 14T). The Californian *Pinus*

radiata (19T) is the only North American species in this list (Fig. 22.3). The only non-woody invader on this list is the stoloniferous grass *Pennisetum clandestinum* (18) from East Africa. Alien annual herbs such as *Anagallis arvensis*, *Briza maxima*, *Bromus hordeaceus* and *Medicago polymorpha* are widespread in the fynbos biome (Richardson *et al.* 1992) and are especially common in renosterveld (e.g. Vlok 1988). The extent of such invasions in unmodified vegetation has not been documented.

22.3.1.2 *The grassland biome*
This biome, like the fynbos biome, is naturally depauperate in tree species (Chap. 10, this volume). As in the fynbos, most of the important invaders are Australian tree species (the fact that most of them occur in less than 10% of the reserves reflects the poor reserve coverage of this biome, particularly in the Transvaal grasslands;

Chap. 23, this volume): *Acacia baileyana* (T), *A. dealbata* (T), *A. decurrens* (T), *A. mearnsii* (49T), *A. podalyriifolia* (T), and *Eucalyptus globulus* (14). Other trees and shrubs are *Pinus patula* (35T), *Pyracantha angustifolia* (T) and *Rubus cuneifolius* (27T). Two grass species, *Nassella tenuissima* (T) and *N. trichotoma* (T) (formerly *Stipa* spp.) were also rated as transformer species in the grassland biome (Wells *et al.* 1986a).

22.3.1.3 *The forest biome*
Most of the important invaders of forests are trees, mainly fast-growing species of the forest fringe: *Casuarina equisetifolia* (16), *Cestrum laevigatum* (27T), *Eucalyptus grandis* (14T), *Litsea glutinosa* (= *L. sebifera*) (11), *Melia azedarach* (55T), *Morus alba* (16), *Pinus elliottii* (23), *Psidium guajava* (41T), *Ricinus communis* (39), *Schinus terebinthifolius* (16), *Senna didymobotrya* (formerly *Cassia*) (11) and *Solanum mauritianum* (48T). Several other important invaders are creepers and climbers: *Cardiospermum grandiflorum* (11), *Chromolaena odorata* (61T), *Lantana camara* (50T), *Passiflora edulis* (14) and *Pereskia aculeata* (27) (Fig. 22.4). The only herbaceous species recorded as most frequent in this biome is the ubiquitous 'Black Jack' forb, *Bidens pilosa* (23). Other important invaders are the trees *Acacia melanoxylon*, *A. mearnsii* and the climbing shrub *Caesalpinia decapetala* (L. Henderson, unpubl. data).

22.3.1.4 *The savanna biome*
In contrast to the previous three biomes, most of the widespread invaders in savanna are herbs, the majority of them from South America: *Alternanthera pungens* (17), *Bidens bipinnata* (16), *Chenopodium album* (13), *Conyza alba* (= *C. floribunda*) (10), *Datura ferox* (13), *D. stramonium* (19), *Schkuhria pinnata* (16), *Solanum elaeagnifolium* (10), *Tagetes minuta* (29), *Verbena bonariensis* (10) and *Zinnia peruviana* (13). Several succulents from the Americas are also important: *Agave sisalana* (16), *Cereus jamacaru* (*C. peruvianus* formerly misapplied in southern Africa) (T) and *Opuntia ficus-indica* (59T) (Fig. 22.5). Several other woody species from South America are also widespread: *Caesalpinia decapetala* (11), *Jacaranda mimosifolia* (10T) and *Sesbania punicea* (21). The succulent tree *Euphorbia tirucalli* from Central Africa was also listed as a transforming alien invader in this biome (Wells *et al.* 1986b), although its alien status is questionable (Leach 1973). *Chromolaena odorata* and *Melia azedarach* are also widespread (Henderson & Wells 1986).

22.3.1.5 *The karoo biomes*
Most of the important alien invaders of the Nama- and succulent karoo biomes are from the New World. These are succulents, *Agave americana* (13), *Opuntia aurantiaca* (29T) and *O. imbricata* (33); tree and shrub invaders of watercourses, *Nicotiana glauca* (54T), *Prosopis* spp. (Fig. 22.6; a complex mixture of up to five taxa, mainly *P. glandulosa* and *P. velutina*) (33T); or forbs, *Argemone ochroleuca* (= *A. subfusiformis*) (25), *Polygonum aviculare* (13), *Xanthium spinosum* (46) and *X. strumarium* (25). There are also a few widespread invaders from Eurasia: the tree *Salix babylonica* (17T) and the forbs *Cirsium vulgare* (38) and *Salsola kali* (21).

Most of the alien invaders of the karoo invade both the Nama- and succulent karoo biomes. There are, however, a few trends that appear to differentiate the two

Figure 22.4 ***Pereskia aculeata*** **(Cactaceae), from South and Central America and the West Indies, was widely used as a barrier plant in southern Africa. Seed dispersal by birds has enabled this thorny climber to spread over large areas, particularly in the eastern coastal regions where it forms dense mats that smother the indigenous vegetation** (Photo: J.H. Hoffmann).

Figure 22.5 *Opuntia ficus-indica* (Cactaceae) from Central America was well-established in southern Africa 245 years ago. The dense stands that developed over large parts of the country, especially in the grassland, karoo and savanna biomes, were reduced to occasional, scattered plants by a phycitid moth, *Cactoblastis cactorum*, and a cochineal bug, *Dactylopius opuntiae*, which were introduced for biological control of the weed during the 1930s. Thickets of the weed, such as this one near Uitenhage, persist in the coastal regions of the Eastern Cape, where the biological control agents are less effective and chemical control measures are needed to contain infestations
(Photos: J.H. Hoffmann).

Figure 22.6 An extensive dense stand of mesquite (*Prosopis* spp.; Fabaceae) near VanWyksvlei. This weed, a complex mixture of hybrids of several *Prosopis* species (mainly *P. glandulosa* and *P. velutina*) has invaded large parts of arid southern Africa, especially the Nama-karoo biome. Most dispersal of the weed occurs when animals, including livestock, feed on the seed-pods and discharge the undigested seed.
(Photo: J.H. Hoffmann).

biomes. For example, the succulent *Opuntia* species appear to be more invasive in the succulent karoo biome than in the Nama-karoo. More species seem to invade in sites other than water courses in the succulent karoo. *Nerium oleander* only invades water courses in a relatively narrow band of the succulent karoo biome where this abuts the fynbos biome (Fig. 22.7). This species is not invasive in the fynbos, nor in the Nama-karoo, although it has been widely planted as an ornamental and barrier plant in both biomes. There are fewer invasive alien species in the Nama-karoo, and most species that do invade

are confined to water courses, drainage lines and pans. *Prosopis* species, which are known to tap water at considerable depth, show most extensive invasions in the Nama-karoo.

22.3.1.6 *The desert biome*
Almost all the alien plant invaders in this biome are restricted to the ephemeral river beds that traverse the desert. Important species are *Datura innoxia*, *Nicotiana glauca*, *Prosopis* spp., *Ricinus communis* and *Argemone* spp. (Brown, Macdonald & Brown 1985; Brown & Gubb 1986;

Figure 22.7 *Nerium oleander* (Apocynaceae), a native of the Mediterranean Basin, was introduced to southern Africa as an ornamental plant during the 19th century. It invades water courses in a narrow band where the succulent karoo biome abuts on the fynbos biome *(Photo: D.M. Richardson).*

Boyer & Boyer 1989; Chap. 9, this volume). The upland portions of this biome impose such stringent environmental constraints on plant growth that species not pre-adapted to it appear to be completely unable to invade. There has been no incentive to introduce suitably adapted plants from similar areas in other parts of the world. In contrast, the seasonal watercourses provide an environment that is basically analogous to that in many other parts of the world. These watercourses are consequently invaded by a range of alien species, a situation that is replicated in other arid parts of the world (Loope *et al.* 1988).

22.3.1.7 *Aquatic ecosystems*
Several alien vascular plants have invaded the freshwater ecosystems of southern Africa. The most important and widespread of these are the free-floating macrophytes, *Azolla filiculoides* (T), *Eichhornia crassipes* (T) and *Salvinia molesta* (Fig. 22.8; not included in the list of transformer species by Wells *et al.* 1986b; but see Ashton, Appleton & Jackson 1986), the submerged hydrophyte, *Myriophyllum aquaticum* (T) and the emergent Spanish Reed *Arundo donax* (T).

Because aquatic ecosystems are poorly represented in the nature reserves of southern Africa and because most of the above-mentioned species are still patchily distributed in the region (Ashton *et al.* 1986; Fig. 22.9), none of these species was recorded from more than 10% of the reserves in any one biome and only *E. crassipes* reached 10% in any biome – in this case it was the savanna biome (Macdonald 1991).

There are no records of introduced seaweeds in the southern African marine flora; this is in contrast to many other regions, particularly in the North Atlantic, where there are many well-documented examples of seaweed invasions (see Chap. 15, this volume). There has been no attempt to assess the status of alien plants in southern African estuaries, but the following alien species were recorded from a survey of five systems in the Western Cape: *Anagallis arvensis, Apium graveolens, Aster subulatus, Bromus inermis* (= *B. japonicus*), *Lolium rigidum* (= *L. loliaceum*), *L. perenne, Parapholis incurva, Polypogon monspeliensis, Sonchus oleraceus, Spergularia media* and *S. rubra* (M. O'Callaghan, unpubl. data). *Spartina maritima* is widespread, but its status as an alien is debatable. Its distribution has been expanding over the past few decades. If it is an alien, it is probably having a major impact on the natural functioning of estuaries (Pierce 1982).

22.3.2 Characteristics of invasive alien species
The 84 important environmental weeds in Appendix 22.1 represent 28 families and 54 genera. Fabaceae (*sensu lato,* including Mimosaceae) with 15 species and Asteraceae (ten species) contribute the most species, followed by Myrtaceae (seven species), Solanaceae (seven species), Cactaceae (five species), Pinaceae (five species) and Poaceae (four species); these seven families together contribute 63% of the species. *Acacia,* with ten species, is the most important genus, followed by *Eucalyptus* and *Pinus* with five species each. When compared with Holm *et al.*'s (1979) list of 7411 of the world's worst weeds (which includes some non-alien weeds), the Myrtaceae, Solana-

(a)

Figure 22.8 *Salvinia molesta* (Pteridophyta: Salvinaceae) is one of the most widespread invader species of freshwater systems in southern Africa and many other parts of the world. (*a*) A section of the Olifants River near Witbank completely covered with mats of *S. molesta* in 1991. (*b*) The same site in 1995 after biological control has almost eradicated the plant. *Salvinia molesta* is one of seven alien plant species in southern Africa considered to be under complete biological control (see text) *(Photos: C.J. Cilliers).*

ceae, Cactaceae and Pinaceae are clearly over-represented in the southern African list of environmental weeds, whereas the Poaceae is under-represented. Families with over 100 species listed by Holm *et al.* (1979), but which are not represented in the list of southern Africa's major environmental weeds, are the Apiaceae, Brassicaceae, Convolvulaceae, Cyperaceae, Lamiaceae, Malvaceae and Scrophulariaceae. Of the genera listed by Holm *et al.* (1979) that contain ten or more weedy species, the following feature in the list in Appendix 22.1: *Acacia*, *Bidens*, *Chenopodium*, *Opuntia*, *Pinus*, *Rubus*, *Senna* (formerly *Cassia*) and *Solanum*. These 'typically weedy' taxa make up 30% of the list.

Sixty-three per cent of the major environmental weeds are trees or shrubs, and 79% of species are perennials. There are only four grass species in the list, and only 14 annuals. The species in the list display a wide range of functional attributes: there is clearly no single set of features that has guaranteed success as an invader [but see Dean, Holmes & Weiss (1986) and Knight (1986) for accounts of aspects of the reproductive ecology, and Whiting, Bate & Erasmus (1986) for ecophysiological characteristics that distinguish some major invaders]. The under-representation of alien pasture plants and the marked over-representation of trees and shrubs in the environmental weed flora of southern Africa is striking (see also Kruger, Richardson & Van Wilgen 1986). The need to introduce woody plants to a region where native timber resources were scarce certainly motivated the importation of more trees and shrubs than was the case in many other parts of the world [see e.g. Kruger *et al.* (1989) for a comparison between the fynbos and other regions with a mediterranean-type climate]. Whether this bias alone can explain the inordinate success of alien woody plants as invaders in the region is not clear. Certainly, the fynbos biome appears to be highly vulnerable to invasion by a suite of trees and shrubs that do not feature as prominently in the lists of invaders in

(b)

Figure 22.8 (*cont.*).

other parts of the world. For example, this is demonstrated by the relative ease with which *Pinus* species invade this biome compared to other biomes in southern Africa and abroad (Richardson & Bond 1991; Richardson *et al.* 1994b; see 22.5.4). But why are there so few grasses in particular, and herbs in general, on the list when these growth forms dominate weed lists in most parts of the world? One possible explanation is that, although hundreds of herbaceous species were introduced to southern Africa, most of these were domesticated varieties specially selected for garden conditions and which are often sterile or nearly so (see Kruger *et al.* 1986 for further discussion).

Almost two-thirds of the 84 species in Appendix 22.1 originate from South America (39%) or Australia (24%), but Europe and Asia (20%) and North America (6%) are also important sources of environmental weeds. It is interesting to compare these statistics with those for the much larger list of invasive alien plants from the *Catalogue of Problem Plants in Southern Africa* of Wells *et al.* (1986a). More than 70% of the 530 species in the latter list originated in Europe and Asia (45.8%) or South America (26.2%) – the continents from which most pre-1800 introductions were made (Table 22.1). It is notable that these continents yielded only 15.5% of the 'transformer'

species. More than a third of the 47 transformer species are of Australian origin, although this continent is the source of only 7.2% of all alien plant species in southern Africa. Almost half of the alien species from Australia are transformer species. The proportional contribution of Australian species to the invasive woody floras is highest in the temperate parts of southern Africa, notably in the fynbos biome. The lowest contribution of Australian taxa is in the eastern coastal areas of the savanna biome, in the two karoo biomes and in the subtropical thicket of the Eastern Cape. Macdonald (1985) argued that Australia appears to have provided a unique evolutionary environment for fire-adapted trees and shrubs suited to temperate climatic conditions and nutrient-poor soils. He suggested that the poor representation of Australian taxa in the alien floras of the four areas mentioned above may be because the Australian flora evolved with few defenses against the heavy defoliation and trampling by the mammalian megafauna that exert a major pressure on the vegetation of these regions (Chap. 17, this volume).

The three major invaders of aquatic systems in southern Africa (*Azolla filiculoides*, *Eichhornia crassipes* and *Salvinia molesta*) originate in South America, whereas another important invader, *Arundo donax*, is indigenous

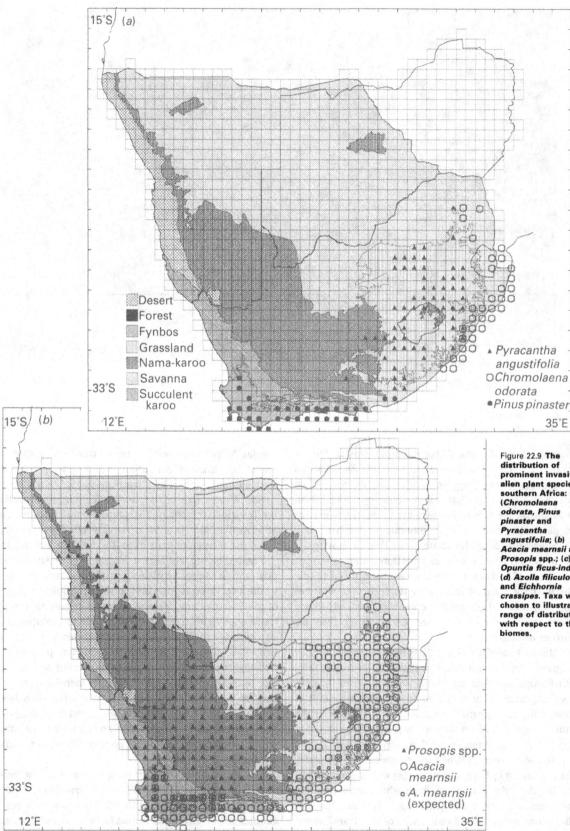

Figure 22.9 **The distribution of prominent invasive alien plant species southern Africa: (a) (*Chromolaena odorata*, *Pinus pinaster* and *Pyracantha angustifolia*; (b) *Acacia mearnsii* and *Prosopis* spp.; (c) *Opuntia ficus-indica*; (d) *Azolla filiculoides* and *Eichhornia crassipes*. Taxa were chosen to illustrate range of distribution with respect to the biomes.**

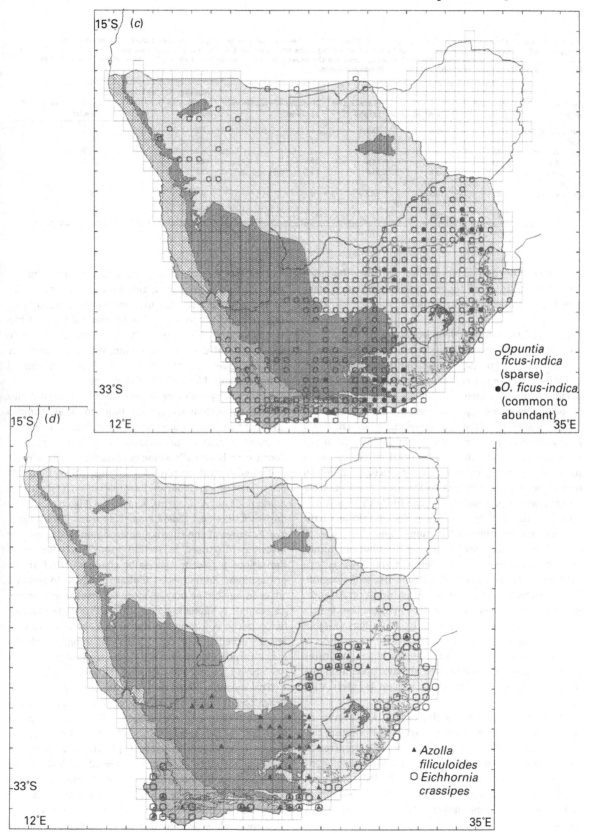

15°S (c)

○*Opuntia
ficus-indica*
(sparse)
●*O. ficus-indica*
(common to
abundant)

33°S

12°E   35°E

15°S (d)

33°S

12°E   35°E

▲ *Azolla
filiculoides*
○ *Eichhornia
crassipes*

Table 22.1 **The origin of 530 alien plant species in southern Africa ('species of foreign origin, but established and reproducing itself as though a native') and of transformer species (a subset of alien plants that 'change the character, condition, form or nature of a natural ecosystem over a substantial area'. The 530 alien species are the most prominent species in a list of 789 taxa given by Wells *et al.* (1986a) in their *Catalogue of Problem Plants in Southern Africa***

Region of origin	Number of species	% of species	% trees and shrubs	Number of transformer species	Number of transformer species as % of alien species	% of all transformer species ($n = 47$)
Europe and Asia	243	45.8	13.6	8	3.3	17.0
South America	139	26.2	27.3	17	12.2	36.2
North America	46	8.7	21.7	4	8.7	8.5
Australia	38	7.2	73.6	17	44.7	36.2
Elsewhere in Africa	20	3.8	20.0	1	5.0	2.1
Other regions	44	8.3	18.2	0	0	0
Total	530	100		47		100

to Europe. The first three species are also the most important invaders of aquatic systems in many other countries (Ashton & Mitchell 1989, p. 120).

22.3.3 Differences between biomes

The foregoing sections showed that each biome has a characteristic set of major invaders, some of which are virtually exclusive to that biome (see also Fig. 22.9). Some invaders occur widely in more than one biome (Appendix 22.1). Examples of species that are confined to one biome or habitat type (or virtually so) are *Hakea sericea* and *Pinus pinaster* (in fynbos) and *Pyracantha angustifolia* (in grassland). There are several possible reasons for a species being confined to a particular biome. In some cases this may be due to features of the environment, whereas in others it is more likely to be a consequence of the history of planting. Not all invaders have reached equilibrium distributions, and the role of environmental features in limiting distribution cannot be accurately assessed. Some patterns are, however, evident. For example, it is thought that *Pyracantha angustifolia* is virtually restricted to high-altitude grasslands, because it needs very cold winters to trigger seed germination (Henderson 1989). Similarly, conditions suitable for the proliferation of *Hakea sericea* (see Richardson & Cowling 1992) exist only in the fynbos biome. A clearer picture will emerge when data are available from the recently initiated Southern African Plant Invaders Atlas project.

Confinement to a particular biome is, however, in some cases undoubtedly an artifact of the planting history of a species. For example, *Acacia dealbata* is currently most widespread in the grassland and savanna biomes, where it has been widely planted; it does, however, spread freely around planting sites in the fynbos biome and is likely to become much more widespread in this biome.

One species that clearly transgresses biome boundaries is *Opuntia ficus-indica*; this species was very widespread in the grassland, karoo and savanna biomes (Fig. 22.9) until its range was markedly reduced through biological control. The species is now most abundant in moist parts of the savanna biome, where climatic conditions hinder the effectiveness of biological control (Zimmermann, Moran & Hoffmann 1986).

Several very widespread invaders in southern Africa owe their extensive distribution to opportunities afforded by rivers. As a result, the distribution of a suite of invaders that occurs mainly along watercourses is poorly correlated with biome boundaries or with any climatic, ecological or physical features other than the occurrence of watercourses. Examples of such species are *Acacia dealbata*, *A. decurrens*, *A. mearnsii* (Fig. 22.10), *Melia azedarach*, *Nicotiana glauca*, *Prosopis* spp. and *Sesbania punicea*. Features of the riparian environment that promote invasions (discussed in section 22.5.1.) override many limitations placed on species by characteristic features of the biome. Some species that are confined to riparian zones in one biome may be more widespread elsewhere; an example is *Nicotiana glauca*, which is restricted to river banks in the Western Cape, but which spreads into various disturbed sites in other parts of the region. All the major invaders in the desert biome are riparian species (see also 22.3.1 and Chap. 9, this volume).

The savanna biome is home to more of southern Africa's most important environmental terrestrial weeds (44; 55.7%) than any other biome, followed by the grassland (40; 50.1%), forest (36; 45.6%), fynbos (27; 34.2%), karoo (19; 24.1%) and desert (5; 6.3%). The savanna biome shares many major environmental weeds with the forest and grassland biomes (coefficient of community values of 44.2 and 47.0, respectively; Fig. 22.11). Other biome pairs with relatively high CC values were forest–grassland and fynbos–savanna. The desert biome was clearly

Figure 22.10 *Acacia mearnsii* (Fabaceae) debris deposited along the Kouga River in the Uniondale District of the Eastern Cape after a flood in 1991. Extensive stands along many watercourses make *A. mearnsii* one of the most widespread invasive alien trees in southern Africa. Thickets of woody alien invaders in riparian habitats destabilize stream banks and enhance erosion *(Photo: D.B. Versfeld).*

the most distinct biome in terms of its environmental weed flora.

Table 22.2 shows the distribution of major plant growth forms for the most important environmental weeds in terrestrial biomes. The list is dominated by woody plants (51 species) and herbs (21 species). Woody plants are best represented in the fynbos (25 species; 92.6% of species in that biome), grassland (22 species; 57.9%), savanna (24 species; 55.8%) and forest (25 species; 73.5%) biomes. Although only six and three woody species occur in the karoo and desert biomes, respectively, these biomes have far fewer major invaders than the other biomes, and these numbers constitute large proportions of the total number of major environmental weeds (35.3% and 60.0%, respectively).

22.4 The extent of invasions

22.4.1 What proportion of each biome is invaded?
Comparing statistics on the extent of alien invasions is complicated by the variety of measures used in different surveys. Even apparently simple estimates of the absolute extent of invasions (e.g. number of hectares invaded) are sometimes made meaningless by the absence of a definition of how an area was rated as 'invaded' as distinct from 'uninvaded'. This section should be read with this in mind. We could find no data on the extent of invasion of aquatic ecosystems [but Wells *et al.* (1983) concluded that these systems were the most severely invaded of all habitat/area classes, with streambanks

second and the winter-rainfall region of the Cape – essentially the fynbos biome – third].

22.4.1.1 *The fynbos biome*
The most comprehensive survey of the extent of alien infestations in any southern African biome was that of the major tree and shrub invasions of the fynbos biome (Macdonald *et al.* 1985). In this exercise a large group of local experts mapped the extent of invasions throughout the biome based on survey data and local knowledge. The minimum density of mature alien plants for an area to be rated as 'invaded' was c. 10 plants km^{-2}.

The combined extent of infestations of *Hakea* and *Pinus* species in 1984 was estimated to be 6641 km^2 which is 8.6% of the biome and 12% of the area of natural vegetation remaining at that time. If the area that has been successfully cleared of these species (1579 km^2) is added, the percentage of the area currently under natural vegetation that had ever been infested is 14%. This figure is similar to the estimate of the peak extent of *Hakea* species invasion of mountain fynbos (14.2%; Macdonald 1984). However, *Hakea* and *Pinus* are mainly invasive in mountain fynbos and if the extent of their combined invasions in this vegetation type is calculated (they usually occur in mixed stands), the percentage of remaining areas of natural vegetation ever invaded rises to 26%, which accords with Richardson's (1984) estimate of 25%.

The combined extent of other thicket-forming alien tree/shrub invasions (mainly Australian species such as *Acacia cyclops, A. longifolia, A. saligna, Eucalyptus* spp., *Leptospermum laevigatum* and *Paraserianthes lophantha*) in 1984 was 11 907 km^2. Of this total, 3815 km^2 was in areas

Table 22.2 **The distribution of major growth forms of 79 important environmental weeds (terrestrial species in Appendix 22.1) in southern African biomes. Numbers in round brackets are the numbers of species in each major growth-form category as a percentage of the total number of species in each biome. Numbers in square brackets in columns 2–7 are the numbers of species in each major growth-form category for each biome as a percentage of the total number of species in that category**

Growth forms	Fynbos	Grassland	Forest	Savanna	Karoo	Desert
Herbs						
Annuals {14}	0	9	5	11	5	1
Perennials {5}	1	5	1	3	1	0
Biennials {1}	0	1	1	1	1	0
Annuals/biennials {1}	0	0	0	0	1	1
Sub-total {21}	1 (3.7)[4.8]	15 (39.5)[71.4]	7 (20.1)[33.3]	15 (34.9)[71.4]	8 (47.1)[38.1]	2 (40.0)[9.5]
Woody plants						
Trees {31}	17	15	13	14	2	1
Trees/shrubs {9}	5	2	5	3	0	0
Shrubs {7}	2	3	3	4	4	2
Shrubs/climbers {4}	1	2	4	3	0	0
Sub-total {51}	25 (92.6)[49.0]	22 (57.5)[43.1]	25 (73.5)[49.0]	24 (55.8)[47.1]	6 (35.3)[11.8]	3 (60.0)[5.9]
Other						
Non-woody climbers {2}	0	0	1	0	0	0
Tall perennial succulents {4}	1	1	1	3	2	0
Short perennial succulents {1}	0	0	0	1	1	0
Sub-total {7}	1 (3.7)[14.3]	1 (2.6)[14.3]	2 (5.9)[28.6]	4 (9.3)[57.1]	3 (17.6)[42.9]	0 (0)[0]
Total {79}	27	40	36	44	19	5

Figure 22.11 **Constellation diagram showing the degree of similarity of the major environmental weed floras in six southern African biomes (data from Appendix 22.1). Floras were compared with the use of Sorensen's coefficients of community (CC). CC = [(2 × c)/(s$_1$ + s$_2$)] × 100, where c is the number of species common to both biomes and s$_1$ and s$_2$ are the number of species in the two biomes. Degrees of similarity between biomes are directly proportional to the thickness of lines. Values of CC ranged from 47.0 for grassland–savanna and 44.2 for forest–savanna to less than 2.0 for forest–desert.**

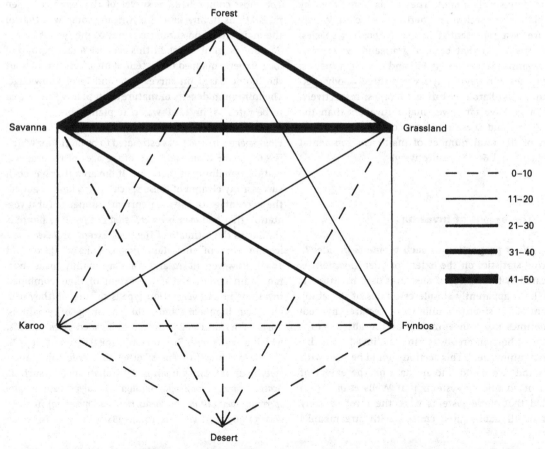

Table 22.3 The extent of dense (>25% canopy cover) and light (<25% canopy cover) stands of alien trees and shrubs in 15 vegetation types on the Cape Peninsula in 1994, and the number of endemic, rare and threatened plant taxa affected by such stands (Richardson et al. 1996). 'Eucalyptus spp.' comprises 5–10 species of this genus (all species other than E. lehmannii)

Vegetation type	Area remaining (ha)	Area invaded (ha) Dense (%)	Area invaded (ha) Light (%)	Endemic and threatened taxa Dense	Endemic and threatened taxa Light	Endemic and threatened taxa Uninvaded	Major alien species (in order of importance)
1. Forest and thicket (FOR)	1 107	181 (16.4)	720 (65.0)	1	20	4	Acacia cyclops, Pinus radiata, Eucalyptus spp., P. pinaster, E. lehmannii
2. Dune asteraceous fynbos (DUN)	2 235	733 (32.8)	358 (16.0)	5	9	21	Acacia cyclops, A. saligna, Eucalyptus spp., E. lehmannii, Pinus radiata
3. Coastal scree asteraceous fynbos (CSA)	237	109 (46.0)	57 (24.1)	2	5	1	Acacia cyclops
4. Wet restioid fynbos (WRF)	3 151	16 (0.5)	445 (14.1)	2	14	36	Eucalyptus spp., E. lehmannii, Acacia longifolia, Pinus pinaster, P. pinea
5. Ericaceous fynbos (ERI)	1 342	0.2 (0.0)	814 (60.7)	0	34	17	Pinus radiata, P. pinaster, Eucalyptus spp.
6. Sandplain proteoid fynbos (SND)	2 433	464 (19.1)	875 (36.0)	4	9	18	Acacia cyclops, Eucalyptus spp., Leptospermum laevigatum, Populus spp., A. saligna
7. Mesic oligotrophic proteoid fynbos (MOP)	8 813	468 (5.3)	1601 (18.2)	19	31	99	Pinus pinaster, Acacia cyclops, A. longifolia, Eucalyptus spp., P. pinea
8. Mesic mesotrophic proteoid fynbos (MMP)	7 018	1013 (14.4)	3062 (43.6)	23	54	45	Acacia cyclops, Eucalyptus spp., A. saligna, Pinus radiata, P. pinaster
9. Undifferentiated cliff communities (CLF)	722	35 (4.8)	447 (62.0)	2	19	10	Acacia cyclops, Pinus pinaster, Eucalyptus spp.
10. Wetlands (WET)	351	33 (9.4)	43 (12.3)	0	0	1	Acacia cyclops, Eucalyptus spp.
11. Wet oligotrophic proteoid fynbos (WOP)	1 030	0.4 (0.0)	408 (39.6)	0	14	21	Acacia saligna, Pinus radiata, Eucalyptus spp., A. cyclops, P. pinaster
12. Wet mesotrophic proteoid fynbos (WMP)	1 234	154 (12.5)	568 (46.0)	4	11	13	Pinus radiata, Eucalyptus spp., Acacia saligna, Pinus pinaster, A. cyclops, Paraserianthes lophantha
13. Renosterveld and grassland (REN)	989	107 (10.8)	637 (64.4)	1	15	2	Eucalyptus spp., Acacia cyclops, Pinus pinaster, P. pinea
14. Upland restioid fynbos (URF)	149	0 (0)	149 (100)	0	15	0	—
15. Vleis (VLE)	144	0 (0)	0 (0)	0	0	2	—
Total	30 955	3313.6	10184				

mapped by Moll & Bossi (1984) as being transformed, giving the infestation in remaining areas of natural vegetation as 8092 km² (Macdonald *et al.* 1985). This is 14% of the total area currently under natural vegetation. If the area considered to have been successfully cleared (870 km²) is added, the percentage ever invaded rises to 16% – 11.6% of the total area of the biome. Most of these invasions occur in lowland fynbos. If this vegetation type is considered separately, then the percentage of remaining natural vegetation ever invaded by this group of species is 68%. This is currently the most extensively invaded vegetation type in South Africa, with important implications for land-use management and biodiversity conservation (see below).

In the regional survey of alien invasions of nature reserves (where the extent of invasions in 1983–1985 was defined as km-square grid cells having at least one mature individual; Macdonald 1991), *Acacia cyclops* showed the most extensive invasions in fynbos reserves, with 52.6% of grid cells invaded (averaged over 26 reserves reporting this statistic). *Acacia saligna* was second with 37% (*n* = 31), *Pinus pinaster* had 35.9% (*n* = 23), *Hakea sericea* had 31.2% (*n* = 20), *A. longifolia* had 21% (*n* = 13) and *A. mearnsii* had 15.5 % (*n* = 23) (see Richardson *et al.* 1992 for additional analyses of these data and distribution maps for these species).

The best data on the extent of woody plant invasions for any area in southern Africa is available for the Cape Peninsula in the Western Cape (Richardson *et al.* 1996). A thorough survey of the distribution of alien plants in 1994 showed that 10.7% of the area not affected by urbanization and agriculture is currently under dense alien stands (> 25% canopy cover) and another 32.9% is lightly invaded. Dense stands of *Acacia cyclops*, the most widespread invader, cover 2510 ha, 75% of the total area under dense alien stands. Table 22.3 shows the extent of invasion in 15 vegetation types and the major alien plants in each type.

The biome-level extent of non-woody invasions of fynbos has never been assessed, but many alien grass and forb species (predominantly from the Mediterranean Basin) are widespread (Macdonald 1991; Richardson *et al.* 1992).

22.4.1.2 *The grassland biome*

The extent of invasions in this 349 174-km² biome is poorly known (Henderson & Wells 1986). At the scale of quarter-degrees, *Acacia mearnsii* (86 grid cells) has the most extensive distribution in the biome. *Acacia dealbata* (80) was the second most widespread, *Salix babylonica* (67) third, *Populus* × *canescens* (63) fourth, *Opuntia aurantiaca* (52) fifth and *Melia azedarach* (50) sixth (Henderson & Wells 1986). *Nassella trichotoma* recorded from only 12

quarter-degree cells, all in the grassland biome, was estimated to have invaded about 600 km² (0.2% of the biome) by 1983 (Henderson & Wells 1986). The most widespread invaders of riparian habitats in this biome are *Salix babylonica*, *A. dealbata*, *Populus X canescens*, *A. mearnsii* and *Salix fragilis*; these species were recorded at 52%, 22%, 19%, 14% and 7% of 1005 watercourse crossings across the biome in a survey between 1979 and 1982 (Henderson 1991a).

From the results of the regional survey of nature reserves (Macdonald 1991), it appears that, of the species reported from more than nine grassland reserves, *Pinus patula* had the highest average extent (16.5% of km-square grid cells invaded; *n* = 22 cells). *Acacia mearnsii* was second (14.9 %; *n* = 22) and *Rubus cuneifolius* third (10.8%; *n* = 9). If the total number of km-square grid cells invaded in grassland reserves is computed, *R. cuneifolius* had the greatest extent (712 km²), *A. mearnsii* was second (480 km²), *Opuntia ficus-indica* third (273 km²), *Solanum mauritianum* fourth (202 km²) and *Pinus patula* fifth (176 km²). From the known statistics of the proportion of the biome included within reserves and the proportion of the biome transformed, and the assumption that an equal proportion of the area inside and outside reserves was invaded, *O. ficus-indica* has a total extent of invasion in the grassland biome of 4882 km-square grid cells.

22.4.1.3 *The forest biome*

There are few published data on the extent of invasions in the small forest biome (closed forests cover no more than 3000 km²) (Geldenhuys *et al.* 1986; Chap. 12, this volume). Almost all invaders are scattered along forest edges and in open patches (Geldenhuys *et al.* 1986), making it particularly difficult to quantify the extent of invasions. The regional survey of alien invasions of nature reserves (Macdonald 1991) provides the best available information on the extent of invasions in this biome. Of the species that had their known extent reported from ten or more forest reserves, *Chromolaena odorata* had the most extensive infestations (49.9% of km-square grid cells invaded, averaged over 15 reserves). *Lantana camara* (44.1%; *n* = 14) was second, *Melia azedarach* (31.1%; *n* = 14) third, *Ricinus communis* (22%; *n* = 12) fourth and *Acacia mearnsii* (10.8%; *n* = 10) fifth. The estimated total extent of *C. odorata* infestations in forest nature reserves was 400 km². This survey allowed an estimate of *C. odorata* infestations in and out of reserves to be made: the total extent of 3249 km² of natural vegetation invaded (Macdonald 1991) is considerably less than the 8000 km² estimated to be invaded by this species in KwaZulu-Natal (Liggitt 1983). However, the latter estimate was based on roadside surveys and is likely to have overestimated the extent of infestations away from roads; it probably also included transformed areas, which were excluded from the former estimate.

22.4.1.4 *The savanna biome*

Savanna covers some 1.4 million km² or 53.7% of southern Africa (Chap. 5, this volume). Quantitative data on the extent of invasion of savanna refer only to South African savannas that cover some 959 000 km². At the scale of quarter-degree cells, *Opuntia ficus-indica* has the most extensive infestation in these savannas (151 grid cells), *Melia azedarach* is second (113), *Lantana camara* third (110) and *Opuntia aurantiaca* fourth (61) (Henderson & Wells 1986). In the regional nature reserve survey (Macdonald 1991), only two species were reported to extend over more than ten savanna reserves: *O. ficus-indica* had the highest average extent (38.7% of km-square grid cells, averaged over 21 reserves), and *M. azedarach* with an average of 13.1% cells over 14 reserves.

In 1939–1941, dense and light stands of *Opuntia ficus-indica* covered at least 8745 km² and 2937 km², respectively (Du Toit 1942). Following a successful biocontrol programme it was stated in the early 1980s that 'it now infests less than 350 000 ha' (Zimmermann & Moran 1982). This implies a reduction in its range of more than 70% over 40 years. However, the known extent of infestations inside nature reserves as reported in the mid-1980s was 2672 km² in savanna reserves alone and, if all biomes were included, this figure rose to 3703 km² (Macdonald 1991). With the extent of infestations inside nature reserves used to estimate the total extent of invasions throughout the biomes, the areas invaded within the savanna biome alone were estimated to be 13 874 km² (Macdonald 1991). This finding agrees with those of a series of road transect surveys conducted in the Transvaal over the period 1979–1982 (Henderson & Musil 1984). In these surveys *O. ficus-indica* was found to be the most frequent alien species, occurring in 72% of all 'bushveld' type road segments and in 62% of all segments sampled over all vegetation types in the province. Furthermore, in the northern Cape Province in 1989, *O. ficus-indica* was also the most frequently recorded alien in savanna vegetation types (56.8% of road segments; *n* = 139; average length of road segment 8.9 km; Henderson 1991b). It is obvious that the minimum density considered to constitute an infestation varies markedly between these different surveys.

Opuntia aurantiaca, considered to be 'South Africa's most costly invasive weed' (Zimmermann *et al.* 1986), has been variously considered to have infested 15 000 km² at its peak (Zimmermann & Moran 1982), and 8300 km² in the mid-1980s following a moderately successful integrated control programme (Zimmermann *et al.* 1986).

Most of this infestation is in the savanna biome, but it also invades karoo and grassland.

Melia azedarach is primarily an invader of riparian vegetation within the savanna biome (Macdonald 1983, 1988, 1991). The extent of its invasion is therefore difficult to quantify. In the Transvaal savanna, *M. azedarach* was the most frequently recorded alien species in the stream-bank habitat, being recorded at 38% of all survey points (*n* = 166; Henderson & Musil 1984). In KwaZulu-Natal and the north-eastern Free State in 1986/7, *M. azedarach* was the second most frequently recorded species on stream banks in savanna habitats (after *Ricinus communis*), being recorded at 39% of survey points (*n* = 115; Henderson 1989).

22.4.1.5 *The karoo biomes*

In the Nama-karoo (here defined as roughly equivalent to the central Cape Province or Great Karoo, i.e. south of the Orange River, bounded on the west by the succulent karoo and Namaqualand, in the east by the Free State and the Eastern Cape, and in the south by the fynbos biome and the Swartberg range) the following species have the most extensive ranges: *Prosopis* spp. (recorded in 103 quarter-degree cells; heavily invasive/very abundant in 27), *Opuntia ficus-indica* (in 56 cells), *Nicotiana glauca* (40 cells), *Atriplex nummularia* (44 cells, abundant in 3), *Opuntia robusta* cultivars (23 cells) and *Tamarix* sp. cf. *ramosissima* (19 cells).

In the succulent karoo (Namaqualand and the Little Karoo) the ranges are: *Nicotiana glauca* (65 cells, abundant in 1), *Opuntia ficus-indica* (44 cells), *Atriplex nummularia* (38 cells, abundant in 3), *Prosopis* spp. (37 cells, abundant in 3), *Acacia cyclops* (34 cells, abundant in 7), *A. saligna* (25 cells, abundant in 9) and *Eucalyptus camaldulensis* (17 cells, abundant in 4) (L. Henderson, unpubl. data).

22.4.1.6 *The desert biome*

The best data on the extent of invasion of desert ecosystems in southern Africa by alien plants were provided by Macdonald & Nott (1987). Here we summarize their data for two of the nine vegetation types surveyed in that study, namely the Northern Namib and the Central Namib, both desert types. Data in brackets are the percentage frequency of species in sampled 10-km sections and the mean abundance ratings (1 = 1 plant in a 10-km section; 9 = a virtually continuous stand at least 1 ha in extent). In the Northern Namib, the only alien plant noted was *Ricinus communis* (3; 1). In the Central Namib, *Nicotiana glauca* (4; 7), *Argemone ochroleuca* (4; 5), *Datura innoxia* (4; 3) and *R. communis* (2; 3) were noted. In another vegetation type, 'semi-desert and savanna transition', the following taxa had mean abundance ratings of 3 or more: *Asclepias fruticosa* (20; 3.4), *Prosopis* spp. (12; 3.3),

Argemone ochroleuca (12; 3), *Datura innoxia* (4; 3), *D. stramonium* (4; 3) and *R. communis* (12; 3).

22.5 Patterns and processes of invasion

Ashton & Mitchell (1989) provided a thorough review of the patterns and processes of invasion of aquatic habitats in southern Africa by alien plants. There have, however, been few similarly detailed studies for major environmental weeds of terrestrial habitats other than those on the set of highly successful woody invaders of fynbos. Rather than catalogue the studies, we selected and summarized three prominent categories: riparian habitats, serotinous trees and shrubs in fynbos and grassland, and acacias in fynbos. We consider the role of disturbance in invasions by comparing its importance in the invasion of different southern African biomes by *Pinus* species.

22.5.1 Invaders of riparian habitats

Features of the riparian environment that promote invasions include the easier access to moisture (which reduces any drought stresses imposed by prevailing features that delimit the biomes), and periodic disturbances in the form of floods that disperse seeds, prepare them for germination, provide seed beds, and remove competing plants. The processes of invasion of riparian zones by hard-seeded legumes were generalized by Hoffmann & Moran (1988), based on their observations of *Sesbania punicea*. They suggest that dense stands in a river obstruct the flow of water, especially during flooding. This leads to erosion of the watercourses and to the conversion of well-defined rivers into diffuse systems of shallow streamlets and trickles. The consequent sedimentation and widening of the stream bed create ideal substrata for expansion of the stand. Essentially the same processes account for the widespread occurrence of *Salix babylonica* in riparian zones of the grassland biome. The violent thunderstorms that are characteristic of this biome create powerful spate flows during which the force of the water breaks off branches and deposits them downstream (Henderson 1991a). Floods are also important for the dispersal of acorns of *Quercus robur* which invades riparian habitats in parts of the fynbos biome (Knight 1985).

22.5.2 Serotinous trees and shrubs in fynbos and grassland

Serotinous trees and shrubs of the genera *Hakea* and *Pinus* have invaded large parts of the natural vegetation in the fynbos biome (see 22.4.1). The analysis of sequen-

tial aerial photographs and autecological studies of pines at several sites have shown that invasion by these serotinous species is a two-phase process. Firstly, satellite foci establish when the winged seeds are dispersed by wind over several kilometres, most seeds being released en masse when fire kills the adult plants. The satellite foci grow and coalesce to form dense even-aged stands (see review in Richardson *et al.* 1992). In detailing the elements of the 'invasion window' for serotinous trees and shrubs in fynbos, Richardson & Cowling (1992) suggested that the alien species behave essentially like the dominant native shrubs (which show marked spatial and temporal fluctuations in population sizes after fires), but that the former eclipse the latter in two key facets of demography: seed dispersal and fire-resilience. The first faculty improves vagility and facilitates spread. Once establishment barriers are overcome, the fynbos is transformed (often over 2–4 fire cycles) into a forest of alien trees. The superior fire resilience of the alien populations buffers them against local extinction, thus disrupting the prevailing non-equilibrium system.

Fire drives these invasions, and fire frequency is of cardinal importance. Conditions in fynbos are such that fires occur at intervals of 4–40 years, but mostly at intervals of 8–20 years (Chap. 6, this volume). These serotinous species do not resprout after fire and are totally reliant on seed banks accumulated between fires for their persistence. A fire-free interval longer than the juvenile period (the time taken to produce the first viable seeds) is thus an important element of the 'invasion window'. Fire cycles are much shorter in grassland (Chap. 18, this volume) and these species invade only in areas where fire frequency has been reduced (Richardson *et al.* 1994b).

22.5.3 *Acacia* species in fynbos

The three most widespread *Acacia* species in the fynbos biome, *A. cyclops*, *A. longifolia* and *A. saligna* have a range of strategies that enable them to deal with the fynbos environment, and this has resulted in different invasion patterns (Richardson *et al.* 1992). *Acacia cyclops*, with its bright red aril, is primarily bird-dispersed, and invasion commences with the deposition of seed under perching sites – typically in clumps associated with indigenous fruit-bearing plants (Glyphis, Milton & Siegfried 1981). As the initial colonists mature, these attract vertebrate dispersers that deposit seeds of various species and disperse seeds outwards. As the complexity (number of fruiting species) of the clump increases, so long-distance dispersal becomes more effective (greater numbers of more species of dispersers are attracted and more seeds are dispersed). Also, as seed production increases, so the level of short-distance dispersal by vertebrates, ants and wind increases. Dispersal by vertebrates is especially important in creating new satellite populations, whereas dispersal by ants and wind leads to lateral expansion of foci. Both types of dispersal lead to the coalescence of foci to form continuous stands. *Acacia cyclops* has a much smaller persistent seed bank than *A. longifolia* and *A. saligna* (Holmes 1989), and germination is not tightly coupled with fire (Jeffrey, Holmes & Rebelo 1988; Richardson *et al.* 1992).

Although *A. longifolia* lacks the brightly coloured arils of *A. cyclops*, it is also dispersed by birds to some extent. Its seeds and those of *A. saligna* are, however, primarily adapted for dispersal by water and ants. Both species accumulate large, persistent seed banks rapidly in the soil (because of the high-percentage viability and water-impermeable dormancy), and germination is cued by fire. Similarities and differences in elements of the ecology of these species, and the influence of these on invasion patterns, are described by Richardson *et al.* (1992).

22.5.4 The role of disturbance: *Pinus* as a case study

It is generally accepted that some form of disturbance is required to allow the invasion of natural vegetation by alien plants (Richardson & Cowling 1992; Richardson *et al.* 1992). Sections 22.5.1 to 22.5.3 show that disturbances that create openings in plant cover are important drivers of alien plant invasions of riparian habitats and fynbos. But how important is disturbance in other southern African biomes, and does southern African vegetation differ in any way in this respect from that in other parts of the world?

In the absence of replicated experiments to test the importance of different forms of disturbance on invasions in different biomes, we must look to natural experiments to provide evidence. Ideally, one would like to compare invasions of the same species, introduced at the same time, propagated to the same extent and thus afforded the same opportunities to spread into natural vegetation in different biomes, with a range of disturbance types and intensities in each. Those responsible for introducing and planting alien plants in southern Africa did not have such experiments in mind and we must deduce what we can form imperfectly designed natural experiments.

One of the most useful natural experiments that sheds light on the role of disturbance in plant invasions in southern African biomes is the widespread planting of several *Pinus* species and the subsequent spread of these species into natural or semi-natural karoo, forest, fynbos and grassland. The same species that invade in southern Africa were also planted, and some have invaded, in other southern hemisphere regions. The differential invasion of habitats provides useful clues to factors that control the invasions.

At least nine *Pinus* species are currently invasive (i.e. regenerate naturally and recruit seedlings more than 100 m from parent plants) in southern Africa (Richardson *et al.* 1994b). The invasive species (and invaded biomes) are *P. canariensis* (fynbos, forest), *P. elliottii* (grassland), *P. halepensis* (fynbos), *P. patula* (grassland, forest), *P. pinaster* (fynbos; Fig. 22.1), *P. pinea* (fynbos), *P. radiata* (fynbos), *P. roxburghii* (karoo) and *P. taeda* (grassland). Six of these species (all except *P. canariensis*, *P. pinea* and *P. roxburghii*) also invade in other parts of the southern hemisphere (Richardson *et al.* 1994b).

Major vegetation types (more precisely ground-cover categories) in the southern hemisphere can be crudely ranked according to their vulnerability to invasion by pines, other factors being equal, as follows: forest << shrubland < grassland << vegetated dunes < bare ground. However, the vegetation or ground cover at a site is determined by various separate (though interacting) factors, all of which influence invadability to some extent. These factors include climate, vegetation structure, species composition, the natural disturbance regime and land use.

Most of the pine invasion events reviewed by Richardson *et al.* (1994b) were in areas where the natural disturbance regime has been noticeably altered by humans, e.g. by increased or decreased herbivore pressure (grazing, browsing and trampling are important), altered fire frequency, or mechanical clearing of vegetation. All records of pine invasions into forests show that severe disturbance is required to facilitate seedling recruitment and population growth (see also Geldenhuys *et al.* 1986, p. 125). Invaded grasslands in Australia, New Zealand and South Africa *invariably* have a history of perturbation (i.e. deviation from the natural disturbance regime). The same applies to most shrubland sites. The ultimate cause of increased invadability in these cases is clearly release from competitive inhibition from resident plants through reduction in ground cover which creates an opportunity for invasion (Richardson *et al.* 1994b). Grazing and browsing animals, and other forms of mechanical disturbance, also play a role in 'planting' the seeds in some cases (Richardson & Bond 1991).

Invasions of *P. halepensis*, *P. pinaster* and *P. radiata* in fynbos stand out from those in other regions. In this biome, the only disturbance required to sustain invasions over large invaded areas is the natural occurrence of fire. Intense fires and the dearth of vigorous plants in the first year or two after fire combine to create opportunities for establishment and subsequent invasion. In general, intermediate levels of perturbation make sites more vulnerable to invasion by pines, whereas severe perturbation prevents invasion.

Pine invasions in forests and grasslands of southern Africa follow very similar patterns to those recorded in other parts of the southern hemisphere (Richardson *et al.* 1994b). In all cases the commencement and continuation of invasions can be clearly attributed to disturbances that create openings in the ground cover. The ease with which pines invade fynbos where there is no evidence of disturbance other than the naturally occurring fire regime suggests that this vegetation type is inherently highly susceptible to invasion by pines. Indeed, pine invasions in fynbos proceeded at similar rates to the spread of pines across deglaciated landscapes of the northern hemisphere, i.e. several kilometres per generation. The elements of this 'invasion window' are described by Richardson & Cowling (1992).

22.6 Consequences of plant invasions

Cronk & Fuller (1995), in their review of the threats posed by invasive alien plants to natural ecosystems worldwide, described seven major categories of threats: replacement of diverse systems with single- (or mixed) species stands of aliens; alterations of soil chemistry; alteration of geomorphological processes; alteration of fire regimes; alteration of hydrology; invasions leading to plant extinction; and those that pose a direct threat to the native fauna. To this list we may add another type of impact which warrants special mention for southern Africa, namely the destruction of riparian habitats. There are examples of all the above-mentioned threats in southern Africa (references in the reviews cited in 22.1), but their magnitude and the processes leading to impacts have not been studied outside the fynbos biome. Table 22.4 summarizes the range of impacts that invasive alien plants are known to cause in fynbos.

22.6.1 Negative economic impacts
All the ecological impacts mentioned above have economic impacts but these are more clearly manifested for some invaders than for others. For example, invasion of grassland, karoo and savanna biomes by alien plants causes the degradation of semi-natural rangelands with obvious economic implications as a result of loss of grazing (through replacement of palatable species or reduced access) or injury to the feet and mouths of livestock in the case of *Opuntia aurantiaca*. More than R10 million is spent annually on combating *O. aurantiaca* alone. Invasive alien plants have diminished the capital value of the

agricultural section in southern Africa by several billion Rand (Huntley, Siegfried & Sunter 1989).

Another impact is the alteration of catchment hydrology caused by dense stands of *Acacia*, *Hakea* and *Pinus* species in the fynbos biome. Runoff from catchments with dense stands of aliens is between 30% and 70% lower than for uninvaded fynbos, depending on the annual rainfall and the age and density of the alien stand. During the dry summer months, when water needs are greatest, runoff in invaded catchments may be reduced to zero, converting perennial streams to seasonal ones. Besides the obviously detrimental impacts to aquatic biota, the reduced streamflow has serious implications for water production in the Western Cape, where the shortage of this resource is already limiting development. Catchments between the Berg and Breede Rivers yield about two-thirds of the region's water requirements and more than 90% of the water supply of the Greater Cape Town Metropolitan Area. Large parts of these catchments are covered by dense alien stands. Recent cost–benefit analyses show that the major costs of clearing alien trees and shrubs is warranted (Van Wilgen, Cowling & Burgers 1996).

The major impacts of invasive aquatic plants arise from river blockages and interrupted water flows, evapotranspirational losses, difficulties in the treatment of water to attain potable standards, development of increased breeding sites for vectors and intermediate hosts of human diseases such as malaria and bilharzia, and inhibition of recreational uses of water bodies (Mitchell 1974). To these can be added the cost of actions taken to ameliorate the ecological effects of the invasion, and of control measures. Ashton *et al.* (1986) present detailed analyses of the economic impacts of invasions of *Eichhornia crassipes*, *Myriophyllum aquaticum* and *Salvinia molesta*.

22.6.2 Positive economic impacts
Several invasive alien plants in southern Africa serve a useful purpose in part of their range in the region. Among these are the pines which, together with *Eucalyptus* spp., form the foundation of the commercial forestry enterprise. *Acacia mearnsii* and *A. melanoxylon* are also important forestry species, the former being cultivated for its bark, which is used in tanning leather, and the timber from the latter usually making up more than half the total amount of timber harvested from southern Cape forests every year (Geldenhuys *et al.* 1986). *Acacia cyclops* and *A. saligna* are harvested for fire-wood throughout their range, but especially in the southwestern fynbos biome, where there is a substantial semi-formal industry based on this resource (Azorin 1992). *Opuntia*

Table 22.4 **Major threats to natural ecosystems in the fynbos biome posed by invasive alien plants**

Threat	Fynbos examples	Key references
Replacement of diverse systems with single- (or mixed) species stands of aliens	Large parts of the fynbos biome are covered by impenetrable thickets of alien trees and shrubs, mainly species of the genera *Acacia*, *Hakea* and *Pinus*	Macdonald & Richardson (1986); Richardson *et al.* (1992); Richardson *et al.* (1996)
Invasions leading to plant extinction	Dense stands of trees and shrubs suppress most native plant species, leading to a local extinction of many species and a marked reduction in biodiversity	Richardson, Macdonald & Forsyth (1989); Richardson *et al.* (1996); see also Table 22.3
Alteration of soil chemistry	The greater productivity of alien stands relative to fynbos increases biomass which changes litterfall dynamics and thus the input of organic matter and nutrients to the soil. Impacts on soil-nutrient budgets depend on soil type and the invading species involved. Major impacts include increased concentrations of nitrogen and other nutrients under *Acacia* stands (mainly on clay-rich soils)	See Chap. 6, this volume
Alteration of geomorphological processes	The stabilization of dunes by alien *Acacia* species has disrupted the natural succession of native colonizers on coastal dunes. This has curtailed the supply of sand to beaches in several areas, preventing the maintenance of a prograding shoreline and causing beach erosion	Lubke (1985)
Alteration of fire regimes	Invasion of fynbos by *Acacia*, *Hakea* and *Pinus* species changes the structure of the fuel bed. Invasion of mountain fynbos by *H. sericea* can increase fuel loads (dead material plus live material <6 mm) by 60%, and lower the moisture content of live foliage, thus increasing fire hazard. Invasion of pines increase biomass in fynbos by up to 300%; the fuel bed in such stands is effectively confined to the litter layer, under all but severe conditions of fire hazard. This causes fires to spread slowly and burn at low intensities under moderate weather conditions. Under extreme weather conditions, fires in such invaded stands burn through live canopies. The increased energy released by such fires makes them self-sustaining, difficult to control and potentially more damaging to ecosystems than fires in natural vegetation. The fire regime under such conditions is likely to change. Fires will become less frequent and more intense	Van Wilgen & Richardson (1985)
Alteration of hydrology	Inference from catchment experiments shows that dense stands of alien trees can reduce runoff by between 30% and 70%, depending on the annual rainfall and the age and density of the alien stand. Recent estimates suggest that invasion could result in an average decrease in water production from fynbos catchments of 347 m^3 water ha^{-1} yr^{-1} over 100 years, resulting in an average loss of more than 30% of the water supply to the city of Cape Town	Le Maitre *et al.* (1996); see also Chap. 6, this volume
Destruction of riparian habitat	Communities of river bank grasses and sedges are replaced by woody aliens (e.g. *Acacia* spp.), changing the habitat from a stable community dominated by herbaceous species to an unstable community dominated by trees and shrubs. The greater exposure of soils leads to increased erosion, often resulting in the formation of levees and hence altered flow patterns	D.M. Richardson pers. obs.

ficus-indica is harvested in the Eastern Cape for various purposes (Brutsch & Zimmermann 1993). Inevitably, the reliance of a proportion of the human population on invasive alien plants which are deemed to be totally undesirable by another section of the population leads to conflicts of interest [see e.g. Geldenhuys (1986) for *A. melanoxylon*; and Armstrong (1992), Cowling (1992) and Donnelly & Morris (1992) for a range of views on the need for control of *A. saligna*].

22.7 Control

Most of the major environmental weeds in southern Africa have been subjected to some form of management. The first campaign against an invasive alien plant was probably that against *Xanthium spinosum* which was initiated in 1860 (Henderson & Wells 1986). The first attempts to 'eradicate' *Opuntia ficus-indica* were launched

in 1883 (Pettey 1947). Without exception, initial control efforts were haphazard; they were applied on a trial-and-error basis without appropriate consideration of economic and sociological factors, or interactions between the plant and its environment at both an individual and a population level. Early control attempts were seldom, if ever, followed up, and in some cases control efforts made the problem worse. This is clearly illustrated for *Hakea sericea*, where attempts to deal with the invader without applying prescribed burning and follow-up measures were useless (Richardson 1985).

Early attempts at controlling invasive alien plants in southern Africa were, like those elsewhere, aimed at eradication rather than at reducing population sizes to below a nuisance level. Control efforts were often focused on inappropriate aspects of the plant's biology. The application of legislation (firstly the Weeds Act of 1943, and later the Conservation of Agricultural Resources Act of 1984) has been totally ineffective, and subsidization schemes (e.g. for herbicides to combat

Opuntia aurantiaca; Zimmermann & Moran 1982) have thwarted integrated control.

It is beyond the scope of this chapter to provide details on control efforts throughout the region [there have been detailed reviews of the history of control operations in several nature reserves: Macdonald (1983) for Hluhluwe/Umfolozi; Macdonald, Clark & Taylor (1989) for the Cape of Good Hope Nature Reserve; Macdonald (1988) for the Kruger National Park]. The extraordinary scale of mechanical control methods over 1.15 million ha in fynbos catchments must, however, be emphasized. In this biome, an average of 57 851 ha is treated each year (data for 1987–1993) by felling dense stands of (mainly) *Hakea* and *Pinus* species using mechanical brush-cutters and chain saws (Cape Nature Conservation, unpubl. data). The felled plants are left lying for 12–18 months to allow the serotinous cones to open and the seeds to germinate. The area is then burnt to kill regenerating seedlings. Regular follow-up surveys are done to remove seedlings that established after the fire. The average annual cost of these operations is R1.32 million (this amount excludes many actual costs; if these are taken into account, costs rise to around R7 million, about 17% of the total budget for conservation in the Western Cape). This is probably the most extensive and expensive campaign against environmental weeds anywhere in the world.

22.7.1 Biological control

Biological control (i.e. the use of natural enemies to regulate troublesome organisms) has been widely used against 38 alien plant species in South Africa.

Seventy-three species of herbivorous natural enemies have been introduced into South Africa for biological control. Almost two-thirds (44 or 60%) of these have become established on the target plants. This proportion may increase when the outcome is known of five introductions that have not yet been assessed, but already it matches the levels recorded in other countries that are active in biological weed control (Julien 1992).

Almost all (93%) of the agents that have been used for biological control of alien plants in South Africa have been herbivorous insects (Fig. 22.12*a*). Of these, Coleoptera, 34 species (47% of the 73 species), and Lepidoptera, 16 species (22%), have been the most utilized taxa. There is increasing emphasis towards the use of microbial pathogens, particularly fungi, as biological control agents, either utilized classically (i.e. released and left to proliferate on the weeds) or formulated and applied regularly as mycoherbicides. A quarter of the agents introduced to South Africa are species that primarily destroy the reproductive parts of the weeds (i.e. flowers, fruits or seeds) (Fig. 22.12*b*). The insect herbivores that

damage vegetative parts include representatives of all the major phytophagous feeding guilds for insects (Fig. 22.13).

The rate at which agents have been released in South Africa has increased steadily since the first introduction in 1913 (Fig. 22.12*c*). Initially, the biological control programmes relied heavily on the use of agents that had previously been used successfully against weeds in other countries. Although some recent programmes still fall into this category, since 1970 an increasing proportion of programmes in southern Africa have used agents that have never been utilized elsewhere (Fig. 22.12*c*), often on weeds that are peculiar to South Africa.

At least six of the introduced agents have colonized alien plant species other than their target species. These occurrences were not unexpected, because some of the alien species are closely related, and some of the herbivorous insect agents are oligophagous (i.e. they develop on two or more closely related plant species). Most prominent among the herbivore species that have colonized more than one alien plant species is *Cactoblastis cactorum*, which feeds on, and causes considerable damage to, at least seven alien cactaceous species in South Africa.

The 38 target weed species embrace a variety of growth forms (Fig. 22.12*d*) in 12 different families, predominantly the Cactaceae (15 species), the Fabaceae (seven species) and the Asteraceae (five species).

Of the 38 alien plant species targeted for biological control, the agents have: (1) completely controlled seven species (*Harrisia martinii*, *Hypericum perforatum*, *Opuntia leptocaulis*, *O. vulgaris*, *Pistia stratiotes*, *Salvinia molesta* and *Sesbania punicea* – 18% of those targeted for biological control) (i.e. the weed populations have been sustained at acceptably low densities by the agents alone); (2) substantially reduced the density and invasiveness of 11 species (29%) so that additional control measures have been attenuated; (3) caused negligible levels of damage on five species (13%) and contributed little or nothing to control of these weeds; (4) not been fully evaluated for 13 species (34%), mostly because releases of the agents have been too recent for meaningful assessments to be made; and (5) failed so far on two species (5%) (Fig. 22.12*e*).

The most intensive biological control programme has been that against *Lantana camara*, on which 17 insect species have been released. Six of the agents have become established in South Africa and contribute substantially to control of the weed. However, *L. camara* comprises a complex mixture of 50 genetically distinct taxa (Spies & Stirton 1982) and it remains a problem in many areas.

The other group of plants that has received a dispro-

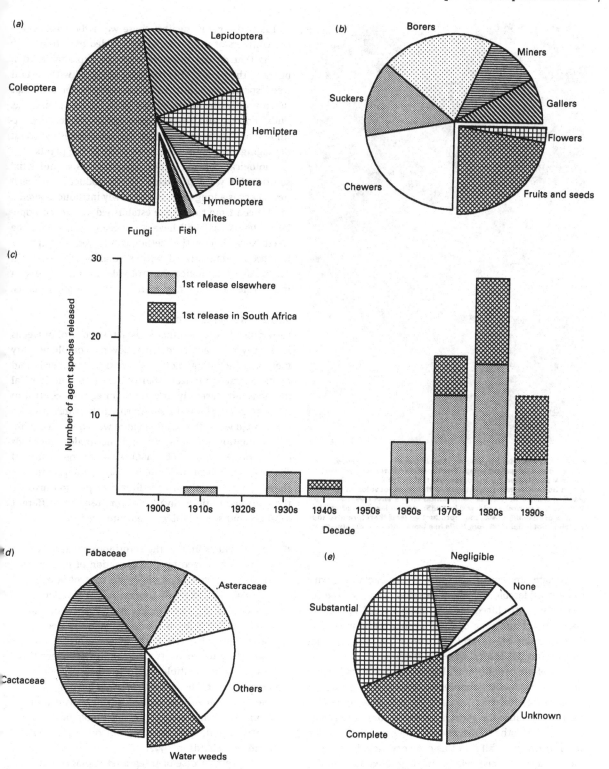

Figure 22.12 **Biological control of weeds in southern Africa.** (*a*) The taxa of biological control agents that have been released. (*b*) The feeding guilds of the biological control agents. (*c*) Releases of biological weed control agents by decade, showing the proportion of releases that were made with agent species that had already been used succesfully in at least one other country (stippled bars), as opposed to agents that were first released in southern Africa (cross-hatched bars). (*d*) The groups (see text) of weeds targeted for biological control. (*e*) The outcome of the biological weed control programmes to date (see text for definition of categories).

Figure 22.13 **Abundant galling on *Acacia longifolia* (Fabaceae) caused by a bud-galling wasp, *Trichilogaster acaciaelongifoliae*, which was released for biological control in 1982. The galls, which house the immature stages of the wasp, develop in the place of normal flowers. In conjunction with a snout beetle, *Melanterius ventralis* (released in 1985), whose larvae feed on the residue of ripening seed, the reproductive (and therefore also the invasive) potential of *A. longifolia* has been curtailed drastically** *(Photo: J.H. Hoffmann).*

portionate amount of attention is the Cactaceae, with some considerable success. Altogether, 14 agent species have been introduced onto 15 cactus weed species in South Africa and nine of these agents are now responsible for complete control of three of the cactus species and substantial control of seven species.

One of the benefits of biological control is that weeds can be permanently suppressed over much, or all, of their range. However, sometimes selective control is required, because the plant is a weed in some situations, but useful in others. The invasive alien plant species in South Africa that fall into this category include the Australian *Acacia* species which are exploited for tannin, timber, fire-wood and fodder, and the North American *Prosopis* species, which produce large quantities of nutritious pods, a valuable fodder, and provide shade and fire-wood in otherwise treeless areas. Biological con-

trol is still possible against these weeds because potential conflicts of interests can be avoided by introducing agents that do not interfere with the exploitable attributes of the plants. This has been achieved with several weed species in South Africa by using herbivores that destroy only the reproductive parts of the plants (i.e. buds, flowers and seeds). These agents reduce the fecundity of the weed and curb its invasiveness without damaging the useful vegetative parts of the plants.

Biological control is not advocated as a 'quick-fix' solution for weed problems. The introduction of each new agent is a long process and newly introduced agents often need time to become established and reach population levels that will have a noticeable impact on the weed. Nevertheless, the method can be very effective, as has been demonstrated with several weed species in South Africa. Biological control will continue to play a major role in the management of invasive alien plants.

22.7.2 Integrated control

Experience has shown that satisfactory control of weeds is usually only achieved when several complementary methods, including biological control, improved land management practices, herbicides and mechanical methods, are carefully integrated. Integrated control of invasive alien plants in southern Africa dates back to about 1940 when *Opuntia ficus-indica* was controlled with a combination of biological and mechanical methods (Pettey 1943). Kluge *et al.* (1986) reviewed the history of the development of integrated control strategies in southern Africa. Here we provide four examples to illustrate the various options available for use on different invaders and some problems encountered.

1   In the fynbos biome, the serotinous shrub *Hakea sericea* is managed by a combination of mechanical clearing, burning and biological control (Kluge *et al.* 1986). The wood of another major invader of fynbos, *Acacia cyclops*, makes good fuelwood, and attempts are being made to manage the weed in parts of its adventive range by encouraging harvesting under controlled conditions where this practice is compatible with the other aims of management. Mechanical clearing, burning and chemical control are applied where harvesting is impractical, and several potential biological control agents have been identified (Van Wilgen, Bond & Richardson 1992).

2   An interesting case of integrated management occurred with the control of *Opuntia lindheimeri* (= *O. tardospina*). Biological control alone was ineffective, because the control agent, *Dactylopius opuntiae*, a cochineal species, was preyed on and

kept at low population levels by predatory ladybird beetles (Coccinellidae). Applications of insecticide sprayed on the cactus in low concentrations killed the ladybirds but did not affect the cochineal, allowing populations to increase and effectively control the cactus (Annecke, Karny & Burger 1969).

3 Mechanical and chemical control of *Sesbania punicea* has become much more effective since the introduction of biological control agents that destroy the flowers and seeds of the weed. The fecundity of *S. punicea* has declined to the extent that there are very few viable seeds in the soil (Hoffmann & Moran 1991) so that there is little or no recolonization by the weed in areas where stands have been cleared (J.H. Hoffmann, unpubl. data).

4 The utilization of *Prosopis* pods for fodder should be managed so that the pods are left for up to six months after falling to the ground, to allow populations of the introduced seed-feeding beetles to build up sufficiently to destroy most of the seeds. This practice reduces the number of viable seeds that are dispersed in the dung of livestock and thereby limits recruitment of the weed (H.G. Zimmermann, unpubl. data).

Certain control measures applied to some weeds are not feasible under southern African conditions. For example, invasive pines are controlled in New Zealand grasslands by manipulating grazing intensity (references in Richardson *et al.* 1994b). This strategy is not feasible in the fynbos biome, where pine invasions are most prevalent, because the fynbos has very low forage value. Furthermore, conditions are such that the fire frequency cannot be manipulated to the detriment of invading pines, which produce seeds before sufficient fuel has accumulated (Richardson & Cowling 1994). Forestry authorities are reluctant to sanction research on biological control that uses seed-attacking insects, as there is an export market for pine seeds. The only control option that remains is mechanical clearing combined with prescribed burning.

22.8 The future

Devising effective methods for managing alien plants is a major challenge facing reserve managers and planners in the next few decades. Attention must be given to reducing the deleterious effects of current invaders, and to reducing the number of future invaders.

22.8.1 Dealing with current invaders

In this section we discuss some of the most important problems that are impeding the effective control of current invaders and suggest how the situation may be improved.

Biological control holds great promise for reducing the density of populations of many major environmental weeds in southern Africa in the long term (i.e. over several decades), and this form of control should be the foundation of all strategies. Mechanical control is, however, the only practical option for clearing existing dense stands in many instances. For example, mechanical clearing is the only way of removing dense stands of alien acacias, hakeas and pines in the fynbos biome, but the method is very expensive and there are increasing demands on the state funds previously allocated for this purpose. Dense stands already cover large areas of fynbos, threatening many native species with extinction and affecting many ecosystem services. In many cases the dense stands have been present for less than 30 years, and although many native organisms are affected, it appears that restoration is still (but may not be for much longer) feasible (P.M. Holmes, unpubl. data). Every effort must be made to clear these stands as soon as possible.

Increasing disturbance of natural areas, for example through altered fire regimes and increased grazing pressure, increases the susceptibility of vegetation to invasion. Disturbance-prone habitats, such as river banks and fire-prone systems such as fynbos, are a special problem because it is impossible to eliminate the major driving force in invasions in these habitats. Sustained management of invasive alien plants is required in these cases. Increasing afforestation with aggressive invaders such as pines increases the seed pool of these species and also the interface between seed sources and areas of natural vegetation.

What can be done to reduce the extent and impacts of invasive alien plants? Legislation aimed at forcing landowners to clear alien plants has been a dismal failure in the past. It is very unlikely that any form of legislation aimed at private landowners and farmers will help curb invasions at this stage in southern Africa's political history. Where new legislation is urgently required is in the case of large-scale landowners whose commercial activities clearly contribute to the increased invasion of natural vegetation, notably the large privately owned forestry companies. These organizations should be obliged to control the spread of forestry trees from plantations. Levies on forest products could be used to fund control and research on biological control options that are compatible with forestry. Inter-organization liaison must be improved. In all cases there should be greater emphasis on developing integrated control strategies. Good progress has been made in the fynbos biome, where rule-based models, using geographic information systems, have been implemented for integrating alien plant control measures with other facets of extensive land management, notably prescribed burning in mountain catchments (Richardson *et al.* 1994a). The ecology of many of the major environmental weeds is poorly understood, and further autecological studies are needed.

22.8.2 New invaders

Although the rate of new arrivals in the region has slowed, there are likely to be many new invaders for several reasons. Firstly, history has shown that there are sometimes time lags between the arrival of an alien and the first invasions; there are therefore probably many 'time bombs'. Secondly, the slow, natural spread of species that currently occupy suboptimal areas to more favourable sites will almost certainly lead to new widespread invasions. Thirdly, although fewer in number, many of the new introductions are made after more detailed site matching (e.g. for forestry; Richardson & McMahon 1992); it is therefore likely that a greater proportion of these introduced species will invade than in the past, when far less attention was given to matching (Fig. 22.14).

It is curious that no invasive alien seaweeds have been reported from southern African waters, and the situation should be carefully monitored to prevent the establishment of species such as *Caulerpa taxiflora*, *Colpomenia peregrina* and *Sargassum muticum* which have caused major problems in other regions. Epiphytes on cultured oysters, on the hulls and in the ballast water of ships, and aquarium escapees are potential invaders of southern Africa's marine waters (see Chap. 15, this volume).

Can we predict which species could invade in the future? This was one of the main questions to emerge from the SCOPE programme (Mooney & Drake 1989). The general consensus at the end of the SCOPE programme was that invasions are highly idiosyncratic and that it is unlikely that useful predictions will be forthcoming (references in Richardson & Cowling 1992; see also Lodge 1993). However, global lists of invaders (e.g. Binggeli 1996; Cronk & Fuller 1995) identify species to watch carefully; many species are invaders in more than one region, and it is clearly appropriate to err on the side of caution. Invaders of some habitats (e.g. riparian zones) are successful across very large geographical areas. All important invaders of such habitats anywhere in the world should be regarded as 'high-risk' introductions.

Global perspectives of the performance of related

(a)

(b)

Figure 22.14 **Several species of *Banksia* (Proteaceae) from Australia were recently introduced to the fynbos biome where they are planted for their showy flowers. Many banksias have similar traits to the major invaders of fynbos (i.e. tall serotinous shrubs with high fecundity, short juvenile periods and low fire-tolerance as adults). One species with these attributes is *B. ericifolia* [(a) shows it growing on the Agulhas Plain near Napier]. Eight-year-old shrubs here held 12 times the number of viable seeds reported for 25-year-old shrubs in Australia. This species has already started to display invasive tendencies in fynbos. Screening procedures have been developed to identify other species that are likely to invade in fynbos. These studies suggest that *Banksia grandis*, a species grown mainly for its attractive cones [(b) shows a plantation of this species near Napier], is unlikely to invade fynbos** *(Photos: D.M. Richardson).*

taxa as invaders can be used to identify other taxa that are inherently predisposed to become invasive. For example, invasiveness of *Pinus* species (derived from a global review) was found to be negatively correlated with features of their life history – their mean seed mass, minimum juvenile period and the mean interval between large seed crops. A robust discriminant function combining these three variables can be used, not only for detection of other invasive pines, but also for preliminary screening of invasive woody species in other groups of seed plants (Rejmánek & Richardson 1996).

The history of introductions and subsequent invasions for a particular biome provides a valuable natural experiment (e.g. Richardson, Cowling & Le Maitre 1990). Although new invaders are also likely to come from source areas other than those of the current invaders, careful analysis of the fate of a range of introduced taxa with reasonably similar histories of dissemination certainly enables one to quantify, to some degree, the 'invasion windows' and the traits that a plant needs to invade a particular environment. This was done for the fynbos biome (Richardson *et al.* 1990; Richardson & Cowling 1992), and facilitated the development of an expert system for screening potential introductions (Tucker & Richardson 1995). This approach requires a detailed understanding of ecosystem processes in the target environment (see Richardson & Cowling 1992), and is firmly based on the 'routes to success' taken by current invaders. This perspective is useful for identifying 'high-risk' species, and could, if rigorously applied, reduce the number of new invasive alien species. Horticulturalists, foresters and nurserymen have a special obligation in this regard.

22.9   Acknowledgements

We thank Steve Higgins and Pat Holmes for their perceptive reviews of the manuscript, Toni Milewski for his help in compiling Appendix 22.1 and Jeanne Hurford for help with Fig. 22.9.

Appendix 22.1

Species	Family	Origin	Growth-form	Functional classification[a]						Seed categories[b]	Seed size[c]	Main dispersal agent[d]	Vegetative propagation[e]	Biome(s) invaded[f]
				S	Lo	H	P	W	Lf					
Acacia baileyana	Fabaceae	Australia	Tree	t	p	u	e	w	se	3	m	wt, an	?	gr
Acacia cyclops	Fabaceae	Australia	Tree	t	p	u	e	w	se	3	m	ml, bd	x	fy, fo, sa
Acacia dealbata	Fabaceae	Australia	Tree	t	p	u	e	w	se	3	m	wt	c	gr
Acacia decurrens	Fabaceae	Australia	Tree	t	p	u	e	w	se	3	m	wt	c	gr
Acacia longifolia	Fabaceae	Australia	Tree	t	p	u	e	w	se	3	m	wt, bd, an	x	fy, sa
Acacia mearnsii	Fabaceae	Australia	Tree	t	p	u	e	w	se	3	m	wt	c	gr, fy, fo, sa
Acacia melanoxylon	Fabaceae	Australia	Tree	t	p	u	e	w	se	3	m	bd, wt	c	fy, gr, fo
Acacia podalyriifolia	Fabaceae	Australia	Tree	t	p	u	e	w	se	3	m	wt, an	?	gr
Acacia pycnantha	Fabaceae	Australia	Tree	t	p	u	e	w	se	3	m	wt, ml, bd?	x?	fy
Acacia saligna	Fabaceae	Australia	Tree	t	p	u	e	w	se	3	m	wt, ml, an	c	fy, fo, sa
Agave americana	Agavaceae	C. America	Shrub	t	p	u	e	w	su	3	m	wd	s	ka
Agave sisalana	Agavaceae	C. America	Shrub	t	p	c	s	s	su	6	—	wt; hu	s, d	sa, fo, gr, ka
Alternanthera pungens	Amaranthaceae	S. America	Herb	t	a	u	s	n	n	3	m	ml	c	sa, gr, fo
Argemone ochroleuca (= A. subfusiformis)	Papaveraceae	S. & C. America	Herb	t	a	u	gm	w	n	2	s?	wd, wt	x?	ka, sa, fo, gr, de
Arundo donax	Poaceae	Europe, Asia	Emergent reed	a	p	u	e	w	n	?	m	wt	?	aq
Azolla filiculoides	Salviniaceae	S. America	Free-floating macrophyte	a	v	f(f)	e	n	n	2	vs	wt, bd	d	aq
Bidens bipinnata	Asteraceae	Europe, Asia	Herb	t	a	u	gm	n	n	2	s	w, ml	c	sa, gr
Bidens pilosa	Asteraceae	S. America	Herb	t	a	u	gm	n	n	2	s	w, ml	c	fo, gr, sa
Caesalpinia decapetala	Fabaceae	Asia	Shrub/climber	t	a	u	e	w	n	3	m	wt, ml	c	sa, gr, fo
Cardiospermum grandiflorum	Sapindaceae	S. America	Climber	t	v	u	v	s	n	?	m	wd, wt	?	fo
Casuarina equisetifolia	Casuarinaceae	Uncertain (pantropical)	Tree	t	p	u	e	w	se	3	m	wd, wt	x	fo, sa
Cereus jemacaru (C. peruvianus formerly misapplied)	Cactaceae	S. America	Tall perennial succulent	a	p	u	e	w	su	4	s	bd	d	sa
Cestrum laevigatum	Solanaceae	S. America	Shrub	t	p	u	e	w	n	4	s	bd	c	fo
Chenopodium album	Chenopodiaceae	Europe, Asia	Herb	t	a	u	gm	n	su	2	s	wd, wt	?	sa, gr
Chromolaena odorata	Asteraceae	S. America, West Indies	Shrub/climber	t	p	u	e	w	n	2	s	wd, wt	c, s	fo, sa
Cirsium vulgare	Asteraceae	Europe, Asia	Herb	t	b	u	gm	n	n	2	s	wd, ml	?	ka, gr, sa, fo
Conyza alba (= C. floribunda)	Asteraceae	S. America	Herb	t	a	u	gm	n	n	2	s	wd?	?	sa, gr
Datura ferox	Solanaceae	Asia	Herb	t	a	u	v	w	n	2	s	wd	?	sa, gr
Datura innoxia	Solanaceae	S. & C. America, West Indies	Herb	t	v	u	v	w	se	2	s	wd	?	de, ka
Datura stramonium	Solanaceae	S. & C. America	Herb	t	a	u	v	w	n	2	s	wd	?	sa, ka, gr, fo
Eichhornia crassipes	Pontederiaceae	S. & C. America	Free-floating macrophyte	a	p	f(f)	e	w	n	6	—	(wt)	d, r	aq
Eucalyptus cladocalyx	Myrtaceae	Australia	Tree	t	p	u	e	w	se	2	s	wd	c	fy
Eucalyptus diversicolor	Myrtaceae	Australia	Tree	t	p	u	e	w	se	2	s	wd	c	fy
Eucalyptus globulus	Myrtaceae	Australia	Tree	t	p	u	e	w	se	2	s	wd	c	gr, fy, sa, fo
Eucalyptus grandis	Myrtaceae	Australia	Tree	t	p	u	e	w	se	2	s	wd	c	fo, gr, sa
Eucalyptus lehmannii	Myrtaceae	Australia	Tree	t	p	u	e	w	se	1	s	wd	c	fy
Euphorbia tirucalli	Euphorbiaceae	Africa	Tall perennial succulent	t	p	u	e	s	su	3	m	hu	d	sa
Hakea gibbosa	Proteaceae	Australia	Tree/Shrub	t	p	u	e	w	sc	1	m	wd	x	fy
Hakea sericea	Proteaceae	Australia	Tree/Shrub	t	p	u	e	w	sc	1	m	wd	x	fy, fo
Hakea drupacea (= H. suaveolens)	Proteaceae	Australia	Tree/Shrub	t	p	u	e	w	sc	1	m	wd	x	fy
Jacaranda mimosifolia	Bignoniaceae	S. America	Tree	t	p	u	l	w	n	5	m	wd	c, r	sa, gr, fo
Lantana camara	Verbenaceae	S. & C. America	Shrub/climber	t	p	u	e/l	w	n	4	m	bd	c, r	fo, sa, fy, gr

Appendix 22.1 Characteristics of 84 important environmental weeds in southern African biomes. The list of species was derived by combining the 47 'transformer' species listed by Wells *et al*. (1986b) and the 67 taxa recorded from at least 10% of reserves in any one biome and which were classified as 'widespread' in a survey of South African nature reserves (Macdonald 1991) (see text)

Species	Family	Origin	Growth form											
Leptospermum laevigatum	Myrtaceae	Australia	Tree/Shrub	t	p	u	e	w	se	2	s	wd, wt	x	fy
Litsea glutinosa (= *L. sebifera*)	Lauraceae	Asia	Tree/Shrub	t	p	u	e	w	n	?	m	bd	?	fo
Melia azedarach	Meliaceae	Asia	Tree	t	p	u	—	w	n	5	m	bd, wt	c	fo, sa, gr
Morus alba	Moraceae	Asia	Tree	t	p	u	—	w	n	4	s	bd	c?	fo, sa
Myriophyllum aquaticum	Haloragaceae	S. America	Floating (attached)	a	p	f(a)	e	n	su	—	wt	d	aq	
Nassella tenuissima	Poaceae	S. America	Herb	t	p	u	—	c	n	2	s	wd, wt	d	gr
Nassella trichotoma	Poaceae	S. America	Herb	t	p	u	—	c	n	2	s	wd, wt	d	gr
Nicotiana glauca	Solanaceae	S. America	Tree/Shrub	t	p	u	e	w	n	2	vs	wd, wt	c	ka, de, sa, fy
Opuntia aurantiaca	Cactaceae	S. America	Short perennial succulent	t	p	u	e	n	su	—	—	ml, wt	d	ka, sa
Opuntia ficus-indica	Cactaceae	C. America	Tall perennial succulent	t	p	u	e	s	su	4	m	ml, bd	d	sa, ka, gr, fo, fy
Opuntia imbricata	Cactaceae	C. America	Tall perennial succulent	t	p	u	e	s	su	4	m	ml, bd	d	ka
Paraserianthes lophantha	Fabaceae	Australia	Tree	t	p	u	e	w	n	3	m	wt	x	fy
Passiflora edulis	Passifloraceae	S. America	Climber	t	p	c	e	n	n	4	s	bd; ml	?	fo
Pennisetum clandestinum	Poaceae	Africa	Herb	t	p	c	—	n	n	6/2	s	hu	r	fy, sa, gr, ka
Pereskia aculeata	Cactaceae	S. & C. America, West Indies	Shrub/climber	t	p	u	e	s	n	4	s	bd	d	fo
Pinus elliottii	Pinaceae	N. America	Tree	t	p	u	e	w	se	1	m	wd	x	fo
Pinus patula	Pinaceae	C. America	Tree	t	p	u	e	w	se	1	m	wd	x	gr
Pinus pinaster	Pinaceae	Mediterranean Basin	Tree	t	p	u	e	w	se	1	m	wd	x	fy, fo, sa
Pinus pinea	Pinaceae	Mediterranean Basin	Tree	t	p	u	e	w	se	1	m	ml	x	fy
Pinus radiata	Pinaceae	N. America	Tree	t	p	u	e	w	se	1	m	wd	x	fy
Polygonum aviculare	Polygonaceae	Europe, Asia	Herb	t	a	u/c	gm	n	n	3	?	?	x?	ka
Populus alba	Salicaceae	Europe, Asia	Tree	t	p	u	—	c	n	6		(wt)	s, d	fy, gr, sa
Populus canescens	Salicaceae	Europe, Asia	Tree	t	p	u	e/l	w	n	6		wt	s, d	fy, gr, sa
Prosopis spp.	Fabaceae	N. & S. America	Tree	t	p	u	e/l	w	n	3	m	ml	c	ka, de, sa
Psidium guajava	Myrtaceae	West Indies	Tree/Shrub	t	p	u	e	w	se	4	m	bd, ml	c	fo, sa, gr
Pyracantha angustifolia	Rosaceae	Europe, Asia	Shrub	t	p	u	e	w	n	4	s	bd	c?	gr
Ricinus communis	Euphorbiaceae	Africa	Shrub	t	p	u	v	w	n	3	m	hu, wt	c	de, fo, sa, ka, fy, gr
Rubus cuneifolius	Rosaceae	N. America	Shrub	t	p	u	e/l	w	n	4	s	bd	c, s	gr, fo
Salix babylonica	Salicaceae	Asia	Tree	t	p	u	—	w	n	6		hu	d	ka, gr
Salsola kali	Chenopodiaceae	Asia	Herb	t	a	u	gm	s	n	3	s	wd	?	ka
Salvinia molesta	Salviniaceae	S. America	Free-floating macrophyte	a	p	f(f)	e	n	n	6		(wt)	d	aq
Schinus terebinthifolius	Anacardiaceae	S. America	Tree	t	p	u	e	w	n	3	m	bd	c	fo, sa
Schkuhria pinnata	Asteraceae	S. America	Herb	t	a	u	gm	c	n	2	s	wd?	?	sa, gr
Senna didymobotrya	Fabaceae	Africa	Tree/Shrub	t	p	u	e	w	n	3	m	wt	?	fo, sa
Sesbania punicea	Fabaceae	S. America	Tree	t	p	u	e/l	w	n	5	m	wt	x	sa, fo, fy, gr
Solanum elaeagnifolium	Solanaceae	N. & S. America	Shrub	t	p	u	gm?	w	n	4	s	?	r	sa

Appendix 22.1. (contd).

Species	Family	Origin	Growth-form	Functional classification[a]						Seed categories[b]	Seed size[c]	Main dispersal agent[d]	Vegetative propagation[d]	Biome(s) invaded[f]
				S	Lo	H	P	W	Lf					
Solanum mauritianum	Solanaceae	S. America	Tree/Shrub	t	p	u	e	w	n	4	s	bd	d	fo, gr, sa, fy
Tagetes minuta	Asteraceae	S. America	Herb	t	a	u	gm	n	n	2	s	wd	?	sa, gr, ka, fo
Verbena bonariensis	Verbenaceae	S. America	Herb	t	p	u	s	n	n	2	s	wt	?	sa, gr
Xanthium spinosum	Asteraceae	S. America?	Herb	t	a	u	gm	n	n	2	?	ml, wt	?	ka, sa, gr
Xanthium strumarium	Asteraceae	C. & S. America?	Herb	t	a	u	gm	n	n	2	s	wt, ml	?	ka, sa
Zinnia peruviana	Asteraceae	S. America	Herb	t	a	u	gm	n	n	2	s	wd	?	sa

[a] Substrate (epiphyte, terrestrial or aquatic); Longevity (annual, perennial (>5 years) or variable); Habit (upright, scandent (lax or drooping, but not creeping), creeping, floating (free), or floating (attached)); Perennation (evergreen, leaf deciduous, stem deciduous, geophytic, germinative (annual, ephemeral or biennial) or variable); Woodiness (woody, semi-woody (stems peripherally woody, with a soft pith) or non-woody); Leaf texture (sclerophyllous, semi-sclerophyllous, normal, fleshy or succulent) (includes leaf-like structures such as cladodes and phyllodes).

[b] 1, canopy stored seed; 2, small dry seed: soil-stored; 3, medium sized dry seeds: soil-stored; 4, seed embedded in soft fruit; 5, short-lived seed; 6, seedless invaders (principal source of data: Dean *et al.* 1986).

[c] m, >1000 seeds 100 g⁻¹; s, >10 000 seeds 100 g⁻¹; vs, >100 000 seeds 100 g⁻¹ (principal source of data: Dean *et al.* 1986). Parentheses indicate that taxa are sterile in southern Africa.

[d] an, ants; bd, birds; hu, humans; ml, mammals; wd, wind; wt, water (principal source of data: Dean *et al.* 1986).

[e] X, none; c, coppice; d, division or rooting of vegetative organ; s, sucker; r, runner (principal source of data: Dean *et al.* 1986).

[f] aq, aquatic systems; de, desert; fo, forest; fy, fynbos; gr, grassland; ka, karoo (Nama and succulent karoo biomes are not differentiated); sa, savanna. The biome listed first is where the densest infestations occur; other biomes listed are where ≥5% of reserves in those biomes were invaded (Macdonald 1991) or, in the case of the desert biome, where ≥5% of 10 km sections in desert vegetation types listed by Macdonald & Nott (1987; see text) had density ratings of 3 or more.

.10 References

amson, R.S. (1938). *The Vegetation of South Africa*. London: British Empire Vegetation Commission.

anecke, D.P., Karny, M. & Burger, W.A. (1969). Improved biological control of the prickly pear, *Opuntia megacantha* Salm-Dyck, in South Africa through the use of an insecticide. *Phytophylactica*, 1, 9–13.

mstrong, G. (1992). A necessary evil? *Veld & Flora*, **78**, 10.

shton, P.J., Appleton, C.C. & Jackson, P.B.N. (1986). Ecological impacts and economic consequences of alien invasive organisms in southern Africa. In *The Ecology and Management of Biological Invasions in Southern Africa*, ed. I.A.W. Macdonald, F.J. Kruger & A.A. Ferrar, pp. 247–57. Cape Town: Oxford University Press.

shton, P.J. & Mitchell, D.S. (1989). Aquatic plants: patterns and modes of invasion, attributes of invading species and assessment of control programmes. In *Biological Invasions: A Global Perspective*, ed., J.A. Drake, H.A. Mooney, F. Di Castri, R.H. Groves, F.J. Kruger, M. Rejmánek & M. Williamson, pp. 111–54. Chichester: Wiley.

zorín, E.J. (1992). *The Potential of Alien Acacias as a Woodfuel Resource in the South Western Cape*. Cape Town: National Energy Council.

ews, J.W. (1916). An account of the chief types of vegetation in South Africa, with notes on the plant succession. *Journal of Ecology*, 4, 129–59.

nggeli, P. (1996). A taxonomic, biogeographical and ecological overview of invasive woody plants. *Journal of Vegetation Science*, 7, 121–4.

olus, H. (1886). Sketch of the flora of South Africa. In *Official Handbook of the Cape of Good Hope*, ed. J. Noble, pp. 288–319. Cape Town: Solomon.

oyer, D.C. & Boyer, H.J. (1989). The status of invasive alien plants in the major rivers of the Namib Naukluft Park. *Madoqua*, 16, 51–8.

Brown, C.J. & Gubb, A.A. (1986). Invasive alien organisms in the Namib Desert, Upper Karoo and the arid and semi-arid savannas of western southern Africa. In *The Ecology and Management of Biological Invasions in Southern Africa*, ed. I.A.W. Macdonald, F.J. Kruger & A.A. Ferrar, pp. 93–108. Cape Town: Oxford University Press.

Brown, C.J., Macdonald, I.A.W. & Brown S.E. (ed.) (1985). *Invasive Alien Organisms in South West Africa/Namibia*. South African National Scientific Programmes Report 119. Pretoria: CSIR.

Brutsch, M.O. & Zimmermann, H.G. (1993). The Prickly Pear (*Opuntia ficus-indica* [Cactaceae]) in South Africa: Utilization of the naturalized weed, and of cultivated plants. *Economic Botany*, 47, 154–62.

Cowling, R.M. (1992). Definitely an unnecessary evil. *Veld & Flora*, 78, 63–4.

Cronk, Q.C.B. & Fuller, J.L. (1995). *Plant Invaders. The Threat to Natural Ecosystems*. London: Chapman & Hall.

Deacon, J. (1986). Human settlement in South Africa and archaeological evidence for alien plants and animals. In *The Ecology and Management of Biological Invasions in Southern Africa*, ed. I.A.W. Macdonald, F.J. Kruger & A.A. Ferrar, pp. 3–19. Cape Town: Oxford University Press.

Dean, S.J., Holmes, P.M. & Weiss, P.J. (1986). Seed biology of invasive alien plants in South Africa and South West Africa/Namibia. In *The Ecology and Management of Biological Invasions in Southern Africa*, ed. I.A.W. Macdonald, F.J. Kruger & A.A. Ferrar, pp. 157–70. Cape Town: Oxford University Press.

Donnelly, D. & Morris, M.J. (1992). Port Jackson biocontrol: dispelling the myths. *Veld & Flora*, 78, 62.

Drake, J.A., Mooney, H.A., Di Castri, F., Groves, R.H., Kruger, F.J., Rejmánek, M. & Williamson, M. (eds) (1989). *Biological Invasions: A Global Perspective*. Chichester: Wiley.

du Toit, R. (1942). The spread of prickly pear in the Union. *Farming in South Africa*, 16, 300–1.

Elton, C.S. (1958). *The Ecology of Invasions by Animals and Plants*. London: Metheun.

Fenn, J.A. (1980). Control of hakea in the Western Cape. In *Proceedings of the Third National Weed. Conference*, ed. S. Neser & A.L.P. Cairns, pp. 167–73. Cape Town: Balkema.

Ferrar, A.A. & Kruger, F.J. (1983). *South African Programme for the SCOPE Project on the Ecology of Biological Invasions*. South African National Scientific Programmes Report 72. Pretoria: CSIR.

Geldenhuys, C.J. (1986). Costs and benefits of the Australian blackwood *Acacia melanoxylon* in South African forestry. In *The Ecology and Management of Biological Invasions in Southern Africa*, ed. I.A.W. Macdonald, F.J. Kruger & A.A. Ferrar, pp. 275–84. Cape Town: Oxford University Press.

Geldenhuys, C.J., le Roux, P.J. & Cooper, K.H. (1986). Alien invasions in indigenous evergreen forest. In *The Ecology and Management of Biological Invasions in Southern Africa*, ed. I.A.W. Macdonald, F.J. Kruger & A.A. Ferrar, pp. 119–31. Cape Town: Oxford University Press.

Glyphis, J.P., Milton, S.J. & Siegfried, W.R. (1981). Dispersal of *Acacia cyclops* by birds. *Oecologia*, 48, 138–41.

Henderson, L. (1989). Invasive alien woody plants of Natal and the North-eastern Orange Free State. *Bothalia*, 19, 237–61.

Henderson, L. (1991a). Alien invasive *Salix* spp. (willows) in the grassland biome of South Africa. *South African Forestry Journal*, 157, 91–5.

Henderson, L. (1991b). Invasive alien woody plants of the Northern Cape. *Bothalia*, 21, 177–89.

Henderson, L. (1992). Invasive alien woody plants of the eastern Cape. *Bothalia*, 22, 119–43.

Henderson, L. & Musil, K.J. (1984). Exotic woody plant invaders of the Transvaal. *Bothalia*, 15, 297–313.

Henderson, L. & Wells, M.J. (1986). Alien plant invasions in the grassland and savanna biomes. In *The Ecology and Management of Biological Invasions in Southern Africa*, ed. I.A.W. Macdonald, F.J. Kruger & A.A. Ferrar, pp. 109–18. Cape Town: Oxford University Press.

Hoffmann, J.H. & Moran, V.C. (1988). The invasive weed *Sesbania punicea* in South Africa and prospects for its biological control. *South African Journal of Science*, 84, 740–2.

Hoffmann, J.H. & Moran, V.C. (1991). Biological control of *Sesbania punicea* (Fabaceae) in South Africa. *Agriculture, Ecosystems and Environment*, 37, 157–73.

Holm, L., Pancho, J.V., Herberger, J.P. & Plucknett, D.L. (1979). *A Geographical Atlas of World Weeds*. New York: Wiley.

Holm, L.G., Plucknett, D.L., Pancho, J.V. & Herberger, J.P. (1977). *The World's Worst Weeds*. Honululu: University Press Hawaii.

Holmes, P.M. (1989). Decay rates in buried alien *Acacia* seed populations of different density. *South African Journal of Botany*, 55, 299–303.

Holzner, W. & Humata, N. (ed.) (1982). *Biology and Ecology of Weeds*. The Hague: Junk.

Huntley, B.J., Siegfried, W.R. & Sunter, C. (1989). *South African Environments into the 21st Century*. Cape Town: Human & Rousseau Tafelberg.

Jeffrey, D.J., Holmes, P.M. & Rebelo, A.G. (1988). Effects of dry heat on seed germination in selected indigenous and alien legume species in South Africa. *South African Journal of Botany*, 54, 28–34.

Julien M.H. (1992). *Biological Control of Weeds: A World Catalogue of Agents and Their Target Weed.*. Wallingford: CAB International.

Kluge, R.L., Zimmermann, H.G., Cilliers, C.J. & Harding, G.B. (1986). Integrated control for invasive alien weed.. In *The Ecology and Management of Biological Invasions in Southern Africa*, ed. I.A.W. Macdonald, F.J. Kruger & A.A. Ferrar, pp. 295–304. Cape Town: Oxford University Press.

Knight, R.S. (1985). A model of episodic, abiotic dispersal for oaks *(Quercus robur)*. *South African Journal of Botany*, 51, 265–9.

Knight, R.S. (1986). A comparative analysis of fleshy fruit displays in alien and indigenous plants. In *The Ecology and Management of Biological Invasions in Southern Africa*, ed. I.A.W. Macdonald, F.J. Kruger & A.A. Ferrar, pp. 171–8. Cape Town: Oxford University Press.

Kruger, F.J., Breytenbach, G.J., Macdonald, I.A.W. & Richardson, D.M. (1989). The characteristics of invaded mediterranean-climate regions. In *Biological Invasions: A Global Perspective*, ed., J.A. Drake, H.A. Mooney, F. Di Castri, R.H. Groves, F.J. Kruger, M. Rejmánek & M. Williamson, pp. 181–213. Chichester: Wiley.

Kruger, F.J., Richardson, D.M. & Van Wilgen, B.W. (1986). Processes of invasion by alien plants. In *The Ecology and Management of Biological Invasions in Southern Africa*, ed. I.A.W. Macdonald, F.J. Kruger & A.A. Ferrar, pp. 145–55. Cape Town: Oxford University Press.

Leach, L.C. (1973). *Euphorbia tirucalli* L. its typification, synonymy and relationships with notes on 'Almeidina' and 'Cassoneira'' *Kirkia*, 9, 69–86.

Le Maitre, D.C., Van Wilgen, B.W., Chapman, R.A. & McKelly, D.H. (1996). Invasive plants and water resources in the Western Cape Province, South Africa: modelling the consequences of a lack of management. *Journal of Applied Ecology*, 33, 161–72.

Liggitt, B. (1983). *The Invasive Alien Plant Chromolaena odorata, With Regard to its Status and Control in Natal*. Pietermaritzburg: Institute of Natural Resources, University of Natal.

Lodge, D.M. (1993). Biological invasions: Lessons for ecology. *Trends in Ecology and Evolution*, 8, 133–7.

Loope, L.L., Sanchez, P.G., Tarr, P.W., Loope, W.L. & Anderson, R.L. (1988). Biological invasions of arid land nature reserves. *Biological Conservation*, 44, 95–118.

Lubke, R.A. (1985). Erosion of the beach at St. Francis Bay, Eastern Cape, South Africa. *Biological Conservation*, 32, 99–127.

Macdonald, I.A.W. (1983). Alien trees, shrubs and creepers invading indigenous vegetation in the Hluhluwe-Umfolozi Game Reserve Complex in Natal. *Bothalia*, 14, 949–59.

Macdonald, I.A.W. (1984). Is the fynbos biome especially susceptible to invasion by alien plants? A re-analysis of available data. *South African Journal of Science*, 80, 369–77.

Macdonald, I.A.W. (1985). The Australian contribution to southern Africa's invasive flora: an ecological analysis. *Proceedings of the Ecological Society of Australia*, 14, 225–36.

Macdonald, I.A.W. (1986). Invasive alien plants and their control in southern African nature reserves. In *Proceedings of Conference on Science in the National Parks*, pp. 63–79. Fort Collins, Colorado: Colorado State University.

Macdonald, I.A.W. (1988). A list of alien plants in the Kruger National Park. *Koedoe*, 31, 137–50.

Macdonald, I.A.W. (1991). *Conservation Implications of the Invasion of Southern Africa by Alien Organisms*. PhD Thesis. Cape Town: University of Cape Town.

Macdonald, I.A.W., Clark, D.L. & Taylor, H.C. (1987). The alien flora of the Cape of Good Hope Nature Reserve. *South African Journal of Botany*, 53, 398–404.

Macdonald, I.A.W., Clark, D.L. & Taylor, H.C. (1989). The history and effects of alien plant control in the Cape of Good Hope Nature Reserve, 1941–1987. *South African Journal of Botany*, 55, 56–75.

Macdonald, I.A.W., Graber, D.M., DeBenedetti, S., Groves, R.H. & Fuentes, E.R. (1988). Introduced species in nature reserves in mediterranean climatic regions of the world. *Biological Conservation*, 44, 37–66.

Macdonald, I.A.W., Jarman, M.L. & Beeston, P.M. (eds) (1985). *Management of Invasive Alien Plants in the Fynbos Biome. South African National Scientific Programmes Report 111*. Pretoria: CSIR.

Macdonald, I.A.W., Kruger, F.J. & Ferrar, A.A. (eds) (1986). *The Ecology and Management of Biological Invasions in Southern Africa*. Cape Town: Oxford University Press.

Macdonald, I.A.W. & Nott, T.B. (1987). Invasive alien organisms in central South West Africa/Namibia: Results of a reconnaisance survey conducted in November 1984. *Madoqua*, 15, 21–34.

cdonald, I.A.W. & Richardson, D.M. (1986). Alien species in terrestrial ecosystems of the fynbos biome. In *The Ecology and Management of Biological Invasions in Southern Africa*, ed. I.A.W. Macdonald, F.J. Kruger & A.A. Ferrar, pp. 77–91. Cape Town: Oxford University Press.

tchell, D.S. (ed.) (1974). *Aquatic Vegetation and its Use and Control*. Paris: UNESCO.

ll, E.J. & Bossi, L. (1984). Assessment of the extent of the natural vegetation of the fynbos biome of South Africa. *South African Journal of Science*, **80**, 355–8.

oney, H.A. & Drake, J.A. (1989). Biological invasions: a SCOPE Program Overview. In *Biological Invasions: A Global Perspective*, ed. J.A. Drake, H.A. Mooney, F. Di Castri, R.H. Groves, F.J. Kruger, M. Rejmánek & M. Williamson, pp. 491–506. Chichester: Wiley.

ran, V.C. & Annecke (1979). Critical reviews of biological pest control in South Africa. 3. The jointed cactus, *Opuntia aurantiaca* Lindley. *Journal of the Entomological Society of Southern Africa*, **42**, 299–329.

ran, V.C. & Zimmermann, H.G. (1991). Biological control of jointed cactus, *Opuntia aurantiaca* (Cactaceae), in Southern Africa. *Agriculture, Ecosystems and Environment*, **37**, 5–27.

tey, F.W. (1943). Prickly-pear eradication by insects and felling of plants. *Farming in South Africa*, **18**, 743–6.

tey, F.W. (1947). *The Biological Control of Prickly Pears in South Africa*. Entomology Series 22. Pretoria: Department of Agriculture.

illips, E.P. (1938). The naturalized species of *Hakea*. *Farming in South Africa*, **13**, 424.

illips, J. (1963). *The Forests of George, Knysna and the Zitzikama – A Brief History of Their Management 1778–1939*. Pretoria: Government Printer.

rce, S.M. (1982). What is *Spartina* doing in our estuaries? *South African Journal of Science*, **78**, 229–30.

ynton, R.J. (1979a). *Tree Planting in Southern Africa*. Volume 1. The Pines. Pretoria: Department of Forestry.

Poynton, R.J. (1979b). *Tree Planting in Southern Africa*. Volume 2. The Eucalypts. Pretoria: Department of Forestry.

Poynton, R.J. (1984). *Characteristics and Uses of Selected Trees and Shrubs Cultivated in South Africa*. Bulletin No. 39. 4th Edition, revised. Pretoria: Government Printer.

Rejmánek, M. & Richardson, D.M. (1996). What attributes make some plant species more invasive? *Ecology*, **77**, 1655–61.

Richardson, D.M. (1984). A cartographic analysis of physiographic factors influencing the distribution of *Hakea* spp. in the S.W. Cape Province. *South African Forestry Journal*, **128**, 36–40.

Richardson, D.M. (1985). *Studies on Aspects of the Integrated Control of Hakea sericea in the South-western Cape Province, South Africa*. MSc Thesis. Cape Town: University of Cape Town.

Richardson, D.M. & Bond, W.J. (1991). Determinants of plant distribution: evidence from pine invasions. *American Naturalist*, **137**, 639–68.

Richardson, D.M. & Cowling, R.M. (1992). Why is mountain fynbos invasible and which species invade? In *Fire in South African Mountain Fynbos*, ed. B.W. van Wilgen, D.M. Richardson, F.J. Kruger & H.J. van Hensbergen, pp. 161–81. Berlin: Springer-Verlag.

Richardson, D.M. & Cowling, R.M. (1994). The ecology of invasive aliens pines (*Pinus* spp.) in the Jonkershoek Valley, Stellenbosch, South Africa. *Bontebok*, **9**, 1–10.

Richardson, D.M., Cowling, R.M. & Le Maitre, D.C. (1990). Assessing the risk of invasive success in *Pinus* and *Banksia* in South African Mountain Fynbos. *Journal of Vegetation Science*, **1**, 629–42.

Richardson, D.M., Macdonald, I.A.W. & Forsyth, G.G. (1989). Reductions in plant species richness under stands of alien trees and shrubs in the fynbos biome. *South African Forestry Journal*, **49**, 1–8.

Richardson, D.M., Macdonald, I.A.W., Holmes, P.M. & Cowling, R.M. (1992). Plant and animal invasions. In *The Ecology of Fynbos. Nutrients, Fire and Diversity*, ed. R.M. Cowling, pp. 271–308. Cape Town: Oxford University Press.

Richardson, D.M. & McMahon, J.P. (1992). A bioclimatic analysis of *Eucalyptus nitens* to identify potential planting regions in southern Africa. *South African Journal of Science*, **88**, 380–7.

Richardson, D.M., van Wilgen, B.W., Higgins, S.I., Trinder-Smith, T.H., Cowling, R.M. & McKelly, D.H. (1996). Current and future threats to plant biodiversity on the Cape Peninsula, South Africa. *Biodiversity & Conservation*, **5**, 607–47.

Richardson, D.M., Van Wilgen, B.W., Le Maitre, D.C., Higgins, K.B. & Forsyth, G.G. (1994a). A computer-based system for fire management in the mountains of the Cape Province, South Africa. *International Journal of Wildland Fire*, **4**, 17–32.

Richardson, D.M., Williams, P.A. & Hobbs, R.J. (1994b). Pine invasions in the Southern hemisphere: Determinants of spread and invadability. *Journal of Biogeography*, **21**, 511–27.

Shaughnessy, G. (1986). A case study of some woody plant introductions to the Cape Town area. In *The Ecology and Management of Biological Invasions in Southern Africa*, ed. I.A.W. Macdonald, F.J. Kruger & A.A. Ferrar, pp. 37–43. Cape Town: Oxford University Press.

Sim, T.R. (1905). *Tree Planting in Natal*. Pietermaritzburg: Department of Agriculture.

Sim, T.R. (1927). Some effects of man's influence on the South African flora. *South African Journal of Science*, **23**, 492–507.

Spies, J.J. & Stirton, C.H. (1982). Meiotic studies of some South African cultivars of *Lantana camara* (Verbenaceae). *Bothalia*, **14**, 101–11.

Stirton, C.H. (ed.) (1978). *Plant Invaders: Beautiful but Dangerous*. Cape Town: Department of Nature and Environmental Conservation of the Cape Provincial Administration.

Swarbrick, J.T. (1991). Towards a rating scheme for environmental weeds. *Plant Protection Quarterly*, **6**, 185.

Taylor, H.C. (1969). Pest-plants and nature conservation in the winter rainfall region. *Journal of the Botanical Society of South Africa*, **55**, 32–8.

Tucker, K. & Richardson, D.M. (1995). An expert system for screening potentially invasive alien plants in South African fynbos. *Journal of Environmental Management*, 44, 309–38.

Usher, M.B., Kruger, F.J., Macdonald, I.A.W., Loope, L.L. & Brockie, R.E. (1988). The ecology of biological invasions into nature reserves: an introduction. *Biological Conservation*, 44, 1–8.

van Wilgen, B.W., Bond, W.J. & Richardson, D.M. (1992). Ecosystem management. In *The Ecology of Fynbos. Nutrients, Fire and Diversity*, ed. R.M. Cowling, pp. 345–71. Cape Town: Oxford University Press.

van Wilgen, B.W., Cowling, R.M. & Burgers, C.J. (1996). Valuation of ecosystem services: a case study from South African fynbos. *BioScience*, 46, 184–9.

van Wilgen, B.W. & Richardson, D.M. (1985). The effects of alien shrub invasions on vegetation structure and fire behaviour in South African fynbos shrublands: a simulation study. *Journal of Applied Ecology*, 22, 955–66.

Vlok, J.H.J. (1988). Alpha diversity of lowland fynbos herbs at various levels of infestation by alien annuals. *South African Journal of Botany*, 54, 623–7.

Wells, M.J., Balsinhas, A.A., Joffe, H., Engelbrecht, V.M., Harding, G. & Stirton, C.H. (1986a). A catalogue of problem plants in southern Africa. *Memoirs of the Botanical Survey of South Africa*, 53, 1–658.

Wells, M.J., Engelbrecht, V.M., Balsinhas, A.A. & Stirton, C.H. (1983). Weed flora of South Africa 3 : More power shifts in the veld. *Bothalia*, 14, 967–70.

Wells, M.J., Poynton, R.J., Balsinhas, A.A., Musil, C.F., Joffe, H., van Hoepen, E. & Abbott, S.K. (1986b). The history of introduction of invasive alien plants to southern Africa. In *The Ecology and Management of Biological Invasions in Southern Africa*, ed. I.A.W. Macdonald, F.J. Kruger & A.A. Ferrar, pp. 21–35. Cape Town: Oxford University Press.

Whiting, B.H., Bate, G.C. & Erasmus, D.J. (1986). Photosynthesis and carbon allocation in invasive plants of South African biomes. In *The Ecology and Management of Biological Invasions in Southern Africa*, ed. I.A.W. Macdonald, F.J. Kruger & A.A. Ferrar, pp. 179–88. Cape Town: Oxford University Press.

Wicht, C.L. (1945). *Preservation of the Vegetation of the South-western Cape*. Cape Town: Special Publication of the Royal Society of South Africa.

Woods, D.H. (1950). The menace of the pine. *Journal of the Mountain Club of South Africa*, 53, 78–87.

Zimmermann, H.G. & Moran, V.C. (1982). Ecology and management of cactus weed. in South Africa. *South African Journal of Science*, 78, 314–20.

Zimmermann, H.G., Moran, V.C. & Hoffmann, J.H. (1986). Insect herbivores as determinants of the present distribution and abundance of invasive cacti in South Africa. In *The Ecology and Management of Biological Invasions in Southern Africa*, ed. I.A.W. Macdonald, F.J. Kruger & A.A. Ferrar, pp. 269–74. Cape Town: Oxford University Press.

Conservation

A.G. Rebelo

23.1 Introduction

South Africa was one of the first countries in the world to produce an assessment of the adequacy of the existing nature reserve system (Edwards 1974). This assessment called for 5% of each vegetation type to be conserved in formal nature reserves. By 1980 there was an international call that each country should strive for a conservation level of 10% (Myers 1979). The adoption of this apparently arbitrary level was justified by the prediction from island biogeography theory that 10% of an area should protect about 50% of species (Diamond 1976; Shafer 1990; RSA 1991). The 1992 Caracas Action Plan of the World Parks Congress adopted 10% as the minimum goal for each of the world's major biomes (World Resources Institute 1994).

Despite early initiatives, little action has been forthcoming. Detailed surveys, such as those of the lowlands of the fynbos (Jarman 1986) and the succulent karoo biomes (Hilton-Taylor & Le Roux 1989), have explored options and proposed suitable sites for conservation action, but there has been little progress over the past two decades since the identification of shortcomings in 1974 (Greyling & Huntley 1984; Siegfried 1989; Rebelo 1992a).

The future of conservation in southern Africa is uncertain. Land reform proposals for South Africa, which seek to redistribute under-utilized, privately owned land and to make state land available to small-scale and subsistence farmers, could potentially conflict with conservation. It is generally agreed that ours is the first, and possibly the last generation, in the position to decide the fate of thousands or millions of species on earth (World Resources Institute 1992). In South Africa, options for the next few decades are currently being decided. It is not true that development must occur at the expense of conservation, although it is the easy option which could lead to the 'wasteland' scenario of conservation planners (Huntley, Siegfried & Sunter 1989). However, because ecotourism is potentially a major source of revenue in southern Africa, maintaining and improving the conservation estate would be a sound investment (Cowling 1991, 1993).

In this chapter the effectiveness of the conservation estate in southern Africa as an efficient conservation network is assessed. General (political) details of the conservation estate are dealt with elsewhere (Greyling & Huntley 1984; Siegfried 1989; World Resources Institute 1994). However, some of the assumptions implicit in the goals set for levels of conservation in the region are queried. Also discussed is the need for an action plan to negate a continuous refinement of requirements and politically expedient reserve proclamation at the expense of meaningful progress.

23.2 A brief history of conservation planning

The conservation estate in southern Africa developed during the colonial era in response to the realization that over-hunting, rinderpest and methods of tsetse fly control were resulting in the loss of animal species (Anderson & Grove 1987). Little concern was given to floristic, ecosystem or social issues. Instead, the focus, determined by the hunting associations, was the preservation of the large mammal species (Pringle 1982). Such reserves, proclaimed towards the end of the nineteenth century, were sited in marginal areas, unsuitable for agriculture and with no known mining potential. Management involved predator elimination, total fire protection, provision of artificial water supplies, 'veld improvement' and maintaining 'the balance of nature' (Huntley 1978; Rebelo 1992a,b).

At the same time, foresters in the Cape Colony were expressing alarm at the rapid loss of indigenous forests in the coastal regions (see Greyling & Huntley 1984). This resulted in the establishment of forest reserves, with a view to future afforestation.

During the first half of the twentieth century, conservation actions were increasingly guided by ecosystem-level considerations. This shift in approach had its roots in the London Convention of 1933, which coincided in South Africa with the proclamation of the largest proportion of the current conservation estate: the water catchments (Fig. 23.1). In one of the few published cases of holistic conservation planning in South Africa, Wicht (1945) reviewed the effects of fire regime, pasturing, ero-

sion, invasive alien organisms, conversion and overharvesting on the vegetation of the fynbos biome. He advocated that a minimum of five reserves of national status should be established in the region. All were in the mountains and the emphasis would be on preserving the flora rather than the fauna. Local and educational reserves were also proposed, but their siting was limited to a comment that every town in the region should have an adjacent reserve. At this same time (1945), a survey was initiated to describe and map the vegetation types of South Africa – resulting in the publication of Acocks' *Veld Types of South Africa* in 1953 (Acocks 1953).

The 1966 meeting of the Association pour l'Etude Taxonomique de la Flore d'Afrique Tropicale (AETFAT) on 'Conservation of vegetation in Africa South of the Sahara' focused on the conservation of representative units of vegetation (Hedberg & Hedberg 1968). In 1968, a survey was initiated in South Africa to outline the conservation status of the 70 Acocks' veld or vegetation types (units of the same farming potential). This was reported in Section CT of the International Biological

Figure 23.1 **The development of the conservation estate in southern Africa according to year of proclamation.**

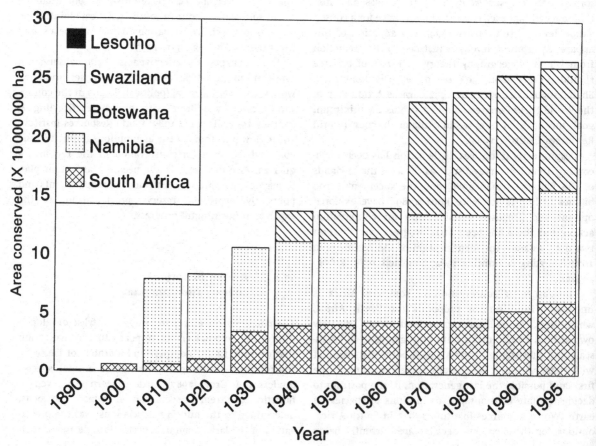

Programme, and taken up by the National Committee for Nature Conservation (NAKOR), which instituted a 'national plan' (Scheepers 1983).

During the 1970s the international focus shifted to rare species and Red Data Book taxa and representative ecosystems. In southern Africa, a shift towards game utilization, chiefly in the form of recreational hunting, resulted in large areas of the savanna biome being managed in a manner compatible with ecosystem conservation (Huntley 1978). However, in some regions conservation of large game was to the detriment of vegetation, largely through 'veld-improvement' and the provision of water sources (Rebelo 1992b). The Wilderness Area Concept resulted in the creation of wilderness areas and hiking trails within the mountain catchments of South Africa, chiefly by the proclamation of old forestry reserves as Wilderness Areas (Ackerman 1972).

The 1980s saw a large increase in literature concerning conservation: Red Data Books (Ferrar 1989; Hall 1989), conservation assessments (Jarman 1986, Cowan 1987), theoretical issues (Hall 1984) and regional syntheses (Huntley 1989), which reflected an international explosion in publications (Huntley 1994).

Today the trend is for the biodiversity of the region to be represented optimally by the reserve network (Huntley 1994). Most southern African states have ratified or signed the Biodiversity Treaty, which requires the formulation of national plans for the conservation of representative systems (World Resources Institute 1994). There is also an increased perception that conservation must involve sustainable utilization, rather than the setting aside of land without benefit to the local people (Anderson & Grove 1987; Stuart & Adams 1990). In southern Africa, partly as a result of the perceived enforcement of colonial and apartheid doctrines, there has been a backlash, and existing conservation areas have to be defended and justified as beneficial to local communities (Huntley 1978; Stuart & Adams 1990; Cowling 1991; West & Brechin 1992).

Recently, there has been an increasing realization that if adequate representation of all ecosystems and species is to be achieved, then areas outside the nature reserve estate also need to be managed for conservation (Western 1989). This approach has its roots in the following assumptions: some forms of land use are compatible with conservation goals; human communities are unwilling to vacate or sell prime conservation land; there are inadequate funds for reserve acquisition; and, for some ecosystems, tracts of untransformed land suitable for reserves do not exist. Privately owned conservation areas in South Africa often have some semi-formal recognition, such as the Langebaan, Knysna and the Lakes Contractual National Parks (under the National

Parks Act 57 of 1976), the Cape Mountain Catchment Areas (under the Water Conservation Act 63 of 1970), the KwaZulu-Natal Conservancies (Grobler 1990), and the Natural Heritage Programme (M. Wahl, pers. comm.). Throughout southern Africa, community-based conservation schemes, usually based on tourism, are becoming increasingly popular (e.g. Jacobsohn 1991).

Most of the subcontinent is too arid for intensive transformation (Chap.2, this volume), so that effectively, preservation is ensured under all except the most drastic farming practices (e.g. planting with alien *Atriplex* spp. and heavy overgrazing). Nevertheless, many of these arid systems are over-utilized relative to their carrying capacity (see Chaps 7, 8, 9, 10 and 11, this volume). Insufficient attention is being given to conservation of land with high transformation potential in exactly those regions where conservation options are rapidly diminishing (Rebelo 1994a). The dilemma of evaluating conservation priorities in systems endangered with radical transformation and in those being slowly degraded will be discussed later.

23.3 The conservation estate

Of the five countries in southern Africa, two have achieved the goal of protecting 10% of their vegetation: Botswana (17.9%) and Namibia (12.6%) (Table 23.1, Fig. 23.2). South Africa is halfway to achieving the goal (5.1%), whereas Swaziland (3.2%) and Lesotho (0.2%) are well below the levels attained internationally (5.9%) or for the African continent (4.6%) (World Resources Institute 1994).

Three of the seven southern African biomes have more than 10% of their area conserved (Table 23.1) – the desert, fynbos and savanna biomes, with the forest biome approaching 9%. In contrast, the Nama-karoo, grassland and succulent karoo biomes have less than 3% of their area conserved. The greater part of these biomes falls largely within South Africa, hence their conservation in the subcontinent is almost entirely this country's responsibility.

In South Africa, 14 of the 70 vegetation types (Veld Types) have more than 10% of their area conserved (Fig. 23.3). This has increased by 10 since a decade ago, but most of these are forest types (Table 23.2), which occupy relatively small areas of the country. Five vegetation types (Fynbos, *Themeda–Festuca* Alpine Veld, Lowveld, Arid Lowveld and Mopane Veld) have over one-fifth of their area conserved, chiefly in the mountain catchments of the fynbos biome and the Drakensberg (first two mentioned), and in the Kruger National Park (last three

Table 23.1 **Current conservation status of the biomes of southern Africa**

Biome[a]	Area				Index	
	Total (× 10^6 ha)[b]	Conserved (%)[b]	Transformed (%)[c]	Required (% gc)[d]	Species (n)[e]	RDB spp. (n)[f]
Fynbos (South Africa)	4.835	14.89	32	90	7316	1326
Succulent karoo	19.720	2.12	<<26	52	2125	558
(Namibia)[g]	3.920	1.17				
(South Africa)	15.800	2.36				
Nama-karoo	41.128	0.25	?	17	2147	67
(Namibia)[g]	19.600	0.00				
(South Africa)	24.529	0.45				
Desert (Namibia)[g]	7.840	85.00	?	12	497	?
Grassland	32.309	1.33	66	31	3788	?
(Lesotho)	3.034	0.21				
(South Africa)	28.929	1.38				
(Swaziland)[g]	0.346	6.50				
Savanna	147.435	11.60	<<48	15	5788	?
(Botswana)	58.173	17.88				
(Namibia)[g]	50.960	5.74				
(South Africa)	37.633	9.96				
(Swaziland)[g]	0.672	4.96				
Forest	8.401	8.56	?	?	1000?	?
(South Africa)	7.681	9.36				
(Swaziland)[g]	0.720	0.00				
Total	261.668	9.89	??	22	21000	2000?

[a] Biomes follow Rutherford & Westfall (1986), but boundaries differ substantially, because Acocks' Veld Types were allocated to biomes as follows: forest = 1–9; fynbos = 46, 47, 69, 70; grassland = 44, 45, 48–68; Nama-karoo = 27, 29, 30, 32, 35–38, 40–42; savanna = 10–23; succulent karoo = 24–26, 28, 31, 33, 34, 39, 43. For Veld Types corresponding to numbers, see Table 23.2.
[b] Data are from sources in Table 23.2 and thus differ from Siegfried (1989), most specifically in that Siegfried (loc. cit.) has an additional 11 000 ha not accounted for in the National Register of protected areas.
[c] Area transformed is based on subsamples from Macdonald (1989), and is biased towards parts of South Africa, where transformation is more extensive than in other countries.
[d] Required index is the number of grid cells (gc) required to represent each plant species once as a percentage of the total number of grid squares in the biome (see text).
[e] Species index (total number of species from core grid cells) from Gibbs Russell (1987), for forest from Geldenhuys & MacDevette (1989).
[f] Red Data Book (RDB) Index (total number of species in the Red Data Book) from C. Hilton-Taylor (unpubl. data).
[g] Biome areas for Swaziland and Namibia are estimates.

mentioned). The number of vegetation types lacking any conservation area has remained unchanged at around 17 since 1974, and amounts to one-quarter of the total. Although the current number of vegetation types with less than 1% conserved (36) is a quarter less than it was eight years ago (45), it still amounts to 51% of South Africa's vegetation types.

Of the vegetation types with no conservation status in South Africa (Fig. 23.3), seven occur in Nama-karoo, six in grassland, two in succulent karoo and two in savanna. This reflects the overall conservation status within the biomes in southern Africa. Even the best conserved biomes have major inadequacies. Despite the recognition that the critically endangered fynbos biome vegetation types, Coastal Fynbos and Coastal Renosterveld, are in urgent need of conservation (Greyling & Huntley 1984; Jarman 1986), little progress has been achieved (Table 23.2). The same is true of the succulent karoo and

grassland biomes, where the area under conservation has remained static, despite the recommendations of Edwards (1974) and Greyling & Huntley (1984).

The conservation status of the major centres of endemism (see Chap. 3, this volume) has been dealt with by Davis, Heywood & Hamilton (1994). The succulent karoo (Hilton-Taylor 1994a) and fynbos biomes (Rebelo 1994b) are recognized as major centres of endemism, and the conservation status of their subcentres of endemism are outlined below. Three lesser centres, the Kaokoveld, Pondoland and Drakensberg Alpine regions are mentioned briefly.

Of the 19 centres of endemism for the Proteaceae in the fynbos, ten have more than half of their area formally conserved or in mountain catchment areas, two have less than 5% conserved and three less than 1%. Three of these poorly conserved centres of endemism are in the lowlands, where land transformation is severe,

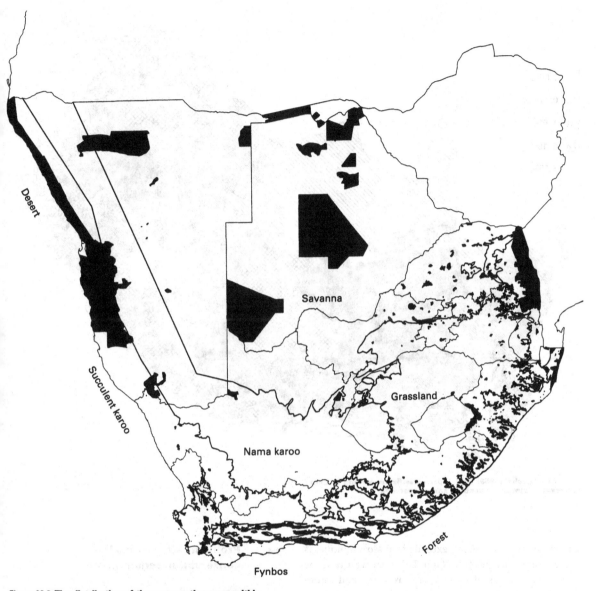

Figure 23.2 **The distribution of the conservation areas within southern Africa (shaded) in relation to the biomes (heavy lines) defined in terms of Veld Types (see Table 23.2). Also shown are international boundaries and the provincial boundaries within South Africa. Boundaries of conservation areas from Department of Environment Affairs and Tourism, and World Conservation Monitoring Centre.**

and two are in the dry inland mountains, which have little water catchment potential. With the exception of the Cape Agulhas region, most of the species-rich districts are well conserved (Rebelo 1994b).

In the succulent karoo, five centres of endemism are recognized (Hilton-Taylor 1994a). The Gariep Centre is the richest of these, and is relatively well conserved by the conservation areas in the Richtersveld. There are no conservation areas in the Kamiesberg and Vanrhynsdorp centres. There are several conservation areas in the West-

ern Mountain (7.7% by area) and Little Karoo centres of endemism, but there are no analyses of the efficiency or adequacy of the reserve system.

The Kaokoveld, a largely desert and Nama-karoo centre of endemism, which extends into southern Angola, is well conserved, but the formally conserved area is confined to the coastal desert region. However, much of the area is managed as a game conservation area, and most of the Kaokoveld in Angola has formal, if ineffective, conservation (Hilton-Taylor 1994b). The

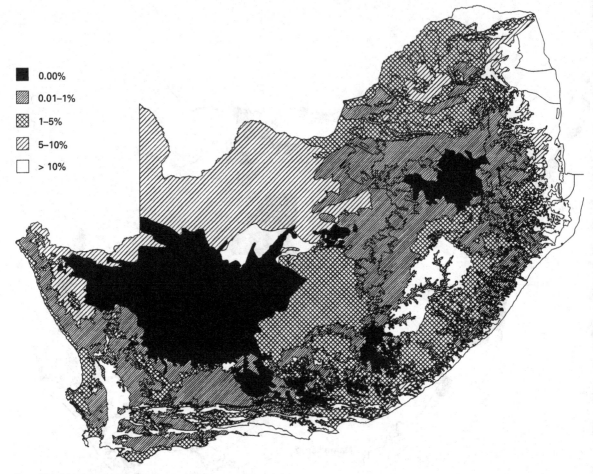

0.00%

0.01–1%

1–5%

5–10%

> 10%

Figure 23.3 **Acocks' (1953) Veld Types classified according to their conservation status (percentage area conserved).**

Maputaland Centre, which extends into Mozambique, is well conserved in KwaZulu-Natal. This is the result of the area being a focus of early 'game' reserves and, more recently, involvement of local people in managing local resources. However, the human population growth in this region is very rapid, exerting extreme pressure on conservation areas (Van Wyk 1994).

About 7% of the Pondoland Centre is conserved. Parts of the region are considered to have the highest conservation priority in southern Africa (Van Wyk 1994). For the Drakensberg Alpine Region, 7% is conserved, mainly in KwaZulu-Natal. However, the alpine and subalpine belts in Lesotho are poorly conserved (Killick 1994).

23.4 Problems with assessing conservation requirements

23.4.1 Scale and representativeness

It is obvious from the above synopsis that the assessment of conservation status is profoundly influenced by the scale of resolution adopted for analysis. Applying the standard prescription of protecting 10% of the subcontinental, national, or biome will not provide for adequate conservation (Slatyer 1975). Thus, although almost 12% of savanna vegetation in the subcontinent has been conserved, and almost 10% of savanna in South Africa, the coverage can hardly be considered adequate. Four of the 13 savanna vegetation types in South Africa have less than 1.0% of their area conserved, whereas only four have more than 10% conserved. The problem of scale extends below the lowest resolution of vegetation types

Table 23.2 **Conservation status of vegetation types (Acocks' Veld Types) in South Africa. The totals in each case are estimates, as detailed analysis of vegetation types within reserves are usually carried out after reserves are proclaimed. Interim assessments usually designate an entire reserve to one vegetation type (see text for details). The transfer of State Forest to provincial conservation authorities still requires confirmation – this applies especially to the tropical forest Veld Types. Sources: Edwards 1974; Scheepers 1983; Cowan 1987; Department of Environment Affairs 1995 – South African Register of Protected Areas (M. Wahl, pers. comm.)**

Vegetation type (Veld Type number)	Area conserved (%)				Area transformed[b] (%)
	1974[a]	1983	1987	1995	
Coastal Tropical Forest	**1.66**	**1.52**	**1.52**	**9.01**	
1 Coastal Forest & Thornveld	2.80	1.30	1.63	12.19	43
2 Alexandria Forest	0.03	0.07	0.11	10.04	[c]
3 Pondoland Sourveld	2.05	4.08	3.91	11.43	94
4 Knysna Forest	0.88	3.46	4.28	16.28	24
5 'Ngongoni Veld	0.09	0.93	0.60	0.23	87
6 Zululand Thornveld	4.89	5.95	5.45	11.06	54
7 Eastern Province Thornveld	0.04	0.03	0.04	5.92	[c]
Inland Tropical Forest	**3.86**	**5.02**	**6.05**	**10.26**	**71**
8 NE Mountain Sourveld	1.56	3.21	7.61	11.29	65
9 Lowland Sour Bushveld	5.83	6.46	4.81	9.45	76
Tropical Bush & Savanna	**7.09**	**9.09**	**10.00**	**10.65**	
10 Lowveld	27.05	22.22	23.65	24.82	36
11 Arid Lowveld	19.76	31.09	31.77	35.17	29
12 Springbok Flats Turfveld	0.00	0.00	0.00	0.28	92
13 Other Turfveld	0.00	0.04	0.05	0.84	60
14 Arid Sweet Bushveld	0.08	0.26	0.37	1.39	27
15 Mopane Veld	25.67	40.87	42.46	41.12	8
16 Kalahari Thornveld	6.86	7.00	8.68	7.26	55
17 Karroid Kalahari Thornveld	0.00	0.04	0.04	10.94	[c]
18 Mixed Bushveld	0.71	1.59	2.14	3.00	55
19 Sourish Mixed Bushveld	0.17	0.44	0.93	1.82	64
20 Sour Bushveld	0.16	2.27	1.29	6.96	28
False Bushveld	**0.00**	**0.00**	**0.00**	**0.00**	
21 False Thornveld	0.00	0.01	0.01	0.00	[c]
22 Thorny Grassland	0.00	0.00	0.00	0.00	[c]
Karoo	**0.07**	**0.09**	**0.60**	**1.57**	
23 Valley Bushveld	0.40	1.09	1.40	4.37	51
24 Noorsveld	0.00	0.00	0.00	0.00	[c]
25 Spekboomveld	0.00	0.74	0.73	2.91	[c]
26 Karroid Broken Veld	0.08	0.14	0.85	0.52	[c]
27 Central Upper Karoo	0.00	0.02	0.02	0.00	[c]
28 Western Mountain Karoo	0.00	0.08	0.09	0.98	[c]
29 Arid Karoo	0.00	0.01	0.01	0.00	[c]
30 Central Lower Karoo	0.00	0.00	0.00	0.00	[c]
31 Succulent Karoo	0.00	0.01	0.41	0.37	[c]
32 Orange River Broken Veld	0.11	0.25	0.83	0.00	[c]
33 Namaqualand Broken Veld	0.18	0.16	1.91	9.13	[c]
34 Strandveld	0.06	0.45	0.51	3.19	24
False Karoo	**0.33**	**1.31**	**1.40**	**0.96**	
35 False Arid Karoo	0.00	0.00	0.00	0.00	[c]
36 False Upper Karoo	0.47	1.06	1.70	1.15	[c]
37 False Karroid Broken Veld	0.70	1.04	1.23	1.08	[c]
38 False Central Lower Karoo	0.00	0.00	0.48	3.36	[c]
39 False Succulent Karoo	0.21	0.00	0.00	0.00	[c]
40 False Orange R B Veld	0.00	0.00	5.66	0.00	[c]
41 Karroid Pan Turf Veld	0.00	0.00	0.00	0.00	[c]
42 Karroid Mountain Veld	0.00	1.26	8.43	10.99	[c]
43 Mountain Renosterveld	0.00	12.10	1.46	2.09[d]	27
Temperate and Transitional	**0.63**	**2.89**	**3.48**	**2.10**	**66**
44 Highland/Dohne Sourveld	0.80	3.19	4.30	1.21	60
45 Mist Belt/'Ngongoni Veld	0.00	0.15	0.30	0.60	89
46 Coastal Renosterveld	0.34	0.59	0.48	6.52[d]	85
47 Coastal Fynbos (Macchia)	0.65	2.42	1.76	4.96	47

Table 23.2 (cont.)

Vegetation type (Veld Type number)	Area conserved (%)				Area transformed[b] (%)
	1974[a]	1983	1987	1995	
Pure Grassveld	**0.18**	**0.49**	**0.58**	**1.66**	
48 Cymbopogon–Themeda Veld	0.00	0.02	0.06	0.10	72
49 Transitional Type 48	0.79	0.43	0.81	0.53	c
50 Dry Cymbopogon–Themeda Veld	0.11	0.29	0.60	0.44	67
51 Pan Turf Veld	0.00	0.00	0.00	0.00	c
52 Themeda Veld	0.00	0.00	0.00	0.00	79
53 Patchy Highveld Transition	0.00	0.00	0.00	0.00	59
54 Turf Highveld Transition	0.00	0.13	0.13	0.26	64
55 Bankenveld Transition	0.00	0.00	0.00	0.00	82
56 Highland Sourveld Transition	0.11	0.12	0.12	1.11	92
57 NE Sandy Highveld	0.00	0.68	0.46	0.67	49
58 Themeda–Festuca Alpine Veld	1.73	6.48	5.32	25.10	32
59 Stormberg Plateau Sweetveld	0.00	0.00	0.00	0.00	c
60 Karroid Mountain Veld	0.00	0.08	0.08	0.13	c
False Grassveld	**0.02**	**0.53**	**0.98**	**0.78**	
61 Bankenveld	0.00	0.57	1.36	0.76	65
62 Sour Sandveld Transition	0.00	0.00	0.34	0.39	41
63 Piet Retief Sourveld	0.00	0.77	0.45	0.45	43
64 Northern Tall Grassveld	0.00	2.11	3.03	3.07	69
65 Southern Tall Grassveld	0.10	0.21	0.31	0.35	78
66 Natal Sour Grassveld	0.00	0.00	0.70	0.77	90
67 Pietersburg False Grassveld	0.00	0.00	0.00	0.62	81
68 Eastern Province Grassveld	0.00	0.00	0.00	0.00	c
Sclerophyllous Bush	**0.78**	**33.55**	**54.97**	**21.59**	**11**
69 Fynbos	0.78	33.55	54.97	21.59	11
False Sclerophyllous Bush	**0.00**	**2.15**	**2.04**	**13.44**	**3**
70 False Fynbos	0.00[e]	2.15[e]	2.04[e]	13.44	3
Total	**2.40**	**3.73**	**4.55**	**5.08**	

[a] The data for 1974 excludes areas of state forest, which were de facto conservation areas.

[b] Area transformed is based on subsamples of vegetation types (Macdonald 1989) – these are based on data for 1979–1984.

[c] No data available.

[d] (Rebelo 1995) has shown the figures to be 0.22% for Coastal Renosterveld and 1.75% for Mountain Renosterveld.

[e] In all but the 1995 assessment, False Fynbos was classified as Fynbos in error.

considered above – where should a reserve be placed within a particular under-represented vegetation type? Without further, finer-scale data, inadequacies similar to the biome versus vegetation type misrepresentation may become apparent.

An excellent example of this problem is the Etosha National Park in Namibia which conserves 2.2×10^6 ha of the savanna biome. However, the Etosha Pan, a saline pan hardly representative of a savanna ecosystem (Chap. 5, this volume), occupies almost one-third of this area. This does not detract from this reserve preserving almost one-third of the salt pan surface area in southern Africa – it emphasizes its uniqueness (and importance), but not as a typical savanna habitat.

This lack of representativeness has been exacerbated by attempts to increase conservation levels by adding to existing reserves rather than creating new, smaller, isolated reserves. While reserve expansion is an ecologically sound approach for conservation management of large game animals (Shafer 1990), it results in increased levels

of conservation of the better conserved vegetation types. Furthermore, since larger reserves are more easily managed, agglomeration has been favoured over increasing the number of smaller reserves. However, large reserves will not suffice if the ecosystems within a Veld Type vary considerably over the Veld Type – this is known as the sample effect (Shafer 1990). In order to reconcile the requirements of many large animal species for large reserves and the need to conserve vegetation diversity, emphasis should be placed on creating new, large reserves in areas where future additions will result in the incorporation of new vegetation types at minimal cost. This strategy does not preclude the need for smaller reserves targeted at species and localized communities, or the need to take into account abiotic factors, such as landscape structure and dynamics (Shafer 1990).

In South Africa, the Red Data Books have provided the focus for fine-scale conservation (Ferrar 1989). However, the effectiveness of this strategy within the framework of regional conservation schemes has never been assessed. Such an assessment is urgently required, given the following: the very high number of threatened plant species – 1326 taxa in fynbos, 558 in succulent karoo (Hilton-Taylor & Le Roux 1989); the lack of detailed distributional records; and competition between plant and animal conservation needs (Ferrar 1989). It appears that, in South Africa, Red Data Book plants have been the focus of conservation research and monitoring, but that they have not played a major role in reserve acquisition and regional conservation planning to date [but see Willis, Cowling & Lombard (1996) and Trinder-Smith, Lombard & Picker (1996)]. Some of the problems with conservation planning based on naturally rare species are dealt with in section 23.5.4.

Other countries in the subcontinent do not have Red Data Books (Hilton-Taylor 1994c), and, interestingly, have very few small nature reserves (World Resources Institute 1994).

23.4.2 Representativeness and parochial interests

Not only is representativeness inadequate in the savanna biome, but the best represented vegetation types in South Africa (Mopane Veld, Arid Lowveld, and Lowveld – Table 23.2) are those that are widespread outside South Africa. Savanna vegetation types largely confined to South Africa are especially poorly conserved (3.1, 0–11% mean and range); versus those that are also widespread outside the country (18.4, 1–41%). Thus, savanna biome vegetation types that cannot be conserved elsewhere are poorly conserved in South Africa. As a general rule it can be stated that the vegetation types shared with other countries ('peripheral rarity' *sensu* Ferrar 1989) are the best conserved, whereas the endemic vegetation types – with the exception of fynbos – are the most poorly conserved.

There is the possibility that this pattern of 'over-conservation' of peripheral vegetation types may in part be due to incompatible vegetation type categories and classifications between neighbouring countries or, alternatively, a lack of studies between countries at similar or compatible scales. However, the establishment of the conservation estate in South Africa reflects expedience. Large conservation areas were established in peripheral areas. These had been perceived to be of low agricultural and mining potential at the time of their creation, and in nutrient-rich savannas that still contained game (Chaps 11 and 17, this volume). Their establishment pre-dated any comprehensive vegetation maps.

The lack of fine-scale classifications of compatible vegetation types across political boundaries hinders a subcontinental perspective to conservation assessment and planning. At present, White's (1983) vegetation map of Africa, which is intermediate in scale between the biome and the Veld Type (Acocks 1953), is the only subcontinental appraisal available. International cooperation in determining vegetation type units should be regarded as a priority for future evaluation and planning at the subcontinental scale. In view of the fact that cross-border conservation areas are being mooted (between Mozambique, South Africa and Zimbabwe, and between Botswana, South Africa and Zimbabwe), these conservation areas and their boundaries should be incorporated into a subcontinental conservation framework. However, it appears that these schemes are centred on large mammal species and scant attention has been given to considerations of plant biodiversity.

Pertinent here is the issue of setting goals at local versus national versus international scales. The goal of representing a Veld Type may already have been achieved at an international and national level. Nonetheless, a local conservation department may aim to preserve a representative proportion of the type within its area of jurisdiction. There is an example of this in Kwa-Zulu-Natal, where conservation authorities aim to conserve all 6186 plant species in the province, even though many of these are adequately preserved elsewhere. This requires that adequate conservation must be maintained in 85% of grid cells as opposed to only 56% if a national strategy for South Africa were adopted (Fig. 23.4).

The strategy of each region attempting to preserve its own biota should be encouraged internationally. It alleviates a far more serious problem for conservation, namely countries abrogating their responsibility. It is in those regions where resources are not available for conserving an endemic vegetation type that international pressure and financial support is most strongly needed.

23.4.3 Biases and responsibilities in reporting

A major problem with conservation evaluation in South Africa is the categorization of older reserves according to outdated vegetation classifications. Although these assessments are later adjusted, they result in large-scale fluctuations in conservation status of vegetation types (see Table 23.2). This is especially true of the rarer vegetation types. This was not a major problem in the past. However, with increasingly frequent evaluation and the possibility of decision-making based on up-to-date data, such categorizations can negatively influence conservation decisions.

An example of the above is the conservation status of Coastal and Mountain Renosterveld in the fynbos biome. These are given in the National Register (Table 23.2) as 6.52% and 2.09% of the total area of the vegetation type, respectively. However, careful scrutiny of the data do not support this – only 0.22% of Coastal Renosterveld and

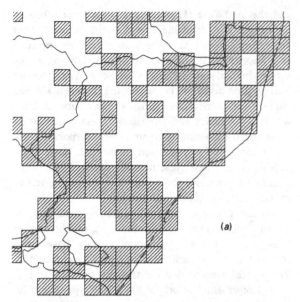

Figure 23.4 **The quarter-degree grid cells (25 km × 25 km) required to represent all 6186 plant species in KwaZulu-Natal assuming that conservation planning is done: (a) on a national scale for South Africa – which requires conservation areas in 56% of the 165 grid cells versus (b) for KwaZulu-Natal only – which requires** conservation areas in 85% of grid cells. (From an unpublished report presented at a workshop to assess the Natal Park Board's proposed scientific programme towards a regional strategy for the conservation of biodiversity in KwaZulu-Natal in 1993).

1.75% of Mountain Renosterveld are currently conserved (Rebelo 1995). Much of this discrepancy arises because, although the reserves are classified as conserving Renosterveld, this type comprises only a small part of the reserves (e.g. Bontebok National Park, Gamka Nature Reserve). Discrepancies also arise from outdated data (areas deproclaimed, e.g. Hartebeest Nature Reserve), old vegetation type delineations [e.g. Acocks (1953) versus Moll & Bossi (1984)] and the inclusion of degraded areas (e.g. old farmlands, submerged areas in reserves fringing dams).

These local problems become more apparent at larger scales. There are major differences between the various estimates for the conservation estate in South Africa. The South African National Register of protected areas (Table 23.2) gives 5.1% of South Africa under conservation, compared to Siegfried's (1989) estimate of 5.8% and the World Resources Institute's (1994) estimate of 6.1%. The number of nature reserves also varies, with 377, 582 and 235 quoted in the above papers, respectively.

In South Africa, the National Register of protected areas is maintained by the Department of Environment Affairs, which is responsible for reporting to international agencies as required by international treatises. However, the department does not have the means to check the data in the register and relies on local conservation authorities to maintain its accuracy (M. Wahl, pers. comm.). Procedures for checking, circulating and

reviewing data are needed for effective national and international conservation planning.

23.5 Designing an efficient conservation network

23.5.1 Evaluation of a levels approach
The advantage of a levels approach is that it ensures international fairness, requiring each nation to strive for the same goal (i.e. to protect 10% of its surface area). An approach with different goals for different countries would be fraught with problems, especially in the case of nations that are required to set aside proportionally more land for conservation. The levels approach also allows for an efficient evaluation of progress of individual countries in their efforts towards achieving the agreed international goal.

The levels approach implicitly assumes that each nation/biome/vegetation unit is a uniform entity and that a random area within the unit will effectively preserve all the components and interactions within the system (cf. Chap. 19, this volume). Alternatively, it assumes that where there is heterogeneity within the unit, this heterogeneity is random. In most cases this is clearly incorrect, especially at larger scales (Shafer 1990): non-random heterogeneity is universally appreciated among conservation biologists. However, any attempt by an international agency to suggest that a nation has preserved the wrong 10%, whether by design or accident, is unlikely to result in remedial action.

The only way an efficient, representative level can be achieved is through detailed local studies. Even at the local scale of vegetation types, few are uniform in terms of species composition and ecosystem processes over their entire extent. The problem with most assessments of this kind is that it is relatively easy to identify a vegetation type (or other unit) requiring urgent conservation action. However, it is often impossible, without further study, to identify representative conservation sites explicitly within the unit.

More importantly, the levels approach is also independent of scale, being downwardly and upwardly compatible within the unit hierarchy. Therefore, the 10% goal applies to both the extensive grassland biome and one of its smaller vegetation types (e.g. *Themeda–Festuca* Alpine Veld). It thus applies equally to each of the 23 grassland biome Veld Types and to the four fynbos biome Veld Types. As a consequence, it neatly sidesteps the major problem with the current classification of vegetation units in South Africa [Acocks' (1953) Veld Types]: a classification that is not hierarchically defined, and comprises many more subdivisions in agriculturally important areas. A case in point is Macchia (fynbos), Veld Type no. 69, which Acocks (1953) admitted was inadequately subdivided. If future studies show fynbos vegetation to comprise *n* subunits, the 10% goal will still apply to each, irrespective of their hierarchical status. Furthermore, the levels approach would apply equally well, irrespective of how the units are characterized.

However, a levels approach assumes that each unit is equal in terms of its conservation requirements. Thus, in South Africa, a level of 10% is targeted for both the fynbos and savanna biomes, even though the fynbos has a quarter more species and occupies only 3% of the area of savanna (Table 23.1). It also assumes that ecosystem processes within the units are equivalent, so that 10% of the area is equally adequate for fynbos, in which large mammal species are generally scarce (Rebelo 1992b), and savanna, which is dominated by large mammal species that require large home ranges (Chap. 11, this volume). This is unlikely to be the case, although where ignorance prevails, this may be the only defensible approach.

Scheepers (1983) suggested that 'special plant communities' may form the focus of conservation attention in South Africa. However, no such assessments have been published on the subcontinent, although the status of wetlands (Breen & Begg 1989) and rivers (O'Keeffe *et al.* 1989) has been reviewed.

Lastly, a levels approach says nothing about conser-

vation urgency. Units are usually evaluated without regard to the factors affecting them. It is obvious that a vegetation type with 5% conserved and a transformation rate of 30% of the remaining vegetation per annum requires more urgent action than another with no conserved areas and a transformation rate of 5% per annum. Unfortunately, data on rates of transformation are not routinely available. Furthermore, transformation can vary from minor degradation to complete conversion. Evaluation is not straightforward, given that the same agricultural practice may be compatible with conservation in one vegetation type, but disastrous for conservation in another. Data for evaluating the impact of transformation and its effect on conservation options are seldom available at a national scale. Furthermore, data are not available for determining the conservation importance of agricultural lands. Thus, although much of the Nama-karoo outside reserves may appear to be suitable for plant conservation, the impacts of insecticides used against locust outbreaks and the elimination of potential livestock predators are poorly understood (Hilton-Taylor & Le Roux 1989). Maverick farming practices, such as one-off planting of virgin veld with specialized cash crops, are also difficult to foresee and control, and the effects of stock grazing can be difficult to detect and predict (see Chap. 21, this volume).

23.5.2 How much should be conserved?

Assuming that 10% of each vegetation type in South Africa should be conserved, how much more needs to be conserved in order to reach the desired level? Given the current reserve configuration in Table 23.2, an additional 7.9×10^6 ha is required – one-third more again than the current conservation estate. The total required conservation estate of 14.0×10^6 ha equals some 12% of South Africa's area. Seven vegetation types still require more than 300×10^3 ha (one-third the size of the Kalahari Gemsbok Park in South Africa or six times the size of Umfolozi Game Reserve) for adequate conservation. These are:

1 False Upper Karoo (36 – Acocks' Veld Type number) – Nama-karoo biome;
2 Arid Karoo (29) – Nama-karoo biome;
3 Karroid Broken Veld (26) – succulent karoo biome;
4 Succulent Karoo (31) – succulent karoo biome;
5 *Cymbopogon–Themeda* Veld (48) – grassland biome;
6 Dry *Cymbopogon–Themeda* Veld (50) – grassland biome, and;
7 Kalahari Thornveld (16) – savanna biome.

However, there are other factors affecting the area (or levels) of vegetation types that need to be conserved.

These include ecosystem viability and heterogeneity limitations. With regard to these factors, what levels are adequate for conservation relative to the 10% level proposed internationally?

23.5.2.1 Ecosystem approaches

Ecosystem requirements dictate that areas should be large enough to maintain viable ecosystems without outside interference and minimum internal management (Shafer 1990). However, most southern African systems are considerably transformed: almost all large predators, scavengers and herbivores are extinct in the fynbos, Nama-karoo and succulent karoo biomes and most of the grassland biome vegetation types, although selected large herbivores have been reintroduced to some reserves (Macdonald 1989). This transformation is offset to some degree by the replacement of larger indigenous ungulates with cattle, sheep and goats, but the degree to which this mimics natural grazing patterns and pressures, especially in the karoo biomes, is poorly understood (Hilton-Taylor & Le Roux 1989; Chap. 17, this volume).

A theoretical minimum size for a normally functioning ecosystem should be one that: includes minimum viable populations of keystone species such as elephant (*Loxodonta africana*), rhinoceros (*Ceratotherium simum* and *Diceros bicornis*) and aardvark (*Orycteropus afer*), which disperse seeds, control woody plant encroachment and maintain other ecosystem processes (Chaps 11 and 17, this volume); and which supports a viable population of top carnivores (Kruger 1977; Shafer 1990; Soulé & Mills 1992). Of course, systems smaller than ideal may function normally if such processes are maintained artificially. In South Africa, as early as 1974, it was suggested that reserves below 100 ha in size had doubtful viability (Edwards 1974). However, adequate conservation of the fynbos biome lowlands will require the creation of a network of small 'witness stands' (Huntley 1978). The idea that small reserves are doomed and thus useless (but see below), not to mention problematic in terms of management, still pervades conservation in the fynbos biome today (pers. obs.) and the concept of 'witness stands' has not been implemented.

Kruger (1977) calculated that a minimum viable area in fynbos vegetation should be between 10 000 and 100 000 ha. Rebelo (1992a) suggested that reserves of between 100 000 and 1 000 000 ha might be required to preserve large mammals. Provided movement of top predators (leopards; *Panthera pardus*) between reserves could be ensured, however, this could be reduced to between 2200 and 16 000 ha. In southern Cape fynbos, Bond, Midgley & Vlok (1988) suggested that areas of 600 ha are sufficient to maintain plant species diversity,

as they found no effects of insularization in fynbos 'islands' exceeding this size within a 'sea' of Afromontane forest. In an acid-sand fynbos landscape, Cowling & Bond (1991) indicated that relative to extensive areas of limestone habitat, limestone fynbos fragments as small as 5 ha did not lose species. Provided that major ecosystem processes (recurrent fires, water table levels) and biotic interactions (pollination and seed dispersal) are maintained, fynbos reserves of a very small size may be viable. However, the cost, technical requirements and social implications of maintaining fire regimes and biotic services in such remnants has not yet been determined.

Cowling (1986) suggested that the minimum area of 15 000 ha would be required to re-establish pre-settlement animal communities in the Nama-karoo biome, where migratory herds followed rain patterns (Chaps 8 and 17, this volume). However, given rainfall dynamics and the historical episodic mass migrations of game (Pringle 1982), much larger areas would probably be required.

No similar analyses have been carried out for other vegetation types in southern Africa. However, it might be predicted that the succulent karoo biome should approach fynbos in its minimum requirements (based on the large number of local endemic plant species and the lack of large mammalian herbivores). It might further be predicted that requirements for the Nama-karoo and grasslands would be intermediate between those of the fynbos and succulent karoo, and those of the savanna biome. The requirements of the forest biome may be determined more by the regeneration requirements of canopy trees, than by the requirements of large herbivores (Chap. 12, this volume) – however, large herbivores are required for regeneration of these species, suggesting that forest needs to be conserved within its savanna or grassland matrix (Stuart-Hill 1991).

Furthermore, the extent of areas required for maintaining functioning ecosystems may vary from region to region. In fynbos, leopards require seven times the area for home ranges in the wetter south than in the drier northwest (Norton & Henley 1987). If related to carrying capacity of herbivores, this requirement may be mirrored by other ecosystem processes. Afromontane forest may plausibly require smaller areas in the southern Cape than in the larger Afromontane regions, because of greater fragmentation over historical time (Chap. 4, this volume). The degree to which these vegetation units require servicing by animals from the surrounding units is, however, unknown. Also unstudied is the migration of pollinators, seed-dispersers and herbivores between adjacent vegetation types.

The problem of resolving the large areas required by

large mammal species with vegetation-based conservation goals may be solved by agglomerating reserves over the boundaries of two, three or more vegetation types. Few large herbivores and carnivores spend their entire lives within one vegetation type. For some highly mobile top predators (e.g. birds and leopards) and herbivores (birds), conservation areas need not be contiguous. Furthermore, some vegetation types may be effectively conserved under grazing by livestock, so that these considerations may not be relevant.

At present in southern Africa there is no evidence to suggest that an area of 10% of any vegetation unit would be inadequate for its effective conservation as a functioning ecosystem. The problem is that few vegetation types are likely to be preserved in contiguous areas of this size. Clearly, more research is required on the minimum viable area for reserves in different vegetation units.

23.5.2.2 *Species approaches*

Data on the spatial heterogeneity of species distributions within most of Acocks' Veld Types are not available; this prevents a more detailed assessment of conservation requirements based on more detailed community types or associations. However, as pointed out by Edwards (1974), the Veld Type concept is based on a large number of extensive vegetation samples, recording species composition and density. The Veld Type is thus a floristically defined vegetation class at the landscape level. This should allow the use of species distributions as a surrogate for heterogeneity within Veld Types.

The species approach embodies the 'fine-scale' approach to determining conservation priorities. This is in contrast to the 'coarse-scale' approach, where the units of interest are biomes, vegetation types, or community types – an approach that is preferred by national and international planners (Noss 1987). However, where data on species distributions do exist, there is no reason why they should not be incorporated into planning at the national and international scales. Such an approach is only now becoming feasible as technology and software develop to allow the manipulation of the vast amount of available data. Two such approaches are Gap Analysis (Karieva 1993) and Iterative Procedures (Pressey *et al.* 1993; Rebelo 1994c,d). Gap analysis has not been used in southern Africa for planning vegetation conservation, although iterative procedures have. Rebelo (1994c) gives a synopsis of the methodology and philosophy behind iterative procedures as used in southern Africa.

Iterative procedures have been used to determine conservation requirements of vegetation types or land units in Australia (Pressey & Nicholls 1989; Pressey, John-son & Wilson 1994). However, this is possible only where land units suitable for acquisition have been determined and the size of land units is in the same order of magnitude as the vegetation units (where units for acquisition are far smaller than the vegetation units, the exercise is trivial). In southern Africa, this is only readily possible in the drier savanna biome – especially Namibia and Botswana. Such iterative procedures are limited in the same way as conventional area assessments in that they assume that the units are homogeneous. An advantage is that they can incorporate additional species data where these exist.

Rebelo (1994d) has determined the minimum conservation requirements for southern Africa. An iterative approach was used to represent each species at least once in a configuration of 25×25 km^2 grid cells occupying a minimal area and taking into account the existing conservation areas (Fig. 23.5). The results mirror the assessment of the biome and vegetation type patterns in some respects: namely, that the top conservation priorities are in the succulent karoo, the grasslands, the escarpment foothills (VorNamib) of Namibia and the fynbos biome lowlands.

However, Rebelo's (1994d) results do suggest where reserves should be located within vegetation types based on species turnover and endemicity – the two major factors influencing the spatial configuration of an ideal reserve network (one that conserves all species as frequently as possible in a minimum area). Several caveats must be borne in mind before the results are interpreted from the perspective of vegetation type. Ecotones tend to have more species, so that Rebelo's approach would select contact zones between vegetation types rather than typical examples of vegetation types. This further exacerbates the difficulty in assigning the 25×25 km grid cells to vegetation types, which results in the biasing of one vegetation type at the expense of its neighbour. Lastly, the results do not imply an aerial requirement for conservation, only the number or proportion of grid cells that probably require a reserve of some size within them. Interpreting these results in terms of area is thus problematic.

In order to represent all species in southern Africa in a single reserve, markedly different proportions of the biomes must be preserved (Table 23.1). The fynbos biome has the highest aerial requirement – 90% of grid cells must be represented. Those biomes with the next highest aerial requirements are succulent karoo (52%) and grassland (31%), with the other biomes requiring less than the subcontinental mean of 22%.

Although aerial prescriptions are not possible, it can be stated that 90% of grid cells in the fynbos biome will require a reserve to preserve all the biome's plant spec-

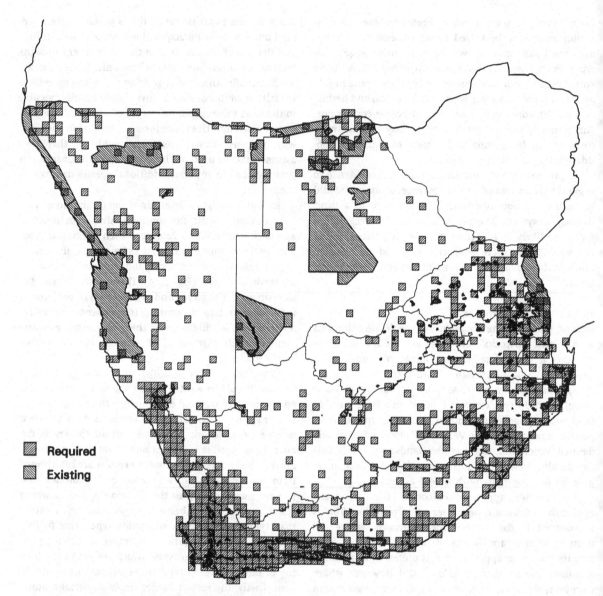

Figure 23.5 **The existing reserve network (shaded polygons) in southern Africa and the distribution of reserves required per quarter-degree grid cell (shaded squares), with areas of overlap (cross-hatched), based on iterative procedures to conserve each plant species at least once (Rebelo 1994c).**

ies, whereas in the savanna biome, only 15% of grid cells will require a reserve for the same purpose (Table 23.1). In absolute terms, the fynbos and grassland biomes both require reserves in about 150 grid cells, the Nama-karoo and succulent karoo biomes in about 110 grid cells, the savanna biome in about 300 grid cells and the desert biome in about 15 grid cells. Thus, the savanna biome requires twice as many grid cells as the fynbos biome, but the fynbos biome is only 3% of its size. These results are similar to those predicted from the relationship

between cumulative plant species richness and area (25 km × 25 km grid units) in the biomes of southern Africa (Cowling *et al.* 1986).

The area of grid cells that contain reserves relative to the area conserved in southern Africa has a ratio of 2 : 1. This ratio implies that the extent of a Veld Type which should be reserved for effective conservation should be 45% for the fynbos biome, 26% for succulent karoo, 15% for grassland, 9% for Nama-karoo, 8% for savanna and 6% for desert. This suggests that the existing conservation area may be 33% of that required for the fynbos biome, 8% for succulent karoo and grassland biomes and 3% of that required for the Nama-karoo biome. The desert and savanna biomes are adequately conserved at a biome scale. Given these guidelines, the total area required for the adequate conservation of the biomes and their plant species would be about three times (20 × 10^6 ha) that of the existing conservation system in South Africa and one and a half times (40 × 10^6 ha) that of the southern African conservation estate.

Although these figures are rough estimates, they do emphasize the importance of South Africa's role in the conservation of the flora of the subcontinent. If these figures are in any way realistic, it bodes ill for the realization of currently stated biodiversity goals. Urgent research is required to determine what data are needed to develop and implement a realistic conservation action plan for southern Africa.

23.5.3 What are the conservation priorities?

The top conservation priorities in southern Africa are poorly conserved vegetation types that are most at risk of being destroyed. Data on the proportion of vegetation left (Tables 23.1 and 23.2) suggest that the grassland biome is most in need of urgent conservation – seven grassland Veld Types have less than a quarter of their original extent untransformed. Although data are lacking for the two karoo biomes, few of these Veld Types are likely to be heavily transformed. The top priorities for conservation in South Africa are the following six vegetation types, which have barely sufficient untransformed vegetation remaining in order to achieve a 10% conservation level. Ranked in order of priority these are:

1 Springbok Flats Turfveld (12 – Acocks' Veld Type number) – savanna biome;
2 Highland Sourveld Transition (56) – grassland biome;
3 Natal Sour Grassveld (66) – grassland biome;
4 Mist Belt-'Ngongoni Veld (45) – grassland biome;
5 'Ngongoni Veld (5) – forest biome; and
6 Coastal Renosterveld (46) – fynbos biome.

A major fault with the above analysis is that it dichotomizes transformation into transformed and untransformed, whereas in reality all vegetation types are transformed to some degree. However, it is easy to use remote sensing techniques to detect total transformation (such as mining or conversion to agricultural crops), whereas it is difficult to detect or quantify degradation caused by factors such as selective grazing, wildflower harvesting and alien plant invasion, all of which alter species composition and (in some cases) ecosystem func-

tioning. Furthermore, degraded systems are generally easier to restore than totally transformed systems, especially when (as is mainly the case) relatively intact portions of vegetation types are scattered throughout the landscape. Until a more detailed assessment of the transformation status of South Africa's vegetation types is available, the priorities listed above will remain the best approximation.

In reality, conservation of these critically endangered vegetation types is not politically expedient. Far greater political returns can be obtained by purchasing land in agriculturally marginal and sparsely populated vegetation types. Reserves in threatened vegetation types will also tend to be more fragmented, and thus less attractive for conservation, with inherent problems of small core areas, high extinction rates and high management costs. Large contiguous reserves can be created only by including transformed land at extremely high costs. The potential for, and costs of, restoration of transformed land are currently unknown.

The correct strategy for acquiring conservation estate depends primarily on the goal (which must be realistic) and the time span to achieve that goal. The most economically sound strategy is to ignore the very expensive sites, urgently acquire sites in those Veld Types likely to increase in value in the near future and defer buying the cheapest sites for as long as possible. This strategy is likely to be unacceptable to many conservationists, who argue that the most endangered vegetation types should be obtained first, and that through greater lobbying, an attempt should be made to ensure that ecologically sound goals are met, irrespective of the economical implications. Similarly, many politicians will want the largest gains achievable within a term of office at the lowest cost and effort.

In the end the problem becomes one of jurisdiction. Local conservation authorities are least likely to have funds for acquiring and managing the many, expensive, small reserves needed for the most threatened vegetation types. National conservation institutions are more likely to have the resources for such acquisitions and management, but are unlikely to sanction them, because of their small size, and parochial nature. International agencies are even less likely to become embroiled in local affairs. It is politically expedient for conservation authorities to ignore the most threatened vegetation types and leave these for the attention of concerned non-governmental organizations.

If a national conservation goal is to be set in accordance with international prescriptions, it is vital that local and national obligations be clearly outlined in any action plans, and that relevant agencies be obliged to meet realistic goals. International obligations to conservation in southern Africa (e.g. the fynbos biome as one of the world's six Floral Kingdoms) need to be considered, and funding should be forthcoming where local or national resources prove to be inadequate for achieving international prescriptions.

23.5.4 Conservation research needs

A major problem with the above analysis of conservation priorities is that Acocks' system of Veld Types has no formal hierarchy. Thus, although succulent karoo has only 2.12% of its area conserved, no succulent karoo vegetation types feature in the top ten conservation priorities. If Acocks' Veld Types had a hierarchical structure, it might be possible to offset under-representation within a vegetation unit with more reserves in closely related units. It would also be possible to rank units across biomes, so that the relative importance of Springbok Flats Turfveld (0.28% conserved, 92% transformed) in the savanna biome (11.6% conserved) can be assessed against the various vegetation units in succulent karoo (2.12% conserved). Thus, it is crucial for conservation that a more objective, hierarchical and detailed vegetation map be developed for southern Africa.

The major threat to conservation is ecosystem transformation. This may vary from a reduction in the abundance of palatable plant species at one extreme to the total destruction of the entire ecosystem at the other extreme. In order to compile conservation priorities and needs objectively, what is needed is a scheme to evaluate the types and degree of transformation, and the implications for conservation. Just how this should be done is unclear. Nonetheless, it is absurd that 90% of succulent karoo should have to be totally transformed before it becomes a national conservation priority. An evaluation as outlined would allow the timely acquisition of land through alternative conservation options. Where alien plants are the major threat to ecosystem functioning and species richness, clearing programmes and biocontrol could be implemented. In cases of overgrazing, strict law enforcement or tax incentives might alleviate the loss of palatable species. Furthermore, the effectiveness of different land uses should be evaluated to guide conservation action – from benign neglect, through subsidization to law enforcement and land acquisition. It is important that statutory conservation areas should strive to meet the criteria for the conservation of their vegetation units. To maintain scientific and political credibility in the evaluation of conservation priorities, it is essential that a vegetation type, not threatened by any transformation, should be considered to be effectively conserved under its current land use.

As shown above, an iterative analysis of the conservation requirements of plant species indicates that the required extent of conservation area (as a proportion of the biome area) for the fynbos biome may be six times higher than for the savanna biome. Further analysis of the local implications of a subcontinental conservation strategy for the region's flora is required. The need for ecosystem preservation (large representative landscapes) should be balanced with rare plant conservation (species-specific conservation) in a matrix of different land uses, where not every plant species is threatened (e.g. some species may benefit from overgrazing or alien encroachment). This would ensure that the maximum conservation benefit could be achieved with the minimum expenditure of resources. The implementation of any such conservation plan would be dependent on the identification of target species. The focus on target species allows the explicit definition of management plans and monitoring requirements, even where the autecology of the target species is currently poorly known.

Because of southern Africa's floristic wealth, there are a large number of localized and threatened species. Any holistic conservation programme is bound to focus on these Red Data Book taxa, and even accord them the highest conservation priority. The iteratively determined conservation plan for the flora of southern Africa determines the location of conservation areas primarily by the most localized species. There are no data on the minimum areas required to preserve naturally rare species: there are many thousands of these localized species in the flora of southern Africa – over 1300 species in the fynbos biome alone. In addition, some widespread species are found as populations occupying a few square metres, and separated from neighbouring populations by tens or hundreds of kilometres of unfavourable terrain. How do these taxa persist? There seems to be a strong imperative, based on empirical data (Cowling & Bond 1991), to establish many small reserves in a region with a flora characterized by extreme habitat specificity and high turnover (Chaps 3 and 19, this volume). Although reserve locations may be selected and even management goals and monitoring requirements defined according to the need to preserve these rarer species, such management requirements remain undeveloped. Owing to the current lack of data, it is probably prudent to implement as many of these 'voucher reserves' as soon as possible and empirically determine the problems associated with this approach. Although many 'voucher reserves' may prove to be unsuccessful, the cost of 'benign neglect' will probably be far higher in terms of species losses and will yield no insights into how the rapidly increasing number of threatened species may be protected.

23.6   The future

The existing reserve system in southern Africa protects 74% of all vascular plants (Siegfried 1989) and covers more than 10% of three of the seven biomes and 14 of the 70 vegetation types (South Africa only). However, is it capable of conserving these species and ecosystems in the longer term?

Any conservation strategy in the future will have to consider the effects of global climate change on the vegetation of southern Africa. Several scenarios for southern Africa have been developed, but all are too generalized for conservation planning, even at the scale of vegetation types and biomes, let alone the smaller scale of nature reserves (Ellery, Scholes & Mentis 1991; Bond 1995; Euston-Brown 1995). The changes in two major variables – rainfall and temperature – are poorly understood on a local and seasonal basis, and it is unlikely that accurate predictions of changes in species distributions will be possible.

Given that conservation is necessarily long-term, will the vegetation units delimited today be valid under a warmer climate? Is it possible that species associations change significantly so that our perception of types changes markedly? Which vegetation types are most likely to be transformed or lost? Which vegetation changes are likely to be subtle, and which obvious in character? Which changes will be slow and which will be rapid or catastrophic? Will vegetation types delimited more by geology than climate be relatively immune to climate change (Euston-Brown 1995)? These issues need to be explored, as they are essential in determining the parameters of monitoring programmes designed to provide the baseline of 'normal' ecosystem functioning. They are also required to plan possible conservation corridors, genetic and species transfer programmes and options for changing reserve configurations.

One research avenue that has not yet received the attention it deserves is the study of plant species refugia. Far greater and perhaps more rapid climatic change occurred during the Pleistocene than is predicted over the next few decades (Chap. 4, this volume). There is a possibility that some of these past refugia may act as a source or sink for plant populations in the likely event of climate change. Given the uncertainty of such climate change, a study of refugia may provide far more information for conservation planning than predictions from models of climate change. Refugia should form the core of an efficient conservation network.

Although the spectre of climate change may seem remote, there is one aspect in which it will dramatically affect conservation planning. Much land today is considered *de facto* conserved, because current economic land uses are compatible with conservation. Changes in carrying capacity and crop suitability may well change the compatibility of agricultural activity with conservation in some vegetation types. However, it is only with well-defined, attainable goals, committed conservation agencies and forward planning that conservation will in any way withstand such climate-induced changes.

23.7   Acknowledgements

I thank M. Wahl (National Register of Protected Areas, Department of Environment Affairs), C. Hilton-Taylor (Plant Red Data Book, National Botanical Institute) for access to current departmental records, and P.M. Holmes and R.M. Cowling for comments on various drafts of the manuscript.

.8   References

kerman, D.P. (1972). The proclamation of wilderness areas by the department of forestry. *South African Forestry Journal*, **82**, 19–21.

ocks, J.P.H. (1953). Veld Types of South Africa. *Memoirs of the Botanical Survey of South Africa*, **28**, 1–128.

Anderson, D. & Grove, R. (1987). The scramble for Eden: past, present and future in African conservation. In *Conservation in Africa. People, Politics and Practice*, ed. D. Anderson & R. Grove, pp. 1–12. Cambridge: Cambridge University Press.

Bond, W.J. (1995). Do mutualisms matter? Assessing the impact of pollinator and disperser disruption on plant extinction. *Philosophical Transactions of the Royal Society of London*, **344**, 83–90.

Bond, W.J., Midgley, J.J. & Vlok, J. (1988). When is an island not an island? Insular effects and their causes in fynbos shrublands. *Oecologia*, **77**, 515–21.

Breen, C.M. & Begg, G.W. (1989). Conservation status of southern African wetlands. In *Biotic Diversity in Southern Africa. Concepts and Conservation*, ed. B.J. Huntley, pp. 254–64. Cape Town: Oxford University Press.

Cowan, G.I. (1987). *SA Plan for Nature Conservation. Annual Report (April 1986 – March 1987*. Internal report. Pretoria: Department Environmental Affairs.

Cowling, R.M. (1986). A description of the karoo biome project. *South African National Scientific Programmes Report 122*. Pretoria: CSIR.

Cowling, R.M. (1991). Options for rural land use in southern Africa: an ecological perspective. In *A Harvest of Discontent. The Land Question in South Africa*, ed. M. de Klerk, pp. 11–22. Cape Town: IDASA.

Cowling, R.M. (1993). Ecotourism: what is it and what can it mean for conservation? *Veld & Flora*, **79**, 3–5.

Cowling, R.M. & Bond, W.J. (1991). How small can reserves be? An empirical approach in the Cape Fynbos. *Biological Conservation*, **58**, 243–56.

Cowling, R.M., Gibbs Russell, G.E., Hoffman, M.T. & Hilton-Taylor, C. (1989). Patterns of plant species diversity in southern Africa. In *Biotic Diversity in Southern Africa. Concepts and Conservation*, ed. B.J. Huntley, pp. 19–50. Cape Town: Oxford University Press.

Davis, S.D., Heywood, V.H. & Hamilton, A.C. (eds) (1994). *Centres of Plant Diversity – a Guide and Strategy for their Conservation. Volume 1. Europe, Africa, South West Asia and the Middle East*. Cambridge: WWF & IUCN.

Diamond, J.M. (1976). Island biogeography and conservation: strategy and limitations. *Science*, **193**, 1027–9.

Edwards, D. (1974). Survey to determine the adequacy of existing conserved areas in relation to vegetation types: a preliminary report. *Koedoe*, **17**, 2–37.

Ellery, W.N., Scholes, R.J. & Mentis, M.T. (1991). An initial approach to predicting the sensitivity of the South African grassland biome to climate change. *South African Journal of Science*, **87**, 499–503.

Euston-Brown, D. (1995). *Environmental and Dynamic Determinants of Vegetation Distribution in the Kouga and Baviaanskloof Mountains, Eastern Cape*. MSc Thesis. Cape Town: University of Cape Town.

Ferrar, A.A. (1989). The role of Red Data Books in conserving biodiversity. In *Biotic Diversity in Southern Africa. Concepts and Conservation*, ed. B.J. Huntley, pp. 136–47. Cape Town: Oxford University Press.

Geldenhuys, C.J. & MacDevette, D.R. (1989). Conservation status of coastal and montane evergreen forest. In *Biotic Diversity in Southern Africa. Concepts and Conservation*, ed. B.J. Huntley, pp. 224–38. Cape Town: Oxford University Press.

Gibbs Russell, G.E. (1987). Preliminary floristic analysis of the major biomes in southern Africa. *Bothalia*, **17**, 213–27.

Greyling, T. & Huntley, B.J. (1984). Directory of southern African conservation areas. *South African National Scientific Programmes Report 98*. Pretoria: CSIR.

Grobler, H. (1990). The private landowner and conservation. *Natal*, **15**, 1–2.

Hall, A.V. (ed.) (1984). Conservation of threatened natural habitats. *South African National Scientific Programmes Report 92*. Pretoria: CSIR.

Hall, A.V. (1989). Rare plant surveys and atlases. In *Biotic Diversity in Southern Africa: Concepts and Conservation*, ed. B.J. Huntley, pp. 148–56. Cape Town: Oxford University Press.

Hedberg, I. & Hedberg, O. (eds) (1968). Conservation of vegetation in Africa south of the Sahara. *Acta Phytogeographica Suecica*, **54**, 1–320.

Hilton-Taylor, C. (1994a). Western Cape Domain (Succulent Karoo) Republic of South Africa and Namibia. In *Centres of Plant Diversity. A Guide and Strategy for Their Conservation. Volume 1. Europe, Africa, South West Asia and the Middle East*, ed. S.D. Davis, V.H. Heywood & A.C. Hamilton, pp. 204–17. Cambridge: WWF & IUCN.

Hilton-Taylor, C. (1994b). The Koakoveld–Namibia and Angola. In *Centres of Plant Diversity. A Guide and Strategy for their Conservation. Volume 1. Europe, Africa, South West Asia and the Middle East*, ed. S.D. Davis, V.H. Heywood & A.C. Hamilton, pp. 201–3. Cambridge: WWF & IUCN.

Hilton-Taylor, C. (1994c). Species level issues and actions. In *Botanical Diversity in Southern Africa*, ed. B.J. Huntley. *Strelitzia*, **1**, 372–80.

Hilton-Taylor, C. & Le Roux, A. (1989). Conservation status of the fynbos and karoo biomes. In *Biotic Diversity in Southern Africa. Concepts and Conservation*, ed. B.J. Huntley, pp. 202–23. Cape Town: Oxford University Press.

Huntley, B.J. (1978). Ecosystem conservation in southern Africa. In *Biogeography and Ecology of Southern Africa*, ed. M.J.A. Werger, pp. 1333–84. The Hague: Junk.

Huntley, B.J. (ed.) (1989). *Biotic Diversity in southern Africa: Concepts and Conservation*, Cape Town: Oxford University Press.

Huntley, B.J. (ed.) (1994). *Botanical Diversity in Southern Africa. Strelitzia*, **1**. Pretoria: National Botanical Institute.

Huntley, B., Siegfried, R. & Sunter, C. (1989). *South African Environments into the 21st Century*. Cape Town: Human & Rousseau.

Jacobsohn, M. (1991). Koakoveld: making conservation work. *New Ground*, **6**, 2–5.

Jarman, M.L. (1986). Conservation priorities in lowland regions of the fynbos biome. *South African National Scientific Programmes Report 87*. Pretoria: CSIR.

Karieva, P. (1993). No shortcuts in new maps. *Nature*, **365**, 292–3.

Killick, D.J.B. (1994). Drakensberg Alpine Region. Lesotho and South Africa. In *Centres of Plant Diversity. A Guide and Strategy for Their Conservation. Volume 1. Europe, Africa, South West Asia and the Middle East*, ed. S.D. Davis, V.H. Heywood & A.C. Hamilton, pp. 257–60. Cambridge: WWF & IUCN.

Kruger, F.J. (1977). Ecological reserves in the Cape Fynbos. Toward a strategy for conservation. *South African Journal of Science*, **73**, 81–5.

Macdonald, I.A.W. (1989). Man's role in changing the face of southern Africa. In *Biotic Diversity in Southern Africa. Concepts and Conservation*, ed. B.J. Huntley, pp. 224–38. Cape Town: Oxford University Press.

Moll, E.J. & Bossi, L. (1984). Assessment of the natural vegetation of the fynbos biome of south Africa. *South African Journal of Science*, **80**, 355–8.

Myers, N. (1979). *The Sinking Ark – A New Look at the Problem of Disappearing Species*. Oxford: Pergamon Press.

orton, P.M. & Henley, S.R. (1987). Home range and movements of male leopards in the Cederberg Wilderness Area, Cape Province. *South African Journal of Wildlife Research*, 17, 41–8.

oss, R.F. (1987). From plant communities to landscapes in conservation inventories: a look at the nature conservancy (USA). *Biological Conservation*, 41, 11–37.

Keeffe, J.H., Davis, B.R., King, J.M. & Skelton, P.H. (1989). Conservation status of southern African rivers. In *Biotic Diversity in Southern Africa: Concepts and Conservation*, ed. B.J. Huntley, pp. 266–89. Cape Town: Oxford University Press.

essey, R.L., Humphries, C.J., Vane-Wright, R.I. & Williams, P.H. (1993). Beyond opportunism – key principles for systematic reserve selection. *Trends in Ecology and Evolution*, 8, 124–8.

essey, R.L., Johnson, I.R. & Wilson, P.D. (1994). Shades of irreplaceability: towards a measure of the contribution of sites to a reservation goal. *Biodiversity and Conservation*, 3, 242–62.

essey, R.L. & Nicholls, A.O. (1989). Application of a numerical algorithm to the selection of reserves in semi-arid New South Wales. *Biological Conservation*, 50, 263–78.

ingle, J.A. (1982). *The Conservationists and the Killers*. Cape Town: TV Bulpin.

belo, A.G. (1992a). Preservation of biotic diversity. In *The Ecology of Fynbos. Nutrients, Fire and Diversity*, ed. R.M. Cowling, pp. 309–44. Cape Town: Oxford University Press.

belo, A.G. (1992b). Red Data Book species in the Cape Floristic Region: threats, priorities and target species. *Transactions of the Royal Society of South Africa*, 48, 55–86.

belo, A.G. (1994a). Ecosystem level issues and actions – designing a conservation network for preserving ecosystem and plant biodiversity in southern Africa. In *Botanical Diversity in Southern Africa. Concepts and Conservation*, ed. B.J. Huntley. *Strelitzia*, 1, 381–6.

Rebelo, A.G. (1994b). Cape Floristic Region. Republic of South Africa. In *Centres of Plant Diversity. Guide and Strategy for their Conservation. Volume 1. Europe, Africa, South West Asia and the Middle East*, ed. S.D. Davis, V.H. Heywood & A.C. Hamilton, pp. 218–24. Cambridge: WWF & IUCN.

Rebelo, A.G. (1994c). Iterative selection procedures: centres of endemism and optimal placement of reserves. In *Botanical Diversity in Southern Africa. Concepts and Conservation*, ed. B.J. Huntley. *Strelitzia*, 1, 231–57.

Rebelo, A.G. (1994d). Using the Proteaceae to design a nature reserve network and determine conservation priorities for the Cape Floristic Region. In *Systematics and Conservation Evaluation*, ed. P.L. Forey, C.J. Humphries & R.I. Vane-Wright, pp. 375–96. Oxford: Clarendon Press.

Rebelo, A.G. (1995). Renosterveld: conservation and research. In *The Sustainable Use and Management of Renosterveld Remnants in the Cape Floristic Region – Proceedings of a Symposium*, ed. A.B. Low & F.E. Jones, pp. 32–42. Cape Town: Flora Conservation Committee.

RSA (1991). *Report of the Three Committees of the President's Council on a National Environmental Management System*. Cape Town: Government Printer.

RSA (1994). *White Paper RDP*. Cape Town: Government Printer.

Rutherford, M.C. & Westfall, R.H. (1986). Biomes of southern Africa: an objective characterization. *Memoirs of the Botanical Survey of South Africa*, 54, 1–98.

Scheepers, J.C. (1983). The present status of vegetation conservation in South Africa. *Bothalia*, 14, 991–5.

Shafer, C.L. (1990). *Nature Reserves – Island Theory and Conservation Practice*. Washington: Smithsonian Institute Press.

Siegfried, W.R. (1989). Preservation of species in south African nature reserves. In *Biotic Diversity in Southern Africa. Concepts and Conservation*, ed. B.J. Huntley, pp. 186–201. Cape Town: Oxford University Press.

Slatyer, R.O. (1975). Ecological reserves: size, structure and management. In *A National System of Ecological Reserves in Australia No 19: Report to the National Academy of Science*, ed. F. Fenner, pp. 22–38. Netley: Griffen Press.

Soulé, M.E. & Mills, L.S. (1992). Conservation genetics and conservation biology: a troubled marriage. In *Conservation for Biodiversity and Sustainable Development*, ed. O.T. Sandlund, K. Hindar & A.H.D. Brown, pp. 55–69. Oslo: Scandinavian University Press.

Stuart, S.N. & Adams, R.J. (1990). Biodiversity in sub-saharan Africa and its islands: conservation, management and sustainable use. *Occasional Papers of the IUCN Species Survival Commission* 6. Gland, Switzerland: IUCN.

Stuart-Hill, G.C. (1991). Elephants the rightful conservators of the valley bushveld. *Veld & Flora*, 77, 9–11.

Trinder-Smith, T.H., Lombard, A.T. & Picker, M.D. (1996). Reserve scenarios for the Cape Peninsula: high, middle and low road options for conserving the remaining biodiversity. *Biodiversity and Conservation*, 5, 649–69.

van Wyk, A.E. (1994). Maputaland–Pondoland Region. South Africa, Swaziland and Mozambique. In *Centres of Plant Diversity. A Guide and Strategy for their Conservation. Volume 1. Europe, Africa, South West Asia and the Middle East*, ed. S.D. Davis, V.H. Heywood & A.C. Hamilton, pp. 227–35. Cambridge: WWF & IUCN.

West, P.C. & Brechin, S.R. (1992). *Resident Peoples and Parks. Social Dilemmas and Strategies in International Conservation*. Tucson: University of Arizona Press.

Western, D. (1989). Conservation without parks: wildlife in the rural landscape. In *Conservation for the Twenty-first Century*, ed. D. Western & M. Pearl, pp. 158–65. New York: Oxford University Press.

White, F. (1983). *The Vegetation of Africa*. Paris: Unesco.

Wicht, C.L. (1945). *Report of the Committee on the Preservation of the Vegetation of the South Western Cape*. Cape Town: Royal Society of South Africa.

Willis, C.K., Lombard, A.T., Cowling, R.M., Heydenrych B.J. & Burgers, C.J. (1996). Reserve systems for limestone endemic flora of the Cape lowland fynbos: iterative versus linear programming. *Biological Conservation*, 77, 53–62.

World Resources Institute (1992). *Global Biodiversity Strategy: Guidelines for Action to Save, Study, and Use Earth's Biotic Wealth Sustainably and Equitably.* Oxford: Oxford University Press.

World Resources Institute. (1994). *World Resources 1994–1995 – A Guide to the Global Environment.* Oxford: Oxford University Press.

Glossary

alien In this volume, referring to organisms that reached southern Africa as a consequence of the activities of humans.

Bantu Refers to the family of related languages spoken by the indigenous farming peoples of southern Africa.

-berg/e Afrikaans word for a hill or mountain. Often used as suffix, e.g. Drakensberg ('dragon mountain').

bergwind The warm dry wind, especially frequent in winter, that blows when the plateau of the interior of southern Africa is covered by a strong high-pressure system and pressure is low over the ocean. The winds blow outwards from the interior, affecting the coastal areas, and become heated by descent. Temperatures may rise above 36 °C and temporarily exceed those of midsummer. Bergwinds may continue for two or three days, significantly increasing the probability of fires, especially on the southern Cape coast.

boer/e Afrikaans word for a farmer; also used with reference to any South African of Dutch descent (see also *trekboer*).

-bos/bosch Afrikaans/Dutch word for forest or 'bush' and widely used as a suffix in place names.

bosveld (see *bushveld*)

broken veld A term applied by South African botanists to a landscape dominated by dwarf shrubs, grasses and succulents, with scattered larger bushes and occasional small trees, usually occurring on flattish, gravelly plains or rugged, rocky mountains.

bush encroachment The expansion of woody plants into vegetation previously dominated by non-woody plants or where woody and non-woody elements were co-dominant, resulting in reduced carrying capacity for domestic livestock (grazers). The phenomenon is very widespread in southern Africa and is generally attributed to the removal of megaherbivores, the alteration of fire regimes, the reduced use of trees, overgrazing or a combination of these factors.

bushveld	Vernacular name, loosely applied in South Africa, for *savanna* vegetation, usually where the canopies of moderately tall (5–10 m) trees frequently touch each other.
dambo	Vernacular term in use in south-central Africa to describe a seasonally wet grassland that occupies the valley bottoms in many moist savanna landscapes; see also *vlei*.
decreaser species	Those plant species that dominate in veld that is well managed, and decrease when veld is either over- or under-grazed.
environmental weed	The small proportion of *alien* plants that invade natural vegetation, often causing a marked change in the composition and structure of communities (see Chap. 22 for further explanation).
ericoid	Referring to plants or plant parts resembling species of the Ericaceae, i.e. small shrubs with minute, rolled-under leaves. One of three characteristic elements of *fynbos* vegetation (see also *restioid*, *proteoid*).
escarpment	The long steep face of a plateau. In southern Africa, usually refers to the eastern escarpment which forms the edge of the inland plateau or *Highveld* (see also *Great Escarpment*).
fairy rings	Circular patches, devoid of vegetation, in the eastern Namib, from Angola to Namaqualand. See section 9.3.3.6. in Chap. 9 for discussion of the various hypotheses suggested to explain their origin.
fynbos	The characteristic shrubland vegetation of the southern and southwestern Cape of South Africa which comprises three major growth forms (*proteoid*, *ericoid* and *restioid*).
gilgai	A component of the distinctive striped clumping of the woody vegetation in savannas, sometimes visible from the air. Gilgai are the whorled, fingerprint-like patterns, the result of the micro-relief variation found on swelling clays (see Chap. 11 for further discussion).
Great Karoo	The basin-like region of the Karoo, bounded in the north by the *Great Escarpment* and in the south by the Cape Fold Mountains.
Great Escarpment	The generally precipitous, horseshoe-shaped transition between the high interior plateau of southern Africa and the lower-lying plains of the coastal hinterland.
heuweltjies	Slightly raised mounds of soil, circular in shape with diameters of 5–35 m, and regularly spaced across the landscape. Similar structures in North America are known as Mima mounds.
Highveld	The interior plateau of southern Africa, at an altitude of above c. 1600 m, mostly covered with grassland.
homeland	Areas of land demarcated under the apartheid regime in South Africa for self-rule by various ethnically defined African/Black populations. Although many of these areas had previously been communal lands, the entrenchment of 'separate development' led to the forced removal of millions of people from areas

designated for whites into these 'homelands', resulting in massive overcrowding and severe shortages of natural resources.

increaser species Those plant species that occur naturally in the vegetation but which increase in abundance with over- or under-grazing or selective grazing.

Karoo The vast arid plain occupying most of the interior of the old Cape Province of South Africa, roughly between the Orange River in the north, Oudtshoorn in the south, the Cape west coast and Somerset East in the east. In this volume, divided into two biomes (succulent and Nama-karoo) on the basis of climatic parameters and growth-form composition, the latter also including parts outside what is typically known as Karoo. The word 'karoo' comes from a *Khoikhoi* word for 'dry'.

Khoikhoi Refers to people who were accepted as belonging to a community where a dialect of the Khoikhoi language was spoken and where pastoralism was the preferred form of economy. Also spelt Khoekhoe (cf. *Khoisan*).

Khoisan A *Khoikhoi* word referring to a cluster of indigenous south African families distinct from *Bantu*-speaking peoples. Khoisan is also used loosely to refer to a genetic population, although not all Khoisan speakers were genetically Khoisan (e.g. the 'negroid' Damara of Namibia speak Nama, a *Khoikhoi* language).

klipveld The vast area underlain by dolomite in the North-West Province of South Africa, thus named because of the abundance of surface rock ('klip' in Afrikaans).

-kloof The Afrikaans word for a gully, ravine or valley, usually with steeply inclined or rocky sides and often well wooded. Often used as a suffix in place names, e.g. Baviaanskloof ('baboon valley').

koppie A small hill, often with a rocky summit.

Little Karoo The region between the Langeberge and the Groot Swartberge in the Western Cape.

Lowveld Term applied to the eastern part of southern Africa lying between c. 100 and 900 m above sea level and mostly covered with savanna.

Mima-like mounds (see *heuweltjies*)

miombo A broad-leaved savanna, dominated by species of *Brachystegia* and *Julbernardia*, occurring in moist, infertile conditions; the dominant vegetation of large areas of south-central Africa.

native In this volume, referring to organisms that evolved in southern Africa, or that arrived in the region before the beginning of the neolithic period or that arrived here since that time by a method entirely independent of human activity.

proteoid Referring to plants resembling many species of the Proteaceae; i.e. evergreen shrubs with hard, leathery leaves. One of three characteristic elements of *fynbos* vegetation (see also *ericoid*, *restioid*).

renosterveld   Vernacular name for the fire-prone shrublands in the fynbos biome which superficially resemble *fynbos*, but which grow on more fertile, finer textured, clay-rich soils. Renosterveld also lacks the *restioid* element and *proteoids* are rare. Renosterveld takes its name from the renosterbos, *Elytropappus rhinocerotis* (Asteraceae), which is the most abundant and conspicuous shrub species in the vegetation type.

restioid   Referring to the grass-like plants of the Cyperaceae and Restionaceae in *fynbos* (also known as the 'Cape reeds'). One of three characteristic elements of *fynbos* vegetation (see also *ericoid, proteoid*).

San   A term, quoted from the older literature, and originally a *Khoikhoi* word which was used dismissively to describe people living by hunting and gathering, denoting low social status and wealth, relative to the ideals of the *Khoikhoi* society. As such, both indigenous hunter-gatherers and impoverished fragments of *Khoikhoi* society were included in this category. The use of the word San is currently viewed as controversial and is avoided.

sandveld   A vernacular name used to describe two distinct and widely separated vegetation types in southern Africa. (1) The low-lying coastal plain, overlain by acid sands, in the Western Cape of South Africa, roughly between Cape Town and Lamberts Bay and bounded on its inland margin by renosterveld on clay-rich soils and on the coastal margin by *strandveld* on calcareous sands. Sandveld vegetation is predominantly *proteoid fynbos* (in the south) or *restioid fynbos* (in the north). (2) An open, broad-leaved savanna on sand or sandy loam, usually dominated by species of the Combretaceae, or sometimes Fabaceae (notably *Burkea africana*) (see also Table 11.3).

savanna   A tropical and subtropical vegetation type co-dominated by woody plants and C_4 grasses.

sourveld   Rangeland (typically short grassland) on acidic substrata with low nutrient status and mesic conditions. Occurs mainly at high elevation. Sourveld provides palatable material only in the growing season (cf. *sweetveld*).

strandveld   Vernacular name for the complex mosaic of *subtropical thicket*, *fynbos* and karroid vegetation along the south and west coasts of the Western Cape.

subtropical thicket   A dense, often impenetrable shrubland or low forest comprising species with subtropical affinities. It occurs on relatively nutrient-rich soils, and in areas that are protected from fire, such as river valleys, rock screes and outcrops, termite mounds or *heuweltjies*, and on coastal dunes, especially in the fynbos biome. Subtropical thicket with a high cover of succulents, especially arborescent *Euphorbia* and *Aloe* species, is often referred to as *valley bushveld*.

sweetveld   Usually grassland on basic, enriched substrata, but with moderate to low rainfall and/or high radiation conditions (north-facing slopes). Sweetveld remains palatable and nutritious when mature (cf. *sourveld*).

thornveld   A southern African colloquial term for savannas dominated by thorny trees or shrubs, typically *Acacia* spp.

top-kill	Killing, by fire, of plant canopy tissue including shoot buds. Plants resprout from the base.
trekboer	A nomadic grazier moving with his flocks and travelling by ox-wagon.
valley bushveld	see *subtropical thicket*.
veld	A term loosely applied to open country.
Veld Type	A term, coined by J.P.H. Acocks in his classic 1953 memoir on *Veld Types of South Africa* to describe a unit of vegetation whose range of variation is small enough to permit the whole of it to have the same farming potential. Although clearly an agricultural concept (designed as a practical guide to farmers), Acocks' map of Veld Types has served as the principal scheme for classifying the vegetation of the region.
vlei	Vernacular term used in South Africa for a seasonally wet grassland; see also *dambo*.

Subject index

Index of biota and taxa

Note: References in bold indicate illustrations. There may also be textual references on these pages.

Printed in the United States
By Bookmasters